Third Edition

Insect Physiology and Biochemistry

Third Edition

Insect Physiology and Biochemistry

James L. Nation, Sr.
UNIVERSITY OF FLORIDA, GAINESVILLE, USA

CRC Press
Taylor & Francis Group
Boca Raton London New York

CRC Press is an imprint of the
Taylor & Francis Group, an **informa** business

CRC Press
Taylor & Francis Group
6000 Broken Sound Parkway NW, Suite 300
Boca Raton, FL 33487-2742

© 2016 by Taylor & Francis Group, LLC
CRC Press is an imprint of Taylor & Francis Group, an Informa business

No claim to original U.S. Government works

Printed on acid-free paper
Version Date: 20150622

International Standard Book Number-13: 978-1-4822-4758-9 (Hardback)

This book contains information obtained from authentic and highly regarded sources. Reasonable efforts have been made to publish reliable data and information, but the author and publisher cannot assume responsibility for the validity of all materials or the consequences of their use. The authors and publishers have attempted to trace the copyright holders of all material reproduced in this publication and apologize to copyright holders if permission to publish in this form has not been obtained. If any copyright material has not been acknowledged please write and let us know so we may rectify in any future reprint.

Except as permitted under U.S. Copyright Law, no part of this book may be reprinted, reproduced, transmitted, or utilized in any form by any electronic, mechanical, or other means, now known or hereafter invented, including photocopying, microfilming, and recording, or in any information storage or retrieval system, without written permission from the publishers.

For permission to photocopy or use material electronically from this work, please access www.copyright.com (http://www.copyright.com/) or contact the Copyright Clearance Center, Inc. (CCC), 222 Rosewood Drive, Danvers, MA 01923, 978-750-8400. CCC is a not-for-profit organization that provides licenses and registration for a variety of users. For organizations that have been granted a photocopy license by the CCC, a separate system of payment has been arranged.

Trademark Notice: Product or corporate names may be trademarks or registered trademarks, and are used only for identification and explanation without intent to infringe.

Visit the Taylor & Francis Web site at
http://www.taylorandfrancis.com

and the CRC Press Web site at
http://www.crcpress.com

Printed and Bound in the United States of America by Edwards Brothers Malloy on sustainably sourced paper.

Contents

Preface xix
Author xxi

Chapter 1
Embryogenesis 1

Preview 1
1.1 Introduction 2
1.2 Morphogenesis 3
    1.2.1 Egg, Fertilization, and Zygote Formation 3
    1.2.2 Variations in Zygotic Nucleus Cleavage, Formation of Energids,
           and Blastoderm Formation 4
           1.2.2.1 Apterygota 5
           1.2.2.2 Hemimetabola 7
           1.2.2.3 Holometabola 8
    1.2.3 Formation of the Germ Band 9
    1.2.4 Gastrulation 9
    1.2.5 Germ Band Elongation 10
    1.2.6 Blastokinesis and Extraembryonic Membranes 11
1.3 Genetic Control of Embryogenesis 15
    1.3.1 Development of a Model for Patterning 16
           1.3.1.1 *Bicoid* Gene and Anterior Determination in *Drosophila* 16
           1.3.1.2 Posterior Group Genes and Posterior Pattern Formation 18
           1.3.1.3 Genes Required in the Acron and Telson 18
           1.3.1.4 Dorsal–Ventral Axis 19
1.4 Segmentation Genes 19
1.5 Homeotic Genes 20
    1.5.1 Homeobox 21
1.6 Organogenesis 21
    1.6.1 Neurogenesis 21
    1.6.2 Development of the Gut 22
    1.6.3 Malpighian Tubules 23
    1.6.4 Tracheal System 23
    1.6.5 Oenocytes 23
    1.6.6 Cuticle Secretion in the Embryo 23
    1.6.7 Cell Movements during Embryogenesis 23
    1.6.8 Programmed Cell Death: Apoptosis 24
1.7 Hatching 24
1.8 Imaginal Discs 24
1.9 Review and Self-Study Questions 28
References 28

Chapter 2
Digestion 33

Preview...33
2.1 Introduction..33
2.2 Relationships between Food Habits and Gut Structure and Function 34
    2.2.1 Plant versus Animal Origin: Solid versus Liquid Diet 34
2.3 Major Structural Regions of the Gut..36
    2.3.1 Foregut...36
    2.3.2 Midgut..39
    2.3.3 Hindgut 39
2.4 Midgut Cell Types..40
    2.4.1 Columnar Cells 40
    2.4.2 Regenerative Cells 40
    2.4.3 Goblet Cells...41
2.5 Microvilli or Brush Border of Midgut Cells 44
2.6 Glycocalyx 45
2.7 Peritrophic Matrix...45
    2.7.1 Functions of the Peritrophic Matrix..46
2.8 Digestive Enzymes...47
    2.8.1 Carbohydrate-Digesting Enzymes...48
    2.8.2 Lipid-Digesting Enzymes 49
    2.8.3 Protein-Digesting Enzymes...49
    2.8.4 Do Proteinase Inhibitors in the Food Influence the Evolution of Proteinase Secreted? 51
2.9 Hormonal Influence on Midgut..51
2.10 Countercurrent Circulation of Midgut Contents and Absorption of Digested Products.......53
2.11 Transepithelial and Oxidation–Reduction Potential of the Gut...55
2.12 Gut pH..55
2.13 Hematophagy: Feeding on Vertebrate Blood...57
2.14 Digestive System Morphology and Physiology in Major Insect Orders 58
    2.14.1 Orthoptera..58
    2.14.2 Dictyoptera...59
    2.14.3 Isoptera...59
    2.14.4 Hemiptera...60
    2.14.5 Homoptera..61
    2.14.6 Coleoptera..61
    2.14.7 Hymenoptera..62
    2.14.8 Diptera..62
    2.14.9 Lepidoptera 63
2.15 Insect Gut as a Potential Target for Population Management and Control of the Spread of Plant and Animal Disease Organisms..64
2.16 Review and Self-Study Questions 65
References...65

Chapter 3
Nutrition..75

Preview...75
3.1 Introduction..75
3.2 Importance of Balance in Nutritional Components 76

3.3	Ability of Insects to Self-Select Nutritional Components	77
3.4	Requirements for Specific Nutrients	77
	3.4.1 Nitrogen Source: Proteins and Amino Acids	78
	3.4.2 Essential Amino Acids	79
	3.4.3 Carbohydrates	81
	3.4.4 Lipids	82
	3.4.5 Sterols	82
	3.4.6 Polyunsaturated Fatty Acids	83
	3.4.7 Vitamins	84
	3.4.8 Minerals	85
3.5	Techniques and Dietary Terms Used in Insect Nutrition Studies	87
3.6	Criteria for Evaluating Nutritional Quality of a Diet	87
3.7	Measures of Food Intake and Utilization	88
3.8	Phagostimulants	90
3.9	Feeding Deterrents	92
3.10	Review and Self-Study Questions	92
References		93

Chapter 4
Integument and Molting 99

Preview		99
4.1	Introduction	100
4.2	Structure of the Integument	100
	4.2.1 Cuticulin Envelope	101
	4.2.2 Epicuticle	101
	4.2.3 Procuticle	102
	4.2.4 Pore Canals and Wax Channels	103
	4.2.5 Epidermal Cells	103
4.3	Molting and Formation of New Cuticle	106
	4.3.1 Apolysial Space	108
	4.3.2 Molting Fluid Secretion	108
	4.3.3 New Cuticle Formation	108
	4.3.4 Reabsorption of Molting Fluid	109
4.4	Ecdysis	109
	4.4.1 Shedding the Old Cuticle: Ecdysis	112
	4.4.2 Post-Ecdysis Wing Expansion and Waterproofing the New Cuticle	113
	4.4.3 Sclerotization of Cuticle	114
4.5	Chemical Composition of Cuticle	116
	4.5.1 Chitin	116
	4.5.2 Biosynthesis of Chitin	121
	4.5.3 Cuticular Proteins	122
	4.5.4 Resilin	125
	4.5.5 Stage-Specific Differences in Cuticle Proteins	126
	4.5.6 Protective Functions of Cuticle Proteins	126
	4.5.7 Cuticular Lipids	126
4.6	Mineralization of Insect Cuticles	129
4.7	Capture of Atmospheric Water on Cuticular Surfaces	129
4.8	Review and Self-Study Questions	129
References		130

Chapter 5
Hormones and Development 135

Preview 135
5.1 Introduction 136
5.2 Historical Beginnings for the Concept of Hormonal Control of Molting
    and Metamorphosis 136
5.3 Interplay of PTTH, Ecdysteroids, and Juvenile Hormone Control Development 137
5.4 Brain Neurosecretory Cells and Prothoracicotropic Hormone 141
    5.4.1   Source and Chemistry 141
    5.4.2   Bioassay for PTTH Activity 141
    5.4.3   Stimuli for the Secretion of PTTH 143
    5.4.4   PTTH Secretion after Brain Activation by Stretch Receptors 143
    5.4.5   Gated PTTH Secretion in Tobacco Hornworm 144
    5.4.6   Secretion of PTTH after Brain Activation by Cold Exposure 144
    5.4.7   Regulation of Tissue and Hemolymph Levels of PTTH 144
    5.4.8   Mode of Action of PTTH 145
5.5 Prothoracic Glands and Ecdysteroids 146
    5.5.1   Biosynthesis of Ecdysone 146
    5.5.2   Conversion of Ecdysone into 20-Hydroxyecdysone 148
    5.5.3   Molecular Diversity in the Structure of the Molting Hormone 148
    5.5.4   *Calliphora* Assay for Ecdysteroids 151
    5.5.5   Radioimmunoassay for Ecdysone and Related Ecdysteroids 151
    5.5.6   Assay by Physicochemical Techniques 152
    5.5.7   Tissues and Cell Cultures Used in Assays 153
    5.5.8   Degradation of Ecdysone 153
    5.5.9   Virus Degradation of Host Ecdysteroids 155
    5.5.10 Dependence of Some Parasitoids on Host Ecdysteroids 155
5.6 Corpora Allata and Juvenile Hormones 155
    5.6.1   Glandular Source and Chemistry 155
    5.6.2   Assays for JH Activity 157
    5.6.3   Regulation of the Tissue and Hemolymph Levels of JH 157
    5.6.4   Insect Growth Regulators and Compounds That Are Cytotoxic
            to the Corpora Allata 161
    5.6.5   Cellular Mode of Action and Receptors for JH 162
    5.6.6   Downstream Transcription Factors 163
5.7 Mode of Action of Ecdysteroids at the Gene Level 164
    5.7.1   Chromosomal Puffs 164
    5.7.2   Isolation of an Ecdysteroid Receptor 166
    5.7.3   Differential Tissue and Cell Response to Ecdysteroids 168
5.8 Possible Timer Gene in the Molting Process 170
5.9 Ecdysone–Gene Interaction Ideas Stimulated Vertebrate Work 170
5.10 Review and Self-Study Questions 171
References 172

Chapter 6
Biological Rhythms 181

Preview 181
6.1 Introduction 182
6.2 Characteristics of Circadian and Photoperiodic Rhythms 182

6.3	Molecular Basis for the Circadian Clock	183
6.4	Evidence for Clock Genes in Many Insects	186
6.5	Examples of Circadian Functions in Insects	186
	6.5.1 Circadian Regulation of Hormone Secretion	186
	6.5.2 Circadian Clock Influence in Peripheral Organs and Tissues	188
	6.5.3 Circadian Clock Influence in Social Behavior of Honeybees	189
	6.5.4 Circadian Clock Influence in Reproduction	190
6.6	Photoperiodic Response: One Clock, Two Clocks, or Multiple Clocks?	193
6.7	Clock Models Based on Experimental Responses of Insects to Varying Light/Dark Regimes	195
	6.7.1 Hourglass Model	196
	6.7.2 External Coincidence Model	197
	6.7.3 Internal Coincidence Model	197
	6.7.4 Resonance Model	197
	6.7.5 Summary Results from Model Experiments	198
6.8	Conclusions	198
6.9	Review and Self-Study Questions	198
References		199

Chapter 7
Diapause 207

Preview 207

7.1	Introduction	207
7.2	Diapause: A Survival Strategy	208
7.3	Phases of Diapause	210
	7.3.1 Prediapause: Induction and Preparation	210
	7.3.2 Diapause: Initiation and Maintenance	211
	7.3.3 Diapause Termination	212
7.4	Hormonal Control of Diapause	212
	7.4.1 Embryonic Diapause	212
	7.4.2 Larval Diapause	213
	7.4.3 Pupal Diapause	213
	7.4.4 Adult Diapause	214
7.5	Role of Daily and Seasonal Biological Clocks in Diapause	215
7.6	Diapause and Gene Expression	215
7.7	Nutrient Accumulation for Diapause and the Storage and Conservation of Nutrients during Diapause	216
7.8	Molecular Studies of Diapause	217
7.9	Review and Self-Study Questions	218
References		219

Chapter 8
Intermediary Metabolism 223

Preview 223

8.1	Introduction: Nutrient Stores—The Fat Body		224
8.2	Energy Demands for Insect Flight		225
8.3	Metabolic Stores		226
	8.3.1 Carbohydrate Resources		226
		8.3.1.1 Trehalose Resources	226
		8.3.1.2 Glycogen: Storage and Synthesis	229

8.4	Hormones Controlling Carbohydrate Metabolism	230
8.5	Pathways of Metabolism Supporting Intense Muscular Activity, Such as Flight	230
	8.5.1 Glycolysis	230
	8.5.1.1 Glycerol-3-Phosphate Shuttle and Regeneration of NAD^+	232
	8.5.1.2 Significance and Control of the Glycerol-3-Phosphate Shuttle	235
	8.5.2 Krebs Cycle	235
	8.5.2.1 Control of Krebs Cycle Metabolism and the Regulation of Carbohydrate Metabolism in Flight Muscles	237
	8.5.3 Electron Transport System	237
	8.5.4 Proline as a Fuel for Flight	240
	8.5.5 Mobilization and the Use of Lipids for Flight Energy	243
	8.5.5.1 Transport of Lipids by Lipophorin	247
	8.5.5.2 Activation of Fatty Acids, Entry into Mitochondria, and β-Oxidation	247
8.6	Review and Self-Study Questions	249
References		249

Chapter 9
Neuroanatomy 255

Preview		255
9.1	Introduction	256
9.2	Central Nervous System	256
9.3	Brain	258
	9.3.1 Protocerebrum	259
	9.3.2 Deutocerebrum	260
	9.3.2.1 Antennal Mechanosensory and Motor Center Neuropil	261
	9.3.2.2 Antennal Lobe	261
	9.3.3 Tritocerebrum	263
9.4	Ventral Ganglia	264
	9.4.1 Abdominal Ganglia	265
	9.4.2 Lateral Nerves	266
9.5	Oxygen and Glucose Supply to the Brain and Ganglia	266
9.6	Neuropil	267
9.7	Hemolymph–Brain (CNS) Barrier	268
9.8	Neurons: Building Blocks of a Nervous System	269
	9.8.1 Afferent or Sensory Neurons	270
	9.8.2 Efferent or Motor Neurons	270
	9.8.3 Interneurons	271
	9.8.4 Glial Cells	272
9.9	Giant Axons in the Insect Central Nervous System	273
9.10	Nervous System Control of Behavior: Motor Programs	274
	9.10.1 Motor Program That Controls Walking	274
	9.10.2 Motor Pattern for Rhythmic Breathing	274
9.11	Neurosecretory Cells and Neurosecretion Products from the CNS	276
	9.11.1 Neurosecretory Cells	276
	9.11.2 Adipokinetic Hormone	278
	9.11.3 Proctolin	278
	9.11.4 FMRFamide-Related Peptides	278
	9.11.5 Tachykinins: Locustatachykinins and Leucokinins	279
	9.11.6 Pigment-Dispersing Factors	279

           9.11.7   Vasopressin-Like Peptide (Locust F2 Peptide) 279
           9.11.8   Allatotropins and Allatostatins 280
           9.11.9   Crustacean Cardioactive Peptide 280
           9.11.10   Pheromone Biosynthesis Activating Neuropeptide 280
Review and Self-Study Questions 281
References 281

Chapter 10
Neurophysiology 285

Preview 285
10.1   Introduction 286
10.2   Nerve Cell Responses to Stimuli 286
           10.2.1   Graded Responses 287
           10.2.2   Spike Potentials 288
10.3   Physiological Basis for Neuronal Responses to Stimuli 289
           10.3.1   Membrane Ion Channels: Bioelectric Potentials 289
           10.3.2   Resting Potential 292
           10.3.3   The Action Potential: Sodium Activation 294
           10.3.4   Sodium Inactivation and Repolarization 296
           10.3.5   Measurement of Ion Fluxes: Voltage Clamp Technique 297
10.4   Conduction of the Action Potential: Local Circuit Theory 298
10.5   Physiology and Biochemistry at the Synapse: Excitatory and Inhibitory Postsynaptic
           Potentials 299
10.6   Acetylcholine-Mediated Synapses 301
           10.6.1   Action of Acetylcholine at the Synapse 301
           10.6.2   Nicotinic and Muscarinic Cholinergic Receptors in Insects 302
           10.6.3   Acetylcholine Receptor Structure 303
10.7   Electric Transmission across Synapses 304
10.8   Neuromuscular Junctions 304
10.9   Review and Self-Study Questions 304
References 305

Chapter 11
Muscles 309

Preview 309
11.1   Introduction 310
11.2   Basic Muscle Structure and Function 310
           11.2.1   Macro- and Microstructure of Muscle 310
           11.2.2   Muscle Attachments to the Exoskeleton 313
           11.2.3   Skeletal Muscle 314
           11.2.4   Polyneuronal Innervation and Multiterminal Nerve Contacts 314
           11.2.5   Transmitter Chemical at Nerve–Muscle Junctions 316
11.3   Synchronous and Asynchronous Muscles 317
11.4   Muscle Proteins and Physiology of Contraction 320
           11.4.1   Active State: Binding of Myosin Heads to Actin and the Sliding of Filaments 322
           11.4.2   Release of Myosin Heads from Actin 322
11.5   Muscles Involved in General Locomotion, Running, and Jumping 323
           11.5.1   Adaptations for Running, Walking, and Survival 323
           11.5.2   Adaptations for Jumping 325

11.6	Sound Production: Tymbal and Stridulatory Muscle	327
	11.6.1 Tymbal Morphology and Physiology	327
	11.6.2 Stridulatory Muscle Physiology	328
11.7	Morphology and Physiology of Nonskeletal Muscle	329
	11.7.1 Visceral Muscles	329
	11.7.2 Heart Muscle	329
	11.7.3 Alary Muscles	330
11.8	Review and Self-Study Questions	330
References		330

Chapter 12
Flight 335

Preview	335
12.1 Introduction	335
12.2 Thoracic Structure, Wing Hinges, and Muscle Groups Involved in Flight	336
12.3 Wing Strokes	338
12.4 Multiple Contractions from Each Volley of Nerve Impulses to Asynchronous Muscles	339
12.5 Flight in Dragonflies and Damselflies	340
12.6 Aerodynamics of Lift and Drag Forces Produced by Wings	341
12.6.1 Lift Forces Generated by Clap and Fling Wing Movements	342
12.6.2 Lift Forces Derived from Drag and Delayed Stall	343
12.7 Hovering Flight	344
12.8 Control of Pitch and Twisting of Wings	346
12.9 Power Output of Flight Muscles	346
12.10 Metabolic Activity of Wing Muscles	347
12.11 Flight Behavior	348
12.12 Review and Self-Study Questions	348
References	349

Chapter 13
Sensory Systems 353

Preview	353
13.1 Introduction	353
13.2 External and Internal Receptors Monitoring the Environment	355
13.3 General Functional Classification of Sensory Receptors	355
13.3.1 Receptors with Multiple Pores	356
13.3.2 Receptors with a Single Pore	357
13.3.3 Receptors without Pores	357
13.4 Mechanoreceptors	357
13.4.1 Structure of a Simple Tactile Hair: A Mechanoreceptor Sensillum	357
13.4.2 Hair Plates	358
13.4.3 Chordotonal Sensilla	358
13.4.4 Subgenual Organs	359
13.4.5 Tympanal Organs: Specialized Organs for Airborne Sounds	360
13.4.6 Johnston's Organ	364
13.4.7 Simple Chordotonal Organs	364
13.4.8 Thermoreceptors and Hygroreceptors	365
13.4.9 Infrared Reception	366

13.5	Chemoreceptors	368
13.5.1	Olfactory Sensilla: Dendritic Fine Structure	368
13.5.2	Contact Chemoreceptors–Gustatory Receptors	369
13.5.3	Specialists versus Generalists among Chemoreceptors	371
13.5.4	Stimulus–Receptor Excitation Coupling	371
13.6	Review and Self-Study Questions	372
References		372

Chapter 14
Vision 377

Preview		377
14.1	Introduction	377
14.2	Compound Eye Structure	379
14.3	Dioptric Structures	381
14.4	Corneal Layering	383
14.5	Retinula Cells	383
14.6	Rhabdomeres	384
14.7	Electrical Activity of Retinula Cells	384
14.8	Neural Connections in the Optic Lobe	384
14.9	Ocelli	385
14.10	Larval Eyes: Stemmata	386
14.11	Dermal Light Sense	387
14.12	Chemistry of Insect Vision	388
14.13	Visual Cascade	389
14.14	Regulation of the Visual Cascade	390
14.15	Color Vision	391
14.16	Vision Is Important in Behavior	394
14.17	Nutritional Need for Carotenoids in Insects	394
14.18	Detection of Plane-Polarized Light	395
14.19	Visual Acuity	397
14.20	Review and Self-Study Questions	400
References		400

Chapter 15
Circulatory System 405

Preview		405
15.1	Introduction: Embryonic Development of the Circulatory System and Hemocytes	405
15.2	Dorsal Vessel: Heart and Aorta	406
15.2.1	Alary Muscles	409
15.2.2	Ostia	410
15.2.3	Heartbeat	410
15.2.4	Ionic Influences on Heartbeat	411
15.2.5	Nerve Supply to the Heart	411
15.2.6	Cardioactive Secretions	412
15.3	Accessory Pulsatile Hearts	413
15.4	Hemocytes	414
15.4.1	Functions of Hemocytes	416
15.4.2	Hemocytopoietic Tissues and Origin of Hemocytes	417
15.4.3	Number of Circulating Hemocytes	418

15.5 Hemolymph 420
    15.5.1 Functions of Hemolymph and Circulation 420
    15.5.2 Hemolymph Volume 422
    15.5.3 Coagulation of Hemolymph 422
    15.5.4 Hemolymph pH and Hemolymph Buffers 424
    15.5.5 Chemical Composition of Hemolymph 425
        15.5.5.1 Inorganic Ions 426
        15.5.5.2 Free Amino Acids 426
        15.5.5.3 Proteins 427
        15.5.5.4 Other Organic Constituents 427
15.6 Rate of Circulation 428
15.7 Hemoglobin 428
15.8 Review and Self-Study Questions 429
References 429

Chapter 16
Immunity 433

Preview 433
16.1 Introduction 433
16.2 Physical Barriers to Invasion 435
16.3 Cellular Immune Reactions 436
16.4 Recognition of Nonself 437
16.5 Synthesis of Antifungal and Antibacterial Peptides 438
16.6 Toll Pathway for the Synthesis of Antimicrobial Peptides 439
16.7 IMD Pathway for the Synthesis of Antimicrobial Peptides 442
16.8 C-Type Lectins 443
16.9 Serpins 443
16.10 Ecology, Behavior, and Immunity 444
16.11 Cost of Defense 445
16.12 Coevolutionary Race between Parasitoid Escape Mechanisms and Host Defense Mechanisms 446
16.13 Autoimmune Consequences of Some Defense Reactions 447
16.14 Gender Differences in Immune Responses 447
16.15 Conclusions 447
16.16 Review and Self-Study Questions 448
References 448

Chapter 17
Respiration 455

Preview 455
17.1 Introduction 455
17.2 Structure of the Tracheal System 457
    17.2.1 Tracheae and Tracheole Structure 457
    17.2.2 Spiracle Structure and Function 459
    17.2.3 Tracheal Epithelium 462
    17.2.4 Development of New Tracheoles 462
    17.2.5 Air Sacs 464
    17.2.6 Molting of Tracheae 464
17.3 Tracheal Supply to Tissues and Organs 464
    17.3.1 Adaptations of Tracheae to Supply Flight Muscles 465

17.4 Ventilation and Diffusion of Gases within the System 466
    17.4.1 Simple Diffusion Is Usually Not Adequate 467
    17.4.2 Active Ventilation of Tracheae 467
    17.4.3 Diffusion from Tracheoles to Mitochondria 471
17.5 Discontinuous Gas Exchange 471
17.6 Water Balance during Flight 475
17.7 Gas Exchange in Aquatic Insects 475
    17.7.1 Compressible Gas Gills 476
    17.7.2 Incompressible Gas Gills: A Plastron 477
    17.7.3 Use of Aquatic Plants as Air Source 477
    17.7.4 Cutaneous Respiration: Closed Tracheal System in Some Aquatic Insects 478
17.8 Respiration in Endoparasitic Insects 481
17.9 Respiratory Pigments 481
17.10 Respiration in Eggs and Developing Embryos 481
17.11 Nonrespiratory Functions of the Tracheal System 482
17.12 Review and Self-Study Questions 483
References 484

Chapter 18
Excretion 489

Preview 489
18.1 Introduction 489
18.2 Malpighian Tubules 490
18.3 Malpighian Tubule Cells 493
18.4 Formation of Primary Urine in Malpighian Tubules 494
18.5 Proton Pump as a Driving Mechanism for Urine Formation and Homeostasis 495
18.6 Selective Reabsorption in the Hindgut 499
    18.6.1 Anatomical Specialization of Hindgut Epithelial Cells 499
    18.6.2 Secretion and Reabsorption in the Ileum 501
    18.6.3 Reabsorption in the Rectum 501
18.7 Role of the Excretory System in Maintaining Homeostasis 502
    18.7.1 Electrolyte Homeostasis 502
    18.7.2 Water Homeostasis 503
        18.7.2.1 Diuretic Hormones 504
        18.7.2.2 Antidiuretic Hormones 506
    18.7.3 Acid–Base Homeostasis 506
    18.7.4 Nitrogen Homeostasis 507
        18.7.4.1 Ammonia Excretion 508
        18.7.4.2 Uric Acid Synthesis and Excretion 509
18.8 Cryptonephridial Systems 511
18.9 Self-Study and Review Questions 513
References 514

Chapter 19
Semiochemicals 523

Preview 523
19.1 Introduction 524
19.2 Classes of Semiochemicals 524
19.3 Importance of the Olfactory Sense in Insects 525

19.4	Active Space Concept	528
19.5	Pheromones Classified according to Behavior Elicited	528
19.6	Pheromone Parsimony	529
19.7	Chemical Characteristics of Semiochemicals	529
19.8	Insect Receptors and the Detection Process	533
	19.8.1 Pheromone-Binding Proteins	533
	19.8.2 Signal Transduction and Receptor Response	535
	19.8.3 Pheromone Inactivation and Clearing of the Receptor	538
	19.8.4 Do Insects Smell the Blend or Just the Major Components?	538
19.9	Information Coding and Processing	539
	19.9.1 Structure of Odor Plumes	539
	19.9.2 Pheromone Signal Processing	541
19.10	Hormonal Control of Pheromone Synthesis and Release	543
	19.10.1 Mode of Action of PBAN	544
19.11	Biosynthesis of Pheromones	545
19.12	Geographical and Population Differences and Evolution of Pheromone Blends	547
19.13	Practical Applications of Pheromones	548
	19.13.1 Mechanisms Operating in Mating Disruption	549
	19.13.1.1 Sensory Fatigue	549
	19.13.1.2 False Trail Following	549
	19.13.1.3 Camouflage of Natural Pheromone Plume	550
	19.13.1.4 Pheromone Antagonists and Imbalanced Blends	550
19.14	Review and Self-Study Questions	550
References		551

Chapter 20
Reproduction 561

Preview		561
20.1	Introduction	562
20.2	Female Reproductive System	562
	20.2.1 Structure of Ovaries	562
	20.2.1.1 Panoistic Ovarioles	564
	20.2.1.2 Telotrophic Ovarioles	564
	20.2.1.3 Polytrophic Ovarioles	565
	20.2.1.4 Oviposition	565
	20.2.2 Nutrients for Oogenesis	566
	20.2.3 Hormonal Regulation of Ovary Development and Synthesis of Egg Proteins	567
20.3	Vitellogenins and Yolk Proteins	572
	20.3.1 Biochemical Characteristics of Vitellogenins and Yolk Proteins	572
	20.3.2 Yolk Proteins of Higher Diptera	573
20.4	Sequestering of Vitellogenins and Yolk Proteins by Oocytes	574
	20.4.1 Patency of Follicular Cells	574
	20.4.2 Egg Proteins Produced by Follicular Cells	576
	20.4.3 Proteins in Addition to Vitellogenin and Yolk Proteins in the Egg	576
20.5	Formation of the Vitelline Membrane	576
20.6	Chorion	576
20.7	Gas Exchange in Eggs	577

20.8 Male Reproductive System 577
    20.8.1 Apyrene and Eupyrene Sperm of Lepidoptera 579
    20.8.2 Male Accessory Glands 580
    20.8.3 Transfer of Sperm 580
20.9 Gender Determination...581
20.10 Review and Self-Study Questions 582
References..583

Chapter 21
Insect Symbioses 591

Preview..591
21.1 Introduction...592
21.2 Symbioses among Leaf-Cutting Ants, Fungi, and Bacteria 593
21.3 Biology of Termites...595
    21.3.1 Symbionts in Termites 596
    21.3.2 Lignocellulose Structure 598
    21.3.3 Nitrogen Metabolism 599
    21.3.4 Fungal Culture 599
21.4 Bark and Ambrosia Beetles and Their Symbionts..599
    21.4.1 Ambrosia Beetles 600
    21.4.2 Bark Beetles 603
    21.4.3 Fungal Role in Supplementing Limited Nutrients in Wood and Phloem 603
    21.4.4 Evolution of Fungal Feeding in Bark Beetles 603
    21.4.5 Bacteria as Part of the Bark Beetle Holobiont 604
    21.4.6 Anthropogenic Effects upon Bark Beetles and Their Symbionts........................605
21.5 *Buchnera* in Aphids 605
21.6 Tsetse Fly Symbionts 606
21.7 *Wolbachia* 607
    21.7.1 Cytoplasmic Incompatibility Inducing the Effect of *Wolbachia* 608
    21.7.2 Parthenogenesis Inducing the Effect of *Wolbachia* 609
    21.7.3 Feminizing Strains of *Wolbachia* 610
21.8 Review and Self-Study Questions 611
References..611

Index 623

Preface

Research work and literature in physiology and biochemistry of insects continues to expand in an explosive and exponential fashion. New journals continue to appear; the literature is very widespread. Experienced and established scientists are likely to find time constraints that make it difficult or impossible to read the general field of insect physiology, and the problem is even more acute for graduate students or undergraduates who are just getting acquainted with the physiology of insects. Even though the Internet and online journals and books offer wonderful advantages, considerable time is still required. My aim in this third edition of *Insect Physiology and Biochemistry* is to produce a book that will be useful to advanced undergraduate and graduate students who may have to take a course in insect physiology, and I hope the book will aid working scientists in a variety of disciplines who conduct research with insects but may have limited time to read general insect physiology. All the topics included in the last edition have been retained, with the addition of recent references to each chapter, more than 500 additional references to those previous chapters. Two new chapters—Biological Rhythms and Insect Symbioses—have been added, with an additional 300+ references, so the book now includes about 2600 references to the literature. Another new feature in this edition is the inclusion of Review and Self-Study Questions at the end of each chapter. I have found these helpful, both in classroom teaching and in a distance education approach with students when I could only communicate with them by e-mail.

I thank Dr. N. Krishnan for suggesting the topic of biological rhythms and for helpful comments on an early version of the chapter and pointing me to some important resources. I am indebted to Dr. Guy Bloch, who read a later version of the chapter and provided very helpful comments. The chapter on insect symbioses was read by Dr. Jiri Hulcr and Dr. Kirsten S. Pelz-Stelinski, both of whom provided very helpful suggestions. I thank again those who read and advised me on the previous editions—Glenn Hall, Marie Nation Becker, Jon Harrison, Tom Miller, and anonymous reviewers.

I am indebted to the many persons who sent photographs to use in the book: Ring Cardé, Herb Oberlander, D.L. Musolin, Jiri Hulcr, M. T. Kasson, Lyle Buss, Andrei Sourakov, Gretert Montano, Marie Nation Becker, Rochelle Nation, Hanife Genc, Maria and Tom Eisner, J. N. Holland, W.B. Hunter, Jerry Butler, Ethel Villabos and Todd Shelly, Ritsuo Nishida, Clay Smith, V. Leclerc and J.-M Reichhart, Alexandra Shapiro, Jimmy Becnel, K. Tomioka, N. Peschel and C. Helfrich-Förster, Thomas Chouvenc, Guy Bloch, D. L. Musolin, C. Hermann, Brian Forschler, Jarrod Scott, Y. Hongoh, M.-Y. Choi, Peter Teal, Paul Shirk, Al Handler, Coby Schal, D.O. Deonier, Robin Giblin-Davis, C. Bordereau, and C.R. Currie.

I thank John Sulzycki, CRC Press, Taylor & Francis Group, for requesting a revision and for offering 24 pages of color illustrations. David Fausel was initially helpful in handling questions and the manuscript. I thank Todd Perry for managing the book project, Gopinath Chandrasekaran for excellent copy editing, and the illustrations editor for arranging the color illustrations. I thank Kathy Milne, who constructed many of the line drawings on a computer. I thank the Department of Entomology and Nematology, University of Florida, Gainesville, for allowing me to hold the title of emeritus professor after retirement and for giving me an office and computer facilities. Finally, I thank my wife, Dorothy, who has been patient as I waded through a thousand research papers and numerous books in order to update this edition.

James L. Nation, Sr.
Professor Emeritus
Entomology and Nematology
University of Florida
Gainesville, Florida

Author

James L. Nation, Sr., PhD, is professor emeritus of entomology at the University of Florida, Gainesville. He holds a BSc in entomology from Mississippi State University, Starkville, and a PhD in entomology from Cornell University, Ithaca, New York. His special interest in entomology is the physiology and biochemistry of insects. Prior to retirement in 2003, he taught and conducted research at the University of Florida for 43 years. Although he sometimes taught other courses, he introduced a graduate-level course and mainly taught graduate-level insect physiology. Research activities included work in insect excretion, pheromones, cuticular hydrocarbons, and insect nutrition. He served as an associate editor of *Florida Entomologist* from 1967 to 1969, as an editor of the *Journal of Chemical Ecology* from 1994 to 2000, and as editor of *Florida Entomologist* from 2004 to 2010. He has continued to teach a course in the Undergraduate Honors Program called *Global Environmental Issues* each fall term since retirement and introduced a graduate-level course in insect physiology at Florida A&M University in Tallahassee, Florida, in 2006 (taught by interactive television). He wrote both the first (2002) and the second (2008) edition of *Insect Physiology and Biochemistry*. He is active in volunteer activities and gardening.

CHAPTER 1

Embryogenesis

PREVIEW

Insect eggs have a central yolk surrounded by a layer of cytoplasm. A proteinaceous chorion put on the egg while it is in the ovary provides a protective covering for the egg. Sperm released from the spermatheca of the female passes through the micropyle, a narrow channel through the chorion, as the egg passes down the oviduct on its way to be deposited in the environment. Usually, the egg nucleus is diploid until the entry of the sperm stimulates meiotic division leading to the haploid egg nucleus. The union of a sperm nucleus with the egg nucleus produces the zygote and stimulates the zygote to begin divisions. Complete cleavage of the zygotic yolk and cytoplasm occurs in eggs of some species during the first few divisions, but yolk cleavage ceases after a few divisions. In most species, cleavage of yolk and cytoplasm is incomplete from the beginning. Ultimately, zygotic divisions in all insect eggs produce large numbers of nuclei lacking cell membranes but each surrounded with a small field of cytoplasm. These nuclei and associated cytoplasm are called energids. Energids gradually migrate into a single layer near the periphery of the egg, forming the blastoderm. Cell membranes become complete after blastoderm formation. A few cells, the pole cells, aggregate at the posterior end of the egg and are the first to become committed to a future developmental track; they will become the gametes of the adult. Cells on the ventral side of the blastoderm enlarge and become committed as the germ band, the cells that will become the embryo. Subsequent development of the germ band is controlled by maternal and zygotic genes. Maternal genes are present and active in the nurse cells of the mother during oogenesis. The mother's nurse cells pass maternal gene transcripts (mRNAs) into the developing oocyte in the ovary, and these begin to function in the zygote. The maternal gene transcripts are translated into proteins in the zygote, and one of the earliest actions of these proteins is control of anterior–posterior and dorsal–ventral axes orientation of the embryo. Later, acting zygotic genes include gap genes that divide the embryo into large domains, pair-rule genes that divide the domains into parasegments, and finally segment polarity genes that control formation of true segments. Homeotic genes begin to function during parasegment formation to give each segment its characteristic identity. Organogenesis leads to formation of the organ systems of the embryo. Insects with complete development retain within the larval body small embryonic clusters of cells called imaginal discs that divide, differentiate, and grow into adult structures during pupation.

1.1 INTRODUCTION

The three major divisions in the Insecta, the Apterygota, Hemimetabola, and Holometabola, are not directly ancestral to each other, and consequently embryological developments in the groups, although similar in some respects, often are divergent. The Apterygota (Protura, Collembola, Diplura, and Thysanura) never evolved wings and lack metamorphosis. The immatures look just like small versions of the adults. The Hemimetabola (Orthoptera, Hemiptera, Heteroptera, and others) evolved wings and have gradual metamorphosis. Immatures have some adult features but lack wings. The Holometabola (Coleoptera, Hymenoptera, Lepidoptera, Diptera, and others) evolved wings and have complete metamorphosis. Immature forms are typically wormlike and thus look very different from the adults. Wingless adults occur in the Hemimetabola and Holometabola, but the wingless condition evolved secondarily from winged forms.

The goals for this chapter are to describe the morphogenetic events and the action of some genes in formation of the embryo. The work by Johannsen and Butt (1941) is still a very valuable source for understanding variations in morphogenesis, as are more recent reviews by Anderson (1972a,b), Jura (1972), Sander et al. (1985), Campos-Ortega and Hartenstein (1985), Panfilio et al. (2006), and Panfilio (2008). A review of the morphology of embryogenesis in the silkworm, *Bombyx mori*, has been provided by Miya (1985), and the early stages of embryogenesis are described for several species of fireflies by Kobayashi and Ando (1985).

More details of genetic control of insect embryogenesis are available for *Drosophila melanogaster* than for any other insect, and a timeline for some of the major morphogenetic events may be helpful (Table 1.1), but one should keep in mind that *Drosophila* is a fast-developing insect, and many other insects do not develop so rapidly. Good reviews of the development and genetics of *Drosophila* are provided by Gehring and Hiromi (1986), Gehring (1987), French (1988),

Table 1.1 Developmental Stages of *Drosophila* Embryogenesis

		Morphological Events (25°C)	Hours[a]
Stage 1	25 min	Cleavage divisions 1 and 2	0:25
Stage 2	40 min	Divisions 3–8 occur	1:05
Stage 3	15 min	Pole bud formation, division 9 occurs	1:20
Stage 4	50 min	Final four divisions, syncytial blastoderm formed, stage 4 ends at beginning of cellularization	2:10
Stage 5	40 min	Cellularization occurs	2:50
Stage 6	10 min	Early stages of gastrulation	3:00
Stage 7	10 min	Gastrulation complete	3:10
Stage 8	30 min	Formation of amnioproctodeal invagination and rapid germ band elongation	3:40
Stage 9	40 min	Transient segmentation of mesodermal layer, stomodeal invagination	4:20
Stage 10	60 min	Stomodeum invaginates, germ band growth continues	5:20
Stage 11	120 min	Growth stage, no major morphogenetic changes, parasegmental furrows develop	7:20
Stage 12	60 min	Germ band shortens	9:20
Stage 13	60 min	Germ band shortening complete, head involution begins	10:20
Stage 14	60 min	Closure of midgut, dorsal closure	11:20
Stage 15	30 min	Gut forms complete tube and encloses yolk sac	13:00
Stage 16	3 h	Intersegmental grooves evident, shortening of ventral nerve cord	16:00
Stage 17		Stage 17 extends to hatching	

Source: Data from Campos-Ortega, J.A. and Hartenstein, V., *The Embryonic Development of Drosophila melanogaster*, Springer-Verlag, Berlin, Germany, 1985, 227pp.
Note: Times and stages probably will be different in other species of insects.
[a] The time is elapsed time after the egg has been laid in hours.

Nüsslein-Volhard (1991), Lawrence (1992), and Bate and Martinez-Arias (1993). Melton (1991) provides a good comparative review of certain aspects of animal development.

1.2 MORPHOGENESIS

1.2.1 Egg, Fertilization, and Zygote Formation

Insect eggs are **centrolecithal**, which means that the eggs have a central yolk surrounded by a layer of cytoplasm. The yolk is a nutrient source to be used by the developing embryo. A **vitelline membrane** surrounds the peripheral cytoplasm (sometimes called the **periplasm**), and a proteinaceous **chorion** provides a protective cover for the egg contents (Figure 1.1). The cytoplasm is a layer of variable thickness in eggs of different groups. In some, there is so little cytoplasm that it is not visually obvious, as for example, in eggs of the Apterygota. The egg nucleus may lie at the periphery of the egg, on top of the yolk and surrounding cytoplasm, or it may lie in the cytoplasm. When an egg is laid, the nucleus usually is still in the diploid state. The entry of sperm as the egg passes down the oviduct of the female often initiates maturation divisions. The first maturation division divides the chromosomes equally, but the nuclear plasma is divided unequally, resulting in a large egg nucleus and a small **polar body** (Figure 1.2). The egg nucleus divides once more to become the haploid female gamete, with production of another small polar body. The first polar body may or may not divide again. If it does divide, two more polar bodies are produced; in any case, polar bodies eventually are reabsorbed into the yolk. The haploid female nucleus usually migrates toward the center of the egg and unites on the way with the sperm nucleus; the developing organism is then called the **zygote**.

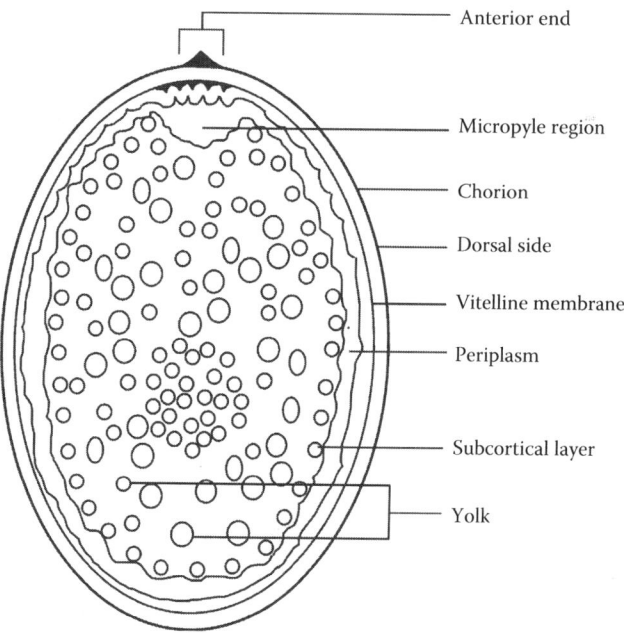

Figure 1.1 Diagram of egg structure.

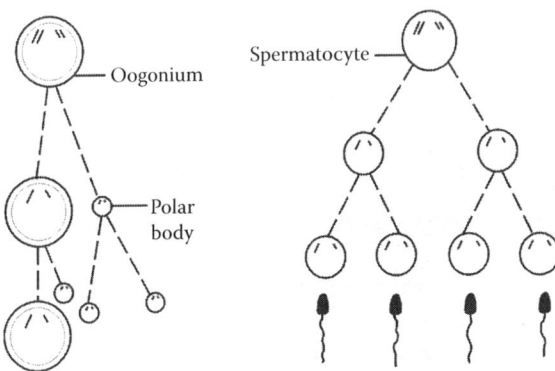

Figure 1.2 Maturation divisions of oocyte and sperm. Oogonia in the germarial region of an ovary divide by meiosis to produce an oocyte and a polar body. A second meiotic division, which may not occur until the oocyte is united with the sperm, produces the final oocyte. The polar bodies are reabsorbed as food for the developing oocyte. Spermatocytes in the germarium of the testes give rise to mature spermatozoa by meiotic divisions. The union of a sperm and egg produces the zygote.

1.2.2 Variations in Zygotic Nucleus Cleavage, Formation of Energids, and Blastoderm Formation

Zygotic nucleus divisions are influenced by the quantity of yolk and cytoplasm. The division in eggs with little yolk, such as in the collembolan *Tetrodontophora bielanensis* (Apterygota), partitions the yolk in a few early divisions (Figure 1.3), but not after the eight-cell stage. In the great majority of insect groups, the zygotic nuclei divide from the beginning without cleavage of the yolk and without formation of cell membranes between nuclei. Repeated nuclear divisions produce

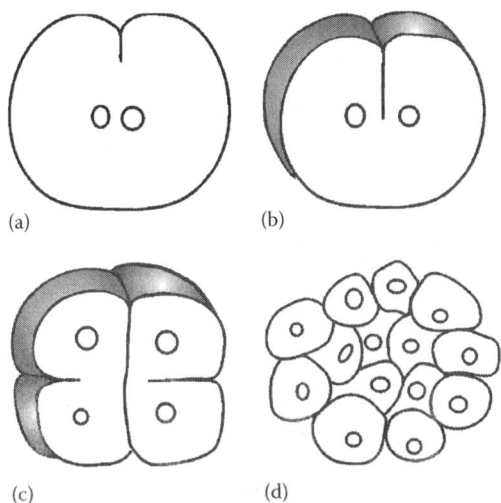

Figure 1.3 The first few cleavages of the yolk may be complete, as in some Collembola, but complete cleavage ceases after a few divisions. (a) The first division is depicted as beginning after the nucleus has divided by mitosis; (b) division into two cells is illustrated; (c) cleavage into four cells is underway, and the four may divide into eight cells, after which, the yolk usually is not cleaved equally with subsequent nuclear divisions; (d) a ball of cells has formed with yolk that has not been partitioned accumulated in the center (not shown) of the mass of cells. (Stages modified from Jura, C., Development of apterygote insects, in J. Counce and C.H. Waddington (eds.), *Developmental Systems: Insects*, vol. 1, Academic Press, New York, 1972, pp. 49–94.)

EMBRYOGENESIS

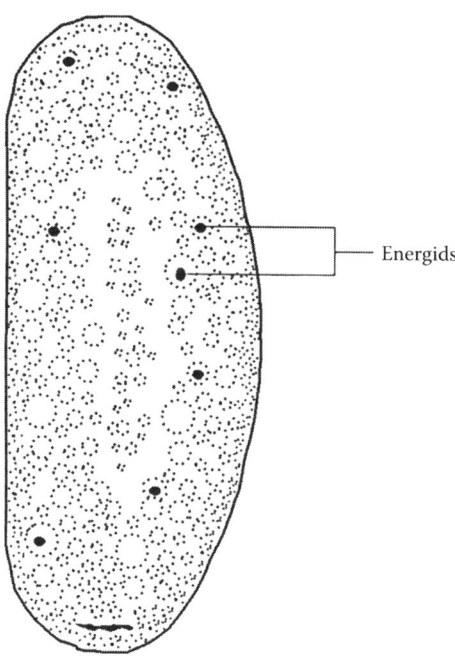

Figure 1.4 An example of an egg in which the yolk is not partitioned and cleavage nuclei (energids) are produced and surrounded by a small amount of cytoplasm. Yolk remains in the interior of the egg. (Redrawn and modified from Johannsen, O.A. and Butt, F.H., *Embryology of Insects and Myriapods*, McGraw-Hill, New York, 1941.)

thousands of nuclei, each surrounded by a small island of cytoplasm. Each nucleus with its island of cytoplasm is called an **energid** (Figure 1.4). Energids migrate toward the periphery, and when a few thousand nuclei have been formed, they distribute themselves in a single layer around the perimeter. Some energids remain in the yolk and become **vitellophages** that digest (liquify) the yolk and make the nutrients available to the developing embryo. Cytoplasmic strands extend from the **blastomeres**, as the energids are now usually called, into the yolk as a route for nutrient uptake. Eventually, cell membranes become complete, the cytoplasmic strands disappear, and the layer of cells is called the **blastoderm** (Figure 1.5). There are numerous differences in the way the blastoderm forms and in subsequent morphogenetic movements among the different groups of insects. A brief summary of major differences is given in the following text; the reviews and reference works cited in the introduction should be consulted if more details about specific groups are desired.

1.2.2.1 Apterygota

Apterygotes are small, wingless insects with ametabolous development (no metamorphosis and no major changes in morphology between immatures and adults) and include the orders Protura (small insects in soil and leaf litter), Collembola (commonly called springtails), Diplura (called bristletails), Archaeognatha (also called bristletails), and Thysanura (some bristletails, silverfish, and firebrats) (Romoser and Stoffolano, 1998). Details and variations in development of the Apterygota have been reviewed by Jura (1972). Even in the Apterygota, the processes of division and cleavage are not the same in all members of the group. In some, the yolk is cleaved at each division, but in others, nuclear division occurs without yolk cleavage. Division continues to make many small blastomeres that move toward the periphery of the egg and gradually align themselves in a single layer around the perimeter of the egg to form the blastoderm. At one pole of the blastoderm, blastomeres

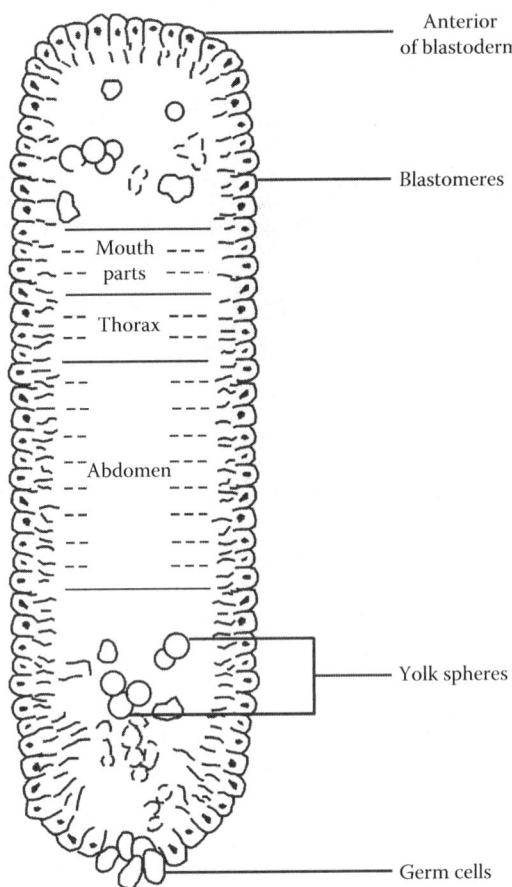

Figure 1.5 An illustration of the blastoderm stage in development. Energids gradually migrate to the periphery to form a single layer of blastomeres around the periphery of the egg. Cell membranes are incomplete in an early blastoderm, and cytoplasmic strands from blastomeres extend into the yolk, but later the cell membranes become complete and junctions develop between cells to hold them together. The dotted lines across the blastoderm indicate diagrammatically a fate map based on the development of *Drosophila melanogaster*, which has a determinate-type egg. Even at this early stage, blastoderm cells in *Drosophila* are known (from marking experiments) to be committed to a specific path of development. The broken lines indicate regions developing into mouth parts, thorax, or abdomen. This first evidence of developmental commitment marks the parasegments, developmental units within which genetic action leads to the final segmentation pattern of the larva and adult.

become increasingly columnar as the **dorsal organ** forms. At the other pole (usually the ventral side), they are smaller but represent the cells that will form the future embryonic rudiment (= germ band) and embryonic membranes (Figure 1.6).

The exact function of the **dorsal organ** is not clear; it may be secretory. If the cells that form it are damaged or destroyed, the embryo does not develop normally and does not hatch. The dorsal organ cells invaginate into the underlying yolk and take the shape of a mushroom with a stalk, and tendrils grow out and contact the developing germ band (the embryo) after gastrulation has occurred. A dorsal organ does not develop in a recognizable form in Hemi- and Holometabola.

In *Japyx solifugus* (Diplura), cleavage is superficial from the start. The blastomeres migrate toward the periphery of the egg at about the 64-cell stage, and after additional divisions, the blastoderm is formed. A dorsal organ is present and behaves much like that in Collembola. Cleavages of the zygote of Thysanura (silverfish) are superficial, and the yolk is not cleaved. Synchrony is

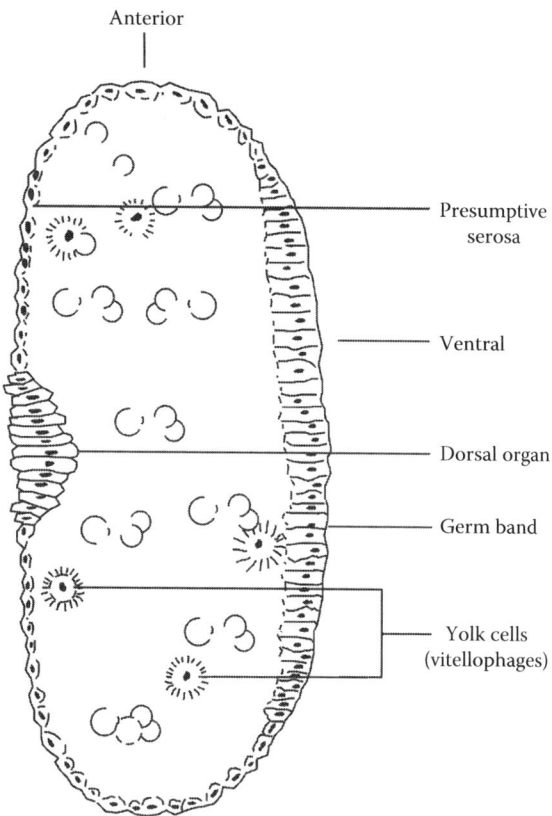

Figure 1.6 A late blastoderm stage with germ band formed on ventral side of egg and dorsal organ on the dorsal side (Modified from Johannsen, O.A. and Butt, F.H., *Embryology of Insects and Myriapods*, McGraw-Hill, New York, 1941). The germ band will subsequently grow into the embryo. The function of the dorsal organ, not present in the blastoderm of some species, is not known in detail.

lost after a few divisions. Some cleavage nuclei migrate to the periphery, while some remain in the yolk, functioning as yolk nuclei, later to become vitellophages. The germ anlage or germ band forms at the posterior pole of the blastoderm. A few blastomeres at the anterior pole of the blastoderm may form a dorsal organ, but some workers have questioned whether a dorsal organ is present. After gastrulation, a part of the serosal membrane sinks into the yolk to form a secondary dorsal organ, but it soon degenerates and its function is unknown. The yolk cells in Thysanura are true vitellophages that digest the yolk; some yolk cells later disintegrate and contribute to the formation of the midgut epithelium. The blastoderm stage exists only briefly in Apterygota and is followed by gastrulation.

1.2.2.2 Hemimetabola

Embryogenesis of the Hemimetabola has been reviewed by Anderson (1972a). In general, eggs of hemimetabolous insects develop slowly, taking weeks or months to hatch. When the egg is released from the ovary, the oocyte is in metaphase of the first maturation division. The nucleus with a small amount of cytoplasm lies at the periphery of the egg, where it stays as maturation divisions produce three polar nuclei and one haploid female pronucleus. The female pronucleus migrates to the interior, while the polar nuclei stay at the periphery. The union with the male pronucleus occurs near the middle of the egg. The three polar nuclei and any unsuccessful male pronuclei are reabsorbed during early cleavage. The eggs are relatively rich in yolk. The zygotic nucleus undergoes divisions without yolk cleavage, and energids are formed. Division is usually synchronous until the blastoderm is formed. The rate of division varies a great deal among the Hemimetabola, but none divides as fast as in the Holometabola. Energids gradually move toward the periphery of the egg and form the blastoderm. The number of nuclear divisions and the number of energids that form the blastoderm vary with species.

Energids that remain in the yolk mass become primary **vitellophages** and continue to divide and produce more vitellophages. In Dictyoptera (cockroaches), Plecoptera (stone flies), and Gryllotalpidae (mole crickets), there are no primary vitellophages, but some secondary ones develop from energids that migrate from the periphery back into the yolk. The final position of the germ band is ventral and usually posterior, but the position along the anterior/posterior axis is somewhat variable in different species. The germ band may lie on the surface of the yolk mass, or it may grow into the interior of the yolk in some groups.

1.2.2.3 Holometabola

The eggs of the Holometabola have only a small amount of yolk and a relatively large peripheral periplasm (= cytoplasm). The outermost part of the periplasm is called the egg cortex. Eggs typically are small, 1 mm or less in length. Eggs of Lepidoptera tend to be round to ovoid in shape, while those of Diptera and Hymenoptera are usually elongated. Typically the egg is in the metaphase of the first maturation division when released from the ovary. Maturation division results in three polar nuclei and the female pronucleus, which migrates into the interior of the egg. The yolk is not cleaved and zygotic nuclear divisions produce energids. Divisions are typically synchronous through 8 to 10 or even more divisions, but synchrony is lost in various Holometabola at different times after about the eighth division. The rate of division is also variable, with higher Diptera (the Cyclorrhapha) having the fastest division rate. The number of cells in the blastoderm varies in the Holometabola from about 500 to 8000. Some of the blastoderm cells may migrate back into the yolk as secondary vitellophages. Although the nuclei of the vitellophages cease dividing and remain in the central yolk region, their DNA replicates and they become polyploid. The pole buds and syncytial blastoderm nuclei continue to divide independently of each other. The zygote usually lies centrally but may be displaced toward either end. Cell division and growth of the embryo are rapid, and eggs usually hatch in a few days in most cases.

In *D. melanogaster*, the morphogenetic events, as well as their genetic control, have been extensively studied, and there typically are 13 synchronous division cycles before cell boundaries are established between nuclei. After the first seven synchronous divisions, there are 128 nuclei arranged in an ellipsoid shape around the central yolk (Zalokar and Erk, 1976). Most of the nuclei begin to migrate to the periphery of the embryo, but about 26 nuclei stay near the yolk in the center of the egg after the seventh division and become vitellophages. The vitellophages and all the other nuclei undergo the eighth nuclear division together. At this time, the first cells become determined (committed to a particular developmental fate), and a few nuclei are incorporated into the posterior pole plasma to become the polar buds or future germ cells. These will give rise to the gametes (reproductive cells) of the adult insect. The remaining energids are destined to become somatic cells of the embryo.

The vitellophages, polar buds, and somatic nuclei divide in synchrony two more times, making a total of 10 divisions for the somatic nuclei. In the 8th, 9th, and 10th divisions, the somatic nuclei progressively move toward the surface of the egg, forming a single layer of nuclei around the perimeter of the egg, the syncytial blastoderm (syncytial denotes the lack of cell boundaries) (Foe and Alberts, 1983).

After 13 divisions of the somatic nuclei during the first 3 h of embryo life in *D. melanogaster*, there are about 5000 syncytial blastoderm nuclei layered around the periphery of the egg (Chan and

Gehring, 1971). Cell membranes begin to form, and **desmosomes** form between cells to hold them together (Mahowald, 1963). The cytoplasmic strands that reach into the yolk gradually disappear as cell membranes are completed. The time at which cells become committed to the formation of specific structures is variable in different insects, but in *D. melanogaster*, formerly totipotent energids become determined in the blastoderm stage, after which they can only develop into certain body segments (Simcox and Sang, 1983; Gehring, 1987). The ultimate development of blastoderm cells in *Drosophila* has been ascertained by marking cells to note their future fate, and the representation of the commitment of blastoderm cells as done diagrammatically in Figure 1.5 is called a fate map (Campos-Ortega and Hartenstein, 1985).

1.2.3 Formation of the Germ Band

Initially, the cells of the blastula are uniform in size and shape, but along the ventral side of the blastula, the cells rapidly thicken and enlarge into the **germ band**, the cells destined to give rise to the embryo. In *D. melanogaster* and other Diptera with large amounts of cytoplasm, nearly the entire cell number of the blastoderm becomes the germ band, leaving only a few cells to build the extraembryonic membranes. In other insects, a variable number of cells along the ventral side of the blastoderm enlarge and become more columnar in shape, while lateral and dorsal to the ventral region, the cells become more flattened and squamous and are destined to form the **extraembryonic membranes** called the **amnion** and **serosa**.

The initial size of the germ band varies in different groups of insects, and three major types occur, characterized as **short germ band**, **long germ band**, and **intermediate germ band** (reviewed by Sander et al., 1985). Short germ band eggs tend to be indeterminate, large eggs from panoistic ovaries, contain a large yolk with relatively little cytoplasm, have a relatively small portion of the blastoderm that becomes the germ band, and develop slowly over days, weeks, or months. Long germ band eggs usually are smaller eggs that come from meroistic ovaries. They tend to have a relatively large amount of cytoplasm and a small amount of yolk. The germ band initially covers a large portion of the blastoderm, and development to hatching is rapid, often hours to days. Long germ band eggs tend to be determinate. **Indeterminate** and **determinate** refer to how soon the blastoderm cells become committed to a specific developmental fate. In determinate eggs, the blastoderm cells become committed very early to a specific developmental pathway. Regardless of the size of the germ band initially, elongation and growth occur as development continues. Short germ band eggs tend to be characteristic of the Orthoptera and Odonata (insects with panoistic ovaries), while long germ band eggs tend to be produced by Lepidoptera, Coleoptera, Diptera, and Hymenoptera. However, there are some groups that do not fit easily into one category, so the correlation between taxon and egg type is not strong (Sander et al., 1985). Evolutionary selection may have led to long germ band eggs as an adaptation to use rapidly decaying vegetable, fruit, or dead animal hosts, as well as reduced exposure of a relatively immobile stage (Sander et al., 1985). Some insects have an intermediate germ band egg in which segmentation in the gnathocephalon and thorax occurs relatively rapidly, but the abdominal portion of the germ band grows slowly and segmentation takes longer to occur. Eggs of the cricket *Acheta domesticus* are intermediate germ band type (Sander et al., 1985).

1.2.4 Gastrulation

During **gastrulation**, part of the germ band sinks into the ball of blastoderm cells (Figure 1.7a), and germ layers are formed that will give rise to different organs and tissues. Immediately after gastrulation occurs, the structure is sometimes referred to as a **gastrula**. Gastrulation is highly variable and so different in some insects from the process in other animals that some embryologists have questioned whether the events occurring are really gastrulation in the classic sense.

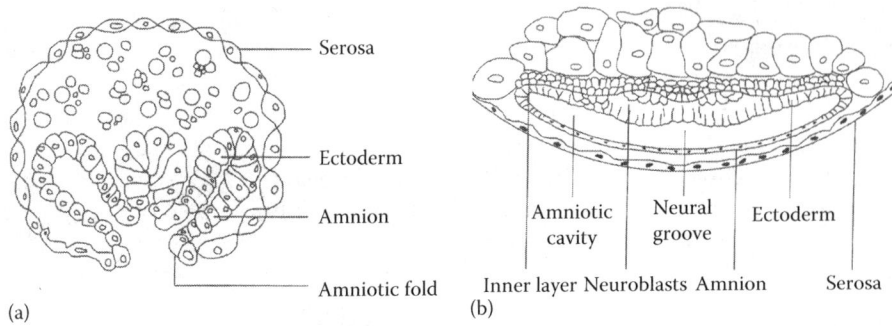

Figure 1.7 (a) Schematic representation of an early stage in gastrulation during which the germ band invaginates. (b) A later stage in which germ layers have formed. (Redrawn and modified from Johannsen, O.A. and Butt, F.H., *Embryology of Insects and Myriapods*, McGraw-Hill, New York, 1941.)

Deep invagination of the germ band, so characteristic of many other types of organisms, does not occur in insects (Johannsen and Butt, 1941). In other organisms, gastrulation results in an outer layer (the ectoderm), an inner layer (the endoderm), and a middle layer (the mesoderm). Typically in insects at the end of gastrulation, there is an outer **ectodermal** layer of cells and an inner layer of cells termed the **mesendodermal** layer (Figure 1.7b). Most of the controversy about gastrulation has been focused on the formation, or some argue the lack of formation, of a classic **endoderm** (Johannsen and Butt, 1941). The only structure formed from the endoderm, providing it is accepted as a germ layer, is the midgut. The outer layer of **ectoderm** gives rise to the nervous system, the tracheal system, the fore- and hindgut, and the integument. Formation of the mesendoderm layer varies with different insect groups. In Coleoptera, invaginated cells along the ventral midline fold into a tube that subsequently unfolds to become an irregular inner layer of cells. In honeybees, *Apis mellifera*, a ventral plate of cells sinks inward and is overgrown by the remaining lateral plates of germ band, and a somewhat similar formation of an inner layer of cells occurs in Orthoptera. The **mesendoderm** gives rise to muscles, circulatory system, fat body, hemocytes, and the midgut (but see the note on controversy about separate endoderm and origin of midgut [Johannsen and Butt, 1941]).

During gastrulation, the ectoderm and mesendoderm become overgrown by some of the remaining surface cells, eventually enclosing the developing embryo, an **amnionic cavity**, and remaining yolk. The layer of squamous cells lining the ventral portion of the amnionic cavity is called the **amnion**, and a thin layer of cells on the outside of the **gastrula** becomes the **serosa**, the two extra-embryonic membranes of the embryo (Figure 1.7b). In *Drosophila*, the two membranes fuse into the **amnioserosa**.

1.2.5 Germ Band Elongation

The germ band grows and elongates in all insects regardless of its size initially. The anterior part of the germ band, the **protocephalon**, includes the antennal segments, intercalary segments (which give rise to the tritocerebrum and parts of the head capsule), and three gnathal segments (primordia of the mandible, maxilla, and labium). The protocephalon is bilaterally widened at the anterior end (like a double-headed hammer), with a "fingerlike" tail. In *D. melanogaster*, the posterior tail of the germ band, as well as the procephalon, is fully formed at the blastoderm stage, and segmentation can proceed at once, usually occurring within hours of formation of the blastoderm and while gastrulation is in progress.

The tail portion of the germ band grows at variable rates in different groups. Dorsoventral (DV) furrows rapidly appear behind the protocephalon in *Drosophila* embryos as segmentation is initiated. These first segments are called **parasegments**, and they are slightly out of register with the final segmentation pattern that will develop. Nevertheless, they represent the first evidence of **metamerization** in the embryo.

Six segments fuse into the head. The thorax consists of three segments. The number of abdominal segments is variable in different insects, but 11, or 12 if the terminal telson is counted, is the primitive number. In rapidly developing Holometabola, body appendages soon appear as bilateral evaginations or small cellular buds from the ectoderm. Buds from the protocephalon form the antennae and mouth parts, and buds from the thoracic segments give rise to the legs and wings. Bilateral outgrowths of buds appear on the abdominal segments but are later reabsorbed in segments 1–7 and 10. In some insects, abdominal buds on segments 8 and 9 continue to develop into the external genitalia, and those on segment 11 form cerci. The exact form that the abdominal limb buds might take, if they were not reabsorbed, is unknown, but some have interpreted them as gill flaps in an ancient insect ancestor (Wigglesworth, 1972).

1.2.6 Blastokinesis and Extraembryonic Membranes

Blastokinesis refers to movements and rotations of the embryo, processes that are variable in different insect groups. Sometimes, blastokinesis is divided into two phases, **anatrepsis** and **katatrepsis**. Various degrees of anatrepsis and katatrepsis occur in different insects. Embryological development in *D. melanogaster* has been well studied, but it is not typical of most other insects. In *Drosophila*, the gene *zerknüllt* (*zen*) specifies a single extraembryonic membrane, the **amnioserosa**. Zen encodes a homeobox transcription factor and is one of the genes influenced by maternal and zygotic DV morphogen gradients. Maternal NF-kB/Dorsal protein gradient represses *zen* in the ventral half of the embryo, resulting in *zen* transcription to a broad dorsal region.

Most of the Hemimetabola insects, those with incomplete metamorphosis, have eggs with a large amount of central yolk, and blastokinesis is divided into two phases, anatrepsis and katatrepsis, with two extraembryonic membranes being formed, the **serosa** and the **amnion**. Panfilio et al. (2006) and Panfilio (2008) described blastokinesis in the milkweed bug, *Oncopeltus fasciatus*, as an example for many insects, particularly the Hemimetabola. *Oncopeltus* has a short germ band, in which a relatively small portion of the single layer of cells in the blastoderm gives rise to the embryo. Blastokinesis means, as the term implies, that movements of the blastoderm cells take place. Briefly stated, the embryo first invaginates into the large amount of yolk typical of eggs of many Hemimetabola, grows in length, and then emerges from the yolk and comes to rest on the ventral side of the egg where subsequent development occurs. The following description of blastokinesis in *O. fasciatus* is based on Panfilio's (2008) review (Figures 1.8 and 1.9).

At about 22% into the embryological development, the first blastokinetic movement, **anatrepsis**, occurs as the germ rudiment invaginates into (becomes immersed in) the yolk at the posterior pole of the egg, illustrated in Figure 1.8a. Panfilio (2008) describes the process like a "sock being pulled through its own opening." It is important to note that the part of the germ rudiment that will eventually give rise to the head of the insect remains oriented near the posterior pole of the egg (see Figures 1.8b and 1.9a and b). Some blastoderm cells separate from the germ rudiment as it invaginates and remain on the surface to become the **serosa**, an extraembryonic membrane that spreads to fully enclose the yolk and the germ rudiment. While immersed in the yolk, the germ rudiment differentiates into the embryo proper and a second extraembryonic membrane, the **amnion** that covers the ventral side of the embryo. The embryo is upside down at this time with its ventral surface

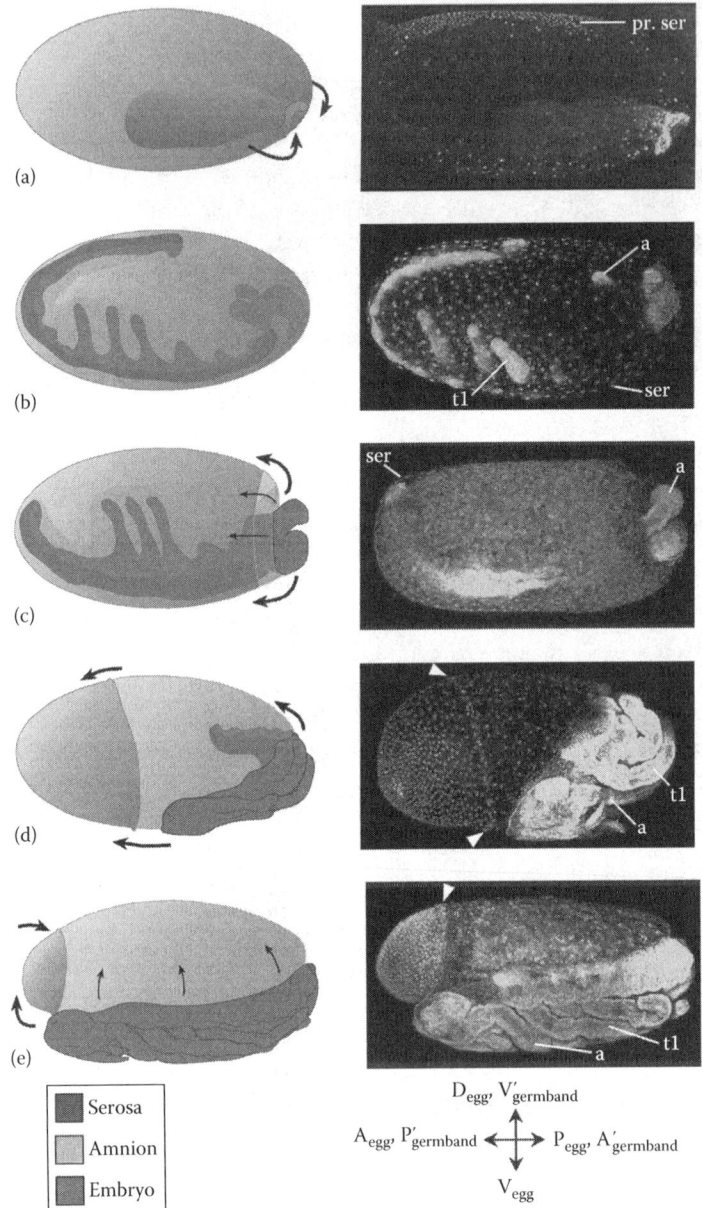

Figure 1.8 **(See color insert.)** Blastokinesis during *Oncopeltus fasciatus* development illustrated in drawings at the left and light micrographs of nuclear staining of dechorionated eggs on the right. (a) Immersion anatrepsis with the embryo invaginating into the yolk; (b) extended germ band stage (see Figure 1.9a for labeling); (c) early katatrepsis with head region emerging from the yolk; (d) midkatatrepsis, with embryo about half out of the yolk; (e) late katatrepsis with embryo fully out of the yolk and now lying on the ventral side of the egg. The serosa was mostly removed in the light micrograph (a) so that the yolk nuclei and germ rudiment are visible. The embryo is partially visible in micrograph panels (b) and (c) in places where it is not so deep in the yolk. Egg orientation is anterior to the left and dorsal up. Color-coding: serosa (blue); amnion (orange); embryo (gray). Black arrows in the drawings indicate the direction of movement; white arrowheads in the micrographs indicate the amnion–serosa boundary. *Abbreviations*: A, P, D, and V denote anterior, posterior, dorsal, and ventral axes of the egg, respectively; A′, P′, and V′ are sides of the germ band stage embryo; a, antennae; pr. ser, presumptive serosa; t1, thoracic segment/leg 1. (From Panfilio, K.A., *Dev. Biol.*, 313, 471, 2008. With permission.)

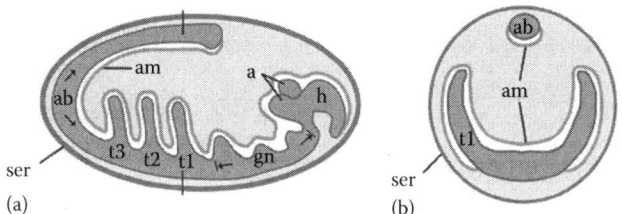

Figure 1.9 (See color insert.) Illustrations of a germ band stage embryo modeled on the milkweed bug *Oncopeltus fasciatus* with the extraembryonic membranes of serosa (blue) and amnion (orange), embryo (gray), and yolk (blue). The amnionic cavity (white) lies between the amnion and the ventral surface of the embryo. (a) Midsagittal section view. (b) Transverse section view approximately across the middle of (a) showing that the germ band has spread laterally over the ventral side of the egg. The portion that will become the abdomen extends around the anterior end of the egg (behind the plane of the paper), and just the tip shows on the dorsal side of the egg in this cross section. The posterior pole of the egg is facing the viewer in the cross section. The anterior end of the egg is at the left in the sagittal section, and egg dorsal is up in both views. *Abbreviations*: a, antennae; am, amnion; ab, abdominal region; gn, gnathal (mouthpart) region; h, head; ser, serosa; t1–3, thoracic segments/legs 1–3. The sagittal section is a labeled view of Figure 1.8b. (From Panfilio, K.A., *Dev. Biol.*, 313, 471, 2008. With permission.)

directed upward toward the dorsal side of the egg and the cells that will become the head oriented toward the posterior pole of the egg (Figures 1.8b and 1.9a). The embryo stays in this orientation for an additional 25% of the development time as it grows in length, undergoes some segmentation, and beginning development of appendages.

About halfway through embryological development of the milkweed bug embryo, the second blastokinetic movement, **katatrepsis**, begins as the germ rudiment starts to evert or turn outward. The amnion turns inside out and begins to fuse with the serosa over the future head of the embryo, but the two subsequently rupture as the embryo continues to evert and emerge from the yolk, resulting in the freeing of the head and antennal region from the membranes as it pushes out of the yolk (Figure 1.8c). The serosa contracts toward the anterior pole of the egg, contracting to a small anterior-pole cap, and eventually, when katatrepsis is over, forming the dorsal organ and then disappearing completely. The internal amnion and embryo continue to emerge from the yolk, with the head out first, then the thorax, and last the abdominal region. As these actions occur, the emerging embryo bends backward over the posterior pole of the egg (somewhat like a "U" lying to the left on its side, as in Figure 1.8d as it emerges from the yolk) and migrates toward the ventral surface of the egg, coming to rest with the head oriented toward the anterior pole of the egg and with the ventral surface of the developing embryo oriented to the ventral surface of the egg (Figure 1.8e). The embryo grows dorsally and laterally until it replaces the amnion and dorsal closure occurs forming a cavity with all the yolk confined within the embryo. The embryonic gut forms around the yolk and organogenesis continues.

Blastokinesis varies widely among insect groups, and the two extraembryonic membranes are not always distinct (Panfilio, 2008). Anatrepsis and katatrepsis are especially prominent in those insects without a pupal stage, the Hemimetabola, that have gradual metamorphosis, although variations exist in these groups. Some Hemimetabola, specifically oviparous Blattaria, Mantodea, and some species of Hemiptera (Hemiptera are now classified as Heteroptera) lack anatrepsis and katatrepsis or have no definable blastokinesis movements. The blastokinetic events in most Holometabola insects, those with a pupal stage, are typically reduced, although some do have distinct movements and extraembryonic membranes. For example, the red flour beetle, *Tribolium castaneum*, has complete development, like *Drosophila*, but it has a short-germ-band-type development in which it produces the main body segments in succession—head first, then thoracic segments, and finally abdominal segments, and it has well-developed extraembryonic membranes, the amnion and serosa (Handel et al., 2000). *Tribolium* has two *zen* homologs—*Tc-zen1* and *Tc-zen2* (*Tc* in the name comes from the first letters of the scientific name, *T. castaneum*, of the beetle). The serosa is specified by *Tc-zen1* in early development. The amnion covers the embryo at the dorsal side, and later *Tc-zen2* promotes fusion of the amnion and serosa required for dorsal closure. Certain Lepidoptera (some moths and butterflies) also have well-defined anatrepsis and katatrepsis. Dorsal closure is a characteristic of all insects.

The review of Panfilio (2008) should be consulted for more detailed descriptions of variations in blastokinesis in a number of insect groups.

A number of possible functions have been suggested for the serosa and amnion, but few of the suggestions have been substantiated with experimental evidence. Suggested functions for the serosa and/or amnion listed by Panfilio (2008) are as follows:

1. General protective function.
2. Cushioning of embryo from physical shock.
3. The serosa may be a source of innate immune response.
4. Detoxification of environmental poisons.
5. Desiccation tolerance due to formation of a serosal cuticle, and in some insects, a cuticle formed by the amnion.
6. Possible role in water regulation in the embryo.
7. Aid in yolk metabolism.
8. Sequestering of metabolic wastes within the amniotic cavity or space external to serosa.
9. Aminoserosa in *Drosophila* is required for germ band elongation, retraction, and dorsal closure (cited in Goltsev et al., 2009).

The main evidence for the possible role of the serosa in forming a protective cuticle, in water regulation, and in aiding in mounting an immune response to challenge comes from work with mosquitoes and the red flour beetle. Mosquitoes are in the order Diptera, but they represent a different branch from *Drosophila*, and they typically show distinct anatrepsis and katatrepsis. Many species of mosquitoes are disease vectors for humans and other mammals, lending impetus to detailed studies. *Aedes aegypti* is one of the main vectors of dengue, a serious debilitating and painful disease that humans can contract from the bite of the female mosquito. According to Rezende et al. (2008), embryos of *A. aegypti* are capable of surviving for several months under arid conditions. The developing eggs form a serosal cuticle, and the onset of desiccation resistance is correlated in time with serosal cuticle formation, leading the authors to suggest that the serosal cuticle is important in desiccation resistance of the eggs. They confirmed that the serosal cuticle contained chitin, a component of the adult cuticle that is important in the adult as a water barrier.

Goltsev et al. (2009) also found that the serosa secreted an embryological cuticle in *Anopheles gambiae*, a serious vector of malaria in sub-Saharan Africa, and that the eggs show desiccation resistance during the dry months when the mosquitoes continue to actively transmit malaria in dry areas and dry months. Goltsev et al. (2009) suggest that the serosal cuticle produced by the serosa is mainly responsible for desiccation resistance of the early embryo. Resistance is not obtained until 14 h after fertilization, and it is at this time that the serosal cuticle is formed. They also observed upregulation of an aquaporin gene that enables the early embryo to retain water more effectively.

Vargas et al. (2014) compared egg desiccation resistance in *A. aegypti*, *Anopheles aquasalis*, and *Culex quinquefasciatus*, all disease vectors, and found that desiccation resistance occurred concomitant with formation of the serosal cuticle. Although perhaps not definitive simply because desiccation resistance in eggs shows up with serosal cuticle formation, the occurrence is suggestive of a function for the serosa in resisting drying of the embryo in these mosquitoes.

In *T. castaneum*, the red flour beetle, there are two homologs of the gene *zerknüllt* (*zen*), *Tc-zen1* and *Tc-zen2* (van der Zee et al., 2005). The serosa is specified by *Tc-zen1* in early development.

The amnion covers the embryo at the dorsal side, and later *Tc-zen2* promotes fusion of the amnion and serosa required for dorsal closure. The use of *Tc-zen2* RNAi results in a failure of normal dorsal closure, and the embryo closes ventrally with an everted (inside out) form in which appendages start to develop inside the embryo and internal organs start to develop outside. Jacobs and van der Zee (2013) used parental *Tc-zerknüllt1* (*Tc-zen1*) RNAi to suppress serosa formation in embryos of the beetles with the results showing that only the amnion formed in the treated embryos, covering the yolk dorsally but not enveloping the embryo. When normal embryos and those treated with RNAi were challenged with both gram-positive and gram-negative bacteria, the serosa-deficient embryos could not upregulate any of a number of tested immune genes, while normal embryos upregulated immune-deficient genes, attacins, cecropins, and defensins after challenge. The authors concluded that the serosa had an immune response function.

1.3 GENETIC CONTROL OF EMBRYOGENESIS

What causes cells to differentiate and become committed to one pathway as opposed to another? A simple answer to this question is not possible, but two principal mechanisms have been identified for determining cell fate during development. One mechanism is **cell-to-cell interaction** in which one cell influences or induces its neighbor to follow a certain developmental pathway. A second mechanism is that of **regional localization** of molecules that provides information to nuclei or cells in contact with the molecule(s) as to a pathway for development. A molecule whose concentration influences a local pattern of determination is called a **morphogen** (Slack, 1987). These two systems influencing development are not mutually exclusive and both seem to work in many systems. Relatively few genes in organisms as diverse as invertebrates and vertebrates may specify cell fates during development (Melton, 1991). Apparent differences in development appear to be much more similar at the molecular and genetic level than the phenotypic and organismal level might suggest.

A very large amount of information is now available about genetic control of development in *D. melanogaster* (Bate and Martinez-Arias, 1993). The examples and explanations of genetic control to follow are based upon *Drosophila* unless otherwise stated. Although there are important implications for all organisms in the large body of genetic information developed from studies of *Drosophila*, it should not be considered representative of all insects. Many genes expressed in the embryo as mutated genes result in death of the embryo prior to hatching, but it is still possible in many cases to determine how the embryo develops differently from normal ones up to the point of death, and thus identify genes that have specific functions.

More than 70 genes are involved in embryonic development of *D. melanogaster*, and most of them have been characterized. They usually are classified broadly into (1) **maternal genes** and (2) **zygotic genes**. Maternal genes are present in nurse cells of the maternal ovary, and gene products (tRNA and mRNA) are transferred to the oocyte by nurse cells while it is developing in the ovary. These gene products begin to function in the oocyte during its growth in the ovary, and some continue to function in the egg for several hours after the egg is laid. Zygotic genes begin to function in the zygote, and some maternal and some zygotic genes function simultaneously and interactively. The early functioning zygotic genes are divided into **segmentation genes** and **homeotic genes**. Segmentation genes specify number and polarity of segments (Nüsslein-Volhard and Wieschaus, 1980). Homeotic genes regulate development after segmentation by determining identity and sequence of body segments (Gehring and Hiromi, 1986). In addition to gene transcripts, the mother primes the egg for development with mitochondria, ribosomes, and food (yolk). Entry of a sperm into the egg sets some developmental events in motion, and cascades of genetic actions are initiated.

1.3.1 Development of a Model for Patterning

The earliest genetically controlled events are to order the axes of the egg and thereby determine the axes of the embryo that will develop. Some pre-*Drosophila* work on formation of the anterior–posterior axis (Sander, 1960, 1976) gave rise to a general model for anterior–posterior development. Sander initially demonstrated a posterior activity center in *Euscelis* (a leafhopper) eggs by ligating the egg into two parts. Neither half could form a perfect embryo after the ligature, but the anterior half of the egg could form a complete, but small embryo, when cytoplasm from the posterior pole was transferred into the anterior half of the egg. Sander inferred from these experiments that both a posterior and an anterior activity center existed in the egg and that diffusion gradients spread to other parts of the egg from these centers. Similar ligation experiments and genetic analysis with eggs of *Drosophila* verified Sander's analysis. This work led to a model in which it was proposed that anterior (A) and posterior (P) factors (**morphogens**) diffuse through the egg from the initial site of deposition (Nüsslein-Volhard et al., 1987). The A gradient is highest at the anterior end and becomes progressively less concentrated as it diffuses toward the posterior end. The direction of the P gradient is just the opposite, highest near the posterior end and lowest at the anterior end of the embryo. The concentration ratio A/P varies continuously throughout the egg and influences some segmentation genes to respond by initiating a **cascade of gene action**. A cascade of action occurs when one or more genes are activated, and they in turn active other genes, and those active still others, and so on. The location of a particular cell within the gradient ratio determines which genes respond, and what the cell will become, whether part of the head, thorax, or abdomen. The A/P ratio may be a general model for insects, and the major aspects of it have been verified in the development of *Drosophila*.

In *D. melanogaster*, both maternal effect and zygotic genes are involved in establishing the anterior–posterior and dorsal–ventral axes (Nüsslein-Volhard, 1979; Anderson, 1987; Lehmann, 1988). Patterning along the anterior to posterior axis is controlled by three systems of genes: the (1) anterior system, (2) the posterior system, and (3) the terminal system. Each system requires the action of some maternal genes and some zygotic genes. The anterior system is responsible for development in the segmented region of the head and thorax, and the posterior system determines segmentation in the abdomen. The terminal system controls development in the nonsegmented **acron** at the anterior end (along with the anterior system which has some control over the acron) and the **telson** at the posterior end.

1.3.1.1 Bicoid *Gene and Anterior Determination in* Drosophila

The anterior pole of the egg is determined by a **morphogen** called **Bicoid protein**. **Transcripts** (mRNA) of the maternal gene *bicoid* in nurse cells are passed to the oocyte through cytoplasmic strands called ring canals (Figure 1.10) connecting nurse cells and oocytes (Bopp et al., 1986; Frigerio et al., 1986). Several maternal genes, including *exuperantia*, *staufen*, and *swallow,* help localize *bicoid* transcript, possibly through binding action between their gene products and the *bicoid* transcript. In embryos with mutations of one or more of these genes, *bicoid* transcript is not well localized, and the development of the anterior region of the embryo is abnormal. The *bicoid*-transcribed mRNA is not translated into protein (the actual morphogen) until shortly after the egg is laid, and then a diffusion gradient of Bicoid protein (Figure 1.11) is established over the anterior half of the egg (Driever and Nüsslein-Volhard, 1988a). Factors promoting the gradient include (1) synthesis at the site of localization of *bicoid* transcripts, (2) diffusion of the Bicoid protein away from the site (cellularization of the blastoderm is not complete yet, so diffusion is not inhibited by cell membranes), and (3) constant rate of proteolytic degradation of Bicoid protein throughout the embryo. The *bicoid* transcripts soon disappear from the egg, but Bicoid protein persists for about 1 h after the mRNA disappears.

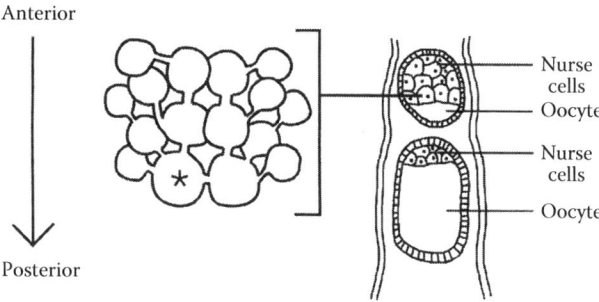

Figure 1.10 Nurse cells in the follicle of an insect with meroistic ovarioles (such as *Drosophila*) with interconnecting ring canals on the left. The 16 cells are diploid and represent four mitotic divisions of an oogonium. The cell marked by the asterisk is, for this example, assumed to have become the oocyte in the ovarian follicle diagrammed on the right. The remaining 15 cells serve as nurse cells, supplying nutrients and gene transcripts to the developing oocyte. The oocyte remains diploid, like the nurse cells, while it grows in the ovarian follicle.

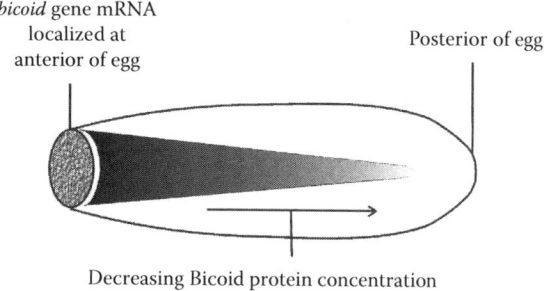

Figure 1.11 An illustration of the gradient established by diffusion of Bicoid protein translated from *bicoid* transcript localized at the anterior end of the egg. The concentration of Bicoid protein, necessary for head development, is greatest at the anterior end, and the concentration decreases toward the posterior.

Overall, Bicoid protein is present for about 4 h in the early life of the embryo (until gastrulation is underway), and then, its job completed, it disappears.

Bicoid protein has a **homeodomain,** and it is a **transcription factor** that binds to DNA, thus initiating gene cascades. Local concentrations of Bicoid are believed to be important to its ability to bind to high-affinity and low-affinity binding sites at the promoter region of target genes. At least three groups of zygotic genes are target genes for Bicoid cascade action, including (1) the gap genes *hunchback* and *Krüppel*, (2) several genes involved in formation of the head including *tailless*, *giant*, and *deformed* (Driever and Nüsslein-Volhard, 1988b; Melton, 1991), and (3) the pair-rule gene *even stripe* 2. Bicoid activates some genes, such as *hunchback*, and sets sharp borders (sometimes only across a few nuclei) where the gene will be expressed. In other cases, it only activates a particular gene, and the borders for that gene action are set by other transcription factors. High concentrations of Bicoid protein are needed at the anterior end to specify head development, where it acts as a positive transcription factor that binds to and activates zygotic *hunchback*. Bicoid also activates a cascade of gap genes (segmentation genes, see in the following text) that help specify the parasegmentation pattern (Driever and Nüsslein-Volhard, 1988b; French, 1988). Both positive and negative interactions between genes and their products are common. For example, *Krüppel*, which promotes posterior thoracic and abdomen development, is inhibited from expression in the anterior embryo by high concentration of

Bicoid, but it can be expressed and influence development in the posterior part of the thorax and posterior embryo because the gradient of Bicoid protein is much lower (Gaul and Jäckle, 1987).

Mothers with a mutant *bicoid* gene may produce an embryo in which the head and thorax are missing and/or sometimes with anterior parts partially replaced by a second abdomen (because Bicoid protein is not present to inhibit *Krüppel*). An egg containing a mutant *bicoid* gene can be (partially) rescued experimentally by injecting cytoplasm from the anterior pole of a wild-type (normal) egg, resulting in normal thoracic segments and a nearly complete head (Frohnhöfer and Nüsslein-Volhard, 1986). A mutation called *dicephalic* allows the nurse cells in an ovarian follicle to be split into two groups, with the developing oocyte sandwiched between two groups of nurse cells. Some embryos from this mutant are abnormal, and start to form an anterior parasegment at each end of the embryo (Lohs-Schardin, 1982), apparently because *bicoid* mRNA is transferred to both ends of the egg (French, 1988).

1.3.1.2 Posterior Group Genes and Posterior Pattern Formation

Posterior development is influenced by the maternal genes *oskar, staufen, tudor, valois, vasa, nanos, pumilio, hunchback*, and the zygotic gene *knirps*. The maternal gene **nanos (nos)** is the principal controlling gene in the posterior group (French, 1988; Nüsslein-Volhard, 1991). The mRNA that *nanos* specifies is localized at the posterior end of the oocyte (Lehmann and Sander, 1988). Research indicates that *nos*-dependent activity spreads over the posterior half of the developing embryo and that the maternal gene *pumilio* is necessary for the spread. One of the critical functions of *nos* is to clear from the posterior of the embryo the transcript of maternal *hunchback* (Lawrence, 1992). The gene **hunchback** is functional both as a maternal gene (in the nurse cells) and as a zygotic gene in the embryo. Although zygotic *hunchback* transcripts are made only at the anterior end (controlled by Bicoid concentration), maternal *hunchback* transcripts are uniformly distributed in the egg. The maternal *hunchback* transcripts repress the zygotic gene *knirps*, whose activity is essential to the formation of posterior structures. Additional genes that work in conjunction with *nanos* and influence posterior development function in localization, packaging, and deployment of *nanos* mRNA posteriorly (Lawrence, 1992). Mutants in which maternal *hunchback* transcripts are not present also do not require the activity of *nanos*; thus, *nanos* is described as a permissive gene for posterior development, in that it allows posterior development by destroying transcripts of maternal *hunchback*, but *nanos* function itself is not determinative and not a transcription factor (Lawrence, 1992). In the absence of maternal *hunchback* transcripts, other gene(s) promote posterior development. Development of germ cells (pole cells) is also influenced by *nanos*.

1.3.1.3 Genes Required in the Acron and Telson

The terminal group maternal genes are needed to specify normal development of the nonsegmented ends of the embryo. In *D. melanogaster*, the **acron** at the anterior includes the labrum and dorsal bridge, while the **telson** at the posterior end includes the anal pads, anal tuft, Fitzkorper, and anal spiracle (Lehmann, 1988). Embryos from mothers that have a mutation in the maternal gene *torso* fail to develop the acron and telson, and sometimes part of the abdominal structures posterior to the seventh abdominal segment, including hindgut and posterior midgut (Denglemann et al., 1986; Schüpbach and Wieschaus, 1986; French, 1988). The gene *torso* appears to be activated by a **ligand** (a molecule that binds to it), and the ligand is probably only present at the extreme ends of the embryo. There is some evidence to suggest that the ligand may be the product of the maternal gene *torsolike*. When activated, one of the roles for *torso* is to coordinate the activities of two (zygotic) gap genes, *tailless* involved in the development of the telson and *huckebein* involved in the development of the midgut (Weigel et al., 1990).

1.3.1.4 Dorsal–Ventral Axis

The dorsal–ventral axis of a *D. melanogaster* egg is established by the concerted action of at least 18 genes (Chasan and Anderson, 1993). One of the principal genes is the maternal gene *dorsal*. The *dorsal* transcript (mRNA) is uniformly distributed in the ooplasm and cytoplasm of the egg and is detectable for about 2 h after the egg is laid. Translation of the message results in **Dorsal protein**. The protein becomes localized in blastomere nuclei on the ventral side of the egg (Steward, 1987), where it acts as a morphogen and influences the zygotic genes *snail* and *twist*, as well as other genes. Mutations in which Dorsal protein is abnormally localized in blastomere nuclei result in abnormal dorsalization and eventual death of the embryo. The cascade of gene action involved in correct localization of Dorsal protein in blastomeres on the ventral side of the egg is complex. A number of maternal genes are involved. The maternal gene *Toll* encodes a receptor protein that probably acts to bind Dorsal. The maternal genes *easter*, *snake*, and possibly others may be expressed in follicle cells touching the part of the oocyte that is destined to become the ventral side and encode gene products that promote binding Toll protein to a particular region of the blastoderm.

1.4 SEGMENTATION GENES

Segmentation genes are zygotic genes, and a cascade action by segmentation genes divides the embryo into broad domains, and then into smaller regions, resulting finally in parasegments. **Parasegments** are the first evidence of **metamerization** (segmentation) and are the site and boundaries of future gene action (Martinez-Arias and Lawrence, 1985; Lawrence 1988). A parasegment includes the posterior one-fourth of a segment and the anterior three-fourths of the segment behind, with "segment" used here to mean the final segmentation pattern of the adult insect. Thus, parasegments are a little out of register with the final segmentation pattern that will ultimately exist. The parasegments are important because genetic analysis has shown that specific gene control for a body region occurs within a parasegment. Thus, parasegments narrow down the region in which specific genes are responsible for coordinating events.

Drosophila segmentation genes have been divided into three classes (Nüsslein-Volhard and Wieschaus, 1980; Howard, 1988). These are (1) **gap genes** that divide the embryo into a series of major domains, (2) **pair-rule genes** that further divide the major domains into parasegments, and (3) **segment polarity genes** that specify the pattern within each segment (Gehring, 1987; Ingham and Gergen, 1988). Gap genes, which are expressed prior to the pair-rule and segment polarity genes (Jäckle et al., 1986), are required in broad aperiodic regions of the blastoderm in order to divide the embryo into broad domains; mutations in the gap genes result in embryos that lack segments where particular gap genes should function. Pair-rule genes function with two-segment periodicity; mutants fail to develop structures in alternate segments. Segment polarity genes are required in every segment (Howard, 1988).

Examples of gap genes are zygotic *hunchback*, *Krüppel*, *Knirps*, and *giant*. Pair-rule genes include *hairy*, *runt*, *even-skipped*, *fushi tarazu*, *paired*, *odd paired*, and *sloppy-paired*. Segment polarity genes include *engrailed*, *wingless*, *patched*, *hedgehog*, *dishevelled*, and *fused*. In general, the segmentation genes are expressed very early in the development of the embryo. Segmentation genes have been characterized by the defects resulting from mutations in the genes, which typically cause the deletion of segments or some alteration in the polarity of a segment. Mutants of *fushi tarazu* (ftz^-), for example, have only about half the normal number of segments (Wakimoto and Kaufman, 1981). Mutants of *even-skipped*, that is, (eve^-), lack all segmentation in the middle region of the embryo and also show altered patterns of expression of two other segmentation genes (pair-rule *fushi tarazu* and *engrailed*), which contributes to the evidence that the expression of pair-rule segmentation genes probably involves cross-regulatory interactions.

Most of the genes involved in development do not act independently of each other, and many appear to be influenced by the action of other genes (see Howard, 1988 for review). Gap genes are regulated by maternal genes and by other gap genes. Pair-rule gene expression is variable in embryos with different types of maternal and gap gene mutations. Expression of some pair-rule genes is dependent on expression of other pair-rule genes (Harding et al., 1986; Ingham and Gergen, 1988), and segment-polarity genes are regulated by pair-rule genes. Thus, gene interactions and gene hierarchies provide instructions for successively dividing the embryo into smaller units and regulate development within these small units.

1.5 HOMEOTIC GENES

Homeotic genes give each segment its own identity and control the proper sequence for the development of segments so that, for example, the three thoracic segments are in the proper sequence, as opposed to being scattered among abdominal segments (reviewed by Gehring and Hiromi, 1986). The name "homeotic" comes from the observation that mutant alleles of these genes alter expression of some feature of a segment so that the expressed feature looks like that of another segment. For example, a homeotic gene that should control development of an antenna might mutate (e.g., mutant *antennapedia*) to cause development of a leg at the normal site of an antenna. By studying such mutants, and many can be induced in *D. melanogaster* by various treatments, it has been possible to identify a family of homeotic genes and to specify much of their control. The homeotic genes cluster in two gene complexes, the **antennapedia complex (ANT-C)** and **bithorax complex (BX-C)**, collectively called the **homeotic complex (HOM-C)** located on the right arm of chromosome 3. The genes are arranged in a linear sequence on the chromosome correlated with the linear axis of the embryo. In *Tribolium*, ANT-C and BX-C are adjacent to each other on the same chromosome, but in *Drosophila*, ANT-C, which controls parasegments in the head and first two thoracic segments, is located proximally on chromosome 3, while BX-C, which controls the third thoracic segment and abdominal segments, is located distally on chromosome 3 (Lewis, 1978). The linear arrangement may mean that each parasegment requires the cumulative activity of genes anterior to it (Lewis, 1978), but there is also evidence that the homeotic genes have regulatory interactive effects on each other (Struhl, 1982). Examples of homeotic genes are *proboscipedia, deformed, sex combs reduced, Antennapedia,* and *Ultrabithorax*.

The same homeotic genes are expressed in more than one segment (Gehring, 1987), but mutations of a gene seem to be preferentially expressed in particular segments, which suggests that the normal gene also has its principal role in that segment. For example, deletion of *Antennapedia* (*Antp*) affects each of the three thoracic segments, but the main effect is to cause the second thoracic segment to look more like the first thoracic segment. This is interpreted to mean that the main role for *Antp* is to control development of segment 2 of the thorax (Gehring, 1987). Expression of *Antp* where it is not supposed to be expressed causes the embryo to grow, or try to grow, an antenna in the wrong place. Homeotic genes regulate, at least in part, the activities of each other. There is evidence that those genes that act more posteriorly may inhibit the expression of more anterior genes. Gehring (1987) has suggested that one mechanism by which homeotic genes may regulate each other is through competition (by their gene products) for the same binding sites on DNA, which could thus control expression of a particular gene or genes.

1.5.1 Homeobox

An important functional part of homeotic genes is the **homeobox**, a particular DNA segment about 180 base pairs in length discovered by characterizing the total nucleotide sequence of the *Antp* gene of *D. melanogaster*. The homeobox sequence is highly conserved and has since been found in every homeotic gene examined, and in a few nonhomeotic but developmental genes. The homeotic genes are transcription factors and the homeobox codes for the DNA-binding sequence. Differences in the homeobox sequence of nucleotides may alter the transcription factor and thus alter the final product (leg, antenna, or bristles) that the homeotic gene controls. The homeobox sequence has been found in vertebrate (including human) and plant developmental genes. Homeotic genes, and the proteins they encode, may have other highly conserved sequences. The sequence of amino acids encoded by the *engrailed* gene from *Drosophila*, honeybee (*A. mellifera*), and a mouse, organisms representing an evolutionary span covering greater than 500 million years, was conserved near the carboxyl terminal of the protein. The *Deformed* gene-encoding protein from *Drosophila*, frogs, and humans contains conserved sequences of amino acids near the amino terminal. These conserved functions suggest that the homeotic genes have fundamental roles in the development of insects and vertebrates (Gehring, 1987).

1.6 ORGANOGENESIS

1.6.1 Neurogenesis

Neurogenesis has been recently reviewed by Campos-Ortega (1994). The development of the nervous system begins in the early germ band stage and typically is the first tissue to differentiate. In each parasegment, ectodermal cells differentiate into three types of nervous system precursor cells, enlarged **neuroblasts** (NBs), **midline precursor cells** (MPCs), and **nonneuronal cells** (NNCs). The NBs divide repeatedly and produce a chain of **ganglion mother cells** (GMCs). The GMCs and the MPCs each divide only once producing pairs of progeny cells.

Initially, any cell in the neuroectoderm can become an NB, but once cells differentiate into NBs, they inhibit neighboring cells from also becoming NBs and promote their ultimate differentiation into NNCs. A number of genes are involved in neurogenesis, including many segmentation genes that are expressed in the developing central nervous system (CNS). The *Drosophila* gene *Notch*, expressed in the neuroectoderm influences whether cells become NBs or NNCs. By 8–9 h after development starts in *D. melanogaster*, there are about 250 neurons in each parasegment, and the segmentation gene *fushi tarazu* (*ftz*) is expressed in a segmentally repeated pattern in about 30 of the neurons in each segment (Doe et al., 1988), although the significance of *ftz* to the developing nervous system is not yet clear. MPCs briefly express *ftz* (during hours 8–9 in different MPCs), in GMCs, and in cells that later become glial cells.

Ganglionic masses of cells become differentiated and separated as segmentation occurs in the embryo. Three bilaterally paired groups of NBs in the protocephalon will give rise to the protocerebrum, deutocerebrum, and tritocerebrum. Three ganglionic masses in the gnathal segments fuse into the subesophageal ganglion. Each thoracic segment and the 11 abdominal segments initially have paired ganglionic masses, but fusion of some of the abdominal ganglia always occurs, and in some insects, all abdominal ganglia and thoracic ganglia fuse into a single thoracic ganglion, or there may be two thoracic ganglia. The stomatogastric ganglia that send nerves to the foregut are derived from ectoderm associated with the stomodeum. Sensory organs are derived from modifications of ectodermal cells in localized parts of the body where they occur. The optic lobes, a part of

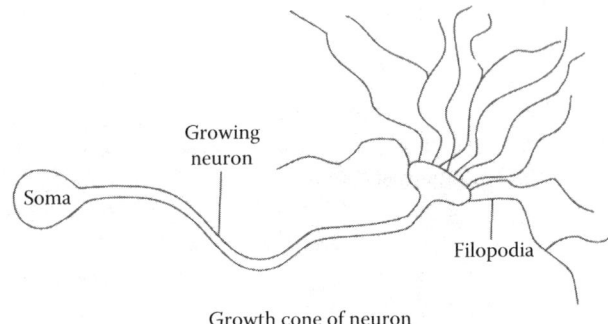

Figure 1.12 Diagrammatic representation of a growing neuron with its growth cone and filopodia. (Redrawn and modified from Goodman, C.S. and Bastiani, M.J., *Sci. Am.*, 251(6), 58, 1984.)

the protocerebrum, do not arise from NBs, but develop from ectodermal cells, and like many other aspects of embryogenesis, they arise differently in different groups.

When nerve cells have located and attached to each other to form a ganglion, they send out axonal processes to make contact with neurons in other ganglia, and/or with effector organs such as muscles and glands. The transcription factor Even-skipped transcribed by the gene *even-skipped* (*eve*) is critical for axon guidance of motoneurons that project to dorsal muscles and for the acquisition of electrical properties in motoneurons and also possibly regulates later aspects of neuron development (Pym et al., 2006). Dendritic processes grow toward the CNS from peripheral sensory cells. A growing neuron exhibits a **growth cone** (Figure 1.12). The growth cone or leading edge of an growing neuronal process contains "fingerlike filopodia" that are constantly extending and retracting, exploring the environment (Goodman and Bastiani, 1984; Harrelson and Goodman, 1988). The filopodia may recognize certain chemical gradients, some of which may attract the growing tip, while others repel it and influence "guide cells" that provide a pathway to follow. A guide cell may be another axon that already has a connection, and the growing neuronal process partially envelops the guide and grows along it. In this way, multiple neurons going to the same general location would aggregate into larger nerves. The cellular and molecular mechanisms that guide neuronal growth cones in insects and vertebrates appear to be very similar and highly conserved. Two glycoprotein **cell adhesion molecules (CAMs)**, fasciclin II and amalgam, expressed on the surface of certain developing neurons, function in specific adhesion and nerve cell recognition in both vertebrates and insects.

The growth cone regions of developing neurons contain high concentrations of actin that is involved in the neuron's ability to move and respond to growth gradients from its target. These actin filaments have the same basic molecular structure as those in muscle cells. Myosin is also present in the growth cone lamellae. Smith (1988) has proposed that ATP provides the energy for actin polymerization, which enables the growth cone to send out filopodia, while retraction of the filopodia is also energized by ATP and possibly involves an interaction between actin and myosin, which is also present in growth cone lamellae.

1.6.2 Development of the Gut

Soon after germ band elongation begins, a group of ectoderm cells at the anterior tip of the embryo invaginate to form the stomodeum, hypopharynx, and other parts of the gnathal segments. Similar invagination of ectodermal cells from the posterior ectoderm indicates the beginnings of the proctodeum. The midgut is derived from multiplication of cells at each end of the invaginating tissue. The three segments of the gut at first develop independently, and the complete alimentary canal is formed when plugs of cells at the end of the foregut, each end of the midgut, and at the end of the hindgut die and the three gut segments unite. As the anterior and posterior midgut primordia come together, they enclose the remaining yolk sac within the midgut. In bees and wasps (Hymenoptera), the foregut is open to the midgut before hatching, but the midgut is not open to the hindgut until just before pupation of mature larvae, so undigested food (such as pollen shells) accumulates in the midgut. Prior to pupation, cells plugging the posterior end of the midgut and anterior end of the hindgut die, opening the connection, and the larva empties the gut in the comb cell. Adult bees clean the excreta from the cell.

1.6.3 Malpighian Tubules

The Malpighian tubules develop from evaginations of the anterior proctodeum and mark the junction between the midgut and hindgut. Although they are derived from the proctodeum, which has a cuticular lining, the tubules themselves are not lined by cuticle and have a cellular morphology more similar to midgut cells than to hindgut cells. In gryllid crickets and mole crickets (Gryllotalpidae), a cuticle-lined excretory tube several millimeters long leads from the gut to a cuticle-lined bladder from which many Malpighian tubules arise. The tubules do not have a cuticular lining.

1.6.4 Tracheal System

The tracheal system develops from ectoderm. Bilaterally paired tracheal pits appear on most segments, and the pits are connected to short tubes shaped like an upside down "T." Eventually, the pieces fuse into continuous longitudinal tracheae with segmentally arranged spiracular openings.

1.6.5 Oenocytes

Oenocytes are large cells in most larval and adult insects that stain evenly pink with eosin. In the embryo, they are derived from ectoderm in abdominal segments. They usually appear as isolated cells scattered here and there at various places in the body. They often occur between epidermal cells beneath the cuticle, and they also commonly lie between fat body cells. Their function is not clearly defined, but they have smooth endoplasmic reticulum suggesting lipid synthesis, and they have been implicated as possible sources of ecdysteroids.

1.6.6 Cuticle Secretion in the Embryo

Several cuticles are secreted and shed by the embryos of some apterygote insects. Embryonic epidermal cells in hemimetabolous insects and some holometabolous ones secrete a cuticle after blastokinesis. The serosa, an extraembryonic membrane, secretes cuticle in some insects (see Section 1.2.6). Some insects including some acridids, *Dysdercus* spp. (Hemiptera), and *Hyalophora cecropia* (Lepidoptera) secrete two embryonic cuticles. The first one is shed, and the second one becomes the cuticle of the first instar (Mueller, 1963).

1.6.7 Cell Movements during Embryogenesis

Many cells move in amoeba-like fashion from their site of origin to another point in the embryo where they join with like cells and become functionally active. The processes by which functionally similar cells find each other and form an organ have fascinated embryologists for decades. A current model for cell migration suggests that gradients of diffusing chemicals, CAMs, and tactile cues guide cells and allow cells of similar function to aggregate into tissues and organs. CAMs are molecules in cell membranes that recognize another CAM of like or unlike molecular structure in another cell membrane. About 10 CAMs currently have been identified, but many more are believed to exist. When two cells with homophilic binding CAMs contact each other, the CAMs bind the two cells together. As additional cells bind, gradually, a tissue or organ is built. Some heterophilic CAMs may exist also.

1.6.8 Programmed Cell Death: Apoptosis

Cell death is a normal part of embryogenesis in *Drosophila* (Abrams et al., 1993) and in all other multicellular organisms. Programmed cell death is called **apoptosis**. Death of cells depends in some cases upon hormonal cues and, in others, on cell-to-cell interactions (Kimura and Truman, 1990; Wolff and Ready, 1991; Campos et al., 1992). A gene, ***reaper (rpr)***, plays a major role in control of apoptosis in *Drosophila* (White et al., 1994), and embryos that are homozygous for a small deletion that includes the *reaper* gene exhibit no apoptosis and contain many extra cells that should have died. These embryos fail to survive. Although the mechanisms involved in the functioning of *reaper* are not yet clear, *reaper* mRNA is expressed in cells that are destined for apoptosis.

1.7 HATCHING

Development to hatching takes days, weeks, or even months in the Hemimetabola but is much faster in the Holometabola, usually a matter of a few days in most. Faster development may have been an evolutionary process driven by selection for ability to take advantage of rapidly decaying resources (as in decaying fruit or dung breeders) and rapidly changing plant growth resources.

1.8 IMAGINAL DISCS

The larva of holometabolous insects have within their body undifferentiated groups of cells that will grow into adult tissues under appropriate hormonal controls during pupation. These embryonic cells, derived from ectoderm, are grouped into **imaginal discs** in various places in the body of the larva (Figure 1.13). The small clumps of imaginal disc cells typically begin to divide and increase in number during the late part of larval development, but the timing of cell increase is highly variable in different groups of insects. When the insect pupates, the imaginal discs provide the cells to make the adult structures. Imaginal discs in Diptera include discs for wings, legs, halteres, compound eye, antennae, genital structures, and some mouthparts. Isolated small groups of abdominal histoblasts scattered in the larval abdomen give rise to abdominal structures in the adult, and groups of imaginal cells in the larval salivary glands and proventriculus give rise to the corresponding adult structures. Imaginal discs have been most intensively studied in Diptera and to some extent in Lepidoptera. Reviews of imaginal disc structure and development (primarily in *Drosophila*) have been provided by Madhavan and Schneiderman (1977), Oberlander (1985), and Larsen-Rapport (1986). Anderson (1963a,b; 1964a,b) described the origin and development of imaginal discs in a tephritid fly, *Bactrocera* (formerly *Dacus*) *tryoni*, and van Ruiten and Sprey (1974) described the development of a leg disc in the blowfly, *Calliphora erythrocephala*.

During larval life, the discs grow by mitotic cell division. Ecdysteroid secretion in the third instar (the last instar) of *Drosophila* signals the imaginal discs to begin rapid differentiation into adult structures during the pupal stage. Discrete discs are first evident in a larva as a thickening of the epidermis in an early instar. The discs separate from the epidermis and migrate to new locations, but they remain in contact with the epidermis by a stalk. The time of appearance and growth rate of different discs in *Drosophila* and other Diptera that have been studied are variable.

In Lepidoptera, the eye, leg, and antennal discs arise late in larval development from fields of diploid cells, the primordia, which retain the embryonic potential to form imaginal discs. In the last

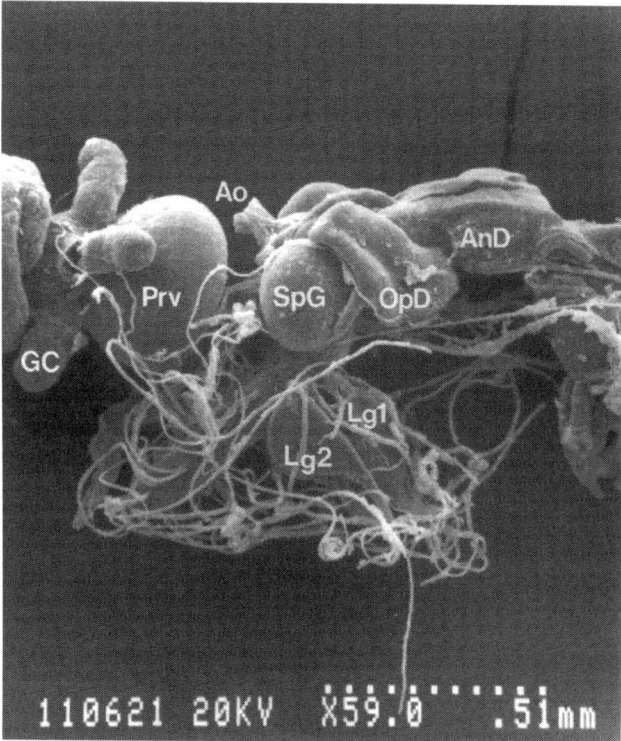

Figure 1.13 Some of the imaginal discs in a third instar of the tephritid fruit fly *Anastrepha suspensa*. AnD, antennal disc; Ao, aorta; Gc, gastric caeca; L1 and L2, leg discs for the prothoracic and mesothoracic legs; OpD, optic lobe and compound eye disc; Prv, proventriculus; SpG, supraesophageal or brain imaginal disc. The short, fingerlike ventral nerve cord to which L1 and L2 leg discs are attached contains the imaginal disc cells for developing the adult ventral nerve cord and nerves. The leg disc for the metathoracic pair of legs is not attached to the nerve cord but is located adjacent to the wing discs near the spiracle and not shown in the photograph. The image is from a late third instar that was almost ready to pupariate (SEM photograph by the author).

instar, disc formation is dependent upon adequate nutrition and the normal falling level of juvenile hormone (JH) that will allow pupation instead of another larval molt. Recently, Truman et al. (2006) showed that the discs failed to develop in late instars that were experimentally starved, an action that also elevated JH levels. These workers showed in further experiments that it was the elevated level of JH and not the lack of nutrition in starved larvae that stopped disc development in late instars. Higher levels of JH, which is normal in earlier instars, suppressed disc formation regardless of adequate feeding or starvation. Even JH mimic applied to feeding last instars failed to stop disc formation. They postulate the release of a metamorphosis initiating factor in the well fed last instar that overrides (experimentally induced) JH suppression of disc development in late last instar *Manduca sexta*. Truman et al. (2006) conclude that JH plays a role in the intermolt periods of early instars in promoting isomorphic growth of primordia and suppressing morphogenetic signals that would turn primordia into imaginal discs too early. Allee et al. (2006) also found the presence of JH could repress development in cultured eye primordial of the lepidopteran *M. sexta*, but JH mimics applied *in vivo* failed to affect early eye imaginal disc development. These authors also suggest that some nutrient-dependent hormonal factor is important in the necessary process of JH repression for metamorphosis to occur.

As a disc grows, it assumes a pocket- or tubelike structure. The pocket or cavity in the disc is called the peripodial cavity. As more cells are produced by mitosis, the tube-like growth folds upon

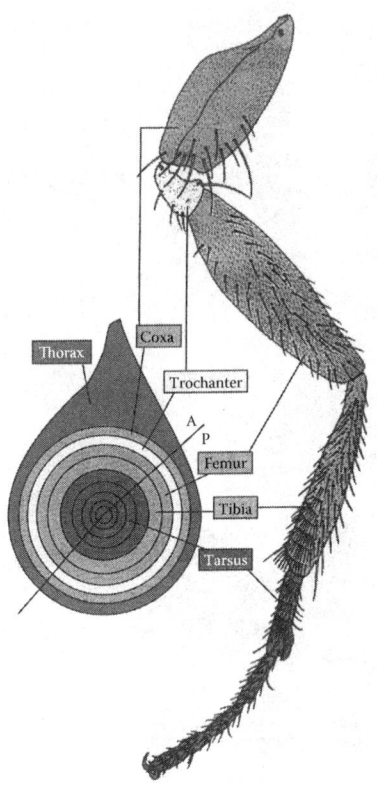

Figure 1.14 **(See color insert.)** An imaginal leg disc of *Drosophila melanogaster* showing regions of the disk that will give rise to parts of the leg and some other parts of the body during pupation. (From Bryant, P.J., *Science*, 259, 471, 1993. With permission.)

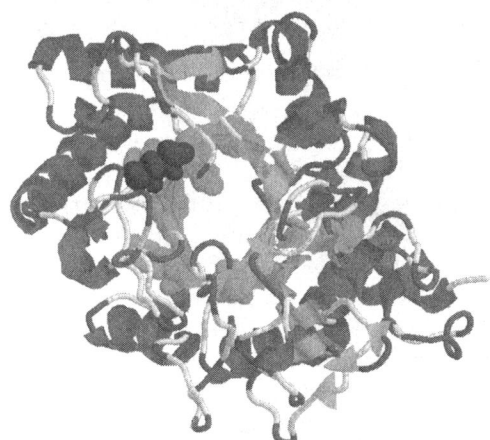

Figure 1.15 **(See color insert.)** A basic view of the growth-promoting glycoprotein (IDGF-DRW protein 3-D structure) from *Diaprepes* root weevil. The structure was predicted with ROBETTA, a full-chain protein prediction server. The beta barrel motif is shown in yellow in the center. The conserved residues in green (in space fill) are apparently essential to maintain the barrel folding as predicted by interface alanine scanning. Residues are Gly109, Gly110, Asp152, Gly153, Leu218, Asp242, Lys295, Gly412, and Asp421. The residue at position 159 is Glu (in brown space fill), which is replaced by Gln in other known IDGY proteins. Structure key: alpha helix (pink), beta strand (yellow), turn (blue), and other (gray). (From Huang, Z. et al., *Florida Entomol.*, 89, 223, 2006. With permission; Photo courtesy of Dr. Wayne B. Hunter, USDA, Ft. Pierce, FL.)

itself to form concentric layers. During metamorphosis, the disc unfolds and elongates, and cells differentiate into an adult structure. In the case of a *Drosophila* leg disc, the innermost part of the disc becomes the most distal part of the leg, the tarsal segments, while the peripheral part becomes the most proximal leg structure, the coxa (Figure 1.14) (Bryant, 1993). The other parts of the leg are sequentially layered within the disc. Similarly, various cells in a wing disc give rise to structures of the thorax as well as wings. A model of an imaginal disc growth factor (IDGF), a protein, isolated from the citrus root weevil, *Diaprepes abbreviatus*, is illustrated in Figure 1.15, and a phylogenetic tree of IDGF proteins is shown in Figure 1.16 (Kawamura et al., 1999; Huang et al., 2006). Bryant (1975, 1993) suggests that many of the same genes, gene products, and pathways functioning in the embryo control imaginal disc development. Patterning of *Drosophila* discs is under control of some segment polarity genes necessary for embryonic development, with at least four of the *wingless* subclass of segment polarity genes required for the development of normal limb pattern (Hatano, 1991; Peifer et al., 1991). Discs do not form in embryos mutant for *wingless* (Simcox et al., 1989).

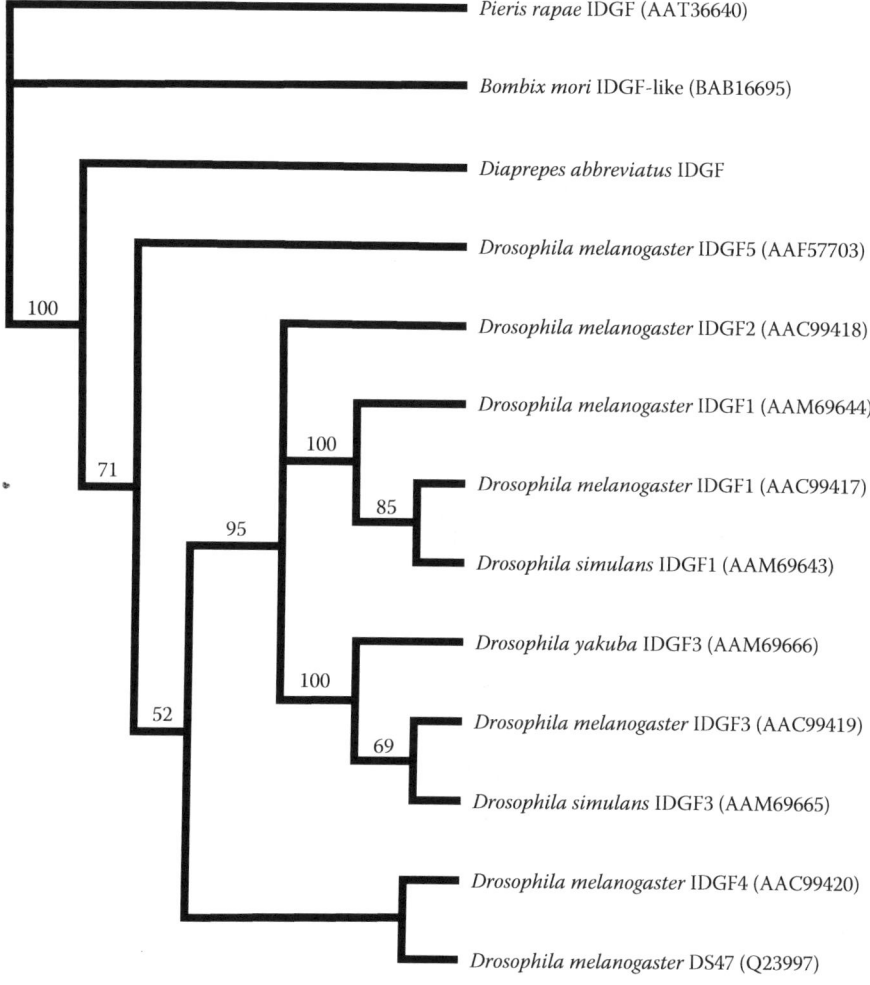

Figure 1.16 A phylogenetic tree of the imaginal disc growth factor (IDGF) family. The accession number of sequences is shown inside parentheses. The tree was generated by bootstrap neighbor joining, 1000x, PAUP 4.0, unrooted. The *Diaprepes abbreviatus* accession number is IDGF_DRW accession n.. AAV68692.1. (From Huang, Z. et al., *Florida Entomol.*, 89, 223, 2006, USDA. With permission.)

Couso et al. (1993) have shown that *wingless* provides information in a polar coordinate system, a system earlier postulated for development of imaginal discs and regenerating limbs (French et al., 1976).

In conclusion, insect embryogenesis is an active and fertile field of research with important results for all of biology. Investigations of gene control in the embryo of *D. melanogaster* have often pointed the way for vertebrate studies. For example, the homeobox in *Drosophila* now appears to be universal in developmental genes. Although there clearly are major differences between some gene functions in insects and vertebrates, there are also fascinating similarities.

1.9 REVIEW AND SELF-STUDY QUESTIONS

1. Define morphogenesis.
2. After a sperm and egg unite, what is the organism or structure called?
3. When an oogonium (in the germarium of the female ovary) divides, what cells result?
4. What are energids?
5. What are energids that remain within the yolk of the developing zygote and digest the yolk called?
6. How is the blastoderm formed?
7. Which cells in the developing zygote are the first to become committed to a future development path?
8. Describe the sequence of blastokinesis in the milkweed bug, *Oncopeltus fasciatus*, including anatrepsis and katatrepsis.
9. What do the term "determinate" and "indeterminate" refer to in the blastoderm?
10. What are some of the larval and adult structures that are derived from the ectoderm?
11. What are parasegments?
12. What is the meaning of the word "metamerization" as applied to insect development?
13. What is the name for a molecule that controls a local or regional pattern of differentiation?
14. What is the name of the principal gene that controls anterior determination?
15. What are maternal genes? Do maternal genes act upon the developing embryo primarily before or after the egg is laid?
16. Which of the following is a morphogen: *bicoid* or Bicoid protein?
17. What are ring canals? Where are they located? What function do they serve?
18. Does Dorsal protein become localized and function on the dorsal or ventral side of the blastoderm?
19. Are segmentation genes maternal genes or zygotic genes?
20. What do homeotic genes do? What is the homeobox?
21. What are imaginal discs?
22. What is apoptosis, and how is it important in development?

REFERENCES

Abrams, J.M., J. White, L.I. Fessler, and H. Steller. 1993. Programmed cell death during *Drosophila* embryogenesis. *Development* 117: 29–43.

Allee, J.P., C.L. Pelletier, E.K. Fergusson, and D.T. Champlin. 2006. Early events in adult eye development of the moth, *Manduca sexta*. *J. Insect Physiol*. 52: 450–460.

Anderson, D.T. 1963a. The embryology of *Dacus tryoni*. 2. Development of imaginal discs in the embryo. *J. Embryol. Exp. Morphol*. 11(Part 2): 339–351.

Anderson, D.T. 1963b. The larval development of *Dacus tryoni* (Frogg.) (Diptera: Trypetidae). I. Larval instars, imaginal discs, and haemocytes. *Aust. J. Zool*. 11: 202–218.

Anderson, D.T. 1964a. The embryology of *Dacus tryoni* (Diptera). 3. Origins of imaginal rudiments other than principal discs. *J. Embryol. Exp. Morphol*. 12(Part 1): 65–75.

Anderson, D.T. 1964b. The larval development of *Dacus tryoni* (Frogg.) (Diptera: Trypetidae). II. Development of imaginal rudiments other than the principal discs. *Aust. J. Zool*. 12: 1–8.

Anderson, D.T. 1972a. The development of hemimetabolous insects, in J. Counce and C.H. Waddington (eds.), *Developmental Systems: Insects*, vol. 1. Academic Press, New York, pp. 95–163.

Anderson, D.T. 1972b. The development of holometabolous insects, in J. Counce and C.H. Waddington (eds.), *Developmental Systems: Insects*, vol. 1. Academic Press, New York, pp. 165–242.

Anderson, K.V. 1987. Dorso-ventral embryonic pattern genes of *Drosophila*. *Trends Genet.* 3: 91–96.

Bate, M. and A. Martinez-Arias (eds.). 1993. *The Development of Drosophila melanogaster*, vols. I and II. Cold Spring Harbor Laboratory Press, Cold Spring Harbor, NY.

Bopp, D., M. Burri, S. Baumgartner, G. Frigerio, and M. Noll. 1986. Conservation of a large protein domain in the segmentation gene paired and in functionally related genes of *Drosophila*. *Cell* 47: 1033–1040.

Bryant, P.J. 1975. Pattern formation in the imaginal wing disc of *Drosophila melanogaster*: Fate map, regeneration, and duplication. *J. Exp. Zool.* 193: 49–78.

Bryant, P.J. 1993. The polar coordinate model goes molecular. *Science* 259: 471–472.

Campos, A.R., K.-F. Fischbach, and H. Steller. 1992. Survival of photoreceptor neurons in the compound eye of *Drosophila* depends on connections with the optic ganglia. *Development* 114: 355–366.

Campos-Ortega, J.A. 1994. Genetic mechanisms of early neurogenesis in *Drosophila melanogaster*. *Adv. Insect Physiol.* 25: 75–103.

Campos-Ortega, J.A. and V. Hartenstein. 1985. *The Embryonic Development of Drosophila melanogaster*. Springer-Verlag, Berlin, Germany, 227pp.

Chan, L.-N. and W. Gehring. 1971. Determination of blastoderm cells in *Drosophila melanogaster*. *Proc. Natl. Acad. Sci. USA* 68: 2217–2221.

Chasan, R. and K.V. Anderson. 1993. Maternal control of dorsal-ventral polarity and pattern in the embryo, in M. Bate and A. Martínez-Arias (eds.), *The Development of Drosophila melanogaster*, vol. 1. Cold Spring Harbor Laboratory Press, Cold Spring Harbor, NY, pp. 387–424.

Couso, J.P., M. Bate, and A. Martínez-Arias. 1993. A *wingless*-dependent polar coordinate system in *Drosophila* imaginal discs. *Science* 259: 484–489.

Denglemann, A., A. Hardy, N. Perrimon, and A. Mahowald. 1986. Developmental analysis of the torso-like phenotype in *Drosophila* produced by a maternal-effect locus. *Dev. Biol.* 115: 479–489.

Doe, C.Q., Y. Hiromi, W.J. Gehring, and C.S. Goodman. 1988. Expression and function of the segmentation gene *fushi tarazu* during *Drosophila* neurogenesis. *Science* 239: 170–175.

Driever, W. and C. Nüsslein-Volhard. 1988a. A gradient of *bicoid* protein in *Drosophila* embryos. *Cell* 54: 83–93.

Driever, W. and C. Nüsslein-Volhard. 1988b. The *bicoid* protein determines position in the *Drosophila* embryo in a concentration- dependent manner. *Cell* 54: 95–104.

Foe, V.A. and B.M. Alberts. 1983. Studies of nuclear and cytoplasmic behaviour during five mitotic cycles that precede gastrulation in *Drosophila* embryogenesis. *J. Cell Sci.* 61: 31–70.

French, V. 1988. Gradients and insect segmentation. *Development (Supplement)* 104: 3–16.

French, V., P.J. Bryant, and S.V. Bryant. 1976. Pattern regeneration in epimorphic fields. *Science* 193: 969–981.

Frigerio, G., M. Burri, D. Bopp, S. Baumgartner, and M. Noll. 1986. Structure of the segmentation gene *paired* and the *Drosophila* PRD gene set as part of a gene network. *Cell* 47: 735–746.

Frohnhöfer, H.G. and C. Nüsslein-Volhard. 1986. Organization of anterior pattern in the *Drosophila* embryo by the maternal gene *bicoid*. *Nature (Lond.)* 324: 120–125.

Gaul, U. and H. Jäckle. 1987. Pole region-dependent repression of the *Drosophila* gap gene *Krüppel* by maternal gene products. *Cell* 51: 549–555.

Gehring, W.J. 1987. Homeo boxes in the study of development. *Science* 236: 1245–1252.

Gehring, W.J. and Y. Hiromi. 1986. Homeotic genes and the homeobox. *Annu. Rev. Genet.* 20: 147–173.

Goltsev, Y., G.L. Rezende, K. Vranizan, G. Lanzaro, D. Valle, and M. Levine. 2009. Developmental and evolutionary basis for drought tolerance of the *Anopheles gambiae* embryo. *Dev. Biol.* 330: 462–470.

Goodman, C.S. and M.J. Bastiani. 1984. How embryonic nerve cells recognize one another. *Sci. Am.* 251(6): 58–66.

Handel, K., C.G. Grünfelder, S. Roth, and K. Sander. 2000. *Tribolium* embryogenesis: A SEM study of cell shapes and movements from blastoderm to serosal closure. *Dev. Genes Evol.* 210: 167–179.

Harding, K., C. Rushlow, H.J. Doyle, T. Hoey, and M. Levine. 1986. Cross-regulatory interactions among pair-rule genes in *Drosophila*. 1986. *Science* 233: 953–959.

Harrelson, A.L. and C.S. Goodman. 1988. Growth cone guidance in insects: Fasciclin II is a member of the immunoglobulin superfamily. *Science* 242: 700–708.

Hatano, Y. 1991. Molecular cloning and analysis of *forked* locus in *Drosophila ananassae*. *Mol. Gen. Genet.* 226: 17–23.

Howard, K. 1988. The generation of periodic pattern during early *Drosophila* embryogenesis. *Development* 104(Suppl.): 35–50.

Huang, Z., W.B. Hunter, C.A. Cleland, M. Wolinsky, S.L. Lapointe, and C.A. Powell. 2006. A new member of the growth-promoting glycoproteins from *Diaprepes* root weevil (Coleoptera: Curculionidae). *Florida Entomol.* 89: 223–232.

Ingham, P. and P. Gergen. 1988. Interactions between the pair-rule genes *runt*, *hairy*, *even-skipped* and *fushi tarazu* and the establishment of periodic pattern in the *Drosophila* embryo. *Development* 104(Suppl.): 51–60.

Jäckle, H., D. Tautz, T. Schuh, E. Seifert, and R. Lehmann. 1986. Cross regulatory interactions among gap genes of *Drosophila*. *Nature (Lond.)* 324: 668–670.

Jacobs, C.G.C. and M. van der Zee. 2013. Immune competence in insect eggs depends on the extraembryonic serosa. *Dev. Comp. Immunol.* 41: 263–269.

Johannsen, O.A. and F.H. Butt. 1941. *Embryology of Insects and Myriapods*. McGraw-Hill, New York.

Jura, C. 1972. Development of apterygote insects, in J. Counce and C.H. Waddington (eds.), *Developmental Systems: Insects*, vol. 1. Academic Press, New York, pp. 49–94.

Kawamura, K., T. Shibata, O. Saget, D. Peel, and P.J. Bryant. 1999. A new family of growth factors produced by the fat body and active on *Drosophila* imaginal disc cells. *Development* 126: 211–219.

Kimura, K.-I. and J.W. Truman. 1990. Postmetamorphic cell death in the nervous and muscular systems of *Drosophila melanogaster*. *J. Neurosci.* 10: 403–411.

Kobayashi, H. and H. Ando. 1985. Early embryogenesis of fireflies, *Luciola cruciata*, *L. lateralis* and *Hotaria parvula* (Coleoptera, Lampyridae), in H. Ando and K. Miya (eds.), *Recent Advances in Insect Embryology in Japan*. ISEBU Co. Ltd., Tsukuba, Japan, pp. 157–169.

Larsen-Rapport, E.W. 1986. Imaginal disc determination: Molecular and cellular correlates. *Annu. Rev. Entomol.* 31: 145–175.

Lawrence, P.A. 1988. The present status of the parasegment. *Development* 104(Suppl.): 61–65.

Lawrence, P.A. 1992. *The Making of a Fly: The Genetics of Animal Design*. Blackwell Scientific Publications, Oxford, U.K., 228pp.

Lehmann, R. 1988. Phenotypic comparison between maternal and zygotic genes controlling the segmental pattern of the *Drosophila* embryo. *Development* 104(Suppl.): 17–27.

Lehmann, R. and K. Sander. 1988. *Drosophila* nurse cells produce a posterior signal required for embryonic segmentation and polarity. *Nature (Lond.)* 335: 68–70.

Lewis, E.B. 1978. A gene complex controlling segmentation in *Drosophila*. *Nature (Lond.)* 276: 565–570.

Lohs-Schardin, M. 1982. Dicephalic—A *Drosophila* mutant affecting polarity in follicle organization and embryonic patterning. *Wilhelm Roux Arch. Dev. Biol.* 191: 28–36.

Madhavan, M.M. and H.A. Schneiderman. 1977. Histological analysis of the dynamics of growth of imaginal discs and histoblasts nests during the larval development of *Drosophila melanogaster*. *Wilhelm Roux's Arch. Dev. Biol.* 183: 269–305.

Mahowald, A.P. 1963. Electron microscopy of the formation of the cellular blastoderm in *Drosophila melanogaster* embryo. *Exp. Cell. Res.* 32: 457–468.

Martínez-Arias, A. and P.A. Lawrence. 1985. Parasegments and compartments in the *Drosophila* embryo. *Nature (Lond.)* 313: 639–642.

Melton, D.A. 1991. Pattern formation during animal development. *Science* 252: 234–241.

Miya, K. 1985. Determination and formation of the basic body pattern in embryo of the domesticated silkmoth, *Bombyx mori* (Lepidoptera, Bombycidae), in H. Ando and K. Miya (eds.), *Recent Advances in Insect Embryology in Japan*. ISEBU Co. Ltd., Tsukuba, Japan, pp. 107–123.

Mueller, N.S. 1963. An experimental analysis of molting in embryos of *Melanoplus differentialis*. *Dev. Biol.* 8: 222–240.

Nüsslein-Volhard, C. 1979. Maternal effect mutations that alter the spatial coordinates of the embryo of *Drosophila melanogaster*, in S. Subtelny and I.R. Königsberg (eds.), *Determinants of Spatial Organization*. Academic Press, New York, pp. 185–211.

Nüsslein-Volhard, C. 1991. Determination of the embryonic axes of *Drosophila*. *Dev.* 1991(Suppl. 1): 1–10.

Nüsslein-Volhard, C. and E. Wieschaus. 1980. Mutations affecting segment number and polarity in *Drosophila*. *Nature (Lond.)* 287: 795–801.

Nüsslein-Volhard, C., H.G. Frohnhöfer, and R. Lehmann. 1987. Determination of anteroposterior polarity in *Drosophila*. *Science* 238: 1675–1681.

Oberlander, H. 1985. The imaginal discs, in G.A. Kerkut and L.I. Gilbert (eds.), *Comprehensive Insect Physiology, Biochemistry and Pharmacology*, vol. 1, *Embryogenesis and Reproduction*. Pergamon Press, Oxford, U.K., pp. 151–182.

Panfilio, K.A. 2008. Extraembryonic development in insects and the acrobatics of blastokinesis. *Dev. Biol.* 313: 471–491.

Panfilio, K.A., P.Z. Liu, M. Akram, and T.C. Kaufman. 2006. *Oncopeltus fasciatus zen* is essential for serosal tissue function in katatrepsis. *Dev. Biol.* 292: 226–243.

Peifer, M., C. Rauskolb, M. Williams, B. Riggleman, and E. Weischaus. 1991. The segment polarity gene *armadillo* interacts with the *wingless* signaling pathway in both embryonic and adult pattern formation. *Development* 111: 1029–1045.

Pym, E.C.G., T.D. Southall, C.J. Mee, A.H. Brand, and R.A. Baines. 2006. The homeobox transcription factor Even-skipped regulates acquisition of electrical properties in *Drosophila* neurons. *Neural Dev.* 1: 3. doi:10.1186/1749-8104-1-3.

Rezende, G.L., A.J. Martins, C. Gentile, L.C. Farnesi, M. Pelajo-Machado, A.A. Peixoto, and D. Valle. 2008. Embryonic desiccation resistance in *Aedes aegypti*: Presumptive role of the chitinized serosal cuticle. *BMC Dev. Biol.* 8: 82. doi:10.1186/1471-213X-8-82.

Romoser, W.S. and J.G. Stoffolano, Jr. 1998. *The Science of Entomology*, 4th ed. WCB/McGraw-Hill, Dubuque, IA, 605pp.

Sander, K. 1960. Analyse des ooplasmatischen Reakionssystems von *Eucelis plebejus* Fall. (Circadina) durch isolieren und Kombinieren von Keimteilen. II. Die Differenzierungsleistungen nach Verlagern von Hinterpolmaterial. *Wilhelm Roux Arch. EntwMech. Org.* 151: 660–707.

Sander, K. 1976. Specification of the basic body pattern in insect embryogenesis. *Adv. Insect Physiol.* 12: 125–238.

Sander, K., J.O. Gutzeit, and H. Jäckle. 1985. Insect embryogenesis: Morphology, physiology, genetical and molecular aspects, in G.A. Kerkut and L.I. Gilbert (eds.), *Comprehensive Insect Physiology, Biochemistry and Pharmacology*, vol. 1, *Embryogenesis and Reproduction*. Pergamon Press, Oxford, U.K., pp. 319–385.

Schüpbach, T. and E. Wieschaus. 1986. Maternal-effect mutations altering the anterior-posterior pattern of the *Drosophila* embryo. *Wilhelm Roux Arch. Dev. Biol.* 195: 302–317.

Simcox, A.A., I.J.H. Roberts, E. Hersperger, M.C. Gribbin, A. Shearn, and J.R.S. Whittle. 1989. Imaginal discs can be recovered from cultured embryos mutant for the segment-polarity genes *engrailed*, *naked* and *patched* but not from *wingless*. *Development* 107: 715–722.

Simcox, A.A. and J.H. Sang. 1983. When does determination occur in *Drosophila* embryos? *Dev. Biol.* 97: 212–215.

Slack, J.M.W. 1987. Morphogenetic gradients-past and present. *Trends Biochem. Sci.* 12: 200–204.

Smith, S.J. 1988. Neuronal cytomechanics: The actin-based motility of growth cones. *Science* 242: 708–715.

Steward, R. 1987. *Dorsal*, an embryonic polarity gene in *Drosophila*, is homologous to the vertebrate proto-oncogene, c-*rel*. *Science* 238: 692–694.

Struhl, G. 1982. Genes controlling segmental specification in the *Drosophila* thorax. *Proc. Natl. Acad. Sci. USA* 79: 7380–7384.

Truman, J.W., K. Hiruma, J.P. Allee, S.G.B. MacWhinnie, D.T. Champlin, and L.M. Riddiford. 2006. Juvenile hormone is required to couple imaginal disc formation with nutrition in insects. *Science* 312: 1385–1388.

van der Zee, M., N. Berns, and S. Roth. 2005. Distinct functions of the *Tribolium zerknüllt* genes in serosa specification and dorsal closure. *Curr. Biol.* 15: 624–636.

van Ruiten, Th.M. and Th.E. Sprey. 1974. The ultrastructure of the developing leg disk of *Calliphora erythrocephala*. *Z. Zellforsch.* 147: 373–400.

Vargas, H.C.M., L.C. Farnesi, A.J. Martins, D. Valle, and G.C. Rezende. 2014. Serosal cuticle formation and distinct degrees of desiccation resistance in embryos of the mosquito vectors *Aedes aegypti*, *Anopheles aquasalis*, and *Culex quinquefasciatus*. *J. Insect Physiol.* 62: 54–60.

Wakimoto, B.T. and T.C. Kaufman. 1981. Analysis of larval segmentation in lethal genotypes associated with the *Antennapedia* gene complex in *Drosophila melanogaster*. *Dev. Biol.* 81: 51–64.

Weigel, D., G. Jürgens, M. Klingler, and H. Jäckle. 1990. Two gap genes mediate maternal terminal pattern information in *Drosophila*. *Science* 248: 495–498.

White, K., M.E. Grether, J.M. Abrams, L. Young, K. Farrell, and H. Steller. 1994. Genetic control of programmed cell death in *Drosophila. Science* 264: 677–683.

Wigglesworth, V.B. 1972. *The Principles of Insect Physiology*, 7th ed. Chapman & Hall, New York.

Wolff, T. and D.F. Ready. 1991. Cell death in normal and rough eye mutants of *Drosophila. Development* 113: 825–839.

Zalokar, M. and I. Erk. 1976. Division and migration of nuclei during early embryogenesis of *Drosophila melanogaster. J. Microsc. Biol. Cell* 25: 97–106.

CHAPTER 2

Digestion

PREVIEW

The structure and function of the alimentary canal, usually called the gut, coevolved with the food habits of insects. Both insects and their alimentary canals are diverse, and the alimentary canal is modified in special ways for solid vs. liquid food and animal vs. plant food. Insects often have to adapt to a dwindling food supply or change in the quality of their food. Food eaten by larval and adult insects is often quite different as well as different according to the sex, for example, adult female mosquitoes and other insects that must feed on vertebrate blood to mature their eggs. Nevertheless, the basic evolutionary plan for three major divisions of the gut—the foregut, midgut, and hindgut—has been retained in all insects. The midgut is the principal site for the secretion of digestive enzymes, digestion of food, and absorption of nutrients in most insects, although some insects display significant or major digestion in the foregut or hindgut. Both are lined with a cuticular intima on the surface of the epithelial cells that is shed with each molt. The midgut does not have an attached cuticular intima, but in many insects the midgut cells secrete a detached peritrophic matrix, an envelope that encloses the food and within which most of the digestion occurs. The peritrophic matrix is not universally present in all insects. Several types of midgut epithelial cells occur in various species; the principal cells have microvilli at the gut lumen surface, a modification providing an extensive surface area for secretion of digestive enzymes and for absorption. Many insects have protein-digesting enzymes with the general characteristics of trypsin and chymotrypsin, and some have protein-digesting enzymes, the cathepsins, that function at an acid pH. A wide variety of carbohydrate-digesting enzymes and general lipases that digest lipids have been identified. Some insects can secrete a complete complement of the three principal enzymes needed to digest cellulose, but also may use cellulose-digesting enzymes made by their gut symbionts. This chapter describes the principles of gut function and structure, and briefly reviews gut modifications and function in major insect orders.

2.1 INTRODUCTION

The evolutionary success and diversity of insects have been driven by their ability to occupy many ecological niches and utilize many different sources of food. Usually, newly hatched insects must obtain food soon or die. Some newly hatched insects eat the eggshell and the small amount of yolk left in the shell after hatching, or begin eating the food on or in which the egg was laid; however, some have to search for food. With suitable food, a larva may successfully grow, molt, and eventually become an adult. Thus, processing of food and functioning of the alimentary canal are activities critical to life. Insects are extraordinarily diverse in food and feeding habits, and have correspondingly high diversity in gut structure and function. Major changes in gut structure and

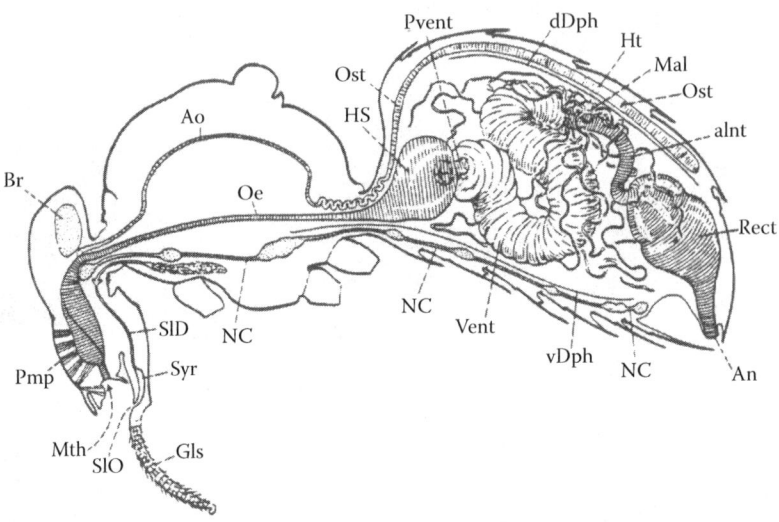

Figure 2.1 The body outline of a honeybee showing gut structure, dorsal vessel, and ventral nerve cord with ganglia. alnt, anterior intestine (= ileum); An, anus; Ao, aorta; Br, brain; dDph, dorsal diaphragm; Gls, tongue; HS, honey stomach (= crop); Ht, heart; Mal, Malpighian tubules; Mth, mouth; NC, nerve cord; Oe, esophagus; Ost, ostia in the heart; Pmp, pharyngeal pump; Pvent, proventriculus; Rect, rectum; SlD, salivary duct; SlO, salivary orifice; Syr, salivary syringe; Vent, ventriculus (= midgut); vDph, ventral diaphragm. (From Dadant & Sons, (eds.). 1975. The Hive and the Honey Bee, Revised edition, Dadant & Sons, Publishers, Hamilton, IL, p. 112. Permission granted by citation and acknowledgment.)

function almost always occur in those insects in which the larval and adult food is different and in those insects with complete metamorphosis. There is, then, no "typical" insect gut, but the less specialized gut of a honeybee (Figure 2.1) or an orthopteroid (Figure 2.2) can be used to illustrate major gut structures. Terra and Ferreira (2012) present a general review of insect digestion.

2.2 RELATIONSHIPS BETWEEN FOOD HABITS AND GUT STRUCTURE AND FUNCTION

2.2.1 Plant versus Animal Origin: Solid versus Liquid Diet

The most likely ancestral-type feeding behavior was probably that of a **general scavenger**, similar to the present-day cockroach, followed by later evolution toward more specialized **phytophagous** or **carnivorous** feeding (Southwood, 1973; Dow, 1986). The gut in such a generalist feeder was probably fairly simple, not much convoluted, and not much longer than the body, conditions that prevail today in generalist feeders. As insects evolved and adapted to new foods, there was concomitant evolution in gut structure and function.

The food that insects consume can be roughly divided into broad categories of **solid food** vs. **liquid food** and of **plant** or **animal** origin. In some insects, solid food is broken up mechanically with the mandibles and with a grinding action by a muscular **proventriculus**. The gut of solid feeders tends to be a relatively straight tube, not much, if any, longer than the body, possibly because solid food does not easily pass through a very convoluted gut. Caterpillars (Lepidoptera), for example, have a simple, straight-through type of gut. They are, for the most part, phytophagous and often have an abundant source of food. They tend to feed frequently and some almost continuously. Cellulose in the plant food is not digested by caterpillars or by most other phytophagous insects, and the incompletely digested food passes rapidly through the gut.

DIGESTION

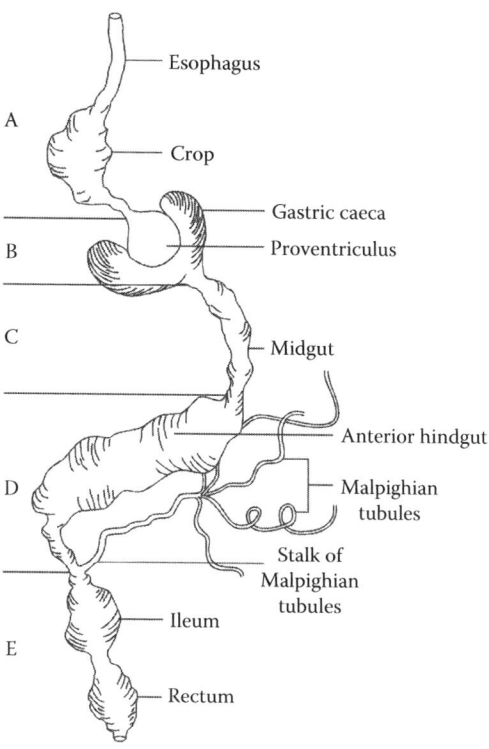

Figure 2.2 Gut structure in a generalized feeder, such as the cricket *Gryllus rubens*. (A) Foregut, including the proventriculus. (B) Two large gastric caeca cupped around the proventriculus. The gastric caeca are part of the midgut. (C) The short and relatively unspecialized midgut. (D) The hindgut is divided into an anterior portion that has a cuticular lining on the surface of the cells (From Nation, J.L., *Int. J. Insect Morphol. Embryol.*, 12, 201, 1983). The Malpighian tubules do not originate at the junction of the midgut and hindgut in gryllid crickets, but arise from a cuticular lined stalk. The stalk arises near the junction of the anterior and posterior hindgut. (E) The posterior hindgut consisting of the ileum and rectum.

A liquid diet more easily passes through a convoluted gut, which many liquid feeders have, but, in some cases, the diet may be so dilute that it presents new problems, such as how to get rid of so much water and, sometimes, other components in excess, most notably sugar. Insects that ingest a dilute liquid food have evolved specialized adaptations to deal with the excess water and other components (for example, salts or sugars). *Rhodnius prolixus* (Hemiptera: Reduviidae), which takes one large blood meal each instar, and *Dysdercus fasciatus* (Reduviidae), which feeds upon the phloem sap of plants, use hormonal controls (diuretic hormones) that regulate rapid water excretion by Malpighian tubules. Homoptera, also plant sap feeders, have morphological modifications of the gut called "filter chambers" in which a loop of the hindgut is in close contact with the foregut to allow a large volume of water to bypass the midgut, thus moving fluids directly from foregut to hindgut. This causes some nutrient loss as well, primarily sugars, but there is an excess of sugar in the phloem sap. Amino acids are present in plant sap in low concentrations, and in order to extract the quantity of amino acids needed for growth and development from the limited volume of fluid that actually goes through the midgut, homopterans have to feed voraciously and excrete a large quantity of honeydew.

The slightly lower nutrient quality of plant food, as compared to animal sources, requires large intake and results in a steady elimination of frass droppings or, if phloem or xylem sap is the food source, elimination of excess fluid. Plants generally supply sufficient carbohydrates, lipids, and phytosterols for good insect nutrition. Plant tissues usually contain lower levels of amino acids than animal tissues and some amino acids may be critically low or absent. A number of amino acids, as well as vitamins and other important dietary components, often are supplied by symbionts (see Chapter 21) and do not always have to come from the diet.

Animal feeders feed less frequently and at irregular intervals, as opportunity affords, so they tend to have a gut specialized for storage (for instance, the large crop in a praying mantis) that enables them to feast when food is available and to hold a large meal for digestion over time. Insects that take food of animal origin generally obtain a better balance of amino acids than those that feed on plants. In addition, animal tissues are a rich source of carbohydrates, cholesterol, and other lipids.

2.3 MAJOR STRUCTURAL REGIONS OF THE GUT

2.3.1 Foregut

In spite of the diversity mentioned in the previous section, three basic divisions—the **foregut**, **midgut**, and **hindgut**—can be recognized on embryological, morphological, and physiological grounds in all insects. During embryonic development, the foregut develops from invaginating ectodermal tissue at the anterior end of the body. The foregut epithelial cells secrete a cuticular lining that is attached to the surface of the cells on the lumen (apical) side. This lining contains both chitin and proteins and is essentially the same as the epicuticle and endocuticle on the body surface. Heavily sclerotized regions of the foregut lining, such as in the proventriculus of some insects, contain hard exocuticle. The foregut can be divided into a buccal cavity (mouth), pharynx, esophagus, crop, proventriculus, and esophageal invagination, and any part may be highly modified. At each molt, the old cuticular lining from the foregut is sloughed off into the gut and any undigested residue is excreted with the feces. Epithelial cells in the foregut are usually flattened squamous cells that do not secrete digestive enzymes into the lumen of the foregut.

The mouth or buccal cavity is usually just an enlarged opening that receives the slightly chewed food in mandibulate insects or fluid ingested by insects with piercing and sucking mouth parts. Powerful muscles in the wall of the pharynx pump fluid into the buccal cavity and aid swallowing in blood feeders and xylem and phloem feeders (Stoffolano and Haselton, 2013). Salivary glands are diverticula from the anterior part of the foregut that secrete fluid and carbohydrate-digesting enzymes (mostly amylases) into the buccal cavity. Salivary secretions contain amylase, lubricate the food, and contribute to digestion in the crop.

The pharynx passes food to the esophagus, which may be a simple tube that continues to the proventriculus at the end of the foregut in some insects. Alternatively, the foregut may expand into a much enlarged and dilated **crop**, or the crop may be a diverticulum from the main part of the foregut, as in some Diptera (Stoffolano et al., 2010; Solari et al., 2013; Stoffolano and Haselton, 2013). Substantial digestion occurs in the crop of some insects, for example, in many Orthoptera and some Coleoptera (Dow, 1986), but the enzymes (except for salivary enzymes) come from the midgut and are present in fluid that is regurgitated from the midgut into the crop. The foregut cuticle is impermeable and little or no absorption occurs from the foregut.

The crop periodically releases some of its contents to pass through the proventriculus and into the midgut. In the cockroach, *Leucophaea maderae*, crop emptying is inversely related to the concentration of the crop contents and to hemolymph levels of some nutrients. Thus, the crop releases materials to the midgut at a rate that allows it to digest and absorb nutrients more efficiently (Englemann, 1968). The contraction of the crop is under hormone control (serotonin) in *Phormia regina*, a dipteran (Liscia et al., 2012). Gut function is probably influenced by hormones in all insects, and serotonin, crustacean cardioactive peptide, allatostatin-A, and neuropeptide F have been specifically implicated in one or more insects (Blenau and Thamm, 2011; Falibene et al., 2012; Mikani et al., 2012; Matsui et al., 2013; French et al., 2014).

Extraoral digestion occurs in many insects (Cohen, 1995). By injecting hydrolytic enzymes into the food source (animal or plant material) and then sucking back the digested products, insects utilize very high percentages of the nutrient value of the food source. Typically in predacious bugs, high concentrations of digestive enzymes occur in the saliva (Zeng and Cohen, 2000; Zibaee et al., 2012), whereas in intermittently feeding heteropterans such as the leaf-footed bug *Leptoglossus zonatus* and other similar feeders, salivary concentrations of digestive enzymes are relatively low, and most of the digestion takes place in the midgut (Augusti and Cohen, 2000; Walker and Allen, 2010; Fialho et al., 2012; Mehrabadi et al., 2012; Rocha et al., 2014). Some insects reflux enzyme secretions and partially digested products by repeatedly sucking up and reinjecting the liquefied juices into the food. Refluxing mixes the secretions and fluids and extends the effective life of the digestive enzymes. Refluxing is particularly effective when the food contains a limiting boundary, such as the shell of a seed or the cuticle of an insect that acts as a container for the liquefying body contents. When larval carabids, which normally feed extraorally upon small arthropods, were allowed to feed upon a large portion of meat, only about 50% of the proteins available, including the digestive enzymes of the larvae could be recovered, apparently because the enzymes and some digested products diffused into the piece of meat (Cheeseman and Gillot, 1987).

The **proventriculus** at the end of the foregut may be very muscular and contain heavily sclerotized teeth, ridges, and spines (Figure 2.3) for further grinding and tearing of the food, or it may be reduced to a simple valve at the entry to the midgut. The proventriculus of worker honeybees, called the **honey stopper** (Figure 2.4), consists of four converging fingers projecting anteriorly into the crop. The fingers, each bearing intermeshing spines, open and close rhythmically to capture pollen grains and sweep them from the crop into a bolus that later enters the midgut for digestion. Nectar is strained through the interlocking spines and retained in the crop for later deposition in the honeycomb. Flap-like or valve-like extensions of the proventriculus sometimes project into the midgut, forming the **esophageal valves** or **cardiac** sphincter as the junction between foregut and midgut. There is a great

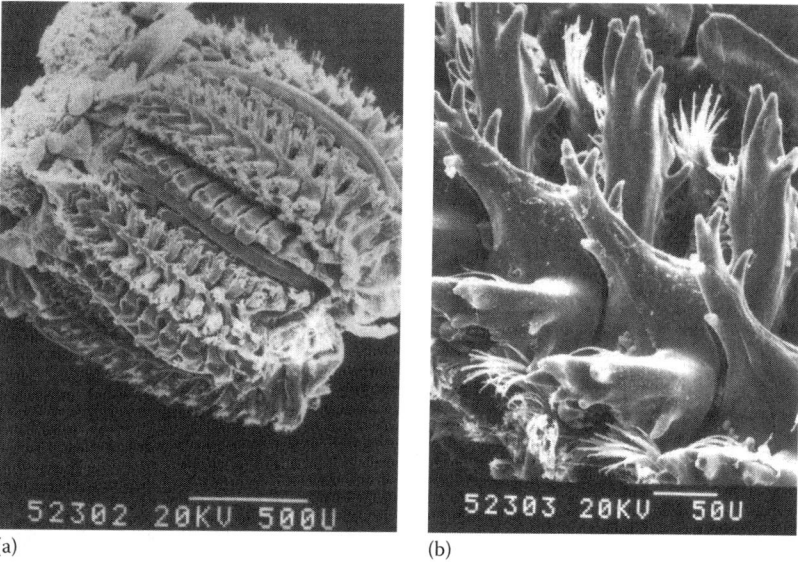

(a)   (b)

Figure 2.3 Proventriculus from a bush cricket. (a) The barrel-shaped proventriculus has been cut longitudinally and turned inside out, so that (b) the heavily sclerotized ridges and "teeth" on the internal surface can be observed. (SEM by the author.)

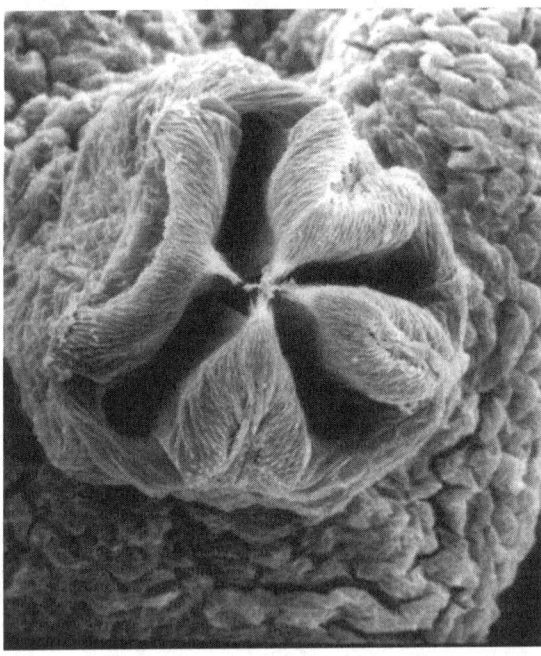

Figure 2.4 The proventriculus (also called the honey stopper) of an adult worker honeybee. The view is from the crop looking toward the midgut. The fingerlike proventricular flaps containing setae strain pollen grains from the nectar in the crop and pass the pollen into the midgut. Most of the nectar can be left in the crop for honey production in the hive. (SEM by the author.)

deal of variability in structure at the **junction** and in the depth of its invagination into the midgut. Wigglesworth (1961) suggested that the main function of the invagination is to channel food entering the midgut into the peritrophic matrix. In mole crickets (Gryllotalpidae), four large esophageal valves (Figure 2.5) channel food and sand grains often present because of their feeding habits past two large, cup-shaped gastric caeca, protecting the delicate microvilli on the lumen surface of the gastric caecal cells from abrasive food particles and sand grains.

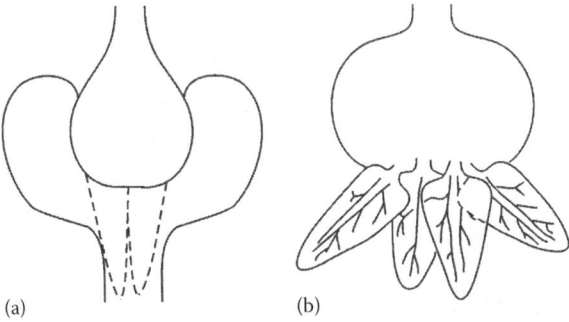

Figure 2.5 Proventricular valve flaps in a mole cricket, *Scapteriscus abbreviatus*. The long valve flaps prevent sand (which is common in the gut of mole crickets) and other rough food particles from entering the delicate gastric caeca just posterior to the proventriculus. (a) The view shows the position of two of the valves (dotted lines) on the inside of the gut. (b) The gut has been dissected and the four valve flaps spread apart. The valve flaps are cusp shaped and project about 1.2 mm past the opening to the gastric caeca. The cells of the gastric caeca protected by the flaps are the only cells in the gut of this mole cricket that have microvilli on the lumenal surface; the remainder of the gut contains a cuticular lining on the surface of the cells.

2.3.2 Midgut

The midgut is the principal site for the secretion of digestive enzymes and for digestion and absorption in most insects (Hakim et al., 2010). In many insects, **gastric caeca** arise at or near the origin of the midgut, but they may be located at various points along the midgut. In some insects, the gastric caeca are a major site of absorption of digestion products, and they also produce digestive enzymes. The role of gastric caeca in absorption, especially if the gastric caeca are near the origin of the midgut, depends upon a countercurrent flow (see Section 2.10) that brings midgut contents back to the gastric caeca.

The origin of the midgut in insects is a controversial topic. Dow (1986) reviewed some of the recent literature regarding whether the midgut develops from endodermal tissue (Richards and Richards, 1977; McFarlane, 1985) or whether it develops from buds of tissue at the invaginated ends of the fore- and hindgut, in which case it would be derived from ectodermal tissue. Dow concludes that, in some insects at least, a case can be made that the midgut may be derived from ectodermal tissue, as are the fore- and hindguts. The midgut does not have an attached cuticular lining on the surface of the cells, but midgut cells in the majority of insects can secrete a chitin and protein-containing membrane, the peritrophic matrix, that surrounds the food and shields the delicate midgut cells from contact with potentially rough and abrasive food particles.

2.3.3 Hindgut

The hindgut, like the foregut, develops in the embryo from ectodermal tissue and, consequently, hindgut cells have an attached cuticular lining on their surface (Figure 2.6). The Malpighian tubules usually mark the beginning of the hindgut (see Chapter 18, for some exceptions). The junction between the mid- and hindgut has been called the pylorus by some authors, and a valvular structure may occur here. The part of the hindgut immediately past the Malpighian tubules has been called the ileum. Sometimes the ileum simply grades into the rectum, but in some insects there is a distinct middle region called the colon by some authors. The terminal part of the hindgut is the rectum.

Both circular and longitudinal muscles lie on the outer or hemolymph side of the hindgut. The arrangement of these muscle bands varies in different insects and even within the same insect at different points along the hindgut as to which band of muscle is outermost; circular or longitudinal

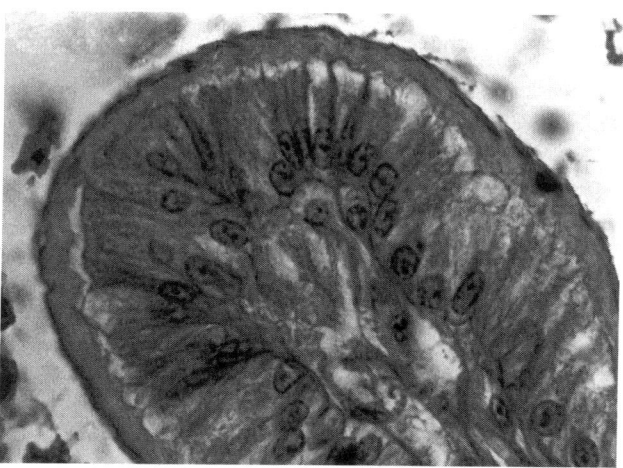

Figure 2.6 The thick cuticular layer on the lumen surface of cells in the hindgut of a mole cricket, *Scapteriscus borelli*. (Photo by the author.)

may be outermost (Gupta and Berridge, 1966; Hopkins, 1967). The entire hindgut has a chitinous lining (sometimes referred to as an intima) on the lumen surface of the cells. Typically, the cells of the hindgut wall are arranged in a single layer of irregularly shaped epithelial cells. Numerous septate desmosomes connect the lateral borders of cells and serve to hold cells together and present a barrier to fluid and molecules that might otherwise pass between adjacent cells.

The contents of the hindgut are generally fluid as they pass into the rectum. The rectum plays a critically important role in the reabsorption of water, ions, and dissolved substances (including some nutrients) from the primary urine flushed into the hindgut by the Malpighian tubules. Specialized cells in the rectal papillae and rectal pads in the rectum of many insects have characteristic ultrastructure and physiological mechanisms that promote reabsorption. Water recovery by the rectum results in the relatively dry frass or fecal pellets characteristic of many terrestrial insects. The cuticular lining on the hindgut cells is thinner and has larger pores in it than the lining in the foregut, and numerous substances can be absorbed from the lumen. More details on the anatomy of hindgut and its role in secretion, reabsorption, and water conservation are described in Chapter 18.

The hindgut is most specialized in those insects that digest cellulose, such as termites. In termites, the hindgut is usually divided into several chambers harboring either bacteria or protozoa that digest cellulose. Glucose is the principal carbohydrate liberated from cellulose digestion and the resident microorganisms usually ferment it, with the end products being short-chain fatty acids (principally acetic acid) that can be absorbed by the termite and used as an energy source. (See Section 2.14.3 for additional details.)

2.4 MIDGUT CELL TYPES

2.4.1 Columnar Cells

Columnar cells, the most numerous cells in the midgut, conduct most of the absorption of digested products and secretion of enzymes. "Columnar" refers to the tall shape of the cells, but some insects have more than one morphological type. The cells have microvilli on the apical or lumen surface and extensive invaginations of the basal cell membrane (Figure 2.7). The extensive membrane infolding at the basal face and microvilli at the apical face of midgut cells are adaptations to present a large membrane surface for metabolic functions, such as absorption and secretion. Midgut cells exhibit other characteristics typical of secretory epithelium, such as rough endoplasmic reticulum, a large Golgi complex, and secretory vesicles.

2.4.2 Regenerative Cells

Midgut cells wear out rapidly and are replaced by new cells that grow from small **regenerative cells** lying randomly near the base of mature cells in larval Diptera and Lepidoptera, or as small cell clusters called nidi (nests) (Figure 2.8) in Orthoptera and Odonata, and at the apex of crypts or caeca projecting through the gut muscle layers in Coleoptera. Regenerative cells grow into mature cells gradually and replace cells lost through age, wear, and loss through apocrine and holocrine secretion. House (1974) reported that midgut cells in *Periplaneta americana* are replaced about every 40–120 h.

Damage to the midgut regenerative cells after irradiation for inducing sterility in insects for population control (**sterile insect technique, SIT**) has been one of the limiting factors in the use of the technique, particularly in the boll weevil, *Anthonomus grandis*. Irradiated boll weevils soon die from midgut disruption because regenerative cells are unable to successfully divide and replace normal loss of midgut cells after irradiation (Riemann and Flint, 1967).

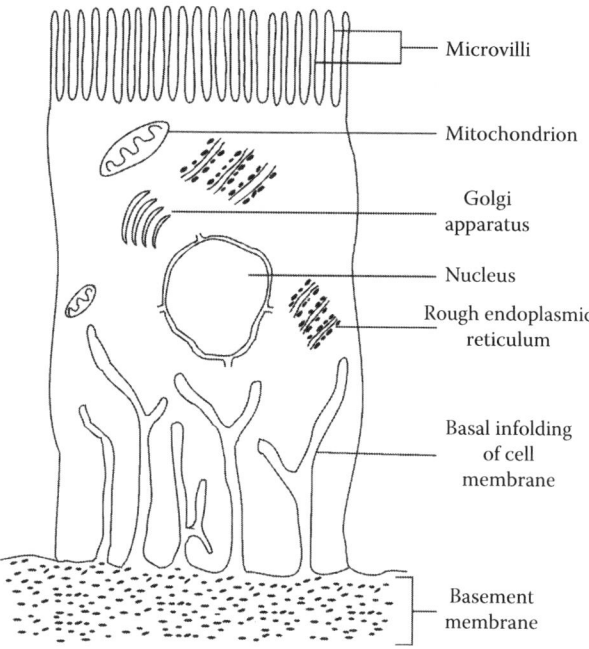

Figure 2.7 Diagrammatic drawing of the main ultrastructural features of a midgut cell. Basal infoldings of the cell membrane project into the cell. Microvilli occur on the gut lumen side. Mitochondria are numerous.

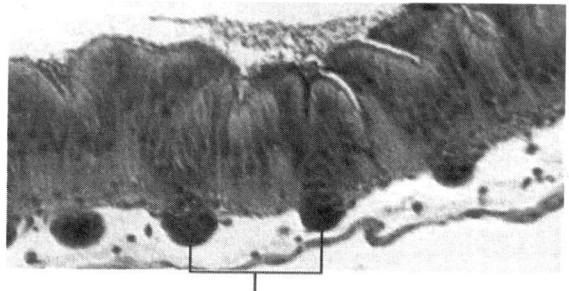

Nidi of regenerative cells in midgut

Figure 2.8 Regenerative cells occur in the midgut of most insects and gradually grow into mature cells to replace worn out cells. The anatomical arrangement of regenerative cells varies in different species; nidi or nests of regenerative cells are shown in this illustration from gastric caeca of a mole cricket, *Scapteriscus vicinus*. (Photo by the author.)

2.4.3 Goblet Cells

Goblet cells are, indeed, somewhat goblet shaped with a large central cavity lined with microvilli (Figure 2.9). The sides of a goblet cell curve around to enclose the cavity and, at the apex, the apical lips have interdigitating microvilli that control fluid exchange between cavity and midgut lumen. Goblet cells are interspersed among midgut epithelial cells in lepidopterous larvae and in Ephemeroptera, Plecoptera, and Trichoptera. The apical membrane facing the goblet cavity houses

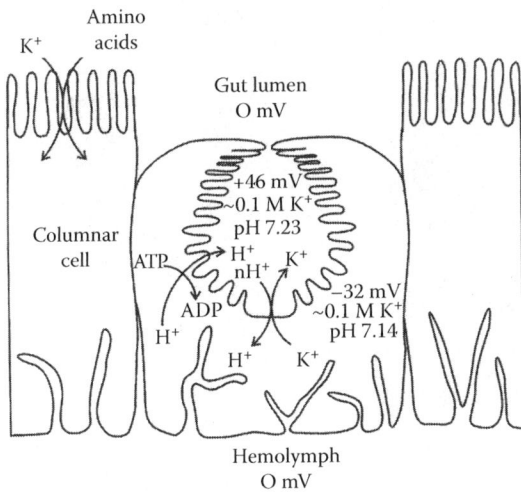

Figure 2.9 Structure and function of goblet cells from the midgut of a lepidopteran. A proton pump actively secretes protons (H^+) into the goblet cavity and an antiporter mechanism in the goblet cell membrane transports K^+ into the goblet cell cavity in exchange for H^+. Goblet cavity contents are eventually emptied into the midgut lumen, creating the strongly alkaline midgut of Lepidoptera.

a vacuolar-type H^+-ATPase (**proton ATPase pump**) (Figure 2.10) that establishes a voltage across the apical membrane and pumps H^+ into the goblet cavity (Chao et al., 1991). Vacuolar proton pumps are now known to occur in many secretory tissues (Huss et al., 2011). A proton pump occurs in the apical membrane of insect salivary glands (Figure 2.11) (Bauumann and Walz, 2012) and in Malpighian tubule cells and drives the formation of fluid (urine) in Malpighian tubules. In goblet cells, a K^+/nH^+ **antiporter mechanism** (Wieczorek et al., 1989, 1991, 2000) exchanges H^+ for K^+ in the goblet cavity. The pump is strongly electrogenic and creates transmembrane voltages that can exceed 240 mV and transmembrane pH gradients that may exceed 4 pH units (Harvey, 1992). The transmembrane voltage created by the pump enables a cotransporter mechanism in the apical membrane of columnar cells to reabsorb K^+ and amino acids from protein digestion (Giordana et al., 1989; Feldman et al., 2000).

The pump consists of a complex of protein subunits with the V_0 base embedded in the goblet cell apical membrane (the surface facing the goblet cavity) and the V_1 head piece projecting into the goblet cell cytoplasm (Merzendorfer et al., 1997; Wieczorek et al., 2000; Baumann and Walz, 2012). The V_1 complex uses energy from adenosine triphosphate (ATP) breakdown to drive protons into the goblet cavity through the V_0 transmembrane complex. The pH of the goblet cavity remains near neutral (at 7.23 ± 0.11) because of the rapid exchange of K^+ for H^+, probably with two or more H^+ per K^+ exchanged (Chao et al., 1991). Potassium ions enter goblet cells from the hemolymph at the basal side through K^+ channels located in the basal membrane (Zeiske et al., 1986; Moffett and Koch, 1988a,b). Electrical coupling of the basal and apical membranes may occur involving anion transport at the basolateral membrane (with concomitant cation movement) as has been demonstrated in Malpighian tubules (Beyenbach, 1995; Beyenbach et al., 2000). Electrical coupling, if it occurs in the goblet cells, would permit the electrical driving forces at the basolateral and apical membranes to rise and fall in parallel so that cation entry from the hemolymph matches cation extrusion into the goblet cavity (Beyenbach et al., 2000). Ultimately goblet cavity contents are emptied into the lumen of the midgut. Several different anions, including bicarbonate HCO_3^-, may be secreted into the lumen with potassium, giving rise to the high pH characteristic of the larval midgut in many Lepidoptera (Dow, 1984; Dow and O'Donnell, 1990). Goblet cells appear to metabolize amino acids preferentially to support pump activity,

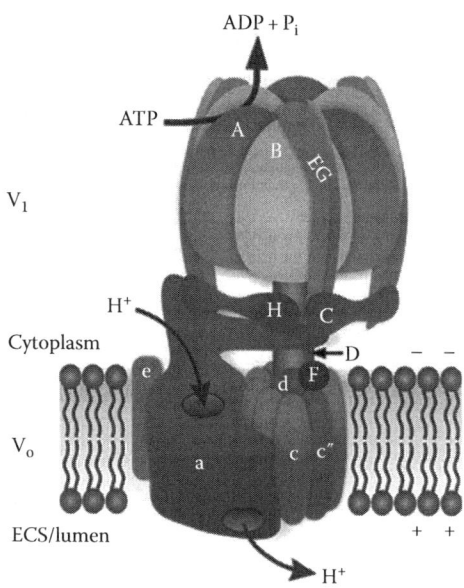

Figure 2.10 (See color insert.) Model of the proton pump, a V-ATPase. The V-ATPase holoenzyme consists of two domains, the V_o complex in the membrane and the V_1 complex inside the cell. The V_o complex is composed of at least five subunits (labeled here as a,c,c″,d,e) and it translocates protons from the cytosol to the extracellular space (ECS) or to the lumen of an organelle, respectively. The peripheral V_1 complex is composed of eight subunits (A–H) and is responsible for ATP hydrolysis, supplying energy to the pump. Proton pumps occur in numerous cells. With respect to goblet cells, V_1 is located within the goblet cell cytoplasm and the pump base, V_o, forms a transmembrane channel through which protons are pumped into the goblet cell cavity. ATP hydrolysis in the V_1 part of the pump provides the energy for the pumping of protons across the cell membrane. A separate mechanism in the goblet cell membrane that has not been elucidated exchanges K⁺ for H⁺ in the goblet cell cavity. The proton pump is the driving force for the concentration of K⁺ against a concentration gradient in the goblet cell cavity. Eventually the K⁺ is released into the gut lumen, creating high pH in the lumen. (From Baumann, O. and Walz, B., *J. Insect Physiol.*, 58, 450, 2012. With permission.)

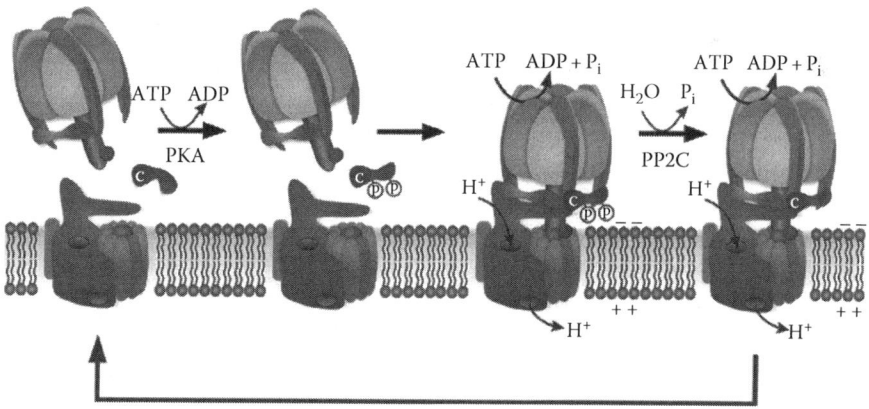

Figure 2.11 (See color insert.) A schematic model of the proton pump illustrating the reversible assembly of the pump. The functioning pump is shown as the third model from the left, while the completely disassembled pump is shown on the extreme left. Subunit C at left, must be phosphorylated (C attached to 2 Ps) in the two middle diagrams, in order for it to promote assembly of V_1 and V_o holoenzymes. Dephosphorylation of subunit C (extreme right model) leads to disassembly of the pump to conserve energy when it is not specifically needed. (From Baumann, O. and Walz, B. *J. Insect Physiol.*, 58, 450, 2012. With permission.)

and L-alanine, L-glutamine, L-glutamate, and L-malate experimentally support pump activity and maintain the transepithelial potential, but glucose is ineffective (Parenti et al., 1985). Only apical (goblet cavity facing) membranes of goblet cells contain significant amounts of V-ATPase proteins, and other plasma membranes and endomembranes of goblet and columnar cells contain little or no pump proteins (Klein et al., 1991). Part of the regulatory process involves dissociation/reassociation of the V_1 cytoplasmic complex from the V_o transmembrane complex (Kane and Parra, 2000; Beyenbach and Wieczorek, 2006; Lafourcade et al., 2008; Voss et al., 2009; Baumann and Walz, 2012).

The high midgut pH may help protect against tannins that are common in the plant hosts of Lepidoptera larvae. Tannins can interact with the insect's own enzymes and proteins in the food and may result in reduced digestion of proteins. Although other factors also may aid in reducing the impact of tannins, there is less formation of nonsoluble protein–tannin complexes at high pH.

2.5 MICROVILLI OR BRUSH BORDER OF MIDGUT CELLS

Microvilli at the apical surface of midgut cells greatly increase the surface area for enzyme secretion and for absorption of digested products (Figure 2.12). Center-to-center spacing (= width) of microvilli in midgut cells of various insects ranges from about 150 to 200 nm (Richards and Richards, 1977), hence, the difficulty of resolving them clearly with light microcopy. Microvilli have been described variously as a **brush border** or a **striated border**. The brush border gives the appearance of many very fine, closely spaced, and relatively short hairs, whereas the striated border gives the appearance of less numerous "sticklike" extensions from the apical surface. Although the terms "brush border" and "striated border" frequently are used in the literature, both are borders of microvilli. The microvilli on midgut cells of the mosquito *Aedes aegypti* are covered by a network of fine strands called the microvilli-associated network (Zieler et al., 2000). A major function of this dense network may be to protect the midgut microvilli from phagocytes and other cells in the blood meal until digestion begins.

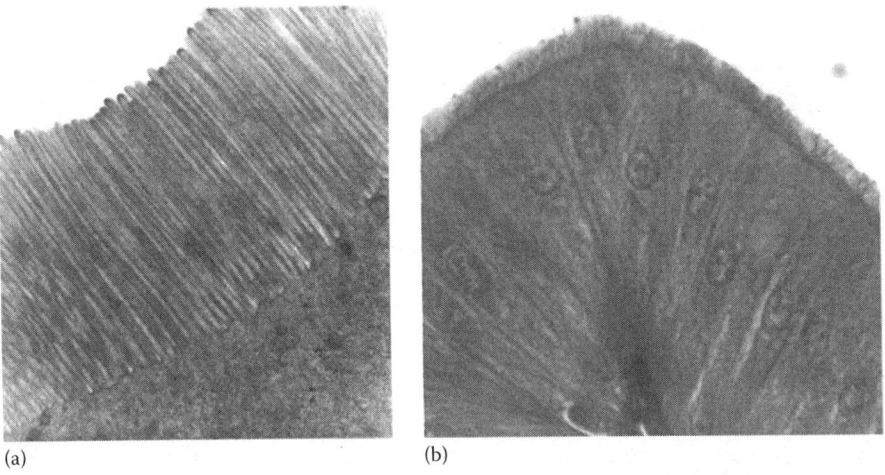

Figure 2.12 (a) Transmission electron micrograph of midgut microvilli from a chironomid larva. (b) The brush border on gastric caeca cells from a mole cricket, *Scapteriscus vicinus* (oil immersion, light microscope). (Photo by the author.)

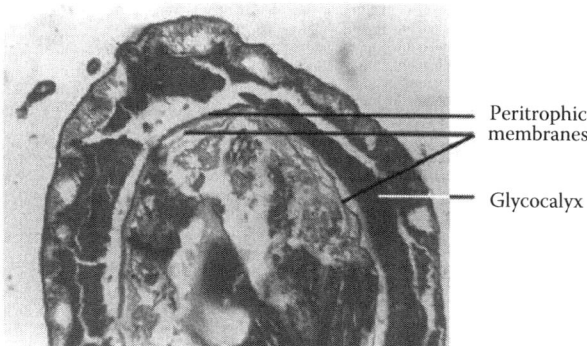

Figure 2.13 Photo of the peritrophic matrix and the dark layer of glycocalyx material between the peritrophic matrix and the surface of the cells in the midgut of a gryllid cricket. (Photo by the author.)

2.6 GLYCOCALYX

Most insects do not have a mucus lining in the midgut that is directly comparable to the mucus lining of various parts of the digestive system of vertebrates, but often there is a viscous secretion consisting of protein and carbohydrates on the surface of, and between, microvilli called the **glycocalyx** (Figure 2.13). The viscous glycocalyx traps and concentrates secreted enzymes and products of digestion (Santos and Terra, 1984; Santos et al., 1986).

2.7 PERITROPHIC MATRIX

The **peritrophic matrix (PM)** (also called the **peritrophic membrane**) surrounds the food in the midgut and serves as a shield to protect microvilli from direct physical contact with food particles (Lehane, 1997) (Figure 2.13). Wigglesworth (1961) characterized the PM as **Type I** when it is secreted as a continuous delamination all along the length of the midgut and **Type II** when it is secreted from a ring of cells at the anterior margin of the midgut. No particular cell type has been identified as secreting the Type I PM. The Type II PM is secreted continuously, similar to a stocking, as food pushes into it from the foregut. Most insects that produce a PM produce the Type I PM, but Dermaptera and larvae of Diptera produce a Type II PM. Mosquito larvae form a Type II PM, but adult mosquitoes secrete a Type I PM. Some insects, including adult mosquitoes, secrete a PM only after taking food into the gut. Stretching of the gut, rather than a secretogogue mechanism, seems to be involved in the case of mosquitoes because an enema of saline can induce PM production. Not all insects fit into the Type I/Type II model, and some do not form a PM at all. *Ptinus* spp. beetles secrete a PM starting only some distance along the middle third region of the midgut. Members of two weevil genera (*Cionus* and *Cleopus*) secrete a PM only toward the posterior of the midgut (Rudall and Kenchington, 1973).

Hemiptera and Homoptera as a group appear not to form a PM; at least a PM has not been unequivocally identified in any of them. In some Hemiptera, a perimicrovillar membrane, a thin membrane over the microvilli, has been described from electron microscope (EM) studies. Gryllid crickets have a PM, but several reports indicate that mole crickets (Gryllotalpidae) do not form a PM. Some adult lepidopterans and some adult tabanids have a PM while others do not; differences occur even within the same family (Waterhouse, 1953). *Drosophila* embryos and newly hatched *Aedes aegypti* mosquito larvae have a PM, but newly hatched honeybee larvae do not acquire one until several days after hatching. Although attempts have been made to relate the presence or absence of a PM to diet and to phylogeny, too many exceptions occur for any satisfactory relationship.

The peritrophic matrix contains chitin, proteins, glycoproteins, and proteoglycans, with chitin making up from 4% to about 20%, and protein composing up to 40% in various insects. Other components that have been reported are acid mucopolysaccharides, neutral polysaccharides, mucins, hyaluronic acid, hexosamine, glucose, and glucuronic acid. Chitin occurs in its α, β, and γ forms (see Chapter 4) in the PM of various insects, but α-chitin is most common. Kato et al. (2006) present evidence that chitin for the PM is synthesized *de novo* in adult *Aedes aegypti* mosquitoes (Culicidae) in response to ingesting a blood meal, but data are not available on anopheline mosquitoes. There appear to be at least two genes involved in chitin synthesis: one controlling chitin synthesis in the cuticle and one for chitin synthesis in the peritrophic matrix (Arakane et al., 2004; Hogenkamp et al., 2005).

Enzymes must pass through the PM to get at the food and small molecules resulting from digestion must pass out, therefore, the PM is porous. Reported pore sizes vary, perhaps with mode of estimation and with species tested. Santos and Terra (1986) reported the presence of 7- to 7.5-nm diameter pores in the PM of the sphingid caterpillar, *Erinnyis ello*. Pore size has been estimated at 200 nm in *Locusta* (Baines, 1978) and 150 nm in some cockroaches (Skaer, 1981). Using the permeability of fluorescently labeled dextrans, Edwards and Jacobs-Lorena (2000) determined that the main part of the PM of two mosquito larvae was permeable to 148 kDa or smaller particles, but that part in the gastric caeca was only permeable to 19.5 kDa or smaller particles. Mechanical damage and possible attack by protein-digesting enzymes of the gut probably act to shorten the life of the PM, and may cause some breakup of it in the posterior midgut and/or hindgut. Perhaps to counter such destructive action, some insects produce several separate peritrophic matrices several times per day, each encasing the one before it. Often multiple layers in the PM are observable with the transmission electron microscope; five layers occur in a dipteran, *Calliphora calcitrans*, but overall the PM is thin, varying from 0.13 to about 0.4 μm thick (Lehane, 1976).

2.7.1 Functions of the Peritrophic Matrix

A great deal of discussion in the literature has been devoted to possible functions of the PM. Suggested functions include

1. Protection of the delicate microvilli on the surface of midgut cells from contact with rough food particles
2. A barrier against entry of viruses, bacteria, or other parasites that would be too large to pass through the unbroken PM
3. An aid in preventing the rapid excretion of digestive enzymes
4. Compartmentalization of digestion within the midgut
5. Prevention of nonspecific binding of undigested materials or plant allelochemicals to midgut microvillar surfaces and/or binding to transport proteins at the midgut surface (reviewed by Terra, 1990; Lehane, 1997)

Although protection of the delicate midgut tissue from rough food particles is the most often mentioned function of the peritrophic matrix in the literature, Lehane (1997) believes that protection from pathogens ingested with the food is probably the most important function.

The PM may protect some phytophagous insects from toxic effects of ingested phenolic compounds, which are common in many plants, by preventing passage of the phenolics through the matrix and/or by complexing the substance within the PM (Bernays and Chamberlain, 1980).

The PM of the grasshopper, *Melanoplus sanguinipes* (Orthoptera: Acrididae), however, allowed some gallotannins to penetrate and adsorbed less than 1% of the tested tannins (Barbehenn et al., 1996).

A PM is present in *Tomocerus minor* (Humbert, 1979), a collembolan and a generalist scavenger and representative of very early evolution of insects. The PM probably evolved very early in a generalist scavenger feeder, in which protection of midgut microvillar surfaces from food particles, sand, or other hard substances coincidentally ingested was likely to be important. The PM has been conserved over long evolutionary time, even if some of its supposed functions are no longer important in a particular insect. A PM is present in many insects that do not feed upon rough or solid food, such as some blood feeders, and in adult lepidopterans that take flower and plant nectars. In these cases, it may be an evolutionary relic, but it may also serve some or all of the other protective functions enumerated earlier. Other insects seem to get along just fine without a PM at all.

2.8 DIGESTIVE ENZYMES

The secretion of digestive enzymes into the gut lumen is characterized as **constitutive secretion** when the enzymes are released from the cells as soon as they are synthesized and as **regulated secretion** when the enzyme is synthesized and stored, often as a **zymogen** (protein containing a peptide sequence that prevents enzymatic activity until the sequence is removed), until a signal to release it is received (Lehane et al., 1995). Most insects studied to date utilize constitutive secretion rather than regulated secretion. Two well-studied features of vertebrate digestive systems—storage of enzymes as inactive zymogens and stimulation of enzyme secretion by the food itself—occur in insects (Blakemore et al., 1995; Moffatt et al., 1995).

Signals to secrete digestive enzymes may come from stimulation of ingested food, in which case it is called **prandial control**, from **hormonal stimulation**, and from **paracrine control** (release of factors from putative endocrine cells in the gut) (Lehane et al., 1995). Clear-cut distinctions between these mechanisms of enzyme control are not always obvious from experimental analyses and subtle overlap of mechanisms may occur (Lehane et al., 1995). In general, paracrine and prandial control mechanisms of enzyme secretion are most common in insects. Proteins in the food are stimulants for digestive enzyme secretion in many insects (Blakemore et al., 1995, and references therein). Whether these act directly on enzyme secreting cells (prandial mechanism) or act through the putative endocrine cells (paracrine mechanism) present in the gut of many insects is not well established.

Midgut cells secrete enzymes in three ways. In the most common type of enzyme secretion, called **merocrine secretion** and also called **exocytosis**, enzymes are processed in the Golgi complex of the columnar cells and enclosed in small vesicles. These enzyme-containing vesicles fuse with the cell plasma membrane, and the enzymes are released to the gut lumen. In another, probably more costly and, thus, less common form of secretion, the entire midgut cell breaks down and the cytoplasmic contents are discharged into the lumen of the gut. This is called **holocrine secretion**. In a variation of this, called **apocrine secretion**, only parts of the cell, typically just the microvillar membranes, fragment and disintegrate into the gut lumen. A further variation on apocrine secretion is **microapocrine secretion** in which small single- or double-membrane vesicles are pinched off from the cell microvilli (Silva et al., 2013). Apocrine and microapocrine secretion are typical of the anterior part of the midgut, while exocytosis most often occurs in the posterior midgut. The mechanism of enzyme secretion may be related to the region of the gut and its particular function. For example, the anterior part of the midgut is often involved with the absorption of digested products, and apocrine or microapocrine secretion in this region may be an adaptation to promote dispersion of secretory vesicle contents into the midgut lumen in a region undergoing absorption processes (Cristofoletti et al., 2000).

In midgut cells of Lepidoptera, trypsin is incorporated into the membrane of small vesicles within the midgut cells. The vesicles migrate to the microvilli of the columnar cells, where trypsin is processed to become soluble within the vesicles. Through an exocytotic process, the vesicles bud from the microvilli as double membrane vesicles as they are released into the gut lumen. Trypsin is released into the gut lumen as the inner vesicle membrane fuses with the outer membrane and/or as the vesicles disintegrate due to the high pH in the lumen (Santos and Terra, 1984; Santos et al., 1986). A similar process occurs in *Aedes aegypti* (Graf et al., 1986), in which the vesicles containing soluble trypsin fuse with the membranes of the microvilli, releasing trypsin by exocytosis. Jordao et al. (1996) found that trypsin in larval midgut cells of the housefly, *Musca domestica*, is initially bound to membranes by a small peptide anchor, processed in the Golgi complex and enclosed in the membrane-bound form in secretory vesicles. These vesicles fuse with the plasma membrane at the gut lumen interface, and the trypsin, thus exposed to the gut pH (near neutral), is released by a conformation change of the anchoring peptide. Enzyme processing and secretion is very likely a costly process in any case, but holocrine and apocrine secretion cause the loss of all or parts of cells, necessitating extensive repair or replacement. Replacement cells grow in from regenerative cells.

2.8.1 Carbohydrate-Digesting Enzymes

Carbohydrate-digesting enzymes are secreted by the salivary glands as well as by the midgut epithelium. Dietary starch is the typical nutritive complex carbohydrate (excepting cellulose, which most insects cannot digest) ingested by phytophagous insects, and glycogen is a complex carbohydrate ingested by carnivorous insects. α-**Amylase**, acting upon starch and glycogen, is a common digestive enzyme in insects (Ramzi and Hosseininaveh, 2010; Vatanparast and Hosseininaveh, 2010; Razavi et al., 2011; Ghamari et al., 2014). It attacks interior glucosidic linkages of starch and glycogen, thus giving rise to a mixture of shorter dextrins. α-**Glucosidase** and oligo-1,6-glucosidase (**isomaltase**) assist in digesting the smaller dextrins, releasing glucose. Many insects also have one or more α- or β-**glycosidases** that digest a broad range of small carbohydrates. α-Glucosidase hydrolyzes maltose, sucrose, trehalose, melezitose, raffinose, and stachyose. α-Galactosidase hydrolyzes melibiose, raffinose, and stachyose. An α, α-**trehalase** in the gut of some insects digests trehalose that occurs in the body of other insects preyed upon as food. β-**Glucosidase** attacks cellobiose, gentiobiose, and methyl-β-glycosides. Lactose is hydrolyzed to glucose and galactose by β-**galactosidase**, and β-**fructofuranosidase** acts upon sucrose and raffinose to release simple sugars. **Chitinase** occurs in the gut of some insects (Souza-Neto et al., 2003; Genta et al., 2006, and references therein). A chitinase purified from the midgut of *Tenebrio molitor* has special properties, including lack of a chitin-binding domain in its structure, which may enable it to aid in digesting chitin-rich food without damaging the peritrophic matrix (Genta et al., 2006). How widespread and how effective midgut chitinase may be in the nutrition of insects is not clear, but many predatory insects, particularly chewing insects, probably ingest chitin. In contrast to a possible nutritional role for chitinases, some work has shown that adding chitinase to the diet of insects or feeding them transgenic plants that express chitinase can impair growth and development (Ding et al., 1998; Fitches et al., 2004), probably by damaging the gut structure. Pechan et al. (2002) showed that certain plants produce a chitin-binding cysteine proteinase that attacks the peritrophic matrix when insects feed on the plants.

An insect usually has only a few of the carbohydrate-digesting enzymes, depending upon the food it eats. Honeybees, *Apis mellifera*, have several α-glucosidases or sucrases that act rapidly upon sucrose, usually the principal carbohydrate in the nectar taken by these insects. They utilize the resulting glucose and fructose for an immediate energy source and for making honey. Termites feed upon and digest cellulose, as do some beetles, a few cockroaches, and woodwasps in the family Siricidae. Cellulose cannot be completely digested by one cellulase enzyme; the crystalline structure and the β-1,4 linkage of glucose units in cellulose make it difficult to hydrolyze, and complete digestion requires a complement of three enzymes, endoglucanases (EC 3.2.1.4), exoglucanases (EC 3.2.1.91), and cellobiases (EC 3.2.1.21) acting in sequential attacks. Recent evidence indicates that some insects are able to produce all the needed enzymes (Wei et al., 2006 and references therein). Some insects depend on protozoa or bacterial symbionts (some termites, beetles, and cockroaches) or fungi ingested with the food (some fungus-culturing termites, some beetle larvae, and woodwasp larvae) for some or all of the necessary cellulases. Endoglucanases and exoglucanases disrupt the crystalline structure of cellulose and digest long chains of cellulose into shorter cellobiose chains. β-Glucosidase (a cellobiase) releases glucose from cellobiose.

2.8.2 Lipid-Digesting Enzymes

Most of the fat eaten by an insect consists of **triacylglycerols**. **Lipases** secreted from the midgut and in some insects probably from symbionts as well, release fatty acids and glycerol from triacylglycerols. In the few insects studied, hydrolysis of triacylglycerols seems to proceed slowly. Slow hydrolysis may be caused by the choice of the substrate tested; natural triacylglycerols are often complex mixtures of different chain-length fatty acids esterified with glycerol, whereas test substrates are often triolein or tripalmitin in which all three fatty acids are oleic or palmitic, respectively. Emulsifying agents that enable hydrophilic enzymes to contact the hydrophobic surface of the triacylglycerol also are likely to be important in the digestion of lipids; natural emulsifying agents are largely unknown in the insect gut, but amino acids, proteins, and fatty acylamino complexes act as emulsifiers in some insects. Components in the glycocalyx layer in the gut may aid in emulsifying fats and in promoting contact between lipases and triacylglycerols.

2.8.3 Protein-Digesting Enzymes

Proteinases are classified as **serine**, **cysteine**, **aspartic acid**, and **metalloproteinases** depending upon the amino acid or metal at the active site of the enzyme (Barrett and Rawlings, 1991). The enzymes are further characterized by their sensitivity to specific inhibitors that act upon the amino acid(s) at the active site, by use of specific substrates that are attacked only by certain types of proteinases and by the pH for optimum activity. Some proteinases act optimally at alkaline pH while others have maximum activity at acid pH.

Serine proteinases include trypsin- and chymotrypsin-like proteinases and elastases that work at alkaline pH. These endoproteinases have been demonstrated in the midgut of many insects. Cysteine and aspartic acid proteinases have mildly acid pH optima. Proteinases with acid pH optima also have been called **cathepsins**. Trypsin, chymotrypsin, and aminopeptidases are common in many insects, but there are no reports of secretion in the same insect of proteinases active at both acid and alkaline pHs. The type of proteinase secreted and gut pH, obviously, must be coordinated in order for effective digestion to occur. The members of a taxonomic group, however, may not all have the same type of proteinases. Many beetles have cysteine proteinases most active at slightly acid pH (Thie and Houseman, 1990; Wolfson and Murdock, 1990). Some members of the family Scarabaeidae have serine proteinases that act at the high midgut pH typical of these insects, and they have no detectable cysteine proteinases (McGhie et al., 1995). Lepidoptera typically secrete trypsin-like enzymes (Valaitis, 1995), which are most active at alkaline pH and, thus, have a generally favorable pH in the midgut due to goblet cell secretion of potassium into the gut lumen.

Some of the enzymes that digest proteins are free in the gut lumen while others are membrane-bound. **Endoproteases** attack large proteins internally at the linkage between certain amino acids, thus, breaking the protein into smaller polypeptides, while **exopeptidases** attack the smaller pieces by cutting off the terminal amino acid. The presence of different types of proteinase inhibitors in the food eaten, especially plant-derived foods, may have promoted evolution of a variety of protein-digesting enzymes so that some enzyme molecules will escape inhibition. **Trypsin** and **chymotrypsin** are two major protein-digesting enzymes produced in the gut of insects (Figure 2.14). Trypsin, an endoproteinase, is a common component of midgut secretions in many insects. It attacks a protein at peptide bonds in which the carbonyl function comes from lysine or arginine. **Chymotrypsin-like endoproteinases** have been found in several insects, including a cockroach, some beetles, some mosquito larvae, and in some wasps and hornets. Chymotrypsin can attack at phenylalanine, tryptophan, and tyrosine residues. The larvae of the Mediterranean fruit fly, *Ceratitis capitata*, for example, have both trypsin and chymotrypsin serine proteinases, while the adult flies have only chymotrypsin-type proteinases (Silva et al., 2006).

Evidence for influence of food on the type of proteinase secreted is weak (Thie and Houseman, 1990; Wolfson and Murdock, 1990). Blood-feeding insects may have either serine proteinases or cathepsins. For example, the major proteolytic enzyme in the midgut of female *A. aegypti* is trypsin-like (Borovsky and Schlein, 1988), as is also the case in *Stomoxys calcitrans*, two blood-feeding dipterans. *Rhodnius prolixus* (Hemiptera), also a blood feeder, secretes cathepsins active at acid pH, but has no trypsin-like enzymes. In *S. calcitrans* and *R. prolixus*, the blood meal acts as a stimulus for secretion of the extracellular proteinases, but not for levels of membrane-bound aminopeptidases (Houseman et al., 1985).

The endoproteinases and some exoproteinases must pass through the peritrophic matrix to promote digestion of the large proteins. Smaller proteins and peptides released by the digestive processes may diffuse out of the peritrophic matrix through pores of the PM, with final digestion taking place at the surface of the microvilli, where there are both free and cell membrane–bound aminopeptidases. **Aminopeptidases** (also called **exopeptidases**) are exoenzymes that remove one amino acid after another from the end of a small peptide. Cell membrane–bound aminopeptidase activity has been found in *R. prolixus* (Hemiptera), *S. calcitrans*, and some other Diptera, some Lepidoptera, and some Coleoptera. A carabid beetle, *Pheropsophus aequinoctialis*, has both free and cell membrane–bound aminopeptidases. When the exopeptidase is bound at the microvillar surface, the amino acids cut from the end of a polypeptide are already at the site, the microvilli, for absorption.

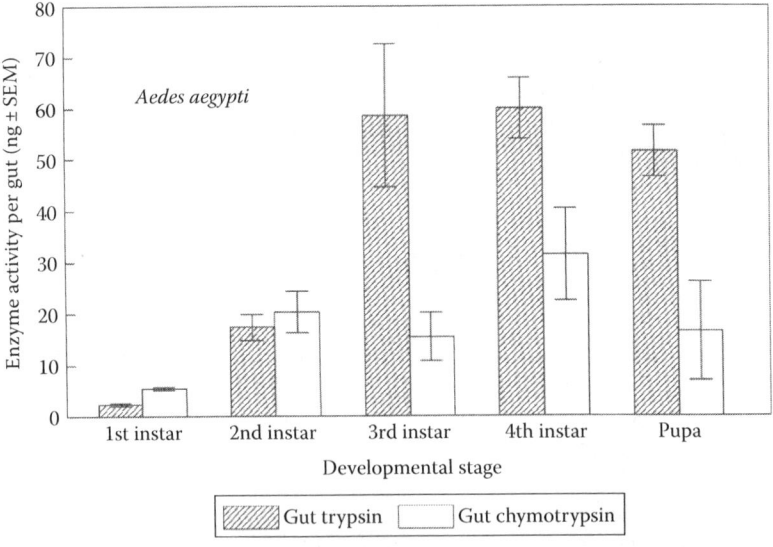

Figure 2.14 Synthesis of trypsin- and chymotrypsin-like enzymes by *Aedes aegypti* larvae and pupae. (Author constructed histogram from data by Borovsky, D., *Pestycydy*, 3, 71, 2005.)

2.8.4 Do Proteinase Inhibitors in the Food Influence the Evolution of Proteinase Secreted?

There may be a relationship between the type of proteinase inhibitors in food and the evolutionary selection for type and number of proteinase secreted by insects (Terra, 1988), but only a few insects have been studied. Proteinase inhibitors typically are small proteins that are known to occur in more than 100 plants. Some of the proteins inhibit serine proteinases, while others act upon cysteine proteinases. Some insects respond to consumption of a trypsin inhibitor by secreting additional trypsin-like enzyme(s) that allow differential susceptibility to the inhibitor and/or hyperproducing enzymes to compensate for inhibition (Broadway, 1995; Broadway and Villani, 1995). Bayés et al. (2006) identified and purified a carboxypeptidase B enzyme in the gut of larval corn earworm, *Helicoverpa zea*, that is not inhibited by potato carboxypeptidase inhibitor, which typically inhibits carboxypeptidases from insects; carboxypeptidase B was not inhibited in several other lepidopteran species surveyed. Inhibitors of α-amylase occur in Triticale seeds (Mehrabadi et al., 2012).

Many insects secrete more than one proteinase, including multiple molecular forms of some proteinases (Moffatt et al., 1995 and references therein). At least 20 trypsin isozyme bands were detected by isoelectric focusing in midgut homogenates of *A. aegypti*, with 5 of them representing major bands (Graf and Briegel, 1985). Lopes et al. (2006) found evidence from structural analyses that insect trypsin-active sites have become more hydrophobic in the course of evolution, apparently as an adaptation to resist inhibition by plant protein inhibitors that typically have polar amino acid (hydrophilic) residues at their active binding sites. Multiple isozymes and hyperproduction of some enzymes may enable some measure of escape from ingested inhibitors. Transgenic plants containing proteinase inhibitors have been tested and proven to have adverse effects upon the growth of some insects (McManus et al., 1994). Nevertheless, it remains to be seen if this technique can be successfully used in insect control and, if so, for how long before insects develop coping mechanisms.

2.9 HORMONAL INFLUENCE ON MIDGUT

Hormonal control of digestive enzyme secretion does not appear to be a widespread mechanism among insect groups, but a well-defined case occurs in *A. aegypti* female mosquitoes. The females must have a series of blood meals to mature successive batches of eggs. The terminal oocytes in each of the numerous ovarioles of both ovaries mature together, and this set of mature eggs must be laid to make room for the growth and maturation of a second set of eggs. A decapeptide hormone (**trypsin-modulating oostatic factor, TMOF**; amino acid sequence YDPAPPPPPP; see Table 2.1 for single-letter notation for amino acid identity), synthesized by ovarian follicular epithelium cells, is released 24–42 h post feeding and transported by the hemolymph to receptors on the hemolymph side (basal side) of midgut cells (Borovsky et al., 1994a,b). The hormone, possibly through a second messenger, signals midgut cells to cease producing late trypsin (see next paragraph) that is necessary to finish digesting the blood meal. Younger or secondary oocytes cease growing, apparently from lack of nutrients, until the inhibition of digestion is released after the female lays the mature eggs (Borovsky et al. 1990, 1993). It still remains to be determined how TMOF acts upon midgut cells and how the inhibition is released so that the next batch of eggs can be nourished and matured. The overall process is highly adaptive for the mosquito since the abdomen of the female is not large enough to hold continuously maturing eggs.

Two forms of trypsin are secreted in *A. aegypti* and both forms appear to be critical to the ultimate formation of eggs. An **early trypsin** (Graf and Briegel, 1989) is regulated at the transcriptional level (Lu et al., 2006) and is present in the midgut during the first 4 h after feeding (Noreiga et al., 1996; Lu et al., 2006). Transcription of the early trypsin gene is controlled by juvenile hormone (JH) level and occurs in the midgut after adult emergence (Noriega et al., 2001). Translation of mRNA occurs in the midgut cells prior to feeding (Felix et al., 1991). Early trypsin begins the process of

Table 2.1 Twenty Amino Acids Found in Proteins

Amino Acid	Three-Letter Designation	Single-Letter Designation	Molecular Weight
Alanine	Ala	A	89
Arginine	Arg	R	174
Asparagine	Asn	N	132
Aspartic acid	Asp	D	133
Cysteine	Cys	C	121
Glutamic acid	Glu	E	147
Glutamine	Gln	Q	146
Glycine	Gly	G	75
Histidine	His	H	155
Isoleucine	Ile	I	131
Leucine	Leu	L	131
Lysine	Lys	K	146
Methionine	Met	M	149
Phenylalanine	Phe	F	165
Proline	Pro	P	115
Serine	Ser	S	105
Threonine	Thr	T	119
Tryptophan	Trp	W	204
Tyrosine	Tyr	Y	181
Valine	Val	V	117

blood digestion. **Late trypsin** is regulated at the transcriptional level (Lu et al., 2006) and peaks 18–24 h after the first blood meal. It is the major endoproteinase in the midgut and it is necessary to finish the blood meal digestion. Although early evidence suggested that early trypsin action on the blood meal provided the necessary stimulus for inducing gene transcription of late trypsin (Barillas-Mury et al., 1995), more recent evidence indicates that early trypsin action on the blood meal may not be necessary for late trypsin transcription; application of ribonucleic acid interference (RNAi) to reduce the level of early trypsin expression did not stop the transcription of late trypsin (Lu et al., 2006). Late trypsin synthesis is the form primarily influenced by TMOF. Sufficient amino acids result from early trypsin action to promote maturation of a batch of eggs before TMOF stops late trypsin synthesis. TMOF, which has been synthesized, can survive potential protease degradation and be absorbed in an active form from the midgut when administered to mosquitoes in a blood meal, offering hope that the synthetic hormone may have potential for population control (Borovsky and Mahmood, 1995). Borovsky (2007) recently reviewed the biology, chemistry, and applications of TMOF. One notable recent discovery is that although the natural hormone is an unblocked decapeptide with the structure of NH_2-YDPAPPPPPP-COOH, the amino terminal end of only four amino acid residues (NH_2-YDPA) had 95% as much activity as the decapeptide, and also indicated that this is the part of the molecule that binds to the midgut receptor (Borovsky, 2005).

A similar hormonal mechanism involving TMOF also has been demonstrated in the grey fleshfly *Neobellieria bullata*, except that the neuropeptide is a hexapeptide (amino acid sequence NPTNLH), called **Neb-TMOF** (Borovsky et al., 1996). Neb-TMOF also has been isolated from larvae of the blue blowfly, *Calliphora vicina*, and it seems to have dual functions, acting in larvae as an **ecdysiostatin** that inhibits the synthesis of ecdysteroid, while inhibiting the synthesis of trypsin in the midgut of adult females (Hua et al., 1994; Hua and Koolman, 1995).

The midgut of many insects appears to be a rich source of (putative) endocrine secretions (Sehnal and Žitňan, 1990), but identity of the secretory products and their functions are largely unknown. The midgut in liver-fed black blowflies, *Phormia regina*, releases a hormone targeted for the brain about 4–8 h after the meal. In the brain, it stimulates median neurosecretory cells to initiate the neuroendocrine cascade (see Chapter 20) leading to ovary and egg development (Yin et al., 1993, 1994). Additional research is needed to determine further details and whether similar hormones occur in other insects. Midgut cells may be involved in secreting neuropeptides that mediate many aspects of digestion and/or enzyme secretion, but few details are available.

2.10 COUNTERCURRENT CIRCULATION OF MIDGUT CONTENTS AND ABSORPTION OF DIGESTED PRODUCTS

Berridge (1970) proposed that insects might have a **countercurrent circulation** of fluid contents in the midgut, called the **endoectoperitrophic countercurrent flow**. Such a process could (1) serve to increase digestive efficiency, (2) conserve nutrients that might be lost by a rapid passage through a short midgut, (3) conserve and reuse enzymes that would otherwise be excreted rapidly with the bulk of food moving through the gut, and/or (4) allow absorption of digested products along the entire length of the midgut and/or absorption in the gastric caeca by permitting fluid containing digested products to flow forward. Dow (1986) suggested that the evolutionary driving force for a countercurrent circulation of midgut contents may have been nutrient conservation.

In a countercurrent circulation, food and gut contents within the peritrophic matrix move posteriorly after having entered the midgut from the foregut, while fluid containing partially or completely digested food materials outside the peritrophic matrix moves anteriorly between the midgut cell surfaces and the peritrophic matrix (Figure 2.15). The anterior movement of fluids outside the

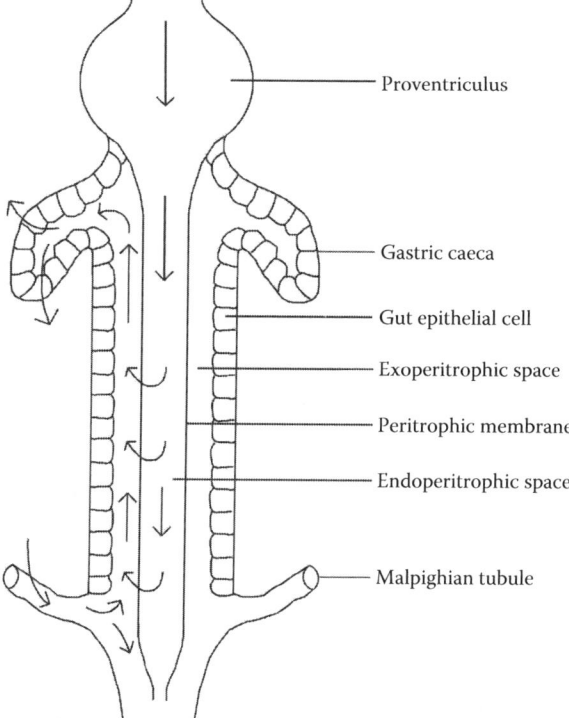

Figure 2.15 Diagrammatic illustration of the countercurrent flow that occurs in the midgut of some insects. Fluid may be passed forward from the Malpighian tubules to help create the forward flow that carries small end products of digestion to the gastric caeca, which are very efficient in the absorption of fluid and nutrients.

peritrophic matrix is promoted by fluid secreted into the ectoperitrophic space by the posterior midgut or, in some cases, by the Malpighian tubules (Dow, 1981). Fluids and dissolved digestion products may be absorbed all along the midgut by the columnar epithelial cells, but in some insects the gastric caeca rapidly absorb fluids and dissolved nutrients brought to them by the countercurrent flow. Only a few species have been studied critically with respect to this mechanism and it is not clear how common it is in insects. Dow (1981) suggested four criteria for determining whether an endoectoperitrophic flow occurred in insects:

1. A posterior region of the gut (could include Malpighian tubules) should be specialized for secretion of fluid into the posterior midgut.
2. An anterior region of the midgut should be specialized for absorption.
3. There should be a concentration gradient of small molecules between the point of fluid secretion and that of absorption.
4. Metabolites from a meal should stay in the gut longer than the bulk of the solid food originally ingested.

Absorption of amino acids from the midgut has been studied in only a few insects, but in several lepidopterans (*M. sexta*, *Philosamia cynthia*, and *Bombyx mori*), amino acids are actively absorbed by transport proteins specific to particular amino acids (Dow, 1986). Several specific transport proteins have been isolated and identified (Giordana et al., 1989). The driving energy for amino acid absorption from the gut in these lepidopterans is the proton-ATPase pump described in goblet cells. The pump creates the high K^+ concentration in the gut lumen and the high transepithelial potential across the gut wall, both shown to be important for amino acid absorption by columnar cells in lepidopterans. There are at least six transport systems, including

1. A transporter of neutral amino acids
2. A specific system for proline
3. A specific system for glycine
4. A specific system for L-lysine
5. A specific transporter for glutamic acid
6. A transporter that is very stereospecific for D-alanine (Giordana et al., 1989)

Sodium ions are an efficient experimental substitution for K^+ in the neutral transport system and in some other transporter systems (Reuveni and Dunn, 1990, 1993; Hennigan et al., 1993a,b), but, in the lepidopteran gut, K^+ is probably the major ion involved because it is present in high concentration, whereas Na^+ is not. Transport proteins and systems for leucine and tyrosine also have been demonstrated in midgut tissue of the Colorado potato beetle, *Leptinotarsa decemlineata* (Reuveni et al., 1993; Hong et al., 1995). Blocking the absorption of amino acids and disruption of their transport systems have been suggested as potential targets for insect control (Hong et al., 1995).

Glucose from digestion of carbohydrates is rapidly absorbed passively by a process known as **facilitated diffusion** (Treherne, 1957, 1958, 1959; Wyatt, 1967). Fat body cells, which can be found attached in small groups on the hemolymph side of the gut, rapidly synthesize the absorbed glucose into the disaccharide trehalose, keeping the hemolymph concentration of glucose low in most insects. Consequently, even low concentrations of glucose in the gut can continue to be absorbed passively.

Fatty acids undergo esterification as they traverse the gut cells and are released into the hemolymph as diacylglycerols for transport to fat body cells (Weintraub and Tietz, 1973, 1978; Turunen, 1975; Turunen and Chippendale, 1977; Chino and Downer, 1979; Thomas, 1984). The diacylglycerols are picked up at the hemolymph side of the gut by lipoprotein complexes, **lipophorins**, that enable transport of the absorbed lipids through the aqueous hemolymph. Lipids are mainly stored in fat body cells as triacylglycerols. After lipophorin delivers the absorbed lipid to the fat body, it can recirculate to load and transport additional lipids absorbed (Chino and Kitazawa, 1981; Chino, 1985; Surholt et al., 1991; Van Heusden et al., 1991; Gondim et al., 1992; Blacklock and Ryan, 1994).

2.11 TRANSEPITHELIAL AND OXIDATION–REDUCTION POTENTIAL OF THE GUT

There is a **transepithelial potential** (TEP) across the gut wall and an oxidation–reduction or **redox potential** within the lumen of various parts of the gut. In most insects, it is not known how these potentials are created or controlled. TEP values ranging from lumen negative to lumen positive occur in different insects. In *P. americana* midgut the TEP varied from −8 to −26 mV (lumen negative to hemolymph), while TEP in the hindgut ranged from E_o = −84 to −240 mV (Bignell, 1981). High negative values, such as those in the hindgut of insects (termites) with large populations of microorganisms that digest cellulose indicate anaerobic conditions, while positive or slightly negative values indicate aerobic conditions. Insects, such as clothes moths, dermestid beetles, and bird lice that digest keratin, the major protein of wool, fur, and feathers, have strongly negative redox (reducing) potentials in the midgut. The reducing conditions facilitate the breaking of disulfide bonds in the keratin molecule, making keratin more digestible.

Redox potential in the various regions of the gut may play a large role in the digestion and assimilation of food materials, in detoxification reactions, and in the production of toxic metabolites from ingested food materials. Ability to adjust gut redox potentials in plant-feeding insects may be one of the ways that insects adapt in the coevolutionary race with plants in combating the wide range of allelochemicals common in many plants. Appel and Martin (1990) found reducing conditions in the midguts of two lepidopterans, *M. sexta* and *Polia latex*, which they suggest would make ingested phenolic compounds less likely to be oxidized to highly toxic and reactive quinones than the oxidizing potentials prevailing in the midgut of several other lepidopterans studied. Oxygen levels in the foregut and midgut lumens of 10 species of caterpillars and 3 species of grasshoppers were generally very low, indicating nearly anoxic conditions (O_2 equal to less than 7.3 mm Hg) and the gut was able to deplete oxygen caused by swallowing oxygen with the food or feeding an artificial diet that increased oxygen tension in the gut in some cases (Johnson and Barbehenn, 2000). The authors suggest that low oxygen tension in the gut may be very common in herbivorous insects, and that it is an adaptation to reduce the rate of oxidation of ingested plant allelochemicals that may be more toxic when oxidized.

2.12 Gut pH

Few generalizations about insect **gut pH** can be made, except that it is highly variable in different insects (Table 2.2). The pH of a gut segment greatly influences the action of any enzymes secreted into or carried with the food into that segment. In addition, gut pH may influence solubility of ingested components, toxicity of some potential toxins, and the population of gut microorganisms. Cathepsin- and trypsin-like enzymes attacking proteins work optimally at acid and alkaline pHs, respectively. Carbohydrate-digesting enzymes usually work best at near neutral pH or under slightly acid conditions. Lipases that digest triglycerides and other esters work best at alkaline pHs near 8.

The crop tends to be slightly acidic in most insects, with little or no presence of buffering agents that could alter pH due to organic acids produced when digestion occurs in the crop, as it does in Orthoptera, Dictyoptera, and some other insects. When proteins are the primary food material for the cockroach, *P. americana*, the crop is slightly acidic at pH 6.3, but when sugars (maltose, lactose, sucrose, or glucose) are eaten, the crop has a pH of 4.5–5.8 because of the glycolytic cycle acids produced.

Larvae of Lepidoptera and Trichoptera tend to have a very high midgut pH, varying from about 8 to 10, promoted by goblet cells that secrete potassium bicarbonate into the lumen of the midgut. The tobacco hornworm larva (*Manduca sexta*) has a midgut pH of 10–12 in different parts of the midgut, which is promoted by the active secretion of K^+ into the midgut in exchange for H^+ by the

Table 2.2 pH in Various Parts of the Gut of Selected Insects

Insect	Foregut	Midgut	Hindgut	Ref.[a]
Orthoptera				
Melanoplus sanguinipes, grasshopper (Acrididae)	5.52	6.75	6.80	1
Photaliotes nebrascensis, grasshopper (Acrididae)	6.03	7.12	6.11	1
Schistocerca gregaria, desert locust (Acrididae)	5.5	6.7–7.0		4
Schistocerca gregaria		5.3		5
Gryllus rubens, cricket (Gryllidae)	5.8–6.0	7.4–7.6	7.6–7.8 (anterior hindgut)	7
Gryllus bimaculatus, cricket (Gryllidae)	5.84 (crop)	8.07	8.50–7.59 (illeum, rectum)	19
Scapteriscus borelli, mole cricket (Gryllotalpidae)	5–7	6–8 (gastric caeca)	7–8 (anterior hindgut)	7
Periplaneta americana, cockroach (Blattidae)		6.3		8
Leucophaea madeirae, cockroach (Blattidae)		9.5 (posterior midgut)		9
Coleoptera				
Popillia japonica larvae, Japanese beetle (Scarabaeidae)		8.5		2
Exomala orientalis larvae, oriental beetle (Scarabaeidae)		8.5–9.0		2
Rhizotrogus majalis larvae, European chafer (Scarabaeidae)		9.0–9.5		2
Maladera castanea larvae, Asiatic garden beetle (Scarabaeidae)		8.5		2
Lichnanthe vulpina larvae, cranberry root grub (Scarabaeidae)		8.5		2
Phyllophaga anixia (Scarabaeidae)		8.5–9.0		2
Oryctes nasicornis (Scarabaeidae)		12.2		14
Epilachna varivestis larvae, Mexican bean beetle (Chrysomelidae)		5.8		3
Anthonomus grandis larvae, boll weevil (Curculionidae)		4.6–5.6		3
Tribolium castaneum, red flour beetle (Tenebrionidae)		6.0		18
Lepidoptera				
Agrotis ipsilon, black cutworm (Noctuidae)		8.5–9.0		2
Manduca sexta, tobacco hornworm (Sphingidae)		9.5–9.7		5
Manduca sexta		6.4 apical folds, ant. MG 8.2 basal folds, ant. MG 7.2 apical folds, post MG		6

(Continued)

Table 2.2 (*Continued*)   pH in Various Parts of the Gut of Selected Insects

Insect	Foregut	Midgut	Hindgut	Ref.[a]
Diptera				
Simulium vitatum, blackfly (Simuliidae)	11.4			15
Tipula abdominalis, cranefly (Tipulidae)	11.6			16
Lucilia cuprina larvae, blowfly (Calliphoridae)		7.4–8 anterior MG 3.3 middle MG 7.4–8 posterior MG		17
Tricoptera				
Caddisfly larvae	7			10
Plecoptera				
Stonefly larvae (Pteronarcyidae)	7			11
Isoptera				
Termites		>10 anterior midgut		12
Five species of soil-feeding termites (Termitidae: Termitinae)	Slightly acid	Slightly acid to slightly alkaline in different species	11–12.5 most anterior hindgut, >10 in second dilation of hindgut, slightly >7 approaching rectum, 4.8–6 in rectum	13

[a] References for data in Table 2.2; additional data and references can be found in Berenbaum (1980). 1. Barbehann et al. (1996), 2. Broadway and Villani (1995), 3. Murdock et al. (1987), 4. Evans and Payne (1964), 5. Martin et al. (1987), 6. Dow and O'Donnell (1990), 7. Thomas and Nation (1984), 8. O'Riordan (1969), 9. Engelmann and Geraerts (1980), 10. Martin et al. (1981a), 11. Martin et al. (1981b), 12. Bignell and Anderson (1980), 13. Brune and Kühl (1996), 14. Bayon (1980), 15. Undeen (1979), 16. Martin et al. (1980), 17. Waterhouse and Stay (1955), 18. Krishna and Saxena (1962), 19. Teo (1997).

proton ATPase pump (see Section 2.4.3) and by transport of ammonia from the gut lumen into the hemolymph (Weihrauch, 2006). Weihrauch suggests that the midgut columnar cells are the likely site of ammonia transport, which aids midgut alkalinization. Lepidoptera caterpillars are predominately phytophagous insects, and Berenbaum (1980) concluded from a survey of published pH values for 60 species in 20 families that midgut pH was related to host plant chemistry. Those larvae that fed upon leaves of trees, which typically contain larger quantities of tannins, had an average midgut pH of 8.67, while those that fed mostly upon herbs and forbs had an average pH of 8.29 in the midgut. The higher midgut pH in those feeding on tannin-rich food may have evolved as a protective mechanism to reduce the toxicity of tannins, which tend to complex with proteins, but do so less readily at higher pH.

Very acid conditions prevail in the special hindgut regions of some termites, crickets, and possibly other insects that have hindgut fauna to digest cellulose. The acid conditions are caused by the anaerobic fermentation of glucose from cellulose digestion, resulting in production of short-chain fatty acids including, acetic, propionic, and butyric acids.

2.13 HEMATOPHAGY: FEEDING ON VERTEBRATE BLOOD

Blood feeding has evolved independently several times in the course of insect evolution. Feeding upon vertebrate blood is of particular concern to humans because of disease transmission from vector to human (Lehane, 2005). Nearly 14,000 species of arthropods in 400 different genera evolved the ability to feed upon vertebrate blood (GraHa-Souza et al., 2006), which might at first seem like a very large number, but actually when one considers that more than one million species of insects are known and, perhaps, millions more exist, the number that can feed upon vertebrate blood is very small. GraHa-Souza et al. (2006) suggest that the blood feeding habit may have evolved about the time that the soft-skinned mammals and birds began to expand with the demise of dinosaurs, and that toxic properties of heme that can generate reactive oxygen radicals leading to the oxidation of lipids has played an important role in the evolution of blood-feeding arthropods. Hemoglobin degradation in the gut of blood feeders releases large quantities of heme and it can potentially alter membrane permeability. Insects and other blood-sucking arthropods have evolved various mechanisms to counter the toxic properties of heme, including the formation of aggregates of heme that take it out of solution, antioxidant properties in the hemolymph of some insects, heme-binding proteins, and metabolic conversion of heme breakdown products into more hydrophilic conjugates that facilitate excretion.

Some insects have the problem of getting rid of excess water and salts after a meal (for example, blood feeders such as mosquitoes, and the tropical hemipteran, *Rhodnius prolixus*) and it is important for them to transport H^+, Na^+, and K^+, and water rapidly from the midgut into the circulating hemolymph and subsequently into Malpighian tubules for excretion (Staniscuaski et al., 2013; Pacey and O'Donnell, 2014). *Rhodnius* takes only one blood meal at the beginning of each instar; the blood may be up to 10 times its unfed weight, ballooning its abdomen, and it must rapidly excrete the excess water and salts. Water movement from the midgut is facilitated by **aquaporins** (Gonen and Walz, 2006), which allow water to move very fast across membranes. Aquaporins are protein-based membrane channels and are widely distributed in all living organisms, including plants (Staniscuaski et al., 2013 and references therein). Some of the channels, called **aquaglyceroporins,** allow glycerol and other small neutral compounds to cross membranes rapidly. In contrast to those insects that need to excrete excess water, most insects need to conserve water, and aquaporins and aquaglyceroporins in phytophagous caterpillars aid them in extracting and conserving all the water from the green leafy food of their dietary plants (Kataoka et al., 2009). Aquaporins and aquaglyceroporins have a major role in freeze-protection of the Antarctic chironomid midge, *Belgica antarctica*, which survives freezing for long months in the Antarctic. Blocking the aquaporins with mercuric chloride decreased the ability of midgut and Malpighian tubules to tolerate freezing without damage (Yi et al., 2011). Functioning aquaporins are important to prevent freezing damage in the tephritid gall fly, *Eurosta solidaginis* (Philip and Lee, 2010; Philip et al., 2011).

2.14 DIGESTIVE SYSTEM MORPHOLOGY AND PHYSIOLOGY IN MAJOR INSECT ORDERS

Detailed studies of digestion have been made in only a relatively few insects within the major insect orders. Because of the diversity of insects, it is always risky to generalize from studies on a few insects, and the reader should keep in mind that there may be exceptions in minor, or even major, details from these limited studies. More extensive details on food habits and related gut structure and function can be found in reviews by Dow (1986), Terra (1990), and Billingsley (1990).

2.14.1 Orthoptera

The crop is a major site of digestion in locusts, grasshoppers, and crickets, and possibly in other Orthoptera. Starch digestion is accomplished in the crop of crickets with enzymes secreted forward from the midgut. Salivary enzymes play only a minor role in digestion. Midgut caeca located at the anterior end of the midgut in locusts and crickets rapidly absorb fluids and dissolved nutrients as these enter from the crop. A Type I peritrophic matrix is secreted in the midgut. Cellulase activity is found in the midgut of some grasshoppers, but its origin and the extent of its function are uncertain. Orthoptera may not generally have the endoectoperitrophic countercurrent flow in the midgut, although in *Schistocerca gregaria*, countercurrent flow occurs in starved locusts, but not in constantly feeding locusts. Terra (1990) has suggested that the countercurrent flow in a starved individual may be an adaptation to keep food and digestive enzymes in the midgut longer for more complete digestion. This might represent a tradeoff of averting starvation vs. keeping potential alleochemicals in the gut longer. The gut of the praying mantis, *Tenodora sinensis*, exhibits extreme modifications to serve the predatory habits and intermittent feeding of this carnivorous insect. The foregut, especially the crop, is long and wide, and occupies nearly the entire length of the body, apparently as an adaptation for the storage of opportunistically available prey (Dow, 1986). The midgut, eight gastric caeca, and the hindgut are shortened and compressed into the last three abdominal segments.

2.14.2 Dictyoptera

Cockroaches are scavengers and opportunistic feeders. The crop is large in cockroaches and is a major organ of digestion. Amylase from the salivary glands and protease, lipase, and carbohydrate-digesting enzymes secreted forward from the midgut contribute to digestion in the crop. Crop emptying is gradual and is regulated by the osmotic pressure created by the small molecules resulting from digestion of crop contents (Englemann, 1968). The higher the osmotic pressure in the crop, the slower the crop empties into the midgut, functionally preventing oversaturation of absorption by gastric caeca and possible loss of poorly absorbed nutrients. Gastric caeca located at the anterior of the midgut are the main sites for absorption. A Type I peritrophic matrix is present and there may be some countercurrent endoectoperitrophic flow. Some final digestion probably occurs in the ectoperitrophic space on the surface of the midgut cells. *Periplaneta americana* incorporates ^{14}C into hemolymph trehalose from labeled cellulose (Bignell, 1977), but the digestion of the cellulose occurs in the hindgut (colon) with the aid of cellulases from bacteria located on the gut luminal wall. The redox potential of the hindgut favors the action of these anaerobic bacteria by varying from -84 to -240 mV, values that are indicative of an anaerobic gut segment (Bignell, 1981). Short-chain fatty acids are produced by bacterial fermentation of glucose liberated from cellulose in the colon. The fatty acids are absorbed through the hindgut wall.

2.14.3 Isoptera

The gut of termites is highly specialized for housing gut microbiota that aid in digestion of cellulose that they obtain from wood, fungi, or other sources depending upon their lifestyle. Gut variation exits among the castes in a colony; for example, soldiers in the family Rhinotermitidae are fed liquid food by the worker caste and do not have to digest cellulose, therefore, they have reduced gut structure. The workers are the social caste responsible for colony construction and nutrition, and they have highly evolved gut chambers to hold various types of microbiota. Termites hatch without their gut microbiota, but soon receive them by feeding upon fluid and excreta from the proctodeum of older nymphs. They lose most of their gut symbionts at each molt, and become reinfected by proctodeal feeding.

The so-called lower termites have flagellate protozoans as well as bacteria in the hindgut, and they get cellulase(s) from their symbionts, and some are known to secrete their own cellulases from the salivary glands (Fujita et al., 2010). Those termites belonging to the "higher termites" in the family Termitidae lack symbiotic protozoa, but have symbiotic bacteria in the hindgut, which is divided into five segments. Many of the higher termites feed upon fungi (Anklin-Mühlemann et al., 1995) or on conidiophores of a fungus growing in their nests which digest cellulose, and some secrete from the salivary glands their own cellulases.

Spirochetes are present in the hindgut of many termites, but their role in digestion is uncertain. They may help to recycle nitrogen, synthesize amino acids, assist in maintaining a low redox potential, and protect from various pathogens (Breznak, 1982; Boucias et al., 1996). The diet of termites tends to be low in protein content, and they acquire protein from the bacterial cells in feces by trophyllaxis and by feeding upon the fecal wastes of each other. Symbionts in the hindgut also can fix atmospheric nitrogen and synthesize proteins from the fixed nitrogen. A peritrophic matrix is usually present in termites (Noirot and Noirot-Timothee, 1969). Protein is digested in the midgut and some termites probably have an endoectoperitrophic flow of fluids and enzymes.

Wood-feeding termites tend to have acetogenic bacteria that ferment glucose to acetate, while some fungus-growing and soil-feeding termites evolve methane from anaerobic fermentation (Brauman et al., 1992). The principal metabolite from cellulose digestion is glucose, which is fermented to acetate that is actively absorbed by the hindgut cells through an energy-requiring mechanism. For example, cellulolytic organisms in the hindgut of *Reticulitermes flavipes* ferment glucose to acetate, CO_2, and H_2. Then CO_2-reducing acetogenic bacteria convert the free H_2 and CO_2 to an additional acetate. Termites absorb the acetate from the hindgut and utilize it for an energy source by metabolism in the Krebs cycle. Some fungus-growing termites convert the hydrogen and carbon dioxide from the initial fermentation of glucose into methane (CH_4) rather than into another acetate molecule. Some investigators have suggested that termites are a significant environmental source of methane, a greenhouse gas, but Brauman et al. (1992) caution that much more data are needed about the distribution of the two processes among termites to make accurate estimates. More details on termite fungal feeding and symbionts in termites can be found in Chapter 21.

2.14.4 Hemiptera

Hemiptera have been reclassified into the order Heteroptra, but the use of the term Hemiptera is retained here because the literature cited uses that term. *Rhodnius prolixus* (family Reduviidae), the "kissing bug," has been a favorite model hemipteran for the study of digestion. Each instar takes just one blood meal (if allowed to feed to repletion), digests it slowly, utilizes the nutrients to support growth and molting to the next instar, and after molting takes another blood meal. The midgut is divided into two major divisions: the anterior midgut and the posterior midgut. Columnar, cuboidal, regenerative cells, and endocrine cells occur in the midgut. The columnar and cuboidal cells have microvilli on the apical surface and basal infoldings. There is no peritrophic matrix, but a peculiar perimicrovillar membrane composed of two trilaminar membranes forms continuously over the microvilli. The two membranes are held very close, but at a constant distance apart, by structural columns or pegs (composition unknown). Little is known about the origin, ultimate fate, and function of the membranes. The anterior midgut functions in carbohydrate and lipid processing, among other functions, but does not secrete enzymes for protein digestion. The posterior midgut has separate functionalities in its anterior and posterior parts. The first part of the posterior midgut secretes cathepsins B and D (active at acid pH), aminopeptidase, and carboxypeptidase, among other enzymes, and protein digestion is initiated. Digestion is completed in the posterior midgut where additional protein-digesting enzymes are secreted and absorption of products occurs. Endocrine cells are concentrated in this region of the midgut, but practically nothing is known of their function. The excellent review by Billingsley (1990) should be consulted for more details on gut structure and function in *Rhodnius*.

Little is known about digestion in other predatory reduviid bugs, but they probably secrete a complete set of digestive enzymes with their saliva as it is injected into the body of prey. Some additional digestion likely occurs in the midgut as partially digested food is swallowed. Many hemipterans are phytophagous and take plant sap or liquefied plant tissues. Seed-feeding hemipterans secrete enzymes into the seed, where a large amount of digestion takes place extraorally, but final digestion occurs in the midgut.

Unequivocal evidence of a peritrophic matrix has not been demonstrated in Hemiptera, most of whom take liquid or semiliquid food not likely to contain rough particles that could abrade the midgut microvilli. If the peritrophic matrix does protect from invading microorganisms, it raises the question of whether hemipterans have other protective means.

2.14.5 Homoptera

Homoptera take xylem or phloem sap, both of which are poor in amino acids and protein, but usually rich in sucrose (150 to >700 mM). Homoptera typically excrete a copious, dilute fluid, and, in some, such as aphids, the fluid contains so much sugar that it is called honeydew. They have to ingest large volumes of fluid to get the amino acids and then they have to get rid of the excess water and sucrose. A characteristic evolutionary feature of the gut in Homoptera is the filter chamber in which a loop of the hindgut is in intimate contact with part of the foregut and some fluid passes directly into the hindgut from the foregut without passing through the midgut. The filter chamber is able to concentrate gut fluid up to tenfold in some xylem feeders (Cicadoidea and Cercopoidea), but only about 2.5-fold in members of the Cicadelloidea, which are phloem feeders. Xylem feeders probably need to concentrate xylem fluids more because of the lower amino acid content (xylem, 3–10 mM amino acids) than do phloem feeders (phloem, 15–65 mM amino acids).

Heteroptera and Fulguroidea (Homoptera) secrete a lipid "membrane" that does not form a distinct sac like the peritrophic matrix, but, nevertheless, it does create a perimicrovillar space between the food mass and the microvillar surface of the cells.

2.14.6 Coleoptera

The crop is often absent or only slightly developed in beetle larvae and in adults of the Polyphaga, but usually present in adult Adephaga. The crop is a site of considerable digestion in the lower (Adephaga) and some higher (Polyphaga) coleopterans by action of enzymes secreted forward from the midgut (Terra et al., 1985). Pre-oral digestion occurs in many of the predacious beetles, and predacious Carabidae complete the process of digestion in the crop by action of enzymes passed forward from the midgut.

Recent research has shown that a number of insects secrete cellulose enzymes from their own tissues and do not depend exclusively on symbionts for cellulose-digesting enzymes (Fujita et al., 2010; Pauchet et al., 2010; Kirsch et al., 2012; Tokuda et al., 2012; Shelomi et al., 2013, 2014). Scarabaeid larvae, some of which feed upon food containing cellulose, probably digest the cellulose with the aid of bacterial derived cellulases, and sometimes they ingest the simpler breakdown products from fungal-digested cellulose. Adult coccinellids may have their own cellulases. Cerambycid larvae live in logs and other wood and digest cellulose by ingesting fungal cellulases from their fungus-infected wood habitat. Their nutrition is marginal, and most have a slow growth pattern and long larval development time. Some Coleoptera (Polyphaga) have no, or a reduced, crop and have cathepsin-like proteinases rather than trypsin-like ones. This might be an evolved adaptation to the presence of trypsin inhibitors in some foods.

Tenebrio molitor has been a favorite coleopteran model insect for digestion studies as well as other physiological experiments because of its size and ease of rearing. It should not be assumed to be typical of beetles, however. The larvae produce amylase, cellobiase, and trehalase from the anterior part of the midgut, and trypsin from the posterior of the midgut. These enzymes are secreted by exocytosis into the gut lumen. Less than 5% of the total of some of the major digestive enzymes are excreted by larvae with the feces and other undigested food, which suggests that the larvae probably have an endoectoperitrophic countercurrent flow of food and digestion products. The majority of digestion seems to occur within the peritrophic matrix, although an aminopeptidase is bound to the microvillar surface, so some final digestion of smaller polypeptides likely occurs at the microvillar surface (Terra et al., 1985; Ferreira et al., 1990).

2.14.7 Hymenoptera

The crop is not a major organ of digestion in Hymenoptera and is reduced in size in many larvae; many hymenopterans also have lost the anterior midgut caeca characteristic of many other groups of insects. The midgut is closed off from the hindgut by a plug of cellular tissue in larval Apocrita (bees and wasps) and the connection does not open until just before pupation. Any undigested residue (e.g., the shell of pollen grains in bees) can then be passed into the hindgut and voided with fecal material so that the gut is cleared before pupation. Larvae of the woodwasp (genus *Sirex*, Symphyta, Siricidae) acquire cellulase and xylanase from fungi ingested with the wood on which they feed. Pollen grains ingested by adult bees are not crushed or cracked by mouth or gut action, but the nutrients inside the pollen grains are dissolved and leached from the grains. The nearly empty pollen grain shells are accumulated in the hindgut and are excreted (only) during flight. There may be an endoectoperitrophic circulation within the midgut, but evidence is not conclusive.

2.14.8 Diptera

There is a prominent esophageal invagination into the midgut in larval mosquitoes, and midgut cells in a ring between the walls of the invagination secrete a Type II peritrophic matrix. The esophageal invagination, thus, acts like a chute to channel food into the stocking-like peritrophic matrix. Continued entry of food from the foregut seems necessary to force the lengthening of the PM. There are some differences in the formation of the PM in anopheline and culicine mosquito adults, but in all of them the peritrophic matrix is secreted only after a blood meal is ingested. A PM (or evidence of its formation) may be present as soon as 30 minutes after a blood meal or only after several hours.

Caeca at the anterior end of the midgut are believed to be the major sites of absorption. Malpighian tubules transfer fluids to the midgut (Stobbart, 1971) and help create a countercurrent endoectoperitrophic flow. Adult mosquitoes have an immature midgut upon emergence and do not usually feed for some period of time. Both males and females take nectar and females (only) also take blood meals for egg maturation.

Nectar taken by male and female mosquitoes is stored in a large, sac-like crop that is a diverticulum from the foregut, but blood meals taken by the females are passed directly into the midgut for the beginning of digestion. The midgut is differentiated functionally into an anterior and a posterior region. The anterior part secretes carbohydrate-digesting enzymes, and nectar components are digested as fluid from the crop and passed into the anterior midgut. The arrangement keeps possible trypsin inhibitors that may be present in nectar away from the site of protein digestion, which occurs in the posterior midgut. Simple sugars resulting from digestion, or those already in the nectar, are absorbed in the anterior midgut.

The posterior midgut cells secrete trypsin-like enzymes and protein (blood) digestion and absorption occur in the posterior midgut. The posterior midgut cells, more so than anterior midgut cells, have extensive microvilli and basal infoldings characteristic of secretion and absorptive processes. The midgut cells in this region get stretched by the large volume of blood that a mosquito takes if it is allowed to feed to repletion. Consequently, the cells have several types of connecting structures between them to help hold them together and prevent excessive leaking of materials in or out between cells while they are stretched. For example, *Anopheles* species have septate desmosomes connecting the apical (nearest the gut lumen) side of adjacent cells. Culicine females have zonula continua attachments between adjacent cells near the apical apex and desmosomes between cells in the basal region. Regenerative cells are common in both anterior and posterior midgut regions.

Cells believed to have endocrine function(s) are common in the posterior midgut. *Aedes aegypti* adults have about 500 such cells concentrated toward the posterior of the midgut (Billingsley, 1990), which, if they are endocrine cells, would make the midgut the largest endocrine organ of adult mosquitoes. Multiple cell types exist, suggesting the possibility of several functions and/or hormone products, but none has been identified as of yet.

Most larvae of the cyclorrhaphous flies (higher Diptera, including the housefly, *Drosophila* spp., and tephritid fruit flies) are saprophagous. The adults feed mostly on liquids or substances they can solubilize by regurgitating a droplet of fluid on the substance. Nectar, oozing fruit juices, sap, bird droppings, and honeydew on leaves or other substrate are utilized. Such nutrient-rich sources are likely to also contain bacteria, yeast, and possibly other microorganisms or fungi from environmental contamination, and these may be ingested with the fluid content as an additional source of nutrients.

Starch digestion occurs in the crop of houseflies by action of salivary amylase. Final carbohydrate digestion may occur on the midgut cell surfaces by action of a membrane-bound maltase. Bacteria in the ingested food are likely to be killed by the low pH of the midgut and the action of lysozyme in the gut. Trypsin acts on proteins in the midgut, but final amino acids are liberated at the midgut cell surface by membrane-bound aminopeptidases.

The midgut of *S. calcitrans* (stable fly, a blood feeder) is divided functionally into three parts. The blood meal is stored temporarily in an anterior region of the midgut where no digestion seems to occur because the blood retains its bright red color. As the blood passes into a middle region of the midgut, known as the opaque zone, it changes color and becomes dark red or brown as it encounters the action of the trypsin-like enzyme. Midgut cells in the opaque zone synthesize trypsin-like enzymes as a zymogen (Moffatt and Lehane, 1990) and store it as granules that are released in part by an apocrine mechanism into the gut lumen. The zymogen is converted to the active enzyme when blood enters the opaque zone, but details of the conversion process have not been elucidated. Possible advantages of storing the enzyme as an inactive zymogen may be that the active enzyme can be made available quickly, and autodigestion of the insect's own midgut cells may be reduced when no blood is present. Digestion of the blood meal is completed and absorption occurs from a posterior region of the midgut.

2.14.9 Lepidoptera

Larvae of Lepidoptera have a very short foregut; a large, long, relatively straight midgut; and a short hindgut. There is no storage or digestion in the short, nearly vestigial foregut. Nearly all lepidopterous larvae are phytophagous feeders, and the gut modifications appear to be an adaptation to pass food quickly into the long midgut so that digestion can begin. Feeding is nearly continuous when plenty of food is available, and larvae may ingest more than their body weight in food daily. Food moves rapidly through the relatively straight gut and frass droppings are frequent in phytophagous caterpillars. A Type I peritrophic matrix is present in the midgut of larvae. Larvae do not have gastric caeca. Digestion and absorption occur along the length of the midgut, with columnar cells secreting the enzymes and performing absorption. Goblet cells secrete K^+ into the midgut lumen, but do not seem to be involved in other gut functions. The presence of an endoectoperitrophic countercurrent flow has been suggested on the basis that digestive enzymes are not rapidly excreted with the steady flow of frass droppings. Clear-cut evidence of such a flow is not available, and a countercurrent flow is to some extent counterintuitive to the observed rapid movement of food through the gut.

Because the larval and adult forms of Lepidoptera have very different life histories and food habits, the adult gut is quite different from that of the larva. Many adult Lepidoptera feed only upon nectar, which is stored in the crop and slowly released into the midgut for digestion to simple sugars. Some adult Lepidoptera have vestigial mouthparts and do not feed at all; they survive and (females) produce eggs at the expense of body substance and generally live only a few days. In addition to nectar, adults of *Heliconius* butterflies feed upon pollen, bird droppings, and other food sources.

In *Erinnyis ello* caterpillars, initial digestion occurs in the endoperitrophic space, with final digestion occurring at the midgut cell surface by membrane-bound enzymes. Digestion does not occur in the foregut.

An unusual food utilized by *Tineola bisselliella* larvae (clothes moth) is wool, and larvae have a very strong reducing action in the midgut that reduces disulfide bonds to sulfhydryl bonds, which facilitates further protein digestion by proteinases (Hughes and Vogler, 2006).

2.15 INSECT GUT AS A POTENTIAL TARGET FOR POPULATION MANAGEMENT AND CONTROL OF THE SPREAD OF PLANT AND ANIMAL DISEASE ORGANISMS

The gut, and particularly the midgut, has been recognized by many entomologists and disease vector specialists as a potential attack point for insect population control and/or control of transmission of the disease organism. The midgut is one of the principal points of entry for toxins, viruses, hormones, bacteria, and other potential agents that might be introduced into insects for population control. For example, a toxin that acts upon the midgut is produced by a family of bacteria *Bacillus thuringiensis* (Bt) (reviewed by Federici, 1993, 1999; Knowles, 1994). Different strains of the bacteria have variable toxicity levels for different insects, and some insects, including many beneficial ones, are not attacked by Bt. Particularly virulent strains have been discovered that are useful for biological control of some Lepidoptera, Diptera, and Coleoptera. The protoxin consists of a mixture of crystalline proteins, the δ-endotoxins. The δ-endotoxin crystals dissolve in the midgut of susceptible insects, releasing proteinacious toxins that range in size from 27 to 140 kDa. These are further broken into smaller toxic polypeptides by the insect's own protein-digesting enzymes. Thus, by its own digestive action, the insect exposes itself to a wide variety of toxins. Some of the toxins bind to the brush border membrane-bound aminopeptidase in the midgut of the gypsy moth (Valaitis et al., 1995). Midgut proteases in the tobacco budworm, *Heliothis virescens*, digest the Cry1Ac Bt protoxin to a 60 kDa toxin that passes through the peritrophic matrix to contact brush border microvilli, where it is further enzymatically cleaved after binding to the extracellular portion of a cadherin transmembrane receptor (Krishnamoorthy et al., 2007). The processed toxin reorganizes into tetramers and binds to second receptors that are bound to the surface of the microvilli, including aminopeptidase and alkaline phosphatase. The tetramers then insert into the midgut epithelial cell membrane, forming pores that allow disruptions in cell osmotic balance, swelling, eventual lysis, and death of the caterpillar. In the pink bollworm, *Pectinophora gossypiella*, Cry1Ac protoxin and activated toxin bind to multiple extracellular sites of cadherin receptor, raising the possibility that activation of the protoxin may occur either before or after binding to the cadherin (Fabrick and Tabashnik, 2007). Some susceptible insects have shown the ability to develop resistance to Bt, however, and strategies are being explored to minimize resistance (McGaughey and Whalon, 1992).

The midgut also plays a role in the transmission of *Leishmania* parasites to humans. *Leishmania* parasites are passed from an infected host to a biting insect, and taken into the midgut of the insect with a blood meal where conditions may be favorable for the parasite to initiate rapid cell division into an early developmental stage, the promastigotes. If the insect is not a suitable host, the promastigotes soon die and are excreted with the fecal wastes of the insect. In susceptible hosts, such as sandflies, specific developmental changes in sugar residues on the surface of the promastigotes enable them to bind to midgut microvilli. As the attached parasites go through further developmental stages, including changes in the surface sugar residues, they are released again into the midgut, where they may be passed to a new host, possibly a human, by regurgitation during a sandfly bite (Pimenta et al., 1992). Agents that could prevent binding of the promastigotes to microvillar surfaces might break the transmission cycle to humans.

Introduction of proteinase inhibitors into commercial plants that would be specific against a particular pest are being investigated. While this probably is technically feasible, a great deal of research will have to be done to determine if significant control can be achieved and whether resistance will develop rapidly.

2.16 REVIEW AND SELF-STUDY QUESTIONS

1. Give examples that show the evolved relationship between food habits and the structure of the alimentary canal in insects.
2. The *Diaprepes* weevil larva feeds upon the roots of citrus and other trees or plants. What sort of gut structure can you expect to find in a *Diaprepes* weevil?
3. What are the three principal morphological and physiological regions of the insect gut?
4. Do most seed feeders that feed on the seeds of legumes (such as peas, soybeans, and other beans) eat the whole seed? How do they get any nutrients out of such seeds?
5. What is the function of the proventriculus in many insects that have this structure as a part of the gut?
6. What are the gastric caeca? What function do they have?
7. Do all parts of the insect gut have a cuticular lining (also called a cuticular intima)? Explain.
8. What is the function of regenerative cells in the midgut?
9. Explain the structure and function of the proton pump that operates in the goblet cells of lepidopterous larvae.
10. What are microvilli (as a part of midgut cells)? What is their function?
11. What function does the glycocalyx play in the digestive system of some insects?
12. What is the peritrophic matrix? Where (in the gut) could you expect to find the peritrophic matrix?
13. Describe functions that have been suggested for the peritrophic matrix.
14. Is cellulose digested by most insects that are plant feeders?
15. What is another name for a triacylglycerol molecule?
16. What is the difference between an endoproteinase and an exoproteinase in the way they digest a protein?
17. How does trypsin-modulating oostatic factor (TMOF) regulate digestion of blood in a female mosquito? In what way is the action of TMOF an adaptive benefit to female mosquitoes?
18. How is gut pH related to digestion in insects?
19. Is gut pH pretty much the same in all insects? Briefly discuss gut pH.
20. How does *Bacillus thuringiensis* (Bt) kill many types of insects?

REFERENCES

Agusti, N. and A. Cohen. 2000. *Lygus hesperus* and *L. liniolaris* (Hemiptera: Miridae): The salivary and midgut enzymes. *J. Entomol. Sci.* 35: 175–182.

Anklin-Mühlemann, R., D.E. Bignell, P.C. Vievers, R.H. Leuthold, and M. Slaytor. 1995. Morphological and biochemical studies of the gut flora in the fungus-growing termite, *Macrotermes subhyalinus*. *J. Insect Physiol.* 41: 929–940.

Appel, H.M. and M.M. Martin. 1990. Gut redox conditions in herbivorous lepidopteran larvae. *J. Chem. Ecol.* 16: 3277–3290.

Arakane, Y., D.G. Hogenkamp, Y.C. Zhu, K.J. Kramer, C.A. Specht, R.W. Beeman, M.R. Kanost, and S. Muthukrishnan. 2004. Characterization of two chitin synthase genes of the red flour beetle, *Tribolium castaneum*, and alternate exon usage in one of the genes during development. *Insect Biochem. Mol. Biol.* 34: 291–304.

Baines, D.M. 1978. Observations on the peritrophic membrane of *Locusta migratoria* migratorioides (R. & F) nymphs. *Acridia* 7: 11–22.

Barbehenn, R.V., M.M. Martin, and A.E. Hagerman. 1996. Reassessment of the roles of the peritrophic envelope and hydrolysis in protecting polyphagous grasshoppers from ingested hydrolyzable tannins. *J. Chem. Ecol.* 22: 1911–1929.

Barillas-Mury, C.V., F.G. Noreiga, and M.A. Wells. 1995. Early trypsin activity is part of the signal transduction system that activates transcription of the late trypsin gene in the midgut of the mosquito *Aedes aegypti*. *Insect Biochem. Mol. Biol.* 25: 242–246.

Barrett, A.J. and N.D. Rawlings. 1991. Proteinases: Types and families of endopeptidases. *Biochem. Soc. Trans.* 19: 707–715.

Baumann, O. and B. Walz. 2012. The blowfly salivary gland—A model system for analyzing the regulation of plasma membrane V-ATPase. *J. Insect Physiol.* 58: 450–458.

Bayés, A., M.R. de la Vega, J. Vendrell, F.X. Aviles, M.A. Jongsma, and J. Beekwilder. 2006. Response of the digestive system of *Helicoverpa zea* to ingestion of potato carboxypeptidase inhibitor and characterization of an uninhibited carboxypeptidase B. *Insect Biochem. Mol. Biol.* 36: 654–664.

Bayon, C. 1980. Volatile fatty acids and methane production in relation to anaerobic carbohydrate fermentation in *Oryctes nasicornis* larvae (Coleoptera: Scarabaeidae). *J. Insect Physiol.* 26: 819–828.

Berenbaum, M. 1980. Adaptive significance of midgut pH in larval Lepidoptera. *Am. Nat.* 115: 138–146.

Bernays, E.A. and D.J. Chamberlain. 1980. A study of tolerance of ingested tannin in *Schistocerca gregaria*. *J. Insect Physiol.* 26: 415–420.

Berridge, M.J. 1970. A structural analysis of intestinal absorption. *Symp. R. Entomol. Soc. Lond.* 5: 135–150.

Beyenbach, K.W. 1995. Mechanisms and regulation of epithelial transport across Malpighian tubules. *J. Insect Physiol.* 41: 197–207.

Beyenbach, K.W., T.L. Pannabecker, and W. Nagel. 2000. Central role of the apical membrane H^+-ATPase in electrogenesis and epithelial transport in Malpighian tubules. *J. Exp. Biol.* 203: 1459–1468.

Beyenbach, K.W. and H. Wieczorek. 2006. The V-type H^+ ATPase: Molecular structure and function, physiological roles and regulation. *J. Exp. Biol.* 209: 577–589.

Bignell, D.E. 1977. An experimental study of cellulose and hemicellulose degradation in the alimentary canal of the American cockroach. *Can. J. Zool.* 55: 579–589.

Bignell, D.E. 1981. Nutrition and digestion, in K.G. Adiyodi and W.J. Bell (eds.), *The American Cockroach*. Chapman & Hall, London, U.K., pp. 57–86.

Bignell, D.E. and J.M. Anderson. 1980. Determination of pH and oxygen status in the guts of lower and higher termites. *J. Insect Physiol.* 26: 183–188.

Billingsley, P.F. 1990. The midgut ultrastructure of hematophagous insects. *Annu. Rev. Entomol.* 35: 219–248.

Blacklock, B.J. and R.O. Ryan. 1994. Hemolymph lipid transport. *Insect Biochem. Mol. Biol.* 24: 855–873.

Blakemore, D., S. Williams, and M.J. Lehane. 1995. Protein stimulation of trypsin secretion from the opaque zone midgut cells of *Stomoxys calcitrans*. *Comp. Biochem. Physiol.* 110B: 301–307.

Blenau, W. and M. Thamm. 2011. Distribution of serotonin (5-HT) and its receptors in the insect brain with focus on the mushroom bodies. Lessons from *Drosophila melanogaster* and *Apis mellifera*. *Arthropod Struct. Dev.* 40: 381–394.

Borovsky, D. 2005. Control of mosquito larval trypsin with *Aea*-trypsin modulating oostatic factor (TMOF) and its analogues. *Pestycydy (Poland)* 3: 71–77.

Borovsky, D. 2007. Trypsin modulating oostatic factor for developing resistant crops, in I. Ishaaya, R. Nauen, and A.R. Horowitz (eds.), *Insecticides Design Using Advanced Technologies*. Springer-Verlag, Berlin, Germany, pp. 135–149.

Borovsky, D., D.A. Carlson, P.R. Griffin, J. Shabanowitz, and D.F. Hunt. 1990. Mosquito oostatic factor: A novel decapeptide modulating trypsin-like enzyme biosynthesis in the midgut. *FASEB J.* 4: 3015–3020.

Borovsky, D., D.A. Carlson, P.R. Griffin, J. Shabanowitz, and D.F. Hunt. 1993. Mass-spectrometry and characterization of *Aedes aegypti* trypsin modulating oostatic factor (TMOF) and its analogs. *Insect Biochem. Mol. Biol.* 23: 703–712.

Borovsky, D., I. Janssen, J. Vanden Broeck, R. Huybrechts, P. Verhaert, H.L. De Bobdt, D. Bylemans, and A. De Loof. 1996. Molecular sequencing and modeling of *Neobellieria bullata* trypsin. Evidence for translational control by *Neobellieria* trypsin-modulating oostatic factor. *Eur. J. Biochem.* 237: 279–287.

Borovsky, D. and F. Mahmood. 1995. Feeding the mosquito *Aedes aegypti* with TMOF and its analogs; effect on trypsin biosynthesis and egg development. *Regul. Pept.* 57: 273–281.

Borovsky, D., C.A. Powell, J.K. Nayar, J.E. Blalock, and T.K. Hayes. 1994b. Characterization and localization of mosquito gut receptors for trypsin modulating oostatic factor using a complementary peptide and immunocytochemistry. *FASEB J.* 8: 350–355.

Borovsky, D. and Y. Schlein. 1988. Quantitative determination of trypsin-like and chymotrypsin-like enzymes in insects. *Arch. Insect Biochem. Physiol.* 8: 249–260.

Borovsky, D., Q. Song, M.C. Ma, and D. Carlson. 1994a. Biosynthesis, secretion, and immunocytochemistry of trypsin modulating oostatic factor of *Aedes aegypti*. *Arch. Insect Biochem. Physiol.* 27: 27–38.

Boucias, D.G., C. Stokes, G. Storey, and J.C. Pendland. 1996. The effects of imidacloprid on the termite *Reticulitemes flavipes* and its interaction with the mycopathogen *Beauveria bassiana*. *Pfanzenschutz-Nachrichten Bayer* 49: 103–144.

Brauman, A., M.D. Kane, M. Labat, and J.A. Breznak. 1992. Genesis of acetate and methane by gut bacteria of nutritionally diverse termites. *Science* 257: 1384–1387.

Breznak, J.A. 1982. Intestinal microbiota of termites and other xylophagous insects. *Annu. Rev. Microbiol.* 36: 323–343.

Broadway, R.M. 1995. Are insects resistant to plant proteinase inhibitors? *J. Insect Physiol.* 41: 107–116.

Broadway, R.M. and M.G. Villani. 1995. Does host range influence susceptibility of herbivorous insects to non-host plant proteinase inhibitors? *Entomol. Exp. Appl.* 76: 303–312.

Brune, A. and M. Kuhl. 1996. pH profiles of the extremely alkaline hindguts of soil-feeding termites (Isoptera: Termitidae): Determined with microelectrodes. *J. Insect Physiol.* 42: 1121–1127.

Chao, A.C., D.F. Moffett, and A. Koch. 1991. Cytoplasmic pH and goblet cavity pH in the posterior midgut of the tobacco hornworm *Manduca sexta*. *J. Exp. Biol.* 155: 403–414.

Cheeseman, M.T. and C. Gillot. 1987. Organization of protein digestion in *Calosoma calidum* (Coleoptera: Carabidae). *J. Insect Physiol.* 33: 1–18.

Chino, H. 1985. Lipid transport: Biochemistry of hemolymph lipophorin, in G.A. Kerkut and L.I. Gilbert (eds.), *Comprehensive Insect Physiology, Biochemistry and Pharmacology*, vol. 10. Pergamon Press, Oxford, U.K., pp. 115–135.

Chino, H. and R.G. Downer. 1979. The role of diacylglycerol absorption of dietary glyceride in the American cockroach. *Insect Biochem.* 9: 379–382.

Chino, H. and K. Kitazawa. 1981. Diacylglycerol-carrying lipoprotein of hemolymph of the locust and some insects. *J. Lipid Res.* 22: 1042–1052.

Cohen, A.C. 1995. Extra-oral digestion in predaceous terrestrial Arthropoda. *Annu. Rev. Entomol.* 40: 85–103.

Cristofoletti, P.T., A.F. Ribeiro, and W.A. Terra. 2000. Apocrine secretion of amylase and exocytosis of trypsin along the midgut of *Tenebrio molitor* larvae. *J. Insect Physiol.* 47: 143–155.

Dadant & Sons (eds.). 1975. *The Hive and the Honey Bee*, Revised Edition. Dadant & Sons, Publishers, Hamilton, IL, p. 112.

Ding, X., B. Gopalakrishnan, L.B. Johnson, F.F. White, X. Wang, T.D. Morgan, K.J. Kramer, and S. Muthukrishnan. 1998. Resistance to tobacco budworm feeding by transgenic plants expressing an insect chitinase gene. *Transgenic Res.* 7: 77–84.

Dow, J.A.T. 1981. Countercurrent flows, water movements and nutrient absorption in the locust midgut. *J. Insect Physiol.* 27: 579–585.

Dow, J.A.T. 1984. Extremely high pH in biological systems: A model for carbonate transport. *Am. J. Physiol.* 246: R633–R635.

Dow, J.A.T. 1986. Insect midgut function. *Adv. Insect Physiol.* 19: 187–328.

Dow, J.A.T. and M.J. O'Donnell. 1990. Reversible alkalinization by *Manduca sexta* midgut. *J. Exp. Biol.* 150: 247–256.

Edwards, M.J. and M. Jacobs-Lorena. 2000. Permeability and disruption of the peritrophic matrix and caecal membrane from *Aedes aegypti* and *Anopheles gambiae* mosquito larvae. *J. Insect Physiol.* 46: 1313–1320.

Englemann, F. 1968. Feeding and crop emptying in the cockroach *Leucophaea maderae*. *J. Insect Physiol.* 14: 1525–1531.

Engelmann, F. and P.M. Geraerts. 1980. The proteases and the protease inhibitor in the midgut of *Leucophaea maderae*. *J. Insect Physiol.* 26: 703–710.

Evans, W.A.L. and D.W. Payne. 1964. Carbohydrases of the alimentary tract of the desert locust, *Schistocerca gregaria* Forsk. *J. Insect Physiol.* 10: 657–674.

Fabrick, J.A. and B.E. Tabashnik. 2007. Binding of *Bacillus thuringiensis* toxin Cry1Ac to multiple sites of cadherin in pink bollworm. *Insect Biochem. Mol. Biol.* 37: 97–106.

Falibene, A., W. Rössler, and R. Josens. 2012. Serotonin depresses feeding behaviour in ants. *J. Insect Physiol.* 58: 7–17.

Federici, B.A. 1993. Insecticidal bacterial proteins identify the midgut epithelium as a source of novel target sites for insect control. *Arch. Insect Biochem. Physiol.* 22: 357–371.

Federici, B.A. 1999. *Bacillus thuringiensis* in biological control, in T.S. Bellows and T.W. Fisher (eds.), *Handbook of Biological Control*. Academic Press, New York, pp. 575–593.

Feldman, D.H., W.R. Harvey, and B.R. Stevens. 2000. A novel electrogenic amino acid transporter is activated by K^+ or Na^+, is alkaline pH-dependent, and is Cl^- independent. *J. Biol. Chem.* 275: 24518–24526.

Felix, C.R., B. Betschart, P.F. Billingsley, and T.A. Freyvogel. 1991. Post-feeding induction of trypsin in the midgut of *Aedes aegypti* L. (Diptera: Culicidae) is separable into two cellular phases. *Insect Biochem.* 21: 197–203.

Ferreira, C., G.L. Bellinello, A.F. Ribeiro, and W.R. Terra. 1990. Digestive enzymes associated with the glycocalyx, microvillar membranes and secretory vesicles from midgut cells of *Tenebrio molitor* larvae. *Insect Biochem.* 20: 839–847.

Fialho, M.C.Q., N.R. Moreira, J.C. Zununcio, A.F. Ribeiro, W.R. Terra, and J.C. Serrão. 2012. Prey digestion in the midgut of the predatory bug *Podisus nigrispinus* (Hemiptera: Pentatomidae). *J. Insect Physiol.* 85: 850–856.

Fitches, E., H. Wilkerson, H. Bell, D.P. Brown, J.A. Gatehouse, and J.P. Edwards. 2004. Cloning, expression and functional characterization of chitinase from larvae of tomato moth (*Lacanobia oleracea*): A demonstration of the insecticidal activity of insect chitinase. *Insect Biochem. Mol. Biol.* 34: 1037–1050.

French, A.S., K.L. Simcock, D. Rolke, S.E. Gartside, W. Blenau, and G.A. Wright. 2014. The role of serotonin in feeding and gut contractions in the honeybee. *J. Insect Physiol.* 61: 8–15.

Fujita, A., M. Hojo, T. Aoyagi, Y. Hayashi, G. Arakawa, G. Tokuda, and H. Watanabe. 2010. Details of the digestive system in the midgut of *Coptotermes formosanus* Shiraki. *J. Wood Sci.* 56: 222–226.

Genta, F.A., L. Blanes, P.T. Cristofoletti, C.L. do Lago, W.R. Terra, and C. Ferreira. 2006. Purification, characterization and molecular cloning of the major chitinase from *Tenebrio molitor* larval midgut. *Insect Biochem. Mol. Biol.* 36: 789–800.

Ghamari, M., V. Hosseininaveh, A. Darvishzadeh, and N.P. Chougule. 2014. Carbohydrases in the digestive system of the spined soldier bug, *Podisus maculiventris* (Say) (Hemiptera: Pentatomidae). *Arch. Insect Biochem. Physiol.* 85: 195–215.

Giordana, B., V.F. Sacchi, P. Parenti, and G.M. Hanozet. 1989. Amino acid transport systems in intestinal brush-border membranes from lepidopteran larvae. *Am. J. Physiol.* 257: R494–R500.

Gondim, K.C., G.C. Atella, J.H. Kawooya, and H. Masuda. 1992. Role of phospholipids in the lipophorin particles of *Rhodnius prolixus*. *Arch. Insect Biochem. Physiol.* 20: 303–314.

Gonen, T. and T. Walz. 2006. The structure of aquaporins. *Q. Rev. Biophys.* 39: 361–396.

Graf, R. and H. Briegel. 1985. Isolation of trypsin isozymes from the mosquito *Aedes aegypti* (L.). *Insect Biochem.* 15: 611–618.

Graf, R. and H. Briegel. 1989. The synthetic pathway of trypsin in the mosquito *Aedes aegypti* L. (Diptera: Culicidae) and in vitro stimulation in isolated midguts. *Insect Biochem.* 19: 129–137.

Graf, R., A.S. Raikhel, M.R. Brown, A.O. Lea, and H. Briegel. 1986. Mosquito trypsin: Immunocytochemical localization in the midgut of blood-fed *Aedes aegypti* (L.). *Cell Tissue Res.* 245: 19–27.

GraHa-Souza, A.V., C. Maya-Monteiro, G.O. Paiva-Silva, G.R.C. Braz, M.C. Paes, M.H.F. Sorgine, M.F. Oliveira, and P.L. Oliveira. 2006. Adaptations against heme toxicity in blood-feeding arthropods. *Insect Biochem. Mol. Biol.* 36: 322–335.

Gupta, B.J. and M.J. Berridge. 1966. Fine structural organization of the rectum in the blowfly, *Calliphora erythrocephala* (Meig.) with special reference to connective tissue, tracheae and neurosecretory innervation in the rectal papillae. *J. Morphol.* 120: 23–82.

Hakim, R.S., K. Baldwin, and G. Smagghe. 2010. Regulation of midgut growth, development, and metamorphosis. *Annu. Rev. Entomol.* 55: 593–608.

Harvey, W.R. 1992. The physiology of V-ATPases. *J. Exp. Biol.* 172: 1–17.

Hennigan, B.B., M.G. Wolfersberger, and W.R. Harvey. 1993b. Neutral amino acid symport in larval *Manduca sexta* midgut brush-border membrane vesicles deduced from cation-dependent uptake of leucine, alanine, and phenylalanine. *Biochim. Biophys. Acta* 1148: 216–222.

Hennigan, B.B., M.G. Wolfersberger, R. Parthasarathy, and W.R. Harvey. 1993a. Cation-dependent leucine, alanine, and phenylalanine uptake at pH 10 in brush-border membrane vesicles from larval *Manduca sexta* midgut. *Biochim. Biophys. Acta* 1148: 209–215.

Hogenkamp, D.G., Y. Arakane, L. Zimoch, H. Merzendorfer, K.J. Kramer, R.W. Beeman, M.R. Kanost, C.A. Specht, and S. Muthukrishnan. 2005. Chitin synthase genes in *Manduca sexta*: Characterization of a gut-specific transcript and differential tissue expression of alternately spliced mRNAs during development. *Insect Biochem. Mol. Biol.* 35: 529–540.

Hong, Y.S., M. Reuveni, and J.J. Neal. 1995. A sodium- and potassium-stimulated tyrosine transporter from *Lepitinotarsa decemlineata* midguts. *J. Insect Physiol.* 41: 527–533.

Hopkins, C.R. 1967. The fine-structural changes observed in the rectal papillae of the mosquito *Aedes aegypti*, L. and their relation to the epithelial transport of water and inorganic ions. *J. R. Microsc. Soc.* 86: 235–252.

House, H.L. 1974. Digestion, in M. Rockstein (ed.), *Physiology of the Insecta*, 2nd ed. Academic Press, New York, pp. 63–117.

Houseman, J.G., A.E.R. Downe, and P.E. Morrison. 1985. Similarities in digestive proteinase production in *Rhodnius prolixus* (Hemiptera: Reduviidae) and *Stomoxys calcitrans* (Diptera: Muscidae). *Insect Biochem.* 15: 471–474.

Hua, Y.-J., D. Bylemans, A. DeLoof, and J. Koolman. 1994. Inhibition of ecdysone biosynthesis in flies by a hexapeptide isolated from vitellogenic ovaries. *Mol. Cell. Endocrinol.* 104: R1–R4.

Hua, Y.-J. and J. Koolman. 1995. An ecdysiostatin from flies. *Regul. Pept.* 57: 263–271.

Hughes, J. and A.P. Vogler. 2006. Gene expression in the gut of keratin-feeding clothes moths (*Tineola*) and keratin beetles (*Trox*) revealed by subtracted cDNA libraries. *Insect Biochem. Mol. Biol.* 36: 584–592.

Humbert, W. 1979. The midgut of *Tomocerus minor* Lubbock (Insecta, Collembola): Ultrastructure, cytochemistry, ageing and renewal during a moulting cycle. *Cell Tissue Res.* 196: 39–57.

Huss, M., O. Vitavska, A. Albertmelcher, S. Bockelmann, C. Nardmann, K. Tabke, F. Tiburcy, and H. Wieczorek. 2011. Vacuolar H^+-ATPases: Intra- and intermolecular interactions. *Eur. J. Cell Biol.* 90: 688–695.

Johnson, K.S. and R.V. Barbehenn. 2000. Oxygen levels in the gut lumens of herbivorous insects. *J. Insect Physiol.* 46: 897–903.

Jordao, B.P., W.R. Terra, A.F. Ribero, M.J. Lehane, and C. Ferreira. 1996. Trypsin secretion in *Musca domestica* larval midguts: A biochemical and immunocytochemical study. *Insect Biochem. Mol. Biol.* 26: 337–346.

Kane, P.M. and K.J. Parra. 2000. Assembly and regulation of the yeast vacuolar H^+-ATPase. *J. Exp. Biol.* 203: 81–87.

Kataoka, N., S. Miyake, and M. Azuma. 2009. Aquaporin and aquaglyceroporin in silkworms, differently expressed in the hindgut and midgut of *Bombyx mori*. *Insect Mol. Biol.* 18: 303–314.

Kato, N., C.R. Mueller, J.F. Fuchs, V. Wessely, Q. Lan, and B.M. Christensen. 2006. Regulatory mechanisms of chitin biosynthesis and roles of chitin in periotrophic matrix formation in the midgut of adult *Aedes aegypti*. *Insect Biochem. Mol. Biol.* 36: 1–9.

Kirsch, R., N. Wielsch, H. Vogel, A. Svatos, D.G. Heckel, and Y. Pauchet. 2012. Combining proteomics and transcriptome sequencing to identify active plant-cell-wall-degrading enzymes in a leaf beetle. *BMC Genomics* 13: 587.

Klein, U., G. Löffelmann, and H. Wieczorek. 1991. The midgut as a model system for insect K^+-transporting epithelia: Immunocytochemical localization of a vacuolar-type H^+ pump. *J. Exp. Biol.* 161: 61–75.

Knowles, B.H. 1994. Mechanism of action of *Bacillus thuringiensis* insecticidal δ-endotoxins. *Adv. Insect Physiol.* 24: 275–308.

Krishna, S.S. and K.N. Saxena. 1962. Digestion and absorption of food in *Tribolium castaneum* (Herbst.). *Physiol. Zool.* 35: 66–78.

Krishnamoorthy, M., J.L. Jurat-Fuentes, R.J. McNall, T. Andacht, and M.J. Adang. 2007. Identification of novel Cry1Ac binding proteins in midgut membranes from *Heliothis virescens* using proteomic analyses. *Insect Biochem. Mol. Biol.* 37: 189–201.

Lafourcade, C., K. Sobo, S. Kieffer-Jaquinod, J. Garin, and F.G. van der Goot. 2008. Regulation of the V-ATPase along the endocytic pathway occurs through reversible subunit association and membrane localization. *PLoS One* 3: e2758.

Lehane, M.J. 1976. Formation and histochemical structure of the peritrophic membrane in the stablefly, *Stomoxys calcitrans*. *J. Insect Physiol.* 22: 1551–1557.

Lehane, M.J. 1997. Peritrophic matrix structure and function. *Annu. Rev. Entomol.* 42: 525–550.

Lehane, M.J. 2005. *The Biology of Blood-Sucking Insects*, 2nd ed. Cambridge University Press, Cambridge, U.K.

Lehane, M.J., D. Blakemore, S. Williams, and M.R. Moffatt. 1995. Regulation of digestive enzyme levels in insects. *Comp. Biochem. Physiol.* 110B: 285–289.

Liscia, A., P. Solari, S.T. Gibbons, A. Gelperin, and J.G. Stoffolano, Jr. 2012. Effect of serotonin and calcium on the supercontractile muscles of the adult blowfly crop. *J. Insect Physiol.* 58: 356–366.

Lopes, A.R., M.A. Juliano, S.R. Marana, L. Juliano, and W.R. Terra. 2006. Substrate specificity of insect trypsins and the role of their subsites in catalysis. *Insect Biochem. Mol. Biol.* 36: 130–140.

Lu, S.J., J.E. Pennington, A.R. Stonehouse, M.M. Mobula, and M.A. Wells. 2006. Reevaluation of the role of early trypsin activity in the transcriptional activation of the late trypsin gene in the mosquito *Aedes aegypti*. *Insect Biochem. Mol. Biol.* 36: 336–343.

Martin, J.S., M.M. Martin, and E.A. Bernays. 1987. Failure of tannic acid to inhibit digestion or reduce digestibility of plant protein in gut fluids of insect herbivores: Implications for theories of plant defense. *J. Chem. Ecol.* 13: 605–621.

Martin, M.M., J.J. Kukor, J.S. Martin, D.L. Lawson, and R.W. Merritt. 1981a. Digestive enzymes of larvae of three species of caddisflies (Trichoptera). *Insect Biochem.* 11: 501–505.

Martin, M.M., J.S. Martin, J.J. Kukor, and R.W. Merritt. 1980. The digestion of protein and carbohydrate by the stream detritivore, *Tipula abdominalis* (Diptera, Tipulidae). *Oecologia* 46: 360–364.

Martin, M.M., J.S. Martin, J.J. Kukor, and R.W. Merritt. 1981b. The digestive enzymes of detritus-feeding stonefly nymphs (Plecoptera; Pteronarcyidae). *Can. J. Zool.* 59: 1947–1951.

Matsui, T., T. Sakai, H. Satake, and M. Takeda. 2013. The pars intercerebralis affect digestive activities of the American cockroach *Periplaneta americana*, via crustacean cardioactive peptide and allostatin-A. *J. Insect Physiol.* 59: 33–37.

McFarlane, J.E. 1985. Nutrition and digestive organs, in M.S. Blum (ed.), *Fundamentals of Insect Physiology*. John Wiley & Sons, New York, pp. 59–89.

McGaughey, W.H. and M.E. Whalon. 1992. Managing insect resistance to *Bacillus thuringiensis* toxins. *Science* 258: 1451–1455.

McGhie, T.K., J.T. Christeller, R. Ford, and P.G. Allsopp. 1995. Characterization of midgut proteinase activities of white grubs: *Lepidiota noxia*, *Lepidiota negatoria*, and *Antitrogus consanguineus* (Scarabeidae, Melolonthini). *Arch. Insect Biochem. Physiol.* 28: 351–363.

McManus, M.T., D.W.R. White, and P.G. McGregor. 1994. Accumulation of a chymotrypsin inhibitor in transgenic tobacco can affect the growth of insect pests. *Transgenic Res.* 3: 50–58.

Mehrabadi, M., A.R. Bandani, R. Mehrabadi, and H. Alizadeh. 2012. Inhibitory activity of proteinaceous α-amylase inhibitors from Triticale seeds against *Eurygaster integriceps* salivary α-amylase: Interactions of the inhibitors and the insect digestive enzymes. *Pestic. Biochem. Physiol.* 102: 220–228.

Merzendorfer, H., R. Gräf, M. Huss, W.R. Harvey, and H. Wieczorek. 1997. Regulation of proton-translocating V-ATPases. *J. Exp. Biol.* 200: 225–235.

Mikani, A., Q.-S. Wang, and M. Takeda. 2012. Brain-midgut short neuropeptide F mechanism that inhibits digestive activity of the American cockroach *Periplaneta americana* upon starvation. *Peptides* 34: 135–144.

Moffatt, M.R., D. Blakemore, and M.J. Lehane. 1995. Studies on the synthesis and secretion of digestive trypsin in *Stomoxys calcitrans* (Insecta-Diptera). *Comp. Biochem. Physiol.* 110B: 291–300.

Moffatt, M.R. and M.J. Lehane. 1990. Trypsin is stored as an inactive zymogen in the midgut of *Stomoxys calcitrans*. *Insect Biochem.* 20: 719–723.

Moffett, D.F. and A.R. Koch. 1988a. Electrophysiology of K^+ transport by midgut epithelium of lepidopteran insect larvae I. The transbasal electrochemical gradient. *J. Exp. Biol.* 135: 25–38.

Moffett, D.F. and A.R. Koch. 1988b. Electrophysiology of K^+ transport by midgut epithelium of lepidopteran insect larvae II. The transapical electrochemical gradients. *J. Exp. Biol.* 135: 39–49.

Murdock, L.L., G. Brookhart, P.E. Dunn, D.E. Foard, S. Kelley, L. Kitch, R.E. Shade, R.H. Shukle, and J.L. Wolfson. 1987. Cysteine digestive proteinases in Coleoptera. *Comp. Biochem. Physiol.* 87B: 783–787.

Nation, J.L. 1983. Specialization in the alimentary canal of some mole crickets (Orthoptera: Gryllotalpidae). *Int. J. Insect Morphol. Embryol.* 12: 201–210.

Noirot, C. and C. Noirot-Timothee. 1969. The digestive system, in K. Krishna and R.M. Wheeler (eds.), *Biology of Termites*, vol. I. Academic Press, New York, pp. 49–88.

Noriega, F.G., K.A. Edgar, W.G. Goodman, D.K. Shah, and M.A. Wells. 2001. Neuroendocrine factors affecting the steady-state levels of early trypsin mRNA in *Aedes aegypti*. *J. Insect Physiol.* 47: 515–522.

Noreiga, F.G., X.-Y. Wang, J.E. Pennington, C.V. Barillas-Mury, and M.A. Wells. 1996. Early trypsin, a female-specific midgut protease in *Aedes aegypti*: Isolation, amino-terminal sequence determination, and cloning and sequencing of the gene. *Insect Biochem. Mol. Biol.* 26: 119–126.

O'Riordan, A.M. 1969. Electrolyte movement in the isolated midgut of the cockroach (*Periplaneta americana*). *J. Exp. Biol.* 51: 699–714.

Pacey, E.K. and M.J. O'Donnell. 2014. Transport of H^+, Na^+ and K^+ across the posterior midgut of blood-fed mosquitoes (*Aedes aegypti*). *J. Insect Physiol.* 61: 42–50.

Parenti, P., B. Giordana, V.F. Sacchi, G.M. Hanozet, and A. Guerritore. 1985. Metabolic activity related to the potassium pump in the midgut of *Bombyx mori* larvae. *J. Exp. Biol.* 116: 69–78.

Pauchet, Y., P. Wilkinson, R. Chauhan, and R.H. ffrench-Constant. 2010. Diversity of beetle genes encoding novel plant cell wall degrading enzymes. *PLoS One* 5(12): e15635.

Pechan, T., A. Cohen, W.P. Williams, and D.S. Luthe. 2002. Insect feeding mobilizes a unique defense protease that disrupts the peritrophic matrix of caterpillars. *Proc. Natl. Acad. Sci. USA* 99: 13319–13323.

Philip, B.N., A.J. Kiss, and R.E. Lee Jr. 2011. The protective role of aquaporins in the freeze-tolerant insect *Eurosta solidaginis*: Functional characterization and tissue abundance of EsAQP1. *J. Exp. Biol.* 214: 848–857.

Philip, B.N. and R.E. Lee. 2010. Changes in abundance of aquaporin-like proteins occur concomitantly with seasonal acquisition of freeze tolerance in the goldenrod gall fly, *Eurosta solidaginis*. *J. Insect Physiol.* 56: 679–685.

Pimenta, P.F.P., S.T. Turco, M.J. McConville, P.G. Lawyer, P.V. Perkins, and D.L. Sacks. 1992. Stage-specific adhesion of *Leishmania* promastigotes to the sandfly midgut. *Science* 256: 1812–1815.

Ramzi, S. and V. Hosseininaveh. 2010. Biochemical characterization of digestive alpha-amylase, alpha-glucosidase and beta-glucosidase in pistachio green stink bug, *Brachynema germari* Kolenati (Hemiptera: Pentatomidae). *J. Asia Pacific Entomol.* 13: 215–219.

Razavi Tabatabaei, P., H. Hosseininaveh, S.H. Goldansaz, and Kh. Talebi. 2011. Biochemical characterization of digestive proteases and carbohydrases of the carob moth, *Ectomyelois ceratoniae* (Zeller) (Lepidoptera: Pyralidae). *J. Asia Pacific Entomol.* 14: 187–191.

Reuveni, M. and P. Dunn. 1990. The use of membrane vesicles as a tool for investigating known and potential pesticides, in D.L. Weigmann (ed.), *Pesticides in the Next Decade: The Challenge Ahead*. Virginia Water Research Center Press, Richmond, VA, pp. 523–531.

Reuveni, M. and P. Dunn. 1993. Absorption pathways of amino acids in the midgut of *Manduca sexta* larvae. *Insect Biochem. Mol. Biol.* 23: 959–966.

Reuveni, M., Y.S. Hong, P.E. Dunn, and J.J. Neal. 1993. Leucine transport into brush border membrane vesicles from guts of *Leptinotarsa decemlineata* and *Manduca sexta*. *Comp. Biochem. Physiol.* 104A: 267–272.

Richards, A.G. and P.A. Richards. 1977. The peritrophic membranes of insects. *Annu. Rev. Entomol.* 22: 219–240.

Riemann, J.G. and H.M. Flint. 1967. Irradiation effects on midguts and testes of the adult boll weevil, *Anthonomus grandis*, determined by histological and shielding studies. *Annu. Entomol. Soc. Am.* 60: 298–308.

Rocha, A.A., C.J.C. Pinto, R.I. Samuels, D. Alexandre, and C.P. Silva. 2014. Digestion in adult females of the leaf-footed bug *Leptoglossus zonatus* (Hemiptera: Coreidae) with emphasis on the glycoside hydrolases α-amylase, α-galactosidase, and α-glucosidase. *Arch. Insect Biochem. Physiol.* 85: 152–163.

Rudall, K.M. and W. Kenchington. 1973. The chitin system. *Biol. Rev. Camb. Philos. Soc.* 48: 597–636.

Santos, C.D., A.F. Ribeiro, and W.R. Terra. 1986. Differential centrifugation, calcium precipitation and ultrasonic disruption of midgut cells of *Erinnyis ello* caterpillars. Purification of cell microvilli and inferences concerning secretory mechanisms. *Can J. Zool.* 64: 490–500.

Santos, C.D. and W.R. Terra. 1984. Plasma membrane associated amylase and trypsin: Intracellular distribution of digestive enzymes in the midgut of the cassava hornworm *Erinnyis ello*. *Insect Biochem.* 14: 587–595.

Santos, C.D. and W.R. Terra. 1986. Midgut alpha-glucosidase and beta-fructosidase from *Erinnyis ello* larvae and imagoes. *Insect Biochem.* 16: 819–824.

Sehnal, F. and D. Žitňan. 1990. Endocrines of insect gut, in A. Epple, C.G. Scanes, and M.H. Stetson (eds.), *Progress in Comparative Endocrinology*. Wiley-Liss, New York, pp. 510–515.

Shelomi, M., W.-S. Lo, L.S. Kimsey, and C.-H. Kuo. 2013. Analysis of the gut microbiota of walking sticks (Phasmatodea). *BMC Res. Notes* 6: 368.

Shelomi, M., H. Watanabe, and G. Arakawa. 2014. Endogenous cellulose enzymes in the stick insect (Phasmatodea) gut. *J. Insect Physiol.* 60: 25–30.

Silva, F.C.B.L., A. Alcazar, L.L.P. Macedo, A.S. Oliveira, F.P. Macedo, L.R.D. Abreu, E.A. Santos, and M.O. Sales. 2006. Digestive enzymes during development of *Ceratitis capitata* (Diptera: Tehpritidae) and effects of SBTI on its digestive serine proteinase targets. *Insect Biochem. Mol. Biol.* 36: 561–569.

Silva, W., C. Cardoso, A.F. Ribeiro, W.R. Terra, and C. Ferreira. 2013. Midgut proteins released by microapocrine secretion in *Spodoptera frugiperda*. *J. Insect Physiol.* 59: 70–80.

Skaer, R.J. 1981. Cellular sieving by a natural, high-flux membrane. *J. Microsc.* 124: 331–333.

Solari, P., J.G. Stoffolano, Jr., J. Fitzpatrick, A. Gelperin, A. Thomson, G. Talani, E. Sanna, and A. Liscia. 2013. Regulatory mechanism and the role of calcium and potassium channels controlling supercontractile crop muscles in adult *Phorma regina*. *J. Insect Physiol.* 59: 942–952.

Southwood, T.R.E. 1973. The insect/plant relationship—An evolutionary perspective. *Symp. R. Entomol. Soc. Lond.* 6: 3–30.

Souza-Neto, J.A., D.S. Gusmao, and F.J.A. Lemos. 2003. Chitinolytic activities in the gut of *Aedes aegypti* (Diptera: Culicidae) larvae and their role in digestion of chitin-rich structures. *Comp. Biochem. Physiol.* 136A: 717–724.

Staniscuaski, F., J.-P. Paluzzi, R. Real-Guerra, C.R. Carlini, and I. Orchard. 2013. Expression analysis and molecular characterization of aquaporins in *Rhodnius prolixus*. *J. Insect Physiol.* 59: 1140–1150.

Stobbart, R.H. 1971. Factors affecting the control of body volume in the larvae of the mosquitoes *Aedes aegypti* (L.) and *Aedes detritus* Edw. *J. Exp. Biol.* 54: 67–82.

Stoffolano, Jr., J.G., L. Guerra, M. Carcupino, G. Gambellini, and A.M. Fausto. 2010. The diverticulated crop of adult *Phormia regina*. *Arthropod Struct. Dev.* 39: 251–260.

Stoffolano, Jr., J.G. and A.T. Haselton. 2013. The adult dipteran crop: A unique and overlooked organ. *Annu. Rev. Entomol.* 58: 205–225.

Surholt, B., J. Goldberg, T.K.F. Schulz, A.M.Th. Beenakkers, and D.J. Van der Horst. 1991. Lipoproteins act as a reusable shuttle for lipid transport in the flying death's-head hawkmoth *Acherontia atropos*. *Biochem. Biophys. Acta* 1086: 15–21.

Teo, L.H. 1997. Tryptic and chymotryptic activities in different parts of the gut of the field cricket *Gryllus bimaculatus* (Orthoptera: Gryllidae). *Annu. Entomol. Soc. Am.* 90: 69–74.

Terra, W.R. 1988. Physiology and biochemistry of insect digestion: An evolutionary perspective. *Brazil J. Med. Biol. Res.* 21: 675–734.

Terra, W.R. 1990. Evolution of digestive systems of insects. *Annu. Rev. Entomol.* 35: 181–200.

Terra, W.R., C. Ferreira, and F. Bastos. 1985. Phylogenetic considerations of insect digestion—Disaccharidases and the spatial organization of digestion in the *Tenebrio molitor* larvae. *Insect Biochem.* 15: 443–449.

Terra, W.T. and C. Ferreira. 2012. Biochemistry and molecular biology of digestion, in L.I. Gilbert (ed.), *Insect Molecular Biology and Biochemistry*. Academic Press/Elsevier, London, U.K., pp. 365–418.

Thie, N.M.R. and J.G. Houseman. 1990. Identification of cathepsin B, D and H in the larval midgut of Colorado potato beetle, *Leptinotarsa decemlineata* Say (Coleoptera: Chrysomelidae). *Insect Biochem.* 20: 313–318.

Thomas, K.K. 1984. Studies on the absorption of lipid from the gut of desert locust, *Schistocerca gregaria*. *Comp. Biochem. Physiol.* 77A: 707–712.

Thomas, K.K. and J.L. Nation. 1984. Protease, amylase and lipase activities in the midgut and hindgut of the cricket, *Gryllus rubens* and mole cricket, *Scapteriscus acletus*. *Comp. Biochem. Physiol.* 79A: 297–304.

Tokuda, G., H. Watanabe, M. Hojo, A. Fujita, H. Makiya, M. Miyagi, G. Arakawa, and M. Arioka. 2012. Cellulolytic environment in the midgut of the wood-feeding higher termite *Nasutitermes takasagoensis*. *J. Insect Physiol.* 58: 147–154.

Treherne, J.E. 1957. Glucose absorption in the cockroach. *J. Exp. Biol.* 34: 478–485.

Treherne, J.E. 1958. The absorption of glucose from the alimentary canal of the locust, *Schistocerca gregaria* (Forsk.). *J. Exp. Biol.* 35: 297–306.

Treherne, J.E. 1959. Gut absorption. *Annu. Rev. Entomol.* 12: 43–58.

Turunen, S. 1975. Absorption and transport of dietary lipid in *Pieris brassicae*. *J. Insect Physiol.* 21: 1521–1529.

Turunen, S. and G.M. Chippendale. 1977. Lipid absorption and transport: Sectional analysis of the larval midgut of the corn borer, *Diatraea grandiosella*. *Insect Biochem.* 7: 203–208.

Undeen, A.H. 1979. Simuliid larval midgut pH and its implications for control. *Mosquito News* 39: 391–393.

Valaitis, A.P. 1995. Gypsy moth midgut proteinases: Purification and characterization of luminal trypsin, elastase and the brush border membrane leucine aminopeptidase. *Insect Biochem. Mol. Biol.* 25: 139–149.

Valaitis, A.P., M.K. Lee, R. Rajamohan, and D.D. Dean. 1995. Brush border membrane aminopeptidase-N in the midgut of gypsy moth serves as the receptor for the CryIA(c) δ-endotoxin of *B. thuringiensis*. *Insect Biochem. Mol. Biol.* 25: 1143–1151.

Van Heusden, M.C., D.J. Van der Horst, J.K. Kawooya, and J.H. Law. 1991. In vivo and in vitro loading of lipid by artificially lipid-depleted lipophorins: Evidence for the role of lipophorin as a reusable lipid shuttle. *J. Lipid Res.* 32: 1789–1794.

Vatanparast, M. and V. Hosseininaveh. 2010. Digestive amylase and pectinase activity in the larvae of alfalfa weevil *Hypera postica* (Coleoptera: Curculionidae. *Entomol. Res.* 40: 328–335.

Voss, M., W. Blenau, B. Walz, and O. Baumann. 2009. V-ATPase deactivation in blowfly salivary glands is mediated by protein phosphatase 2C. *Arch. Insect Biochem. Physiol.* 71: 130–138.

Walker, W. and M.L. Allen. 2010. Expression and RNA interference of salivary polygalacturonase genes in the tarnished plant but *Lygus liniolaris*. *J. Insect Sci.* 10: 173. doi:10.1673/031.010.14133.

Waterhouse, D.F. 1953. The occurrence and significance of the peritrophic membrane with special reference to adult Lepidoptera and Diptera. *Aust. J. Zool.* 1: 299–318.

Waterhouse, D.F. and B. Stay. 1955. Functional differentiation in the midgut epithelium of blowfly larvae as revealed by histochemical tests. *Aust. J. Biol. Sci.* 8: 253–277.

Wei, Y.D., K.S. Lee, A.Z. Gui, H.J. Yoon, I. Kim, Y.H. Je, S.M. Lee et al. 2006. N-linked glcosylation of a beetle (*Apriona germari*) cellulose AG-Egase II is necessary for enzymatic activity. *Insect Biochem. Mol. Biol.* 36: 435–441.

Weihrauch, D. 2006. Active ammonia absorption in the midgut of the tobacco hornworm *Manduca sexta* L.: Transport studies and mRNA expression analysis of a Rhesus-like ammonia transporter. *Insect Biochem. Mol. Biol.* 36: 808–821.

Weintraub, H. and A. Tietz. 1973. Triglyceride digestion and absorption in the locust, *Locusta migratoria*. *Biochem. Biophys. Acta* 306: 31–41.

Weintraub, H. and A. Tietz. 1978. Lipid absorption by isolated intestinal preparation. *Insect Biochem.* 8: 267–274.

Wieczorek, H., G. Grüber, W.R. Harvey, M. Huss, H. Merzendorfer, and W. Zeiske. 2000. Structure and regulation of insect plasma membrane H^+ V-ATPase. *J. Exp. Biol.* 203: 127–135.

Wieczorek, H., M. Putzenlechner, W. Zeiske, and U. Klein. 1991. A vacuolar-type proton pump energizes K^+/H^+ antiport in an animal plasma membrane. *J. Biol. Chem.* 266: 15340–15347.

Wieczorek, H., S. Weerth, M. Schindlbeck, and U. Klein. 1989. A vacuolar-type proton pump in a vesicle fraction enriched with potassium transporting plasma membranes from tobacco hornworm midgut. *J. Biol. Chem.* 264: 11143–11148.

Wigglesworth, V.B. 1961. *The Principles of Insect Physiology*. Methuen & Co. Ltd., London, U.K.

Wolfson, J.L. and L.L. Murdock. 1990. Diversity of digestive proteinase activity among insects. *J. Chem. Ecol.* 16: 1089–1102.

Wyatt, G.R. 1967. The biochemistry of sugars and polysaccharides in insects. *Adv. Insect Phsyiol.* 4: 287–360.

Yi, S.-X., J.B. Benoit, M.A. Elnitsky, N. Kaufmann, J.L. Brodsky, M.L. Zeidel, D.L. Denlinger, and R.E. Lee Jr. 2011. Function and immune-localization of aquaporins in the Antarctic midge *Belgica antarctica*. *J. Insect Physiol.* 57: 1096–1105.

Yin, C.-M., H. Duan, and J.G. Stoffolano, Jr. 1993. Hormonal stimulation of the brain for its control of oogenesis in *Phormia regina* (Meigen). *J. Insect Physiol.* 39: 165–171.

Yin, C.-M., B.-X. Zou, M.-F. Li, and J.G. Stoffolano, Jr. 1994. Discovery of a midgut peptide hormone which activates the endocrine cascade leading to oogenesis in *Phormia regina* (Meigen). *J. Insect Physiol.* 40: 1–9.

Zeiske, W., W. Van Driessche, and R. Zeigler. 1986. Current-noise analysis of the basolateral route for K^+ ions across a K^+-secreting insect midgut epithelium (*Manduca sexta*). *Pflugers Arch.* 407: 657–663.

Zeng, F. and A.C. Cohen. 2000. Comparison of α-amylase and protease activities of a zoophytophagous and two phytozoophagous Heteroptera. *Comp. Biochem. Physiol. A* 126: 101–106.

Zibaee, A., H. Hoda, and M. Fazeli-Dinan. 2012. Role of proteases in extra-oral digestion of a predatory bug, *Andrallus spinidens*. *J. Insect Sci.* 12: 51. doi:10.1673/031.012.5101.

Zieler, H., C.F. Garon, E.R. Fischer, and M. Shahabuddin. 2000. A tubular network associated with the brush-border surface of the *Aedes aegypti* midgut: Implications for pathogen transmission by mosquitoes. *J. Exp. Biol.* 203: 1599–1611.

CHAPTER 3

Nutrition

PREVIEW

Insects need the same basic nutritional components that larger animals need. Balance of nutrients is very critical to most insects studied. Experimentally, some insects are able to self-select among multiple choices of artificial diet formulations in order to compensate for single-diet deficiencies. This suggests that some oligophagous insects may do the same thing in nature. Immature insects often have different nutritional requirements from adults. Some adults (e.g., some Lepidoptera) do not feed as adults and acquire all the nutritional components needed for development of ovaries and eggs during larval life. Most adults need a nitrogen source to mature ovaries and eggs and a carbohydrate source for energy. Although slight variability is known among different insects, the majority studied need dietary arginine, histidine, isoleucine, leucine, lysine, methionine, phenylalanine, threonine, tryptophan, and valine, the same 10 essential amino acids required by larger animals. Some insects need a carbohydrate source for complete development, while others do not. Many adult insects need carbohydrate as an energy source. Insects cannot synthesize sterols, and thus immature insects need a dietary sterol as a precursor that can be transformed into the molting hormone, which has a sterol structure. Eggs contain sterols and the first instar may be able to molt without a dietary source, but subsequent molts may be impossible if dietary sterol is not present. Some adult insects need dietary sterol in order to produce the normal number and/or hatching of eggs. Immatures of some groups need polyunsaturated fatty acids for normal development. Insects generally need B vitamins, vitamin A (or a carotenoid), and ascorbic acid, and some may need small amounts of other vitamins. Insects do not need vitamin D, and probably not vitamin K. Vitamins, some essential amino acids, and sterols may be supplied by symbionts in the gut or fat body. The development of artificial diets for culture of insects has been stimulated by desire to learn and compare nutritional requirements as well as the need to rear large numbers of insects efficiently and economically for commercial and scientific purposes. Procedures to measure growth, digestibility, and conversion of food into body weight and tissues have been devised to evaluate growth and development of insects on artificial diets. Feeding stimulants and deterrents are important in the feeding behavior of insects in their natural environment. Frequently, the presence of natural or similar feeding stimulants and absence of feeding deterrents are important factors in getting insects to eat artificial diets.

3.1 INTRODUCTION

The basic nutritional requirements for growth and reproduction of insects are well known and were largely determined in the middle decades of the twentieth century. Insects generally have about the same basic nutritional needs as large animals (Dadd, 1985), although minor variations in both qualitative and quantitative requirements are known in some insects. In general, insects need the 10 essential amino acids required by the rat, a model animal for larger vertebrates. One notable difference between vertebrates and insects is the insect requirement for a dietary source of sterol; although some can synthesize squalene (the hydrocarbon backbone needed for the ring structure of a sterol), they cannot form the rings.

Research on insect nutrition stems from

1. Comparative scientific interest
2. Efforts to achieve increased productivity from desirable insects, such as silkworms, honeybees, pollinators, and experimental insects
3. Mass production of parasites, predators, or insects for sterile-insect release programs
4. Development of control strategies that might exploit nutritional requirements
5. Understanding metabolic pathways related to nutritional requirements
6. Understanding nutritional influence on polymorphism

The development of insect diets for ease of rearing and for mass rearing has been an especially active field (Singh, 1974, 1976, 1977; Anderson and Leppla, 1992).

The literature on insect nutrition, diet development, and rearing on artificial or semiartificial diets is extensive. There are reviews, including but not limited to House (1965a, 1974), Davis (1968), House et al. (1971), Gordon (1972), Hsiao (1972), Schoonhoven (1972), Dadd (1973, 1985), Vanderzant (1974), Scriber and Slansky (1981), Slansky (1982), Reinecke (1985), Slansky and Scriber (1985), Waldbauer and Friedman (1991), Anderson and Leppla (1992), Locke and Nichol (1992), Simpson and Raubenheimer (1995), Simpson et al. (1995), and Cohen (2004).

Currently, many insects (largely those of some economic importance) can be reared on a synthetic or semisynthetic diet, including some endoparasitoids (Bracken, 1966; Yazgan, 1972), but few representatives of some groups have been reared on synthetic diets. For example, it appears that only three or possibly four butterflies have been reared on synthetic diets, although many moths are reared relatively easily.

3.2 IMPORTANCE OF BALANCE IN NUTRITIONAL COMPONENTS

One of the strongest principles to come from insect diet studies from the middle of the last century is that **balance of nutrients** is important to effective growth, development, and reproduction. Gordon (1959) stressed that balance of nutrients is the most dominant quantitative factor in a diet. Sang (1956) presented detailed quantitative data to document the adverse effect of nutrient imbalance upon growth of *Drosophila melanogaster* larvae, and numerous authors have found similar effects with other insects. House (1965a,b, 1966a, 1969, 1974) found that nutrient imbalance resulted in reduced food intake and that optimum growth depended, among other things, on a proper ratio of amino acids to minerals. In one experiment, 67% of the larvae of *Agria affinis* selected a complete, balanced diet from among choices including a diet deficient in essential amino acid, a complete but imbalanced diet (improper proportion of essential to nonessential amino acids), and agar (House, 1967). About 45% of the larvae reached maturity in the normal time of 6 days on the complete, balanced diet, while those selecting the other diets remained in the first instar or died. The stress of excreting excess nutrients may be detrimental and wasteful of energy. Optimal balance, however, frequently changes with species, sex, and age or stage of development. Bauerfeind et al. (2007) attributed the increased egg-laying potential to nutrient balance in tropical fruit-feeding butterflies, *Bicyclus anynana* (Nymphalidae), that were allowed to feed on fresh or fermenting banana fruit compared with sugar or sugar supplemented with lipids, yeast, or ethanol. Allocation of resources (Fischer et al., 2004; Hahn, 2005; Boggs, 2009), stress resistance (Sisodia and Singh, 2012), reproduction (Telang and Wells, 2004; Boggs, 2009; Dmitriew and Rowe, 2011; Lyimo et al., 2012; Richards et al., 2012; Aguila et al., 2013; Pan et al., 2014), immunity (Telang et al., 2011; Bauerfeind and Fischer, 2014), length of life (Grandison et al., 2009; Joy et al., 2010), and adult morphology, fitness, and activity (Boggs and Freeman, 2005; Prenter et al., 2013; Johnson et al., 2014) are influenced by nutrition.

3.3 ABILITY OF INSECTS TO SELF-SELECT NUTRITIONAL COMPONENTS

Many animals, including some insects, demonstrably **self-select dietary components** both from natural foods and from defined diets. The criteria of self-selection are that (1) there is nonrandomness of choice; (2) a uniform cohort of individuals tend to select nutrients, at least the major ones, in consistent proportions; and (3) individuals having a choice to self-select do as well or better than if self-selection is not possible (see review by Waldbauer and Friedman, 1991). Confused flour beetles, *Tribolium confusum*, given a choice of 1:1:1 mix of particles of germ, bran, and endosperm selected 81% germ, 2% bran, and 17% endosperm that provided a protein/carbohydrate ratio of 57:43, close to the optimum of 50:50 for these immature beetle larvae (Waldbauer and Bhattacharya, 1973). Corn earworms, *Helicoverpa zea*, self-selected portions from defined diets providing a protein/carbohydrate ratio of 79:21, a ratio almost identical to the 80:20 ratio of protein/carbohydrate shown to be optimal for growth of tobacco hornworm larvae, *Manduca sexta* (Waldbauer et al., 1984; Cohen et al., 1987b). Nymphal brown-banded cockroaches, *Supella longipalpa*, self-selected a ratio of 16:84 protein/carbohydrate when given a choice of diets (Cohen et al., 1987a). In an attempt to rationalize the very different ratios of protein/carbohydrate selected by the two lepidopterans and the cockroach, Waldbauer and Friedman (1991) observed that these particular (and indeed most) lepidopterans have relatively short life cycles, and the high proportion of protein in the self-selected diet of larvae permits rapid growth to the pupal stage, with little expenditure of energy in searching for food. Longer lived cockroaches, on the other hand, are genetically programmed to grow more slowly (about 256 days required to reach the adult stage), and a high-protein diet does not speed up growth appreciably. During its long developmental period, however, it needs carbohydrates to provide the fuel for foraging. Food requirements and habits, of necessity, are correlated with life history.

The mechanisms by which insects self-select dietary components are not known. Changes in chemoreceptor sensitivity that correlate with feeding behavior have been observed in locusts (Simpson et al., 1990) and some lepidopterous caterpillars (Schoonhoven et al., 1991), leading to the postulate that changes in peripheral taste receptor sensitivity regulates self-selection through feedback from the metabolic and physiological state of various tissues (Simpson and Simpson, 1990; Schoonhoven et al., 1991). Associative learning, association of a specific stimulus with a reward, such as associating a chemical component with a food that promotes growth, also may play a role (Simpson and White, 1990). In an attempt to obtain some evidence for a peripheral receptor mechanism, Ahmad et al. (1993) maxillectomized third instars of tobacco hornworm, *M. sexta*, a procedure known to alter their ability to discriminate among host plants and gave them a choice of defined diets lacking carbohydrate or protein. The larvae still self-selected from both diet formulations to obtain a protein/carbohydrate ratio equal to that of sham-operated insects. Thus, the mechanism involved in self-selection of diet by insects is still unknown, and direct evidence for feedback that changes peripheral receptor sensitivity is lacking.

3.4 REQUIREMENTS FOR SPECIFIC NUTRIENTS

In order to determine the nutrient requirements of insects, it is most desirable and often necessary (1) to rear the insects through multiple generations and (2) to rear them under **aseptic** or **axenic** conditions. The reason for the first practice is that small insects require only small amounts of some nutrients, and slight contamination of dietary components with traces of those nutrients and/or carryover in the egg and body of the insects may be sufficient for several generations. Some examples will be enumerated in the following account of specific nutrients. The second practice, rearing under axenic conditions, is necessary because many, and perhaps all, insects contain microfauna and microflora in the gut or in special bodies called mycetomes or as bacteroids scattered among fat body cells, and these symbionts usually supply some nutrients to their host (Singh and House, 1970). Blood-feeding insects, phloem and xylem sap feeders, stored-products insects, and cockroaches and termites have symbionts that are known to, or may, supply some vitamins, essential amino acids, and sterols (Dadd, 1985).

3.4.1 Nitrogen Source: Proteins and Amino Acids

Most insects obtain amino acids from their foods by ingesting proteins. An interesting case among butterflies is pollen feeding by adult *Heliconius* and *Laparus doris* butterflies, the only butterflies known to feed upon pollen. These adult butterflies collect a ball of pollen on the proboscis and process it by repeatedly coiling and uncoiling the proboscis, a procedure that enable them to extract essential amino acids and other nutrients from the pollen grains (Krenn et al., 2009, and references therein). The extracted ball of pollen with many ruptured and deformed grains is subsequently discarded.

Purified proteins, such as casein from milk, gluten from wheat, albumin from eggs, and sometimes soybean and peanut protein preparations, have been used in artificial diets. A product called wheast, prepared from milk whey and yeast used in the brewery industry, also is available. In artificial diet formulations, investigators commonly add one or more of these protein sources. Unfortunately, no single purified protein is an entirely satisfactory source of amino acids of balanced proportions for all insects. Casein and egg albumen have a good balance of most amino acids and have been the most widely used in insect diets. Each, however, is relatively low in histidine and tryptophan, two of the essential amino acids. Casein is also relatively low in cystine and glycine, although these are not essential for insects. Most insects probably have an optimum level of protein required in the diet for best growth, but this varies widely for different species. Restricting the dietary protein of several species of cockroaches retarded growth, but prolonged longevity. From 22% to 24% protein in the diet gave fastest growth and lowest nymphal mortality of German and Oriental cockroaches, but 11% protein promoted maximum longevity. The American cockroach grew fastest on 49%–78% protein, but survived longest on 22%–24% protein (Haydak, 1953).

Many adult female insects require a source of protein in order to mature their ovaries and eggs. Many species of female mosquitoes require a blood meal to supply proteins for metabolism and incorporation of yolk into eggs (Zhou and Miesfeld, 2009; Lyimo et al., 2012; Yamany et al., 2012; Farjana and Tuno, 2013). Some species of mosquitoes, said to be autogenous, can lay one batch of eggs without a blood meal by using proteins incorporated from larval development, but they need a blood meal thereafter for each batch of eggs. Rearing mosquitoes in the laboratory for various research programs often requires keeping some suitable animals for the mosquitoes to feed upon, although some adult female mosquitoes will feed upon vertebrate blood enclosed in a membrane. Pitts (2014) reported rearing *Aedes albopictus*, an important vector of several arboviruses, in the laboratory with bovine serum albumin as the only protein source, and the females actually produced more eggs than blood-fed ones.

Protein deprivation may manifest itself in failure to secrete juvenile hormone (JH), which is needed for ovary and egg development, but even if JH or an analog, such as methoprene, is administered to protein-starved insects, they do not produce the normal complement of eggs simply because they do not have enough protein reserves in the body. Male insects usually do not require protein as adults in order to mature sperm.

These examples of sex and developmental requirements illustrate the generalization that optimal nutritional requirements frequently differ with age, sex, and physiological stress. Any attempt to state optimal requirements for protein or amino acids must include a definition of the evaluation criteria.

3.4.2 Essential Amino Acids

The classic method for determining amino acid requirements has been deletion of one amino acid at a time from a diet fed to a group of insects. This is obviously quite time consuming, requiring rearing many insects on a large series of diets, and it also presupposes that the insects can be grown on a synthetic diet of known composition. Some knowledge of the amino acid composition of the protein sources commonly eaten by a particular insect may be helpful in formulating an amino acid mixture on which the insect can survive. For example, Vanderzant (1958) determined that an amino acid mixture characteristic of proteins from cotton supported growth and the development of the pink bollworm, *Pectinophora gossypiella*, better than a mixture based on casein. Achieving a suitable balance of essential and nonessential amino acids (and sometimes their relationship to other nutrients) is often critical to successful rearing (House, 1965b, 1966a).

The **essential amino acids** of a number of insects from different orders have been demonstrated to be **arginine, histidine, isoleucine, leucine, lysine, methionine, phenylalanine, threonine, tryptophan**, and **valine** (all in the L-form). These are the same essential amino acids required by vertebrates. In the absence of one of these essential amino acids, growth and development ceases in the insects listed in Table 3.1. In some cases, nonessential amino acids stimulate growth, and this may be related to the optimization of nutrient balance and the efficiency of the biochemical pathways involved in synthesis of the nonessential amino acids. Although some species can be reared on a diet

Table 3.1 Insects Known to Require the 10 Essential L-Amino Acids: Arginine, Histidine, Isoleucine, Leucine, Lysine, Methionine, Phenylalanine, Threonine, Tryptophan, and Valine

Species	Ref.[a]
Aedes aegypti	4
Agria affinis	5
Cochliomyia hominivorax	18
Culex pipiens	18
Drosophila melanogaster	6
Phormia regina	8
Stegobium paniceum	19
Bombyx mori	18
Myzus persicae	17
Ctenicera destructor	16
Agrotis orthogonia	10
Tribolium confusum	2
Trogoderma granarium	3
Hylemya antiqua	7
Chilo suppressalis	11
Pectinophora gossypiella	12
Helicoverpa zea	13
Attagenus sp.	1
Argyrotaenia velutinana	14
Anthonomus grandis	15
Apis mellifera	9

[a] Reference guide: 1. Moore (1946), 2. Lemonde and Bernard (1951), 3. Pant et al. (1958), 4. Singh and Brown (1957), 5. House (1954), 6. Hinton et al. (1951), 7. Friend (1958), 8. Kasting and McGinnis (1960), 9. DeGroot (1952), 10. Kasting and McGinnis (1962), 11. Ishii and Hirano (1955), 12. Vanderzant (1958), 13. Rock and Hodgson (1971), 14. Rock and King (1968), 15. Vanderzant (1973), 16. Kasting et al. (1962), 17. Dadd and Krieger (1968), Mittler (1971), 18. Dadd (1985), 19. Pant et al. (1960).

in which the 10 essential amino acids are the only amino acid source, others cannot be reared with only the 10 essential ones. Most insects that have been studied actually grow better with a balance of essential and nonessential amino acids. Such a mixture undoubtedly saves energy that otherwise must be expended in synthesizing the nonessential amino acids from the essential ones.

The classic deletion method for determining the essential amino acids cannot be used, of course, if the insect cannot be reared on a defined diet. Kasting and McGinnis (1958) demonstrated the value of an indirect method for defining the essential amino acids for such insects based on injecting or feeding the insects with glucose-U-^{14}C (uniformly labeled glucose) or another suitable general precursor compound (Table 3.2). Essential amino acids are expected to have no ^{14}C label, since they are not supposed to be synthesized, and nonessential amino acids should be labeled because they are synthesized. In practice, low label incorporation is often found in some amino acids known to be essential from deletion studies in cases where both methods have been compared. Evidently, slight synthesis of some essential amino acids may occur, but the rate is too low to eliminate the need for a dietary supply. The method was compared with the classic deletion procedure in a test with the blowfly, *Phormia regina* (Kasting and McGinnis, 1960). Third instars were injected with 3–6 μL of glucose-U-^{14}C containing 5000 counts/min/μL. Sixty-eight hours later, the larvae were homogenized and their body proteins were hydrolyzed to yield free amino acids, which were then separated by ion exchange column chromatography. Those amino acids synthesized from the radioactive glucose were expected to contain significant ^{14}C label and these would be considered nonessential. The essential amino acids, which the blowfly larvae could only get from their food, should contain very little or no label. The results suggested several discrepancies between the deletion and radiolabel techniques. For example, the deletion study indicated that proline was essential, while the label incorporation data indicated that proline was synthesized, that is, it was nonessential. This was later clarified by the finding that some strains of *P. regina* need proline, while other strains do not. Also, the label was not incorporated into tyrosine, suggesting it was essential, while the deletion study showed it to be nonessential. In other insects, tyrosine also has been shown to be nonessential. It is probable that phenylalanine is the precursor

Table 3.2 Determination of Essential and Nonessential Amino Acids by Administration of Glucose-^{14}C to Prairie Grain Wireworms and Two-Spotted Spider Mites

	Prairie Grain Wireworm[a]		Two-Spotted Spider Mite[b]	
	c/min/μM	Requirement	c/min/μM	Requirement
Arginine	0.6	+	19	+
Histidine	0.5	+	25	+
Isoleucine	0.11	+	57	+
Leucine	0.2	+	19	+
Lysine	0.01	+	11	+
Methionine	3.5	+	285	+
Phenylalanine	0.7	+	17	+
Threonine	?	?	2989	−
Valine	0.5	+	94	+
Proline	22		605	−
Alanine	170	−	7247	−
Aspartic acid	37	−	5500	−
Serine	79	−	1350	−
Glycine	38	−	1256	−
Glutamic acid	24	−	2671	−
Tyrosine	1.3	−	50	+

[a] Data from Kasting and McGinnis (1962).
[b] Data from Rodriquez and Hampton (1966).

of tyrosine in the blowfly, as has been demonstrated in a number of other insects. If so, no label would be expected in either phenylalanine or tyrosine. Finally, the labeling technique indicated that both methionine and cystine were essential, whereas the deletion study had indicated neither was essential. Subsequent deletion studies showed that when both methionine and cystine were simultaneously deleted, at least one of them (either one) was essential. Overall, the labeling technique gave results that agreed well with deletion studies for *P. regina* after certain ambiguities were clarified by additional research.

The isotope technique has been used with the prairie grain wireworm, *Agrotis othogonia* (Kasting and McGinnis, 1962), two-spotted spider mites, *Tetranychus urticae* (Rodriquez and Hampton, 1966), and corn earworm, *Helicoverpa zea* (Rock and Hodgson, 1971), who compared the labeling technique with the deletion method for *H. zea*. The deletion method indicated that arginine, histidine, isoleucine, leucine, lysine, methionine, phenylalanine, threonine, tryptophan, and valine were essential. These amino acids had a low label content, but tryptophan and phenylalanine were labeled strongly enough to be inconclusive. Other amino acids were highly labeled and, in agreement with the deletion method, they were considered nonessential.

The isotope labeling technique is clearly useful when a defined diet is not available, and it is much faster in producing results than the slower deletion method that requires many trials. The previous results indicate, however, that the results have to be viewed with caution and confirmed, when possible, with deletion techniques. Additionally, neither method can account for the possible contribution of symbionts in synthesis of amino acids.

It is a common observation that insects often eat their molted exuviae, and Mira (2000) has suggested that one possible benefit from such behavior may be acquisition of a protein meal. In experiments, Mira found that larval cockroaches (*Periplaneta americana*) usually ate the exuviae during larval life, and adult females ate the exuviae more often than males, possibly because of a greater need for nitrogen for reproduction. Cockroaches reared on high-protein diets most often did not eat the exuviae, while those rendered aposymbiotic always ate their exuviae, both of which tend to support a nutritional role for the exuviae, although other explanations also are possible.

3.4.3 Carbohydrates

Most insects do not have an absolute growth requirement for a specific carbohydrate in the diet, although carbohydrates are a major energy source for most insects. Generally, insects can synthesize carbohydrates from amino acids and from lipids. Species in the genera *Tenebrio* (meal worm), *Ephestia* (flour moth), and *Oryzaephilus* (saw-toothed grain beetle) need a carbohydrate source to reach maturity. Other stored grain insects, such as species of *Tribolium* (flour beetles), *Lasioderma* (cigarette beetle), and *Ptinus* (powder post beetles), can be reared to maturity on diets lacking carbohydrates. The adults of the dipterans *Calliphora erythrocephala*, *Lucilia cuprina*, and *Anastrepha suspensa*, some other tephritid fruit flies, and probably many adult dipterans require carbohydrates (typically satisfied by sucrose) for an energy source and continued survival. Worker honeybees have a requirement for carbohydrates at the time of pupation. Worker larvae can be reared in the laboratory on worker jelly, but they fail to pupate on the worker jelly (Shuel and Dixon, 1968). Worker larvae fed royal jelly, the food normally fed to developing queen larvae, pupate normally in the laboratory. Worker larvae seem to have a sugar requirement for pupation, and worker jelly, with only 4% carbohydrate content, has too little. Royal jelly contains about 12% carbohydrate content, so the pupation requirement must lie between 4% and 12% carbohydrate. Shuel and Dixon showed that the addition of 40 mg glucose and 40 mg fructose per gram worker jelly (i.e., 8% additional sugar) allowed worker larvae to pupate in the laboratory when fed the altered worker jelly. Worker jelly is a glandular secretion produced by worker bees and fed to worker larvae for the first 3 days by adult bees. In a honeybee colony, the adult bees feed older worker larvae on a modified worker food containing honey and some pollen, so the carbohydrate content of their natural food is high as they approach pupation.

3.4.4 Lipids

"Lipid" is a broad term that includes biological molecules that are soluble in such organic solvents as ether, alcohol, and similar solvents. Typical lipids in biological organisms include free and bound fatty acids; short- and long-chain alcohols; tri-, di-, and monoacylglycerols; steroids and their esters; phospholipids; and several other groups of compounds. Most insects have the necessary metabolic machinery to convert carbohydrates into lipids, and many insects synthesize lipids and store them in the fat body tissue. A specific lipid required in the diet is sterol and some insects require polyunsaturated fatty acids. Some adult insects do not feed (e.g., some lepidopterans) as adults and do not have mouthparts adapted for feeding. They survive and reproduce by using nutrients accumulated as larvae and stored in the body during pupation. During pupation, *D. melanogaster* (and probably many other dipterans) conserve fat body cells from the larval stage, and the adult uses the lipids and other components stored in these cells as a resource during a short, nonfeeding period until additional food resources are found (Aguila et al., 2007). Similarly, small clumps of larval/pupal fat body cells can be identified by their morphological appearance in newly emerged adult Caribbean fruit flies (Tephritidae) for 2–3 days before they are used up.

3.4.5 Sterols

Because of their inability to synthesize sterols, insects must obtain sterol(s) from their food and/or from their symbionts. They use sterol as a precursor for synthesis of the ecdysteroid molting hormone (see Chapter 5) and as a component of all cell membranes. Cholesterol and sparing sterols, such as cholestanol, are incorporated into all tissues of *Eurycotis floridana*, a cockroach. A **sparing sterol** is one that can be incorporated into cell membranes, but cannot be used to synthesize the molting hormone. The requirement for a dietary sterol was first noted by Hobson (1935) in larvae of the blowfly, *Lucilia sericata*, and has since been verified in many different insects. The firebrat *Ctenolepisma* sp., representative of a very primitive group of insects, can synthesize some sterol, but probably not enough for its needs. There is a recent review of the role of sterols in nutrition and physiology of insects (Behmer and Nes, 2003).

Cholesterol usually satisfies the sterol requirement. Frequently, several different sterols can replace or spare cholesterol, probably serving in a relatively nonspecific capacity as a cell membrane component in place of cholesterol. In only a few insects is it known that the ecdysteroid molting hormone can be synthesized from a sterol other than cholesterol. Although cholesterol has been detected in small quantity in some plants, phytophagous insects normally ingest β-sitosterol and stigmasterol as major plant sterols and other sterols that occur in plants. Biochemical pathways for conversion of plant sterols into cholesterol have been demonstrated in many phytophagous insects and some nonphytophagous ones (Svoboda et al., 1975). Cholesterol or β-sitosterol satisfies the sterol requirement in most of 18 species studied and ergosterol (or 7-dehydrocholesterol) can be utilized by about three-fourths of them (see Chapter 5 for more details on specific sterol use for synthesis of the molting hormone and for conversion pathways).

A few species are known to have a requirement for a very specific dietary sterol. *Drosophila pachea* breeds only in the senita cactus, *Lophocereus schottii*, in the Sonoran Desert of the southwestern United States, and it requires 7-stigmasten-3β-ol, an uncommon sterol found in senita cactus. Only 7-cholesten-3β-ol and 5,7-cholestadien-3β-ol can substitute for the cactus sterol (Kircher et al., 1967). *Xyleborus ferrugineus*, a scolytid beetle, will use cholesterol and lanosterol for egg production and hatching, but larvae fail to pupate unless ergosterol or 7-dehydrocholesterol is available in the diet, presumably indicating a critical need for either of these two sterols for synthesis of its molting hormone. The larvae normally obtain ergosterol from the symbiotic fungus *Fusarium solani* growing on dead trees that are hosts for the scolytid larvae (Norris and Baker, 1967). *Lobesia botrana*, the grape berry moth, has a mutualistic relationship with the fungus *Botrytis cinerea*, and the moth grows faster, survives longer, and has greater fecundity when grown on an artificial diet containing mycelium or purified sterols from the fungus (Mondy and Corio-Costet, 2000).

Sterol deficiency may be manifested in any of the stages of an insect. Newly hatched larvae that lack dietary sterol usually die in the first or second instar because they exhaust the sterol received in the egg from the mother. The lack of a sterol by adult female houseflies results in 80% reduction in egg hatch, although the number of eggs laid is not affected. Adult female boll weevils fed a diet in which cholestanol replaces half the cholesterol requirement cannot maintain normal egg production, but eggs that are laid do hatch. Replacement of more than one-half the cholesterol requirement with cholestanol, however, results in eggs that fail to hatch (Earle et al., 1967).

No general rule can be drawn as to the quantity of sterol needed in an artificial diet. Wide variation in required quantity exists, apparently related to species differences. Generally, 0.1% or less of cholesterol is considered to be satisfactory in artificial diets. A few insects, such as *Musca vicina*, *Dermestes vulpinus*, and *Attagenus piceus*, need as little as 0.01% sterol by weight of diet, while others need about 0.1% by weight.

The ability to synthesize a sterol is mixed among other invertebrates. Some marine annelids can synthesize sterols, but the earthworm, *Lumbricus terrestris*, cannot. Other invertebrates that cannot synthesize a sterol include the crabs *Astacus astucus* and *Cancer pagurus*, sea urchin *Paracentrotus lividus*, oyster *Ostrea gryphaea*, the mollusk *Mytilus californians*, the tapeworm *Spirometra mansonoides*, and the nematodes *Caenorhabditis briggsae*, *Turbatrix aceti*, and *Panagrellus redivivus*.

3.4.6 Polyunsaturated Fatty Acids

About 50 species of insects from five orders have been shown to require a dietary source of **polyunsaturated fatty acids** (Dadd, 1961; Chippendale et al., 1964; Dadd, 1985). Linoleic acids—(Z,Z)-9,12 octadecadienoic acid and (Z,Z,Z)-6,9,12-octadecatrienoic acid (Figure 3.1)—are effective in relieving the symptoms of deficiency; in some species, one of these is more effective than the other.

The requirement for the polyunsaturated fatty acids was initially discovered in lepidopterans, in which a deficiency is dramatically displayed in failure of pupal or adult ecdysis. Individuals that successfully or partially ecdyse are likely to have misformed wings and lack normal scales on the body. Some hymenopterans show similar difficulty in ecdysis when linoleic and linolenic acids are absent or very low in the diet. Acridid grasshoppers also tend to produce deformed adults on fatty acid-deficient diets. Coleopterans show slowed growth and decreased adult fecundity in response to deficiency in polyunsaturated fatty acids. The growth of some insects is improved by adding polyunsaturated fatty acids to the diet even though an absolute requirement has not been demonstrated. Possibly so little is required in the diet that traces in the test diet, or carryover from the egg, may contain enough to support some insects through a generation. It may be necessary to

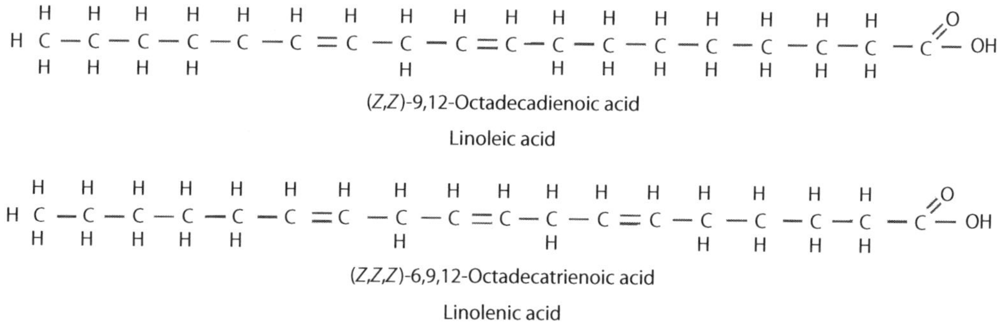

(Z,Z)-9,12-Octadecadienoic acid

Linoleic acid

(Z,Z,Z)-6,9,12-Octadecatrienoic acid

Linolenic acid

Figure 3.1 The structures of the two polyunsaturated fatty acids essential for growth and development of some insects.

test for a requirement through more than one generation, especially if no apparent defect is noted, and if adult performance is a criterion evaluated (Dadd, 1985).

In general, it appears that dipterans may not have an absolute requirement for polyunsaturated fatty acids, even though they are unable to synthesize them. Some dipterans, however, show improved growth when polyunsaturated fatty acids are added to a diet. A few insects, including the cricket *Acheta domesticus*, American cockroach *P. americana*, and the termite *Zootermopsis angusticollis*, are able to synthesize polyunsaturated fatty acids.

3.4.7 Vitamins

Studies of insect vitamin requirements especially are subject to ambiguity if axenic conditions are not maintained. In some early work with stored product insects reared on very dry diets seemed to avoid the interfering effect of microorganisms, which can synthesize many of the vitamins that insects then utilize. These studies showed that insects need thiamine, riboflavin, pyridoxine, niacinamide, pantothenic acid, biotin, folic acid, and choline. Carnitine, also called vitamin B_T, is a requirement for *Tenebrio molitor* (Leclercq, 1950) and *Tribolium obscurus*, *T. confusum*, and *T. castaneum*. Different strains of these insects show variable requirements. One of the critical roles for carnitine is as a participant in the passage of fatty acids across mitochondrial membranes in insects and vertebrates.

Houseflies and blowflies are able to use β-methylcholine and γ-butyrobetaine to reduce the need for choline. When these compounds are fed to flies, the phospholipids of most tissues contain β-methylcholine, but acetylcholine in central nervous system tissue is not substituted. *T. molitor* larvae also incorporate β-methylcholine into body phospholipids, which spares the choline requirement.

There is demonstrated requirement for water-soluble vitamins in the nutrition of a few insects. Ascorbic acid is required for normal growth and development of some insects and seems particularly needed by phytophagous insects (Vanderzant and Richardson, 1963; Beck et al., 1968). Boll weevils, *Anthonomus grandis*, grown under aseptic conditions require inositol for normal growth and development, as do *Blattella germanica* and *P. americana*, German and American cockroaches, respectively. Inositol often seems to improve the growth of many insects, but it has not been demonstrated to be essential in most insects.

Carotene and/or vitamin A are required by insects for normal pigmentation and eye function. *Schistocerca gregaria* needs β-carotene for normal body coloration. Vitamin A is required by houseflies, *Musca domestica*, and tobacco hornworms, *M. sexta*, for normal structure of the eye. Because so little carotene or vitamin A is required (and/or small amounts were contaminating the "purified" diet), houseflies had to be reared for 15 generations on a diet lacking carotenoids and vitamin A to demonstrate conclusively that the vitamin is needed. The 12th and 13th generations had about the same sensitivity in the compound eyes, measured by electroretinograms in response to 340 and 500 nm light, as the first generation. The response from eyes of deficient flies was 2 log units (100×) less sensitive than from eyes of normal flies (Goldsmith et al., 1964). There were changes in rhabdom structure, loss of basement membrane in some places, and degeneration of nervous tissue in the eyes of *M. sexta* reared for several generations on a diet deficient in vitamin A or β-carotene. Moths from the deficient diet showed abnormal orientation to light, and the eyes failed to adapt to the dark (Carlson et al., 1967). Vitamin A accelerates growth of the fly *A. affinis* and the silkworm *Bombyx mori*, but it is not clear that it has a metabolic function in growth apart from its visual function.

Vitamin B_{12} stimulates the growth of some insects, but a clear-cut requirement for growth has not been shown. Possibly, the small amount that may satisfy an insect requirement can occur as a contaminant of other nutrients or be provided by symbionts. Omission of the vitamin from the diet of *B. germanica* results in nonviable eggs, so possibly, it plays a biochemical role in at least some insects.

NUTRITION

Vitamin E is necessary to a beetle, *Cryptolaemus montrouzieri*, in order for adult females to mature and oviposit eggs. The vitamin also is required for spermatogenesis in male house crickets, *A. domesticus*. The parasitoid *A. affinis* needs vitamin E in the larval diet for adult females to produce viable offspring. The vitamin also stimulates growth and development of larvae (House, 1966b). There is no evidence that vitamin D is required by insects. Vitamin K in its several forms has been tested on some insects, usually without any observable effects, but it may have some positive benefit (mechanism unknown) on crickets and may act as a phagostimulant for adult worker honeybees (Dadd, 1985).

3.4.8 Minerals

In general, only major mineral requirements are known for a few insects. Contamination of other food materials with small amounts of minerals, as well as formulation and chemical interactions when minerals are added to a synthetic diet, makes determination of trace element requirements very difficult (Dadd, 1968). It stands to reason that insects need small amounts of many minerals because metal ions are required as enzyme cofactors and as constituents of metalloenzymes. For example, molybdenum (Mo) is a part of the xanthine dehydrogenase involved in purine metabolism of insects. Thus, it seems reasonable to conclude that insects will need traces of Mo in their diet. Insects and vertebrates clearly have some notable differences in quantitative requirements for certain minerals. Vertebrates need large quantities of iron and calcium for hemoglobin and bone formation, respectively. Insects use these elements for neither of these functions (except a few species of insects that do have iron incorporated into hemoglobin) and require only trace amounts of iron and calcium (see review by Locke and Nichol, 1992). Many phytophagous insects need relatively large quantities of potassium and only trace amounts of sodium, while vertebrates need these elements in the reverse order.

Just how much sodium an insect needs is not known, but some Lepidoptera seem to have a need that is met by puddling or drinking at standing water, usually on the ground (Figure 3.2). Puddling behavior has been observed primarily in male Lepidoptera, but females also are known to puddle. Smedley and Eisner (1995) experimentally evaluated puddling behavior in male *Gluphisia septentrionis* moths (Notodontidae). The adult moth has its mouthparts modified in a

Figure 3.2 **(See color insert.)** Butterflies in the family Lycaenidae, Genus *Agrodiaetus*, puddling and taking moisture and salts from the damp soil at the Kaçkar Mountains in Eastern Turkey. The slightly larger and lighter colored individual in the middle of the group is in the genus *Meleageria*. (Photo courtesy of Andrei Sourakov, University of Florida and McGuire Center, Gainesville, FL, 2001. With permission.)

way that seems to facilitate rapid sucking up of puddle water while straining out debris that might be present (Figure 3.3). A male moth may imbibe so much water at natural puddles that it ejects an average of 8 µL of fluid from the anus about every 3 s (Figure 3.4); such behavior was observed to continue for more than 200 min in some individuals. A maximum excretion of fluid equivalent to 600 times the body weight of a moth was observed. The authors showed by quantitative analyses of imbibed fluid and excreted fluid that there is specifically a gain in body sodium by the male moth without a necessary gain of potassium, magnesium, or calcium in test solutions imbibed. The time spent in puddling and volume of fluid excreted is inversely related to solutions containing 0.01, 0.1, and 1 mM Na. Males transfer sodium acquired by puddling to females at mating, and females incorporate it into eggs.

Aphids, which feed upon a liquid diet that can be highly purified, have proven useful in mineral studies. Two aphids, *Myzus persicae* and *Aphis fabae*, require trace amounts of Fe, Mn, Zn, and Cu, as well as major quantities of K, Mg, and phosphate (Dadd, 1967, 1968). Potassium and magnesium are major requirements of *D. melanogaster* (Sang, 1956). Zinc is necessary to *T. molitor* and about 6 µg/g diet satisfies the requirement. Cockroaches grown on artificial diets

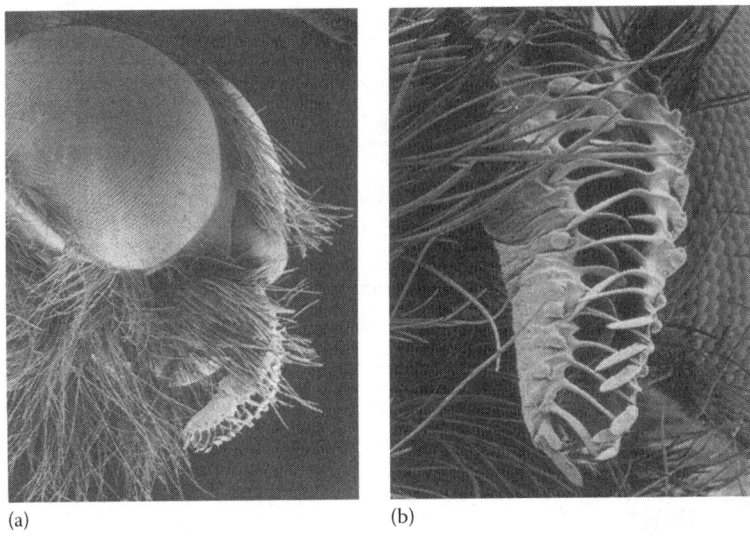

Figure 3.3 (a) Head of male moth *Gluphisia septentrionis* showing the stubby, highly modified proboscis that enables the moth to suck up puddle water while straining out debris. (b) An enlarged view of the male proboscis illustrating the oral cleft and sieving apparatus. (Photos courtesy of Maria and Tom Eisner; From Smedley, S.R. and Eisner, T., *Science*, 270, 1816, 1995. With permission.)

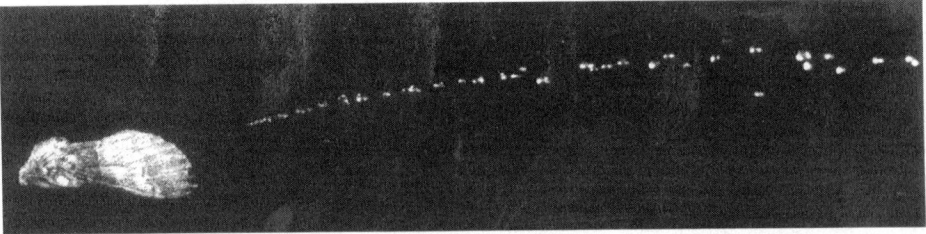

Figure 3.4 A male *Gluphisia septentrionis* forcefully excreting excess fluid during puddling behavior. Some males may eject fluid equal to 600 times the body weight, excreting an average of 8 µL every 3 s for up to 200 min. (Photos courtesy of Tom Eisner and Scott Smedley; From Smedley, S.R. and Eisner, T., *Science*, 270, 1816, 1995. With permission.)

with very low levels of manganese and zinc tend to lose symbionts from their mycetocytes, but the mechanism for this interaction is unknown (Brooks, 1960).

The balance of minerals in a salt mixture and the proportion of minerals to other groups of nutrients are important. Wesson's salt mixture is designed for vertebrates (Osborne and Mendel, 1932) and is high in Ca, Fe, and Na. Wesson's salt mix (often called Salt Mix W) is not adequate to support the development of the European corn borer on an artificial diet. The Ca level in Wesson's salt mixture is toxic to *B. germanica* (Gordon, 1959). A mixture of salts based upon a successful formula for confused flour beetles, *T. confusum* (Medici and Taylor, 1966), supports development of corn borers (Beck et al., 1968).

3.5 TECHNIQUES AND DIETARY TERMS USED IN INSECT NUTRITION STUDIES

Diets for rearing insects are important not only as a way to study insect nutritional requirements but also in mass rearing programs for sterile releases and augmentation of natural parasites and predators. **Holidic** diets consist of chemicals that have a precisely known chemical structure before the various chemicals are mixed. Holidic diets are sometimes referred to as chemically defined diets, although chemical components may react upon mixing to produce new chemical compounds that may not be known. Holidic diets are important in the study of nutritional requirements. **Meridic** diets contain a holidic base with addition of one (or possibly a few) unknown or poorly defined substance(s). **Oligidic** diets contain complex organic material, such as lettuce for grasshoppers, dog food or chick mash for crickets and cockroaches, or ground pinto beans for some lepidopteran larvae. When insects are reared so that no other species (no bacteria, no fungi, no internal symbionts) are present, the culture is called **axenic** culture. Axenic rearing is quite difficult to achieve, but precise definition of nutritional requirements for insects demands axenic rearing, and it has been successfully accomplished in a few cases. **Gnotobiotic** culture is one in which all the species existing are known; such a culture may or may not be axenic, depending upon how many species exist in the culture. For practical purposes of maintaining laboratory cultures and mass production, insects are usually grown in xenic cultures (unknown number of organisms in the culture) on oligidic diets.

3.6 CRITERIA FOR EVALUATING NUTRITIONAL QUALITY OF A DIET

The measurement of growth rate has frequently been used to determine nutritional quality of diets fed to immature stages. Measurements of weight gains, time between molts or to pupation, or time to adult emergence have been used. The percent of successful pupation or emergence of adults may be used. Adult diets can be evaluated by number of eggs laid, percent hatch of eggs, longevity of adults, time to sexual maturity, or other physiological parameters the investigator believes to be influenced by nutrition. It is usually desirable to have more than one criterion for evaluating a diet. Nutritional quality may have little or no effect upon one criterion, while causing great changes in another.

Requirements for some nutrients may not be manifest in the first generation; nutrient reserves are frequently stored in body tissues or egg yolk. Hence, in the absence of any effects of an experimental deficiency being tested, several generations should be reared. Axenic rearing conditions should be maintained when possible. Microorganisms present in the gut or in mycetomes frequently contribute to digestion, availability of nutrients, and biosynthesis of some nutrients, particularly vitamins, and sometimes sterols and some essential amino acids.

The purity of dietary components becomes crucial in some experiments. Traces of sterols frequently occur in protein sources, such as casein or egg albumin. In general, requirements for trace elements, such as Na, Zn, Fe, Mn, and Cu, have not been established for insects because other dietary components contain these elements in sufficient quantity as contaminants. Removal of contaminants, whether sterols, vitamins, or trace minerals, may be tedious and costly.

3.7 MEASURES OF FOOD INTAKE AND UTILIZATION

Animal breeders have been very successful in breeding and selecting animals that maximize weight gains per unit of food consumed. Entomologists also have become more concerned with such factors as efficiency of utilization of food by insects because of the increasing costs of large mass rearing programs. Several procedures for measuring the efficiency of food utilization by insects have been developed, including the following (Waldbauer, 1964; Slansky and Scriber, 1985):

1. **Relative growth rate (R.G.R.)**

    R.G.R. = (dry weight gained)/(feeding days × mean dry weight)

2. **Approximate digestibility (A.D.)**

    A.D. = [(dry weight of food ingested

    − dry weight of feces)/(dry weight of food ingested)] × 100

3. **Efficiency of conversion (E.C.I.)** of ingested food to body matter

    E.C.I. = (weight gained/dry weight of food ingested) × 100

4. **Efficiency with which digested food is converted to body matter (E.C.D.)**

    E.C.D. = [(weight gained)/(dry weight of food ingested − dry weight of feces)] × 100

Experimental measurements of A.D., E.C.I., or E.C.D. require quantitative data for food ingested and weight of feces excreted. In some cases, weighing the food remaining after the insects have ceased feeding, or at the end of a chosen interval, may be satisfactory. Feces of some insects can be manually separated from uneaten food and weighed.

The addition of chromic oxide to the food ingested and subsequent chemical analysis of the amount of chromic oxide in the feces has been used in vertebrates and, in some insects, to indicate the amount of food consumed (McGinnis and Kasting, 1964). The method works if (1) the chromic oxide is uniformly distributed in the food, (2) it has no toxic effects nor alters digestion or physiology of the animal, and (3) it is not absorbed from the gut. In the chromic oxide method, the percentage of ingested food that is utilized is given by the formula

1 − (weight of chromic oxide/unit dry weight of food)/(weight of chromic oxide/unit dry weight of feces) × 100

Utilizing the chromic oxide method with fifth instars of the pale western cutworm *Agrotis orthogonia*, McGinnis and Kasting (1964) found that 41% of the sprouts of Thatcher wheat were utilized by the larvae, but only 21% of a mixture of equal parts of sprouts and powdered cellulose and 16% of the pith from Rescue wheat were utilized. The method requires that a sample of feces must be separated from the uneaten food, the dry weight obtained, and the concentration of chromic oxide determined. If only a sample of feces, and not the total quantity of feces, is collected for analysis, possible error may occur if the concentration of chromic oxide is not reasonably the same in feces excreted at different ages or time of day.

The uric acid produced by insects from protein and purine catabolism also has been used as an indicator of food utilization (Bhattacharya and Waldbauer, 1970, 1972). Uric acid, which does not occur in most foods, especially not in stored grains, is easily determined quantitatively. A small sample of mixed food and feces must be carefully separated manually, the feces weighed, and the quantity of uric acid determined per unit weight of feces. Then, the uric acid content of a larger unseparated sample of food–feces is determined. For example, if a carefully separated sample of feces contained 10% uric acid, while a much larger weighed mixture of food and feces contained 1% uric acid, one could estimate that the mixture was 90% uneaten food and 10% feces. This method is subject to error if the quantity of uric acid per unit weight of feces varies with age or stage of development of the insect (and it often does). The potential errors in the chromic oxide and uric acid methods should be investigated before indiscriminate use of either method in a particular insect. The use of the uric acid method gave good agreement with the more laborious technique of manually separating all the uneaten food and feces in *T. molitor*, *T. confusum*, *Argyrotaenia velutinana*, and *Heliothis virescens*. Chow et al. (1973) compared the uric acid method with the manual separation and gravimetric analysis of feces and uneaten food in a study of approximate digestibility in two lepidopterans, and there was no difference in results between the two methods (Table 3.3) and no differences in the results from a comparison of female and male *A. velutinana* (Table 3.4).

Even though digestibility of a particular food may be good, it may not be readily converted into body substance. Bhattacharya and Waldbauer (1970) found that *T. confusum* larvae digested more than 50% of cracked wheat, wheat germ, ground wheat, and ground wheat supplemented with 5% brewer's yeast, but they converted 15% or less of the digested food into body weight (Figure 3.5). An imbalance in nutrients resulting from digestion may be one factor that prevents efficient conversion into body substance. About 45% of the food ingested goes into net weight gain of honeybee worker larvae when they are fed royal jelly, but this is not their normal food. An interesting technique that made use of a radioactive nuclide, ^{32}P, was devised to study food consumption in honeybee larvae. The ^{32}P was mixed with royal jelly and the labeled jelly was hand-fed to larvae growing in incubators. Because honeybee larvae do not have a complete gut until just prior to pupation, larvae only void feces just before they pupate. Thus, the ^{32}P ingested each day accumulated in the body. By knowing the specific activity of the food, it was possible to calculate the rate and cumulative food intake after removing a larva from the food and obtaining a total body count. For ^{32}P with a half-life of approximately 14 days, some correction should be made if the experiment runs for the 5-day life of a honeybee larva.

Table 3.3 Approximate Digestibility Data from Larval to Pupal Ecdysis of *Heliothis virescens* and *Argyrotaenia velutinana*

Insect	Method for Approximate Digestibility	
	Manual Separation	Uric Acid Procedure
H. virescens	56.0 ± 1.6	55.4 ± 1.6
A. velutinana	51.0 ± 1.8	50.7 ± 1.8

Source: From Chow, Y.M. et al., *Ann. Entomol. Soc. Am.*, 66, 627, 1973. With permission.
Note: Twenty insects were used in each test.

Table 3.4 Comparisons between Male and Female *Argyrotaenia velutinana* in Food Utilization Determined by the Uric Acid Procedure

	A.D.	E.C.I.	E.C.D.
Male	50.5 ± 3.4	12.2 ± 0.8	26.4 ± 3.2
Female	50.5 ± 2.5	13.4 ± 0.8	27.1 ± 3.0

Source: From Chow, Y.M. et al., *Ann. Entomol. Soc. Am.*, 66, 627, 1973. With permission.

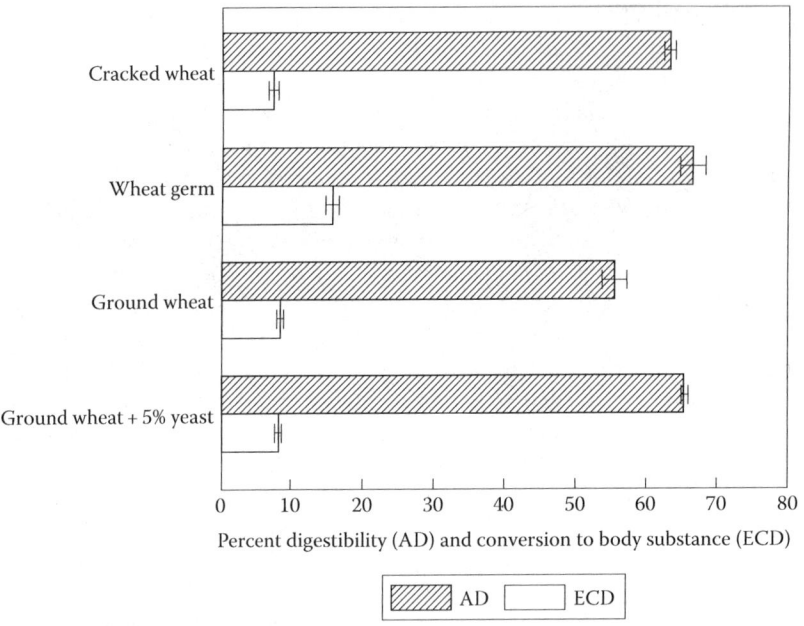

Figure 3.5 The effect of diet on approximate digestibility and efficiency of conversion of digested food to body weight by *Tribolium confusum* larvae. (Author constructed histogram with data from Bhattacharya, A.K. and Waldbauer, G.P., *J. Insect Physiol.*, 16, 1983, 1970.)

3.8 PHAGOSTIMULANTS

Phagostimulants are chemical compounds that induce feeding. Insects are induced to feed by chemosensory stimulation from components in or on their food (Thorsteinson, 1960; Chapman, 1995). Gustatory sensilla (Figure 3.6) are located on the mouthparts, tarsi, and antennae (Schoonhoven et al., 1991; Chyb et al., 1995). Adult western corn rootworms show strong phagostimulatory responses to several L-amino acids (Kim and Mullin, 1998), to γ-aminobutyric acid, and to as little as 0.1 μM of cucurbitacin B, a bitter (to humans) substance in cucurbits. Caterpillars of the tobacco hornworm *M. sexta* have two pairs of myo-inositol-sensitive gustatory sensilla, and the caterpillars respond strongly to inositol by feeding. The compound also counteracts inhibitory effects of some aversive stimuli (Glendinning et al., 2000).

In a few cases, an insect's behavior in the presence of phagostimulants has been divided into a biting response and a swallowing response. Phagostimulants are important to normal insect feeding, and they are very useful in insect diet, nutrition, and mass rearing studies. One commonly encountered frustration suffered by many entomologists has been the finding that a diet formulated upon the principles of good nutrition is refused by insects brought in from the field. Part of the problem may be absence of a normal phagostimulant.

If food recognition mechanisms are acted upon by natural selection, then one might expect a mechanism responsive to specific nutritional requirements to evolve. Major nutritional food materials, such as carbohydrates, proteins, and lipids, are logical indicators in plant and animal tissues of food, and in fact, virtually all the major nutritional substances normally required by animals serve as phagostimulants for one species of insect or another. Many insects have evolved sensory mechanisms responsive to nonnutritive substances as food recognition signals, apparently as a result of the coevolution of plants and insects.

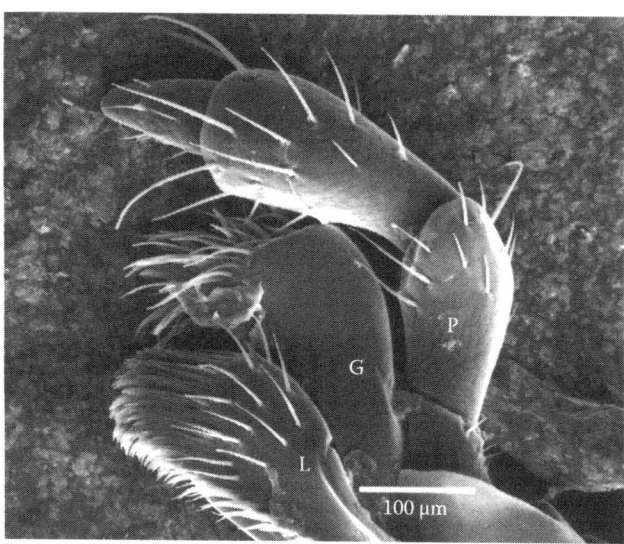

Figure 3.6 A scanning electron micrograph of the maxilla of an adult *Diabrotica virgifera virgifera* illustrating the maxillary palp (P), galea (G), and lacinea (L). The galea contains chemosensilla and (putative) mechanosensilla. (Photo courtesy of Chris Mullin; From Chyb, S. et al., *J. Chem. Ecol.*, 21, 313, 1995. With permission.)

Among primary nutrients, sugars, and sucrose especially, are phagostimulants (e.g., Detrain and Prieur, 2014, and references therein). Glucose is a strong phagostimulant for cockroaches, but a strain of German cockroaches has been found with an aversion to glucose, but not to other sugars (Wada-Katsumata et al., 2011). Glucosides, which are combinations of glucose with a nonsugar molecule, serve as both phagostimulants and deterrents to feeding in different insects. The aglycone or nonsugar moiety seems to control the role of glucosides.

Combinations of nutrients are important in many cases. Whole proteins, such as wheat gluten for confused flour beetles, are sometimes phagostimulants, but more frequently amino acids induce feeding. Leucine, methionine, lysine, and isoleucine in phosphate buffer were effective feeding stimulants for female houseflies. In this case, the presence of phosphate ions was important to the phagostimulant activity of the amino acids, but in some insects, amino acids may act alone or in combination with other amino acids or derivatives of amino acids (Chen and Henderson, 1996; Hollister and Mullin, 1998). Houseflies are also induced to feed by casein and yeast hydrolysates and guanosine monophosphate in phosphate buffers.

Reduced glutathione, a tripeptide composed of glutamic acid, cysteine, and glycine, is a phagostimulant for some and, perhaps all, ticks. Adenosine triphosphate in the presence of sodium ions is a phagostimulant for the adult mosquito *Aedes aegypti*. Lipids frequently are phagostimulants. Phospholipids, triglycerides, sterols, sterol esters, and fatty acids are important phagostimulants for various insects.

Many secondary plant substances, sometimes referred to as "token factors or stimulants," induce insects to feed. These substances are usually present in small amounts and frequently restricted to a group of related plants. Insects have no absolute metabolic or nutritional requirement for these token stimulants, so far as is known; they appear to serve as indicators of appropriate food. Combinations of secondary plant substances are frequently more effective stimulants than single compounds. The boll weevil *A. grandis* is stimulated to feed by a mixture of numerous compounds known to be present in cotton plants, including gossypol. Gossypol by itself, however, is only a weak stimulant to feeding. Secondary plant substances may also signal females to lay eggs. Sinigrin, a mustard oil glucoside, induces the butterfly *Pieris brassicae* females to oviposit even on foreign or nutritionally sterile substrates.

3.9 FEEDING DETERRENTS

Many substances are known that have an inhibitory effect upon insect feeding, and chemoreceptors responsive to specific compounds are involved, as in phagostimulation. For example, (−)-β-hydrastine and strychnine HCL are powerful feeding deterrents for adult western corn beetles mediated through the response of chemoreceptors (Chyb et al., 1995). Cantharidin from the blood of blister beetles is effective at 10^{-5} M as a feeding deterrent to carabid beetles, *Calosoma prominens* (Carrel and Eisner, 1974). Ammonium nitrate inhibits feeding of the sweet clover weevil, *Sitona cylindricollis*. Feeding by the American cockroach, *P. americana*, is inhibited by 1,4-naphthoquinone. Food intake by the desert locust, *S. gregaria*, is decreased by injection of Lom-sulfakinin, a neuropeptide found in the corpus cardiacum of locusts. Although the mechanism has not been elucidated, authors suggest the sulfakinins may reduce the sensitivity of taste receptors (Wei et al., 2000). For more details and a list of feeding deterrents, see Schoonhoven (1969, 1972) and Dethier (1980).

3.10 REVIEW AND SELF-STUDY QUESTIONS

1. What does balance of nutrients mean with respect to insect nutrition?
2. What physiological explanation can be given to show why the balance of nutrients is so important to insects?
3. Why is it very difficult to rear insects under axenic conditions?
4. If cockroaches are reared on dog biscuits, is the diet considered to be holidic, meridic, or oligidic? Support your choice.
5. List the 10 essential amino acids required by insects? What does it mean to say that these are "essential" amino acids?
6. Suppose you wanted to determine whether tyrosine (which is not normally an essential AA for insects) is an essential amino acid for a particular insect you are studying. You have a holidic diet formula for the insect that contains all the amino acids. How could you plan to determine if tyrosine is essential?
7. Suppose you wanted to determine if the larva of the Asian long-horned beetle (this beetle is an exotic beetle accidentally introduced into New York State) has an essential requirement for lysine. This beetle is in the family Cerambycidae, and the larvae feed upon the inner bark of hardwood trees. The life cycle is long, maybe 1 year or more. How might you conduct this experiment?
8. Are insects known generally to have specific requirements for a carbohydrate (say, glucose, fructose, or melezitose) in their diet? Justify your answer.
9. Do insect usually need a sterol in their diet? Why or why not?
10. What is a typical symptom of a deficiency of a polyunsaturated fatty acid (such as linoleic acid) in the diet?
11. Why are axenic conditions absolutely necessary to determine the vitamin requirements of insects?
12. What dietary and physiological reason can be given for the fact that most Lepidoptera have a relatively high value for potassium (K^+) in the hemolymph?
13. What is a phagostimulant? What kinds of compounds have been shown to be phagostimulants for insects?
14. What sugar does a predatory (e.g., a reduviid) insect most likely get when it sticks its beak into a soft caterpillar and sucks out the hemolymph and dissolved body contents?
15. Is plant starch a linear polymer (long chain of glucose units, like beads on a string) or is it branched? Write the names of the linkages of glucose in starch.
16. Glucose is linked into polymers (long chains, complex organization) by both alpha linkages and beta linkages between glucose units. Which type of linkage is most easily and most commonly digested by insects to release individual glucose molecules for energy?
17. What is the chemical name for fat that is stored in the body of a feeding caterpillar and which will be used as an energy source and for formation of adult tissues in the nonfeeding pupal stage?
18. What molecules are linked together by peptide bonds?
19. Why is it advantageous for an insect to have both endo- and exoproteinases in its midgut?
20. What enzymes digest fat molecules?

REFERENCES

Aguila, J.R., J. Suszko, A.G. Gibbs, and D.K. Hoshizaki. 2007. The role of larval fat cells in adult *Drosophila melanogaster*. *J. Exp. Biol.* 210: 956–963.

Aguila, J.R., D.K. Hoshizaki, and A.G. Gibbs. 2013. Contribution of larval nutrition to adult reproduction in *Drosophila melanogaster*. *J. Exp. Biol.* 216: 399–406.

Ahmad, I., G.P. Waldbauer, and S. Friedman. 1993. Maxillectomy does not disrupt self-selection by larvae of *Manduca sexta* (Lepidoptera: Sphingidae). *Ann. Entomol. Soc. Am.* 86: 458–463.

Anderson, T.E. and N.C. Leppla. 1992. *Advances in Insect Rearing for Research and Pest Management*. Westview Press, Boulder, CO.

Bauerfeind, S.S. and K. Fischer. 2014. Integrating temperature and nutrition—Environmental impacts on an insect immune system. *J. Insect Physiol.* 64: 14–20.

Bauerfeind, S.S., K. Fischer, S. Hartstein, S. Janowitz, and D. Martin-Creuzburg. 2007. Effects of adult nutrition on female reproduction in a fruit-feeding butterfly: The role of fruit decay and dietary lipids. *J. Insect Physiol.* 53: 964–973.

Beck, S.D., G.M. Chippendale, and D.E. Swinton. 1968. Nutrition of the European corn borer, *Ostrinia nubilalis*. VI. A larval rearing medium without crude plant fractions. *Ann. Entomol. Soc. Am.* 61: 459–462.

Behmer, S.T. and W.D. Nes. 2003. Insect sterol nutrition and physiology: A global overview. *Adv. Insect Physiol.* 31: 1–72.

Bhattacharya, A.K. and G.P. Waldbauer. 1970. Use of faecal uric acid method in measuring the utilization of food by *Tribolium confusum*. *J. Insect Physiol.* 16: 1983–1990.

Bhattacharya, A.K. and G.P. Waldbauer. 1972. The effect of diet on the nitrogenous end products excreted by larval *Tribolium confusum* with notes on correction of A.D. and E.C.D. for fecal urine. *Entmol. Exp. Appl.* 15: 238–247.

Boggs, C.L. and K.D. Freeman. 2005. Larval food limitation in butterflies: Effects on adult resource allocation and fitness. *Oecologia* 144: 353–361.

Boggs, C.L. 2009. Understanding insect life histories and senescence through a resource allocation lens. *Funct. Ecol.* 23: 27–37.

Bracken, G.K. 1966. Role of ten dietary vitamins on fecundity of the parasitoid *Exeristes comstockii* (Cresson) (Hymenoptera: Ichneumonidae). *Can. Entomol.* 98: 918–922.

Brooks, M.A. 1960. Some dietary factors that affect ovarial transmission of symbiotes. *Proceedings: Helminthology Society of Washington* 27: 212–220.

Carlson, S.D., H.R. Steeves III, J.S. VandeBerg, and W.E. Robbins. 1967. Vitamin A deficiency: Effects on retinal structure of a moth *Manduca sexta*. *Science* 158: 268–270.

Carrel, J.E. and T. Eisner. 1974. Cantharidin: Potent feeding deterrent to insects. *Science* 183: 755–757.

Chapman, R.F. 1995. Chemosensory regulation of feeding, in R.F. Chapman and G. de Boer (eds.), *Regulatory Mechanisms in Insect Feeding*. Chapman & Hall, New York, pp. 101–136.

Chen, J. and G. Henderson. 1996. Determination of feeding preference of Formosan subterranean termite (*Coptotermes formosanus* Shiraki) for some amino acid additives. *J. Chem. Ecol.* 22: 2359–2369.

Chippendale, G.M., S.D. Beck, and F.M. Strong. 1964. Methyl linolenate as an essential nutrient for the cabbage looper, *Trichoplusia ni* Hübner. *Nature (London)* 204: 710–711.

Chow, Y.M., G.C. Rock, and E. Hodgson. 1973. Consumption and utilization of chemically defined diets by *Argyrotaenia velutinana* and *Heliothis virescens*. *Ann. Entomol. Soc. Am.* 66: 627–632.

Chyb, S., H. Eichenseer, B. Hollister, C.A. Mullin, and J.L. Frazier. 1995. Identification of sensilla involved in taste mediation in adult western corn rootworm (*Diabrotica virgifera virgifera* LeConte). *J. Chem. Ecol.* 21: 313–329.

Cohen, A.C. 2004. *Insect Diets, Science and Technology*. CRC Press, Boca Raton, FL, 324pp.

Cohen, R.W., S.L. Heydon, G.P. Waldbauer, and S. Friedman. 1987a. Nutrient self-selection by the omnivorous cockroach *Supella longipalpa*. *J. Insect Physiol.* 33: 77–82.

Cohen, R.W., G.P. Waldbauer, S. Friedman, and N.M. Schiff. 1987b. Nutrient self-selection by *Heliothis zea* larvae: A time-lapse film study. *Entomol. Exp. Appl.* 44: 65–73.

Dadd, R.H. 1961. The nutritional requirements of locusts—V. Observations on essential fatty acids, chlorophyll, nutritional salt mixtures, and the protein or amino acid components of synthetic diets. *J. Insect Physiol.* 6: 126–145.

Dadd, R.H. 1967. Improvement of synthetic diet for the aphid *Myzus persicae* using plant juices, nucleic acids, or trace metals. *J. Insect Physiol.* 13: 763–778.

Dadd, R.H. 1968. Problems connected with inorganic components of aqueous diets. *Bull. Entomol. Soc. Am.* 14: 22–26.

Dadd, R.H. 1973. Insect nutrition: Current developments and metabolic implications. *Annu. Rev. Entomol.* 18: 381–420.

Dadd, R.H. 1985. Nutrition: Organisms, in G.A. Kerkut and L.I. Gilbert (eds.), *Comprehensive Insect Physiology, Biochemistry and Pharmacology*, vol. 4. Pergamon Press, Oxford, U.K., pp. 313–390.

Dadd, R.H. and D.L. Krieger. 1968. Dietary amino acid requirements of the aphid, *Myzus persicae*. *J. Insect Physiol.* 14: 741–764.

Davis, G.R.F. 1968. Phagostimulation and consideration of its role in artificial diets. *Bull. Entomol. Soc. Am.* 14: 27–29.

De Groot, A.P. 1952. Amino acid requirements for growth of the honeybee (*Apis mellifica* L.). *Experientia* 8: 192–193.

Dethier, V.G. 1980. Evolution of receptor sensitivity to secondary plant substances with special reference to deterrents. *Am. Nat.* 115: 45–66.

Detrain, C. and J. Prieur. 2014. Sensitivity and feeding efficiency of the black garden ant *Lasius niger* to sugar resources. *J. Insect Physiol.* 64: 74–80.

Dmitriew, C. and L. Rowe. 2011. The effects of larval nutrition on reproductive performance in a food-limited adult environment *PLoS One* 6: e17399.

Earle, N.W., A.B. Walker, M.L. Burks, and B.H. Slatten. 1967. Sparing of cholesterol by cholestanol in the diet of the boll weevil *Anthonomus grandis* (Coleoptera: Curculionidae). *Ann. Entomol. Soc. Am.* 60: 599–603.

Farjana, T. and N. Tuno. 2013. Multiple blood feeding and host-seeking behavior in *Aedes aegypti* and *Aedes albopictus* (Diptera: Culicidae). *J. Med. Entomol.* 50: 838–846.

Fischer, K., D.M. O'Brien, and C. Boggs. 2004. Allocation of larval and adult resources to reproduction in a fruit-feeding butterfly. *Funct. Ecol.* 18: 656–663.

Friend, W. 1958. Nutritional requirements of phytophagous insects. *Annu. Rev. Entomol.* 3: 57–74.

Glendinning, J.I., N.M. Nelson, and E.A. Bernays. 2000. How do inositol and glucose modulate feeding in *Manduca sexta* caterpillars? *J. Exp. Biol.* 203: 1299–1315.

Goldsmith, T.H., R.J. Barker, and C.F. Cohen. 1964. Sensitivity of visual receptors of carotenoid-depleted flies: A vitamin A deficiency in an invertebrate. *Science* 146: 65–67.

Gordon, H.T. 1959. Minimal nutritional requirements of the German roach, *Blattella germanica* L. *Ann. N. Y. Acad. Sci.* 77: 290–315.

Gordon, H.T. 1972. Interpretation of insect quantitative nutrition, in J.G. Rodriquez (ed.), *Insect and Mite Nutrition: Significance and Implications in Ecology and Pest Management*. Elsevier-North Holland, Amsterdam, the Netherlands, pp. 73–105.

Grandison, R.C., M.D. Piper, and L. Partridge. 2009. Amino-acid imbalance explains extension of lifespan by dietary restriction in *Drosophila*. *Nature* 462: 1061–1065.

Hahn, D.A. 2005. Larval nutrition affects lipids storage and growth, but not protein or carbohydrate storage in newly enclosed adults of the grasshopper *Schistocerca americana*. *J. Insect Physiol.* 51: 1210–1219.

Haydak, M.H. 1953. Influence of the protein level of the diet on the longevity of cockroaches. *Ann. Entomol. Soc. Am.* 46: 547–560.

Hinton, T., D.T. Noyes, and J. Ellis. 1951. Amino acids and growth factors in a chemically defined medium for *Drosophila*. *Physiol. Zool.* 24: 335–353.

Hobson, R.P. 1935. On a fat-soluble growth factor required by blowfly larvae. II. Identity of the growth factor with cholesterol. *Biochem. J.* 29: 2023–2026.

Hollister, B. and C.A. Mullin. 1998. Behavioral and electrophysiological dose-response relationships in adult western corn rootworm (*Diabrotica virgifera virgifera* LeConte) for host pollen amino acids. *J. Insect Physiol.* 44: 463–470.

House, H.L. 1954. Nutritional studies with *Pseudosarcophaga affinis* (Fall), a dipterous parasite of the spruce budworm, *Choristoneura fumiferana* (Clem). I. A chemically defined medium and aseptic culture technique. *Can. J. Zool.* 32: 331–341.

House, H.L. 1965a. Insect nutrition, in M. Rockstein (ed.), *Physiology of Insecta*, 1st ed., vol. 2. Academic Press, New York, pp. 769–813.

House, H.L. 1965b. Effects of low levels of the nutrient content of a food and of nutrient imbalance on the feeding and the nutrition of a phytophagous larva, *Celerio euphorbiae* (Linnaeus) (Lepidoptera: Sphingidae). *Can. Entomol.* 97: 62–68.

House, H.L. 1966a. Effects of varying the ratio between the amino acids and the other nutrients in conjunction with salt mixture on the fly *Agria affinis* [Fall.]. *J. Insect Physiol.* 12: 299–310.

House, H.L. 1966b. Effects of vitamins E and A on growth and development and the necessity of vitamin E for reproduction in the parasitoid *Agria affinis* [Fall.]. *J. Insect Physiol.* 12: 409–418.

House, H.L. 1967. The role of nutritional factors in food selection and preference as related to larval nutrition of an insect, *Pseudosarcophaga affinis* (Diptera, Sarcophagidae) on synthetic diets. *Can. Entomol.* 99: 1310–1321.

House, H.L. 1969. Effects of different proportions of nutrients on insects. *Entmol. Exp. Appl.* 12: 651–669.

House, H.L. 1974. Nutrition, in M. Rockstein (ed.), *The Physiology of Insects*, 2nd ed. Academic Press, New York, pp. 1–62.

House, H.L., P. Singh, and W.W. Batsch. 1971. Lepidoptera, in *Artificial Diets for Insects: A Compilation of References with Abstracts*. Information Bulletin No. 7, Research Institute, Canada Department of Agriculture, Belleville, Ontario, Canda, pp. 55–115.

Hsiao, T.H. 1972. Chemical feeding requirements of oligophagous insects, in J.G. Rodriquez (ed.), *Insect and Mite Nutrition: Significance and Implications in Ecology and Pest Management*. Elsevier-North Holland, Amsterdam, the Netherlands, pp. 225–240.

Ishii, S. and C. Hirano. 1955. Qualitative studies on the essential amino acids for the growth of the larva of the rice stem borer, *Chilo simplex* (Butler), under aseptic conditions. *Bull. Natl. Inst. Agric. Sci. (Japan) Ser. C.* 5: 35–48.

Johnson, H., M.J. Solensky, D.A. Satterfield, and A.K. Davis. 2014. Does skipping a meal matter to a butterfly's appearance? Effects of larval food stress on wing morphology and color in monarch butterflies. *PLoS One* 9(4): e93492.

Joy, T.K., A.J. Arik, V. Corby-Harris, A.A. Johnson, and M.A. Riehle. 2010. The impact of larval and adult dietary restriction on lifespan, reproduction and growth in the mosquito *Aedes aegypti*. *Exp. Gerontol.* 45: 685–690.

Kasting, R. and A.J. McGinnis. 1958. Use of glucose labelled with carbon-14 to determine the amino acids essential to an insect. *Nature (London)* 182: 1380–1381.

Kasting, R. and A.J. McGinnis. 1960. Use of glutamic acid-U-C^{14} to determine nutritionally essential amino acids for larvae of the blowfly, *Phormia regina*. *Can. J. Biochem. Physiol.* 38: 1229–1234.

Kasting, R. and A.J. McGinnis. 1962. Nutrition of the pale western cutworm, *Agrotis orthogonia* Morr.—IV. Amino acid requirements determined with glucose-U-C^{14}. *J. Insect Physiol.* 8: 97–103.

Kasting, R., G.R.F. Davis, and A.J. McGinnis. 1962. Nutritionally essential and non-essential amino acids for the prairie grain wireworm, *Ctenicera destructor* Brown, determined with glucose-U-C^{14}. *J. Insect Physiol.* 8: 589–596.

Kim, J.H. and C.A. Mullin. 1998. Structure-phagostimulatory relationships for amino acids in adult western corn rootworm, *Diabrotica virgifera virgifera*. *J. Chem. Ecol.* 24: 1499–1511.

Kircher, H.W., W.B. Heed, J.S. Russell, and J. Groove. 1967. Senita cactus alkaloids: Their significance to Sonoran desert *Drosophila* ecology. *J. Insect Physiol.* 13: 1869–1874.

Krenn, H.W., M.J.B. Eberhard, S.H. Eberhard, A.-L. Hikl, W. Huber, and L.E. Gilbert. 2009. Mechanical damage to pollen aids nutrient acquisition in *Heliconius* butterflies (Nymphalidae). *Arthropod Plant Interact.* 32: 203–208.

Leclercq, J. 1950. La vitamine T, facteur de croissance pour les larves de "*Tenebrio molitor*". *Arch. Int. Physiol. LVII* 30: 350–352.

Lemonde, A. and R. Bernard. 1951. Nutrition des larves de *Tribolium confusum* (Duval). II. Importance des acides amine. *Can. J. Zool.* 29: 80–83.

Locke, M. and H. Nichol. 1992. Iron economy in insects: Transport, metabolism, and storage. *Annu. Rev. Entomol.* 37: 195–215.

Lyimo, I.N., S.P. Keegan, L.C. Ranford-Cartwright, and H.M. Ferguson. 2012. The impact of uniform and mixed species blood meals on the fitness of the mosquito vector *Anopheles gambiae* s.s: Does a specialist pay for diversifying its host species diet? *J. Evol. Biol.* 25: 452–460.

McGinnis, A.J. and R. Kasting. 1964. Chromic oxide indicator method for measuring food utilization in a plant-feeding insect. *Science* 144: 1464–1465.

Medici, J.C. and M.W. Taylor. 1966. Mineral requirements of the confused flour beetle, *Tribolium confusum* (Duval). *J. Nutr.* 88: 181–186.

Mira, A. 2000. Exuviae eating: A nitrogen meal? *J. Insect Physiol.* 46: 605–610.

Mittler, T.E. 1971. Dietary amino acid requirements of the aphid *Myzus persicae* affected by antibiotic uptake. *J. Nutr.* 101: 1023–1028.

Mondy, N. and M.-F. Corio-Costet. 2000. The response of the grape berry moth (*Lobesia botrana*) to a dietary phytopathogenic fungus (*Botrytis cinerea*): The significance of fungus sterols. *J. Insect Physiol.* 46: 1557–1564.

Moore, W. 1946. Nutrition of *Attagenus* (?) sp. II. (Coleoptera: Dermestidae). *Ann. Entomol. Soc. Am.* 39: 513–521.

Norris, D.M. and J.K. Baker. 1967. Symbiosis: Effects of a mutualistic fungus upon the growth and reproduction of *Xyleborus ferrugineus*. *Science* 156: 1120–1122.

Osborn, T.B. and L.B. Mendel. 1932. A modification of the Osborne–Mendel salt mixture containing only inorganic constituents. *Science* 75: 339–340.

Pan, X., K. Lu, S. Qi, and Q. Zhou. 2014. The content of amino acids in artificial diet influences the development and reproduction of brown planthopper, *Nilaparvata lugens* (Stål). *Arch. Insect Biochem. Physiol.* 86: 75–84.

Pant, N.C., B. Gupta, and J.K. Nayar. 1960. Physiology of intracellular symbiotes of *Stegobium paniceum* L. with special reference to amino acid requirements of the host. *Experientia* 16: 311–312.

Pant, N.C., J.K. Nayar, and P. Gupta. 1958. On the significance of amino acids in the larval development of Khapra-beetle, *Trogoderma granarium* Everts. (Coleoptera: Dermestidae). *Experientia* 14: 176–177.

Pitts, R.J. 2014. A blood-free protein meal supporting oogenesis in the Asian tiger mosquito, *Aedes albopictus* (Skuse). *J. Insect Physiol.* 64: 1–6.

Prenter, J., C.W. Weldon, and P.W. Taylor. 2013. Age-related activity patterns are moderated by diet in Queensland fruit flies *Bactrocera tryoni*. *Physiol. Entomol.* 38: 260–267.

Reinecke, J.P. 1985. Nutrition: Artificial diets, in G.A. Kerkut and L.I. Gilbert (eds.), *Comprehensive Insect Physiology, Biochemistry and Pharmacology*, vol. 4. Pergamon Press, Oxford, U.K., pp. 391–419.

Richards, S.L., S.L. Anderson, and S.A. Yost. 2012. Effects of blood meal source on the reproduction of *Culex pipiens quinquefasciatus* (Diptera: Culicidae). *J. Vector Ecol.* 37: 1–7.

Rock, G.C. and K.W. King. 1968. Amino acid synthesis from glucose-U-^{14}C in *Argyrotaenia velutinana* (Lepidoptera: Tortricidae) larvae. *J. Nutr.* 95: 369–373.

Rock, G.C. and E. Hodgson. 1971. Dietary amino requirements for *Heliothis zea* determined by dietary deletion and radiometric techniques. *J. Insect Physiol.* 17: 1087–1097.

Rodriquez, J.G. and R.E. Hampton. 1966. Essential amino acids determined in the two-spotted spider mite, *Tetranychus urticae* (Acarina: Tetranychidae), with glucose-U-^{14}C. *J. Insect Physiol.* 12: 1209–1216.

Sang, J.H. 1956. The quantitative nutritional requirements of *Drosophila melanogaster*. *J. Exp. Biol.* 35: 45–72.

Schoonhoven, L.M. 1969. Gustation and foodplant selection in some lepidopterous larvae. *Entmol. Exp. Appl.* 12: 555–564.

Schoonhoven, L.M. 1972. Secondary plant substances and insects, in V.C. Runeckles and T.C. Tso (eds.), *Structural and Functional Aspects of Phytochemistry, Recent Advance in Phytochemistry*. vol. 5, Academic Press, NY, pp. 197–224.

Schoonhoven, L.M., M.S.J. Simmonds, and W.M. Blaney. 1991. Changes in the responsiveness of the maxillary styloconic sensilla of *Spodoptera littoralis* to inositol and sinigrin correlate with feeding behavior during the final larval stadium. *J. Insect Physiol.* 37: 261–268.

Scriber, J.M. and F. Slansky, Jr. 1981. The nutritional ecology of immature insects. *Annu. Rev. Entomol.* 26: 183–211.

Shuel, R. and S. Dixon. 1968. The importance of sugar for the pupation of the worker honeybee. *J. Apicult. Res.* 7: 109–112.

Simpson, C.L., S. Chyb, and S.J. Simpson. 1990. Changes in chemoreceptor sensitivity in relation to dietary selection by adult *Locusta migratoria*. *Ent. Exp. Appl.* 56: 259–268.

Simpson, S.J. and C.L. Simpson. 1990. The mechanisms of nutritional compensation by phytophagous insects, in E.E. Bernays (ed.), *Insect–Plant Interactions*, vol. 2. CRC Press, Boca Raton, FL, pp. 111–160.

Simpson, S.J. and D. Raubenheimer. 1995. The geometric analysis of feeding and nutrition: A user's guide. *J. Insect Physiol.* 41: 545–553.

Simpson, S.J. and P.R. White. 1990. Associative learning and locust feeding: Evidence for a "learned hunger" for protein. *Anim. Behav.* 40: 506–513.

Simpson, S.J., D. Raubenheimer, and P.G. Chambers. 1995. The mechanisms of nutritional homeostasis, in R.F. Chapman and G. Boer (eds.), *Regulatory Mechanisms in Insect Feeding*. Chapman & Hall, New York, pp. 251–278.

Singh, K.R.P. and A.W.A. Brown. 1957. Nutritional requirements of *Aedes aegypti* L. *J. Insect Physiol.* 1: 199–220.

Singh, P. 1974. Artificial diets for insects: A compilation of references with abstracts (1970–72). *N. Z. Dept. Sci. Ind. Res. Bull.* 214: 96.

Singh, P. 1976. Synthetic diets for insects and mites, in M. Recheigl (ed.), *Handbook of Nutrition and Food*, vol. 2, CRC Press, Cleveland, OH, pp. 131–250.

Singh, P. 1977. *Artificial Diets for Insects, Mites, and Spiders*. IFI/Plenum, New York.

Singh, P. and H.L. House. 1970. Antimicrobials: "Safe" levels in a synthetic diet of an insect *Agria affinis*. *J. Insect Physiol.* 16: 1769–1782.

Sisodia, S. and B.N. Singh. 2012. Experimental evidence for nutrition regulated stress resistance in *Drosophila ananassae*. *PLoS One* 7: e46131.

Slansky, F. Jr. 1982. Insect nutrition: An adaptationist's perspective. *Fla Entomol.* 65: 45–71.

Slansky, F. Jr. and M. Scriber. 1985. Food consumption and utilization, in G.A. Kerkut and L.I. Gilbert (eds.), *Comprehensive Insect Physiology, Biochemistry and Pharmacology*. Pergamon Press, Oxford, U.K., pp. 87–163.

Smedley, S.R. and T. Eisner. 1995. Sodium uptake by puddling in a moth. *Science* 270: 1816–1818.

Svoboda, J.A., J.N. Kaplanis, W.E. Robbins, and M.J. Thompson. 1975. Recent developments in insect steroid metabolism. *Annu. Rev. Entomol.* 20: 205–220.

Telang, A., A.A. Qayum, A. Parker, B.R. Sacchetta, and G.R. Byenes. 2011. Larval nutritional stress affects vector immune traits in adult yellow fever mosquito *Aedes aegypti* (*Stegomyia aegypti*). *Med. Vet. Entomol.* 26: 271–281.

Telang, A. and M.A. Wells. 2004. The effect of larval and adult nutrition on successful autogenous egg production by a mosquito. *J. Insect Physiol.* 50: 677–685.

Thorsteinson, A.J. 1960. Host selection in phytophagous insects. *Annu. Rev. Entomol.* 5: 193–218.

Vanderzant, E.S. 1958. The amino acid requirements of the pink bollworm. *J. Econ. Entomol.* 51: 309–311.

Vanderzant, E.S. 1973. Axenic rearing of larvae and adults of the boll weevil on defined diets: Additional tests with amino acids and vitamins. *Ann. Entomol. Soc. Am.* 66: 1184–1186.

Vanderzant, E.S. 1974. Development, significance, and application of artificial diets for insects. *Annu. Rev. Entomol.* 19: 139–160.

Vanderzant, E.S. and C.D. Richardson. 1963. Ascorbic acid in the nutrition of plant-feeding insects. *Science* 140: 989–991.

Wada-Katsumata, A., J. Silverman, and C. Schal. 2011. Differential inputs from chemosensory appendages mediate feeding responses to glucose in wild-type and glucose-averse German cockroaches, *Blattella germanica*. *Chem. Senses* 36: 589–600.

Waldbauer, G.P. 1964. The consumption, digestion and utilization of solanaceous and non-solanaceous plants by larvae of the tobacco hornworm, *Protoparce sexta* (Johan.) (Lepidoptera: Sphingidae). *Ent. Exp. Appl.* 7: 253–269.

Waldbauer, G.P. and A.K. Bhattacharya. 1973. Self-selection of an optimum diet from a mixture of wheat fractions by the larvae of *Tribolium confusum*. *J. Insect Physiol.* 19: 407–418.

Waldbauer, G.P. and S. Friedman. 1991. Self-selection of optimal diets by insects. *Annu. Rev. Entomol.* 36: 43–63.

Waldbauer, G.P., R.W. Cohen, and S. Friedman. 1984. Self-selection of an optimal nutrient mix from defined diets by larvae of the corn earworm, *Heliothis zea* (Broddie). *Physiol. Zool.* 57: 590–597.

Wei, Z., G. Baggerman, R.J. Nachman, G. Goldsworthy, P. Verhaert, A. De Loof, and L. Schoofs. 2000. Sulfakinins reduce food intake in the desert locust, *Schistocerca gregaria*. *J. Insect Physiol.* 46: 1259–1265.

Yamany, A.S., H. Mehlhorn, and F.K. Adham. 2012. Yolk protein uptake in the oocyte of the Asian tiger mosquito *Aedes albopictus* (Skuse) (Diptera: Culicidae). *Parasitol. Res.* 111: 1315–1324.

Yazgan, S. 1972. A chemically defined synthetic diet and larval nutritional requirements of the endoparasitoid *Itoplectis conquisitor* (Hymenoptera). *J. Insect Physiol.* 18: 2123–2141.

Zhou, G. and R. Miesfeld. 2009. Differential utilization of blood meal amino acids in mosquitoes. *OAIP* 1–12.

CHAPTER 4

Integument and Molting

PREVIEW

The integument of insects comprises the cuticle and the epidermal cells beneath that secrete the cuticle. The cuticle is the skeleton of insects and skeletal muscles are attached to it. The cuticle may be hard and rigid, as that of adult beetles, or soft and pliable, as in the case of many immature insects and some adults. The head capsule and the thorax, which supports the leg and wing attachments, are usually the most heavily sclerotized (hardened) parts of the body. All stages of insects contain an epicuticular layer that waterproofs the body. This layer is heavily sclerotized, but does not contain chitin. Beneath the epicuticle layer, many insects have an exocuticle layer (hard, sclerotized cuticle) and a layer of soft, relatively unsclerotized endocuticle next to the epidermal cells. Either of the latter two layers may be reduced or absent; there is no endocuticle in very hard cuticle, such as the elytra of beetles, and no exocuticle in very soft-bodied insects, such as some larval diptera. In all immature insects, the exoskeleton gets too small as the insect grows and the cuticle must be molted. Periodically, under hormonal regulation, the old cuticle separates from the epidermal cells, a process called apolysis. Parts of the old cuticle are digested by molting enzymes, and reabsorbed components are used in the synthesis of new cuticle. Secretion of a new cuticle begins beneath the old cuticle even as it is being digested. The first part of the new cuticle to be secreted is the outermost layer, the cuticulin layer of the epicuticle. Additional cuticle, usually called procuticle because it is not sclerotized at this stage regardless of its future destiny, is secreted underneath the cuticulin layer. Cuticle is secreted in thin sheets, with successive sheets pushed up from below. The sheets of cuticle contain a protein matrix with chitin rods embedded in the matrix. Often each successive sheet is rotated slightly with respect to the long axis of the previous sheet, and successive layers give rise to a helicoid appearance in cross section. For some time interval, lasting from hours to days in different insects, an insect has two cuticular coverings: the old and the new. Muscle attachments to the old cuticle are at last severed, freeing the old cuticle to be shed, and the muscles rapidly attach to the new cuticle. Eclosion, or shedding of the old cuticle, and especially eclosion of adults from the pupal stage, is regulated by a complex of neuropeptide hormones. A period of quiescence is necessary in most cases for the cuticle to harden sufficiently to withstand the strain imposed by muscle contraction. Cuticle sclerotization is regulated by a neurohormone, bursicon, secreted from the nervous system. The cuticle contains chitin, a polysaccharide polymer of N-acetylglucosamine, and proteins, lipids that function in waterproofing, and phenols and quinones that are important in sclerotization, the hardening process in the cuticle.

Sclerotization of the cuticle over most of the body is accompanied by darkening, the formation of brown to black melanin pigments, but cuticle can sclerotize without darkening. For example, the compound eyes usually are covered by relatively clear cuticle, and some insects have transparent cuticle over some or even most of the body. The melanin pigments are formed from chemical changes in the polyphenols in the cuticle. Many different proteins have been detected in cuticle prior to sclerotization, but once these proteins are cross-linked, they usually cannot be dissolved from the cuticle. There are some differences, however, in the proteins comprising soft versus hard regions of cuticle. The hardness of cuticle is a function of sclerotization, not chitin content. Some of the hardest parts of cuticle do not contain chitin. Not only are proteins cross-linked to each other during sclerotization, but also the multiple thin sheets of cuticle are cross-linked to each other, giving the cuticle great strength. Lipids on the surface of the epicuticle and within the layers give cuticle excellent waterproofing properties, an important function for nearly all insects because of their large surface area to volume ratio. Aquatic insects benefit from water-impermeability of the cuticle by not absorbing large quantities of water by osmosis.

4.1 INTRODUCTION

The integument (Moussian, 2010) is composed of the **cuticle** and the underlying **epidermal cells** that secrete the cuticle. The cuticle serves as the exoskeleton of the insect; the site for muscle attachment; the first line of defense from fungi, bacteria, predators, parasites, and environmental chemicals including pesticides; and in prevention of desiccation. The serosal cuticle that forms around the embryo in particular is important in preventing desiccation of the embryo (Vargas et al., 2014) as it develops. The integument functions in some or all insects in locomotion, breathing and respiration, feeding, excretion, protection from desiccation, behavior, osmoregulation, water control, and as a food reserve. The many roles played by the integumentary covering of insects is, in part, reflected in the complexity of its structure and chemistry (Vincent and Wegst, 2004) and in the special ways it is adapted to function in the ecology of its owner. The surface morphology of the external cuticle is extraordinarily varied, reflecting species specificity and diversity. The beauty, color, shapes, and intricate sculpturing on the surface of insects attracts amateurs and professionals alike to collect and study insects. Moreover, taxonomists and systematists traditionally have used the surface sculpturing, setae, and sutures on the cuticle in classification of insect species.

Despite many species-specific features, there are certain common features found in the integument. There is always a single layer of epidermal cells lying immediately beneath the cuticle. These cells secrete the new cuticle at molting and, in some insects at least, continue to secrete cuticle even in the adult. In all insects there is a thin layer of cuticle with special properties called the epicuticle at the surface of the insect, and beneath this there is additional cuticle that can be divided into several layers depending primarily upon the degree of sclerotization or cross-linking of the molecules of protein and chitin. This chapter will explore the physiology and biochemistry of cuticle and relate these to species similarities and differences in the integument.

4.2 STRUCTURE OF THE INTEGUMENT

The integument includes the cuticle on the surface of the body and the single layer of epidermal cells beneath the cuticle (Figure 4.1). Locke (2001) has proposed that the cuticle should be described by three primary layers—the cuticulin envelope, the epicuticle, and the procuticle—an updated system of terminology that brings insects in line with other systematic groups. The epidermal cells secrete new cuticle at each molt. There is always a cuticulin or envelope layer in all insects and always an epicuticle layer. Beneath the epicuticle is the procuticle, and its chemical composition and degree of sclerotization varies greatly among different groups of insects and even in different stages of the same insect. When the outermost part of the procuticle is heavily sclerotized (cross-linked hard cuticle), it is called exocuticle. Not all insects have a hard exocuticle.

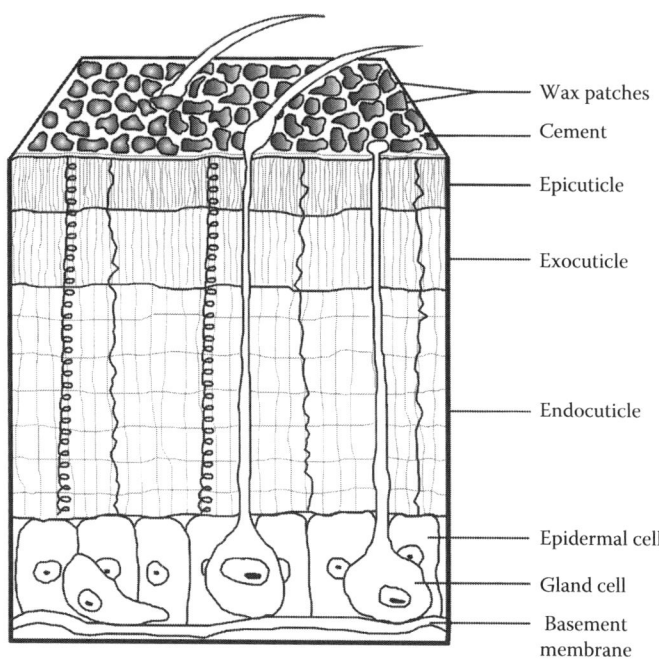

Figure 4.1 Diagrammatic representation of a cross-sectional area of the integument illustrating its major layers.

For example, most soft bodied larvae and other soft parts of insects contain little or no exocuticle in parts of the body that are soft. The envelope, epicuticle, and exocuticle (if any) are not digested prior to a molt, and it is these parts of the cuticle that are shed at the molt. The part of the procuticle that is only lightly cross-linked is the endocuticle. Endocuticle may be greatly reduced or absent in particular parts of the cuticle of the same or different insects, as, for example, in the hard outer wing covers, the elytra, of scarab beetles. Electron micrographs of cuticle cross sections often reveal numerous layers of differing electron density in the cuticle, but these usually have not been given names. The epidermal cells secrete the cuticle, the lipids (waxes), cement, and often many additional chemical components that occur on or in the cuticle layers. When a new cuticle is secreted at molting, the envelope is secreted first, and the epicuticle is secreted on its inner surface. Procuticle is soft at first, but varying degrees of sclerotization occur in different insects soon after the cuticle is secreted.

4.2.1 Cuticulin Envelope

The cuticulin envelope is from 10 to 30 nm thick (Locke, 2001) and is formed at the external surface of the epidermal cell plasma membrane. It separates from the plasma membrane and is pushed upward as new epicuticle, and then procuticle is secreted beneath it. Because it is so thin, its chemical composition is poorly understood, although sclerotized or cross-linked protein is probably one of the main components. Neither the cuticulin envelope nor the epicuticle layer contains chitin.

4.2.2 Epicuticle

The epicuticle layer typically is from 1 to 4 μm in thickness and, like the envelope, its detailed chemical structure has been difficult to discern, but it is known to contain sclerotized proteins impregnated with lipoproteins, lipids, waxes, cement, and minor amounts of various minerals and

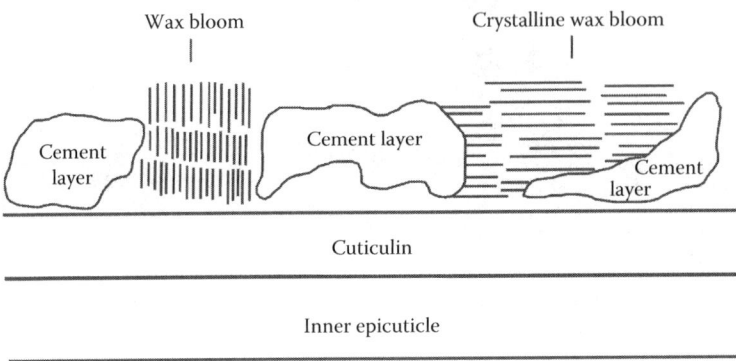

Figure 4.2 Diagrammatic illustration of the mosaic of wax (lipids) and cement on the surface of the epicuticle of some insects. (Drawing modified from Locke, M., *Science*, 147, 295, 1965.)

other chemical components. It does not contain chitin, a major structural carbohydrate in the procuticle. The proteins and some of the lipids appear to be covalently linked, and the proteins are tanned or sclerotized by phenolic compounds and their oxidized products, quinones. **Sclerotization**, the cross-linking of molecules, gives the epicuticle strength, hardness, and low water permeability. Lipids and the cement layer on the surface also provide reduced permeability to water.

Cement, often described as a shellac-like substance at the air interface with the cuticle, is secreted by specialized epidermal cells called dermal glands and transported to the surface of the cuticle. A traditional view is that the cement layer is the outermost layer on the cuticle with a lipid layer just beneath the cement. Probably in some insects this view is valid, but Locke (1965) suggested on the basis of electron microscope studies of the cuticle of the lepidopteran caterpillar, *Calpodes ethlius*, that the cement layer is not continuous, but is broken by patches of lipids, called **wax blooms**, at the surface (Figure 4.2). It seems likely in many insects that a mosaic patchwork of cement and lipid exists at the surface of the epicuticle. Some examples of insects that have much or even all the body covered with lipid at the surface will be described in later sections of this chapter. Despite the thin nature of the epicuticle, it nevertheless is extremely important to surface pattern and features, to permeability of the cuticle, and it represents a limitation on expansion of the cuticle in immature insects, necessitating molting during growth.

4.2.3 Procuticle

The procuticle, containing both chitin and protein, lies just beneath the epicuticle (Neville, 1986). Parts of the procuticle (typically the outer part nearest the epicuticle) may be highly sclerotized and, therefore, is hard and rigid and called the **exocuticle**. Lamellae or layers within this exocuticle may refract light in such a way to produce structural colors in some insects. Many of the iridescent greens and blues of insects are structural colors due to refracted light rather than to pigments. The thickness of the exocuticle is variable and species specific. Adult insects generally have a thicker and more sclerotized exocuticle than larval insects. In particular, the thorax in flying insects has heavily sclerotized exocuticle to support the strong flight muscles. Many larvae have a soft flexible cuticle with little or no exocuticle. As in so many cases with insects, exceptions exist. There are larval insects with hard sclerotized exocuticle and soft-bodied adults with little or none. The harder the cuticle, the greater the degree of sclerotization. The content of chitin does not control hardness of the cuticle, but sclerotization does. Because of the sclerotization, little or none of the exocuticle is digested by molting fluid, and it is shed, along with the epicuticle, at molting.

In some insects, the highly sclerotized exocuticle grades into less sclerotized cuticle, called **mesocuticle** in the earlier literature, or there may be a rapid change to soft, little sclerotized cuticle called the **endocuticle**. In classical histological sections stained with dyes, mesocuticle stains red with Mallory's triple stain, while endocuticle stains blue. Exocuticle and epicuticle do not usually stain with Mallory's stain.

The endocuticle is soft, flexible cuticle containing both chitin and proteins. It has little sclerotization, which is why it is soft and flexible. Stabilization of the proteins and chitin in the endocuticle occurs through some covalent bonds, hydrogen bonds, and probably occasional quinone cross-links. Generally soft-bodied insects have relatively thick endocuticle and thin or no exocuticle. There is, however, always an envelope and epicuticle at the surface of soft-bodied insects.

4.2.4 Pore Canals and Wax Channels

Pore canals are passageways from 0.1 to 0.15 μm in diameter, extending from the epidermal cells through the procuticle, but terminating at the interface between procuticle and epicuticle. Larger canals are often flattened, ribbon-like, and may be twisted or straight. Pore canals transport lipids and cement, and, sometimes, additional chemical components. Formation of the pore canals has not been entirely resolved. Some research has suggested that the passageways are the result of cytoplasmic extensions of the epidermal cells present during cuticle secretion. Usually after a new cuticle has been secreted, the cell extensions are withdrawn leaving the open canals. Some workers who have studied cuticle formation do not accept this mode of formation because cell extensions are not always evident even during new cuticle secretion, but no competing alternative idea has been advanced. The pore canals can be very numerous; for example, as many as $1.2 \times 10^6/mm^2$ in the procuticle of some cockroaches or as few as $15,000/mm^2$ (about 50–70 per epidermal cell) in a sarcophagid (flesh fly) larva.

Although pore canals do not penetrate the epicuticle, there are smaller passageways through the epicuticle called **wax channels** (about 0.006–0.013 μm in diameter). The wax channels are 10–20 times smaller than pore canals (Locke, 1965). Wax channels stain with osmium tetroxide, a stain for unsaturated lipids, and this is taken as evidence that the channels continue the transport of lipids, and probably other materials, to the surface. There is no evidence of 1:1 correspondence of pore canals to wax channels at the junction of the procuticle and epicuticle.

Chemicals passing through the pore canals probably diffuse out of the pore canals laterally to some extent all along their length, and also at the epicuticular interface and, thus, impregnate the entire cuticle. Some of the lipids find their way to the surface of the insect where they provide waterproofing and, perhaps, other ecological and behavioral functions. Continued secretion of cuticular lipids in many insects is a dynamic process, replacing epicuticular lipids that are volatilized (as semiochemicals, for example), lipids that no doubt rub or wear off the cuticle in longer-lived insects, and lipids lost with each molt.

4.2.5 Epidermal Cells

The cells underlying the cuticle are arranged in a single layer (Figure 4.3). They usually are simply called the epidermal cells, but sometimes referred to as the **cuticular epithelium**, the **epidermis**, and (in older literature) the **hypodermis**. All epidermal cells probably secrete chitin and proteins, and some may secrete lipids. Specially modified glandular epidermal cells may secrete cement. There frequently are specialized glandular cells in the cell layer, either in small groups or as scattered, isolated glandular cells that secrete special products (Noirot and Quennedey, 1974). For example, sex pheromones in most female lepidoptera are secreted by small patches of tall, columnar epidermal cells located beneath the cuticle of the ventral intersegmental membrane between the eighth to ninth segments of the abdomen.

Epidermal cells are separated from the circulating hemolymph by a **basement membrane**, a layer of poorly defined chemical composition, but with pores large enough to permit passage of larger hemolymph proteins and other molecules from hemolymph into the epidermal cells. Small tracer

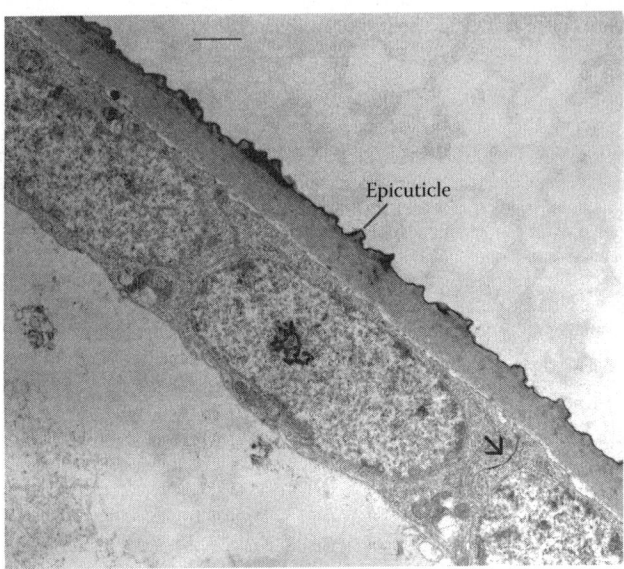

Figure 4.3 Epidermal cells and cuticle from a larva of the mosquito *Ochlerotatus* (formerly *Aedes*) *triseriatus*. The arrow points to a row of septate junctions between the membranes of two adjacent cells. The scale bar at the upper left represents 1 μm. (Photo from the author, Jimmy Becnel and Alexandra Shapiro, USDA, Gainesville, FL.)

particles of gold or ferritin pass from the hemolymph through the basement membrane, but larger particles are stopped on the hemolymph side. The origin of the basement membrane is not defined for most insects; some evidence supports secretion by hemocytes, but some evidence suggests that the epidermal cells themselves secrete it. Tracheae, tracheoles, and nerves pass through the basement membrane to reach the epidermal cells. The basement membrane of epidermal cells may be smooth or may contain many and deep infoldings, depending upon the function and stage in which the insect exists. The membrane changes in appearance, and undoubtedly in function, as the insect develops, and there are marked changes in preparation for and during a molt. **Hemidesmosomes** hold the basement membrane to the epidermal cells.

In the intermolt period, epidermal cells typically have a regular polygonal outline and form a sheet of cells beneath the cuticle. Contiguous epidermal cells, as well as gut and Malpighian tubule epithelium, follicular cells in the ovary, and other cells in insects, are held together by various types of **junctional contacts** (Lane and Skaer, 1980) (Figure 4.4). These junctional contacts not only hold cells together, but also, depending upon the type of contact, may have other functions. **Septate junctions** are close together and numerous, giving a ladder-like appearance between cells (Figure 4.5). Septate junctions tend to occur between the lateral faces of epidermal cells, particularly toward the apical (cuticular) surface and play an important role in preventing inward and outward movement of materials between the cells. **Gap junctions** also occur between the lateral faces of cells. The two cell membranes are very close together (a gap of 2–4 nm) at gap junctions, and for this reason they also have been called **close junctions**. Gap junctions appear to confer electrical coupling upon cells and can make cells function like a syncytium, and play a role in cell-to-cell communication. They also have been called *macula communicans* because of their apparent or suspected role in communication. Close junctions also act as sieves, allowing certain sizes of molecules to pass through while excluding others, and may function in controlling the speed of entry of some types of molecules. **Tight junctions** occur in some insect tissues, such as between cells in the compound eyes, testes, rectal pads, and between perineural cells forming part of the blood–brain barrier in the central nervous system. Where cells are bonded by tight junctions, the adjacent cell membranes appear to contain rows of

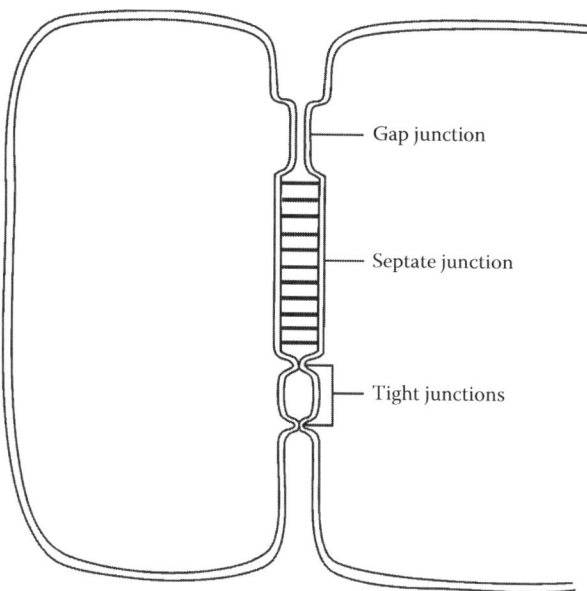

Figure 4.4 Examples of the most common cell-to-cell structures holding cells together and, in some cases, preventing passage of chemical substances between cells. Insect cells typically have tight junctions near the basal cell surface. The complex interdigitating membranes between adjacent cells probably, in many cases, aid in making tight junctions resist the passage of substances between cells.

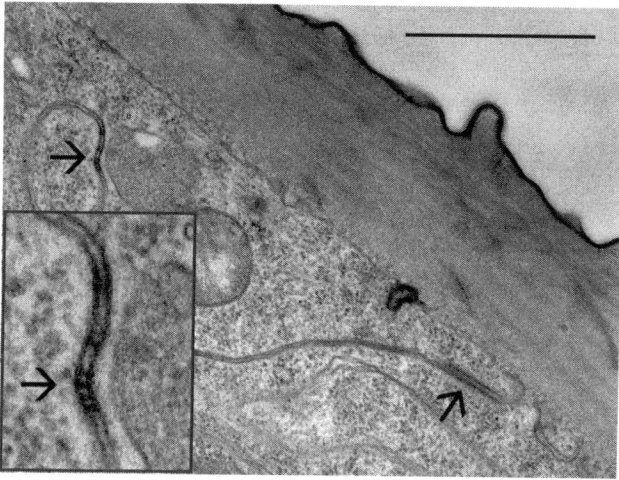

Figure 4.5 Septate junctions between adjacent cell membranes in the epidermal cells of a larva of the mosquito *Ochlerotatus* (formerly *Aedes*) *triseriatus*. The arrows in the lower magnification view point to septate junctions, and the inset shows a higher magnification view of the junctions above the inset box. The scale bar in the upper right corner represents 1 µm. (Photo from the author, Jimmy Becnel and Alexandra Shapiro, USDA, Gainesville, FL.)

particles tightly packed in ridges that make contact or even fuse the adjacent membranes together, obliterating any intercellular space. Tight junctions seal the passageway between adjacent cells and present a barrier to passage of substances. Depending on cell function and physiological condition of the insect, there may be sinuses with varying width from time to time between adjacent epidermal cells, with cells being held together mainly at basal and apical surfaces. Cationic ferritin does not penetrate these spaces, an indication of the general lack of permeable pathways between epidermal cells.

Under the influence of hormones secreted in preparation for a molt, epidermal cells generally enlarge and change shape to become more columnar. They begin to divide by mitosis and move or expand into new space and, sometimes, new shapes that will become the body outline of the next instar as they secrete the new cuticle. Epidermal cells secrete cuticle at the apical face. The cuticle is attached to the apical face of cells by a very large 2.5 MDa extracellular matrix protein (Wilkin et al., 2000). The protein is encoded in *Drosophila melanogaster* by the *dumpy* (*dp*) gene, a complex locus in excess of 100 kb. The gene is expressed at numerous locations including epidermal cell–cuticle attachment sites, at muscle–tonofibrillae attachment sites, and where a cuticle intima (lining) is attached to cells, such as in the tracheae and fore- and hindgut sites. Dumpy protein comprises 308 epidermal growth factor (EGF) modules and 185 of a new class of modules named DPY. Near the carboxy terminal, the protein has a cross-linking *zona pellucida* domain and a transmembrane anchoring sequence. The DPY module forms a β sheet motif that is stabilized by covalent disulfide bonds, and linked end-to-end with EGF modules to form a fiber that may be as much as 0.8 μm in length. Functionally, the protein provides a strong anchor for tissue–cuticle connections, permitting mechanical tension without allowing the tissue to tear away from the cuticle. Such tension sites will occur at many places in the exoskeleton and gut, and most notably at muscle attachment sites.

Epidermal cells have extensive **rough endoplasmic reticulum** (**RER**) where protein synthesis occurs and areas of **smooth endoplasmic reticulum** (**SER**) for lipid synthesis. Cell nuclei are often polyploid, with multiple nucleoli. During development, nucleoli enlarge and develop multiple lobes when there is evidence of new synthesis of ribonucleic acid (RNA) and ribosomes. The **Golgi complex** is prominent in epidermal cells (Locke, 1984) and probably serves several functions including

1. Processing of secretory substances necessary to synthesize cuticle
2. Production of material for the plasma membrane of the cell
3. Modification of newly synthesized proteins
4. Packaging of cellular components in isolation envelopes for later autophagy
5. Processing and packaging of lysozymes needed for autophagy and heterophagy

Epidermal cells are involved in wound repair and can move from an area of undamaged cells into an area of damaged or destroyed cells. Cells at the leading edge spread over the wound until they cover it and establish contact with another epidermal cell. The population of cells in the peripheral zone around a wound is temporarily reduced as cells migrate toward the wound area, but cell divisions soon repopulate the area.

Oenocytes are large, prominent cells scattered among the epidermal cells and also clustered at spiracles near the origin of larger tracheae and scattered among fat body cells. Oenocytes located among the epidermal cells are considered a type of epidermal cell and are differentiated from epidermal tissues during embryogenesis as well as later in development. These cells are usually large, polyploid, and always have extensive tubular smooth endoplasmic reticulum and a well-developed plasma membrane reticular system. Their function is variable, but one function is lipid secretion and lipid metabolism (Makki et al., 2014).

4.3 MOLTING AND FORMATION OF NEW CUTICLE

The external skeleton gradually becomes too small for the growing body tissues of an immature insect and it must **molt** its cuticle (Figure 4.6). Molting is a vulnerable time for insects; they are easy prey for predators and subject to environmental hazards, particularly desiccation. The muscles that move the body must be detached from the old cuticle, but they are detached only immediately before the ecdysis and new muscle attachments are made quickly to the new epicuticle. The new cuticle

INTEGUMENT AND MOLTING

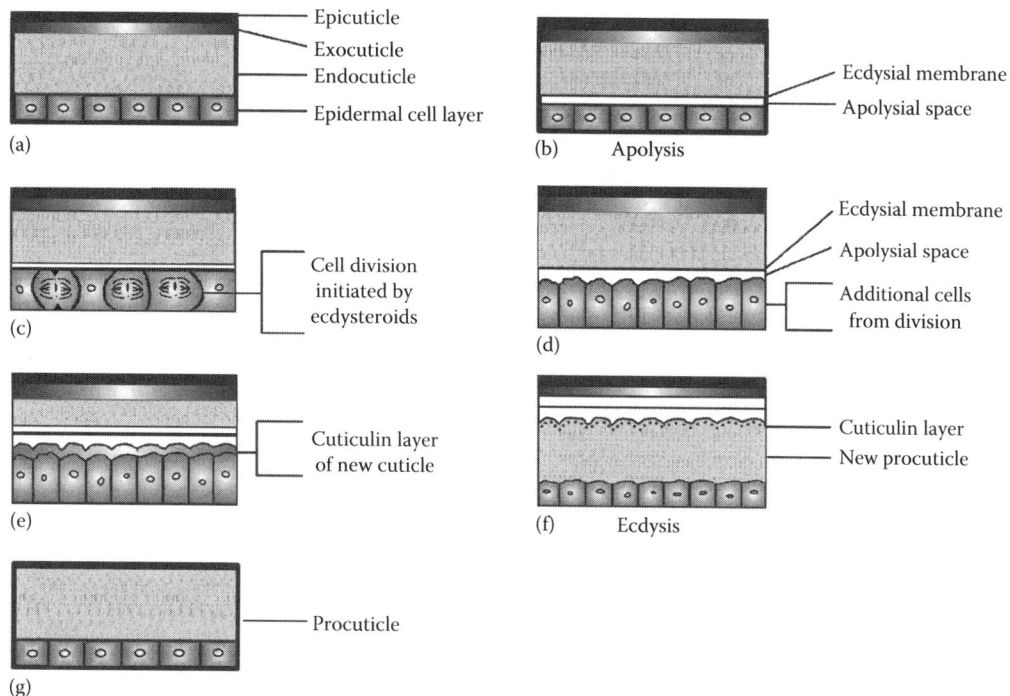

Figure 4.6 Diagrammatic illustration of the process of apolysis, secretion of new cuticle, and ecdysis of the old cuticle. (a) Old cuticle just before molting begins. (b) Formation of the ecdysial membrane and apolysial space. (c) Initiation of cell division in epidermal layer in response to molting hormone. (d) New epidermal cells, usually developing an irregular apical surface. (e) New cuticle secretion begins with the secretion of cuticulin layer. Digestion of the old endocuticle continues. (f) New unsclerotized procuticle is formed. (g) The old cuticle shell has been ecdysed and the new cuticle will be covered with a wax and cement layer, and some of the procuticle may be sclerotized into exocuticle, depending upon the insect and location on the body.

must harden sufficiently to resist the pull of the musculature or muscle action can cause skeletal deformation and result in permanent restriction of movement, and especially failure of flight ability.

Preparation for molting is under endocrine and nervous control (Shaik et al., 2014). Although covered in more detail in Chapter 5, a brief summary of hormonal control of molting in a tobacco hornworm larva is provided here. The events leading to molting in the tobacco hornworm larva represent a described scenario, but this pattern is not typical of all insects. The evidence suggests that each instar of the tobacco hornworm grows until it reaches a certain size and/or weight. Stretch receptors in the body probably are stimulated by the increasing growth of body tissues and, when it attains the critical size, the brain secretes prothoracicotropic hormone (PTTH). PTTH is released into the circulating hemolymph, circulates around the body, and binds specific receptor proteins on the surface of prothoracic gland (PGL) cells. Multiple biochemical reactions are initiated that result in the synthesis of ecdysone by the PGL. Ecdysone is released into the circulating hemolymph and converted into 20-hydroxyecdysone by epidermal cells as well as other tissues. Epidermal cells respond to 20-hydroxyecdysone by separating from the old cuticle (apolysis) (Figure 4.6a and b) and by mitotic activity that produces new cells to spread over the larger body surface that must be enclosed within the new epidermal cell layer and new cuticle (Figure 4.6c). **Apolysis**, or separation of the epidermal cells from the old cuticle (Jenkin and Hinton, 1966), marks the beginning of a molt and of a new instar, and the animal within the loosened, but not yet shed, cuticle is the **pharate** next instar (or stage, if the next form is the pupa or adult).

4.3.1 Apolysial Space

The **apolysial space** is at first a minute space created by the separation of the epidermal cells from the old cuticle. Soon, molting fluid is secreted into the space and activated and, later, as the molting fluid digests some of the old cuticle, the space widens. Typically, there is the discharge of discontinuous patches of membrane-bound secretion into the apolysial space. The vesicles of secretion, often called the apolysial droplets, are secreted by exocytosis from the plasma membrane of epidermal cells. The presence of apolysial droplets seems to precede apolysis in *Calpodes ethlius* and some other insects and follows apolysis in others, such as *Galleria mellonella* and *Hyalophora cecropia*. Although it is still open to experimental analysis, it seems reasonable that some secretion, even though not directly visible, may be involved in dissolving the attachments of the old cuticle to the epidermal cells.

4.3.2 Molting Fluid Secretion

Molting fluid is first evident as osmiophilic droplets (i.e., droplets likely rich in polar lipids) secreted by the epidermal cells into the apolysial space. An **ecdysial membrane** soon appears and can be observed in histological sections. It may result from coalescence of the droplets in some insects, as reported from both *Rhodnius prolixus* and *Calpodes ethlius*, or may be formed from inner layers of the old endocuticle as reported in *Schistocerca gregaria*. The ecdysial membrane persists through the premolt period and, later, is shed with the old exuvium.

Molting fluid contains both **proteinases** and **chitinases** that digest the proteins and chitin, respectively, in old endocuticle (and possibly some of the less heavily sclerotized mesocuticle in some insects). The chitin-digesting enzymes have received more detailed attention so far. **Chitinase** (EC 3.2.1.14) is an endoenzyme and attacks a chitin chain at random by internal hydrolysis. It produces smaller, soluble oligosaccharides that are attacked at the ends by ***N*-acetyl-β-D-glucosaminidase** (EC 3.2.1.30), an exoenzyme yielding free ***N*-acetylglucosamine**. Chitinase isolated from a *Drosophila* cell line has a pH optimum of about 6. Injection of 20-hydroxyecdysone into fifth instars of the silkworm, *Bombyx mori* and *Manduca sexta*, causes secretion of chitinase and *N*-acetyl-β-D-glucosaminidase, but induction of chitinase requires higher levels of hormone than induction of *N*-acetyl-β-D-glucosaminidase. Chitinase exists in more than one molecular size (88 and 65 kDa), and may exist as a **zymogen** (215 kDa) that is converted to the active enzyme at the proper time in *B. mori*.

At least 10 proteases occur in the molting fluid from tobacco hornworm pharate pupae (Brookhart and Kramer, 1990). Both endo- and exocleaving proteolytic enzymes occur and are most active in the neutral-to-alkaline pH range. Some of the proteases have trypsin-like and chymotrypsin-like activity, but they differ from similar gut enzymes in that they are not affected by some inhibitors of gut trypsin and chymotrypsin. None of the enzymes detected is a sulfhydryl or carboxyl protease. At least some of the enzymes are secreted in a zymogen form.

In general, it seems likely that digestion of the old cuticle starts with proteolytic enzymes acting on the proteins of the cuticle and, thereby, exposing chitin rods or crystallites embedded in the protein matrix. Perhaps there also are places where chitin can be attacked without prior release from surrounding protein.

4.3.3 New Cuticle Formation

New cuticle secretion begins soon after the ecdysial space opens. During new cuticle secretion, an epidermal cell has a series of ridges or knobby projections on its apical face where the proteins and fibers of chitin are secreted. Locke (1984) has called these projections **plasma membrane plaques**. At the initiation of new cuticle synthesis, cuticulin, the new envelope, is the first new secretion that begins to form on these knobby plaques (Figure 4.6d and e). It is secreted initially as small discontinuous patches over the plasma membrane plaques, but the patches enlarge and eventually form a continuous layer of cuticulin. There also may be long microvilli on the apical face of epidermal cells during the molting process. Molting fluid digests the old endocuticle and the products of digestion are reabsorbed by epidermal cells and used in the synthesis of new cuticle. The manner in which the new cuticulin is protected from digestion is not known; one suggestion is that the ecdysial membrane that is formed represents a barrier between the molting fluid and the new cuticulin layer being secreted just below the membrane. It may be that the cuticulin envelope is simply not digestible by the chitinases and proteases that are secreted in the molting fluid. As old endocuticle is digested, new procuticle (unsclerotized) is secreted below the cuticulin envelope, thus pushing it up and outward. The new cuticle secreted at this stage is unsclerotized and many authors have called it **procuticle** (Figure 4.6f and g).

Later, after ecdysis, some or much of the new cuticle, depending upon the insect and body part involved, may become heavily sclerotized and very hard (exocuticle), while other layers might be less sclerotized (mesocuticle) or very lightly sclerotized (endocuticle) (Shaik et al., 2014). The cuticle layers can be described in terms of the envelope (the cuticulin layer), the epicuticle, and procuticle, each of which is secreted in a different process (Locke, 2001). The envelope is laid down at the plasma membrane surface, usually on plaques at the tips of microvilli, as noted earlier. The newly impermeable envelope then protects the epidermal cell surface from digestive enzymes in the molting fluid, but allows the digestion of the old endocuticle, so that the amino acids from protein digestion and glucose from chitin digestion can be reassembled into new procuticle. The epicuticle is secreted on the inner face of the envelope and, with the envelope, forms the outer boundary of the cuticle compartment. Procuticle is then formed at the cell surface until it fills the cuticle compartment. Locke notes that some type of limiting boundary—an envelope (but not a cuticulin envelope)—covers the cells in most invertebrate phyla, including bacteria, protozoa, trematodes, nematodes, mollusks, and arthropods. The envelope, according to Locke, provides a mechanism for extending metabolic control of the extracellular compartment (i.e., the cuticle) that limits size; provides protection from environmental chemicals, bacteria, and fungi; regulates permeability of the cuticular compartment; and is involved in surface reflectivity and color.

4.3.4 Reabsorption of Molting Fluid

The molting fluid that accumulates in the apolysial space disappears shortly before ecdysis. Most of it appears to be reabsorbed by the insect; at this critical stage, preventing loss of fluid volume by terrestrial insects may be vital. Some of the molting fluid may be reabsorbed through the epidermal cells, but in *M. sexta* pharate pupae (Cornell and Pan, 1983) and in pharate pupae of the skipper *Calpodes ethlius* (Yarema et al., 2000), it primarily flows beneath the old and new cuticles to the mouth and anal openings and is accumulated in the midgut. Similar swallowing of molting fluid through the mouth occurs in the pharate adult *M. sexta* (Miles and Booker, 1998). The fluid probably contains protein (enzymes) and other potentially useful nutrients, and the water may be useful in helping to flush the gut and/or Malpighian tubules of accumulated waste products, particularly in newly eclosed adults, many of which excrete the **meconium** as the accumulated waste from molting. Excretion of any excess water also lightens the body for flight.

4.4 ECDYSIS

To facilitate **ecdysis** of the old cuticle, some insects swallow air, or swallow water if they are aquatic insects, to expand the gut and split the old cuticle. Prior to emerging Diptera expand the soft cuticle of the frons (a part of the head) to make a balloon-like sac called the **ptilinum** (Figure 4.7) that ruptures the puparium and provides an exit out of which the adult wiggles free. Soon after

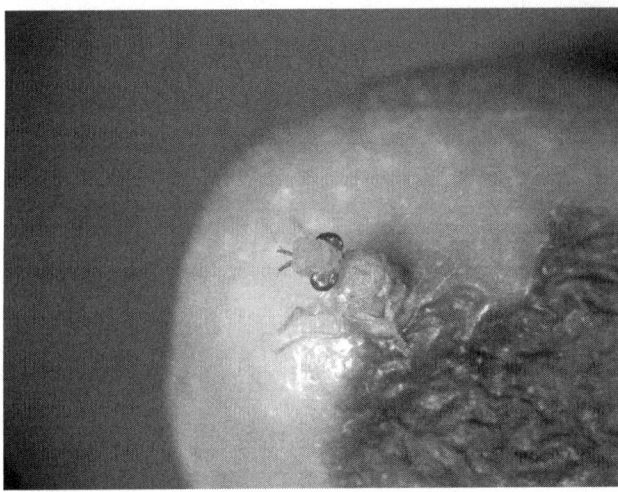

Figure 4.7 **(See color insert.)** An olive fruit fly emerging with the ptilinum expanded to aid it in breaking out of the puparial cuticle. (Photo courtesy of Dr. Hanife Genc, Çanakkale Onsekiz Mart Üniversitesi, Çanakkale, Turkey.)

emergence, the ptilinum is retracted by muscles that atrophy, and the cuticular surface hardens into the ptilinum suture. The muscular actions involved in ecdysis are controlled by nervous motor programs. Ecdysis-related motor programs have been identified in a number of insects, and some insects have sequential stages or sequential motor programs that must occur in proper sequence for molting to be successful. Events leading up to and completion of ecdysis in insects illustrate interactions between the endocrine and nervous systems, for, although the nervous system controls muscular movements in all insects, the motor program is initiated by hormonal action. The sequence of nervous and hormonal controls were initially documented in the lepidopterans *Manduca sexta* and *Bombyx mori* (Truman, 1978; Horodyski, 1996), but many details have now been extended to many holometabolous and hemimetabolous insects (Roller et al., 2010; Lee et al., 2013, and references in both).

On the basis of distinct behaviors and physiology, ecdysis has been divided into phases: **pre-ecdysis I, pre-ecdysis II, ecdysis (= eclosion)**, and **post–ecdysis** behavior and physiology (Reynolds, 1980). The process of molting from one stage to another and eventual metamorphosis to the adult form is coordinated by a suite of neuropeptides: **corazonin, pre-ecdysis-triggering hormone (PETH), e**dysis-triggering hormone (ETH) eclosion hormone (EH) and CCAP and bursicon (Truman and Riddiford, 1970; Žitňan et al., 1996, 2007; Žitňan and Adams, 2000), and the hormones **PTTH** (also a neurohormonee), **ecdysone** or a related ecdysteroid, and **juvenile hormone (JH)**.

Several days prior to beginning of ecdysis, brain neurosecretory cells release **PTTH** that targets the prothoracic gland cells, inducing synthesis and release of **ecdysone** into the hemolymph. Ecdysone is converted to 20-hydroxyecdysone which then programs the epidermal cells to get ready to make a new cuticle; **juvenile hormone** is also secreted from the corpora allata, and it will determine whether the new cuticle produced will be larval, pupal, or adult (more details about these three hormones that initiate molting can be found in Chapter 5, **Hormones and Development**). The high level of 20-hydroxyecdysone in the hemolymph induces receptors in the central nervous system for the ecdysis-triggering neuropeptides (**PETH and ETH**) and stimulates production of ETH in the **Inka cells**. About 30 min before the visible signs of ecdysis begin in *Manduca sexta*, the neuropeptide **corazonin** is produced in brain neurosecretory cells and released into the hemolymph from axons leading to the corpora cardiaca–corpora allata glands (Kim et al., 2004). Corazonin in the hemolymph binds to a receptor in Inka cells and causes initial and low-level secretion of PETH and ETH neuropeptides into the hemolymph. These neuropeptides cause **pre-ecdysis I** contractions. Pre-ecdysis dorsoventral contractions occur synchronously in thoracic and abdominal segments. These pre-ecdysis contractions occur every 10–12 s with a contraction lasting 5–7 s. Figure 4.8 shows a sequence of shedding the larval cuticle in the last instar of a monarch butterfly caterpillar. A video slightly less than 8 min in duration (real time) of the shedding of the cuticle by a monarch butterfly is present on YouTube at https://www.youtube.com/watch?v=MeaNCvIXhyM&list=PLa-k6B3eFyuauSWnMCVM8ZxlbKvYG10rB&index=1 (last accessed April 8, 2015).

Initially, a monarch caterpillar ready to pupate attaches itself to some substrate by hooking its cremaster on the last abdominal segment into a silk pad previously secreted from the mouth, and then hangs down in a "j" shape. At the beginning of the final ecdysis process it hangs straight down and within a few seconds the dorsal cuticle splits just behind the head. The green pupa then begins to emerge, and the larval cuticle is retracted posteriorly by intersegmental muscles. When fully retracted to the posterior end of the body, it is dislodged from the body by vigorous wiggling of the pupa, which then becomes quiet and soon takes the characteristic shape and color of a monarch pupa.

Isolated ventral nerve cords (VNC) of larvae of *Hyalophora cecropia*, a moth, are capable of generating the motor program indicating that the nervous activity is centered in the nervous system (Truman and Riddiford, 1970). At first, each ganglion along the VNC produces bursts of action

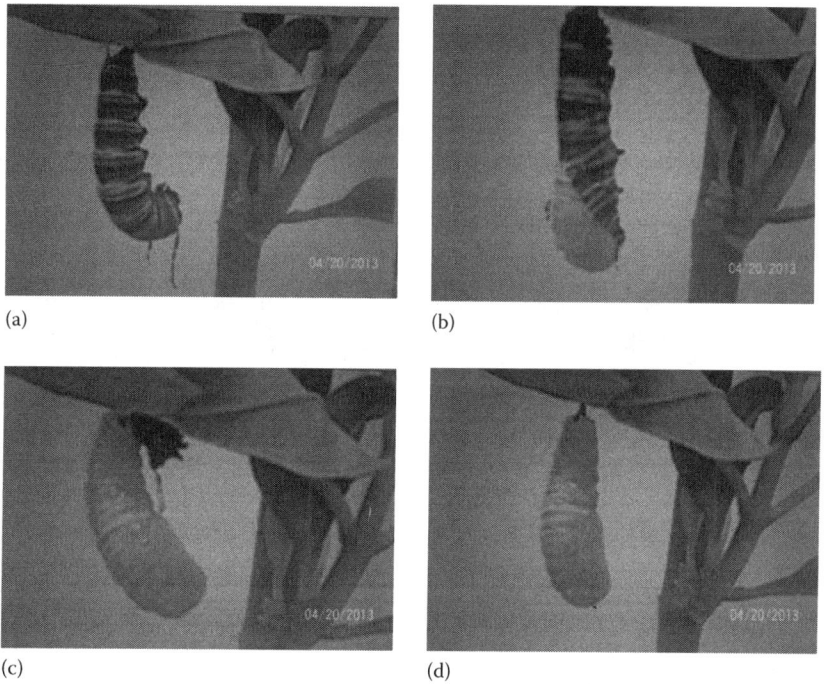

Figure 4.8 (See color insert.) A sequence showing molting of the last instar cuticle by a larval monarch butterfly caterpillar as it hangs by its cremaster from a silk pad. (a) The larva is still in the "j" shape; the pharate pupa is inside the larval integument but cannot be seen at this point. (b) The larva now hangs straight down and the larval integument has split dorsally, beginning at the head of the larva, and the green pupa can be seen as the larval integument retracts toward the posterior. (c) The larval integument has been retracted to the posterior end of the pupa by contractions of intersegmental muscles attached to the old integument. (d) The newly emerged pupa has vigorously wiggled until the old larval integument was dislodged and fell away from the pupa. The pupa will continue to shorten in length and soon take on the typical green appearance of a monarch pupa. (Photos by the author.)

potentials alternately from the right and left sides. In an intact animal, these nerve impulses would initiate muscle action in each segment that could cause wiggling and rotatory movements of the abdomen. Slowly PETH and ETH accumulate in the hemolymph, leading to **pre-ecdysis II** with more vigorous and frequent contractions. The increased level of ETH activates ventromedial cells in the brain to release **eclosion hormone (EH)** about 40 min before ecdysis in *Drosophila*, and this results in an increase of **cyclic GMP (cGMP)** in tracheae and associated neurons, increase in heart rate, and filling of new tracheae with air (Baker et al., 1999). The gradual build up of EH in the hemolymph leads to massive release of PETH and ETH from Inka cells and the transition to ecdysis (Kim et al., 2004).

4.4.1 Shedding the Old Cuticle: Ecdysis

The final ecdysis contractions are a series of peristaltic waves of contractions that originate in the most posterior segment and pass anteriorly for about 10 min until ecdysis is complete. Monitoring of nervous activity in the lateral nerves indicates that synchronous bursts of action potentials occur from both sides of each ganglion, but alternating in time between successive ganglia. These action potentials cause the peristaltic muscle contractions necessary to push the adult out of the old cuticle. Some minutes after the adult is out of the pupal cuticle, all ganglia simultaneously begin to produce prolonged bursts of action potentials resulting in a steady tonic contraction of the abdomen that aids the pumping of hemolymph into the wings to inflate them (Figure 4.9a through d).

Figure 4.9 **(See color insert.)** (a) Ecdysis of the adult cicada *Neocicada hieroglyphica* from the last nymphal cuticle. (b) The adult with partially expanded wings. (c) The adult *Neocicada hieroglyphica* with fully expanded wings, but cuticle not yet hardeneded and darkened. (d) Adult *Neocicada hieroglyphica* with sclerotized and darkened cuticle. Note how it blends with the tree trunk on which it is resting, waiting for its wings to fully harden. (Photos courtesy of Lyle Buss, University of Florida, Entomology & Nematology, Gainesville, FL.)

Eclosion hormone(s) probably exists in most or possibly all insects (Truman and Riddiford, 1970; Truman et al., 1981), but behavioral and biochemical assays for eclosion hormone are most thorough in Lepidoptera.

4.4.2 Post-Ecdysis Wing Expansion and Waterproofing the New Cuticle

Hardening (sclerotization) and darkening of the cuticle and wing expansion (Figure 4.9d) are under control of the tanning hormone, **bursicon** (Cottrell, 1962; Fraenkel and Hsiao, 1962; Luan et al., 2006; Loveall and Deitcher, 2010). Bursicon is a very large protein of about 30,000 Da and was difficult to sequence, but some peptide sequences were obtained, enabling a search for similar fragments in the *Drosophila* genome that led to identification of the *bursicon* (*burs*) gene, and subsequently to a second gene *partner of bursicon* (*pburs*). It turns out that the peptide encoded by either of these two genes is not active by itself, but a heterodimer with each peptide is active and stimulates synthesis of cAMP as a necessary second messenger for bursicon function (Baker and Truman, 2002; Luo et al., 2005). Mutant flies that lacked the *burs* gene could not carry out post-ecdysial tanning and lacked bursicon in the central nervous system (Dewey et al., 2004). The receptor for bursicon is encoded by the gene, *rickets*, in *Drosophila* (Baker and Truman, 2002).

There is evidence that **crustacean cardioactive peptide (CCAP)** is important in the final phase of ecdysis (Gammie and Truman, 1997, 1999; Žitňan and Adams, 2000), probably in regulating the release of bursicon. In *Drosophila* larvae group of neurons that express CCAP also synthesize bursicon, the tanning hormone (Park et al., 2003; Luo et al., 2005). Luan et al. (2006) were able to show that the CCAP-expressing neurons can be identified as two distinct groups, with one releasing bursicon and the other group releasing CAAP that regulates bursicon release. Genetic mutants of *Drosophila* missing CCAP and *pburs* genes suffered severe abnormalities at pupation (Lahr et al., 2012), while *Tribolium castaneum* failed to ecdyse when RNAi was used to reduce CCAP and its receptor levels (Arakane et al., 2008; Li et al., 2011). Lee et al. (2013) showed that the CCAP-signaling pathway is important to ecdysis in the hemimetabolous *Rhodnius prolixus*. Thus, CCAP seems to be important in ecdysis of a broad group of insects.

Following actual shedding of the old cuticle, lipids and cement are secreted onto the new epicuticle, muscle reattachments become firmly fixed to the epicuticle, and the cuticle begins to sclerotize. In *Calpodes* larvae, wax secretion is under control of the corpora allata/corpora cardiaca complex. Formation of layers of endocuticle continues in many insects during intermolt periods and even during adult life. In some insects, there are two distinctive layers formed daily—one during the day and another at night—so that growth rings can be observed and counted in a cross section of cuticle. These growth layers are present in certain grasshoppers and some cockroaches (*Periplaneta*) and are present, but not so clearly demarcated, in milkweed bugs for the first 8 days of adult life.

The role of corazonin in an early initial phase of ecdysis seems well established for *M. sexta* and perhaps *B. mori*, but it has not been established that it has the same role in other insects (Žitňan and Adams, 2012), although corazonin and peptides similar to corazonin have been identified in a number of pterygote and apterygote insects (Roller et al., 2003). A corazonin receptor has been identified in *Drosophila melanogaster* (Park et al., 2002). In *M. sexta* corazonin is localized in the same cells that express the gene *period*, which is involved in circadian rhythm behavior (Wise et al., 2002). Thus, corazonin may play a role in the gated eclosion of larvae to the pupal stage (Truman and Riddiford, 1970; Truman, 1972) and possibly has a role in stress behaviors of insects (Zhao et al., 2010).

Peptide hormones similar to PETH and ETH in Lepidoptera (described as ETH1 and ETH2 in other insects) are members of a conserved family of peptide hormones that occur in many other insects that have been studied, and in ticks and other arthropods (Roller et al., 2010).

EH, a neurosecretory polypeptide of 62 amino acids, is secreted in *M. sexta* by ventrally located neurosecretory cells in the tritocerebrum in response to falling ecdysteroid titer (Truman, 1985; Kingan and Adams, 2000). Its release can be inhibited by injecting 20-hydroxyecdysone, thus preventing the acquisition of competence in the Inka cells (Kingan and Adams, 2000). Under falling ecdysteroid levels, it is released from the proctodeal nerve, arising from the terminal abdominal ganglion, during larval and pupal ecdysis. The cells secreting EH in the pharate adult are modified during adult development so that the cells send axons to the corpora cardiaca (CC). EH is transferred through this axon for storage in the CC and released into the circulating hemolymph for pharate adult eclosion. Expression of the basic leucine zipper gene *cryptocephal* (*crc*) is required for expression of ETH in Inka cells (Gauthier and Hewes, 2006).

4.4.3 Sclerotization of Cuticle

Sclerotization is the process of cross-linking protein-to-protein chains, chitin-to-chitin chains, and possibly protein-to-chitin chains (see reviews by Sugumaran, 1988; Hopkins and Kramer, 1992). Sclerotization is also called **tanning** and sometimes simply described as hardening of the cuticle. Tanning refers to the cross-linking process itself and not to a color change, although sclerotization often is accompanied by tan, brown, or black colors. The colors are created by a variety of pigments, including melanin. The phenols associated with sclerotization easily undergo autoxidation (phenols to quinones) and quinones readily polymerize, processes usually leading to melanin and tan or brown to black colors. Hardening (sclerotization/tanning) and darkening are two different processes, and cuticle can become sclerotized without darkening, for example, over the compound eyes.

Only protein-to-protein sclerotization occurs in the epicuticle because no chitin occurs there, but in other layers of the cuticle all the combinations may exist. Sclerotization gives strength and rigidity to the cuticle. Apodemes, the elytra of beetles, and the mandibles of chewing insects are examples of heavily sclerotized cuticle. Intersegmental membranes and the cuticle of soft larvae are lightly sclerotized. Hardness of the cuticle is a function of the degree of sclerotization and is not indicative of the content of chitin in the cuticle, as once thought.

The cross-linking or sclerotizing agents are phenols and their oxidation products, quinones. A number of phenols and quinones exist in various insect cuticles and all probably participate to some extent as sclerotizing agents, but the chemistry has been best elucidated for production of **N-acetyldopamine**, a common and major sclerotizing agent in many cuticles. N-acetyldopamine is formed from tyrosine by a number of enzymatically controlled steps (Figure 4.10). Early instars of Diptera, in which the process has been studied in some detail, metabolize tyrosine to N-acetyltryamine and p-hydroxyphenyl propionic acid, which are not involved in sclerotization. Only late in the last instar is there a switch in synthesis to N-acetyldopamine under the influence of the molting hormone.

N-β-alanyldopamine has been implicated as the principal tanning agent in the pupal cuticle of *M. sexta*, in which it increases as much as 800-fold during the tanning of the pupa. It is a major cuticular constituent of a number of insects in various orders and may be the typical sclerotizing agent in pupae since the pupal o-diphenol oxidase oxidizes it most readily among a variety of potential substrates (Hopkins et al., 1982).

Quinones cross-link protein chains by reacting with free amino groups, such as those of lysine, tryptophan, arginine, histidine, and the terminal amino group at one end of a protein. Chitin chains are also linked to each other and possibly to protein chains through the amino group of N-acetyl-glucosamine. The sulfhydryl group (–SH) of the amino acid cysteine may also participate in cross-linking protein chains through formation of disulfide linkages (–S–S–).

When protein chains are linked to the ring of the phenolic cross-linking agent, the process is called **quinone tanning** or **quinone sclerotization**. Proteins also may be linked to the β-carbon (the

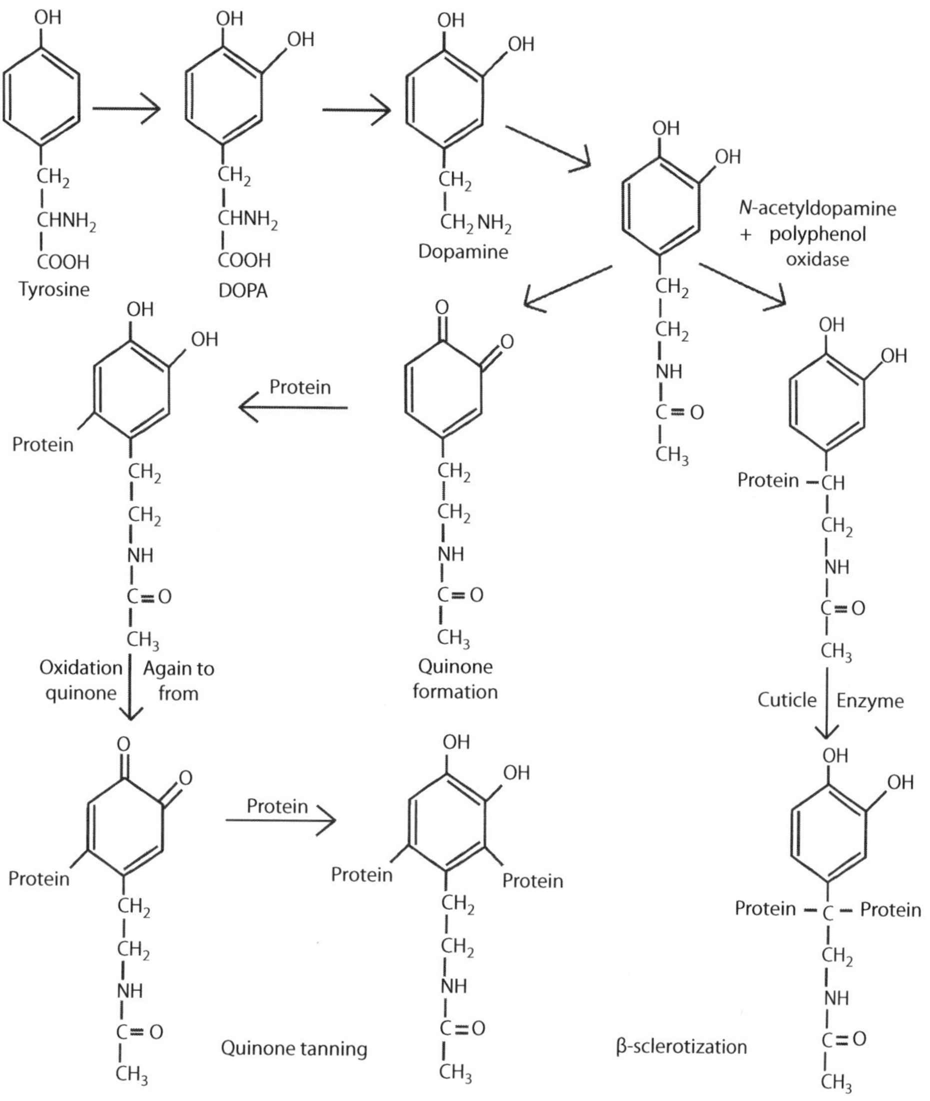

Figure 4.10 A generalized biosynthetic pathway for metabolism of tyrosine to N-acetyldopamine, a common sclerotizing agent, and the linking of proteins to the phenolic ring in either quinone tanning or to the beta carbon in β-sclerotization.

carbon nearest the ring in the side chain) of N-acetyldopamine, a process called **β-sclerotization** to distinguish it from protein attachment to the ring. Quinones are involved in both types of sclerotization. How an insect controls the type of sclerotization that occurs is unknown. Some evidence indicates that both quinone tanning and β-sclerotization can occur in the same small region of cuticle. It has been suggested by some workers that β-sclerotization can harden cuticle to produce lighter colored or transparent cuticle, although how this might be controlled is unknown. Transparent cuticle is important in the covering of the compound eyes, for example, and many insects have lightly colored patches of cuticle elsewhere on the body. An important area of integumentary physiology still to be elucidated is the production of transparent cuticle.

The tanning of the ootheca of American cockroaches, *Periplaneta americana,* is a model for sclerotization in the cuticle (Figure 4.11). The ootheca of cockroaches contains proteins but no chitin.

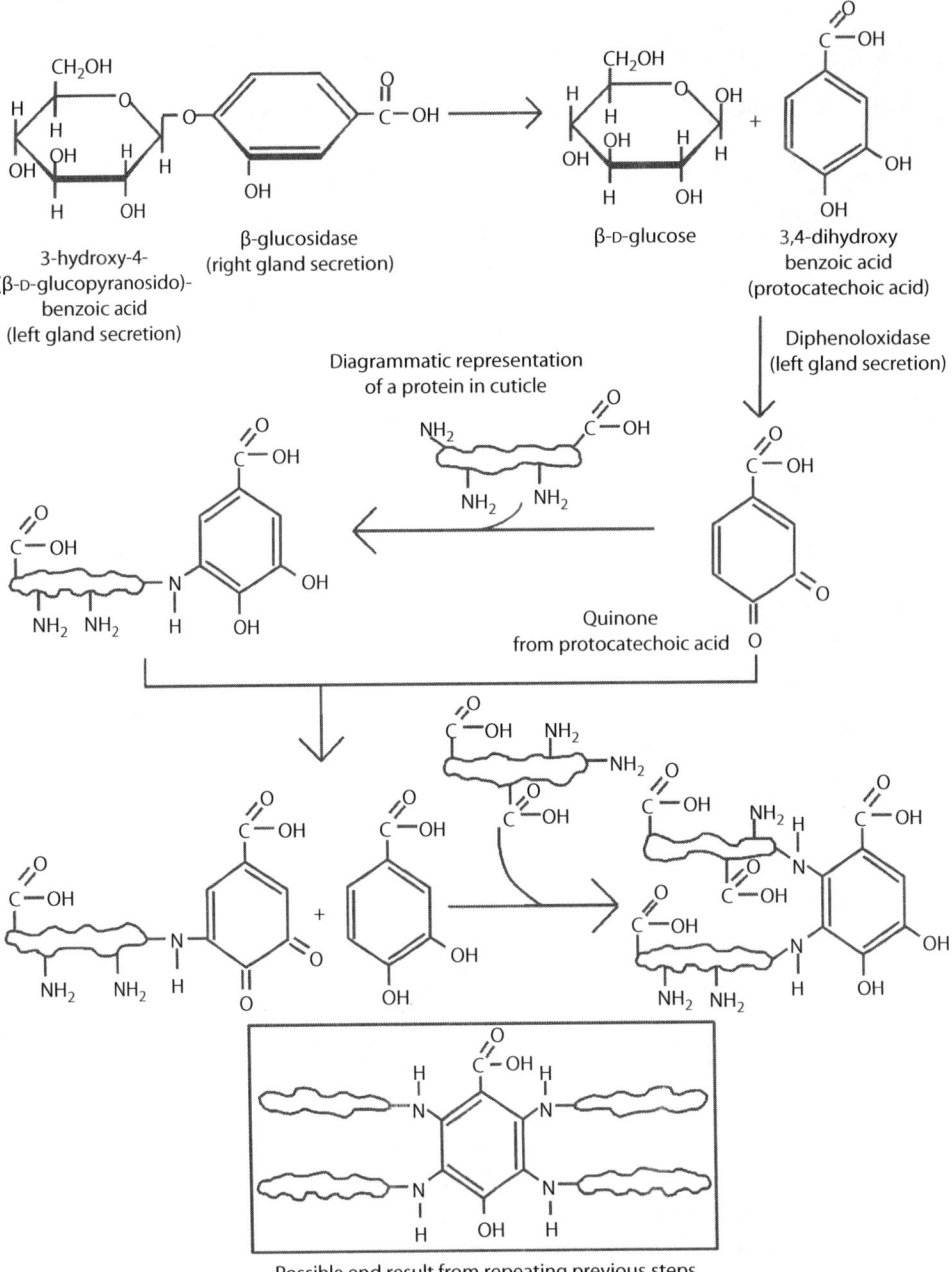

Figure 4.11 A general sclerotization model based on sclerotization of the ootheca in the American cockroach, *Periplaneta americana*.

When first formed, the ootheca is white and soft, but it soon sclerotizes and darkens to a hard covering for the developing eggs and embryos. The sclerotizing process involves secretions from two accessory or collateral glands that are part of the reproductive tract in female cockroaches. The left gland contains the enzyme **diphenoloxidase**, and the glucoside of 3,4-dihydroxybenzoic acid and of 3,4-dihydroxybenzyl alcohol. The right gland contains **β-glucosidase**. When the secretions from the two glands are poured over the newly formed ootheca, β-glucosidase cleaves the two glycosides to free glucose and the diphenol, **3,4-dihydroxy benzoic acid**. Diphenoloxidase oxidizes the phenolic acid to its quinone form. The quinone reacts without enzymatic help with a free amino group from a protein, hooking the protein to the ring of the compound (i.e., quinone sclerotization) and simultaneously becoming reduced again to the phenol form. In the presence of excess free quinone, the protein–phenol complex is again oxidized to a quinone, which may then react with another free amino group of a protein, with this protein also becoming hooked to the phenolic ring. These reactions may be repeated until several protein chains have been linked to the phenolic ring in quinone sclerotization. In the process, the ootheca becomes a hard, tough, waterproof case covering the eggs and developing embryos.

Bursicon is a **neuropeptide** that promotes sclerotization and specifies how much crosslinking of molecules will occur (Loveall and Deitcher, 2010). The hormone is secreted by the nervous system. It has not been isolated in pure enough form for a complete amino acid determination, but it is a small polypeptide of about 40,000 Da. It has been found in various ganglia of the central nervous system of many insects and is now believed to occur in most or possibly all insects.

It was first discovered in a newly formed adult fly (Cottrell, 1962; Frankel and Hsiao, 1962, 1963; Frankel et al., 1966). Soon after an adult fly emerges from the puparium, its peripheral nervous system sends signals to the brain to secrete bursicon (from Greek *bursikos*, tanning or pertaining to tanning). Bursicon is released from neurosecretory cells (NSC) in the pars intercerebralis of the brain and from NSC in the large combined abdominal and thoracic ganglion of cyclorrhapha dipterans, in which bursicon is present at even higher concentrations than in the brain. Bursicon also has been demonstrated in the nervous system and corpora cardiaca of the cockroach, *P. americana*. In ways that have not been elucidated in detail, bursicon promotes hardening or sclerotization of the cuticle. Bursicon may promote production of some of the sclerotizing enzymes, control access to quinone precursors, or control penetration and permeability of the cuticle to phenols and quinones.

4.5 CHEMICAL COMPOSITION OF CUTICLE

4.5.1 Chitin

Chitin, one of the most widely occurring polysaccharides in nature, is found in the cuticle of crustaceans and insects, in many other invertebrates, in nematode eggs, and as a structural cell wall component of fungi. As previously noted, chitin does not occur in the epicuticular layer of cuticle, and it may not be the major constituent in other parts of the cuticle. In some cuticles protein is present in greater percentage by weight than chitin (Figure 4.12).

Chitin is a polymer of **N-acetyl-β-D-glucosamine** (2-acetamido-2-deoxy-β-D-glucose) residues held together by **β-(1-4)-glycosidic linkages** (Figure 4.13). Enzymatic hydrolysis of chitin with chitinase and chitobiase releases *N*-acetylglucosamine as well as some free glucosamine. From these and other experiments, it has been suggested that every sixth or seventh residue in chitin may be glucosamine, with the remainder being *N*-acetyl-glucosamine. Chitin cannot be extracted directly from cuticles with any solvent, but it is left behind when the proteins, mineral deposits (if present), lipids, and other chemicals have been removed. The procedures that remove the other

118   INSECT PHYSIOLOGY AND BIOCHEMISTRY

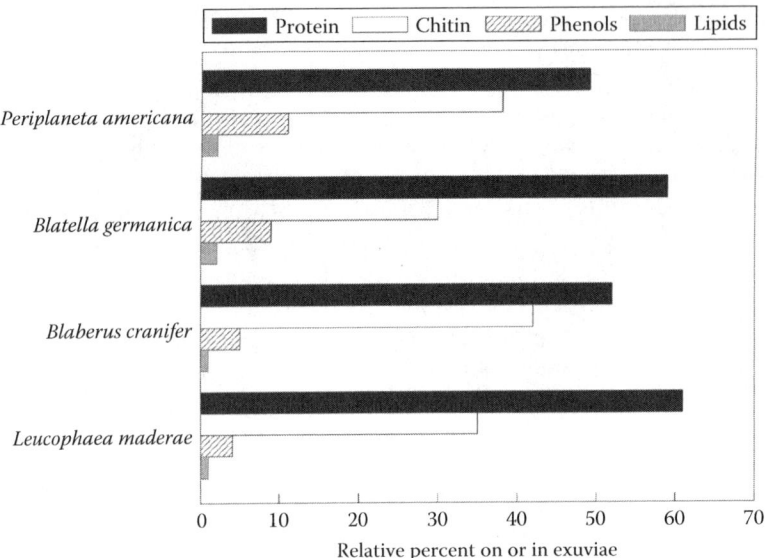

Figure 4.12 Determination by ^{13}C-NMR of the proportion of proteins, chitin, phenolic compounds, and lipids in or on the exuviae of selected species of cockroaches. (Data modified from Kramer, K.J. et al., *Insect Biochem.*, 21, 149, 1991.)

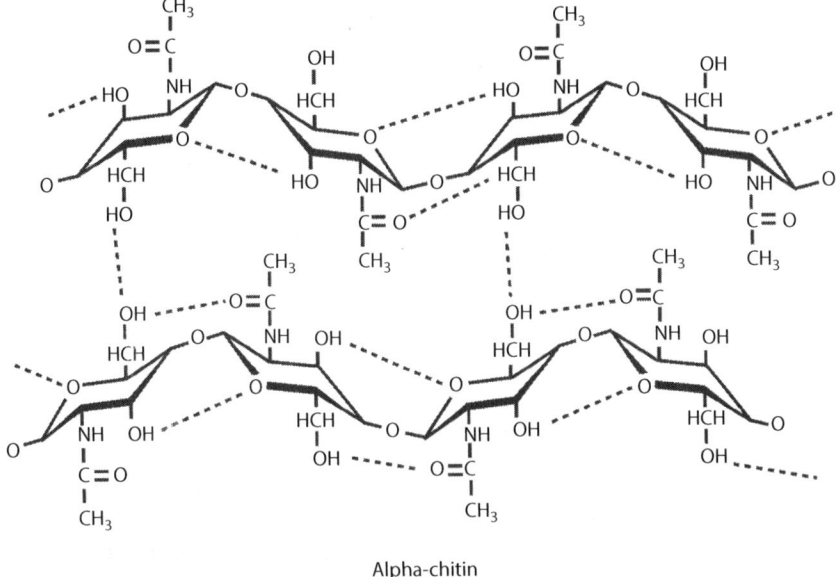

Figure 4.13 An illustration of the chemical structure of α-chitin chains and the antiparallel arrangement of (two) chains that is typical in insect cuticles. Hydrogen bonds, some of which are intrachain and some interchain bonds, are represented by the dotted lines. Not shown are hydrogen bonds between chains in adjacent layers of chitin.

constituents almost always cause varying degrees of degradation of the chitin that is left, however. For example, treatment with hot alkali (KOH) removes protein from insect cuticles. Cold alkali does the same thing, but requires a longer time period; certain cuticular structures, such as genitalia, are "cleared" for taxonomic purposes by allowing the tissue to stay in cold alkali for several days to weeks. Cold and hot alkali remove some of the acetyl groups from chitin, leaving a less acetylated product called chitosan. Transparent and flexible, chitosan reacts with iodine (van Wisslingh's test) to give a dark purple color, and this often has been used to test for the presence of chitin, although it may not be infallible. In light of the vigorous procedures required to purify chitin from insect cuticles, it is not surprising that isolated chitin does not show all the same properties of the original insect cuticle.

The inertness and insolubility of chitin in the cuticle serve insects well, but make it difficult to characterize the molecular arrangement of chitin within cuticle. Studies with x-ray diffraction indicate that chitin exhibits a crystalline structure, but unit crystal dimensions and number of chitin chains per unit crystal vary with the source of the cuticle. Three types of chitin, named α-, β-, and γ-chitin, have been described, and all three occur in some form in insects. The three types of chitin have differences in crystal cell size, number of chitin chains per unit cell, and degree of hydration (Rudall, 1963). Differences in orientation of the three chitin chains in chitinized structures lead to differences in physical packing of chains and in overall physiological properties of structures. In **α-chitin** adjacent chains run antiparallel to each other, which allows them to pack closer together and maximizes the number of within- and between-chain hydrogen bonds. Typically 18–20 α-chitin chains are packed together in a roughly circular (about 3 nm diameter) rod or **crystallite** that is embedded in a protein matrix (Figure 4.14). The extensive **hydrogen bonding** within chains, between chains, and between adjacent sheets of cuticle contributes to the rigidity, strength, and waterproofing of the cuticle, leaving few hydration sites for water. Water molecules form hydrogen bonds with appropriate partners, and most of the potential sites for

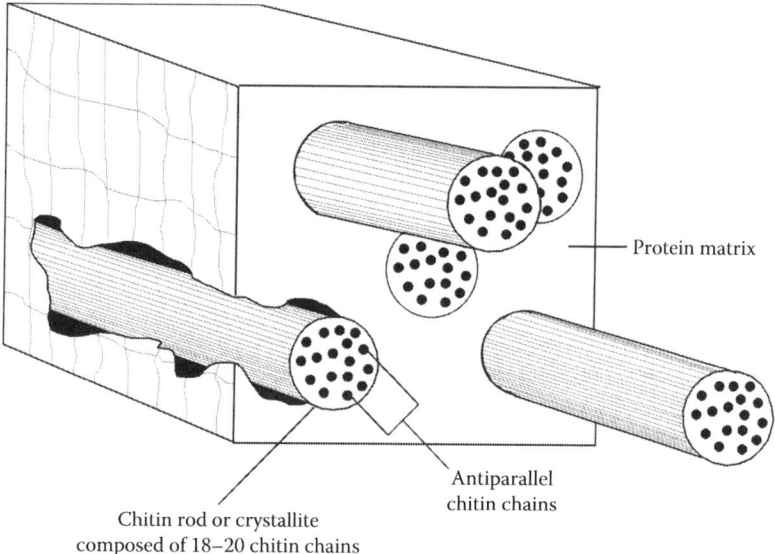

Figure 4.14 Chitin chains (about 20 typically in antiparallel arrangement) held together by hydrogen bonds to form chitin rods or crystallites embedded in a protein matrix in cuticle. (Drawing modified from Giraud-Guille, M.M. and Bouligand, Y., Chitin-protein molecular organization in arthropods, in R. Muzzarelli, Ch. Jeuniaux, and G.W. Gooday (eds.), *Chitin in Nature and Technology*, Plenum Press, New York, 1986, pp. 29–35.)

hydration in α-chitin are already involved in hydrogen bonds. High content of water of hydration not only weakens the tightness of chain packing, but it allows more water movement across the cuticle, disrupting the high degree of impermeability typical of the cuticle. The tight packing and intra- and interchain hydrogen bonds provide strength, stability, and contribute to the impermeability of the cuticle to water.

Adjacent chains in **β-chitin** run parallel to each other, and the relatively large N-acetyl groups projecting from the chain act like spacers holding chains farther apart, reducing tightness of packing and the number of hydrogen bonds that can form between chains. Chains in **γ-chitin** may be oriented in various ways, but one common orientation is a repeating pattern of two parallel chains adjacent to an antiparallel chain, again reducing packing tightness and interchain hydrogen bonds. Chains in β- and γ-chitin have more free groups that can form hydrogen bonds with water of hydration. β-Chitin and γ-chitin occur in cocoons of some beetles, and both are found in some other noninsect invertebrates. γ-Chitin has been identified in the peritrophic membrane of some insects. Greater hydration and less packing of chains allow chitinous structures with large amounts of β- and γ-chitins to be flexible and soft.

Cuticle is secreted as thin lamellae or sheets, like sheets of paper stacked one on top of each other. The rods or crystallites of chitin are embedded in the protein matrix of a sheet (Giraud-Guille and Bouligand, 1986) and provide strengthening in much the same way as that provided by steel rods (reinforcement rods commonly called "rebars" in the construction industry) that are embedded in concrete columns and walls. Adjacent sheets of cuticle are stabilized by quinone tanning agents and by hydrogen bonds between chitin rods in adjacent cuticle sheets when the rods are near the surface of each sheet. It is still uncertain if quinone tanning agents directly link chitin rods to protein.

In successive sheets of cuticle, the chitin rods often are shifted slightly in orientation relative to the sheet above it (the older sheet), and this gives rise to **bouligand helicoids** (Figure 4.15) in thin transmission electron micrographs sections. One structural model suggests that chitin rods are embedded parallel to each other in the protein matrix in a plane or sheet of cuticle only a few nanometers thick (Bouligand, 1972), and in each successive sheet of cuticle the model

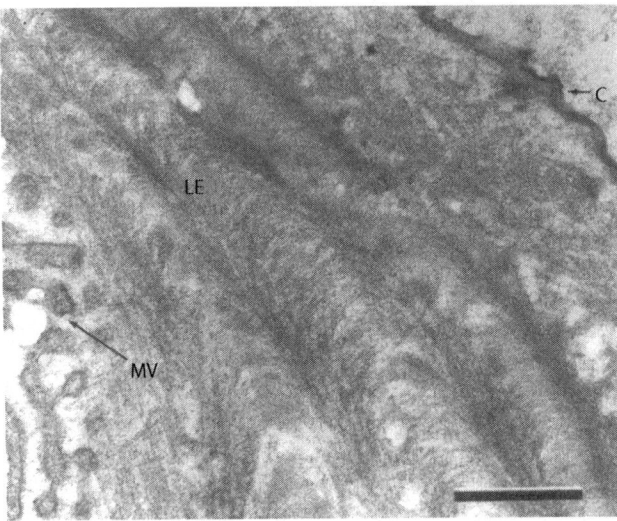

Figure 4.15 Bouligand helicoids in the cuticle of wing discs cultured *in vitro* from the Indian meal moth, *Plodia interpunctella*. C, cuticulin layer; LE, lamellate endocuticle showing Bouligand helicoid patterns; MV, microvillae of underlying epidermal cell. (Photo courtesy of Herbert Oberlander, USDA, Gainesville, FL [Retired].)

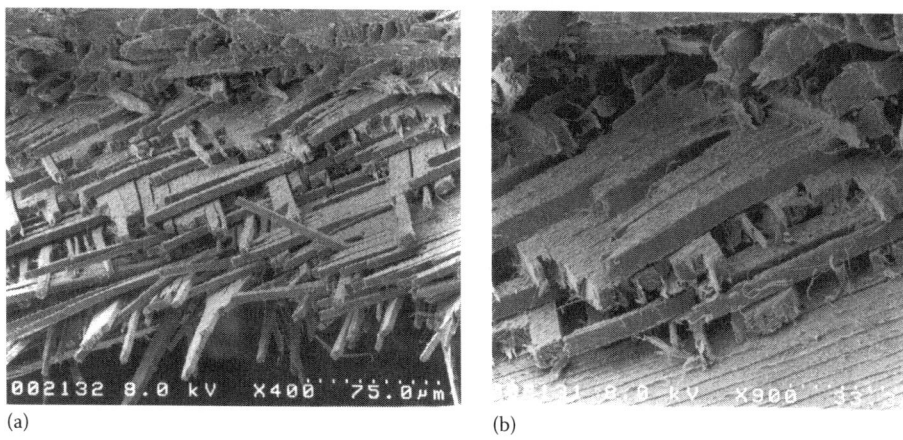

Figure 4.16 Freeze-fractured break in the thoracic cuticle of the weevil, *Rhynchophorus cruentatus*, showing plywood-like arrangement of cuticle layers that gives the cuticle added strength (a: 400×, b: 900×). (Photos courtesy of Robin Giblin-Davis, professor, Department of Entomology & Nematology, University of Florida, Research and Education Center, Ft. Lauderdale, FL.)

suggests that rods are reoriented slightly through a small angle relative to rods in the plane lying above. The shift in orientation of the chitin rods produces a helicoid pattern when oblique sections are cut through the cuticle. When the rotation has passed through 180°, the result is a lamella of cuticle.

Cross sections and freeze-fracture of insect cuticles often show a "plywood-like" arrangement (Figure 4.16a and b). The plywood structure occurs when chitin rods do not shift during the formation of many overlying layers of cuticle, then they shift rapidly through 90° in a few thin lamella, and finally do not shift again while another thick layer of cuticle is laid down. Manufactured plywood sheets are designed in this way for added strength, and one can assume that similar strength is imparted to cuticle with this arrangement.

4.5.2 Biosynthesis of Chitin

Information as to how chitin is synthesized has been obtained from a variety of invertebrates, including insects, and from fungi. The process is still not completely understood in all of these organisms. The starting point for synthesis of chitin is β-D-glucose (Figure 4.17), although the β-D-glucose may come from a storage form, such as trehalose or glycogen. The latter two compounds are formed from α-D-glucose, but a rapid and dynamic isomerization of glucose in aqueous solution occurs so that at any given moment there is 37% α-D-glucose and 63% β-D-glucose. Any removal of one form, for example, by synthesis into a polysaccharide, rapidly leads to a new equilibrium. A general review of biosynthesis has been presented by Cohen (1987). Initially, glucose is phosphorylated at the expense of adenosine triphosphate (ATP) (Figure 4.18) and then isomerized to fructose-6-phosphate. An amino group is transferred from glutamine to fructose-6-phosphate to form glucosamine-6-phosphate. The latter molecule is acetylated, probably with acetyl CoA contributing the acetyl group, to form *N*-acetyl-glucosamine-6-phosphate. A transfer of the phosphate group from carbon 6 to carbon 1 is necessary to form *N*-acetyl-glucosamine-1-phosphate, which reacts with uridine triphosphate and forms uridine diphospho-*N*-acetyl-glucosamine (UDPGlcNAc). The detailed steps (likely there are several) between UDPGlcNAc and the linking of *N*-acetyl-β-D-glucosamine units together with β-1,4 linkages by the enzyme chitin synthetase are poorly clarified. Difficulties have been encountered in purifying and maintaining the stability of the enzymes needed in the final polymerization step.

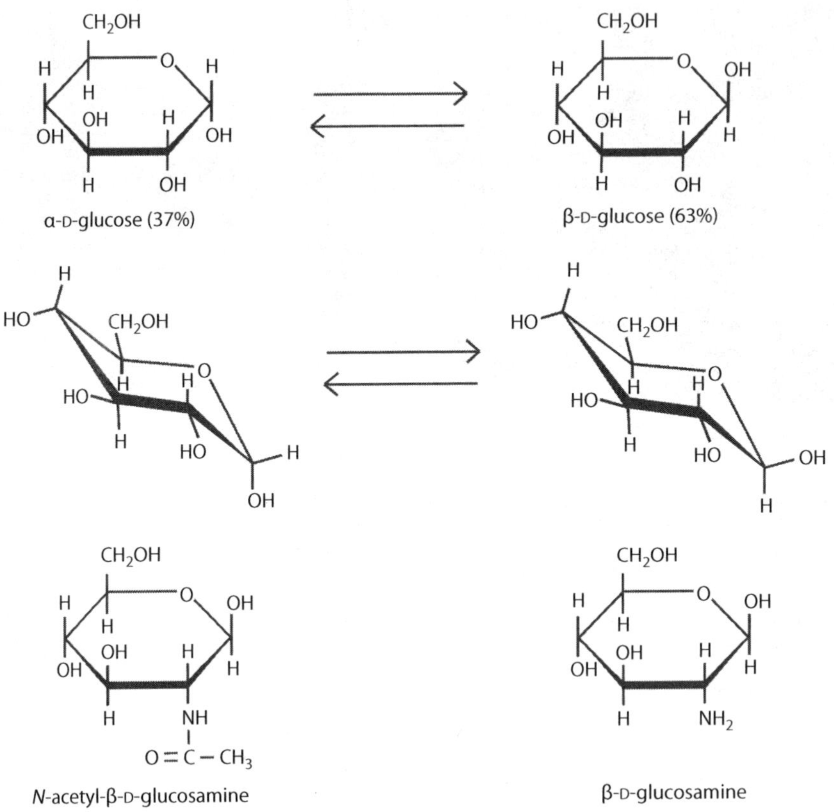

α-D-glucose (37%)   β-D-glucose (63%)

N-acetyl-β-D-glucosamine   β-D-glucosamine

Figure 4.17 An illustration of two ways of representing the structure of α-D-glucose and β-D-glucose, the proportions of each in the dynamic equilibrium that always exists in aqueous solution, and the structures of N-acetyl-β-D-glucosamine and β-D-glucosamine, all precursors of chitin.

The steps up to and including the formation of UDPGlcNAc have been verified in a number of insects (Kramer and Koga, 1986) and are similar in crustaceans, insects, and fungi. Little is known about the specifics of the final steps in chitin synthesis in any organism. Probably one intermediate step involves transfer of the N-acetyl-glucosamine residue to a lipid to form dolicyl-diphosphate N-acetyl-glucosamine (Turnbull and Howells, 1982, 1983), as has been described in integumentary tissue from the sheep blowfly, *Lucilia cuprina*. A similar process has also been described from crustaceans. Chitin synthase, necessary for the final step(s) has been prepared from insects as a large complex of proteins, some of which are very unstable in isolated form. The enzyme is present in microsomal preparations from epidermal cells, integument, and gut; multiple forms of chitin synthase may occur in different tissues and insects (Kramer and Koga, 1986). It may exist as a zymogen that requires activation when chitin synthesis is underway (Mayer et al., 1980).

4.5.3 Cuticular Proteins

Cuticular proteins, as well as chitin, are synthesized in the epidermal cells and released as an amorphous secretion around the apical microvilli of the epidermal cells, whereupon the chitin and proteins appear to self-assemble into fibrils, with proteins filling in the matrix around chitin rods (crystallites). In hard cuticles, the proteins are stabilized (sclerotized, cross-linked) by phenolic and quinone compounds that form covalent bonds and cross-link proteins to each other (and possibly to chitin),

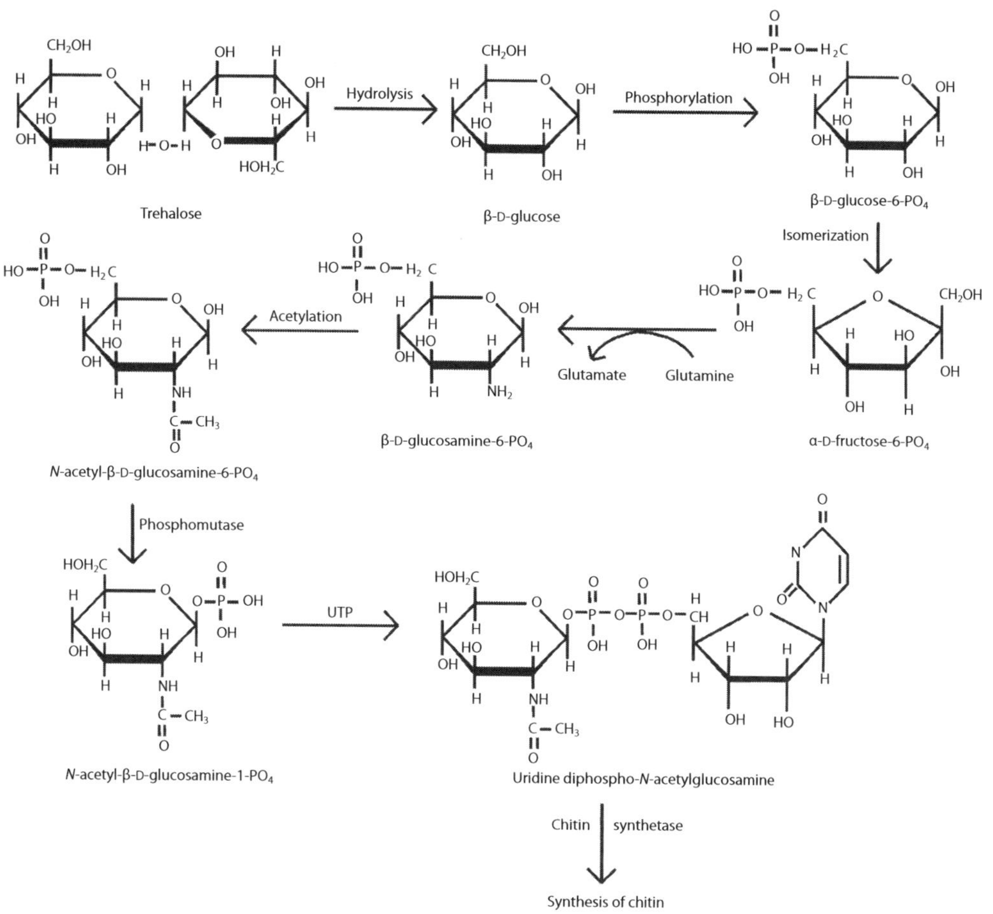

Figure 4.18 A general and tentative scheme for the biosynthesis of chitin.

forming a very hard, rigid structure. Even in very soft cuticles there is some degree of stabilization, but probably relatively few cross-links. Cuticle proteins in proximity to chitin rods may be bound to chitin by quinones or hydrogen bonds or both. Protein and chitin bound together have never been extracted from cuticle, however, because the degradative procedures necessary to extract proteins or chitin from cuticle break any potential protein–chitin bonds. After the cuticular proteins are sclerotized, they are difficult or impossible to extract without extensive degradation resulting from treatments needed to break the phenolic cross-links and breaking of peptide bonds within proteins.

Current research clearly indicates that there are many cuticular proteins. A good review of cuticular proteins with many amino acid sequences of cuticle proteins known at the time of the review was made available by Andersen et al. (1995). Many more proteins and sequences have been identified in the intervening years (Andersen 1998, 2000), and a cuticular protein database sequences is available at http://bioinformatics2.biol.uoa.gr/cuticleDB/index.jsp.

Additional protein sequences are available in the Prosite database at http://www.expasy.ch/prosite/ as "insect cuticle proteins signature" (PS00233) and in the Blocks database http://blocks.fhcrc.org/ as "insect_cuticle" (IPB000618) (Rebers and Willis, 2001). At least 100 electrophoretically separable proteins can be extracted from the cuticle of newly eclosed migratory locusts, *Locusta migratoria*, before the cuticle becomes sclerotized (Andersen et al., 1986). Proteins make up to

70% of the cuticle dry weight in locusts and about 90% of the proteins can be extracted prior to sclerotization, although only a few proteins can be extracted from hard or sclerotized cuticle. Using proteomic analyses, He et al. (2007) identified 295 putative cuticular peptides from the malaria mosquito, *Anopheles gambiae*. Kucharski et al., (2007) used the genomic database for honey-bees (http://racerx00.tamu.edu/bee_resources.html) to characterize three genes (*apd* 1, 2, and 3) expressing three proteins (APD 1, 2, and 3) named apidermins in which five amino acids comprise 74%–86% of their amino acid content. Apidermin 1 is found in the exoskeleton cuticle only during the late pupal stage and early adult when the cuticle is darkening. Apidermin 2 is found in the cuticle of tracheae, foregut, midgut, and in the embryo. Apidermin 3 overlaps the locations of APD 1 and 2, but particularly occurs in nonpigmented cuticle of the compound eyes and external cuticle of white pupae (early pupae before sclerotization and darkening of the cuticle).

Three genes (*BmGRP* 1, 2, and 3) that are induced by a pulse of 20-hydroxyecdysone were identified from *Bombyx mori* that encode three glycine-rich proteins in wing cuticle, and may contribute to cuticle of larvae, pupae, and adults (Zhong et al., 2006). Additional cuticle protein genes (the *BMWCPs*, Noji et al., 2003; and *BmCPG*1, Suzuki et al., 2002) have been identified that are inducible by 20-hydroxyecdysone and are expressed in different body regions and at different times in *B. mori*.

A conserved amino acid domain in seven cuticular proteins was noted by Rebers and Ridderford (1988) and has been found in many additional arthropod proteins as a 35- to 36-amino acid motif that is now generally described as the R&R consensus (Rebers and Willis, 2001). Andersen (1998) found that the consensus motif varied slightly in soft versus hard cuticles in the desert locust and proposed an RR-1 and RR-2 sequence for soft and hard cuticle, respectively, and the two groups are now often called the extended R&R consensus (Iconomidou et al., 1999). Andersen (1999) proposed a model in which the R&R consensus served to tie the proteins to chitin, that is, a chitin-binding domain. Rebers and Willis (2001) provided the first experimental evidence for binding by the RR-2 extended form of the R&R consensus to hard cuticle with the use of fusion proteins from *Anopheles gambiae* expressed in *Escherichia coli*, with the protein binding to chitin beads. Togawa et al. (2004) demonstrated that BMCP30, a purified protein from *B. mori* that has the RR-1 consensus sequence, bound chitin in a chitin-affinity assay. Thus, the evidence indicates that the extended R&R consensus is involved in protein to chitin binding, with RR-1 proteins typical of soft cuticle and RR-2 proteins typical in hard cuticle. However, Suderman et al. (2006) found an RR-1 cuticular protein in very hard, highly sclerotized and nonpigmented cuticle from *M. sexta*. Iconomidou et al. (2005) provide supportive evidence for the probable folding of the extended consensus region into an antiparallel β-pleated sheet that is bound to chitin.

Some studies show that certain cuticle proteins are specific to anatomical structures or body regions, some are specific to certain stages of development, and some are even specific to age within a stage. Proteins of soft cuticles often are different from proteins in hard cuticles in ways other than the degree of sclerotization. According to Andersen (1998), the same proteins occur in cuticle laid down after ecdysis (i.e., the procuticle) in the thorax of *Schistocerca gregaria* as in the intersegmental membranes of males and females, but the thoracic cuticle is hard and stiff to support the flight musculature, while the cuticle of intersegmental membranes has different properties. In males, the intersegmental membranes are tough, flexible, and not extensible, but in females the intersegmental membranes are viscoelastic and stretchable up to 10 times their original length. The viscoelasticity of these membranes is critical to egg laying behavior in locusts, which work the abdomen into the soil several centimeters, stretching the abdomen, before laying a pod of eggs. Proteins from soft cuticles frequently have a high content of polar amino acids, while proteins in the harder or more sclerotized cuticles have a higher content of hydrophobic amino acids. Locust proteins from regions of flexible cuticle have isoelectric points between 4.4 and 5.0, and these

INTEGUMENT AND MOLTING

proteins are missing from harder cuticle (Cox and Willis, 1985). In contrast, only minor differences in the proteins of segmental (hard) cuticle and intersegmental (flexible) cuticle occur in the silkmoth, *Antheraea polyphemus* (Sridhara, 1983).

4.5.4 Resilin

Resilin is an important structural protein of cuticle that is rubber-like, colorless, transparent, and insoluble in water. It has remarkable properties of elasticity, like rubber, but shows less deformation upon prolonged stretching than rubber. A stretched rubber band gets longer upon prolonged stretching, but, even after extended periods of stretching, resilin returns to about 97% of its original length. In order to be flexible, resilin obviously cannot be very sclerotized or have many cross-links with other protein chains, but dityrosine and trityrosine residues (Figure 4.19) provide a few internal cross-links, giving resilin chains some stability, yet allowing high elasticity. Resilin occurs in the wing hinges of some insects and in the hinge mechanism of the jumping leg of fleas. The prealar arm connecting the mesotergum to the first basalar sclerite of the thoracic wall in *S. gregaria* wings contains about 50 μg resilin and 15 μg chitin. The elastic tendon connecting the pleuro-subalar muscle to the ventral wall in dragonflies contains resilin. In large *Aeshna* spp., the tendon is about 0.7 mm long by 0.15 mm wide and contains 5–7 μg resilin. The main hinge ligament of the forewings of locusts located between the mesopleural wing process and the second axillary wing sclerite contains about 100 μg resilin and 20 μg chitin. Like such structural proteins as collagen, elastin, and silk fibroin, resilin is rich in glycine and proline, but contains no hydroxyproline, hydroxylysine, tryptophan, or sulfur-containing amino acids. Biosynthesis of resilin has not been extensively studied, probably because of the rather small areas in which it occurs, but it is believed to be secreted by specialized epidermal cells.

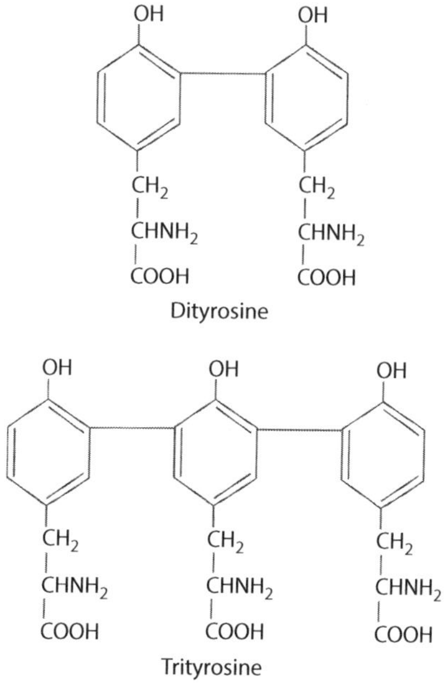

Figure 4.19 The chemical structure of dityrosine and trityrosine, two unique amino acids in resilin that are believed to be important in holding the protein chains together, but allowing for elasticity.

4.5.5 Stage-Specific Differences in Cuticle Proteins

Developmental biologists have frequently raised the question whether each stage in the development of an insect is controlled by specific genes or gene sets. There might be a larval set of genes, a pupal set, and an adult set. Somehow each set would be activated and then at the appropriate time deactivated. The identification of specific cuticular proteins at a particular stage of development might provide an opportunity for isolating and identifying the gene that controlled the proteins and, ultimately, of resolving the questions whether there are stage-specific genes. Regional specific, stage-specific, and temporal differences in the proteins of *Hyalophora cecropia* (Cox and Willis, 1985, 1987) have been found, but a *cecropia* silk moth larva also conserves proteins from some stages to the next as well as synthesizes new proteins specific to its current stage of development. A protein band has been found in the extracts of head and thoracic cuticle of *Drosophila* that does not occur in abdominal cuticle (Chihara et al., 1982).

Stage-specific proteins occur in *Drosophila melanogaster, Manduca sexta, Antheraea polyphemus, Bombyx mori,* and *Tenebrio molitor.* cDNA, mRNA, and electrophoretic analyses have been used to determine that there are different proteins in pupal and adult wing cuticle of *Antheraea polyphemus* (Sridhara, 1983, 1985). Cuticular proteins from larval, pupal, and adult cuticle of *Tenebrio* are specific for each stage (Lemoine and Delachambre, 1986). Electrophoretic and immunoblot analyses have been used to demonstrate different cuticular proteins in pupal and larval cuticles of the silkworm, *Bombyx mori* (Nakato et al., 1990).

In an apparent contradiction to much of the given evidence of stage-specificity of proteins, the cotton boll weevil, *Anthonomus grandis*, has a high degree of antigenic similarity among cuticular proteins of various stages (Stiles and Leopold, 1990). Many of the proteins are glycosylated, however, which may promote antigenic cross-reactivity and lead to false indication of similarity (Stiles, 1991). Thus, it is still not clear whether the proteins in the boll weevil are stage-specific.

In summary, some proteins in the cuticle appear to be stage specific, but also there is evidence that some proteins are common to several stages. The data do not cast much light upon the original question of whether there are specific gene sets governing each developmental stage. The same protein in more than one stage could simply mean that the gene coding for it is present in one or more sets. If there were no proteins that were the same, in say, a larva and the pupa, then there would be a stronger case for two sets of genes.

4.5.6 Protective Functions of Cuticle Proteins

Cuticular proteins in some or all insects may function in protection against invading bacteria or other organisms (Marmaras et al., 1993). When the surface of the epicuticle of silkworm larvae is lightly abraded and treated with bacteria (*Bacillus lichenformis* or *Enterobacter cloacae*) or bacterial cell wall compounds, mRNAs are produced that code for cecropin, a large protein with antibacterial properties. The response is systemic and cecropin appears in epidermal cells and fat body cells remote from the abrasion (Brey et al., 1993). Subsequently and as early as 8 h post abrasion, antibacterial activity appears in the matrix of the abraded cuticle, but not in unabraded cuticle. More intense abrasion of cuticle results in a wider distribution of the antibacterial activity including activity in fat body and hemolymph. It is not clear how surface abrasion of the cuticle and the introduction of bacteria into the abrasion is communicated to epidermal and fat body cells. See Chapter 16 Immunity for more details.

4.5.7 Cuticular Lipids

The lipids on the cuticle of insects have received a great deal of attention, not only because they promote water conservation, but because often they have additional behavioral, pheromonal, ecological, and taxonomic significance (Howard and Blomquist, 1982; Liepert and Dettner, 1996;

Doi et al., 1997; Darrouzet et al., 2014; Jennings et al., 2014). Among the lipids on the cuticle are true chemical waxes, hydrocarbons, alcohols, fatty acids, glycerides, sterols, ketones, aldehydes, and esters. Cuticular lipids are relatively easy to separate, identify, and quantitatively measure with gas chromatography combined with mass spectrometry (Bagnères and Morgan, 1990).

The lipid layer on the cuticle has often been described as a wax layer. In most insects only a small percentage of the lipids are true waxes, which are defined chemically as esters between long-chain alcohols and long-chain fatty acids. In most cases "wax layer" is clearly a misnomer. The quantity of waxes on the cuticle varies among insects and stages of development (Lockey, 1988). Only about 3% of the cuticular lipids of a weevil, *Ceutorrhynchus assimilis*, are true wax esters, but up to 74% of the lipids from the cuticle of the burrowing cockroach, *Arenivaga investigata,* are wax esters. Wax esters make up 24.4%, 29.2%, and 3.9% of the cuticular lipids of the larva, pupa, and adult, respectively, of the bean beetle, *Epilachna varivestis*. Honeybees have wax esters (34% of total lipids) on their cuticle and secrete several types of wax esters to make beeswax, which typically contains about 50% wax esters.

The lipid layer may lie beneath the cement layer, as many introductory entomology books suggest, but this is not always true. There is a mosaic pattern consisting of patches of lipids and cement on the cuticle of *Calpodes ethlius* (Locke, 1965). A tenebrionid beetle, *Cryptoflossus verrucosa*, that lives in the Sonoran Desert in the American southwest has a "wax bloom" on its surface. The wax bloom serves as camouflage, reduces water evaporation, and provides some thermal protection by restricting air movement at the cuticular surface (Hadley, 1979). The beetles change color depending upon how much wax is secreted to the surface. The wax is secreted mainly at low humidity from the tips of miniature tubercles at the cuticle surface, and the beetles are blue in low humidity and black at high humidity (Figure 4.20). The filaments of wax spread over the cuticle to form a fibrous meshwork about 20 μm thick and light reflected from the surface makes the beetles appear light bluish-white in color. Under high humidity conditions, the lipid filaments are not secreted and the beetles are black. The layering of the filaments and their thickness and the boundary air layer trapped between the meshwork and the surface of the cuticle probably retard transcuticular water loss and possibly reduces the rate of body heating by acting as a reflective shield. Bluish beetles lose 0.109 ± 0.032 mg/cm^2/h at 40°C and 0% relative

Figure 4.20 Color phases of the beetle *Cryptoglossa verrucosa* (Le Conte) (Coleoptera: Tenebrionidae) in response to humidity. The beetle is dark, nearly black (right) at higher humidity and blue (left) at low humidity. Wax filaments secreted during low humidity by tubercles on the elytra disperse the light and give the beetles a beautiful blue color. (The photos were taken in Bahia de Kino, Sonora, Mexico, and are courtesy of J. Nathaniel Holland, Department of Ecology and Evolutionary Biology, University of Arizona, Tucson, AZ.)

humidity (RH), while black beetles lose 0.140 ± 0.026 mg/cm^2/h (Hadley, 1979). Another insect with lipid at the surface of the cuticle is the eri silkworm, which covers itself with a white powder composed of two long-chain alcohols (Bowers and Thompson, 1965).

The type and quantity of lipids on the cuticle seem likely to be one of the many ways insects are adapted for the environment in which they live. For example, young larvae of *Sarcophaga bullata* (Diptera:Sarcophagidae), which live in very wet environments, have only small quantities of cuticular hydrocarbons on the cuticle. Pupae, representing a closed system to water, and adults subject to the drying influence of the air have greater quantities of surface lipids (Armold and Regnier, 1975). Diapausing pupae of *M. sexta* have a thicker lipid layer on the cuticle than non-diapausing ones (Bell et al., 1975), apparently providing greater protection from desiccation during the long diapause. Nymphs of the desert cicada, *Diceroprocta apache*, live underground and have smaller quantities of cuticular hydrocarbons than adults, which fly and live at a temperature close to 50°C (Hadley, 1980). The underground burrows or chambers in which nymphs live have lower and more stable temperatures, and probably high humidity, thus affording more protection from water transpiration than the air in which adults live. Season, and apparently temperature are correlated with the type and quantity of hydrocarbons on the cuticle of the desert tenebrionid, *Eleodes armata* (Hadley, 1977). Summer beetles have a greater quantity of hydrocarbons and more long-chain ones than winter beetles, but longer-chain molecules can be induced by holding winter beetles at 35°C for 5–10 weeks.

Hydrocarbons often are the major components on the cuticle, and many different compounds are usually present including straight-chain saturated compounds (alkanes), olefins (unsaturated alkenes), and alkanes with methyl branches (2-methyl, 3-methyl, and various internally branched alkanes are common) (Lockey, 1988). Molecules with complex branching and multiple double bonds occur on some cuticles. In many insects, cuticular hydrocarbons are species-specific, leading to the use of cuticular hydrocarbon analyses in taxonomy and systematics (Lockey, 1988, 1991; Page et al., 1990, 1997; Haverty et al., 1996, 1997; Caputo et al., 2007). Some caution must be exercised in interpreting hydrocarbon composition as species indicators, however, because there may be sexual, seasonal, stage-specific, or dietary-related variations in the cuticular hydrocarbons (Espelie et al., 1994). Cuticular lipids are "worn off" the cuticle and periodically replaced, especially in long-lived insects, and are lost with the epicuticle at each molt. At ecdysis, new lipids are secreted again just before and/or just after ecdysis of the old cuticle. Hydrocarbons are synthesized by abdominal oenocytes and/or epidermal cells in female *Blatella germanica* cockroaches and transported by lipophorins in the hemolymph to distribution sites in the body including the cuticle (Gu et al., 1995).

Insects have a large surface-to-volume ratio, and desiccation is a potential hazard for many insects. One of the major functions of the cuticle is to protect insects from losing excess water from the body by transevaporation across the cuticle and from imbibing water and flooding the body in the case of aquatic insects (Hadley, 1984). All parts of the integument, including the epidermal cells, are important in maintaining the impermeable nature of the cuticle, but the lipids and wax bloom at the surface of the cuticle are especially important in providing waterproofing (Noble-Nesbitt, 1991). Although the cuticle is relatively impermeable to water, 85% or more of the water lost from the body is lost through the cuticle.

Insects experience sudden, rapid water loss through the cuticle at a critical temperature (Ramsay, 1935; Wigglesworth, 1945), a "transition temperature" that varies with the species. The mechanisms involved in this sudden shift in water permeability are not yet understood, but it may be due in part to a reorientation of the lipid molecules on the cuticle surface. Although this phenomenon has been suspected to be related to changes in the lipids on the cuticle, experimental details have, until recently, been lacking. It has now been documented that increasing temperature does indeed modify the physical state (induces melting) of epicuticular lipids and disrupts packing of molecules on the cuticle as a waterproof barrier (Gibbs, 1995, 1998, 2011; Gibbs and Pomonis, 1995; Rouke and Gibbs, 1999). The critical transition temperature T_m was defined by Gibbs (1995) as the temperature causing 50% melting of cuticular hydrocarbons and subsequent disruption of molecular orientation. Melting and orientation depend upon carbon chain length, degree of branching, location of branching, and saturation (alkanes) or unsaturation (alkenes) in the mixture of hydrocarbons. Longer chain alkanes melt at higher temperatures than alkenes of the same carbon number, and *n*-alkanes (straight chains) melt at higher temperatures than branched chains with the same number of carbons. Internally branched hydrocarbons melt at lower temperatures than terminally branched alkanes (2- or 3-methylalkanes) (Gibbs, 1995; Gibbs and Pomonis, 1995). Rourke and Gibbs (1999) determined that the transition temperature for rapid water loss from the cuticle of the grasshopper, *Melanoplus sanguinipes,* occurs when the cuticular lipids are about 30% melted, and Rourke (2000) found that increased quantity and higher melting points of cuticular lipids, rather than loss of water from the tracheae, correlate with lower rates of body water loss. Increased water loss through the cuticle of the German cockroach, *Blattella germanica* L., occurs even when cuticular hydrocarbons are only about 5% melted (Young et al., 2000). Although the T_m does not correlate well with typical environmental temperature exposure of some insects, the type and packing of hydrocarbons on the cuticle does influence water loss and, in some cases, possibly at only a few degrees higher than the typical environmental exposure (Young et al., 2000).

4.6 MINERALIZATION OF INSECT CUTICLES

Insect cuticles generally do not incorporate significant quantities of minerals into their cuticles. One insect that does is the face fly, *Musca autumnalis*. About 63% of its puparium dry weight is ash, and the major mineral ions are calcium, phosphorus, and magnesium. The related housefly, *M. domestica*, and the stable fly, *Stomoxys calcitrans*, have only 3.65% and 2.31% ash, respectively, as dry weight of their puparia, but the major ions are still calcium, phosphorus, and magnesium, though in much lower quantity than in the face fly (Roseland et al., 1985).

4.7 CAPTURE OF ATMOSPHERIC WATER ON CUTICULAR SURFACES

A few insects have specialized regions of cuticle upon which water condensation from the atmosphere can occur, with the result that the insect can absorb the water into the body. The desert cockroach, *Arenivaga investigata*, can capture water vapor on the cuticular lining in the mouth, and the mealworm, *Tenebrio molitor,* and the thysanuran firebrat, *Thermobia domestica*, can absorb water across the rectal cuticle from moist air. *Tenebrio molitor* can absorb water across its rectal cuticle at a rate of 0.4 mg/cm^2/h in high humidities, but larvae lose water no faster than 0.08 mg/cm^2/h even at low humidity. A Namib desert tenebrionid, *Onymacris unguicularis*, captures water over its entire body surface, as it orients in a head-down position, at or near the top of a sand dune where water condenses on its body from wind-driven fogs at night. The vertical orientation of the body causes the water to run in droplets toward its mouth, and the beetle ingests the water.

4.8 REVIEW AND SELF-STUDY QUESTIONS

1. What are the different divisions of the cuticle that can be discerned with staining and other histological techniques?
2. What is the typical thickness of the epicuticle?
3. Describe the cuticle in (1) a beetle, (2) a caterpillar, and (3) a *Drosophila* larva.
4. What cells secrete the cuticle at molting?
5. What is the apolysial space? When is it formed?

6. What is the chemical composition of the molting fluid?
7. What is procuticle?
8. What is the wax layer? What is its chemical composition?
9. What does sclerotization mean? What occurs during sclerotization of cuticle?
10. What are the differences in structure and function between wax channels and pore canals?
11. What are the different roles for N-acetyldopamine and N-acetylglucosamine in the cuticle?
12. What is bursicon and what role does it have in the cuticle?
13. What are the differences between quinone tanning and beta-sclerotization?
14. What is diphenoloxidase?
15. What linkage occurs in chitin chains? What is the building block for chitin?
16. Are hydrogen bonds important in the cuticle? Support your answer.
17. Is chitin always the greatest component in the cuticle?
18. What is resilin and why is it important in the cuticle?
19. Describe the chemical nature of the cuticular lipids on and in the cuticle.

REFERENCES

Chihara, C.J., D.J. Silvert, and J.W. Fristrom. 1982. The cuticle proteins of *Drosophila melanogaster*: Stage specificity. *Dev. Biol.* 89: 379–388.

Cohen, E. 1987. Chitin biochemistry: Synthesis and inhibition. *Annu. Rev. Entomol.* 32: 71–93.

Cornell, J.C. and M.L. Pan. 1983. The disappearance of moulting fluid in the tobacco hornworm, *Manduca sexta*. *J. Exp. Biol.* 107: 501–504.

Cottrell, C.B. 1962. The imaginal ecdysis of blowflies. Detection of the blood-borne darkening factor and determination of some of its properties. *J. Exp. Biol.* 39: 413–430.

Cox, D.L. and J.H. Willis. 1985. The cuticular proteins of *Hyalophora cecropia* from different anatomical regions and metamorphic stages. *Insect Biochem.* 15(3): 349–362.

Cox, D.L. and J.H. Willis. 1987. Analysis of the cuticular proteins of *Hyalophora cecropia* with two dimensional electrophoresis. *Insect Biochem.* 17: 457–486.

Darrouzet, E., M. Labédan, X. Landré, E. Perdereau, J.P. Christidès, and A.G. Bagnères. 2014. Endocrine control of cuticular hydrocarbon profiles during worker-to-soldier differentiation in the termite *Reticulitermes flavipes*. *J. Insect Physiol.* 61: 25–33.

Dewey, E.M., S.L. McNabb, J. Ewer, G.R. Kuo, C.L. Takanishi, J.W. Truman, and H.-W. Honegger. 2004. Identification of the gene encoding bursion, an insect neuropeptide responsible for cuticle sclerotization and wing spreading. *Curr. Biol.* 14: 1208–1213.

Doi, M., T. Nemoto, H. Nakanishi, Y. Kuwahara, and Y. Oguma. 1997. Behavioral response of males to major sex pheromone component, (Z,Z)-5,25-Hentriacontadiene, of *Drosophila ananassae* females. *J. Chem. Ecol.* 23: 2067–2078.

Espelie, K., R.F. Chapman, and G.A. Sword. 1994. Variation in the surface lipids of the grasshopper, *Schistocerca americana* (Drury). *Biochem. Syst. Ecol.* 22: 563–575.

Frankel, G. and C. Hsiao. 1962. Hormonal and nervous control of tanning in the fly. *Science* 138: 27–29.

Frankel, G. and C. Hsiao. 1963. Tanning in the adult fly: A new function of neurosecretion in the brain. *Science* 141: 1057–1058.

Frankel, G., C. Hsiao, and M. Seligman. 1966. Properties of bursicon: An insect protein hormone that control cuticular tanning. *Science* 151: 91–93.

Gammie, S.C. and J.W. Truman. 1997. Neuropeptide hierarchies and the activation of sequential motor behaviours in the hawkmoth, *Manduca sexta*. *J. Neurosci.* 17: 4389–4397.

Gammie, S.C. and J.W. Truman. 1999. Eclosion hormone provides a link between ecdysis-triggering hormone and crustacean cardioactive peptide in the neuroendocrine cascade that controls ecdysis behaviour. *J. Exp. Biol.* 202: 343–352.

Gauthier, S.A. and R.S. Hewes. 2006. Transcriptional regulation of neuropeptide and peptide hormone expression by the *Drosophila dimmed* and *cryptocephal* genes. *J. Exp. Biol.* 209: 1803–1815.

Gibbs, A. 1995. Physical properties of insect cuticular hydrocarbons: Model mixtures and lipid interactions. *Comp. Biochem. Physiol.* 112B: 667–672.

Gibbs, A.G. 1998. Waterproofing properties of cuticular lipids. *Am. Zool.* 38: 471–482.
Gibbs, A.G. 2011. Thermodynamics of cuticular transpiration. *J. Insect Physiol.* 57: 1066–1069.
Gibbs, A.G. and J.G. Pomonis. 1995. Physical properties of insect cuticular hydrocarbons: Model mixtures and interactions. *Comp. Biochem. Physiol.* 112B: 243–249.
Giraud-Guille, M.M. and Y. Bouligand. 1986. Chitin-protein molecular organization in arthropods, in R. Muzzarelli, Ch. Jeuniaux, and G.W. Gooday (eds.), *Chitin in Nature and Technology.* Plenum Press, New York, pp. 29–35.
Gu, X., D. Quilici, P. Juarez, G.J. Blomquist, and C. Schal. 1995. Biosynthesis of hydrocarbons and contact sex pheromone and their transport by lipophorin in females of the German cockroach (*Blatella germanica*). *J. Insect Physiol.* 41: 257–267.
Hadley, N.F. 1977. Epicuticular lipids of the desert tenebrionid beetle, *Eleodes armata*: Seasonal and acclimatory effects on composition. *Insect Biochem.* 7: 277–283.
Hadley, N.F. 1979. Wax secretion and color phase of the desert tenebrionid beetle *Cryptoglossa verrucosa* (LeConte). *Science* 203: 367–369.
Hadley, N.F. 1980. Cuticular lipids of adult and nymphal exuviae of the desert cicada, *Diceroprocta apache* (Homoptera, Cicadidae). *Comp. Biochem. Physiol. B* 65: 549–553.
Hadley, N.F. 1984. Cuticle: Ecological significance, in J. Bereiter-Hahn, A.G. Matoltsy, and K. Sylvia Richards (eds.), *Biology of the Integument 1. Invertebrates.* Springer-Verlag, New York, pp. 685–693.
Haverty, M.I., M.S. Collins, L.J. Nelson, and B.L. Thorne. 1997. Cuticular hydrocarbons of termites of the British Virgin Islands. *J. Chem. Ecol.* 23: 927–964.
Haverty, M.I., B.L. Thorne, and L.J. Nelson. 1996. Hydrocarbons of *Naustitermes acajutlae* and comparison of methodologies for sampling cuticular hudrocarbons of Caribbean termites for taxonomic and ecological studies. *J. Chem. Ecol.* 22: 2081–2109.
He, N., J.M.C. Botelho, R.J. McNall, V. Belozerov, W.A. Dunn, T. Mize, R. Orlando, and J.H. Willis. 2007. Proteomic analysis of cast cuticles from *Anopheles gambiae* by tandem mass spectrometry. *Insect Biochem. Mol. Biol.* 37: 135–146.
Hopkins, T.L. and K.J. Kramer. 1992. Insect cuticle sclerotization. *Annu. Rev. Entomol.* 37: 273–302.
Hopkins, T.L., T.D. Morgan, Y. Aso, and K.J. Kramer. 1982. N-β-alanyldopamine: Major role in insect cuticle tanning. *Science* 217: 364–366.
Horodyski, F.M. 1996. Neuroendocrine control of insect ecdysis by eclosion hormone (mini-review). *J. Insect Physiol.* 42: 917–924.
Howard, R.W. and G.J. Blomquist. 1982. Chemical ecology and biochemistry of insect hydrocarbons. *Annu. Rev. Entomol.* 27: 149–172.
Iconomidou, V.A., J.H. Willis, and S.J. Hamodrakas. 1999. Is β-pleated sheet the molecular conformation, which dictates formation of helicoidal cuticle? *Insect Biochem. Mol. Biol.* 29: 285–292.
Iconomidou, V.A., J.H. Willis, and S.J. Hamodrakas. 2005. Unique features of the structural model of 'hard' cuticle proteins: Implications for chitin-protein interactions and cross-linking in cuticle. *Insect Biochem. Mol. Biol.* 35: 553–560.
Jenkin, P.M. and H.E. Hinton. 1966. Apolysis in arthropod moulting cycles. *Nature (Lond.)* 211: 871.
Jennings, J.H., W.J. Etges, T. Schmitt, and A. Hoikkala. 2014. Cuticular hydrocarbons of *Drosophila montana*: Geographic variation, sexual dimorphism and potential roles as pheromones. *J. Insect Physiol.* 61: 16–24.
Kim, Y.-J., I. Spalovská-Valachová, K.-H. Cho, I. Žitňanová, Y. Park, M.E. Adams, and D. Žitňan. 2004. Corazonin receptor signaling in ecdysis initiation. *Proc. Natl. Acad. Sci. USA* 101: 6704–6709.
Kingan, T.G. and M.E. Adams. 2000. Ecdysteroids regulate secretory competence in Inka cells. *J. Exp. Biol.* 203: 3011–3018.
Kramer, K.J., A.M. Christensen, T.D. Morgan, J. Schaefer, T.H. Czapla, and T.L. Hopkins. 1991. Analysis of cockroach oothecae and exuviae by solid-state ^{13}C-NMR spectroscopy. *Insect Biochem.* 21: 149–156.
Kramer, K.J. and D. Koga. 1986. Insect chitin, physical state, synthesis, degradation and metabolic regulation. *Insect Biochem.* 16: 851–877.
Kucharski, R., J. Maleszka, and R. Maleszka. 2007. Novel cuticular proteins revealed by the honey bee genome. *Insect Biochem. Mol. Biol.* 37: 128–134.
Lahr, E.C., D. Dean, and J. Ewer. 2012. Genetic analysis of ecdysis behavior in *Drosophila* reveals partially overlapping functions of two unrelated neuropeptides. *J. Neurosci.* 32: 6819–6829.
Lane, N.J. and H.B. Skaer. 1980. Intercellular junctions in insect tissues. *Adv. Insect Physiol.* 15: 35–213.
Lee, D., I. Orchard, and A.B. Lange. 2013. Evidence for a conserved CCAP-signaling pathway controlling ecdysis in a hemimetabolous insect, *Rhodnius prolixus*. *Front. Neurosci.* 7: 1–9.

Lemoine, A. and J. Delachambre. 1986. A water-soluble protein specific to the adult cuticle in *Tenebrio*. Its use as a marker of a new programme expressed by epidermal cells. *Insect Biochem.* 16(3): 483–489.

Li, B., R.W. Beeman, and Y. Park. 2011. Functions of duplicated genes encoding CCAP receptors in the red flour beetle, *Tribolium castaneum*. *J. Insect Physiol.* 57: 1190–1197.

Liepert, C. and K. Dettner. 1996. Role of cuticular hydrocarbons of aphid parasitoids in their relationship to aphid-attending ants. *J. Chem. Ecol.* 22: 695–706.

Locke, M. 1965. Permeability of insect cuticle to water and lipids. *Science* 147: 295–298.

Locke, M. 1984. Epidermal cells, in J. Bereiter-Hahn, A.G. Matoltsy, and K.S. Richards (eds.), *Biology of the Integument 1. Invertebrates*. Springer-Verlag, New York, pp. 502–522.

Locke, M. 2001. The Wigglesworth lecture: Insects for studying fundamental problems in biology. *J. Insect Physiol.* 47: 495–507.

Lockey, K.H. 1988. Lipids of insect cuticle: Origin, composition and function. *Comp. Biochem. Physiol.* 89B: 595–645.

Lockey, K.H. 1991. Insect hydrocarbon classes: Implications for chemotaxonomy. *Insect Biochem.* 21: 91–97.

Loveall, B.J. and D.L. Deitcher. 2010. The essential role of bursicon during *Drosophila* development. *BMC Dev. Biol.* 10: 92

Luan, H., W.C. Lemon, N.C. Peabody, J.B. Pohl, P.K. Zelensky, D. Wang, M.N. Nitabach, T.C. Holmes, and B.H. White. 2006. Functional dissection of a neuronal network required for cuticle tanning and wing expansion in *Drosophila*. *J. Neurosci.* 26: 573–584.

Luo, C.-W., E.M. Dewey, S. Sudo, J. Ewer, A.Y. Hsu, H.-W. Honegger, and A.J.W. Hsueh. 2005. Bursicon, the insect cuticle-hardening hormone, is a heterodimeric cysteine knot protein that activates G protein-coupled receptor LGR2. *Proc. Natl. Acad. Sci. USA* 102: 2820–2825.

Makki, R., E. Cinnamon, and A.P. Gould. 2014. The development and functions of oenocytes. *Annu. Rev. Entomol.* 59: 405–425.

Marmaras, V.J., S.N. Bournazos, P.G. Katsoris, and M. Lambropoulou. 1993. Defense mechanisms in insects: Certain integumental proteins and tyrosinase are responsible for nonself-recognition and immobilization of *Escherichia coli* in the cuticle of developing *Ceratitis capitata*. *Arch. Insect Biochem. Physiol.* 23: 169–180.

Mayer, R.T., A.C. Chen, and J.R. DeLoach. 1980. Characterization of a chitin synthase from the stable fly, *Stomoxys calcitrans* (L.). *Insect Biochem.* 10: 549–556.

Miles, C.I. and R. Booker. 1998. The role of the frontal ganglion in the feeding and eclosion behavior of the moth *Manduca sexta*. *J. Exp. Biol.* 20: 1785–1798.

Moussian, B. 2010. Recent advances in understanding mechanisms of insect cuticle differentiation. *Insect Biochem. Mol. Biol.* 40: 363–375.

Nakato, H., M. Toriyama, S. Izumi, and S. Tomino. 1990. Structure and expression of mRNA for pupal cuticle protein of the silkworm, *Bombyx mori*. *Insect Biochem.* 20: 667–678.

Neville, C. 1986. The biology of the arthropod cuticle. *Carolina Biol. Reader* 103: 16.

Noble-Nesbitt, J. 1991. Cuticular permeability and its control, in K. Binnington and A. Retnakaran (eds.), *Physiology of the Insect Epidermis*. CSIRO Publications, Melbourne, Victoria, Australia, pp. 252–283.

Noirot, C. and A. Quennedey. 1974. Fine structure of insect epidermal glands. *Annu. Rev. Entomol.* 19: 61–80.

Noji, T., M. Ote, M. Taketa, K. Mita, T. Shimata, and H. Kawasaki. 2003. Isolation and comparison of different ecdysone-responsive cuticle protein genes in wing discs of *Bombyx mori*. *Insect Biochem. Molec. Biol.* 33: 671–679.

Page, M., L.J. Nelson, G.J. Blomquist, and S.J. Seybold. 1997. Cuticular hydrocarbons as chemotaxonomic characters of pine engraver beetles (*Ips* spp.) in the *grandicollis* subgeneric group. *J. Chem. Ecol.* 23: 1053–1099.

Page, M., L.J. Nelson, M.I. Haverty, and G.J. Blomquist. 1990. Cuticular hydrocarbons as chemotaxonomic characters for bark beetles: *Dendroctonus ponderosae, D. jeffreyi, D. brevicomin,* and *D. frontalis* (Coleoptera: Scolytidae). *Ann. Entomol. Soc. Am.* 83: 892–901.

Park, J.H., A.J. Schroeder, C. Helfrich-Förster, F.R. Jackson, and J. Ewer. 2003. Targeted ablation of CCAP neuropeptide-containing neurons of *Drosophila* causes specific defects in execution and circadian timing of ecdysis behavior. *Development* 130: 2645–2656.

Park, Y., Y.-S. Kim, and M.E. Adams. 2002. Identification of G protein-coupled receptors for *Drosophila* PRXaminde peptides, CCAP, corazonin, and AKH supports a theory of ligand-receptor coevolution. *Proc. Natl. Acad. Sci. USA* 99: 11423–11428.

Ramsay, J.A. 1935. The evaporation of water from the cockroach. *J. Exp. Biol.* 12: 373–383.

Rebers, J.E. and L.M. Riddiford. 1988. Structure and expression of a *Manduca sexta* larval cuticle gene homologous to *Drosophila* cuticle genes. *J. Mol. Biol.* 203: 411–423.

Rebers, J.E. and J.H. Willis. 2001. A conserved domain in arthropod cuticular proteins binds chitin. *Insect Biochem. Mol. Biol.* 31: 1083–1093.

Reynolds, S.E. 1980. Integration of behavior and physiology in ecdysis. *Adv. Insect Physiol.* 15: 475–595.

Roller, L., Y. Tanaka, and S. Tanaka. 2003. Corazonin and corasonin-like substances in the central nervous system of the Pterygote and Apterygote insects. *Cell Tissue Res.* 312: 393–406.

Roller, L., I. Žitňanová, L. Dai, L. Šimo, Y. Park, H. Satake, Y. Tanaka, M.E. Adams, and D. Žitňan. 2010. Ecdysis triggering hormone signaling in arthropods. *Peptides* 31: 429–441.

Roseland, C.R., M.J. Grodowitz, K.J. Kramer, T.L. Hopkins, and A.B. Broce. 1985. Stabilization of mineralized and sclerotized puparial cuticle of muscid flies. *Insect Biochem.* 15: 521–528.

Rourke, B.C. 2000. Geographic and altitudinal variation in water balance and metabolic rate in a California grasshopper, *Melanoplus sanguinipes*. *J. Exp. Biol.* 203: 2699–2712.

Rourke, B.C. and A.G. Gibbs. 1999. Effects of lipid phase transitions on cuticular permeability: Model-membrane and in situ studies. *J. Exp. Biol.* 202: 3255–3262.

Rudall, K.M. 1963. Chitin protein complexes of insect cuticle. *Adv. Insect Physiol.* 1: 257–311.

Shaik, K.S., Y. Wang, L. Aravind, and B. Moussain. 2014. The Knickkopf DOMON domain is essential for cuticle differentiation in *Drosophila melanogaster*. *Arch. Insect Biochem. Physiol.* 86: 100–106.

Sridhara, S. 1983. Cuticular proteins of the silkmoth, *Antheraea polyphemus*. *Insect Biochem.* 13: 665–676.

Sridhara, S. 1985. Evidence that pupal and adult cuticular proteins are coded by different genes in the silkmoth, *Antheraea polyphemus*. *Insect Biochem.* 15: 333–339.

Stiles, B. 1991. Cuticle proteins of the boll weevil, *Anthonomus grandis*, abdomen: Structural similarities and glycosylation. *Insect Biochem.* 21: 249–258.

Stiles, B. and R.A. Leopold. 1990. Cuticle proteins from the *Anthonomus grandis* abdomen: Stage specificity and immunological relatedness. *Insect Biochem.* 20: 113–125.

Suderman, R.J., N.T. Dittmer, M.R. Kanost, and K.J. Kramer. 2006. Model reactions for insect cuticle sclerotization: Cross-linking of recombinant cuticle proteins upon their laccase-catalyzed oxidative conjugation with cetechols. *Insect Biochem. Mol. Biol.* 36: 353–365.

Sugumaran, M. 1988. Molecular mechanisms for cuticle sclerotization. *Adv. Insect Physiol.* 21: 179–231.

Suzuki, Y., T. Matsuoka, Y. Iimura, and H. Fujiwara. 2002. Ecdysteroid dependent expression of a novel cuticle protein gene *BMCPG1* in the silkworm, *Bombyx mori*. *Insect Biochem. Mol. Biol.* 32: 599–607.

Togawa, T., H. Nakato, and S. Izumi. 2004. Analysis of the chitin recognition mechanism of cuticle proteins from the soft cuticle of the silkworm, *Bombyx mori*. *Insect Biochem. Mol. Biol.* 34: 1059–1067.

Truman, J.W. 1972. Physiology of Insect Rhythms. II. The silkmoth brain as the location of the biological clock controlling eclosion. *J. Comp. Physiol.* 81: 99–114.

Truman, J.W. 1978. Hormonal release of stereotyped motor programmes from the isolated nervous system of the *Cecropia* silkmoth. *J. Exp. Biol.* 74: 151–174.

Truman, J.W. 1985. Hormonal control of ecdysis, in G.A. Kerkut and L.I. Gilbert (eds.), *Comprehensive Insect Physiology, Biochemistry and Pharmacology*. Pergamon Press, Oxford, U.K., pp. 109–151.

Truman, J.W. and L.M. Riddiford. 1970. Neuroendocrine control of ecdysis in silkmoths. *Science* 167: 1624–1626.

Truman, J.W., P.H. Taghert, P.F. Copenhaver, N.J. Tublitz, and L.M. Schwartz. 1981. Eclosion hormone may control all ecdyses in insects. *Nature* 291: 70–71.

Turnbull, I.F. and A.J. Howells. 1982. Effects of several larvicidal compounds on chitin biosynthesis by isolated larval integuments of the sheep blow fly *Lucilia cuprina*. *Aust. J. Biol. Sci.* 35: 491–503.

Turnbull, I.F. and A.J. Howells. 1983. Integumental chitin synthase activity in cell-free extracts of larvae of the Australian sheep blow-fly, *Lucilia cuprina* and two other species of Diptera. *Aust. J. Biol. Sci.* 36: 251–262.

Vargas, H.C.M., L.C. Farnesi, A.J. Martins, D. Valle, and G.L. Rezende. 2014. Serosal cuticle formation and distinct degrees of desiccation resistance in embryos of the mosquito vectors *Aedes aegypti*, *Anopheles aquasalis*, and *Culex quinquefasciatus*. *J. Insect Physiol.* 62: 54–60.

Vincent, J.F.V. and U.G.K. Wegst. 2004. Design and mechanical properties of insect cuticle. *Arthropod Struct. Dev.* 33: 187–199.

Wigglesworth, V.B. 1945. Transpiration through the cuticle of insects. *J. Exp. Biol.* 21: 97–114.

Wilkin, M.B., M.N. Becker, D. Mulvey, I. Phan, A. Chao, K. Cooper, H.-J. Chung, I.D. Campbell, M. Baron, and R. MacIntyre. 2000. *Drosophila* dumpy is a gigantic extracellular protein required to maintain tension at epidermal-cuticle attachment sites. *Curr. Biol.* 10: 559–567.

Wise, S., N.T. Davis, E. Tyndale, J. Noveral, M.G. Folwell, V. Bedian, I.F. Emery, and K.K. Siwicki. 2002. Neuroanatomical studies of *period* gene in the hawkmoth, *Manduca sexta*. *J. Comp. Neurol.* 447: 366–380.

Yarema, C., H. McLean, and S. Caveney. 2000. L-Glutamate retrieved with the moulting fluid is processed by a gluamine syntetase in the pupal midgut of *Calpodes ethlius*. *J. Insect Physiol.* 46: 1497–1507.

Young, H.P., J.K. Larabee, A.G. Gibbs, and C. Schal. 2000. Relationship between tissue-specific hydrocarbon profiles and lipid melting temperatures in the cockroach *Blattella germanica*. *J. Chem. Ecol.* 26: 1245–1263.

Zhao, Y., C.A. Bretz, S.A. Hawksworth, J. Hirsh, and E.C. Johnson. 2010. Corazonin neurons function in sexually dimorphic circuitry that shape behavioral responses to stress in *Drosophila*. *PLoS One* 5:e9141.

Zhong, Y.-S., K. Mita, T. Shimada, and H. Kawasaki. 2006. Glycine-rich protein genes, which encode a major component of the cuticle, have different developmental profiles from other cuticle protein genes in *Bombyx mori*. *Insect Biochem. Mol. Biol.* 36: 99–110.

Žitňan, D. and M.E. Adams. 2000. Excitatory and inhibitory roles of central ganglia in initiation of the insect ecdysis behavioural sequence. *J. Exp. Biol.* 203: 1329–1340.

Žitňan, D. and M.E. Adams. 2012. Neuroendocrine regulation of ecdysis, in L.I. Gilbert (ed.), *Insect Ecology*. Elsevier, Amsterdam, the Netherlands, pp. 253–309.

Žitňan, D., Y.-J. Kim, I. Žitňanová, L. Roller, and M.E. Adams. 2007. Complex steroid-peptide-receptor cascade controls insect ecdysis. *Gen. Comp. Endocrinol.* 153: 88–96.

Žitňan, D., T.G. Kingan, J.L. Hermesman, and M.E. Adams. 1996. Identification of ecdysis-triggering hormone from an epitracheal endocrine system. *Science* 271: 88–91.

CHAPTER 5

Hormones and Development

PREVIEW

Insects have an external skeleton and, as they grow, it becomes too small. Consequently, all insects periodically secrete a new, more flexible exoskeleton that they can "grow into" inside the old one and then shed (molt, ecdysis) the old skeleton. The majority of insects also metamorphose into an adult form at the last molt. Molting and metamorphosis are under the control of hormones, with the brain as a master control gland. A few brain neurosecretory cells (NSCs) secrete prothoracicotropic hormone (PTTH) at an appropriate time in each instar to set in motion further hormonal and physiological events necessary for molting. The cues that stimulate NSC to secrete PTTH have been identified in only a few insects, but at least three types of stimuli are known in different insects, including an environmental stimulus (cold exposure in a diapausing insect), attaining a certain weight or size, and stretching of the abdomen in response to a large blood meal. PTTH passes down the axons of the secreting NSC to the corpora cardiaca (CC), small paired (and sometimes fused) masses of tissue of ectodermal origin just behind the brain. In the tobacco hornworm and possibly in other Lepidoptera, NSC axons terminate in the corpora allata (CA), small paired structures just posterior to the CC. PTTH is released from the CC (or CA in some or possibly all Lepidoptera) and is picked up by the circulating hemolymph. The CC and CA are neurohemal organs where neurosecretions are passed into the hemolymph. PTTH binds to receptors on the outer cell membrane of the prothoracic glands, and adenylate cyclase on the inner side of the cell membrane is activated. Adenylate cyclase converts adenosine triphosphate (ATP) into cyclic adenosine monophosphate (cAMP), the second messenger that sets in motion the cascade of reactions resulting in synthesis of ecdysone from cholesterol or one of the C-28 or C-29 plant sterols. Ecdysone is not stored in the prothoracic glands (PGL) but is secreted into the hemolymph as it is produced. It generally is considered a prohormone, and a 20-monooxygenase enzyme (present in many tissues, but not in the prothoracic gland cells) requiring cytochrome P450 rapidly converts ecdysone into the active hormone 20-hydroxyecdysone by adding the hydroxyl group at the C-20 position in the β-configuration (which is the rationale for the older name of β-ecdysone for 20-hydroxyecdysone). 20-Hydroxyecdysone is the molting hormone, although it cannot be absolutely concluded that ecdysone does not have hormonal activity itself. Several molecular structures similar to ecdysone and to 20-hydroxyecdysone with hormonal activity are known from different insects, and frequently, all the steroid hormones are described as ecdysteroids. Receptors on the epidermal cells are the targets for ecdysteroids in immature insects.

A number of actions are initiated by ecdysteroids, including mitosis and cell division of epidermal cells, apolysis (separation) of the old cuticle from the cells, secretion of molting fluid, and secretion of a new cuticle. Later, when holometabolous insects are about to pupate, many tissues express ecdysteroid receptors and become targets for reorganization into pupal and, finally, adult structures. Juvenile hormone (JH) is secreted in each instar prior to the peak of ecdysteroid secretion and modifies cuticle secreted so that an additional juvenile-type cuticle is secreted. When the insect is large enough to pupate, JH is present only in very small quantity, and the ecdysteroid molting hormone then causes a pupal cuticle to be secreted, with appropriate changes in various internal tissues as well. Subsequently, ecdysteroid secretion with little or no JH allows the epidermal cells to secrete an adult cuticle, and internal organs and tissues are also reorganized to reflect the adult stage. Each time ecdysteroids are secreted, PTTH is secreted first, and, in the immature stages, JH is also secreted ahead of the ecdysteroid peak. The exact stimuli and controls upon secretion of JH are still not well-defined, but nervous control is believed to be a major influence. Ecdysteroids act at the gene level by regulating or modifying expression of genes. The hormone binds to a receptor in the nucleus, and zinc fingers on the receptor bind the receptor–hormone complex to DNA. Several different receptor isoforms are known, and expression and number of receptors on the cell surface may be one of the ways that some cells respond to ecdysteroid, while others do not respond, or respond only at certain times, such as at pupation. Recent evidence indicates that in some of its functions, JH also acts at the gene level.

5.1 INTRODUCTION

Two critical and important physiological events in the life of insects are **molting** and **metamorphosis**. All insects molt periodically in order to grow, and all but a very few go through either gradual (no pupal stage) or complete (with pupal stage) metamorphosis to become an adult. How are these events in the life of all insects controlled? In many insects, perhaps in all, there is a circadian rhythm that initiates the hormonal and other physiological changes that occur in molting and metamorphosis (Schiesari et al., 2011). Molting and metamorphosis are not rapid changes in the same sense that response to many other daily encountered stimuli causes rapid movement away from, or attraction to, the source of the stimulus. The nervous system controls the latter type of rapid responses, but the hormonal system is better suited to control the slower physiological and biochemical changes requiring sustained stimulation needed in molting and metamorphosis. As in vertebrates, so in insects, the nervous system exerts control by secretion of neurohormones (Gilbert et al., 1988) and by nervous feedback over many and possibly all endocrine functions. Thus, nervous control of endocrine function antedates the split between vertebrate and invertebrate lines of evolution.

This chapter describes endocrine controls of growth, molting, and metamorphosis. Many other functions are under endocrine regulation and will be described in the appropriate subject chapters. Additional details on developmental hormones and hormones regulating other functions in insects can be found in books by Nijhout (1994), Gilbert et al. (1996b), and Gilbert (2012) and in recent reviews by Riddiford and Truman (1993), Bernardo and Dubrovsky (2012), Riddiford (2007, 2012), Hill et al. (2013), Jindra et al. (2013), Gilbert et al. (2000), Truman and Riddiford (2002), and Yamanaka et al. (2013).

5.2 HISTORICAL BEGINNINGS FOR THE CONCEPT OF HORMONAL CONTROL OF MOLTING AND METAMORPHOSIS

Stefan Kopeč (1917) published that the brain of gypsy moth caterpillars is necessary for successful pupation. His experiments involved surgically removing the brain from some larvae while performing sham surgery (incision made, but the brain not removed) on control larvae. A high percentage of the sham-operated larvae pupated, while brainless larvae usually failed to form pupae, although they continued to live. Kopec found that he also could isolate the posterior body region from brain influence by tying a silk ligature tightly around the body at various points posterior to the head. Regions posterior to a ligature failed to show the cuticular changes associated with pupation, while anterior to the ligature the cuticle changed to look more like pupal cuticle. If he removed the brain late in last instars, the brainless larvae pupated anyway, leading Kopec to suggest that the brain was necessary for successful pupation for only a short period of time. The latter experiments led to the concept of a **critical period**, a time period when the brain is necessary for its hormonal influence to be exerted.

Thus, although the idea that the brain controlled metamorphosis was current during the 1920s and 1930s, not much attention was given to this concept. Experiments by Fukuda (1940, 1944) on *Bombyx mori* led him to the conclusion that a secretion from the prothoracic region was necessary for pupation. Kopec and Fukuda were each partially correct; both brain and a gland in the prothorax are now known to be necessary for molting from one instar to the next and for successful pupation and transformation to the adult.

Kopec and Fukuda were each looking at different halves of a two-step endocrine mechanism regulating molting and metamorphosis. However, it turns out that a third critical step also is involved—the secretion of a third hormone, the **juvenile hormone (JH)**, from glands in the head modifies the type of molt. Identification of the **CA** as the source of this third hormone stems from classical extirpation and reimplantation experiments conducted by V.B. Wigglesworth on the reduviid blood-feeding bug *Rhodnius prolixus* in the 1930s. Wigglesworth (1936) first called the hormone from the CA an inhibitory hormone. When he implanted multiple CA into last instars of *Rhodnius*, the bugs molted into **supernumerary larvae** rather than changing into adults as expected. In this sense, it did inhibit metamorphosis. Later, Wigglesworth (1940) called the hormone the JH as it became clearer that it functioned in other insects and generally had a juvenilizing effect rather than a strictly inhibitory effect.

Finally, dichotomy surrounding the roles of the brain and prothoracic glands was resolved by Carroll Williams at Harvard University in a series of experiments. Williams (1947) designed experiments to test the idea that the **brain hormone** might activate the **prothoracic glands** to produce a **molting hormone**. Williams used pupae of a native silk moth, *Hyalophora cecropia*, for his experiments. These large pupae have an obligatory pupal diapause in which they survive the winter in the soil and leaf litter. Following a period of cold exposure at 5°C–10°C for at least 6 weeks, pupae will molt into adults after a few weeks at warm temperatures. Williams found that diapausing pupae could be induced to complete development even without chilling when an "active" brain (the brain from a pupa that had been chilled) was implanted. Williams was able to slice these large pupae in half, seal the abdominal half with a glass coverslip and wax, and implant either an active brain, bits of prothoracic gland, or both into the abdomen. Only when both brain and prothoracic gland tissue were implanted did these isolated abdomens metamorphose into adult abdomens with moth scales and adult reproductive structures. Thus, Williams (1947) established that both brain and prothoracic glands were necessary for adult development and that the secretion from the brain activated the prothoracic glands.

5.3 INTERPLAY OF PTTH, ECDYSTEROIDS, AND JUVENILE HORMONE CONTROL DEVELOPMENT

The necessity to grow a new cuticle, periodically shed the old cuticle, organize changes in internal organs, and finally metamorphose into an adult involves a surprising number of hormones, listed in Table 5.1. Each molt is under both nervous and hormonal control. Overall the brain is in control and initiates each molt by secreting **PTTH** from **brain NSCs** (Smith and Rybcynski, 2012). PTTH targets cells in the prothoracic gland and activates them to synthesize and secrete **ecdysone** or a closely related ecdysteroid molecule. **JH** synthesized in the CA is released in molts of early instars

Table 5.1 Hormones Involved in Molting and Metamorphosis

Hormone	Description
Prothoracicotropic hormone, PTTH	A neuropeptide secreted by brain neurosecretory cells. The target is prothoracic gland cells; it stimulates secretion of ecdysone or related ecdysteroids.
Ecdysone, ecdysteroids	The steroid molting hormones. The main target is epidermal cells; it has several molecular forms and promotes synthesis of new cuticle.
Juvenile hormone, JH	"Status quo" hormone that determines the type of cuticle secreted by epidermal cells and keeps ecdysone from forcing precocious metamorphosis; it has several molecular forms.
Corazonin	Neuropeptide from brain important in molting that initiates ecdysis action in moths and maybe other insects.
Preecdysis-triggering hormone, PETH	From Inka cells on tracheae, with ecdysis-triggering hormone (ETH) that initiates preecdysis I nervous and muscular action.
Ecdysis-triggering hormone, ETH	Secreted from Inka cells, with PETH that triggers preecdysis I; ETH stimulates EH from brain cells.
Eclosion hormone, EH	EH released into hemolymph; stimulates Inka cells to make massive secretion of PETH and ETH, leading to preecdysis II and ecdysis.
Crustacean cardioactive peptide, CCAP	Neuropeptide from nervous system, important in shedding old cuticle.
Bursicon	The tanning hormone, a neuropeptide from nervous system that promotes cuticle sclerotization.

Note: The hormones are listed in the approximate sequence in which they come into play in molting and ecdysis.

and modifies the type of cuticle secreted. There are molecular variants of PTTH, ecdysone, and JH in various insects, and more details will be given in the sections that follow.

Ecdysteroids combine with a receptor protein in the nucleus of cells, and the ecdysteroid-receptor complex binds to DNA and induces transcription of a few master genes. Transcripts from these few genes turn on a cascade of gene activity ultimately resulting in cell division in epidermal cells, secretion of molting fluid, and secretion of a new cuticle and (depending upon the stage) may result in numerous structural changes in morphology and physiology of internal organs, such as the nervous system, gut, and reproductive organs. The timing of secretion and quantity of JH at target cells modulates the action of ecdysteroid by influencing the nature of the molt, whether larval–larval, larval–pupal, or pupal–adult (Gilbert et al., 1996a). JH also determines whether major changes will occur in internal organs; usually little or no changes in internal morphology occur between larval molts, but major changes occur during transformation into pupa or adult.

JH is secreted in advance of the rise of ecdysteroid secretion in early instars of hemimetabolous insects, such as *Nauphoeta cinerea* (Figure 5.1a), and falls toward the end of the instar, allowing secretion of another nymphal cuticle. JH titers are very low or not measurable in the last instar, and this appears to be important in allowing the molt into an adult. In locusts small, premolt pulses of ecdysteroid secretion near the end of the last instar are important in inducing mitosis in wing pads, initiating growth of future flight muscles, and starting changes in male accessory glands before the molt actually starts. Subsequently, a major pulse of ecdysteroid in the continued absence of JH causes secretion of an adult cuticle with concomitant changes in internal structure and physiology characteristic of adult locusts.

JH also tends to be high, relative to ecdysteroid, during the early part of an instar in holometabolous insects, such as *Manduca sexta*, and falls only moderately just prior to each molt (Figure 5.1b). As a consequence of this hormonal interplay, the epidermal cells secrete another larval cuticle in the early instars. In the last instar, however, JH falls to a low level before the molt, partly in response to a decline in synthesis due to a decreased level of methyl transferase (Bhaskaran et al., 1986), the enzyme that adds the methyl group to JH acid, and to an increase in JH esterase (JHE) (Roe and Venkatesh, 1990), the enzyme that hydrolyzes JH. As a consequence, the JH level drops below detectable levels early in the last instar.

In *M. sexta*, there is only one peak of PTTH secretion in the penultimate instar but two peaks in the last instar. The first peak of PTTH induces a small peak of ecdysteroid that reprograms the

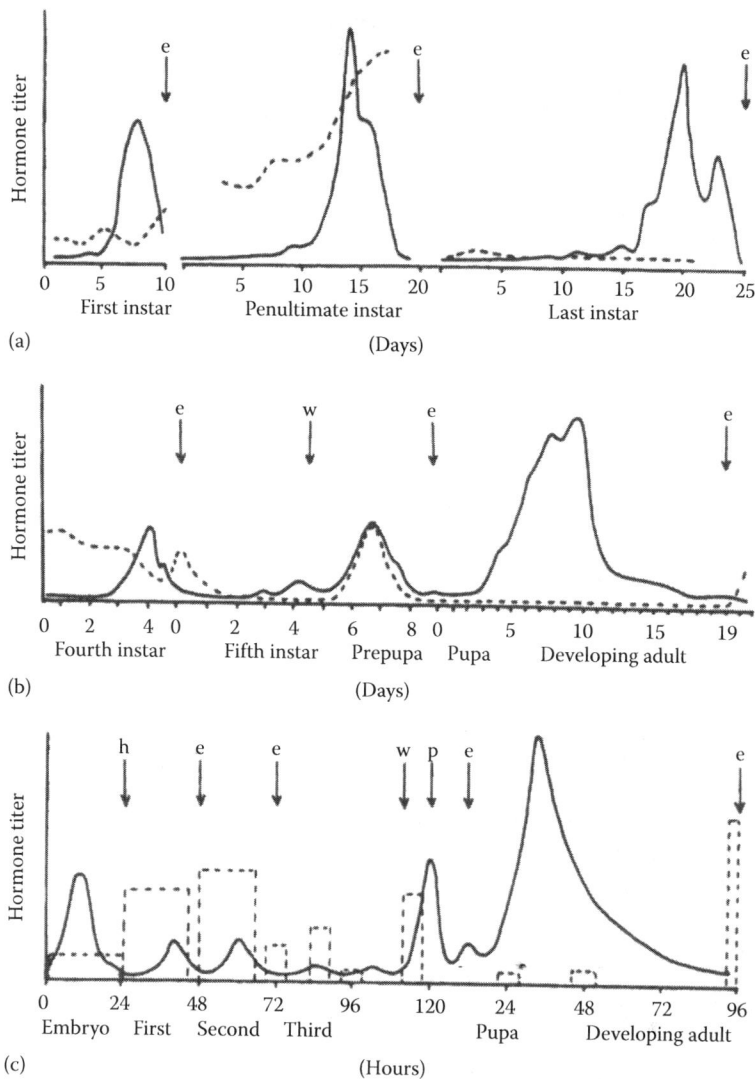

Figure 5.1 Ecdysteroid and JH titers during development of model insects. (a) The cockroach, *Nauphoeta cinerea*, with gradual metamorphosis. (b) *Manduca sexta*, the tobacco hornworm, with complete metamorphosis. (c) *D. melanogaster*, also with complete metamorphosis. In each case, the solid line indicates ecdysteroid titer and the broken line indicates JH titer. Key: e, ecdysis; w, wandering larva; p, pupation. (From Riddiford, L.M., *Adv. Insect Physiol.*, 24, 213, 1994. With permission.)

larval tissues and causes the epidermal cells to become committed to pupal development. This reprogramming peak of ecdysteroid can occur only when JH titer falls below a critical level because JH appears to act directly upon the brain to prevent PTTH secretion (Nijhout and Williams, 1974; Roundtree and Bollenbacher, 1986). Physiological changes in the nervous system cause cessation of feeding and induce wandering behavior, and metabolic changes occur in the fat body. When a larva finds a suitable pupation site, wandering behavior ceases, and a large release of ecdysteroid (in response to another release of PTTH) and a rise in JH cause the epidermal cells to secrete a pupal cuticle. Later, a rise in the level of JHE removes JH, allowing the large pupal pulse of ecdysteroid to promote adult development. Each level of ecdysteroid—the rising phase, peak, and falling phase—may be important to the responsiveness of certain cells and tissues in some or even all insects. For example, the DOPA (dihydroxyphenylalanine) decarboxylase gene in *Manduca* is regulated by decreasing ecdysteroid titer (Hiruma and Riddiford, 1993), and low ecdysteroid levels may play a role in decreasing the titer of JH at metamorphosis (Gu and Chow, 1993). Experimental elevation of ecdysteroid levels in some insects at times when the titer should be low or falling has detrimental effects.

There are six clearly defined pulses of ecdysteroid secretion during development of *Drosophila melanogaster* (Figure 5.1c) (Handler, 1982), each preceded by a pulse of **prothoracicotropic hormone (PTTH)** secretion. JH is secreted in conjunction with each peak of ecdysteroid except the last one in which the pupa molts to the adult. The first pulse of ecdysteroid in *D. melanogaster* occurs about 10 h into embryonic development. A second pulse of ecdysteroid during the first instar initiates secretion of a larval cuticle and the molt into the second instar. The third pulse of ecdysteroid causes the secretion of another larval cuticle and molting into the third (the last) instar. In contrast to the situation in a hemimetabolous insect, JH is produced in *Drosophila* in the early part of the last instar, but it falls late in the instar to a very low or nondetectable level. As part of many physiological and morphological changes induced by the fourth secretory pulse of ecdysteroid in the last instar, body shortening and pupariation to a prepupa are the most obvious. About 12 h after pupariation, a fifth pulse of ecdysone induces secretion of a pupal cuticle. Finally, a sixth and broad pulse of ecdysteroid is secreted in the presence of very low levels of JH in the pupa and adult development begins.

Exogenous JH administered prior to a critical period late in the penultimate or final instar of Hemiptera, Coleoptera, and Lepidoptera usually results in one or more supernumerary molts (Riddiford, 1996). For example, in *Tenebrio molitor*, two gene-encoding proteins specific to the adult cuticle are expressed normally in the pupal stage with the falling ecdysteroid titer. JH application to the newly ecdysed pupa prevents the appearance of the cuticle-specific proteins and a second pupa is formed instead of an adult (Riddiford, 1996). Some insects, such as Diptera and Siphonaptera, are unable to undergo supernumerary molts, and exogenous JH applied at a critical time late in the last instar usually results in death of the individual. Not only does JH interact in some unknown way to modify the type of cuticle secreted, there is evidence that JH can act directly on the brain to inhibit secretion of PTTH (Nijhout and Williams, 1974; Roundtree and Bollenbacher, 1986), and depending upon the stage of *M. sexta*, JH can both stimulate and inhibit synthesis of ecdysteroids by the prothoracic glands (Gilbert et al., 1996a). For example, in the lepidopteran *M. sexta*, there are two pulses of PTTH in the last instar: the first a small pulse and the second a larger, more sustained pulse. The first small peak of PTTH can occur only when JH titer falls below a critical level. A number of ideas have been advanced as to how JH may specify the nature of a molt, and more than one may be correct. For example, one possibility is that JH and JH receptor may interact with some transcription factors important in the regulation of continuing larval gene transcription during the intermolt. Another possibility is that JH might be involved in protein–protein interactions that stabilize chromatin configuration. When JH is low or absent, ecdysteroid may be able to destabilize the chromatin and open regions for new gene expression. Not all of JH action may be nuclear; JH might act on posttranslational processing or on translation (Nijhout, 1994; Riddiford, 1994, 1996).

Although the interplay of ecdysteroid and JH holds the fate of cells and tissues, not all cells respond in the same way or at the same time. Imaginal discs that give rise to adult structures do not grow in synchrony (Riddiford, 1994). In *Drosophila* and in tephritid fruit flies, the eye discs that make the compound eyes of adults are large and contain many cells even in the first instar, whereas the leg discs are too small to be located easily in early instars, but the latter grow rapidly in the last instar. Some cells and tissues, such as those in larval organs, are destined to die as the imaginal discs replace them with adult structures. Application of exogenous JH within the critical period, when some cells and tissues are still sensitive to its influence while others are not, can cause mosaic insects (Wigglesworth, 1940) that have a mixture of morphological and physiological characteristics representing two stages.

5.4 BRAIN NEUROSECRETORY CELLS AND PROTHORACICOTROPIC HORMONE

5.4.1 Source and Chemistry

Although the hormone from the brain that induces prothoracic glands to secrete molting hormone was initially called **brain hormone**, the more appropriate name is the **prothoraciotropic hormone (PTTH)**. This avoids confusion because, in addition to PTTH, the brain produces a number of hormones that regulate other functions in insects. PTTH is produced by **NSCs** in the brain (Figure 5.2a through c). In some insects, only a few large NSCs in the brain are involved in producing PTTH; only two NSCs in each hemisphere of the protocerebrum of *B. mori* were pinpointed as producers with immunofluorescent antibodies to PTTH. PTTH is not released directly from the brain, but passes down monopolar axons (Figure 5.2b) as electron-dense granules to a **neurohemal organ**, the general name for a structure from which neurosecretory hormones are released into the circulating hemolymph. The neurohemal organs that release PTTH in most insects are the bilaterally paired **CC**, although in some insects these organs are fused. In *M. sexta*, and perhaps in other or even all Lepidoptera, PTTH is released from the paired **CA** (Agui et al., 1980). The CA are paired bilaterally in Lepidoptera and some other insects but are fused in some groups of insects. Serving as a release site for hormones made elsewhere does not preclude a neurohemal organ from also synthesizing hormones. The CA, for example, release PTTH made in the brain, and parts of the gland also synthesize and release JH.

PTTH occurs in two different molecular sizes, a relatively small polypeptide of about 4.4 kDa and a large form of about 30 kDa. These often are designated as **4 K PTTH** and **30 K PTTH**. The 30 K PTTH was first thought to be about 22 kDa and was so described in some early reports. The 4 K PTTH can be further separated by electrophoresis into three species designated as **PTTH-I**, **PTTH-II**, and **PTTH-III**. Isolation of PTTH and identification of its polypeptide molecular structure occupied a number of laboratories for more than a decade before it was successfully isolated and identified from 648,000 heads of adult male *Bombyx* moths (Nagasawa et al., 1984). The 4 K PTTH currently is known as **bombyxin** because it was isolated from *B. mori*. It is part of a family of insulin-like-related peptides coded by genes expressed in four pairs of medial brain NSC whose axons terminate in the CA. Bombyxin is a dimer bonded by disulfide bonds and is related to the insulin family of polypeptides. Bombyxin does not have activity on *B. mori* or *M. sexta* but does have molt-inducing activity in the bioassay with *Samia cynthia ricini*. *Samia*, for several practical reasons, was usually used as the bioassay animal in the isolation of PTTH from *B. mori* extracts.

The 30 K PTTH is now established as the hormone that induces the prothoracic glands to produce ecdysteroid hormone. It exists as a homodimer with the two monomer units **glycosylated** and held together by inter- and intramonomer **disulfide bonds** (Kawakami et al., 1990). It has been isolated from the brain of *B. mori* and *M. sexta*. The two hormones are not particularly similar to each other, and the *B. mori* hormone has no hormonal activity in *M. sexta* or *D. melanogaster* (Gilbert et al., 1996a). There may be other species-specific PTTHs.

5.4.2 Bioassay for PTTH Activity

Each step in any successful isolation and identification of a natural product must be monitored with an assay. Initially at least, a biological response assay, or **bioassay**, is used because the chemistry is usually unknown. Even when chemical identity is known, the biological response usually is more sensitive than a chemical assay. The bioassay used in isolating PTTH was induction of adult development in a brainless (**Dauer**) pupa of another silk moth, *S. cynthia ricini,* after injection of active material. Dauer pupae were prepared by surgical removal of the brain from newly pupated *S. cynthia ricini*. This rendered them incapable of producing PTTH, of course, but left them alive

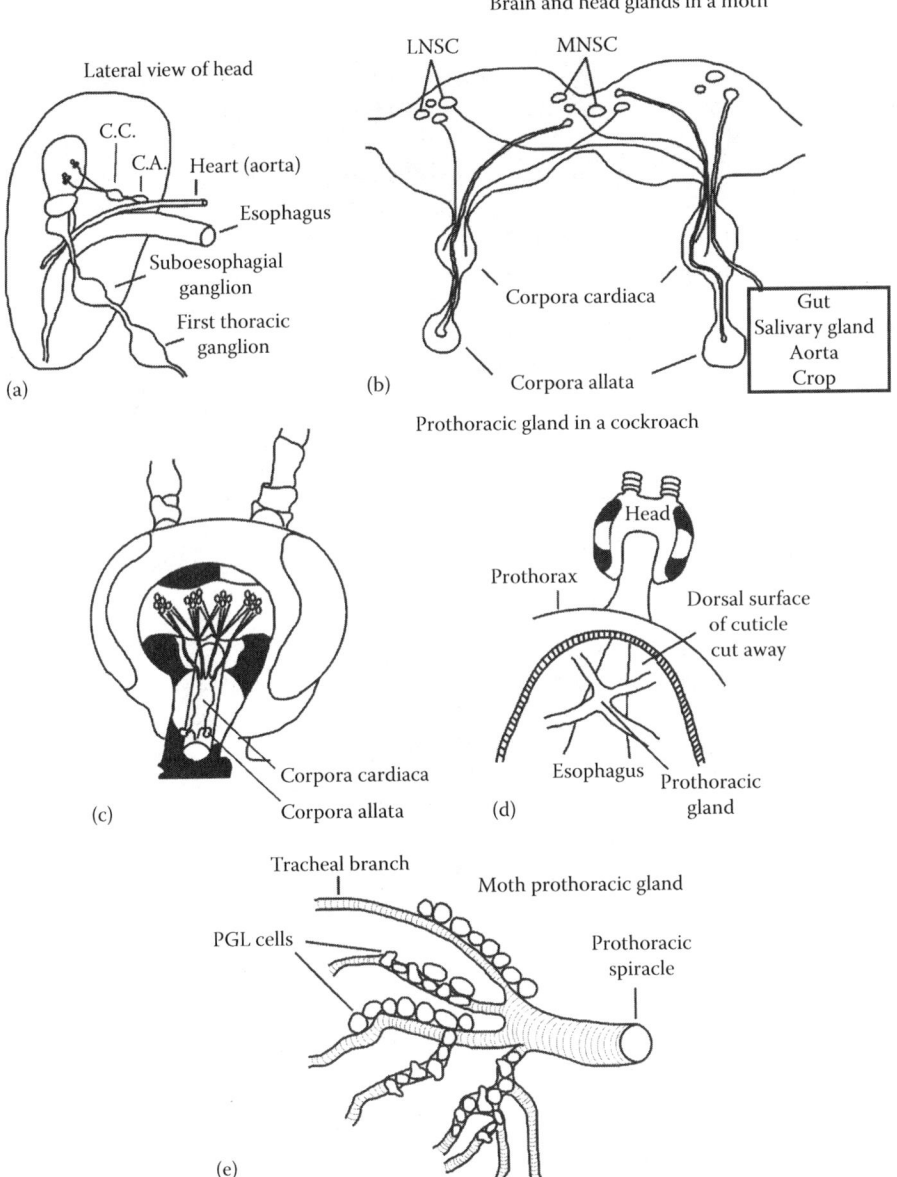

Figure 5.2 Representations of neuroendocrine structures in different insect groups. (a) Generalized concept of a lateral view of the three regions of the brain (protocerebrum, deutocerebrum, and tritocerebrum) with NSCs in the protocerebrum terminating in the CC and CA. (b) Cross-sectional view of the protocerebrum showing general positioning of medial and lateral NSCs and their axons to CC and CA typical of Lepidoptera larvae. (c) Cutaway view of the head of an American cockroach to show the fused CC and lateral CA lying on top of the aorta, with diagrammatic representation of neurosecretory cells in the protocerebrum of the brain with axons leading to CC and CA. (d) The "X"-shaped, nearly transparent prothoracic gland typical in an American cockroach nymph. (e) Prothoracic glands cells clustered around the larger tracheal branches near the prothoracic spiracle in a Lepidoptera larva.

for long periods of time in a suspended state of development ("Dauer" is a German word that roughly translates as "continuing without change"). These pupae needed a stimulus to the prothoracic glands to cause them to produce ecdysone in order to continue toward adult development. Even with exogenously added PTTH to stimulate the prothoracic glands, these pupae could not produce a new brain in the developing adult, so, of course, they would not be normal adults. Often they could successfully metamorphose into an adult, sometimes needing a little help from the investigator to completely get loose from the pupal cuticle. For an assay of PTTH, however, complete development is not necessary and the wait is undesirable; it is enough to see the beginning secretion of an adult cuticle, and particularly wing cuticle, as an indication that a preparation contains PTTH activity. In another assay, the sample to be assayed is added to a culture of *Samia* prothoracic glands. About 1×10^{-11} M PTTH, equal to about 40 ng of PTTH, activates the glands to produce ecdysone, which can be measured in several ways.

5.4.3 Stimuli for the Secretion of PTTH

What stimuli induce the brain to secrete PTTH and initiate the molting process? The answer is only partially known for a few insects, but those examples illustrate the diversity, adaptiveness, and complexity that characterize insect biology. In addition, they serve to stimulate ideas for investigating other possible control mechanisms in insects. In the known examples, the brain gets its cue (1) from the filled gut activating **stretch receptors** that send a message to the brain, (2) from some measurement of having attained a **critical size**, and (3) from cold exposure, and probably other **environmental stimuli** that can stimulate the brain. An example of each of these three mechanisms will be described in the next sections.

5.4.4 PTTH Secretion after Brain Activation by Stretch Receptors

Activation of stretch receptors in the gut and abdomen as a stimulus for PTTH secretion was first defined from *R. prolixus*, a blood-feeding reduviid bug. *Rhodnius* typically takes one large blood meal (if allowed to feed successfully to repletion) in each instar. It spends the rest of the instar digesting the meal and getting ready to molt into the next instar. By decapitating insects at various times after a blood meal and sealing the wound with wax, Wigglesworth and subsequent researchers showed that *Rhodnius* needed to retain its head (i.e., the brain) for about 3 days after feeding in the fourth instar and for about 5–7 days in the last (fifth) instar for a successful molt to be initiated. Thus, PTTH secretion was not an instantaneous event, but one that had to continue for a critical period of time. Apparently, the prothoracic glands need a sustained stimulus in order to secrete enough ecdysone to cause epidermal cells to begin to produce a new cuticle. The period during which the brain must secrete PTTH is called the **critical period** and was defined by Nijhout (1994) as "the point in time at which half the animals are able to complete a normal molt in the absence" of the brain. The duration of the critical period probably varies in different insects and, as evidence from *Rhodnius* suggests, in different stages of the same species. Clearly, the critical period cannot be as long in housefly pupae, which emerge as adults in about 3 days, as it is in *Rhodnius*. That the stimulus is stretching, and not nutritional, was shown by giving a nonfed *Rhodnius* an enema of saline. The saline enema stretched the gut and initiated the process of molting, while an intermittent series of small blood meals that never really stretched the abdomen did not allow molting to occur. The stretch receptors are **tonic receptors** located in one of the abdominal nerves. These stretch receptors do not adapt, but continue to fire bursts of nerve impulses while pressed against the abdominal wall by the filled gut. Abdominal stretching also is the activator of the brain in another reduviid bug, *Dipetalogaster maximus*, and the receptors are located in abdominal nerves that became stretched up to one and a half times the normal length during feeding (Nijhout, 1984). Stretching is important in the milkweed bug,

Oncopeltus fasciatus, but other undefined factors also influence molting. An injected saline enema can induce molting in milkweed bugs only after a sharply defined **critical weight** is attained (Nijhout, 1979).

5.4.5 Gated PTTH Secretion in Tobacco Hornworm

Attainment of a **critical body mass** or weight by *M. sexta* larvae is necessary for PTTH secretion. Secretion of PTTH occurs only during a well-defined temporal window, or gate, near the beginning of **scotophase** (dark period) after a larva reaches a critical weight. About one-third of fourth instars reared at 25°C with a 12:12 L:D cycle reach the critical size necessary to secrete PTTH about 36 h into the fourth instar. Secretion of PTTH in these larvae begins in the first hours after the beginning of the scotophase; they are designated as **Gate I larvae**. The two-thirds of the larvae that have not grown to the critical size to make Gate I must wait 24 h even if they reach the critical size before the next scotophase. They cannot begin to secrete PTTH until the scotophase, and these larvae are described as **Gate II larvae**.

The stimulus for PTTH secretion is probably a nervous system stimulus and may be due to stretching of some parts of the body, although this has not been proven. The secretion of PTTH is clearly related to internal physiology and the growth rhythm of the larva. Release of PTTH activates the prothoracic glands to produce ecdysone, and larvae molt into the fifth instar at the beginning of a photophase about 50 h after Gate I or II. During the 50 h between PTTH secretion and ecdysis, the epidermal cells are active in cell division, apolysis occurs, and new cuticle secretion begins as some of the old cuticle is digested.

Stretching is not indicated in *Drosophila* larvae, but growth of the prothoracic gland cells is mediated through **insulin/insulin-like growth factor**, and a critical gland weight must be achieved before metamorphosis can be initiated in *D. melanogaster* (Oldham and Hafen, 2003; Mirth et al., 2005).

5.4.6 Secretion of PTTH after Brain Activation by Cold Exposure

The brain is "made ready" to secrete PTTH by **exposure to low temperature** for a required minimum of days in a diapausing pupa of the *cecropia* silk moth, *H. cecropia*, but secretion of PTTH begins only some weeks after return of pupae to room temperature. How the cold activates or primes the brain has not been explained. Carroll Williams (who discovered this process in *cecropia*), along with his students, took advantage of the need for chilling by keeping *cecropia* diapausing pupae that were collected in the late summer in the refrigerator until needed for experiments. The requirement for a prolonged period of chilling to activate the brain to produce PTTH, followed by a period of warm exposure, is clearly adaptive for *H. cecropia*, which must get through the winter as an inactive pupa in the soil or leaf litter and emerge as an adult with the coming warm weather of late spring. Activation of the brain after only a few nights of cold exposure, followed by a few warm days, could have the moths emerging in the late fall, long before there would be any new leaves on the host trees for its caterpillars. Probably many other insects that must pass a period of dormancy may experience similar environmentally induced stimuli.

5.4.7 Regulation of Tissue and Hemolymph Levels of PTTH

Tissue and hemolymph levels of PTTH currently are difficult to measure. Availability of very sensitive radioimmunoassay (RIA) techniques should allow more data to be accumulated on regulation, but at present little is known about how the hemolymph and tissue levels of PTTH are regulated. Probably there are enzymes, such as proteinases, that degrade PTTH that is not bound to a receptor at the target tissue (prothoracic glands), and tissue sequestration and excretion may occur.

5.4.8 Mode of Action of PTTH

PTTH binds at the outer surface of **prothoracic gland cells** with Torso (target of rapamycin [TOR]), a transmembrane receptor (a tyrosine kinase) (Figure 5.3), initiating a complex network of molecular actions (Rewitz et al., 2009, 2013; Gu et al., 2013; Hsieh et al., 2013, 2014; Smith and Rybczynski, 2012). There is an increase in calcium ions, reactive oxygen species, phospholipase C (PLC), cAMP, and a number of downstream kinases. **PLC** and Ca^{2+} are involved in reactive oxygen stimulation (**ROS**) (Hsieh et al., 2013, 2014). **cAMP** in the prothoracic gland cells, produced by action of adenylate cyclase upon ATP, is a **second messenger**. At least part of the function of cAMP is to regulate Ca^{2+} ions in the prothoracic glands. A series of downstream **kinases**

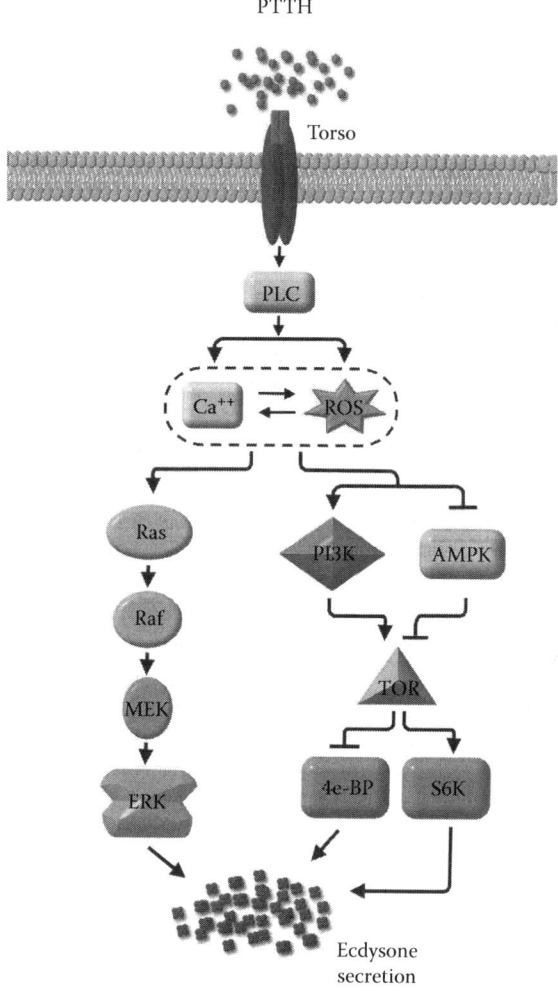

Figure 5.3 (See color insert.) A conceptional presentation of the signaling network involved in PTTH-stimulated ecdysteroidogenesis in prothoracic gland cells, involving reactive oxygen species, calcium ions, and a number of proteins. PTTH, prothoracicotropic hormone; torso, receptor protein for PTTH; Ras, Ras GTP-binding protein; ERK, extracellular signal-regulated kinase; cAMP, cyclic adenosine monophosphate; PI3K, phosphatidylinositol 3-kinase; TOR, target of rapamycin; 4E-BP, eIF4E-binding protein; S6K, p70 ribosomal protein S6 kinase; AMPK, adenosine 5/-monophosphate-activated protein kinase. (From Hsieh, Y.-C. et al., *J. Insect Physiol.*, 63, 32, 2014. With permission.)

is set in motion, ultimately resulting in synthesis of ecdysone (Gu et al., 2010, 2011, 2012, 2013; Marchal et al., 2010; Lin and Gu, 2011; Hsieh et al., 2014). **Kinases** are enzymes that are involved in phosphorylation reactions, a process that often activates proteins and enzymes. There are both stimulatory and inhibitory controls of some of the components in the pathway (Truman, 2006). Inactive cellular protein kinases are converted into active protein kinases (Smith et al., 1984, 1985, 1986; Smith 1993), reaching a maximum in about 5 min (Smith et al., 1986) and correlating well with rapid generation of cAMP. There is some evidence that PTTH may modulate or influence growth of the PGLs and may play a role in regulating ecdysone synthesizing enzymes (Gilbert et al., 1996a).

5.5 PROTHORACIC GLANDS AND ECDYSTEROIDS

The **prothoracic glands** secrete ecdysone or a closely related compound. The glands were described by Pierre Lyonet (1706–1789) in 1762, who apparently had no idea they were involved in molting or metamorphosis. Toyama (1902) redescribed the glands in silkworm larvae, and Ke (1930) suggested the name prothoracic glands. Cells of the prothoracic glands are derived from ectodermal tissue in the embryo. Prothoracic glands have a variety of shapes and names in different insects, and they have been called thoracic glands, or peritracheal glands in some insects and ventral glands in Ephemeroptera and Odonata. The glands in Lepidoptera consist of loose clusters of cells that are widely scattered in the prothorax but may even reach into the head region. In *H. cecropia* larvae, the glands consist of loose clusters of large cells (about 47×22 μm in fourth instars) scattered along the major tracheal branches near the prothoracic spiracle (Figure 5.2e). The cells receive neurons from the prothoracic ganglion and also from the subesophageal ganglion. There are 220 cells in the prothoracic glands of *M. sexta*. In cockroaches, the glands form the figure of an "X" in the prothoracic segment (Figure 5.2d). When activated by PTTH, the prothoracic glands secrete ecdysone or a closely related ecdysteroid. The prothoracic glands in the hemipteran, *R. prolixus*, have an endogenous photosensitive circadian oscillator that regulates ecdysteroid synthesis after a blood meal causes the release of PTTH, which acts as an entraining agent (Vafopoulou and Steel, 1999). The prothoracic gland cells of the cockroach, *P. americana*, do not have an endogenous circadian oscillator, but rhythmicity of ecdysteroid secretion during the photophase is controlled by secretion of PTTH during the scotophase (Richter, 2001). The brain in larvae of the blowfly, *Calliphora vicina*, contains extractable prothoracicotropic and prothoracicostatic compounds, and ecdysteroid synthesis seems likely to be under control of a complex of several factors with interacting and opposing activity (Hua et al., 1997). The prothoracic glands of most insects degenerate during or soon after metamorphosis, and possibly they, like some other tissues, are JH dependent in larval life (Gilbert et al., 1996a). Ecdysteroids are produced by adult female insects in the follicular epithelial cells of the ovary.

5.5.1 Biosynthesis of Ecdysone

The prothoracic glands sequester **cholesterol** from the circulating hemolymph and convert it into **ecdysone** or a closely related ecdysteroid. One pathway that has been described is shown in Figure 5.4, and a recent review by Gilbert et al. (2002) provides additional details. Cholesterol must be obtained from the diet; insects cannot synthesize it. The first step in synthesis of ecdysone is the conversion of cholesterol to 5,7-dehydrocholesterol. In *Locusta migratoria* and *M. sexta*, the enzyme responsible is a microsomal cytochrome P-450 monooxygenase requiring nicotinamide adenine dinucleotide phosphate (NADPH) (Kappler et al., 1988; Grieneisen et al., 1993). The 7-dehydrocholesterol is shuttled from the cytoplasm into the mitochondria where oxidative and hydroxylation steps occur that produce a diketol or a ketodiol, depending upon the sequence of

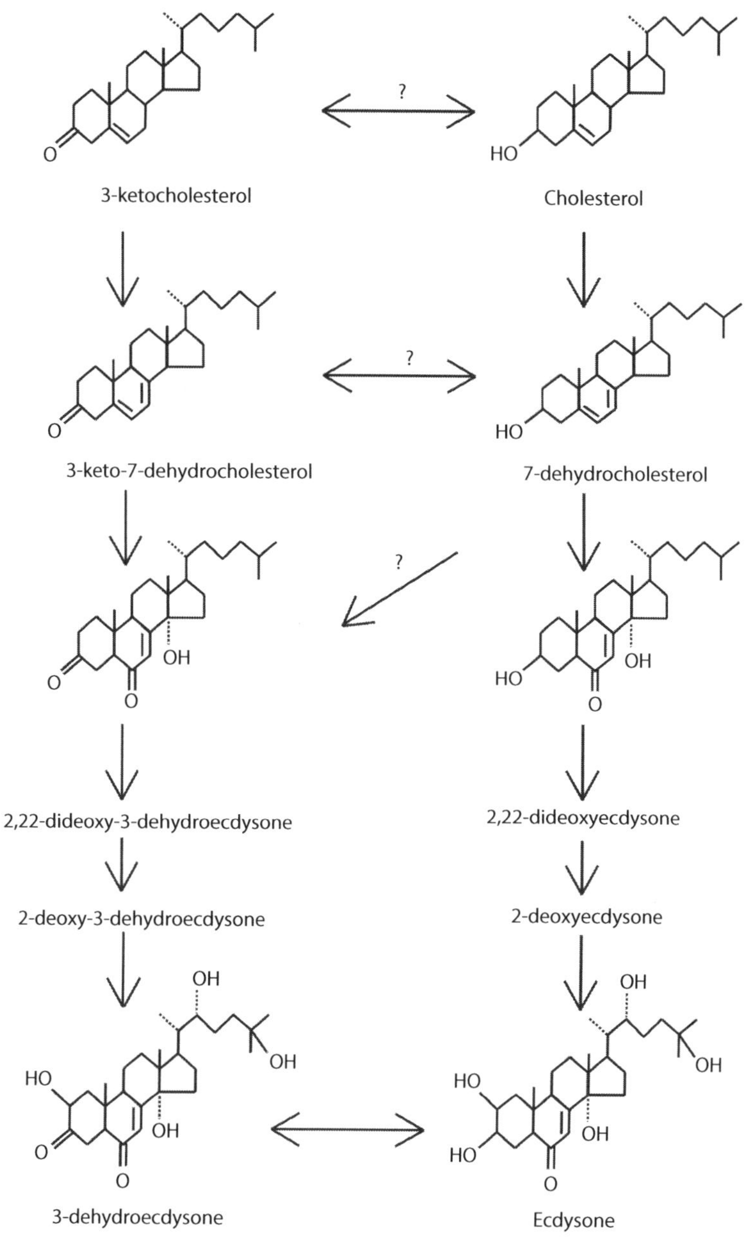

Figure 5.4 Possible biosynthetic routes to ecdysone and analogs in the prothoracic glands. (From Grieneisen, M.L. et al., *Insect Biochem. Mol. Biol.*, 23, 13, 1993. With permission.)

reactions at carbon 3. A mitochondrial membrane shuttle then moves the resulting trideoxyecdysteroid back to the endoplasmic reticulum for hydroxylation at carbon 25 by a microsomal cytochrome P450 enzyme. Finally, the shuttle returns the steroid to the mitochondrial compartment where the C-22 and C-2 hydroxyl groups are added by mitochondrial cytochrome P450 enzymes (Kappler et al., 1988; Grieneisen et al., 1993). The frequent shuttling between the endoplasmic reticulum and the mitochondria requires expenditure of ATP energy and is facilitated by a sterol carrier protein (Grieneisen et al., 1993). The end products, depending on the chemistry at C-3, are either 3-dehydroecdysone or ecdysone. Both compounds have been found in the prothoracic glands of several lepidopterans and in the Y-organ of some crustaceans. 3-Dehydroecdysone can be converted to ecdysone by reduction of the 3-keto group. In *M. sexta*, 3-dehydroecdysone is released from the prothoracic glands (Warren et al., 1988a,b), and it is reduced in the hemolymph to ecdysone by a ketone reductase (Sakurai et al., 1989).

5.5.2 Conversion of Ecdysone into 20-Hydroxyecdysone

The prothoracic glands do not store ecdysone, but secrete it into the hemolymph as it is made. Ecdysone seems to have low hormonal activity itself, although this is hard to gauge because it is rapidly converted to **20-hydroxyecdysone** (Figure 5.5; note that ecdysone and 20-hydroxyecdysone each have 27 carbons in their structure) by the **enzyme 20-hydroxymonooxygenase** present in most insect tissues. The Malpighian tubules, gut, and fat body are especially rich in 20-hydroxymonooxygenase. The enzyme, however, is not present in the prothoracic glands. In older literature, 20-hydroxyecdysone is also known as β-**ecdysone, ecdysterone**, and **crustecdysone**. 20-Hydroxyecdysone was first called β-ecdysone because the hydroxyl group at carbon 20 is in the β-configuration (β designation means that the OH group projects from the plane of the paper toward the reader). The name "crustecdysone" came from its isolation and description from crustaceans at a time when it was not known that its structure was identical to 20-hydroxyecdysone.

5.5.3 Molecular Diversity in the Structure of the Molting Hormone

Although cholesterol is the sterol obtained from the food by carnivorous insects, about half of all insects are plant feeders and, thus, have access to plant sterols that typically have 28 or 29 carbons in the molecule, while cholesterol has 27 (Figure 5.6). Phytophagous insects that have been studied usually dealkylate plant sterols to cholesterol. A few phytophagous insects (only a few have been studied) do not dealkylate the plant sterols but synthesize a C-28 or a C-29 ecdysteroid molting hormone. **Makisterone A** or **20-hydroxy-24-α-methyl ecdysone** (Figure 5.7) is the principal molting hormone of honeybees, and radiolabeled tracers have shown that they synthesize it from the plant sterol **campesterol**, a C-28 sterol with the C-24 methyl in the α-configuration. Makisterone A has also been detected in embryos of the milkweed bug, *O. fasciatus*, and in other hemipterans. A sensitive enzyme immunoassay with a detection limit of about 3 picograms (pg) for Makisterone A has been developed (Royer et al., 1993).

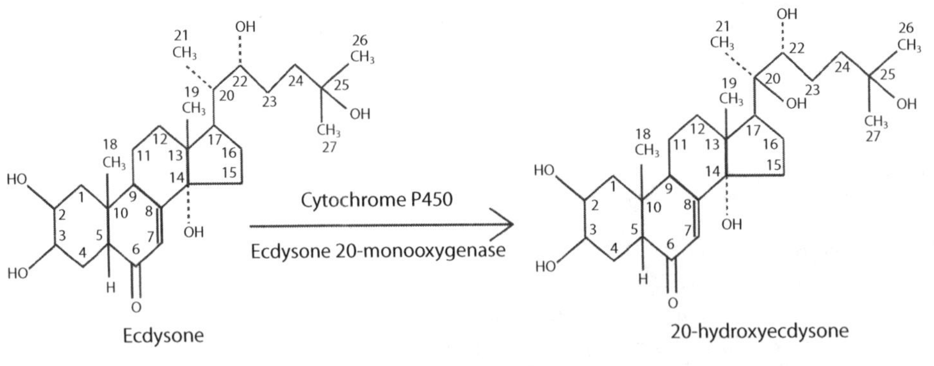

Figure 5.5 Structures of ecdysone, the principal prohormone produced by the prothoracic glands, and 20-hydroxyecdysone, the active hormone. Ecdysone is converted to 20-hydroxyecdysone by the enzyme ecdysone 20-monooxygenase, with participation of cytochrome P450. The reaction is characteristic of many tissues, but does not occur in prothoracic gland cells.

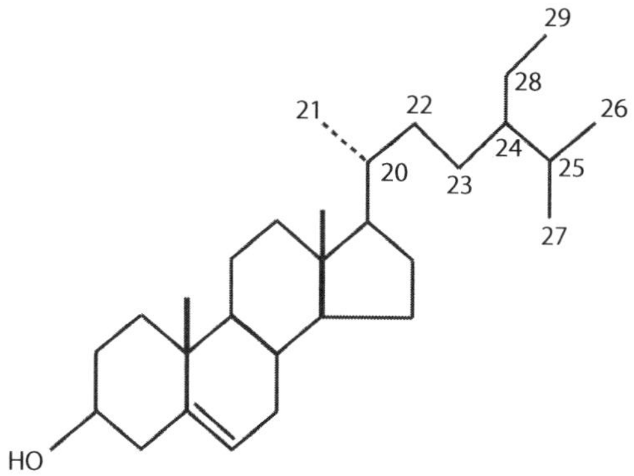

HORMONES AND DEVELOPMENT

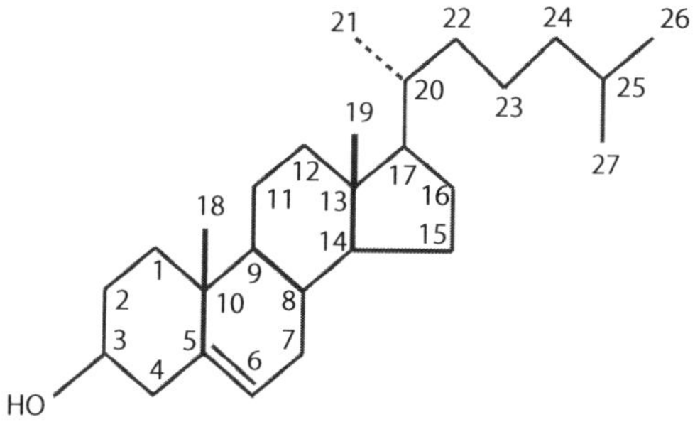

Figure 5.6 Two typical plant sterols—β-sitosterol (a 29-carbon sterol) and campesterol (a 28-carbon sterol)—and the typical animal sterol (cholesterol) with 27 carbons.

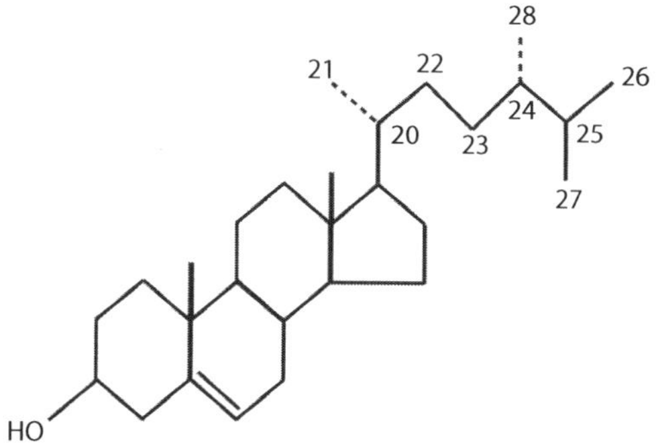

Leaf-cutting ants, *Acromyrmex octospinosus*, make **24-epi-Makisterone A (20-hydroxy-24-β-methyl ecdysone)** (Figure 5.8) from a sterol in a fungus they farm. The ants cut leaves, take them into their underground nest, and eat the fungus that grows on the leaves. The fungus contains a C-28 sterol with the C-24 methyl group in the β-configuration, from which the ants synthesize 24-epi-Makisterone A.

The cotton stainer bug, *Dysdercus fasciatus*, feeds upon plant sap and ingests the plant sterol, **sitosterol**, a C-29 sterol that it uses to make **Makisterone C**, a C-29 ecdysteroid, as its molting hormone (Figure 5.9).

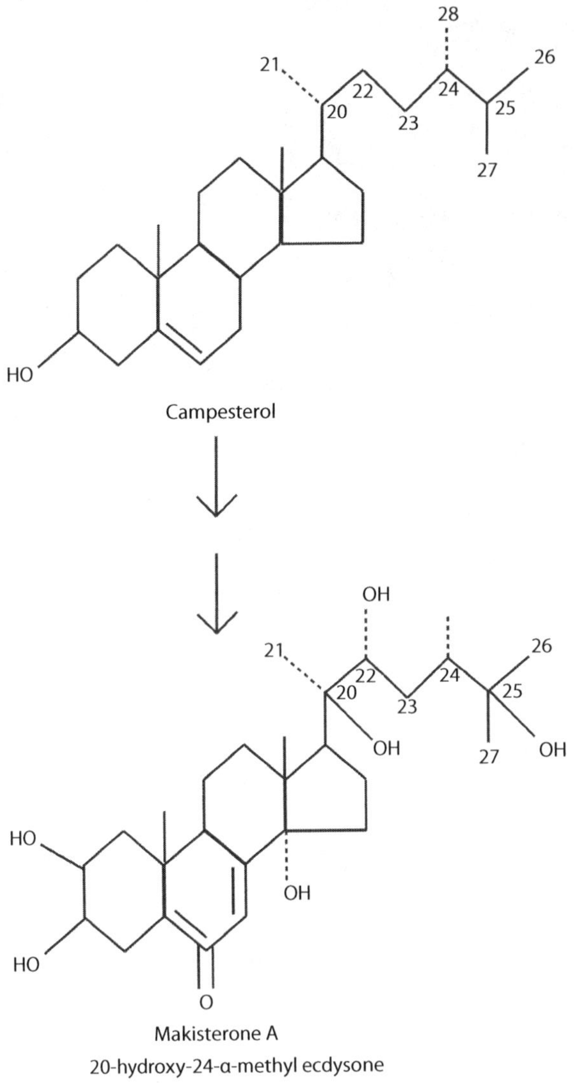

Figure 5.7 Conversion of campesterol to form the molting hormone, 20-hydroxy-24-α-methyl ecdysone, or Makisterone A, in honeybees, *Apis mellifera* L.

D. melanogaster can convert campesterol and sitosterol into cholesterol when reared aseptically on defined diets (3.3% and 8.1% cholesterol in tissues of insects raised on campesterol and sitosterol, respectively). Thus, the biochemical machinery to make the conversion of plant sterols may be widespread in insects (Feldlaufer et al., 1995). *D. melanogaster* contains small amounts of 3-dehydro-20-hydroxyecdysone and 3-dehydroecdysone, but the principal ecdysteroids in the flies are ecdysone and 20-hydroxyecdysone. A small amount of Makisterone A, probably formed from campesterol, is present in pupae. *M. sexta* embryos have small amounts of 20,26-hydroxyecdysone and 26-hydroxyecdysone. There is evidence of synthesis of ecdysteroids in isolated abdomens of *M. sexta* pupae.

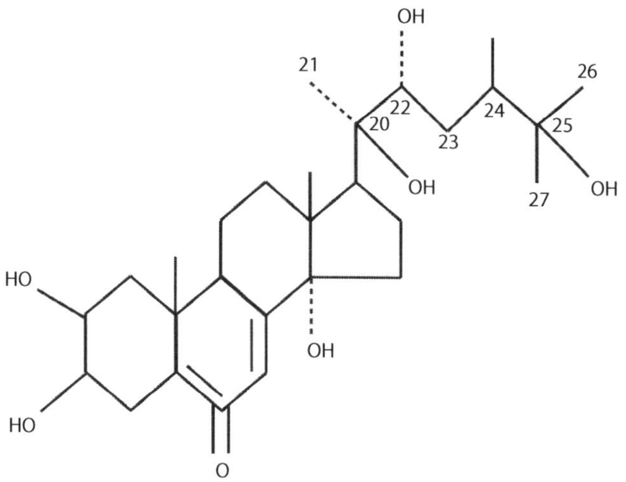

24-epi-Makisterone A

20-hydroxy-24-β-methyl ecdysone

Figure 5.8 24-epi-Makisterone A (20-hydroxy-24-β-methyl ecdysone), the molting hormone of leaf-cutting ants, *Acromyrmex octospinosus*.

5.5.4 *Calliphora* Assay for Ecdysteroids

The oldest bioassay for ecdysteroids is the ***Calliphora* assay** developed by Fraenkel (1935). In this assay, with late third instars of *Calliphora erythrocephala*, a blowfly, a ligature is placed around the body about one-third of the body length from the anterior end, so that the brain is isolated from chemical communication with the posterior part of the body. The ligature must be placed before the brain secretes PTTH, or at least before the ring gland secretes the molting hormone. The **ring gland** is a tissue in dipterous larvae that contains both ecdysone-secreting and JH-secreting cells (Figure 5.10). If larvae are successfully ligatured before secretion of ecdysone, then the portion of the larva posterior to the ligature will not shorten and form a puparium since it receives no ecdysteroid, whereas the anterior region proceeds to form a modified puparial cuticle because of ecdysteroid secreted by the ring gland.

The body region posterior to the ligature can be caused to form puparial cuticle by dipping the abdomen in a solution of ecdysone, by topical application of ecdysone, or by injecting ecdysone. In a typical assay, 20–30 larvae are treated as a group, and the percentage forming the puparial cuticle posterior to the ligature is recorded. One *Calliphora* unit was originally defined as that amount of hormone that would cause 50%–70% of the treated abdomens to pupariate, but after ecdysone and 20-hydroxyecdysone were isolated as crystals by Butenandt and Karlson (1954), a *Calliphora* unit was redefined as 0.01 µg pure ecdysone.

5.5.5 Radioimmunoassay for Ecdysone and Related Ecdysteroids

The most widely used method for determining ecdysteroid levels in tissue and hemolymph is a **radioimmunoassay (RIA)** (Borst and O'Connor, 1972). Improved RIA methods for ecdysone are sensitive to about 1–10 pg of ecdysone, or about 1000-fold the sensitivity of the *Calliphora* assay. Although the antibody molecules used in the original RIA method were not completely specific for either ecdysone or 20-hydroxyecdysone, more specific antibody preparations and refined procedures have increased

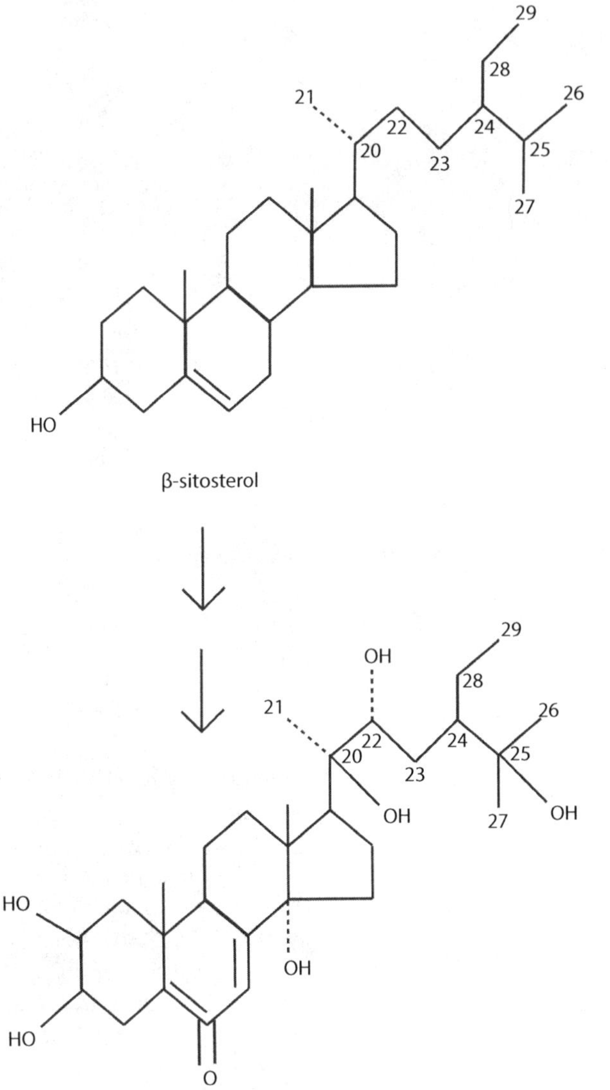

β-sitosterol

Makisterone C

20-hydroxy-24-ethyl ecdysone

Figure 5.9 Conversion of β-sitosterol to the molting hormone, 20-hydroxy-24-β-ethyl ecdysone, or Makisterone C, in the cotton stainer bug, *Dysdercus fasciatus*.

specificity for particular ecdysteroids (Porcheron et al., 1989; Royer et al., 1993). Nevertheless, because there are several steroidal molecules with hormonal activity in insects, analyses are often reported as total ecdysteroids detected rather than attempting to attribute activity to specific ecdysteroids.

5.5.6 Assay by Physicochemical Techniques

Thin-layer chromatography (**TLC**), **high-performance liquid chromatography** (**HPLC**), and **gas chromatography** coupled with **mass spectroscopy** (**GC–MS**) are methods that have the advantage of allowing separation of the various ecdysteroids present in a biological sample

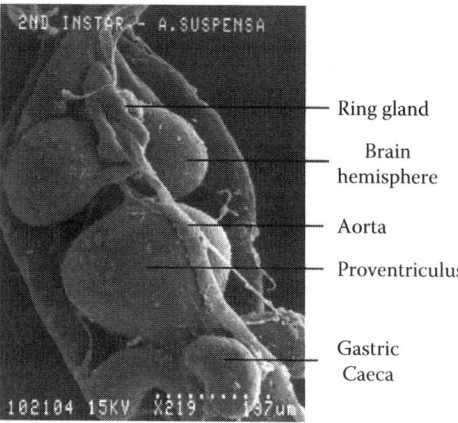

Figure 5.10 A scanning electron micrograph of the larval ring gland and brain in a second instar of the tephritid fruit fly, *Anastrepha suspensa*. Note the aorta passing through the ring gland. (SEM by the author.)

before quantitation. TLC is currently used mainly as a preliminary cleanup method because its ability to separate closely related steroids is limited. HPLC is probably the most widely used of the physicochemical methods (Lafont and Wilson, 1990; Wainwright et al., 1997; Louden et al., 2002), providing good separation of the various ecdysteroids and quantitation. The most definitive method for identification of ecdysteroids is the use of GC–MS or HPLC–MS (Wainwright et al., 1997; Louden et al., 2002). The ecdysteroids are too polar to chromatograph directly by GC, however, and suitable derivatives must be made. Several problems can be encountered in derivatization, including difficulty of reaction, uncontrolled chemical changes in the ecdysteroids, and very large molecular weight compounds after derivatization that require high temperatures for successful elution. Successful separation of derivatized ecdysteroids by GC coupled to a mass spectrometer, however, then allows characterization and quantitation from the mass spectrum. A further refinement of MS known as **coupled MS–MS** involving two tandemly operated mass spectrometers gives even greater resolution. This technique can be coupled with GC or HPLC (e.g., HPLC–MS–MS). Mauchamp et al. (1993) used GC–MS–MS and HPLC–MS–MS to demonstrate the application of the technique to ecdysteroid analysis and found Makisterone A and Makisterone C in the eggs of *D. fasciatus*, confirming a similar analysis by Royer et al. (1993), who used enzyme immunoassay. The most obvious disadvantage of the physiochemical techniques is the cost of the equipment.

5.5.7 Tissues and Cell Cultures Used in Assays

Cultured tissues and cells have been used for many years to assay activity of ecdysteroids (reviewed by Oberlander and Ferkovich, 1994) and were very important in elucidation of an ecdysone receptor (see Section 5.7.2). Cell lines from several lepidopterans (Sohi et al., 1995) have been used as model systems to test the ecdysone agonists RH-5849 and RH-5992, two nonsteroidal compounds that are capable of binding to the ecdysone receptor and can induce precocious and incomplete molting when added exogenously to some insects.

5.5.8 Degradation of Ecdysone

There are mechanisms to destroy and remove ecdysteroids rapidly after the hormone has accomplished its intended function. The half-life of most ecdysteroids is very short, a few hours at most. Excretion is one route to remove ecdysteroids from the body, and ecdysteroids have been detected in the fecal wastes as **conjugates** with glucose, glucuronic acid, and sulfates. These conjugates render the molecule physiologically inactive, and even though the ecdysteroids are remarkably water soluble (a necessity for excretion via the Malpighian tubules), the conjugates probably further facilitate excretion. Oxidation or modification of some of the functional groups on the molecule also reduce or destroy biological activity. An active enzyme in the larval midgut of *M. sexta* converts ecdysteroids to inactive compounds by C-3 epimerization, that is, the conversion of the 3β-hydroxy-group to a 3-oxo-group or to a 3α-hydroxy-group (Weirich et al., 1993). Conversion of 20-hydroxyecdysone to 20,26-hydroxyecdysone is also a degradation pathway. These conversion products have only very low molting hormone activity. Remarkably enough, the high-level pulses of molting hormone secretion characteristic of immature insects in preparation for each molt may induce the enzymes necessary for hormone removal, a kind of self-destruct mechanism. There is recent evidence for this in the cotton leafworm, *Spodoptera littoralis*, in which treatment of last instars with ecdysone, 20-hydroxyecdysone, or the ecdysone agonist RH 5849 induced 26-hydroxylase activity (Chen et al., 1994). The induction required new mRNA (gene transcription) and new protein synthesis (gene translation). The degradation pathway (Figure 5.11) is 20-hydroxyecdysone → 20,26-hydroxyecdysone → 20-hydroxyecdysone-26-aldehyde → 20-hydroxyecdysone-26-oic acid (Chen et al., 1994).

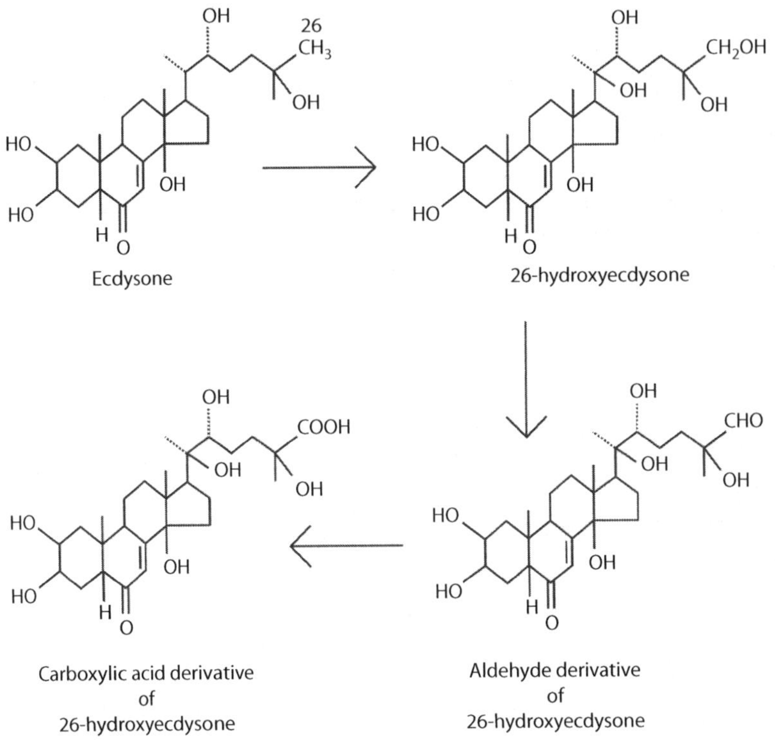

Figure 5.11 One of the ways that the molting hormone is inactivated by the cotton leafworm, *Spodoptera littoralis*. In this scheme, ecdysone is converted to the molting hormone 20-hydroxyecdysone. Hormone not bound to receptors is converted rapidly to 20,26-hydroxyecdysone with low hormonal activity; 20,26-hydroxyecdysone is further metabolized to the 26-aldehyde derivative, with the aldehyde group subsequently oxidized to a carboxyl group. The latter two compounds do not have hormonal activity and are rapidly excreted. (Scheme modified from Chen, J. et al., *Biochem. J.*, 301, 89, 1994.)

5.5.9 Virus Degradation of Host Ecdysteroids

A remarkable adaptation has evolved in a baculovirus, *Autographa californica* nuclear polyhedrosis virus (AcMNPV), that enables the virus to produce a **uridine 5′-diphosphate (UDP)-glucosyl transferase** that catalyzes the transfer of glucose from UDP-glucose to ecdysteroids, thus inactivating the hormone (O'Reilly and Miller, 1989). Although enzymes of this type typically transfer glucuronic acid to various drugs and carcinogens in vertebrates, glucose is normally transferred in insects. Virus expression of the *egt* gene that controls this enzyme allows the virus to interfere with normal larval development and prevents molting of the host, the fall armyworm, *Spodoptera frugiperda*. The evolutionary advantage gained by the virus is not entirely clear, but as a model it introduces the fascinating possibility that other pathogens and parasites may manipulate the hormonal milieu of their host. Indeed, Palli et al. (2000) demonstrate another case of virus modulation of its host's hormonal titers: Larvae of the spruce budworm, *Choristoneura fumiferana*, infected with an entomopoxvirus (*C. fumiferana* entomopoxvirus) feed and grow in size, but eventually die without metamorphosing, apparently because the infected larvae have increased JH titer and decreased ecdysteroid titer.

5.5.10 Dependence of Some Parasitoids on Host Ecdysteroids

Some parasitoids depend on the host's hormones for their own development and pupation. The parasitoid *Biosteres longicaudatus*, an endoparasite of larvae of the Caribbean fruit fly (a tephritid), typically oviposits into early third instar larvae where the parasitoid first instar feeds on host tissues and grows to a critical size ready to molt (Lawrence, 1986). The parasitoid does not produce its own molting hormone, but depends upon its host's hormone. The parasitoid does not molt until the host secretes ecdysone to initiate its own larval to pupal transformation. The parasitoid then molts to the next instar and continues to feed upon the pupa of the fly. Eventually the parasitoid kills the host by feeding upon critical tissues and then pupates and completes its transformation into an adult wasp inside the puparial shell of the host. *Diapetimorpha introita* is an ichneumonid ectoparasitoid of the fall armyworm, *S. frugiperda*, and is capable of producing its own ecdysteroids when grown on an artificial diet. Hemolymph ecdysteroid titers were higher, however, in host-reared than in diet-reared individuals, suggesting a significant nutritional role for most effective ecdysteroid biosynthesis in these ectoparasitoids (Gelman et al., 2000).

5.6 CORPORA ALLATA AND JUVENILE HORMONES

5.6.1 Glandular Source and Chemistry

JHs are **sesquiterpenoid** compounds produced in the cells of the **CA**, which are bilaterally paired structures in Lepidoptera, but are often fused into one mass of tissue in other groups of insects. The CA, derived embryologically from ectodermal tissue, lie beneath the aorta and posterior to the brain and CC. A nerve tract connects the CA with the CC and brain and brain NSCs. In Orthoptera, Thysanura, and Ephemeroptera, there is a nerve tract to the subesophageal ganglion. Five JH variants (six, if methyl farnesoate is counted; it is the immediate precursor of the JH molecules and is secreted by some insects and has JH activity on its own) are known. The structures of **JH I, JH II, JH III, JH 0**, and **iso JH 0** (also called **4-methyl JH I**), and **JH bisepoxide** (**JHB3**) are shown in Figure 5.12. Although JH III has most often been found as the principal or only JH molecule in many insects, more detailed analyses with GC–MS (Bergot et al., 1981) have shown multiple JHs in some insects, particularly Lepidoptera. JH III is only detectable as a trace or sometimes not at all in some Lepidoptera (Edwards et al., 1995; Ramaswamy et al., 1997;

Juvenile hormone molecules

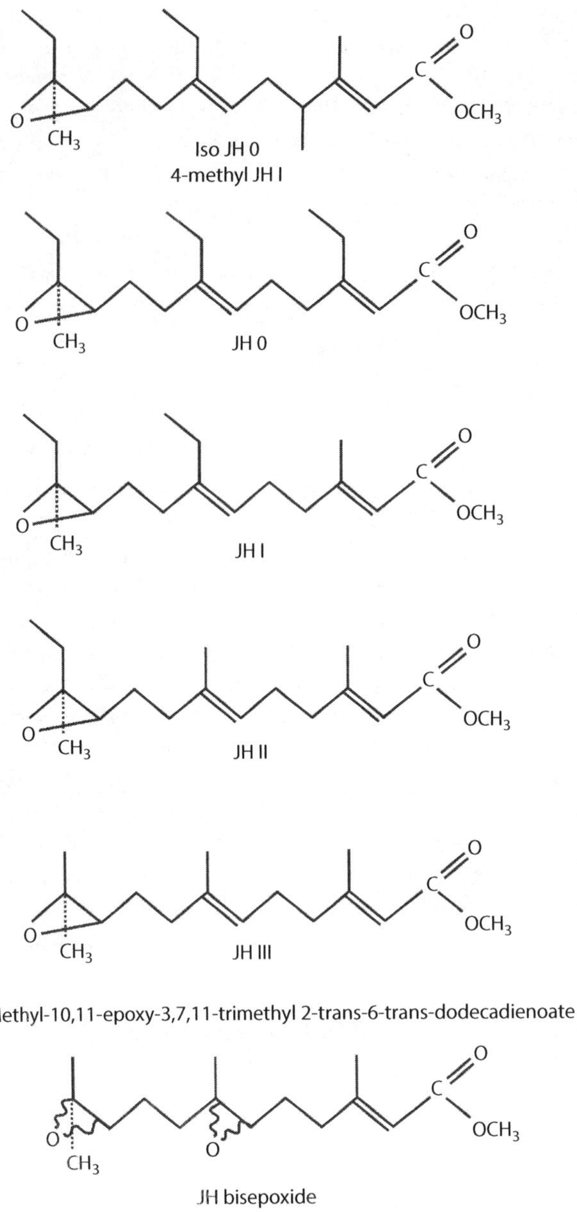

Iso JH 0
4-methyl JH I

JH 0

JH I

JH II

JH III

Methyl-10,11-epoxy-3,7,11-trimethyl 2-trans-6-trans-dodecadienoate

JH bisepoxide

Figure 5.12 Naturally occurring molecules with JH activity. JH III may be the most common JH of insects. JH I, JH II, JH III, JH 0, and iso JH 0 have been found in some Lepidoptera. JH bisepoxide is synthesized by the ring gland of some Diptera.

Park et al., 1998), and JH I and JH II often are the principal JHs in larvae. JH I, JH II, and JH III are released from isolated CA of 10-day-old *Actebia fennica* moth females, but JH II is the principal JH (Everaerts et al., 2000). JH II is the principal JH in the fourth and fifth instars of the tomato moth larvae, *Lacanobia oleracea*, and in the pupa, along with some JH I (Edwards et al., 1995). However, Audsley et al. (2000) found that 90% of the JH released by isolated CA is JH II. In addition,

JH II acid and **JH I acid** also are released from the CA, but Audsley et al. (2000) attribute the acid analogs to the presence of JHE activity in the CA and breakdown of already formed JH, rather than failure of methylation of farnesoic acid in biosynthesis. It has not been determined if the multiple forms of JH have unique functional roles in all the insects where they have been found, but in some cases, they do have specific functional roles (Gilbert et al., 2000). Synthesis of JH is regulated by neuropeptides and biogenic amines, and probably by nervous control.

Although JH I is the principal JH of early instars of *M. sexta* (Schooley et al., 1984), production of the JH acids seems to be the normal process in the last instar of *M. sexta*, which loses the enzymatic ability to methylate the final step in synthesis of the various JHs (Bhaskaran et al., 1986; Baker et al., 1987). JH acid production continues into the pupal and adult stages. The JH acids have little biological activity, but they can be methylated slowly in imaginal disc tissues (Sparagana et al., 1985), and the slow methylation may be important to overall function of JH and metamorphosis in *M. sexta*, and possibly other Lepidoptera (Gilbert et al., 1996a). Enough evidence has accumulated, however, to show that the JH acids themselves have hormonal activity in some insects (Gilbert et al., 2000).

JHB3 is the principal JH of *D. melanogaster* (Richard et al., 1989a, 1989b; Yin et al., 1995) and occurs also in *Phormia regina* (Richard et al., 1989a), *Calliphora vomitoria* (Cusson et al., 1991) and, the sheep blowfly, *Lucilia cuprina* (Lefevere et al., 1993) and in four species of mosquitoes (Borovsky et al., 1994). In addition to JHB3, *L. cuprina* also produces some JH III (East et al., 1997). **JHSB3, skipped bisepoxide**, another recently discovered variation in JH structure, is the natural JH of the heteropteran, *Plautia stali* (Kotaki et al., 2009, 2011).

The JH molecules have chiral centers, and some evidence indicates that the natural enantiomers may be more active and less rapidly degraded than unnatural enantiomers (Peter et al., 1979; King and Tobe, 1993). JH III has a chiral center at carbon 10 and the other JH molecules have chiral centers at C-10 and C-11. JHB3 has chiral centers at C-6, C-7, and C-10.

5.6.2 Assays for JH Activity

The oldest assay is the ***Galleria* wax test** in which a small hole about 1 mm² is cut through the cuticle of a newly molted pupa, a test material is applied, and the wound is sealed with molten wax. It is more effective to administer samples potentially containing JH in an agent, such as peanut oil, that protects it from JHEs (Gilbert and Schneiderman, 1960). If JH is present in the sample applied to the pupal wound, the adult that emerges about 10 days later has a small patch of pupal cuticle devoid of scales, or with few scales, at the wound site. The size and quality of the pupal patch is scored to give a rough measure of JH present in the test solution. The *Galleria* assay has the disadvantages that it is slow, requiring about 2 weeks to determine if a sample has JH activity, considerable experience is required to obtain reliable data, and biological variability requires that a large sample of pupae must be tested. Currently, most of the determinations of JH activity are done by selective ion monitoring with GC–MS. It is highly specific for the different JH molecules (Figure 5.13; Table 5.2) and is a good quantitative method (Bergot et al., 1976, 1981; Teal et al., 2000). An RIA technique for JH also is available (Hunnicutt et al., 1989; Huang et al., 1994; Borst et al., 2000).

A very sensitive method for measuring biosynthesis of JH is based on transfer of the radiolabeled methyl group from methionine to JH during the biosynthesis of JH. This method, however, is not suitable for the measurement of total tissue or hemolymph titer of JH.

5.6.3 Regulation of the Tissue and Hemolymph Levels of JH

The titer of JH in the hemolymph and tissues is a function of biosynthesis by the CA balanced against degradation in the tissues. JH is secreted continuously by CA during larval life, but there are stage-specific fluctuations (Tobe and Stay, 1985). JH is not stored in the CA or in

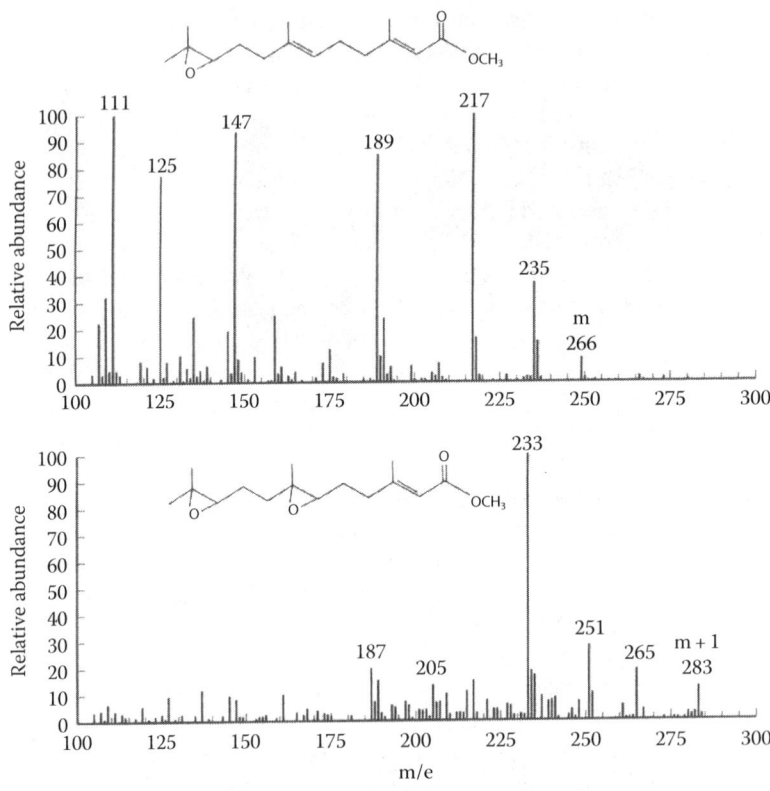

Figure 5.13 Mass spectra of JH III (top) and JH III bisepoxide (bottom) extracted from the hemolymph of sexually mature, 12-day-old male tephritid fruit fly, *Anastrepha suspensa*. Samples were analyzed in a Finnigan Magnum Ion Trap mass spectrometer interfaced with a Varian Star 3400 gas chromatograph (GC) containing a 30 m × 0.25 mm × 0.1 μm film thickness J&W DB5-MS analytical column. The ion trap was operated in chemical ionization mode with isobutane as the reagent gas. A 10 m × 0.25 mm uncoated, deactivated, fused silica retention gap column in the GC and a 10 cm × 0.5 mm length of uncoated, deactivated, fused silica column in the injector allowed large volumes of sample to be injected without loss of resolution. The conditions of chromatography were as follows: initial injector temperature of 40°C for 30 s, injector temperature increased at 170°C/min to 270°C, initial column temperature 40°C for 5 min, and column temperature increased at 5°C/min to 210°C. Helium (He) carrier gas linear flow velocity equal to 24 cm/s. GC-MS transfer line temperature was 230°C. Ion trap–operating conditions were as follows: multiplier voltage, 1900 V; manifold temperature, 130°C; emission current, 16 μamps; mass acquisition range, 60–350 amu; 1 scan/s. (Mass spectrometry record courtesy of Peter Teal, USDA, Gainesville, FL.)

other tissues. In some insects, there is evidence that nervous connections from the brain influence biosynthesis of JH in the CA, either through stimulation or inhibition. These examples generally come from studies of endocrine control of reproduction in adults. Nervous suppression of JH synthesis is found in virgin female cockroaches, *Diploptera punctata*, which normally do not mature oocytes or ovary until mating releases the inhibition. Mating often stimulates the CA to begin JH production, possibly by relieving the inhibition that the nervous system may have on CA activity (Tobe and Feyereisen, 1983). Experimentally severing nerve connections between the brain and CA allows oocytes to mature and promotes a cycle of JH secretion without mating. In *Schistocerca* sp., however, severing the nerve connections to the CA causes a decline in JH production.

Table 5.2 Description of Cleavage Assignments Resulting in Diagnostic Ions Used for Quantitation of JH Compounds

Ion No.		Mass to Charge (m/e)	
		JH III	JH III Bisepoxide
1	M + 1		283
	Ion 1 − HOH (from ring cleavage of epoxide)		265
2	Ion 1 − CH_3OH (from methyl ester)	235	251
3	Ion 2 − HOH (from ring cleavage of epoxide)	217[a]	233[a]
4	Ion 3 − CO (from methyl ester)	189	205
5	Ion 4 − HOH (from ring cleavage of second epoxide)		187
6	Ion 1 − $C_2H_4O_2$ (from methyl ester) − C_3H_8O (from epoxy terminus)	147	
7	M − $C_8H_{13}O_2$ (cleavage between C-6 and C-7)	125	
8	$C_7H_{11}O$ (scission between C-6 and C-7 after loss of CH_3OH)	111	

Source: Data courtesy of Peter Teal, USDA, Gainesville, FL.
[a] Parent ion.

The brain clearly has some control over the secretion of JH by the CA (reviewed by Stay et al., 1994). There are numerous reports of changes in the CA and JH titers in reaction to severing nervous connections between the CA and brain and after total removal of the brain. Some of the brain's control is likely to be through the intact nerve connections and nerve impulses, but some studies have demonstrated also that extracts of the brain influence CA activity and JH levels. A number of **allatotropins (ATs)** and **allatostatins (ASTs)**, peptides from the nervous system of several different species and from several orders of insects, have been isolated and bioassayed for effect on JH biosynthesis. ATs stimulate JH synthesis, while ASTs inhibit synthesis. In general, these peptides have mainly been studied in adult insects in which JH is necessary in most insects for female reproduction (reviewed by Stay et al., 1994). Audsley et al. (2000) found, however, that Mas-AT (isolated from *M. sexta* and the only AT that has been completely sequenced) seems to be the principal regulator of CA activity in the larval tomato moth, *Lacanobia oleracea*. These authors suggest that the larval CA are activated by Mas-AT as needed and that removal of the peptide when JH is not needed stops JH synthesis with no or little need for Mas-AST to turn off synthesis.

The insulin signaling pathway may influence JH production (Tu et al., 2005). Although insulin-like peptides are known from a number of insects, manipulation of genes in *Drosophila* has been important in describing in more detail the production and role of insulin peptides (Tatar et al., 2001, 2003). The genome of *D. melanogaster* encodes seven insulin-like peptides (Brogiolo et al., 2001), with cells in the **pars intercerebralis** of the brain producing the peptides (Cao and Brown, 2001). The neuronal processes from the cells innervate the CC and the heart, and the heart probably acts as a neurohemal organ where the insulin-like peptides are released and enter the circulation. Tu et al. (2005) found that AT-positive axons in the brain and CC/CA complex of mutants for the insulin receptor (mutant *InR*) in *Drosophila* were less immunoreactive, leading them to conclude that insulin signaling could influence JH synthesis by controlling regulatory neuropeptides.

Additional agents have been shown to stimulate or inhibit CA in some insects, including neurotransmitters, such as dopamine and octopamine. Second messenger systems, including cAMP, inositol trisphosphate, diacylglycerol, and calcium signaling; location, kind, and number of potential receptors; and ecdysteroid action are known in various insects to have stimulatory or inhibitory effects on the CA (Woodring and Hoffmann, 1994; Granger et al., 1996, 2000; Gilbert et al., 2000).

The main degradative pathways for JH involve specific and nonspecific **JHEs** described from numerous insects and **JH epoxide hydrolases** (**JHEHs**) reported from some insects. The esterases attack the ester linkage, while epoxide hydrolase opens the epoxide ring and creates a diol (Figure 5.14). Only one action, or both, may occur in some insects. The metabolic changes not only eliminate all or most of the hormonal activity of the molecules, but the molecules become more water soluble and can be excreted by the Malpighian tubules.

JH is transported through the hemolymph bound to a protective lipoprotein. The proteins are known as **JH-binding proteins** (**JHBPs**), and proteins with high and low specific binding have been isolated. Lipophorins are primary JH carriers in hemimetabolous insects (Kanost et al., 1990). In two cockroach species, lipophorin and vitellogenin are the two principal JHBPs (Engelmann and Mala, 2000). In *M. sexta*, 32 kDa, JHBP has been purified. JH binds in a hydrophobic pocket in the protein, so it is well protected from esterase attack (Touhara and Prestwich, 1992). A major factor contributing to degradation of JH in *M. sexta* is a JH-specific esterase that is synthesized in large amounts during the latter part of each larval molt and in the final instar. Although the JHBPs protect JH from esterase attack, the high titer of esterase just prior to a molt destroys any JH that dissociates. Hydrolysis of undissociated JH continues to pull the equilibrium of bound JH toward dissociation and degradation, allowing the molt to the next instar (Riddiford, 1996). In *L. migratoria*, JHBP protects the natural enantiomer, (10*R*)-JH III, from hemolymph esterase activity better than it protects the unnatural enantiomer, (10*S*)-JH III (Peter et al., 1979, 1983). In *N. cinerea* cockroaches, however, natural (10*R*)-JH III was degraded by hemolymph esterase more rapidly than a racemic mixture of the enantiomers (Lanzrein et al., 1993). Thus, fluctuations in the level of JHBP in the hemolymph at different

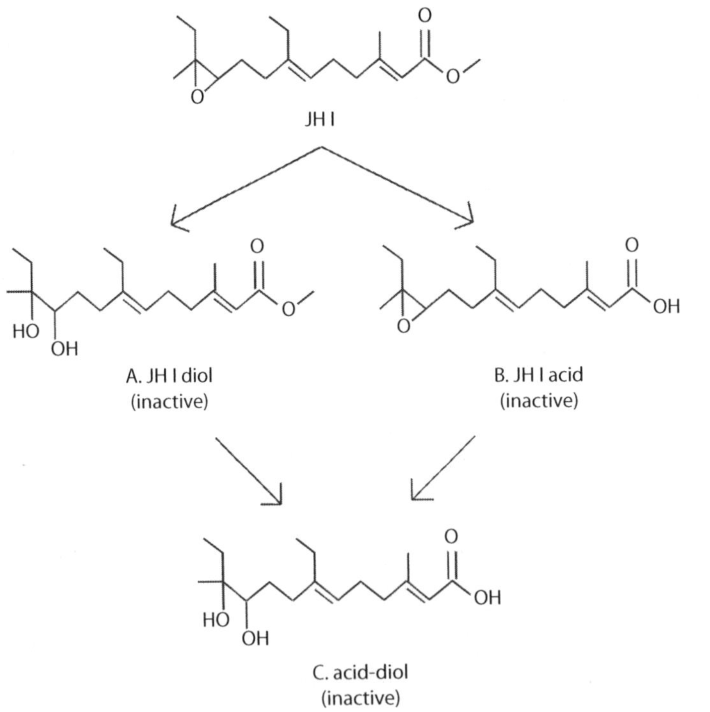

Figure 5.14 Metabolic pathways for the degradation of JH. The epoxide ring may be opened with hydrolysis and production of two hydroxyl groups, as in A, or the ester group may be hydrolyzed to the free acid. Both A and B are inactive, and either or both may occur in most insects.

developmental times are a factor in JH titer. Therefore, although JH titer is influenced by (1) the rate of synthesis, (2) rate of breakdown, (3) sequestering in some target tissues, (4) presence of JHBP, and (5) excretion, the two main processes appear to be rate of synthesis and rate of breakdown.

JH biosynthesis occurs through an isoprenoid pathway in which acetyl CoA units are used to build farnesyl pyrophosphate and, finally, methyl farnesoate. In the final enzymatically controlled step occurring in CA, methyl farnesoate epoxidase, a cytochrome P450 containing monooxygenase, epoxidizes the 10,11-double bond. Photoaffinity labels have been used to label a protein (presumably methyl farnesoate epoxidase) of about 55 kDa in the CA of the cockroach, *D. punctata* (Andersen et al., 1995). The label binds to the heme iron of cytochrome P450 and to the hydrophobic substrate-binding pocket of the enzyme. These or similar probes can be useful in defining the exact chemistry of the epoxidase enzyme and the final epoxidation step, as well as demonstrating where exactly within the CA cells the reaction occurs.

5.6.4 Insect Growth Regulators and Compounds That Are Cytotoxic to the Corpora Allata

For reasons still not understood, JH is an extremely easy molecule to mimic in all or part of its physiological functions. Thousands of compounds (about 5000) and extracts of many tissues, including those of vertebrates, have JH activity to varying degrees. Compounds with JH activity generally are called insect growth regulators (IGRs). A few IGRs have enough JH activity (examples are kinoprene, hydroprene, and methoprene) to be used commercially as insecticides. Methoprene is very effective against the late larval stage of mosquitoes and fleas. Kinoprene is most active against Lepidoptera and hydroprene against Orthoptera.

Some plants contain substances that are active against the CA. For example, chromene compounds given the trivial names **precocene I** and **precocene II** (Figure 5.15) occur in various vegetative parts of *Ageratum houstonianum*, a common bedding ornamental (Bowers et al., 1976). These compounds have a powerful effect upon milkweed bugs, *O. fasciatus*, and numerous other insects, but they are usually most effective on Hemiptera. The compounds cause second and third instars of milkweed bugs to precociously metamorphose into small, imperfectly formed adults. The small adults are incapable of reproducing because the ovaries do not develop. Considerable excitement ensued from this discovery, and scientists in many laboratories around the world synthesized derivatives of the precocenes and/or started searches for other naturally occurring compounds with similar hormonal effects. Neither the precocenes nor derivatives, and no other naturally occurring compounds with similar action, have been commercialized at the present time.

The precocenes do not directly antagonize the action of JH at target sites. Treated insects can sometimes be rescued with large doses of JH or an effective JH analog. The compounds have a cytotoxic effect upon the cells of the CA, causing them to atrophy and fail to produce JH. Lack of JH allows ecdysteroids to catapult young instars into precocious adult development.

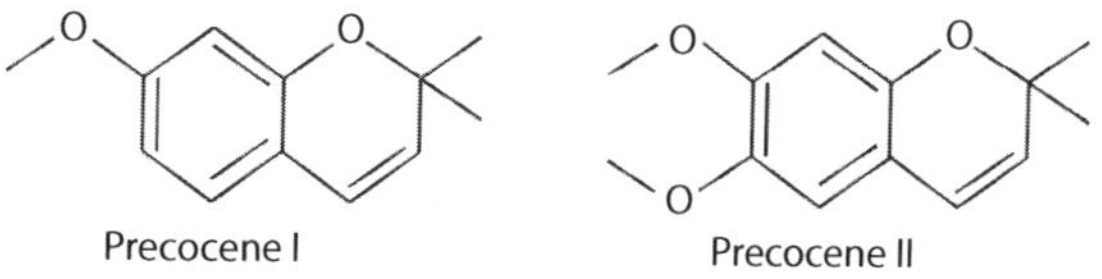

Figure 5.15 Naturally occurring compounds precocene I and precocene II discovered in *Ageratum houstonianum* ornamental bedding plants. The compounds cause precocious molting in some insects, most notably Hemiptera.

5.6.5 Cellular Mode of Action and Receptors for JH

Intensive searches for cellular and nuclear protein receptors for JH have been conducted by numerous investigators. In addition to its role in molting and metamorphosis, JH has pleiotropic action, influencing sexual behavior, caste determination in some insects, pheromone biosynthesis, diapause, synthesis of egg yolk proteins, and migration in some insects, and its mode of action may differ in some of these roles, sometimes requiring gene activation, while other actions likely are not dependent upon transcription of genes.

Because of the large amount of genetic information and many mutants available, *Drosophila* was an obvious choice for studies of JH receptors. JH is essential for head eversion in *Drosophila* and for normal development in the fat body of prepupae, and elimination of JH by allatectomy results in death of prepupae (Liu et al., 2009; Riddiford et al., 2010). A potential cellular JH receptor was found in certain *D. melanogaster* mutants by Wilson and Fabian (1986), who found that some larvae in induced mutagenic strains of *Drosophila* were able to tolerate the exogenous addition of methoprene, a JH mimic, which would be expected to cause several types of metamorphic defects during pupal formation. This led to the discovery of the *Methoprene-tolerant* gene (*Met*) and to the suggestion that it encodes a JH receptor protein (MET) in *Drosophila* (Figure 5.16). The discovery, however, that *Met*-null mutants of *Drosophila* were able to metamorphose to the adult stage and showed only relatively minor defects cast some doubt upon MET as a JH receptor; how had the larvae withstood the adverse effect of added methoprene or JH if a receptor protein were not present? This small mystery was solved when an additional gene, *germ cell expressed* (*gce*), which encodes the protein GCE, was discovered. This gene probably evolved from gene duplication in a *Drosophila* ancestor (Baumann et al., 2010a,b), also binds JH and acts as a backup to bind JH and allow metamorphosis; both *Met* and *gce* had to be absent for failure to metamorphose (Abdou et al., 2011a,b). Mutants deprived of both *Met* and *gce*, however, prematurely activated the gene *broad*, required for metamorphosis (Riddiford et al., 2010; Abdou et al., 2011a), and showed some differentiation of the optic lobes, an early indication of metamorphosis preparation (Riddiford et al., 2010). Miura et al. (2005) showed that recombinant *Drosophila* MET protein bound JH, and Charles et al. (2011) confirmed binding in *Drosophila* and various insects, including the primitive firebrat, *Thermobia domestica*.

Genetic research with the beetle *Tribolium castaneum*, following elucidation of its genome, helped confirm MET as a JH receptor. Although *Tribolium* does not have the *Drosophila Met* and

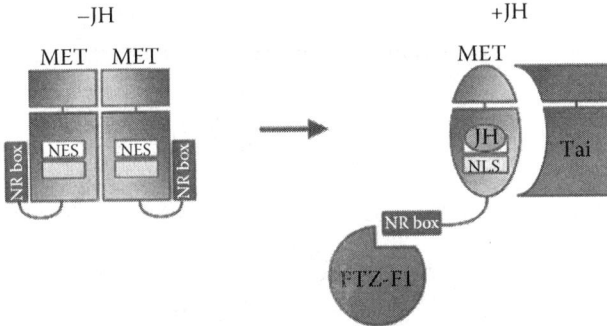

Figure 5.16 (See color insert.) A model for activation of MET protein by JH-dependent conformational changes. At the left when JH is absent, MET exists as a homodimer by interactions in the bHLH and PAS domains (gray in diagram). Binding of JH (right) stimulates an active confirmation and promotes an interaction with Taiman/AaFISC/TcSRC with the bHLH-PAS region and with FTZ-F1 at a C-terminal NR box. Possibly JH may influence the activity of nuclear export and import signals in the ligand-binding domain, promoting nuclear localization. (From Bernado, T.J. and Dubrovsky, E.B., *Insects*, 3, 324, 2012. With permission.)

gce genes, it has a gene similar to the *Met/gce* genes of *Drosophila*, usually referred to in the literature as a *Met* gene. *Tribolium* larvae deprived genetically or surgically of JH can precociously molt into small pupae and adults, a convenient test for a role of JH in metamorphosis when *Tribolium* is deprived of its JH receptor. Konopova and Jindra (2007) used *Met* RNAi to reduce *Met* activity in third and fourth instar *Tribolium* larvae, which precociously metamorphosed as fifth and sixth instars into small adults. Similarly, Met RNAi also allowed the hemipteran, the linden bug *Pyrrhocoris apterus*, to precociously metamorphose into an adult (Konopova et al., 2011). Thus, inhibiting the ability of JH to bind to MET protein prevented JH from exerting its juvenilizing effect in a holometabolous insect (*Tribolium*) and in a hemimetabolous one (*Pyrrhocoris*). MET proteins from *Drosophila* and from *Tribolium* have high binding of JH III; K_d is 5.3 nM for *Drosophila* (Miura et al., 2005) and 3.9 nM for *Tribolium* (Charles et al., 2011).

MET and GCE are members of a family of proteins that form dimers, in this case as MET–MET or MET–GCE. When JH binds to the dimers (Godlewski et al., 2006), the two proteins separate, leaving JH bound to MET, and JH–MET then associates with the protein Taiman in *Drosophila* (Bai et al., 2000). In the mosquito *Aedes aegypti*, JH–MET associates with a protein named βFtz-F1 (Zhu et al., 2006; Li et al., 2011) and JH–MET associates with the protein SRC in *Tribolium* (Zhang et al., 2011). Because these three proteins, Taiman, βFtz-F1, and SRC, interact with the ecdysone receptor, as well as JH–MET, they are referred to as steroid coactivators.

JH also interacts with several nuclear receptors in the ecdysone pathway, for example, methoprene or JH III acting with the nuclear receptor βFtz-F1- and Gce-induced *E75A*, one of the early genes in the ecdysone cascade, but Met was not required (Dubrovsky et al., 2011). Ultraspiracle (USP) is another nuclear receptor in the ecdysone cascade, and it may also be a receptor for JH (Riddiford, 2012). JH III binds USP from *Drosophila*, albeit only moderately so ($K_d = 2 \times 10^{-7}$ M), but USP binds methyl farnesoate at the high nanomolar level more typical of a hormone receptor (Jones et al., 2006).

From an evolutionary perspective, MET proteins are conserved across a number of insect orders (Konopova et al., 2011), and a non-MET protein that binds JH III has been found in the firebrat, *T. domestica* (Charles et al., 2011). The firebrat, an early insect to evolve, molts a number of times, but does not undergo metamorphosis. Juveniles simply look like small adults. JH, however, is important for reproduction, leading Sehnal et al. (1996) to suggest that the original function of JH was in reproduction and only later did it come to have a major role in metamorphosis.

5.6.6 Downstream Transcription Factors

Two additional genes—*Krüppel homolog 1* (*Kr-h1*) and *broad* (*br*)—are necessary players in molting and metamorphic processes. JH is the main activator of *Kr-h1*, which encodes a putative transcription factor with DNA-binding zinc fingers. *Kr-h1* is widespread across insect orders, tending to retain developmental "status quo." The amino acid composition of its protein Kr-h1 and regulation of its expression are highly conserved (Jindra et al., 2013). Ecdysone regulates *broad*, and it is one of the early genes initiated by ecdysone at the start of the molting process. Broad protein, a member of a family of transcription factors, promotes metamorphosis in an ecdysone-dependent manner. Although *broad* is initially activated by ecdysone, there are a number of examples showing JH modulation of broad expression and expression of some of the other early genes activated by ecdysone, mediated probably through JH-dependent Kr-h1 (Zhou and Riddiford, 2002; Berger and Dubrovsky, 2005; Minakuchi et al., 2008; Jindra et al., 2013).

In *Drosophila* broad protein appears in epidermal cells during the final larval instar (Bayer et al., 1996). The JH mimic pyriproxyfen added at pupariation prevented the normal disappearance of broad in abdominal imaginal epidermis, and as the next pulse of ecdysone occurred, normally leading to the pupal-to-adult molt, the epidermal cells again secreted a pupal cuticle (Zhou and Riddiford, 2002). In *Tribolium br*, mRNA is present throughout most of larval life at a low level, increases in the final larval instar, reaches its highest level in the prepupal stage, and is not detectable in the early pupa. The results indicate that *br* activity is necessary for larval to larval molts and for the larval to pupal molt, but *br* must be turned off for the pupal-to-adult molt. Continued or reexpression of *broad* from added exogenous JH or JH mimic at the beginning of adult development causes a second pupa (Konopova and Jindra, 2007; Minakuchi et al., 2008; Suzuki et al., 2008). The first pulse of ecdysteroid is sufficient to turn *broad* on, and the elevated second pulse, in the absence of JH, causes the cessation of *broad* action and allows pupal-to-adult metamorphosis. Broad is detectable in epidermal cells after the first small peak of ecdysone, which also causes cessation of feeding and wandering in *M. sexta*. Broad signals the epidermal cells for pupal differentiation (Zhou et al., 1998; Hiruma and Riddiford, 2010; Riddiford, 2012). Broad reaches a peak in epidermal cells during the second large peak of ecdysone and then falls rapidly to undetectable levels by day 4 of pupal development. When pupae were treated with the JH mimic pyriproxyfen, broad failed to disappear, and second pupal cuticle was formed. The results again suggest that ecdysone can turn on *broad* and turn it off.

In two hemimetabolous insects that do not have a pupal stage, *O. fasciatus*, the milkweed bug, and *Blattella germanica*, the German cockroach, expression of *broad* is necessary for normal nymphal development (Piulachs et al., 2010). In *Oncopeltus* broad expression disappears in the final nymphal instar, allowing metamorphosis to the adult (Erezyilmaz et al., 2006). Application of pyriproxyfen allowed continuation of *broad* expression, and a supernumerary nymph formed. Treatment of third instars of *Oncopeltus* with precocene, an agent that prevents the CA from being able to synthesize JH, allowed *broad* mRNA to disappear in the fourth instar, and small precocious adults were formed.

The data from several hemimetabolous and holometabolous insects suggest that *broad* expression is necessary for embryonic and nymphal development in hemimetabolous insects and for pupal specification in holometabolous insects and that JH prevents precocious adult development through Kr-h1 (Jindra et al., 2013).

5.7 MODE OF ACTION OF ECDYSTEROIDS AT THE GENE LEVEL

5.7.1 Chromosomal Puffs

The **polytene chromosomes** in both *Chironomus tentans*, a midge (Chironomidae, Diptera), and *D. melanogaster* have been important for understanding the mode of action of ecdysteroids. Polytene chromosomes are the result of chromosomal replication without mitosis, and polytene chromosomes in a single cell may consist of up to 2^{13} chromatids (Lezzi, 1996). Alternating bands of condensed and decondensed chromatin occur along the length of the polytene chromosomes, producing the characteristic banding pattern observed in the microscope. Work with *Drosophila* chromosomes was aided greatly by the extensive genetic and developmental background for *D. melanogaster*. The polytene chromosomes (Figure 5.17) in salivary glands of *D. melanogaster* and *C. tentans* show characteristic **puff patterns** (Figure 5.18) that enlarge and regress during development (Becker, 1959; Clever and Karlson, 1960). The puffs are sections of the DNA, that is, genes, in the act of transcribing the genetic code to new mRNA. Clever and Karlson (1960) took advantage of the availability of isolated and identified molting hormone, 20-hydroxyecdysone (Butenandt and Karlson, 1954), to show that the hormone injected into last instars of *C. tentans* induces chromosomal puffing within 2 h and the puff regions are the same as those observed during normal initiation of pupariation; they ventured the seminal suggestion that "the primary effect of ecdysone is to alter the activity of specific genes." Numerous studies of the effect of 20-hydroxyecdysone on chromosomes of *C. tentans* and *D. melanogaster* were soon underway, with significant improvement in developing and standardizing the use of organ

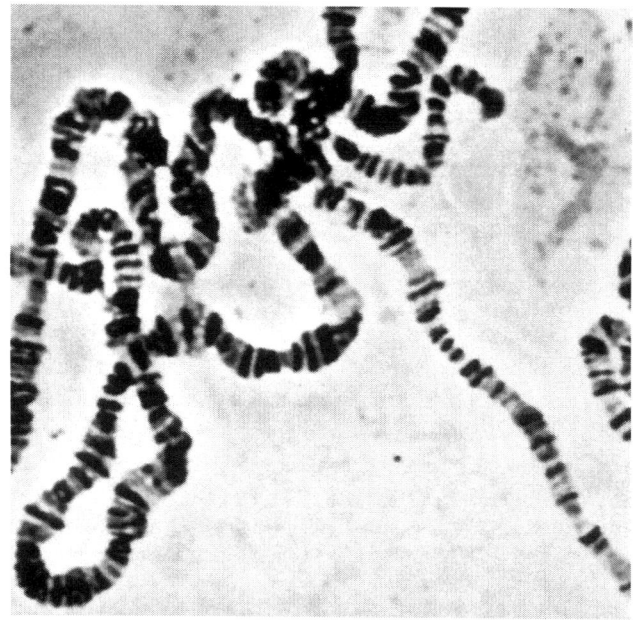

Figure 5.17 A polytene chromosome from *D. melanogaster*. (Micrograph courtesy of Marie Nation Becker, PhD Scientist, Science Writer, Tallahassee, FL.)

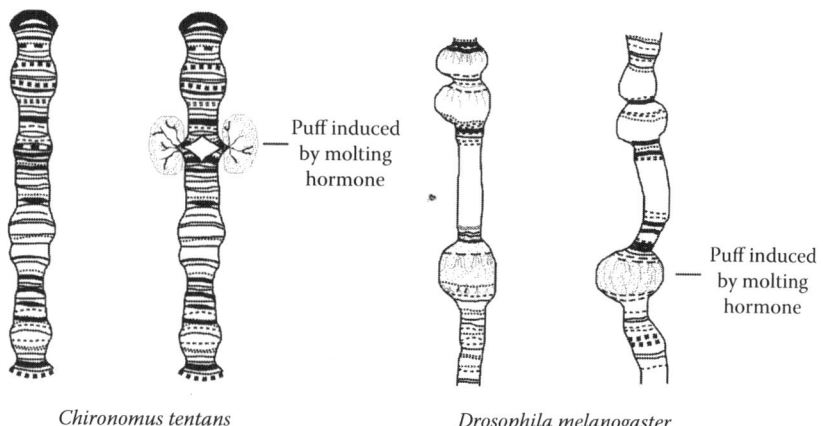

Figure 5.18 A drawing of a chromosomal puff from a polytene chromosome of *D. melanogaster* induced by molting hormone. (Redrawn and modified from Becker, H.J., *Chromosoma*, 10, 654, 1959.)

cultures of *Drosophila* salivary glands and a testable model scheme for how the genes worked (Ashburner, 1972). Within a few minutes after addition of 20-hydroxyecdysone to a culture medium containing isolated third instar *Drosophila* salivary glands, **six "early" puffs** develop at sites designated as 22B4-5, 23E, 63F, 74EF, 75B, and 74C (Ashburner, 1972). The early puffs grow larger over a period of about 1 h, persist for about 4 h, and then begin to regress and disappear in about 6 h. The six puff regions are the same ones that occur at the initiation of pupariation of late third instars, thus providing satisfying evidence that isolated salivary glands respond in the same way as intact glands in whole insects. A few hours after the early puffs regress, up to 100 puffs can be observed in other chromosomal regions; these are called **"late" puffs** and have been divided into "early–late" and "late–late" puffs in reference to their time of puffing.

The puffed regions of polytene chromosomes are indicative of a loosening of the DNA strands of polytene chromosomes and **transcription** of genes into mRNA. Experiments showed that the six early puffs appeared when cycloheximide, a general inhibitor of protein synthesis, was added with 20-hydroxyecdysone to the culture medium. These results indicated that no new protein synthesis was required to induce the early genes to transcribe new mRNA. Ashburner and Richards (1976) suggested that 20-hydroxyecdysone acted directly on the early genes, probably, they speculated, after combining with a receptor protein.

The early puffs begin to regress after about 4 h and many new puffs appear. These actions occur even in the continued presence of 20-hydroxyecdysone, but regression of early puffs and development of late puffs can be prevented in the presence of cycloheximide. This indicated that new protein synthesis was necessary for gene regression and appearance of late puffs. Utilizing the idea of negative feedback, Ashburner postulated that when enough of some new gene product(s) had been made, it (they) acted to inhibit the early genes and induce the late genes.

Inhibition of early genes and induction of late genes requires exposure of the salivary gland chromosomes to hormone for a critical period of time, as shown by washing the added 20-hydroxyecdysone from gland cultures after various exposure periods. After only short exposure to 20-hydroxyecdysone, early puff regression and late puff development are aborted. Ashburner and Richards (1976) suggested that the early gene transcription products have to be made in sufficient quantity to compete with the 20-hydroxyecdysone–receptor complex for binding sites on the DNA.

5.7.2 Isolation of an Ecdysteroid Receptor

The search for an **ecdysone receptor** led to attempts to demonstrate whether ^{14}C- or ^{3}H-labeled ecdysone accumulated at the site of chromosomal puffs. While some workers found tantalizing suggestions that ecdysone accumulated in the nucleus and at puffs, the miniscule amount of ecdysone at a puff site and the lack of very high specific radiolabeling of ecdysone with ^{14}C or ^{3}H were drawbacks to conclusive proof. A major development occurred with the discovery that an analog of ecdysone, ponasterone, has good hormonal activity after iodination to form 26-[^{125}I]-iodoponasterone (Figure 5.19). The radioactive iodine gives iodoponasterone very high specific labeling, thus facilitating detection in the nucleus. Labeled iodoponasterone shows powerful hormonal effects upon cultured *Drosophila* Kc cells, an embryonic cell line derived from *D. melanogaster* and *in vivo* activity, and it is bound in the nucleus to DNA (Cherbas et al., 1988, 1991).

A gene, *EcR*, from *D. melanogaster* that codes a protein with ecdysteroid-binding properties has been isolated and sequenced (Koelle et al., 1991). The receptor protein that it encodes, designated **EcR**, shows high specific binding to DNA and to labeled ecdysteroids. EcR is localized in the nucleus and can be detected with anti-EcR monoclonal and polyclonal antibodies in a variety of *Drosophila* tissues including imaginal discs, fat body, tracheae, salivary glands, central nervous system (CNS) tissue, gut, ring gland, and epidermal cells and in 13- to 16-h-old embryos, older embryos, and late third instars.

The EcR protein contains two highly conserved domain characteristic of vertebrate steroid receptors. The N-terminal portion of the molecule contains the **DNA-binding domain**, while the **ecdysteroid-binding domain** is near the C-terminus where there is a dimerization sequence, again a similarity to previously discovered vertebrate steroid receptors. EcR forms dimers that are stable only when binding between protein and ecdysteroid occurs; in contrast, some of the vertebrate receptors form stable dimers even when not bound to hormone (Ozyhar et al., 1991).

When the sequence of amino acids in the hormone-binding domain of the EcR protein is compared with sequences of amino acids in vertebrate steroid receptors, EcR is most similar to a

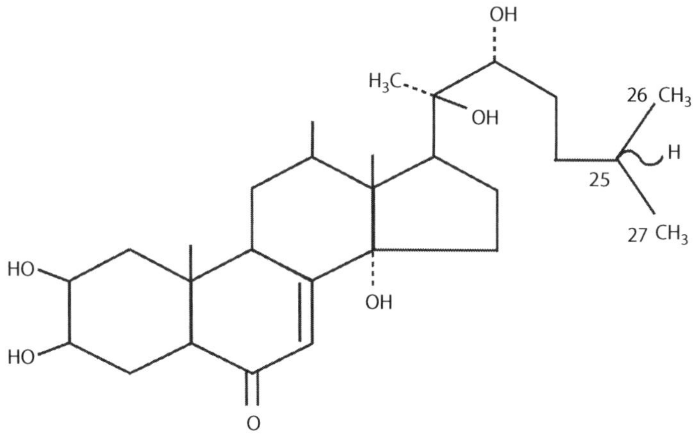

HORMONES AND DEVELOPMENT

Figure 5.19 Structure of ponasterone, a synthetic ecdysteroid with low hormonal activity *in vivo* and in *Drosophila* cell lines and after iodination with ^{125}I to form 26-iodoponasterone, which has relatively high hormonal activity and high specific activity due to the ^{125}I.

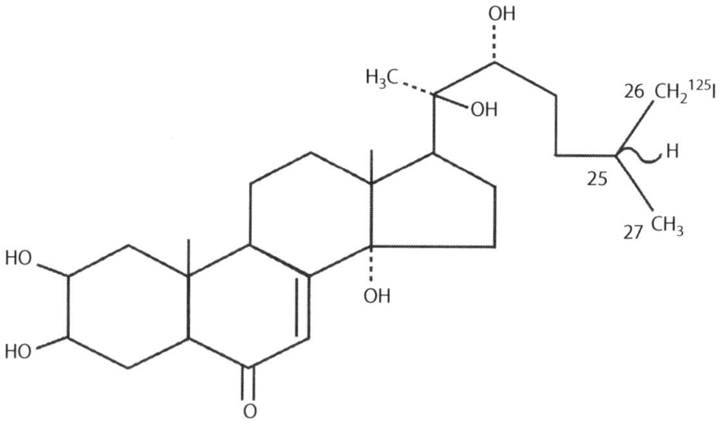

subfamily of steroid receptors that include the human thyroid receptor, human vitamin D receptor, and human retinoic acid receptor as well as some other hormone receptors. All of the most highly conserved amino acids in the DNA-binding domain of the vertebrate steroid receptors are present in EcR, including **nine cysteine residues**. Eight of the cysteine residues coordinate two zinc atoms, as in the vertebrate receptors, resulting in the folding of the receptor chain into **two zinc fingers** (Figure 5.20). The zinc fingers are involved in binding the hormone-receptor complex to DNA at base sequences known as **ecdysone response elements** (**EREs**). In the case of vertebrates in which there are a number of different steroid hormones, the binding response elements are called simply hormone response elements (HREs).

EcR binds to the heat shock protein gene *hsp27*, a gene that at present is not known to have a defined role in molting. It codes for heat shock protein 27 (MSP27) and is activated both by heat shock and by 20-hydroxyecdysone. EcR specifically binds to the *hsp27* gene promoter region at the EREs with the **imperfect palindromic base sequence GGTTCA–TGCACT** that resembles the known vertebrate steroidal HREs (Ozyhar et al., 1991). The *hsp27*-DNA-bound EcR has a molecular mass of about 270 kDa (Ozyhar et al., 1991).

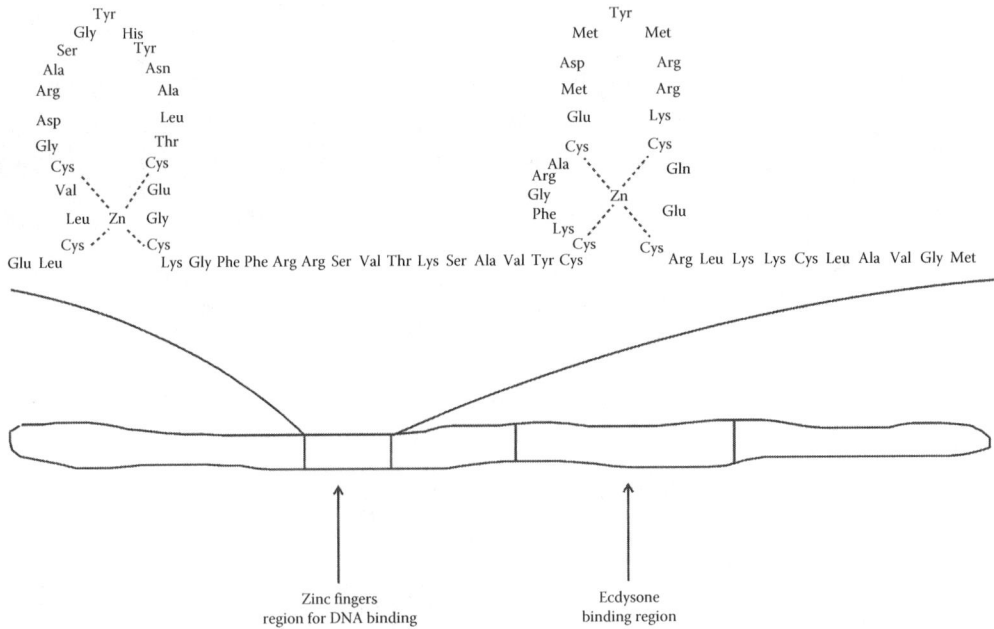

Figure 5.20 Representation of the amino acid sequence and Zn atoms in the two zinc fingers in the DNA-binding region of the ecdysone receptor isolated from *Drosophila*-cultured K_c cells. Additional portions of the receptor molecule (bottom diagram) are involved with actual binding of the ecdysteroid hormone and possibly with transactivation after binding to the DNA.

5.7.3 Differential Tissue and Cell Response to Ecdysteroids

A major challenge is to explain the basis for differential cell and tissue response to ecdysone. Ecdysteroid action on genes may involve activation of some genes, repression of some genes, and indirect action through transcription products. Clearly the interaction with JH also must be considered, but its interaction with ecdysteroid hormone to control development is much less clear than the present state of ecdysteroid action. Not all tissues and cells respond to ecdysone and JH in the same way, yet presumably all are exposed to the same hormonal stimulation at each molt. For example, epidermal cells respond at each molt and secrete a cuticle that may be larval, pupal, or adult in structure, depending upon the interaction of ecdysteroids and JH. On the other hand, some tissues, such as nervous system and imaginal discs, change little or not at all during some of the ecdysteroid pulses but may respond later in development, or even during adult life (e.g., reproductive organs). How can these tissues, all exposed to the same hormonal stimuli, respond so differently? The answer, at least in part, must lie in (1) the presence of different molecular forms of the ecdysteroid receptor, (2) different number and/or combinations of receptors in different tissues, (3) dimerization and/or heterodimerization of the receptor, and (4) the presence and interaction of specific tissue factors and hormone-induced transcription factors acting in conjunction with the hormone-receptor complex.

Multiple ecdysone receptors are known to exist, but tissue distribution and numbers of the different ecdysone receptors remain uncertain. A single gene has been identified and cloned from *D. melanogaster* that codes for three different ecdysteroid receptors (Talbot et al., 1993). Only the B1 receptor is predominant in larvae. The imaginal discs mainly contain the A form (Talbot et al., 1993). The three receptor forms, EcR-A, EcR-B1, and EcR-B2, have common domains for binding DNA and ecdysteroids, but each has its own unique N-terminal domain, the region believed to

HORMONES AND DEVELOPMENT

direct or modify the type of response a cell makes to a steroid hormone. Thus, one way that differential cell and tissue response to ecdysteroid secretion can be mediated is through the number, type, and distribution of different receptors.

During larval development of *Drosophila* and *Manduca*, the CNS contains little ecdysone receptor, but at pupariation all cells have high levels of EcR, and in *Drosophila* the receptor is the B1 form. The B1 receptor disappears after pupariation, and in adult development, the type and amount of EcR varies with whether a particular cell is a new cell or a cell from the larva that has metamorphosed (Truman et al., 1994). Application of JH at the appropriate time can prevent the appearance of EcR in cells of the CNS of *Manduca* (Riddiford, 1996), but JH does not prevent the appearance of 20-hydroxyecdysone-induced EcR mRNA in epidermal cells during larval or pupal molt in *Manduca*. In the latter, tissue EcR is present in larval epidermis throughout larval life with higher levels occurring at molts.

Another possible mechanism for differential response is in the **dimerization** of the ecdysteroid receptor. Steroid receptors in vertebrates, and apparently in insects, typically form dimers before they bind to the HREs of DNA (Figure 5.21). Both homodimers and heterodimers are known in vertebrates. In **homodimers**, two receptor molecules bind together with each also binding a steroid hormone molecule. **Heterodimers** are composed of one molecule of a receptor combined with a different receptor with each receptor binding a steroid molecule. Heterodimers and homodimers may confer upon the receptor complex different or variable gene regulatory properties. The nature of the dimerization may be one way that tissues respond differentially to ecdysone exposure and the way that the same tissue responds differently to ecdysone at different times. An important role for heterodimer formation in vertebrate hormone signaling has been demonstrated, but its importance remains to be shown in insects. *Drosophila* EcR, however, is known to form a heterodimer with the *Drosophila* protein Ultraspiracle (USP), a product of the *ultraspiracle* (*usp*) gene locus (Yao et al., 1992; Thomas et al., 1993). Heterodimerization conferred DNA binding and functional activity on the complex in the presence of 20-hydroxyecdysone in cotransfected CV1 monkey kidney cells.

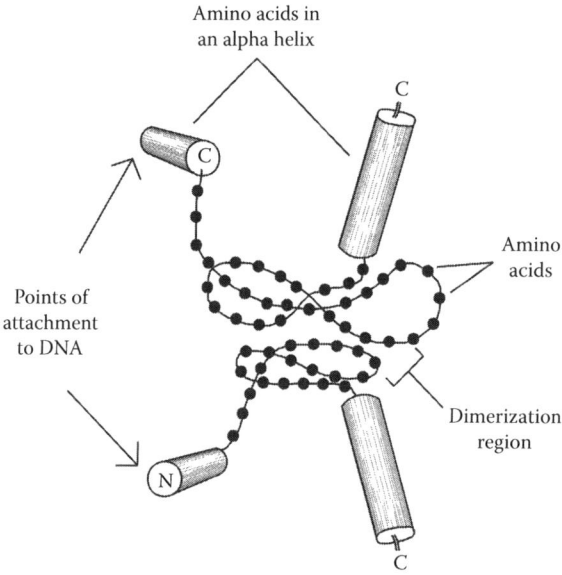

Figure 5.21 A model representing the potential binding of a dimer composed of two vertebrate steroid receptor molecules. The zinc fingers, the DNA-binding region of each receptor, may bind to HREs situated at adjacent major grooves located 34 Å apart in the DNA helix. (Redrawn and modified from Schwabe, J.W.R. and Rhodes, D., *Trends Biochem. Sci.*, 16, 291, 1991.)

Tissue and cell response may be related also to the presence of **tissue-specific factors** and **transcription products** that can modify hormone-receptor action. For example, the transcription factor designated E74 occurs in *D. melanogaster* in two forms appearing sequentially during pupariation. The two forms may occur in response to different transcription times and possibly in response to different levels of 20-hydroxyecdysone in the tissues in the time leading up to and during pupariation (Thummel, 1990; Thummel et al., 1990). When details of these and perhaps other control mechanisms are elucidated, it may be clearer why some late third instar tissues exposed to the same strong pulse of ecdysone undergo histolysis (e.g., gut, salivary glands), while others (imaginal discs) grow and form adult structures.

Finally, all DNA-binding sites for ecdysteroid-receptor complex may not function as response elements (i.e., all may not allow the hormone-receptor complex to promote, inhibit, or modify gene action). Additional transcription factors, possibly specific to particular cells, may be required to turn a binding site into a response element. Since it is believed that the ecdysone receptor must form a dimer in order to bind to DNA, as all vertebrate steroid receptors do, then formation of and interactions between homo- and heterodimers may influence how the hormonal message is applied in each cell and tissue.

5.8 POSSIBLE TIMER GENE IN THE MOLTING PROCESS

The early gene *E74A* unit in *D. melanogaster* is large, consisting of 60 kb, and it takes about 1 h to transcribe the gene, with elongation of the RNA transcript taking place at a rate of about 1.1 kb/min (Thummel et al., 1990). In agreement with this time frame, *E74A* mRNA is not detectable in the cell cytoplasm sooner than about 60 min after gene puffing action. These data provide an explanation why *E74* puffs continue to expand during the first hour after exposure to ecdysone; they expand because new transcripts of mRNA, beginning at the 5′ end, are being formed at the puff site; as more transcript is made, the puff gets bigger. After about an hour, the first complete transcripts are released from the end of the 60 kb unit, and the rate of new transcript formation reaches an equilibrium with the release of transcripts. This agrees with observations that the puff slows in expansion after about an hour and reaches a stable size until it begins to regress at about 4 h posttreatment. Regression in size after about 4 h presumably reflects the time it takes for enough of the new mRNA transcript to be translated into protein products or for some other product(s) of transcription to be made, which act to repress the *E74A* promoter. The half-life of *E74A* mRNA appears to be about 1 h (Thummel et al., 1990).

At least three of the early puff genes, *E74A*, *E75*, and *2B5*, are large and also might serve as timer genes. If the genes were small, on the order of only a few kb in size, then transcripts would be pushed off the end in a few minutes, and the genes likely would be repressed by the products of their transcripts in a much shorter time than the 4–6 h that actually occurs. Whether the large size of several of the early genes is indeed related to a significant timing of events is not known, but the **timer gene concept** is important as a model. One can now imagine that some physiological, morphogenetic, or biochemical event may be timed by the size of a gene unit and the time required for transcripts to be made.

5.9 ECDYSONE–GENE INTERACTION IDEAS STIMULATED VERTEBRATE WORK

It took about 30 years of work to prove the astute guess made by Clever and Karlson in 1960 that ecdysone alters "the activity of specific genes." The idea that steroid hormones might work at the gene level was actively pursued also by vertebrate biologists and geneticists, and they were successful much sooner than insect biologists in demonstrating that vertebrate steroid hormones bind to DNA and work at the gene level. There are many more details available at the molecular level on the receptors and binding of vertebrate steroid hormones than is known about ecdysteroid function in insects and other invertebrates. One important aspect of the comparative vertebrate and invertebrate work is the revelation that the basic mechanism of steroid hormone action has been very conserved, and some functional aspects of steroid hormone action clearly predates the separation of vertebrate and invertebrate lines of evolution.

Vertebrate steroid hormones first bind either to a cytoplasmic receptor (the glucocorticoid steroids) or a nuclear receptor (estrogen). The receptors are proteins. After binding occurs, the hormone-receptor complex translocates to the nucleus (if the receptor is not already located in the nucleus), where binding to nuclear DNA occurs. Genes may be turned on or off by the action of the bound hormone receptor and transcription of messenger RNA is regulated.

Vertebrate steroid receptors bind the steroid hormone near the carboxy terminal end of the protein molecule. Near the amino terminal end, the vertebrate steroid receptors contain a highly conserved region with a sequence of amino acids that recognizes and binds to a specific sequence of bases in DNA, the HRE. The sequence of amino acids in several vertebrate steroid receptors has been determined, and they share a common feature in which the DNA-binding region of the protein chain is folded into two zinc fingers. Each zinc finger structure is maintained by a single zinc atom that forms coordinate bonds to four cysteine residues.

The hormone-receptor complex typically binds to DNA as a dimer, and it is the carboxy terminal region of the receptor that contains the dimerization sequence. Both ends of the receptor protein contain regions that have transactivation and possibly transcription regulatory action. Further discussions of vertebrate mechanisms are not appropriate here, but much more extensive reviews can be found in Carson-Jurica et al. (1990) and Schwabe and Rhodes (1991).

5.10 REVIEW AND SELF-STUDY QUESTIONS

1. What does the acronym PTTH stand for?
2. PTTH is produced or synthesized in one organ in the insect and released into the hemolymph from a different organ. The release site is different in some insects. Name the synthesis site and the release sites that are known.
3. What is a neurohemal organ? What is its function? Name one in insects.
4. What is the definition of the "critical period" as it applies to insect hormones?
5. What organ (gland) synthesizes JH?
6. What are the three mechanisms that have been described as the stimulus for secretion of PTTH?
7. Why is a sterol a dietary requirement for insects?
8. What is the biochemical difference between ecdysone and 20-hydroxyecdysone?
9. Does the designation of the hydroxyl group in 20-hydroxyecdysone as "beta" mean that the hydroxyl group projects out toward the reader or back into the plane of the paper in a diagram of β-hydroxyecdysone on the page of a book?
10. What is one of the best ways to detect JH in insect hemolymph or insect tissues?
11. Why is it important that mechanisms for degradation of JH and ecdysone occur in insect tissues?
12. What rather unusual chemical group occurs in all of the naturally occurring JH molecules?
13. What are ATs and ASTs? Name at least one function of each?
14. Why are JHBPs in the hemolymph important?
15. What is the name of some IGRs? Do they have hormonal activity?
16. What are polytene chromosomes? How are they formed? What are chromosomal puffs?
17. Define transcription and translation as these terms are applied to gene and ecdysteroid action.
18. What sort of molecule is the ecdysone receptor? Where is it located in cells?
19. How was 26-iodoponasterone important in working out how ecdysone acts at the gene level?

20. What are some of the ways that scientists have tried to explain the observed fact that different tissues respond differently (or even not at all) to developmental hormones such as ecdysteroids at different times in the life of an immature insect?
21. What is the nature of possible dimers in the function of the ecdysone receptor?
22. When the ecdysone receptor binds ecdysone, what happens next?
23. Describe the biochemical or molecular mechanism for the action of ecdysone in insect development. Explain in some detail how it works.

REFERENCES

Abdou, M., C. Peng, J. Huang, O. Zyaan, S. Wang, S. Li, and J. Wang. 2011b. Wnt signaling cross-talks with JH signaling by suppressing *Met* and Gce expression. *PLoS One* 6: e26772.

Abdou, M.A., Q. He, D. Wen, O. Zyaan, J. Wang, J. Xu, A.A. Baumann et al. 2011a. *Drosophila* Met and Gce are partially redundant in transducing juvenile hormone action. *Insect Biochem. Mol. Biol.* 41: 938–945.

Agui, N., W. Bollenbacher, N. Granger, and L.I. Gilbert. 1980. Corpus allatum is release site for the insect prothoracicotropic hormone. *Nature (Lond.)* 285: 669–670.

Andersen, J.F., M. Ceruso, G.C. Unnithan, E. Kuwano, G.D. Prestwich, and R. Feyereisen. 1995. Photoaffinity labeling of methyl farnesoate epoxidase in cockroach corpora allata. *Insect Biochem. Mol. Biol.* 25: 713–719.

Ashburner, M. 1972. Patterns of puffing activity in the salivary gland chromosomes of *Drosophila*. *Chromosoma* 38: 255–281.

Ashburner, M. and G. Richards. 1976. Sequential gene activation of ecdysone in polytene chromosomes of *Drosophila melanogaster*. *Develop. Biol.* 54: 241–255.

Audsley, N., R.J. Weaver, and J.P. Edwards. 2000. Juvenile hormone biosynthesis by corpora allata of larval tomato moth, *Lacanobia oleracea*, and regulation by *Manduca sexta* allatostatin and allatotropin. *Insect Biochem. Mol. Biol.* 30: 681–689.

Bai, J., Y. Uehara, and D.J. Montell. 2000. Regulation of invasive cell behavior by Taiman, a *Drosophila* protein related to AIB1, a steroid receptor coactivator amplified in breast cancer. *Cell* 103: 1047–1058.

Baker, F.C., L.W. Tsai, C.C. Reuter, and D.A. Schooley, 1987. In vivo fluctuation of JH, JH acid, and ecdysteroid titer, and JH esterase activity, during development of fifth stadium *Manduca sexta*. *Insect Biochem.* 17: 989–996.

Baumann, A., J. Barry, S. Wang, Y. Fujiwara, and T.G. Wilson. 2010a. Paralogous genes involved in juvenile hormone action in *Drosophila melanogaster*. *Genetics* 185: 1327–1336.

Baumann, A., Y. Fujiwara, and T.G. Wilson. 2010b. Evolutionary divergence of the paralogs Methoprene tolerant *(Met)* and germ cell expressed *(gce)* within the genus *Drosophila*. *J. Insect Physiol.* 56: 1445–1455.

Bayer, C., L. von Kalm, and J.W. Fristrom. 1996. Gene regulation in imaginal disc and salivary gland development during *Drosophila* metamorphosis, in L.I. Gilbert, J.R. Tata, and D.G. Atkinson (eds.), *Metamorphosis, Postembryonic Reprogramming of Gene Expression in Amphibian and Insect Cells*. Academic Press, San Diego, CA, pp. 321–361.

Becker, H.J. 1959. Die Puffs Der Speicheldrusenchromosomen von *Drosophila melanogaster*. *Chromosoma* 10: 654–678.

Berger, E.M. and E.B. Dubrovsky. 2005. Juvenile hormone molecular actions and interactions during development of *Drosophila melanogaster*. *Vitam. Horm.* 73: 175–215.

Bergot, B.J., M. Ratcliff, and D.A. Schooley. 1981. Method for quantitative determination of the four known juvenile hormones in insect tissue using gas chromatography-mass spectroscopy. *J. Chromatogr.* 204: 231–244.

Bergot, B.J., D.A. Schooley, G.M. Chippendale, and C.-M. Yin. 1976. Juvenile hormone titer determinations in the southwestern corn borer, *Diatraea grandiosella* by electron capture-gas chromatography. *Life Sci.* 18: 811–820.

Bernado, T.J. and E.B. Dubrovsky. 2012. Molecular mechanisms of transcription activation by juvenile hormone: A critical role for bHLH-PAS and nuclear receptor proteins. *Insects* 3: 324–338.

Bhaskaran, G., S.P. Sparagana, P. Barrera, and K.H. Dahm. 1986. Change in corpus allatum function during metamorphosis of the tobacco hornworm *Manduca sexta*. Regulation at the terminal step in juvenile hormone biosynthesis. *Arch. Insect Biochem. Physiol.* 3: 321–338.

Borovsky, D., D.A. Carlson, R.G. Hancock, H. Rembold, and E. Van Handel. 1994. *De novo* biosynthesis of juvenile hormone III and I by the accessory glands of the male mosquito. *Insect Biochem. Mol. Biol.* 24: 437–444.

Borst, D.W., M.R. Eskew, S.J. Wagner, K. Shores, J. Hunter, L. Luker, J.D. Hatle, and L.B. Hecht. 2000. Quantification of juvenile hormone III, vitellogenin, and vitellogenin-mRNA during the oviposition cycle of the lubber grasshopper. *Insect Biochem. Mol. Biol.* 30: 813–819.

Borst, D.W. and J.D. O'Connor. 1972. Arthropod molting hormone: Radioimmune assay. *Science* 178: 418–419.

Bowers, W.J., T. Ohta, J.S. Cleere, and P.A. Marsella. 1976. Discovery of insect anti-juvenile hormones in plants. *Science* 193: 542–547.

Brogiolo, W., H. Stocker, T. Ikeya, F. Rintelen, R. Fernandez, and E. Hafen. 2001. An evolutionarily conserved function of the *Drosophila* insulin receptor and insulin-like peptides in growth control. *Curr. Biol.* 11: 213–221.

Butenandt, A. and P. Karlson. 1954. Über die isolierung eines Metamorphose-Hormone der Insekten in kristallierten Form. *Z. Naturforsch.* 9b: 389–391.

Cao, C. and M.R. Brown. 2001. Localization of an insulin-like peptide in brains of two flies. *Cell Tissue Res.* 304: 317–321.

Carson-Jurica, M.A., W.T. Schrader, and B.W. O'Malley. 1990. Steroid receptor family: Structure and functions. *Endocr. Rev.* 11: 210–220.

Charles, J.P., T. Iwema, V.C. Epa, K. Takaki, J. Rynes, and M. Jindra. 2011. Ligand-binding properties of a juvenile hormone receptor, Methoprene-tolerant. *Proc. Natl. Acad. Sci. USA* 108: 21128–21133.

Chen, J., M. Kabbouh, M.J. Fisher, and H.H. Rees. 1994. Induction of an inactivation pathway for ecdysteroid in larvae of the cotton leafworm, *Spodoptera littoralis*. *Biochem. J.* 301: 89–95.

Cherbas, L., K. Lee, and P. Cherbas. 1991. Identification of ecdysone response elements by analysis of the *Drosophila Eip28/29* gene. *Genes Dev.* 5: 120–131.

Cherbas, P., L. Cherbas, S. Lee, and K. Nakanishi. 1988. 26-[^{125}I]Iodoponasterone A is a potent ecdysone and a sensitive radioligand for ecdysone receptors. *Proc. Natl. Acad. Sci. USA* 85: 2096–2100.

Clever, U. and P. Karlson. 1960. Induktion Von Puff-Veranderugen In den Speicheldrusenchromosomen von *Chironomus tentans* durch Ecdyson. *Exp. Cell Res.* 20: 623–626.

Cusson, M., K.J. Yagi, Q. Ding, H. Duve, A. Thorpe, J.N. McNeil, and S.S. Tobe. 1991. Biosynthesis and release of juvenile hormone and its precursors in insects and crustaceans: The search for a unifying arthropod endocrinology. *Insect Biochem.* 21: 1–6.

Dubrovsky, E.B., V.A. Dubrovskaya, T. Bernardo, V. Otte, R. DiFilippo, and H. Bryan. 2011. The *Drosophila* FTZ-F1 nuclear- receptor mediates juvenile hormone activation of *E754* gene expression through an intracellular pathway. *J. Biol. Chem.* 286: 33689–33700.

East, P.D., T.D. Sutherland, S.C. Trowell, A.J. Herlt, and R.W. Richards. 1997. Juvenile hormone synthesis by ring glands of the blowfly *Lucilia cuprina*. *Arch. Insect Biochem. Physiol.* 34: 239–253.

Edwards, J.P., T.S. Corbitt, H.F. McArdle, J.E. Short, and R.J. Weaver. 1995. Endogenous levels of insect juvenile hormones in larval, pupal, and adult stages of the tomato moth, *Lacanobia oleracea*. *J. Insect Physiol.* 41: 641–651.

Engelmann, F. and J. Mala. 2000. The interactions between juvenile hormone (JH), lipophorin, vitellogenin, and JH esterases in two cockroach species. *Insect Biochem. Mol. Biol.* 30: 793–803.

Erezyilmaz, D.F., L.M. Riddiford, and J.W. Truman. 2006. The pupal specifier *broad* directs progressive morphogenesis in a direct developing insect. *Proc. Natl. Acad. Sci. USA* 103: 6925–6930.

Everaerts, C., M. Cusson, and J.N. McNeil. 2000. The influence of smoke volatiles on sexual maturation and juvenile hormone biosynthesis in the black army cutworm, *Actevia fennica* (Lepidoptera: Noctuidae). *Insect Biochem. Mol. Biol.* 30: 855–862.

Feldlaufer, M.F., G.F. Weirich, R.B. Imberski, and J.A. Svoboda. 1995. Ecdysteroid production in *Drosophila melanogaster* reared on defined diets. *Insect Biochem. Mol. Biol.* 25: 709–712.

Fraenkel, G. 1935. A hormone causing pupation in the blowfly *Calliphora erythrocephala*. *Proc. R. Soc. Lond., Ser. B* 118: 1–12.

Fukuda, S. 1940. Hormonal control of moulting and pupation in the silkworm. *Proc. Imp. Acad. Jpn.* 16: 417–420.

Fukuda, S. 1944. The hormonal mechanism of larval molting and metamorphosis in the silkworm. *J. Fac. Sci. Tokyo Univ.* 4: 477–532.

Gelman, D.B., J.E. Carpenter, and P.D. Greany. 2000. Ecdysteroid levels/profiles of the parasitoid wasp, *Diapetimorpha introita*, reared on its host, *Spodoptera frugiperda* and on an artificial diet. *J. Insect Physiol.* 46: 457–465.

Gilbert, L.I. (ed.). 2012. *Insect Endocrinology*. Elsevier, Amsterdam, the Netherlands.

Gilbert, L.I., W.L. Combest, W.A. Smith, and V.H. Meller. 1988. Neuropeptides, second messengers and insect molting. *BioEssays* 8: 153–157.

Gilbert, L.I., N.A. Granger, and R.M. Roe. 2000. The juvenile hormones: Historical facts and speculations on future research directions. *Insect Biochem. Mol. Biol.* 30: 617–644.

Gilbert, L.I., R. Rybczynski, and S.S. Tobe. 1996a. Endocrine cascade in insect metamorphosis, in L.I. Gilbert, J.R. Tata, and B.G. Atkinson (eds.), *Metamorphosis: Postembryonic Reprogramming of Gene Expression in Amphibian and Insect Cells*. Academic Press, New York, pp. 59–107.

Gilbert, L.I., R. Rybczynski, and J.T. Warren. 2002. Control and biochemical nature of the ecdysteroidogenic pathway. *Annu. Rev. Entomol.* 47: 883–916.

Gilbert, L.I. and H. Schneiderman. 1960. The development of a bioassay for the juvenile hormone of insects. *Trans. Am. Micros. Soc.* 79: 38–67.

Gilbert, L.I., J.R. Tata, and B.G. Atkinson (eds.). 1996b. *Metamorphosis: Postembryonic Reprogramming of Gene Expression in Amphibian and Insect Cells*. Academic Press, New York.

Godlewski, J., S. Wang, and T.G. Wilson. 2006. Interaction of bHLH-PAS proteins involved in juvenile hormone reception in *Drosophila*. *Biochem. Biophys. Res. Commun.* 342: 1305–1311.

Granger, N.A., R. Ebersohl, and T.C. Sparks. 2000. Pharmacological characterization of dopamine receptors in the corpus allatum of *Manduca sexta* larvae. *Insect Biochem. Mol. Biol.* 30: 755–766.

Granger, N.A., S.L. Sturgis, R. Ebersohl, C. Geng, and T.C. Sparks. 1996. Dopaminergic control of corpora allata activity in the larval tobacco hornworm, *Manduca sexta*. *Arch. Insect Biochem. Physiol.* 32: 449–466.

Grieneisen, M.L., J.T. Warren, and L.I. Gilbert. 1993. Early steps in ecdysteroid biosynthesis: Evidence for the involvement of cytochrome P-450 enzymes. *Insect Biochem. Mol. Biol.* 23: 13–23.

Gu, S.-H. and Y.-S. Chou. 1993. Role of low ecdysteroid levels in the early last larval instar of *Bombyx mori*. *Experientia* 49: 806–809.

Gu, S.-H., Y.-C. Hsieh, S.-C. Young, and P.-L. Lin. 2013. Involvement of phosphorylation of adenosine 5′-monophosphate-activated protein kinase in PTTH-stimulated ecdysteroidogenesis in prothoracic glands of the silkworm, *Bombyx mori*. *PLoS One* 8: e63102.

Gu, S.-H., J.-L. Lin, and P.-L. Lin. 2010. PTTH-stimulated ERK phosphorylation in prothoracic glands of the silkworm, *Bombyx mori*: Role of Ca^{2+}/calmodulin and receptor tyrosine kinase. *J. Insect Physiol.* 56: 93–101.

Gu, S.-H., W.-I. Yeh, S.-C. Young, P.-L. Lin, and S. Li. 2012. TOR signaling is involved in PTTH-stimulated ecdysteroidogenesis by prothoracic glands in the silkworm, *Bombyx mori*. *Insect Biochem. Mol. Biol.* 42: 296–303.

Gu, S.-H., S.-C. Young, J.-L. Lin, and P.-L. Lin. 2011. Involvement of PI3K/Akt signaling in PTTH-stimulated ecdysteroidogenesis by prothoracic glands in the silkworm, *Bombyx mori*. *Insect Biochem. Mol. Biol.* 41: 197–202.

Handler, A.M. 1982. Ecdysteroid titers during pupal and adult development in *Drosophila melanogaster*. *Dev. Biol.* 93: 73–82.

Hill, R.J., I.M.L. Billas, F. Bonneton, L.D. Graham, and M.C. Lawrence. 2013. Ecdysone receptors: From the Ashburner model to structural biology. *Annu. Rev. Entomol.* 58: 251–271.

Hiruma, K. and L.M. Riddiford. 1993. Molecular mechanisms of cuticular melanization in the tobacco hornworm, *Manduca sexta* (L.) (Lepidoptera: Sphingidae). *Int. J. Insect Morphol. Emryol.* 22: 103–117.

Hiruma, K. and L.M. Riddiford. 2010. Developmental expression of mRNAs for epidermal and fat body proteins and hormonally regulated transcription factors in the tobacco hornworm, *Manduca sexta*. *J. Insect Physiol.* 56: 1390–1395.

Hsieh, Y.-C., S.-L. Hsu, and S.-H. Gu. 2013. Involvement of reactive oxygen species in PTTH-stimulated ecdysteroidogenesis in prothoracic glands of the silkworm, *Bombyx mori*. *Insect Biochem. Mol. Biol.* 43: 859–866.

Hsieh, Y.-C., P.-L. Lin, and S.-H. Gu. 2014. Signaling of reactive oxygen species in PTTH-stimulated ecdysteroidogenesis in prothoracic glands in the silkworm, *Bombyx mori*. *J. Insect Physiol.* 63: 32–39.

Hua, Y.-J., R.-J. Jiang, and J. Koolman. 1997. Multiple control of ecdysone biosynthesis in blowfly larvae: Interaction of ecdysiotropins and ecdysiostatins. *Arch. Insect Biochem. Physiol.* 35: 125–134.

Huang, Z.-H., G.E. Robinson, and D.W. Borst. 1994. Physiological correlates of division of labor among similarly aged honeybees. *J. Comp. Physiol. A* 174: 731–739.

Hunnicutt, D., Y.C. Toong, and D.W. Borst. 1989. A chiral specific antiserum for juvenile hormone. *Am. Zool.* 29: 48a.

Jindra, M., S.R. Palli, and L.M. Riddiford. 2013. The juvenile hormone signaling pathway in insect development. *Annu. Rev. Entomol.* 58: 181–204.

Jones, G., D. Jones, P. Teal, A. Sapa, and M. Wozniak. 2006. The retinoid-X receptor ortholog, ultraspiracle, binds with nanomolar affinity to an endogenous morphogenetic ligand. *FEBS J.* 273: 4983–4996.

Kanost, M.R., J.K. Kawooya, J.H. Law, R.O. Ryan, M.C. Van Heusden, and R. Ziegler. 1990. Insect hemolymph proteins. *Adv. Insect Physiol.* 22: 299–396.

Kappler, C., M. Kabbouh, C. Hetru, F. Durst, and J.A. Hoffmann. 1988. Characterization of three hydroxylases involved in the final steps of biosynthesis of the steroid hormone ecdysone in *Locusta migratoria* (Insecta, Orthoptera). *J. Steroid Biochem.* 31: 891–898.

Kawakami, A., H. Kataoka, T. Oka, A. Mizoguchi, M. Kimura-Kawakami, T. Adachi, M. Iwami, H. Nagasawa, A. Suzuki, and H. Ishizaki. 1990. Molecular cloning of the *Bombyx mori* prothoracicotropic hormone. *Science* 247: 1333–1335.

Ke, O. 1930. Morphological variation of the prothoracic gland in the domestic and the wild silkworm (In Japanese). *Bult. Sci. Fak. Terkult. Kjusu Imp. Univ.* 4: 12–21.

King, L.E. and S.S. Tobe. 1993. Changes in the titre of a juvenile hormone III binding lipophorin in the haemolymph of *Diploptera punctata* during development and reproduction: Functional significance. *J. Insect Physiol.* 39: 241–251.

Koelle, M.R., W.S. Talbot, W.A. Segraves, M.T. Bender, P. Cherbas, and D.S. Hogness. 1991. The *Drosophila* EcR gene encodes an ecdysone receptor, a new member of the steroid receptor super family. *Cell* 67: 59–77.

Konopova, B. and M. Jindra. 2007. Juvenile hormone resistance gene Methoprene-tolerant controls entry into metamorphosis in the beetle *Tribolium castaneum*. *Proc. Natl. Acad. Sci. USA* 104: 10488–10493.

Konopova, B., V. Smykal, and M. Jindra. 2011. Common and distinct roles of juvenile hormone signaling genes in metamorphosis of holometabolous and hemimetabolous insects. *PLoS One* 6: e28728.

Kopeč, S. 1917. Experiments on metamorphosis of insects. *Bull. Int. Acad. Cracovie* (B): 57–60.

Kotaki, T., T. Shinada, K. Kaihara, Y. Ohfune, and H. Numata. 2009. Structure determination of a new juvenile hormone from a heteropteran insect. *Org. Lett.* 11: 5234–5237.

Kotaki, T., T. Shinada, K. Kaihara, Y. Ohfune, and H. Numata. 2011. Biological activities of juvenile hormone III skipped bisepoxide in last instar nymphs and adults of a stink bug, *Plautia stall*. *J. Insect Physiol.* 57: 147–152.

Lafont, R. and I.D. Wilson. 1990. Advances in ecdysteroid high performance liquid chromatography, in A.R. McCaffery and I.D. Wilson (eds.), *Chromatography and Isolation of Insect Hormones and Pheromones*. Plenum Press, New York, pp. 79–94.

Lanzrein, B., R. Wilhelm, and R. Reichsteiner. 1993. Differential degradation of racemic and 10R-juvenile hormone-III by cockroach (*Nauphoeta cinerea*) haemolymph and the use of lipophorin for long-term culturing of corpora allata. *J. Insect Physiol.* 39: 53–63.

Lawrence, P.O. 1986. The role of 20-hydroxyecdysone in the moulting of *Biosteres longicaudatus*, a parasite of the Caribbean fruit fly, *Anastrepha suspensa*. *J. Insect Physiol.* 32: 329–337.

Lefevere, K.S., M.J. Lacey, P.H. Smith, and B. Roberts. 1993. Identification and quantification of juvenile hormone biosynthesized by larval and adult Australian sheep blowfly *Lucilia cuprina* (Diptera: Calliphoridae). *Insect Biochem. Mol. Biol.* 23: 713–720.

Lezzi, M. 1996. Chromosome puffing: Supramolecular aspects of ecdysone action, in L.I. Gilbert, J.R. Tata, and B.G. Atkinson (eds.), *Metamorphosis: Postembryonic Reprogramming of Gene Expression in Amphibian and Insect Cells*. Academic Press, New York, pp. 145–173.

Li, M., F.A. Mead, and J. Zhu. 2011. Heterodimer of two bHLH-PAS proteins mediates juvenile hormone-induced gene expression. *Proc. Natl. Acad. Sci. USA* 108: 638–643.

Lin, J.-L. and S.-H. Gu. 2011. Prothoracicotropic hormone induces tyrosine phosphorylation in prothoracic glands of the silkworm, *Bombyx mori*. *Arch. Insect Biochem. Physiol.* 78: 144–155.

Liu, Y., Z. Sheng, H. Liu, D. Wen, Q. He, S. Wang, W. Shao et al. 2009. Juvenile hormone counteracts the bHLH-PAS transcription factors MET and GCE to prevent caspase-dependent programmed cell death in *Drosophila*. *Development* 136: 2015–2025.

Louden, D., A. Handley, R. Lefont, S. Taylor, I. Sinclair, E. Lenz, T. Orton, and I.D. Wilson. 2002. HPLC analysis of ecdysteroids in plant extracts using superheated deuterium oxide with multiple on-line spectroscopic analysis (UV, IR, ^{1}H NMR, and MS). *Anal. Chem.* 74: 288–294.

Marchal, E., H.P. Vandersmissen, L. Badisco, S. Van de Velde, H. Verlinden, M. Iga, P. Van Wielendaele et al. 2010. Control of ecdysteroidogenesis in prothoracic glands of insects: A review. *Peptides* 31: 506–519.

Mauchamp, B., C. Royer, L. Kerhoas, and J. Einhorn. 1993. MS/MS analyses of ecdysteroids in developing eggs of *Dysdercus fasciatus*. *Insect Biochem. Mol. Biol.* 23: 199–205.

Minakuchi, C., X. Zhou, and L.M. Riddiford. 2008. Krüppel homolog 1 (*Kr-h1*) mediates juvenile hormone action during metamorphosis of *Drosophila melanogaster*. *Mech. Dev.* 125: 91–105.

Mirth, C., J.W. Truman, and L.M. Riddiford. 2005. The role of the prothoracic gland in determining critical weight for metamorphosis in *Drosophila melanogaster*. *Current Biol.* 15: 1796–1807.

Miura, K., M. Oda, S. Makita, and Y. Chinzei Y. 2005. Characterization of the *Drosophila Methoprene-tolerant* gene product. *FEBS J.* 272: 1169–1178.

Nagasawa, H., H. Kataoka, A. Isogai, S. Tamura, and A. Suzuki. 1984. Amino-terminal amino acid sequence of the silkworm prothoracicotropic hormone: Homology with insulin. *Science* 226: 1344–1345.

Nijhout, H.F. 1979. Stretch-induced moulting in *Oncopeltus fasciatus*. *J. Insect Physiol.* 25: 277–281.

Nijhout, H.F. 1984. Abdominal stretch reception in *Dipetalogaster maximus* (Hemiptera: Reduviidae). *J. Insect Physiol.* 30: 629–633.

Nijhout, H.F. 1994. *Insect Hormones*. Princeton University Press, Princeton, NJ.

Nijhout, H.F. and C.M. Williams. 1974. Control of moulting and metamorphosis in the tobacco hornworm *Manduca sexta* (L.): Cessation of juvenile hormone secretion as a trigger for pupation. *J. Exp. Biol.* 61: 493–501.

Oberlander, H. and S.M. Ferkovich. 1994. Physiological and developmental capacities of insect cell lines, in K. Maramorosch and A.H. McIntosh (eds.), *Insect Cell Technology*. CRC Press, Boca Raton, FL, pp. 127–140.

Oldham, S. and E. Hafen. 2003. Insulin/IGF and target of rapamycin signaling: A TOR de force in growth control. *Trends in Cell Biol.* 13: 79–85.

O'Reilly, D.R. and L.K. Miller. 1989. A baculovirus blocks insect molting by producing ecdysteroid UDP-glucosyl transferase. *Science* 245: 110–112.

Ozyhar, A., M. Strangmann-Diekmann, H. Kiltz, and O. Pongs. 1991. Characterization of a specific ecdysteroid receptor-DNA complex reveals common properties for invertebrate and vertebrate hormone-receptor/DNA interactions. *Eur. J. Biochem.* 200: 329–335.

Palli, S.R., T.R. Ladd, W.L. Tomkins, S. Shu, S.B. Ramaswamy, Y. Tanaka, B. Arif, and A. Retnakaran. 2000. *Choristoneura fumiferana* entomopoxvirus prevents metamorphosis and modulates juvenile hormone and ecdysteroid titers. *Insect Biochem. Mol. Biol.* 30: 869–876.

Park, Y.I., S. Shu, S.B. Ramaswamy, and A. Srinivasan. 1998. Mating in *Heliothis virescens*: Transfer of juvenile hormone during copulation by male to female and stimulation of biosynthesis of endogenous juvenile hormone. *Arch. Insect Biochem. Physiol.* 38: 100–107.

Peter, M.G., S. Gunawan, G. Gellisen, and H. Emmerich. 1979. Differences in hydrolysis and binding of homologous juvenile hormones in *Locusta migratoria* haemolymph. *Z. Naturforsch.* 34: 558–598.

Peter, M.G., H.P. Stupp, and K.U. Lentes. 1983. Reversal of the enantioselectivity in the enzymatic hydrolysis of juvenile hormone as a consequence of the protein fractionation. *Angew. Chem.* 95: 773–774.

Piulachs, M.D., V. Pagone, and X. Bellés. 2010. Key roles of the *Broad-Complex* gene in insect embryogenesis. *Insect Biochem. Mol. Biol.* 40: 468–475.

Porcheron, P., M. Moriniere, J. Grassi, and P. Pradelles. 1989. Development of an enzyme immunoassay for ecdysteroids using acetylcholinesterase as label. *Insect Biochem.* 19: 117–122.

Ramaswamy, S.B., S. Shu, Y.I. Park, and F. Zeng. 1997. Dynamics of juvenile hormone-mediated gonadotropism in the Lepidoptera. *Arch. Insect Biochem. Physiol.* 35: 539–558.

Rewitz, K.F., N. Yamanaka, L.I. Gilbert, and M.B. O'Connor. 2009. The insect neuropeptide PTTH activates receptor tyrosine kinase torso to initiate metamorphosis. *Science* 326: 1403–1405.

Rewitz, K.F., N. Yamanaka, and M.B. O'Connor. 2013. Developmental checkpoints and feedback circuits time insect maturation. *Current Topics in Developmental Biology* 103: 1–33.

Richard, D.S., S.W. Applebaum, and L.I. Gilbert. 1989b. Developmental regulation of juvenile hormone biosynthesis by the ring gland of *Drosophila melanogaster*. *J. Comp. Physiol. B* 159: 383–387.

Richard, D.S., S.W. Applebaum, T.J. Sliter, F.C. Baker, D.A. Schooley, C.C. Reuter, V.C. Henrich, and L.I. Gilbert. 1989a. Juvenile hormone bisepoxide biosynthesis in vitro by the ring gland of *Drosophila melanogaster*: A putative juvenile hormone in the higher Diptera. *Proc. Natl. Acad. Sci. USA* 86: 1421–1425.

Richter, K. 2001. Daily changes in neuroendocrine control of moulting hormone secretion in the prothoracic gland of the cockroach *Periplaneta americana* (L.). *J. Insect Physiol.* 47: 333–338.

Riddiford, L.M. 1994. Cellular and molecular actions of juvenile hormone I. General considerations and premetamorphic actions. *Adv. Insect Physiol.* 24: 213–274.

Riddiford, L.M. 1996. Molecular aspects of juvenile hormone action in insect metamorphosis, in L.I. Gilbert, J.R. Tata, and B.G. Atkinson (eds.), *Metamorphosis: Postembryonic Reprogramming of Gene Expression in Amphibian and Insect Cells*. Academic Press, New York, pp. 223–251.

Riddiford, L.M. 2007. Juvenile hormone action: A 2007 perspective. *J. Insect Physiol.* 54: 895–901.

Riddiford, L.M. 2012. How does juvenile hormone control insect metamorphosis and reproduction? *Gen. Comp. Endocrinol.* 179: 477–484.

Riddiford, L.M. and J.W. Truman. 1993. Hormone receptors and the regulation of insect metamorphosis. *Am. Zool.* 33: 340–347.

Riddiford, L.M., J.W. Truman, C.K. Mirth, and Y.C. Shen. 2010. A role for juvenile hormone in the prepupal development of *Drosophila melanogaster*. *Development* 137: 1117–1126.

Roe, R.M. and K. Venkatesh. 1990. Metabolism of juvenile hormones: Degradation and titer regulation, in A.P. Gupta (ed.), *Morphogenetic Hormones of Arthropods*, vol. 1. Rutgers University, New Brunswick, NJ, pp. 125–179.

Roundtree, D.B. and W.E. Bollenbacher. 1986. The release of prothoracicotropic hormone in the tobacco hornworm, *Manduca sexta*, is controlled intrinsically by juvenile hormone. *J. Exp. Biol.* 120: 41–58.

Royer, C., P. Porcheron, P. Pradelles, and B. Mauchamp. 1993. Development and use of an enzymatic tracer for new enzyme immunoassay for Makisterone A. *Insect Biochem. Mol. Biol.* 23: 193–197.

Sakurai, S., J.T. Warren, and L.I. Gilbert. 1989. Mediation of ecdysone synthesis in *Manduca sexta* by a hemolymph enzyme. *Arch. Insect Biochem. Physiol.* 10: 179–197.

Schiesari, L., C.P. Kyriacou, and R. Costa. 2011. The hormonal and circadian basis for insect photoperiodic timing. *FEBS Lett.* 585: 1450–1460.

Schooley, D.A., F.C. Baker, L.W. Tsai, C.A. Miller, and C.G. Jamieson. 1984. Juvenile hormones 0, I, and II exist only in Lepidoptera, in J.A. Hoffmann and M. Porchet (eds.), *Biosynthesis, Metabolism and Mode of Action of Invertebrate Hormones*. Springer-Verlag, Heidelberg, Germany, pp. 373–383.

Schwabe, J.W.R. and D. Rhodes. 1991. Beyond zinc fingers: Steroid hormone receptors have a novel structural motif for DNA recognition. *Trends Biochem. Sci.* 16: 291–296.

Sehnal, F., P. Svacha, and J. Grzavy. 1996. Evolution of insect metamorphosis, in L.I. Gilbert, J.R. Tata, and B.G. Atkinson (eds.), *Metamorphosis: Postembryonic Reprogramming of Gene Expression in Amphibian and Insect Cells*. Academic Press, San Diego, CA, pp. 3–58.

Smith, W.A. 1993. Second messengers and the action of prothoracicotropic hormone in *Manduca sexta*. *Am. Zool.* 33: 330–339.

Smith, W.A., W.L. Combest, and L.I. Gilbert. 1986. Involvement of cyclic AMP-dependent protein kinase in prothoracicotropic hormone-stimulated ecdysone synthesis. *Mol. Cell. Endocrinol.* 47: 25–33.

Smith, W.A., L.I. Gilbert, and W.E. Bollenbacher. 1984. The role of cyclic AMP in the regulation of ecdysone synthesis. *Mol. Cell. Endocrinol.* 37: 285–294.

Smith, W.A., L.I. Gilbert, and W.E. Bollenbacher. 1985. Calcium-cyclic AMP interactions in prothoracicotropic hormone stimulation of ecdysone synthesis. *Mol. Cell. Endocrinol.* 39: 71–78.

Smith, W.A. and R. Rybczynski. 2012. Prothoracicotropic hormone, in L.I. Gilbert (ed.), *Insect Endocrinology*. Academic Press, New York, pp. 1–62.

Sohi, S.S., S.R. Palli, B.J. Cook, and A. Retnakaran. 1995. Forest insect cell lines responsive to 20-hydroxyecdysone and two nonsteroidal ecdysone agonists, RH-5849 and RH-5992. *J. Insect Physiol.* 41: 457–464.

Sparagana, S.P., G. Bhaskaran, and P. Barrera. 1985. Juvenile hormone acid methyltransferase activity in imaginal discs of *Manduca sexta* prepupae. *Arch. Insect Biochem. Physiol.* 2: 191–202.

Stay, B., S.S. Tobe, and W.G. Bendena. 1994. Allatostatins: Identification, primary structures, functions and distribution. *Adv. Insect Physiol.* 25: 267–337.

Suzuki, Y., J.W. Truman, and L.M. Riddiford. 2008. The role of Broad in the development of *Tribolium castaneum*: Implications for the evolution of the holometabolous insect pupa. *Development* 135: 569–577.

Talbolt, W.S., E.A. Swyryd, and D.S. Hogness. 1993. *Drosophila* tissues with different metamorphic response to ecdysone express different ecdysone receptor isoforms. *Cell* 73: 1323–1337.

Tatar, M., A. Bartke, and A. Antebi. 2003. The endocrine regulation of aging by insulin-like signals. *Science* 299: 1346–1351.

Tatar, M., A. Kopelman, D. Epstein, M.-P. Tu, C.-M. Yin, and R.S. Garofalo. 2001. A mutant *Drosophila* insulin receptor homolog that extends life-span and impairs neuroendocrine function. *Science* 292: 107–110.

Teal, P.E.A., Y. Gomez-Simuta, and A.T. Proveaux. 2000. Mating experience and juvenile hormone enhance sexual signaling and mating in male Caribbean fruit flies. *Proc. Natl. Acad. Sci. USA* 97: 3708–3712.

Thomas, H.E., H.G. Stunnenberg, and F.A. Stewart. 1993. Heterodimerization of the *Drosophila* ecdysone receptor with retinoid X receptor and ultraspiracle. *Nature* 362: 471–475.

Thummel, C.S. 1990. Puffs and gene regulation-molecular insights into the *Drosophila* ecdysone regulatory hierarchy. *BioEssays* 12: 561–568.

Thummel, C.S., K.C. Burtis, and D.S. Hogness. 1990. Spatial and temporal patterns of E74 transcription during *Drosophila* development. *Cell* 61: 101–111.

Tobe, S.S. and R. Feyereisen. 1983. Juvenile hormone biosynthesis: Regulation and assay, in R.G.H. Downer and H. Laufer (eds.), *Endocrinology of Insects*. Alan R. Liss, Inc., New York, pp. 161–178.

Tobe, S.S. and B. Stay. 1985. Structure and regulation of the corpus allatum. *Adv. Insect Physiol.* 18: 305–432.

Touhara, K. and G.D. Prestwich. 1992. Binding site mapping of a photoaffinity-labeled juvenile hormone binding protein. *Biochem. Biophys. Res. Commun.* 182: 466–473.

Toyama, K. 1902. Contributions to the study of silk-worms. I. On the embryology of the silk-worm. *Bull. College Agric. Tokyo Imp. Univ.* 6: 73–118.

Truman, J.W. 2006. Steroid hormone secretion in insects comes of age. *Proc. Natl. Acad. Sci. USA* 103: 8909–8910.

Truman, J.W. and L.M. Riddiford. 2002. Endocrine insights into the evolution of metamorphosis in insects. *Annu. Rev. Entomol.* 47: 467–500.

Truman, J.W., W.S. Talbot, S.E. Fahrbach, and D.S. Hogness. 1994. Ecdysone receptor expression in the CNS correlates with stage-specific responses to ecdysteroids during *Drosophila* and *Manduca* development. *Development* 120: 219–234.

Tu, M.-P., C.-M. Yin, and M. Tatar. 2005. Mutations in insulin signaling pathway alter juvenile hormone synthesis in *Drosophila melanogaster*. *Gen. Comp. Endocrinol.* 142: 347–356.

Vafopoulou, X. and C.G.H. Steel. 1999. Daily rhythm of responsiveness to prothoracicotropic hormone in prothoracic glands of *Rhodnius prolixus*. *Arch. Insect Biochem. Physiol.* 41: 117–123.

Wainwright, G., M.C. Prescott, L.L. Lomas, S.G. Webster, and H.H. Rees. 1997. Development of a new high-performance liquid chromatography-mass spectrometric method for the analysis of ecdysteroids in biological extracts. *Arch. Insect Biochem. Physiol.* 35: 21–31.

Warren, J.T., S. Sakurai, D.B. Roundtree, and L.I. Gilbert. 1988a. Synthesis and secretion in vitro of ecdysteroids by the prothoracic gland of *Manduca sexta*. *J. Insect Physiol.* 34: 571–576.

Warren, J.T., S. Sakurai, D.B. Roundtree, L.I. Gilbert, S.-S. Lee, and K. Nakanishi. 1988b. Regulation of the ecdysteroid titer of *Manduca sexta:* Reappraisal of the role of the prothoracic glands. *Proc. Natl. Acad. Sci. USA* 85: 958–962.

Weirich, G.F., M.F. Feldlaufer, and J.A. Svoboda. 1993. Ecdysone oxidase and 3-oxoecdysteroid reductases in *Manduca sexta*: Developmental changes and tissue distribution. *Arch. Insect Biochem. Physiol.* 23: 199–211.

Wigglesworth, V.B. 1936. The function of the corpus allatum in the growth and reproduction of *Rhodnius prolixus* (Hemiptera). *Q. J. Microscop. Sci.* 79: 91–121.

Wigglesworth, V.B. 1940. The determination of characters at metamorphosis in *Rhodnius prolixus*. *J. Exp. Biol.* 17: 201–222.

Williams, C.M. 1947. Physiology of insect diapause. II. Interactions between the pupal brain and prothoracic glands in the metamorphosis of the giant silkworm, *Platysamia cecropia*. *Biol. Bull. Woods Hole* 93: 86–98.

Wilson, T.G. and J. Fabian. 1986. A *Drosophila melanogaster* mutant resistant to a chemical analog of juvenile hormone. *Dev. Biol.* 118: 190–201.

Woodring, J. and K.H. Hoffmann. 1994. The effects of octopamine, dopamine and serotonin on juvenile hormone synthesis, in vitro, in the cricket *Gryllus bimaculatus*. *J. Insect Physiol.* 40: 797–802.

Yamanaka, N., K.F. Rewitz, and M.B. O'Connor. 2013. Ecdysone control of developmental transitions: Lessons from *Drosophila* research. *Annu. Rev. Entomol.* 58: 497–516.

Yao, T., W.A. Segraves, A.E. Oro, M. McKeown, and R.M. Evans. 1992. *Drosophila* ultraspiracle modulates ecdysone receptor function via heterodimer formation. *Cell* 71: 63–72.

Yin, C.-M., B.-X. Zou, M. Jiang, M.-F. Li, W. Qin, T.L. Potter, and J.G. Stoffolano, Jr. 1995. Identification of juvenile hormone III bisepoxide (JHB_3), juvenile hormone III and methyl farnesoate secreted by the corpus allatum of *Phormia regina* (Meigen), in vitro and function of JHB_3 either applied alone or as a part of a juvenoid blend. *J. Insect Physiol.* 41: 473–479.

Zhang, Z., J. Xu, Z. Sheng, Y. Sui, and S.R. Palli. 2011. Steroid receptor co-activator is required for juvenile hormone signal transduction through a bHLH-PAS transcription factor, Methoprene tolerant. *J. Biol. Chem.* 286: 8437–8447.

Zhou, B., K. Hiruma, T. Shinoda, and L.M. Riddiford. 1998. Juvenile hormone prevents ecdysteroid-induced expression of Broad Complex RNAs in the epidermis of the tobacco hornworm. *Manduca sexta. Dev. Biol.* 203: 233–244.

Zhou, X. and L.M. Riddiford. 2002. Broad specifies pupal development and mediates the "status quo" action of juvenile hormone on the pupal-adult transformation in *Drosophila* and *Manduca*. *Development* 129: 2259–2269.

Zhu, J., L. Chen, G. Sun, and A.S. Raikhel. 2006. The competence factor βFtz-F1 potentiates ecdysone receptor activity via a p160/SRC coactivator. *Mol. Cell. Biol.* 26: 9402–9412.

CHAPTER **6**

Biological Rhythms

PREVIEW

Since the beginning of life on earth, organisms have been subjected to alternating light and dark (L/D) fluctuations as the earth rotates on its axis and to differing day and night lengths and temperatures as the earth moves around the sun. Circadian rhythms are internally generated by autonomous clocks that are entrained by environmental cycles. Biological rhythms must have developed in the earliest organisms in response to the cycling environment in which they evolved. Rhythms benefit living organisms by saving energy, reducing competition for food, defining habitat niche, regulating physiological and biochemical machinery, and coordinating mating behavior. The brain may contain a master clock, but that clock does not necessarily directly control peripheral tissue clocks. The products (mRNA and proteins) from the clock genes oscillate in response to regulation in feedback loops, resulting in temperature-compensated (within certain biological limits for life) rhythms that entrain to approximately a 24 h period in response to L/D cycles and temperature. Do circadian rhythms and photoperiodic rhythms operate with the same genes and clock mechanism? That is a question in chronobiology that has occupied thinking and research for nearly 100 years, and still without an unequivocal answer that all accept. The list of activities and physiological process that insects (and other animals) undergo during their development from immature to adult stage that are known to be regulated by circadian clock genes comprises just about every body function and activity. Light typically entrains or resets the clock each day, and many of the genes and molecular aspects of the clock have been identified, especially in *Drosophila melanogaster*, and it is likely that the basic mechanism of circadian rhythm function in *Drosophila* applies to most if not all insects, and in many cases, to plants and higher animals, including humans, likely with some variations. The photoperiodic response is highly adaptive for insects, especially those that moved into temperate and more polar regions. They were able, unconsciously of course, to anticipate climate changes that were going to be unfavorable for continued development, and so they prepared to diapause, migrate, or hibernate in preparation for the coming adverse conditions. It would be detrimental to development for one or two cool nights or shorter days to send them into a photoperiodic response, so a counter evolved to keep track of the number of increasingly shorter days, and once the critical limit has been recorded, a photoperiodic response is initiated. Once initiated, the photoperiodic response (diapause, migration, production of winged forms in aphids) is not reversed by the continuation of some favorable days of warm weather. Plant and animal scientists devised experiments to understand biological responses to daily and seasonal L/D and temperature regimes, and many models were proposed. Several of the major ones are described in this chapter. Today, of course, the most enlightening model is a molecular one, also described herein.

6.1 INTRODUCTION

With the exception of a few animals that live in caves or underground for long periods of time, animals and plants are affected by the natural cycles they experience as the earth rotates on its axis every 24 h and moves around the sun on a yearly basis. **Circadian rhythms** are internal responses entrained by daily cycles of light and darkness and temperature, while responses to the annual movement of the earth around the sun are typically called **photoperiodic responses**, or sometimes **seasonal responses** (Nelson et al., 2010). By timing activity and behavior to different times of the day or night, biological rhythms aid insects in reducing competition for resources, promote physiological and behavioral adaptation, and influence evolution.

Responses to day and night and to annual cycles have no doubt always been practiced by humans (perhaps unconsciously). The earliest humans must have discovered that certain animals were best hunted for food at certain times of the day or night, and when agriculture began some 10,000 years ago, those first farmers discovered that certain plants could not be grown all year round. As scientist began to try to understand their own rhythms and those of plants and animals, they began studies on flowering plants that were to have far-reaching consequences into the twenty-first century. Those early scientists included Garner and Allard (1920), Bünning (1936), Bünsow (1953), and Nanda and Hamner (1958). Studies on insects (and mites) soon followed, including major work by Lees (1973), Pittendrigh and Minis (1964), Pittendrigh (1966), Pittendrigh et al. (1970), Beck (1980), Saunders (1971, 1973a), and additional experimental data cited by these and other authors; in fact, the literature is extraordinarily voluminous and scattered in many different journals and books. The field has been extensively reviewed (Saunders, 2002, 2005, 2010, 2011, 2013; Danks, 2005; Bradshaw and Holzapfel, 2007a,b, 2010; Zhang and Kay, 2010; Koštál, 2011; Saunders and Bertossa, 2011; Goto, 2013).

6.2 CHARACTERISTICS OF CIRCADIAN AND PHOTOPERIODIC RHYTHMS

Circadian rhythms are internally generated by autonomous clocks that are entrained by environmental cycles to regulate activity, physiological, and behavioral rhythms. Circadian rhythms have the following characteristics: (1) they are entrained by environmental cues, mainly light and/or temperature; (2) they are temperature compensated, that is, they demonstrate a stable period length over a range of physiological temperatures; and (3) they free run under constant light or constant dark conditions with a period length equal to or close to that in alternating L/D conditions (Bloch et al., 2013). **Biological rhythms** must have developed in the earliest insects in response to the cycling environment in which they evolved. Rhythms benefit insects by saving energy, reducing competition for food, defining habitat niche, regulating physiological and biochemical machinery, and coordinating mating behavior.

Saunders (2009) suggested that insects probably evolved in warm tropical climates, and their daily activity and behavior clocks evolved with two half cycles in response to 12 h of light or **photophase** followed by 12 h of darkness or **scotophase** typical of day and night length at the equator. The circadian clock is reset daily by light although, in most insects that have been studied, the clock measures the dark period (the scotophase) rather than the light period (the photophase). The clock in *Drosophila*, and probably in many other organisms, can be reset by temperature if light is held constant (Glaser and Stanewsky, 2007). Whereas many biochemical processes, such as enzyme and metabolic reactions, proceed faster or slower, depending upon the temperature, the clock is temperature compensated within relevant physiological temperatures, and little or no changes in rhythms occur. An **entrained circadian rhythm** free runs (continues its normal periodicity) under constant environmental conditions (such as those induced experimentally), including total darkness or continuous light, although there may be some drift in timing of phases and damping (and even cessation) of the rhythm over time.

As insects expanded their populations into temperate climates, they experienced seasons of longer summer days and shorter winter days. Their life cycles were influenced by the changing seasons and a photoperiodic clock, along with appropriate survival mechanisms, evolved in those insects that were successful in colonizing temperate habitats. Once accumulated night measurements were sufficient, a seasonally appropriate response was initiated, and the photoperiodic clock was not reset even though favorable conditions might continue for some time before the unfavorable environmental conditions occurred.

Thus, the **photoperiodic clock** enables insects to prepare for impending harsh environmental conditions before those conditions arrive and to make avoidance behaviors such as diapause, migration, and/or production of seasonal morphs more adapted to escape or survive unfavorable environmental conditions to come. The photoperiodic clock, like the circadian clock, seems in most cases to measure the duration of the dark period. Much work has been done, and continues to be directed, at defining how the clocks work and how their messages are translated into behavioral, physiological, and biochemical actions.

Are circadian and photoperiodic responses governed by the same clock mechanism, or are there different clocks? The short answer is that this issue has not been resolved to the satisfaction of all insect biologists who work on rhythms. What are the molecular and genetic components of the clock(s)? Current research has shown that at least some of the components are the same or similar. There is not just a single circadian clock and a single photoperiodic clock; there are numerous cellular clocks, but whether they all operate on the same genetic and molecular basis is still open in the minds of some scientists.

6.3 MOLECULAR BASIS FOR THE CIRCADIAN CLOCK

Elucidation of the circadian clock molecular mechanisms began with studies of *D. melanogaster* and especially its eclosion and locomotor rhythms (Konopka and Benzer, 1971; Hardin, 2005, 2009; Sandrelli et al., 2008; Allada and Chung, 2010; Peschel and Helfrich-Förster, 2011; De et al., 2012; Tomioka et al., 2012). Self-regulating oscillatory action in the *Drosophila* clock model is kept in motion by daily light entrainment of **clock genes** and their protein products, and the clock comprises several interacting feedback loops (Figure 6.1), as well as posttranslational regulation (Weber et al., 2011). The first clock gene identified was *period* (*per*) (Konopka and Benzer, 1971; Bargiello et al., 1984; Reddy et al., 1984; Zehring et al., 1984), followed shortly by identification

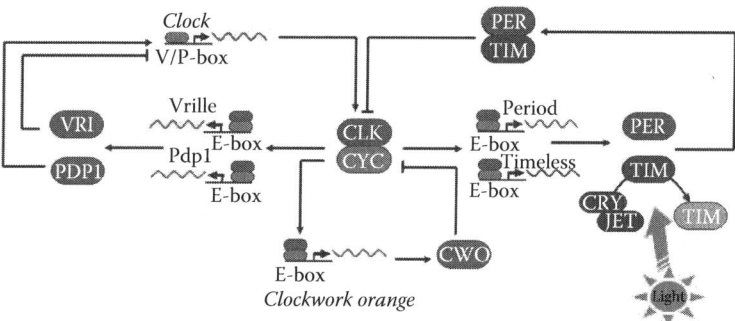

Figure 6.1 **(See color insert.)** A model depicting the molecular oscillatory mechanism of the *Drosophila* circadian clock. The clock comprises three interlocking transcriptional–translational loops. Light acts to degrade TIM through Cry, and then JETLAG (JET) resets the clock in a time-of-day–dependent manner. Lines ending in an arrow head are stimulatory; those ending in a negative sign are inhibitory. CLK, clock protein; CRY, cryptochrome protein; CWO, clockwork orange protein; CYC, cycle protein; PER, period protein; PDP1, Par domain protein 1; TIM, timeless protein; VRI, vrille protein. (From Tomioka, K. et al., *J. Comp. Physiol. B*, 182, 729, 2012. With permission.)

of *timeless* (*tim*) (Sehgal et al., 1994; Myers et al., 1995). The genes *per* and *tim* that encode the proteins **PERIOD** (PER) and **TIMELESS** (TIM), respectively, along with the genes *cycle* (*cyc*) and *clock* (*clk*) that encode **CYCLE** (CYC) and **CLOCK** (CLK), respectively, comprise the major feedback loop in the clock. PER and TIM accumulate in the cytoplasm during scotophase (Siwicki et al., 1988; Gekakis et al., 1995) and form the heterodimer PER/TIM, which enters the nucleus in the latter part of the night and suppresses the action of the **dimer** CYC/CLK. The measurement of mRNA has shown that *per* and *tim* expression is low during the light phase (photophase) and high during the early part of the dark phase (scotophase), with proteins PER and TIM delayed a few hours as expected for their synthesis, but clearly tracking levels of *per* and *tim* mRNA. CYC/CLK dimer acts as a positive regulator of transcription of the genes *per* and *tim* (Allada et al., 1998; Darlington et al., 1998; Rutila et al., 1998), but the protein product, PER/TIM, suppresses CYC/CLK. Thus, *per* and *tim* actually regulate their own production in *Drosophila* through this feedback action. Another feedback loop influences rhythmical production of CLK by action of the genes *vrille* (*vri*) and *PAR domain protein 1ε* (*Pdp1ε*) (Cyran et al., 2003; Glossop et al., 2003), and CLK/CYC dimer promotes transcription of both *vri* and *Pdp1ε*. VRI protein binds upstream to *clk* and suppresses its transcription, while PDP1E protein stimulates transcription of *clk*, although there is some evidence that it may only have a minor role in stimulating *clk* (Benito et al., 2007). Still, another loop with feedback mechanics regulates rhythmical expression of the gene *clockwork orange* (*cwo*) (Matsumoto et al., 2007). **CLOCKWORK ORANGE** protein influences the amplitude of *per* and *tim* oscillations by negative action on the function of the protein CYC (Kadener et al., 2007; Lim et al., 2007; Richier et al., 2008).

The rhythm is entrained (reset) by the effect of light on the response of the photopigment **CRYPTOCHROME** (CRY) encoded by *cyptopchrome-1* (*cry-1* = *cry-d*) (Emery et al., 2000). CRY is a flavin containing protein and shows maximum sensitivity to blue light at 400–500 nm. Light exposure enables CRY to bind to TIM, degrading it, which also allows PER to degrade (Emery et al., 1998; Stanewsky et al., 1998; Ceriani et al., 1999; Busza et al., 2004). Thus, the clock starts a new cycle.

The *Drosophila* clock neurons are sensitive to low light levels. A subset of pacemaker neurons in *D. melanogaster* responded to illumination equivalent to one-quarter moonlight intensity by increased activity and shifted morning and evening activity peaks into the night (Bachleitner et al., 2007). The detection of such low light levels depends upon the photopigments rhodopsin-6 in cell R8 and rhodopsin-1 in cells R1 to R6 in the compound eyes (Schlichting et al., 2014). Kistenpfennig et al. (2012) showed that a 1 h exposure to 1000-lux light pulses given to cry^{01} mutants of *Drosophila* caused the flies to phase shift their activity rhythm in both morning and evening (but with reduced magnitude).

In summary, CYC/CLK dimer stimulates production of PER/TIM, and PER/TIM feedback inhibits CYC/CLK, but neither of these actions is instantaneous, and the period of the clock correlates approximately with the length of the day of an entrained circadian rhythm. Light entrains and resets the rhythm each day through its action on CRY, which degrades TIM and PER (releasing inhibition of CYC/CLK), and the protein JETLAG resets the clock for the next 24 h cycle, as CYC/CLK renews the production of PER/TIM (Koh et al., 2006).

Clock mRNA and clock proteins are present in about 150 neurons in the brain of *D. melanogaster* (Helfrich-Förster, 2006; Peschel and Helfrich-Förster, 2011; Hermann et al., 2012), and they oscillate with a period of about 24 h and even in constant environmental conditions of continuing darkness (Figure 6.2). The brain clock, often described as a "**master clock**" (Hermann et al., 2012), controls behaviors and in some cases, including *Drosophila*, exerts an influence on **peripheral clocks**. The clock neurons in the brain are organized into networks (Figure 6.3) with different neurochemically controlled subsets of neurons controlling different functions, including the morning and evening locomotor rhythms (Pittendrigh and Daan, 1976; Helfrich-Förster, 2004, 2006; Peschel and Helfrich-Förster, 2011; Hermann et al., 2012).

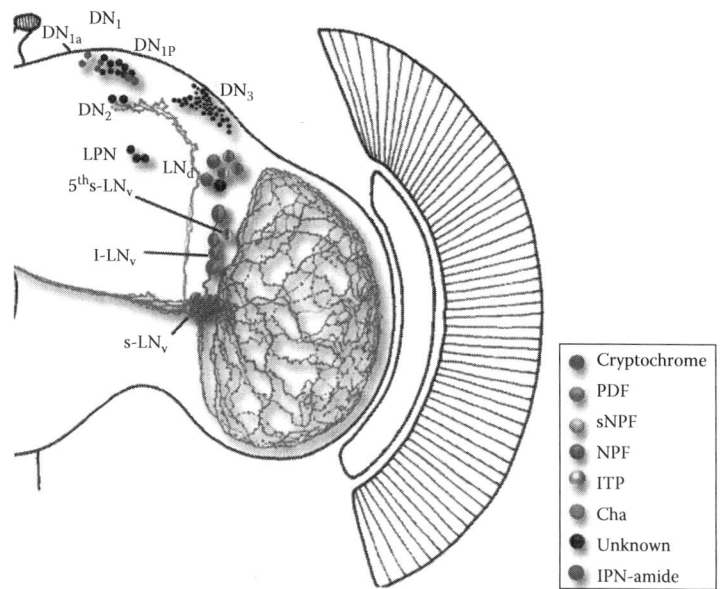

Figure 6.2 (See color insert.) A basic overview of clock-gene-expressing neurons and their neurochemical characterization. The different colors of the neurons indicate the peptides/proteins that are expressed in those cells. Cryptochrome, yellow; pigment dispersing factor (PDF), green; short neuropeptide F, blue; neuropeptide F, red; ion transport peptide, gray; choline acetyltransferase, purple; IPN-amide, orange. Cells with unknown peptidergic content are colored black. Arborization of PDF is indicated in green. DN, Dorsal neurons LN, Lateral N\neurons. (From Peschel, N. and Helfrich-Förster, C., *FEBS Lett.*, 585, 1435; *Physiology*, 57, 935, 2011. With permission.)

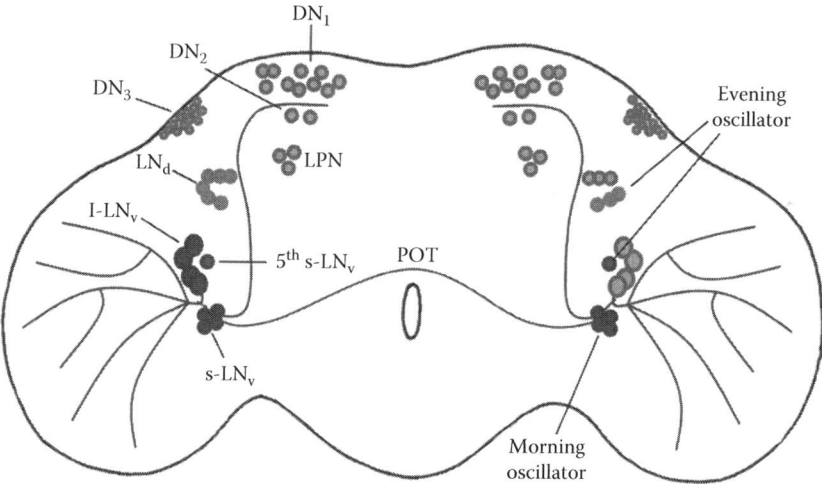

Figure 6.3 (See color insert.) The neuronal network of the "master" clock in the adult *Drosophila melanogaster* brain. The dorsally located clock neurons (DNs) consist of three different groups, the DN_1s, DN_2s, and DN_3s. The lateral clock neurons (LNs) can be divided into a group of six dorsal lateral neurons (LN_ds, left side, yellow) and the more ventrally located LN_vs (left side, blue). There are four LN_vs per hemisphere with small cell bodies, the s-LN_vs, and four to six neurons with large soma, the l-LN_vs; both express the neuropeptide PDF. Usually located among the l-LN_vs is another single neuron with a smaller cell body, the fifth s-LNv (red), which is PDF-negative. Another group of clock-gene-expressing neurons, the LPNs, is located in the lateral posterior brain. According to the dual oscillator model, three of the six LN_ds and the fifth s-LNv were identified to be evening neurons (right side, yellow, and red), whereas the s-LN_vs were found to be morning neurons (right side, blue). (From Hermann, C. et al., *J. Comp. Neurol.*, 520, 970, 2012. With permission.)

6.4 EVIDENCE FOR CLOCK GENES IN MANY INSECTS

Evolution of insects led to two *cry* genes, *cry-d* (named for discovery in *Drosophila*) and *cry-m* (discovered in mammals) and designated as *cry-1* and *cry-2*, respectively, in some reports. *D. melanogaster* and *Drosophila pseudoobscura* are believed to have lost *cry-2* (*cry-m*) and have the derived clock with only *cry-d* (Yuan et al., 2007; Sandrelli et al., 2008; Merlin and Reppert, 2010; Tomioka and Massumoto, 2010; Saunders, 2012). Although some species-specific differences were found in clock mechanisms in 10 *Drosophila* species, Hermann et al. (2013) concluded that the general structure of the circadian clock network is highly conserved in the *Drosophila* genus.

Clock genes and clock proteins are present in a variety of insects. For example, Yuan et al. (2007) found CRY-m protein to be a core component of the circadian clock of *Danaus plexippus*, the monarch butterfly. The clock appears to be important in **migration behavior** and orientation of monarch butterflies (Sauman et al., 2005; Reppert, 2007; Zhu et al., 2008; Goto, 2013). A variation of the clock mechanism in the monarch butterfly is that TIM/PER dimer binds to CRY-m, and this stable complex represses CYC-CLK transcription. An additional feedback loop allows CRY-m to negatively regulate PER and TIM.

The **clock genes** *cry-d* and *tim* have been lost in the course of evolution of honeybees, *Apis mellifera* and *Nasonia vitripennis*, which have *cry-m* (Rubin et al., 2006; Schurko et al., 2010). The genes *per*, *cry-d*, *cry-m*, *tim-1* (*tim-1* = *timeless*), and *tim-2* (*tim-2* = *timeout*) have been found in at least 18 insect species, including *Bombyx mori*, *D. plexippus*, *Anopheles gambiae*, *Aedes aegypti*, *Culex quinquefasciatus*, and the aphid *Acyrthosiphon pisum*, but *cry-d* and *tim-1* are not present in five species of Hymenoptera (Saunders, 2012). The Madeira cockroach *Rhyparobia* (*Leucophaea*) *maderae* contains the genes *per*, *tim-1*, and *cry-2*, and their expression showed circadian oscillation, peaking in the first half of the scotophase (Werckenthin et al., 2012). The clock gene *cyc* is present in the cricket *Gryllus bimaculatus* (Uryu and Tomioka, 2010; Uryu et al., 2013).

The details of the clock mechanism known thus far have been worked out mainly in *Drosophila*, aided by the availability of the complete genome. Reports of clock genes in other insects are well established, but a complete clock mechanism comparable to that in *Drosophila* has not been established. It is of some interest whether the clock genes evolved with the earliest insects or whether some of the genes evolved with the increasing diversity and complexity of insects. Kamae et al. (2010) and Kamae and Tomioka (2012) investigated the gene *tim* in the firebrat, *Thermobia domestica*, an early-to-evolve apterygote insect that molts but does not undergo metamorphosis. Young firebrats look like small copies of the adults. The gene *tim* was present and functional in this primitive insect, but there were some differences from *Drosophila*, a highly evolved insect with complete metamorphosis. These investigations showed that at least one of the principal genes in the *Drosophila* clock evolved with the earliest insects, but many genes have pleiotropic effects, and the presence of a gene may not always mean that it is functioning as part of a circadian (or photoperiodic) clock.

6.5 EXAMPLES OF CIRCADIAN FUNCTIONS IN INSECTS

6.5.1 Circadian Regulation of Hormone Secretion

Molting and metamorphosis have been shown to be under circadian control in at least some insects (Truman, 1972; Bloch et al., 2013; Di Cara and King-Jones, 2013, and references therein). In immature *Rhodnius prolixus*, a circadian rhythm regulates release of prothoracicotropic hormone (PTTH) from the brain after the bug takes a blood meal. Clock neurons project to the

PTTH-producing cells (Vafopoulou et al., 2007, 2010). **PTTH** is bound on the cell surface of the prothoracic gland cells by receptor proteins that reach peak density concomitantly with the rise of circulating PTTH (Vafopoulou and Steel, 1999). As *Rhodnius* molts and metamorphoses into adults, most of the earlier clock network is retained, but new clock cells grow into place with an increase in axonal arborizations invading the mushroom bodies in the brain, corpus allatum, and ventral nerve cord, all orchestrated with each characteristic pulse of ecdysteroid during the molting and metamorphosis process (Vafopoulou and Steel, 1991, 2012). *Rhodnius* adults continue a daily rhythmic synthesis and release of PTTH in the scotophase during egg development, although its exact function in *Rhodnius* is not clear (Vafopoulou et al., 2012). In some adult females, however, it is known that the ovaries produce ecdysone, which is important in egg maturation, in response to PTTH secretion, so probably the PTTH in adult females of *Rhodnius* stimulates the ovaries to produce ecdysone.

Juvenile hormone (JH) is under circadian regulation in some insects. In honeybees, Bloch et al. (2013) suggest that JH biosynthesis is regulated in many, and perhaps all, insects in a circadian manner by direct or indirect neuronal pathways from cells in the circadian network (Figure 6.4). JH regulates photoperiodism in honeybees and other social insects (Bloch et al., 2013).

Their model assumes that the circadian system influences the expression or activity of enzymes of the JH pathway. In a very different insect, the pentatomid bug *Plautia stali*, levels of JH rise within 4 days after transferring the bugs from a short-day regime to a long-day program, suggesting clock influence (Matsumoto et al., 2013). JH biosynthesis is under circadian influence in

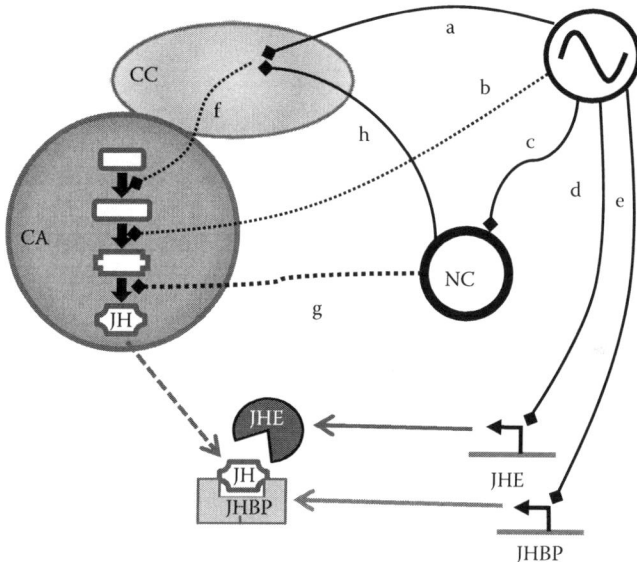

Figure 6.4 (See color insert.) A working model for circadian modulation of JH titers. The model assumes that JH biosynthesis is regulated in many, and perhaps all, insects in a circadian manner by direct or indirect neuronal pathways from cells in the circadian network (represented in the model by the circle with a sine symbol in the upper right corner). Continuous lines ending in a diamond head depict neuronal pathways. Dotted lines show pathways innervating the corpora allata (CA) and are assumed to contain allatoactive neuropeptides such as allatotropins and allatostatins. The shapes and arrows in the circle representing the CA depict chemical intermediates and enzymes, respectively, in the JH biosynthetic pathway. The model assumes that the circadian system influences the expression or activity of enzymes of the JH pathway. Pathways *d* and *e* show circadian influences on the transcription or activity of proteins that degrade (e.g., JH esterase) or bind (e.g., JH binding protein) JH, respectively. CA, corpora allata; CC, corpora cardiaca; JH, juvenile hormone; JHBP, juvenile hormone binding protein; JHE, JH esterase; NC, neurosecretory cell. (From Bloch, G. et al., *J. Insect Physiol.*, 59, 56, 2013. With permission.)

long-winged *Gryllus firmus* crickets that can fly, but not in short-winged ones that cannot fly; Stay and Zera (2010) used immunostaining of cells in the **corpora allata** for **allatostatin** to show that staining was inversely correlated with the rate of JH biosynthesis in the crickets, suggesting that a circadian role in JH biosynthesis and release likely is due to clock controlling an allatostatin synthesis or its movement into the corpora allata.

Many insects produce a multicomponent **sex pheromone** to attract a mate, and the production and release of the pheromone components typically occur only during certain times of the day or night. In Lepidoptera, mating in many species occurs in the early dark period, and pheromone production and release is controlled by the neuropeptide pheromone biosynthesis-activating neuropeptide (**PBAN**). A circadian rhythm regulates PBAN production and release, courtship, and mating behavior in both female and male moths (Závodská et al., 2009; Bober and Rafaeli, 2010; Bober et al., 2010).

6.5.2 Circadian Clock Influence in Peripheral Organs and Tissues

Largely independent of a central pacemaker, circadian clocks seem to be widely dispersed in cells throughout the body of insects. Koštál (2011) reasoned that various physiological functions in different organs could be controlled by peripheral oscillatory clocks rather than by a central clock because light, as a reset mechanism, probably can penetrate the cuticle and reach tissues in most parts of the insect body, and evidence for peripheral clocks has been found in a number of tissues of insects.

There probably are many oscillating clocks in insects, but more details are available for clock oscillations in compound eyes of *Drosophila*, some cockroaches, and some moths; antennae of fruit flies, cockroaches, moths, and some butterflies; prothoracic gland of fruit flies and some moths; Malpighian tubules of fruit flies; testes of fruit flies, crickets, and moths; digestive system of crickets; and the terminal abdominal ganglion of crickets. Circadian oscillations of PER and TIM occur in the compound eyes and ocelli of *Drosophila* (Siwicki et al., 1988; Zerr et al., 1990; Cheng and Hardin, 1998; Houl et al., 2006; Barth et al., 2010), but cyclic expression of *per* in isolated body parts of *Drosophila* suggests that autonomous circadian pacemakers may be widespread in the body (Plautz et al., 1997). In houseflies, *Musca domestica*, *D. melanogaster*, and the blowfly *Calliphora vicina* (Pyza and Meinertzhagen, 1995, 1999; Pyza and Cymborowski, 2001), there is a rhythm in size of interneurons in the visual system. Peripheral pacemakers are part of chemoreceptor cells in *D. melanogaster*, and Krishnan et al. (1999) and Tanoue et al. (2004) suggest that they may function to modulate sensitivity of chemoreceptor cells on a circadian basis. A circadian rhythm in function of chemosensory receptors in the antennae is further supported by an electroantennogram (EAG) rhythm in the cockroach *Periplaneta americana* and *D. melanogaster*. In the cockroach, the rhythm is driven by a central clock in the optic lobe of the brain, and although the EAG rhythm is lost after removal of the optic lobe (Page and Koelling, 2003), individual olfactory neurons in the antennae maintain rhythmic response to odors (Saifullah and Page, 2009) indicating some level of independence of a brain clock. In *Drosophila*, the EAG rhythm is driven by a clock in olfactory neurons (Tanoue et al., 2004) but is abolished in *per* null mutants (Krishnan et al., 1999). In monarch butterflies *Danaus plexippus* and the hawk moth *Manduca sexta*, *per* is expressed in a rhythmical pattern in the antennae (Schukel et al., 2007), and the antennal clock appears to be involved in the time-compensated sun-compass orientation of the monarch butterfly, which is lost if the antennae are excised or the antennae are painted to prevent light reception (Merlin et al., 2009). The compass orientation is used by monarchs in their migration to and from Mexico (Goto, 2013).

Taste receptors are under circadian control by a clock in gustatory receptor neurons in *Drosophila* (Chatterjee et al., 2010). Disabling the clock causes the flies to become hyperactive and search for food, typical behavior of flies that are hungry. The sense of taste in normal flies is most sensitive in the early morning, characteristic of the time when their activity rhythm also is at a peak (Chatterjee and Hardin, 2010).

The genes *per* and *tim* are expressed in a rhythmical manner in **Malpighian tubules** of *Drosophila*, and the rhythmical activity continues in cultured tubules removed from the flies (Hege et al., 1997; Giebultowicz et al., 2000) and even in those tubules transplanted into adults entrained to a different rhythm. A function for the rhythm of *per* and *tim* in Malpighian tubules, however, has not been demonstrated. Bioluminescent glowworms, the larval form of a fly in New Zealand and Australia, produce light in cells at the tip of the Malpighian tubules as a way to lure prey into its web. Merritt and Aotani (2008) showed that when the glowworms (as larvae) were placed in constant darkness, they continued a free-running cyclical light production for 28 days, indicative of a circadian mechanism in control of **bioluminescence**. Genes responsible for bioluminescence were not identified in the glowworms.

A clock in the **cuticular epidermis** that controls cuticle layer deposition in adult *D. melanogaster* is entrained by light and temperature. The clock maintains cuticle layer deposition under constant conditions, even in cultured tissues (Ito et al., 2008). The heteropteran *Riptortus pedestris* also has a clock that functions in cuticle deposition, and the use of RNA interface (RNAi) to reduce expression of the genes *per* and *cyc* abolished the rhythm (Ikeno et al., 2010, 2011).

Xu et al. (2011) reported a very extensive table of cyclically expressed genes in *Drosophila* fat body controlling metabolism, detoxification, immune response, and steroid hormone processes and found that restricting food to the flies desynchronized various rhythms by altering the phase of rhythmic gene expression independent of the brain clock. They suggested a role for a peripheral clock in the fat body that is driven independently of the brain, especially as influenced by restricted feeding.

6.5.3 Circadian Clock Influence in Social Behavior of Honeybees

Social insects including ants, termites, and bees have evolved highly structured societies with division of labor and function with the colony. The circadian clock network is important in the complex behaviors of individuals in these social colonies. Perhaps because they have been cultured by humans for so long, honeybees have been studied intensively with respect to biological rhythms. Honeybees have been shown to have **time-linked memory**, **sun-compass orientation**, and **dance communication** (Shemesh et al., 2007, 2010; Bloch, 2009, 2010). Some components of the clock in honeybees are different from those in *Drosophila*. Specifically in honeybees, cryptochrome is encoded only by *cry-m*, rather than by *cry-d* as in *Drosophila*, and the honeybee cryptochrome is not sensitive to light and is not the oscillator for the clock. Honeybees do not have the gene *tim-1*, but the genome includes the gene for *tim-2* and the genes *per*, *cyc*, *clk*, *vri*, *Par domain protein 1* (*Pdp1*), and *cwo*, all genes involved in the clock mechanism. Forager honeybees have a strong circadian rhythm and are able to encode a time-related single foraging (feeding) experience. The time-memory experience free runs under constant conditions and can be phase shifted, aspects typical of a circadian rhythm. In foragers, the brain levels of *cry-m* and *per* mRNA oscillate with L/D entrainment and free run under constant darkness. Nurse bees, which typically spend their entire life inside the dark colony tending (feeding) the brood of larval bees, do not exhibit a circadian rhythm in locomotor activity (Bloch, 2009, 2010), although some transcripts in the brain of nurse bees oscillate ($n = 160$), likely indicating that some processes probably do cycle even though there is an absence of circadian rhythms in behavior or clock gene expression (Rodriguez-Zas et al., 2012). Foragers show oscillations in as many as 541 transcripts (Rogriguez-Zas et al., 2012) as they carry out the work of collecting and storing nectar and pollen for the colony. Nurse bees can show a strong circadian rhythm when transferred from brood feeding to cages under constant laboratory conditions, and old forager bees, which have strong circadian rhythms, sometimes take on brood feeding, whereupon they show attenuated oscillations of clock mRNA. Shemesh et al. (2007, 2010), Bloch (2010), and Rodriguez-Zas et al. (2012) suggest that **social organization** with division of labor modulates circadian rhythms in honeybees and

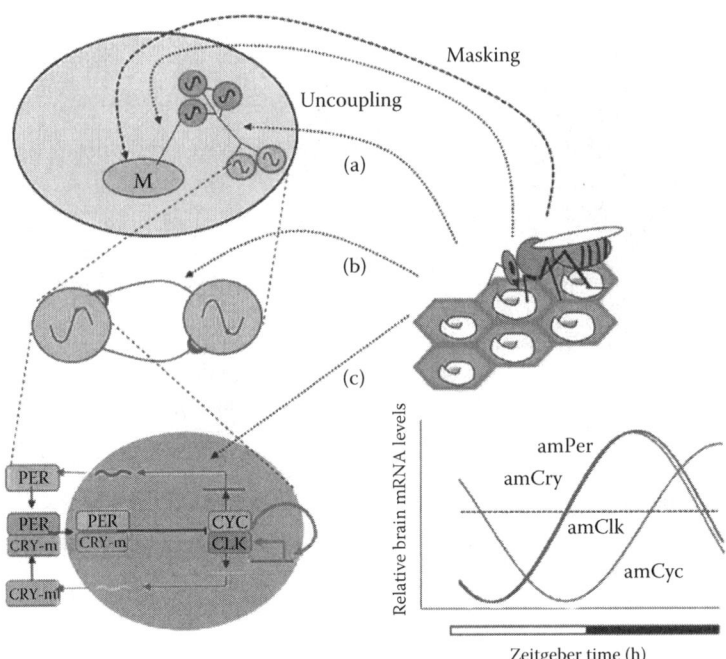

Figure 6.5 (See color insert.) A multilevel regulation of circadian rhythms in the brain and central nervous system and their possible social modulation in honeybees, *Apis mellifera*. Top: Pacemaker cells and other neuronal circuits interact to produce overt circadian rhythms. The circadian network is composed of clusters of coupled pacemaker cells (circles with sinusoidal wave lines). Left middle: Interactions between cells in the same cluster are needed to generate circadian rhythms in behavior. Left bottom: Circadian rhythms are generated autonomously within pacemaker cells in the honeybee brain. (a) The dotted line indicates modulation of the interaction between clusters of pacemakers. (b) The dotted line suggests interactions between pacemakers within the same cluster. (c) The dotted line indicates possible modulation within pacemaker cells. Honeybee on brood comb: Direct interaction between nurse bees and the brood may influence circadian rhythms in several ways and at several levels, as indicated by the dotted lines. Lower right: Gene mRNA cycling (with the prefix *am* for *Apis mellifera*, as follows: for Period [amPer], Cryptochrome-m [amCry], and Cycle [amCyc]). In the honeybee, the phase of amCyc transcript is almost in antiphase to that of amPer and amCry. The straight dashed line represents clock mRNA, which does not vary through the day. Not shown: the mRNA pattern for *Timeout* (*amTim2*), which showed variation over time as influenced by the environment. M, motor control center; "masking," modulation of motor controlling centers without affecting the circadian system; uncoupling, modulating the interaction between the circadian system and motor controlling centers. (From Bloch, G., *J. Biol. Rhythms*, 25, 307, 2010. With permission.)

that interaction with the brood modulates the circadian system (Figure 6.5), resulting in social modulation and **"task-related plasticity"** in the circadian system of honeybee. Similar socially related plasticity has been shown in other social insects, including ants and bumblebees.

6.5.4 Circadian Clock Influence in Reproduction

A rhythm regulating sperm release in *Drosophila* males has been demonstrated (Giebultowicz et al., 1989; Bebas et al., 2001), and RNAi against *per* decreases sperm release (Gvakharia et al., 2000; Kotwica et al., 2009a). Rhythmical release of sperm bundles by *Spodoptera littoralis* males occurs in the early evening hours and continues when the reproductive system is isolated from the body, suggesting a role for a peripheral clock mechanism. In cyst cells and barrier cells in the testes, PER protein showed a rhythm with sperm release. The use of RNA interference (double-stranded fragments of *per* mRNA) lowered *per* mRNA and protein in cyst and barrier cells and delayed (but did not stop) sperm release (Kotwica et al., 2009a). The authors suggested a role for a molecular oscillator and *per* gene in sperm release in *S. littoralis*.

Rhythmical mating activity that is dependent mainly upon female behavior in *Drosophila* has been shown by Sakai and Ishida (2001). A circadian rhythm in egg-laying behavior in *D. melanogaster* is independent of the clock neurons in the brain that control the locomotor and eclosion rhythms. Beaver et al. (2002) found in *Drosophila* that clock genes were expressed in a rhythmical pattern in testes and seminal vesicles, and arrhythmic clock mutants for *per*, *tim*, *cyc*, and *clk* laid fewer eggs and more unfertilized eggs, and numbers of sperm passed to the seminal vesicles at mating were reduced. Tobback et al. (2011) observed reduced progeny production in female *Schistocerca gregaria* desert locusts after using RNAi to reduce clock gene expression. Kotwica et al. (2009b) showed that PER protein is expressed uniformly in the cytoplasm of previtellogenic follicle cells but is located mainly in the nucleus of follicle cells during early vitellogenesis (Figure 6.6).

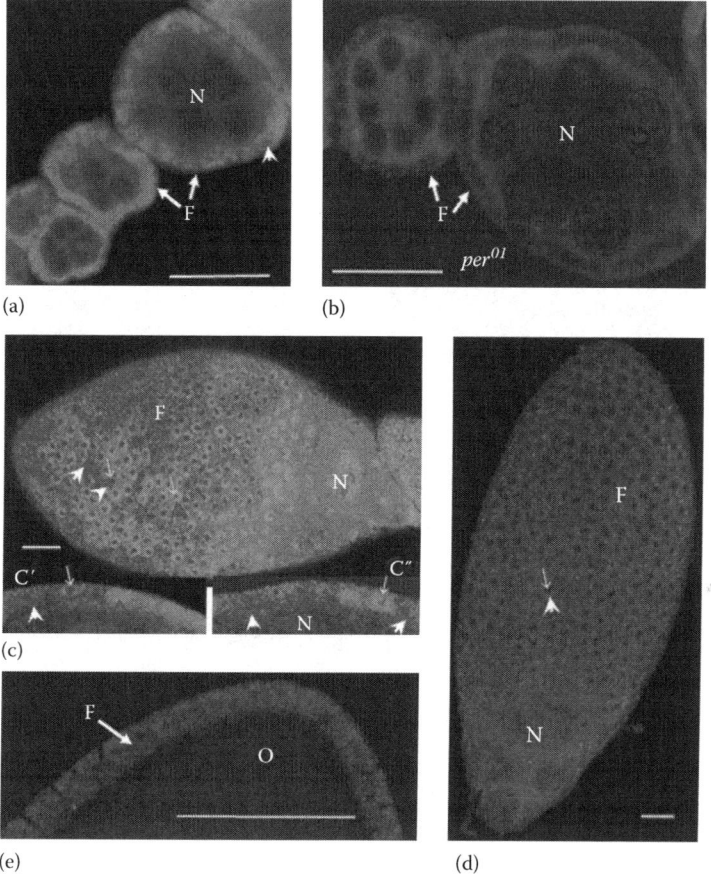

Figure 6.6 **(See color insert.)** Expression of PER protein in ovaries of *D. melanogaster*. (a) Transverse section in which PER is cytoplasmic in young previtellogenic follicles. (b) Young previtellogenic follicles in *per^{01}* flies—negative control. (c) An image near the surface of an early previtellogenic follicle. Insets C′ and C″ show transverse sections of enlarged follicular cells. PER is in nuclei of some cells (red arrowheads) but absent in other nuclei (white arrowheads). Yellow arrows in C and D point to large dark nucleoli. Patchy nuclear pattern was observed in all stage 9 follicles. (d) PER was not detected in older vitellogenic follicular cells in stage 10; specks of green are nonspecific and occur in *per^{01}* flies (not shown). (e) Anterior fragment of egg chamber with yolk and follicular cells (stage 10, transverse section). PER protein stained green, nuclei stained blue. Bar = 20 μm. N, nurse cells; F, follicular cells; O, oocyte. (From Kotwica et al., 2009b. With permission.)

The authors suggest that PER is randomly distributed and its translocation to the nucleus is not synchronous because the follicle cells do not exhibit a clock entrained by cryptochrome. They further suggest that DOUBLETIME protein (Figure 6.7), a kinase that can phosphorylate PER, may contribute to its movement into the nucleus by phosphorylating it.

Merlin et al., (2007) demonstrated a circadian clock and rhythm controlling olfactory **pheromone reception** in the antennae of male *S. littoralis* moths. The clock genes *per*, *cry-1*, and *cry-2* were identified in the antennae, and their transcripts fluctuated in a cyclical manner in both antennae and brain. The response of the male's antenna, determined by **EAG** recording, showed a circadian rhythm in response to female sex pheromone, and an odorant-degrading enzyme transcript in the antennae cycled rhythmically.

Larsdotter-Mellström et al. (2012) found that the dynamics of a male-produced sex pheromone in the butterfly *Pieris napi* differed when the butterfly was in **diapause**. A summer generation that develops rapidly in Sweden requires approximately 1 day after adult emergence to synthesize its pheromone comprised of geranial and neral (1:1), but those individuals that spend several months in diapause emerge with the pheromone components already synthesized prior to emergence. No specific circadian or photoperiodic clock functions were investigated in this study.

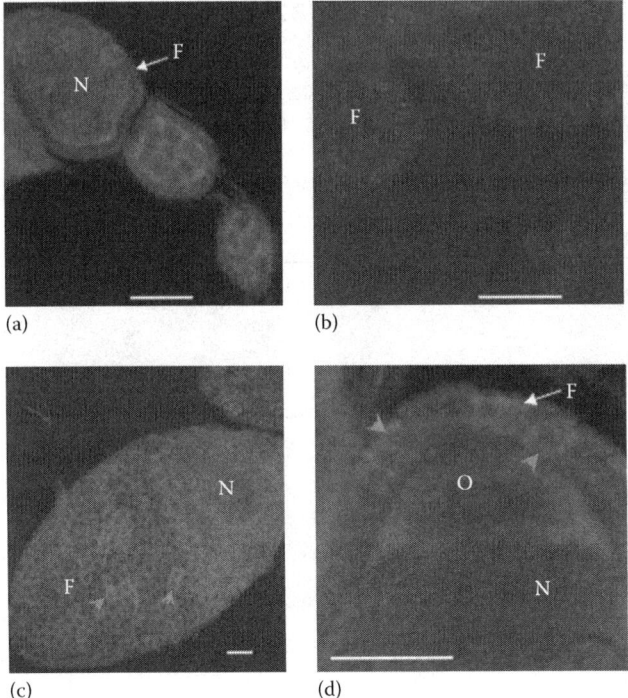

Figure 6.7 **(See color insert.)** Expression of DOUBLETIME (DBT) protein in ovaries of *D. melanogaster*. (a) DBT is present in nurse cells of young previtellogenic oocytes in this transverse section. (b) DBT is absent in follicular cells of previtellogenic oocytes (stages 3–7) in this image near the surface. (c) DBT shows patchy cytoplasmic pattern of expression (green arrowheads) in stage 9 vitellogenic egg chamber in this surface image. (d) Transverse section of anterior fragment of egg chamber showing cytoplasmic DBT in some follicular cells (green arrowheads). The pattern of DBT distribution is similar for ovaries collected from different time points (data not shown). DBT is stained red; nuclei are blue. Bar = 20 μm. N, nurse cells; F, follicular cells; O, oocyte. (From Kotwica et al., 2009b. With permission.)

Levi-Zada et al. (2014) found that male *Aphomia* (*Arenipses*) *sabella* moths release volatile components in a circadian rhythm during the night between 03:00 and 05:00 h. The compounds are released from androconium glands in the forewings of males only. The authors did not investigate the role of any particular clock genes.

Gaten et al. (2012) were able to alter the circadian rhythm of phase-dependent polyphenism and flight in *S. gregaria* by manipulating rearing density of the locusts.

The malaria mosquito, *A. gambiae*, shows circadian influence on insecticide resistance and in metabolic detoxification of DDT (Balmert et al., 2014). The circadian clock is important in *D. melanogaster* in combating reactive oxygen species (Krishnan, et al., 2008; Subramanian et al., 2014) and pathogenic infections (Lee and Edery, 2008).

6.6 PHOTOPERIODIC RESPONSE: ONE CLOCK, TWO CLOCKS, OR MULTIPLE CLOCKS?

The photoperiodic response enables insects to detect approaching environmental conditions that may be unfavorable for continuous activity and development (Denlinger, 1986; Denlinger, 2002; Schiesari et al., 2011; Denlinger and Armbruster, 2014) and to avoid the unfavorable environment by diapausing, hibernating, or migrating. Typically, insects measure the increasing long dark night of fall as indicators of the harsher climate of the coming winter. They use the information to make seasonal adjustments in behavior and physiology, including **migration**, **diapause**, and in some insects, development of **winged morphs** (Hardie, 1987) that can fly away from the impending unfavorable conditions. Bünning (1936) ignited a continuing controversy with his suggestion that the circadian clock was the basic mechanism for photoperiodic responses in bean plants. The translation into English of his title is "The **endogenous daily rhythm** as the foundation for the photoperiodic reaction." An endogenous daily rhythm is called a circadian rhythm in current literature. Many experiments have been conducted in an attempt to verify, clarify, and/or refute Bunning's suggestion. A common experiment has been to measure the percent of a population of insects entering diapause after exposure to a regime of decreasing day length and increasing night length. Such an L/D regime is a characteristic of approaching winter in temperate and polar climates. The threshold for a positive diapause response typically was indicated by 50% of the population entering diapause. With identification of clock genes, however, another approach is to look for the participation of these genes in the photoperiodic response. Ikeno et al. (2010, 2011) recently used this approach to show that the clock genes *per* and *cyc* regulate both the circadian rhythm of cuticle deposition in the bean bug *R. pedestris* and diapause in response to long nights. These authors used RNAi against *per* and *cyc* to show that cuticle deposition was disrupted. Application of *per* RNAi caused the bugs to avoid diapause even during long night exposures, and *cyc* RNAi induced diapause even when the bugs were exposed to long days and short nights. Thus, *per* and *cyc* were clearly involved in both the circadian rhythm of cuticle deposition and the photoperiodic response in *R. pedestris*, providing some of the most straightforward evidence that the circadian clock genes function in photoperiodic responses.

Interpretation of the photoperiodic responses result can be complicated by additional factors that influence diapause in a population, including temperature, quantity and quality of diet, and various biotic interactions (Saunders, 2002). The temperature at which experiments are conducted can mimic the clock system in complete darkness (Saunders, 1973a) and in continuous light, but it is not known if the temperature receptors and photoreceptors are the same, or if different, whether they share some data input.

How much the circadian clock and the seasonal or photoperiodic clock share the same or similar biochemical and genetic machinery is a matter that currently is debated by researchers.

Koštál (2011) and Saunders (2013) expressed (in slightly different form) three possible interpretations for the relationship between the circadian and photoperiodic clock, summarized as follows:

1. The circadian timing system is causally involved in photoperiodic measurements (i.e., there is basically one time measuring system).
2. The circadian system is not part of the core functioning of the photoperiodic system, but the two systems functionally cooperate.
3. The two clocks evolved independently in function and physical location in insects.

Saunders (2013) and Goto (2013) support, with their data, the idea that there is only one clock system serving daily and photoperiodic responses, although they acknowledge that there may be some differences in the two systems. Koštál (2011) tends to support the second idea that the clocks are independent but cooperate in their functions. Danks (2005) and Bradshaw and Holzapfel (2007a,b, 2010) believe that the daily and photoperiodic timing mechanisms evolved independently.

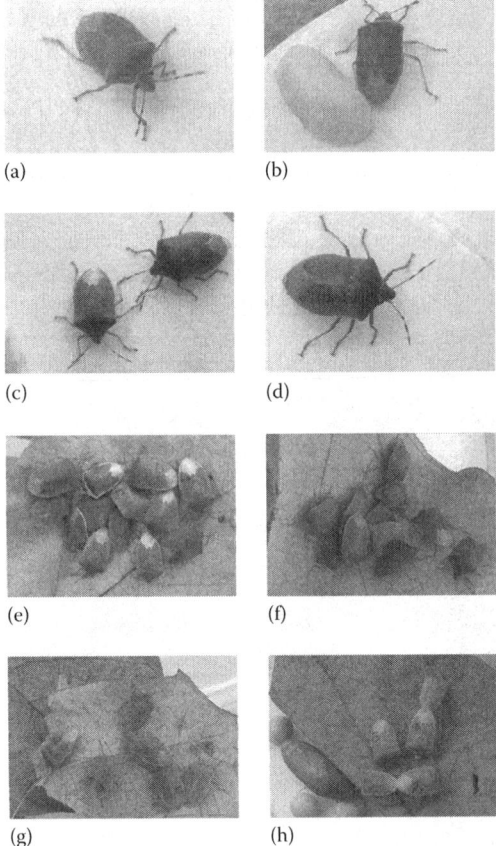

Figure 6.8 (See color insert.) Diapause-associated color changes in the heteropteran stinkbug *Nezara viridula* reared under laboratory or quasi-natural outdoor conditions. Laboratory rearing occurred in Kyoto, Japan. (a) reproductive, nondiapausing green adult; (b) and (c) diapausing adults of intermediate color; (d) russet brown diapausing adult; (e) and (f) mostly intermediate colored and russet overwintering adults in early stage of diapause after outdoor rearing in Kyoto (November to December); (g) adults in March of different color grades in the last stage of diapause; (h) adults in April that have terminated diapause, changed color, and are beginning post-diapause reproduction. (From Musolin, D.L., *Physiol. Entomol.*, 37, 309, 2012. With permission.) (Photograph courtesy of Prof. Dr. D.L. Musolin.)

Experimental results from various insects have been used to support each of the three ideas. Because all insect evolution is not considered to be monophyletic (Grimaldi and Engel, 2005), the circadian clock may have evolved into the photoperiodic clock in some insects, while in others, a variety of timing mechanisms may have evolved.

Manifestly, both circadian clocks and photoperiodic clocks depend upon (1) the input of light and alternating darkness, although in most insects the dark period is the parameter actually measured, (2) a timing mechanism to measure the advent of L/D periods, and (3) response of effector nerves, hormones, and muscles to bring about appropriate responses. The circadian clock is reset (entrained) each day by the dawn light. The photoperiodic clock measures length of the dark periods over time and (4) has a counter that accumulates the number of successively longer nights until a threshold is reached for seasonal response(s), such as diapause, migration, development of a seasonal morph, or combination of these responses (e.g., monarch butterflies go into reproductive **diapause** and **migrate** south in response to short days and long nights). Many other diapausing insects simply go into a semidormant state where they have passed the summer. For example, the green stinkbug *Nezara viridula*, an important agricultural pest, diapauses under leaf litter, in cracks and crevices, within the crown of dense trees, and other suitable sheltered places such as under the roof of buildings (Musolin, 2012). There are a number of color morphs of this heteropteran, some of which are genetically controlled and some are photoperiodically controlled (Figure 6.8).

Photoperiodic responses often occur only once in the life of an insect (such as diapause in many temperate climate insects, migration in a number of insects, or formation of a winged morph in some aphids) and typically in a species-specific stage, either egg, larval, pupal, or adult. A few insects, however, particularly some living in extremely harsh environments like the arctic with very short favorable seasons for development, experience a seasonal diapause several times because they mature so slowly, requiring several years to mature in the successive short summers. Even if environmental conditions are not yet life threatening to the insect, photoperiodic behavioral response is set in motion when the counter has accumulated sufficient data and the nervous/endocrine effectors have been activated. The response is not immediately reversible even if environmental conditions continue for some time favoring further development.

6.7 CLOCK MODELS BASED ON EXPERIMENTAL RESPONSES OF INSECTS TO VARYING LIGHT/DARK REGIMES

A direct way to solve the issue whether there is one clock or more than one system for time measurement in insects might be to determine the genes and their protein products in both circadian and photoperiodic systems. Only relatively recently have molecular data for the circadian clock become available in *Drosophila*. Although the genome is now known for a number of insects, most of those do not have a strong photoperiodic response, and *D. melanogaster*, whose genome is known, has only a weak reproductive diapause in response to long nights, so it is not so easy to use *Drosophila* for studies of the photoperiodic clock and its molecular machinery.

During most of the twentieth century, insect researchers interested in clock mechanisms conducted experiments with insects that readily went into diapause when exposed to short days/long nights and constructed models to explain the photoperiodic responses. Because the circadian clock was known to be free running under constant conditions, the basic idea was to induce a mixed-age population of insects to make a photoperiodic response (e.g., enter diapause when exposed to successive short days, long nights) and then hold that population under constant dark conditions and look for any evidence of a rhythm near or at multiples of the 24 h circadian rhythm in the portion of the population entering diapause. Some experimenters also incorporated light pulses of various duration and intensity at intervals during the constant dark (DD) conditions to try to see if a reversal of response or a phase shift occurred. If such a rhythm was detected, it was taken as evidence that the circadian clock was at least somehow involved in the photoperiodic response. This type of approach was pioneered by plant scientists (Bünsow, 1953, 1960; Nanda and Hamner, 1958) and by insect scientists (Pittendrigh and Minis, 1964; Pittendrigh, 1966, 1972; Lees, 1973, 1986; Saunders, 1973a, 1974; Danks, 2005).

Numerous models were constructed to explain photoperiodic responses. Vaz Nunes and Saunders (1999) listed 13 of the most robust models that came out of these experiments, and they briefly described the main features of each model with supporting references (Table 1 in Vaz Nunes and Saunders, 1999). Most of the 13 models can be grouped into four groups: the hourglass model, resonance model, external coincidence model, and internal coincidence model. In some of these groups, there are models that show variations within the group. The experiments and models were intended to support or refute the idea that the circadian clock was the basis for the photoperiodic clock. The four major groups of models listed earlier are briefly described here.

6.7.1 Hourglass Model

The **hourglass model** was developed by Lees (1973, 1986, and references therein) working in England in a temperate climate with distinct summer and winter seasons. His work was mainly with **green vetch aphids**, *Megoura viciae*, that formed seasonal morphs with wings in the late fall. In the long days of summer, the aphids reproduced parthenogenetically as **virginoparae**, producing wingless individuals. In the fall, sexual and egg-laying **oviparae** were produced and eggs were laid that hatched into a winged morph. The adaptive benefit was that winged individual might fly to a more favorable location or habitat as the environment became increasingly less favorable to their host and to themselves. Lees suggested a night length timer measured the increasingly longer nights of fall, and he proposed the analogy of an hourglass that would have to be "turned over" for continuous development to continue. As slightly shorter days began in the early fall, the hourglass would be turned over each dawn, but as enough increasingly longer scotophases were "counted," the hourglass would not be turned over resulting in the development of winged morphs. The timer was not dependent on a circadian mechanism and was not free running; it would not reset itself in continuous darkness. The critical night length for inducing formation of sexual forms and eggs in *Megoura* proved to be about 9.5 h; a night that long was sufficient to prevent the hourglass from being turned over. Lees, using pulses of light spanning the dark period, found two points, however, during the critical night length that application of light could reverse the decision to produce oviparae. The first point occurred during the first 2–3 h of the critical night length, when pulses of blue light could reverse the timing mechanisms and virginoparae would continue to be produced. From hours 3–4 in the critical night phase, the system could not be altered by short light pulses, and oviparae were produced. A second reversal of oviparae production occurred from hour 5 of darkness to just before the critical night length of 9.5 h. Lees interpreted the data as two light-sensitive periods in the physiology of the aphids: the first reversal in the laboratory at 3–4 h into the night phase as indicative of the situation in the field in summer when the natural long-day light would be available to illuminate the early part of the (subjective) night phase, and virginoparae would continue to be produced, and again the early dawn light of a long day would illuminate the late sensitive phase just before the critical night length, and virginoparae also would continue to be produced. In longer nights of late fall, both the first light-sensitive phase and the late sensitive phase would be in the dark after the short day and long night, respectively, and oviparae would be produced. He concluded that there was no indication of a circadian rhythm or cycling in the photoperiodic response of the aphids. Saunders and Lewis (1987) suggested that a heavily damped circadian mechanism might still be involved but so damped down that there is no obvious cycling, and Saunders (1973a) interpreted his own work with *Sarcophaga argyrostoma*, which showed similarities to that with the green vetch aphid, as heavily damped oscillations.

6.7.2 External Coincidence Model

The **external coincidence model** grew out of research on the **eclosion rhythm** of *D. pseudoobscura* (Pittendrigh and Minnis, 1964; Pittendrigh, 1966, 1972). Pittendrigh called the response the external coincidence model because, in Pittendrigh's words (1972), "...photoperiodic induction occurs only when some phase of the circadian rhythm coincides in time with a phase (e.g., light) in the external daily cycle of environmental change." Light in the external coincidence model has two roles, entrainment of the daily rhythm and photoinduction of an appropriate response (in the case of *D. pseudoobscura*, the eclosion rhythm). Hence, the term external coincidence means "coincidence of a phase in some internal rhythm with a phase of an external cycle." The external coincidence model has been applied to explain the response of a number of other insects to a seasonal change in the environment. The model assumes that there is a single circadian oscillator that is entrained by the L/D cycle. Similar to the work of Lees, Pittendrigh proposed a hypothetical light-sensitive period or photoinducible phase (symbolized by φ_i) positioned at or near the end of the critical dark phase. This hypothetically positioned φ_i does not move, and in long nights of fall, it is positioned during the late part of the dark period (i.e., assuming a 12L/12D ideal cycle) well before the long night comes to an end, and the clock signals that fall and winter months are approaching when conditions for continuing development will be very unfavorable. In many temperate climate insects, diapause is the response. During spring or early summer when the photophase is increasing in length and the scotophase is becoming relatively short, the light-sensitive period falls in the early dawn light after the short night, and the clock is reset for the next cycle and continued development is signaled. The external coincidence model did not fit responses of most other insects that were being tested by various experimenters, but one case in which it did fit well was diapause in the flesh fly *S. argyrostoma* (Saunders, 1973a).

6.7.3 Internal Coincidence Model

This model also was proposed by Pittendrigh (1972), who postulated that "only under some photoperiods would critical phase–points of two (internal) oscillations coincide" to produce an internal coincidence. Light has only the single role of entrainment in the **internal coincidence model**; light is not the direct promoter of the **photoinducible response** that occurs when the internal oscillations are in phase and coincide. This model proposes that there are two or more circadian oscillators at work, with some oscillators being cued by the light-on phase, while others are cued by the light-off (dark) phase. The photoinduction role that may signal diapause or continued development is determined by the different phase relationships between the oscillators (lengthening interval between the oscillators in spring and summer with increasingly longer days and shorter nights [greater deviation from a 12L/12D cycle] and decreasing interval between oscillators in fall as days shorten and nights lengthen [approaching the 12L/12D ideal]). Data from relatively few insects have been found to fit the internal coincidence model, but data from diapause induction in photoperiodic experiments with the parasitic wasp *N. vitripennis* (Walker) fit the internal coincidence model well (Saunders, 1973b, 1974). Repeating peaks and valleys in percent of larvae diapausing occurred after adult *N. vitripennis* were exposed to L/D cycles of light from 4 to 28 h light and variable scotophases totaling combined L/D cycles up to 72 h in duration. These repeating peaks and valleys in percent diapausing were interpreted by Saunders (1974) as evidence for dawn and dusk oscillators and as evidence for a circadian process involved in the induction of diapause (Saunders, 1974).

6.7.4 Resonance Model

In the **resonance model**, suggested by Pittendrigh (1972), light entrains several oscillators involved in L/D measurement, but light may not have a direct role in photoperiodic (diapause) induction. At least two (but possibly more) oscillators occur with at least one measuring lights on and another measuring lights off. Lights on, of course, would reset the circadian clock each morning. A photoperiodic response was visualized as occurring when the light-on/light-off phases were close together in length, and thus in resonance. For example, as summer came to an end and fall began with shorter days and longer nights, the summer 14L/10D cycle would begin shifting more toward the circadian period of 12L/12D with more resonance of phases occurring, and a seasonal response could occur. The L/D cycle did not necessarily have to reach perfect resonance of a 12L/12D phase before a seasonal response could occur.

6.7.5 Summary Results from Model Experiments

Vaz Nunes and Saunders (1999) interpret the resonance model, external coincidence models, and internal coincidence models as indicating a role for the circadian clock system, albeit with a number of variations. They present a table showing insects (and the mites *Tetranychus urticae* and *Amblyseius potentillae*) in which experimental data are interpreted as indicating circadian involvement in a photoperiodic response induced by long night exposures in 13 species of insects in 5 orders and 2 mites. Negative responses, suggesting lack of circadian involvement in photoperiodic induction, were found in 11 insect species in four orders and in the red spider mite *Metatetranychus ulmi*, although they argue that a negative response does not have to mean lack of participation of the circadian clock (i.e., perhaps the experimental procedure just did not show a circadian clock involvement, or apparent lack of circadian influence might be due to heavily damped circadian oscillators).

6.8 CONCLUSIONS

As Zhang and Kay (2010) so aptly stated "Clocks [are] not winding down...." In fact, the importance of biological clocks in all of biology is increasing, with discovery of significant roles for rhythms in medicine, drug therapy, intermediary metabolism, agriculture, and human behavior. Probably, most of the advances in the future will be made with genetic and molecular tools. No doubt, more clock genes will be identified. Probably, clock details evolved as insects and other organisms evolved, and differences in the clock mechanism(s) may be defined in many different organisms. The increasingly successful elaboration of the complete genome of insects and other organisms will enable determination of similarities and differences in the clock genes. Perhaps, soon there can be definitive genetic and molecular evidence that everyone will accept to show whether the circadian and photoperiodic clocks of insects are the same or different.

6.9 REVIEW AND SELF-STUDY QUESTIONS

1. Describe the differences between a circadian rhythm and a photoperiodic rhythm.
2. What are the principal characteristics of a circadian rhythm?
3. What part of the 24 h cycle does the circadian clock measure?
4. What does it mean to say that a circadian rhythm is "entrained"?
5. What is CHRYPTOCHROME and what is its role in the clock mechanism?
6. Be able to name the principal clock genes and their protein products and explain the role of each.
7. What are the presently described variations in the *cry* gene?
8. What evidence is there for circadian influence on hormone secretion?
9. Describe some of the evidence for peripheral clocks.
10. What are at least three variations of ideas on whether the circadian and photoperiodic clocks are the same or different?

11. Design an experiment in which a group of insects is subjected to an L/D regime to determine what regime causes at least 50% of them to diapause.
12. How did plant scientists studying flowering in several different plants in the early part of the twentieth century influence biological rhythm research for the entire century?
13. What are the differences in the external coincidence model and the internal coincidence model for response of insects subjected to various L/D regimes?
14. Who developed the hourglass model for response of an insect to L/D regimes, and what insect was primarily used in those experiments?

REFERENCES

Allada, R. and B.Y. Chung. 2010. Circadian organization of behavior and physiology in *Drosophila*. *Annu. Rev. Physiol.* 72: 605–624.

Allalda, R., N.E. White, W.V. So, J.C. Hall, and M. Rosbash. 1998. A mutant *Drosophila* homolog of mammalian *Clock* disrupts circadian rhythms and transcription of *period* and *timeless*. *Cell* 93: 791–804.

Bachleitner, W., L. Kempinger, C. Wülbeck, D. Rieger, and C. Helfrich-Förster. 2007. Moonlight shifts the endogenous clock of *Drosophila melanogaster*. *Proc. Natl. Acad. Sci. USA* 104: 3538–3543.

Balmert, N.J., S.S.C. Rund, J.P. Ghazi, P. Zhou, and G.E. Duffield. 2014. Time-of-day specific changes in metabolic detoxification and insecticide resistance in the malaria mosquito *Anopheles gambiae*. *J. Insect Physiol.* 64: 30–39.

Bargiello, T.A., F.R. Jackson, and M.W. Young. 1984. Restoration of circadian behavioral rhythms by gene transfer in *Drosophila melanogaster*. *Nature* 312: 752–754.

Barth, M., M. Schultze, C.M. Schuster, and R. Strauss. 2010. Circadian plasticity in photoreceptor cells control visual coding efficiency in *Drosophila melanogaster*. *PLoS One* 5: e9217.

Beaver, L.M., B.L. Rush, B.O. Gvakharia, and J.M. Giebultowicz. 2002. Loss of circadian clock function decreases reproductive fitness in males of *Drosophila melanogaster*. *Proc. Natl. Acad. Sci. USA* 99: 2134–2139.

Bebas, P., B. Cymborowski, and J.M. Giebultowicz. 2001. Circadian rhythm of sperm release in males of the cotton leafworm, *Spodoptera littoralis:* In vivo and in vitro study. *J. Insect Physiol.* 47: 859–866.

Beck, S.D. 1980. *Photoperiodism*. Academic Press, New York.

Benito, J., H. Zheng, and P.E. Hardin. 2007. PDP1ε functions downstream of the circadian oscillator to mediate behavioral rhythms. *J. Neurosci.* 27: 2539–2547.

Bloch, G. 2009. Plasticity in the circadian clock and the temporal organization of insect societies, in J. Gadau, J. Fewell, and E.O. Wilson (eds.), *Organization of Insect Societies: From Genome to Sociocomplexity*. Harvard University Press, Cambridge, MA, pp. 402–431.

Bloch, G. 2010. The social clock of the honeybee. *J. Biol. Rhythms* 25: 307–317.

Bloch, G., E. Hazan, and A. Rafaeli. 2013. Circadian rhythms and endocrine functions in adult insects. *J. Insect Physiol.* 59: 56–69.

Bober, R., A. Azielli, and A. Rafaeli. 2010. Developmental regulation of the pheromone biosynthesis activating neuropeptide-receptor (PBAN-R): Reevaluating the role of juvenile hormone. *Insect Mol. Biol.* 19: 77–86.

Bober, R. and A. Rafaeli. 2010. Gene-silencing reveals the functional significance of pheromone biosynthesis activating neuropeptide receptor (PBAN-R) in a male moth. *Proc. Natl. Acad. Sci. USA* 107: 16858–16862.

Bradshaw, W.E. and C.M. Holzapfel. 2007a. Evolution of animal photoperiodism. *Annu. Rev. Ecol. Syst.* 38: 1–25.

Bradshaw, W.E. and C.M. Holzapfel. 2007b. Tantalizing *timeless*. *Science* 316: 1851–1852.

Bradshaw, W.E. and C.M. Holzapfel. 2010. What season is it anyway? Circadian tracking vs. photoperiodic anticipation in insects. *J. Biol. Rhythms* 25: 155–165.

Bünning, E. 1936. Die Endonome Tagesrhythmik als Grundlage der Photoperiodischen Reaktion. *Berichte der deutschen botanischen Geschellschaft* 53: 590–607.

Bünsow, R.C. 1953. Uber Tages- und Jahresrhythmische Änderungen der Photoperiodischen Lichtempfindlichkeit bei *Kalanchoë blossfeldiana* und ihre Beziehungen zur endogenen Tagesrhythmik. *Zeitschrift für Botanik* 41: 257–276.

Bünsow, R.C. 1960. The circadian rhythm of photoperiodic responsiveness in Kalanchoe. *Cold Spr. Harb. Symp. Quant. Bol.* 25: 257–260.

Busza, A., M. Emery-Le, M. Rosbash, and P. Emery. 2004. Roles of two *Drosophila* CRYPTOCHROME structural domains in circadian photoreceptors. *Science* 304: 1503–1506.

Ceriani, M.F., T.K. Darlington, D. Staknis, P. Mas, A.A. Petti, C.J. Weitz, and S.A. Kay. 1999. Light-dependent sequestration of TIMELESS by CRYPTOCHROME. *Science* 285: 553–556.

Chatterjee A., S. Tanoue, J.H. Houl, and P. Hardin. 2010. Regulation of gustatory physiology and appetitive behavior by the *Drosophila* circadian clock. *Curr. Biol.* 20: 300–309.

Chauerjee, A. and P.E. Hardin. 2010. Time to taste: Circadian clock function in the *Drosophila* gustatory system. *Fly* 4: 283–287.

Cheng, Y. and P.E. Hardin. 1998. *Drosophila* photoreceptors contain an autonomous circadian oscillator that can function without *period* gene. *J. Neurosci.* 18: 741–750.

Cyran, S.A., A.M. Buchsbaum, K.L. Reddy, M.-C. Lin, N.R.J. Glossop, P.E. Hardin, M.W. Young, R.V. Storti, and J. Blau. 2003. *vrille*, *Pdp1* and *dClock* form a second feedback loop in the *Drosophila* circadian clock. *Cell* 112: 329–341.

Danks, H.V. 2005. How similar are daily and seasonal biological clocks? *J. Insect Physiol.* 51: 609–619.

Darlington, D.K., K. Wager-Smith, M.F. Ceriani, D. Staknis, N. Gekakis, T.D.L. Steeves, C.J. Weitz, J.S. Takahashi, and A. Kay. 1998. Closing the circadian loop: CLOCK-induced transcription of its own inhibitors *per* and *tim*. *Science* 280: 1599–1603.

De, J., V. Varma, and V.K. Sharma. 2012. Adult emergence rhythm of fruit flies *Drosophila melanogaster* under seminatural conditions. *J. Biol. Rhythms* 27: 280–286.

Denlinger, D.L. 1986. Dormancy in tropical insects. *Annu. Rev. Entomol.* 31: 239–264.

Denlinger, D.L. 2002. Regulation of diapause. *Annu. Rev. Entomol.* 47: 93–122.

Denlinger, D.L. and P.A. Armbruster. 2014. Mosquito diapause. *Annu. Rev. Entomol.* 59: 73–93.

Di Cara, F. and K. King-Jones. 2013. How clocks and hormones act in concert to control the timing of insect development, in *Current Topics in Developmental Biology*, vol. 105, edited by Paul Wassarman, Elsevier, Inc., Amsterdam, the Netherlands.

Emery, P., W.V. So, M. Kaneko, J.C. Hall, and M. Rosbach. 1998. CRY, a *Drosophila* clock and light-regulated cryptochrome, is a major contributor to circadian rhythm resetting and photosensitivity. *Cell* 95: 669–679.

Emery, P., R. Stanewsky, C. Helfrich-Forster, M. Emery-Le, J.C. Hall, and M. Rosbash. 2000. *Drosophila* CRY is a deep-brain circadian photoreceptor. *Neuron* 26: 493–504.

Garner, W.W. and H.A. Allard. 1920. Flowering and fruiting of plants as controlled by the length of day, in *Yearbook of the U.S. Department of Agriculture*, United States Dept. of Agriculture. pp. 377–400.

Gaten, E., S.J. Huston, H.B. Dowse, and T. Matheson. 2012. Solitary and gregarious locusts differ in circadian rhythmicity of a visual output neuron. *J. Biol. Rhythms* 27: 196–205.

Gekakis, N., L. Saez, A.M. Delahaye-Brown, M.P. Myers, A. Sehgal, M.W. Young, and C.J. Weitz. 1995. Isolation of timeless by PER protein interaction: defective interaction between timeless protein and long-period mutant PER[1]. *Science* 270: 811–815.

Giebultowicz, J.M., J.G. Riemann, A.K. Raina, and R.L. Ridgway. 1989. Circadian system controlling release of sperm in the insect testes. *Science* 245: 1098–1100.

Giebultowicz, J.M., R. Stanewsky, J.C. Hall, and D.M. Hege. 2000. Transplanted *Drosophila* excretory tubules maintain circadian clock cycling out of phase with the host. *Curr. Biol.* 10: 107–110.

Glaser, F.T. and R. Stanewsky. 2007. Synchronization of the *Drosophila* circadian clock by temperature cycles. *Cold Spring Harb. Symp. Quant. Biol.* 72: 233.

Glossop, N.R., J.H. Houl, H. Zheng, F.S. Ng, S.M. Dudek, and P.E. Hardin. 2003. VRILLE feeds back to control circadian transcription of *Clock* in the *Drosophila* circadian oscillator. *Neuron* 37: 249–261.

Goto, S.G. 2013. Roles of circadian clock genes in insect photoperiodism. *Entomol. Sci.* 16: 1–16.

Grimaldi, D. and M.S. Engel. 2005. *Evolution of the Insects*. Cambridge University Press, Cambridge, U.K., 755pp.

Gvakharia, B.O., J.A. Kilgore, P. Bebas, and J.M. Giebultowicz. 2000. Temporal and spatial expression of the period gene in the reproductive system of the codling moth. *J. Biol. Rhythms* 15: 27–35.

Hardie, J. 1987. The photoperiodic control of wing development in the black bean aphid, *Aphis fabae*. *J. Insect Physiol.* 33: 543–549.

Hardin, P. 2005. The circadian timekeeping system of *Drosophila*. *Curr. Biol.* 15: R714–R722.

Hardin, P. 2009. Molecular mechanisms of circadian timekeeping in *Drosophila*. *Sleep Biol. Rhythms* 7: 235–242.

Hege, D.M., R. Stanewsky, J.C. Hall, and J.M. Giebultowicz. 1997. Rhythmic expression of a PER-reporter in the Malpighian tubules of decapitated *Drosophila:* Evidence for a brain-independent circadian clock. *J. Biol. Rhythms* 12: 300–308.

Helfrich-Förster, C. 2004. The circadian clock in the brain: a structural and functional comparison between mammals and insects. *J. Comp. Physiol.* 190: 601–613.

Helfrich-Förster, C. 2006. The neural basis of Drosophila's circadian clock. *Sleep Biol. Rhythms* 4: 224–234.

Hermann, C., R. Saccon, P.R. Senthilan, L. Domnik, H. Dircksen, T. Yoshii, and C. Helfrich-Förster. 2013. The circadian clock network in the brain of different *Drosophila* species. *J. Comp. Neurol.* 521: 367–388.

Hermann, C., T. Yoshii, V. Dusik, and C. Helfrich-Förster. 2012. Neuropeptide F immunoreactive clock neurons modify evening locomotor activity and free-running period in *Drosophila melanogaster. J. Comp. Neurol.* 520: 970–987.

Houl, J.H., W. Yu, S.M. Dudek, and P.E. Hardin. 2006. *Drosophila* CLOCK is constitutively expressed in circadian oscillator and non-oscillator cells. *J. Biol. Rhythms* 21: 93–103.

Ikeno, T., H. Numata, and S.G. Goto. 2011. Circadian clock genes *period* and *cycle* regulate photoperiodic diapause in the bean bug *Riptortus pedestris* males. *J. Insect Physiol.* 57: 935–938.

Ikeno, T., S.I. Tanaka, H. Numata, and S.G. Goto. 2010. Photoperiodic diapause under the control of circadian clock genes in an insect. *BMC Biol* 8: 116.

Ito, C., S.G. Goto, S. Ship, K. Tomioka, and H. Numala. 2008. Peripheral circadian clock for the cuticle deposition rhythm in *Drosophila melanogaster. Proc. Natl. Acad. Sci. USA* 105: 8446–8451.

Kadener, S., D. Stoleru, M. McDonald, P. Nawathean, and M. Rosbash. 2007. *Clockwork orange* is a transcriptional repressor and a new *Drosophila* circadian pacemaker component. *Genes Dev.* 21: 1675–1686.

Kamae, Y., F. Tanaka, and K. Tomioka. 2010. Molecular cloning and functional analysis of the clock genes, *Clock* and cycle, in firebrat *Thermobia domestica. J. Insect Physiol.* 56: 1291–1299.

Kamae, Y. and K. Tomioka. 2012. timeless is an essential component of the circadian clock in a primitive insect, the firebrat *Thermobia domestica. J. Biol. Rhythms* 27: 126–134.

Kistenpfennig, C., J. Hirsh, T. Yoshii, and C. Helfrich-Förster. 2012. Phase-shifting the fruit fly clock without cryptochrome. *J. Biol. Rhythms* 27: 117–125.

Koh, K., X. Zheng, and A. Sehgal. 2006. JETLAG resets the *Drosophila* circadian clock by promoting light-induced degradation of TIMELESS. *Science* 312: 1809–1812.

Konopka, R.J. and S. Benzer. 1971. Clock mutants of *Drosophila melanogaster. Proc. Natl. Acad. Sci. USA* 68: 2112–2116.

Koštál, V. 2011. Insect photoperiodic calendar and circadian clock: Independence, cooperation, or unity? *J. Insect Physiol.* 57: 538–556.

Kotwica, J., P. Bebas, B.O. Gvakharia, and J.M. Giebultowicz. 2009a. RNA interference of the *period* gene affects the rhythm of sperm release in moths. *J. Biol. Rhythms* 24: 25–34.

Kotwica, J., M.K. Larson, P. Bebas, and J.M. Giebultowicz. 2009b. Developmental profiles of PERIOD and DOUBLETIME in *Drosophila melanogaster* ovary. *J. Insect Physiol.* 55: 419–425.

Krishnan, B., S.E. Dryer, and P.E. Hardin. 1999. Circadian rhythms in olfactory responses of *Drosophila melanogaster. Nature* 400: 375–378.

Krishnan, N., A.J. Davis, and J.M. Giebultowicz. 2008. Circadian regulation of response to oxidative stress in *Drosophila melanogaster. Biochem. Biophys. Res. Commun.* 374: 299–303.

Larsdotter-Mellström, H., R. Murtazina, A.-K. Borg-Karlson, and C. Wiklund. 2012. Timing of male sex pheromone biosynthesis in a butterfly—Different dynamics under direct or diapause development. *J. Chem. Ecol.* 38: 584–591.

Lee, J.E. and I. Edery. 2008. Circadian regulation in the ability of *Drosophila* to combat pathogenic infections. *Curr. Biol.* 18: 195–199.

Lees, A.D. 1973. Photoperiodic time measurement in the aphid *Megoura viciae. J. Insect Physiol.* 19: 2279–2316.

Lees, A.D. 1986. Some effects of temperature on the hour glass photoperiod timer in the aphid *Megoura viciae. J. Insect Physiol.* 32: 79–89.

Levi-Zada, A., M. David, D. Fefer, V. Seplyarsky, A. Sadowsky, S. Dobrinin, T. Ticuchinski, D. Harari, D. Blumberg, and E. Dunkelblum. 2014. Circadian release of male-specific components of the greater date moth, *Aphomia (Arenipses) sabella,* using sequential SPME/GC/MS analysis. *J. Chem. Ecol.* 40: 236–243.

Lim, C., B.Y. Chung, J.L. Pitman, J.J. McGill, S. Pradham, J. Lee, K.P. Keegan, J. Choe, and J. Allada. 2007. Clockwork orange excodes a transcriptional repressor important for circadian-clock amplitude in *Drosophila*. *Curr. Biol.* 17: 1082–1089.

Matsumoto, A., M. Ukai-Tadenuma, R.G. Yamada, J. Houl, K.D. Uno, T. Kasukawa, B. Dauwalder et al. 2007. A functional genomics strategy reveals *clockwork orange* as a transcriptional regulator in the *Drosophila* circadian clock. *Gene Dev.* 21: 1687–1700.

Matsumoto, K., H. Numata, and S. Shiga. 2013. Role of the brain in photoperiodic regulation of juvenile hormone biosynthesis in the brown-winged green bug *Plautia stali*. *J. Insect Physiol.* 59: 387–393.

Merlin, C., M.C. François, I. Queguiner, M. Maibèche-Coisné, and E. Jacquin-Joly. 2007. Evidence for a putative antennal clock in *Mamestra brassicae*: Molecular cloning and characterization of two clock genes—*Period* and *cryptochrome*—In antennae. *Insect Mol. Biol.* 15: 137–145.

Merlin, C., R.J. Gegear, and S.M. Reppert. 2009. Antennal circadian clocks coordinate sun compass orientation in migratory monarch butterflies. *Science* 325: 1700–1704.

Merlin, C. and S.M. Reppert. 2010. Lepidopteran circadian clocks. From molecules to behavior, in M.R. Goldsmith and F. Marec (eds.), *Molecular Biology and Genetics of the Lepidoptera*. CRC Press, Boca Raton, FL, pp. 137–152.

Merritt, D.J. and S. Aotani. 2008. Circadian regulation of bioluminescence in the prey-luring glowworm, *Arachnocampa flava*. *J. Biol. Rhythms* 23: 319–329.

Musolin, D.L. 2012. Surviving winter: Diapause syndrome in the southern green stink bug *Nezara viridula* in the laboratory, in the field, and under climate change conditions. *Physiol. Entomol.* 37: 309–322.

Myers, M.P., K. Wager-Smith, C.S. Wesley, M.W. Young, and A. Sehgal. 1995. Positional cloning and sequence analysis of the *Drosophila* clock gene, *timeless*. *Science* 270: 805–808.

Nanda, K.K. and K.C. Hamner. 1958. Studies on the nature of the endogenous rhythm affecting photoperiodic responses of Biloxi soybean. *Botan. Gaz.* 120: 14–25.

Nelson, R.J., D.L. Denlinger, and D.E. Somers (eds.). 2010. *Photoperiodism: The Biological Calendar*. Oxford University Press, Oxford, U.K., 581pp.

Page, T.L. and E. Koelling. 2003. Circadian rhythm in olfactory response in the antennae controlled by the optic lobe in the cockroach. *J. Insect Physiol.* 49: 697–707.

Peschel, N. and C. Helfrich-Förster. 2011. Setting the clock—By-nature: Circadian rhythm in the fruitfly *Drosophila melanogaster*. *FEBS Lett.* 585: 1435–1442; *Physiology* 57: 935–938.

Pittendrigh, C.S. 1966. The circadian oscillation in *Drosophila pseudoobscura* pupae: A model for the photoperiodic clock. *Z. für Pflanzenphysiologie* 54: 275–307.

Pittendrigh, C.S. 1972. Circadian surfaces and the diversity of possible roles of circadian organization in photoperiodic induction. *Proc. Natl. Acad. Sci. USA* 69: 2734–2737.

Pittendrigh, C.S. and S. Daan. 1976. A functional analysis of circadian pacemakers in nocturnal rodents. V. Pacemaker structure: A clock for all seasons. *J. Comp. Physiol. A* 106: 333–355.

Pittendrigh, C.S., J.H. Eichhorn, D.H. Minis, and V.G. Bruce. 1970. Circadian systems VI. Photoperiodic time measurement in *Pectinophora gossypiella*. *Proc. Natl. Acad. Sci. USA* 66: 758–764.

Pittendrigh, C.S. and D.H. Minis. 1964. The entrainment of circadian oscillations by light and their role as photoperiodic clocks. *Am. Naturalist* 98: 261–294

Plautz, J.D., M. Kaneko, J.C. Hall, and S.A. Kay. 1997. Independent photoreceptive circadian clocks throughout *Drosophila*. *Science* 278: 1632–1635.

Pyza, E. and B. Cymborowski. 2001. Circadian rhythms in behaviour and in the visual system of the blowfly *Calliphora vicina*. *J. Insect Physiol.* 47: 897–904.

Pyza, E. and A. Meinertzhagen. 1995. Monopolar cell axons in the first optic neuropil of the housefly, *Musca domestica* L. undergo daily fluctuations in diameter that have a circadian basis. *J. Neurosci.* 15: 407–418.

Pyza, E. and I.A. Meinertzhagen. 1999. Daily rhythmic changes of cell size and shape in the first optic neuropil in *Drosophila melanogaster*. *J. Neurobiol.* 40: 77–88.

Reddy, P., W.A. Zehring, D.A. Wheeler, V. Pirrotta, C. Hadfield, J.S. Hall, and M. Rosbash. 1984. Molecular analysis of the period locus in *Drosophila melanogaster* and identification of a transcript involved in biological rhythms. *Cell* 38: 701–710.

Reppert, S.M. 2007. The ancestral circadian clock of monarch butterflies: Role in time-compensated sun compass orientation. *Cold Spring Harb. Symp. Quant. Biol.* 72: 113–118.

Richier, B., C. Michard-Vanhée, A. Lamouroux, C. Papin, and F. Rouyer. 2008. The clockwork orange *Drosophila* protein functions as both an activator and a repressor of clock gene expression. *J. Biol. Rhythms* 23: 103–116.

Rodriguez-Zas, S.L., B.R. Southey, Y. Shemesh, E.B. Rubin, M. Cohen, G.E. Robinson, and G. Bloch. 2012. Microarray analysis of naturally socially regulated plasticity in circadian rhythms of honey bees. *J. Biol. Rhythms* 27(1): 12–24.

Rubin, E.B., Y. Shemesh, M. Cohen, S. Elgavish, H.M. Robertson, and G. Bloch. 2006. Molecular and phylogenetic analyses reveal mammalian-like clockwork in the honey bee (*Apis mellifera*) and shed new light of the molecular evolution of the circadian clock. *Genome Res.* 16: 1352–165.

Rutila, J.E., V. Suri, M. Le, W.V. So, M. Rosbash, and J.C. Hall. 1998. CYCLE is a second bHLH-PAS clock protein essential for circadian rhythmicity and transcription of *Drosophila period* and *timeless*. *Cell* 93: 805–814.

Saifullah, A.S.M. and T.L. Page. 2009. Circadian regulation of olfactory neurons in the cockroach antenna. *J. Biol Rhythms* 24: 144–152.

Sakai, T. and N. Ishida. 2001. Circadian rhythms of female mating activity governed by clock genes in *Drosophila*. *Proc. Natl. Acad. Sci. USA* 98: 9221–9225.

Sandrelli, F., R. Costa, C.P. Kyriacou, and E. Rosato. 2008. Comparative analysis of circadian clock genes in insects. *Insect Mol. Biol.* 17: 447–463.

Sauman, I.A.D. Briscoe, H. Zhu, D. Shi, O. Froy, J. Stalleicken, Q. Yuan, A. Casselman, and S.M. Reppert. 2005. Connecting the navigational clock to sun compass input in monarch butterfly brain. *Neuron* 46: 457–467.

Saunders, D.S. 1971. The temperature-compensated photoperiodic clock "programming" development and pupal diapause in the flesh-fly, *Sarcophaga argyrostoma*. *J. Insect Physiol.* 17: 801–812.

Saunders, D.S. 1973a. The photoperiodic clock in the flesh-fly, *Sarcophaga argyrostoma*. *J. Insect Physiol.* 19: 1941–1954.

Saunders, D.S. 1973b. Thermoperiodic control of diapause in an insect: theory of internal coincidence. *Science* 181: 358–360.

Saunders, D.S. 1974. Evidence for "dawn" and "dusk" oscillators in the *Nasonia* photoperiodic clock. *J. Insect Physiol.* 20: 77–88.

Saunders, D.S. 2002. *Insect Clocks*, 3rd edn. Elsevier Science, Amsterdam, the Netherlands, 560pp.

Saunders, D.S. 2005. Erwin Bünning and Tony Lees, two giants of chronobiology, and the problem of time measurement in insect photoperiodism. *J. Insect Physiol.* 51: 599–608.

Saunders, D.S. 2009. Circadian rhythms and the evolution of photoperiodic timing in insects. *Physiol. Entomol.* 34: 301–308.

Saunders, D.S. 2010. Controversial aspects of photoperiodism in insects and mites. *J. Insect Physiol.* 56: 1491–1502.

Saunders, D.S. 2011. Unity and diversity in the insect photoperiodic mechanism. *Entomol. Sci.* 14: 235–244.

Saunders, D.S. 2012. Insect photoperiodism: Seeing the light. *Physiol. Entomol.* 37: 207–218.

Saunders, D.S. 2013. Insect photoperiodism: Measuring the night. *J. Ins. Physiol.* 59: 1–10.

Saunders, D.S. and R.C. Bertossa. 2011. Deciphering time measurement: The role of circadian 'clock' genes and formal experimentation in insect photoperiodism. *J. Insect Physiol.* 57: 557–566.

Saunders, D.S. and R.D. Lewis. 1987. A damped circadian oscillator model of an insect photoperiodic clock. III. Circadian and "hourglass" responses. *J. Theoretical Biol.* 128: 73–85.

Schiesari, L., C.P. Kyriacou, and R. Costa. 2011. The hormonal and circadian basis for insect photoperiodic timing. *FEBS Lett.* 585: 1450–1460.

Schlichting, M., R. Grebler, N. Peschel, T. Yoshii, and C. Helfrich-Förster. 2014. Moonlight detection by *Drosophila*'s endogenous clock depends on multiple photopigments in the compound eyes. *J. Biol. Rhythms* 29: 75–86.

Schuckel, J., K.K. Siwicki, and M. Stengl. 2007. Putative circadian pacemaker cells in the antenna of the hawkmoth *Manduca sexta*. *Cell Tissue Res.* 330: 271–278.

Schurko, A.M., D.J. Mazur, and J.M. Logsdon, Jr. 2010. Inventory and phylogenetic distribution of meiotic genes in *Nasonia vitripennis* and among diverse arthropods. *Insect Mol. Biol.* 19: 165–180.

Sehgal, A., J.L. Price, B. Man, and M.W. Young. 1994. Loss of circadian behavioral rhythms and per mRNA oscillations in the *Drosophila* mutant timeless. *Science* 263: 1603–1606.

Shemesh, Y., M. Cohen, and G. Bloch. 2007. Natural plasticity in circadian rhythms is mediated by reorganization in the molecular clockwork in honeybees. *FASEB J.* 21: 2304–2311.

Shemesh, Y., A. Eban-Rothschild, M. Cohen, and G. Bloch. 2010. Molecular dynamics and social regulation of context-dependent plasticity in the circadian clockwork of the honey bee. *J. Neurosci.* 30: 12517–12525.

Siwicki, K.K., C. Eastman, G. Petersen, M. Rosbash, and J.C. Hall. 1988. Antibodies to the *period* gene product of *Drosophila* reveal diverse tissue distribution and rhythmic changes in the visual system. *Neuron* 1: 141–150.

Stanewsky, R., M. Kaneko, P. Emery, B. Beretta, K. Wager-Smith, S.A. Kay, M. Rosbash, and J.C. Hall. 1998. The cry^b mutation identifies cryptochrome as a circadian photoreceptor in *Drosophila*. *Cell* 95: 681–692.

Stay, B. and A.J. Zera. 2010. Morph-specific diurnal variation in allatostatin immunostaining in the corpora allata of *Gryllus firmus*: implications for the regulation of a morph-specific circadian rhythm for JH biosynthetic rate. *J. Insect Physiol.* 56: 266–270.

Subramanian, P., V. Prasanna, J.J. Jayapalan, P.S.A. Rahman, and O.H. Hashim. 2014. Role of *Bacopa monnieri* in the temporal regulation of oxidative stress in clock mutant (cry^b) of *Drosophila melanogaster*. *J. Insect Physiol.* 65: 37–44.

Tanoue, S., P. Krishnan, B. Krishnan, S.E. Dryer, and P.E. Hardin. 2004. Circadian clocks in antennal neurons are necessary and sufficient for olfaction rhythms in *Drosophila*. *Curr. Biol.* 14: 638–649.

Tobback, J., B. Boerjan, H.P. Vandersmissen, and R. Huybrechts. 2011. The circadian clock genes affect reproductive capacity in the desert locust *Schistocerca gregaria*. *Insect Biochem. Mol. Biol.* 41: 313–321.

Tomioka, K. and A. Matsumoto. 2010. A comparative view of insect clock systems. *Cell. Mol. Life Sci.* 67: 1397–1406.

Tomioka, K., O. Uryu, Y. Kamae, Y. Umezaki, and T. Yoshii. 2012. Peripheral circadian rhythms and their regulatory mechanism in insects and some other arthropods: A review. *J. Comp. Physiol. B* 182: 729–740.

Truman, J.W. 1972. Physiology of insect rhythms I. Circadian organization of the endocrine events underlying the moulting cycle of larval tobacco hornworms. *J. Exp. Biol.* 57: 805–820.

Uryu, O. and K. Tomioka. 2010. Circadian oscillations outside the optic lobe in the cricket *Gryllus bimaculatus*. *J. Insect Physiol.* 56: 1284–1290.

Uryu, O., S.G. Karpova, and K. Tomioka. 2013. The clock gene *cycle* plays an important role in the circadian clock of the cricket *Gryllus bimaculatus*. *J. Insect Physiol.* 59: 697–704.

Vafopoulou, X., M. Cardinal-Aucoin, and C.G.H. Steel. 2012. Rhythmic release of prothoracicotropic hormone from the brain of an adult insect during egg development. *Comp. Biochem. Physiol. Part A* 161: 193–200.

Vafopoulou, X. and C.G.H. Steel. 1991. Circadian regulation of synthesis of ecdysteroids by prothoracic glands of the insect *Rhodnius prolixus*: Evidence of a dual oscillator system. *Gen. Comp. Endocrinol.* 83: 27–34.

Vafopoulou, X. and C.G.H. Steel. 1999. Daily rhythm of responsiveness to prothoracicotropic hormone in prothoracic glands of *Rhodnius prolixus*. *Arch. Insect Biochem. Physiol.* 41: 117–123.

Vafopoulou, X. and C.G.H. Steel. 2012. Metamorphosis of a clock: Remodeling of the circadian timing system in the brain of *Rhodnius prolixus* (Hemiptera) during larval-adult development. *J. Comp. Neurol.* 520: 1146–1154.

Vafopoulou, X., C.G.H. Steel, and K.L. Terry. 2007. Neuroanatomical relations of prothoracicotropic hormone neurons with the circadian timekeeping system in the brain of larval and adult *Rhodnius prolixus* (Hemiptera). *J. Comp. Neurol.* 503: 511–524.

Vafopoulou, X., K.L. Terry, and C.G.H. Steel. 2010. The circadian timing system in the brain of fifth larval instar of *Rhodnius prolixus* (Hemiptera). *J. Comp. Neurol.* 518: 1264–1282.

Vaz Nunes, M. and D. Saunders. 1999. Photoperiodic time measurement in insects: A review of clock models. *J. Biol. Rhythms* 14: 84–104.

Weber, F., D. Zorn, C. Rademacher, and H.-C. Hung. 2011. Post-translational timing mechanisms of the *Drosophila* clock. *FEBS Letters* 585: 1443–1449.

Werckenthin, A., C. Derst, and M. Stengl. 2012. Sequence and expression of *per*, *tim 1*, and *cry 2* genes in the Madeira cockroach *Rhyparobia maderae*. *J. Biol. Rhythms* 27: 453–466.

Xu, K., J.R. DiAngelo, M.E. Hughes, J.B. Hogenesch, and A. Sehgal. 2011. The circadian clock interacts with metabolic physiology to influence reproductive fitness. *Cell Metabolism* 13: 639–654.

Yuan, Q., D. Metterville, A.D. Briscoe, and S.M. Reppert. 2007. Insect cryptochromes: Gene duplication and loss define diverse ways to construct insect circadian clocks. *Mol. Biol. Evol.* 24: 948–955.

Závodská, R., G. von Wowern, C. Löfstedt, W.-Q. Rosén, and I. Sauman. 2009. The release of a pheromonotropic neuropeptide, PBAN, in the turnip moth *Agrotis segetum*, exhibits a circadian rhythm. *J. Insect Physiol.* 55: 435–440.

Zehring, D.A., D.A. Wheeler, P. Reddy, R.J. Konopka, C.P. Kyriacou, M. Rosbash, and J.C. Hall. 1984. P-element transformation with period locus DNA restores rhythmicity to mutant, arrhythmic *Drosophila melanogaster*. *Cell* 39: 369–376.

Zerr, D.M., J.C. Hall, M. Rosbach, and K.K. Siwicki. 1990. Circadian fluctuations of period protein immunoreactivity in the CNS and the visual system of *Drosophila*. *J. Neurosci.* 10: 2749–2762.

Zhang, E.E. and S.A. Kay. 2010. Clocks not winding down: Unravelling circadian networks. *Nat. Rev. Mol. Cell Biol.* 11: 764–776.

Zhu, H., I. Sauman, Q. Yuan, A. Casselman, M. Emery-Le, P. Emery, and S.M. Reppert. 2008. Cryptochromes define a novel circadian clock mechanism in monarch butterflies that may underlie sun compass navigation. *PLoS Biol.* 6: e4.

CHAPTER 7

Diapause

PREVIEW

Diapause is a genetically regulated program of altered development. Typically, during diapause, development is slowed or stopped in immature insects and reproduction is repressed in adult insects. For some insects, diapause is obligatory and for others it can be facultative depending on environmental conditions. Some insects never diapause. In those insects that can, during any developmental stage or period in the life of the insect, it may enter diapause. Diapause evolved as a survival strategy to avoid adverse environmental conditions unfavorable to continued development, activity, or reproduction. Environmental cues are the principal inducers of diapause. Insects enter and pass through several phases in the diapause process, including initiation, preparation, diapause, termination, and sometimes a postdiapause phase. Several different hormones play a role in one or more phases of diapause; in some cases, diapause is regulated by the secretion of one or more hormones, and sometimes diapause is regulated by the lack of one or more hormones.

Circadian (daily) and seasonal (photoperiod) biological clocks influence many aspects of insect life, including diapause. The genetic basis for diapause is emerging as an active area of investigation. Some genes are known to be upregulated in diapause, and others are downregulated. A complete genetic explanation for diapause is not yet known. Prior to entering diapause, insects usually, but not invariably, accumulate lipid stores in the body that will serve their slowed metabolic needs for the long period of nonfeeding that usually occurs in diapausing immature insects. Some immatures and some adults in diapause continue to feed and may remain active. Another active area of investigation is the molecular basis of diapause and diapause termination. Although several hundred species of insects are known to diapause, most of the present knowledge is based on the in-depth study of only a few species.

7.1 INTRODUCTION

Insects have evolved a number of adaptive mechanisms to deal with an environment that can be unfavorable for activity and development temporarily or for long periods. Three of the ways that insects adapt to an unfavorable environment include **dormancy**, **quiescence**, and **diapause**. Dormancy is a broadly encompassing term that has been used to describe any state of suppressed development that is ecologically or evolutionarily meaningful, usually accompanied by suppressed metabolism. In this broad sense, dormancy encompasses two very different states: quiescence and diapause (Koštál, 2006). Quiescence is an immediate response to temporarily unfavorable environmental conditions, such as a few days of unseasonably cold weather in the life of an insect; the insect typically becomes quiet and physiological responses slow down. This quiescent state is not genetically programmed or regulated, and with the return of favorable environmental conditions, the insect responds immediately with resumption of normal activity. Diapause is a genetically regulated process that "represents an alternative developmental pathway prompted by unique patterns of gene expression that result in the sequestration of nutrient reserves, suppression of metabolism, a halt or slowing of development, and the acquisition of increased tolerance to environmental stresses" (Rinehart et al., 2001). Diapause often begins while favorable environmental conditions still exist and does not end with temporary return of favorable conditions. Insects are able to colonize environments that may not be suitable for continuous activity and development, maximize use of seasonally fluctuating resources, diversify into niches in the environment, and colonize temperate and polar habitats by diapausing during unfavorable conditions (Koštál, 2006). Insects sense, and apparently measure, predictably changing environments (such as approaching winter), and they prepare for, and enter, diapause well before conditions become unfavorable for continued activity and development and often continue in diapause for some time after favorable conditions return. Once entered, diapause typically continues for months, and in a few cases, for a year or more. Diapausing insects usually show increased resistance to harsh environmental conditions, including increased cold hardiness and resistance to desiccation (Denlinger, 2002). Metabolism typically is depressed, so they use less energy, and nutrient reserves in the body are conserved (reviewed by Hahn and Denlinger, 2007).

Species are known to diapause in the embryonic stage, as larvae, as pupae, or as adults. Usually an insect does not diapause in more than one stage, but exceptions have been described. Diapause is **obligatory** for some insects, typically those with a **univoltine** life cycle, so that every generation enters diapause during the same point in the life cycle. In most insects, diapause is **facultative**, and as long as favorable conditions exist, generations continue without diapause, but when unfavorable conditions are eminent, diapause is entered.

The environmental token cues that cause insects to enter diapause have been described for a great many insects, and those cues include drought, desiccation, low moisture content of the food, scarcity of food, high temperature, low temperature, critical day length, crowding, maternal diet, maternal age at oviposition, and maternal exposure to certain environmental conditions at particular periods in its life. The environmental conditions that induce diapause in the greatest number of insects, especially those in temperate and arctic regions, are seasonal alternation of day/night length and decreasing temperature as reliable indicators of approaching winter.

Diapause occurs in many invertebrates in addition to insects and has been described in more than 500 species in 17 orders (Nishizuka et al., 1998; Koštál, 2006, and references therein); so the literature on diapause is extensive and scattered in many sources. Recent reviews of diapause, primarily in insects, include those by Denlinger (1985, 2002), Danks (1987), Saunders et al. (2002), Koštál (2006), Hahn and Denlinger (2007), MacMillan and Sinclair (2011), Sláma and Denlinger (2013), and Denlinger and Armbruster (2014).

7.2 DIAPAUSE: A SURVIVAL STRATEGY

Insects benefit in several ways from the capacity to diapause during part of their life cycle. For insects that live in temperate and polar climates, the major benefit of diapause is a way to survive the harsh cold of winter. Some insects benefit by synchronizing adult emergence from the immature insects that are not synchronized in their development. Conversely, synchrony of developing stages can be eliminated when some individuals enter diapause while others do not, thus staggering adult development and eclosion in a bet-hedging strategy so that all adults do not emerge at a given time when adverse survival conditions might occur. Staggered adult emergence also allows dispersal and reduces sibling mating. A reproductive diapause in adults can allow nutrients to be redirected from egg development to migratory flight. For example, monarch butterflies, *Danaus plexippus*, go through several generations as they migrate northward in the spring and early summer across the United States, finally arriving in southern Ontario, Canada. The adult generation emerging in late summer enters a reproductive diapause and shifts their energy resources to support migratory flight as they make their way to wintering grounds in the mountains near Mexico City. These adults stay in reproductive diapause in Mexico, but on warm days they fly in search of water and nectar from nearby plants (Figure 7.1). In March of each year, they begin to migrate northward again, the ovaries develop mature eggs and they lay them on milkweed plants in northern Mexico and southern Texas. Successive generations continue without diapause as they move farther north, until in late summer, adults from all over the North American continent, that have never migrated to Mexico, will enter reproductive diapause and repeat the migration.

How polar invertebrates survive harsh temperature is particularly interesting (Everatt et al., 2013). An insect with an especially fascinating life cycle is the arctic Woolybear, *Gynaephora groenlandica*. It lives "on the edge" in the arctic and survives harsh winters by diapausing. It requires 14 years to complete its life cycle at Alexandra Fiord, Ellesmere Island, Canada, and possibly as long as 20 years farther north at Lake Hazen also on Ellesmere Island (Bennett et al., 1999, and references therein). In June of each year, it spends a lot of time basking in the sun to warm up because the mean temperature in June is only about 10°C. The larva can rapidly increase its metabolic rate when favorable conditions occur and it becomes active and feeds upon its primary host, the arctic willow, *Salix arctica*. Metabolic rate and activity drop rapidly to conserve energy during brief spells of unfavorable conditions. It spends years as a larva, feeding during a brief interval in the arctic summer, and diapausing year after year to survive the very harsh winter. In July, before the arctic summer has peaked, it ceases to feed, spins a hibernacula in a concealed and somewhat protected place, and diapauses as a larva throughout the long arctic winter. During the winter, it freezes at approximately −8°C to −10°C without detrimental effects on cell structure, and it can then tolerate temperatures as low as −70°C. Even in the summer, its tissues can tolerate sudden cold spells and it survives freezing temperatures down to −15°C in short cold periods.

Another freeze-tolerant lepidopteran, *Pringleophaga marioni* (Tineidae), lives on Marion Island, a sub-Antarctic island that is cold, wet, and now experiencing climate change. These numerous and large caterpillars are considered a keystone decomposer species on the island. Their life

(a)

(b)

Figure 7.1 (**See color insert.**) (a) An adult monarch butterfly taking nectar from a flower on a sunny day in mid-January at the El Rosario colony of diapausing and overwintering monarchs in the mountains near Mexico City. (b) Several diapausing monarchs feeding from a flower in the Sierra Chincua colony in the mountains near Mexico City on a warm day in mid-January. (Photos courtesy of Rochelle Carlson Nation.)

cycle requires several years of larval development because they are subjected to repeated bouts of cold exposure, which limits their feeding and growth (Sinclair and Chown, 2005, and references therein). Sinclair and Chown found that repeated exposure to −5°C did not cause mortality, but did cause weight loss due to the cessation of feeding.

7.3 PHASES OF DIAPAUSE

Insects programmed to diapause pass through successive phases, including prediapause induction and preparation, initiation of diapause, continuing maintenance, termination of diapause, and postdiapause physiology and behavior (Koštál, 2006). Many different descriptive terms have been used in the literature to characterize the stages of diapause. Koštál has attempted to standardize these terms. Table 7.1 summarizes the descriptive diapause terms used by Koštál, with a brief characterization of the stages of diapause.

7.3.1 Prediapause: Induction and Preparation

The **prediapause** program typically begins well before environmental conditions become adverse for further development or survival. Entry into diapause is not a sudden, triggered event, but typically a slow process, with changes occurring gradually. The principal token cues inducing diapause are environmental changes in day length and temperature, but other factors also influence the decision to diapause in some species. These include inadequate nutritional resources, low water content and/or senescence and quality of food resources, excessive crowding, drought conditions and desiccation, the mother's diet, the mother's age at oviposition, and exposure of the mother to certain environmental conditions during development or during oogenesis. Some parasitoids enter diapause when their host enters diapause and then they depend on the host's endocrine system to terminate diapause. Sometimes the presence of a parasitoid in a host can cause the host to fail to enter diapause when it otherwise would have done so (Tauber et al., 1986; Denlinger, 2002). Many papers and reviews have been published describing cues leading to the induction of diapause in a variety of insects, including, but not limited to, De Wilde (1962), Danilevskii (1965), Beck (1980, 1983), Tauber et al. (1986), Danks (1987), and Denlinger (2002).

Table 7.1 Stages of Insect Diapause as Characterized by Koštál

Prediapause	Continuing morphogenesis, environmental signals condition the insect to diapause
Induction phase	Token stimuli (cues) from environment are transduced and switch direct development into diapause program
Preparation phase	Behavioral and physiological preparation for diapause can occur; some direct development may continue
Diapause	Diapause occurs and is maintained
Initiation	Energy reserves accumulated, morphogenesis stops, metabolism suppressed, some feeding may continue, may seek microhabitat for diapause, diapause intensity increases
Maintenance	Arrested development continues, token stimuli aid maintenance of diapause, gradual decrease in diapause intensity, greater sensitivity to diapause termination conditions
Termination	Environmental signals are transduced and cause decrease in diapause intensity; members of a population may become synchronized for further development; direct development may resume if environment suitable
Postdiapause quiescence	Unfavorable conditions for immediate development may result in a period of quiescence until favorable conditions occur

Source: Koštál, V., *J. Insect Physiol.*, 52, 113, 2006.

Overwintering insects (Koštál et al., 2007) that diapause in order to survive cold temperatures typically synthesize low molecular weight **polyhydric alcohols**, such as glycerol, trehalose, and sorbitol, and accumulate energy reserves as lipids that are stored in the fat body as **triacylglycerols**. The polyols may serve as cryoprotectants, or in some cases, they may have a role in the control of the diapause program (Horie et al., 2000). For many insects, diapause is often a long period without feeding or taking water, and accumulation of reserves before diapause and conservation of resources during diapause are important to survival and to the fitness of the insects when diapause is terminated. The literature on energy resources, conservation, and ultimate effects upon fitness when diapause is terminated has been reviewed by Hahn and Denlinger (2007).

The heteropteran, *Pyrrhocoris apterus*, showed nearly doubling of free amino acids, with particular increases in proline and α-alanine, in preparation for overwintering, and changes in cell membrane lipids including sterols and tocopherols as winter approached (Koštál et al., 2011, 2013). The bark beetle, *Pityogenes chalcographus*, enters a photoperiodically induced diapause as winter approaches with concomitant increases in carbohydrate reserves of glycogen and trehalose and converts some of the carbohydrates into glycerol during the peak of winter as a cryoprotectant against ice crystals (Koštál et al., 2014). *Drosophila melanogaster* adults survived long-term cold treatment better when exposed to low acclimation temperature. The low temperature acclimation was accompanied by slight changes in membrane glycerophospholipids (Overgaard et al., 2008). False codling moth larvae, *Thaumatotibia leucotreta*, survive better when exposed to fluctuating stressors of temperature and dehydration regimes than when exposed to constant stress conditions in preparation for winter survival (Boardman et al., 2013). The fruit fly species, *Drosophila montana*, a member of the *Drosophila virilis* group, is a northern species that has adapted to various climatological environments. In Finland, the northern populations are univoltine, and reproductive diapause of females strongly depends upon exposure to short day length (Tyukmaeva et al., 2011) after eclosion of the adults (Salminen et al., 2012), and exposure to 3 short days is sufficient to cause females to enter reproductive diapause (Salminen and Hoikkala, 2013). Sláma and Denlinger (2013) and MacMillan and Sinclair (2011) reviewed physiological and metabolic mechanisms that enable insects to survive cold exposure and diapause.

7.3.2 Diapause: Initiation and Maintenance

Initiation of diapause may or may not be easy to determine. Koštál (2006) defines initiation at that point in time when direct development ceases, as, for example, when the insect molts into a specific diapause stage that has characteristic color or morphological features. Sometimes determining that diapause has been initiated requires dissection of insects to detect internal changes in tissue, such as the failure of ovaries to grow and eggs to develop when diapause occurs. Another characteristic of initiation of diapause is the decrease in metabolic rate as measured by oxygen consumption, but the change in metabolic rate may be gradual in those species that remain mobile in diapause. Metabolic depression that usually occurs during diapause in most insects is an aid in conserving both energy reserves and water (Williams and Lee, 2005; Hahn and Denlinger, 2007) and some insects secrete additional hydrocarbons on the cuticle to help minimize water loss during a long diapause (Yoder and Denlinger, 1991; Benoit and Denlinger, 2007). Some lepidopterous larvae continue to be active and feed and may experience molting without growth during diapause. Adults in diapause may continue with intense physical and metabolic activity supporting migratory flight. There is evidence (see, for example, Lee and Denlinger, 1997) that diapause somehow involves the hormones of development (prothoracicotropic hormone [PTTH], ecdysteroids, juvenile hormone [JH] (Eizaguirre et al., 2005), and possibly others), but the physiological determinants that alter hormonal mechanisms are relatively obscure in diapausing insects. In a recent study, Kang et al. (2014) found that knockdown of allatotropin mRNA by application of RNAi in *Culex pipiens* mosquito females that were programmed for nondiapause development resulted in ovary suppression similar to that in diapause, again reaffirming that the lack of JH is at least one factor in reproductive diapause.

Once diapause is initiated, it is usually maintained for weeks or months even when environmental conditions may be suitable for development. Beyond the obvious changes in activity and reduced metabolism that can be measured in diapausing insects, relatively little is known about the internal physiology and biochemistry of diapausing insects, although currently this is an area of intense investigation.

7.3.3 Diapause Termination

Diapausing insects gradually become more sensitive to diapause-terminating conditions as diapause intensity decreases, but like the initiation of diapause, this is a gradual process. Termination usually depends on the reception of **token stimuli** concerning the environment. The physiology of termination often commences long before the insect manifests behavior that clearly means diapause is over. Factors that lead to diapause termination usually are environmental conditions, but the transduction mechanisms by which signals are converted into physiological and biochemical processes leading to diapause termination and resumption of development in most species are not resolved. Although it is often difficult to pinpoint the specific factors that are central to diapause termination, numerous physiological and biochemical processes occur in diapausing insects as they near the minimum exposure period for diapause termination.

7.4 HORMONAL CONTROL OF DIAPAUSE

Virtually all aspects of insect life are influenced by hormones and diapause is no exception. The hormonal changes associated with diapause in the various life stages have been characterized by numerous investigators and the literature has been reviewed by Chippendale (1977) and Denlinger (1985, 2002).

7.4.1 Embryonic Diapause

Many temperate-zone insects diapause in the egg stage or embryonic stage. Diapause of *Bombyx mori* in the embryonic stage has been studied extensively and is reasonably well understood mechanistically. Diapause in *B. mori* is facultative, but a univoltine strain is known in which diapause is obligatory. Whether an embryo is destined to diapause depends on the photoperiod to which the mother was exposed during her development as a larva. Female silkworm moths that grow up in the early spring under short-day conditions lay eggs that do not diapause, but hatch into the summer generation. A diapause hormone produced by females growing up under long-day summer conditions is secreted from the subesophageal glands and transferred to her eggs, which then enter diapause about 2 days after being laid. The hormone is a neuropeptide comprising 24 amino acids (Nakagaki et al., 1991; Yamashita, 1996). The adaptive benefit for embryos in eggs laid in late summer to diapause is that they would not have time to grow to maturity and pupate before the arrival of winter; thus, they survive the winter in diapause. A period of prolonged cold exposure in the winter followed by a return of springtime temperatures terminates diapause, and the larvae become the spring generation growing under short-day conditions. This spring generation of adults will again lay nondiapausing eggs, and the summer generation of adults will lay eggs in which embryos will spend the winter in diapause.

The *B. mori* model is based on embryonic diapause just after the eggs are laid and before the neuroendocrine system of the embryo has developed. Species in which the diapause occurs in older embryos may have different regulatory controls than the maternal one in *B. mori*, but little research has been done on such species. A species with a late embryo diapause is the grasshopper, *Melanoplus sanguinipes*. It is distributed over most of the midwestern and western United States, and northward into western Canada and Alaska. Eggs are laid in late summer or early fall and the embryo usually spends the winter in diapause after completing about 89% of embryonic development. The embryonic diapause may be obligatory or facultative in different parts of its range, and embryos may enter diapause at different ages (Fielding, 2006). For example, in Idaho, *M. sanguinipes* is univoltine, and eggs are laid in the late summer. Embryonic development continues until diapause or cold temperatures force a cessation of development. Holding early prediapause embryos from the Idaho population at 5°C for up to 90 days seems to allow them to meet some minimum exposure requirement and avert going into full diapause, which would normally last longer than 90 days. In subarctic Alaska, however, diapause of *M. sanguinipes* embryos is obligatory, and they always enter diapause even if given chilling conditions like those in Idaho populations (Fielding, 2006).

7.4.2 Larval Diapause

Larval diapause is common among species of Lepidoptera, Diptera, Hymenoptera, Coleoptera, Neuroptera, Odonata, Orthoptera, Homoptera, Hemiptera, and Plecoptera (Denlinger, 1985). Any instar may diapause, but diapause of last instars is more common. Diapausing larvae may display active movement and feeding. Some diapausing larvae molt "in place," that is, they molt without growing into a larger instar. Those that have stationary molts usually lose weight because they use body reserves without replacement during the diapause. **Hormonal regulation** of larval diapause is consistent in many species with a high titer of JH (Yin and Chippendale, 1976, 1979; Chippendale, 1977, and references therein) and relative inactivity of the ecdysteroid-producing prothoracic glands or their failure to produce enough ecdysteroid to counter the level of JH present. There are some species, however, in which there is no apparent role for JH in larval diapause, and a hormonal role, if any, is unknown.

7.4.3 Pupal Diapause

Pupal diapause is a very common overwintering strategy. In his studies of diapausing *Hyalophora cecropia* pupae, Carroll Williams (1952) discovered the interactions between brain hormone (as it was then known, but now known as PTTH), prothoracic glands, and ecdysone in insect development. *H. cecropia* is a univoltine species that overwinters as a diapausing pupa, entering diapause shortly after pupation. It does not break diapause in the northeastern United States until late in the spring or early summer, depending upon the location, when mulberry leaves are available on which females lay eggs and larvae feed. Williams found that a period of cold exposure was necessary for diapausing *H. cecropia* pupae to break the diapause, and he found that keeping them in a refrigerator at about 5°C for at least 6 weeks and then returning them to room temperature allowed adult development in a few weeks. He also found that he could prolong diapause by holding pupae at 5°C for many months and then return them to room temperature. Manipulating the temperature to maintain a population in diapause, and using surgical techniques, Williams demonstrated that brain hormone (later named PTTH, the prothoracicotropic hormone produced in a few large neurosecretory cells in the brain) was not secreted until the brain had been chilled for a certain period of time; without PTTH, the prothoracic glands were not stimulated to secrete ecdysone, and without ecdysone, development of the adult could not occur. The relationship between PTTH secretion and ecdysone was extremely important at the time because, prior to Williams' work, both the brain and a thoracic center were known to have some influence on pupation, but how the two did this and that PTTH drives the prothoracic glands to produce ecdysone was not known.

The flesh flies, *Sarcophaga crassipalpis* and *Sarcophaga bullata,* have a facultative pupal diapause. Development is continuous under long-day conditions, but exposure of larvae to short-day conditions (12 h of light) at 25°C causes pupae to enter diapause. Typically *S. crassipalpis* pupae remain in diapause several months, but break diapause during the mid-winter, when it is still cold, so they stay in a postdiapause quiescent stage until warm weather allows adults to emerge (Hahn, personal communication). During diapause, the cells in the brain of *S. crassipalpis* are arrested in development in the C0/G1 phase of the cell cycle (Tammariello and Denlinger, 1998). When diapause was broken by treating the pupae with hexane (see Section 7.8), brain cells began to develop and cycle into growth phases within 12 h. The gene *PCNA* encoding the proliferating cell nuclear antigen (PCNA) was expressed after the termination of diapause, but not during diapause. Some other genes (*cyclin E, p21,* and *p53*) were expressed during and after diapause about equally. Tammariello and Denlinger (1988) concluded that PCNA was likely important as a regulator of cell cycle arrest during diapause.

Denlinger (1985) notes that the unifying mechanism in insects that diapause as pupae is the lack of sufficient ecdysteroid to stimulate adult development, but he suggests that the mechanisms that cause the lack of ecdysteroid may be diverse in different species. Some species that have been studied depend on an intact brain only briefly at the beginning of pupal diapause, whereas *H. crecropia* requires that the brain must be chilled for weeks before it becomes competent to secrete PTTH. There is no evidence in *H. cecropia* that JH is involved in diapause, but in diapausing *S. crassipalpis,* JH may have a role in initiating, maintaining, and terminating pupal diapause (Denlinger, 1985).

7.4.4 Adult Diapause

Adult diapause is common in species of Coleoptera, Lepidoptera, Diptera, Heteroptera, Orthoptera, Neuroptera, and Trichoptera (Denlinger, 1985). Several hormones are known to be required for reproduction in female insects (see Chapter 20 for more details), but it is generally believed that low titer or lack of JH is crucial in regulating adult diapause, although Denlinger (2002) cautions that it may not be the only hormone involved.

One of the most intensively studied species with adult diapause is the Colorado potato beetle, *Leptinotarsa decemlineata* (De Wilde et al., 1959; De Wilde and de Boer, 1961, 1969; de Kort, 1990; Noronha and Cloutier, 2006; Doležal et al., 2007). Adults emerging in summer lay eggs for several months and development is continuous under long-day conditions (but 20%–30% enter diapause no matter what photoperiod exposure they receive) (De Wilde et al., 1959). Exposure to short-day photoperiods in early autumn induces adults to enter diapause after a short period of feeding. They dig into the soil, histolyze their wing muscles (as a source of nutrient reserves), and overwinter in adult diapause. Diapausing adults in the soil can survive subzero temperatures, and they may emerge briefly to feed. The great majority emerges the following spring and begin to feed and lay eggs, but some stay in diapause for 2 years, a few for 3–7 years, and one is known to have been in diapause for 10 years (Tauber and Tauber, 2002).

Adult males and females of *Chrysopa carnea* (Neuroptera) diapause in response to short-day exposure during their development (MacLeod, 1967; Tauber and Tauber, 1969). Under constant temperature conditions and a 16:8 long-day (LD) photoperiod, development and reproduction are continuous, but if exposed first to a 16:8 LD photoperiod and then are transferred to a 12:12 LD cycle, adults enter a reproductive diapause lasting about 3 months. Tauber and Tauber (1970) subsequently showed that the 12:12 LD cycle is not in itself the determining factor in diapause, but rather the previous photoperiod history to which the insects had been exposed. The 12:12 LD cycle could induce diapause, prevent diapause, or terminate diapause, depending on the photoperiod history of the insects prior to the short-day cycle. They concluded that *C. carnea* is able to perceive and respond to decreasing and increasing day lengths that do not cross the critical photoperiod for diapause.

7.5 ROLE OF DAILY AND SEASONAL BIOLOGICAL CLOCKS IN DIAPAUSE

Daily (**circadian**) and **seasonal** (photoperiod) **biological clocks** influence daily and seasonal rhythms in insects (Tauber and Tauber, 1970; Denlinger, 1986, 2002; Koštál et al., 2000; Saunders et al., 2002; Danks, 2003, 2005; Hua et al., 2005; Goto et al., 2006). Daily repeating **circadian rhythms** are set by light and dark cues, giving an insect a reference point for a particular time of the 24 h cycle. For many insects, circadian rhythms program such behaviors as pupation, eclosion from eggs, and from the pupal stage, and time internal physiological events (such as pheromone secretion and release). Seasonal activity in many temperate insects is based on monitoring the **photoperiod**, with insects somehow calculating the duration and accumulation of daily changes in day or night length over a period of time. Daily clocks measure the time of day while seasonal clocks monitor the duration of each day and the number of days of a given length, as well as additional environmental factors; although similar, daily and seasonal biological clocks are different and function differently (Danks, 2005). Kumar et al. (2007) demonstrated that circadian clocks in *D. melanogaster* are heritable and populations can be selected with different circadian rhythms.

The receptors for circadian rhythms and for seasonal activity are located in the brain of some insects, in the compound eyes of others, and, in some cases, in both brain and compound eyes (Saunders and Cymborowski, 1996; Morita and Numata, 1997, 1999; Nakamura and Hodkova, 1998). Cryptochrome, a light sensitive pigment encoded by the gene *cryptochrome* (Hall, 2000), contains the vitamin riboflavin and a protein. It is mainly involved in sensing circadian light information (Ishikawa et al., 1999; Sancar, 2000; Ivanchenko et al., 2001). Cryptochrome is located in both brain tissue and compound eyes of *D. melanogaster* (Holfrich-Förster et al., 2001; Rieger et al., 2003). In the brain it regulates the morning activity rhythm of *Drosophila*, while in the compound eyes it controls the evening peak of activity in *Drosophila*. The chromophore receptor that responds to seasonal changes contains carotinoids or vitamin A (Veerman, 2001), but little is known about how it functions in relation to seasonal changes.

7.6 DIAPAUSE AND GENE EXPRESSION

The ultimate regulation of diapause lies in genes that regulate the hormonal and metabolic changes associated with entry into diapause, diapause maintenance, and diapause termination and return to normal activity. The genes that are directly involved in diapause, however, remain to be identified. A number of **clock** and **timing genes** have been identified, including *period, timeless, dClock, cycle, doubletime,* and *vrille* (Dunlop, 1999; Schotland and Sehgal, 2001; Denlinger, 2002), but none of these is known to directly control diapause. Although *period* (*per*) has been studied the most and tested in *Drosophila* with respect to diapause, null mutants for *per* diapause just as the wild type does, but the null mutants show a lack of circadian rhythms and altered singing behavior (Saunders et al., 1989). A mutant allele of *per* in *Chymomyza costata* (a drosophilid species) does not prevent diapause (Shimada, 1999).

Numerous genes are upregulated, or downregulated or are intermittently expressed during diapause (reviewed by Denlinger, 2002). Joplin et al. (1990) found that more than 300 proteins (representing gene expression) occurred in brain tissue of nondiapausing *S. crassipalpis,* but only about 180 were detectable in the brain of diapausing pupae. They concluded that about 40% of the genes active in the brain of nondiapausing pupae were silenced during diapause, and that about 10% of the genes expressed during diapause are only expressed at that time. Flannagan et al. (1998) found that some genes in the brain of diapausing *S. crassipalpis* are expressed throughout diapause, some are expressed only during the early part of diapause, some late in diapause, and some

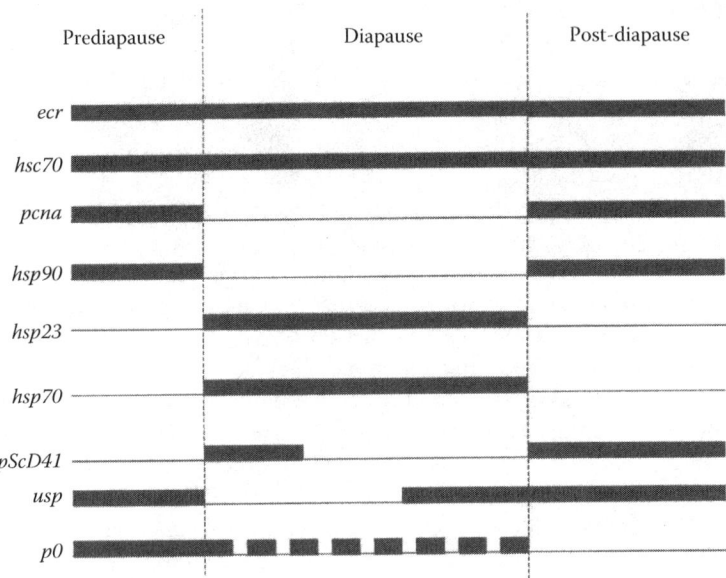

Figure 7.2 Patterns of gene expression in the flesh fly, *S. crassipalpis,* before, during, and after diapause. Proteins encoded by genes are named on the left of the diagram. Some genes are not influenced by diapause (*ecr* and *hsc70*), some are downregulated in diapause (*pcna* and *hsp90*), and some are upregulated throughout diapause (*hsp23* and *hsp70*). There are early diapause genes (*pScD41*), late diapause genes (*usp*), and genes expressed intermittently during diapause (*p0*). (From Denlinger, D.L., *Annu. Rev. Entomol.*, 47, 93, 2002. With permission.)

intermittently (Denlinger, 2002) (Figure 7.2). Denlinger (2002) concludes that many of the genes downregulated during diapause are genes one would expect to be silenced because of the reduction in metabolic and physical activity. They are likely a consequence of diapause rather than a cause of diapause and probably do not play a controlling role in diapause. HSP-90, the 90 kDa heat shock protein, is downregulated during diapause and upregulated when diapause is terminated, presumably a consequence of 20-hydroxyecdysone control of the gene *hsp90*. Ecdysteroids upregulate *hsp90*, and the gene is downregulated when ecdysteroids are absent. Even when downregulated, however, *hsp90* remains responsive to environmental conditions. Because pupal diapause may be, in part at least, a consequence of ecdysteroids deficiency, it is possible that *hsp90* and ecdysteroid are linked in important physiological ways to diapause, possibly with *hsp90* serving in some way in the functioning of the EcR/USP dimer that binds 20-hydroxyecdysone and then binds DNA in the nucleus, allowing transcription of many genes (Arbeitman and Hogness, 2000).

7.7 NUTRIENT ACCUMULATION FOR DIAPAUSE AND THE STORAGE AND CONSERVATION OF NUTRIENTS DURING DIAPAUSE

Most diapausing insects feed only sparingly or not at all (Hahn and Denlinger, 2007); certainly diapausing pupae and embryos have no opportunity to feed. Prior to entering diapause, insects typically feed vigorously and synthesize and store nutrients, primarily as lipids and proteins. Depressed basal metabolism, reduced activity or quiescence, and low temperatures common to many diapausing insects aid in making the accumulated reserves last through the diapause. Hahn and Denlinger (2007) review data showing that nutrient storage and metabolic activity can influence the decision to diapause and the duration of diapause in some species. They also stress that fitness of the insect coming out of diapause, particularly if it cannot feed immediately, is related to the availability of conserved nutrients. Lipids, the richest energy source—more than twice the calories from a gram of lipid compared to carbohydrate or protein (see Chapter 8)—usually are stored as triacylglycerols in the fat body. Lipases hydrolyze fatty acids from the triacylglycerols, and metabolism of fatty acids provides energy for diapause maintenance and for postdiapause until feeding resumes.

Many insects accumulate proteins, especially proteins called **hexamerins**, prior to diapause (Koopmanschap et al., 1995; Wheeler et al., 2000; Denlinger, 2002, and references therein). The proteins are called hexamerins because they typically are made of six equal size subunits, although a few variations in subunits have been described. The proteins usually, but not always, have a high content of aromatic amino acid residues in their structure, and frequently are referred to as **arylphorins** or **storage proteins**. It was thought at one time that these were specific to diapause, and they were called diapause proteins in some early literature. However, it is now recognized that many insects synthesize hexamerins, but never diapause. They are synthesized by the fat body and released into the hemolymph. Generally they are resequestered by fat body cells and stored until used in building new tissues, generally during pupation and formation of adult tissues. In nondiapausing insects, they usually do not persist in the body for long because they are used in metamorphosis and growth of new tissues. They do persist at high concentration in the hemolymph of diapausing insects, usually until diapause ends, presumably because diapausing insects are not building new tissues. They disappear soon after diapause ends, probably again being used as a source of amino acids for tissue construction in renewed development.

Other proteins may be important as factors in diapause. Lee et al. (1998) found that actin is one of two major proteins in the central nervous system (CNS) of gypsy moths that is no longer synthesized in pharate larvae that enter diapause near the end of embryonic development. Actin synthesis does not begin again in the CNS until diapause is broken. Actin functions in the central nervous system in axonal transport of synaptic vesicles and in accumulation and release of neuropeptides and neurotransmitters. These authors conclude that actin in gypsy moth larvae is involved in critical processes in development, especially in postdiapause larvae. In contrast to the downregulation of actin in gypsy moth, *actin* 1 and 2 genes in *C. pipiens* L. mosquitoes are upregulated during adult diapause (Kim et al., 2006; Robich et al., 2006). The two genes are expressed throughout diapause, but expressed more highly early in diapause. Kim et al. (2006) suggest that one function of the actin is to strengthen the cytoskeleton for withstanding the long winter diapause.

7.8 MOLECULAR STUDIES OF DIAPAUSE

The flesh fly, *S. crassipalpis*, has been a favorite model for study of diapause by a number of researchers. The flies diapause in the pupal stage when exposed to short-day conditions (12 h of light) at 25°C. Ecdysone is absent or at very low levels in diapausing pupae, but diapause can be broken by injection of ecdysone, implicating its absence as an important factor in diapause. Rinehart et al. (2001) found that the expression of ecdysone receptor (EcR) remained detectable and was not downregulated during pupal diapause, but transcripts of the *ultraspiracle* (*usp*) gene were downregulated and only began rising near the end of normal diapause, rising rapidly upon the termination of diapause. They suggested that the availability of **Ultraspiracle** (USP) may be a major factor in diapause termination in *S. crassipalpis*. USP protein is the partner needed by the EcR protein in order to bind 20-hydroxyecdysone and form a functional transcription factor for genes involved in development.

An interesting way to break pupal diapause in *S. crassipalpis* (and *Manduca sexta*) is to treat diapausing pupae with a few microliters of a nonpolar organic solvent, such as hexane (Zdarek and Denlinger, 1975; Denlinger et al., 1980). Fujiwara and Denlinger (2007) recently showed that within 10 min after administering hexane, phosphorylation of **extracellular signal-regulated**

kinases (ERKs) occurs in brain tissue and in various peripheral tissues, including epidermis and fat body cells (but not in ring gland cells). ERKs and **mitogen-activated protein kinases** (MAPKs, also called **MAP kinases**) are enzymes that participate in amplification cascades, and, typically, are activated by phosphorylation. They, in turn, phosphorylate other molecules, including members of the MAPK family. ERK and MAPKs have been implicated in signal transduction cascades and in termination of diapause in several different insects (Rybcznski et al., 2001; Iwata et al., 2005b; Fujiwara and Shiomi, 2006; Fujiwara et al., 2006a,b; Kidokoro et al., 2006a,b; Fujiwara and Denlinger, 2007). The authors speculate that p-ERK (phosphorylated-ERK) in the brain of *S. crassipalpis* leads to secretion of PTTH, which could then activate the prothoracic glands to produce ecdysteroids to reinitiate development.

ERKs and P38 MAPKs have been implicated in the termination of silkworm embryo diapause (Fujiwara and Shiomi, 2006). Iwata et al. (2005b) found that phosphorylation of ERK in yolk cells increased in diapausing *Bombyx* embryos that were chilled at 5°C for 45–60 days (conditions that break diapause) and then returned to 25°C. Phosphorylation of ERK (p-ERK) is regulated by MAPK–ERK (MEK) and possibly by p38 MAPK (Fujiwara et al., 2006b). p-ERK regulates diapause termination by activating transcription of genes controlling enzymes needed to free bound ecdysteroids in the yolk (Sonobe and Yamada, 2004) and convert sorbitol to glycogen in *Bombyx*.

Diapausing embryos of *B. mori* accumulate **sorbitol** and **glycerol** in the eggs, and although they may serve as antifreeze compounds as often suggested, more recent evidence suggests that sorbitol may have a much more central role as a controlling factor in maintaining diapause (Iwata et al., 2005a). Embryos from diapausing eggs continue to develop and do not enter diapause if they are removed from the eggs (called denuded embryos) and cultured *in vitro* without sorbitol in the medium; however, additions of sorbitol or trehalose to the incubation medium inhibits development and sends embryos into diapause. Measurement of sorbitol and **trehalose** in diapausing eggs indicates that the normal level of trehalose in nondiapausing eggs is too low to inhibit development, but the sorbitol concentration is close to the level shown by experimental addition of sorbitol to inhibit development. Sorbitol and glycerol are synthesized at the expense of glycogen (Horie et al., 2000). Based on experiments in which diapause could be broken in embryos removed from eggs and incubated in a medium devoid of sorbitol or trehalose, Horie et al. (2000) concluded that sorbitol is an **arresting factor** of development in diapausing eggs, and not simply an antifreeze compound as formerly thought.

7.9 REVIEW AND SELF-STUDY QUESTIONS

1. What is a physiological definition of diapause?
2. What is the difference between obligatory and facultative diapause?
3. Explain how diapause is a survival strategy in insects?
4. Does diapause occur in any animals other than insects? If so what?
5. What are some of the most common "token" environmental cues that induce diapause?
6. Do any insects diapause more than once in a single life-time? Elaborate briefly.
7. What are the stages of diapause as defined by Koštál?
8. Do any diapausing insects continue to feed and stay active? Explain with some examples.
9. Is diapause initiation and the breaking of diapause a trigger reaction or a slow process?
10. Briefly describe the hormonal control of embryonic diapause in the race of *B. mori* that always diapauses.
11. Can a diapausing insect molt? Does it grow larger?
12. What is the common or unifying theme in insects that diapause as pupae?
13. How was the diapause of *H. cecropia* involved in the discovery of the brain and the prothoracic glands in molting?
14. Describe diapause in the Colorado potato beetle.
15. What are the different roles for circadian and seasonal biological clocks?

16. What is the present state of knowledge about the genetics of diapause?
17. What are hexamerins and what function(s) do they seem to have in insects, including diapausing ones?
18. What do these acronyms stand for? (1) USP, (2) ERKs, (3) MAPK, and (4) p-ERK.
19. What is the evidence to support the idea that sorbitol has a central role in diapause of silkworm embryos.

REFERENCES

Arbeitman, M.N. and D.S. Hogness. 2000. Molecular chaperones activate the *Drosophila* ecdysone receptor, an RXR heterodimer. *Cell* 101: 67–77.
Beck, S.D. 1980. *Insect Photoperiodism*, 2nd edn. Academic Press, New York.
Beck, S.D. 1983. Insect thermoperiodism. *Annu. Rev. Entomol.* 28: 91–108.
Bennett, V.A., O. Kukal, and R.E. Lee, Jr. 1999. Metabolic opportunists: Feeding and temperature influence the rate and pattern of respiration in the high Arctic woollybear caterpillar *Gynaephora groenlandica* (Lymanitriidae). *J. Exp. Biol.* 202: 47–53.
Benoit, J.B. and D.L. Denlinger. 2007. Suppression of water loss during adult diapause in the northern house mosquito, *Culex pipiens. J. Exp. Biol.* 210: 217–226.
Boardman, L., J.G. Sørensen, and J.S. Terblanche. 2013. Physiological responses to fluctuating thermal and hydration regimes in the chill susceptible insect, *Thaumatotibia leucotreta. J. Insect Physiol.* 59: 781–794.
Chippendale, G.M. 1977. Hormonal regulation of larval diapause. *Annu. Rev. Entomol.* 22: 121–138.
Danilevskii, A.S. 1965. *Photoperiodism and Seasonal Development of Insects.* Oliver & Boyd, Edinburgh, Scotland.
Danks, H.V. 1987. *Insect Dormancy: An Ecological Perspective.* Ottawa Biological Survey, Ottawa, Ontario, Canada.
Danks, H.V. 2003. Studying insect photoperiodism and rhymicity: Components, approaches and lessons. *Eur. J. Entomol.* 100: 209–221.
Danks, H.V. 2005. How similar are daily and seasonal biological clocks? *J. Insect Physiol.* 51: 609–619.
de Kort, C.A.D. 1990. Thirty-five years of diapause research with the Colorado potato beetle. *Entomol. Exp. Applicata* 56: 1–13.
De Wilde, J. 1962. Photoperiodism in insects and mites. *Annu. Rev. Entomol.* 7: 1–26.
De Wilde, J. and J.A. de Boer. 1961. Physiology of diapause in the adult Colorado beetle—II. Diapause as a case of pseudo-allatectomy. *J. Insect Physiol.* 6: 152–161.
De Wilde, J. and J.A. de Boer. 1969. Humoral and nervous pathways in photoperiodic induction of diapause in *Leptinotarsa decemlineata. J. Insect Physiol.* 661–664.
De Wilde, J., C.S. Duintjer, and L. Mook. 1959. Physiology of diapause in the adult Colorado beetle (*Leptinotarsa decemlineata* Say)—I. The photoperiod as a controlling factor. *J. Insect Physiol.* 3: 75–80.
Denlinger, D.L. 1985. Hormonal control of diapause, in G.A. Kekut and L.I. Gilbert (eds.), *Comprehensive Insect Physiology, Biochemistry and Pharmacology.* Pergamon, Oxford, U.K., pp. 353–412.
Denlinger, D.L. 1986. Dormancy in tropical insects. *Annu. Rev. Entomol.* 31: 239–264.
Denlinger, D.L. 2002. Regulation of diapause. *Annu. Rev. Entomol.* 47: 93–122.
Denlinger, D.L. and P.A. Armbruster. 2014. Mosquito diapause. *Annu. Rev. Entomol.* 59: 73–93.
Denlinger, D.L., J.J. Campbell, and J.Y. Bradfield. 1980. Stimulatory effect of organic solvents on initiating development in diapausing pupae of the flesh fly, *Sarcophaga crassipalpis*, and the tobacco hornworm, *Manduca sexta. Physiol. Entomol.* 5: 7–15.
Doležal, P., O. Habuštová, and F. Sehnal. 2007. Effects of photoperiod and temperature on the rate of larval development, food conversion efficiency, and the imaginal diapause in *Leptinotarsa decemlineata. J. Insect Physiol.* 53: 849–857.
Dunlop, J.C. 1999. Molecular bases for circadian clocks. *Cell* 96: 271–290.
Eizaguirre, M., C. Schafellner, C. Lopez, and F. Sehnal. 2005. Relationship between an increase of juvenile hormone titer in early instars and the induction of diapause in fully grown larvae of *Sesamia nonagrioides. J. Insect Physiol.* 51: 1127–1134.
Everatt, M.J., J.S. Bale, P. Convey, M.R. Worland, and S.A.L. Hayward. 2013. The effect of acclimation temperature on thermal activity thresholds in polar terrestrial invertebrates. *J. Insect Physiol.* 59: 1057–1064.

Fielding, D. 2006. Optimal diapause strategies of a grasshopper *Melanoplus sanguinipes*. *J. Insect Sci.* 6: 1–16. (Available online: insectscience.org/6.02, Accessed on April 20, 2015.)

Flannagan, R.D., S.P. Tammariello, K.H. Joplin, R.A. Cikra-Ireland, G.D. Yocum, and D.L. Denlinger. 1998. Diapause-specific gene expression in pupae of the flesh fly *Sarcophaga crassipalpis*. *Proc. Natl. Acad. Sci. USA* 95: 5616–5620.

Fujiwara, Y. and D.L. Denlinger. 2007. High temperature and hexane break pupal diapause in the flesh fly, *Sarcophaga crassipalpis*, by activating ERK/MAPK. *J. Insect Physiol.* 53: 1276–1282.

Fujiwara, Y. and K. Shiomi. 2006. Distinct effects of different temperatures on diapause termination, yolk morphology and MAPK phosphorylation in the silkworm, *Bombyx mori*. *J. Insect Physiol.* 52: 1194–1201.

Fujiwara, Y., C. Shindome, M. Takeda, and K. Shiomi. 2006a. The roles of ERK and P38 MAPK signaling cascades on embryonic diapause initiation and termination of the silkworm, *Bombyx mori*. *Insect Biochem. Mol. Biol.* 36: 47–53.

Fujiwara, Y., Y. Tanaka, K. Iwata, R.O. Rubio, T. Yaginuma, O. Yamashita, and K. Shiomi. 2006b. ERK/MAPK regulates ecdysteroid and sorbitol metabolism for embryonic diapause termination of the silkworm, *Bombyx mori*. *J. Insect Physiol.* 52: 569–575.

Goto, S.G., B. Han, and D.L. Denlinger. 2006. A nondiapausing variant of the flesh fly, *Sarcophaga bullata*, that shows arrhythmic adult eclosion and elevated expression of two circadian clock genes, *period* and *timeless*. *J. Insect Physiol.* 52: 1213–1218.

Hahn, D.A. and D.L. Denlinger. 2007. Meeting the energetic demands of insect diapause: Nutrient storage and utilization. *J. Insect Physiol.* 53: 760–773.

Hahn, D. Personal communication, August 2007.

Hall, J.C. 2000. Cryptochromes: Sensory reception, transduction, and clock functions subserving circadian systems. *Curr. Opin. Neurobiol.* 10: 456–486.

Holfrich-Förster, C., C. Winter, A. Hofbauer, J.C. Hall, and R. Stanewsky. 2001. The circadian clock of fruit flies is blind after elimination of all known photoreceptors. *Neuron* 30: 249–261.

Horie, Y., T. Kanda, and Y. Mochida. 2000. Sorbitol as an arrester of embryonic development in diapausing eggs of the silkworm, *Bombyx mori*. *J. Insect Physiol.* 46: 1009–1016.

Hua, A., D. Yang, S. Wu, and F. Xue. 2005. Photoperiodic control of diapause in *Pseudopidorus fasciata* (Lepidoptera: Zygaenidae) based on a qualitative time measurement. *J. Insect Physiol.* 51: 1261–1267.

Ishikawa, T., A. Matsumoto, T. Kato, Jr., S. Togashi, H. Ryo, M. Ikenaga, T. Todo, R. Ueda, and T. Tanimura. 1999. DCRY is a *Drosophila* photoreceptor protein implicated in light entrainment of circadian rhythm. *Genes Cells* 4: 57–65.

Ivanchenko, M., T. Stanewsky, and J. Giebultowicz. 2001. Circadian photoreception in *Drosophila*: Functions of cryptochrome in peripheral and central clocks. *J. Biol. Rhythms* 16: 205–215.

Iwata, K., Y. Fujiwara, and M. Takeda. 2005a. Effects of temperature, sorbitol, alanine and diapause hormone on the embryonic development in *Bombyx mori*: In vitro tests of old hypotheses. *Physiol. Entomol.* 30: 317–323.

Iwata, K., C. Shindome, Y. Kobayashi, M. Takeda, O. Yamashita, K. Shiomi, and Y. Fujiwara. 2005b. Temperature-dependent activation of ERK/MAPK in yolk cells and its role in embryonic diapause termination in the silkworm *Bombyx mori*. *J. Insect Physiol.* 51: 1306–1312.

Joplin, K.H., G.D. Yocum, and D.L. Denlinger. 1990. Diapause specific proteins expressed by the brain during pupal diapause of the flesh fly, *Sarcophaga crassipalpis*. *J. Insect Physiol.* 36: 775–783.

Kang, D.S., D.L. Denlinger, and C. Sim. 2014. Suppression of allatotropin simulates reproductive diapause in the mosquito *Culex pipiens*. *J. Insect Physiol.* 64: 48–53.

Kidokoro, K., K. Iwata, Y. Fujiwara, and M. Takeda. 2006a. Effects of juvenile hormone analogs and 20-hydroxyecdysone on diapause termination in eggs of *Locusta migratoria* and *Oxya yezoenisi*. *J. Insect Physiol.* 52: 473–479.

Kidokoro, K., K. Iwata, Y. Fujiwara, and M. Takeda. 2006b. Involvement of ERK/MAPK in regulation of diapause intensity in the false melon beetle, *Atrachya menetriesi*. *J. Insect Physiol.* 52: 1189–1193.

Kim, M., R.M. Robich, J.P. Rinehart, and D.L. Denlinger. 2006. Upregulation of two actin genes and redistribution of actin during diapause and cold stress in the northern house mosquito, *Culex pipiens*. *J. Insect Physiol.* 52: 1226–1233.

Koopmanschap, A.B., J.H.M. Lammers, and C.A.D. de Kort. 1995. The structure of the gene encoding diapause protein 1 of the Colorado potato beetle (*Leptinotarsa decemlineata*). *J. Insect Physiol.* 41: 509–518.

Koštál, V. 2006. Eco-physiological phases of insect diapause. *J. Insect Physiol.* 52: 113–127.

Koštál, V., B. Miklas, P. Doležal, J. Rozsypal, and H. Zahradníčková. 2014. Physiology of cold tolerance in the bark beetle, *Pityogenes chalcographus* and its overwintering in spruce stands. *J. Insect Physiol.* 63: 62–70.

Koštál, V., H. Noguchi, K. Shimada, and Y. Hayakawa. 2000. Circadian component influences the photoperiodic induction of diapause in a drosophilid fly, *Chymomyza costata*. *J. Insect Physiol.* 46: 887–896.

Koštál, V., D. Renault, and J. Rozsypal. 2011. Seasonal changes of free amino acids and thermal hysteresis in overwintering heteropteran insect *Pyrrhocoris apterus*. *Comp. Biochem. Physiol. A* 160: 245–251.

Koštál, V., T. Urban, L. Řimnáčová, P. Berková, and P. Šimek. 2013. Seasonal changes in minor membrane phospholipid classes, sterols and tocopherols in overwintering, *Pyrrhocoris apterus*. *J. Insect Physiol.* 59: 934–941.

Koštál, V., H. ZahradniČková, P. Šimek, and J. Zelený. 2007. Multiple component system of sugars and polyols in the overwintering spruce bark beetle, *Ips typographus*. *J. Insect Physiol.* 2007: 580–586.

Kumar, S., D. Kumar, D.A. Paranjpe, C.R. Akarsh, and V.K. Sharma. 2007. Selection on the timing of adult emergence results in altered circadian clocks in fruit flies *Drosophila melanogaster*. *J. Exp. Biol.* 210: 906–918.

Lee, K.-Y. and D.L. Denlinger. 1997. A role for ecdysteroids in the induction and maintenance of the pharate first instar diapause of the gypsy moth, *Lymantria dispar*. *J. Insect Physiol.* 43: 289–296.

Lee, K.-Y., S. Hiremath, and D.L. Denlinger. 1998. Expression of actin in the central nervous system is switched off during diapause in the gypsy moth, *Lymantria dispar*. *J. Insect Physiol.* 44: 221–226.

MacLeod, D.G.R. 1967. Experimental induction and elimination of adult diapause and autumnal coloration in *Chrysopa carnea* (Neuroptera). *J. Insect Physiol.* 13: 1343–1349.

MacMillan, H.A. and B.J. Sinclair. 2011. Mechanisms underlying insect chill-coma. *J. Insect Physiol.* 57: 12–20.

Morita, A. and H. Numata, 1997. Distribution of photoperiodic receptors in the compound eyes of the bean bug, *Riptortus clavatus*. *J. Comp. Physiol. A* 180: 181–185.

Morita, A. and H. Numata, 1999. Localization of the photoreceptor for photoperiodism in the stink bug, *Plautia crossota stali*. *Physiol. Entomol.* 24: 189–195.

Nakagaki, M., R. Takei, E. Nagashima, and T. Yaganuma. 1991. Cell cycles in embryos of the silkworm, *Bombyx mori*: G2-arrest at diapause stage. *Roux's Arch. Dev. Biol.* 200: 223–229.

Nakamura, K. and M. Hodkova. 1998. Photoreception in entrainment of rhythms and photoperiodic regulation of diapause in a hemipteran, *Graphosoma lineatum*. *J. Biol. Rhythms* 13: 159–166.

Nishizuka, M., A. Azuma, and S. Masaki. 1998. Diapause response to photoperiod and temperature in *Lepisma saccharina* Linnaeus (Thysanura: Lepismatidae). *Entomol. Sci.* 1: 7–14.

Noronha, C. and C. Cloutier. 2006. Effects of potato foliage age and temperature regime on prediapause Colorado potato beetle *Leptinotarsa decemlineata* (Coleoptera: Chrysomelidae). *Physiol. Ecol.* 35: 590–599.

Overgaard, J., A. Tomčala, J.G. Sørensen, M. Holmstrup, P.H. Krogh, P. Šimek, and V. Koštál. 2008. Effects of acclimation temperature on thermal tolerance and membrane phospholipids composition in the fruit fly *Drosophila melanogaster*. *J. Insect Physiol.* 54: 619–629.

Rieger, D., R. Stanewsky, and C. Helfrich-Foerster. 2003. Cryptochrome, compound eyes, Hofbauer-Buchner eyelets, and ocelli play different roles in the entrainment and masking pathway of the locomotor activity rhythm in the fruit fly *Drosophila melanogaster*. *J. Biol. Rhythms* 18: 377–391.

Rinehart, J.P., R.A. Cikra-Ireland, R.D. Flannagan, and D.L. Denlinger. 2001. Expression of ecdysone receptor is unaffected by pupal diapause in the flesh fly, *Sarcophaga crassipalpis*, while its dimerization partner, USP, is downregulated. *J. Insect Physiol.* 47: 915–921.

Robich, R.M., J.P. Rinehart, L.J. Kitchen, and D.L. Denlinger. 2006. Diapause-specific gene expression in the northern mosquito, *Culex pipiens* L., identified by suppressive substractive hydridization. *J. Insect Physiol.* 53: 235–245.

Rybcznski, R., S.C. Bell, and L.I. Gilbert. 2001. Activation of an extracellular signal-regulated kinase (ERK) by the insect prothoracicotropic hormone. *Mol. Cell. Endocrinol.* 184: 1–11.

Salminen, T.S. and A. Hoikkala. 2013. Effect of temperature on the duration of sensitive period and on the number of photoperiodic cycles required for the induction of reproductive diapause in *Drosophila montana*. *J. Insect Physiol.* 59: 450–457.

Salminen, T.S., L. Vesala, and A. Hoikkala. 2012. Photoperiodic regulation of life-history traits before and after eclosion: Egg-to-adult development time, juvenile body mass and reprodutive diapause in *Drosophila monatana*. *J. Insect Physiol.* 58: 1541–1547.

Sancar, A. 2000. Cryptochrome: The second photoactive pigment in the eye and its role in circadian photoreception. *Annu. Rev. Biochem.* 69: 31–67.

Saunders, D.S. and B. Cymborowski. 1996. Removal of optic lobes of adult blowflies (*Calliphora vicina*) leaves photoperiodic induction of larval diapause intact. *J. Insect Physiol.* 42: 807–811.

Saunders, D.S., V.C. Henrich, and L.I. Gilbert. 1989. Induction of diapause in *Drosophila melanogaster*: Photoperiodic regulation and impact of arrhythmic clock mutations on time measurement. *Proc. Natl. Acad. Sci. USA* 86: 3748–3752.

Saunders, D.S., C.G.H. Steel, X. Vafopoulou, and R.D. Lewis. 2002. *Insect Clocks*, 3rd edn. Elsevier, Amsterdam, the Netherlands.

Schotland, P. and A. Sehgal. 2001. Molecular control of *Drosophila* circadian rhythms, in D.L. Denlinger, J.M. Giebultowicz, and D.S. Saunders (eds.), *Insect Timing: Circadian Rhythmicity to Seasonality*. Elsevier, Amsterdam, the Netherlands, pp. 15–30.

Shimada, K. 1999. Genetic linkage analysis of photoperiodic clock genes in *Chymomyza costata* (Diptera: Drosophilidae). *Entomol. Sci.* 2: 575–578.

Sinclair, B.J. and S.L. Chown. 2005. Deleterious effects of repeated cold exposure in a freeze-tolerant sub-Antarctic caterpillar. *J. Exp. Biol.* 208: 869–879.

Sláma, K. and D.L. Denlinger. 2013.Transitions in the heartbeat pattern during pupal diapsuse and adult development in the flesh fly, *Sarcophaga crassipalpis*. *J. Insect Physiol.* 59: 767–780.

Sonobe, H. and R. Yamada. 2004. Ecdysteroids during early embryonic development in silkworm *Bombyx mori*: Metabolism and functions. *Zool. Sci.* 21: 503–516.

Tammariello, S.P. and D.L. Denlinger. 1998. G0/G1 cell arrest in the brain of *Sarcophaga crassipalpis* during pupal diapause and the expression pattern of the cell cycle regulator, proliferating cell nuclear antigen. *Insect Biochem. Mol. Biol.* 28: 83–89.

Tauber, M.J. and C.A. Tauber. 1969. Diapause in *Chrysopa carnea* (Neuroptera:Chrysopidae)—I. Effect of photoperiod on reproductively active adults. *Can. Entomol.* 101: 364–370.

Tauber, M.J. and C.A. Tauber. 1970. Photoperiodic induction of diapause in an insect: Response to changing day lengths. *Science* 167: 170.

Tauber, M.J. and C.A. Tauber. 2002. Prolonged dormancy in *Leptinotarsa decemlineata* (Coleoptera: Chrysomelidae): A ten-year field study with implications for crop rotation. *Environ. Entomol.* 31: 499–504.

Tauber, M.J., C.A. Tauber, and S. Masaki. 1986. *Seasonal Adaptations of Insects*. Oxford University Press, New York.

Tyukmaeva, V.I., T.S. Salminen, M. Kankare, K.E. Knott, and A. Hoikkala. 2011. Adaptation to a seasonally varying environment: A strong latitudinal cline in reproductive diapause combined with high gene flow in *Drosophila montana*. *Ecol. Evol.* 1: 160–168.

Veerman, A. 2001. Photoperiodic time measurement in insects and mites: A critical evaluation of the oscillator-clock hypothesis. *J. Insect Physiol.* 47: 1097–1109.

Wheeler, D.E., I. Tuchinskaya, N.A. Buck, and B.E. Tabashnik. 2000. Hexameric storage proteins during metamorphosis and egg production in the diamondback moth, *Plutella xylostella* (Lepidoptera*)*. *J. Insect Physiol.* 46: 951–958.

Williams, C.M. 1952. Physiology of insect diapause. IV. The brain and prothoracic glands as an endocrine system in the cecropia silkworm. *Biol. Bull.* 103: 120–138.

Williams, J.B. and R.E. Lee, Jr. 2005. Plant senescence cues entry into diapause in the gall fly *Eurosta solidaginis*: Resulting metabolic depression is critical for water conservation. *J. Exp. Biol.* 208: 4437–4444.

Yamashita, O. 1996. Diapause hormone of the silkworm, *Bombyx mori*: Structure, gene expression and function. *J. Insect Physiol.* 42: 669–679.

Yin, C.-M. and G.M. Chippendale. 1976. Hormonal control of larval diapause and metamorphosis of the southwestern corn borer *Diatraea grandiosella*. *J. Exp. Biol.* 64: 303–310.

Yin, C.-M. and G.M. Chippendale. 1979. Diapause of the southwestern corn borer, *Diatraea grandiosella*: Further evidence showing juvenile hormone to be the regulator. *J. Insect Physiol.* 25: 513–523.

Yoder, J.A. and D.L. Denlinger. 1991. Water balance in flesh fly pupae and water vapor absorption associated with diapause. *J. Exp. Biol.* 157: 273–286.

Zdarek, J. and D.L. Denlinger. 1975. Action of ecdysteroids, juvenoids, and non-hormonal agents on termination of pupal diapause in the flesh fly. *J. Insect Physiol.* 21: 1193–1202.

CHAPTER 8

Intermediary Metabolism

PREVIEW

Metabolism is the sum of all chemical reactions occurring in an organism. This chapter deals primarily with catabolism, reactions that break down molecules to release energy. The most intense and rapid energy demands made by insects come with flight. Within seconds, the available high-energy phosphates in the body are used in flight, and energy must be made available rapidly and for long periods for flight to continue. The tracheal system of insects is able to supply oxygen to mitochondria, even during flight, and insects do not incur an oxygen debt in flight. As a consequence of the efficiency of the tracheal system and a fast glycerol-3-phosphate shuttle that regenerates NAD$^+$ for use in glycolysis, virtually all metabolic glucose can go directly to pyruvate, and pyruvate can go directly into mitochondria for metabolism to release large amounts of energy. Flight muscle mitochondria are highly specialized to support a high rate of metabolism. They are extraordinarily large (up to 4 μm long) and irregular in shape. The cristae of flight muscle mitochondria are numerous, like the pages of a book, and there is relatively little open (matrix) space within flight muscle mitochondria. Up to 40% of the wet weight of flight muscle from *Phormia regina*, a blowfly, is mitochondrial mass, and half the muscle protein is mitochondrial protein. One microgram of flight muscle from a blowfly may contain 1.1×10^8 mitochondria. Biochemist Albert Lehninger estimated that flight muscle mitochondria have as much as 400 m^2 surface/g mitochondrial protein. By way of comparison, rat liver mitochondria have about 40 m^2/g protein. The outer membrane of insect mitochondria is permeable to most soluble components, but the inner membrane is very selectively permeable. Cytochrome *c* reductase and hexokinase, among other enzymes, are located on the outer membrane. The space between the outer and inner membrane contains adenylate kinase and nucleoside diphosphokinase activities. The outer surface of the inner membrane contains glycerol-3-phosphate dehydrogenase, proline dehydrogenase, and trehalase. The inner membrane contains the respiratory chain enzymes, adenosine triphosphate (ATP)-synthesizing enzymes, and α-ketoglutarate dehydrogenase. The inner side of the inner membrane contains succinic dehydrogenase and nicotinamide adenine dinucleotide (reduced form) (NADH) dehydrogenase. The matrix contains citrate synthetase, aconitase, isocitrate dehydrogenase, fumarase, malate dehydrogenase, alanine and aspartate aminotransferase, and carnitine, acetyl, and palmitoyl transferases. Most of the Krebs cycle intermediates do not readily cross the inner membrane and usually are not metabolized when added exogenously to isolated mitochondria. Knob-like structures about 8–9.5 nm in diameter are connected to the cristae by stalks that are 3–4 nm in diameter and 4–5 nm in length, and it is within these knob-like structures that ATP is actually synthesized by a chemiosmotic gradient.

Some groups (Lepidoptera, Orthoptera, and some others) burn lipids (fatty acids) as flight fuels. Fatty acids, which must be metabolized within the mitochondria and, hence, require availability of oxygen, release large amounts of energy per unit weight of substrate metabolized. The ability to mobilize and transport lipids from fat body rapidly and the availability of oxygen from the tracheal system are major adaptations in those insects that burn fatty acids for flight. Some insects that metabolize lipids are able to fly continuously for hours and undertake long-distance migration. A few insects use proline as flight fuel. Its complete metabolism yields much less energy per unit weight metabolized, and only a few insects have evolved to depend upon it as a major flight fuel.

8.1 INTRODUCTION: NUTRIENT STORES—THE FAT BODY

The day-to-day activities of an insect require a constant supply of energy. Many larval insects feed almost continuously, accumulating reserves that will be used during pupation, for diapause in some, and to support adult activities. Most adults need an intake of food to support activities, such as dispersal, reproduction, and flight, but a few adult insects do not feed as adults and live their life out in a few days, using stored nutrients from larval feeding. The fat body is a major organ for storing accumulating reserves (reviewed by Arrese and Soulages, 2010; Costa-Leonardo et al., 2013). It is also a major organ of metabolism, and it is important in detoxification. It is functionally divided into several physiological processes (Haunerland and Shirk, 1995). The fat body is not so much a "body" as it is a loosely organized collection of tissue in the head, thorax, and abdomen, although the largest amount of the fat body is usually in the abdomen. Larvae of most Lepidoptera typically have extensive and large fat body tissue as they approach pupation. Newly emerged adult Lepidoptera often still have a large fat body and mostly feed upon sugars in nectar or oozing fruit juices, while adult Diptera and Hymenoptera tend to have only small amounts of fat body after emerging as adults. The latter insects feed as adults to support reproductive processes. The distribution of the fat body within the body puts the tissue near most other organs, and nutrients from the fat body can be delivered rapidly to other organs and tissues.

The fat body is a major storage site of lipids (fats), carbohydrates, and a major site of protein synthesis. Triacylglycerols (TAGs) constitute about 90% of the lipids in the fat body of most insects. TAGs are stored in lipid droplets surrounded by various proteins to make a lipoprotein complex (Arrese et al., 2008a,b; Bickel et al., 2009). Lipolysis, the process mobilizing the lipids as energy sources (Chaves et al., 2011) involves multiple enzymatic steps. The short explanation is that the TAG is converted to a diacylglycerol (DAG) by removing one fatty acid, and the DAG is transported by the hemolymph as a lipoprotein complex to other organs (Weers and Ryan, 2006). In flight muscles of Lepidoptera, the two remaining fatty acids are released for metabolism and the glycerol is recycled to the fat body where it again is used as a building block for TAG synthesis. Major uses for fatty acid metabolism are pupation, diapause (Denlinger, 2005; Hahn and Denlinger, 2007) and, in Lepidoptera and some other insects, fuel for flight. The main enzyme acting upon the TAGs in the fat body is active phospholipase a (Arrese et al., 2006). Diptera and Hymenoptera, however, do not use fatty acids as a flight fuel; instead they use glucose (Kaufmann and Brown, 2008).

The fat body is a major site for synthesis of proteins, including most of the proteins in the hemolymph, vitellogenin (Tufail and Takeda, 2009), that will eventually be incorporated into eggs and other proteins (such as diapause proteins).

Glucose absorbed from the midgut during active feeding is rapidly converted into trehalose, the major hemolymph sugar (Thompson, 2003), and it circulates in high concentration in the hemolymph of many insects, and if not immediately needed for metabolic activity it is stored in the fat body. Glycogen is another main carbohydrate in insects and it is synthesized from glucose-6-phosphate and stored mainly in the fat body (see Section 8.3.1.2), although small amounts also occur in muscles and gut tissues. Li et al. (2014) recently presented experimental evidence that an important part of the circadian clock mechanism, PAR domain protein 1 (PDP1), has a function in mobilization of energy in the fat body, linking mechanistic target of rapamycin (mTOR) in circadian function with cellular metabolism.

Several different types of cells occur in the fat body, although the major ones, and the only ones in many insects, are adipocytes (trophocytes or fat cells). Other cells that occur in the fat body of some insects include urate cells that accumulate uric acid (particularly common in cockroaches and a few other groups of insects), mycetocytes that house microorganisms, and oenocytes (Makki et al., 2014). Hemoglobin cells that synthesize hemoglobin are found in the larvae of bot flies (*Gastrophilus* sp.) and in backswimmers in the genera *Anisops* and *Buenoa*.

Metabolism involves all the biochemical reactions occurring in an organism, coverage that clearly is not possible in one chapter. Thus, those metabolic reactions most directly involved in mobilizing stored energy reserves and the metabolic cycles for releasing energy for the various activities of insects are the subject of this chapter. Additional details of mobilization of stored energy reserves from the fat body and their metabolism to release energy will be given in the sections that follow.

8.2 ENERGY DEMANDS FOR INSECT FLIGHT

Only birds, insects, and bats fly with their own muscle power. Flight enables insects to disperse rapidly and widely and seek new areas to colonize. It also enables them to seek new and/or sparsely distributed food resources, locate potential mates, and search for oviposition sites. In some insects, such as blowflies (Diptera) and some Hymenoptera, flight is the most energy-intensive biological process known per unit weight of tissue. Perhaps because of the unique position it holds, flight metabolism has been studied extensively and many reviews are available, including Sacktor (1974), Bailey (1975), Candy (1985), Friedman (1985), Downer (1985), and Beenakkers et al. (1986). Blacklock and Ryan (1994) present an excellent review of lipid transport and metabolism.

A honeybee in continuous flight burns up to 2400 cal/g muscle/h (Weis-Fogh, 1952). Contrast this with the recorded metabolic rate of 215 cal/g muscle/h for hummingbirds during hovering flight, (Hainsworth, 1981), one of the highest rates of metabolism known among vertebrates. The mass-specific metabolic rates for flying honeybees are about 3 times greater than those measured for hovering hummingbirds, and 30 times those of human athletes in maximum exercise activity (Suarez et al., 2000, and references therein). Not only do some flying insects have high oxygen and calorie consumption values, but they can reach these high metabolic rates within a few seconds after taking flight. Upon cessation of flight, metabolic rate returns almost instantly to a low, "resting" rate. An oxygen debt does not have to be paid after intense activity because flight metabolism in insects is aerobic in contrast to largely anaerobic work accomplished in vertebrate muscles during intense muscular activity.

How insects control the rapid "turn-on" and "turn-off" of flight metabolism has been of great interest. Biochemists describe the adjustment in metabolic rate from rest to activity as the control value, calculated as the ratio of oxygen consumption rate during intense muscular activity divided by the resting rate. Control implies, of course, that an animal regulates its oxygen consumption and metabolic processes to support the intense activity, and then scales down the processes when the activity ceases. Upon initiation of flight, oxygen consumption rate in many insects jumps within seconds to values as much as 100 times the resting rate. A blowfly, *Lucilia sericata*, consumes 33–50 μL oxygen/min/g tissue while resting, but almost instantly increases that to as much as 1625–3000 μL oxygen/min/g tissue upon taking flight (Davis and Fraenkel, 1940). A simple calculation shows the control value to be at least 50, and possibly as much as 100 times the resting value. A variety of moths, which are not especially fast flyers, have oxygen consumption values of 7–12 μL/min/g muscle at rest and 700–1660 μL/min/g muscle in flight, again yielding control values of approximately 100 (Zebe, 1954).

There are many dynamic changes occurring in an insect upon initiation of flight, including changes in various metabolites and ions, increased nerve firing, flight muscle contractions, mobilization of components from the fat body, transport through the hemolymph, and the release of hormones. All of these events contribute to the physiological control ability of flying insects to achieve the very rapid 50–100-fold increase in metabolic activity and oxygen consumption during flight. Hummingbirds in flight have only about a fivefold control from resting metabolism to flight (Pearson, 1950), and trained, conditioned human sprinters also have control values of about 5 during a sprint.

Insects are able to use several different substrates as fuel during flight. As energy is released, it is trapped in the universal metabolic currency, ATP. As in all other organisms, ATP is present in relatively small amounts in cells and more is synthesized as needed. The ATP concentration in cells is one of the regulators of metabolism, with "large" amounts inhibiting some key enzymes involved in ATP synthesis, while decreased amounts stimulate new synthesis.

Probably in a typical insect, the level of **ATP** in the flight musculature is sufficient for only about 1 s of flight; a **phosphagen** reserve of **arginine phosphate**, sufficient for an additional 2–4 s of flight, can be used to transfer the high energy phosphate group from arginine phosphate to ADP, converting it to ATP (Candy, 1989).

Clearly, metabolism of some additional substrates and synthesis of new ATP must start in the first 1 or 2 s if flight is to continue. All insects appear to initially metabolize carbohydrates and a little bit of proline to "prime the Krebs cycle" upon taking flight. Some, such as Diptera and Hymenoptera, can sustain flight only as long as carbohydrates are available to metabolize, while Lepidoptera, Orthoptera, and a number of other insect groups rapidly switch to metabolism of lipids before their carbohydrates are gone. A few insects metabolize the amino acid proline as a major fuel supply to support flight.

8.3 METABOLIC STORES

8.3.1 Carbohydrate Resources

The two most common carbohydrate-stored reserves of insects are the disaccharide **trehalose** and the polysaccharide **glycogen**. The hemolymph, fat body, and gut tissue are major sources of stored carbohydrates, but small amounts of trehalose and glycogen occur in muscles. Trehalose usually is present in a large quantity in the hemolymph and is rapidly hydrolyzed to two glucose molecules for muscles or other tissues to use. Glycogen stored in fat body cells and gut cells must be hydrolyzed to release glucose units, which are then converted to trehalose and transported by the hemolymph to active tissues. Carbohydrate reserves are sufficient usually in a well-fed dipteran or hymenopteran to support continuous flight for 30 min to perhaps 2 h, depending on the species, size of the insect, size of fat body (which varies considerably in insects), and the trehalose content of the hemolymph (also variable).

8.3.1.1 Trehalose Resources

When energy is needed, trehalose is usually the first metabolite used, and its hydrolysis yields two molecules of glucose for each trehalose molecule hydrolyzed. Trehalose is the principal storage sugar of insects, and from 200 mg to as much as 1.5 g per 100 mL hemolymph occur in the hemolymph of various insect species. Trehalose is a disaccharide (α-D-glucopyranosyl-α-D-glucopyranoside), with the two glucose units linked α-1,1 (Figure 8.1). As a consequence of the 1,1 linkage of the two glucose units, trehalose is a nonreducing sugar, perhaps an important feature

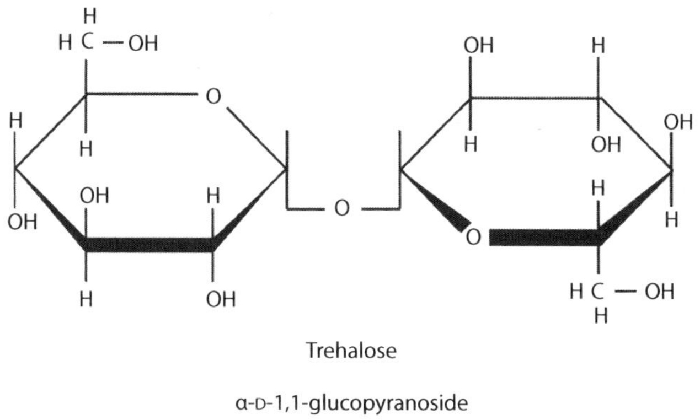

INTERMEDIARY METABOLISM

Figure 8.1 Trehalose (α-D-glucopyranosyl-α-D-glucopyranoside), a principal disaccharide storage and transport sugar in most insects.

because a reducing sugar that occurs in such large concentration as trehalose in the hemolymph might interact with and reduce other components in the hemolymph or tissues. Additional stores of trehalose occur in muscle cells and fat body cells.

Trehalose is synthesized rapidly from glucose as it is absorbed from the midgut. The absorbed glucose may have several fates, as shown in Figure 8.2, but with few exceptions, glucose is not stored as such in insects and is not usually present in any appreciable quantity in the hemolymph. Rapid synthesis of the absorbed glucose into trehalose keeps the hemolymph level of glucose very low, and glucose absorption occurs without an energy-requiring membrane transport mechanism. The process is called **facilitated diffusion** and, even when the gut concentration is low, glucose is still effectively absorbed.

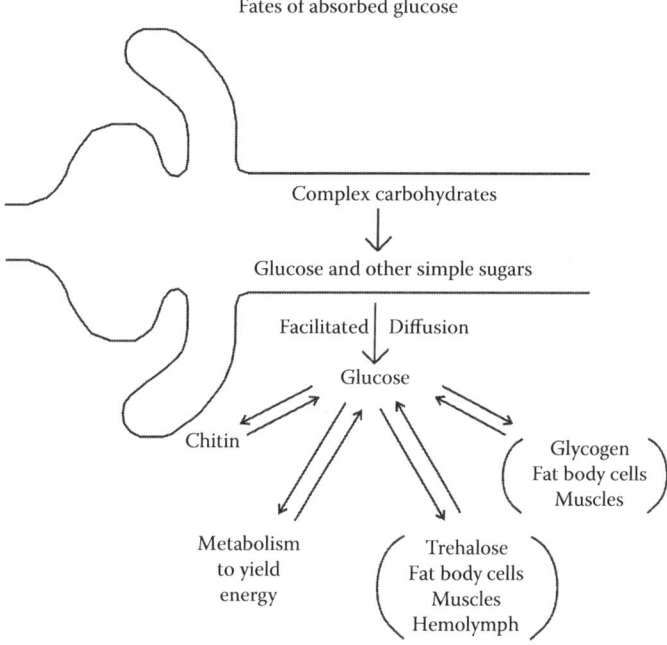

Figure 8.2 Possible fates for glucose absorbed from the gut.

The conversion of [^{14}C]glucose into trehalose has been demonstrated in tissue preparations from a number of insects, including the orthopterans *Schistocerca gregaria*, *Locusta migratoria*; a dipteran, *P. regina*; a dictyopteran (cockroach), *Leucophaea maderae*; and the lepidopterans, *Bombyx mori* and *Hyalophora cecropia*. Most of the synthesis of trehalose occurs in fat body cells, with small amounts synthesized in other tissues, such as gut and muscle cells.

Trehalose is a costly sugar for insects to synthesize, requiring several enzymes, steps, and input of high-energy phosphate from ATP. The pathway for trehalose synthesis in fat body (Figure 8.3) is based upon the work of investigators in several different laboratories. Immediately after absorption from the gut, glucose is converted to glucose-6-phosphate by the enzyme hexokinase with ATP supplying the phosphate group and energy for its transfer to glucose. Two molecules of glucose-6-phosphate are needed for trehalose synthesis, which necessitates investment of two ATPs. No significant energy exchange is required when the phosphate group is transferred to carbon-1 of glucose in a subsequent reaction. The enzyme uridine diphosphate glucose pyrophosphorylase catalyzes the synthesis of uridine diphosphate glucose (UDPG) from glucose-1-phosphate and uridine triphosphate (UTP). Trehalose-6-phosphate synthetase catalyzes the formation of trehalose-6-phosphate

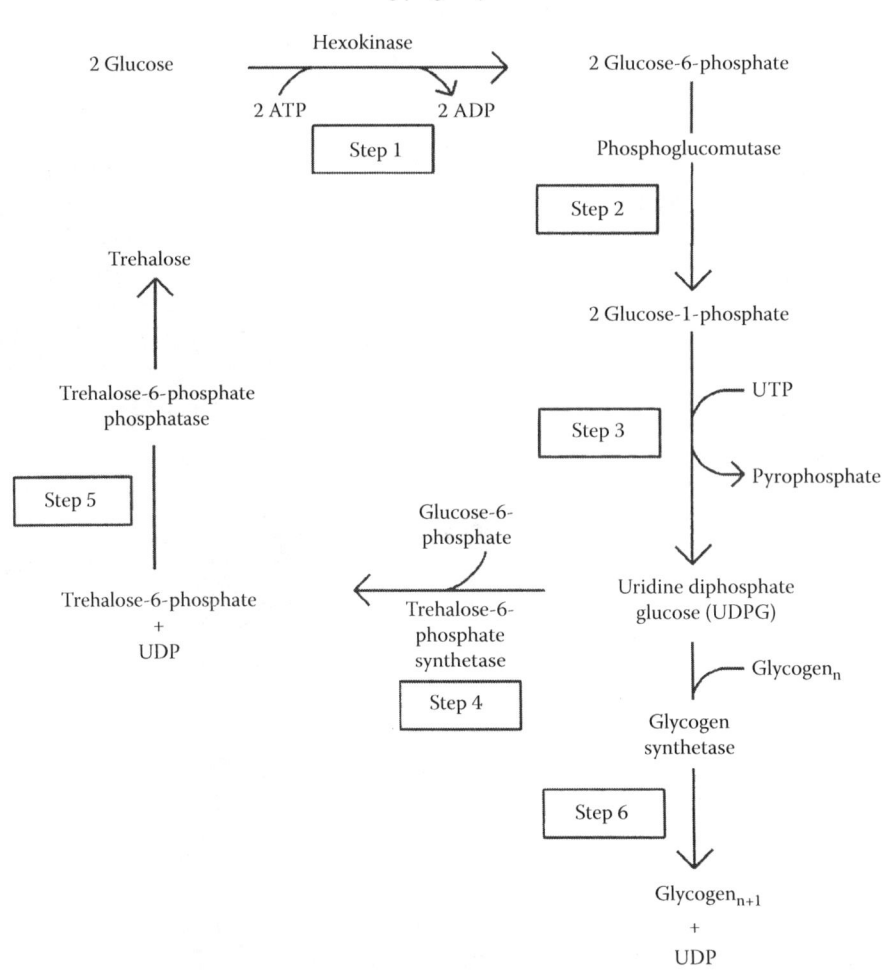

Figure 8.3 Biosynthetic pathway for the formation of trehalose and glycogen from glucose.

from glucose-6-phosphate and UDPG, with release of uridine diphosphate (UDP). The regeneration of UTP for future reactions requires enzymatically catalyzed phosphorylation of UDP by ATP, so one more ATP must be counted in the cost of synthesis of trehalose. Finally, trehalose-6-phosphate synthetase removes the phosphate group from trehalose-6-phosphate to form trehalose.

Synthesis of trehalose, at least in part, is regulated by a **negative feedback mechanism** in which free trehalose inhibits trehalose-6-phosphate synthetase and, thus, acts like a brake on the system. By slowing new synthesis of trehalose when the concentration of trehalose is high and little metabolic need for an energy supply exists, negative feedback presumably helps shift the synthesis of glucose into glycogen. The exact interaction of these two systems for storage of glucose, however, has not been worked out in detail in insects.

Even if no attempt is made to account for the energy required to synthesize the enzymes of the trehalose pathway, or for maintenance of the pathway, at least 3 mol of ATP are required to synthesize 1 mol of trehalose from glucose. Why do insects use a costly storage form for sugar when so many other organisms store sugar as glucose? The answer is not clear, but various guesses have been offered. Possibly one selection pressure during evolution was the value of a large reserve of an immediate energy source in the hemolymph that is available to support the energy demand of sustained flight. The nonreducing nature of trehalose, in contrast to the reducing nature of glucose, may prevent undesirable reactions in the hemolymph, and still provide a ready source of glucose when the enzyme trehalase is activated and releases 2 glucose molecules from trehalose. Moreover, there may have been evolutionary selection to reduce the osmotic effect of dissolved nutrients in the hemolymph. For example, 10 mM of trehalose has the equivalent energy value of 20 mM of glucose, but the trehalose in solution will have only half the osmotic value that 20 mM glucose will have because osmotic pressure is dependent upon the number of chemical particles in solution or suspension and not upon their size or chemical nature.

Most insects have high levels of **trehalase** in the hemolymph and in fat body cells, but the enzyme exists as an inactive proenzyme. The mechanism by which an insect converts the proenzyme to the active form under normal physiological processes has not been elucidated. Trehalase can be activated rapidly by the simple act of wounding and collection of hemolymph for trehalose assay, and by disruption of other tissues during sampling. Thus, to measure the true level of trehalose in hemolymph or tissues requires care to inactivate or minimize trehalase activity during tissue collection and processing.

8.3.1.2 Glycogen: Storage and Synthesis

Glycogen is a second form of energy storage. Insect flight muscles contain glycogen, but most muscles are too small to store very much. Glycogen is present at 10-15 mg/g thorax wet weight, which is mostly muscle tissue, in the blowfly, *P. regina* (Childress et al., 1970), and this is sufficient for a few minutes of flight. In order to sustain flight, additional fuels must be brought to the flight muscles. Glucose can be released from glycogen stored in fat body by **glycogen phosphorylase**. This enzyme is present in muscle tissue and fat body as inactive phosphorylase *b*, and it must be activated to phosphorylase *a*. Activation is under the control of the **hypertrehalosemic hormone (HTH)**, a peptide hormone formerly called the **hyperglycemic hormone (HGH)**. Initiation of flight activates the corpora cardiaca, probably through nervous control, to secrete the hormone (Steele, 1961, 1980, 1985). HTH requires the participation of a second messenger, **cAMP** (cyclic adenosine monophosphate), at the fat body cell membrane surface (Hanaoka and Takahashi, 1977) to activate phosphorylase *b* kinase, which converts inactive phosphorylase *b* to active phosphorylase *a*. Glycogen phosphorylase *a* then cleaves 1 glucose molecule from glycogen. Because glucose-1-phosphate is the form cleaved from glycogen, the investment of 1 ATP to activate free glucose is saved in the initial stages of glycolysis (see Figure 8.3). Free Ca^{2+} at concentrations as low as 10^{-8} M and inorganic phosphate stimulate phosphorylase *b* kinase, and stimulation is near maximum at 10^{-6} M Ca^{2+} (Chaplain, 1967; Hansford and Sacktor, 1970). Both free Ca^{2+} and inorganic PO_4 are increased as a result of the initiation of flight. The reverse reaction that inactivates glycogen phosphorylase by conversion of phosphorylase *a* to phosphorylase *b* is catalyzed by phosphorylase *a* phosphatase, but little is known about how this enzyme functions in insects.

Storage glycogen occurs mainly in fat body cells, but some glycogen is stored in gut epithelial cells and, to a slight extent, in muscle cells. Synthesis of glycogen is catalyzed by the enzyme **UDP-glucose-glycogen transglycosylase**, also known as **glycogen synthetase**. Glycogen synthetase catalyzes the transfer of glucose from UDP-glucose to glycogen, freeing UDP for some further reaction (most probably conversion back to ATP).

The precise regulatory controls determining the synthesis of trehalose vs. glycogen are not clear in insects, but one factor known to stimulate glycogen synthesis in insect tissues is the accumulation of glucose-6-phosphate. Glucose-6-phosphate can accumulate slowly as the rate of trehalose synthesis declines due to the feedback inhibition of free trehalose upon trehalose 6-phosphate synthetase. The declining synthesis of trehalose is likely to shift the UDP-glucose pool toward the synthesis of glycogen.

8.4 HORMONES CONTROLLING CARBOHYDRATE METABOLISM

The principal hormone controlling carbohydrate metabolism is the **hypertrehalosemic hormone (HTH)**. A related peptide hormone, **adipokinetic hormone (AKH)**, may supplant the action of HTH in some insects. For example, *Manduca sexta*, the tobacco hornworm, utilizes AKH for controlling carbohydrate metabolism during larval growth and development, but adults use AKH to mobilize lipids for flight fuel (Zeigler et al., 1990; Nijhout, 1994).

HTH and AKH have been purified and sequenced. The two hormones are closely related chemically and are considered to be members of the same family of peptide hormones (Gäde, 1990; Nijhout, 1994; Kaufmann and Brown, 2008). HTH is a polypeptide of 10 amino acids, and the sequence of amino acids varies slightly from species to species. AKH, isolated from several different insects, may have from 8 to 10 amino acids in its structure.

8.5 PATHWAYS OF METABOLISM SUPPORTING INTENSE MUSCULAR ACTIVITY, SUCH AS FLIGHT

8.5.1 Glycolysis

All insects tested metabolize carbohydrates first upon taking flight. For some insects, such as dipterans and hymenopterans, carbohydrate is the only fuel they can mobilize fast enough to support flight. Other insects metabolize carbohydrate at the initiation of flight, but if flight continues, they switch to another fuel, such as proline or fatty acids. **Glycolysis** (Figure 8.4), the process by which insects start to metabolize glucose, is similar to the process in vertebrates and other organisms, with the exception that glycolysis in insect flight muscle is always aerobic, never anaerobic as in actively working vertebrate muscle. The tracheal supply to insect flight muscle is extensive, and capable of supplying sufficient oxygen for totally aerobic oxidation during flight. The enzymes of glycolysis function equally well under aerobic conditions or anaerobic conditions.

Another specialization in insect flight muscle glycolysis is the way in which NADH in the cytoplasm is oxidized to NAD^+. The quantity of cytoplasmic NAD^+ is limited in the flight muscles of insects, just as it is in vertebrate muscles, and in order for glycolysis to continue, NAD^+ must

INTERMEDIARY METABOLISM

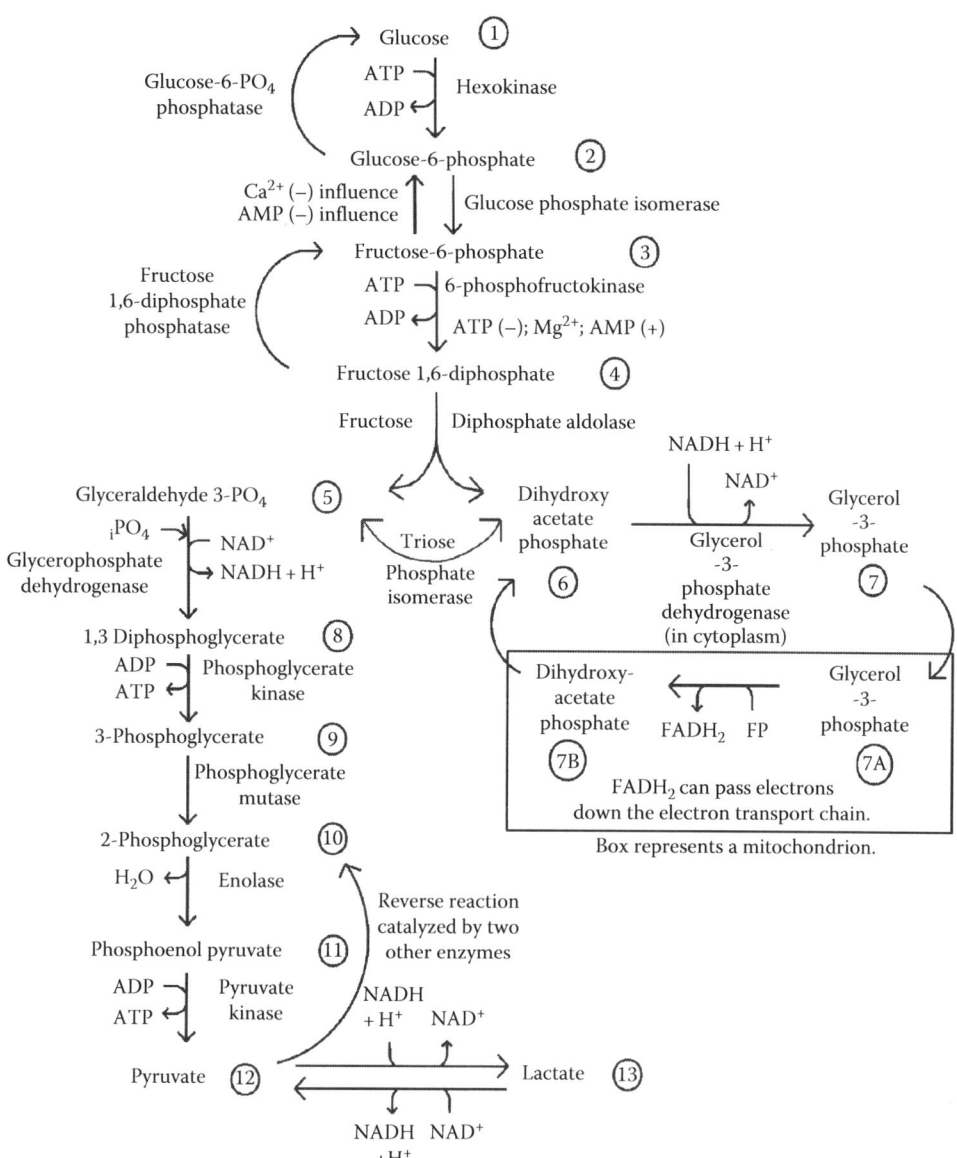

Figure 8.4 The glycolytic pathway for the metabolism of glucose in insects.

be constantly regenerated. Cytoplasmic NAD⁺ in insect flight muscle is regenerated through the **glycerol-3-phosphate shuttle**, not through conversion of pyruvate to lactate as in vertebrates. Some slower working skeletal muscles in insects, such as leg muscles, may oxidize NADH to NAD⁺, however, by the pyruvate to lactate step.

Because glucose in most insects is not present in significant quantities in the cell cytoplasm or hemolymph, glucose entering the glycolytic process will be derived first from the hydrolysis of trehalose and slightly later from glycogen. Glucose derived from trehalose must be phosphorylated with the participation of ATP and hexokinase. This investment of ATP to get the process started must be subtracted later from the total number of ATPs produced as the result of complete glucose metabolism.

Paradoxically for an insect initiating flight and needing energy from glucose, insect muscle hexokinase activity is easily inhibited by the product of its action, glucose-6-phosphate, as it is in other animal systems. The inhibition is countered by other products, however, such as inorganic phosphate that accumulates from the use of ATP to power the sudden intense muscle activity of flight. Initially, the situation is somewhat analogous to driving a car with one foot on the brake and the other on the accelerator. During sustained flight, however, a steady state is soon reached so that glycolysis proceeds smoothly. Glucose is released from glycogen in a phosphorylated state, as glucose-1-phosphate, without the expenditure of a high-energy phosphate, such as ATP, because the phosphate group comes from inorganic phosphate. Thus, when glycogen is the source of glucose for metabolism, there is one less ATP required than when glucose comes from trehalose. The phosphate group in glucose-1-phosphate is moved to carbon-6 by phosphoglucomutase to form glucose-6-phosphate without further expenditure of ATP. An important next step is the conversion of glucose-6-phosphate, a 6-carbon sugar, into fructose-6-phosphate, and this reaction is catalyzed by phosphoglucoisomerase without additional input of ATP. Multiple allele frequencies for phosphoglucoisomerase that are under selection by temperature have been found in a flightless beetle, *Chrysomela aeneicollis*, and may be important in the ability of this montane leaf beetle to adjust to changing climate conditions (Rank et al., 2007).

Conversion of fructose-6-phosphate to fructose-1,6-diphosphate requires another phosphorylation, and this time ATP is required to provide the energy and phosphate group. Thus, depending upon the source of glucose at the start, either one ATP (if the glucose is derived from glycogen as glucose-1-phosphate), or two ATPs (if it is from trehalose), must be invested to get glycolysis underway. This step from fructose-6-phosphate to fructose-1,6-diphosphate is one of the major control points for carbohydrate metabolism in insects, as it is in other organisms. Excess ATP inhibits phosphofructokinase isolated from blowfly insect flight muscle, and acts as a brake on glycolysis when demand for ATP drops. Phosphofructokinase is stimulated by AMP, inorganic phosphate, and cyclic AMP (Walker and Bailey, 1969), products expected to accumulate from the initiation of flight and the use of available ATP in muscle contractions. Although ATP decreased in the blowfly *Phormia* upon initiation of flight, it fell only slightly from 6.9 to 6.2 mM (Sacktor and Hurlbut, 1966), a drop that is unlikely to relieve inhibition of phosphofructokinase because the lower concentration of 6.2 mM ATP still inhibited isolated phosphofructokinase in vitro. There is also a concomitant rise in AMP level from 0.12 mM at rest to 0.30 mM in flight, but again a magnitude of change that seems insufficient to account for the large increase in flight metabolism. Additional factors, perhaps relating to compartmentalization, and other as yet unidentified agents acting upon this control point are probably involved.

8.5.1.1 Glycerol-3-Phosphate Shuttle and Regeneration of NAD⁺

Fructose-1,6-diphosphate is split into two 3-carbon products, glyceraldehyde-3-phosphate and dihydroxyacetone phosphate. These two compounds are interconvertible and the enzyme for conversion is **triosephosphate isomerase**. The oxidation of glyceraldehyde-3-phosphate to 1,3-diphosphoglycerate is a very important step because it is dependent upon the availability of inorganic phosphate and the oxidized form of **nicotinamide adenine dinucleotide** (**NAD$^+$**). Inorganic phosphate (possible forms might be $NaHPO_4$ or $KHPO_4$) is unlikely ever to be a limiting factor in the reaction, but only small amounts of NAD$^+$ are present in the cytoplasm, and the oxidized form must be regenerated as rapidly as it is used in order for this cytoplasmic reaction to continue. In the reaction, two electrons and two protons are removed from glyceraldehyde-3-phosphate, thereby oxidizing it to 1,3-diphosphoglycerate. The two electrons and one proton are accepted by NAD$^+$, reducing it to **NADH** (Figure 8.5), and one proton is buffered by the cytoplasmic medium. If NADH, resulting from this reaction, could get into the mitochondria, which as noted earlier always have available oxygen in flight muscles, it could be reoxidized to NAD$^+$, but flight muscle mitochondria

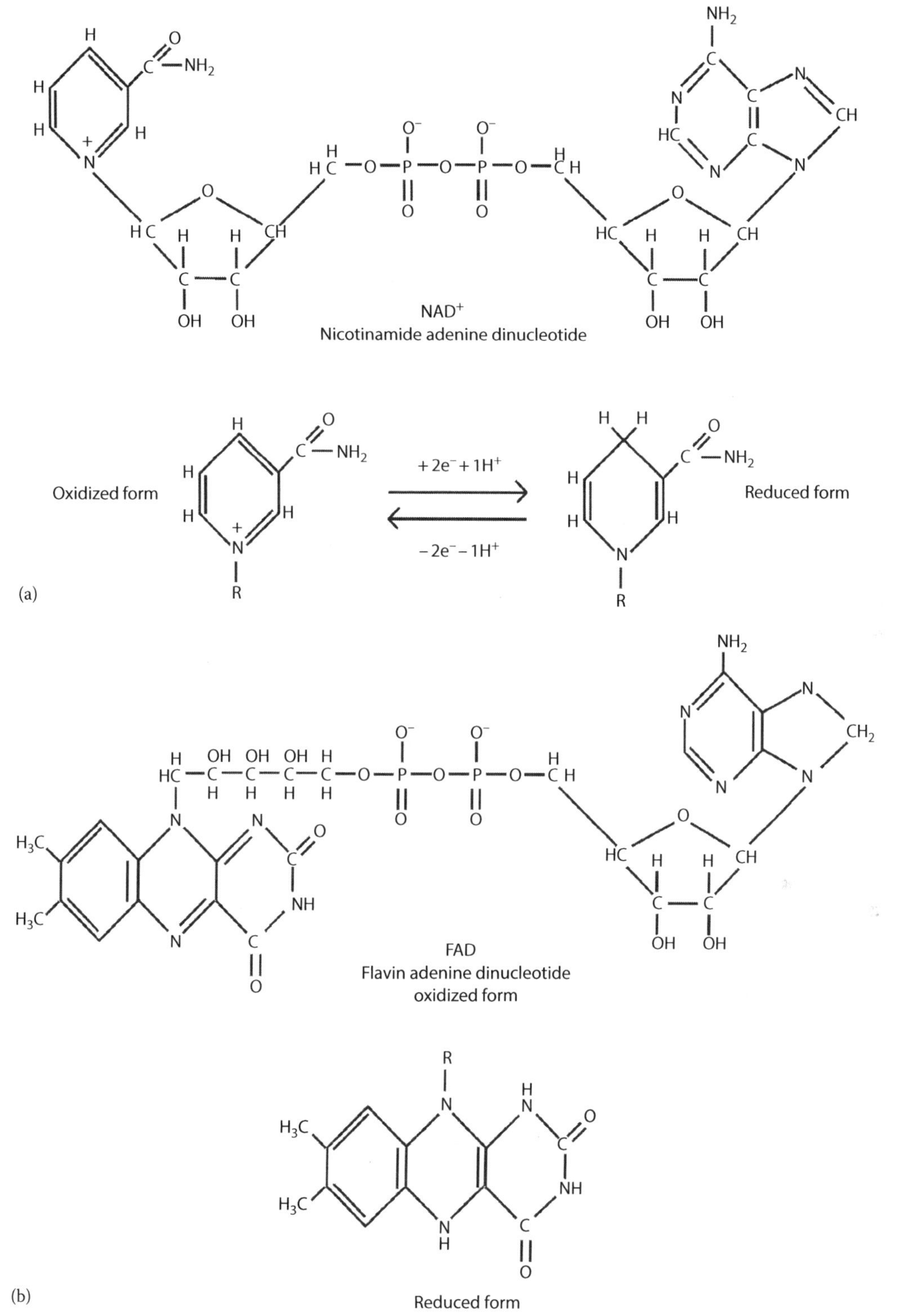

Figure 8.5 Reduced and oxidized forms of (a) nicotinamide adenine dinucleotide (NAD⁺) and (b) flavin adenine dinucleotide (FAD).

are relatively impermeable to NADH, NADPH, NAD^+, and $NADP^+$ (Sacktor, 1961; Sacktor and Dick, 1962). Thus, a cytoplasmic mechanism is necessary to regenerate NAD^+. The common cytoplasmic mechanism for regenerating NAD^+ in working vertebrate muscle is the transfer of the two electrons and one proton from NADH (and a cytoplasmic proton, H^+) to pyruvate, thereby reducing it to lactate. Although lactic dehydrogenase, the catalyst for this reaction, occurs in insect walking leg muscles and other muscles that perform slower movements, its activity in flight muscle is very low and is unable to regenerate NAD^+ fast enough to allow carbohydrate metabolism to continue at a high rate. Flight muscles have high levels of another enzyme, **cytoplasmic glycerol-3-phosphate dehydrogenase** (Table 8.1), that catalyzes the regeneration of NAD^+ in the cytoplasm much faster by the cytoplasmic half of the glycerol-3-phosphate shuttle reactions, in which dihydroxyacetone phosphate is reduced (receiving two protons into its structure) and thereby converted into glycerol-3-phosphate and NAD^+ is regenerated (see reactions 6 and 7, Figure 8.4).

This cytoplasmic reaction is the first of a two-step shuttle for transferring electrons from the cytoplasm to mitochondria. Regeneration of cytoplasmic NAD^+ allows continued oxidation of glyceraldehyde-3-phosphate, with substrate-level production of 1 mol of ATP/3-carbon fragment oxidized to pyruvic acid. Pyruvate rapidly enters mitochondria and leads to further oxidation and production of ATP through the Krebs cycle.

Glycerol-3-phosphate (G-3-P) from the reaction above does not accumulate in the cytoplasm of flight muscle tissues as lactate does in a working vertebrate muscle. About the same concentration, 2 mM, is present in both resting and working flight muscle (Sacktor and Wormser-Shavit, 1966). As fast as it is produced, G-3-P crosses the outer membrane of flight muscle mitochondria and, at the outer surface of the inner mitochondrial membrane, it is rapidly oxidized to dihydroxyacetone phosphate by a flavin adenine dinucleotide (FAD)-linked **mitochondrial glycerol-3-phosphate dehydrogenase** bound to the inner membrane (Note reactions 7A and 7B in the box, Figure 8.4). The flavoprotein (FAD) accepts the two electrons and two protons (becoming designated as $FADH_2$). The protons get transferred rapidly through the electron transport system in mitochondria to molecular oxygen as the final acceptor. Several evolutionary adaptations have made this shuttle possible, including (1) high activity of the cytoplasmic glycerol-3-phosphate dehydrogenase, (2) localization of an active glycerol-3-phosphate dehydrogenase on the outer surface of the inner membrane of flight muscle mitochondria, and (3) availability of oxygen and a fully functional electron transport system in the mitochondria of working muscle. Dihydroxyacetone phosphate at the inner membrane surface rapidly diffuses out of mitochondria into the cytoplasm. The overall process can be succinctly summarized as follows: NADH is oxidized to NAD^+ in the cytoplasm and its electrons and proton (plus another cytoplasmic proton) are shuttled across the outer mitochondrial membrane via a carrier, glycerol-3-phosphate. At the outer surface of the inner membrane, the carrier is oxidized to dihydroxyacetone phosphate, which returns to the

Table 8.1 Glycerol-3-Phosphate Dehydrogenase Activity

Tissue Source	µmol/g wet wt./min
Blowfly flight muscle	1230
Honeybee flight muscle	700
Locust flight muscle	167
Cockroach flight muscle	48
Cockroach leg muscle	32
Locust leg muscle	33
Rat skeletal muscle	50
Beef smooth muscle	0.1

Source: Adapted from Bailey, E., Utilization of fuels by muscle, in: Candy, D.J. and Kilby, B.A. (eds.), *Insect Biochemistry and Function*, Chapman & Hall, London, U.K., 1975, pp. 3–87.

cytoplasm to repeat the process, and the electrons and protons pass down the electron transport chain with the generation of 2 ATP/cytoplasmic NADH oxidized.

Evidence for the importance of the G-3-P shuttle in flight comes from mutants of *Drosophila melanogaster* that are deficient in cytoplasmic glycerol-3-phosphate dehydrogenase (Bewley et al., 1974; Collier et al., 1976) and are incapable of flight, presumably because they have no effective way to rapidly regenerate ATP in the cytoplasm.

8.5.1.2 Significance and Control of the Glycerol-3-Phosphate Shuttle

The significance of the glycerol-3-phosphate shuttle hinges on the assumption that the shuttle is self-generating when catalytic amounts of dihydroxyacetone phosphate are introduced (Sacktor and Dick, 1962). A small number of dihydroxyacetone phosphate (DHAP) molecules converted to G-3-P in the cytoplasm, with subsequent conversion of G-3-P to DHAP in the membranes of mitochondria, may cycle over and over during flight and keep the cytoplasmic level of NAD^+ high. This would allow nearly all the DHAP produced from the splitting of fructose-1,6-diphosphate to be converted to glyceraldehyde-3-phosphate, and ultimately converted to pyruvate. In this scenario, all the initial glucose can be converted to two pyruvate molecules that enter the Krebs cycle; thus, ATP production in glycolysis and in the Krebs cycle is maximized. The shuttle itself should produce 4 mol ATP per mole glucose metabolized (two ATPs for each cytoplasmic NAD^+ regenerated, or said another way, two ATPs for each FADH2 produced within the mitochondria as a result of the shuttle action), and if the shuttle makes it possible for nearly all of the glucose to be converted to pyruvate, then 4 mol ATP moles/mole glucose would be produced by substrate oxidations in the reactions of glycolysis. Thus, in this scenario, as many as 8 mol of ATP could be produced per mole of glucose metabolized during glycolysis. If glucose is derived from trehalose, then two ATP will be required in early phosphorylations (to form glucose-6-phosphate and fructose-1,6-diphosphate), so the net production would be six ATP. If glycogen provides the glucose, then only one ATP is needed in an early phosphorylation, and the net production of ATP/glucose metabolized is seven. The important point is that flight can continue for long periods supported by aerobic metabolism, which provides much more ATP than anaerobic metabolism.

The importance of the shuttle in insect flight muscle suggests that there must be control points in the shuttle mechanism, and indeed there are. Free Ca^{2+}, and possibly Mg^{2+}, are important in stimulating the metabolism of glycerol-3-phosphate. Ethylene diamine tetraacetic acid (EDTA), a sequestering agent for divalent cations, inhibits oxidation of glycerol-3-phosphate, but the inhibition can be reversed by adding additional Ca^{2+} or Mg^{2+}, thus implicating one or both of these ions as potential control factors in the shuttle reactions (Estabrook and Sacktor, 1958). Although Ca^{2+} is bound to the sarcoplasmic reticulum (SR), a network of membranes in muscle tissue, the arrival of nerve impulses causes its release. Under these conditions, concentrations of free Ca^{2+} ions at 10^{-6}–10^{-7} M occur in the sarcoplasm. Ca^{2+} released in muscle tissue stimulates G-3-P dehydrogenase and antagonizes resting inhibition of the dehydrogenase (Sacktor and Wormser-Shavit, 1966), with 10^{-7} M Ca^{2+} stimulating G-3-P dehydrogenase to about half-maximal activity (Hansford and Chappel, 1967; Donnellan and Beechey, 1969; Carafoli and Sacktor, 1972).

8.5.2 Krebs Cycle

The reactions of the **Krebs cycle** are shown in Figure 8.6. The two products from glucose metabolism in the glycolytic pathway, pyruvate and glycerol-3-phosphate, rapidly enter mitochondria (Sacktor and Wormser-Shavit, 1966). Cytoplasmic pyruvate accumulates very briefly for the first few seconds after flight begins, probably due to the need to prime the Krebs cycle with intermediates, particularly oxaloacetate as the acceptor for acetate resulting from the oxidation of pyruvate. Priming of the cycle may result from proline metabolism. Proline decreases initially at the start of

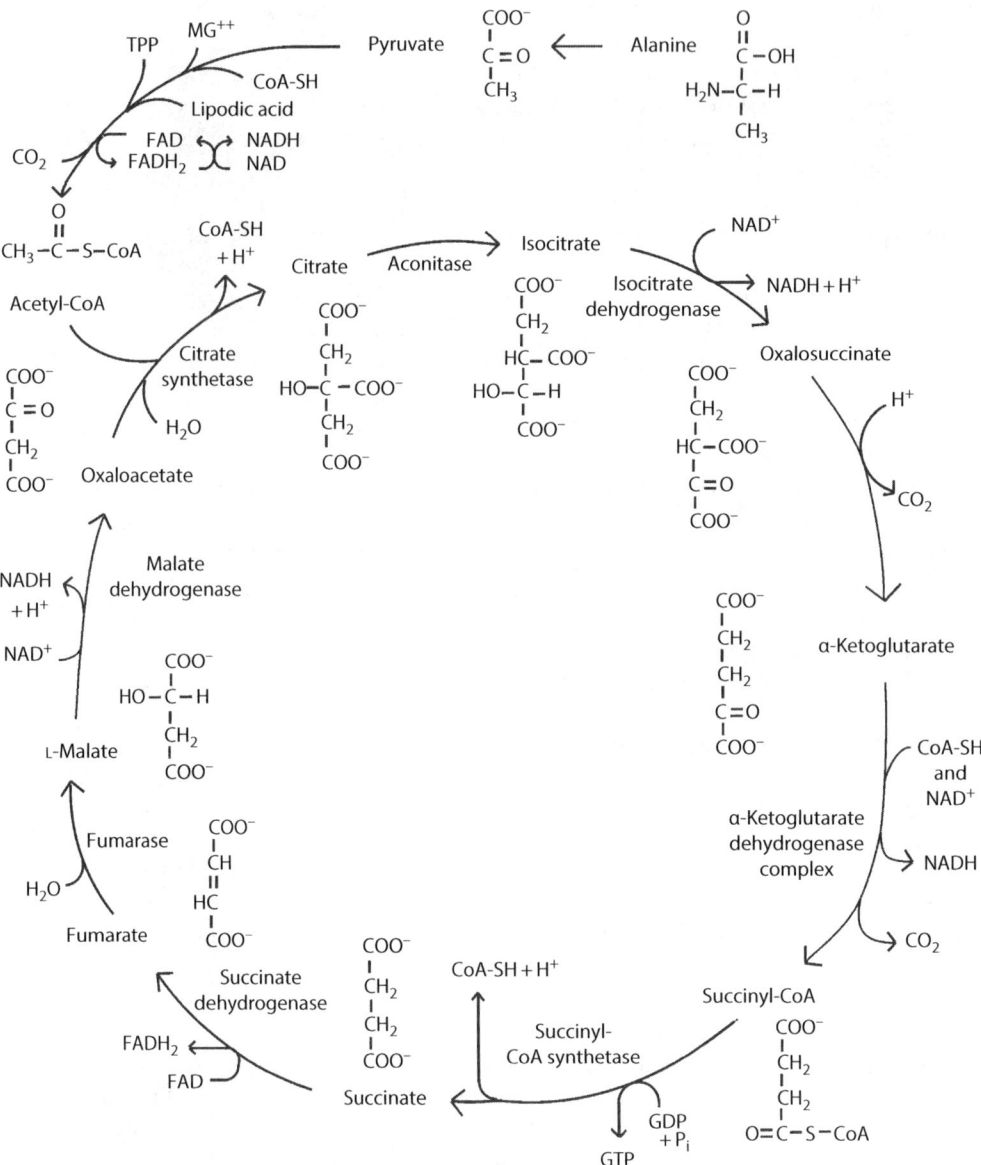

Figure 8.6 The reactions of the Krebs cycle in insect mitochondria. The principal source of pyruvate entering the Krebs cycle is glycolysis, but insects can convert alanine to pyruvate as one way to metabolize amino acids for energy. Amino acids also can enter the Krebs cycle at other points.

flight, and it may be converted by proline dehydrogenase to glutamate, which in turn undergoes transamination with pyruvate to form α-ketoglutarate and alanine. α-Ketoglutarate, a normal component of the Krebs cycle, is oxidized through several steps to oxaloacetate. In any event, the delay in oxidation of pyruvate is rapidly relieved.

Most substrates from the Krebs cycle are not readily oxidized when added exogenously to isolated mitochondria, apparently because they cannot penetrate the mitochondrial membranes (Van den Bergh and Slater, 1962). In preparations of isolated mitochondria from some insects, added succinate is metabolized, apparently after transport by a carrier in the mitochondrial membrane.

The carrier is rapidly saturated by inorganic phosphate buffer, and succinate oxidation is best demonstrated in nonphosphate buffer systems. A few examples of rapid oxidation of other Krebs cycle intermediates have emerged, and it is probable that some variability exists among insect species as to which, if any, of the Krebs cycle intermediates can be metabolized by isolated mitochondria. In intact insects, of course, Krebs cycle intermediates are produced from within, and do not have to penetrate mitochondria. Lack of membrane permeability to the Krebs cycle intermediates serves to keep internal concentrations high, since they do not leak out.

The enzyme **pyruvate dehydrogenase** in insect flight muscle tissue exists in an inactive form (phosphorylated) and an active form (dephosphorylated), with activation controlled by a phosphatase enzyme activated by free Ca^{2+}. Entry of pyruvate into mitochondria and its conversion to acetyl-coenzyme A is likely to involve a multistep sequence in insects, as it does in other organisms. A complex five-step sequence for passage of pyruvate across the mitochondrial membranes and into the mitochondrial matrix has been described for some noninsect mitochondrial preparations (Lehninger, 1975).

8.5.2.1 Control of Krebs Cycle Metabolism and the Regulation of Carbohydrate Metabolism in Flight Muscles

Control points in the Krebs cycle enable insects to rapidly increase metabolic activity upon taking flight. **Isocitrate dehydrogenase**, an NAD-linked enzyme, is a major **control point** (Goebell and Klingenberg, 1964; Hansford, 1972; Zahavi and Tahori, 1972). It is inhibited by high levels of ATP and stimulated by isocitrate, ADP, and inorganic PO_4. The latter two compounds accumulate from the use of ATP to initiate and support flight. The concentrations and relative ratios of ATP, ADP, and AMP play an important role in regulating metabolism in insects, as they do in other organisms. Flight muscle of *P. regina* contains 6.9, 1.5, and 0.13 µmol/g wet weight of ATP, ADP, and AMP, respectively (Sacktor and Hurlbut, 1966). Upon initiation of flight, the level of ATP falls rapidly, while concentrations of ADP, AMP, and inorganic PO_4 increase. Many steps in the mobilization and metabolism of carbohydrates require regulatory mechanisms to enable the rapid increase in the rate of metabolism upon initiation of flight and the equally rapid decrease in metabolism when flight stops.

8.5.3 Electron Transport System

The **electron transport system** in insect mitochondria is similar to that in other animals (Sacktor, 1974). The components of the respiratory chain are arranged in a sequence on the inner mitochondrial membrane so that electrons flow down the chain, as illustrated in Figure 8.7. Electrons pass sequentially to a component with a lower oxidation-reduction potential (a more positive value) until molecular oxygen accepts two electrons and two protons to form water.

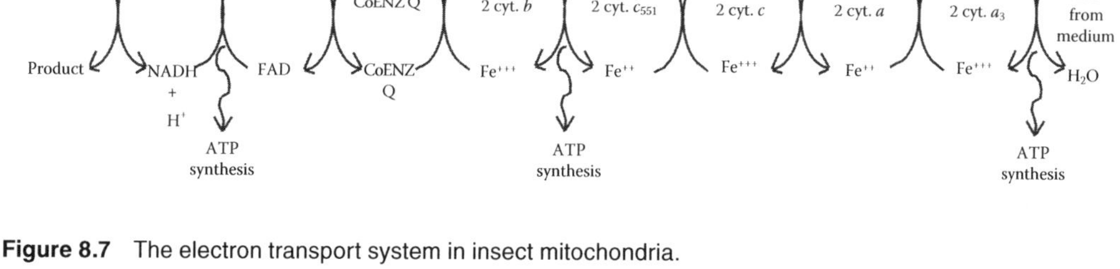

Figure 8.7 The electron transport system in insect mitochondria.

Flight muscle mitochondria are not permeable to NAD^+ or NADH, so cytoplasmic nucleotides do not readily enter mitochondria, nor do those inside the mitochondria pass outside. The dehydrogenase enzyme-NAD^+ complex often is tightly bound to the inner membrane. An exception is the mitochondrial glycerol-3-phosphate dehydrogenase involved in the shuttle described in glycolysis; this dehydrogenase is located on the outer part of the inner membrane and is linked to a flavoprotein other than NAD^+. Insect mitochondria have nonheme iron containing coenzyme Q, also called ubiquinone, between the flavoprotein and cytochrome b in the chain sequence. Like the cytochromes, one molecule of coenzyme Q accepts only one electron and no protons. Coenzyme Q has an isoprene side chain of varying length in different insects.

The **cytochromes** are proteins (enzymes) that contain iron (Fe) held in a heme-porphyrin structure (Figure 8.8). The iron is available from an iron transport protein, transferrin, in the hemolymph and from a storage protein, ferritin (Nichol et al., 2002). One molecule of any of the cytochromes can accept only one electron (or give up one upon oxidation). The atom of iron in the heme structure is the part of the molecule that accepts one electron (thereby becoming reduced Fe^{2+}). Fe^{2+} becomes oxidized to Fe^{3+} when the electron is passed to the next cytochrome in the sequence. No protons can be accepted by the cytochromes; the protons removed from FPH_2 are pumped into the mitochondrial matrix, while the two electrons are passed to two molecules of cytochrome b. Later these protons will return through **ATP synthase complexes** in the inner membrane, enabling the formation of new ATP, and the protons join with two electrons in the structure of H_2O. All the cytochromes have absorption characteristics in the reduced state determined by the heme-porphyrin rings, the Fe, and the protein chain. Thus, cytochrome c^{551} is a characteristic cytochrome of insects and its absorption maximum in the reduced state is at 551 nm. In vertebrates, the cytochrome in this position in the chain is called cytochrome c_1, and its absorption spectrum has a maximum at 554 nm indicating that it is slightly different in some way, perhaps only in the folding of the protein backbone of the cytochrome. Cytochrome c, the next cytochrome in the sequence, has been isolated and purified from many organisms, including different families of insects. The last two cytochromes in the chain make up a unit known as cytochrome oxidase. An electron is transferred from cytochrome a to cytochrome a_3 and subsequently to an atom of oxygen. The two necessary protons for water formation are taken from the mitochondrial matrix. Water formed through transfer of substrate electrons and protons to oxygen is called **metabolic water**. It is a very important source of water for all insects, and especially those that live in very dry environments.

At three points in the electron transfer chain enough energy is released in a single step to enable the synthesis of an ATP molecule from ADP and inorganic phosphate (a minimum of about 7.5 kcal is required) (Figure 8.8). Thus, when two electrons are passed through the complete electron transfer pathway, three ATPs can be formed. The first synthesis of ATP occurs when two electrons are passed from NADH to FAD. If this step is bypassed, as when FAD directly accepts the electrons from substrate oxidation (e.g., in the mitochondrial reaction of glycerol-3-phosphate to dihydroxyacetone as part of the glycerol-3-phosphate shuttle), then the first ATP is not formed and only two remaining ATPs per two electrons transferred are formed. The second ATP is formed when two cytochrome b molecules each transfer one electron to two cytochrome c molecules. The final ATP is formed when two cytochrome a_3 molecules transfer two electrons to molecular oxygen to form water (with the two protons coming from the general buffers of the mitochondrial matrix). In every step where ATP is formed, more energy is released than can be captured in one ATP synthesized, but in this step in particular, a large release of energy occurs, equal to about 23.8 kcal/mol of substrate. This is enough energy to synthesize nearly three ATPs, but only one is actually formed. The remainder of the energy is dissipated as heat. Insects, like other biological organisms, are capable of capturing only about 40% of the energy in a glucose molecule as ATP.

Although fine details of ATP production in insects have not been elucidated, the **chemiosmotic hypothesis** proposed by Mitchell (1979, and additional references therein) seems the likely mechanism. In the chemiosmotic process, energy from the transfer of electrons is used to pump protons

INTERMEDIARY METABOLISM   239

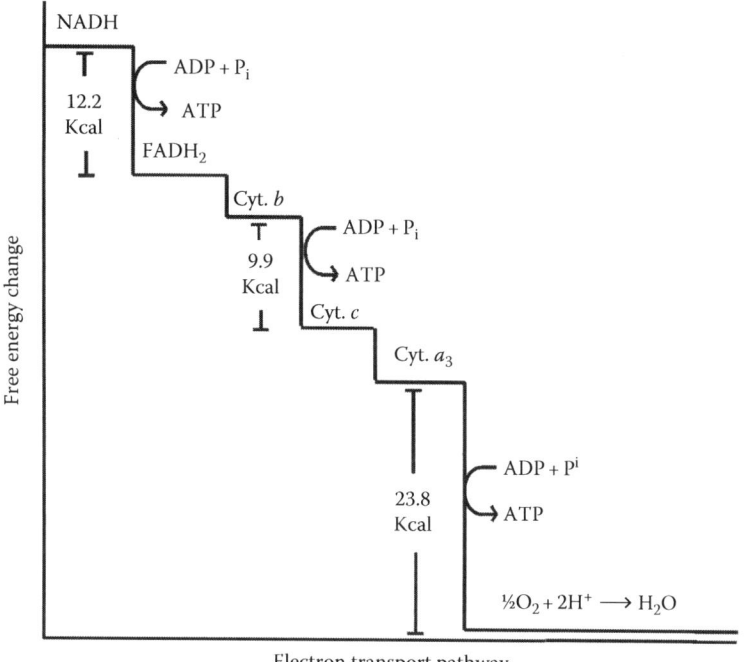

Figure 8.8 Changes in free energy and points at which enough energy is available to enable synthesis of ATP as electrons pass down the transport chain. A cytochrome, illustrated diagrammatically here, is a large enzymatic protein attached to an iron heme group. The iron atom in the cytochromes accepts one electron (becoming reduced) or gives up one electron (becoming oxidized) as electrons flow from component to component in the chain. Electron flow is like a one-way street, with electrons always moving toward the component with the higher redox potential, and finally to molecular oxygen.

into the space between the inner and outer mitochondrial membranes (Figures 8.9 and 8.10). This leaves the inner compartment of the mitochondrion negatively charged relative to the intermembrane space, which has a positive charge because of the accumulating H^+. Thus, a small battery results. Mitochondria utilize the potential difference between the membranes to supply the energy for ATP synthesis from ADP as the protons pass down the chemiosmotic gradient by returning to the inner compartment through F_1 membrane complexes, also called the **F_1 ATP synthase complex**. The complex is a **membrane ion channel** composed of proteins that allow the passage of protons, and ATP synthesizing enzymes are part of the complex.

The energy yield from the metabolism of 1 mol of glucose could be as much as 36 net moles of ATP (Figures 8.11 and 8.12). If glycogen is the source metabolized, the number of ATP could be higher because one less ATP is initially needed to phosphorylate the glucose-1-phosphate resulting from the cleavage of one glucose unit from a glycogen molecule.

8.5.4 Proline as a Fuel for Flight

The amino acid **proline** is a major metabolic fuel for the tsetse fly (Bursell, 1963, 1965, 1966, 1981), for adults of the Colorado potato beetle, *Leptinotarsa decemlineata* (DeKort et al., 1973; Khan and DeKort, 1978; Mordue and DeKort, 1978; Weeda et al., 1980); some beetles in the family Scarabaeidae (Pearson et al., 1979), including *Melolontha melolontha*, *Heliocopris dilloni*, and *Popillia japonica*; and some beetles in the family Cerambycidae (Gäde and Auerswald, 2000). In the South African long-horned beetle, *Phryneta spinator*, about 50% of the carbohydrates and 40% of the proline in the hemolymph were metabolized to support 5 min of flight, and alanine increased. Increase of alanine is expected when proline is metabolized for energy because of the transamination reaction in which the amino group from glutamic acid is transferred to pyruvic acid, creating alanine and α-ketoglutarate (see reactions in Figure 8.13). Complete proline metabolism releases up to 14 mol ATP/mol proline and is a mitochondrial process, so flight and proline metabolism are linked to and dependent on a rich supply of oxygen to flight muscles. The pathway for metabolism of proline, proposed in part by Bursell (1967), is shown in Figure 8.13. Proline readily enters mitochondria and is first oxidized to glutamate (Bursell, 1967) by a very active **proline dehydrogenase** located in tsetse fly flight muscle mitochondria. A flavoprotein accepts the two electrons and two protons that are removed.

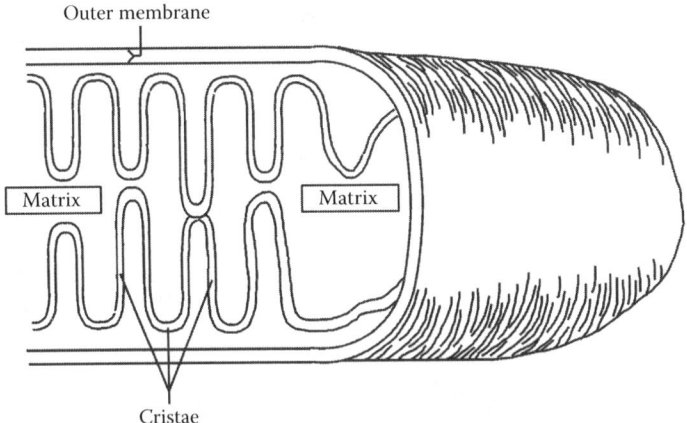

Figure 8.9 A diagrammatic cut-away view of a mitochondrion to illustrate the cristae typical of flight muscle mitochondria where the electron transport reactions occur, and the matrix where most of the citric acid cycle reactions occur. Cristae are much more numerous than this diagram shows, more like the pages in a book.

INTERMEDIARY METABOLISM

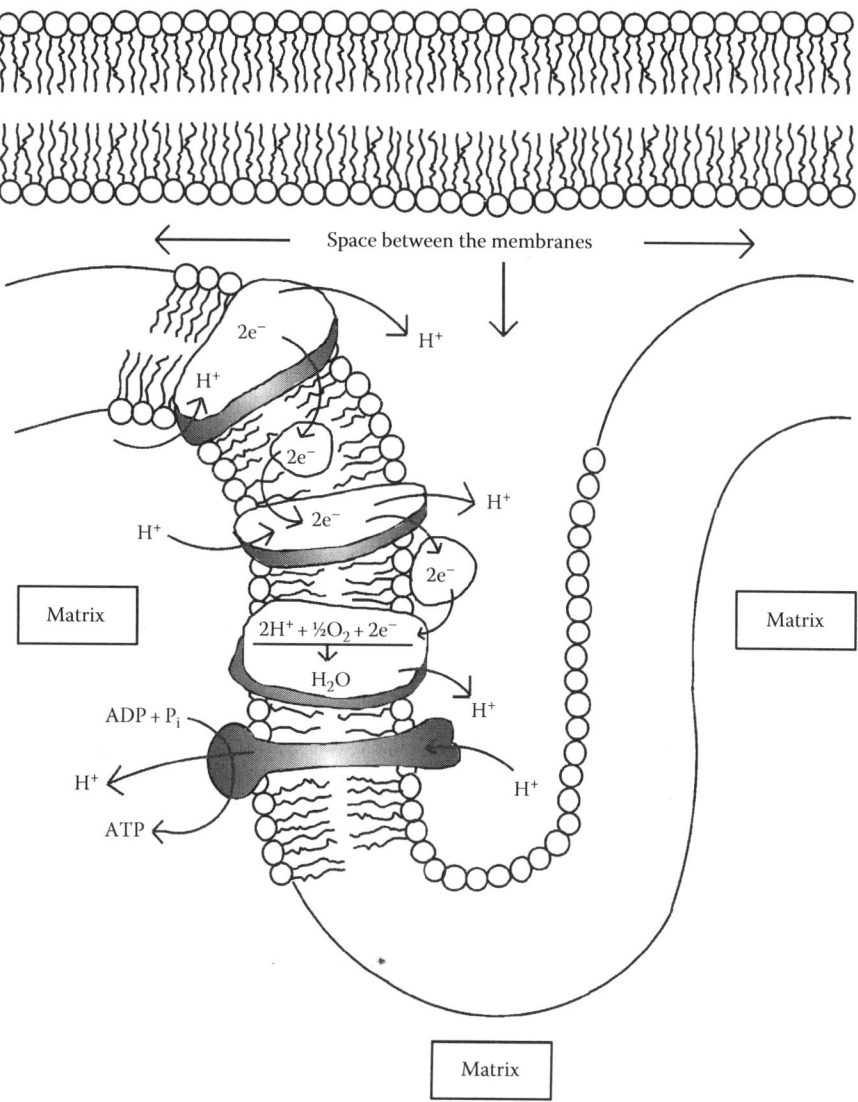

Figure 8.10 An illustration of the steps in transfer of electrons down the electron transfer chain, the pumping of protons into the space between the membranes, and the passage of protons back to the matrix through the ATP synthetase complex causing the formation of ATP.

Glutamate then undergoes a transamination reaction with pyruvate to produce α-ketoglutarate and alanine. The α-ketoglutarate formed is a normal component of the Krebs cycle and it is readily metabolized by the cycle pathway.

One of the products from proline metabolism, alanine, is rapidly removed from the muscle and transported to the fat body where, by addition of a 2-carbon unit derived from fatty acids, it is converted into proline again. It can then be transported to the muscles to repeat the proline cycle reactions. Thus, the proline pathway represents a shuttle to transfer 2-carbon units from the fat body to the muscles for metabolism (Candy, 1989).

No satisfying explanation has been offered as to why some insects evolved metabolic machinery to metabolize proline, while others use glucose or lipid metabolism, both of which produce larger amounts of ATP per mole substrate metabolized than proline. Possibly many insects metabolize

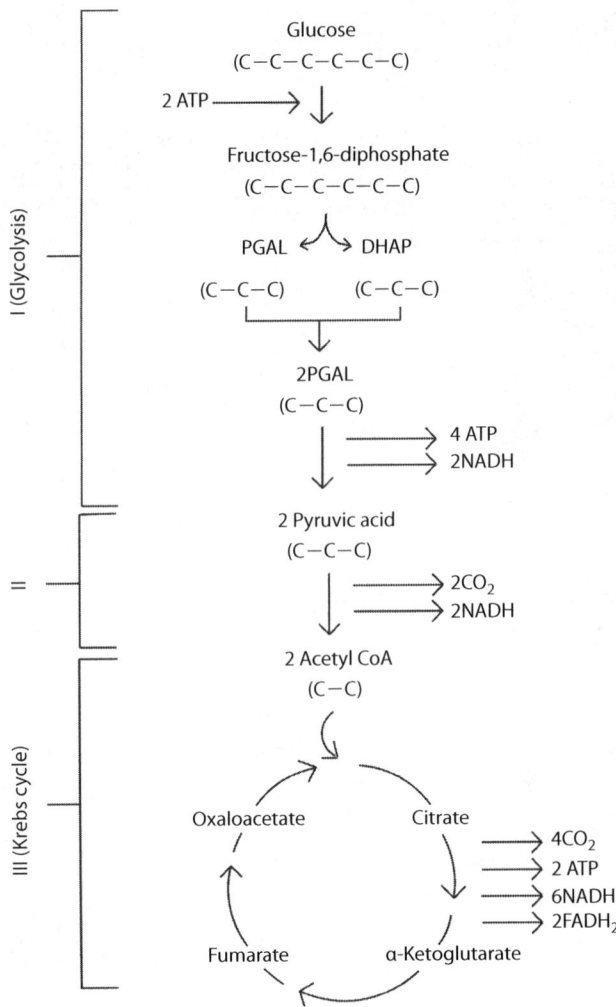

Figure 8.11 A summary of the major stages in the metabolism of glucose in flight muscle, with CO_2, reduced cofactors, and a small amount of substrate level formation of ATP resulting.

small amounts of proline at the beginning of flight to prime the Krebs cycle. Evidence for proline use in this way has been obtained for the housefly *Musca domestica*, the blowflies *P. regina* and *Sarcophaga nodosa*, and the locust *S. gregaria* (Sacktor and Wormser-Shavit, 1966). In these insects, there is an initial disappearance of proline during the first minute of flight, with all of them switching rapidly to carbohydrate, and the locust ultimately switching to lipids if flight continues for an hour or so. Thus, perhaps a sort of preadaptation of the necessary enzymes may have been present in the early evolution of insects. A few insects depend on the proline pathway for energy release, but because the pathway releases much less ATP per mole of initial substrate metabolized, it probably was selected against in very strong and longer-distance fliers. Some attempts to relate proline metabolism to the blood feeding behavior of tsetse flies no longer seem tenable in light of the several beetles now known to utilize proline.

Proline metabolism is subject to control by the cellular level of ADP. An increase in ADP stimulates oxidation of proline, while glutamate resulting from oxidation inhibits through a feedback

INTERMEDIARY METABOLISM

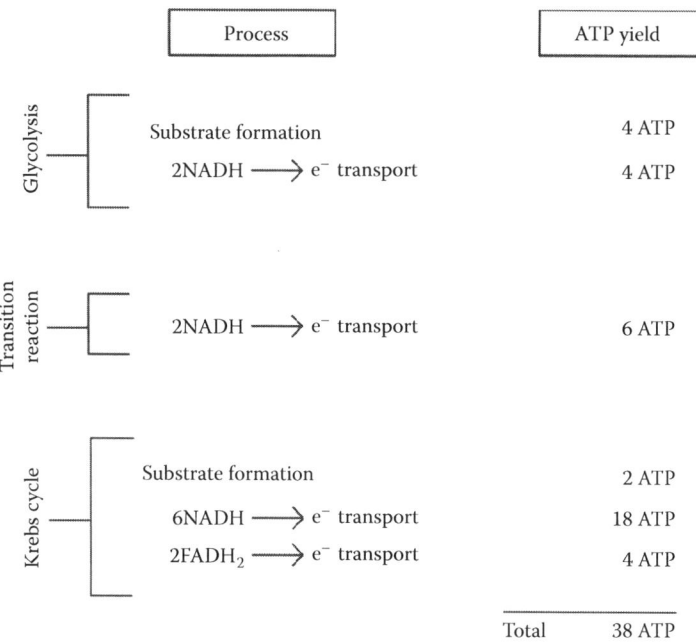

Figure 8.12 A summary of the energy yield in the several stages of glucose metabolism in flight muscle. Electron transport (e⁻ transport) occurs through the electron transport chain.

mechanism. Isolated mitochondria from a blowfly, *P. regina*, are stimulated by the addition of ADP (Hansford and Sacktor, 1970), which allosterically lowers the apparent K_m of proline dehydrogenase for proline from 33 to 6 mM. This is probably significant for the blowfly because the concentration of proline in flight muscle tissue was found to be between 6 and 7 mM (Sacktor and Wormser-Shavit, 1966). Similar effects of ADP upon mitochondrial metabolism of proline were shown for tsetse flies, (another) blowfly, and houseflies, but not for locusts. The stimulatory action of ADP is probably a control point for the oxidation of proline in intact insects, since ADP would be expected to accumulate upon initiation of flight. Glutamate formed from proline oxidation inhibits proline dehydrogenase by negative feedback. ADP can counter the inhibition of glutamate, which aids the tsetse fly in using it as its main flight fuel.

8.5.5 Mobilization and the Use of Lipids for Flight Energy

The fat body is a major storage site of lipids (fats). Triacylglycerols (TAGs), constituting about 90% of the lipids in the fat body of most insects, are stored in lipid droplets surrounded by proteins, some of which are called perilipins, to make a lipoprotein complex (Arrese et al., 2008a,b; Bickel et al., 2009; Kühnlein, 2011). Some perilipins in *Drosophila* seem to have dual roles, some protecting the encased lipids from metabolism and others aiding in mobilizing the lipids (Bi et al., 2012). Oenocytes in the fat body and epidermis function in the accumulation of lipid droplets (Gutierrez et al., 2007), lipid processing, and in detoxification (Martins and Ramalho-Ortigão, 2012). In order for the fat to be used for metabolism by other tissues, the TAG

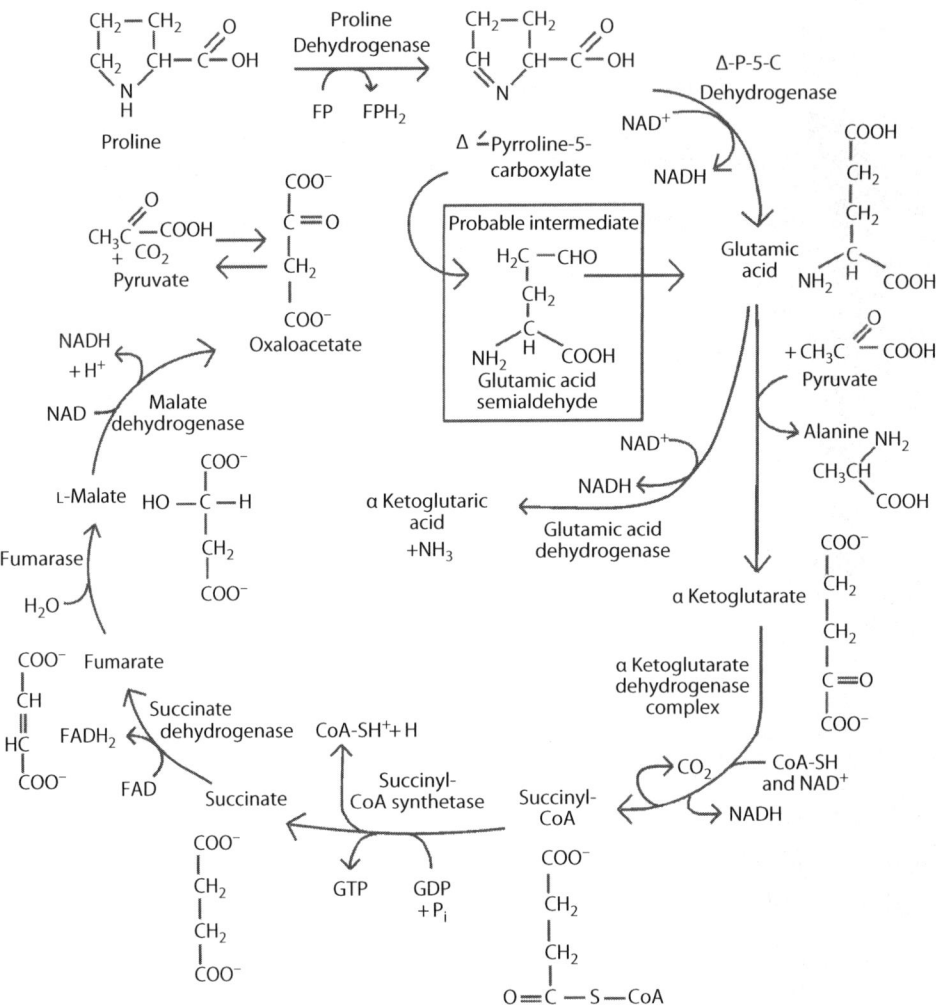

Figure 8.13 The pathway for proline metabolism in mitochondria in support of flight in the tsetse fly and some other insects. Proline enters mitochondria and is metabolized to glutamate and thence to α-ketoglutarate by a transamination reaction. The rest of the pathway is identical with the Krebs cycle. There is little of the enzyme glutamic acid dehydrogenase in tsetse fly mitochondria, so very little of the glutamic acid is converted to α-ketoglutarate in that way.

is converted to a diacylglycerol (DAG) by removing one fatty acid in the fat body. In insects, the DAG is transported by the hemolymph as a lipoprotein complex (Blacklock and Ryan, 1994; Weers and Ryan, 2006; Van der Horst et al., 2009) (Figure 8.14) to other organs, where the two remaining fatty acids are released for metabolism and the glycerol is recycled to the fat body where it again is used as a building block for TAG synthesis. Major uses for fatty acids occur during pupation, diapause (Denlinger, 2005; Hahn and Denlinger, 2007) and, in Lepidoptera and some other insects, during flight (Figure 8.15). Diptera and Hymenoptera, however, do not use fatty acids as a flight fuel; instead they use glucose. The main enzyme acting upon the TAGs in the fat body is active phospholipase *a* (Arrese et al., 2006).

Adult Lepidoptera, some Orthoptera, and some other insects can use fatty acids as a flight fuel, and well-fed ones generally have sufficient lipids in the body to support flight for much longer periods than can be supported by the available supply of carbohydrates. The metabolism

INTERMEDIARY METABOLISM

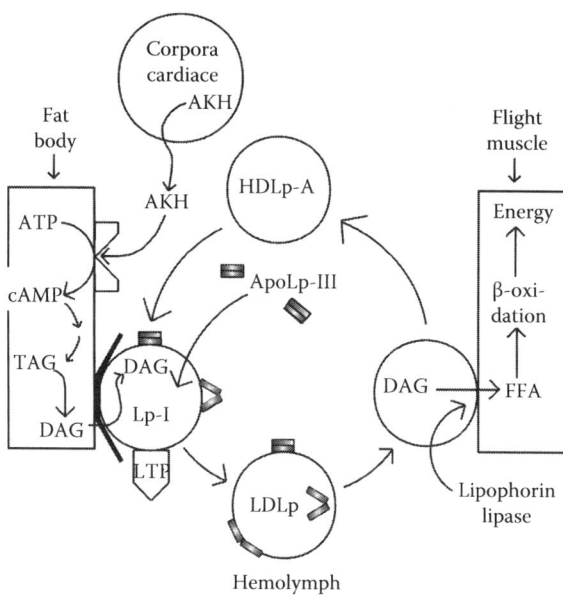

Figure 8.14 A diagram of the role of AKH in promoting DAG release from the fat body, loading of DAG into a lipoprotein transport complex, and delivery of it to muscle cells. At the muscle cell membrane the two fatty acids are removed from DAG by a membrane-bound lipase, and the fatty acids are transported into the muscle cells for metabolism. (Drawing modified from Blacklock, B.J. and Ryan, R.O., *Insect Biochem. Mol. Biol.*, 24, 855, 1994.)

of fatty acids to support flight, however, presents several problems. Nearly all the fatty acids are stored in fat body cells as TAGs and thus before they can be metabolized, they must be released from fat body cells and transported to the muscles. Release of lipid from fat body cells is under control of a peptide hormone, **adipokinetic hormone (AKH)**. AKH is secreted from the corpora cardiac (Beenakkers et al., 1986), circulates through the hemolymph, and binds to its adipokinetic hormone receptor (AKHR) in the fat body (Grönke et al., 2007). AKH action, mediated through participation of cAMP at the fat body cell membrane and subsequent activation of a lipase, causes the release of DAG from fat body cells. Metabolism of fatty acids is a mitochondrial process, and the efficient delivery of oxygen to flight muscles was a preadaptation to evolution of lipid metabolism by flight muscles.

Although AKH is the major hormone involved in lipid mobilization, the hormone has a number of other functions not related to lipid metabolism (Kodrík, 2008; Kodrík et al., 2012; Krishnan and Kodrík, 2012; Bednářová et al., 2013a,b,c; Vinokurov et al., 2014). In its metabolic function for lipids, it promotes rapid release of DAG from fat body cells. The stimulus for the secretion of AKH from the corpora cardiaca is not well defined, but probably nervous control associated with the activation of the thoracic musculature is involved. AKH peptides have been isolated from a number of insects and the amino acid sequence indicates they are all members of the same family of neurohormones, and they generally have cross-reactivity (Nijhout, 1994). The peptides have 8–10 amino acids, depending on the species from which the different AKHs have been isolated. A typical structure, that of Locust AKH-1, is H_2N-Thr-Gly-Trp-Asn-Pro-Thr-Phe-Asn-Leu-PCA.

Pyrrolidone carboxylic acid (PCA), at the carboxy terminal end, is formed from the amino acid glutamine, sometimes also shown as pGLU (cyclized pyroglutamate). Both names and symbols stand for the same chemical structure, and all the AKHs isolated have this modified amino acid at the carboxy terminal end. The amino acid at the amino terminal varies with the species. A specific radioimmunoassay for AKH has been developed (Fox and Reynolds, 1990).

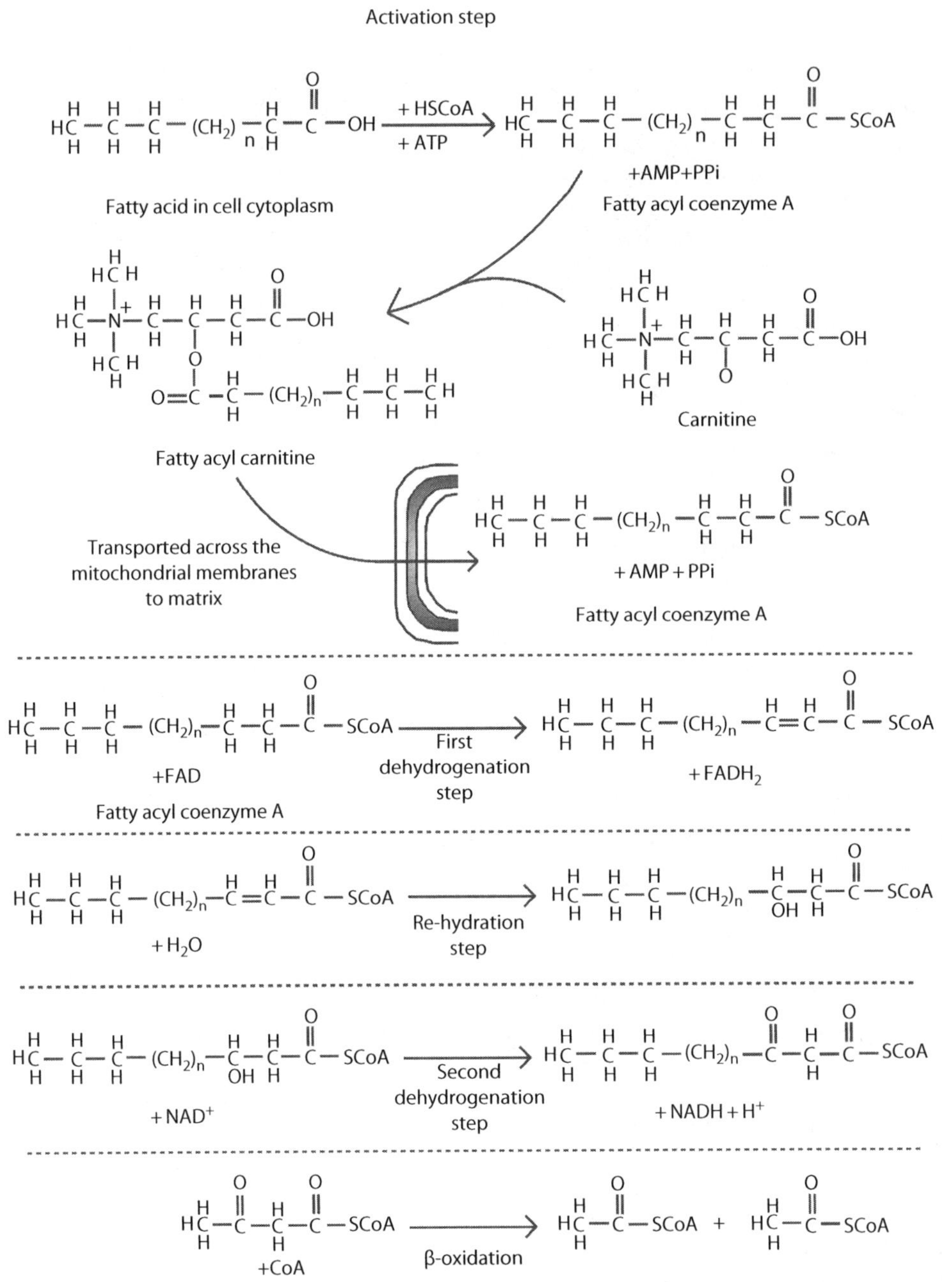

Figure 8.15 A summary of the metabolic reactions leading to β-oxidation of a fatty acid and release of acetyl CoA units within flight muscle mitochondria in insects that utilize lipids for flight energy. Carnitine aids in transferring fatty acids into the mitochondria. Subsequent reactions occur within muscle mitochondria, and the acetyl CoA generated is further metabolized by the Krebs cycle. An even-numbered fatty acid such as C18:3DB (oleic acid with 3 double bonds, which is a common fatty acid in insect lipids) will yield 9 acetyl CoA units, each of which yields 12 ATPs in the Krebs cycle. Palmitic acid (C16) would yield eight acetyl CoA units. Thus, fatty acid metabolism yields a large number of ATPs.

The adipokinetic hormones of *Pyrrhocoris apterus* and *L. migratoria* promote selective mobilization of DAGs in the fat body (Bártů et al., 2010; Tomčala et al., 2010). In both insects, DAGs with C16 and C18:1 (and in *Locusta*, also with C18:2) fatty acids are preferentially mobilized. In *Rhodnius prolixus* lipophorin carrying lipids bound at specific sites in the fat body on different days. A radiolabel in palmitic acid (C16 fatty acid) appeared in 60 min in TAGs in the fat body, indicating rapid movement of the fatty acid and storage in the fat body of *Rhodnius* (Pontes et al., 2008).

8.5.5.1 Transport of Lipids by Lipophorin

Insects transport lipids through the aqueous hemolymph as lipoprotein complexes called **lipophorins**. A variety of lipids have been found in lipophorin particles, including hydrocarbons, phospholipids, and tri- and diacylglycerols. There are three identified proteins associated with lipophorin: apoLp-I, apoLp-II, and apoLp-III. ApoLp-III associates reversibly with DAG-loaded lipophorin and dissociates from lipophorin in the unloaded state. It appears to be necessary to stabilize the loaded lipophorin–DAG particle. ApoLp-I and II remain in the structure of lipophorin in both the loaded and unloaded states.

Although the structure of lipophorins may vary, it appears that, in general, they have a hydrophobic core of hydrocarbons, a middle layer of DAG, and a surface monolayer composed of phospholipids and apolipoproteins. In contrast to the situation in vertebrates, insect lipophorin is not degraded after it delivers a lipid load to muscles (or to other tissues), but circulates in the hemolymph. The unloaded lipophorin may pick up more dietary lipids at the midgut or return to the fat body for a new load of lipids (Chino and Kitazawa, 1981; Chino, 1985; Surholt et al., 1991; Van Heusden et al., 1991; Gondim et al., 1992). Loading of an already existing lipophorin particle, as opposed to having to synthesize a new particle, is one of the adaptations evolved by insects to facilitate rapid mobilization of lipids for flight (Blacklock and Ryan, 1994). Palm et al. (2012) found that the particular lipoproteins associate with lipid transport in *D. melanogaster* varied in tissues-specific ways.

8.5.5.2 Activation of Fatty Acids, Entry into Mitochondria, and β-Oxidation

At the hemolymph–flight muscle cell interface, DAG is unloaded under the influence of AKH (Wheeler, 1989), the apoLp-III protein dissociates from the lipophorin, and the remaining high-density lipoprotein particle (HDLp-A) is shuttled back through the hemolymph to transport DAG again, either from the gut to the fat body or from fat body to the muscles. A membrane-bound lipase in flight muscle tissue has high affinity for DAG, releasing fatty acids and glycerol from DAG bound to lipoprotein A$^+$, the transport lipoprotein complex, very quickly (Wheeler and Goldsworthy, 1985; Van Heusden et al., 1986). Glycerol resulting from the hydrolysis can be phosphorylated and metabolized for energy (as glycerol-3-phosphate) or returned to the fat body where it can be used again to form TAGs. The free fatty acids are bound to an intracellular protein in the locust, *S. gregaria* (Haunerland and Chisholm, 1990), which helps to move them through the aqueous medium of the cell cytoplasm to the site of metabolism, the mitochondria. The fatty acids must be activated to cross mitochondrial membranes, and the reactions can be broken down into the following seven steps:

1. *Activation of fatty acids in the cytoplasm of muscle fibers and entry into mitochondria*: The sequence of reactions leading to β-oxidation within mitochondria is shown in Figure 8.15. Fatty acids in the cytoplasm must be complexed with the vitamin **carnitine** in order to cross mitochondrial membranes. In the cytoplasm, the fatty acid is activated by reaction with coenzyme A (CoA) and ATP to form a fatty acyl CoA derivative. The reaction costs the equivalent of two ATPs per molecule of fatty acid activated because the reaction results in AMP and PPi as by-products rather than ADP. Two ATPs will be needed to phosphorylate AMP in the process of replenishing the supply of ATP for additional reactions.
2. *Carnitine* is complexed in the presence of carnitine acyl tranferase to the activated fatty acyl CoA, with the release of CoA. CoA can now participate in the activation of another fatty acid in the cytoplasm, while the fatty acyl carnitine derivative passes through the outer and inner mitochondrial membranes.

Inside mitochondria at the inner surface of the inner membrane, carnitine is removed and acetyl CoA reacts with fatty acid to reactivate it. Carnitine returns to the cytoplasm and is available to assist in the entry of another molecule of activated fatty acid. The fatty acyl CoA molecule, now in the matrix of the mitochondrion, undergoes further reactions leading to β-**oxidation** and removal of acetyl CoA units that readily enter the oxidative reactions of the Krebs cycle. The following reactions are necessary.

3. *Dehydrogenation with FAD as the cofactor—the first matrix reaction*: The initial matrix reaction is a **dehydrogenation** (an oxidation) in which a double bond is introduced at the β-carbon. FAD is the electron and proton acceptor. The reduced $FADH_2$ transfers its electrons through the electron transport system, which will result in the production of two ATPs/two electrons transferred.
4. *Addition of water to the activated fatty acid molecule*: Water is added across the double bond introduced in Step 3, thereby reducing the fatty acid, but no significant energy input or release occurs.
5. *Dehydrogenation to yield the β-keto form of the molecule*: The activated fatty acid is oxidized again, but this time the two electrons and two protons are removed from the same carbon, carbon-3, leaving a β-keto group at this position (Figure 8.15). NAD^+ participates in this reaction and the resulting NADH passes its two electrons through the electron transport chain with the production of three ATPs.
6. *β-oxidation*: Acetyl CoA is cleaved from the fatty acid chain, with a second CoA participating in the reaction so that the fatty acid (now minus two carbons) is left in an activated form. This new, shorter fatty acyl CoA molecule repeats the reactions in Steps 1–4. The shortened fatty acyl CoA molecule continues to repeat these steps, becoming shorter by two carbons each time as another acetyl CoA is removed from it, until in its final passage through the steps, when only four carbons of the original molecule remain, it will yield two molecules of acetyl CoA.
7. *The energy derived from fatty acid metabolism*: The metabolism of fatty acids yields many more ATPs per mol of fatty acid metabolized than can be derived from metabolism of glucose. Each $FADH_2$ from the oxidation in Step 3 yields two molecules of ATP as the electrons pass down the electron transport chain and, similarly, each NADH from Step 5 yields three ATPs. Each acetyl CoA released from a fatty acid molecule results in 12 ATPs from oxidation through the Krebs cycle. Since a fatty acid must pass successively through Steps 3–6 each time a 2-carbon unit is released by β-oxidation, five ATPs will be produced with each pass. A fatty acid will have to make $(n/2 - 1)$ passes (n = number of carbons in the original fatty acid). The scheme is based on the metabolism of fatty acids with an even number of carbons, but generally these are the ones found in both plant and animal fats. Moreover, each 2-carbon unit passed through the Krebs cycle will produce 12 ATPs. Thus, the total ATPs produced for palmitic acid with 16 carbons, as an example, when it has been completely metabolized to carbon dioxide and water can be calculated from formation, and subsequent passage through the Krebs cycle/and or the electron transport system, of $8CH_3COSCoA + 7FADH_2 + 7NADH$.

When the $7FADH_2$s and 7NADHs transfer their electrons through the electron transport system, 35 ATPs will be produced. Each acetyl CoA metabolized through the Krebs cycle will yield 12 ATPs, for a total of 96 ATPs from 8 acetyl CoA molecules.

The total ATPs derived from complete metabolism of 1 molecule of palmitic acid will then be 35 ATPs + 96 ATPs = 131 ATPs. Net production of ATP will be 129 ATPs because 2 ATPs were needed in Step 1 to activate the fatty acid in the cytoplasm.

ATP produced in these reactions will be used by a flying moth, locust, or beetle to work the flight muscles, to supply energy to the nervous system, and for all the other physiological processes of the body that must go on even in flight. The metabolic water produced as a result of metabolism of just one fatty acid (108 molecules H_2O per molecule of palmitic acid) is indicative of the large amounts of metabolic water that can be produced when an insect is able to metabolize lipid for flight. Metabolic water is a valuable resource to insects. Pupae of many insects contain large amounts of lipids accumulated during larval life, and these large stores of lipids can be metabolized slowly during pupal transformation to the adult to provide energy for new syntheses and cellular changes and also supply water to the closed system. Only some insects, however, can mobilize, transport, and metabolize fatty acids rapidly enough to support flight.

8.6 REVIEW AND SELF-STUDY QUESTIONS

1. Name any five specific activities in insects that require expenditure of energy.
2. What does the term "metabolic control" mean?
3. What substrates can different insects use to support flight?
4. Where are the largest quantities of trehalose and glycogen stored in insects?
5. What hormone influences the release of glucose (as glucose-1-phosphate) from glycogen?
6. What enzyme is required by insects to convert trehalose into two molecules of glucose?
7. Give names in words for these acronyms: ADP, ATP, NAD$^+$, NADH, FAD, FADH2, G-3-P, TAG, DAG.
8. How is cytoplasmic NAD$^+$ regenerated in insect flight muscle?
9. How is mitochondrial NAD$^+$ regenerated in insects? Is there a major difference from how it is regenerated in vertebrate mitochondria?
10. Do typical cells (of any tissue) have too much ATP present at any given time? Explain your answer with a brief explanation why or why not.
11. Do typical cells (of any tissue) have too much NAD$^+$ or FAD present at any given time? Explain.
12. If an insect could only carry out anaerobic metabolism in the glycolytic cycle (glucose to pyruvate), could it be a very successful long-distance migratory insect? Explain the reasoning behind your answer.
13. If an insect could not deliver a rich oxygen supply to the mitochondria, could it metabolize fatty acids or praline during flight? Explain or justify your answer with a physiological or biochemical statement.
14. Is there very much direct formation of ATP molecules in either glycolysis or the Krebs Cycle? If not, in what form does the energy from the original glucose (or fatty acid, proline, or other amino acids) exist within the Krebs Cycle?
15. How much energy (as ATP) can be released from metabolism of myristic acid (a fatty acid with 14 carbons in its structure) by a flying moth?
16. What is the function of the glycerol-3-phosphate shuttle in insects?
17. What is the function of the oxygen that an insect breathes?

REFERENCES

Arrese, E.L., S. Mirza, L. Rivera, A.D. Howard, P.S. Chetty, and J.L. Soulages. 2008a. Expression of lipid storage droplet protein-1 may define the role of AKH as a lipid mobilizing hormone in *Manduca sexta*. *Insect Biochem. Mol. Biol.* 38: 993–1000.

Arrese, E.L., R.T. Patel, and J.L. Soulages. 2006. The main triglyceride-lipase from the insect fat body is an active phospholipase A$_1$: Identification and characterization. *J. Lipid Res.* 47: 2656–2667.

Arrese, E.L., L. Rivera, M. Masakazu, S. Mirza, D.D. Hartson, S. Weintraub, and J.L. Soulages. 2008b. Function and structure of lipid storage droplet protein 1 studied in lipoprotein complexes. *Arch. Biochem. Biophys.* 473: 42–47.

Arrese, E.L. and J.L. Soulages. 2010. Insect fat body: Energy, metabolism, and regulation. *Annu. Rev. Entomol.* 55: 207–225.

Bailey, E. 1975. Utilization of fuels by muscle, in D.J. Candy and B.A. Kilby (eds.), *Insect Biochemistry and Function*. Chapman & Hall, London, U.K., pp. 3–87.

Bártů, I., A. Tomčala, R. Socha, P. Šimek, and D. Kodrík. 2010. Analysis of the lipids mobilized by adipokinetic hormones in the firebug *Pyrrhocoris apterus* (Heteroptera: Pyrrhocoridae). *Eur. J. Entomol.* 107: 509–520.

Bednářová, A., D. Kodrík, and N. Krishnan. 2013a. Adipokinetic hormone exerts its anti-oxidative effects using a conserved signal-transduction mechanism involving both PKC and cAMP by mobilizing extra- and intracellular Ca^{2+} stores. *Comp. Biochem. Physiol. Part C* 158: 142–149.

Bednářová, A., D. Kodrík, and N. Krishnan. 2013b. Unique roles of glucagon and glucagon-like peptides: Parallels in understanding the functions of adipokinetic hormones in stress responses in insects. *Comp. Biochem. Physiol. Part A* 164: 91–100.

Bednářová, A., N. Krishnan, I.-C. Cheng, J. Večeřa, H.-J. Lee, and D. Kodrík. 2013c. Adipokinetic hormone counteracts oxidative stress elicited in insects by hydrogen peroxide: In vivo and in vitro study. *Physiol. Entomol.* 38: 54–62.

Beenakkers, A.M.Th., D.J. Van der Horst, and W.J.A. van Marrewijk. 1986. Insect lipids and lipoproteins, and their role in physiological processes. *Prog. Lipid Res.* 24: 19–67.

Bewley, G.C., J.M. Rawls, Jr., and J.C. Lucchesi. 1974. α-Glycerophosphate dehydrogenase in *Drosophila malanogaster*: Kinetic differences and developmental differentiation of the larval and adult isozymes. *J. Insect Physiol.* 20: 153–165.

Bi, J., Y. Xiang, H. Chen, Z. Liu, S. Grönke, R.P. Kühnlein, and X. Huang. 2012. Opposite and redundant roles of two *Drosophila* perilipins in lipid metabolism. *J. Cell Sci.* 125: 3568–3577.

Bickel, P.E., J.T. Tansey, and M.A. Welte. 2009. PAT proteins, an ancient family of lipid droplet proteins that regulate cellular lipid stores. *Biochim. Biophys. Acta* 1791: 419–440.

Blacklock, B.J. and R.O. Ryan. 1994. Hemolymph lipid transport. *Insect Biochem. Mol. Biol.* 24: 855–873.

Bursell, E. 1963. Aspects of the metabolism of amino acids in the tsetse fly, *Glossina* (Diptera). *J. Insect Physiol.* 9: 439–452.

Bursell, E. 1965. Oxaloacetic carboxylase in flight musculature of the tsetse fly. *Comp. Biochem. Physiol.* 16: 259–266.

Bursell, E. 1966. Aspects of the flight metabolism of tsetse flies (*Glossina*). *Comp. Biochem. Physiol.* 19: 809–818.

Bursell, E. 1967. The conversion of glutamate to alanine in the tsetse fly (*Glossina morsitans*). *Comp. Biochem. Physiol.* 23: 825–829.

Bursell, E. 1981. The role of proline in energy metabolism, in R.G.H. Downer (ed.), *Energy Metabolism in Insects.* Plenum, New York, pp. 135–154.

Candy, D.J. 1985. Intermediary metabolism, in G.A. Kerkut and L.I. Gilbert (eds.), *Comprehensive Insect Physiology, Biochemistry and Pharmacology*, vol. 10. Pergamon Press, Oxford, U.K., pp. 1–41

Candy, D.J. 1989. Utilization of fuels by the flight muscles, in G.J. Goldsworthy and C.H. Wheeler (eds.), *Insect Flight*. CRC Press, Boca Raton, FL, pp. 305–319.

Carafoli, E. and B. Sacktor. 1972. The effects of ruthenium red on reactions of blowfly flight muscle mitochondrial with calcium. *Biochem. Biophys. Res. Commun.* 49: 1498–1503.

Chaplain, R.A. 1967. The effect of Ca^{2+} and fibre elongation on the activation of the contractile mechanism of insect fibrillar flight muscle. *Biochim. Biophys. Acta* 131: 385–392.

Chaves, V.E., D. Frasson, and N.H. Kawashita. 2011. Several agents and pathways regulate lipolysis in adipocytes. *Biochimie* 93: 1631–1640.

Childress, C.C., B. Sacktor, I.W. Grossman, and E. Bueding. 1970. Isolation, ultrastructure, and biochemical characterization of glycogen in insect flight muscle. *J. Cell. Biol.* 45: 83–90.

Chino, H. 1985. Lipid transport: Biochemistry of hemolymph lipophorin, in G.A. Kerkut and L.I. Gilbert (eds.), *Comprehensive Insect Physiology, Biochemistry and Pharmacology*, vol. 10. Pergamon Press, Oxford, U.K., pp. 115–135.

Chino, H. and K. Kitazawa. 1981. Diacylglycerol-carrying lipoprotein of hemolymph of the locust and some insects. *J. Lipid Res.* 22: 1042–1052.

Collier, G.E., D.T. Sullivan, and R.J. MacIntyre. 1976. Purification of α-glycerophosphate dehydrogenase from *Drosophila melanogaster. Biochim. Biophys. Acta* 429: 316–323.

Costa-Leonardo, A.M., L.T. Laranjo, V. Janei, and I. Haifig. 2013. The fat body of termites: Functions and stored materials. *J. Insect Physiol.* 59: 577–587.

Davis, R.A. and G. Fraenkel. 1940. The oxygen consumption of flies during flight. *J. Exp. Biol.* 17: 402–407.

DeKort, C.A.D., A.K.M. Bartelink, and R.R. Schuurmans. 1973. The significance of L-proline for oxidative metabolism in the flight muscles of the Colorado potato beetle, *Leptinotarsa decemlineata*. *Insect Biochem.* 3: 11–17.

Denlinger, D.L. 2005. Diapause in the mosquito *Culex pipiens* evokes a metabolic switch from blood feeding to sugar gluttony. *Proc. Natl. Acad. Sci. USA* 102: 15912–15917.

Donnellan, J.F. and R.B. Beechey. 1969. Factors affecting the oxidation of glycerol-1-phosphate by insect flight-muscle mitochondria. *J. Insect Physiol.* 15: 367–372.

Downer, R.G.H. 1985. Lipid metabolism, in G.A. Kerkut and L.I. Gilbert (eds.), *Comprehensive Insect Physiology, Biochemistry and Pharmacology*, vol. 10. Pergamon Press, Oxford, U.K., pp. 77–113.

Estabrook, R.W. and B. Sacktor. 1958. α-Glycerophosphate oxidase of flight muscle mitochondria. *J. Biol. Chem.* 233: 1014–1019.
Fox, A.M. and S.E. Reynolds. 1990. Quantification of *Manduca* adipokinetic hormone in nervous and endocrine tissue by a specific radioimmunoassay. *J. Insect Physiol.* 36: 683–689.
Friedman, S. 1985. Carbohydrate metabolism, in G.A. Kerkut and L.I. Gilbert (eds.), *Comprehensive Insect Physiology, Biochemistry and Pharmacology*, vol. 10. Pergamon Press, Oxford, U.K., pp. 43–76.
Gäde, G. 1990. The adipokinetic hormone/red pigment-concentrating hormone peptide family: Structures, interrelationships and functions. *J. Insect Physiol.* 36: 1–12.
Gäde, G. and L. Auerswald. 2000. Flight substrates and their regulation by a member of the AKH/RPCH family of neuropeptides in Cerambycidae. *J. Insect Physiol.* 46: 1575–1584.
Goebell, H. and M. Klingenberg. 1964. DPN-spezifische Isocitrat-dehydrogenase der Mitochondrien-I. Kinetische Eigenschaften, Vorkommen und Function der DPN-spezifischen Isocitrat-dehydrogenase. *Biochem. Z.* 340: 441–464.
Gondim, K.C., G.C. Atella, J.H. Kawooya, and H. Masuda. 1992. Role of phospholipids in the lipophorin particles of *Rhodnius prolixus*. *Arch. Insect Biochem. Physiol.* 20: 303–314.
Grönke, S., C. Müller, J. Hirsch, S. Fellert, A. Andreou, T. Haase, H. Jäckle, and R.P. Kühnlein. 2007. Dual lipolytic control of body fat storage and mobilization in *Drosophila*. *PLoS Biol.* 5: e137.
Gutierrez, E., D. Wiggins, B. Fielding, and A.P. Gould. 2007. Specialized hepatocyte-like cells regulate *Drosophila* lipid metabolism. *Nature* 445: 275–280.
Hahn, D.A. and D.L. Denlinger. 2007. Meeting the energetic demands of insect diapause: Nutrient storage and utilization. *J. Insect Physiol.* 53: 760–773.
Hainsworth, F.R. 1981. Energy regulation in hummingbirds. *Am. Sci.* 69: 420–429.
Hanaoka, K. and S.Y. Takahashi. 1977. Adenylate cyclase system and the hyperglycaemic factor in the cockroach, *Periplaneta americana*. *Insect Biochem.* 7: 95–99.
Hansford, R.G. 1972. Some properties of pyruvate and 2-oxoglutarate oxidation by blowfly flight-muscle mitochondria. *Biochem. J.* 127: 271–283.
Hansford, R.G. and J.B. Chappel. 1967. The effect of Ca^{2+} on the oxidation of glycerol phosphate by blowfly flight-muscle mitochondria. *Biochem. Biophys. Res. Commun.* 27: 686–692.
Hansford, R.G. and B. Sacktor. 1970. The control of the oxidation of proline by isolated flight muscle mitochondria. *J. Biol. Chem.* 245: 991–994.
Haunerland, N. and J.M. Chisholm. 1990. Fatty acid binding protein in flight muscle of the locust *Schistocerca gregaria*. *Biochim. Biophys. Acta* 1047: 233–238.
Haunerland, N.H. and P.D. Shirk. 1995. Regional and functional differentiation in the insect fat body. *Annu. Rev. Entomol.* 40: 121–145.
Kaufmann, C. and M.R. Brown. 2008. Regulation of carbohydrate metabolism and flight performance by a hypertrehalosaemic hormone in the mosquito *Anopheles gambiae*. *J. Insect Physiol.* 54: 367–377.
Khan, M.A. and C.A.D. DeKort. 1978. Further evidence for the significance of L-proline for flight in the Colorado potato beetle, *Leptinotarsa decemlineata*. *Comp. Biochem. Physiol. B* 60: 407–411.
Kodrík, D. 2008. Adipokinetic hormone functions that are not associated with insect flight. *Physiol. Entomol.* 33: 171–180.
Kodrík, D., K. Vinokurov, A. Tomčala, and R. Socha. 2012. The effect of adipokinetic hormone on midgut characteristics in *Pyrrhocoris apterus* L. (Heteroptera). *J. Insect Physiol.* 58: 194–204.
Krishnan, N. and D. Kodrík. 2012. Endocrine control of oxidative stress in insects, in A.A. Farooqui, and T. Farooqui (eds.), *Oxidative Stress in Vertebrates and Invertebrates: Molecular Aspects of Cell Signalling*. Wiley-Blackwell, NJ, pp. 261–270.
Kühnlein, R.P. 2011. The contribution of the *Drosophila* model to lipid droplet research. *Prog. Lipid Res.* 50: 348–356.
Lehninger, A.L. 1975. *Biochemistry*, 2nd ed. Worth Publishers, New York.
Li, Z., J. Jiang, Y. Chen, L. You, Y. Huang, and A. Tan. 2014. *PDP1* regulates metabolism through the IIS-TOR pathway in the red flour beetle, *Tribolium castaneum*. *Arch. Insect Biochem. Physiol.* 85: 127–136.
Makki, R., E. Cinnamon, and A.P. Gould. 2014. The development and functions of oenocytes. *Annu. Rev. Entomol.* 59: 405–425.
Martins, G.F. and J.M. Ramalho-Ortigão. 2012. Oenocytes in insects. *Invertebr. Surviv. J.* 9: 139–152.

Mitchell, P. 1979. Keilin's respiratory chain concept and its chemiosmotic consequences. *Science* 206: 1148–1159.

Mordue, W. and C.A.D. DeKort. 1978. Energy substrates for flight in the Colorado potato beetle, *Leptinotarsa decemlineata. J. Insect Physiol.* 24: 221–224.

Nichol, H., J.H. Law, and J.J. Winzerling. 2002. Iron metabolism in insects. *Annu. Rev. Entomol.* 47: 535–559.

Nijhout, H.F. 1994. *Insect Hormones*. Princeton University Press, Princeton, NJ.

Palm, W., J.L. Sampaio, M. Brankatschk, M. Carvalho, A. Mahmoud, A. Shevchenko, and S. Eaton. 2012. Lipoproteins in *Drosophila melanogaster*—Assembly, function, and influence on tissue lipid composition. *PLoS Genet.* 8: e1002828.

Pearson, D.J., M.O. Imbuga, and J.B. Hoek. 1979. Enzyme activities in flight and leg muscles of the dung beetle in relation to proline metabolism. *Insect Biochem.* 9: 461–466.

Pearson, O.P. 1950. The metabolism of hummingbirds. *Condor* 52: 145–152.

Pontes, E.G., P. Leite, D. Majerowicz, G.C. Atella, and K.C. Gondim. 2008. Dynamics of lipid accumulation by the fat body of *Rhodnius prolixus*: The involvement of lipophorin binding sites. *J. Insect Physiol.* 54: 790–797.

Rank, N.E., D.A. Bruce, D.M. McMillan, C. Barclay, and E.P. Dahlhoff. 2007. Phosphoglucose isomerase genotype affects running speed and heat shock protein expression after exposure to extreme temperatures in a montane willow beetle. *J. Exp. Biol.* 210: 750–764.

Sacktor, B. 1961. The role of mitochondria in respiratory metabolism of flight muscle. *Annu. Rev. Entomol.* 6: 103–130.

Sacktor, B. 1974. Biological oxidations and energetics in insect mitochondria, in M. Rockstein (ed.), *Physiology of Insecta*, 2nd ed., vol. 4. Academic Press, New York, pp. 271–353.

Sacktor, B. and A. Dick. 1962. Pathways of hydrogen transport in the oxidation of extramitochondrial reduced diphosphopyridine nucleotide in flight muscle. *J. Biol. Chem.* 237: 3259–3263.

Sacktor, B. and E.C. Hurlbut. 1966. Regulation of metabolism in working muscle in vivo. II. Concentrations of adenine nucleotides, arginine phosphate, and inorganic phosphate in insect flight muscle during flight. *J. Biol. Chem.* 241: 632–634.

Sacktor, B. and E. Wormser-Shavit. 1966. Regulation of metabolism in working muscle in vivo. I. Concentrations of some glycolytic, tricarboxylic acid cycle, and amino acid intermediates in insect flight muscle during flight. *J. Biol. Chem.* 241: 624–631.

Steele, J.E. 1961. Occurrence of a hyperglycaemic factor in the corpus cardiacum of an insect. *Nature* 192: 680–681.

Steele, J.E. 1980. Hormonal modulation of carbohydrate and lipid metabolism in the fat body, in M. Locke and D.S. Smith (eds.), *Insect Biology in the Future*. Academic Press, New York, pp. 253–271.

Steele, J.E. 1985. Control of metabolic processes, in G.A. Kerkut and L.I. Gilbert (eds.), *Comprehensive Insect Physiology, Biochemistry and Pharmacology*. Pergamon Press, New York, pp. 99–145.

Suarez, R.K., J.F. Staples, J.R.B. Lighton, and O. Mathieu-Costello. 2000. Mitochondrial function in flying honeybees (*Apis mellifera*): Respiratory chain enzymes and electron flow from complex III to oxygen. *J. Exp. Biol.* 203: 905–911.

Surholt, B., J. Goldberg, T.K.F. Schulz, A.M.Th. Beenakkers, and D.J. Van der Horst. 1991. Lipoproteins act as a reusable shuttle for lipid transport in the flying death's-head hawkmoth *Acherontia atropos. Biochem. Biophys. Acta* 1086: 15–21.

Thompson, S.N. 2003. Trehalose: The insect 'blood' sugar. *Adv. Insect Physiol.* 31: 205–285.

Tomčala, A., I. Bártů, P. Šimek, and D. Kodrík. 2010. Locust adipokinetic hormones mobilize diacylglycerols selectively. *Comp. Biochem. Physiol., Part B* 156: 26–32.

Tufail, M. and M. Takeda. 2009. Insect vitellogenin/lipophorin receptors: Molecular structures, role in oogenesis, and regulatory mechanisms. *J. Insect Physiol.* 55: 87–103.

Van den Bergh, S.G. and E.C. Slater. 1962. The respiratory activity and permeability of housefly sarcosomes. *Biochem. J.* 82: 362–371.

Van der Horst, D.J., S.D. Roosendaal, and K.W. Rodenburg. 2009. Circulatory lipid transport: Lipoprotein assembly and function from an evolutionary perspective. *Mol. Cell. Biochem.* 326: 105–119.

Van Heusden, M.C., D.J. Van der Horst, J.K. Kawooya, and J.H. Law. 1991. In vivo and in vitro loading of lipid by artificially lipid-depleted lipophorins: Evidence for the role of lipophorin as a reusable lipid shuttle. *J. Lipid Res.* 32: 1789–1794.

Van Heusden, M.C., D.J. Van der Horst, J.M. Van Doorn, J. Wes, and A.M.Th. Beenakkers. 1986. Lipoprotein lipase activity in the flight muscle of *Locusta migratoria* and its specificity for haemolymph lipoproteins. *Insect Biochem.* 16: 517–523.

Vinokurov, K., A. Bednářová, A. Tomčala, T. Stašková, N. Krishnan, and D. Kodrík. 2014. Role of adipokinetic hormone in stimulation of salivary gland activities: The fire bug *Pyrrhocoris apterus* L. (Heteroptera) as a model species. *J. Insect Physiol.* 60: 58–67.

Walker, P.R. and E. Bailey. 1969. A comparison of the properties of the phosphofructokinase of the fat body and flight muscle of the adult male desert locust. *Biochem. J.* 111: 365–369.

Weeda, E., C.A.D. DeKort, and A.M.Th. Beenakkers. 1980. Oxidation of proline and pyruvate by flight muscle mitochondria of the Colorado potato beetle, *Leptinotarsa decemlineata*. *Insect Biochem.* 10: 305–311.

Weers, P.M.M. and R.O. Ryan. 2006. Apoliophorin III: Role model apolipoprotein. *Insect Biochem.Mol. Biol.* 36: 231–240.

Weis-Fogh, T. 1952. Fat combustion and metabolic rate of flying locusts (*Schistocerca gregaria* ForskDl). *Phil. Trans. Roy. Soc. B* 237: 1–36.

Wheeler, C.H. 1989. Mobilization and transport of fuels to the flight muscles, in G.J. Goldsworthy and C.H. Wheeler (eds.), *Insect Flight*. CRC Press, Boca Raton, FL, pp. 273–303.

Wheeler, C.H. and G.J. Goldsworthy. 1985. Specificity and localisation of lipoprotein lipase in the flight muscles of *Locusta migratoria*. *Biol. Chem. Hoppe-Seyler*. 366: 1071–1077.

Zahavi, M. and A.S. Tahori. 1972. Activity of mitochondrial NAD-linked isocitric dehydrogenase in alatiform and apteriform larvae of *Myzus persicae*. *J. Insect Physiol.* 18: 609–614.

Zebe, E. 1954. Über den Stoffwechsel der Lepidopteren. *Z. Vergl. Physiol.* 36: 290–317.

Zeigler, R., K. Eckart, and J.H. Law. 1990. Adipokinetic hormone controls lipid metabolism in adults and carbohydrate metabolism in larvae of *Manduca sexta*. *Peptides* 11: 1037–1040.

CHAPTER 9

Neuroanatomy

PREVIEW

The central nervous system (CNS) comprises the brain, ventral nerve cord, and ventral ganglia. The brain consists of fused ganglia that make up the protocerebrum, deutocerebrum, and tritocerebrum. The protocerebrum is a major integrative center and receives sensory input from the compound eyes. The deutocerebrum receives sensory input from the antennae and sends motor output to the antennae. The tritocerebrum sends motoneurons to muscles in the labrum and pharynx and innervates the stomatogastric (foregut) nervous system controlling foregut muscles. In some insects, sensory axons from sensory receptors on the head terminate in the tritocerebrum, and in some insects, the tritocerebrum receives sensory input from receptors on the mouthparts. The subesophageal ganglion has sensory and motor connections to sensory structures and muscles of the mouthparts, salivary glands, neck receptors (in some insects), and neck muscles. Axons from neurons in the subesophageal ganglion project forward to the brain and posterior to the thoracic ganglia. The subesophageal ganglion has influence over motor patterns involved with walking, flying, and breathing, although those motor patterns originate in other ganglia. Typically, there are three thoracic ganglia, the pro-, meso-, and metathoracic ganglion, located, respectively, in the pro-, meso-, and metathoracic segments. Each thoracic ganglion sends motor axons to the leg muscles of its segment and receives sensory axons from sensory receptors in the tarsi and leg joints. The meso- and metathoracic ganglia send motor nerves to the wing muscles. Although the primitive evolutionary condition seems to have been that each abdominal ganglion innervated and received sensory information from structures in its segment, in all existing insects, fusion of some abdominal ganglia has occurred. Some Apterygota have eight abdominal ganglia, some Odonata larvae have seven, and Orthoptera have five or six. In some highly evolved dipterans and hemipterans, all abdominal ganglia have fused with thoracic ganglia. Nerves radiate from fused ganglia to organs and muscles representing the evolutionary segmental origin of the ganglia that have fused. The central region of all ganglia is an area of synaptic connections called the neuropil, and cell bodies of motor neurons and interneurons tend to be located peripherally.

The cell bodies of sensory neurons are located in peripheral parts of the body near the sensory site, for example, in the antennae or tarsi, and many other internal and cuticular sites. The brain, ventral connectives and ganglia, and large lateral nerves are protected from direct contact with the hemolymph by a selectively permeable barrier, the hemolymph–brain barrier, consisting of an outer acellular layer called the neurolemma and an inner cellular layer, the perineurium. Nerve cells have a high demand for oxygen and nutrients. The CNS receives a rich supply of tracheae delivering oxygen, and hemolymph bathes the nerves and ganglia in the open circulatory system. Neurons are classified as sensory, or afferent, if they deliver signals to the CNS, and motor, or efferent, if they deliver output to muscles, glands, and organs. Interneurons mediate between sensory and motor neurons and, generally, are located within the CNS. Neurosecretory cells (NSC) are present in all ganglia and play a major role in producing neuropeptides that regulate a variety of physiological and behavioral functions. Motor programs originate in various ganglia and control repetitive muscular actions, such as tracheal ventilation, walking, and ecdysis.

9.1 INTRODUCTION

So little was known about the anatomy and function of nervous systems just a little more than a century ago that the prevailing idea then was that the nervous system was composed of anastomosing cells, and individual cells were thought not to exist. The very small size of neurons, their long extended processes, and histological stains and procedures that did not clearly differentiate individual neurons led some early anatomists, including Camillo Golgi, Carl Weigert, and Franz Nissl (all of whom, nevertheless, made major contributions to neuroanatomy), to make or support this erroneous conclusion. Despite the difficulty of studying the detailed internal anatomy of the nervous system, studies of gross anatomy continued to reveal fine details. With the development of better stains, a great deal of progress in defining the anatomy of nervous systems occurred in the last half of the nineteenth century. Golgi developed the staining procedure in 1873 that bears his name and is still used today. It allowed him to see evidence of individual neurons within a ganglion. The Spanish anatomist, Ramon y Cajal, used Golgi staining with improvements he devised. He published (in 1888) his "neuron doctrine," in which he declared that nervous systems of animals were composed of individual cells just as all other organs were. Cajal and Golgi later shared the Nobel Prize for their pioneering work on neuroanatomy. Each advance stimulated another, and many anatomical details of the nervous system in animals, including insects, were published in the late nineteenth and early twentieth centuries. Most of the major details concerning insect nervous system anatomy were known by the early part of the twentieth century. Indeed, as early as 1762, Pierre Lyonet had published a highly detailed and accurate drawing of the entire nervous system of a caterpillar, *Cossus* (cited in Strausfeld, 1976).

9.2 CENTRAL NERVOUS SYSTEM

The central nervous system (CNS) of insects consists of the **brain**, the **ventral ganglia**, and the ventral **nerve cord** (Figures 9.1 and 9.2). In the earliest insects to evolve, a ganglion probably occurred in each segment and nerves from it controlled the muscles and glands in that segment. Aristotle believed that the brain in insects resided somewhere between the head and the tail (cited in Strausfeld, 1976, p. 3); he may have been uncertain because a great deal of autonomy for control resides in each ganglion. The brain is in the head in most insects (but it is located in several body segments posterior to the anterior end of the body in dipterous larvae). Some insects experimentally rendered headless, or merely brainless, can live for long periods and may even walk, mate, and lay eggs. Removal of the brain, entire head, or other parts of the nervous system, while keeping the insect alive, has been a useful approach in the study of hormones produced by the nervous system.

Fusion of ganglia has occurred in all insects, sometimes to the extent that there are no ganglia in the abdominal segments. The ventral nerve connectives between ganglia still show pairing indicative of bilateral symmetry of the system, but ganglia from the two sides of a segment have fused along the midline in all extant insects.

The ventral ganglia and connecting nerve cord usually lie close to the cuticle on the ventral side of the body. Typically, but not invariably, nerves from a ganglion innervate muscles and organs within the segment where the ganglion resides. Nerves from fused ganglia project to the various body segments and structures in the posterior of the body.

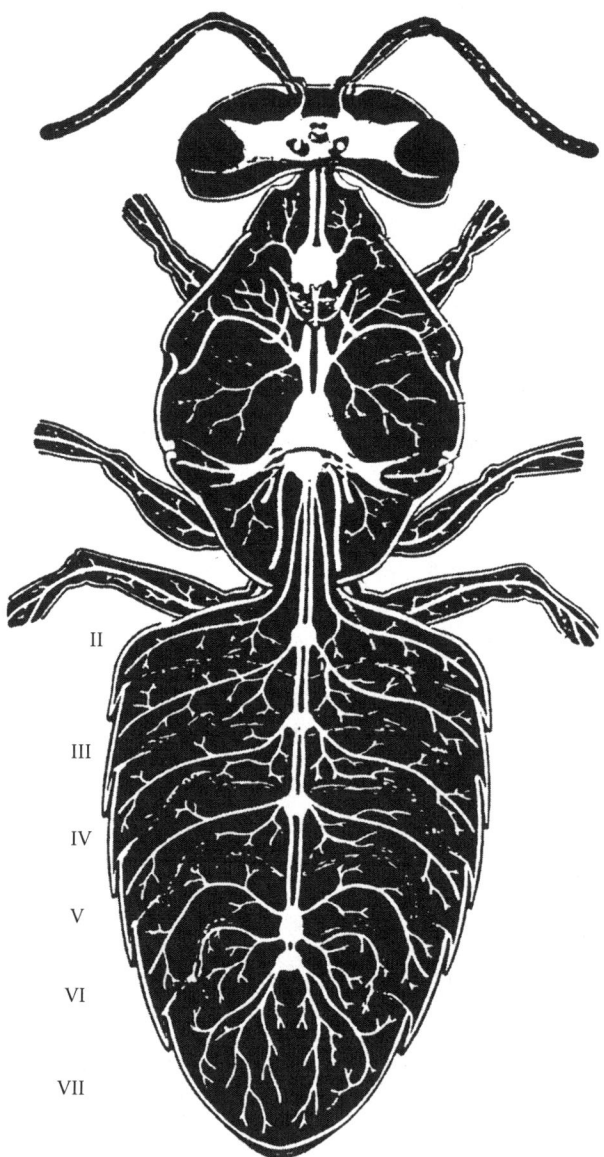

Figure 9.1 A drawing of the brain, ventral ganglia, ventral connectives, and some of the major nerves in a worker honeybee. The ganglia and nerve cord lie beneath the ventral cuticle, so this drawing depicts the system as a dissection from the dorsal surface, with all organs removed except the nervous system. In the earliest insects to evolve, there probably was a ganglion in each segment, but coalescence of ganglia has occurred in all living insects. The bilateral symmetry of the nervous system is still evident in the paired ventral connectives between ganglia, although in some insects the connectives also are fused together. There has been a tendency during the evolution of some insect groups for abdominal ganglia to become fused with thoracic ganglia, and only nerves pass into the abdomen to innervate the various muscles and organs. (From Dadant & Sons (eds.), *The Hive and the Honey Bee*, Revised edn., 1975, Dadant & Sons Publishers, Hamilton, IL. Permission given with acknowledgment.)

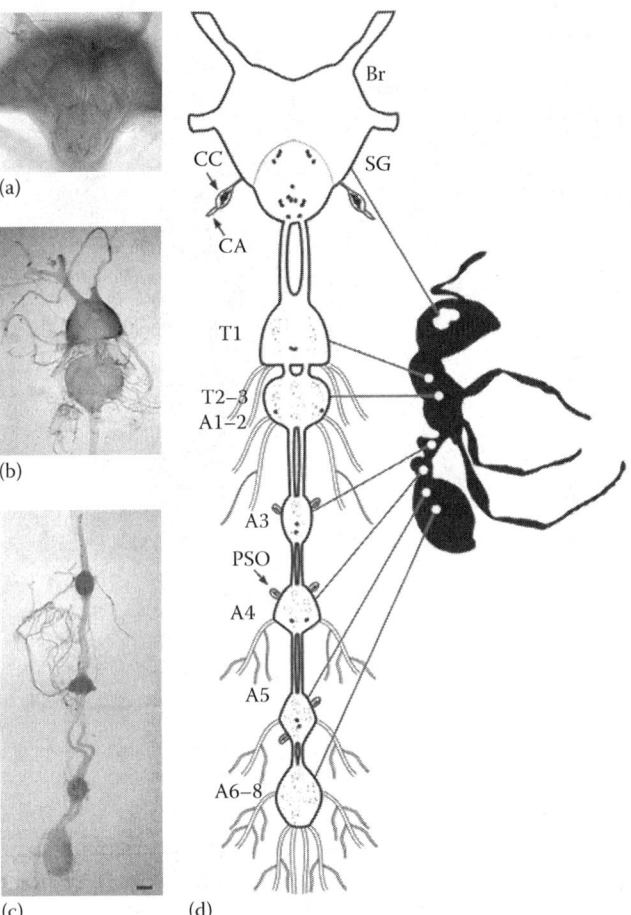

Figure 9.2 The central nervous system (CNS) of an adult fire ant; fusion of ganglia has occurred in the thoracic and abdominal ganglia. Photograph (a) shows the brain, (b) the two large thoracic ganglia, and (c) the four abdominal ganglia. (d) A drawing of the CNS of the ant is shown adjacent to the photographs. Br, brain; SG, suboesophageal ganglion; CC, corpora cardiac; CA, corpora allata; PSO, perisympathetic organ; T1–T3, first to third thoracic ganglia; A1–A8, first to eighth abdominal ganglia. Bar in photograph (c) is 50 μm. Dots of stain in the photograph and drawing show PBAN-like immunoreactivity in the CNS. (From Choi, M.-Y. et al., *Cell Tissue Res.*, 335, 431, 2009. With permission.)

A ganglion typically contains a mass of neuronal cell bodies of inter- and motoneurons at the periphery and a central region, the **neuropil**, where synapses occur. The cell bodies of sensory neurons of insects usually are near the site of sensory stimulus reception, and consequently, many sensory neuron cell bodies are located peripherally in the cuticle and in or on internal organs.

9.3 BRAIN

The brain consists of three fused ganglionic masses, the **protocerebrum**, **deutocerebrum**, and **tritocerebrum** (Figure 9.3). These three ganglionic masses typically rest on top of the esophagus, which passes posteriorly between the connectives to the subesophageal ganglion. Together, the three parts are sometimes called the supraesophageal ganglion. Two excellent books have been written detailing the anatomy and function of the brain and other parts of the CNS in insects; Strausfeld (1976) deals primarily with Diptera, and Burrows (1996) concentrates on locusts.

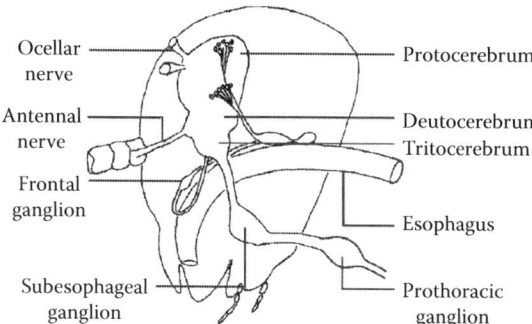

Figure 9.3 A lateral view of the three major parts of the brain (protocerebrum, deutocerebrum, and tritocerebrum) with associated head connections and ventral connectives to the subesophageal ganglion. (Drawing modified from Snodgrass, R.E., *Principles of Insect Morphology*, McGraw-Hill, New York, 1935; Jenkin, P.M., *Animal Hormones: A Comparative Survey*, Pergamon Press, Oxford, U.K., 1962.)

9.3.1 Protocerebrum

The protocerebrum is the site of major integrative centers that process incoming information from many sensory sources. The optic lobes, which process information from the compound eyes, are part of the protocerebrum. The optic lobes contain several neuropil regions related to visual processing. The protocerebrum also receives input from ocelli via the ocellar nerves. Small paired nerves, the *nervus corporis cardiaci I* and *II*, link the protocerebral neurosensory cells (NSC) with the corpora cardiaca (CC) and corpora allata (CA).

The **corpora pedunculata**, the mushroom bodies, are large, bilateral integrative centers in the protocerebrum (Figure 9.4). The size of the protocerebrum dedicated to the mushroom bodies varies in

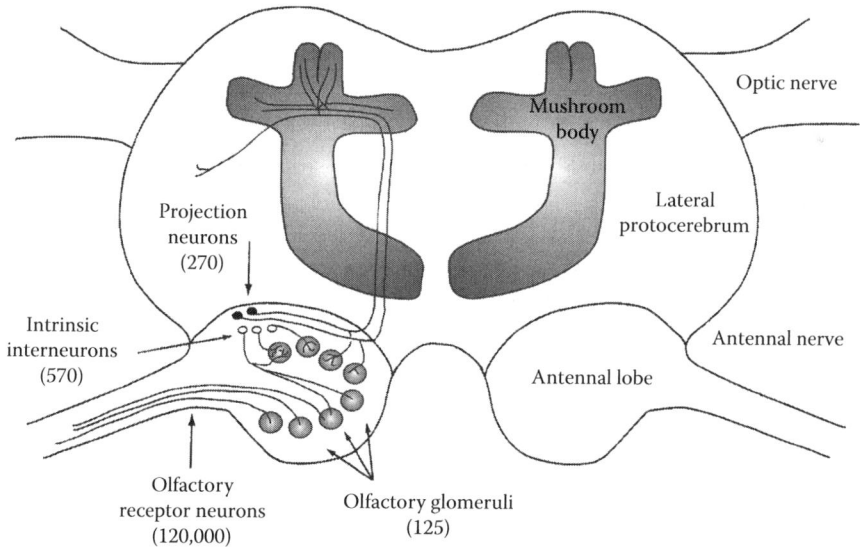

Figure 9.4 A frontal section through the brain of a female cockroach illustrating the antennal lobes and mushroom bodies. Sensory neurons from the antennae pass into the deutocerebrum where they synapse in glomeruli with interneurons projecting to the protocerebrum. The numbers indicate the approximate numbers of neurons in various parts of the olfactory system in a cockroach. (From Lemon, W.C. and Getz, W.M., *Ann. Entomol. Soc. Am.*, 92, 861, 1999. With permission.)

insects, with estimates of about 50,000 cells in locusts and as many as 1.2×10^6 in honeybees. These integrative centers are believed to be involved with olfactory learning through connections and input from the olfactory lobe integrative centers in the deutocerebrum, the region that receives olfactory input from the antennae. The mushroom bodies are divided into the peduncle (or stalk) and the calyx (the cap part of the mushroom shape). The peduncle contains fibers of neurons going to and from the calyx, a synaptic region. The mushroom shape of the corpora pedunculata is not so evident in all insects.

Another major region in the central part of the protocerebrum is the **central body complex** (Pfeiffer and Homberg, 2014), consisting of closely interconnected neuropils located between the bases of the stalks of the mushroom bodies. The protocerebral bridge, an elongated neuropil, that connects the two hemispheres of the protocerebrum is part of the central body. The central complex is connected to various parts of the protocerebrum, but with a few exceptions in several insects, it is only indirectly connected to the deutocerebrum and ventral nerve cord. There are four principal types of neurons based upon arborization patterns associated with the central body: (1) tangential neurons, (2) amacrine neurons, (3) pontine neurons, and (4) columnar neurons. Tangential neurons innervate the protocerebral bridge and various parts of the central complex. Amacrine neurons are not very numerous and have been reported only from the upper portion of the central body (CBU). Pontine neurons occur only in the central body and their somata are located in the pars intercerebralis. Columnar neurons have arborizations in the protocerebral bridge and in the central body, and their somata also are located in the pars intercerebralis. Columnar appear to be the principal output neurons of the central complex.

A number of neurotransmitters and neuromodulators have been described from the different types of neurons in the central body complex, with glutamate and acetylcholine being the major transmitters. The transmitter in many of the tangential neurons that serve the lower division of the central body is γ-aminobutyric acid (GABA), as well as in some of the synapses in the CBU. Some tangential neurons in the CBU have histamine and dopamine as transmitters (Nässel, 1999). Octopamine occurs in some of the neurons coming from the subesophageal ganglion to the central body and in a pair of neurons projecting to the protocerebral bridge (Busch et al., 2009; Homberg et al., 2013; Kahsai et al. 2010, 2012). Kahsai and Winther (2011) mapped receptors to several different neurotransmitters in the central complex of *Drosophila*. A number of neuropeptides have been detected in the brain (Nässel, 1993) and central complex neurons (Nässel, 2002; Nässel and Homberg, 2006). Heuer et al. (2012) and Kahsai and Winther (2011) found the neuropeptide SIFamide in *Manduca sexta* and *Drosophila*, respectively. Nitric oxide (NO) has been detected at synaptic endings in some populations of tangential, pontine, and columnar neurons (Müller, 1997; Kurylas et al., 2005; Wenzel et al., 2005; Siegl et al., 2009). Although specific physiological reactions associated with most of these neurons, neurotransmitters, and neuropeptides are unknown, tachykinin and neuropeptide F have been shown to modulate locomotor behavior in *Drosophila* (Kahsai et al., 2010), and acetylcholine, GABA, and NO are involved in modulation of sound production in grasshoppers (Kunst et al., 2011).

9.3.2 Deutocerebrum

The deutocerebrum receives sensory input from mechano- and chemosensory receptor neurons on the antennae (Homberg et al., 1989; Rodrigues and Pinto, 1989; Hösl, 1990; Stocker, 1994; Hildebrand, 1995, 1996) and sends motor signals to muscles of the antennae. There are separate neuropil regions in the deutocerebrum that process the information from the chemosensory and mechanosensory neurons, that is, the axons from the two types of receptors project (send axons) to separate areas within the deutocerebrum. Chemosensory input goes to the **antennal lobe (AL)** neuropil, while the **antennal mechanosensory and motor center (AMMC)** receives the mechanoreceptor input and sends motor information out. Each of these centers is represented on the left and right sides of the brain, with the AMMC located posterior and ventral to the AL (Marion-Poll and Tobin, 1992; Hildebrand, 1995).

At least in some insects (e.g., Lepidoptera) and possibly in all, the chemosensory inputs are further partitioned into separate synaptic sites within the AL based upon whether the information comes from receptors sensitive to sex pheromone, food or host odors, or carbon dioxide. The antenna is not reproduced in an exact spatial way in the AL (i.e., receptor axons from the distal portion of the antenna may project to the same site as axons from proximal antennal receptors), but directional information is retained in some species by unilateral input from an antenna to the ipsilateral side of the brain.

9.3.2.1 Antennal Mechanosensory and Motor Center Neuropil

Relatively little is known at present about the AMMC, but as the name indicates, it contains arborizations of both motor and sensory neurons. Motor centers controlling muscles and glands of the head are in the deutocerebrum (with some additional ones controlling these structures also in the tritocerebrum). The deutocerebrum sends motor neurons to the antennal muscles and muscles of the labrum.

As a sensory center, the AMMC primarily receives mechanoreceptor neuron terminals from Böhm's organ, Johnston's organ, Janet's organ, and other mechanosensory structures located on the two basal segments of the antenna (the scape and pedicel) of various insects. In addition to terminal arborizations in the AMMC, some arborizations from mechanosensory cells project to the protocerebrum, the subesophageal ganglion, and into thoracic ganglia, indicating widespread distribution of some mechanosensory information.

9.3.2.2 Antennal Lobe

In contrast to the meager information known about the AMMC, a great deal of information is now known about the **AL neuropil** and its interconnections with sensory structures and other parts of the nervous system (Figure 9.5). The AL is the first-order olfactory center in insects. Each AL

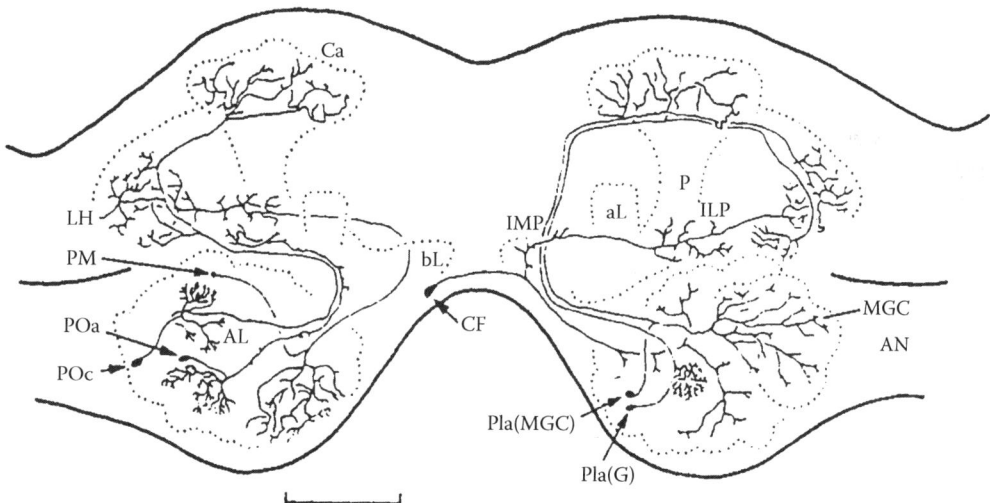

Figure 9.5 Horizontal cross section through the brain in an adult *M. sexta*, the tobacco hornworm moth, showing a variety of neurons from the antennal lobe projecting to parts of the protocerebrum. The antennal lobes receive sensory input from the antennae. Key: AL, antennal lobe; CF, centrifugal neuron; MGC, macroglomerular complex; bL, beta lobe; aL, alpha lobe; ILP, inferior lateral protocerebrum; P, pedunculus of mushroom body; IMP, inferior medial protocerebrum; Ca, calyces of mushroom bodies; LH, lateral horn of protocerebrum; Pla(G), Pla(MGC), POa, POc, various projection neurons. (From Homberg, U. et al., *Annu. Rev. Entomol.*, 34, 477, 1989. With permission.)

receives sensory information from chemoreceptors on the flagellum of the antenna on the ipsilateral (same) side of the body. Within the chemosensory neuropil of the AL, there are two neuropil divisions in some insects: (1) one concerned with food, host, and, perhaps, other general environmental odors and (2) a **macroglomerular complex** (**MGC**) that is sex specific in males of Dictyoptera (cockroaches) and Lepidoptera (moths) as the sensory neuropil for the sex pheromone receptor neurons. Axons from the antennal receptor cells pass through the antennal nerve and terminate in AL neuropil regions in structures called **glomeruli**.

9.3.2.2.1 Organization of Glomeruli in the Antennal Lobe

Glomeruli are somewhat cup-shaped masses of axonal terminals. Typically, a glomerulus is about 50–100 µm in diameter, and each is separated from other glomeruli by layers of glial cell membranes. The number of glomeruli is species specific, with approximately 10 in *Aedes aegypti* mosquitoes, about 200 in *Formica pratensis* (an ant), and approximately 60 in *Manduca sexta*. There is a great deal of convergence of the sensory axons: from 10^5 to 10^8 receptor neurons converge on 10 to a few hundred glomeruli in different insects. The glomeruli house axonal terminals of antennal olfactory receptor cells, neurites of local and output (projection) neurons, and terminals of centrifugal neurons (CNs) projecting to the AL from other sites in the brain. A given glomerulus appears to be associated with a group of axons related to particular odor identification rather than representing a strict morphological array on the antenna, that is, in a given glomerulus, the converging sensory axons may come from various sites on the antenna where there are receptors sensitive to a specific odor.

Inside the glomeruli are the soma of one to several interneurons whose terminals make synaptic connections with the incoming sensory axons. Three classes of interneurons are associated with the glomeruli:

1. Local interneurons (LNs) that interconnect areas of the AL but remain within the AL.
2. Projection neurons (PNs) that have dendrites located in the AL with axons projecting into the mushroom bodies and lateral parts of the protocerebrum to which they relay information.
3. CNs whose axons project into the AL, but cell bodies and dendrites are located outside the AL (e.g., some, but not exclusively, in the protocerebrum).

Somata of the LNs and of most PNs are located in groups in the peripheral part of the AL. Somata of some PNs and of most CNs are located in the protocerebrum instead of in the deutocerebrum.

Fiber tracts connect each AL with the protocerebrum and the subesophageal ganglion. In some Diptera, the ALs on opposite sides of the brain are interconnected by fiber tracts. ALs on each side of the brain can also communicate with each other through commissures in several groups of insects, especially Diptera (commissures are larger pathways between parts of the brain). Through synaptic connections with LNs, PNs, and CNs, whose terminals project to many parts of the brain, sensory information is distributed to higher centers in the brain.

In male *M. sexta*, there are three sexually dimorphic glomeruli comprising the MGC (Rospars and Hildebrand, 1992). These glomeruli receive axonal arborizations from pheromone receptors on male antennae, and in addition to interneuronal connections noted earlier, glomeruli also receive terminals from an identified 5-hydroxytryptamine (5-HT)-immunoreactive neuron (Sun et al., 1993). The magnitude and duration of neuronal potentials from pheromone receptors on the antenna are increased by 5-HT added to the bathing saline, suggesting that 5-HT is a neuromodulator of pheromone signals relayed from the MGC to higher-order integrative centers in the protocerebrum (Kloppenburg and Heinbockel, 2000).

NEUROANATOMY

The projection of axons from olfactory receptors to discrete glomeruli is a characteristic feature of both invertebrate and vertebrate olfactory systems (Hildebrand, 1995), and the organizational pattern may be as much as 500 million years old (Dethier, 1990, cited in Hildebrand, 1995). The role of glomeruli in odor perception has been reviewed by Galizia and Menzel (2001).

9.3.3 Tritocerebrum

The tritocerebrum sends motoneurons to muscles in the labrum and pharynx and innervates the **stomatogastric nervous system**, a system of several small ganglia, including the frontal ganglion, hypocerebral ganglion, and ingluvial ganglia that have control of foregut muscles. In some insects, sensory axons from sensory receptors on the head terminate in the tritocerebrum, and in *M. sexta*, it is known to receive PNs from sensory receptors on the mouthparts (Kent and Hildebrand, 1987). A commissural connective from each side of the tritocerebrum passes around the esophagus and provides cross communication between the two halves, and lateral connectives connect each half of the tritocerebrum with the subesophageal ganglion.

The unpaired frontal ganglion (Figures 9.3 and 9.6), lying on top of the esophagus and anterior to the brain, is connected to the tritocerebrum by lateral connectives. The recurrent nerve and other small nerves arise from the frontal ganglion and carry motoneurons to muscles of the gut wall. Nerves from the frontal ganglion innervate the pharynx. Posteriorly, the median recurrent nerve runs along the surface of the esophagus, passes beneath the brain, and connects behind the brain with the hypocerebral ganglion, also lying on the surface of the esophagus. The small, unpaired **hypocerebral ganglion** innervates the CC. The small, paired **ingluvial ganglia** send nerves to the posterior foregut. The tritocerebrum has lateral nerve cord connections to the **subesophageal ganglion**. The brain typically rests on top of the gut, which passes between the lateral nerve cord connectives to the subesophageal ganglion. The subesophageal ganglion is formed from the fusion of three pairs of ganglia. It has sensory and motor connections to sensory structures and muscles of the mouthparts, salivary glands, neck receptors in some insects, and neck muscles. Axons from neurons in the subesophageal ganglion project forward to the brain and posterior to the thoracic ganglia. The subesophageal ganglion has influence over motor patterns involved with walking, flying, and breathing, although those motor patterns originate in other ganglia.

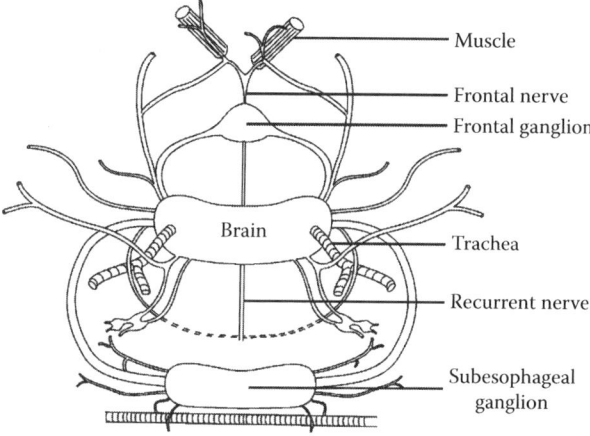

Figure 9.6 A dorsal view of the brain and associated head ganglia and nerves of a larval tomato hornworm *Manduca* spp. Lepidoptera: Sphingidae. Note the large tracheal trunks that penetrate the brain tissue and provide for gas exchange in the brain.

9.4 VENTRAL GANGLIA

Many insects have three thoracic ganglia—the **pro-**, **meso-**, and **metathoracic ganglia**, located in corresponding thoracic segments (Figure 9.7)—and such an arrangement was probably an early evolutionary one. In some insect groups (particularly Hemiptera and Diptera and others), the meso- and metathoracic ganglia have fused, often with fusion of some of the abdominal ganglia into a large thoracic ganglion (Figures 9.2 and 9.8). Each thoracic ganglion sends motor axons to the leg muscles of its segment and receives sensory axons from sensory receptors in the tarsi and leg joints. The meso- and metathoracic ganglia supply motor nerves to the wing muscles, respectively, of the meso- and metathoracic segments.

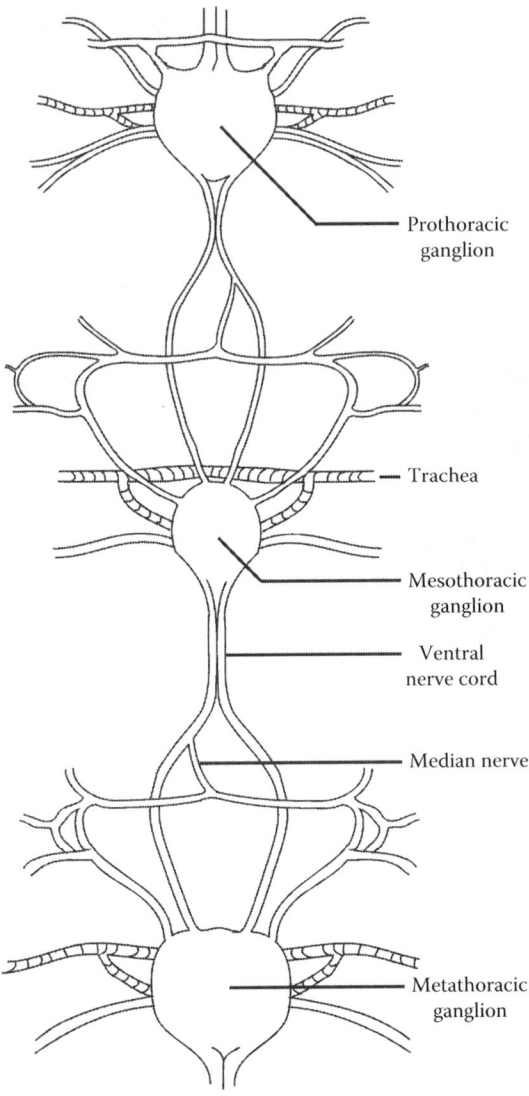

Figure 9.7 A dorsal view of the thoracic ganglia of a larval tomato hornworm. Note the median nerve that arises from the ganglion in front, travels with the connective for a distance and then separates from the ventral connective, and splits into two branches passing to each side of the posterior segment. A branch of the median nerve innervates the spiracular muscles that control opening and closing. Large tracheal trunks supply numerous smaller branches to the ganglia.

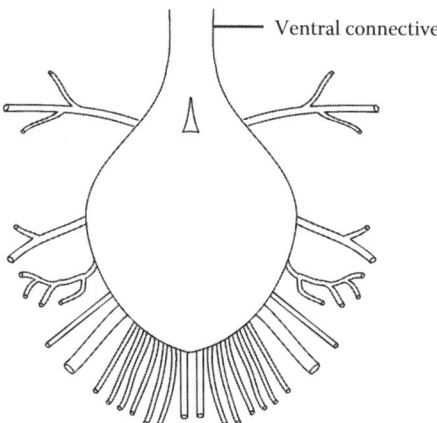

Figure 9.8 The large second thoracic ganglion of the hemipteran, *Oncopeltus fasciatus*, the large milkweed bug. The ganglion contains all the fused neuromeres from the abdomen. Note the many nerves that arise from the ganglion and pass into the abdomen.

9.4.1 Abdominal Ganglia

The number of abdominal ganglia is highly variable in different orders. One ganglion in each segment was the evolutionarily primitive condition, but in all living insects, some fusion of abdominal ganglia has occurred. Some Apterygota still have eight abdominal ganglia, some Odonata larvae have seven, and Orthoptera have five or six. There may be fewer than five in some insects (e.g., Figure 9.2), and in some highly evolved dipterans and hemipterans, all abdominal ganglia have fused with thoracic ganglia into the one, large metathoracic ganglion. In locusts, *Schistocerca gregaria*, for example, the first three abdominal ganglia are fused in the adult with the large metathoracic ganglion. The large sixth abdominal ganglion in the American cockroach is the **terminal abdominal ganglion** (**TAG**) and is a fusion of **neuromeres** (ganglionic masses associated with a particular segment) from the posterior segments (Figure 9.9).

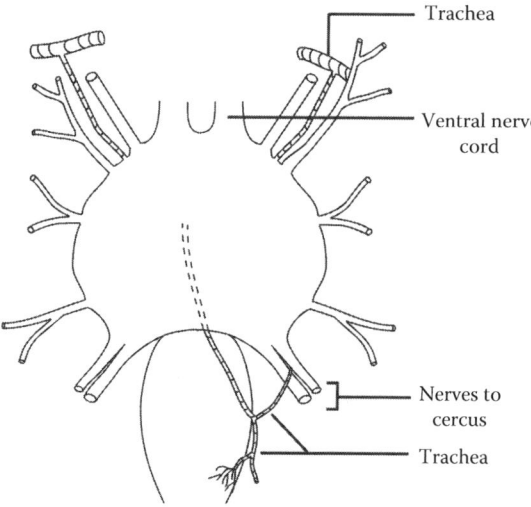

Figure 9.9 The sixth and terminal abdominal ganglion (TAG) of the American cockroach, *Periplaneta americana*. The TAG contains fused neuromeres from several posterior abdominal segments. Nerves from the TAG pass into the cerci, and sensory neurons from the cerci synapse in the TAG with large (giant) axons that project forward to the thoracic ganglia and brain.

Numerous nerves radiate out from it to the posterior structures in the body and it is well supplied with tracheae. Cercal nerves from the cerci carry large numbers of sensory axons into the TAG and synapse with giant axons functioning as a fast escape mechanism when the cockroach is threatened. The giant axons pass forward through abdominal ganglia without synapsing to eventually synapse with interneurons connecting to leg motoneurons in the thoracic ganglia. Thus, information can be sent from the cerci to the legs very rapidly.

Fused ganglionic masses (i.e., fused neuromeres) send nerves to the various muscles and glands of the body segments according to their evolutionary origin and also carry sensory axons back to the fused neuromeres. According to Snodgrass (1935), the best morphological indication of the composition of fused ganglia (or of a ganglion that has migrated out of the primitive segmentation pattern) is the distribution of nerves to various segments and segmental muscles or glands.

Thoracic and abdominal ganglia tend to be divided into three anatomical and functional divisions—a **dorsal motor neuropil**, a **middle integrative neuropil**, and a **ventral sensory neuropil**. Sensory information tends to come into the ventral portion of a ganglion from lateral nerves and from nerve tracts in the ventral nerve cord. Motor output more often occurs from the dorsal portion of a ganglion after associative interneurons allow the two regions to communicate. Internally, the middle layer or associative zone appears to contain the most complex array of variously shaped interneurons. Intersegmental ventral nerve cord connectives also tend to maintain the pattern of dorsal motor nerves and ventral sensory nerves. Burrows (1996) does not attribute any special significance to the anatomical and functional divisions of ganglia.

9.4.2 Lateral Nerves

Ganglia give rise to variable numbers of nerve tracts in different groups of insects. Most of the nerve tracts that arise from the brain or ventral ganglia contain mixed nerves, that is, the tract carries both sensory and motor neurons. An exception is the ocellar nerve connecting the ocelli with the protocerebrum. It is purely sensory, carrying interneurons with sensory information inward to the protocerebrum. The antennal nerves are mixed, carrying sensory fibers from olfactory and mechanoreceptors inward, and motoneurons leading to tentorial ptilinial muscles and tentorial frontal muscles. The labrofrontal nerves and maxillary-labellar nerves are mixed, bringing sensory information from receptors on the mouthparts and carrying motoneurons to the muscles of the mouthparts.

Lateral nerves from the thoracic ganglia innervate the wing musculature and return sensory information from receptors associated with wing orientation. Motor neurons from the thoracic ganglia innervate the musculature of the legs and tarsi, and thoracic ganglia receive mechanosensory information from the legs and probably mechano- and chemosensory information from the tarsi in most insects. Motor neurons and sensory neurons do not directly synapse with each other, but each makes synaptic connections with interneurons (probably with many interneurons). Interneurons make synaptic connections with many other interneurons and, thus, every part of the nervous system is potentially in communication. Incoming sensory information is received in various neuropils, interpreted, and motor commands are formulated to send to muscles or glands.

9.5 OXYGEN AND GLUCOSE SUPPLY TO THE BRAIN AND GANGLIA

Nerve cells typically have a high rate of metabolism requiring a steady supply of nutrients and efficient gas exchange. Glucose is delivered (as trehalose) by the hemolymph, and gas exchange occurs through the tracheal system. Large tracheal trunks penetrate ganglia, branch

NEUROANATOMY

into smaller tracheae and tracheoles, and bring air to within a few micrometer of each neuron or glial cell. Carbon dioxide is delivered back to the outside via the spiracles. Trachea and tracheoles have been estimated to take up 4%–8% of the neuropil volume in the brain of the housefly (Strausfeld, 1976).

9.6 NEUROPIL

The main tissue mass of any ganglion is a central region of synaptic connections between arborizations of sensory, motor, and interneurons called the **neuropil**. The largest and most complex neuropils occur in the brain in such associative centers as the mushroom bodies, the central body, and optic lobes, but all ganglia contain neuropil regions. The neuropil is concentrated in the center of a ganglion and is surrounded with a shell of cell bodies, the **somata** (**perikarya**, by some authors) of motoneurons, interneurons, and glial cells (Figure 9.10). The neuropil continues to grow in size with new synaptic connections in the milkweed bug, *Oncopeltus fasciatus*, and in *Drosophila*

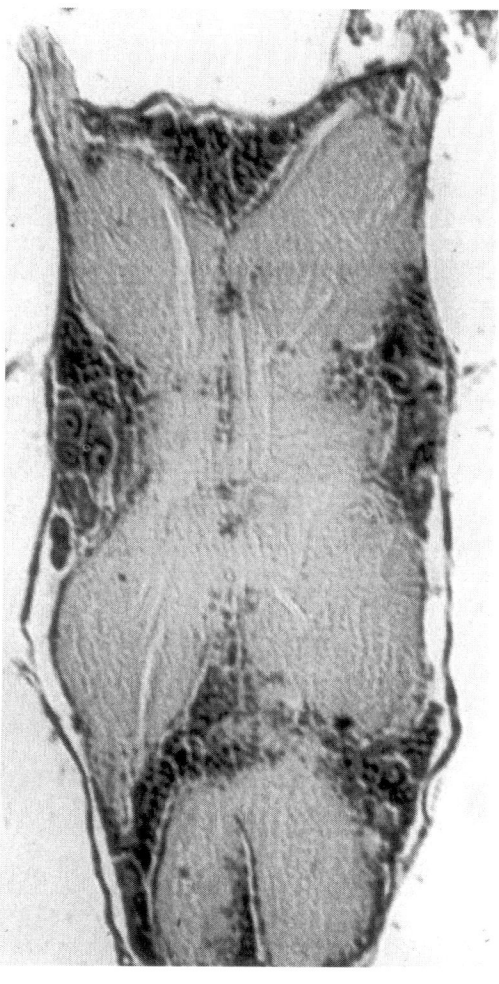

Figure 9.10 A light microscope view of a cross section through the large thoracic ganglion of a dipteran to show the central neuropil and surrounding shell of darkly stained somata. The ganglion contains the fused abdominal ganglia. (Photo by the author.)

melanogaster, and possibly in most or all insects, even after the nervous system as a whole stops growth. This suggests the importance of new information processing and integration that an insect uses to acquire food, mate, lay eggs, and survive in its environment.

9.7 HEMOLYMPH–BRAIN (CNS) BARRIER

At the hemolymph interface, the brain, ventral connectives, ventral ganglia, and large nerves are protected from direct contact with the hemolymph by a selectively permeable barrier, the **hemolymph–CNS barrier**, or **blood–brain barrier** (BBB) (Scharrer, 1939; Strausfeld, 1976; Abbott and Treherne, 1977; Treherne, 1985). The barrier consists of an outer acellular layer called the **neurolemma** (also neural lemma, neural lamella) and an inner cellular layer, the **perineurium**, consisting of the perineural cells (Figure 9.11).

The open circulatory system with high hemolymph concentrations of trehalose and amino acids, some of which may function as neurotransmitters (Iversen et al., 1975), and the high potassium to sodium ratio in some phytophagous insects may have greatly influenced the evolution of the BBB. The BBB protects not only the brain but also all ganglia and major nerves from direct contact with the hemolymph.

The entire acellular and cellular barrier is thin, varying from 7 μm to about 15 μm thick around the brain of the housefly, *Musca domestica*, for example. Some insects have a continuous fatty sheath of variable thickness composed of fat body cells exterior to the neurolemma that covers the entire brain and ventral nerve cord (Treherne, 1985). When present, the fatty envelope also acts as a barrier to ions and osmolytes in the hemolymph.

The acellular neurolemma of some insects is composed of several layers identifiable in the electron microscope and contains fibrils of collagen-like material that are embedded in a matrix of glycosaminoglycans (Ashhurst and Costin, 1971a,b). The fibrous matrix is tough and elastic and is the first barrier that a neurophysiologist must push a microelectrode through in order to record from within the CNS. The neurolemma is relatively permeable to large molecules, such as inulin and methylene blue, and in itself does not appear to be much of a barrier to hemolymph substances. This acellular neurolemma is probably secreted by the outermost layer of cells, sheath cells, located just below the surface of the neurolemma (Burrows, 1996).

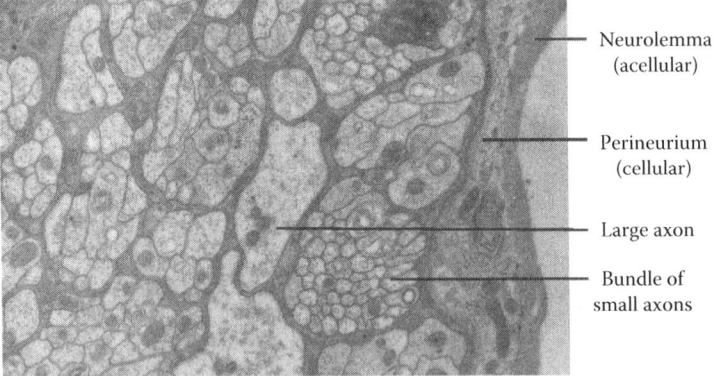

Figure 9.11 A transmission electron micrograph (TEM) of a portion of the ventral nerve cord of a mosquito, *Culex* species. The brain, ventral nerve cord, ganglia, and major nerves are protected from contact with the hemolymph by the hemolymph–brain barrier. The barrier consists of an acellular neurolemma and a cellular perineurium made up of various types of glial cells. The neurolemma is rather permeable and leaky and the perineurium is the main barrier. (TEM by the author.)

NEUROANATOMY

The cellular layers beneath the neurolemma (the perineural cells) constitute the main hemolymph–CNS barrier. The perineural layer is several cells in thickness. Wigglesworth (1959) classified the perineural cells as glial cells, but that is not generally followed now (Strausfeld, 1976). The perineural cells have extensive couplings between the cells via tight and gap junctions (Lane and Skaer, 1980), which probably account for much of the impermeability of the cell layer since large molecules and ions would have great difficulty in passing between the cells because of the between-cell barriers (Strausfeld, 1976). The physiology of these cells is poorly known, but it has been suggested that the perineural cells are nutritive as well as protective. They probably participate in nutrient transfer from the hemolymph to underlying glial and nerve cells.

9.8 NEURONS: BUILDING BLOCKS OF A NERVOUS SYSTEM

Nerve cells (neurons) have many shapes, and no single shape can be said to be characteristic. Neurons have one or more **dendrites**, a cell body called the **soma** (perikaryon, by some authors) that contains the nucleus, and an **axon** (Figure 9.12). Sometimes the axon has collateral branches. Extensive arborization of axonal and dendritic processes is typical. Within the CNS, the arborizations enable synaptic contact with many other neurons, thus making possible communication throughout the nervous system. A dendrite is defined as any process conducting electrical excitation from the site of stimulation toward the soma, and the axon is the process conducting excitation away from the soma. An axon may conduct excitation toward the CNS (sensory neuron) or away from the CNS (motoneuron) toward an effector, such as a muscle or gland.

The use of dendrite as a term for the extensive arborizations of neurons in the neuropil of insects has been criticized as borrowed and defined from vertebrate neurobiology (Burrows, 1996). Some of the arborizations from certain neurons in insects are known to receive input in the neuropil (i.e., act like dendrites), while others deliver output (act like axons), and morphology cannot be used to discriminate between these. A few motoneurons have been identified with input and output synapses on the same nerve branch. As a consequence, the word dendrite for arborizations of central neurons is often avoided, and different authors have used **neurites**, branches,

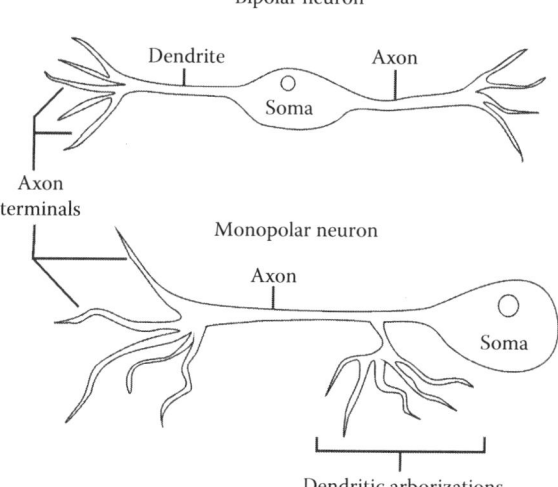

Figure 9.12 A schematic drawing of a sensory and motor neuron to illustrate axon, dendrite, soma, and arborizations.

and arborizations to describe branching without having to specify whether it functions in input or output (Burrows, 1996). When a clearly defined neurite emerges from the neuron in question and transmits spikes along its length, it is called an axon. There are a number of morphological types of cells in the nervous system, but they can be grouped functionally into **sensory neurons**, **motoneurons**, **interneurons**, **NSC**, and **glial cells**.

9.8.1 Afferent or Sensory Neurons

Sensory neurons carry nerve impulses toward the CNS. There is a major morphological difference between insects and vertebrates in the location of the somata (sing., soma) of sensory neurons. In vertebrates, the somata of afferent neurons rest in the paravertebral chain of ganglia near the dorsal spinal cord; the dendrites are long and the axons relatively short. The somata of insect sensory neurons, with few exceptions, are located very near the site of stimulus detection, that is, at the cuticular surface for external receptors and as part of various internal organs. The dendritic processes are very short, and the axonal processes tend to be relatively long. Insect sensory neurons tend to be **bipolar**. The short dendritic processes of insect receptor cells probably do not generate or conduct spikes, as they must in the long dendrites of vertebrates, but develop graded electrical activity that spreads from the site of the stimulus. When the graded electrical activity is strong, it is likely to spread to the region where spikes are generated. The number of sensory neurons may change with molts and metamorphosis. For example, mechanoreceptors (innervated hairs) on the abdominal cerci of *Acheta domesticus* increased from 50 to 750 during growth and molting.

9.8.2 Efferent or Motor Neurons

Motoneurons have their somata located in the peripheral region of a ganglion and are usually monopolar in insects. The somata of motoneurons are usually very large, up to 100 µm in diameter. Motoneurons are usually paired on each side of a ganglion, and neurites and axons from a motoneuron usually remain on the side, or exit in case of an axon, from the side of the ganglion that houses the soma. **Dorsal unpaired median (DUM) motoneurons** are not paired but have neurites in each half of a ganglion and an axon on each side of the ganglion that emerges through a lateral nerve of the ganglion to innervate an effector gland or muscle.

A few motoneurons have branches on both sides of a ganglion, and the axon may even emerge from the contralateral side. Usually the single process that arises from the peripheral soma of a motoneuron passes into the neuropil of its ganglion and branches off a network of arborizations (neurites) that make many synaptic connections with neurites of other interneurons, and the axon exits the ganglion on the ipsilateral side through a large nerve and continues to its target site of gland or muscle. Some axons exit into the ventral connectives between ganglia and terminate in another part of the CNS, or eventually pass out through the lateral nerve of a different ganglion than that of its origin.

Motoneurons, when individually identified, are named (or numbered) after the muscle they innervate, a system started by Snodgrass (1929). The number of motoneurons is a very small percentage of the total nerve cells in an insect. Typically, only about 100 or so metathoracic motoneurons are involved with control of wing and leg muscles and other thoracic musculature on each side of the body in a locust (Burrows, 1996). Often only two or three, and only in a few cases more, motoneurons go to large muscles. For example, the extensor tibiae of the third pair of legs in a locust receives four neurons, including a fast axon, a slow axon, an inhibitory neuron, and a dorsal unpaired median extensor tibiae (DUMETi) neuron to the extensor tibiae muscle.

Axon to soma synaptic connections, a common type of synapse in vertebrate motoneurons, have not been found in insects. All synapses occur within the neuropil between neurites of neurons. The axon conducts a spike, but the soma and fine neurites of a motoneuron do not ordinarily conduct the spike, although somata of DUM neurons can conduct spikes. Under certain experimental conditions, even the soma of a motoneuron that usually does not conduct a spike can be shown to be excitable and to possess voltage-sensitive Na^+ and Ca^{2+} channels that may be sensitive to neuromodulators, which could create new dimensions in integration of signals (Burrows, 1996).

A motoneuron generally is bombarded with synaptic input from the many fine neurites in the neuropil. The electrical activity initiated in the neurites, while spike-like, is not transmitted over the many fine neurites and to the soma without decrement, as it typically is once it enters an axon. There are changes in rise time, amplitude, and duration of the excitability wave as it is conducted toward the soma and to the axon. If the excitation is strong enough, true spikes propagated without decrement will be generated at a zone in the axon sometimes called the axon hillock. Speed of transmission in axons is variable depending upon size of the axon (faster transmission occurs in larger axons, such as the so-called giant axons) and other axonal characteristics.

9.8.3 Interneurons

Interneurons, also called **association neurons** and **internuncials**, may be located entirely within a ganglion or they may send intersegmental (axonal) processes through the ventral nerve cord to make synaptic connections in other ganglia. Interneurons make extensive synaptic connections with other interneurons, with incoming afferent (sensory) neurons, and with motoneurons. With their vast array of neurites and interconnections, interneurons are extremely important in coordinating communication between sensory and motor systems and within the CNS. The somata of interneurons lie in the peripheral region of a ganglion.

Local Interneurons (LNs) are those that make connections within a ganglion and do not exit from it. Interneurons that exit a ganglion to make connections in other ganglia are called **intersegmental interneurons**. LNs are further classified into spiking and nonspiking interneurons, depending on whether the neuron transmits a spike or only graded electrical activity. Spiking LNs are most common in the optic neuropils, mushroom bodies of the protocerebrum, and the ALs of the deutocerebrum, but they can be found in thoracic ganglia and a few in other parts of the nervous system. Generally, LNs are paired in the two halves of a ganglion, but the DUM LNs, as the name indicates, are not paired. The somata of these DUM neurons are located in the periphery of a ganglion near the midline, and extensive neurites from each DUM spread into both halves of the ganglion. DUM neurons may be local, intersegmental, or efferent DUM neurons (Burrows, 1996). Intersegmental DUM neurons have neurites in both halves of a ganglion, and an axon arises from among the neurites on each side of the ganglion and passes through the intersegmental connectives into the next ganglion. (Efferent DUM neurons were described in Section 9.8.2.)

There may be several different neurotransmitters involved with spiking interneurons. Some show inhibitory action and evidence (immunoreactivity) of release of **GABA**, a common neurotransmitter at inhibitory synaptic endings. Other spiking interneurons have excitatory action at the postsynaptic connections, and they must release a stimulatory neurotransmitter, but the chemical nature has not been identified. Efferent DUM interneurons appear to be octopaminergic (releasing octopamine) and probably exert neurohormonal control over muscle or gland action.

Nonspiking interneurons have only a vast array of neurites projecting into the neuropil of a ganglion and no identifiable axons. They are contained within a single ganglion. They release their neurotransmitter at the synaptic junctions without transmitting spikes. The monopolar neurons in the neural cartridges of the lamina ganglionaris in the optic lobe are nonspiking interneurons. They also occur in the neural network controlling walking behavior in cockroaches (Pearson and Fourtner, 1975) and other insects and in the TAG, where they process signals from certain mechanoreceptors on the cerci (Kondoh et al., 1991, 1993). The dynamics of the electrical excitation and its movement over these nonspiking interneurons is not well defined. There is evidence for K^+ and possibly Ca^{2+} ion channels in the membrane (Laurent, 1990, 1991) but no evidence for fast Na^+ currents (Burrows, 1996). The nature of neurotransmitters involved with nonspiking interneurons is not known.

9.8.4 Glial Cells

Glial cells are always present in nervous systems and are in intimate contact with neurons, ensheathing axonal and neurite processes, as well as the somata, of neurons. Glial cells serve neurons and the nervous system in a number of ways: they provide structural support, nutritive and metabolic functions, and protection from outside chemical and ionic influence (Freeman and Doherty, 2006). They may also provide regenerative guidance for regrowth of severed or damaged neuronal processes, repair and maintenance of neurons, and possibly function in signaling and information processing (Burrows, 1996; Kretzschmar and Pflugfelder, 2002; Oland and Tolbert, 2003). Multiple layers of glial cells help provide the BBB that protects the CNS from direct contact with chemical components in the hemolymph.

Typically in insects (and sometimes in vertebrates as well), many axons may be ensheathed by the same glial cell. For example, in *A. domesticus*, the house cricket, from 10 to 35 contiguous small-diameter axons, may share a glial cell in the dorsolateral bundles of the metathoracic nerve as well as in the cercal nerves of adult crickets. Even though a glial cell may wrap several times around an axon, it is not considered among neurophysiologists to be equivalent to the myelin sheath characteristic of vertebrate axons because the wrappings are not as numerous nor as tight in the case of insects. Insects do not have nodes of Ranvier and do not show saltatory conduction. Generally, an axon must be large to have a glial wrapping not shared with other axons. In the house cricket, an axon must have a diameter greater than 1 µm to have its own glial sheath. Glial cells serve protective functions by isolating neurons or their processes from each other and probably have a role in nutritive support of neurons, although direct evidence for this is not so clear in insects. They may also remove and degrade excess neurotransmitter, or transmitter precursors, and possibly other toxins, and store nutritive macromolecules (Smith and Treherne, 1963; Treherne and Pichon, 1972; Strausfeld, 1976).

Great numbers of glial cells are to be found in the peripheral parts of every ganglion where they surround the cell bodies of neurons and the neurites or axons arising from the cell body. Glial cell types have been described primarily on a morphological basis. Wigglesworth (1959) described four types of glia from *Rhodnius* and included the perineural cells that probably secrete the acellular neurolemma of the BBB as one type. Other authors have excluded the perineural cells from their classification of glial cells. Strausfeld (1976) describes four types based on electron microscopy:

1. **Type 1** neuroglia occur just inside the perineural cell layer and form tight junctions with perineural cells. The cellular processes from these neuroglia do not appear to enter the neuropil but wrap around neuronal somata located in the perimeter of the ganglion. Multiple wrappings of motoneuron somata are common, and motoneurons are more heavily wrapped than DUM neurons (Burrows, 1996). Extensive interdigitation occurs between glial cell processes and neuronal processes, especially at points where a neurite branches off from a soma to enter the neuropil.
2. **Type 2** glial cells have somata located interior to the Type 1 cells but still lie mainly at the peripheral edge of the ganglion. They isolate somata of neurons and surround neuronal processes passing into the neuropil.
3. **Type 3** cells are more interior in the ganglion and their somata lie just at the interface of the central neuropil. Type 3 cells send extensive lamellate, mossy, or spinniform processes into the neuropil and mainly insulate neuronal processes within the neuropil. Some Type 3 somata have extensions reaching far back into the periphery of a ganglion where they make tight junctional contacts with perineural cells.
4. **Type 4** neuroglia are the sheath cells that enclose single axons (large or giant axons) or groups of smaller axons in the brain, ganglia, ganglionic connectives, and in lateral nerves. When groups of small axons are enclosed by a single sheath cell, the enclosed axons may lie naked and adjacent to each other, or in other cases, the sheath cell invaginates a little way between the axons, partially isolating them (Strausfeld, 1976).

Synaptic junctions are free from glia, and some electron micrographs have shown indications that not even all neuronal processes in the neuropil may be isolated by glia. The number of glial cells in the nervous system is not known in any insect and may vary widely among insects. A gene, *repo*, is expressed in most glial cells, but not in neurons (Halter et al., 1995) of *D. melanogaster*, and may prove to be a good marker of glial cells and facilitate counting.

9.9 GIANT AXONS IN THE INSECT CENTRAL NERVOUS SYSTEM

In contrast to the very small diameter of most neurons (typical soma diameter is only a few micrometers, and the axon diameter is even less), some insects have **"giant"** axons of varying size running through the abdominal ganglia and into the thoracic ganglia. Table 9.1 shows data on some known giant axons. The CNS giant axons of insects develop by anastomosis of adjacent segmental neurons. The ganglionic junctional synapses are electrical rather than chemical and much faster conducting than chemical synapses. The large diameter of giants also promotes a more rapid rate of impulse conduction. A few such giant axons may be common in many, if not all, insects, but only a few have been studied. In the cockroach, the somata of the giant axons are located in the last abdominal ganglion, and the giants make synaptic contacts within the neuropil of the ganglion with sensory neurons from the paired cerci, the short appendages from the last segment of the insect. The giant axons run without synapsing through the abdominal ganglia and synapse in the thoracic ganglia with motoneurons going to the legs. These axons provide a rapid pathway for sensory information from the cerci to activate the legs in an escape behavior. A puff of air on the cerci causes a cockroach to make rapid escape movements. Some of the smaller giants also pass through the thorax and synapse in the brain (Hess, 1958). The giant axons in the CNS of dragonfly larvae (*Anax* spp.) run from the last abdominal ganglion to the thoracic ganglia, but these giants have synapses in each abdominal ganglion with motoneurons to abdominal muscles as well as with motoneurons in the thoracic ganglia to the legs (Fielden, 1960). The escape reaction in dragonfly larvae involves raising the legs, releasing the hold on the substrate, and simultaneous contracting abdominal muscles to force water out of the large rectum in a jet propulsion mechanism that propels the insect forward. Insect giant axons are hardly "giants" in comparison with the giant fibers of mollusks, some of which may be up to 1 mm in diameter. The very large axons in mollusks were very useful in determining the physiological properties of nerve impulse generation and transmission (Hodgkin and Huxley, 1952).

Giant axons have been found in the functioning of very fast trap jaws of ponerine ants in the genus *Odontomachus* (Gronenberg et al., 1993). These ants capture small prey in their jaws, which they lock open in a cocked position. Each of a number of mechanoreceptor hairs on the inner edges of the mandibles contains a large sensory neuron (15–20 μm in diameter) that passes through the mandibular nerve to the subesophageal ganglion. When the mechanoreceptor trigger hairs are touched by prey, the jaws snap closed in only 0.33–1 ms.

Table 9.1 Characteristics of Giant Axons of Selected Insects

Characteristics	*Periplaneta americana*	*Locusta migratoria*	*Anax* spp.
No. giant axons	6–8	4	6–7
Axon diameters (μm)	20–60	8–15	12–16
Location of somata	TAG[a]	TAG	Not determined
Sensory connections	Cerci	Cerci	Paraprocts
Connections to CNS	Thoracic MN[b]	Thoracic MN	Abdominal and thoracic MN

[a] TAG = terminal abdominal ganglion.
[b] MN = motoneurons.

9.10 NERVOUS SYSTEM CONTROL OF BEHAVIOR: MOTOR PROGRAMS

Motor programs are neural mechanisms for coordinating and regulating repetitive behaviors. Although motor programs control and support many, perhaps all, of the repetitive behavioral actions of insects, few have been defined in any detail, and very few identified neurons in the pathways are known. Several examples of motor programs are described here to illustrate their nature and complexity. Motor programs originating in the CNS allow an animal to conduct rhythmic behavior without requiring timing signals from rhythmically stimulated sensory organs (Delcomyn, 1980). Some motor neurons may function in more than one motor program, for example, in programs controlling walking, running, and posture, all involving the legs. A motor program controlling **ecdysis** has been described in Chapter 4.

9.10.1 Motor Program That Controls Walking

One of the better defined motor programs (Figure 9.13) controls muscles involved in walking by the American cockroach, *Periplaneta americana* (Pearson, 1972, 1976; Pearson and Fourtner, 1975; Fourtner and Pearson, 1976; Delcomyn, 1985; Graham, 1985). There are six control centers in the thoracic ganglia of a cockroach, one for each leg, with a segmental ganglion controlling the pair of legs attached to its segment. A small number of **central command (COM) neurons** provide coordination among the centers and ensure that only one leg at a time is moved. The COM neurons send output to a group of interneurons, the **flexor burst generator (FBG) neurons**. The FBG neurons produce rhythmic **excitatory** output to motoneurons whose axons synapse with flexor leg muscles and **inhibitory** output to neurons whose axons connect with extensor muscles of a leg. Thus, the leg can be flexed for the next step. The flexor muscles bend the leg and swing it forward. As it nears its full forward swing, hair sensilla near the coxal base are activated and send negative feedback to the FBG neurons, which inhibits their output to the flexor motoneurons and reduces inhibition of extensor motoneurons. The hair sensilla also have positive feedback to the extensor motoneurons, thus helping to make them ready for the leg to contact the substrate.

As the leg contacts the surface on which the cockroach is walking, the extensor muscles, now receiving activation from the COM neurons and the hair sensilla, extend the leg, push it backward, and move the body forward. As the leg takes some of the weight of the body, dome sensilla near the femoral–tibial joint are stimulated, and their input provides additional positive feedback to the extensor motor neurons while simultaneously inhibiting the FBG neurons. With the leg in this extended position, the hair sensilla at the coxal joint are not active, so their previous inhibition of the FBG neurons is relieved, and the FBG neurons now send impulses that flex the leg and swing it forward for the next step, and the cycle repeats.

Both hair sensilla and dome sensilla have positive and negative feedback loops so that when activated, they simultaneously stimulate one set of neurons and inhibit another set. Such positive and negative feedback loops are very common in control of antagonistic sets of muscles in all animals.

There are four types of sensory organs on the legs, including a femoral chordotonal organ, campaniform sensilla, hair plates, and hair sensilla (see Chapter 13 for more specific details about these sensory structures). Sensory input from one or more of the leg structures superimposes on the CNS output to provide proper timing of the start of leg movements and stance. Experimentally preventing input from only one of the four structures causes only minor alteration in leg movements, and the absence of all sensory input from the legs makes leg movement abnormal but does not destroy the ability of the insect to walk (Pearson and Iles, 1973).

9.10.2 Motor Pattern for Rhythmic Breathing

Large insects (generally larger than a *Drosophila*) have to have pumping or ventilatory movements of the abdomen in order to force air through the tracheal system and create exchange of tissue gases.

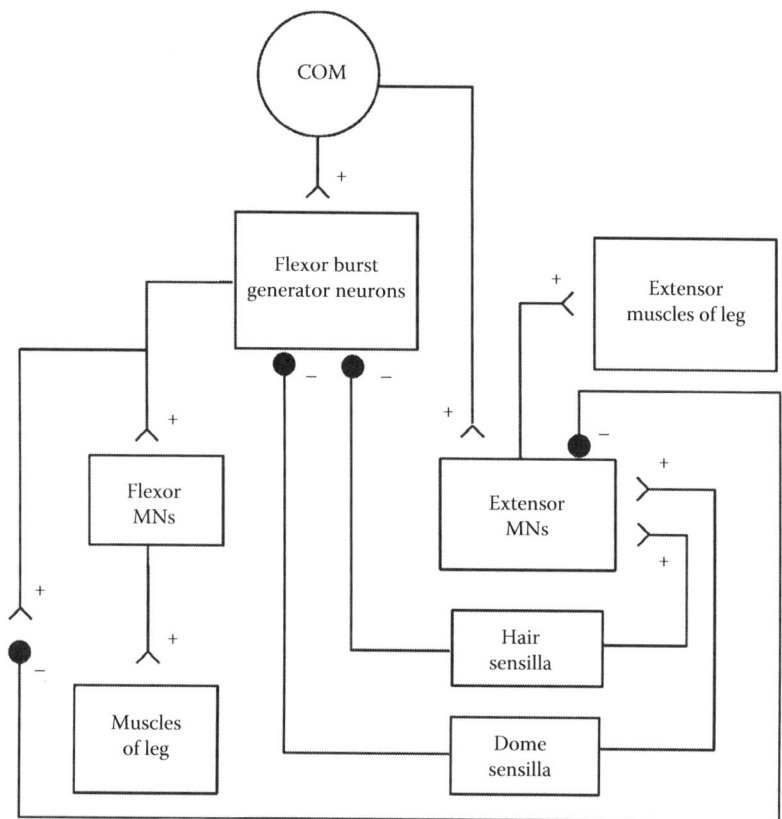

Figure 9.13 A schematic diagram of the major components in a motor program to control walking in the American cockroach, *Periplaneta americana*. (Adapted from Pearson, K.G., *J. Exp. Biol.*, 56, 173, 1972; Pearson, K.G. and Fourtner, C.R., *J. Neurophysiol.*, 38, 33, 1975; Fourtner, C.R., Central nervous control of cockroach walking, in Herman, R.M., Grillner, S., Stein, P.S.G., and Stuart, D.G. (eds.), *Neural Control of Locomotion*, Plenum Press, New York, 1976, pp. 519–537; Fourtner, C.R. and Pearson, K.G., Morphological and physiological properties of motor neurons innervating insect leg muscles, in Hoyle, G. (ed.), *Identified Neurons and Behavior of Arthropods*, Plenum Press, New York, 1976, pp. 87–99.)

A motor pattern has been observed in several large insects, and the pattern in the locusts, *S. gregaria* and *Locusta migratoria*, has been well described.

The driving force for the motor program originates in the metathoracic ganglion. The evidence for this location is that the (experimentally) isolated metathoracic ganglion maintains its rhythmical output of spike activity, whereas no other isolated thoracic or head ganglia display the pattern. Isolated abdominal ganglia have a rhythm, but it is slow and somewhat abnormal. The regularity and frequency of the output from the metathoracic ganglion indicate that it drives the rhythm in the abdominal ganglia. Only abdominal segments 3–8 participate in the ventilatory movements. Segments 1 and 2 have no dorsoventral muscles to function in the inspiratory phase, and segments 9–11 are highly modified to bear the genitalia. In an active, deeply ventilating locust, the participating abdominal segments contract together, but during very shallow ventilation, there is a delay of about 80–400 ms between activation of segments, so that a sort of ripple movement runs from anterior segments toward the posterior of the insect. The thorax, of course, is too rigid to participate in routine ventilatory movements, but in flight, the flight muscle contractions, as well as some movement of the thoracic wall, help to create a faster rate of airflow through the large tracheae.

Airflow through the large, longitudinal tracheal trunks is a directed flow from anterior to posterior. There are two spiracles in the thorax and eight in the abdomen of locusts. Spiracles 1–4 are open during inspiration and the remainder are closed. During expiration, the pattern is reversed, and in shallow ventilation, fewer spiracles are open in inspiration and only spiracle 10 may be open for expiration. Each spiracle opening is guarded by two valves. Some spiracles have both an opener and a closer muscle, while others have only a closer muscle.

Ventilatory movements in each segment involve 13 muscles receiving innervation from two pairs of lateral nerves and a single median nerve from each ganglion. Lateral nerve 1 contains axons of about 30 motoneurons, some of which terminate on dorsal longitudinal muscles. Contraction of these muscles pulls the segments closer together; in stressed ventilation, there is marked telescoping of the abdomen. These movements result in expansion of the abdomen and air is sucked in through the open anterior spiracles. Lateral nerve 2 contains axons of about 13 motoneurons, some of which innervate expiratory dorsoventral muscles. These axons display bursts of spike activity during expiration. Contraction of the dorsoventral muscles lifts the sternites upward and compresses the body cavity, forcing air out through open posterior spiracles. The single median nerve contains axons of four motoneurons that divide and innervate the spiracular valve muscles and muscles on each side of the body involved in inspiration in the next posterior segment. The axons in the median nerve spike during inspiration, indicating that the median nerve output is related to the inspiration phase of the cycle.

Lewis et al. (1973) proposed that an interneuron (IN 1) in the metathoracic ganglion produces a burst of spikes and is the COM neuron. It receives feedback stimulatory input and inhibitory input from receptors responding to carbon dioxide and oxygen and possibly other factors (e.g., neuromodulators) at sites in the CNS. In the model, IN 1 output acts as a brake on two coordinating interneurons (IN 2), one in each ventral nerve cord connective. Axons of IN 2 run the length of the ventral nerve cord and synapse in each ganglion with a small interneuron (IN 3) in each ganglion that controls expiratory motoneurons (MN 4s) for that segment. IN 3 also sends inhibitory input to inspiratory motoneurons (MN 5s) in each ganglion. IN 2s directly send weak inhibitory signals to MN 5s. The inspiratory MN 5s have a spontaneous firing rhythm, and when released from inhibition by IN 3s, they promote inspiration while simultaneously inhibiting the expiratory MN 4s through a sixth interneuron (IN 6). Feedback to IN 1 from CO_2 and O_2 receptors in the tissues determines how much it inhibits IN 2s and, thus, can determine the rate of ventilatory movements.

Coordination of these ventilatory movements by the rhythm driven from the metathoracic ganglion results in alternate contraction and expansion of the abdomen, sucking air in through open anterior spiracles and forcing it out through open posterior ones. Opening and closing of the spiracles is under the control of the median nerve, but the spiracle muscles receive rhythmic nerve input linked to the ventilatory rhythm.

9.11 NEUROSECRETORY CELLS AND NEUROSECRETION PRODUCTS FROM THE CNS

NSC and their secretory products are very important functional parts of the nervous system. The cells occur throughout the CNS. Neurosecretory products serve as hormones, neuromodulators, neurotransmitters, regulators of hormonal secretion, and have a variety of additional functions on many tissues.

9.11.1 Neurosecretory Cells

NSC usually have a very large soma, are usually monopolar, and occur in all ventral ganglia and the brain of insects. The somata are located peripherally in a ganglion. They are usually characterized by their large size and staining properties. Axonal processes from neurosecretory neurons often project to the periphery of the body, and staining suggests that they carry the neurosecretory products to functional sites. Except for a few neurotransmitter and modulating chemicals, most neurosecretory hormones (in both insects and vertebrates) are peptides or small proteins.

Neurosecretion, the secretion and release of products that may function as hormones and as neuromodulators, is one of the major functions of the nervous system. Neurosecretion is ideally suited for control of physiological and biochemical processes in which sustained stimulation is needed, such as secretion of prothoracicotropic hormone (PTTH) over several days in some insects in order to stimulate the prothoracic glands to begin to produce the molting hormone.

All known physiologically active molecules secreted by the nervous system (of all animals) have been peptides or small proteins (except neurotransmitters, such as acetylcholine, GABA, and some biogenic amines), and they are usually called neuropeptides. Immunocytochemistry, in which an internal secretory component reacts with antibodies prepared to identify specific neurosecretory products, has enabled cytologists to rapidly determine the location of cells that secrete specific products. Thus, more than 100 neuropeptide sequences have been described from insects, but few have proven functions because the products have not been isolated and a functional bioassay developed. Usually the neuropeptides found in insects have been small, composed of 10–15 or fewer amino acids. The sequence of a number has been determined and the molecules can be synthesized for bioassay tests. Nevertheless, a clearly defined function has been demonstrated for only a few of the isolated peptides, and more often a described function is tenuous and imprecise, often said to be "adipokinetic hormone (AKH)-like" or "proctolin-like" or having properties similar to some other well-defined neuropeptide.

Immunoreaction to rabbit antiserum is one of the favorite methods of peptide detection. The antiserum reagent often has been used in an Enzyme Linked Immunosorbent Assay (ELISA) reaction or complexed with a fluorescent dye, as is usually the case in immunohistochemistry. Current work to isolate and sequence neuropeptides is a very active research field, and new natural products and synthetically modified peptides based on a natural structure with physiological functions appear in the literature regularly. The greatest knowledge gap in neurosecretion at present is understanding the function of the many neuropeptides and characterization of receptors for the described neuropeptides.

Many regions of the CNS of insects have been mapped for the presence of various neuropeptide-reactive neurons, with detection of neuropeptides in interneurons, motoneurons, and NSC. A few of the major neuropil regions in the insect brain receive neuronal connections from many different neuropeptide-containing neurons. For example, the fan-shaped body in the protocerebrum is innervated by neurons that react to antisera to FaRPs, proctolin, AKH, leucokinins, locustatachykinin, and several other known neuropeptides. The pars intercerebralis in the protocerebrum and the medulla in the optic lobe show similar diversity in contacts with neuropeptide-containing neurons. This diversity of innervation is further argument for great diversity in function and behavior modulating activity of the insect CNS.

The identified peptides have been grouped into families based on similarity of structure. A family does not, however, necessarily indicate similarity of function. There are about 20 such families. Some of the neuropeptides have described functions as hormones (such as PTTH, AKH, eclosion hormone, and diuretic hormone), but it is highly likely that some neuropeptides function as neurotransmitters and as neuromodulators that could modify the input or output from neural connections. There is isolated evidence in insects of the co-localization of neuropeptides and neurotransmitters in the same nerve terminals, and neuropeptides are known in some cases to be released simultaneously with neurotransmitters. If some neuropeptides do indeed work in this way, a single neuron in a network may be able to regulate many variations on a basic behavior by modulation with neuropeptides. As an example, the cardioacceleratory peptides in *M. sexta* modulate four different behaviors at different periods in the life of the tobacco hornworm, with two modifying the feeding, ingestion, and nutrition and two others regulating heart activity in relation to wing inflation and flight (Tublitz et al., 1991). Neuromodulators also might alter response characteristics of neurons, including such activities as feedback, feed-forward, motor output, and muscle or gland response to nervous activity. Neuropeptides also may have roles in embryonic development (Hokfelt, 1991; Li et al., 1991, 1992) and as cytokines in nonself-recognition and response (Scharrer, 1991; Johnson et al., 1992).

Included here are a few of the neurosecretory peptides that have been found in insects to illustrate the diversity and function of neurosecretion. Additional details about the function of some of these compounds may be found in other chapters and in a review by Nässel (1993).

9.11.2 Adipokinetic Hormone

AKH, first isolated from the CC of *L. migratoria* (Stone et al., 1976), is a decapeptide with the amino acid (coded) sequence pQLNFTPNWGTamide. It is now known from a large number of insects and crustaceans and was described by a number of names. Identity of many of these with AKH was only recognized later. For example, substances with AKH functions were described from crustaceans as "red pigment concentrating hormone," and from cockroaches as periplanetins CCI-II, MI, MII, and as neurohormone D. About 20 members in this family have been sequenced from 7 orders of insects. Some are octapeptides instead of decapeptides, but usually they begin with pyroglutamate and have an amide function at the carboxy-terminal end. Locust AKH is synthesized as two inactive prohormones, pro-AKH-I and AKH-II, requiring two different messenger RNAs, followed by complex processing that eventually results in AKH-I and AKH-II and three dimeric peptides with as yet undescribed functions. In various bioassay preparations, AKH has functional activity on mobilization of lipids or carbohydrates, acceleration of heartbeat, myoactivity, and inhibition of fatty acid and protein synthesis (Orchard, 1987; Gäde, 1990). AKH-like immunoreactive (AKH-LI) cells are present in CC, brain, and subesophageal ganglion of many insects. The brain of the blowfly (*Calliphora*) contains about 50 AKH-LI neurons scattered in the proto-, deuto-, and tritocerebrum and several hundred in the medulla of each optic lobe. One very prominent AKH-LI neuron on each side of the protocerebrum of *Calliphora* has arborizations that cover most of the superior protocerebrum on the ipsilateral side.

9.11.3 Proctolin

Proctolin was the first insect neuropeptide to be sequenced (Starrat and Brown, 1975). It contains the amino acid sequence arginine–tyrosine–leucine–proline–threonine (amino acid code RYLPT). Proctolin has action on skeletal, heart, and visceral muscle. In crustaceans, it controls central pattern-generating nerve networks that regulate feeding and ventilatory behaviors. There are about 40 cell bodies that are proctolin immunoreactive in the brains of cockroaches, a similar number in Colorado potato beetle, from 80 to 90 in blowfly brain and about 100 in each lobulus (a specific neuropil region) of the optic lobe in blowflies. Proctolin in the insect brain may act as a neurohormone and as a modulator of responses within central synaptic neuropil (Nässel and O'Shea, 1987). In *M. sexta*, NSC that are proctolin immunoreactive send branches and arborizations to the CA, and in Colorado beetle, such terminals are found in the CC.

9.11.4 FMRFamide-Related Peptides

Neuropeptides with the general structure of **FMRFamides** (see Chapter 2, Table 2.1, for the amino acids identified with single letters) are widely represented throughout the Metazoa and are characterized by the amino acid sequence MRFamide at the carboxy-terminal end. Because of their wide distribution and important actions in many groups, the FaRPs are the best studied of all the neuropeptides. FaRPs have been isolated and sequenced from *Leucophaea cinerea*, *M. sexta*, *S. gregaria*, *Aedes* spp., and *Calliphora erythrocephala*. There is striking diversity in the distribution and number of FMRFamides within insects as well as in other groups. In *Calliphora,* 13 different sequences are found, of which CaliFRMFamide 5 is the most abundant and located in the ventral nerve cord. *D. melanogaster* has 13 known FMRFamides, while a closely related species, *D. virilis*, has only 10 known ones. Five are shared by the two species. One of those in *D. melanogaster*, a heptapeptide is the same as CalliFMRFamide 11 in *Calliphora*. The distribution of neurons with sequences that show a positive FMRFamide immunoreaction is very widespread in the brain of a number of insects, including the Colorado potato beetle, *Drosophila*, a blowfly, *M. sexta*, and the honeybee. About 240 cell bodies that react positive are located in the proto-, deuto-, and tritocerebrum and subesophageal ganglion of *Drosophila*. Few functions of FMRFamide peptides have been determined in insects, but CalliFMRFamide 1, 2, and 3 induce salivation from blowfly salivary glands at nanomolar concentrations. FMRFamide peptides may have multiple physiological effects in different tissues.

9.11.5 Tachykinins: Locustatachykinins and Leucokinins

Tachykinins are a large family of peptides in lower vertebrates and mammals, and now found in some insects. The group in vertebrates is best represented by **Substance P**. There are four *Locusta* tachykinins (**LomTK I, II, II, and IV**) isolated from the brain and CC of *L. migratoria*. There is about 50% homology of the sequence of LomTK I with a vertebrate tachykinin, physalaemin. Another group of insect kinins is the **leucokinins** first discovered in the cockroach, *Leucophea maderae*, but now known also from a cricket, *A. domesticus*, and a locust. The leucokinins show myotropic action on visceral muscle and influence ion transport in Malpighian tubules. Eight known leucokinins are octapeptides (**LK I–VIII**); there are five known **achetakinins** and one **locustakinin**. Neurons that show immunoreactivity for these peptides are present in the brains of the insects indicated by the names of the peptides. It is thought that both tachykinins and kinins act as neuromodulators, and the kinins also may have important roles as neurohormones. In *D. melanogaster*, two different tachykinin receptors have been isolated by recombinant DNA procedures. There is also evidence for leucokinin receptors in the gut of some insects. One tachykinin receptor protein has been identified and cloned from *Drosophila*, and it was expressed when put into mouse NIH-3t3 cells, causing the cells to increase synthesis of inositol trisphosphate (IP3) in response to locustatachykinin II (Monnier et al., 1992).

9.11.6 Pigment-Dispersing Factors

A number of octadecapeptides that have the ability to disperse pigment in chromatophores of some crustaceans have been isolated from insects and crustaceans. Although the bioassay utilizes the pigment-dispersing action of the compounds, pigment dispersing is not the normal physiological role of these peptides in insects because pigment dispersion is not a typical mechanism in insects as it is in some crustaceans. Highly characteristic groups of neurons that are immunoreactive for these peptides are associated with the visual system and seem to be similar in a number of insect species. These peptides may be involved in regulatory activity within the visual system, possibly regulating a circadian pacemaker system. A neurohormonal role cannot be excluded. Neurons with these neuropeptides are not so widely distributed within the nervous system.

9.11.7 Vasopressin-Like Peptide (Locust F2 Peptide)

Vasopressin is a peptide in vertebrates with activity on smooth muscle of blood vessels, and it can elevate blood pressure by causing constriction of the vessels. The role of this neurohormone in insects is not vasoconstriction, however. Two neurons in *L. migratoria* react with antisera against vasopressin, and these neurons have extensive axonal connections throughout the brain and optic lobes. Two locust peptides, F1 and F2, were isolated. F1 is a monomer and is inactive, but F2 is an antiparallel dimer of two F1s and is active in having diuretic activity in the locust assay. Recent reports indicate that F2 neurons have some input from the visual system, and it may indicate that F2 is secreted in response to a light-driven circadian rhythm.

9.11.8 Allatotropins and Allatostatins

Allatotropins (ATs) and **allatostatins (ASTs)** are neuropeptides isolated from nervous and some nonneural tissues (reviewed by Gilbert et al., 2000) that either stimulate or inhibit, respectively, the CA. Although several ATs have been discovered based on bioassays showing stimulation of CA (Gilbert et al., 2000; Stay, 2000), the only one of known structure is Mas-AT from *M. sexta*. Mas-AT has the structure Gly-Phe-Lys-Asn-Val-Glu-Met-Met-Thr-Ala-Arg-Gly-Phe-NH$_2$ (Kataoka et al., 1989). It also has been isolated from the lepidopterans *Spodoptera frugiperda* (fall armyworm) (Oeh et al., 2000) and *Lacanobia oleracea* (tomato moth) (Audsley et al., 2000). Physiological stimulation of juvenile hormone (JH) synthesis by ATs has generally been demonstrated in adult females, but Mas-AT stimulates, and Mas-AST inhibits, the larval CA of the tomato moth (Audsley et al., 2000).

There are three identified ASTs: Mas-AST from *M. sexta* (pGlu-Val-Arg-Phe-Arg-Gln-Cys-Tyr-Phe-Asn-Pro-Ile-Ser-Cys-Phe-OH), Dip-AST from the cockroach *Diploptera punctata* (a pentapeptide with amidated C-terminal sequence and a variable number of amino acids at the N-terminus in different cockroach species), and an AST from the cricket, *Gryllus bimaculatus*, that is similar to the cockroach family of ASTs and also a different AST. Mas-AST and Dip-AST have physiological action on larvae and adults and, in the case of Dip-AST, in the embryo. Action of the cricket AST has been demonstrated only in adults (Stay, 2000, and references therein). The *Diploptera* ASTs are a family of 13 ASTs, each of which has physiological action in inhibiting the CA but with strikingly different effectiveness. Tobe et al. (2000) suggest from experimental studies with mixture of the peptides that they likely act in concert to regulate JH biosynthesis by interacting with receptors in the CA.

ATs and ASTs are widely distributed in various tissues of insects and in other invertebrates, and they have physiological action on tissues other than the CA. For example, they are known to have action on the heart, midgut, hindgut, foregut, oviduct, and fat body in various insects (Stay, 2000), so it seems highly likely that they have fundamental actions unrelated to JH synthesis (Gilbert et al., 2000; Stay, 2000; Truesdell et al., 2000).

9.11.9 Crustacean Cardioactive Peptide

A **cardioactive peptide** with the sequence PFCNAFTGCamide has been isolated from *L. migratoria* and a crab. Similar or possibly identical peptides appear to occur in *Tenebrio molitor* and in *M. sexta*. As the name implies, one action may be to stimulate the heart, but its true function(s) in insects is not known. **Corazonin** is another cardioactive peptide that has been isolated from the cockroach *Periplaneta*. Corazonin antiserum D reacts with lateral NSC in the protocerebrum and in two descending neurons in the blowfly, suggesting the same or a very similar molecule. Additional functions for crustacean cardioactive peptide (CCAP) and corazonin in the molting process can be found in **Chapter 4, Integument**.

9.11.10 Pheromone Biosynthesis Activating Neuropeptide

Pheromone biosynthesis activating neuropeptide (PBAN), a neuropeptide, controls the biosynthesis of the pheromone in glands of some female moths, the best documented of which is *Helicoverpa* (formerly *Heliothis*) *zea* (Raina et al., 1989). PBAN immunoreactive neurons have also been demonstrated in the CNS of several other species (Choi et al., 2001, 2009, 2011; Choi and Vander Meer, 2012). More detailed discussion of PBAN function in pheromone biosynthesis is provided in Chapter 20.

REVIEW AND SELF-STUDY QUESTIONS

1. What parts of the insect nervous system comprise the CNS?
2. What is the neuropil?
3. What are the three major anatomical parts of the brain?
4. What part of the brain receives sensory input from the compound eyes in an adult insect?
5. What are the mushroom bodies? What is their function?
6. What part of the brain receives input information from the antennae?
7. Explain the sexual differences in the brain of insects.
8. What is the structure of a ganglion in the nervous system?
9. Is the ventral nerve cord a single or double cord of interconnections between ganglia?
10. Describe variations in the location of ventral ganglia in insects.
11. What, if anything, is the difference between a nerve and a neuron?
12. What is the name given to the cell body of neurons?
13. How is a dendrite defined? An axon?
14. Will insects typically have long or short dendrites?
15. What is the role or function of interneurons in the nervous system?
16. What is a glial cell and what are its functions?
17. What is a simple definition of a motor program? Describe an example.
18. Are NSC common in insects?
19. In what part of the nervous system are NSC found?
20. Where is the cell body of a sensory neuron found in insects?

REFERENCES

Abbott, N.J. and J.E. Treherne. 1977. Homeostasis of the brain microenvironment: A comparative account, in B.L. Gupta, R.B. Moreton, J.L. Oschman, and B.J. Wall (eds.), *Transport of Ions and Water in Animals*. Academic Press, London, U.K., pp. 481–510.

Ashhurst, D.E. and N.M. Costin. 1971a. Insect mucosubstances. II. The mucosubstances of the central nervous system. *Histochem. J.* 3: 297–310.

Ashhurst, D.E. and N.M. Costin. 1971b. Insect mucosubstances. III. Some mucosubstances of the nervous systems of the wax-moth (*Galleria mellonella*) and the stick insect (*Carausius morosus*). *Histochem. J.* 3: 379–387.

Audsley, N., R.J. Weaver, and J.P. Edwards. 2000. Juvenile hormone biosynthesis by corpora allata of tomato moth, *Lacanobia oleracea*, and regulation by *Manduca sexta* allatostatin and allatotropin. *Insect Biochem. Mol. Biol.* 30: 681–689.

Burrows, M. 1996. *The Neurobiology of an Insect Brain*. Oxford University Press, Oxford, U.K.

Busch, S., M. Selcho, K. Ito, and H. Tanimoto. 2009. A map of octopaminergic neurons in the *Drosophila* brain. *J. Comp. Neurol.* 513: 643–667.

Choi, M.-Y., A. Rafaeli, and R.A. Jurenka. 2001. Pyrokinin/PBAN-like peptides in the central nervous system of *Drosophila melanogaster*. *Cell Tissue Res.* 306: 459–465.

Choi, M.-Y., A. Raina, and R.K. Vander Meer. 2009. PBAN/pyrokinin peptides in the central nervous system of the fire ant, *Solenopsis invicta*. *Cell Tissue Res.* 335: 431–439.

Choi, M.-Y., R.K. Vander Meer, D. Shoemaker, and S.M. Valles. 2011. PBAN gene architecture and expression in the fire ant, *Solenopsis invicta*. *J. Insect Physiol.* 57: 161–165.

Choi, M.-Y. and R.K. Vander Meer. 2012. Molecular structure and diversity of PBAN/pyrokinin family peptides in ants. *Frontiers Endocrinol.* 3: 1–8. Doi:10.3389/fendo.2012.00032.

Dadant & Sons (eds.). 1975. *The Hive and the Honey Bee*, Revised ed. Dadant & Sons Publishers, Hamilton, IL.

Delcomyn, F. 1980. Neural basis of rhythmic behavior in animals. *Science* 210: 492–498.

Delcomyn, F. 1985. Factors regulating insect walking. *Annu. Rev. Entomol.* 30: 239–256.

Dethier, V.G. 1990. Five hundred million years of olfaction, in K. Colbow (ed.), *Frank Allison Linville's R.H. Wright Lectures on Olfactory Research*. Simon Fraser University, Burnaby, British Columbia, Canada, pp. 1–37.

Fielden, A. 1960. Transmission through the last abdominal ganglion of the dragonfly, *Anax imperator*. *J. Exp. Biol.* 37: 832–844.

Fourtner, C.R. 1976. Central nervous control of cockroach walking, in R.M. Herman, S. Grillner, P.S.G. Stein, and D.G. Stuart (eds.), *Neural Control of Locomotion*. Plenum Press, New York, pp. 519–537.

Fourtner, C.R. and K.G. Pearson. 1976. Morphological and physiological properties of motor neurons innervating insect leg muscles, in G. Hoyle (ed.), *Identified Neurons and Behavior of Arthropods*. Plenum Press, New York, pp. 87–99.

Freeman, M.R. and J. Doherty. 2006. Glial cell biology in *Drosophila* and vertebrates. *Trends Neurosci.* 29: 82–90.

Gäde, G. 1990. The adipokinetic hormone/red pigment-concentrating hormone peptide family: Structures, interrelationships and functions. *J. Insect Physiol.* 36: 1–12.

Galizia, C.G. and R. Menzel. 2001. The role of glomeruli in the neural representation of odours: Results from optical recording studies. *J. Insect Physiol.* 47: 115–130.

Gilbert, L.I., N.A. Granger, and R.M. Roe. 2000. The juvenile hormones: Historical facts and speculations on future research directions. *Insect Biochem. Mol. Biol.* 30: 617–644.

Graham, D. 1985. Pattern and the control of walking in insects. *Adv. Insect Physiol.* 18: 31–140.

Gronenberg, W., J. Tautz, and B. Holldobler. 1993. Fast trap jaws and giant neurons in the ant *Odontomachus*. *Science* 262: 561–563.

Halter, D.A., J. Urban, C. Rickert, S.S. Ner, K. Ito, A.A. Travers, and G.M. Technau. 1995. The homeobox gene *repo* is required for the differentiation and maintenance of glia function in the embryonic nervous system of *Drosophila melanogaster*. *Development* 121: 317–332.

Hess, A. 1958. The fine structure of nerve cells and fibres, neuroglia, and sheaths of the ganglion chain in the cockroach (*Periplaneta americana*) *J. Biophys. Biochem. Cytol.* 4: 731–742.

Heuer, C.M., M. Binzer, and J. Schachtner. 2012. SIFamide in the brain of the sphinx moth, *Manduca sexta*. *Acta Biol. Hung.* 63(Suppl. 2): 48–57.

Hildebrand, J.G. 1995. Analysis of chemical signals by nervous systems. *Proc. Natl. Acad. Sci. USA* 92: 67–74.

Hildebrand, J.G. 1996. Olfactory control of behavior in moths: Central processing of odor information and the functional significance of olfactory glomeruli. *J. Comp. Physiol A* 178: 5–19.

Hodgkin, A.L. and A.F. Huxley. 1952. Currents carried by sodium and potassium ions through the membrane of the giant axon of *Loligo*. *J. Physiol. (London)* 116: 449–472.

Hokfelt, T. 1991. Neuropeptides in perspective: The last ten years. *Neuron* 7: 867–879.

Homberg, U., J. Seyfarth, U. Binkle, M. Monastirioti, and M.J. Alkema. 2013. Identification of distinct tyraminergic and octopaminergic neurons innervating the central complex of the desert locust, *Schistocerca gregaria*. *J. Comp. Neurol.* 521: 2025–2041.

Homberg, U., T.A. Christensen, and J.G. Hildebrand. 1989. Structure and function of the deutocerebrum in insects. *Annu. Rev. Entomol.* 34: 477–501.

Hösl, M. 1990. Pheromone-sensitive neurons in the deutocerebrum of *Periplaneta americana*: Receptive fields on the antenna. *J. Comp. Physiol. A* 167: 321–327.

Iversen, L.L., S.D. Iversen, and S.H. Snyder (eds.). 1975. Amino acid transmitters, in *Handbook of Psychopharmacology*, vol. 4. Plenum Press, New York.

Jenkin, P.M. 1962. *Animal Hormones: A Comparative Survey*. Pergamon Press, Oxford, U.K.

Johnson, H.M., M.O. Downs, and C.H. Pontzer. 1992. Neuroendocrine peptide hormone regulation of immunity. *Chem. Immunol.* 52: 49–83.

Kahsai, L., M.A. Carlsson, Å.M.E. Winther, and D.R. Nässel. 2012. Distribution of metabotropic receptors of serotonin, dopamine, GABA, glutamate, and short neuropeptide F in the central complex of *Drosophila*. *Neuroscience* 208: 11–26.

Kahsai, L. and Å.M.E. Winther. 2011. Chemical neuroanatomy of the *Drosophila* central complex: Distribution of multiple neuropeptides in relation to neurotransmitters. *J. Comp. Neurol.* 519: 290–315.

Kahsai, L., J.-R. Martin, and Å.M.E. Winther. 2010. Neuropeptides in the *Drosophila* central complex in modulation of locomotor behavior. *J. Exp. Biol.* 213: 2256–2265.

Kataoka, H., A. Toschi, J.P. Li, R.L. Carney, D.A. Schooley, and S.J. Kramer. 1989. Identification of an allatotropin from adult *Manduca sexta*. *Science* 243: 1481–1483.

Kent, K.S. and J.G. Hildebrand. 1987. Cephalic sensory pathways in the central nervous system of larval *Manduca sexta* (Lepidoptera: Sphingidae). *Phil. Trans. Roy. Soc. London B* 315: 1–36.

Kloppenburg, P. and T. Heinbockel. 2000. 5-Hydroxytryptamine modulates pheromone-evoked local field potentials in the macroglomerular complex of the sphinx moth *Manduca sexta*. *J. Exp. Biol.* 203: 1701–1709.

Kondoh, Y., H. Morishita, T. Arima, J. Okuma, and Y. Hasegawa. 1991. White noise analysis of graded response in a wind-sensitive, nonspiking interneuron of the cockroach. *J. Comp. Physiol.* 168A: 429–443.

Kondoh, Y., T. Arima, J. Okuma, and Y. Hasegawa. 1993. Response dynamics and directional properties of nonspiking local interneurons in the cockroach cercal system. *J. Neurosci.* 13: 2287–2305.

Kretzschmar, D. and G.O. Pflugfelder. 2002. Glia in development, function, and neurodegeneration of the adult insect brain. *Brain Res. Bull.* 57: 121–131.

Kunst, M., R. Pförtner, K. Aschenbrenner, and R. Heinrich. 2011. Neurochemical architecture of the central complex related to its function in the control of grasshopper acoustic communication. *PLoS One* 6: c25613.

Kurylas, A.E., S.R. Ott, J. Schachtner, M.R. Elphick, L. Williams, and U. Homberg. 2005. Localization of nitric oxide synthase in the central complex and surrounding midbrain neuropils of the locust *Schistocerca gregaria*. *J. Comp. Neurol.* 484: 206–223.

Lane, N.J. and H. leB. Skaer. 1980. Intercellular junctions in insect tissues. *Adv. Insect Physiol.* 15: 35–213.

Laurent, G. 1990. Voltage-dependent nonlinearities in the membrane of locust nonspiking local interneurons, and their significance for synaptic integration. *J. Neurosci.* 10: 2268–2280.

Laurent, G. 1991. Evidence for voltage-activated outward currents in the neuropilar membrane of locust nonspiking local interneurons. *J. Neurosci.* 11: 1713–1726.

Lemon, W.C. and W.M. Getz. 1999. Neural coding of general odors in insects. *Ann. Entomol. Soc. Am.* 92: 861–872.

Lewis, G.W., P.L. Miller, and P.S. Mills. 1973. Neuro-muscular mechanisms of abdominal pumping in the locust. *J. Exp. Biol.* 59: 149–168.

Li, X.J., W. Wolfgang, Y.N. Wu, R.A. North, and M. Forte, M. 1991. Cloning, heterologous expression and developmental regulation of a *Drosophila* receptor for tachykinin-like peptide. *EMBO J.* 10: 3221–3229.

Li, X.J., Y.N. Wu, R.A. North, and M. Forte. 1992. Cloning, functional expression and developmental regulation of a neuropeptide Y receptor from *Drosophila melanogaster*. *J. Biol. Chem.* 267: 9–12.

Marion-Poll, F. and T.R. Tobin. 1992. Temporal coding of pheromone pulses and trains in *Manduca sexta*. *J. Comp. Physiol.* 171: 505–512.

Monnier, D.J., F. Cloas, P. Rosay, R. Hen, E. Borelli, and L. Maroteaux. 1992. NKD, a developmentally regulated tachykinin receptor in *Drosophila*. *J. Biol. Chem.* 267: 1298–1302.

Müller, U. 1997. The nitric oxide system in insects. *Prog. Neurobiol.* 51: 363–381.

Nässel, D.R. 1993. Ncuropeptides in the insect brain: A review. *Cell Tissue Res.* 273: 1–29.

Nässel, D.R. 1999. Histamine in the brain of insects: A review. *Microsc. Res. Tech.* 44: 121–136.

Nässel, D.R. 2002. Neuropeptides in the nervous system of *Drosophila* and other insects: Multiple roles as neuromodulators and neurohormones. *Prog. Neurobiol.* 68: 1–84.

Nässel, D.R. and M. O'Shea. 1987. Proctolin-like immunoreactive neurons in the blowfly central nervous system. *J. Comp. Neurol.* 265: 437–454.

Nässel, D.R. and U. Homberg. 2006. Neuropeptides in interneurons of the insect brain. *Cell Tissue Res.* 326: 1–24.

Oeh, U., M.W. Lorenz, H. Dyker, P. Lösel, and K.H. Hoffmann. 2000. Interaction between *Manduca sexta* allatotropin and *Manduca sexta* allatostatin in the fall armyworm *Spodoptera frugiperda*. *Insect Biochem. Mol. Biol.* 30: 719–727.

Oland, L.A. and L.P. Tolbert. 2003. Key interactions between neurons and glial cells during neural development in insects. *Annu. Rev. Entomol.* 48: 89–110.

Orchard, I. 1987. Adipokinetic hormone: An update. *J. Insect Physiol.* 33: 451–463.

Pearson, K. 1976. The control of walking. *Sci. Am.* 235: 72–86.

Pearson, K.G. 1972. Central programming and reflex control of walking in the cockroach. *J. Exp. Biol.* 56: 173–193.

Pearson, K.G. and C.R. Fourtner. 1975. Nonspiking interneurons in the walking system of the cockroach. *J. Neurophysiol.* 38: 33–52.

Pearson, K.G. and J.F. Iles. 1973. Nervous mechanisms underlying intersegmental coordination of leg movements during walking in the cockroach. *J. Exp. Biol.* 58: 725–744.

Pfeiffer, K. and U. Homberg. 2014. Organisation and functional roles of the central complex in the insect brain. *Annu. Rev. Entomol.* 59: 165–184.

Raina, A.K., H. Jaffe, T.G. Kempe, P. Keim, R.W. Blacher, H.M. Fales, C.T. Riley, J.A. Klun, R.L. Ridgway, and D.K. Hayes. 1989. Identification of a neuropeptide hormone that regulates sex pheromone production in female moths. *Science* 244: 796–798.

Rodrigues, V. and L. Pinto. 1989. The antennal glomerulus as a functional unit of odor coding in *Drosophila melanogaster*, in R.N. Singh and N.J. Strausfeld (eds.), *Neurobiology of Sensory Systems*. Plenum Press, New York, pp. 387–396.

Rospars, J.P. and J.G. Hildebrand. 1992. Anatomical identification of glomeruli in the antennal lobes of male sphinx moth *Manduca sexta*. *Cell Tissue Res*. 270: 205–227.

Scharrer, B. 1991. Neuroimmunology: The importance and role of a comparative approach, in G.B. Stefano and E.M. Smith (eds.), *Advances in Neuroimmunology*, vol. 1. Manchester University Press, Manchester, U.K., pp. 1–6.

Scharrer, B.C.J. 1939. The differentiation between neuroglia and connective tissue sheath in the cockroach (*Periplaneta americana*). *J. Comp. Neurol*. 70: 77–88.

Siegl, T., J. Schachtner, G.R. Holstein, and U. Homberg. 2009. NO/cGMP signalling: L-citrulline and cGMP immunostaining in the central complex of the desert locust *Schistocerca gregaria*. *Cell Tissue Res*. 337: 327–340.

Smith, D.S. and J.E. Treherne. 1963. Functional aspects of the organization of the insect nervous system. *Adv. Insect Physiol*. 1: 140–148.

Snodgrass, R.E. 1929. The thoracic mechanism of a grasshopper, and its antecedents. *Smithson. Misc. Collect*. 82: 1–111.

Snodgrass, R.E. 1935. *Principles of Insect Morphology*. McGraw-Hill, New York.

Starrat, A.N. and B.E. Brown. 1975. Structure of the pentapeptide proctolin, a proposed neurotransmitter in insects. *Life Sci*. 17: 1253–1256.

Stay, B. 2000. A review of the role of neurosecretion in the control of juvenile hormone synthesis: A tribute to Berta Scharrer. *Insect Biochem. Mol. Biol*. 30: 653–662.

Stocker, R.F. 1994. The organization of the chemosensory system in *Drosophila melanogaster*: A review. *Cell Tissue Res*. 275: 3–26.

Stone, J.V., W. Mordue, K.E. Betley, and H.R. Morris. 1976. Structure of locust adipokinetic hormone, a neurohormone that regulates lipid utilization during flight. *Nature* 265: 207–211.

Strausfeld, N.J. 1976. *Atlas of an Insect Brain*. Springer-Verlag, New York.

Sun, X.J., L.P. Tolbert, and J.G. Hildebrand. 1993. Ramification pattern and ultrastructural characteristics of the serotonin immunoreactive neuron in the antennal lobe of the moth *Manduca sexta*: A laser scanning confocal and electron microscopic study. *J. Comp. Neurol*. 338: 5–16.

Tobe, S.S., J.R. Zhang, P.R.F. Bowser, B.C. Donly, and W.G. Bendena. 2000. Biological activities of the allatostatin family of peptides in the cockroach, *Diploptera punctata*, and potential interactions with receptors. *J. Insect Physiol*. 46: 231–242.

Treherne, J.E. 1985. Blood-brain barrier, in G.A. Kerkut and L.I. Gilbert (eds.), *Comprehensive Insect Physiology Biochemistry and Pharmacology*, vol. 5. Pergamon Press, Oxford, U.K., pp. 115–137.

Treherne, J.E. and Y. Pichon. 1972. The insect blood-brain barrier. *Adv. Insect Physiol*. 9: 257–313.

Truesdell, P.F., P.M. Koladich, H. Kataoka, K. Kojima, A. Suzuki, J.N. McNeil, A. Mizoguchi, S.S. Tobe, and W.G. Bendena. 2000. Molecular characterization of a cDNA from the true armyworm *Pseudaletia unipuncta* encoding *Manduca sexta* allatotropin peptide. *Insect Biochem. Mol. Biol*. 30: 691–702.

Tublitz, N., D. Brink, K.S. Broadie, P. Loi, and A.W. Sylwester. 1991. From behavior to molecules: An integrated approach to the study of neuropeptides. *Trends Neurosci*. 14: 254–259.

Wenzel, B., M. Kunst, C. Günther, G.K. Ganter, R. Lakes-Harlan, N. Elsner, and R. Heinrich. 2005. Nitric oxide/cyclic guanosine monophosphate signaling in the central complex of the grasshopper brain inhibits singing behavior. *J. Comp. Neurol*. 488: 129–139.

Wigglesworth, V.B. 1959. The histology of the nervous system of an insect, *Rhodnius prolixus* (Hemiptera). II. The central ganglia. *Q. J. Micro. Sci*. 100: 299–313.

CHAPTER 10

Neurophysiology

PREVIEW

Neurons function like batteries, that is, they develop and store a potential difference across the cell membrane. With appropriate stimulation, a neuron discharges a flow of electricity along its axonal or dendritic processes. Afferent axons synapse with interneuronal processes, which enable the stimulation to be passed on to many other neurons, including motor neurons. Axons from motor neurons synapse directly with glands or muscles. In nearly all cases the transmission from neuron–neuron or from neuron–tissue is by chemical transmission. A few electrical synapses occur in the central nervous system (CNS) of insects in which neuronal processes have physically fused so that chemical transmission is not necessary. Graded neuronal responses occur at synapses and receptor neuron endings. Graded responses develop relatively slowly, are not self-propagated, and are proportional in strength to the stimulus intensity. Sufficiently strong-graded potentials typically lead to a generation of spikes or all-or-none potentials at the axon hillock, a region of the axon where spikes can be produced. All-or-none potentials are called "action potentials." Action potentials are not proportional to the stimulus strength, provided the stimulus is above the threshold for spike generation. They rise extremely rapidly, last a few milliseconds, and are propagated along the axon without decrement. The resting potential is the potential difference across the cell membrane when the cell is not being stimulated. Typically in insects, the resting potential across the axon membrane is about 70 mV, inside negative to the outside. The resting potential depends on ion distribution, which is a result of a Na^+–K^+ exchange pump that pumps Na^+ out of the cell and brings K^+ into the cell. The resting membrane is very impermeable to Na^+ reentry. Other ions involved include negatively charged proteins inside the neuron, and Cl^- on both sides of the neuronal membrane. The outside of the neuron refers to the very small space (called the mesaxon) between neuronal membrane and the surrounding glial cell; all neurons are surrounded by glial cell membranes. Thus, it is the distribution of ions between the inside of the neuron and its mesaxon space that determines the resting potential.

Stimulation above the characteristic threshold for the neuron causes an action potential, in which Na^+ channels rapidly open, allowing an influx (picomolar amounts) of Na^+. This influx of positive ions reverses the potential so that the inside of the neuron is briefly positive to the outside. Na^+ channels close in a few milliseconds and an outward flow of K^+ ions (again picomolar amounts) repolarizes the neuron and restores the resting potential. The Na^+–K^+ pump only works in long-term maintenance, not in restoration of the resting potential after a stimulus. Transmission across synapses is by chemical diffusion, a slower process than the flow of electrical current represented by the transmission of an action potential. If the synaptic transmitter chemical is acetylcholine or L-glutamic acid, the synapse is a stimulatory synapse, and the postsynaptic potential is called an **excitatory postsynaptic potential** (**EPSP**). At inhibitory synapses the neurotransmitter is gamma (γ)-aminobutyric acid (GABA) and the potential is called an **inhibitory postsynaptic potential** (**IPSP**). Acetylcholine is the stimulatory neurotransmitter at neuron–neuron synapses in the central nervous system (CNS), and L-glutamic acid and possibly L-aspartic acid are stimulatory transmitters at neuromuscular junctions. The only inhibitory transmitter known from insects is GABA.

10.1 INTRODUCTION

Neurons are composed of a cell body, the **soma**, and **axonal** and **dendritic processes**. **Integration**, the alteration or modification of electrical signals, may occur at a number of levels and sites within a single neuron and in the neuropil of ganglia. The corpora pedunculata (mushroom bodies) and the central body in the protocerebrum are examples of major integrative centers.

This chapter illustrates the functioning of neurons with examples from insect biology whenever possible. The reader should be aware, however, that most of the experiments that first revealed nerve function were conducted on organisms other than insects, mostly on mollusks and some crustaceans because they have very large, giant axons (some approaching 1 mm in diameter) and, in the case of marine mollusks, seawater was a sufficient saline in which to study the properties of neurons. Enough of the major neurophysiological experiments have been repeated on insects to confidently verify that the basic pattern of nerve cell function in insects agrees with principles derived from other groups (Pichon and Ashcroft, 1985). Indeed, experiments on all major groups of animals show that the physiological and biochemical principles of nerve function evolved early in the evolution of animals and have been highly conserved in the course of evolution.

Detailed study of single nerve cells began in the early 1930s, and the three outstanding leaders were Alan L. Hodgkin, Andrew F. Huxley, and Sir John C. Eccles. They and their colleagues conducted innovative experimentation in nerve function, mostly with mollusks and crustaceans, which culminated in the Nobel Prize in Physiology or Medicine awarded to Hodgkin, Huxley, and Eccles in 1963. The physiological model (Eccles, 1964; Hodgkin, 1964; Huxley, 1964) that grew out of these studies came to be called simply the **Hodgkin and Huxley model**, and it is the basis for understanding how neurons function in insects.

10.2 NERVE CELL RESPONSES TO STIMULI

A number of different types of electrical responses (potentials) from a neuron are possible in response to a stimulus. When an experimental electrical stimulus (electrical stimulation is the usual means of stimulating neurons in the laboratory because the stimulus characteristics can be controlled and repeated) is delivered to a nerve cell membrane, even if it is too weak to excite the cell into action, it causes a passive alteration in the membrane potential, called **electrotonus** (Figure 10.1), because cells are conductors of electricity. The electrotonic effect passively spreads along the length of the cell as far as the natural tissue resistance and capacitance will allow. If the tissue is electrically excitable, however, and the electrical stimulation is strong enough, the neuron responds with a **graded membrane response**, often called a **local potential**. An even stronger stimulus can cause the neuron to respond by changing the graded response into an **all-or-none spike**, the **action potential** (Figure 10.1). A graded response

NEUROPHYSIOLOGY

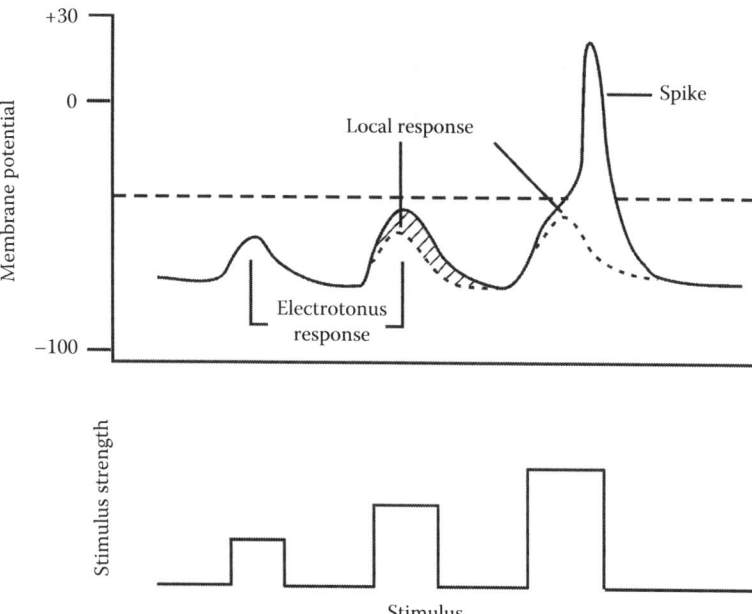

Figure 10.1 A conceptual diagram to illustrate the difference between passive electrotonic response, graded (local) potential response, and spike response of a nerve cell membrane subjected to stimuli by increasingly large square wave current pulses.

always precedes a spike. In some neurons, any part of the neuron may conduct a spike, but typically, in insect neurons, spikes are conducted by axonal processes.

Electrically excitable cells have a characteristic **membrane threshold** that must be exceeded in order to generate a response. Generally, a stimulus must be strong enough to cause a change of about 10–15 mV in the axon membrane potential to exceed the threshold and thereby generate a spike. An oscilloscope or computer usually is used to visualize nerve activity and permanent recordings are usually made on tape for playback and evaluation. Software is currently available that allows a computer to store and process data from stimulation of nerve cells. Action potentials similar to that shown in Figure 10.2 have been recorded from axons of many insects (Pichon and Boistel, 1966; Treherne and Maddrell, 1967; Gwilliam and Burrows, 1980; Tanouye et al., 1981).

10.2.1 Graded Responses

Graded responses are very important responses that nerve cells make, and some neurons (e.g., some interneurons in the ventral nerve cord) only make graded responses. Graded responses are often named after the site of their occurrence, such as synaptic potentials, receptor potentials, pacemaker potentials, or local potentials. Graded potentials always are localized and usually do not spread very far from their origin; they are propagated decrementally, becoming weaker the farther away they travel from their origin, and they soon are extinguished by resistance of the tissue to current flow. Their intensity is proportional to the strength of the stimulus. They rise and fall slowly in comparison with the 1–3 ms time duration of a spike. Because they rise and fall slowly, graded potentials are sometimes simply called **slow potentials**. Some defining characteristics of graded potentials compared with characteristics of spikes are described in Table 10.1.

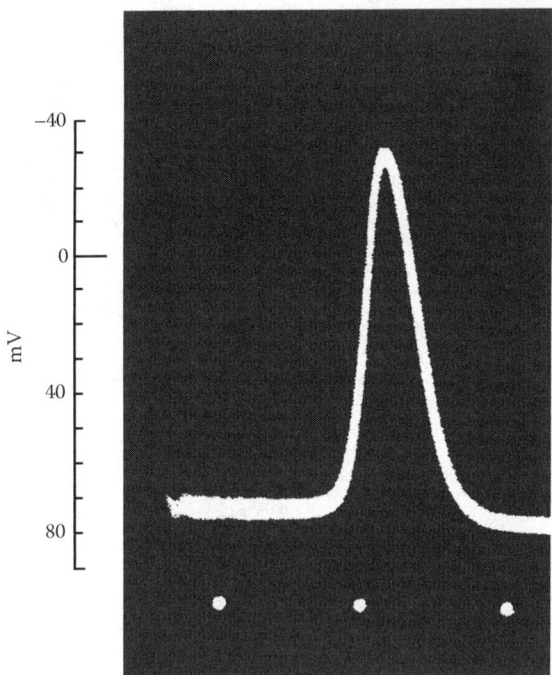

Figure 10.2 An action potential or spike recorded from a cockroach giant axon. The recording and potentials are displayed relative to the outside of the axon; thus, the overshoot, the magnitude by which the outside of the neuron becomes negative to the inside, is indicated by the height of the spike above the "0" potential. At rest, the outside potential of an axon is positive to the inside, which is the negative pole. Transitory reversal of polarity occurs across the axonal membrane during a spike. The dots along the *x*-axis indicate time in milliseconds. (From Yamasaki, T. and Narahashi, T., *J. Insect Physiol.*, 3, 146, 1959. With permission.)

Table 10.1 Comparison of Graded Potential and Spike Potential Characteristics

Response Characteristics	Type of Potential	
	Graded	Spike
1. Rate of rise	Slow, related to stimulus strength	Rapid, not related to stimulus strength[a]
2. Threshold	No threshold	Definite threshold
3. Magnitude	Related to stimulus strength	Characteristic of the neuron; not related to stimulus
4. Propagation	Decremental, usually	Self-generating, over a few mm only, nondecremental
5. Refractory period	None	Definite relative and absolute refractory period
6. Ability to summate	Yes	No

[a] The stimulus is assumed to be above the threshold for the neuron.

10.2.2 Spike Potentials

All-or-none spike or action potentials rise and fall very rapidly, self-generate along the axon, and are propagated without decrement. The size of the spike is not proportional to the stimulus strength, provided that the stimulus exceeds the spike threshold. Within the same neuron over a short period, spikes may be about the same size, but different neurons develop spikes of different sizes. Investigators are often interested in a train or burst of spikes arising from the stimulation of a receptor and can sometimes determine how many neurons may be responding in the receptor by the differential size of the recorded spikes.

10.3 PHYSIOLOGICAL BASIS FOR NEURONAL RESPONSES TO STIMULI

10.3.1 Membrane Ion Channels: Bioelectric Potentials

Ion homeostasis is necessary for neurons to continue to function, but stimuli cause changes in membrane permeability of excitable cells and temporarily alter ionic homeostasis. Neurons are like a charged battery: a potential difference across the cell membrane of about 70 mV is common in insect neurons, but lower and higher values have been recorded. The potential difference usually is written with a negative sign to connote that the inside is negative to the outside of the cell. The potential difference results from the unequal distribution of ions (both inorganic and organic) on the two sides of the membrane. Moreover, there is differential permeability to diffusible ions between the inside and outside.

The major ions involved in the transmembrane potential of nerve cells are K^+, Na^+, Cl^-, Ca^{2+}, and large negatively charged organic ions (largely proteins$^-$). The membrane has different permeabilities to each of these, and the permeabilities are very different in a resting neuron than in one undergoing stimulation. There is virtually no permeability to large negatively charged protein ions at any time in a healthy neuron, and permeability to the other ions is variable and depends on the physiological state and voltage across the membrane. The permeabilities to K^+ and Na^+ in a resting neuron can change rapidly upon stimulation. Chloride ions tend to follow the dictates of the distributions of the positive ions. In nerve cells, calcium is mainly involved with the release of neurotransmitters at the presynaptic terminals, and is considered in that respect later.

The distribution of sodium, potassium, chloride, and negatively charged proteins that influence the membrane potential can be represented as follows:

$$K^+_{inside} + Na^+_{inside} = Protein^-_{inside} + Cl^-_{inside} \tag{10.1}$$

$$K^+_{outside} + Na^+_{outside} = Protein^-_{outside} + Cl^-_{outside} \tag{10.2}$$

It is important to note that even though there is a potential difference *across* the cell membrane, there is electrical neutrality on one side. The total negatively charged ions on a given side equals the total positively charged ions on the same side. The "outside" in Equation 10.2 is the small space, the **mesaxon** between the nerve cell membrane and the membrane of the protective glial cell. The mesaxon space and representative ion concentrations in the axon, mesaxon space, and the surrounding hemolymph of a cockroach neuron are shown in Figure 10.3. Glial cells that surround all neurons play a major role in ion regulation and ion homeostasis (Money et al., 2009; Rodgers et al., 2009; Kocmarek and O'Donnell, 2011; Spong and Robertson, 2013). The concentration of protein is normally very low in the mesaxon space; therefore most of the negatively charged ions in the mesaxon space are chloride ions. Thus, Cl^- concentration is normally greater outside the neuron than inside because the inside negative charge is shared by Cl^- and large nondiffusible negative ions, such as proteins$^-$.

During depolarization and repolarization of a nerve cell, ions move through microscopic **pores** in the membrane. **Transmembrane proteins** (Figure 10.4) control the movement of the ions by forming narrow, hydrophilic channels through the cell membrane. These pores, which are capable of opening and closing rapidly, are frequently described as **gates**, **channels**, **gated channels**, and **ion channels**. These terms are used interchangeably. Some ion channels are very selective to a particular ion, while others are relatively nonselective. Potassium channels and sodium channels in nerve cell membranes generally are very selective for the named ion. Numerous ion channels occur in biological tissues (Stevens, 1984), but the most important ones for nerve cell function (and muscle cells) are channels for Na^+, K^+, Cl^-, and Ca^{2+}.

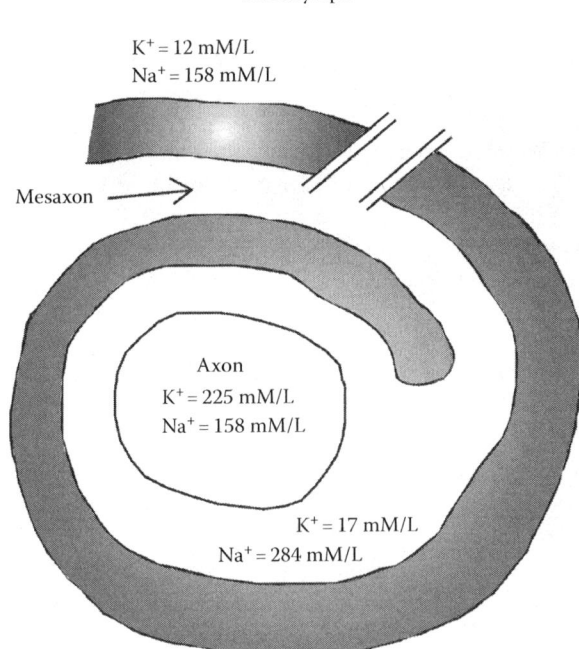

Figure 10.3 The diagram shows the mesaxon and the importance of the glial cell that protects a nerve cell from hemolymph ion concentrations. The sodium–potassium exchange pump maintains the distribution of ions within the mesaxon channel necessary for normal nerve cell function. The concentrations of ions shown come from data in Narahashi and Yamasaki (1960).

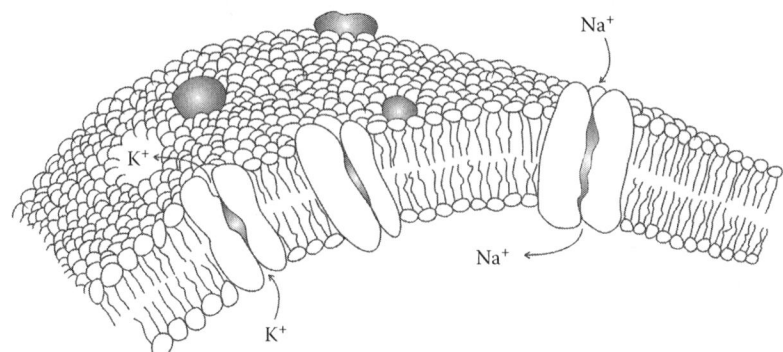

Figure 10.4 A conceptual illustration of transmembrane proteins functioning as ion channels. Some channels are ion selective, while others may allow passage of several different ions. Binding of a ligand by the ion channel protein(s) opens some ion channels, while others may be voltage gated and open in response to changes in the voltage across the cell membrane.

Channels are **ligand gated** (Figure 10.5), if a neurotransmitter or other molecule controls the opening of the gates, or **voltage gated**, if the membrane voltage controls the opening of the channels. Acetylcholine, for example, and other neurotransmitters at nerve and muscle synapses are ligands that bind to one or more of the channel proteins composing the gate and cause the gate to open. Some potassium gates (those involved in membrane repolarization) are voltage gated, but others in *Drosophila melanogaster* muscle are known to be calcium activated (i.e., ligand gated, with calcium as the ligand) (Ganetzky et al., 1993).

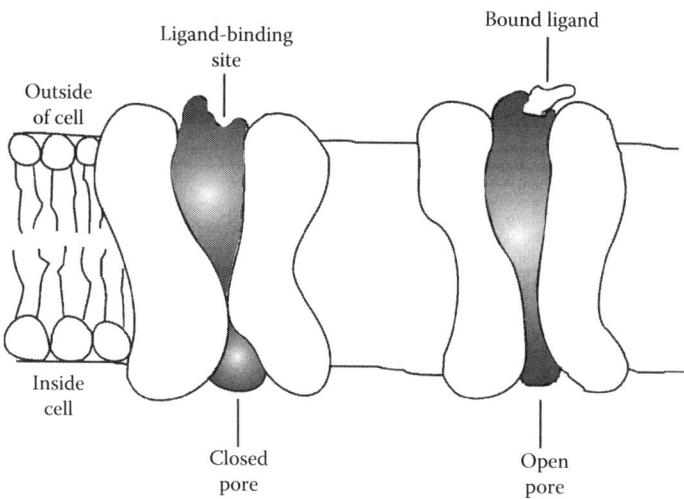

Figure 10.5 An illustration of an ion channel controlled by the binding of a ligand. The hypothetical channel in this diagram is composed of four transmembrane proteins, but only three are shown; the fourth protein forming the front of the channel has been omitted to allow a view of the channel. The ligand is shown being bound to the stippled transmembrane protein and opening the channel.

Sodium channels along an axon are voltage gated, and are opened by the flow of weak **local currents** flowing out from the main site of depolarization. Once an all-or-none depolarization develops, it gives rise to the local currents that depolarize the region ahead of the action potential, thus leading to the "self-generating" movement of the depolarization along the neuron. Sodium channels, as well as other channels, can be demonstrated by use of antibodies to specific peptide sequences in channel proteins. In the thoracic ganglion of a cockroach, an appropriate antibody intensely stains axons in tracts, commissures, and nerves, indicating many voltage-gated sodium channels. There is very little staining of the cell body membranes, indicating few sodium channels in the soma (French et al., 1993). In this particular study of the cockroach, the density of channels in the membrane was about 90 channels per µm^2 and this is similar to the density in some noninsect axons. An analysis of sodium channels in larvae and adult moths of *Heliothis virescens* shows that the channels generally exhibit similar characteristics to sodium channels in vertebrates, but the moth channels are more sensitive to scorpion toxin than vertebrate channels (Lee and Adams, 2000).

The precise mechanisms by which transmembrane proteins control movement of ions through cell membranes are still not well known, but evidence is accumulating that conformational changes in some of the transmembrane proteins making up a channel open a pore between adjacent proteins. A conceptual model (Armstrong and Bezanilla, 1977) for ion channel function is known as the **ball and chain model**. The model consists of several transmembrane protein domains, and an intracellular ball and chain sequence of amino acids. A molecular conformation allows the ball portion to swing in to block the channel or swing out to open it. Channel proteins may be composed of a variable number of transmembrane segments and protein subunits.

Drosophila has been important in the study of potassium ion channels. The *Shaker* (*Sh*) gene codes for the proteins that make the potassium channel in *D. melanogaster*, and it was the first potassium channel gene to be cloned from any biological tissue and studied in detail (Timpe et al., 1988). The *shaker* potassium channel polypeptide includes seven hydrophobic transmembrane segments (Miller, 1991) that loop back and forth across the membrane. Six segments of the molecule exist as alpha helices that span the membrane, while one segment, a beta-hairpin loop within the membrane, forms the pore and is capable of conformational changes that allow or occlude the passage of K$^+$. The ball portion of the model is thought to consist of amino acid residues 1–20 and the chain residues 23–40. The functional gate consists of a tetramer of four polypeptide subunits (Li et al., 1992; Mackinnon et al., 1993).

A number of other potassium channel genes have been cloned from *D. melanogaster*, including three genes, *Shaw*, *Shab*, and *Shal*, that are similar to *Shaker* (Salkoff et al., 1992), all of which code for channel proteins that form voltage-gated channels. Ganetzky et al. (1993) cloned two additional *Drosophila* potassium channel genes, *slowpoke* (*slo*) and *ether à go-go* (*eag*), two genes that code for calcium-activated potassium ion channels in muscle cells. The gene for *slo* is expressed in neurons of the CNS and peripheral system, in muscle cells, in some cells in the midgut, and in tracheal cells of *D. melanogaster* (Becker et al., 1995). It is likely that the gene controls ion movements in cells with very different functionalities. *Drosophila* gene probes have been used to isolate potassium channel genes from other insects and from vertebrates. For example, a homolog to *slo* was isolated from a mosquito, *Aedes* sp., with >90% similarity to the *Drosophila* gene in coding for amino acid sequence. Both *slo* (Ganetzky et al., 1993) and *eag* (Warmke and Ganetzky, 1993) homologs have been cloned from mouse and human tissues. The proteins coded by the mouse and human genes have 71% and 48% identity, respectively, with *Drosophila eag* protein. The *slo* proteins from mice and humans have about 70% identity to the *Drosophila* proteins coded by *slo*. Defining how the different potassium channels function is a major research effort for the future. Sodium and potassium channels close on a time-dependent basis, and neither ligands nor voltage directly control closing. Certain drugs, insecticides, and naturally occurring poisons, however, can block various ion gates open or closed.

10.3.2 Resting Potential

The **resting potential** is the potential difference between the inside and the outside of the cell at rest (the inside of the cell is negative relative to the outside). At rest, the cell membrane is nearly impermeable to sodium ions and to the large, charged protein molecules, but relatively permeable to potassium and chloride ions. The space outside a nerve cell is the mesaxon space between the neuron membrane and the protective glial cell membrane, and potassium ions are typically in low concentration here, but sodium ion concentration is usually high. Thus, both an electrical gradient and a concentration gradient act upon any movement of sodium and potassium ions.

The negative charge on the inside of a resting neuron will attract positively charged ions, such as K^+ and Na^+. The concentration gradient acting on K^+ will tend to force it outward until the attraction of the negative charge (the electrical gradient) on the inside just balances the concentration force. Any change in the membrane potential immediately results in a redistribution of K^+ ions until a new equilibrium is achieved. Although both a concentration gradient and an electrical gradient act upon Na^+ to promote inward movement when a channel is open, the membrane at rest is nearly impermeable to Na^+ entry. In a membrane at rest, Na^+ only very slowly leaks in with time.

An energy-requiring membrane pump, the sodium–potassium exchange pump, works continuously to pump K^+ from the mesaxon space into the nerve cell and to pump Na^+ out of the cell and into the mesaxon space. It is important to understand, however, that this pump serves a long-term maintenance function, and it does not move ions fast enough to cause the repolarization process, which usually requires only a few milliseconds following a stimulus response. Ion pumps in some cases can induce a slow potential change, but the important point here is that repolarization is not a function of the Na^+–K^+ exchange pump. Repolarization is based on different physiological properties of a neuron, which are explained in Section 10.3.4.

The presence and distribution of chloride has little effect on resting or action potentials, generally, and chloride ions tend to distribute according to the dictates of Equations 10.1 and 10.2. The presence of the large, nondiffusible organic ions, mostly proteins, that carry a net negative charge at physiological pH are extremely important to overall nerve cell charge maintenance, enabling the inside to hold its negative charge relative to the outside. The mesaxonal space contains little of these large molecules.

The **Nernst equation**, a physical chemistry model derived from studies with solutions of ions separated by a semipermeable membrane in laboratory situations, gives a fairly accurate prediction of membrane potentials of a neuron. The resting potential, influenced most strongly by potassium distribution inside and outside the cell, is calculated as a potassium equilibrium potential, as follows:

$$E_m = \frac{RT}{\eta F} \ln\left(\frac{[K^+]_o}{[K^+]_i}\right) \quad (10.3)$$

where

E_m is the membrane potential in volts

R is the universal gas constant (0.082 L × atmospheres per mole per degree Kelvin temperature scale or 8.314 joules per mole per degrees Kelvin)

T is the temperature on the Kelvin or absolute scale (°C + 273)

η is the valence of the ion involved (for potassium, the valence is 1)

F is the Faraday unit (96,496 coulombs of electricity per gram-equivalent of ion moving)

ln is the logarithm to the base e (natural log as opposed to log base 10)

$[K^+]_i$ and $[K^+]_o$ refer to ionic activity of potassium on the inside and outside, respectively, of the membrane

The equation asks for ionic activities of potassium (indicated by the brackets in the equation), rather than concentration, but the more conveniently measured concentrations often are used in the equation to obtain a good approximation of the voltage across a nerve cell membrane.

When a neuron membrane is resting and its permeability to Na$^+$ is very low (as is normal), then the magnitude of the resting membrane potential is directly influenced by the K_o^+/K_i^+ ratio, shown in Figure 10.6. The graph shows the magnitude of the resting potential when the external K$^+$ concentration in the saline bathing the nerve is varied. High levels of potassium outside the cell

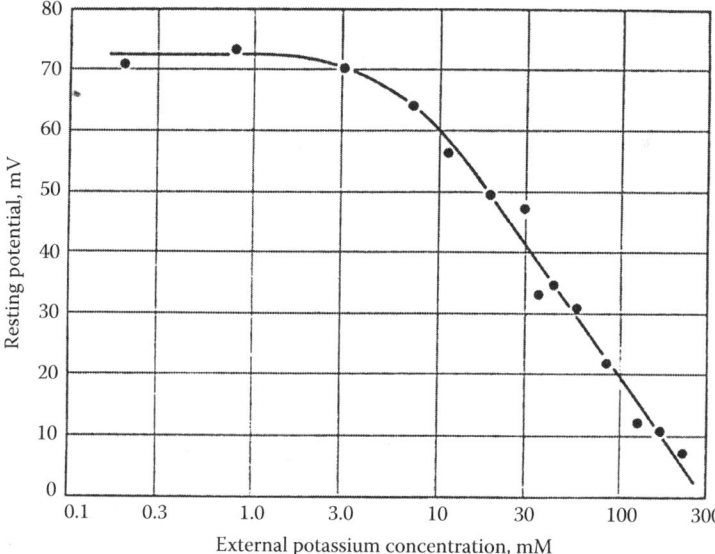

Figure 10.6 The magnitude of the resting membrane potential as a function of the potassium in experimentally controlled external bathing solution. Experimentally raising the external K$^+$ concentration in the bathing solution progressively reduces the magnitude of the resting potential. The normal bathing solution of a neuron is the fluid in the mesaxon channel in which K$^+$ is typically low in concentration. (From Yamasaki, T. and Narahashi, T., *J. Insect Physiol.*, 3, 146, 1959. With permission.)

reduce, and may destroy, the resting membrane potential, as predicted by the Nernst equation. The equation predicts that the membrane resting potential will decrease as the K_o^+/K_i^+ ratio approaches a value of 1, and when the value is 1, there is no transmembrane potential. This experimental effect of manipulating the external potassium concentration reinforces the importance of glial cells, which shield all parts of the nervous system from direct exposure to the hemolymph. Phytophagous insects often have a very high level of potassium ions in the hemolymph and a low level of sodium ions, conditions that would be deleterious to nerve cell function.

The Nernst equation was expanded into the **Goldman constant field equation** to include the three major ions moving during an action potential in an attempt to make even better predictions. The expanded equation is

$$E_m = \frac{RT}{\eta F}\left(\ln\left(\frac{P_k[K^+]_i + P_{Na}[Na^+]_i + P_{Cl}[Cl^{-1}]_o}{P_k[K^+]_o + P_{Na}[Na^+]_o + P_{Cl}[Cl^{-1}]_i}\right)\right) \quad (10.4)$$

where P is the membrane permeability value for each of the major ions that move during an action potential. The Goldman constant field equation gives slightly better agreement with actual observations, but absolute values for P are not available for very many neuronal preparations. *Relative* permeability values for K^+, Na^+, and Cl^-, equal to 1:0.04:0.45, respectively, were used for the squid giant axon to test the equation (Hodgkin and Katz, 1949). The Nernst equation was valuable when it was first applied to neuron physiology because it provided numerous testable predictions that generated many experiments, and it is a classical case of the benefits of mathematical modeling of a biological phenomenon.

10.3.3 The Action Potential: Sodium Activation

Upon stimulation that causes depolarization, a nerve cell undergoes remarkable permeability changes to both Na^+ and K^+. The most dramatic change is that the excited membrane becomes explosively permeable to Na^+, and a small number of sodium ions, acted upon by both electrical and concentration gradients, rush into the membrane causing **depolarization** of the membrane. The sudden increase to Na^+ influx is called **sodium activation**. The rush of Na^+ for 2–3 ms into the neuron when the sodium channels are completely open dominates membrane physiology and its electrical properties. The sodium ions carry an inward current, and this is usually recorded on an oscilloscope or similar device as a **spike** or **action potential**, and described as the **all-or-none response** of a neuron. A partial action potential does not occur (except in special experimental situations controlled by the investigator).

The rate at which the membrane potential changes (i.e., the rise of the spike potential) is very fast. For example, the rise of the spike has been measured at 1370 mV/ms in a giant axon of the cockroach, *Periplaneta americana*. At this rate, it takes much less than 10 µs to depolarize the resting membrane potential from its typical resting value of about −70 mV in a cockroach giant axon.

Relatively few sodium ions move across the membrane to cause the depolarization. Although the number of ions moving per cm² of membrane surface has not been determined for insect axons, experiments with squid giant axons bathed in seawater containing radioactive Na^+ demonstrated that an average of 3.7×10^{-12} mol of Na^+ ions moved across 1 cm² membrane surface with each stimulus (Keyes, 1951). Thus, movement of only a few picomoles of sodium ions causes the spike, and the concentration of sodium ions in the mesaxon space is hardly changed.

The rapid inward movement of sodium ions during an action potential allows the membrane potential to overshoot zero, and the inside becomes positive to the outside over that portion of the membrane surface affected by the action potential. The reversal is called the **overshoot potential**.

In a cockroach, giant axon overshoot potentials of about +35 mV have been recorded (the plus sign indicates that now the inside is positive to the outside). The total magnitude of the action potential is stated as the sum of the absolute values of the resting potential and the overshoot potential. Thus, if a neuron has a resting potential of −70 mV and an overshoot potential of +35 mV, then the action potential is 105 mV.

During the spike, the membrane is absolutely refractory (the **absolute refractory period**) to further stimulation. The sodium channels are open and Na⁺ enters at a maximal rate, and a stimulus of even great magnitude cannot cause more to enter or the gates to open any wider. The membrane is not capable of making any greater response. Thus, a second stimulus delivered within 1–2 ms of the first one will not elicit a response from an axon. Furthermore, in addition to the absolute refractory period, a neuron is partially or **relatively refractory** for a further few to many milliseconds, and only a very strong stimulus will elicit a new response during the relatively refractory period. Thus, although the absolute refractory period and the relative refractory period are short, they set an upper limit (typically about 100 impulses/s) on how many separate nerve impulses a neuron can transmit in 1 s.

The Nernst equation also predicts the magnitude of the overshoot potential, which is a sodium equilibrium potential in which sodium movements dominate the membrane. The form for the equation describing sodium equilibrium potential is

$$E_m = \frac{RT}{\eta F} \ln\left(\frac{[Na]_o}{[Na]_i}\right) \tag{10.5}$$

Sodium concentration in the mesaxonal space is very important to the action potential, and low sodium concentrations cause small action potentials. Although the concentration of sodium ions is relatively high in the mesaxon space, only a small volume of solution containing the ions occurs in this restricted space. The clefts in the mesaxon channel around a cockroach giant axon contain only enough free sodium ions for about 20–30 action potentials if no corrective pump action occurs (Treherne and Schofield, 1981). Thus, without some way to restore the ionic composition of this microchannel around an axon, it would soon fail to fire. In reality, axons are capable of firing continuously for many minutes (Narahashi and Yamasaki, 1960; Parnas et al., 1969) because at the inner membrane surface of the glial cell (i.e., the surface nearest the axon) coupled Na⁺–K⁺ pumps pump Na⁺ from the glial cell into the mesaxon channel and pump K⁺ from the mesaxon channel into the glial cell (Treherne and Schofield, 1981). Accumulation of K⁺ in the mesaxon also would be detrimental to continued function. The axon membrane also has a coupled Na⁺–K⁺ pump that works to pump Na⁺ out of the axon and bring K⁺ back into the axon. It is important to understand that the pumps provide for long-term maintenance by keeping Na⁺ in the mesaxon channel high and the concentration of K⁺ low (Treherne, 1985); **the pumps do not repolarize the membrane after a depolarization**.

Figure 10.7 shows graphically the magnitude of the action and overshoot potential as a function of progressive loss of external sodium concentration in the case of a giant axon of the American cockroach (Yamasaki and Narahashi, 1959). As sodium concentration in the bathing saline is reduced, the action potential is reduced in size and, at very low sodium concentration, the action potential is abolished. Similar experiments with similar results have been conducted on nerves from *Blaberus craniifer* (Pichon and Boistel, 1966), *Carausius morosus* (Treherne and Maddrell, 1967), and *Manduca sexta* (Pichon et al., 1972). The importance of sodium ions to the action potential was further demonstrated when desheathed (removal of the protective fat body and perineural layers) crural nerves of *P. americana* and *Locusta migratoria* bathed in a sodium-free saline failed to develop action potentials (Pichon and Treherne, 1973) because there were no sodium ions to carry an inward depolarizing current.

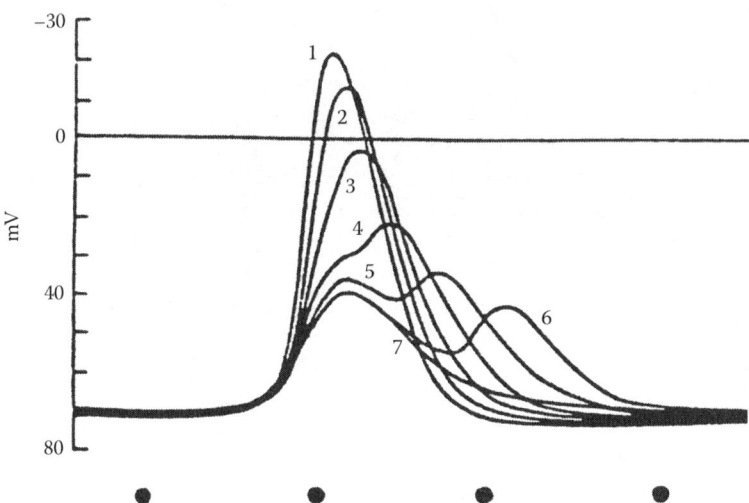

Figure 10.7 The effect of experimentally altering external sodium concentration on the magnitude of the action potential. Experimentally eliminating external Na+, which carries the inward current, progressively reduces the size of the neuron response to stimulation, and finally at very low external sodium concentration, no spike can be developed. The black dots below the figure indicate time in milliseconds. The recording is made relative to the outside of the neuron, so the outside becomes negative to the inside with about 20 mV overshoot potential. (From Yamasaki, T. and Narahashi, T., *J. Insect Physiol.*, 3, 146, 1959. With permission.)

10.3.4 Sodium Inactivation and Repolarization

The excited membrane state normally is a very transitory event and the sodium channels have a time-dependent closing mechanism called **sodium inactivation**. The time required for complete sodium inactivation to occur is variable, but can be from a few to several hundred milliseconds. During sodium inactivation, the sodium channels close. The spike falls rapidly (a falling rate of 640 mV/ms was recorded from a cockroach giant axon), and the rate of fall largely reflects the closing of the sodium channels. Generally, as the spike falls, there is a slow positive after potential (slight **hyperpolarization**) and, shortly thereafter, an even smaller and slower negative after potential. The after potentials are graded or slow potentials and are caused by the transitory displacement of ions in the mesaxon and across the membrane.

Membrane permeability to potassium changes immediately after the spike develops and potassium starts to move outward across the axonal membrane, but it moves slowly at first, and the membrane is dominated by the inwardly directed sodium ion movements for a few milliseconds. Maximum potassium flux was measured at 440 mV/ms in a cockroach giant axon at the peak of the overshoot potential. The outwardly directed potassium current is counter to the inwardly directed current flow carried by sodium. Only when the sodium channels have partially closed, thus restricting the inward flow of Na+, does the outward flow of potassium begin to bring the membrane potential back toward the resting value. Repolarization is a much slower process, relatively speaking, than depolarization, and total recovery of a neuron may take from about 10 ms to many tens of milliseconds, depending on the neuron.

As K+ continues to move out and the sodium channels close, the membrane potential begins to return to its resting condition in which the inside is negative to the outside. The negative pole of this small biological battery now attracts the positively charged potassium ions and slows their outward movement. Net outward flux ceases when the membrane potential has become negative enough to attract potassium and counterbalance the concentration gradient that drives it outward. The neuron has recovered its resting value; it is repolarized and ready to respond to a new stimulus.

The few picomoles of sodium ions that enter a neuron during an action potential **do not** have to be removed from the cell for repolarization to occur. **Repolarization occurs when approximately the equivalent number of positively charged potassium ions exit from the neuron**. Experiments with a squid giant axon that had been injected with radioactive potassium ions demonstrated that an average of 4.3×10^{-12} mol radioactive K^+ ions/cm^2 exited into the bathing saline with each stimulus (Keyes, 1951), approximately equal to the Na^+ ions per cm^2 that entered with each impulse. The Na^+–K^+ exchange pump works to restore the normal distribution of ions, but it **does not** account for repolarization of the neuron membrane. Repolarization and continued nerve cell function without the pump was conclusively shown by selectively poisoning the pump in a squid giant axon, which nevertheless continued to develop spikes and repolarize repeatedly for hours before the redistribution of Na^+ and K^+ became physiologically limiting. The pump serves a long-term maintenance function to keep the Na^+ ions outside and the major quantity of K^+ ions inside. Nerve cell membranes are leaky and allow Na^+ to leak across the membrane, necessitating further action for the membrane pumps. The pumps work slowly and continuously, and require a constant supply of energy. Thus, nervous tissue has high metabolic demands in an insect, as in all other organisms.

10.3.5 Measurement of Ion Fluxes: Voltage Clamp Technique

How can something that happens in 2–3 ms be observed? Even with the oscilloscope it is impossible to follow all the details of the ion fluxes because of the extremely rapid and transitory nature of the nerve response to a stimulus. What is needed is some way to prevent the sudden explosive changes of the action potential—a way to stop an action potential at a given point and measure what ions are moving and in what direction. An ingenious technique, the **voltage clamp**, was devised independently by Cole (1949) and Marmont (1949), and it continues to be a useful tool (Trudeau et al., 1995).

The voltage clamp technique uses a feedback amplifier in the recording circuit to feed into the membrane just enough current in an opposing direction to counter the action of the ion currents induced by a stimulus. With this technique, the membrane could be stabilized (clamped) at any membrane potential desired by the investigator. For example, the membrane potential can be held at -20 mV, inside negative. This would be equivalent to stopping a depolarization about half way, something that does not occur naturally. By measuring the magnitude of the current needed to hold the membrane at a given potential, the investigator can get a measure of the strength of the ion current at that potential. Furthermore, instead of the experiment being over in 1–2 ms, a sustained membrane response is possible. The voltage clamp technique allows the investigator to reconstruct data from an experiment to show the inwardly directed current carried by Na^+ separated from the outwardly directed current carried by K^+. The net sum of the two currents reconstructed over time represents the spike.

By using the voltage clamp technique to hold the membrane potential near, but not exceeding the threshold value for a spike, an investigator can demonstrate that even slight changes in the membrane potential allow a few sodium channels to open. For example, a squid axon voltage clamped at only 8 mV below its resting value for 20 ms did not result in a spike (because the spike generation threshold was not reached), but it did cause as much as 40% reduction in the spike or sodium current on subsequent depolarization. Hodgkin and Huxley explained this experiment as showing that even the slight drop in membrane potential had activated the time-dependent inactivation mechanism controlling the sodium channels. When the membrane was finally depolarized, the timing mechanism did not allow the sodium channels to stay open long enough for the spike to be normal in size. This led directly to the concept of a leaky membrane and helped explain why the action potential is usually not as large as predicted by the Nernst equation. Thus, in a typical neuron, the sodium gates allow some leaking. Upon stimulation of a leaky neuron, the sodium conductance values are lower than expected, and sodium conductance does not continue as long as expected. Conversely, hyperpolarization, even by a few mV (raising the inside potential to a greater negative value, e.g., from −70 to −90 mV), increases the sodium current and the size of the spike upon depolarization. Hyperpolarization also raises the threshold necessary for causing a neuron to fire, and inhibition in the nervous system often works through hyperpolarization of one or more neurons in a circuit, making it more difficult for other stimulating synaptic connections to fire the circuit.

10.4 CONDUCTION OF THE ACTION POTENTIAL: LOCAL CIRCUIT THEORY

After a spike arises, it is self-propagating and travels rapidly along the length of the axon. Local electrical disturbances (Figure 10.8) in the nerve set up a pattern of local currents or local circuits around the excited region of the membrane. The propagated impulse does not go backward over the same length of axon it has just passed because that part of the membrane is still too refractory to be depolarized again by the local currents.

According to the **local circuit theory**, the currents flowing into the axon membrane just ahead of the active region open voltage-gated Na^+ channels leading to depolarization. The depolarization then allows local currents to flow into the next part of the membrane and cause a spike to develop in that new location, and these actions are repeated along the neuron as the depolarized region moves along the axon. The membrane immediately behind the active region also receives the local circuits, but this part of the membrane is in a state of incomplete recovery (in the relative refractory period) and the local circuits are not strong enough to cause a new spike; thus, nerve transmission is normally unidirectional. The local circuit theory of spike propagation is based on experimental evidence from studies on frog sciatic nerve (Hodgkin, 1937a,b). In those experiments, a small section (about 1 mm in length) of the nerve was frozen in order to block transmission across the frozen section because the frozen sodium gates could not open. Local electrical currents, however, could not be stopped by the frozen section because the tissue would still be a conductor of electricity. The intent of the experiment was to freeze a long enough section so that the currents, although attenuated, would flow through the frozen section and do some of the work of lowering the threshold beyond the block. Stimulating electrodes placed just past the frozen section were used to deliver a stimulus just as a wave of depolarization reached the blocked section. The strength of stimulus needed to initiate a new spike just beyond the frozen section was about 10% of normal, and Hodgkin interpreted the experiment as showing that local currents from the wave of depolarization did 90% of the work of exceeding the threshold beyond the block. Without the new stimulus just as the local currents arrived at the post block site, the nerve impulse would

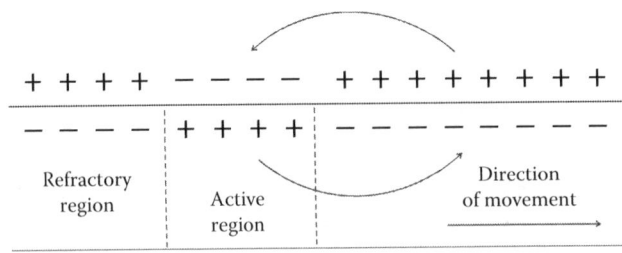

Figure 10.8 Local currents and an illustration of how they function to propagate a spike along the axon. Small currents flowing ahead of the active region open voltage-gated Na^+ channels in the axon. The currents also flow back into the region over which the impulse has just passed, but that region is still partially refractory, and the local currents are not strong enough to open Na^+ channels.

not have continued because the threshold would not have been reached. He reasoned that the local currents should be strong enough in a normal axon with no blocked portion to exceed the threshold and open the sodium channels immediately ahead of the spike. In such a case, the wave of excitation would be self-propagating, as observations indicated. He also concluded from this experiment that the action potential and local currents extend over about 1 mm of surface of an axon. The exact surface of an axon that is excited at any given moment in an insect has not been measured in a similar manner, but the excited state probably spreads over an axon in a larger insect, such as a locust, grasshopper, or cockroach in much the same way.

10.5 PHYSIOLOGY AND BIOCHEMISTRY AT THE SYNAPSE: EXCITATORY AND INHIBITORY POSTSYNAPTIC POTENTIALS

A **synapse** is any site where one nerve cell influences another neuron. In insects, synapses typically occur between the axon of one cell and a neurite of another neuron within the neuropil of a ganglion. A single neuron can have synaptic contacts with many other neurons. The axon-to-muscle contact is sometimes described as a synapse. Some synaptic contacts, in addition to the axon-to-dendrite, are known in other animals, but they are either not known in insects or are very uncommon. For example, axon-to-soma synapses are very common in vertebrates, but none of these has been observed in insects. The transmitter chemical is contained within membrane-bound synaptic vesicles, usually from 200 Å to about 400 Å in diameter, near the presynaptic membrane (Figure 10.9). The arrival of spikes at the presynaptic terminals opens calcium channels and Ca^{2+} enters the presynaptic membrane. The entry of Ca^{2+} promotes the fusion of transmitter vesicle membranes with the presynaptic membrane at the synaptic cleft, thus releasing the transmitter chemical into the synaptic cleft. The released chemical diffuses across the synaptic cleft, a distance of some 100–200 Å, and

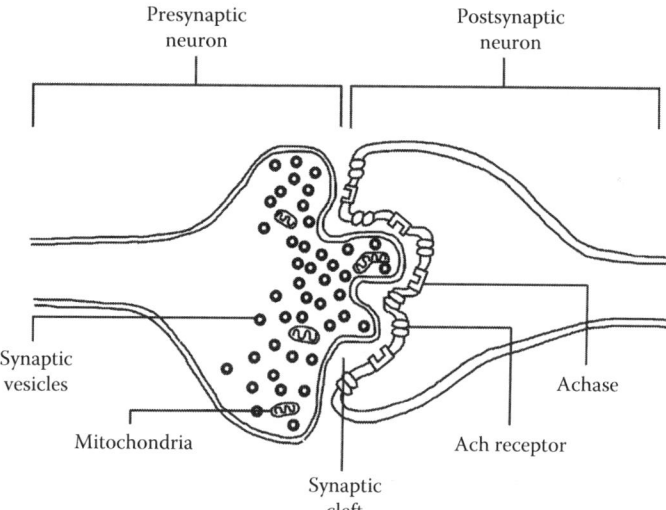

Figure 10.9 A schematic diagram for a synapse in an insect. Synaptic vesicles store the synaptic transmitter near the presynaptic membrane. Arrival of a nerve impulse opens voltage-gated calcium channels; calcium enters the presynaptic membrane and promotes the fusion of synaptic vesicles with the presynaptic membrane at the synaptic cleft region. This part of the membrane has been shown to lengthen in a repeatedly stimulated neuron as the vesicles fuse with the membrane, thereby lengthening it. In a stimulatory neuron in which the transmitter chemical is acetylcholine (ACh), the released ACh diffuses across the synaptic cleft and has about equal probability of encountering an acetylcholine receptor (ACh receptor) or the enzyme acetylcholine esterase (AChase).

binds to highly specific receptor proteins on the postsynaptic membrane surface. The binding of the transmitter to receptor opens ion channels in the postsynaptic membrane and a postsynaptic membrane potential (a graded potential) develops. A particular synapse is stimulatory or inhibitory depending on the neurotransmitter released. Stimulatory synapses give rise to **Excitatory Post Synaptic Potentials (EPSPs)**, while inhibitory synapses generate **Inhibitory Post Synaptic Potentials (IPSPs)**. EPSPs and IPSPs are graded potentials, and their magnitude, rate of rise, duration, and spread will depend on the amount of transmitter chemical released, which in turn might depend on the number of spikes arriving per second at the presynaptic terminals. Larger release of neurotransmitter causes larger postsynaptic potentials. If the postsynaptic potential is a stimulatory potential and if it is strong enough to spread over the postsynaptic membrane surface and reach the spike generation region of the postsynaptic neuron, then spikes may be generated. If the transmitter chemical is an inhibitory chemical, spike generation will be suppressed or prevented in the postsynaptic neuron. This could mean reduction or cessation of spike generation in a postsynaptic neuron that spontaneously is active.

There is evidence that **acetylcholine (ACh)** is a synaptic transmitter at stimulatory synapses in the CNS, and **γ-aminobutyric acid (GABA)** is an inhibitory transmitter in the CNS of insects (Figure 10.10). These two transmitter chemicals are not likely to be the only neurotransmitters in the insect CNS, but others remain to be positively identified. The transmitter at nerve–muscle junctions (neuromuscular synapses) in insects is **L-glutamic acid** and possibly **L-aspartic acid**. There is some evidence from insects for a number of additional putative transmitter chemicals, including 5-hydroxytryptamine, catecholamines, octopamine, and some peptides. Callec (1985) outlined a number of criteria needed to demonstrate a transmitter function for a putative transmitter chemical.

Inhibition is a vital process in nervous systems and acts like a brake on the system. One of the mechanisms of inhibition is hyperpolarization of the cell membrane by increasing permeability to chloride ions. Synapses in the sixth abdominal ganglion of *P. americana*, for example, exhibit a high degree of sensitivity to applied GABA, and 1.05×10^{-13} M is sufficient to hyperpolarize postsynaptic membranes (Kerkut et al., 1969a,b). GABA selectively increases permeability to chloride ions, allowing more of these ions with a negative charge to enter the postsynaptic membrane. This causes the inside negative potential to become slightly more negative (e.g., from −70 mV to as much as −80 mV or even more). A neuron that has been hyperpolarized requires a greater than normal stimulus to change the membrane potential enough to exceed the characteristic membrane threshold for spike generation. GABA channels may be designated in the literature as GABACls. L-Glutamate itself has been shown to have inhibitory action on chloride channels (symbolized as GluCls) in some insects (Janssen et al., 2010; Boumghar et al., 2012; el Hassani et al., 2012; Démares et al., 2013;

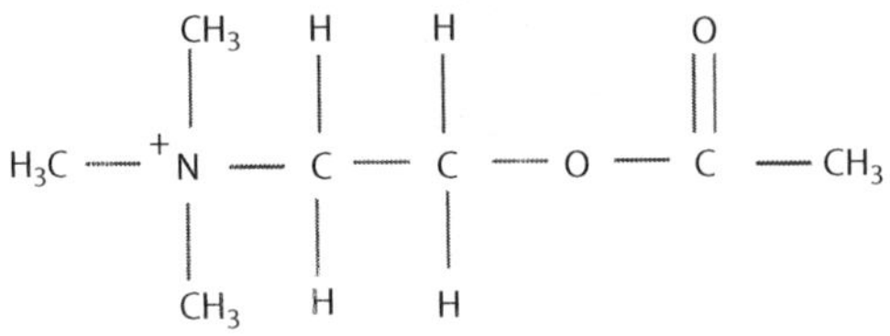

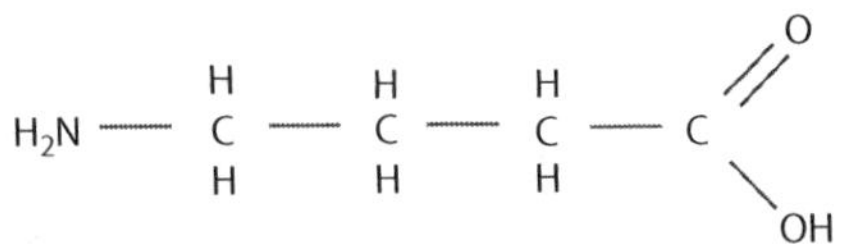

Figure 10.10 The structures for acetylcholine, the stimulatory transmitter at synapses in the central nervous system (CNS) of insects, and for γ-aminobutyric acid (GABA), the inhibitory transmitter in the CNS.

Kita et al., 2013; Liu and Wilson, 2013), and chloride channels have been exploited as safer insecticide targets (Narahashi et al., 2010; Ozoe et al., 2010; Nakao et al., 2013; Ozoe, 2013).

GABA is synthesized in insects from glutamate by action of the enzyme L-glutamic acid decarboxylase, an enzyme widely distributed in high titers in insect nervous tissue. Following secretion at the nerve endings, GABA is probably rendered inactive by uptake at the synaptic terminal and/or by glial cells. It also may be oxidized to succinic acid.

10.6 ACETYLCHOLINE-MEDIATED SYNAPSES

Numerous studies have shown the presence of the necessary components of a cholinergic system in the CNS of insects, that is, acetylcholine (ACh), choline acetyltransferase that synthesizes ACh, **acetylcholinesterase** (**AChase**) that breaks down ACh after its secretion into the synapse, and receptors in the CNS for ACh. Cholinergic receptors are located only within the CNS in insects and are not found at neuromuscular junctions as in vertebrates. Applications of acetylcholine in isolated insect preparations have not always produced spike activity at physiologically relevant concentrations of ACh, possibly because the CNS is protected from ACh applied in a bathing saline by the hemolymph–CNS barrier and fatty sheath surrounding the brain, ventral ganglia, and connectives. A very active acetylcholinesterase acts quickly to destroy applied ACh. However, high sensitivity of insect neurons to ACh is revealed by the application of the ACh with a microsyringe, or iontophoretically, to neurons inside a ganglion (Callec, 1985). For example, the application of 5×10^{-6} M ACh by a microsyringe to the dorsal unpaired median cell bodies in the neuropil of the sixth abdominal ganglion of *P. americana* depolarized the cells and produced a volley of spikes (Callec and Boistel, 1967). Acetylcholine administered iontophoretically into the sixth abdominal ganglion of *P. americana* was stimulatory at a dilution of 1.31×10^{-13} M ACh (Kerkut et al., 1969a,b). Pretreatment of nervous tissue with inhibitors of AChase further increases sensitivity of desheathed ganglia to ACh (Narahashi, 1971; Shankland et al., 1971).

10.6.1 Action of Acetylcholine at the Synapse

A volley of spikes arriving at the presynaptic terminal increases permeability of the presynaptic membrane to calcium, which diffuses into the terminal and, through a second messenger, sets up a cascade of actions that facilitate attachment of synaptic vesicles to the synaptic membrane. The synaptic vesicles fuse with the membrane and release quanta or packets of ACh into the synaptic cleft. Electron micrographs have shown that the presynaptic membrane expands slightly with the incorporation of vesicular membranes. When released into the synaptic cleft, the ACh molecules diffuse randomly, with some contacting and attaching to acetylcholine receptors and others encountering acetylcholinesterase. ACh is rapidly released from the receptor and may randomly collide with another receptor to repeat the process. When ACh is bound to its receptor, Na^+ channels are opened. When large numbers of sodium channels are opened in the postsynaptic membrane, the inward movement of Na^+ depolarizes the postsynaptic membrane with production of an EPSP that is conducted decrementally away from the site of origin. If the stimulation is strong enough, the excitation may spread to the region of the axon that generates spikes.

ACh molecules seem to have about the same probability of encountering the enzyme acetylcholinesterase, which is also bound to the postsynaptic membrane, as they do of encountering a receptor molecule. An ACh molecule encountering acetylcholinesterase is hydrolyzed to acetic acid and choline, neither of which has any physiological action at the synapse. Both breakdown products diffuse out of the synapse and/or are taken up by the presynaptic neuron and may be used to synthesize new ACh through the action of the enzyme choline acetyltransferase. Acetylcholinesterase is protective in function, and poisoning it with molecules such as organophosphate insecticides results in prolonged stimulation at synaptic sites throughout the CNS of insects. Poisoned insects typically show uncontrolled leg tremors, buzzing of the wings without control for flight, and eventual death. Probably many other physiological and biochemical processes are disrupted, such as release of neurohormones and depletion of energy reserves in uncontrolled muscular actions, and all of which contribute to the death of a poisoned insect.

10.6.2 Nicotinic and Muscarinic Cholinergic Receptors in Insects

In insects, as in vertebrates, more than one cholinergic receptor type exists, and the cholinergic receptors in insects have been characterized as **nicotinic**, **muscarinic**, and **mixed** receptors (Callec, 1985). ACh is the neurotransmitter at all of these cholinergic receptors. At very low concentrations, nicotine and muscarine mimic the action of ACh, but at higher concentrations they block the receptor. The Ach-mediated skeletal muscle receptors in vertebrates are of the nicotinic type, while vertebrate heart and gut muscle have muscarinic-type ACh receptors. Brain tissue of vertebrates has both nicotinic and muscarinic receptors, with muscarinic receptors generally outnumbering nicotinic ones in the brain of vertebrates. In contrast, insect central nervous tissue has more nicotinic receptors that muscarinic ones (Figure 10.11) (Breer et al., 1987), and ACh is not the synaptic mediator at insect neuromuscular junctions, so insect muscle tissue has neither type of ACh receptor.

Nicotinic cholinergic receptors in insects, as in vertebrates, are sensitive to the inhibitory action of a very potent toxin, α-bungarotoxin, derived from the venom of snakes of the family Elapidae (snakes in Southeast Asia). The toxin binds irreversibly to nicotinic-type receptors, producing a block of the synapse. It is often used pharmacologically to characterize and study the properties of nicotinic receptors. α-Bungarotoxin binding data indicate that nicotinic ACh receptors are present in the CNS of *D. melanogaster*, *P. americana*, *Musca domestica*, and *M. sexta* (reviewed by Callec, 1985). Membrane-bound nicotinic receptors that have a very high binding capacity (B_{max}) of 8926 fmol/mg protein are highly localized in the neuropil regions of the brain of the American cockroach (Orr et al., 1990).

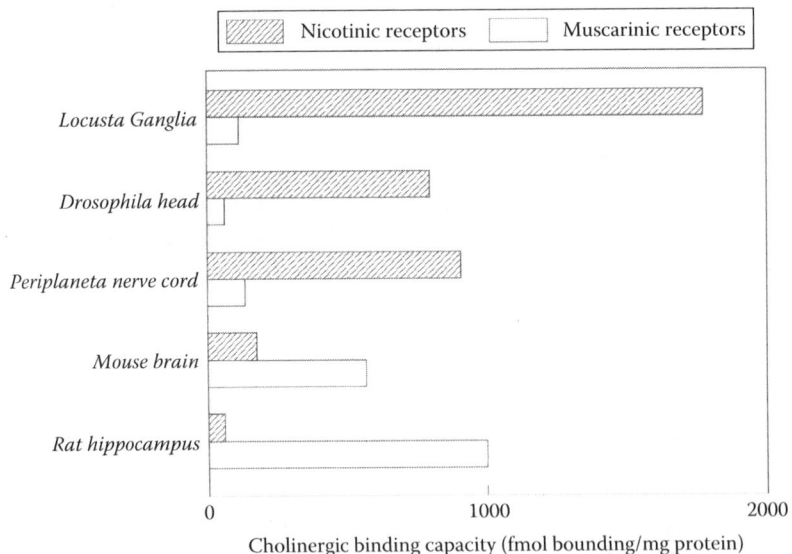

Figure 10.11 The type and binding capacity of cholinergic receptors in the nervous system of selected insects compared with mouse brain and rat hippocampus tissue. (Original histogram by author, with data from Breer, H., et al., Identification, reconstitution and expression of neuronal acetylcholine receptor polypeptides from insects, in: John Law (ed.), *Molecular Entomology*, Alan R. Liss, Inc., New York, 1987, pp. 95–105.)

ACh receptors that specifically bind muscarine, a very potent poison obtained from an *Amanita* spp. of mushroom, are classified as muscarinic-type receptors. Muscarinic receptors have been demonstrated in the head of *D. melanogaster* (Haim et al., 1979) and in the last abdominal ganglion of the cricket, *Acheta domesticus* (Meyer and Edwards, 1980). Insects appear to have a much higher concentration of nicotinic receptor types than muscarinic receptor types (Lummis and Sattelle, 1985; Breer et al., 1987). Pyrrolizidine alkaloids (PAs), present in many plant families, bind to muscarinic cholinergic receptors (Schmeller et al., 1997), and may exert some of their toxicity through this mode of action. Insect nicotinic receptors have become molecular targets for insecticide action (Matsuda et al., 2001; Thany et al., 2006).

10.6.3 Acetylcholine Receptor Structure

The complete structure of insect acetylcholine receptors has not been elucidated, but there is evidence that the ACh receptor comprises part of the sodium channel as it does in vertebrates. In contrast to the vertebrate receptor, however, the nicotinic receptor isolated from locust nervous tissue appears to be composed of identical polypeptide subunits (Breer et al., 1987). The electric fish ACh receptor at the neuromuscular junction is composed of five polypeptide subunits, two alpha, and one each of beta, gamma, and delta polypeptide units. The five subunits form a barrel-shaped transmembrane protein with the sodium channel through the middle (Figure 10.12) (Changeux et al., 1984, 1992; Changeux, 1993). One molecule of Ach binds to each of the two alpha subunits to open the channel to entry of Na^+.

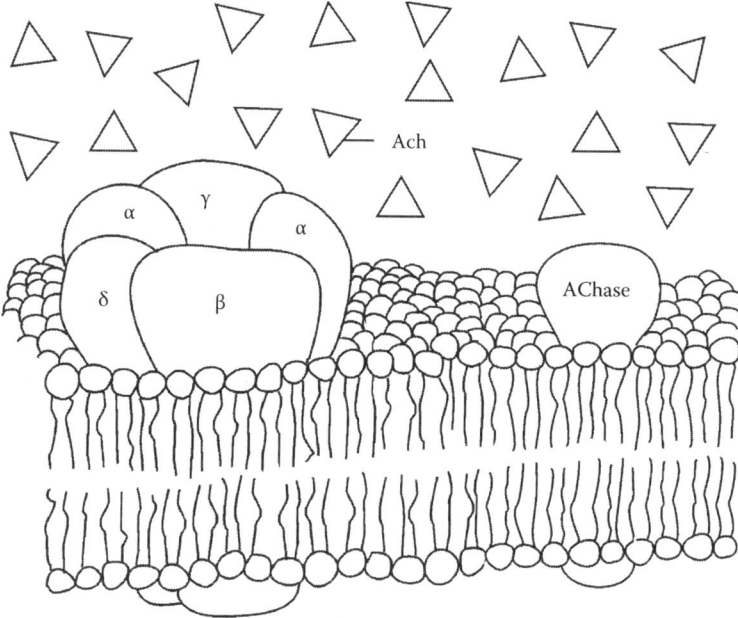

Figure 10.12 A conceptual model for the postsynaptic membrane in which acetylcholine (ACh triangles) may encounter its receptor (the α subunits of the sodium channel, schematically represented at the left) or acetylcholine esterase molecules (AChase, at the right) at the postsynaptic membrane. The ACh receptor modeled here is that shown to occur in a vertebrate and consists of five protein subunits, two α plus one β, δ, and γ. The structure of the receptor in an insect has not been completely elucidated. When Ach binds to each of the α subunits, the channel formed by the proteins opens and sodium ions enter the postsynaptic neuron, resulting in a postsynaptic (graded) potential. The ACh receptor quickly releases a bound ACh molecule and it may bind to another receptor and repeat the action, or it may encounter AChase and be hydrolyzed into acetic acid and choline, both of which are inactive as far as nerve response is concerned.

10.7 ELECTRIC TRANSMISSION ACROSS SYNAPSES

Transmission of impulses across some synapses is electrical. In insects, some or possibly all of the synapses within giant fiber systems of the ventral nerve cord are electrical. The spike crosses the electrical synapse without involvement of a chemical transmitter. Electrical synapses allow faster transmission of spikes (a message) than a network containing several synapses. The cercal nerve giant axon complex provides circuitry for a startle and escape reaction in cockroaches. An escape reaction starts with the reception of stimuli at the mechanoreceptors on the cerci. Strong stimuli result in large receptor potentials and a series of spikes that travel over the cercal nerve to the sixth abdominal ganglion. In the neuropil of the sixth ganglion, acetylcholine is released at synapses with one or more of the giant axons. There is a delay in the giant axons' response of 0.68 ms to the released ACh, followed by a slow rise time of EPSPs over about 2–3 ms. The EPSPs are only about 2–5 mV in amplitude, but they give rise to spikes in the giant axon that do not have to cross additional chemically mediated synapses until they synapse in thoracic ganglia with mononeurons to the leg muscles. Much of the time delay in escape can be attributed to the slowness of chemically mediated synapses in the sixth abdominal ganglion and in the thorax. In general, synaptic transmission, and especially a circuit with multiple synapses, appreciably slows speed of communication within the nervous system. The giant fibers actually represent multiple neurons that have anastomosed together, and the points of fusion are electrical synapses that a spike crosses without a chemical mediator. Probably selection for speed of transmission was a major evolutionary force acting on the development of electrical synapses in the giant fiber system as part of control for an escape mechanism.

10.8 NEUROMUSCULAR JUNCTIONS

The junction between the nerve and muscle is a special type of synapse, usually called the neuromuscular junction. L-Glutamate is the excitatory transmitter chemical at the neuromuscular junction in a few insects studied, and it is generally believed to be the typical insect neuromuscular transmitter at stimulatory junctions. There is some evidence that L-aspartic acid also may act as a transmitter at some neuromuscular synapses. GABA is the transmitter chemical at inhibitory nerve–muscle junctions.

10.9 REVIEW AND SELF-STUDY QUESTIONS

1. What does the word "integration" mean in the nervous system?
2. How is a dendrite different physiologically from an axon?
3. What is a spike potential?
4. What might be some stimuli that would initiate action potentials in an insect?
5. How is the resting potential in a neuron established?
6. What is the typical ionic distribution within and outside of a neuron at rest?
7. What role do proteins play in establishing the resting potential of a neuron?
8. What is the typical construction of an ion channel or ion gate in the nerve cell membrane?
9. What is a ligand? What is a ligand-gated channel?
10. Are sodium channels in an axon voltage gated or ligand gated?
11. Are sodium channels at the synapse voltage gated or ligand gated?
12. How is a neuron repolarized after a single stimulation and spike?
13. What is the function of the sodium–potassium pump in neurons?
14. What is the concept behind the "local-circuit" theory of spike transmission in an axon?
15. Do insect neurons typically have long or short dendrites? Explain why.

16. How is the entry of sodium into a neuron during a spike phase stopped or terminated?
17. What is an excitatory postsynaptic potential (EPSP)?
18. What is the name of one of the best studied stimulatory transmitters at an insect synapse?
19. What is GABA? What words do the letters stand for? What is the role of GABA?
20. Explain the physiology behind poisoning of an insect with an organophosphate, such as Malathion.

REFERENCES

Armstrong, C.M. and F. Bezanilla. 1977. Inactivation of the sodium channel II. Gating current experiments. *J. Gen. Physiol.* 70: 567–590.

Becker, M.N., R. Brenner, and N.S. Atkinson. 1995. Tissue-specific expression of a *Drosophila* calcium-activated potassium channel. *J. Neurosci.* 15(9): 6250–6259.

Boumghar, K., T. Couret-Fauvel, M. Garcia, and C. Armengaud. 2012. Evidence for a role of GABA- and glutamate-gated chloride channels in olfactory memory. *Pharmacol. Biochem. Behav.* 103: 69–75.

Breer, H., D. Benke, W. Hanke, R. Kleene, M. Knipper, and L. Wieczorek. 1987. Identification, reconstitution and expression of neuronal acetylcholine receptor polypeptides from insects, in John Law (ed.), *Molecular Entomology*. Alan R. Liss, Inc., New York, pp. 95–105.

Callec, J.J. 1985. Synaptic transmission in the central nervous system, in G.A. Kerkut and L.I. Gilbert (eds.), *Comprehensive Insect Physiology, Biochemistry and Pharmacology*, vol. 5. Pergamon Press, Oxford, U.K., pp. 139–179.

Callec, J.J. and J. Boistel. 1967. Les effets de l'acetylcholine aux nivaux synaptique et somatique dans le cas du dernier ganglion abdominal de al blatte, *Periplaneta americana* L. *C.R. Soc. Biol.* 161: 442–446.

Changeux, J.-P. 1993. Chemical signaling in the brain. *Sci. Am.* 1993: 58–62.

Changeux, J.-P., A. Devillers-Thiéry, and P. Chemouilli. 1984. Acetylcholine receptor: An allosteric protein. *Science* 225: 1335–1345.

Changeux, J.-P., J.L. Galzi, A. Devillers-Thiéry, and D. Bertrand. 1992. The functional architecture of the acetylcholine nicotinic receptor explored by affinity labelling and site-directed mutagenesis. *Q. Rev. Biophys.* 25: 395–432.

Cole, K.S. 1949. Dynamic electrical characteristics of the squid axon membrane. *Arch. Sci. Physiol.* 3: 253–258.

Démares, F., V. Raymond, and C. Armengaud. 2013. Expression and localization of glutamate-gated chloride channel variants in honeybee brain (*Apis mellifera*). *Insect Biochem. Mol. Biol.* 43: 115–124.

Eccles, J.C. 1964. Ionic mechanism of postsynaptic inhibition. *Science* 145: 1140–1147.

el Hassani, A.K., S. Schuster, Y. Dyck, F. Démares, G. Leboulle, and C. Armengaud. 2012. Identification, localization and function of glutamate-gated chloride channel receptors in the honeybee brain. *Eur. J. Neurosci.* 36: 2409–2420.

French, A.S., E.J. Sanders, E. Duszyk, S. Prasad, P.H. Torkkeli, J. Haskins, and R.A. Murphy. 1993. Immunocytochemical localization of sodium channels in an insect central nervous system using a site-directed antibody. *J. Neurobiol.* 24: 939–948.

Ganetzky, B., J.W. Warmke, G. Robertson, N. Atkinson, and R. Drysdale. 1993. Genetic and molecular analysis of potassium channels in *Drosophila*, in A.B. Borkovec and M.J. Loeb (eds.), *Insect Neurochemistry and Neurophysiology*. CRC Press, Boca Raton, FL, pp. 9–22.

Gwilliam, G.F. and Burrows, M. 1980. Electrical characteristics of the membrane of an identified insect motor neurone. *J. Exp. Biol.* 86: 49–61.

Haim, N., S. Nahum, and Y. Dudai. 1979. Properties of a putative muscarine cholinergic receptor from *Drosophila melanogaster*. *J. Neurochem.* 32: 543–552.

Hodgkin, A.L. 1937a. Evidence for electrical transmission in nerve. Part I. *J. Physiol.* 90: 183–210.

Hodgkin, A.L. 1937b. Evidence for electrical transmission in nerve. Part II. *J. Physiol.* 90: 211–232.

Hodgkin, A.L. 1964. The ionic basis of nerve conduction. *Science* 145: 1148–1154.

Hodgkin, A.L. and B. Katz. 1949. The effect of sodium ions on the electrical activity of the giant axon of the squid. *J. Physiol. (London)* 108: 37–77.

Huxley, A.F. 1964. Excitation and conduction in nerve: Quantitative analysis. *Science* 145: 1154–1159.

Janssen, D., C. Derst, J.-M. Rigo, and E. Van Kerkhove. 2010. Cys-loop ligand-gated chloride channels in dorsal unpaired median neurons of *Locusta migratoria*. *J. Neurophysiol.* 103: 2587–2598.

Kerkut, G.A., R.M. Pitman, and R.J. Walker. 1969a. Sensitivity of neurones of the insect central nervous system to iontophoretically applied acetylcholine or GABA. *Nature* 222: 1075–1076.

Kerkut, G.A., R.M. Pitman, and R.J. Walker. 1969b. Iontophoretic application of acetylcholine and GABA onto insect central neurones. *Comp. Biochem. Physiol.* 31: 611–633.

Keyes, R.D. 1951. The ionic movements during nervous activity. *J. Physiol.* 114: 119–150.

Kita, T., F. Ozoe, M. Azuma, and Y. Ozoe. 2013. Differential distribution of glutamate- and GABA-gated chloride channels in the housefly *Musca domestica*. *J. Insect Physiol.* 59: 887–893.

Kocmarek, A.L. and M.J. O'Donnell. 2011. Potassium fluxes across the blood brain barrier of the cockroach, *Periplaneta americana*. *J. Insect Physiol.* 57: 127–135.

Lee, D. and M.E. Adams. 2000. Sodium channels in central neurons of the tobacco budworm, *Heliothis virescens*: Basic properties and modification by scorpion toxins. *J. Insect Physiol.* 46: 499–508.

Li, M., Y.N. Jan, and L.Y. Jan. 1992. Specification of subunit assembly by the hydrophilic amino-terminal domain of the *Shaker* potassium channel. *Science* 257: 1225–1229.

Liu, W.W. and R.I. Wilson. 2013. Glutamate as an inhibitory transmitter in the *Drosophila* olfactory system. *Proc. Natl. Acad. Sci. USA* 110: 10294–10299.

Lummis, S.C.R. and D.B. Sattelle. 1985. Binding of N-[propionyl-^{3}H] propionylated α-bungarotoxin and L-[benzilic-4,4'-^{3}H] quinuclidinyl benzelate to CNS extracts of the cockroach *Periplaneta americana*. *Comp. Biochem. Physiol.* 80C: 75–83.

Mackinnon, R., R.W. Aldrich, and A.W. Lee. 1993. Functional stoichiometry of *Shaker* potassium channel inactivation. *Science* 262: 757–759.

Marmont, G. 1949. Studies on the axon membrane. I. A new method. *J. Cell. Comp. Physiol.* 34: 351–382.

Matsuda, K., S.D. Buckingham, D. Kleier, J.J. Rauh, M.Grauso, and D.B. Sattelle. 2001. Neonicotinoids: Insecticides acting on insect nicotinic acetylcholine receptors. *Trends Pharmacol. Sci.* 22: 573–580.

Meyer, M.R. and J.S. Edwards. 1980. Muscarinic cholinergic binding sites in an orthopteran central nervous system. *J. Neurobiol.* 11: 215–219.

Miller, C. 1991. 1990: Annus mirabilis of potassium channels. *Science* 252: 1092–1096.

Money, T.G., C.I. Rodgers, S.M. McGregor, and R.M. Robertson. 2009. Loss of potassium homeostasis underlies hyperthermic conduction failure in control and preconditioned locusts. *J. Neurophysiol.* 102: 285–293.

Nakao, T., S. Banba, M. Nomura, and K. Hirase. 2013. Meta-diamide insecticides acting on distinct sites of RDL GABA receptor from those for conventional noncompetitive antagonists. *Insect Biochem. Mol. Biol.* 43: 366–375.

Narahashi, T. 1971. Effects of insecticides on excitable tissues. *Adv. Insect Physiol.* 8: 1–93.

Narahashi, T. and T. Yamasaki. 1960. The mechanism of after-potential production in the giant axons of the cockroach. *J. Physiol. (London)* 151: 75–88.

Narahashi, T., X. Zhao, T. Ikeda, V.L. Salgado, and J.Z. Yeh. 2010. Glutamate-activated chloride channels: Unique fipronil targets present in insects but not in mammals. *Pestic. Biochem. Physiol.* 97: 149–152.

Orr, G.L., N. Orr, and R.M. Hollingworth. 1990. Localization and pharmacological characterization of nicotinic-cholinergic binding sites in cockroach brain using α and neuronal bungarotoxin. *Insect Biochem.* 20: 557–566.

Ozoe, Y. 2013. γ-Aminobutyrate- and glutamate-gated chloride channels as targets of insecticides. *Adv. Insect Physiol.* 44: 211–286.

Ozoe, Y., M. Asahi, F. Ozoe, K. Nakahira, and T. Mita. 2010. The antiparasitic isoxazoline A1443 is a potent blocker of insect ligand-gated chloride channels. *Biochem. Biophy. Res. Commun.* 391: 744–749.

Parnas, L., M.E. Spira, R. Werman, and F. Bergmann. 1969. Non-homogeneous conduction in giant axons of the nerve cord of *Periplaneta americana*. *J. Exp. Biol.* 50: 635–649.

Pichon, Y. and F.M. Ashcroft. 1985. Nerve and muscle: Electrical activity, in G.A. Kerkut and L.I. Gilbert (eds.), *Comprehensive Insect Physiology, Biochemistry and Pharmacology*, vol. 5. Pergamon Press, Oxford, U.K., pp. 85–113.

Pichon, Y. and J. Boistel. 1966. Application aux fibres géantes de Blattes (*Periplaneta americana* L. et *Blabera craniifer* Bürm.) d'une technique permettant l'introduction d'une microélectrode dans le tissu nerveux sans résection préalable de la gaine. *J. Physiol. (Paris)* 58: 592.

Pichon, Y., D.B. Sattelle, and N.J. Lane. 1972. Conduction processes in the nerve cord of the moth, *Manduca sexta*, in relation to its ultra-structure and haemolymph ionic composition. *J. Exp. Biol.* 56: 717–734.

Pichon, Y. and J.E. Treherne. 1973. An electrophysiological study of the sodium and potassium permeabilities of insect peripheral nerves. *J. Exp. Biol.* 59: 447–461.

Rodgers, C.I., J.D. Labrie, and R.M. Robertson. 2009. K⁺ homeostasis and central pattern generation in the metathoracic ganglion of the locust. *J. Insect Physiol.* 55: 599–607.

Salkoff, L., K. Baker, A. Butler, M. Covarrubias, M.D. Pak, and A. Wei. 1992. An essential "set" of K⁺ channels conserved in flies, mice and humans. *Trends Neurosci.* 15: 161–166.

Schmeller, T., A. El-Shazly, and M. Wink. 1997. Binding of pyrrolizidine alkaloids to acetylcholine, serotonin, and dopamine receptors. *J. Chem. Ecol.* 23: 399–416.

Shankland, D.L., J.A. Rose, and C. Donniger. 1971. The cholinergic nature of the cercal nerve-giant fiber synapse in the sixth abdominal ganglion of the American cockroach, *Periplaneta americana* L. *J. Neurobiol.* 2: 247–262.

Spong, K.E. and R.M. Robertson. 2013. Pharmacological blockade of gap junctions induces repetitive surging of extracellular potassium within the locust CNS. *J. Insect Physiol.* 59: 1031–1040.

Stevens, C.F. 1984. Biophysical studies of ion channels. *Science* 225: 1346–1350.

Tanouye, M.A., A. Ferrus, and S.C. Fujita. 1981. Abnormal action potentials associated with the *Shaker* complex locus of *Drosophila*. *Proc. Natl. Acad. Sci. USA* 78: 6548–6552.

Thany, S.H., G. Lenaers, V. Raymond-Delpech, D.B. Sattelle, and B. Lapied. 2006. Exploring the pharmacological properties of insect nicotinic acetlcholine receptors. *Trends Pharmacol. Sci.* 28(1): 14–22. doi: 10.1016/j.tips.2006.11.006.

Timpe, L.C., Y.N. Jan, and L.Y. Jan. 1988. Four cDNA clones from the *Shaker* locus of *Drosophila* induce kinetically distinct A-type potassium currents in *Xenopus oocytes*. *Neuron* 1: 659–667.

Treherne, J.E. 1985. Blood-brain barrier, in G.A. Kerkut and L.I. Gilbert (eds.), *Comprehensive Insect Physiology, Biochemistry and Pharmacology*, vol. 5. Pergamon Press, Oxford, U.K., pp. 115–137.

Treherne, J.E. and S.H.P. Maddrell. 1967. Membrane potentials in the central nervous system of a phytophagous insect, (*Carausius morosus*). *J. Exp. Biol.* 46: 413–421.

Treherne, J.E. and P.K. Schofield. 1981. Mechanisms of ionic homeostasis in the central nervous system of an insect. *J. Exp. Biol.* 95: 61–73.

Trudeau, M.C., J.W. Warmke, B. Ganetzky, and G.A. Robertson. 1995. *HERG*, a human inward rectifier in the voltage-gated potassium channel family. *Science* 269: 92–95.

Warmke, J.W. and B. Ganetzky. 1993. A novel potassium channel gene family: *EAG* homologs in *Drosophila*, mouse and human. *Biophys. J.* 64: A340 (abstract).

Yamasaki, T. and T. Narahashi. 1959. The effects of potassium and sodium ions on the resting and action potentials of the cockroach giant axon. *J. Insect Physiol.* 3: 146–158.

CHAPTER 11

Muscles

PREVIEW

The primary focus in this chapter is on skeletal muscle anatomy and function, with only brief information on flight muscles, gut muscles, and heart muscle for comparison. Insect flight, wing and thoracic structure, flight muscles, and physiology of flight are described in a separate chapter. Insect skeletal muscles are composed of cells that have anastomosed into multinucleate fibers of myofibrils. Myofibrils are divided into sarcomeres, which are the contractile units of muscle. Although wing muscles and jumping leg muscles in some insects are relatively large, muscles in small insects and in small appendages are necessarily small and often are composed of only a few fibers. Skeletal muscles are attached to the cuticle, typically by tonofibrillae that pass through the endo- and exocuticle and attach to the inner layer of the epicuticle by a large protein encoded by the *dumpy* gene.

Mitochondria, originally called sarcosomes because of their large and irregular size, are the powerhouses for muscle function. Only a few motor neurons are allocated to innervate most insect muscles, and typically, there is a fast axon producing a rapid, twitch-like response in the muscle and a slow axon that produces a slower but more sustained contraction. The fast axon innervates each fiber in a muscle, while the slow axon innervates only about 30%–40% of the fibers. Some muscles also receive one or more inhibitory neurons. A few large muscles receive multiple motor neurons. Graded contractions are achieved in some muscles by activating the slow or the fast axon, depending on the degree of muscle action needed, and perhaps by combining these with the action of the inhibitory axon. Each motor nerve breaks into many terminals that make contact with the muscle fibers at intervals of 40–80 μm apart. Generally, an action potential is not conducted by the muscle fiber itself, but contractions occur around the nerve terminals and, thus, sum over the entire muscle. The transmitter chemical at excitatory motor endings is L-glutamic acid or L-aspartic acid, and the transmitter at inhibitory neurons is gamma-aminobutyric acid (GABA).

Most skeletal muscles of insects are synchronous muscles that require nerve input for each contraction, but wing muscles in some insects are asynchronous, and multiple contractions for each motor nerve input can be obtained. The ability of fibrillar muscles to yield multiple contractions is based on anatomical arrangement in the thorax, internal anatomy, and physiology. During the active state induced in muscles by the arrival of nerve impulses, calcium ions bound to the sarcoplasmic reticulum (SR) are released and then bind to a subunit of troponin. This induces a conformation change that pulls tropomyosin away from an active site on actin. Myosin binds to the active site and pulls actin into a new position. Myosin releases from actin when adenosine triphosphate (ATP) binds to it and is split, releasing energy for return to the original state of myosin. Binding, sliding, releasing, and repeating are very rapid events, occurring in about 0.1 ms. Contraction is terminated by rapid binding of calcium to the sarcoplasmic reticulum, necessitating more nerve input to free Ca^{2+} for any additional contraction. SR sequestering of calcium is slower in asynchronous muscle than in synchronous muscle, which is part of the explanation for multiple contractions per nerve input in asynchronous muscle. All muscles in insects are striated, including visceral muscle. Some insects have special adaptations in skeletal anatomy and muscle function for jumping and singing.

11.1 INTRODUCTION

The microstructural units of muscles that are most clearly identifiable with low magnification are muscle fibers. Each muscle fiber in skeletal and wing muscles is composed of many cells that have anastomosed so that cell membranes are no longer distinct. The nuclei from these cells, however, are still evident, and muscle fibers are multinucleate in histological sections. In insect gut muscles, the individual cells are more distinct and uninucleate. All insect muscle is striated, including gut muscle. Typically, the muscle fibers are as long as the muscle itself. In some muscles, fibers are grouped into bundles, while in others, especially fibrillar muscles, they are only loosely held together. Muscles are usually divided into broad categories, such as skeletal, flight, and visceral muscle. Visceral muscles can be further subdivided into muscles associated with the heart and dorsal vessel, alimentary canal, Malpighian tubules, accessory pulsatile organs, various diaphragms, and glandular structures. Skeletal muscles are a mixture of fast and slow-contracting muscles. Fibrillar muscles are exclusively fast-contracting muscles. Heart, alary, and gut muscle are slower-contracting muscles. The muscular system helps support the body, aids in maintaining posture, and enables movement of limbs, antennae, sting apparatus in some insects, ovipositor, viscera, spiracle closure, cibarial and various glandular pumps, and of course, flight and general locomotion. The muscles require large amount of adenosine triphosphate (ATP) energy and produce heat, which may be useful on cool days for those insects that shiver (some lepidopterans) to warm up for flight, but may have to be distributed to the abdomen for cooling the thorax in some insects during flight (some lepidopterans, again). Although wing muscles and jumping leg muscles in some insects are relatively large, muscles in small insects and in small appendages are necessarily small and often are composed of only a few fibers. For example, in the small fruit fly *Drosophila melanogaster*, the dorsal longitudinal flight muscle has only 12 fibers; the dorsoventral flight muscle has 14 fibers, and fibers are only about 1 mm in length and vary from 100 to 200 μm in diameter (Swank, 2012). Larger insects will necessarily have larger muscles with longer and more muscle fibers.

11.2 BASIC MUSCLE STRUCTURE AND FUNCTION

11.2.1 Macro- and Microstructure of Muscle

A muscle can be subdivided into fibers and a fiber into **myofibrils**. Myofibrils are composed of **sarcomeres**, each of which contains **actin** and **myosin** and other proteins involved in the contraction mechanism (Figure 11.1). In addition, muscle contains **Sarcoplasmic Reticulum (SR)**, which is much reduced in fibrillar muscles but well developed in fast synchronous muscles (Figure 11.2). The SR is an extensive network of internal membranes broken into vesicles that run longitudinally on the surface of the muscle fibers. The SR plays a major role in the contraction process as a storehouse of calcium ions. Transverse (T) tubules penetrate the muscle from the outside, originating usually, but not always, at the Z bands. The network of T tubules and SR membranes do not open to each other, but they do intersect at closed junctions believed to be the major sites of calcium storage. These junctions are called **dyad** or **triad** junctions, depending on whether a T tubule intersects with one or two SR vesicles. The T tubules carry the electrical wave of excitation arriving at the surface of a muscle (via a nerve) inward where it also spreads to the SR and releases bound calcium as the free ions necessary for contraction to occur.

Muscles contain abundant and often large, irregularly shaped mitochondria (also called **sarcosomes**, especially in thoracic musculature associated with wing movements), nuclei, and

MUSCLES

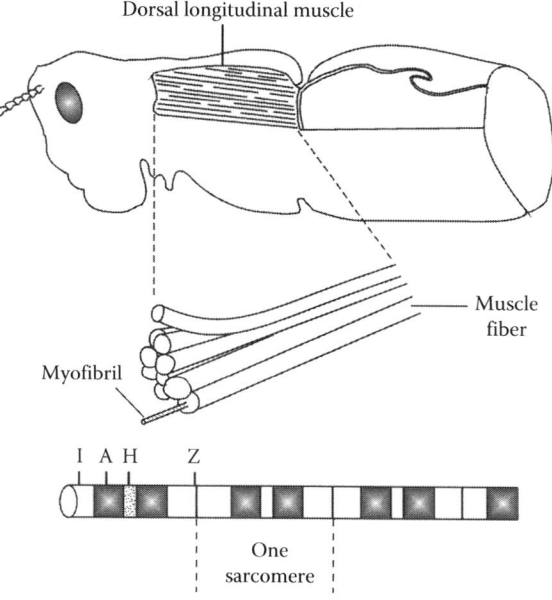

Figure 11.1 An illustration of progressively smaller units that compose muscles. Muscle is composed of muscle fibers that in turn are composed of myofibrils. At still higher magnification, myofibrils can be seen to be composed of sarcomere units, which are the contractile units of the muscle. The distribution of muscle proteins within the sarcomeres, which make the light and dark areas in a sarcomere, is designated as I, A, and H bands.

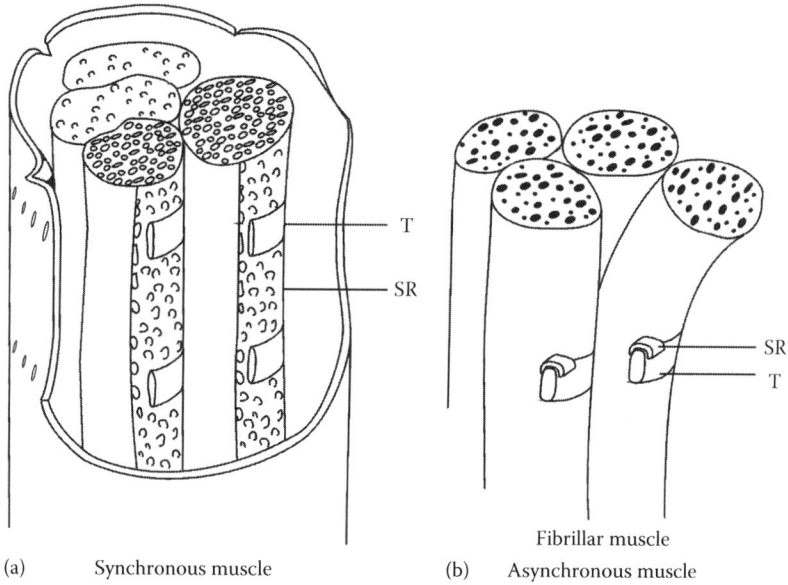

Figure 11.2 Diagrammatic illustration of the sarcoplasmic reticulum and T tubules in synchronous muscle (a) and in asynchronous muscle (b).

intracellular tracheoles (Figure 11.3). Intracellular tracheoles are not really inside the plasma membrane of the muscle, but have merely pushed into the muscle interior, like a finger pushed into a soft balloon.

The myofibrils (called fibrils by some authors) are made up of repeating sarcomere units. A sarcomere, the region between two Z bands, is typically about 2–3 μm long in a muscle at rest, but sarcomeres up to 10 μm long occur in some very slow muscles. Sarcomere length is shorter in fast-contracting muscles. Upon contraction, sarcomere length decreases, as does the entire muscle length. The Z line is a plate-like sheet of protein to which actin (thin filaments, about 5 nm in diameter) and some other muscle proteins are attached. The thin filaments extend on either side of the Z line about two-thirds to nearly the midpoint of a sarcomere. Thick filaments of myosin, about 20 nm in diameter, lie between the thin filaments (Ashhurst, 1967). The thick filaments extend across the middle of a sarcomere, but usually do not extend to the Z line. The various overlapping regions of thick and thin filaments give muscle in thin histological sections a banded appearance (Figure 11.3) as light passes through regions of different density. The A band appears dark because light must pass through the overlapping regions of actin and myosin filaments, while the H zone in the middle of the sarcomere and the I band near the Z line transmit more light because these regions contain only myosin or actin filaments, respectively. The M line across the middle of the H zone is created by cross-links between myosin filaments that help hold the myosin filaments in place. The various zones are of variable length in different muscles, depending on the degree of overlapping of filaments. The width of the H zone bears some relationship to how much the sarcomere (and, thus, the muscle) will shorten upon stimulation and how fast it can accomplish its shortening. Muscles with very narrow H and I zones,

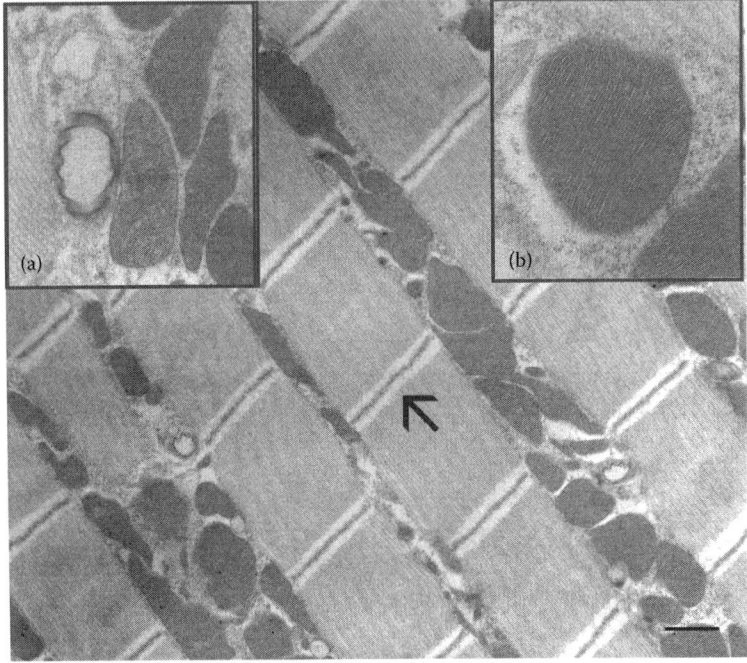

Figure 11.3 A transmission electron micrograph of muscle from *Tachinaephagus zealandicus* (Hymenoptera) showing well-defined Z lines (arrow); numerous, large, irregular mitochondria between the myofibrils; and intracellular tracheoles. (a) Shows an enlarged view of one of the intracellular tracheoles and several mitochondria. (b) Shows an enlargement of a mitochondrion. Membranes of the cristae in mitochondria can be seen in the enlargements. The I bands (the light areas on each side of the dark Z line) are very narrow in this muscle, indicating that the myosin filaments extend nearly to the Z line. There is just a hint of a lighter, narrow H zone at the middle of the sarcomeres. (Photographs by the author, Jimmy Becnel, and Alexandra Shapiro, USDA, Washington, DC.)

such as fibrillar muscles that cause the wing movements in Diptera and Hymenoptera, shorten only a small amount upon contraction (2%–3%), while some muscles may shorten much more. Generally, in skeletal and flight muscles, the striations or bands of adjacent myofibrils are aligned side by side. Thus, the light and dark bands appear evenly lined up in a large section of muscle. However, there are exceptions to this in some insect muscles, and there especially is less alignment in gut muscles.

Cross sections of fibrillar flight muscle myofibrils viewed with the electron microscope typically show each thick filament surrounded by six thin filaments, with the thin filaments positioned about equally between two thick filaments to give a ratio of three thin filaments to one thick filament. Other ratios are found in some muscles, including up to 12 thin filaments arranged in such a way as to present a 6 thin/1 thick filament arrangement in intersegmental muscles of the cockroach, *Periplaneta americana* (Smith, 1966) and *Rhodnius prolixus* (Hemiptera: Reduviidae) (Toselli and Pepe, 1968), and in wing muscles of large saturniid silk moths (Lepidoptera: Saturniidae) that have slow wingbeats of five to six beats per second (Carnevali and Reger, 1982).

The physiology and biochemistry of muscle contraction appear to be essentially the same in insects as in other organisms. The sliding filament theory in which actin and myosin filaments slide over each other, drawing Z bands closer together and, thus, shortening, adequately explains insect muscle contraction (Guschlbauer et al., 2007).

11.2.2 Muscle Attachments to the Exoskeleton

One end of a skeletal or wing muscle is anchored to a relatively nonmoveable part of the exoskeleton. This is called the **origin** of the muscle, and the opposite end that is attached to a moveable part of the body, such as a wing hinge, an appendage, or a part of the exoskeleton, is called the **insertion**. Skeletal muscles are usually anchored to the epicuticle layer of the exoskeleton (Figure 11.4). A large 2.5 MDa

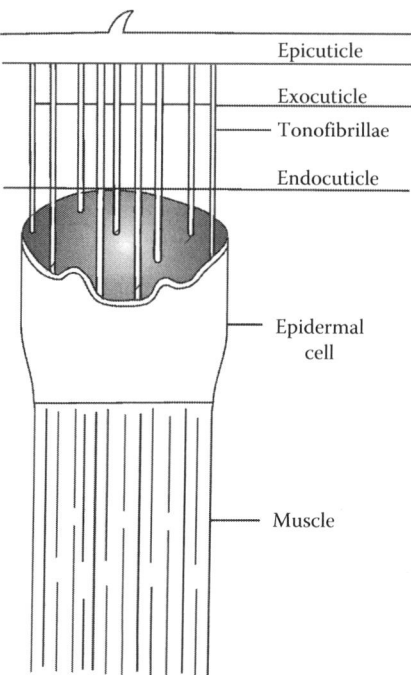

Figure 11.4 A schematic illustration of tonofibrillae connecting muscle fibers with the epicuticle. (Drawing modified from Elder, H.Y., Muscle structure, in P.R.N. Usherwood (ed.), *Insect Muscle*, Academic Press, New York, 1975, pp. 1–74.)

extracellular matrix protein (**Dumpy protein**) encoded by the *dumpy* (*dp*) gene in *D. melanogaster* connects the outer membrane of muscle fibers to the basal membrane of epidermal cells through a cross-linking zona pellucida domain and a transmembrane anchoring sequence (Wilkin et al., 2000). Dumpy protein provides a strong attachment for muscles to cells and cuticle, allowing high mechanical tension and preventing the tearing of muscle away from the cuticle. Bundles of intracellular microtubules originate at the junction with the epidermal cells, pass through the epidermal cells, and, at the cuticle, are cemented to fibers of chitin (**tonofibrillae**) that are formed extracellularly and embedded in the cuticle (Hinton, 1973). Tonofibrillae are very resistant to the action of molting fluid and allow muscles to remain attached to the exoskeleton after apolysis and secretion of new cuticle begins. In some insects, apolysis occurs hours and even days before ecdysis, and movement during the interval between apolysis and ecdysis may be critical to feeding and predator escape. The final factors that enter into the breaking of the tonofibrillae to allow the old cuticle to be ecdysed have not been clarified. After the old attachments of tonofibrillae are dissolved, new attachments to the epicuticle layer occur quickly. It should be remembered that the new cuticle present at ecdysis is largely unsclerotized, and the new epicuticle is probably the most stabilized part of the new cuticle and the best place to anchor the muscles. The new cuticle must sclerotize adequately before the muscles are used or the soft cuticle will be distorted in shape by the pull of powerful muscles, particularly those used in flight and jumping. Most insects rest quietly for some minutes or hours immediately after molting until the cuticle has hardened.

The *dumpy* gene has multiple functions—it is expressed on the inner surface of the micropyle in a mature *Drosophila* egg (Carmon et al., 2007), it interacts with several proteins during embryonic development of the tracheal system, and it is important as a part of the extracellular matrix of the wing (Carmon et al., 2010b)—and has a number of other functions revealed through studies with mutants at the *dumpy* locus (Carmon et al., 2010a).

11.2.3 Skeletal Muscle

Skeletal muscles usually are organized in antagonistic pairs. One muscle of the pair (the **flexor**) bends an appendage, while the second muscle of the pair (the **extensor**) straightens the appendage. A few muscles, for example, the tymbal muscles of cicadas, are not antagonistically paired, but depend upon the natural elasticity of the cuticle to stretch the muscle to the precontraction condition. Skeletal muscles typically are **synchronous muscles** in which the rate of contraction is in 1:1 proportion to the incoming nerve impulses. Peak tension and speed of contraction are influenced by sarcomere length, the degree of overlap of myosin and actin filaments, and degree of development of the SR (Elder, 1975).

11.2.4 Polyneuronal Innervation and Multiterminal Nerve Contacts

Because of the small size of insects, only a few neurons are allocated to control each muscle. Pringle (1939) showed that the leg muscles (and many other muscles, it is now known) typically receive two stimulatory neurons, commonly described as the **slow neuron** and the **fast neuron** (Figure 11.5). At the muscle surface, the two nerves usually share a glial sheath and lie in a shallow groove on the muscle fiber surface, or sometimes, they are invaginated within the muscle fiber outer membrane. Both fast and slow axons break into a number of arborizations that make multiple contacts a few μm apart on muscle fibers. This is called **multiterminal innervation** and it is common in invertebrates. In vertebrates, a neuronal branch makes only one contact with a muscle fiber although the motoneuron typically sends axonal branches to many fibers to form a motor unit.

The fast axon sends multiple terminals to all or most of the muscle fibers in locust jumping leg muscle, but the slow axon makes junctional contacts with only 30%–40% of the muscle fibers (Hoyle, 1955). In muscles of nonjumping legs, the percent of fibers innervated by the slow axon is

MUSCLES

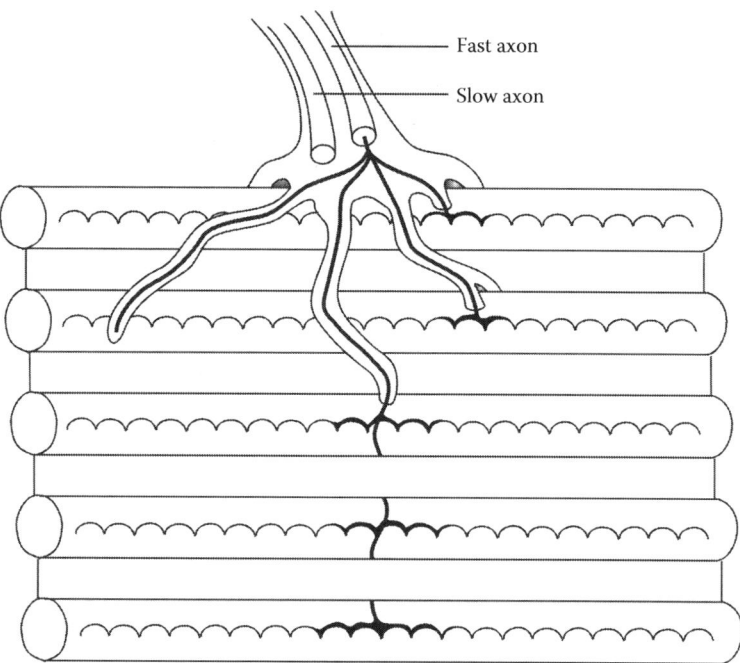

Figure 11.5 Polyneuronal innervation of muscle by fast and slow axons and the multiterminal junctional contacts with the muscle fibers. For clarity, only the contacts from the fast axon, which makes multiple contacts with each muscle fiber, are shown. The slow axon usually makes contact with 30%–40% of the muscle fibers. Both axons may share the same glial sheath and contact points on the muscle fibers.

usually greater than 40%, and in some small muscles, such as some spiracular muscles, every fiber is dually innervated. Junctional locations on the muscle fibers are generally shared by the fast and slow axons. The distance between the multiterminal contacts on a muscle fiber varies with different species, but the contacts are always close together. In flight muscles of *Geotrupes* spp. (a beetle) junctional contacts are about 80 μm apart, in *Musca domestica* 50 μm apart, 40 μm apart in cockroach leg muscle, and about 60 μm apart in locust and grasshopper leg muscle.

Insect skeletal muscle usually does not develop a propagated action potential (an all-or-none response similar to nerve) as vertebrate skeletal muscle does, although resting potentials of 40–60 mV, inside negative, have been recorded from insect muscles (Aidley, 1985, 1989). Instead, graded potentials are produced around each of the junctional endings, and local contraction of the muscle fiber occurs around each ending. Because junctional terminals are very close together, the net result is a nearly simultaneous contraction of the whole fiber without a propagated muscle potential. After very careful dissection, some investigators have reported that jumping leg muscle fibers will give a spike instead of the summed graded responses around end plates. Collet and Belzunces (2007) found that skeletal muscle fibers isolated from the metathoracic tibia of adult honeybees do develop all-or-none action potentials when the membrane potential is held close to the resting value by the voltage clamp procedure. GABA had no effect on the muscle, indicating no inhibitory neurons going to this particular muscle, but L-glutamate induced fast activation of the muscle.

The fast neuron produces a fast, twitch-type response in the muscle, while the slow axon typically produces a much slower, graded response. The slow neuron junctional contacts at the muscle show facilitation, and a characteristic frequency of nerve impulses must arrive at the junction before the muscle contracts. Typically, a muscle innervated by a fast axon contracts rapidly in response to each nerve impulse arriving. The designations "fast" and "slow," indicate the speed of the muscle contraction, not the rate of nerve impulse conduction. Both fast and slow neurons are about the same size and conduct the nerve impulse at the same rate. They usually share the same glial sheath. This dual (or triple if an inhibitory neuron is present) innervation of a muscle in insects is referred to as **polyneuronal innervation**. A third neuron, an inhibitory one that causes hyperpolarization, was described by Hoyle (1955) from the jumping leg muscle of the locust and also from a grasshopper, *Romalea microptera* (Usherwood and Grundfest, 1964; Usherwood, 1968). The same inhibitory neuron typically innervates several muscles, that is, they are considered to be common neurons as opposed to being specific for a particular muscle. However, Bräunig et al. (2006) recently described three inhibitory neurons from *Locusta migratoria*, two of which provide endings to several muscles, but a third is specific to the longitudinal muscle M60 (muscle terminology from Snodgrass, 1929). One of the common neurons innervates intersegmental muscle M59 and dorsal longitudinal muscles (DLM) M81 and M82, while a second common inhibitory neuron also sends a branch to M59 and to the ventral longitudinal muscle M60. Bräunig et al. (2006) claim that the neuron sending terminals only to muscle M60 is the only specific inhibitory neuron (i.e., innervating only one muscle) so far described.

Only a few large muscles get more than two stimulatory neurons. A few cases have been reported in which there are multiple axons to a large muscle. For example, seven to nine axons to the basalar flight muscle of large beetles (Darwin and Pringle, 1959; Ikeda and Boettiger, 1965). The locust DLM, which are large, powerful muscles involved in flight, receive five fast motoneurons (Neville, 1963; Burrows, 1977). The mesothoracic DLM in a katydid, *Neoconocephalus robustus*, receives four fast axons, while DLM in the metathorax receive five fast axons (Josephson and Stokes, 1982). There is no evidence for inhibitory neurons to the DLM in *N. robustus*, and the possibility of slow fibers is uncertain. There are at least five motor neurons to the basalar flight muscle of the scarab beetle, *Cotinis mutabilis* (Josephson et al., 2000a). Thus, although the flight muscles of some insects receive multiple axons and skeletal muscles may receive both slow and fast axons, indirect flight muscles of Hymenoptera and Diptera receive only fast axons. Slow flapping of the wings may be unnecessary and even impossible in some insects. The tymbal muscles of cicadas also receive only fast axons.

A few muscles may receive no nervous connections. For example, the long, thin muscle that spirals round the Malpighian tubules of some insects has its own myogenic rhythm and does not receive a nerve supply. At least some of the accessory heart muscles located at the base of legs and antennae of some insects appear to have no nerve supply, while others receive neurosecretory neurons and may be a neurohemal organ site.

The ability to achieve graded muscle contractions is important to the behavior of all animals, and insects appear to have an adequate repertoire of mechanisms available to them to achieve graded contractions. In addition to the simple arithmetic of how many muscle fibers may be activated depending on the use of the fast or slow axon, **facilitation** and **summation** commonly occur at the junctional endings when the slow axon is activated. In the leg, typically about 15 neuronal impulses per second must arrive at the junctional endings of the slow axon in order to produce contraction. Temporal facilitation and summation can occur when several small volleys of nerve impulses arrive within 1 s. Thus, an insect that has some combination of fast, slow, and inhibitory axons to the same muscle may realize a number of degrees of graded muscle responses from the neuronal system or systems utilized, and response may be further modified by neuropeptides secreted at some nerve endings.

11.2.5 Transmitter Chemical at Nerve–Muscle Junctions

The synaptic cleft between nerve ending and muscle membrane is about 30 nm wide (Aidley, 1989), and a transmitter chemical must diffuse across this gap from axonal ending to muscle sarcolemma. The stimulatory transmitter chemical at fast and slow nerve–muscle junctions is **L-glutamic acid** (Irving and Miller, 1980; Atwood and Cooper, 1996) and possibly **L-aspartate** as well. At inhibitory junctions, it is **GABA** (Figure 11.6). The transmitter chemicals are contained in large numbers of

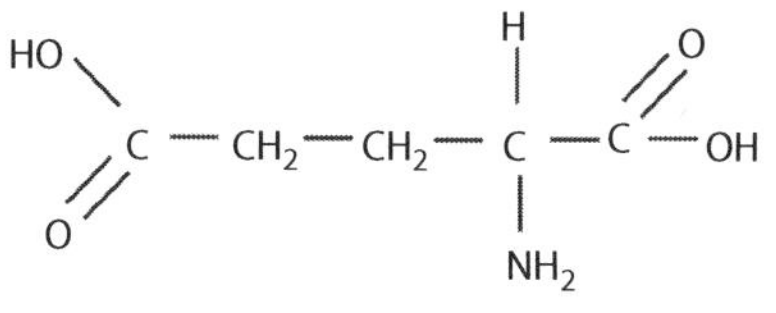

Figure 11.6 The chemical structure of L-glutamic acid, a stimulatory neurotransmitter at nerve–muscle junctions in insects, and γ-aminobutyric acid, an inhibitory transmitter at nerve–muscle junctions.

vesicles that are from 20 to 60 nm in diameter. Although vertebrates utilize GABA as an inhibitory transmitter in the central nervous system (CNS), they use acetylcholine as the stimulatory transmitter at neuromuscular junctions in skeletal muscles.

Several types of glutamate receptors (GluR) have been characterized in insect muscles (and also in the insect CNS) based on binding to indicator compounds, including quisqualate (quisqualate glutamate receptor, qGluR), ibotenate (iGluR), and aspartate (aspGluR), but how these different receptors function in muscle activity is unknown. Multiple receptors may be involved in different degrees of graded responses from the muscle. In *Schistocerca gregaria* and *Tenebrio molitor*, where the most detailed studies have been done, populations of these receptors are present at nerve–muscle junctions as well as at extrajunctional sites (Usherwood, 1994). The fate of L-glutamic acid and GABA at the muscle receptors is unclear, but the neuron may reabsorb the transmitter to deactivate it. The transmitter at visceral muscle junctions has not been definitely determined. There is evidence, however, that some neuropeptides may play a role in gut muscle function, either as neurotransmitters or as neuromodulators that modify the response to a transmitter.

11.3 SYNCHRONOUS AND ASYNCHRONOUS MUSCLES

Muscles of insects can be divided into two groups based on the contraction rate per nerve input. Contraction frequency in **synchronous muscles** is directly controlled in a 1:1 manner by the output of nerve impulses from the CNS. In **asynchronous muscles,** the contraction frequency is a property of the muscle itself and its anatomical arrangement in the musculoskeletal system. Asynchronous wing muscles, the **dorsal longitudinal** and **dorsoventral**, do not attach directly to the wings, but are attached to the thoracic cuticle. Asynchronous muscles oscillate and produce several to many contractions in response to a single neuronal stimulus. The small **direct wing muscles,** called by some authors "control muscles" (Dickinson and Tu, 1997), insert directly on the wing hinge sclerites or on axillary sclerites or movable sclerites of the pleuron of the thorax. These muscles are the **basalar** (wing depressors), **subalar** (wing depressors and twisting or supinations of the wing), and **third axillary** (wing folding) muscles. The control muscles are synchronous muscles and are activated by the nervous system so that one motor spike produces one muscle contraction (Dickinson and Tu, 1997). The indirect flight muscles are called fibrillar muscles because bundles of muscle fibers are unaligned and the much reduced endoplasmic reticulum is located at the periphery (Roy and VijayRaghavan, 2011). The gene *salm* (Schnorrer et al., 2010) encodes a zinc finger transcription factor, Salm, that regulates the differentiation of precursor muscle cells into indirect flight muscles during the development of *Drosophila*, and Salm is expressed in indirect flight muscles of the fly *Calliphora* and the beetle *Tribolium* (Schonbauer et al., 2011). For more details on flight musculature and physiology, see Chapter 12.

General skeletal muscles of all insects, some tymbal and stridulatory muscles, and muscles that move the wings of some insects are synchronous. Synchronous muscles typically have a well-developed SR and T system with regularly occurring triad junctions (see Figure 11.2). Slowly contracting synchronous muscles are an exception, and they have a poorly developed SR. For example, the SR composes only about 1% of the fiber volume in the slow extensor tibia muscle of *S. gregaria* (desert locust: Orthoptera) (Cochrane et al., 1972), whereas the SR fills about 30% of the muscle volume in the fast synchronous tymbal muscle of a cicada, *Platypleura capitata* (Homoptera: Cicadidae) (Josephson and Young, 1985).

Pringle (1957, 1968, 1976) and others who made detailed studies of the function of synchronous muscles suggested that a contraction rate greater than about 100 times/s was unlikely because of the time needed for neuronal repolarization. A firing rate of 100 times/s allows 10 ms for the neuron to repolarize. The contraction frequency of most synchronous muscle is well below 100 Hz (contractions per second), but some stridulatory and tymbal muscles with characteristic internal anatomy of synchronous muscles have rates of contraction two or more times the expected 100 Hz limit. Exactly what is going on in these muscles is not known for sure; they may be synchronous most of the time but capable of going into the oscillatory behavior of asynchronous muscles (Josephson and Young, 1985) (see Section 11.6.1 for more details). Two types of synchronous muscles, **tubular** and **close packed**, have been well characterized and are found in both general skeletal and flight musculature of some insects.

The fibers of tubular muscle are multinucleate, with nuclei located along the center of the fiber (Figure 11.7) and slab-like myofibrils typically radiating from the center like spokes of a wheel. These slab-like myofibrils run the length of the muscle and shorten in the plane of the long axis of the muscle. Numerous large, somewhat elongated mitochondria are arranged radially between myofibrils. Both longitudinal and T tubules of the sarcoplasmic system are well developed in tubular muscles (Smith, 1961). Tubular muscles are the typical skeletal muscles of several orders, including Diptera and Hymenoptera, and are involved in movement of legs and other appendages, spiracle muscle control, and in such activities as compressing the abdomen (the tergosternal muscles) dorsoventrally as part of the breathing rhythm. Tubular muscles (also called radial fibers flight muscle by some authors) are found in the flight musculature of many insect groups, including both direct and indirect flight muscles in some insects. These flight muscles, which are the typical flight musculature in the lower Orthoptera (such as the Blattidae) and in Odonata, may be most similar to the evolutionarily primitive flight muscles.

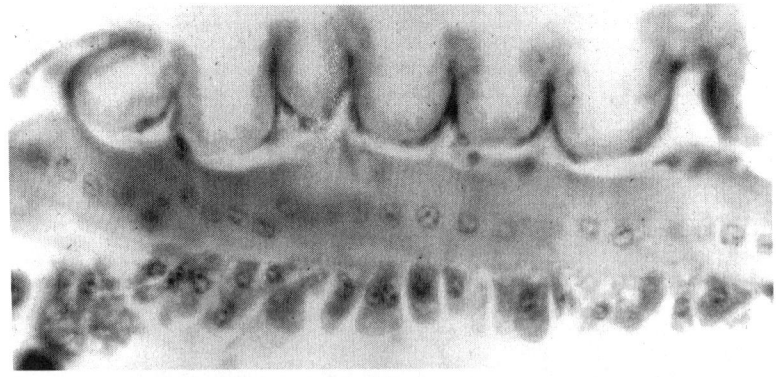

Figure 11.7 Photograph of tubular muscle from a tephritid fruit fly adult (*Anastrepha suspensa*) showing the nuclei lying along the middle of the muscle. (Photograph by the author.)

Other synchronous muscles occurring in the flight musculature of many groups of insects are named **close-packed fibers** (called microfibrillar and mosaic fibers by various authors). The muscle fibers are typically 10–100 μm in diameter. Nuclei are numerous, peripherally located, and somewhat flattened. Small myofibrils of 0.5–1 μm diameter are interspersed with columns of large mitochondria. The muscle fibers may be circular in shape or have a polygon shape with many angles. The flight muscles of some Orthoptera, Trichoptera, and Lepidoptera are close-packed muscle. In some Orthoptera, strap-like, radially arranged myofibrils of close-packed muscle look somewhat like the arrangement in tubular muscle of lower Orthoptera. Both longitudinal and T tubules of the SR are extensively developed.

Asynchronous muscles, also called **fibrillar muscles**, have arisen about 10 times in different groups of insects (Cullen, 1974), and the majority of insects that fly do so with asynchronous muscles. Asynchronous muscles are more efficient than synchronous muscles because they reduce the repetitive cycling of Ca^{2+} and, thus, minimize the energy cost of such cycling (Josephson et al., 2000a). Moreover, asynchronous muscles likely have been selected during evolution of flying insects because they allow a greater power output at high contraction frequencies typical of the flight of many insects (Josephson et al., 2000a,b). Asynchronous flight muscles occur in both direct and indirect muscles associated with wing movements in Diptera, Hymenoptera, Coleoptera, some Hemiptera, and in a number of other insect groups. Asynchronous muscles allow an insect to beat the wings multiple times for each volley of nerve impulses, and wingbeat frequencies greater than 100 Hz are common in insects with asynchronous flight muscles. One very small midge (a dipteran) may beat the wings up to 1000 times/s (Sotavalta, 1953). Although skeletal muscles are not asynchronous muscles, the tymbal muscles of some cicadas are asynchronous. Asynchronous muscles require rhythmical input from the CNS for continued contraction, but the contraction frequency is a property of the muscles and their anatomical orientation and is not proportionally related to the nerve input. There may be three, four, or many repeated contractions for each volley of nerve impulses. Wing loading and thoracic resonance properties in insects with asynchronous wing muscles influence the number of wingbeats per second. Honeybees with naturally caused age-related wing wear experience changes in aerodynamic power production, hovering flight, aerodynamic reserve capacity, load carrying capacity, flight maneuverability, and evasion of predators (Vance and Roberts, 2014). Wingbeat frequency can be increased experimentally by reducing wing loading (by cutting small portions off the wings). In contrast, wing loading does not have much influence on wingbeat frequency in insects with synchronous muscles because the frequency is determined by the rate of nerve impulses coming to the muscles.

Fibrillar muscles are so-named because the muscle fibers separate from each other easily upon even slight shearing or tearing action (Figure 11.8). The muscle fibers are very large, ranging from about 100 μm up to 1 mm in diameter. In larger insects, the fibrillar muscles consist of bundles of fibers separated by tracheae, which often push into the plasma membrane of individual fibers to become intracellular tracheoles. The muscle fibers are cylindrical in shape and multinucleate, with nuclei usually lying peripherally in neat rows. Sarcomeres vary from 1.7 to 2.5 μm in length. The A band covers about 90% of the sarcomere length and I bands are very narrow. These anatomical arrangements mean that maximum contraction can occur with very little shortening of the sarcomere and, consequently, with little shortening of the total muscle length, another property enabling rapid contraction rates. The myofibrils of fibrillar muscles are large, varying from 1 to 5 μm in diameter (as large as muscle fibers in some other types of muscles) and are as long as the muscle itself. The SR is poorly developed and not continuous along the long axis of the muscle fiber so that junctions between the SR and T system of tubules are reduced to dyad junctions (see Figure 11.2b). The T tubules are well developed, penetrating usually at the Z lines and ramifying among myofibrils. Six thin filaments (actin) surround the thick filament (myosin) in a very regular order so that the ratio of thick to thin filaments is 1:3. Large, irregular mitochondria lie between myofibrils and may occupy up to 30% of the volume of the muscle. Glycogen deposits and lipid droplets have been observed around the mitochondria in some insects (Ashurst and Cullen, 1977).

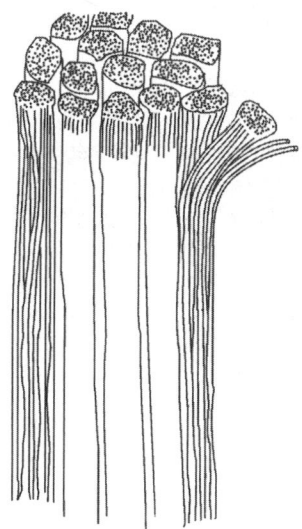

Figure 11.8 A diagrammatic illustration of fibrillar muscle with large myofibrils that tend to separate from each other with slight shearing action.

11.4 MUSCLE PROTEINS AND PHYSIOLOGY OF CONTRACTION

The muscle proteins and contraction physiology are essentially the same in insects as in muscles of other animals (see review by Maruyama, 1985). The major proteins in muscle fibers are **actin**, **tropomyosin**, **troponin** (all part of the thin filaments), and **myosin** (the thick filaments) (Figure 11.9). Shortening of a muscle occurs by the sliding of actin and myosin filaments over each other, pulling the Z bands closer together, and ultimately shortening the muscle toward the point of its fixed attachment, its origin. A single myosin molecule has the shape of a double-headed golf club (Figure 11.9a). Myosin molecules are arranged with their long tails forming the core of the thick filament with heads projecting from the filament. The globular heads have calcium-dependent ATPase activity and contain the binding sites for attachment to actin during contraction. The myosin heads form the structures called **crossbridges** (Squire, 1977).

The smallest unit of actin is a globular polypeptide named **G actin**. G actin subunits are linked together by polypeptide bonds to form a long chain of filament actin (**F actin**) (Figure 11.9c), and a thin filament consists of two chains of F actin twisted around each other in an α-helix. Each G actin subunit in the chain has an active site where a myosin head can attach. Another protein, tropomyosin, is associated with the thin filaments, and it consists of a long filamentous chain running along each of the two grooves created by the F actin helix. Each tropomyosin chain is composed of two α-helical units in a coiled-coil pattern, and the coiled chain runs along the groove created by the F actin helix. Tropomyosin filaments cover the active sites where a myosin head can attach. Troponin is a globular protein composed of three subunits, TnT (**tropomyosin binding**), TnC (**calcium binding**), and TnI (**actin binding**) (Figure 11.9c and d). Troponin functions during nerve stimulation to the muscle by changing shape and pulling tropomyosin away from the myosin-binding sites. A troponin unit is associated with each G actin unit.

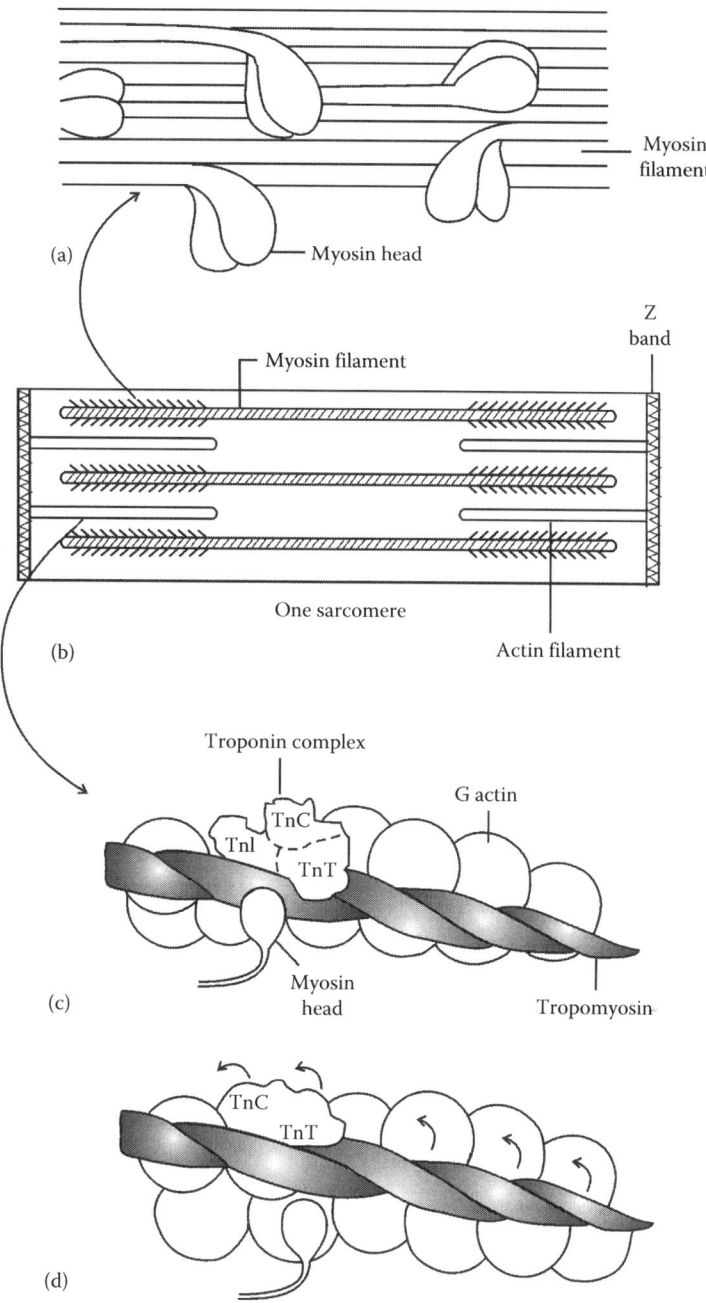

Figure 11.9 The arrangement of the major proteins involved with contraction in each sarcomere of a myofibril. (a) A diagram of myosin molecules to show the double heads of each myosin filament. (b) A sarcomere unit illustrating actin and myosin filaments. (c) Helically entwined strands of F actin with a helical coil of tropomyosin lying in the groove of the F actin molecules. Troponin, consisting of three polypeptide units, is bonded through one polypeptide to a G actin molecule and by another subunit to tropomyosin, leaving one subunit free to bind to calcium ions. (d) When the TnC part of troponin binds calcium ions, a change in shape of troponin occurs, and tropomyosin is pulled away from the active site on G actin, allowing the myosin heads to bind to actin and initiate the power stroke of contraction.

11.4.1 Active State: Binding of Myosin Heads to Actin and the Sliding of Filaments

The signal for initiating a contraction is the arrival of a nerve impulse at the junctional ending on the muscle surface, resulting in the release of neurotransmitters at the polyneuronal neuromuscular junctions. The nerve impulse sets in motion a cascade of reactions mediated by G-proteins and results in the release of Ca^{2+} from binding sites in the SR at the triad (synchronous muscle) or dyad (asynchronous muscle) junctions in the muscle fiber. The threshold of free calcium ions for activation of the contraction process is about 10^{-7} M, with maximum contraction occurring at about 10^{-5} M (Aidley, 1989). The free Ca^{2+} ions in the sarcoplasm and availability of ATP create an **active state** in each myofibril. Once initiated, the active state can persist as long as free Ca^{2+} ions and ATP are available. When the concentration of free Ca^{2+} ions falls below about 10^{-7} M in the sarcoplasm because of the sequestering of Ca^{2+} by the SR, the active state is abolished and a further contraction is not possible until a new volley of nerve impulses releases more Ca^{2+}. The active state persists much longer in asynchronous muscle than in synchronous muscle because the poorly developed SR in asynchronous muscle sequesters calcium ions only slowly.

During the active state, free Ca^{2+} ions bind to the TnC subunit of troponin. This promotes a change in shape of the troponin complex causing the TnT subunit to pull tropomyosin away from the active site where a myosin head can attach to actin. Attachment sites covered by troponin occur on each of the globular or G actin subunits, spaced about 38.5 nm apart. This allows myosin heads to bind to actin at many sites during the active state. When myosin binds to actin, a conformational change occurs in the myosin molecule in a "hinge-like" region of the myosin head and at another hinge site in the arm to which the head is attached, causing the arm to bend similar to the bending of a human arm. These changes in shape create the power stroke that causes sliding of actin over myosin.

11.4.2 Release of Myosin Heads from Actin

The myosin head must release from actin in order to return to its original state and prepare for another power stroke. The conformational change in the myosin molecule associated with the power stroke exposes a site on the head where ATP can bind. ATP binds to this head site, and myosin, which is an ATPase as well as a structural protein (Lymn and Taylor, 1970), splits the bound ATP into ADP and PO_4. The energy liberated enables the myosin head to release from actin and to return to its original shape, ending the power stroke. Further, ADP and inorganic phosphate remain attached to the head until a new attachment to actin occurs, at which time they also are released from the head and a new power stoke begins.

The binding of ATP, detachment of the myosin head from actin, and reattachment occur extremely rapidly and are estimated to require only about 0.1 ms (Lymn and Taylor, 1970). If calcium ions are available to combine with TnC and pull tropomyosin away from the active sites, myosin can bind, accomplish its power stroke, detach, and bind again many times per second. The movement caused by a power stroke has been estimated to be 5–10 nm (Harrington, 1981) and thousands of such movements along the length of the muscle fiber create total fiber shortening. The myosin heads along the length of a myosin filament are not all attached or detached at the same time, and their independent action results in a steady tension exerted to pull the Z bands of a sarcomere closer together, and thus, the entire muscle shortens. After death of an insect, the myosin heads bind to actin, and contraction occurs, but lack of ATP prevents them from detaching from actin, resulting in the condition of rigor mortis as in other animals. However, ATP must be available at all times for normal muscle functions because it is necessary for the SR to actively sequester calcium and for binding to the myosin heads so that the heads detach from actin.

The very large fibrillar muscles of the giant water bug, *Lethocerus* spp. (Hemiptera: Belostomatidae), provided important data for developing the concept of the conformation change in the myosin head region when it is attached to actin and for flexing or movement of the head on its arm.

Electron micrographs of *Lethocerus* muscle in a state of extended contraction (rigor) showed crossbridges (the myosin heads) fixed at an angle, which suggested they had to flex during contraction (Reedy et al., 1965). X-ray diffraction studies of active muscle suggested that the heads actually move during contraction (Tregear and Miller, 1969).

11.5 MUSCLES INVOLVED IN GENERAL LOCOMOTION, RUNNING, AND JUMPING

Muscles, of course, are essential to mobility of insects, enabling them to walk, run, crawl, or fly. There are obvious contrasts in the different modes of locomotion among slow-crawling caterpillars (Trimmer and Issberner, 2007) and insects that jump, run, or fly, but Bejan and Marden (2006) unify all forms of locomotion based on the physical theory that flow systems evolve in such a way that they utilize the minimum useful energy. They compare the running, flying, and swimming of many different animals (including insects) and calculate that speed and body mass (M_b) scale to $M_b^{-1/6}$ for a wide variety of organisms.

Locomotion in a crawling caterpillar presents certain unique features. Unlike annelid worms that use functionally antagonistically arranged muscles to produce extension and contraction of the body, caterpillars (*Manduca sexta*) grip the substrate with their anterior legs and activate dorsal and ventral muscles at the same time in a body segment, typically with two or three abdominal segments moving simultaneously (Simon et al., 2010a). Using phase-contrast synchrotron X-ray imaging and transmission light microscopy, Simon et al. (2010b) discovered that as the terminal prolegs release the substrate and dorsal and ventral muscles are activated, the gut slides forward, resulting in the midgut advancing forward before the body wall and prolegs of the midgut segment move forward. The body wall and associate structures catch up with the next abdominal proleg movement.

Most insect behaviors involve muscle action in some way. For example, the ovipositor opener muscle of *L. migratoria* provides driving force for rhythmic digging behavior that opens a hole in the soil for egg laying (Rose, 2004). Locomotion clearly plays a major role in the evolution and success of all animals, and certainly in insects, enabling them to search for food and mates, escape from danger, search for oviposition sites, migrate into new habitats, and make additional behavioral and lifestyle adaptations.

The crickets, *Gryllus bimaculatus* and *Acheta domesticus*, sometimes autotomize a leg during escape behavior, which not surprisingly affects their locomotion. They ran more slowly, stopped more frequently, and traveled less distance with greater expenditure of energy, had less ability to jump, and were more often caught by lizards and mice than control crickets that had all their legs (Bateman and Fleming, 2005, 2006; Fleming and Bateman, 2007). Fleming and Bateman present a table comparing the energetics of a broad range of arthropods, with references to the literature; in general, the minimum cost of transport (MCOT) (MCOT along a linear distance in J/kg/m) scales negatively with increasing body size (i.e., as body size increases, the MCOT decreases). Lipp et al. (2005) showed that the mean cost of transport in ants (*Camponotus* sp., 130 J/g/km) is small and does not vary appreciably whether running along a level substrate or on slopes with angles up to about 60°. Clusella-Trullas et al. (2010) analyzed locomotion of *Messor capensis*, seed harvester ants, and found that neither active metabolic rate nor cost of transport was affected by acclimation temperature.

11.5.1 Adaptations for Running, Walking, and Survival

Evolution and survival of populations of the chrysomelid willow beetle, *Chrysomela aeneicollis*, in a montane habitat may depend upon allele frequencies in the population for phosphoglucose isomerase (PGI), an important enzyme in carbohydrate metabolism. The gene frequencies may be especially important in the ability of the beetles to survive potential climate changes.

Certain genetic alleles modulate the action of PGI and alter the running speed of *C. aeneicollis*. Running speed is important for male beetles to locate unmated females and for females to find suitable oviposition sites. Both sexes run to escape predators and to gather food. Rank et al. (2007), who documented allele frequency and fitness performance in the beetles, were particularly interested in whether, and how, a population might adapt to changing climate conditions. Beetles in populations in the Sierra Nevada mountains in the western United States living at elevations of 2400–3600 m experience great daily variations in temperature. Allele 1 for PGI is more frequent in populations that live in more northern and colder regions, while allele 4 is more frequent in populations living in warmer areas. The authors concluded that PGI allele frequencies (and upregulation of heat shock protein 70, Hsp 70) are under selection by environmental temperature, and populations of this particular beetle have a range of allele frequencies for PGI that influence fitness and may help them adjust to climate change.

Cataglyphis fortis are desert ants that forage singly over distances of 100 m or more from their nest in the ground. They may meander in many directions over irregular terrain, but when they find food, they return to their nest by the most direct route. They do this by measuring the distance traveled using a stride indicator (a pedometer) and by direction from the nest at any given moment by a celestial compass based on detection of plane polarized light from the sky, a navigation system known as path integration (Wittlinger et al., 2007a). These experimenters caught ants away from their nest and altered the leg length of the ants by surgically shortening the leg so that the ants walked on the stumps or lengthened the legs by gluing pig bristles to the stumps. Ants caught away from the nest and forced to walk home on shortened legs underestimated the nest distance by 30%–40%; conversely, those walking on pig bristle stilts overshot their nest by up to 50% in some trials. Ants traveling outbound from the nest on modified legs (short legs or stilts) calculated the return to the nest correctly. The ants were not using the hair plate mechanoreceptors on the neck region and petiole (narrow connection between thorax and abdomen) to monitor travel over uneven ground (Wittlinger et al., 2007b). Neither shaving the sensory hairs nor immobilizing the joints monitored by the hairs affected the ability of the ants to integrate the correct path and homing distance to their nest. They are affected, however, in their integration of distance traveled if they are denied access to their celestial compass information, as Ronacher et al. (2006) demonstrated by allowing the ants to forage in a Z-shaped experimental system open to the sky but partially covered with Perspex that was opaque to ultraviolet (UV) light. The UV part of the spectrum carries information about the plane of polarized light. Ronacher et al. (2006) discussed and dismissed several possible explanations for the failure of the ants to integrate distance correctly without information from the celestial compass, but the mechanisms by which they integrate all the information remain elusive.

Narendra et al. (2007) studied the Australian desert ant, *Melophorus bagoti*, which also forages in a manner similar to *C. fortis*, but *M. bagoti* typically have more landmarks in their environment and they follow foraging routes marked by low shrubs and other environmental objects. Even in the absence of landmarks, however, they measure distances accurately. Their navigational memory for distance is time limited, however, and ants captured on an outbound trip and held captive for about 24 h revert to landmark-based navigation to find their way home.

In contrast to the navigational systems in the walking desert ants, honeybees, which also have a celestial compass for direction based on detection of plane polarized light, measure distance flown only by the movement of the environment across their visual field (Dacke and Srinivasan, 2007). In experimental situations, bees measure the distance flown along a path only by the total distance flown and are not influenced by the three-dimensional (3D) nature of experimental paths.

The gene *takeout* (*to*) modulates locomotion behavior, feeding, circadian rhythm, and probably juvenile hormone (JH) level in tissues of *D. melanogaster* (Sarov-Blat et al., 2000; Meunier et al., 2007). The protein controlled by the gene, *to*, influences the sensitivity of gustatory neurons and also binds JH. Flies that have a mutated gene do not eat more after having been starved as normal flies do because starvation apparently does not increase sensitivity of taste neurons to sugar in the mutant flies nor do they increase locomotion in search of food. Conversely, mutant flies eat excessively when food is plentiful and become fat. The level of circulating JH influences their locomotary behavior, and the mutant flies apparently have less JH in the body; addition of methoprene, a JH mimic, rescues the mutant flies.

Members of the family Drosophilidae are found at high altitude in the Sierra Nevada mountains in California where they experience extreme temperatures and low oxygen levels. Dillon and Frazier (2006) found that the walking performance of *D. melanogaster* was reduced by low temperature, as might be expected in an ectothermic insect, but was not limited by the low oxygen tension at the higher elevations, indicating the efficiency of the tracheal system in supplying oxygen to the muscles. Flies showed reduced flight performance, however, at the higher elevations and especially at lower temperatures, showing the stress that these environmental parameters put on the muscle and tracheal systems.

11.5.2 Adaptations for Jumping

The ability of some insects to jump many times the body length and height might suggest that insect muscles are stronger than vertebrate muscles. However, actual measurements of the force developed per unit cross-sectional area for insect muscles reveal about the same force per unit cross-sectional area as for other animals. It is anatomy of muscles, leverage, and physiology that enable some insects to jump so far. Jumping muscles tend to be "fast" muscles, while muscles that are used in posture or in walking tend to be "slow" muscles (James et al., 2007). Jumping in a locust occurs by contraction of the large extensor tibiae muscles in the femur of the metathoracic legs. The extensor tibiae muscle is composed of many short muscle fibers with origins on the epicuticle of the femur. The muscle fibers insert on the tendon or apodeme from the tibia that runs the length of the femur (Figure 11.10). Thus, each muscle fiber shortens very little, but the many fibers along the length of the long femur apply tension that straightens the

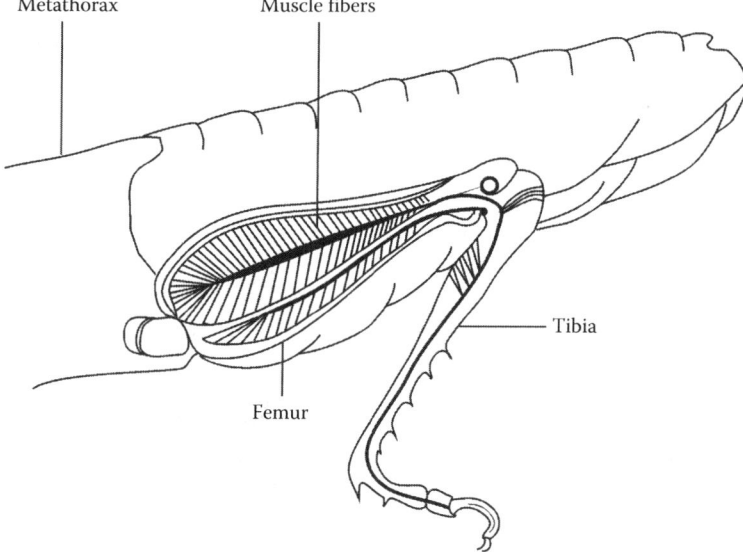

Figure 11.10 The arrangement of muscle fibers in the femur of the jumping leg (the metathoracic leg) of a grasshopper. Many very short muscle fibers are anchored on the epicuticle and insert on long cuticular tendons attached to the tibia in such a way as to flex or extend the leg, depending upon the muscles activated.

leg and propels the locust into the air in a jump. Furthermore, the long femur and tibia of the metathoracic leg (which is much longer than either of the pro- or mesothoracic legs) give the locust something of a "pole-vaulting" advantage, similar to a human using a vaulting pole as an extension of the arms.

Bennet-Clark (1975) found in experiments that more than half of the energy needed for jumping by the locust, *S. gregaria*, is stored just prior to a jump in stress applied to cuticular structures in the leg as the large extensor tibiae muscle contracts nearly isometrically (contraction with little or no shortening). The stress on the semilunar process, a slightly flexible cuticular process located near where the femur meets the tibia, stores 4 mJ (millijoules) of energy, and 3 mJ energy is stored as stress is applied to the cuticular apodeme on which the extensor tibiae inserts. A jump by a male requires 9 mJ, while the female, heavier with eggs and fat, requires 11 mJ. The stored energy in these two structures is released as kinetic energy of the jump and augmented by isotonic (shortening) contraction of the muscle. Storing energy in these cuticular structures just prior to the jump is advantageous in providing power and velocity on jump takeoff that could not be achieved by mere shortening of the muscle fibers in the time frame of a successful jump. In a variety of trials, Bennet-Clark (1975) measured the highest velocity at takeoff to be 3.2 m/s, and great force for the jump could be produced at low environmental temperature because, even though the muscle contracted more slowly at cooler temperatures, the stored energy could be released rapidly.

Some, but not all, fleas are jumpers. Very massive muscles whose fibers must shorten only a small percentage of the muscle length are a common adaptation in fleas that make extraordinary jumps. Jumping fleas have an extraordinarily large coxa, in contrast to most insects, containing the large coxal muscle whose fibers are anchored to the thorax. The coxal muscle inserts on the femur, which is also very large and long (overall the jumping leg is about 80% of the length of the body of a cat flea) and contains large muscles inserting on the tibia. In preparation for a jump, a flea utilizes a cocking mechanism in which the femur is pulled up to overlap the large coxa. In this cocked position, both coxa and femur are perpendicular to the body axis and to the substrate (Rothschild et al., 1988). The cocking action compresses an arch of resilin in the pleural arch near where the coxa is attached to the thorax. The compression of the pad of resilin stores energy for the jump by isometric contraction (i.e., contraction of the muscle with little or no shortening) of the jumping leg muscle (Bennet-Clark and Lucey, 1967). In addition, the cocking action clamps the three thoracic segments together by engagement of "catch" mechanisms between the three thoracic segments. The sternum of the mesothorax, in particular, is rigidly held against the metathorax, and this allows the large muscles involved in raising the femur and locking the catches to relax, resisting tiring and reducing energy requirements. A flea typically remains in this crouched and motionless position, poised for its leap, for up to 0.1 s (Bennet-Clark and Lucey, 1967) with trochanter and tibiotarsal joint and most of the tarsi resting on the substrate. The flea propels itself forcefully into the air by relaxing the levator muscle holding the femur in the perpendicular position and relaxing the ventral longitudinal muscles holding the catches in the cocked position. The stored energy in the compressed resilin, like a compressed spring, is released in 0.7 s, forcefully driving the trochanter against the substrate to provide the initial leverage off the substrate. Powerful muscles extend the joint between the trochanter and the femur and extend the leg providing further thrust as the terminal tarsal segments and tarsal claws press against the substrate. Additional details on the jump of the flea can be found in the illustrated and informative papers by Bennet-Clark and Lucey (1967) and Rothschild et al. (1988). A website that illustrates some aspects of the jump of a flea, among other things, is located at http://www.ftexploring.com/lifetech/flsbws2.html#Science (last accessed April 21, 2015).

11.6 SOUND PRODUCTION: TYMBAL AND STRIDULATORY MUSCLE

11.6.1 Tymbal Morphology and Physiology

Tymbals are thin, often ribbed patches of cuticle in male cicadas (Homoptera), some other homopterans, some Hemiptera, and some Lepidoptera (moths in the families Arctiidae, Ctenuchidae, and Pyralidae, and some nymphalid butterflies) (Ewing, 1989, p. 34) that are used for sound production and communication. The tymbals of cicadas have received the most detailed analysis. Two-paired tymbals, each occurring on the lateral surface of the first abdominal segment of male cicadas, are used to produce both calling and protest songs. The tymbal membrane is a convex layer of cuticle, often bearing a series of ribs, covering an air-filled cavity that acts as a sound enhancer. The large tymbal muscle, the sound-producing muscle, is anchored on the sternal cuticle and inserted dorsolaterally on the tymbal membrane (Figure 11.11), sometimes by means of a thickened rod of chitin. Contraction of the muscle is initially isometric against the resistance of the convex tymbal, but suddenly, the tymbal membrane buckles inward into an unstable, concave shape with a loud clicking sound. The inward buckling of the tymbal releases the load on the tymbal muscle, and it stops developing tension and shortening. The sudden reduction of muscle tension allows the natural elasticity of the tymbal cuticle to click back into the resting convex shape. Sounds may be produced by either, or both, inward or outward movement of the tymbal membrane. Some species have ribs in the convex tymbal surface that buckle progressively inward from posterior to anterior producing a series of sound pulses as the successive ribs buckle even though only a single muscle twitch is involved (Young, 1972; Young and Josephson, 1983b).

The tymbal muscles of some cicadas are synchronous, while others have asynchronous muscles (Pringle, 1981). For example, *Cyclochila australasiae* have synchronous tymbal muscles and *P. capitata* have asynchronous ones, but the flight muscles in both are synchronous

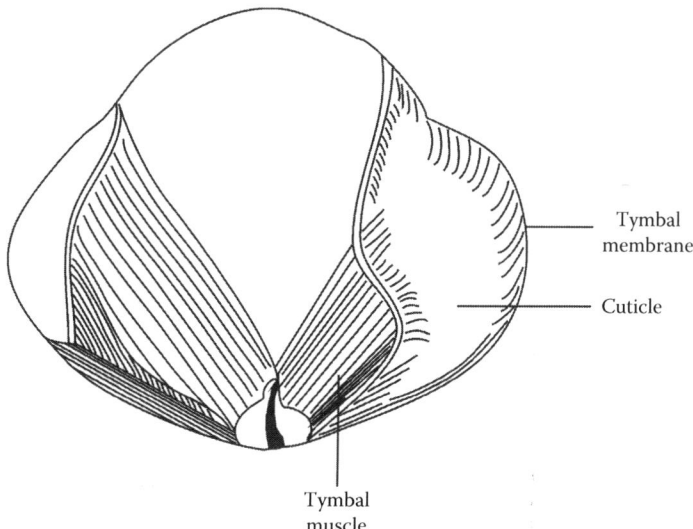

Figure 11.11 A cross-sectional view of the tymbal muscles that contract and flex the tymbal cuticular surface inward to produce the loud singing sounds of a cicada. The muscle on each side is unpaired, and the elastic nature of the cuticle stretches the muscle and returns the tymbal to its resting shape after a contraction, usually producing another sound with the outward snap of the cuticle.

Table 11.1 Comparison of Tymbal Muscle Contraction Frequency, Twitch Duration, and Time from Onset of One Muscle Contraction to the Beginning of the Next Contraction (Cycle Period) during Singing in Selected Cicadas

Cicada Species	Tymbal Muscle Contraction Frequency (Hz)	Twitch Duration (ms)	Cycle Period (ms)
Okanagana vanduzeei	550	~6	—
Psaltoda claripennis	224	6.6	4.46
Psaltoda harrisii	~150	7.7	6.7
Pennahia argentata	192	7.9	5.51
Tamasa tristigma	82	8.7	12.2
Magicicada cassini	~90	9.8	11.1
Magicicada septendecim	~68	11.8	14.7
Abricta curvicosta	72	12.6	13.9
Arunta perulata	31	14.9	32.3
Chloroclysta viridis	56	15.2	17.9
Cystosoma saundersii	~40	22.0	25.0
Cyclochila australasiae	~116	12.3	8.55

Sources: Data from Young, D. and Josephson, R.K., *J. Comp. Physiol.*, 152, 183, 1983a; Josephson, R.K. and Young, D., *J. Exp. Biol.*, 118, 185, 1985.

(Josephson and Young, 1981). Each cicada tymbal muscle is innervated by a single motor axon with multiterminal branches to all the muscle fibers in the muscle (Pringle, 1954a,b; Hagiwara, 1955; Simmons, 1977; Josephson and Young, 1981). A number of cicadas that have synchronous tymbal muscles have contraction frequencies less than 100 Hz (Table 11.1) (Young and Josephson, 1983a) as might be expected of synchronous muscle, but several have high contraction frequencies well above 100 Hz. *Okanagana vanduzeei* has a contraction frequency of 550 Hz. In part, the rapid contractions are enabled by very short twitch duration times (Table 11.1) in the fastest synchronous muscles (Josephson, 1984). However, a contraction rate of 550 Hz in a synchronous muscle means that the nerve must repolarize and be ready to fire again in less than 2 ms. As Table 11.1 shows, only 4–6 ms are available for repolarization in several of the cicadas. In the cicadas with very fast synchronous muscles, the muscle may be delicately balanced on the edge of oscillatory instability and may possibly go into oscillatory behavior (Josephson and Young, 1985). The inward click of the tymbal cuticle releases the load against which the muscle works, and the natural cuticular elasticity that causes it to click back into the convex shape could then stretch and reintroduce the load to the activated muscle, inducing another contraction if the concentration of free calcium ions is still high enough in the cytoplasm. Thus, these tymbal muscles may work like asynchronous wing muscles (see Section 12.4, for details on asynchronous wing muscles).

11.6.2 Stridulatory Muscle Physiology

Another sound-producing mechanism, **stridulation**, involves the rubbing together of two parts of the body. Stridulation is widespread among insect orders, with examples known from Odonata, Orthoptera, Coleoptera, Mecoptera, Lepidoptera, Siphonaptera, and Hymenoptera (Ewing, 1989). As in the tymbal muscles of some cicadas, synchronous muscles involved in stridulation by some katydids (Orthoptera: Tettigoniidae) show unusually high contraction frequencies. Male katydids stridulate to attract females by rubbing their forewings (mesothoracic wings) together. In the mesothoracic segment, the same muscles used to power the downstroke of the wings during flight, the mesothoracic DLM, are used in stridulation. The wing stroke frequency of the DLM in the mesothoracic segment during stridulation of a katydid, *N. robustus* (Orthoptera: Tettigoniidae), has been measured at 200 Hz (Josephson and Stokes, 1982). The DLM in both segments are used in flight, but then the wing stroke frequency is 20 Hz. In a related katydid, *Neoconocephalus triops*, wing stroke frequency during stridulation is 100 Hz, and wing stroke frequency during flight is 22.9 Hz (Josephson, 1984). The tropical bush cricket, *Hexacentrus unicolor* (Orthoptera:Cynocephalidae), produces a song with a stridulatory frequency of 320–415 Hz with synchronous muscles (Heller, 1986). Thus, it is clear that morphologically synchronous muscle can produce contraction frequencies higher than earlier expected. The evolution of asynchronous muscle, which has occurred as many as 10 times in different groups of insects (Cullen, 1974), may be related more to operating efficiency, economy in calcium recycling between the cytoplasm and SR, and overall structural economy than to the potential for high operating frequency (Josephson and Young, 1985).

Many insects and other animals synchronize visual and acoustic calling signals used to attract mates (Fertschai et al., 2007, and references therein). Perfect synchronization does not occur, resulting in "leaders" and "followers." In most cases studied, females tend to prefer the leading signal in arena tests. Thus, female choice would be expected to exert selection pressure on males to avoid being a follower. Fertschai et al. (2007) found that followers in field conditions were more successful in attracting females of the bush cricket, *Mecopoda elongata*, than expected from arena trials and, for those interested in ethology, Fertschai et al. (2007) discuss evolutionary implications for synchronously calling Orthoptera males.

11.7 MORPHOLOGY AND PHYSIOLOGY OF NONSKELETAL MUSCLE

11.7.1 Visceral Muscles

Antagonistic muscle sets in the gut are arranged as bands of longitudinal and circular muscle. They create peristaltic action in the gut that mixes food with enzymes and aids digestion. Peristalsis helps macerate food particles in the powerful proventriculus of some insects. Visceral muscles in the foregut and midgut are innervated by nerves from the stomatogastric nervous system, consisting of several small ganglia lying on top of the esophagus and crop. Some visceral muscles in insects are **myogenic** and may or may not have a nerve supply.

Visceral muscle fibers usually are short and small (1–5 μm in diameter) and, in contrast to skeletal and flight muscle, uninucleate. Long sarcomere lengths (7–10 μm long) are characteristic of, and consistent with, the slow contractions of visceral muscles. Z lines are irregular and often not well lined up in adjacent fibers. In visceral muscles, myosin filaments are surrounded by up to 10–12 actin filaments, whereas in skeletal muscle, the ratio is usually 1 myosin to 6 actin filaments (Elder, 1975). Muscles in the gastric caeca of *Aedes* mosquitoes have eight to nine actin filaments for each myosin filament (Jones and Zeve, 1968). The SR usually is poorly developed, and the T tubules are irregularly located. Mitochondria are small and few in number. The neurotransmitter in gut muscles has not been conclusively identified; neuropeptides are probably important in regulating the rate and force of contraction in gut muscles. Proctolin, a neurohormone from various parts of the nervous system in different insects, has profound action on the muscles in the hindgut.

11.7.2 Heart Muscle

The heartbeat in insects is **myogenic**, but heart rate is influenced by nerves that innervate the heart in most insects (McCann, 1970) and by neurohormones. Sarcomeres tend to be short in heart muscle, with an A band of 1.8 μm, a short I band, and no H band. There are numerous, but small, mitochondria, and extensive tracheal connections, characteristics suggesting moderately high energy demands. Myosin filaments are surrounded by 10–12 actin filaments. The SR is usually not well developed, and T tubules, while well developed, are irregularly spaced. Dyad junctions are common and some triad junctions occur. Intercalated discs, regions of extensive interdigitation of plasma membranes between adjacent fibers, described from heart muscle of *Blattella germanica* and *Hyalophora cecropia* may be important to the spread of the wave of contraction from fiber to fiber (Edwards and Challice, 1960; Sanger and McCann, 1968a).

11.7.3 Alary Muscles

Alary muscles are thin sheets of wing-shaped muscles that help support the heart. The muscle fibers often branch and are variable in length (from 1 to 20 µm long) (Sanger and McCann, 1968b). Sarcomere lengths are long, with the A band measuring about 5.5 µm. Few mitochondria are present, suggesting a very low metabolic activity, and the SR is poorly developed. Myosin filaments are surrounded by 10–12 actin filaments. Some alary fibers have intercalated disc junctions with heart muscle fibers, suggesting intercommunication of alary and heart muscle, but this has not been shown experimentally. Little is known about specific neurotransmitters or neurohormonal effects on alary muscles.

Long sarcomeres, high actin/myosin filament ratio, a poorly developed SR, and T tubule system are general characteristics of tonic or slow-contracting muscles. Visceral, heart, and alary muscles are slow-contracting muscles as opposed to the twitch or rapid contractions more typical of skeletal muscles.

11.8 REVIEW AND SELF-STUDY QUESTIONS

1. Describe the structure of a muscle fiber.
2. Are all muscles in insects striated? Even gut muscles? What causes the appearance of striations?
3. What is the name of the contractile unit of muscle between adjacent Z bands?
4. Is fibrillar muscle synchronous or asynchronous muscle?
5. Physiologically, what are some of the functional differences between synchronous and asynchronous muscles?
6. What is sarcoplasmic reticulum? Is it more extensive in synchronous or asynchronous muscles?
7. What is a dyad or triad junction in muscle?
8. What is another name for the very large mitochondria in flight muscle?
9. Is there more actin or more myosin in muscle tissue?
10. Based upon present evidence, what appears to be the role of Dumpy protein in the muscle system?
11. What are tonofibrillae? To what part of the cuticle are skeletal muscles attached?
12. What action does a flexor muscle perform? An extensor muscle?
13. Describe the general nature of polyterminal or multiterminal nerve contacts on skeletal muscles in insects.
14. What is the stimulatory transmitter chemical at the nerve–muscle junction in insect skeletal muscles?
15. What are the differences between G actin and F actin?
16. What protein binds calcium ions during an active contraction of muscle? Is ATP required for a muscle fiber to relax and return to its resting length? Explain.
17. Does myosin have enzyme activity, that is, can we consider myosin as an enzyme?
18. What are the crossbridges in muscle?

REFERENCES

Aidley, D.J. 1985. Muscle contraction, in G.A. Kerkut and L.I. Gilbert (eds.), *Comprehensive Insect Physiology, Biochemistry and Pharmacology*, vol. 10. Pergamon Press, Oxford, U.K., pp. 407–437.

Aidley, D.J. 1989. Structure and function in flight muscle, in G.J. Goldsworthy and C.H. Wheeler (eds.), *Insect Flight*. CRC Press, Boca Raton, FL, pp. 31–49.

Ashhurst, D.E. 1967. The fibrillar flight muscles of giant water-bugs: An electron-microscope study. *J. Cell Sci.* 2: 435–444.

Ashhurst, D.E. and M.J. Cullen. 1977. The structure of fibrillar flight muscle, in R.T. Tregear (ed.), *Insect Flight Muscle*. North-Holland Publishing Co., New York, pp. 9–14.

Atwood, H.L. and R.L. Cooper. 1996. Assessing ultrastructure of crustacean and insect neuromuscular junctions. *J. Neurosci. Methods* 69: 51–58.

Bateman, M.W. and P.A. Fleming. 2005. Direct and indirect costs of limb autotomy in field crickets *Gryllus bimaculatus*. *Anim. Behav.* 69: 151–159.

Bateman, M.W. and P.A. Fleming. 2006. Increased susceptibility to predation for autotomized house crickets (*Acheta domestica*)[sic]. *Ethology* 112: 670–677.

Bejan, A. and J.H. Marden. 2006. Unifying constructal theory for scale effects in running, swimming and flying. *J. Exp. Biol.* 209: 238–248.

Bennet-Clark, H.C. 1975. The energetics of the jump of the locust *Schistocerca gregaria*. *J. Exp. Biol.* 63: 53–83.

Bennet-Clark, H.C. and E.C.A. Lucey. 1967. The jump of the flea: A study of the energetics and a model of the mechanism. *J. Exp. Biol.* 47: 59–76.

Bräunig, P., M. Schmäh, and H. Wolf. 2006. Common and specific inhibitory motor neurons innervate the intersegmental muscles in the locust thorax. *J. Exp. Biol.* 209: 1827–1836.

Burrows, M. 1977. Flight mechanisms of the locust, in G. Hoyle (ed.), *Identified Neurons and Behaviour of Arthropods*. Plenum Press, New York, pp. 339–356.

Carmon, A., M.J. Guertin, O. Grushko, B. Marshall, and R. MacIntyre. 2010a. A molecular analysis of mutations at the complex *dumpy* locus in *Drosophila melanogaster*. *PLoS One* 5(8): e12319.

Carmon, A., F. Topbas, M. Baron, and R. MacIntyre. 2010b. *dumpy* interacts with a large number of genes in the developing wing of *Drosophila melanogaster*. *Fly* 4(2): 117–127.

Carmon, A., M. Wilkin, J. Hassan, M. Baron, and R. MacIntyre. 2007. Concerted evolution within the *Drosophila dumpy* gene. *Genetics* 176: 309–325.

Carnevali, M.D.C. and J.F. Reger. 1982. Slow-acting flight muscles of saturniid moths. *J. Ultrastruct. Res.* 79: 241–249.

Clusella-Trullas, S., J.A. Terblanche, and S.L. Chown. 2010. Phenotypic plasticity of locomotion performance in the seed harvester *Messor capensis* (Formicidae). *Physiol. Biochem. Zool.* 83: 519–530.

Cochrane, D.G., H.Y. Elder, and P.N.R. Usherwood. 1972. Physiology and ultrastructure of phasic and tonic skeletal muscle fibres in the locust, *Schistocerca gregaria*. *J. Cell. Sci.* 10: 419–441.

Collet, C. and L. Belzunces. 2007. Excitable properties of adult skeletal muscle fibres from the honeybee *Apis mellifera*. *J. Exp. Biol.* 210: 454–464.

Cullen, M.J. 1974. The distribution of asynchronous muscle in insects with particular reference to the Hemiptera: An electron microscope study. *J. Entomol.* (A) 49: 17–41.

Dacke, M. and M.V. Srinivasan. 2007. Honeybee navigation: Distance estimation in the third dimension. *J. Exp. Biol.* 210: 845–853.

Darwin, F.W. and J.W.S. Pringle. 1959. The physiology of insect fibrillar muscle. I. Anatomy and innervation of the basalar muscle of lamellicorn beetles. *Proc. R. Soc. Lond. B* 151: 194–203.

Dickinson, M.H. and M.S. Tu. 1997. The function of dipteran flight muscle. *Comp. Biochem. Physiol.* 116A: 223–238.

Dillon, M.E. and M.R. Frazier. 2006. *Drosophila melanogaster* locomotion in cold thin air. *J. Exp. Biol.* 209: 364–371.

Edwards, G.A. and C.E. Challice. 1960. The ultrastructure of the heart of the cockroach, *Blattella germanica*. *Ann. Entomol. Soc. Am.* 53: 369–383.

Elder, H.Y. 1975. Muscle structure, in P.R.N. Usherwood (ed.), *Insect Muscle*. Academic Press, New York, pp. 1–74.

Ewing, A.W. 1989. *Arthropod Bioacoustics—Neurobiology and Behaviour*. Comstock Publishing Associates, Cornell University Press, Ithaca, NY.

Fertschai, I., J. Stradner, and H. Römer. 2007. Neuroethology of female preference in the synchronously singing bush cricket *Mecopoda elongata* (Tettigoniidae; Orthoptera): Why do followers call at all? *J. Exp. Biol.* 210: 465–476.

Fleming, P.A. and P.W. Bateman. 2007. Just drop it and run: The effect of limb autotomy on running distance and locomotion energetics of field crickets (*Gryllus bimaculatus*). *J. Exp. Biol.* 210: 1446–1454.

Guschlbauer, C., H. Scharstein, and A. Büschges. 2007. The extensor tibiae muscle of the stick insect: Biomechanical properties of an insect walking leg muscle. *J. Exp. Biol.* 210: 1092–1108.

Hagiwara, S. 1955. Neuromuscular mechanism of sound production in the cicada. *Physiol. Comp. Oecol.* 4: 142–153.

Harrington, W.F. 1981. *Muscle Contraction*. Carolina Biological Supply Co., Burlington, NC.

Heller, K.-G. 1986. Warm-up and stridulation in the bushcricket, *Hexacentrus unicolor* serville (Orthoptera, Conocephalidae, Listroscelidinae). *J. Exp. Biol.* 126: 97–109.

Hinton, H.E. 1973. Neglected phases in metamorphosis: A reply to V.B. Wigglesworth. *J. Entomol. (A)* 48: 57–68.

Hoyle, G. 1955. Neuromuscular mechanisms of a locust skeletal muscle. *Proc. R. Soc. B* 143: 343–367.

Ikeda, K. and E.G. Boettiger. 1965. Studies on the flight mechanisms of insects. III. The innervation and the electrical activity of the basalar fibrillar muscles of the beetle, *Oryctus rhinoceros*. *J. Insect Physiol.* 11: 791–802.

Irving, S.N. and T.A. Miller. 1980. Aspartate and glutamate as possible transmitters of the 'slow' and 'fast' neuromuscular junctions of body wall muscles of *Musca* larvae. *J. Comp. Physiol.* 135: 299–314.

James, R.S., C.A. Navas, and A. Herrel. 2007. How important are skeletal muscle mechanics in setting limits on jumping performance? *J. Exp. Biol.* 210: 923–933.

Jones, J.C. and V.H. Zeve. 1968. The fine structure of the gastric caeca of *Aedes aegypti* larvae. *J. Insect Physiol.* 14: 1567–1575.

Josephson, R.K. 1984. Contraction dynamics of flight and stridulatory muscles of tettigoniid insects. *J. Exp. Biol.* 108: 77–96.

Josephson, R.K., J.G. Malamud, and D.A. Stokes. 2000a. Power output by an asynchronous flight muscle from a beetle. *J. Exp. Biol.* 203: 2667–2689.

Josephson, R.K., J.G. Malamud, and D.A. Stokes. 2000b. Asynchronous muscle: A primer. *J. Exp. Biol.* 203: 2713–2722.

Josephson, R.K. and D.R. Stokes. 1982. Electrical properties of fibres from stridulatory and flight muscles of a tettigoniid. *J. Exp. Biol.* 99: 109–125.

Josephson, R.K. and D. Young. 1981. Synchronous and asynchronous muscles in cicadas. *J. Exp. Biol.* 91: 219–237.

Josephson, R.K. and D. Young. 1985. A synchronous insect muscle with an operating frequency greater than 500 Hz. *J. Exp. Biol.* 118: 185–208.

Lipp, A., H. Wolf, and F.-O. Lehmann. 2005. Walking on inclines: Energetics of locomotion in the ant *Camponotus*. *J. Exp. Biol.* 208: 707–719.

Lymn, R.W. and E.W. Taylor. 1970. Transient state phosphate production in the hydrolysis of nucleoside triphosphates by myosin. *Biochemistry* 9: 2975–3983.

Maruyama, K. 1985. Biochemistry of muscle contraction, in G.A. Kerkut and L.I. Gilbert (eds.), *Comprehensive Insect Physiology, Biochemistry and Pharmacology*, vol. 10. Pergamon Press, Oxford, U.K., pp. 487–498.

McCann, F.V. 1970. Physiology of insect hearts. *Annu. Rev. Entomol.* 15: 173–200.

Meunier, N., Y.H. Belgacem, and J.-R. Martin. 2007. Regulation of feeding behaviour and locomotor activity by *takeout* in *Drosophila*. *J. Exp. Biol.* 210: 1424–1434.

Narendra, A., K. Cheng, and R. Wehner. 2007. Acquiring, retaining and integrating memories of the outbound distance in the Australian desert ant *Melophorus bagoti*. *J. Exp. Biol.* 210: 570–577.

Neville, A.C. 1963. Motor unit distribution of the locust dorsal longitudinal flight muscles. *J. Exp. Biol.* 40: 123–136.

Pringle, J.W.S. 1939. The motor mechanism of the insect leg. *J. Exp. Biol.* 16: 220–231.

Pringle, J.W.S. 1954a. A physiological analysis of cicada song. *J. Exp. Biol.* 31: 525–560.

Pringle, J.W.S. 1954b. The mechanism of the myogenic rhythm of certain insect striated muscles. *J. Physiol.* 124: 269–291.

Pringle, J.W.S. 1957. *Insect Flight*. Cambridge University Press, Cambridge, U.K.

Pringle, J.W.S. 1968. Comparative physiology of the flight motor. *Adv. Insect Physiol.* 5: 163–227.

Pringle, J.W.S. 1976. The muscles and sense organs involved in insect flight, in R.C. Rainey (ed.), *Insect Flight*. Royal Entomological Society and Blackwell Science Publications, Oxford, U.K., pp. 3–15.

Pringle, J.W.S. 1981. The evolution of fibrillar muscle in insects. *J. Exp. Biol.* 94: 1–14.

Rank, N.E., D.A. Bruce, D.M. McMillan, C. Barclay, and E.P. Dahlhoff. 2007. Phosphoglucose isomerase genotype affects running speed and heat shock protein expression after exposure to extreme temperatures in a montane willow beetle. *J. Exp. Biol.* 210: 750–764.

Reedy, M.K., K.C. Holmes, and R.T. Tregear. 1965. Induced changes in the orientation of the cross-bridges of glycerinated insect flight muscle. *Nature (Lond.)* 207: 1276–1280.

Ronacher, B., E. Westwig, and R. Wehner. 2006. Integrating two-dimensional paths: Do desert ants process distance information in the absence of celestial compass cures? *J. Exp. Biol.* 209: 3301–3308.

Rose, U. 2004. Morphological and functional maturation of a skeletal muscle regulated by juvenile hormone. *J. Exp. Biol.* 207: 483–495.

Rothschild, M., Y. Schlein, K. Parker, C. Neville, and S. Sternberg. 1988. The flying leap of the flea. *Sci. Am.* 229: 92–100.

Roy, S. and K. VijayRaghavan. 2011. Developmental biology: Taking flight. *Curr. Biol.* 22(2): R64. Doi: 10.1016/j.cub.2011.12.031.

Sanger, J.W. and F.V. McCann. 1968a. Ultrastructure of the myocardium of the moth, *Hyalophora cecropia*. *J. Insect Physiol.* 14: 1105–1111.

Sanger, J.W. and F.V. McCann. 1968b. Ultrastructure of the moth alary muscles and their attachment to the heart wall. *J. Insect Physiol.* 14: 1539–1544.

Sarov-Blat, L., W.V. So, L. Liu, and M. Rosbash. 2000. The *Drosophila takeout* gene is a novel molecular link between circadian rhythms and feeding behavior. *Cell* 101: 647–656.

Schnorrer, F., C. Schönbauer, C.H.C. Langer, G. Dietzl, M. Novatchkova, K. Schernhuber, M. Fellner et al. 2010. Systematic genetic analysis of muscle morphogenesis and function in *Drosophila*. *Nature* 464: 287–291.

Schönbauer, C., J. Distler, N. Jahrling, M. Radolf, H.U. Dodt, M. Frasch, and F. Schnorrer. 2011. Splat mediates an evolutionarily conserved switch to fibrillar muscle fate in insects. *Nature* 479: 406–409.

Simmons, P.J. 1977. Neuronal generation of singing in a cicada. *Nature* 270: 243–245.

Simon, M.A., S.J. Fusillo, K. Colman, and B.A. Trimmer. 2010a. Motor patterns associated with crawling in a soft-bodied arthropod. *J. Exp. Biol.* 213: 2303–2309.

Simon, M.A., W.A. Woods, Jr., Y.V. Serebrenik, S.M. Simon, L.I. van Griethuijsen, J.A. Socha, W.-K. Lee, and B.A. Trimmer. 2010b. Visceral-locomotory pistoning in crawling caterpillars. *Curr. Biol.* 20: 1–6.

Smith, D.S. 1961. The organization of the flight muscle in a dragonfly, *Aeshna* sp. (Odonata). *J. Biochem. Biophy. Cytol.* 11: 119–145.

Smith, D.S. 1966. The structure of intersegmental muscle fibers in an insect, *Periplaneta americana* L. *J. Cell Biol.* 29: 449–459.

Snodgrass, R.E. 1929. *The thoracic mechanism of a grasshopper, and its antecedents. Smithsonian Miscellaneous Collections*, Smithsonian Institution, Washington, DC, U.S. Vol. 82, pp. 1–112.

Sotavalta, O. 1953. Recordings of high wing-stroke and thoracic vibration frequency in some midges. *Biol. Bull.* 104: 439–444.

Squire, J.M. 1977. The structure of insect thick filaments, in R.T. Tregear (ed.), *Insect Flight Muscle*. North-Holland Publishing Co., New York, pp. 91–111.

Swank, D.M. 2012. Mechanical analysis of *Drosophila* indirect flight and jump muscles. *Methods* 56: 69–77.

Toselli, P.A. and F.A. Pepe. 1968. The fine structure of the ventral segmental abdominal muscles of the insect *Rhodnius prolixus* during the molt cycle. I. Muscle structure at molting. *J. Cell Biol.* 37: 445–461.

Tregear, R.T. and A. Miller. 1969. Evidence of cross-bridge movement during contraction of insect flight muscle. *Nature (Lond.)* 222: 1184–1185.

Trimmer, B. and J. Issberner. 2007. Kinematics of soft-bodied, legged locomotion in *Manduca sexta* larvae. *Biol. Bull.* 212: 130–142.

Usherwood, P.N.R. 1968. A critical study of the evidence for peripheral inhibitory axons in insects. *J. Exp. Biol.* 49: 201–222.

Usherwood, P.N.R. 1994. Insect glutamate receptors. *Adv. Insect Physiol.* 24: 309–341.

Usherwood, P.N.R. and H. Grundfest. 1964. Inhibitory postsynaptic potentials in grasshopper muscle. *Science* 143: 817–818.

Vance, J.T. and S.P. Roberts. 2014. The effects of artificial wing wear on the flight capacity of the honey bee *Apis mellifera*. *J. Insect Physiol.* 65: 27–36.

Wilkin, M.B., M.N. Becker, D. Mulvey, I. Phan, A. Chao, K. Cooper, H.-J. Chung, I.D. Campbell, M. Baron, and R. MacIntyre. 2000. *Drosophila* dumpy is a gigantic extracellular protein required to maintain tension at epidermal-cuticle attachment sites. *Curr. Biol.* 10: 559–567.

Wittlinger, M., R. Wehner, and H. Wolf. 2007a. The desert ant odometer: A stride integrator that accounts for stride length and walking speed. *J. Exp. Biol.* 210: 198–207.

Wittlinger, M., R. Wehner, and H. Wolf. 2007b. Hair plate mechanoreceptors associated with body segments are not necessary for three-dimensional path integration in desert ants, *Cataglyphis fortis*. *J. Exp. Biol.* 210: 375–382.

Young, D. 1972. Neuromuscular mechanism of sound production in Australian cicadas. *J. Comp. Physiol.* 79: 343–362.

Young, D. and R.K. Josephson. 1983a. Mechanisms of sound-production and muscle contraction kinetics in cicadas. *J. Comp. Physiol.* 152: 183–195.

Young, D. and R.K. Josephson. 1983b. Pure-tone songs in cicadas with special reference to the genus *Magicicada. J. Comp. Physiol.* 152: 197–207.

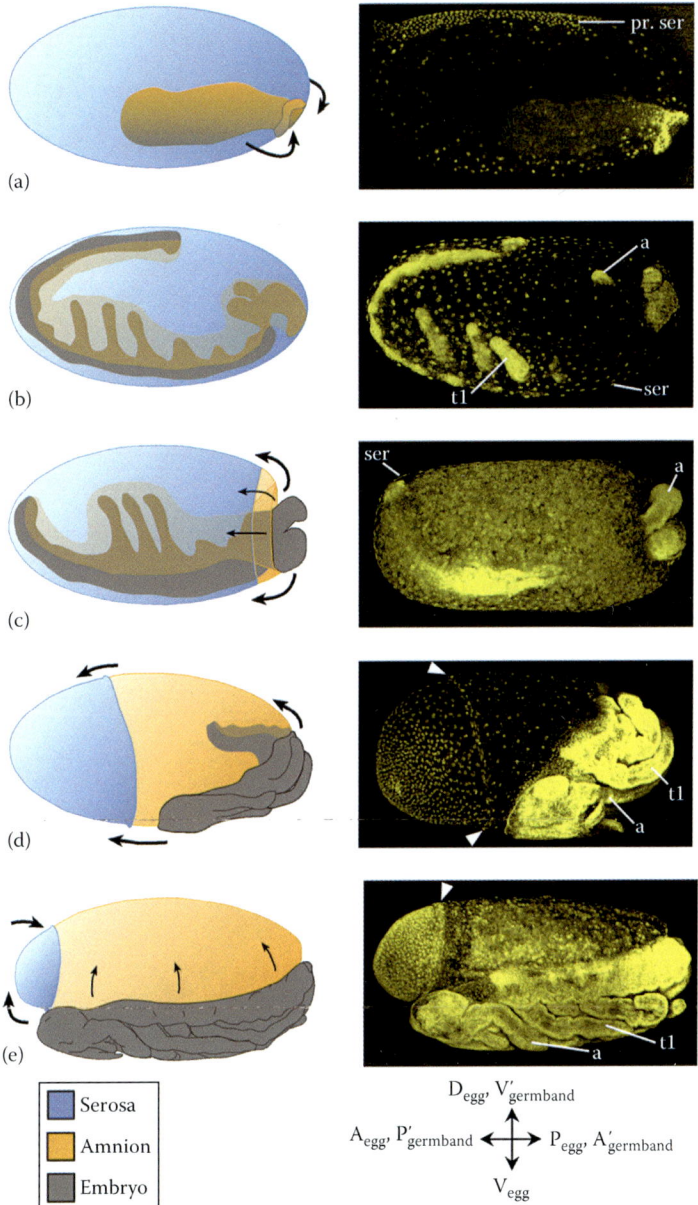

Figure 1.8 Blastokinesis during *Oncopeltus fasciatus* development illustrated in drawings at the left and light micrographs of nuclear staining of dechorionated eggs on the right. (a) Immersion anatrepsis with the embryo invaginating into the yolk; (b) extended germ band stage (see Figure 1.9a for labeling); (c) early katatrepsis with head region emerging from the yolk; (d) midkatatrepsis, with embryo about half out of the yolk; (e) late katatrepsis with embryo fully out of the yolk and now lying on the ventral side of the egg. The serosa was mostly removed in the light micrograph (a) so that the yolk nuclei and germ rudiment are visible. The embryo is partially visible in micrograph panels (b) and (c) in places where it is not so deep in the yolk. Egg orientation is anterior to the left and dorsal up. Color-coding: serosa (blue); amnion (orange); embryo (gray). Black arrows in the drawings indicate the direction of movement; white arrowheads in the micrographs indicate the amnion–serosa boundary. *Abbreviations*: A, P, D, and V denote anterior, posterior, dorsal, and ventral axes of the egg, respectively; A′, P′, and V′ are sides of the germ band stage embryo; a, antennae; pr. ser, presumptive serosa; t1, thoracic segment/leg 1. (From Panfilio, K.A., *Dev. Biol.*, 313, 471, 2008. With permission.)

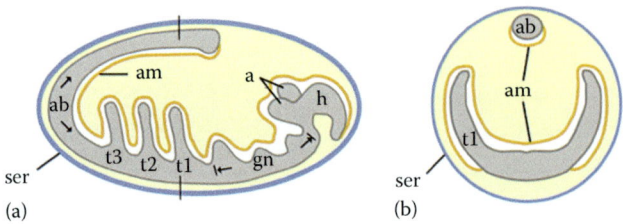

Figure 1.9 Illustrations of a germ band stage embryo modeled on the milkweed bug *Oncopeltus fasciatus* with the extraembryonic membranes of serosa (blue) and amnion (orange), embryo (gray), and yolk (blue). The amnionic cavity (white) lies between the amnion and the ventral surface of the embryo. (a) Midsagittal section view. (b) Transverse section view approximately across the middle of (a) showing that the germ band has spread laterally over the ventral side of the egg. The portion that will become the abdomen extends around the anterior end of the egg (behind the plane of the paper), and just the tip shows on the dorsal side of the egg in this cross section. The posterior pole of the egg is facing the viewer in the cross section. The anterior end of the egg is at the left in the sagittal section, and egg dorsal is up in both views. *Abbreviations*: a, antennae; am, amnion; ab, abdominal region; gn, gnathal (mouthpart) region; h, head; ser, serosa; t1–3, thoracic segments/legs 1–3. The sagittal section is a labeled view of Figure 1.8b. (From Panfilio, K.A., *Dev. Biol.*, 313, 471, 2008. With permission.)

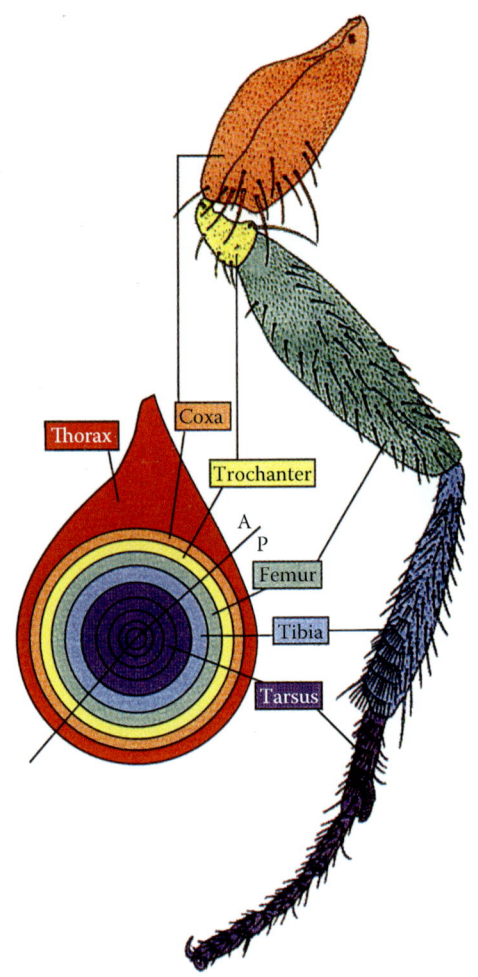

Figure 1.14 An imaginal leg disc of *Drosophila melanogaster* showing regions of the disk that will give rise to parts of the leg and some other parts of the body during pupation. (From Bryant, P.J., *Science*, 259, 471, 1993. With permission.)

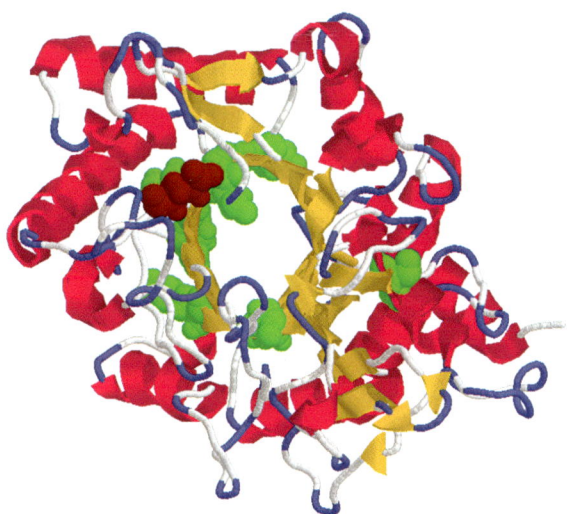

Figure 1.15 A basic view of the growth-promoting glycoprotein (IDGF-DRW protein 3-D structure) from *Diaprepes* root weevil. The structure was predicted with ROBETTA, a full-chain protein prediction server. The beta barrel motif is shown in yellow in the center. The conserved residues in green (in space fill) are apparently essential to maintain the barrel folding as predicted by interface alanine scanning. Residues are Gly109, Gly110, Asp152, Gly153, Leu218, Asp242, Lys295, Gly412, and Asp421. The residue at position 159 is Glu (in brown space fill), which is replaced by Gln in other known IDGY proteins. Structure key: alpha helix (pink), beta strand (yellow), turn (blue), and other (gray). (From Huang, Z. et al., *Florida Entomol.*, 89, 223, 2006. With permission; Photo courtesy of Dr. Wayne B. Hunter, USDA, Ft. Pierce, FL.)

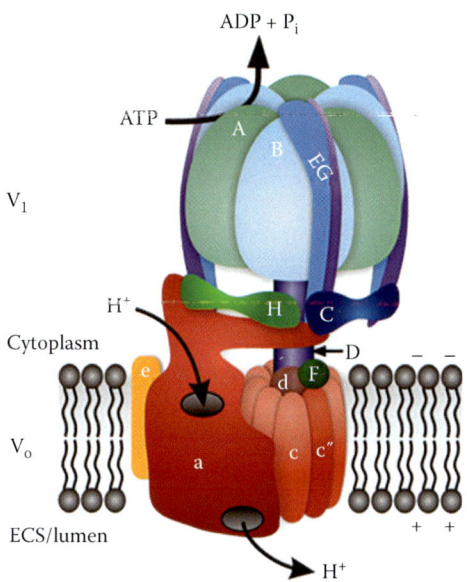

Figure 2.10 Model of the proton pump, a V-ATPase. The V-ATPase holoenzyme consists of two domains, the V_o complex in the membrane and the V_1 complex inside the cell. The V_o complex is composed of at least five subunits (labeled here as a,c,c″,d,e) and it translocates protons from the cytosol to the extracellular space (ECS) or to the lumen of an organelle, respectively. The peripheral V_1 complex is composed of eight subunits (A–H) and is responsible for ATP hydrolysis, supplying energy to the pump. Proton pumps occur in numerous cells. With respect to goblet cells, V_1 is located within the goblet cell cytoplasm and the pump base, V_o, forms a transmembrane channel through which protons are pumped into the goblet cell cavity. ATP hydrolysis in the V_1 part of the pump provides the energy for the pumping of protons across the cell membrane. A separate mechanism in the goblet cell membrane that has not been elucidated exchanges K^+ for H^+ in the goblet cell cavity. The proton pump is the driving force for the concentration of K^+ against a concentration gradient in the goblet cell cavity. Eventually the K^+ is released into the gut lumen, creating high pH in the lumen. (From Baumann, O. and Walz, B., *J. Insect Physiol.*, 58, 450, 2012. With permission.)

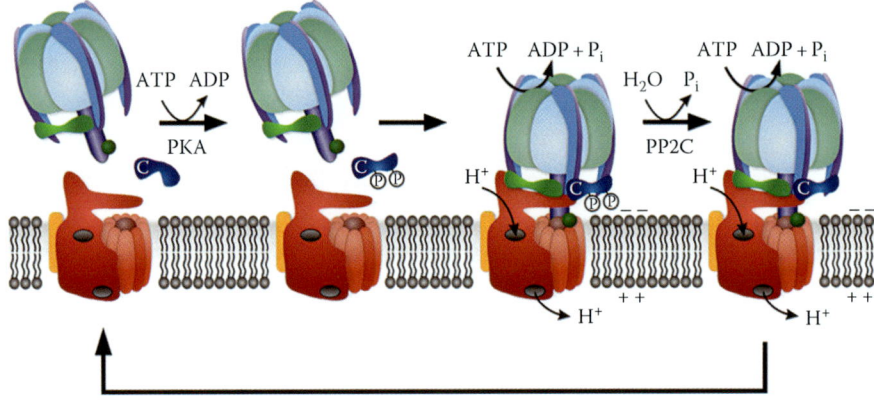

Figure 2.11 A schematic model of the proton pump illustrating the reversible assembly of the pump. The functioning pump is shown as the third model from the left, while the completely disassembled pump is shown on the extreme left. Subunit C at left, must be phosphorylated (C attached to 2 Ps) in the two middle diagrams, in order for it to promote assembly of V_1 and V_o holoenzymes. Dephosphorylation of subunit C (extreme right model) leads to disassembly of the pump to conserve energy when it is not specifically needed. (From Baumann, O. and Walz, B. *J. Insect Physiol.*, 58, 450, 2012. With permission.)

Figure 3.2 Butterflies in the family Lycaenidae, Genus *Agrodiaetus*, puddling and taking moisture and salts from the damp soil at the Kaçkar Mountains in Eastern Turkey. The slightly larger and lighter colored individual in the middle of the group is in the genus *Meleageria*. (Photo courtesy of Andrei Sourakov, University of Florida and McGuire Center, Gainesville, FL, 2001. With permission.)

Figure 4.7 An olive fruit fly emerging with the ptilinum expanded to aid it in breaking out of the puparial cuticle. (Photo courtesy of Dr. Hanife Genc, Çanakkale Onsekiz Mart Üniversitesi, Çanakkale, Turkey.)

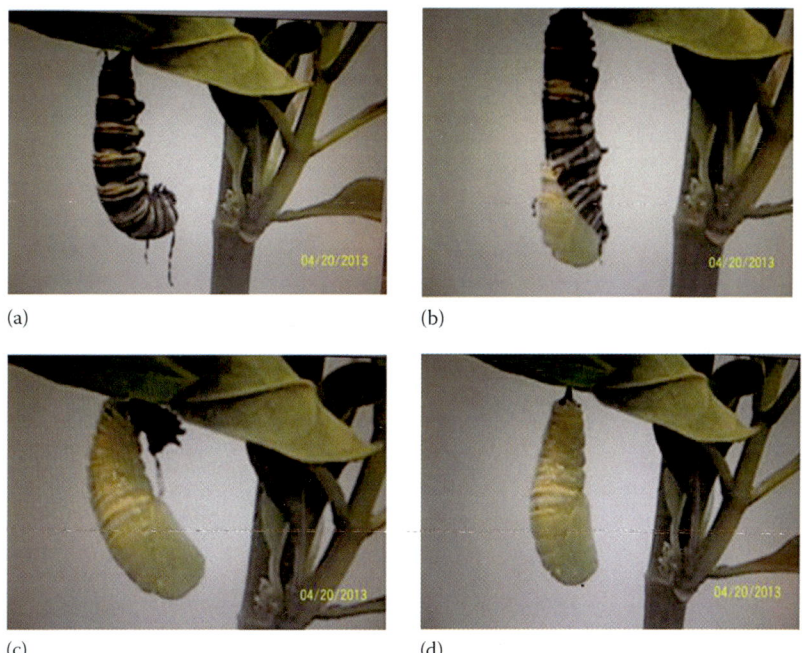

Figure 4.8 A sequence showing molting of the last instar cuticle by a larval monarch butterfly caterpillar as it hangs by its cremaster from a silk pad. (a) The larva is still in the "j" shape; the pharate pupa is inside the larval integument but cannot be seen at this point. (b) The larva now hangs straight down and the larval integument has split dorsally, beginning at the head of the larva, and the green pupa can be seen as the larval integument retracts toward the posterior. (c) The larval integument has been retracted to the posterior end of the pupa by contractions of intersegmental muscles attached to the old integument. (d) The newly emerged pupa has vigorously wiggled until the old larval integument was dislodged and fell away from the pupa. The pupa will continue to shorten in length and soon take on the typical green appearance of a monarch pupa. (Photos by the author.)

Figure 4.9 (a) Ecdysis of the adult cicada *Neocicada hieroglyphica* from the last nymphal cuticle. (b) The adult with partially expanded wings. (c) The adult *Neocicada hieroglyphica* with fully expanded wings, but cuticle not yet hardeneded and darkened. (d) Adult *Neocicada hieroglyphica* with sclerotized and darkened cuticle. Note how it blends with the tree trunk on which it is resting, waiting for its wings to fully harden. (Photos courtesy of Lyle Buss, University of Florida, Entomology & Nematology, Gainesville, FL.)

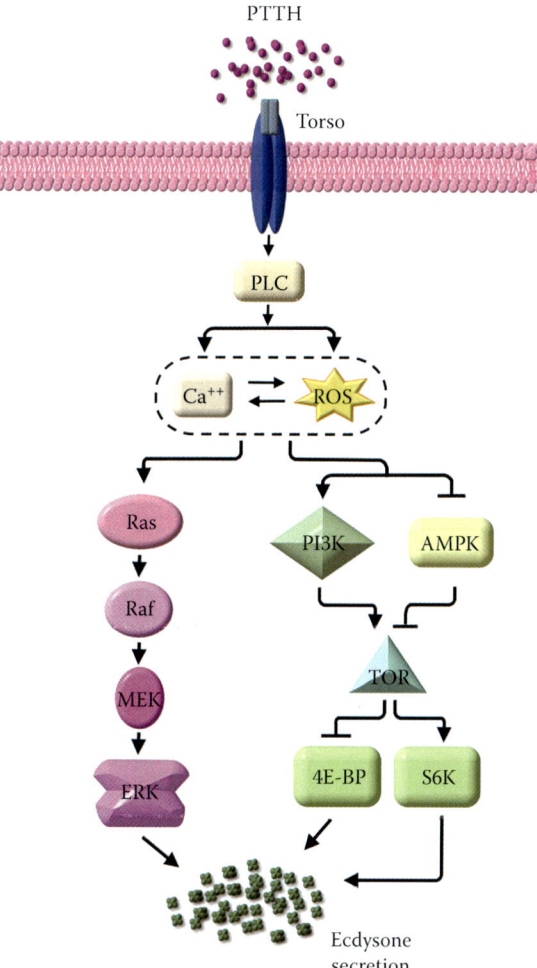

Figure 5.3 A conceptional presentation of the signaling network involved in PTTH-stimulated ecdysteroidogenesis in prothoracic gland cells, involving reactive oxygen species, calcium ions, and a number of proteins. PTTH, prothoracicotropic hormone; torso, receptor protein for PTTH; Ras, Ras GTP-binding protein; ERK, extracellular signal-regulated kinase; cAMP, cyclic adenosine monophosphate; PI3K, phosphatidylinositol 3-kinase; TOR, target of rapamycin; 4E-BP, eIF4E-binding protein; S6K, p70 ribosomal protein S6 kinase; AMPK, adenosine 5/-monophosphate-activated protein kinase. (From Hsieh, Y.-C. et al., *J. Insect Physiol.*, 63, 32, 2014. With permission.)

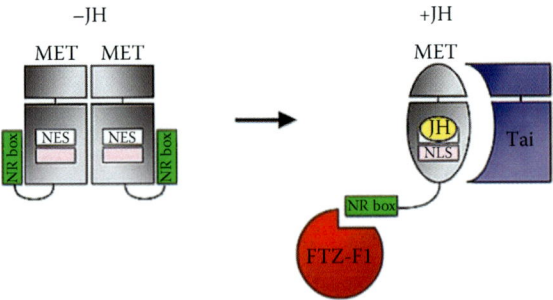

Figure 5.16 A model for activation of MET protein by JH-dependent conformational changes. At the left when JH is absent, MET exists as a homodimer by interactions in the bHLH and PAS domains (gray in diagram). Binding of JH (right) stimulates an active confirmation and promotes an interaction with Taiman/AaFISC/TcSRC with the bHLH-PAS region and with FTZ-F1 at a C-terminal NR box. Possibly JH may influence the activity of nuclear export and import signals in the ligand-binding domain, promoting nuclear localization. (From Bernado, T.J. and Dubrovsky, E.B., *Insects*, 3, 324, 2012. With permission.)

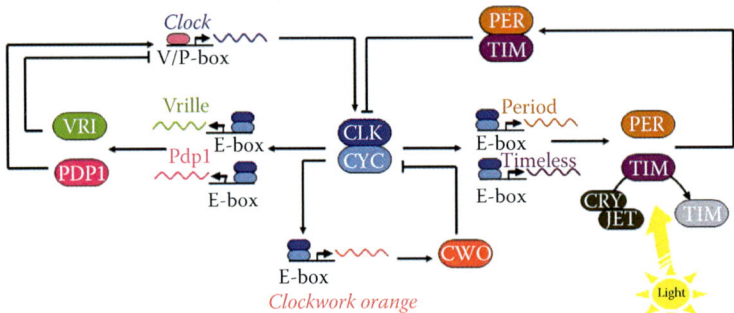

Figure 6.1 A model depicting the molecular oscillatory mechanism of the *Drosophila* circadian clock. The clock comprises three interlocking transcriptional–translational loops. Light acts to degrade TIM through Cry, and then JETLAG (JET) resets the clock in a time-of-day–dependent manner. Lines ending in an arrow head are stimulatory; those ending in a negative sign are inhibitory. CLK, clock protein; CRY, cryptochrome protein; CWO, clockwork orange protein; CYC, cycle protein; PER, period protein; PDP1, Par domain protein 1; TIM, timeless protein; VRI, vrille protein. (From Tomioka, K. et al., *J. Comp. Physiol. B*, 182, 729, 2012. With permission.)

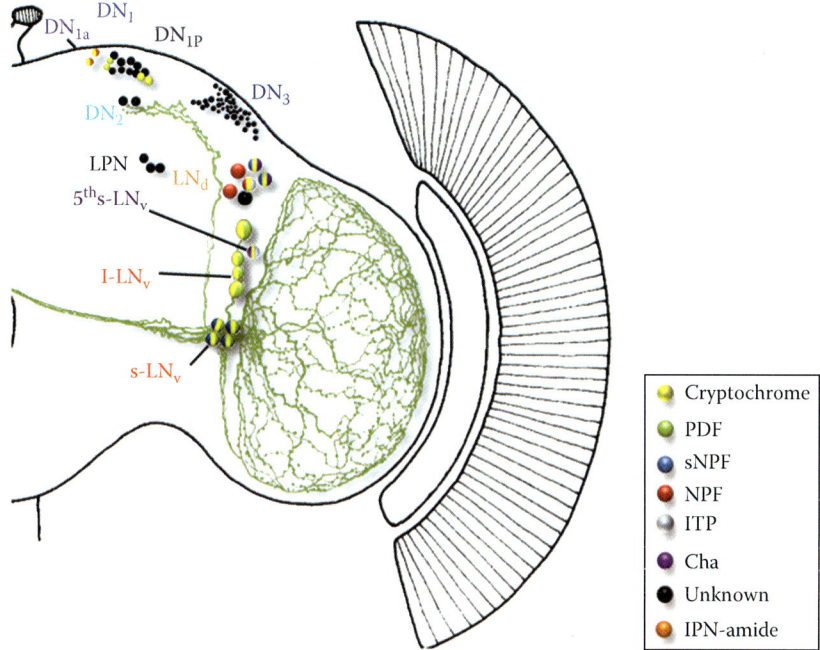

Figure 6.2 A basic overview of clock-gene-expressing neurons and their neurochemical characterization. The different colors of the neurons indicate the peptides/proteins that are expressed in those cells. Cryptochrome, yellow; pigment dispersing factor (PDF), green; short neuropeptide F, blue; neuropeptide F, red; ion transport peptide, gray; choline acetyltransferase, purple; IPN-amide, orange. Cells with unknown peptidergic content are colored black. Arborization of PDF is indicated in green. DN, Dorsal neurons LN, Lateral N\neurons. (From Peschel, N. and Helfrich-Förster, C., *FEBS Lett.*, 585, 1435; *Physiology*, 57, 935, 2011. With permission.)

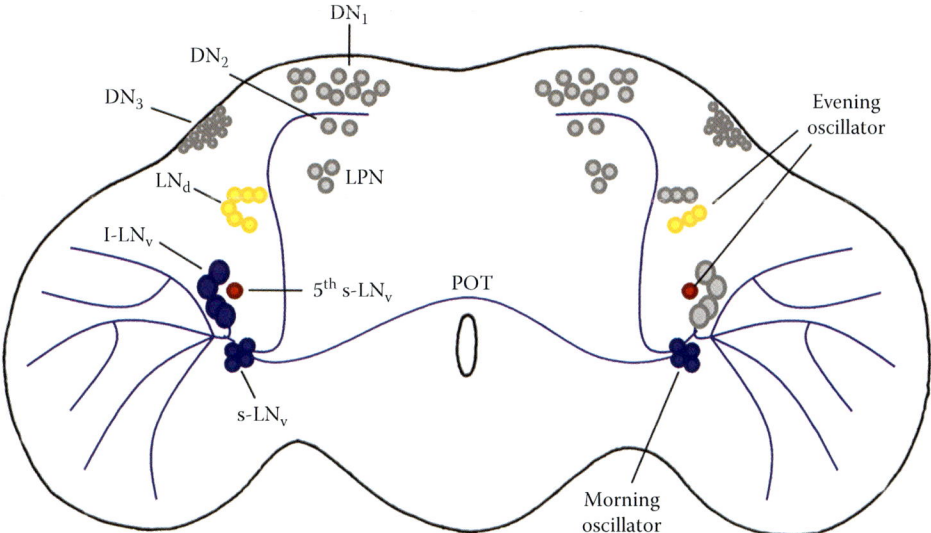

Figure 6.3 The neuronal network of the "master" clock in the adult *Drosophila melanogaster* brain. The dorsally located clock neurons (DNs) consist of three different groups, the DN_1s, DN_2s, and DN_3s. The lateral clock neurons (LNs) can be divided into a group of six dorsal lateral neurons (LN_ds, left side, yellow) and the more ventrally located LN_vs (left side, blue). There are four LN_vs per hemisphere with small cell bodies, the s-LN_vs, and four to six neurons with large soma, the l-LN_vs; both express the neuropeptide PDF. Usually located among the l-LN_vs is another single neuron with a smaller cell body, the fifth s-LNv (red), which is PDF-negative. Another group of clock-gene-expressing neurons, the LPNs, is located in the lateral posterior brain. According to the dual oscillator model, three of the six LN_ds and the fifth s-LNv were identified to be evening neurons (right side, yellow, and red), whereas the s-LN_vs were found to be morning neurons (right side, blue). (From Hermann, C. et al., *J. Comp. Neurol.*, 520, 970, 2012. With permission.)

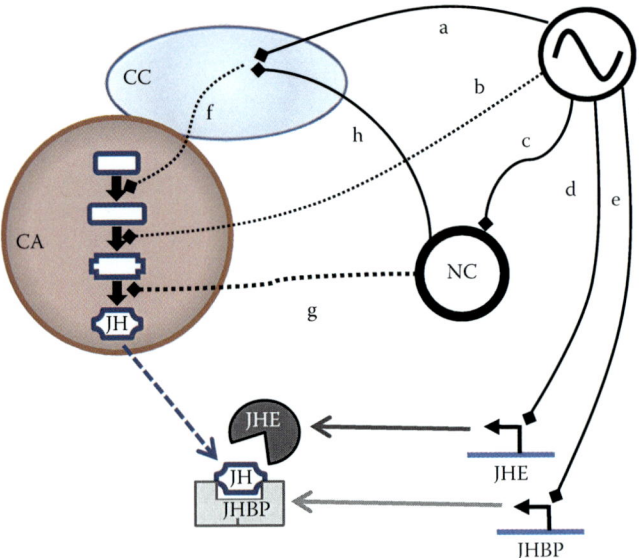

Figure 6.4 A working model for circadian modulation of JH titers. The model assumes that JH biosynthesis is regulated in many, and perhaps all, insects in a circadian manner by direct or indirect neuronal pathways from cells in the circadian network (represented in the model by the circle with a sine symbol in the upper right corner). Continuous lines ending in a diamond head depict neuronal pathways. Dotted lines show pathways innervating the corpora allata (CA) and are assumed to contain allatoactive neuropeptides such as allatotropins and allatostatins. The shapes and arrows in the circle representing the CA depict chemical intermediates and enzymes, respectively, in the JH biosynthetic pathway. The model assumes that the circadian system influences the expression or activity of enzymes of the JH pathway. Pathways *d* and *e* show circadian influences on the transcription or activity of proteins that degrade (e.g., JH esterase) or bind (e.g., JH binding protein) JH, respectively. CA, corpora allata; CC, corpora cardiaca; JH, juvenile hormone; JHBP, juvenile hormone binding protein; JHE, JH esterase; NC, neurosecretory cell. (From Bloch, G. et al., *J. Insect Physiol.*, 59, 56, 2013. With permission.)

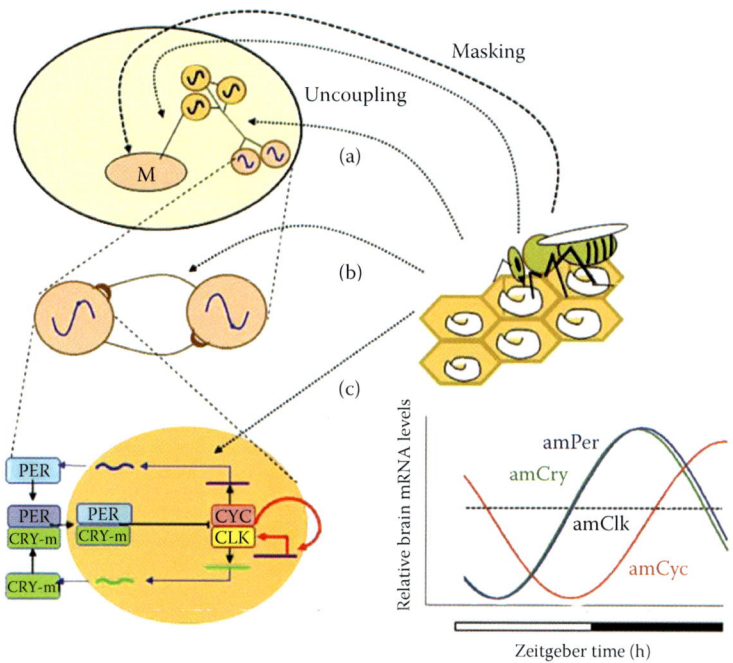

Figure 6.5 A multilevel regulation of circadian rhythms in the brain and central nervous system and their possible social modulation in honeybees, *Apis mellifera*. Top: Pacemaker cells and other neuronal circuits interact to produce overt circadian rhythms. The circadian network is composed of clusters of coupled pacemaker cells (circles with sinusoidal wave lines). Left middle: Interactions between cells in the same cluster are needed to generate circadian rhythms in behavior. Left bottom: Circadian rhythms are generated autonomously within pacemaker cells in the honeybee brain. (a) The dotted line indicates modulation of the interaction between clusters of pacemakers. (b) The dotted line suggests interactions between pacemakers within the same cluster. (c) The dotted line indicates possible modulation within pacemaker cells. Honeybee on brood comb: Direct interaction between nurse bees and the brood may influence circadian rhythms in several ways and at several levels, as indicated by the dotted lines. Lower right: Gene mRNA cycling (with the prefix *am* for *Apis mellifera*, as follows: for Period [amPer], Cryptochrome-m [amCry], and Cycle [amCyc]). In the honeybee, the phase of amCyc transcript is almost in antiphase to that of amPer and amCry. The straight dashed line represents clock mRNA, which does not vary through the day. Not shown: the mRNA pattern for *Timeout* (amTim2), which showed variation over time as influenced by the environment. M, motor control center; "masking," modulation of motor controlling centers without affecting the circadian system; uncoupling, modulating the interaction between the circadian system and motor controlling centers. (From Bloch, G., *J. Biol. Rhythms*, 25, 307, 2010. With permission.)

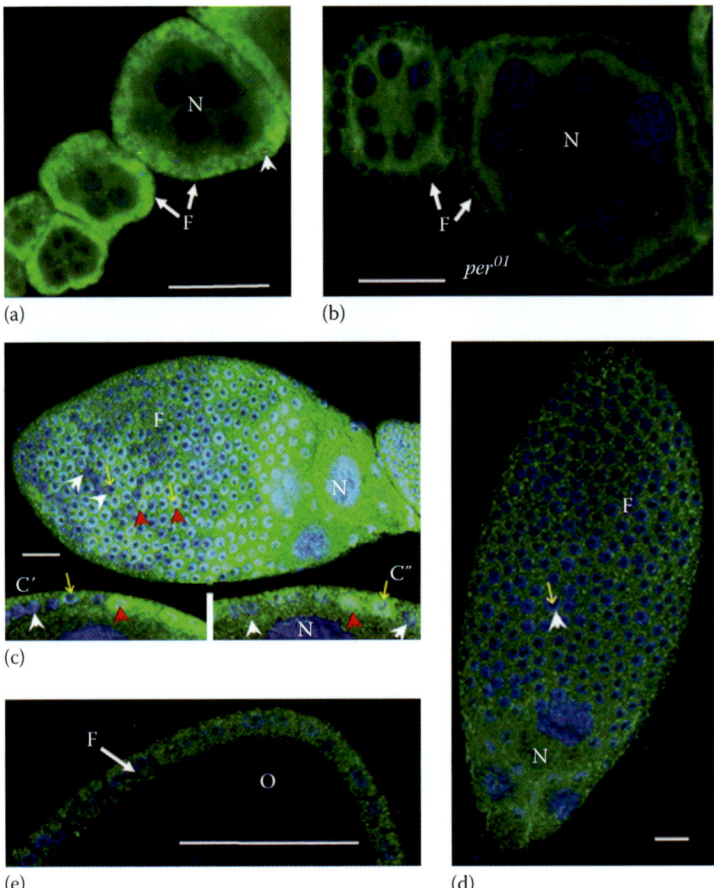

Figure 6.6 Expression of PER protein in ovaries of *D. melanogaster*. (a) Transverse section in which PER is cytoplasmic in young previtellogenic follicles. (b) Young previtellogenic follicles in *per^{01}* flies— negative control. (c) An image near the surface of an early previtellogenic follicle. Insets C′ and C″ show transverse sections of enlarged follicular cells. PER is in nuclei of some cells (red arrowheads) but absent in other nuclei (white arrowheads). Yellow arrows in C and D point to large dark nucleoli. Patchy nuclear pattern was observed in all stage 9 follicles. (d) PER was not detected in older vitellogenic follicular cells in stage 10; specks of green are nonspecific and occur in *per^{01}* flies (not shown). (e) Anterior fragment of egg chamber with yolk and follicular cells (stage 10, transverse section). PER protein stained green, nuclei stained blue. Bar = 20 μm. N, nurse cells; F, follicular cells; O, oocyte. (From Kotwica et al., 2009b. With permission.)

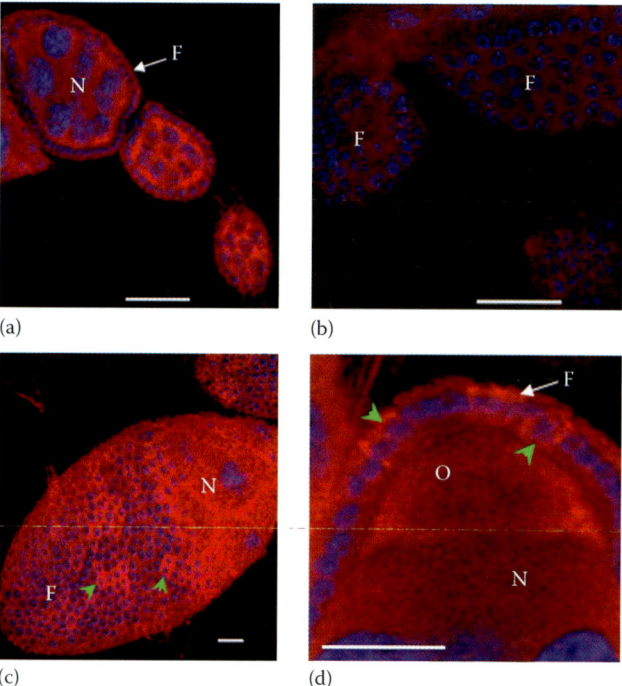

Figure 6.7 Expression of DOUBLETIME (DBT) protein in ovaries of *D. melanogaster*. (a) DBT is present in nurse cells of young previtellogenic oocytes in this transverse section. (b) DBT is absent in follicular cells of previtellogenic oocytes (stages 3–7) in this image near the surface. (c) DBT shows patchy cytoplasmic pattern of expression (green arrowheads) in stage 9 vitellogenic egg chamber in this surface image. (d) Transverse section of anterior fragment of egg chamber showing cytoplasmic DBT in some follicular cells (green arrowheads). The pattern of DBT distribution is similar for ovaries collected from different time points (data not shown). DBT is stained red; nuclei are blue. Bar = 20 μm. N, nurse cells; F, follicular cells; O, oocyte. (From Kotwica et al., 2009b. With permission.)

Figure 6.8 Diapause-associated color changes in the heteropteran stinkbug *Nezara viridula* reared under laboratory or quasi-natural outdoor conditions. Laboratory rearing occurred in Kyoto, Japan. (a) reproductive, nondiapausing green adult; (b) and (c) diapausing adults of intermediate color; (d) russet brown diapausing adult; (e) and (f) mostly intermediate colored and russet overwintering adults in early stage of diapause after outdoor rearing in Kyoto (November to December); (g) adults in March of different color grades in the last stage of diapause; (h) adults in April that have terminated diapause, changed color, and are beginning post-diapause reproduction. (From Musolin, D.L., *Physiol. Entomol.*, 37, 309, 2012. With permission.) (Photograph courtesy of Prof. Dr. D.L. Musolin.)

Figure 7.1 (a) An adult monarch butterfly taking nectar from a flower on a sunny day in mid-January at the El Rosario colony of diapausing and overwintering monarchs in the mountains near Mexico City. (b) Several diapausing monarchs feeding from a flower in the Sierra Chincua colony in the mountains near Mexico City on a warm day in mid-January. (Photos courtesy of Rochelle Carlson Nation.)

Figure 12.6 A male tabanid fly hovering in flight waiting for a female to fly by so that it can give chase. (Photograph courtesy of Jerry Butler, University of Florida, Gainesville, FL, retired.)

Figure 13.5 The polka-dot moth, *Syntomeida epilais*, a wasp mimic and distasteful because it sequesters cardenolides from its larval host plant, *Nerium oleander*. (Photo courtesy of Gretert Montano, Miami-Dade College, Miami, FL.)

Figure 14.4 A tabanid fly with a dark band across the lower part of the compound eyes. (Photo courtesy of Dr. Jerry Butler, University of Florida, Department of Entomology & Nematology, Gainesville, FL, Retired.)

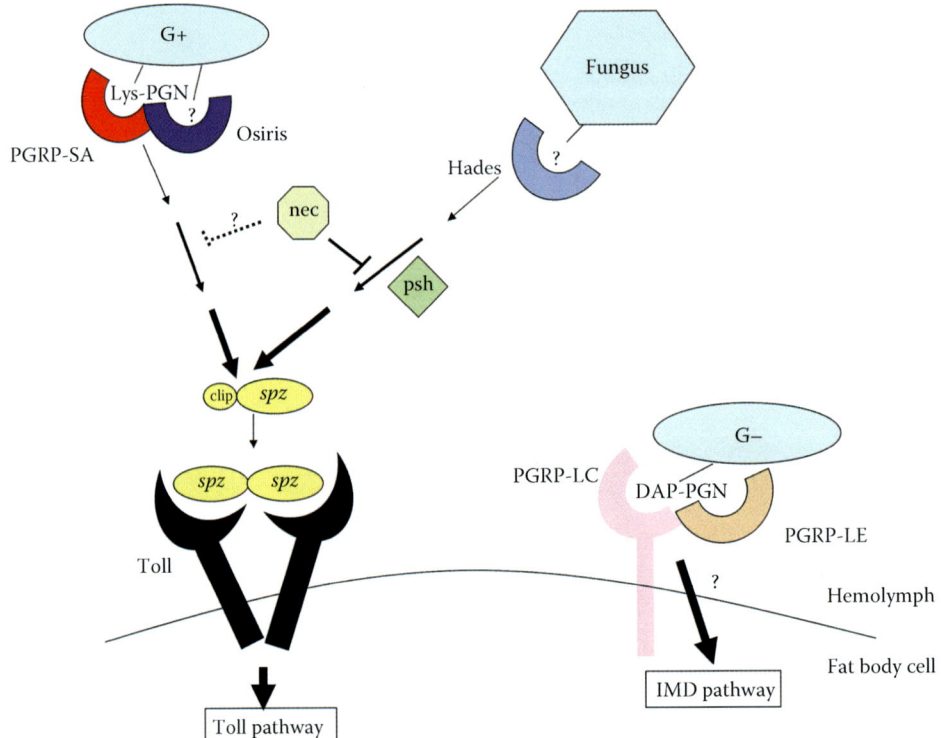

Figure 16.3 The recognition of microbial infection in *Drosophila* leading to activation of the Toll and IMD pathways. Gram (+) bacteria are recognized by PGRP-SA, a peptidoglycan recognition protein that binds the lysine-type peptidoglycan of Gram (+) bacteria. Osiris is another peptidoglycan recognition protein that participates in the initial recognition reaction. A proteolytic kinase cascade is initiated that finally removes a clip domain from the dimer Späetzle, the ligand for Toll, a transmembrane protein in the fat body cell membrane. Fungal invasion also elicits recognition proteins, one of which may be Hades. A proteolytic cascade is activated, and one of the kinases, persephone, proteolytically removes the clip domain from Späetzle, which then binds Toll. Toll dimerizes, perhaps as a result of binding *spz*, and then binds two molecules of *spz*. Gram (−) bacteria, which have diaminopimelic acid (DAP)-type peptidoglycan in the outer coat, is recognized and bound by PGRP-LDC and possibly PGRP-LE, and the immune deficiency (IMD) pathway is activated. (From Leclerc, V. and Reichhart, J.-M., *Immunol. Rev.*, 198, 59, 2004. With permission.)

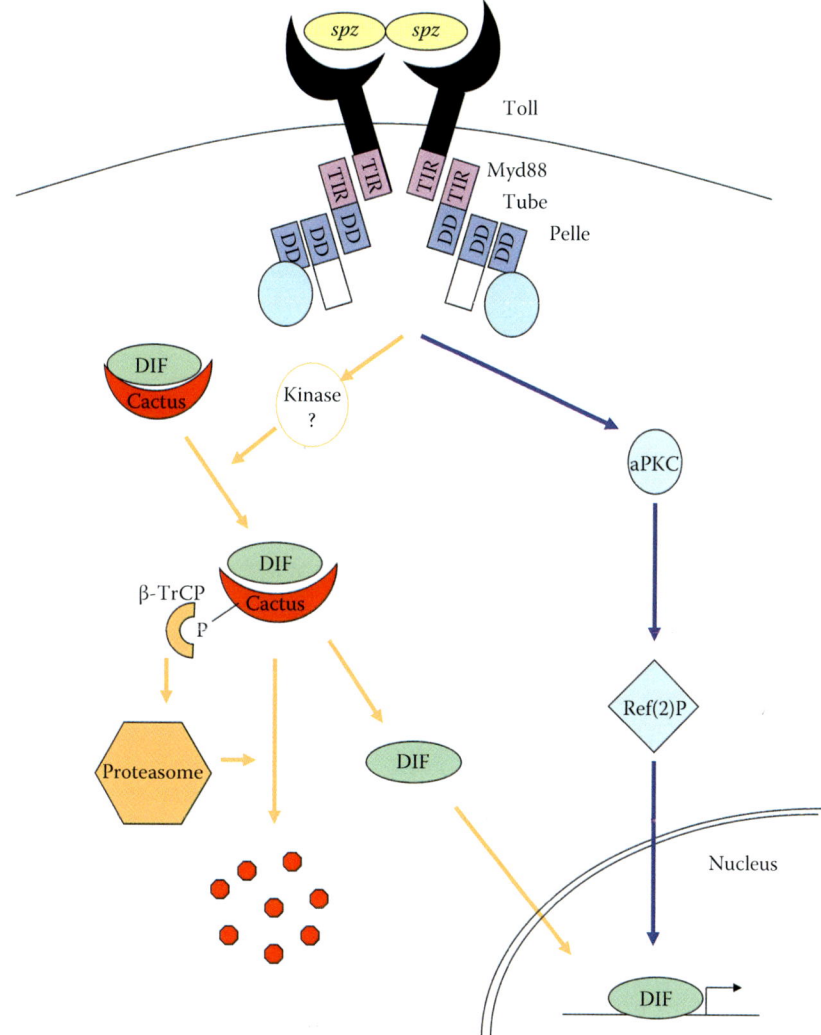

Figure 16.4 An illustration of selected details of the intracellular Toll pathway that culminates in the synthesis of antimicrobial peptide drosomysin. Each of the Toll receptors recruits several additional proteins including MYD88, Tube, and Pelle, and an intracellular proteolytic kinase cascade is activated. At least one of the kinases in the cascade hydrolyzes the Cactus domain from the Cactus-DIF complex in the cell cytoplasm. Cactus in then further degraded by enzymes, and DIF is translocated to the nucleus where it acts as a transcription factor for *drosomyosin*. Expression of *drosomyosin* results in synthesis of the antimicrobial peptide Drosomyosin and its secretion into the circulating hemolymph where it attacks the invading microoganisms. (From Leclerc, V. and Reichhart, J.-M., *Immunol. Rev.*, 198, 59, 2004. With permission.)

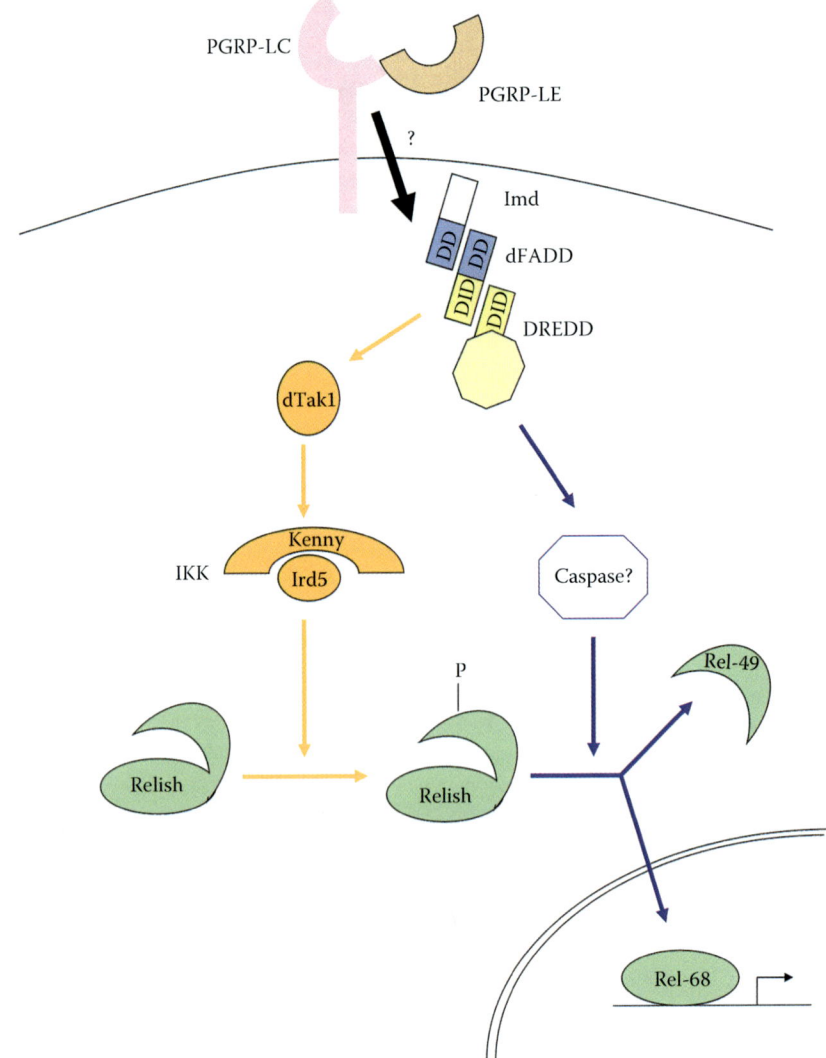

Figure 16.5 The IMD pathway results in the synthesis of an antimicrobial peptide, diptericin, effective against Gram (−) bacteria. Several factors are recruited when the IMD receptor is activated and two intracellular proteolytic cascades are initiated. One cascade results in phosphorylation of Relish. The second cascade cleaves Relish into an inhibitory molecule, Rel-49, which remains in the cytoplasm, while Rel-68 is translocated to the nucleus where it serves as a transcription factor for *diptericin*. The antimicrobial peptide diptericin is then synthesized and secreted into the hemolymph. (From Leclerc, V. and Reichhart, J.-M., *Immunol. Rev.*, 198, 59, 2004. With permission.)

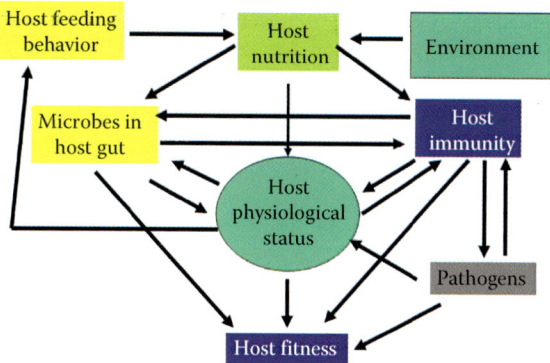

Figure 16.6 A scheme to show some of the factors affecting the host physiological status, host immunity, and host fitness. The interactions are multiple and complex. Environmental influences may act directly upon most of the other components illustrated in this diagram, but to avoid excessive lines, it is shown interacting through host nutrition, which in turn has an influence on all the other components. (Based on Ponton, F. et al., *PloS Pathog.*, 7(12), e1002223, 2011; Ponton, F. et al., *J. Insect Physiol.*, 59, 130, 2013; Chambers, M.C. and Schneider, D.S., *Curr. Opin. Immunol.*, 198, 116, 2012; Bauerfeind, S.S. and Fischer, K., *J. Insect Physiol.*, 64, 14, 2014.)

Figure 19.3 Hair pencils of a male giant danaine butterfly, *Idea leuconoe*. Pheromonal components are released from the hair pencils that influence female response. (A previously unpublished photograph, courtesy of Dr. Ritsuo Nishida, Kyoto, Japan.)

Figure 19.11 A smoke plume released into the air in an orchard in California. The eddies of smoke show the likely eddies in a volatile chemical, such as a pheromone, or another type of semiochemical, released into the air. (Photo courtesy of Dr. Ring Cardé; From *J. Chem. Ecol.* Cover Photo, Vol. 39(9), 2013. With permission.)

Figure 19.13 (a) Males of the Oriental fruit fly, *B. dorsalis*, collect naturally occurring phenylpropanoid compounds from the petals of the Hawaiian lei flower, *Fagraea berteriana*, and use the compounds to make a male-produced pheromone, *trans*-coniferyl alcohol, that attracts female flies. (From Nishida, R. et al., *J. Chem. Ecol.*, 22, 949, 1997. With permission. Color photograph courtesy of Dr. Ritsuo Nishida.) (b) *B. dorsalis* on a flower of the Hawaiian rainbow shower tree, a *Cassia* sp. that releases methyl eugenol, a pheromone precursor for the Oriental fruit fly males. (Photograph courtesy of Ethel Villalobos and Todd Shelly, USDA, Waimanalo, HI.)

Figure 20.3 A female olive fruit fly probing a green olive fruit, perhaps about to lay an egg beneath the skin of the olive fruit. (Photo courtesy of Dr. Hanife Genc, Çanakkale Onsekiz Mart Üniversitesi, Çanakkale, Turkey.)

Figure 21.1 Leaf-cutter ants, nest, and dump. (a and b) Fresh plant material being harvested by foraging workers returning to the nest. (c) An active nest comb with fungal tissue and leaf shown. (d) Older nest comb. (e) Older substrate and spent fungal materials are removed from the bottom of the garden and transported to a refuse dump. Older workers manipulate material on the dump presumably to facilitate degradation of the material. (a, d, e: From Scott, J.J. et al., *PLoS ONE* 5, e9922, 2010. With permission; b, c: From Aylward, F.O. et al., *ISME J.*, 6, 1688, 2012. With permission; all photos courtesy of J. Scott.)

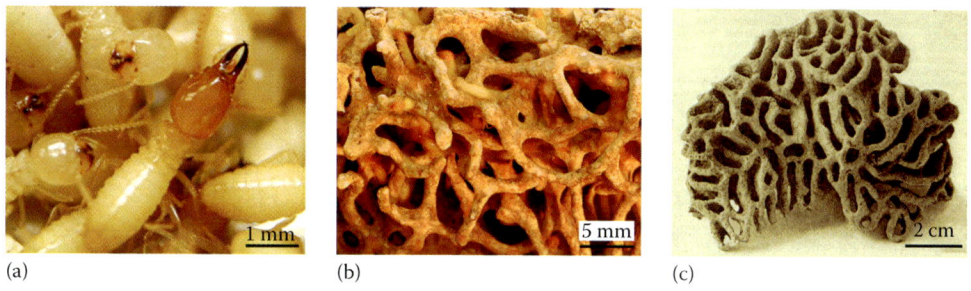

Figure 21.4 (a) The subterranean termite *Coptotermes formosanus*, (b) nest material, and (c) *Termitomyces* fungal comb from the termite *Macrotermes bellicosus*. (From Chouvenc, T. et al., *Proc. R. Soc. Lond. B*, 280, 20131885, 2013. With permission, Photos courtesy of Thomas Chouvenc, University of Florida, Ft. Lauderdale, FL.)

Figure 21.5 (a) The ambrosia beetle *Xylosandrus crassiusculus* in its gallery with fungal garden. (b) *X. crassiusculus* gallery with brood in pupal stage. (c) Characteristic frass appearance of *X. crassiusculus* attack on a tree. All photos courtesy of Jiri Hulcr, University of Florida. (a and c: Unpublished photos; b: From Six, D.L., *Insects*, 3, 339, 2012. With permission.)

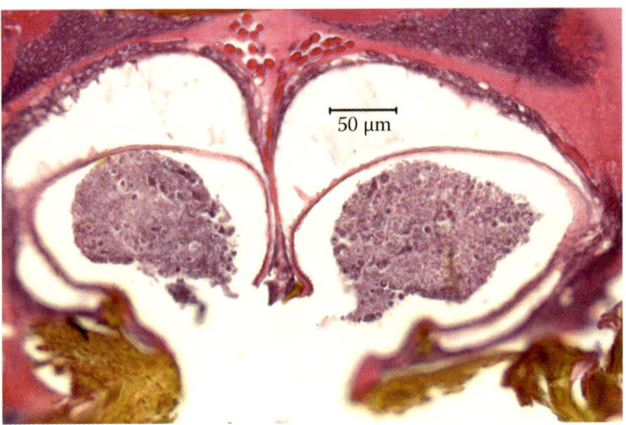

Figure 21.6 Transverse section through the paired mandibular mycangia of a female ambrosia *Euwallacea validus* beetle showing novel *Fusarium* and *Raffaelea* nutritional symbionts. The preparation was paraffin embedded, sectioned, and stained with Harris hematoxylin and eosin–phloxine. (From Kasson, M.T. et al., *Fungal Genet. Biol.*, 56, 147, 2013. With permission, photo courtesy of Mike Kasson.)

Figure 21.7 (a) Multiple small entrances by the mycocleptic beetle *Diuncus duodecimspinatus* around the larger gallery entrance of its host or provider species, *Hadrodemius globus*, in a branch of a *Ficus* sp. in Papua New Guinea. (b) Detrimental effect of mycocleptic beetle *Camptocerus suturalis* on the gallery of its host, *Camptocerus aeneipennis* in Guyana. The bottom half of the host gallery has been destroyed and replaced by a gallery and brood of the mycocleptic *C. suturalis*. Typically, in healthy galleries of *C. aeneipennis*, the larval chambers line both sides of the maternal gallery. Photos in a from Jiri Hulcr, University of Florida, and in b from Sarah M. Smith, Michigan State University. (From Hulcr, J. and Cognato, A.I., *Evolution*, 64, 3205, 2010. With permission.)

Figure 21.8 Feminization of the butterfly *Eurema hecabe* infected with two different strains of *Wolbachia* (*w*HecCI2 and *w*HecFem2). Butterfly caterpillars were infected with the two strains of *Wolbachia*, and then as different instars fed a diet containing an antibiotic to suppress, but not eliminate, the feminization effect of the male genotype. (a and b) Adult *E. hecabe* emerged with deformed wings from infection with the double strain of *Wolbachia*, when fed the antibiotic diet from third to fourth instar. (c) Failure of an adult butterfly infected with the double strain of *Wolbachia* to escape from the pupal case, when fed the antibiotic diet from first to fourth instar. (d and e) Normal adult *E. hecabe* females pale in ground color and without sex brands, from a nontreated line of insects singly infected with *w*HecCI2; (f and g) normal adult male *E. hecabe* bright in color with sex brands (arrow) from a nontreated insect line singly infected with *w*HecCI2. Bars = 10 mm. (From Narita, S. et al., *Appl. Environ. Microbiol.*, 73, 4332, 2007. With permission.)

CHAPTER 12

Flight

PREVIEW

Insects first evolved as flightless animals, and some authorities believe that flight evolved only once in some ancient insect ancestor. How the wings evolved is uncertain, and numerous theories have been proposed. Insects were the first animals to evolve wings sometime between 400 and 300 million years ago. The metabolic activity of flight muscles during flight is intense. Oxygen consumption can go from a resting value to one 50–100 times resting value in seconds. The ability to fly has been a major factor in the success of insects, enabling them to fill many ecological niches, disperse in searching for food and mates, migrate long distances, and escape their enemies. Flight has clearly contributed to the diversification of insects. The thoracic structure of insects that fly has evolved to be relatively rigid and heavily sclerotized to withstand the forces generated by the large flight muscles attached to the cuticle of the thorax. Muscles that enable flight may be attached directly to the wing hinges in Odonata, but, in other insects, the large flight muscles are not attached to the wing hinges at all, but are attached to the thoracic wall in such a way as to deform the shape of the thorax, which is translated into wing movements. Small muscles attach to the wing hinges to aid in steering and turning.

12.1 INTRODUCTION

Insects began as wingless creatures, documented by fossil discoveries dating to nearly 400 million years ago. However, sometime during the next 100 million years, forms with primitive extensions of cuticle from the thorax appeared (Marden and Kramer, 1994). Flight probably evolved only once in some ancient ancestor of modern insects (Brodsky, 1994), and it was probably in an aquatic ancestor.

A number of theories have been advanced to attempt to explain the selection forces acting on the evolution of wings (reviewed by Kingsolver and Koehl, 1994). Apparently, there is some evolutionary plasticity in flight performance with respect to temperature; for example, Frazier et al. (2008) found that *Drosophila melanogaster* reared at colder temperatures (15°C) had better flight performance in cold air than flies reared at 28°C. The cold-reared flies also had an increase in wing area and wing length, and slower wingbeat frequencies than flies reared at warmer temperature.

How the wings evolved in the first place (and probably only once) often has been discussed but not solved. The paranotal lobe theory (reviewed by Wootton, 1976) proposed that wings evolved from rigid extensions of the thoracic terga. Another idea that has received considerable support is that wings evolved from moveable gill flaps whose original function was respiration. Marden and Kramer (1994) have suggested that wing-like appendages were first used in the air by surface-skimming aquatic insects in much the same way that some stone flies (Plecoptera: Taeniopterygidae) and some subadult mayflies (Ephemeroptera) still skim across the water. It is not possible to say with certainty what factors were at work to selectively promote the evolution of wings. It may be that multiple factors were important (Kingsolver and Koehl, 1994). If the wings did evolve from gill flaps, did those gill flaps originally occur only on thoracic segments or also on abdominal segments? Wigglesworth (1976) supported the latter idea, and there may be neurophysiological support (Robertson et al., 1982).

Large interneurons in the meso- and metathoracic ganglia of *Locusta migratoria* have rhythmic nervous output in phase with the motoneuron output to the dorsal longitudinal muscles (DLMs) (the muscles that control the downstroke of the wings) and to motoneurons controlling the dorsoventral muscles (DVMs) that raise the wings. Those particular interneurons, however, have their cell bodies located in the third abdominal ganglion (for DLMs) and first abdominal ganglion (for DVMs), indicative that their original function might have been to control appendages on the abdomen. In the earliest insects, neurons from a segmental ganglion probably controlled the structures within its own segment. Although these three abdominal ganglia are fused with the metathoracic ganglion in modern adult locusts, embryonically, the tissue housing the cell bodies is abdominal ganglionic tissue (Robertson et al., 1982). The question is still open as to what appendages or structures those particular neurons might have controlled early in the evolution of insects, but they might have controlled abdominal gill flaps.

Two reviews (Lehmann, 2008; Lehmann and Schützner, 2010) summarize flight mechanics and oxygen and energy use during flight, particularly in *Drosophila*. The deformation of the wings during flight and the energy loss caused by flapping flight have been analyzed by Lehmann et al. (2011). Card and Dickinson (2008) describe startle flight dynamics vs. voluntary flight in *Drosophila*, noting that the two different flight behaviors involve a trade-off between speed (in startle flight) and stability (in voluntary flight). Utilizing the technique of microtomography with high-speed X-ray images, Walker et al. (2014) and Hedenström (2014) provide new insight into the mechanics of insect flight, and movies from the research of Walker et al. (2014) can be viewed on YouTube from directions in the paper. A different video on movements of the wings by the major groups of muscles can be viewed in Wikipedia at Wikipedia Insect Wing http://en.wikipedia.org/wiki/Insect_flight (last accessed April 21, 2015). Actions of both direct and indirect wing muscles are shown.

12.2 THORACIC STRUCTURE, WING HINGES, AND MUSCLE GROUPS INVOLVED IN FLIGHT

The thorax, although heavily sclerotized to withstand the pull of the flight musculature, is composed of plates joined by sutures that allow some flexibility and movement in several planes relative to the body axis. The wings are hinged to the thoracic plates by a number of small, hard sclerites at the junction between the tergum and the pleuron. The wings pivot up and down over the **pleural wing process**, a heavily sclerotized, fingerlike fulcrum of cuticle that is part of the pleuron (Figure 12.1). In addition, the hinge points let the thorax move inward and outward during a stroke cycle, which aids in snapping the wing rapidly over the pleural wing process.

The forewings attached to the mesothorax are not used in flight by some insects, such as beetles. In other insects in which both fore- and hind wings are used in flight, the wings may beat together, or the fore- and hind wings may be slightly out of phase. When forewings and hind wings are used in flight, the frontal edges of the wings are reinforced with larger, tubular veins for strength. The frontal edge of the wing leads in both the upstroke and the downstroke. Airflow hits the lower side of the wing during the downstroke, which generates the principal lift forces, and hits the upper side during the upstroke, which can also produce lift forces (Nachtigall, 1989). Like a mechanical toggle switch, the wings are only in a stable position when positioned either up

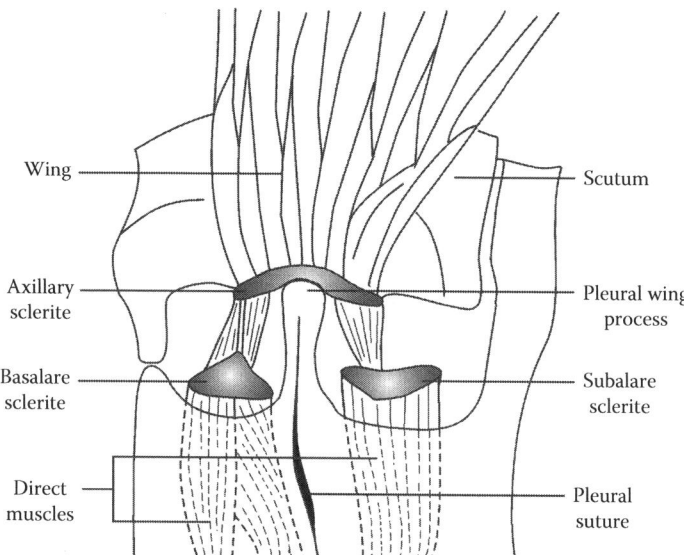

Figure 12.1 Thoracic structure showing the subalar and basalar wing hinge sclerites and heavily sclerotized pleural wing process over which the wing pivots. Subalar and basalar muscles attached to their, respectively, named wing hinges are important in controlling wing orientation movements in all insects, and they also produce the downstroke in dragonflies and damselflies, but not in most other insects.

or down. As every insect collector knows, the horizontal position of the wings of butterflies and moths preserved in an insect collection can be achieved only by pinning the wings in this position until the thorax dries.

At least 10 pairs of muscles are involved in flight, wing orientation, and steering (Pringle, 1976). The **indirect flight muscles** power wing movements by changing the shape of the elastic thorax. These are power muscles and include the DLMs (the wing depressors) that arch the tergum and the **dorsoventral** and **oblique dorsal** muscles (the wing elevators). In *D. melanogaster*, the DLM comprises 12 fibers, with 14 fibers in the dorsal ventral muscle. The fibers are approximately a millimeter in length and vary from 100 to 200 μm in diameter (Swank, 2012). In larger insects, the muscle fibers would be longer and greater in number than in a small fly like *Drosophila*. The DLM fibers and the dorsal ventral fibers do not attach to the wing directly, but attach to the cuticle of the thorax. The indirect flight muscles are fibrillar muscles, so called because the muscle fiber bundles are unaligned and the endoplasmic reticulum is located at the periphery (Roy and VijayRaghavan, 2011). A zinc finger transcription factor, Splat major (Salm), encoded by the gene *salm* (Schnorrer et al., 2010) regulates the differentiation of muscle cells into indirect flight muscles during development of *Drosophila*, and Salm is expressed in indirect flight muscles of the fly *Calliphora* and the beetle *Tribolium* (Schonbauer et al., 2011).

The **direct wing muscles**, called by some authors "control muscles" (Dickinson and Tu, 1997), insert directly on the wing hinge sclerites or on axillary sclerites or movable sclerites of the pleuron of the thorax. These muscles are the **basalar** (wing depressors), **subalar** (wing depressors and twisting or supinations of the wing), and **third axillary** (wing folding) muscles. The control muscles are activated by the nervous system by one-to-one motor spikes (Dickinson and Tu, 1997). A third group of muscles in the thorax are accessory indirect muscles consisting of the pleurosternal, anterior tergopleural, posterior tergopleural, and intersegmental muscles. They modify the way in which the power-producing muscles move the wings and the angle of attack. Contraction of a small muscle, the pleuroalar (also called the pleuroaxillary) muscle, in the locust,

L. migratoria, reduces pronation of the forewings during the downstroke and supination during the upstroke, thus, helping to control the angle of attack and the development of additional lift forces (Wolf, 1990). The muscle inserts on the third pleuroaxillary sclerite and has a broad fan-shaped anchor on the pleural wing process. The angle of attack is important to the generation of lift forces and to speed of flight and hovering (Nachtigall, 1989). The control muscles are important in rapid alteration of wing kinematics needed to allow aerial maneuvers in feeding and escape behavior (Dickinson and Tu, 1997).

12.3 WING STROKES

The wing **downstroke** is produced by contraction of DLMs (in all insects, except dragonflies and damselflies, in which the mechanism is different; see Section 12.5). The DLMs are indirect muscles that do not attach directly to the wing hinge sclerites. These large powerful muscles are attached to the phragma, which are invaginated and hardened cuticular processes at the anterior and posterior of each of the meso- and metathoracic segments (Figures 12.2 and 12.3). When they contract, they shorten the thoracic segments by arching the tergum, slightly lifting the attachment base of the wings at the tergopleural junction and forcing the wings downward. As the wings approach the unstable horizontal position, they suddenly pivot downward over the pleural wing process. This releases the load on the DLMs, and they cease to shorten, but the new position of the wings introduces the load to the antagonistic set of muscles, the DVMs.

The DVMs (Figure 12.2) cause the **upstroke** of the wings in all insects, including the Odonata. These powerful muscles are anchored on the heavily sclerotized, relatively rigid ventral thoracic cuticle. They insert on the dorsum of the thorax. When they contract, they pull on the tergum and reduce the arching of the thorax (Figure 12.3a). This causes the wings to pivot upward, and, when they again reach the unstable position on the pleural wing process, they snap into the up position. This reduces the load on the DVMs, and they cease to shorten.

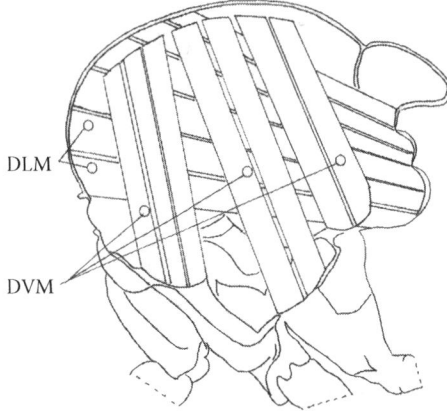

Figure 12.2   A diagrammatic illustration of the dorsal longitudinal muscles (DLMs) and dorsoventral muscles (DVMs), the indirect power flight muscles, in the thorax of *Drosophila repleta*. The illustration shows six DLMs located interior to seven DVMs arranged in three groups. The DLMs are attached to a phragma near the dorsum of the meso- and metathoracic segments in most insects. Upon contraction, these muscles arch the thorax and cause the downstroke of the wings. The DVMs have their origin on heavily sclerotized epicuticle on the ventrum and insert on the thoracic epicuticle. When they contract, they depress or reduce the arching of the thorax and pivot the wings up. (Drawing modified from Williams, C.M. and Williams, M.V., *J. Morphol.*, 72, 589, 1943; Dickinson, M.H. and Tu, M.S., *Comp. Biochem. Physiol.*, 116A, 223, 1997; Hedenström, A., *PLoS Biol.*, 12(3), e1001822, 2014.)

FLIGHT

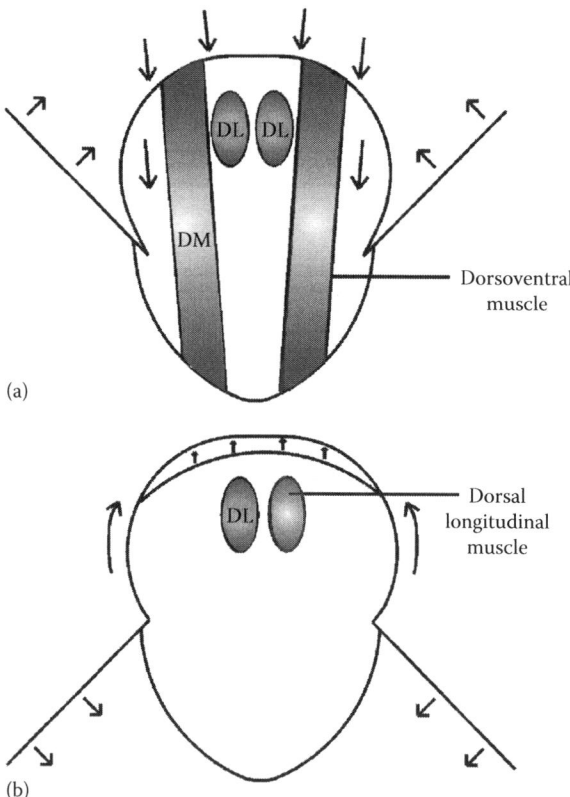

Figure 12.3 A cross section through the thorax to illustrate the indirect antagonistically arranged interior dorsal longitudinal muscles (DLMs) and dorsoventral muscles (DVMs) that produce the downstroke and upstroke, respectively, of the wings in most insects and allow some insects to beat the wings several times for each nerve input. (a) When the DVMs contract, they pull upon the tergum of the thorax and depress the arching of the thorax. This causes the wings to pivot up over the pleural wing process. (b) Contraction of the DLMs, attached to phragma at the front and back of each of the meso- and metathoracic segments, shorten the length of the segments slightly, arch the thorax, and lift the dorsal thoracic region so that the wings pivot downward. Contraction of one pair of muscles stretches the antagonistic pair and serves as a stimulus for a repeated contraction. Thus, several contractions may occur for each nerve impulse received.

In insects with synchronous muscles, nerve impulses must arrive at the DLMs and DVMs to evoke each successive contraction. The rhythm for the repeated nerve impulses and continued contraction of the flight muscles in locusts is based in groups of interneurons in the thoracic ganglia that generate a motor program (Wilson, 1968). Similar motor programs originating in the thorax probably exist for other insects. The wingbeat frequency of selected insects is shown in Table 12.1.

12.4 MULTIPLE CONTRACTIONS FROM EACH VOLLEY OF NERVE IMPULSES TO ASYNCHRONOUS MUSCLES

Insects with asynchronous muscles can achieve several to many contractions of the two antagonistic sets of muscles with one delivery of nerve impulses to each set. The reduced extent of the sarcoplasmic reticulum in asynchronous muscles does not rapidly sequester calcium ions released by the arrival of nerve impulses, and this allows the muscle fibers to remain in the active state

Table 12.1 Selected Insects to Show Representative Wingbeat Frequencies

Insect	Wingbeat Frequency (Hz)	References
Drosophila hydei (Diptera)	170–180	Dickinson and Lighton (1955)
Apis mellifera (Hymenoptera)	180	Pringle (1983)
Anastrepha suspensa (Diptera)	145	Webb et al. (1976)
Forcipomyia sp. (Diptera)	800–950	Sotavalta (1953)
Chironomus sp. (Diptera)	600–650	Sotavalta (1953)
Blowfly (Diptera)	120	Pringle (1983)
Hawk moth (Lepidoptera)	40	Pringle (1983)
Swallowtail butterfly (Lepidoptera)	5	Pringle (1983)
Schistocerca gregaria (Orthoptera)	17	Pringle (1983)
Neoconocephalus robustus (Orthoptera)		Josephson (1984)
Flight	20	
Stridulation	200	
Musca domestica (Diptera)	About 150	Roeder (1951)
Polycanthagyna melanictera (dragonfly)	Flight 33.4, hovering 35	Wang et al. (2003)
Exaerete frontalis (euglossine bee)	110.6	Dillon and Dudley (2004)
Eulaema meriana (euglossine bee)	271.4	Dillon and Dudley (2004)

for a (relatively) long time. Each time the wings snap into the up or down position, the load on one set of antagonistic muscles is released and reintroduced with a stretching stimulus to the other set of muscles. Stretching the muscles acts as a stimulus, and, with Ca^{2+} and adenosine triphosphate (ATP) available, the stretched muscles start to shorten again without another volley of nerve impulses. The time elapsed during a single wing stroke is very short because the muscles shorten only 2%–3% of resting length before the wings snap into a new position. Minimal shortening, a prolonged active state with free calcium ions available, and reintroduction of a load with stretching allow asynchronous muscles to oscillate and produce several contractions for each burst of nerve impulses received. A control center in the thorax sends out periodic nerve impulses to keep a rhythm going.

12.5 FLIGHT IN DRAGONFLIES AND DAMSELFLIES

Members of the order Odonata, the dragonflies and damselflies, represent the most primitive condition with respect to wing movement and flight of all the currently living insects. Dragonflies are able to operate each wing independently with synchronous muscles that attach directly to the wing hinge sclerites. The basalar and subalar muscles produce the downstroke (see Figure 12.1). The basalar muscles insert on the basalar sclerite at the anterior base of the wing in front of the pleural wing process. In the Odonata, they pull the wing down and twist downward, or pronate, the anterior leading edge of the wing. The subalars insert on the subalar sclerite at the base of the wing posterior to the pleural wing process and also pull the posterior edge of the wing downward with some twisting of the trailing edge. The basalar and subalar muscles are large, powerful muscles in dragonflies and damselflies (Nachtigall, 1989), and they work on the short end of the lever to pivot the relatively long wings over the pleural wing process against the resistance of air. They are anchored to parts of the heavily sclerotized ventral and pleural cuticle. This direct muscle arrangement for the downstroke of the wings only occurs in Odonata, according to Nachtigall, but Pringle (1976) states that the basalar and subalar muscles are also power-producing muscles for the wings in some other groups. Pringle (1983) indicates that one basalar muscle group in locusts aids in pronating and pulling the wings down.

Flight in the Odonata has been well studied. Dragonflies hover, fly at speeds up to 10 m/s, accelerate for brief intervals at about 4 g forces, cruise at 2 g (Thomas et al., 2004), and make sharp turns in pursuit of prey, which they capture up to 97% of the time (Olberg et al., 2000). The two pair of wings typically counterstroke when dragonflies are cruising, with the pair of hind wings in the downstroke first, followed by the anterior pair. The phase separation in the movement of the wings varies from 55° to 180°. During flight acceleration and in complicated maneuvers, such as sudden turning, they beat the wings in phase or nearly so, producing more lift and acceleration. A leading-edge vortex (LEV) forms on the wings on each downstroke. Wingbeat frequency, stroke amplitude of the wings, and forward speed cause wake formation over the wing surfaces leading to a LEV during the downstroke. The vortex may extend continuously across the body from one wing to another.

In a study of filmed dragonfly flight, Thomas et al. (2004) found 38 photo frames suitable for detailed analyses, including 28 frames showing counterstroking wingbeats with an LEV on the forewing, 5 frames showing wingbeats involved attached flows over both wings (extending continuously over the thorax and across the wings), and 4 frames showing in-phase wing strokes during periods of acceleration. An important conclusion from the study was that whether the LEV formed on the wing and remained bound (attached over the wing for the duration of the downstroke) or was shed depended upon the angle of attack. Formation, growth, and stabilization of the LEV were caused by the increase in angle of attack, and shedding of the LEV was associated with the decreases in angle of attack. Dragonflies seem to be able to alter angle of attack at any time from zero to high attack angles. Three brief movies of dragonfly flight from the study by Thomas et al. (2004) are presented in supplementary material to their paper at the following URL: http://jeb.biologists.org/cgi/content/full/207/24/4299/DC1 (last accessed April 21, 2015).

Dragonflies glide for long periods and may use gliding to conserve energy and to thermoregulate (Heinrich, 1996; Wakeling and Ellington, 1997a) as convective airflows over the body. In the analyses of free-flying dragonflies, *Sympetrum sanguineum*, and damselflies, *Calopteryx splendens*, Wakeling and Ellington (1997b) found that wingbeat frequency in the damselfly was only half that in *S. sanguineum* even though both are about the same size. The damselfly performed a clap and fling wing movement (see Section 12.6.1), which gave it more lift per wing stroke than the dragonfly, which did not perform the clap and fling maneuver.

12.6 AERODYNAMICS OF LIFT AND DRAG FORCES PRODUCED BY WINGS

To remain airborne, an insect must generate **lift** forces at least equal to its weight, and to move forward, the horizontal thrust vector must exceed the drag of air resisting forward motion (Pringle, 1983). Sane (2003) and Lehmann (2004) provide recent reviews of insect flight, including the mechanisms of aerodynamic force production and lift generation.

The dynamics of flight are complex, and the lift force needed is related to factors such as body weight, wing size, wing shape, speed of air movement over the wings (i.e., wingbeat frequency), and angle of attack of the wings. Smaller insects have to beat their wings faster than larger insects to gain the lift forces to keep them in the air. There is a popular myth, recounted by McMasters (1989), which suggests that someone is supposed to have calculated that the wings of bumblebees are too small to produce enough lift for the insect to fly. The calculations, if they ever really existed, would have been based on steady-state aerodynamic calculations, which predict sufficient lift forces for some insects (e.g., locusts; Jensen, 1956), but not for most insects. **Steady-state aerodynamics** is based largely on calculations derived to explain lift in fixed-wing aircraft. Insects do not have fixed wings, and the flapping of the wings presents special problems, such as changing wing shape during a wing stroke, acceleration, and deceleration of wings as they change direction of movement. Moreover, insect wings do not present a smoothly contoured airfoil typical of an airplane wing (McMasters, 1989). Unsteady-state conditions in which there are momentary very high lift forces followed by lower lift forces, or even negative ones, describe insect flight more adequately than steady-state calculations (Weis-Fogh, 1973).

Numerous researchers have performed experiments and computational investigations with model wings or free-flying *Drosophila* (Lehmann and Dickinson, 1998; Sane and Dickinson, 2001, 2002; Sun and Tang, 2002; Birch and Dickinson, 2003; Fry et al., 2003; Wang et al., 2004; Lehmann and Pick, 2007; Ramamurti and Sandberg, 2007), model wings or dragonflies in free flight (Wang et al., 2003; Maybury and Lehmann, 2004; Sun and Lan, 2004; Thomas et al., 2004; Wang and Sun, 2005), *Manduca sexta* (family Sphingidae) (Ellington et al., 1996; Liu et al., 1998; Hedrick and Daniel, 2006), orchid bees (Dillon and Dudley, 2004), and a variety of other insects (Usherwood and Ellington, 2002; Srygley and Thomas, 2003; Srygley, 2004). Some studies were based on dynamically scaled models of insect wings many times larger than the actual insect wing, and the models were observed in a wind tunnel with smoke or with photography after emersion in a viscous fluid, such as a light oil, while others were based on free-flying insects or tethered insects. Data and computations from model wings have been very useful in gathering data on **unsteady forces** acting on insect wings during flight, and the data generally support the more limited data obtained with tethered insects or free-flying ones.

12.6.1 Lift Forces Generated by Clap and Fling Wing Movements

Weis-Fogh (1973) initially described one example of an **unsteady lift** condition that occurs in very small chalcid wasps as a **clap and fling** wing motion. In this small insect, the wings "clap together" at the top of the upstroke, and then twisting motions controlled by some of the small muscles fling them apart at the start of the downstroke. The rapid flinging motion sets up air movements above the wings that increase the lift force of the downstroke and may aid very small insects to generate enough lift for flight in spite of having little airfoil surface in their tiny wings. *D. melanogaster* uses the clap and fling at the start of the downstroke and at the beginning of the upstroke. Lehmann et al. (2005) suggest that the enhancement of lift from the clap and fling motion in *D. melanogaster* requires an angular separation between the two wings of no more than about 10°–12°. A movie of the clap and fling movement in a simulated wing from the paper by (Lehmann et al., 2005) can be viewed at http://jeb.biologists.org/cgi/content/full/208/16/3075/DCi. last accessed April 21, 2015. *L. migratoria* performs a clap and fling during climbing, but not in horizontal flight. Marden (1987) concluded that insects that have a clap and fling wing motion have as much as 25% more muscle mass-specific lift than insects that do not use a clap and fling maneuver. Wakeling and Ellington (1997b) calculated that the damselfly, *C. splendens*, which uses a clap and fling maneuver, gets 44% more muscle mass-specific lift than the dragonfly, *S. sanguineum*, which does not use clap and fling. Miller and Peskin (2005) presented calculations of lift enhancement from simulations of clap and fling sequences over a range of Reynolds numbers (Re). Re are dimensionless numbers that are proportional to the ratio of inertial to viscous forces acting on an object moving within a fluid medium. At Re values from about 10^3 to 10^4, flow is nearly laminar, but at very high Re values (>10^6, as in an airplane flying through the air), there is great turbulence reflecting inertially driven flows and viscous effects are reduced (Dudley, 2000). Low Re values typically occur in insect flight. Detailed flight analyses from model wings, as well as from free-flying insects, have shown a range of unsteady lift forces operating during flapping flight. These lift-generating forces are clap and fling, LEV, dynamic stall, rotational lift, and wake capture. Insect wings typically bite into the air at high angles of attack, a maneuver that would cause immediate stall in a fixed-wing aircraft. Contrary to the situation in fixed-wing aircraft, drag generated in several complex ways by flapping wings actually contributes to lift in insects.

12.6.2 Lift Forces Derived from Drag and Delayed Stall

Wang (2004) used computational analyses on model wings to show that up to 75% of lift of a hovering dragonfly can be contributed by drag induced by complex airflow over the downstroking wing. Hovering dragonflies and hoverflies have a highly inclined stroke plane of wing movements, an attack angle of 35°–40°, and lift and drag are about equal. Sun and Lan (2004) found similar lift forces due to drag. The hind wings begin the downstroke first, followed by the forepair, so there are two large lift forces in each flapping cycle. The downstroke of the wings produces more lift than the upstroke, which produces more thrust. Each downstroke of the wings produces an LEV ring of air on the anterior edge of the wings associated with the delayed stall of the attack angle. Somps and Luttges (1985) measured large, transient lift forces 15–20 times the body weight of tethered dragonflies, with time-averaged lift values equal to 2–3 times body weight created by the turbulent flow of air generated by the independent movement of the front and rear wings. Similar analyses of unsteady lift forces are described by Brodsky (1994).

Tethered tobacco hornworm moths (*M. sexta*) generate unsteady lift forces equal to at least 1.5 times the body weight during the downstroke of the wings (Ellington et al., 1996). Coincident with the downstroke movement, intense LEVs of low-pressure air above the wings are created by the pronation of the wing (tilting downward of the leading edge) during the downstroke. The vortices form first over the leading edge of the wing, move out toward the wing tips, and finally extend behind the insect in a ring of turbulent air (Figure 12.4) (Alexander, 1996; Ellington et al., 1996). These low-pressure vortices increase the lift of the downstroke.

Figure 12.4 A diagrammatic illustration of vortices above the wings and behind the body created by the downstroke of the wings. The model used to illustrate the vortices is a butterfly, the painted lady *Vanessa cardui*, but the concept is based upon illustrations and data obtained from experiments with the moth *Manduca sexta* flying in a smoke chamber (Alexander, 1996; Ellington et al., 1996). The vortices contribute to the lift generated by the downstroke, the principal lift-generating wing movement. The size, shape, and trailing nature of the vortices may vary with wing shape and flight speed.

The high angle of attack of the wing, a condition that would rapidly create a stall in a fixed-wing aircraft, creates the vortices. This condition of dynamic or delayed stall can be tolerated in an insect for the brief interval of one downstroke, at the end of which the stall conditions are eliminated as the wings change direction. The rotational flip upward of the wings just prior to reaching the full downstroke also allows some of the wake disturbance of the air from the downstroke to be captured as lift.

Using a dynamically scaled model of *D. melanogaster* with built-in sensors, Dickinson et al. (1999) describe three interacting mechanisms that provide the lift forces for flight in the fruit fly, and likely in most other insects, perhaps with some degree of variation in the importance of some of the components. The three mechanisms they describe are (1) the upstroke and downstroke of the wings with a high angle of attack (delayed stall), (2) rotational circulation of air eddies above the wings, and (3) wake capture, with the latter two mechanisms promoted by the pronation and supination of the wings as they rapidly rotate and change direction at the end of or, in some insects, just before the end of each half stroke. Larger insects, such as sphingid moths, generate lift forces from the development of an LEV that produces transient aerodynamic lift forces to keep the insect in the air (Dudley, 1999). The high angle of attack of the wings as they move through the up- and downstroke generates high unsteady lift forces during the fast movement and short stroke amplitude of an insect wing, but come close to stall conditions. Just prior to stall, relief is provided by reversal of wing direction in movement. The rotation of the wings at the end of each half-stroke (pronation at beginning of downstroke and supination at beginning of upstroke) generates an upwardly directed lift force initially, followed by a downward force. Although the turbulence or wake following a moving object in a fluid medium typically produces drag, by rotating the wings at the end of each half-stroke, the wing encounters its own wake in such a way to generate momentary positive lift (Dudley, 1999). The magnitude of the lift generated by wake capture varies from positive to negative depending on exactly when in time the rotation occurs relative to the beginning of the next half-stroke. Wing rotation that is delayed until the start of a new stroke direction produces negative lift. Dickinson et al. (1999) postulate that the small but significant lift forces associated with the rotational nature of the wings are a powerful means that insects use in steering maneuvers during flight.

Dragonfly wings act as ultralight airfoils during gliding flight (Kesel, 2000). In a cross-sectional view, the wings have well-defined corrugations in which rotating vortices develop. The corrugations might be expected to lead to high drag values, but in fact, the drag from air flowing over the wings is low. The vortices filling the profile valleys smooth out the profile geometry, resulting in low drag similar to that of air flowing over flat plates (Kesel, 2000), while lift forces are much higher than expected from flat plates.

Insects have excellent maneuverability and control in flight (Figure 12.5) and often change direction. Most moths are night-flying insects, and they use evasive maneuvers to reduce the chances that a bat will be able to catch them. Anyone who has tried to capture even slow-flying insects, such as butterflies, in a net will appreciate the maneuvering ability of flying insects.

12.7 HOVERING FLIGHT

Many insects are capable of hovering flight. Hovering flight is metabolically very expensive (Willmott and Ellington, 1997) and dynamically complex (Sun and Wang, 2007; Geurten et al., 2010). Male tabanid flies (Figure 12.6) often hover in wait for a female to fly by, which they then chase. Sphingid moths (also called hawk moths and hummingbird moths) feed from flower nectaries while hovering. In an investigation of feeding by hovering *M. sexta*, Sprayberry and Daniel (2007) demonstrated that the moths feed most effectively from stationary flowers, but they have the ability to adjust their position and track the motions of a flower from which they

FLIGHT 345

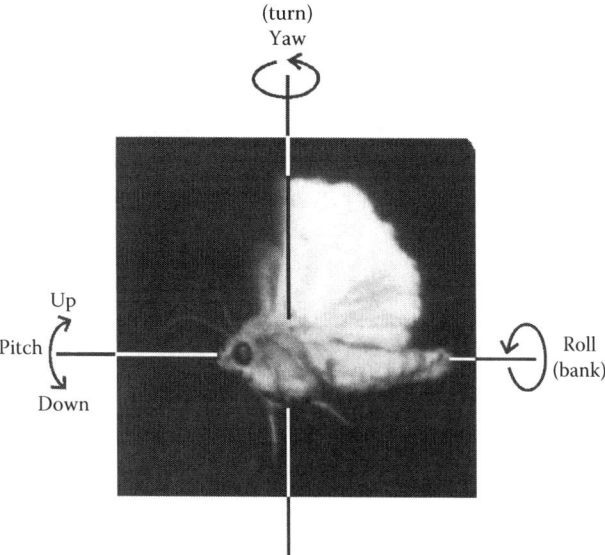

Figure 12.5 Insects in flight use wing movements to maneuver the body in a roll around the long axis of the body, pitch up or down around the perpendicular axis of the body, and yaw or turn the body in flight. The model used in this illustration is a flying moth, *Heliothis virescens*. (Photograph courtesy of Peter Teal, USDA, deceased.)

Figure 12.6 **(See color insert.)** A male tabanid fly hovering in flight waiting for a female to fly by so that it can give chase. (Photograph courtesy of Jerry Butler, University of Florida, Gainesville, FL, retired.)

are feeding. Tracking in hovering flight is very important in keeping the extended proboscis in contact with the nectary, and each movement of the nectary requires some movement in the flight of the moth. Motion of the flower is likely to be induced by the wingbeats of the moth (Sane and Jacobson, 2006) as well as by natural wind flow *M. sexta*, tracked moving flowers and maintained a constant distance from the flower best at about 1 Hz (1 movement/s) when the movement was sideways or up and down relative to the moth in front of the flower. They could track the movement of the nectary at 2 Hz in the horizontal and vertical planes, but with some lag in following the movement, and even though they poorly tracked artificial nectaries moving at 3 Hz in the horizontal and vertical planes, they nevertheless could feed from them. However, they were able to track a flower moving toward them or away from them only at lower frequency and could not feed from a nectary moving at 3 Hz toward or away from them. The authors concluded that the energy cost of tracking was negligible in terms of the volume of nectar that could be consumed during feeding; view at http://faculty.washington.edu/danielt/sprayberry06.mov.

In hovering flight, enough lift must be achieved to counter the body weight while still maintaining balance of aerodynamic forces and body stabilization. Fry et al. (2005) analyzed and photographed hovering flight in *Drosophila* and found that during hovering, the wings move in a U shape and there are large drag forces at the start of a downward movement of each stroke.

12.8 CONTROL OF PITCH AND TWISTING OF WINGS

Pitch and twisting of the leading and trailing edges of the wings during a wing stroke are important in generating lift forces. The basalar and subalar muscles inserted directly on the wing hinges are mainly involved in controlling the wing angle during a wing stroke. The basalar muscles attached to hinge points in front of the pleural process pull down (pronate) the leading edge of the wings in a downstroke, while the subalar muscles, attached to hinge points behind the pleural process, supinate or pull down the posterior edge of the wings and cause the leading edge to tilt upward during an upstroke. These movements cause air to flow faster over the upper surface of the wing than over the lower surface, and a lift force is produced. Thrust forces push the insect forward through the air, and the wing tip tends to describe a figure eight during flight.

In flies, bees, and wasps, a different mechanism for pitch and twist of the wings occurs because of the way in which the thorax is constructed. The tergum of the thorax is divided by the scutal cleft into two plates, the scutum and the scutellum. Sclerites at the posterior of the wing base join to the posterior plate, the scutellum. When the scutellum is pulled slightly forward by the contraction of the DLMs, the wing tilts slightly forward, or pronates, as it pivots down over the pleural process. Relaxation of the DLMs at the end of the downstroke allows the tergal plates to move farther apart, and the wing tilts slightly backward (supinates) as it is forced upward by the contraction of the DVMs.

12.9 POWER OUTPUT OF FLIGHT MUSCLES

The inertia of stopping the movement and changing the up and down direction of the wings is energy demanding, but insects do not have superefficient muscles. Flight, the most energy-intensive activity in the life of a flying insect, may extract near maximum mechanical power output by working flight muscles in some circumstances; however, Tu and Daniel (2004) found that the DLMs, the powerful indirect muscles that cause the downbeat of the wings, generated only 40%–67% of their maximal potential power output during *in vivo* flight conditions. The power output of flight muscle of a katydid, *Neoconocephalus triops* (Orthoptera: Tettigoniidae), is equal to about 37 W/kg muscle when the (synchronous) muscle is stimulated to give a single twitch and maximum output of 76 W/kg muscle during contractions at 25 Hz and at 30°C (Josephson, 1985a). Weis-Fogh (1973, 1977) estimated the mechanical power of flight muscle at 60–360 W/kg based on an assumed 20% metabolic-to-mechanical efficiency of flight muscle. Ellington (1984a,b) estimated the mechanical power output of wing muscle of an insect in hovering flight to be 70–190 W/kg, but questioned whether the metabolic-to-mechanical efficiency was as high as 20%. Metabolic conversion to mechanical muscle efficiency values of only about 10% were calculated for flight muscle of the fruit fly *Drosophila hydei* (Dickinson and Lighton, 1995), and only 3% efficiency of metabolic energy to mechanical power conversion was calculated for *N. triops* during stridulatory singing (Josephson, 1985b). The efficiency of muscles during flight of *S. sanguineum* and *C. splendens* (dragonfly and damselfly, respectively) was estimated at 12.6% for the dragonfly and 8.7% for the damselfly (Wakeling and Ellington, 1997c). Although flight is demanding of energy, it does not appear that evolution has acted to select flight muscles that are superefficient nor do they work at full potential power output during flight.

Dragonflies have one of the highest ratios of flight muscle to body mass of any animal, and they have powerful and efficient flight muscles equipping them to outmaneuver and capture other insects on the wing (Wakeling and Ellington, 1997c). Josephson and Stevenson (1991) measured oxygen consumption and power output together during experiments on the locust, *Schistocerca americana*, and calculated about 6% muscle efficiency in flight. Assuming higher efficiency makes power output calculations higher than they may really be, and, according to Josephson (1985a), the comparative value of power output measurements from muscles of various animals is very limited anyway because most measurements have been made under highly variable conditions, techniques, and assumptions. Marden (1987, 2005) found that takeoff performance in flying insects scales isometrically with flight muscle mass and maximum-specific force output (F_{max}) for flying insects, birds, bats, swimming fish, and running animals scaled to M^1 where M equals motor (working muscle) mass. Weis-Fogh (1964) suggested that insects store kinetic energy as elastic energy during one wing movement (either up or down) and then release the stored energy as kinetic energy when the wing movement reverses. They may store the energy in several different ways, including in the muscle system as in dragonflies (*Aeshna* spp.), in the elasticity of the cuticle of the thorax (sphinx moths), and in the resilin-forming elastic wing hinges (Weis-Fogh, 1964; Dickinson and Lighton, 1995). The natural elasticity of the thoracic cuticle that is distorted by a wing stroke and the compressed resilin in the wing hinge absorb energy and tend to spring back after a wing stroke. Additional reviews and discussion of flight and of wing morphology are presented by Wootton (1992).

12.10 METABOLIC ACTIVITY OF WING MUSCLES

Working insect flight muscles have the highest rate of metabolism per gram of muscle tissue of all biological tissues and the highest control values, ranging from 50 to 100 times resting values. The control value is the ratio of the oxygen consumption in active flight (or other exercise) divided by the oxygen consumption at rest. Intense energy demands of flight are supported by an extensive tracheal and tracheolar network that can supply oxygen to the flight muscles as they need it, so they do not develop an oxygen debt and do not have to resort to anaerobic glycolysis. Within the wing muscles, there are very large and numerous mitochondria (the sarcosomes) with many cristae, like leaves in a book. A large and continuous supply of ATP must be available to flight muscles to support the long periods of time on the wing that many insects demonstrate. Swank et al. (2006) found ATP levels in flight muscles of the fruit fly, *D. melanogaster*, to be as much as seven times the concentration in slower-acting skeletal muscle and in "slow" flight muscle fibers in a mutant fruit fly. Contraction force decreased in the flight muscles of wild flies as inorganic phosphate, Pi, increased because Pi competed with ATP for a binding site on the myosin molecules. Wild-type fruit flies compensated for the competitive binding of ATP and Pi by increasing the concentration of ATP in the muscle fibers. In contrast, as Pi increased in slow muscles, such as skeletal muscles, the maximum contraction force actually increased, indicating some major differences in the physiology and biochemistry of slow muscles and fast flight muscles. A major difference in metabolic performance between Diptera and Hymenoptera and members of the orders Orthoptera, Lepidoptera, and some others is the ability of the latter to deliver lipids to the flight muscles and then metabolize the lipids rapidly to support flight. Diptera and Hymenoptera can only fly on carbohydrates, and their inability to use lipids for flight may be due to lack of hormonal control of lipid storage, difficulties in transport of lipids from fat body to muscles, metabolism at the muscle level, or some combination of these limitations that prevent use of lipids in a timely fashion to support flight.

12.11 FLIGHT BEHAVIOR

Borst et al. (2010) review vision including optic flow, motion detection, and receptive field and describe some of the neural connections in the optic lobe. Many flying insects do not fly in a straight line even when no immediate objects are in the flight path; they typically fly straight for some period of time and then make brief saccadic (saclike) turns. Blowflies (*Lucilia* spp.) were shown to fly in such a meandering pattern even in open space (Kern et al., 2012). In order to navigate while flying, insects must process information on their position, speed, and orientation in space, as well as the position of objects in the environment. They obtain this information primarily from the pattern of images produced on the compound eyes as they fly, using optic flow fields to evaluate self-motion. **Optic flow field** denotes the pattern of apparent motion across the eyes, and there are two components—rotational optic flow caused by rotations about the axes of roll, pitch, and yaw (turning) and translational optic flow derived along the roll, pitch, and yaw axes (Baird et al., 2010). The perceived translational optic flow that the insect receives depends upon the ratio of relative speed and the distance to an environmental object in the field of view (Portelli et al., 2011). Two reflexes in flying *Drosophila*, a turn reflex and an expanding field of view (object in view apparently becoming closer), compete (1) to induce turns and (2) to go forward. The flies balance these reflexes and make forward progress, veering away from impending collisions with objects in their environment (Sugiura and Dickinson, 2009; Reiser and Dickinson, 2010; Straw et al., 2010). Translational optic flow is important as a way to gage the forward speed of the insect in respect to objects on the ground or in the environment. The visual system of *Drosophila*, and other flying insects, is extremely fast, as expected to enable them to avoid objects in their flight path (Theobald et al., 2010, and references therein). Honeybees and *D. melanogaster* maintain constant translational optic flow (i.e., constant ground speed) in the direction of movement, called axial optic flow, resulting in high ground speed when flying in a relatively open area and lower ground speed when flying through dense vegetation (Baird et al., 2010). Clearly, keeping a constant movement of images on the retina of the compound eyes is adaptive and functions as a way to automatically adjust flight speed to avoid collisions with objects in the environment. The insect flies slower in dense foliage and faster in open spaces. When *Drosophila* has to make turn, the flies can make a 120° turn in 80 ms (18 wingbeats) by inducing differences between the beating of the left and right wings (Bergou et al., 2010). Bumblebees (Baird et al., 2010; Dyhr and Higgins, 2010) and honeybees (Portelli et al., 2011) trained to fly through an experimental tunnel regulate ground speed according to whether the tunnel is wide or narrow, with visual texture on the walls to provide strong or weak optic flow information, and whether distances between the walls change abruptly. Return of foraging honeybees to their hive or nest, or return to a rich forage field, is well known, and Dittmar et al. (2010, 2011) and Braun et al. (2012) suggest that they are able to match a current visual scene with a stored pattern of scenery surrounding the forage site or nest site.

12.12 REVIEW AND SELF-STUDY QUESTIONS

1. What is the name of the heavily sclerotized "finger-like" part of the thorax that acts like a fulcrum for the wings and over which they pivot either up or down?
2. Describe the motion of the frontal part of the forewing in both the upstroke and the downstroke of the wings in most insects.
3. What are the names for the two large groups of indirect flight muscles in the thorax of most insects?
4. Describe the relatively small muscles and their functions that are attached to the wing sclerites in most insects.
5. Describe lift production in relations to the wing movements.
6. Describe the muscles and the actions that produce the downstroke and upstroke of the wings.

7. Explain briefly how some insects can beat the wings three or more times for each volley of nerve impulses that come to the muscle.
8. Describe the active state in wing muscles.
9. Why is it beneficial to the rapid wingbeat frequency of many insects that the thoracic indirect wing muscles have to contract only 2%–3% of their length in producing the wing movements?
10. Do dragonflies have asynchronous or synchronous wing muscles? Are the muscles attached directly to the wing sclerites, or are they indirect wing muscles?
11. What is meant by "unsteady lift" as this is applied to the flight of insects?
12. Do smaller insects, which also have smaller wings, have to beat their wings faster than large insects, which have larger wings, in order to remain in flight?
13. Have calculations of metabolic energy (i.e., energy supplied by ATP) converted into mechanical energy of flight shown high, low, or intermediate conversion of energy? Another way of looking at this process is this: Is flight an energetically very costly, or not so costly, way to move about?
14. When, in approximate number of millions of years ago, did insects evolve wings?
15. Do flight muscles of fast-flying insects develop an oxygen debt? Explain and justify your answer.

REFERENCES

Alexander, R.M. 1996. Smokescreen lifted on insect flight. *Nature* 384: 609–610.
Baird, E., T. Kornfeldt, and M. Dacke. 2010. Minimum viewing angle for visually guided ground speed control in bumblebees. *J. Exp. Biol.* 213: 1625–1632.
Bergou, A.J., L. Ristroph, J. Guckenheimer, I. Cohen, and A.J. Wang. 2010. Fruit flies modulate passive wing pitching to generate in-flight turns. *Phys. Rev. Lett.* 104: 1–4. Doi: 10.1103/PhysRevLett.104.148101.
Birch, J.M. and M.H. Dickinson. 2003. The influence of wing-wake interactions on the production of aerodynamic forces in flapping flight. *J. Exp. Biol.* 206: 2257–2272.
Borst, A., J. Haag, and D.F. Reiff. 2010. Fly motion vision. *Annu. Rev. Neurosci.* 33: 49–70.
Braun, E., L. Dittmar, N. Boeddeker, and M. Egelhaaf. 2012. Prototypical components of honeybee homing flight behavior depend on the visual appearance of objects surrounding the goal. *Front. Behav. Neurosci.* 6: 1–16. Doi: 10.3389/fnbeh.2012.00001.
Brodsky, A.K. 1994. *The Evolution of Insect Flight*. Oxford University Press, New York.
Card, G. and M. Dickinson. 2008. Performance trade-offs in the flight initiation of *Drosophila*. *J. Exp. Biol.* 211: 341–353.
Dickinson, M.H., F.-O. Lehmann, and S.P. Sane. 1999. Wing rotation and the aerodynamic basis of insect flight. *Science* 284: 1954–1960.
Dickinson, M.H. and J.R.B. Lighton. 1995. Muscle efficiency and elastic storage in the flight motor of *Drosophila*. *Science* 268: 87–90.
Dickinson, M.H. and M.S. Tu. 1997. The function of dipteran flight muscle. *Comp. Biochem. Physiol.* 116A: 223–238.
Dillon, M.E. and R. Dudley. 2004. Allometry of maximum vertical force production during hovering flight of neotropical orchid bees (Apidae: Euglossini). *J. Exp. Biol.* 207: 417–425.
Dittmar, L., M. Egelhaaf, W. Stürzl, and N. Boeddeker. 2011. The behavioural relevance of landmark texture for honeybee homing. *Front. Behav. Neurosci.* 5:20. Doi: 10.3389/fnbeh.2011.00020.
Dittmar, L., W. Stürzl, E. Baird, N. Boeddeker, and M. Egelhaaf. 2010. Goal seeking in honeybees: Matching of optic flow snapshots? *J. Exp. Biol.* 213: 2912–2923.
Dudley, R. 1999. Unsteady aerodynamics. *Science* 284: 1937–1939.
Dudley, R. 2000. *The Biomechanics of Insect Flight: Form, Function, Evolution*. Princeton University Press, Princeton, NJ.
Dyhr, J.P. and C.M. Higgins. 2010. The spatial frequency tuning of optic-flow-dependent behaviors in the bumblebee *Bombus impatiens*. *J. Exp. Biol.* 213: 1643–1650.
Ellington, C.P. 1984a. The aerodynamics of hovering flight. III. Kinematics. *Phil. Trans. R. Soc. Ser. B* 305: 41–78.
Ellington, C.P. 1984b. The aerodynamics of hovering flight. VI. Lift and power requirements. *Phil. Trans. R. Soc. Ser. B* 305: 145–181.

Ellington, C.P., C. van den Berg, A.P. Willmott, and A.L.R. Thomas. 1996. Leading-edge vortices in insect flight. *Nature* 384: 626–630.

Frazier, M.R., J.F. Harrison, S.D. Kirkton, and S.P. Roberts. 2008. Cold rearing improves cold-flight performance in *Drosophila via* changes in wing morphology. *J. Exp. Biol.* 211: 2116–2122.

Fry, S.N., R. Sayaman, and M.H. Dickinson. 2003. The aerodynamics of free-flight maneuvers in *Drosophila*. *Science* 300: 495–498.

Fry, S.N., R. Sayaman, and M.H. Dickinson. 2005. The aerodynamics of hovering flight in *Drosophila*. *J. Exp. Biol.* 208: 2303–2318.

Guerten, B.R.H., R. Kern, E. Braun, and M. Egelhaaf. 2010. A syntax of hoverfly flight prototypes. *J. Exp. Biol.* 213: 2461–2475.

Hedenström, A. 2014. How insect flight steering muscles work. *PLoS Biol.* 12(3): e1001822.

Hedrick, T.L. and T.L. Daniel. 2006. Flight control in the hawkmoth *Manduca sexta*: The inverse problem of hovering. *J. Exp. Biol.* 209: 3114–3130.

Heinrich, B. 1996. *The Thermal Warriors, Strategies of Insect Survival*. Harvard University Press, Cambridge, MA.

Jensen, M. 1956. Biology and physics of locust flight. III. The aerodynamics of locust flight. *Phil. Trans. R. Soc. Lond. B* 239: 511–552.

Josephson, R.K. 1984. Contraction dynamics of flight and stridulatory muscles of tettigoniid insects. *J. Exp. Biol.* 108: 77–96.

Josephson, R.K. 1985a. Mechanical power output from striated muscle during cyclic contraction. *J. Exp. Biol.* 114: 493–512.

Josephson, R.K. 1985b. The mechanical power output of a tettigoniid wing muscle during singing and flight. *J. Exp. Biol.* 117: 357–368.

Josephson, R.K. and R.D. Stevenson. 1991. The efficiency of a flight muscle from the locust *Schistocerca americana*. *J. Physiol. (Lond.)* 442: 413–429.

Kern, R., N. Boeddeker, L. Dittmar, and M. Egelhaaf. 2012. Blowfly flight characteristics are shaped by environmental features and controlled by optic flow information. *J. Exp. Biol.* 215: 2501–2514.

Kesel, A.B. 2000. Aerodynamic characteristics of dragonfly wing sections compared with technical aerofoils. *J. Exp. Biol.* 203: 3125–3135.

Kingsolver, J.G. and M.A.R. Koehl. 1994. Selective factors in the evolution of insect wings. *Annu. Rev. Entomol.* 39: 425–451.

Lehmann, F.-O. 2004. The mechanism of lift enhancement in insect flight. *Naturwissenschaften* 91: 101–122.

Lehmann, F.-O. 2008. When wings touch wakes: Understanding locomotor force control by wake-wing interface in insect wings. *J. Exp. Biol.* 211: 224–233.

Lehmann, F.-O. and M.H. Dickinson. 1998. The control of wing kinematics and flight forces in fruit flies (*Drosophila* spp.). *J. Exp. Biol.* 201: 385–401.

Lehmann, F.-O., S. Gorb, N. Nasir, and P. Schützner. 2011. Elastic deformation and energy loss of flapping wings. *J. Exp. Biol.* 214: 2949–2961.

Lehmann, F.-O. and S. Pick. 2007. The aerodynamic benefit of wing-wing interaction depends on stroke trajectory in flapping insect wings. *J. Exp. Biol.* 210: 1362–1377.

Lehmann, F.-O., S.P. Sane, and M. Dickinson. 2005. The aerodynamic effects of wing-wing interaction in flapping insect wings. *J. Exp. Biol.* 208: 3075–3092.

Lehmann, F.-O. and P. Schützner. 2010. The respiratory basis of locomotion in *Drosophila*. *J. Insect Physiol.* 56: 543–550.

Liu, H., C. Ellington, K. Kawachi, C. van den Berg, and A.P. Willmott. 1998. A computational fluid dynamic study of hawk moth hovering. *J. Exp. Biol.* 201: 461–477.

Marden, J.H. 1987. Maximum lift production during takeoff in flying animals. *J. Exp. Biol.* 130: 235–258.

Marden, J.H. 2005. Scaling of maximum net force output by motors used for locomotion. *J. Exp. Biol.* 208: 1653–1664.

Marden, J.H. and M.G. Kramer. 1994. Surface-skimming stoneflies: A possible intermediate stage in insect flight evolution. *Science* 266: 427–430.

Maybury, W.J. and F.-O. Lehmann. 2004. The fluid dynamics of flight control by kinematic phase lag variation between two robotic insect wings. *J. Exp. Biol.* 207: 4707–4726.

McMasters, J.H. 1989. The flight of the bumblebee and related myths of entomological engineering. *Am. Sci.* 77: 164–169.

Miller, L.A. and C.S. Peskin. 2005. A computational fluid dynamics of 'clap-and-fling' in the smallest insects. *J. Exp. Biol.* 208: 195–212.

Nachtigall, W. 1989. Mechanics and aerodynamics of flight, in G.J. Goldsworthy and C.H. Wheeler (eds.), *Insect Flight*. CRC Press, Boca Raton, FL, pp. 1–29.

Olberg, R.M., A.H. Worthington, and K.R. Venator. 2000. Prey pursuit and interception in dragonflies. *J. Comp. Physiol. A* 186: 155–162.

Portelli, G., F. Ruffier, F.L. Roubieu, and N. Franceschini. 2011. Honeybees' speed depends on dorsal as well as lateral, ventral and frontal optic flows. *PLoS One* 6(5): e19486.

Pringle, J.W.S. 1976. The muscles and sense organs involved in insect flight, in R.C. Rainey (ed.), *Insect Flight*. Royal Entomological Society and Blackwell Scientific Publications, Oxford, U.K., pp. 3–15.

Pringle, J.W.S. 1983. *Insect Flight*. Carolina Biological Supply Co., Burlington, NC.

Rammaurti, R. and W.C. Sandberg. 2007. A computational investigation of the three-dimensional unsteady aerodynamics of *Drosophila* hovering and maneuvering. *J. Exp. Biol.* 210: 881–896.

Reiser, M.B. and M.H. Dickinson. 2010. *Drosophila* fly straight by fixating objects in the face of expanding optic flow. *J. Exp. Biol.* 213: 1771–1781.

Robertson, R.M., K.G. Pearson, and H. Reichert. 1982. Flight interneurons in the locust and the origin of wings. *Science* 217: 177–179.

Roeder, K.D. 1951. Movements of the thorax and potential changes in the thoracic muscles of insects during flight. *Biol. Bull. Mar. Biol. Lab., Woods Hole* 100: 95–106.

Roy, S. and K. VijayRaghavan. 2011. Developmental biology: Taking flight. *Curr. Biol.* 22(2): R64. Doi: 10.1016/j.cub.2011.12.031.

Sane, S. 2003. The aerodynamics of insect flight. *J. Exp. Biol.* 206: 4191–4208.

Sane, S. and N. Jacobson. 2006. Induced airflow in flying insects. II. Measurements of induced flow. *J. Exp. Biol.* 109: 43–53.

Sane, S.P. and M.H. Dickinson. 2001. The control of flight force by a flapping wing: Lift and drag production. *J. Exp. Biol.* 204: 2607–2626.

Sane, S.P. and M.H. Dickinson. 2002. The aerodynamic effects of wing rotation and a revised quasi-steady model of flapping flight. *J. Exp. Biol.* 205: 1087–1096.

Schnorrer, F., C. Schönbauer, C.C.H. Langer, G. Dietzl, M. Novatchkova, K. Schernhuber, M. Fellner et al. 2010. Systematic genetic analysis of muscle morphogenesis and function in *Drosophila*. *Nature* 464: 287–291.

Schonbauer, C., J. Distler, N. Jahrling, M. Radolf, H.U. Dodt, M. Frasch, and F. Schnorrer. 2011. Splat mediates an evolutionarily conserved switch to fibrillar muscle fate in insects. *Nature* 479: 406–409.

Somps, C. and M. Luttges. 1985. Dragonfly flight: Novel uses of unsteady separated flows. *Science* 228: 1326–1329.

Sotavalta, O. 1953. Recordings of high wing-stroke and thoracic vibration frequency in some midges. *Biol. Bull.* 104: 439–444.

Sprayberry, J.D.H. and T.L. Daniel. 2007. Flower tracking in hawk moths: Behavior and energetics. *J. Exp. Biol.* 210: 37–45.

Srygley, R.B. 2004. The aerodynamic costs of warning signals in palatable mimetic butterflies and their distasteful models. *Proc. R. Soc. Lond. B* 271: 589–594.

Srygley, R.B. and A.L. Thomas. 2003. Unconventional lift-generating mechanisms in free-flying butterflies. *Nature* 420: 660–664.

Straw, A.D., S. Lee, and M.H. Dickinson. 2010. Visual control of altitude in flying *Drosophila*. *Curr. Biol.* 20: 1550–1556.

Sugiura, H. and M.H. Dickinson. 2009. The generation of forces and moments during visual-evoked steering maneuvers in flying *Drosophila*. *PLoS One* 4(2): e4883.

Sun, M. and S.L. Lan. 2004. A computational study of the aerodynamic forces and power requirement of dragonfly (*Aeshna juncea*) hovering. *J. Exp. Biol.* 207: 1887–1901.

Sun, M. and J. Tang. 2002. Unsteady aerodynamic force generation by a model fruit fly wing in flapping motion. *J. Exp. Biol.* 205: 55–70.

Sun, M. and J.K. Wang. 2007. Flight stabilization control of a hovering model insect. *J. Exp. Biol.* 210: 2714–2722.

Swank, D.M. 2012. Mechanical analysis of *Drosophila* indirect flight and jump muscles. *Methods* 56: 69–77.

Swank, D.M., V.K. Vishnudas, and D.W. Maughan. 2006. An exceptionally fast actomyosin reaction powers insect flight muscle. *Proc. Natl. Acad. Sci. USA* 103: 17543–17547.

Theobald, J.C., D.L. Ringach, and M.A. Frye. 2010. Dynamics of optomotor responses in *Drosophila* to perturbations in optic flow. *J. Exp. Biol.* 213: 1366–1375.

Thomas, A.L.R., G.K. Taylor, R.B. Srygley, R.L. Nudds, and R.J. Bomphrey. 2004. Dragonfly flight: Free-flight and tethered flow visualizations reveal a diverse array of unsteady lift-generating mechanisms, controlled primarily via angle of attack. *J. Exp. Biol.* 207: 4299–4323.

Tu, M.S. and T.L. Daniel. 2004. Submaximal power output from the dorsolongitudinal flight muscles of the hawk moth *Manduca sexta*. *J. Exp. Biol.* 207: 4651–4662.

Usherwood, J.R. and C.P. Ellington. 2002. The aerodynamics of revolving wings II. Propeller force coefficients from mayfly to quail. *J. Exp. Biol.* 205: 1565–1576.

Wakeling, J.M. and C.P. Ellington. 1997a. Dragonfly flight. I. Gliding flight and steady-state aerodynamic forces. *J. Exp. Biol.* 200: 543–556.

Wakeling, J.M. and C.P. Ellington. 1997b. Dragonfly flight. II. Velocities, accelerations and kinematics of flapping flight. *J. Exp. Biol.* 200: 557–582.

Wakeling, J.M. and C.P. Ellington. 1997c. Dragonfly flight. III. Lift and power requirements. *J. Exp. Biol.* 200: 583–600.

Walker, S.M., D.A. Schwyn, R. Mokso, M. Wicklein, T. Müller, M. Doube, M. Stampanoni, H.G. Krapp, and G.K. Taylor. 2014. In vivo time-resolved microtomography reveals the mechanics of the blow fly flight motor. *PLoS Biol.* 13(3): e1001823.

Wang, H., L. Zeng, H. Liu, and C. Yin. 2003. Measuring wing kinematics, flight trajectory and body attitude during forward flight and turning maneuvers in dragonflies. *J. Exp. Biol.* 206: 745–757.

Wang, J.K. and M. Sun. 2005. A computational study of the aerodynamics and forewing hindwing interactions of a model dragonfly in forward flight. *J. Exp. Biol.* 208: 3785–3804.

Wang, Z.J. 2004. The role of drag in insect hovering. *J. Exp. Biol.* 207: 4147–4155.

Wang, Z.J., J.M. Birch, and M.H. Dickinson. 2004. Unsteady forces and flows in low Reynolds number hovering flight: Two-dimensional computations vs robotic wing experiments. *J. Exp. Biol.* 2207: 449–460.

Webb, J.C., J.L. Sharp, D.L. Chambers, J.J. McDow, and J.C. Benner. 1976. The analysis and identification of sounds produced by the male Caribbean fruit fly, *Anastrepha suspensa* (Loew). *Ann. Entomol. Soc. Am.* 69: 415–420.

Weis-Fogh, T. 1964. Elasticity and wing movements in insects, in P. Freeman (ed.), *Proceedings XII International Congress of Entomology*, July 8–16, 1965. Royal Entomological Society, London, U.K., pp. 186–188.

Weis-Fogh, T. 1973. Quick estimates of flight fitness in hovering animals, including novel mechanisms for lift production. *J. Exp. Biol.* 59: 169–230.

Weis-Fogh, T. 1977. Dimensional analysis of hovering flight, in T.J. Pedley (ed.), *Scale Effects in Animal Locomotion*. Academic Press, London, U.K., pp. 405–420.

Wigglesworth, V.B. 1976. The evolution of flight, in R.C. Rainey (ed.), *Insect Flight*. Royal Entomological Society and Blackwell Scientific Publications, London, U.K., pp. 255–269.

Williams, C.M. and M.V. Williams. 1943. The flight muscles of *Drosophila repleta*. *J. Morphol.* 72: 589–599.

Willmott, A. and C. Ellington. 1997. The mechanics of flight in the hawk moth *Manduca sexta*. I. Kinematics of hovering and forward flight. *J. Exp. Biol.* 200: 2705–2722.

Wilson, D.M. 1968. The flight-control system of the locust. *Sci. Am.* 218: 83–90.

Wolf, H. 1990. On the function of a locust flight steering muscle and its inhibitory innervation. *J. Exp. Biol.* 150: 55–80.

Wootton, R.J. 1976. The fossil record and insect flight, in R.C. Rainey (ed.), *Insect Flight*. Royal Entomological Society and Blackwell Scientific Publications, London, U.K., pp. 235–254.

Wootton, R.J. 1992. Functional morphology of insects' wings. *Annu. Rev. Entomol.* 37: 113–140.

CHAPTER **13**

Sensory Systems

PREVIEW

Sensory structures transduce many different kinds of internal and external stimuli into electrical signals and feed these signals into the central nervous system (CNS). Sensory receptors are classified in several different ways based on morphology, but morphology is not always a sure indication of physiological function. Sensory receptors on insects are often small, and many receptors have been described from transmission electron microscopy studies without proven physiological functions. A single sensory neuron with its sheath cells is called a sensillum. Frequently, a sensory structure consists of many sensilla, that is, many neurons each enclosed in one or more sheath cells. Mechanoreceptors located at many sites on the body monitor body or appendage orientation in space and serve as wind speed indicators, tympanal organs, simple contact receptors, and environmental vibration receptors. Thermo-, hydro-, and infrared (IR) receptors also are mechanoreceptors. Mechanoreceptors do not have pores opening on the cuticular surface. Proprioceptors located internally are usually mechanoreceptors that monitor stretching, filling of the gut, and other internal movements. Chemoreceptors can be divided into olfactory and gustatory receptors. Olfactory receptors tend to have multiple pores at the cuticular surface, while gustatory receptors tend to have a single pore, usually at the tip of a hair. Olfactory receptors are often concentrated on the antennae, and gustatory receptors are located on the palps, on other mouthparts, and sometimes on the tarsi. Chemoreceptors (probably functioning as contact or gustatory receptors) often are located on the ovipositor of females and enable them to sample an oviposition site. Some olfactory receptors are relatively specialized, as, for example, receptors for the sex pheromone of the species, while other may be responsive to a number of chemicals. Gustatory receptors tend to have varying sensitivity to a number of chemicals, and the firing pattern (number and frequency of action potentials and rate of firing) that several gustatory receptors send into the CNS after exposure to a particular chemical compound has been called across-fiber patterning. To paraphrase the late Vincent Dethier, a noted sensory biologist, across-fiber patterning is the way in which a paucity of receptors can detect a surfeit of stimuli. Considerable data have accumulated in support of the stereochemical theory for the interaction of chemicals at the receptor site. In this theory, the chemical combines with a receptor at the dendritic membrane, and this leads to a receptor potential in the dendritic membrane.

13.1 INTRODUCTION

Insects have a surprisingly diverse array of sensory receptors that feed them information about their internal and external environment. The sensory neurons of insects have their cell bodies, with only rare exceptions, located very near the stimulus site, rather than in or near the central nervous system (CNS) as in vertebrates. Many receptors detect changes occurring at the cuticular surface, and the cell bodies are located peripherally just beneath the cuticle. Most sensory neurons are bipolar, with a few multipolar ones, and the dendritic terminals are usually very short compared to the relatively long axon leading to the CNS. The axons from many sensory neurons pass into the brain prior to synapsing and are classified as **primary** or **type I** sensory neurons. **Secondary** or **type II** sensory neurons synapse prior to entering the brain.

A common characteristic of all types of sensory neurons is that they transduce the stimulus energy, such as light, heat, chemical, or mechanical energy, into a slow, or graded, electrical potential. The receptor process can be divided into three steps: (1) absorption of the stimulus energy, (2) transduction into the receptor potential, and (3) repetitive impulse discharge from the axon portion of the receptor neuron. Repetitive discharge occurs only if the receptor potential is of sufficient magnitude to exceed the threshold for spike generation in the axon.

The input energy may have an excitatory effect upon the receptor neuron (depolarization), or it may have an inhibitory effect (hyperpolarization). Sensory neurons are sensitive to change in a stimulus; thus, a receptor cell will make an initial response (depolarization or hyperpolarization) when the stimulus starts (the "on" response), and then it makes the reverse response when the stimulus ceases (the "off" response). The upward deflection in Figure 13.1 indicates a depolarizing stimulus, and the receptor cell membrane has become less negative on the inside as a receptor potential has been produced.

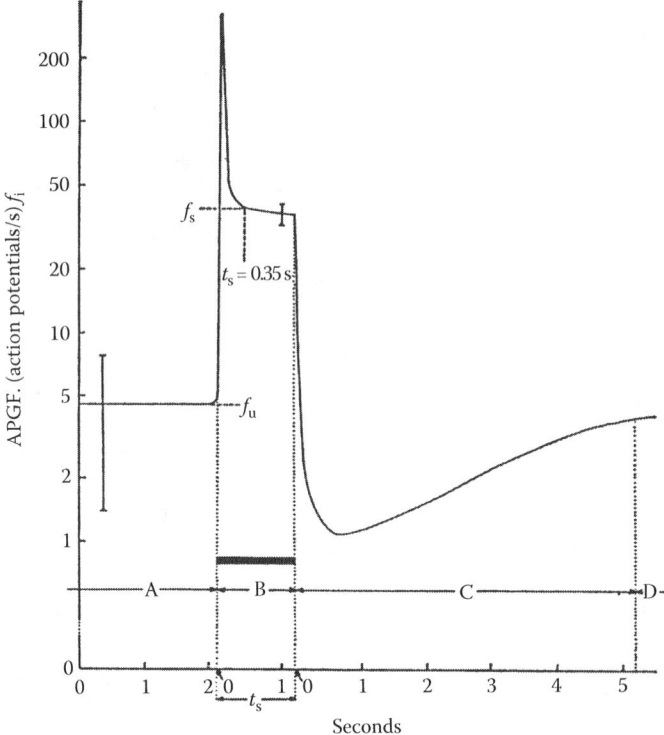

Figure 13.1 The response of a chemoreceptor on the labellum of the blowfly *Phormia regina* to a solution of 0.4 M KCl. The stimulus duration is indicated by the dark bar at B. Periods A and D show the unstimulated spontaneous activity (about 5 spikes/s) of the receptor. The receptor neuron initially makes a strong response to the stimulus by increasing spike output to a rate greater than 200 spikes/s, but adaptation occurs rapidly, and after about 0.35 s, the output rate falls to a tonic output of about 40 impulses/s. The poststimulation response or "off" response is shown in C. (From Rees, C.J.C., *J. Insect Physiol.*, 14, 1331, 1968. With permission.)

Adaptation, illustrated in Figure 13.1, to a steady stimulus is a characteristic feature of many receptor neurons. During adaptation, the receptor potential falls from its initial response level to some lower level or perhaps even to a silent state. Receptors that adapt rapidly to continuing steady stimuli are **phasic receptors**, while those that adapt slowly are **tonic receptors**. The foregut stretch receptor in *Phormia regina* is a tonic receptor that maintains a relatively sustained and uniform rate of firing when a constant stretch is applied (Gelperin, 1967). Two bipolar neurons connect the recurrent nerve with the foregut, and the neurons function as stretch receptors indicating peristalsis and fullness of the gut. Severing the branch of the recurrent nerve carrying the neurons results in failure of a fly to stop feeding and results in hyperphagia and a grossly expanded abdomen (Dethier and Gelperin, 1967).

With appropriate equipment and technique, the receptor potential can be measured, but it sometimes is easier and more convenient to measure the number of spikes produced from the receptor as an indicator of receptor action. A large receptor potential will produce spikes in rapid succession. As the receptor adapts, the frequency of spike generation falls. The chemoreceptor cell in Figure 13.1 shows another feature of many receptors. It is **spontaneously active**, firing about 5 spikes/s in the absence of any applied stimulus. Such spontaneously active receptors probably are never silent. They may play important roles in information coding by reducing spike frequency or increasing it, respectively, in response to inhibitory or stimulatory stimuli. Most sensory neurons in insects are organized into a complex morphological unit containing associated sheath cells, and the whole structure is called a **sensillum** (*pl.*, **sensilla**). Sensory organs, such as the compound eye, tympanum, or Johnston's organ, are composed of many units or sensilla.

13.2 EXTERNAL AND INTERNAL RECEPTORS MONITORING THE ENVIRONMENT

Receptors may be broadly classified as providing information about the external environment or internal environment. Receptors that monitor the external environment are usually given descriptive names as compound eyes, ocelli, tympanum, or Johnston's organ, but simple tactile hairs are also common. Receptors providing information about internal body conditions are called **proprioceptors**. Proprioceptors are present in the connective tissue of the body, among muscles, and along the surface of the alimentary canal of all insects. The sensory neurons are frequently multipolar. Some types of proprioceptors function as mechanoreceptors or stretch receptors to indicate gut filling, muscle tension, and general body movements, while others act as chemoreceptors relaying information about the chemical composition of the body. Proprioceptors generally adapt slowly to a constant stimulus, which clearly is adaptive if they must indicate body orientation, equilibrium, limb positioning, or fullness of the gut.

Many proprioceptors monitor stress and strain in the cuticle and provide information about body and limb movements. For example, head movement relative to the long axis of the body is indicated by movement against hair plates on the back of the head and on the prothorax of some insects. The campaniform sensilla in Table 13.1 are proprioceptors indicating cuticle stresses due to movement, but the only external cuticular evidence of these sensilla is a slightly raised dome of cuticle where they are attached.

13.3 GENERAL FUNCTIONAL CLASSIFICATION OF SENSORY RECEPTORS

Receptors can be classified functionally with respect to the type of energy they transduce. For example, they might be classified as light and/or visual detectors, mechanoreceptors (including tactile, vibration, and sound detectors), chemoreceptors (including contact receptors [gustatory or taste] and olfactory receptors), humidity receptors, temperature receptors, magnetic receptors, and geodetectors.

Table 13.1 Classification Scheme Often Used by Insect Biologists in Describing Sensory Structures

1. *Sensilla trichodea* are sensory hairs or setae ("Sinneshaare") and their elaborations, including hair plates. These sensilla are widely distributed among insects and other arthropods. Many of the hairs on insects are innervated by sensory neurons. The hair may not have any openings (tactile hairs, hair plates) and may be perforated by only one pore (gustatory contact chemoreceptors) or by many pores (olfactory receptors).
2. *Sensilla chaetica* are sensory spines or bristles ("Sinnesborsten") that are stouter than *S. trichodea* and often are located singly rather than in groups. *S. chaetica* consist of an innervated hair in a flexible socket. The hair may have a single pore (gustatory and some mechanoreceptors), or pores may be absent (most mechanoreceptors).
3. *Sensilla squamiformia* are flattened hairs or sensory scales ("Sinnesscchuppen"). They are common on the wings of Lepidoptera, although not every wing scale is innervated as a sensory organ.
4. *Sensilla basiconica* ("Sinneszapfen") are sensory pegs, cones, or stumpy hairs. They may be either thick walled or thin walled and lack pores (thermo- and hygroreceptors) or have a single pore (contact chemoreceptors and gustatory receptors) or have many pores (olfactory receptors). *S. basiconica* may contain only a few or many neurons.
5. *Sensilla coeloconica* ("Grubenkegel") are cones or pegs set in small depressions or pits in the cuticle. Both thick-walled and thin-walled sensilla occur. Some have multiple pores and serve as olfactory receptors, and some have no pores and appear to serve as thermo- and hygroreceptors.
6. *Sensilla styloconica* consist of elevated cones that may be located in pits or at the cuticle surface. *S. styloconica* typically serve a gustatory function and have a single pore at the tip.
7. *Sensilla ampullacea* are sensory tubes ("Sinnesflaschen"). They may be cone shaped and rest on long tubes in sunken pits. They occur on the antennae of bees and other Hymenoptera.
8. *Sensilla placodea* are multiporous plate structures ("Sinnesplatten") with an olfactory function. *S. placodea* have many neurons with dendritic terminals ending on thin plates covering a fluid filled canal. These sensilla occur on antennae of several insect orders.
9. *Sensilla campaniformia* include structures variously called campaniform sensilla, Hick's organs, cupola organs, and sensory pores ("Sinneskuppeln"). *S. campaniformia* are innervated by a single neuron that terminates in a small dome 20–30 µm in diameter. Sometimes, there is a scolopale at the center of the dome where nerve contact occurs. These sensilla may be recessed in cuticular depressions or elevated. Generally, they convey information about mechanical strain in the cuticle, for example, information about joint movements, movements of appendages, or other strain or pressure at the cuticular surface. Campaniform sensilla are often arranged in small groups along the long axis of a limb. It seems likely that they are important in maintaining the stance of an insect, in geotaxis, and in coordinating walking or running.

Anatomical classification of insect receptors has been a common practice (Table 13.1) (Horridge, 1965). Given the diversity of insects, probably all the sensory structures that have been, or will be, described from scanning electron microscope (SEM) studies will not fit the described categories. Unfortunately for physiology, it is much easier to obtain excellent SEM photos of structure than to obtain physiological data, and functional information about many sensilla on the surface of insects is sparse or nonexistent. It is important to remember that it is not possible to assign function with absolute certainty to all insect sensory structures based on morphology.

Altner and Prillinger (1980), Zacharuk (1980), and Frazer (1985) recommend a simplified classification scheme based primarily on the presence or absence of pores, and number of pores in external sensilla, with general functional significance when known. They recommend three major groupings as (1) receptors with multiple pores, (2) receptors with a single pore, and (3) receptors without pores.

13.3.1 Receptors with Multiple Pores

Receptors with multiple pores in the external cuticular structure tend to be **olfactory** receptors that detect airborne chemicals. The external cuticular structure may take the form of a hair (*Sensilla trichodea*), plate (*S. placodea*), peg (*S. basiconica*), or peg in a pit (*S. coeloconica*). Olfactory receptors are probably present on the antennae of all insects and occur elsewhere on the body of some insects. Functionally, the openings or pores through the cuticular covering allow airborne molecules to enter the sensillum. The external structure tends to be thin walled and the cuticular socket inflexible.

13.3.2 Receptors with a Single Pore

Receptors that have a single pore near or at the tip of the cuticular structure usually have a **gustatory** function (i.e., taste receptor) and detect chemical substances in solution. Gustatory receptors also are called contact chemoreceptors. The external appearance may be that of a hair (*Sensilla chaetica* or *S. trichodea*), peg (*S. basiconica*), and dome (*S. styloconica*). The external cuticular structure is thick walled, and sockets may be flexible or inflexible. Gustatory receptors are numerous on the mouthparts as well as on the tarsi and ovipositor of some insects.

13.3.3 Receptors without Pores

The lack of any pore in the external cuticular part of a sensillum is typical of mechanoreceptors that detect vibrations in the air (insect "ears" and vibration receptors), water, or substrate on which an insect rests. Mechanoreceptors may consist of a dome (*Sensilla campaniformia*) or hair (*S. chaetica*) set in a flexible socket. Humidity and temperature receptors, which may have the form of a peg (*S. basiconica*), or peg in a pit (*S. coeloconica*) in an inflexible socket, also usually lack a pore.

Frazer (1985) cautions that although the structural unit is the sensillum, the functional units are the neurons within the sensillum. Multiple neurons in the same sensillum may, and sometimes are known to, serve multiple physiological roles. For example, one neuron may function as a chemoreceptor while another neuron in the sensillum functions as a mechanoreceptor. Dethier (1955) and Hodgson (1956, 1958) described such combinations in the labellar hairs of blowflies, and subsequent studies have shown that multimodal sensilla occur on many appendages and other parts of the body of insects.

13.4 MECHANORECEPTORS

The anatomy and basic physiology of mechanoreceptors were reviewed by Dethier (1963) and Horridge (1965). Mechanoreceptors are involved in detection of airborne sounds, substrate vibrations, appendage movement and orientation (proprioceptors), flight speed, gravity, and, possibly, heat detection. Although all mechanoreceptors contain one or more bipolar sensory neurons and associated sheath cells, there are many anatomical variations. Some are located entirely internally, but many have external components.

13.4.1 Structure of a Simple Tactile Hair: A Mechanoreceptor Sensillum

The simplest mechanoreceptor, a sensory hair, contains a minimum of three cells, all derived from a common epidermal mother cell that divides to give rise to the **trichogen**, the **tormogen**, and the bipolar sensory neuron. The trichogen and tormogen are sheath cells. The trichogen is the inner sheath cell enclosing the soma and parts of the dendrites and axon of the sensory neuron, with the tormogen cell then enclosing both trichogen and neuron—a double sheath arrangement. Sometimes, other sheath or specialized glial cells are present as well, and the inner one in contact with the neuron in some receptors has been called a **thecogen** cell by some authors. As in other parts of the nervous system, the sheath cells insulate the neuron, may provide it with nutrients, and may help control concentrations of ions necessary for nerve function. The elements of a bipolar neuron and sheath cells in a tactile hair are the common elements of all insect sensilla, although some sensilla contain more complicated structures.

Single tactile hairs, and hairs grouped together in a hair plate, are common tactile mechanoreceptors on the body surface of insects. Tactile hairs are numerous on the antennae (especially on those insects that spend part or all of their lives in darkness, such as bees, ants, cockroaches, and cave dwellers) and on the cerci of Orthoptera and Dictyoptera. Cercal receptors detect a range of vibrations in the air and substrate and can act as sound receptors and vibration receptors. In cockroaches, and possibly in other insects, cercal receptors function in an escape mechanism in which the tactile hairs respond to sudden vibrations or loud sounds by sending spikes through the cercal nerve to connect with giant axons at synapses in the sixth abdominal ganglion. The giant axons pass uninterrupted to the thoracic ganglia where synaptic connections are made with motoneurons to the leg muscles. The system results in rapid transmission of stimuli that give rise to escape maneuvers. Many caterpillars also have single hairs that detect air vibrations and/or sounds, and caterpillars make behavioral responses to loud sounds and other airborne vibrations. Hairs are relatively insensitive to sound, however, in contrast to more complex tympanal organs that often are very sensitive.

Depending on the way in which the tactile hair is set in its socket, it may bend in only one direction and, thus, can indicate the direction of the bending energy, while others are omnidirectional. Tactile hairs usually occur in multiples on appendages, however, and directionality is often possible from the combination of stimuli and receptors responding.

13.4.2 Hair Plates

Hair plates are common at leg joints and at points of limb articulations with the body. They respond to touch, bending, and joint flexion with output of nerve spikes. They adapt slowly, a characteristic of static receptors that indicate body orientation. These tactile structures enable an insect to know the position of its limbs with respect to the body, and probably function in locomotion. Greater numbers of tactile hairs typically occur on the coxa and trochanter, probably enabling more precise movement of these large, heavily muscled parts of the leg. Hair plates also are common on sclerites at the back of the head and/or neck and on the anterior parts of the prothorax in mantids, locusts, and bees and act as proprioceptors enabling the insect to know its head orientation with respect to the body. They may also be important in some cases, at least, in flight ability because destroying the hairs on the cervical plates of locusts influences their equilibrium in flight. Sensory neurons, like other parts of the nervous system, have high demand for oxygen; the mechanoreceptors of honeybee cervical hair plates were shown to be very sensitive to oxygen deficiency, with spikes ceasing after 2 min and the receptor potential after 10 min.

13.4.3 Chordotonal Sensilla

A **chordotonal sensillum** is anatomically more complex than a tactile hair. It occurs at most exoskeletal joints, limb joints, and body segment joints. Field and Matheson (1998) presented a comprehensive review of chordotonal sensilla. There are many morphological variations, but basically the sensory neuron is enclosed (ensheathed) within parts of two or three other cells, including a characteristic **scolopale** cell and cap cell (Figure 13.2). There may also be additional sheath cells. Instead of the dendritic terminals being attached directly to the cuticle, they terminate within the cap cell, and the cap cell is attached to some internal structure or to the cuticle. Any stress or pull on the cap cell is then transmitted to the neuronal endings as a stimulus. Often, the cap cell extends well down over the scolopale sheath cell. Authors have not applied the same terminology, unfortunately, and the scolopale cell may be referred to as the sense rod or scolopale body. A single sensory unit with the scolopale and cap structure is called a **scolopidium** or, alternatively, a chordotonal sensillum. Complex chordotonal organs, such as a tympanum or Johnston's organ, contain many scolopidia, and both external and proprioceptors contain scolopidia as the morphological unit.

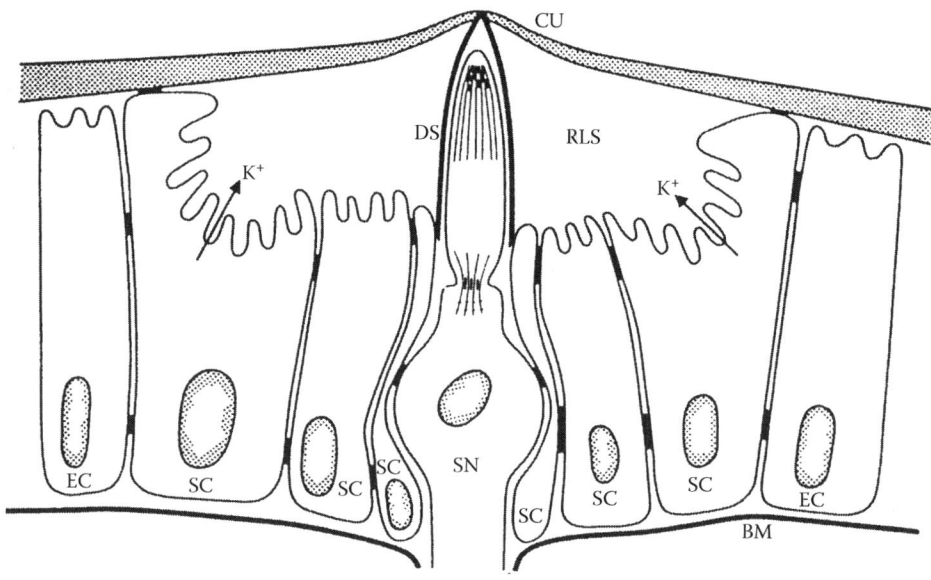

Figure 13.2 Diagram of a hypothetical chordotonal sensillum or mechanoreceptor, showing the scolopale and cap cell. Any stress or strain at the cuticular surface where the scolopale cap cell is attached is transmitted to the sensory neuron beneath. BM, basement membrane; CU, cuticle; DS, dendrite sheath or scolopale; EC, epidermal cells; RLS, receptor lymph space; SC, sheath cells; and SN, sensory neuron. In the diagram, potassium ions are shown being pumped into the receptor lymph space. The short dark bars between adjacent epidermal cells are cell junctions that prevent ion movement between cells and provide high electrical resistance. (From French, A.S., *Annu. Rev. Entomol.*, 33, 39, 1988. With permission.)

Examples of sensory organs containing scolopidia are the following:

1. Subgenual organs found just below the epidermis at the femorotibial joints of most adult insects. These allow an insect to know where its limbs are and whether they are flexed or extended in relation to the body.
2. Tympanal organs involved in detection of substrate vibrations and sound detection.
3. Johnston's organ, also a vibrational/sound detector located in the pedicel of the antennae of most adult insects and in some larvae.
4. Simple structures, each with only a few scolopidia that occur in various parts of the body of larvae and adults (Horridge, 1965).

Sometimes the simple structures of type 4 simply are called chordotonal organs, while the more complex structures, such as subgenual organs or Johnston's organ, have other names, but all are chordontonal organs composed of multiple scolopidia. Even a single sensory (tactile) hair may contain a scolopale, but many of these do not (see earlier).

13.4.4 Subgenual Organs

The **subgenual organ** is a complex chordotonal organ composed of multiple scolopidia. The term "subgenual" means below the knee, from Latin for knee (*genu*), and this complex chordotonal organ usually is located near the joint between the femur and tibia (Figure 13.3). The organ contains as few as three scolopidia in some earwigs (*Forficula* spp.), but contains more in most insects. It acts as a proprioceptor and detects vibrations of the substrate, and it has become specialized as a tympanal organ in some insects. The subgenual organ is especially well developed in crickets (Gryllidae)

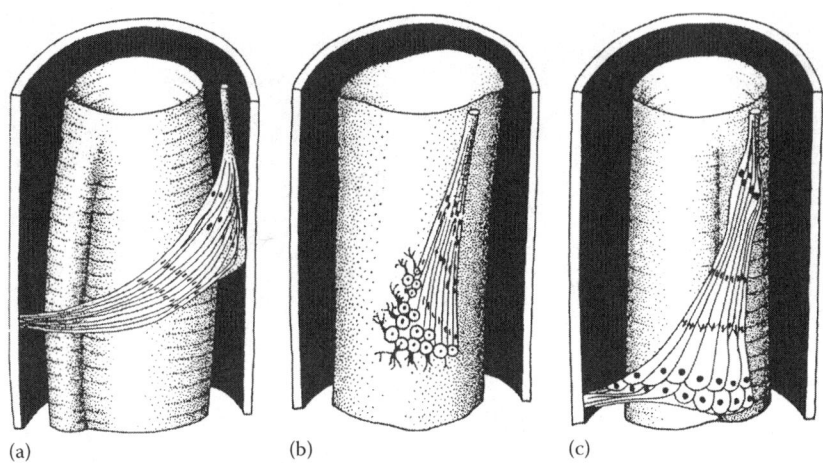

Figure 13.3 Structure of subgenual organs from (a) an orthopteran, (b) a lepidopteran, and (c) a hymenopteran. (From Autrum, H. and Schneider, W., *Z. Vgl. Physiol.*, 31, 77, 1948. With permission.)

and katydids (Tettigoniidae) and is associated with the tympanal organ, with both organs located on the tibia. The two organs have separate innervation, however, and probably have separate functions (Haskell, 1961). In some insects, the scolopidia vary in length, suggesting that different scolopidia might respond to vibrations of different amplitude according to length. The subgenual organ of the American cockroach, *Periplaneta americana*, is sensitive to vibrations that would displace the foot of the insect by as little as 10^{-9}–10^{-7} cm (Autrum and Schneider, 1948).

The subgenual organ is less well developed in Lepidoptera, Hymenoptera, and Hemiptera than in the Orthoptera. Some Hemiptera, Coleoptera, and Diptera do not have subgenual organs and display only low sensitivity to high-frequency substrate vibrations.

Probably all insects have additional chordotonal sensilla on the legs, particularly at or near the leg joints (Haskell, 1961), and some insects lacking a subgenual organ have a similar organ at the distal end of the tibia that may serve much the same function as the subgenual organ (Horridge, 1965).

13.4.5 Tympanal Organs: Specialized Organs for Airborne Sounds

Tympanal organs are chordotonal organs or "insect ears" that are specialized for high-frequency sound detection as opposed to low-frequency vibration detection. Tympanal organs, which have evolved a number of times independently in seven orders of insects, probably evolved from some early form of mechanoreceptor, probably a stretch-registering proprioceptor.

Tympanal organs are located at various places on the body of insects (Figure 13.4) (Yack and Fullard, 1993; Hoy and Robert, 1996). Examples of locations are near the sternum of the first abdominal segment of Acrididae (grasshoppers) and Cicadidae (cicadas), on the tibia of Tettigoniidae (longhorn grasshoppers) and Gryllidae (crickets), on the thorax of Notonectidae (aquatic hemipterans), and on the thorax or abdomen of some Lepidoptera.

Tympanal organs are specialized for airborne sound pressure waves and permit sound detection over a relatively long distance. They are sensitive to a wide range of frequencies from 2 kHz up to about 100 kHz (Hoy and Robert, 1996). Typically in insects, as well as in other animals, tympanal organs are paired. A single pressure receptor is not very efficient at detecting the directionality of the sound source, but two receptors, preferably well separated from each other, can detect directionality by differences in reception at the two locations. Tympanal ears typically have a minimum of three components: (1) a thin cuticular tympanum on the cuticular surface, (2) an

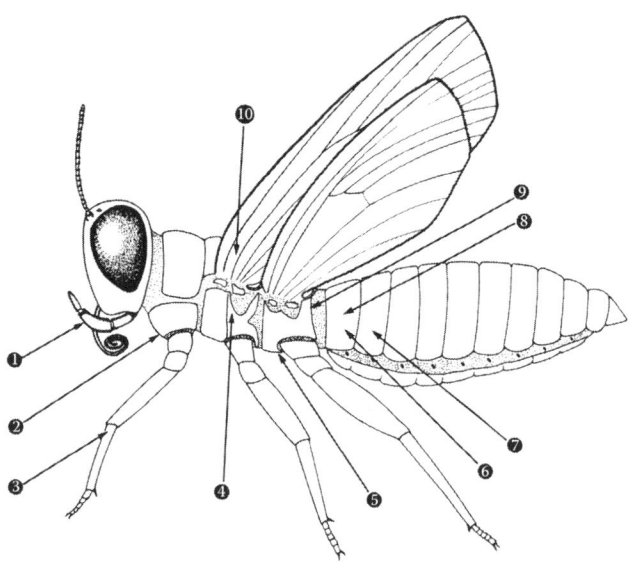

Figure 13.4 Diagram to illustrate the multiple places that tympanal organs (functional ears) have been located on insects. 1, Lepidoptera (Sphingidea); 2, Diptera (Tachinidae); 3, Orthoptera (Ensifera); 4, Hemiptera (Corixidae); 5, Mantodea (Mantidae); 6, Lepidoptera (Geometroidea and Pyraloidea); 7, Hemiptera (Cicadidae); 8, Orthoptera (Acrididae); 9, Lepidoptera (Noctuoidea); 10, Neuroptera (Chrysopidae). (From Yack, J.E. and Fullard, J.H., *Ann. Entomol. Soc. Am.*, 86, 677, 1993. With permission.)

air sac or other tracheal structure behind the tympanum, and (3) sensory neurons organized in scolopidia attached to the tympanal membrane or attached near it, so that they vibrate in response to the vibrations of the tympanum. Airborne sound waves cause the tympanum to vibrate, and sensory neurons enclosed in the scolopale cells detect the vibrations and respond, first, by graded electrical potentials, followed by a burst of spikes in the axon. Experimental examination of the vibration of the tympanum in several adult noctuid moths revealed that ultrasound stimulation caused the tympanum to vibrate with greatest deflection at the location of the receptor neurons, and other parts of the tympanal structure vibrated only weakly (Windmill et al., 2007). The tympanum is tunable in some moths (in this case, *Noctua pronuba*), and stimulation with certain ultrasound frequencies causes the tympanum to tune to higher-frequency levels, thus making it more adaptive for detection of a range of ultrasounds (Windmill et al., 2006). An air cavity or tracheal sac behind the tympanum plays an important role as a resonating chamber and in preventing damping of the sound. Some insects have a tympanum that can respond to sound waves striking it from the inside of the air chamber as well as from outside; such tympanal organs are pressure-difference receivers, and they are especially sensitive to directionality of the sound. Some tympanal organs have scolopidia of different length, suggesting sensitivity to various frequencies, but the function is unproven.

Insects in the orders Dictyoptera, Orthoptera, Neuroptera, Coleoptera, and Lepidoptera have tympanal organs sensitive to ultrasounds and are known to hear the echolocation pulses of insectivorous bats (Hoy, 1989; Fullard and Yack, 1993). One of the main forces promoting evolution of insect tympanal organs may have been foraging of insectivorous bats (Minet and Surlykke, 2003). Nocturnal insects must contend with echolocating insectivorous bats that can locate insects as small as (or smaller than) mosquitoes. Insectivorous bats emit three types of ultrasonic calls—first, general searching calls; next, approach calls as the bat locates a target; and, finally, attack calls when about to attack. Roeder (1965) reported that hunting bats produce pulses of ultrasound repetitively from 10 to 100 times/s, ranging in frequency from about 70 kilocycles/s at the beginning and tapering off to about 35 kilocycles near the end. The duration of cries varies from 1 to 10 ms. Beginning in the early 1960s, studies of moth responses to bat cries were conducted by Griffin et al. (1960), Roeder (1962, 1964, 1966a,b), and others. There are numerous reviews describing these studies for readers to consult including the following: Haskell (1961), Busnel (1963), Sales and Pye (1974), Spangler (1988), Bailey (1991), Fullard and Yack (1993) Hoy and Robert (1996) and Conner and Corcoran (2012). Ewing (1989) provides excellent explanations of the physics of sound and vibration production, transmission, reception, behavioral functions, and evolution of sound in insects.

The acoustical sense in insects also is important in some insects in mate location, courtship, and mate selection. In singing insects, multiple species may be singing at the same time, and there may be other acoustical signals in the environment, so conspecifics need to be able to detect the song of its species. Cicadas, for example, are singing insects that depend on particular acoustical frequencies for mate location. The auditory neurons in the cicada, *Tettigetta josei*, native to Portugal, are tuned to a wide range of low and high frequencies, but especially grouped around the 16 kHz peak of the calling song. Although environmental temperatures caused an upward shift of the acoustic frequency at certain frequencies, the tympanal response was temperature independent in the temperature range of 18°C–35°C, temperatures at which the insects typically call (Fonseca and Correia, 2007). The tympanum in the female cicadas, *Cicadatra atra*, is mechanically tuned to the song of the male, but the male tympanum is only partially tuned to its own song.

The ability to detect and produce acoustic signals, some of which are in the ultrasound range, provided some moths an evolutionary window to use acoustic communication in courtship and mating. The diversity of acoustic signaling in courtship and mating was tabulated by Conner (1999) in moths in the Arctiidae, Noctuidae, Pyralidae, and Sphingidae. As early as 1864, the French scientist Alexandre Laboulbène (cited by Conner, 1999) published a report describing sounds that he could hear coming from the arctiid *Chelonia pudica*, and he suggested that the sounds were important in mating. Subsequent studies on arctiid moths by many authors have confirmed that both pheromone and acoustic signals are important in successful mating (Conner, 1987; Sanderford et al., 1998; Conner, 1999, and references cited therein). The polka-dot moth *Syntomeida epilais* is an example of a highly developed combination of sex pheromone and ultrasound communication in mating (Sanderford and Conner, 1990, 1995; Conner, 1999). The moth (Figure 13.5) is a wasp mimic and distasteful because its larva feeds upon oleander plants, *Nerium oleander*, and sequesters cardenolides. Although a day-flying moth, bat predation may have contributed to evolution of its ultrasound capabilities because it mates in the predawn hours, when it might be exposed to predation by bats. During courtship, both sexes produce sounds in response to each other's presence (Sanderford and Conner, 1990). Males are led to within about 6 m of the female by a sex pheromone, whereupon males begin a series of sound clicks of 4–9 modulation cycles lasting from 1 to several seconds. As the male continues to move closer to the female, she responds with trains of clicks from 1 to more than 25 cycles. The male and female continue in a series of rapid exchanges of clicks as the male searches for the female, followed by very rapid clicks as copulation attempts are made. Both sexes are silent and stationary while mating continues into the morning (Sanderford and Conner, 1990).

Another function of sound production by some insects may be warning signals (aposematic signals) that help deter some predators. For example, Brown et al. (2007) present evidence for trains of clicks at 58 to about 79 dB at 10 cm (audible to a human) produced by silk moth larvae of *Antheraea polyphemus* when attacked by small animals or experimentally pinched with forceps. Sound production in *Antheraea* usually preceded defensive regurgitation, and the regurgitant proved to be somewhat of a deterrent to ants and mice in laboratory tests. Bats learn to avoid some arctiid moths that are distasteful and that respond to bat ultrasound pulses by emitting their own sound pulses (Barber and Conner, 2006, 2007; Fullard et al., 2007; Barber et al., 2009). Arctiid moths generally

Figure 13.5 (See color insert.) The polka-dot moth, *Syntomeida epilais*, is a wasp mimic and distasteful because it sequesters cardenolides from its larval host plant, *Nerium oleander*. (Photo courtesy of Gretert Montano, Miami-Dade College, Miami, FL.)

are distasteful because of pyrrolizidine alkaloids sequestered as larvae from their food plant. The arctiid moth *Cycnia tenera* preferentially responds to bat attack calls as distinct from bat searching calls and then produces ultrasonic clicks that may warn the bat of its distasteful quality as well as having a jamming effect on the bat's sonar receptors (Fullard et al., 2007). The moth uses its ultrasonic response after the bat switches to its attack echolocation calls, and also typically switches into erratic flight, and/or folds the wings and precipitously drops from flight. The authors believe the moth discriminates on the basis of a CNS template that evaluates the pulse period of the bat calls as they change to attack mode as well as by the acoustic power of the calls as the bat gets very close.

Movies simulating the deflections of the tympanum can be viewed as supplementary material in the paper by Seuer et al. (2006) at http://jeb.biologists.org/cgi/content/full/209/20/4115/DCi (Sueur et al., 2006) (last viewed April 22, 2015). In the desert locust, *Schistocerca gregaria*, the tympanum converts acoustic energy into mechanical energy and directs specific vibrational frequencies to different neurons; thus, the tympanum functions to detect sound and is involved in frequency analysis as well (Windmill et al., 2005).

Although most of the energy from an insect that is calling from a perch on a leaf or stem radiates as airborne sound, some is nearly always transmitted to the substrate as a low-frequency vibration (Bailey, 1991). The low-frequency vibrations are not transmitted very far, but insects have receptors capable of transducing both types of energy, and those close to the sound producer may receive both types.

In addition to tympanal organs, some insects also hear some sounds with other organs, including the Johnston's organ, subgenual organs, scattered simple chordotonal sensilla, and simple hair sensilla. Air pressure receptors occur in flies and mosquitoes in which the antennae oscillate and transmit impinging air pressure waves to Johnston's organ (Göpfert and Robert, 2002). Lapshin and Vorontsov (2013) showed that individual auditory receptors in the Johnston organ in some female mosquitoes exhibit frequency tuning, enabling them to use frequency analysis in predator avoidance and mate seeking.

13.4.6 Johnston's Organ

Johnston's organ is a large, complex chordotonal organ that may consist of several groups or a single grouping of scolopidia located between the second (the pedicel) and third joints of each antenna of most adult insects, although some Apterygota (Collembola and Diplura) do not have a Johnston's organ. A simplified form of the organ occurs in some larvae. Johnston's organ responds to several kinds of stimuli in different insects, including acting as a proprioceptor to indicate movement of the antennae, monitoring wingbeat frequency in relation to flight speed in some Diptera, indicating gravity, detecting ripples at the water surface in gyrinid beetles, and receiving sound in mosquitoes and, perhaps, other insects.

With its location in the second antennal segment, Johnston's organ is positioned to monitor movements of the antennal flagellum, whether due to muscles controlled by the insect or displacements of the antennae by wind and flight. There are variable numbers of scolopidia radially arranged and attached to the wall of the pedicel at one end and to the intersegmental membrane between the pedicel and flagellum. Johnston's organ seems to have reached its apex of development in dipterans in the families Chironomidae and Culicidae, in which the pedicel is much enlarged and the organ completely fills it. In these small swarming dipterans, the large organ is directionally sensitive and functions in successful swarming and mating. Frequency of sound is detected by the arista, which vibrates in resonance to the sound of the wings of the female in flight. In addition, the males have numerous long hairs on the antennae, and these vibrate in response to the flight sounds produced by the wings of flying females. Their vibration causes the flagellum (the major portion of the length of the antenna) to vibrate. Males of the mosquito, *Aedes aegypti*, are most sensitive to frequencies from 400 to 650 Hz, corresponding closely to the natural wingbeat frequency of females (Roth, 1948). Johnston's organ functions as a flight speed indicator in adult *Calliphora erythrocephala* (Burkhardt, 1960) and probably also in some other insects, such as the housefly, honeybee, and related insects. It is probably an important gravity indicator for most insects, enabling them to have a sense of their body in relation to horizontal and vertical planes because the weight of the antenna excites scolopidia depending on the pull of gravity relative to the body. Gyrinid water beetles swim at the water surface and avoid colliding with other swimming beetles. They do not crash into other beetles or the sides of a small container because Johnston's organ enables them to detect disturbances and ripples in the water created by other beetles or their own ripples bouncing off the container walls.

13.4.7 Simple Chordotonal Organs

Simple chordotonal structures that have no specific name occur widely over the body of most orders of insects, including adults and larvae. The structures usually consist of only a few scolopidia. There are about 90 such small chordotonal organs arranged along the length of *Drosophila* larvae (Horridge, 1965). Relatively simple chordotonal organs occur in the legs (on the femur and, sometimes, on the tibia), on the wings at the base of the radial and subcostal veins, and within the lumen of the radial vein of many, but not all, insects. Simple chordotonal sensilla occur on the legs in addition to the subgenual organ, and on the antennae, in addition to Johnston's organ, usually at or near the antennal joints, of most insects (Haskell, 1961). Some of the simple chordotonal sensilla respond to certain airborne frequencies (i.e., they are sound receptors), but they are not very sensitive and have a narrow response range. Similar simple chordotonal organs may have been the precursors of tympanal ears in the early evolution of insects (Bailey, 1991).

13.4.8 Thermoreceptors and Hygroreceptors

The literature on insect **thermoreceptors** and **hygroreceptors** has been reviewed by Loftus (1978), Altner and Prillinger (1980), and Altner and Loftus (1985). Experimentally, insects can be shown to respond to warm, moist, and cold air, but conclusive identification of the receptors by which they monitor these environmental changes and even whether they routinely use such information in their behavioral activities is tentative or sparse. Receptors tentatively believed, or in a few cases proven, to function as thermo- and hygroreceptors often occur in the same sensillum on the antennae, most frequently as a triad of three neurons, although these types of receptors are not numerous. It has been estimated that the American cockroach, *P. americana,* has about 1300 sensilla on the antenna that house a thermoreceptor neuron, but this represents only about 0.4% of the receptors on the antennae of a male cockroach (Altner et al., 1983). The most common triad arrangement is that one neuron is sensitive to cold air, one to moist air, and one to dry air. The cold receptor responds to a sharply falling temperature by a rapid rise in firing rate, the moist air receptor fires more frequently when the humidity rises, and the dry air receptor fires in response to falling humidity. Triads have been found on the antennae of the American cockroach, *P. americana*; the migratory locust, *Locusta migratoria*; the European walking stick, *Carausius morosus*; *Triatoma infestans* (Hemiptera); the honeybee, *Apis mellifera*; and a noctuid moth, *Mamestra brassicae.* A warm receptor that fires in response to rising temperature has been found in the same sensillum with a cold receptor on the antennae of the mosquito, *A. aegypti.*

Typically, the cuticular portion of these sensilla has no pore and is set in an inflexible socket; *S. trichodea, S. basiconica, S. coeloconica,* and *S. styloconica,* as well as other morphological structures, are known to house receptors believed to be thermo- and hygroreceptors. A few examples of either a thermoreceptor or a hygroreceptor associated with an olfactory receptor in a sensillum with multiple pores have been described. Thus, as in other types of receptors, it is not possible to identify function with certainty based upon morphology. In the most common triad arrangement (Figure 13.6, a peg in a pit), typically, the dendritic portion of two of the neurons fill the lumen of the peg (or other cuticular arrangement), while the dendrite of the third neuron has short multiple branches, often forming lamellae and ending beneath the peg. The former cells are called type 1 cells and the latter is a type 2 cell. In some cases, a type 3 cell has been found in which the outer dendritic segment is slender, like the cilium portion, and ends much before the outer cuticular structure. A few sensilla have been found with four or five sensory neurons. Although experimental evidence is not conclusive, the arrangement and some data suggest that type 1 cells are actually mechanoreceptors that respond to cuticular distortion due to changes of water content in the cuticular portion of the sensillum. If they are mechanosensitive and can accurately sense cuticular distortions due to water content in the air, they must be well protected from mechanical disturbance, which would create noise in the system. The inflexible socket, short cuticular projection, pit or collar arrangement that is common, and location beneath more massive mechanoreceptors seem likely to protect them from ordinary mechanical disturbance. It should be emphasized, however, that a mechanosensitive functionality is not yet firmly established, and other ways to detect water in the air are possible, for example, by humidity-induced changes in electrolyte concentrations (Loftus, 1976, 1978).

The type 2 cell may be a thermoreceptor. The dendritic portion of the type 2 cell is more variable among different species than that in type 1 cells. Electrophysiological investigations on a cave-dwelling beetle, *Speophyes lucidulus,* showed that a cold and a warm receptor existed in the same sensillum (with a third neuron, possibly also a thermoreceptor). The possibility that the type 2 cell is the thermoreceptor is still quite tentative, but speculation is that the number of lamellae in the distal portion of the dendrite may be correlated with a range of temperatures that can be detected (Corbière-Tichané and Loftus, 1983).

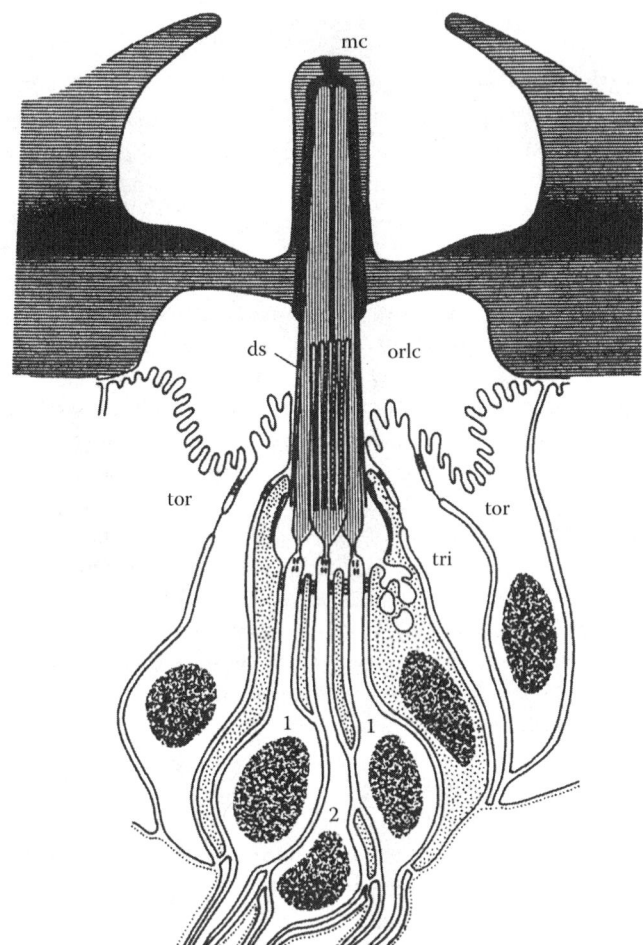

Figure 13.6 The triad arrangement of three sensory neurons and trichogen and tormogen cells in a peg-in-a-pit arrangement of a hygroreceptor. The peg has no pore. The dendritic (1) outer segments of the two type 1 sensory cells enter the lumen of the peg, but (2) the outer segment of the type 2 cell branches into lamellae and ends below the peg. There are three sheath cells, the thecogen (stippled), the trichogen (tri), and the tormogen (tor); ds, dendritic sheath; mc, molting channel of peg; and orlc, outer receptor lymph cavity. (From Altner, H. and Loftus, R., *Annu. Rev. Entomol.*, 30, 273, 1985. With permission.)

The arrangement of pairs of receptors, such as a warm and cold receptor or a moist air receptor with a dry air receptor in the same sensillum, may improve the ability of the system to discriminate changes in environmental conditions. Each will respond in firing rate to a change in temperature or humidity, but the change will be in opposite directions. For example, the warm receptor will respond to rising temperature by increasing its rate of firing, while the cold receptor will respond by decreasing its rate of firing. The reverse will occur during cooling. The integration of such information in the CNS, if indeed used by the CNS, awaits further exploratory research.

13.4.9 Infrared Reception

Many species (about 40; Hart, 1998) of insects have specialized IR **receptors** and are attracted to forest fires, where they lay their eggs in fire-damaged trees and burned-over debris. Most of the insects are beetles. *Melanophila acuminata*, a beetle in the family Buprestidae, has been most intensively studied and is known to be attracted in large numbers to burning forests (Evans, 1962, 1966a; Apel, 1989), where they mate and begin laying eggs in the burned wood. *M. acuminata* has paired pits adjacent to the mesothoracic coxa on the pleuro-ventral thorax (Figure 13.7). The pits are slightly variable in size, 170–320 μm long × 80–150 μm wide and 70–100 μm deep. Within each pit are 50–100 slightly oblong hollow domes and associated multipore wax glands (Evans, 1966b; Vondran et al., 1995; Schmitz et al., 1997). Each dome is innervated by one bipolar neuron and associated sheath cells (Figure 13.8). The cell body lies just beneath the cuticle at the base of the pit, and its axon passes without synapsing into the metathoracic ganglion. The dendrite of the sensory neuron contains neurotubules and has a **ciliary constriction** near its midpoint. The distal tip of the dendrite is attached at the base of the spherical dome. The dome surface is thin and unsclerotized, possibly allowing the hollow sphere beneath the domed cuticle to change its volume due to absorption of IR radiation and, thus, mechanically stretch the dendritic tip of the neuron.

If this is the correct interpretation, then the IR receptor functions like a modified mechanoreceptor. The beetles respond behaviorally to IR radiation at 3 μm wavelength (Evans, 1966a), and IR wavelengths ranging from 2.5 to 4 μm are emitted by intense forest fires. Atmospheric CO_2 and H_2O have narrow bands of strong IR absorption within the same wavelengths, but a window exists at 3.6–4.1 μm in which atmospheric components do not absorb strongly. Another insect demonstrated to use heat, and possibly IR, is *Rhodnius prolixus* (Hemiptera: Reduviidae), a blood-feeding insect that is attracted to warm-blooded animals. They are known to orient and approach thermal sources, and recent work with a pure IR source and a cooled IR transmitting window between the IR source and the bug strongly indicates that they can detect IR radiation (Schmitz et al., 2000). It still is not clear whether they might use a mechanoreceptor that is warmed by absorption of IR, as apparently in *M. acuminata*, or might possibly have nonthermal IR receptors.

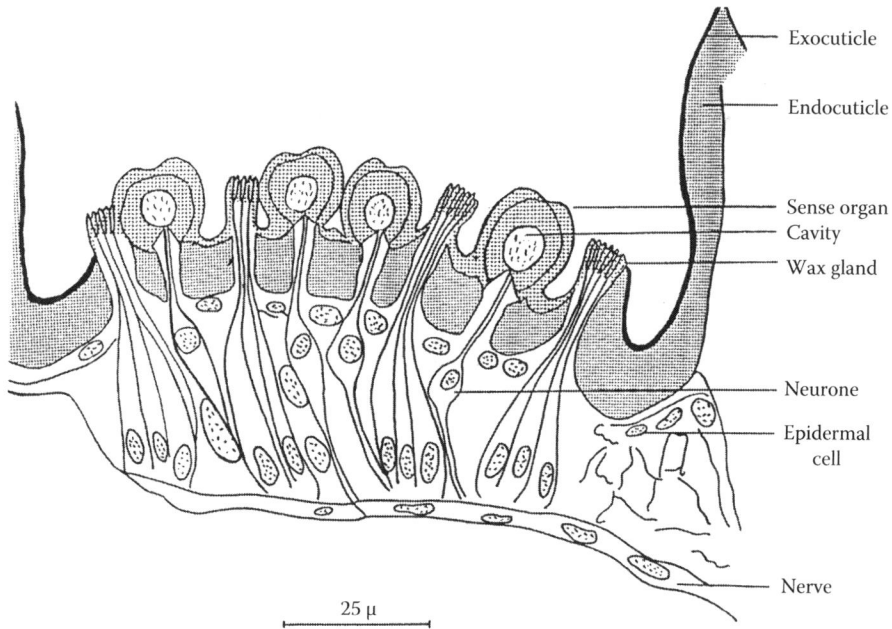

Figure 13.7 Diagram of an infrared (IR) organ with several receptors shown from the beetle *Melanophila acuminata*. The beetle is attracted to the IR radiation from forest fires and thereby locates suitably damaged tree hosts in which to lay its eggs. The receptors probably function as mechanoreceptors responding to distortions induced by the IR radiation in the bulbous cavity of each receptor. (From Evans, W.G., *Ann. Entomol. Soc. Am.*, 59, 873, 1966b. With permission.)

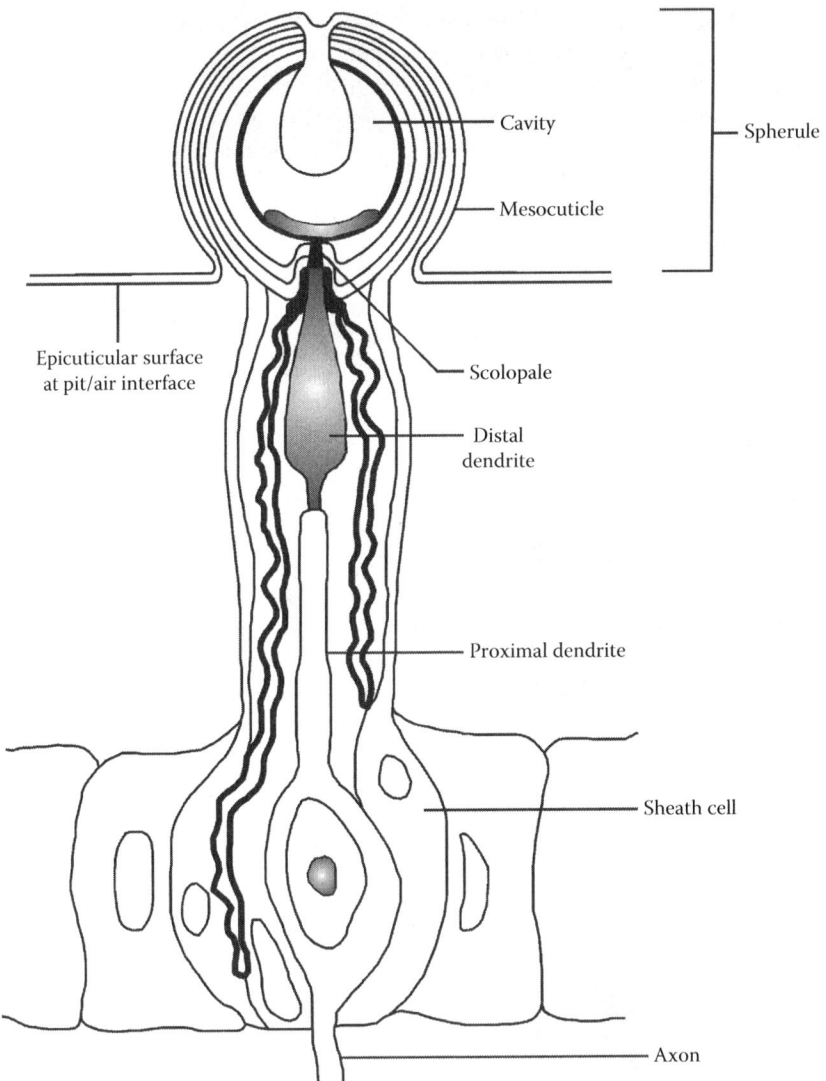

Figure 13.8 A schematic diagram of one of the IR receptors from *Melanophila acuminata*. The spherule is covered by a thin cuticle. The sensory neuron, which has a ciliary constriction, terminates in a scolopale that is attached to the spherule. (Modified from Vondran, T. et al., *Tissue Cell*, 27, 645, 1995.)

13.5 CHEMORECEPTORS

13.5.1 Olfactory Sensilla: Dendritic Fine Structure

Olfaction is a very important to most insects. Males (and sometimes females) depend on detecting and orienting to sex pheromones released by the opposite sex. Both adult sexes search for, and recognize, food by following odor signals. Adult females use olfaction as well as other cues to locate suitable oviposition sites. Larvae of many insects feed on the host where the mother has laid eggs, but more mobile immatures, such as grasshoppers and cockroaches, move frequently in search of food, and olfaction and taste are important to them. Galizia and Rössler (2010) reviewed olfactory systems in insects, and Hansson and Stensmyr (2011) present views on the evolution of olfaction in insects. Although much more is known about the olfactory system in *Drosophila* than in any other insect, an antennal transcriptome is now available for *Manduca sexta* (Grosse-Wilde et al., 2011), and more than 60 olfactory genes and 70 glomeruli in the antennal lobe have been identified (Grosse-Wilde et al., 2010, 2011; Howlett et al., 2012).

Like other sensory neurons, **chemoreceptor neurons** are bipolar, with the cell bodies located peripherally near the stimulus site. Characteristically, the dendrite of an olfactory neuron comprises a relatively large inner segment connected by a narrow ciliary segment to a smaller outer process extending to the tip of the sensillum. In some sensilla, notably *S. basiconica*, there are many branches of the fine terminal process. Occasionally, the ciliary segment is missing. The large inner segment contains many neurotubules and mitochondria, suggesting a high rate of metabolic activity and utilization of oxygen. There may be a need for high metabolic pump activity in the very fine dendritic endings if there is only a small reserve of the ions necessary for nerve function. The function of the ciliary region, when it is present, is not clear. It contains nine pairs of neurotubules attached to the dendritic membrane. Although the two central neurotubules typical of ciliary structures are sometimes present, they are more often absent. Typically, there is a basal body present. Several theories regarding function have been advanced, including a suggestion that the ciliary organelles might function as organizers for regeneration of dendritic endings following molting. The neurotubules extend into the outer dendritic segment. The cuticular walls of olfactory sensilla contain pores varying from 10 to 100 nm in diameter. Microtubules lead from the inner wall of the pore inward and frequently seem to make direct contact with the dendritic endings. The walls of these microtubules are about 3 nm thick. Chemical molecules enter through the pores, are captured by odorant binding proteins, and are transported across the sensillum liquor (an aqueous medium) to the dendritic endings. Olfactory receptors typically give a phasic–tonic response. The axons of olfactory sensory neurons usually are small, measuring from 0.1 to 0.2 nm in diameter, and they usually pass directly into the deutocerebrum without synapsing, though there are examples of axons from antennal receptors that coalesce or merge by synapsing near the base of the antenna.

The role of G-proteins in insect odor receptors is somewhat uncertain. Wicher et al. (2008) found some odorant receptors in *Drosophila* to be ligand gated and G-protein coupled, while others function as cation channels and may be cyclic nucleotide dependent or ligand gated (Sato et al., 2008; Wicher et al., 2008). Yao and Carlson (2010), however, did not find any evidence for G-protein-mediated olfactory function. Although dragonflies and damselflies have large eyes and probably depend more upon vision than olfaction (Olberg, 2012), they nevertheless do respond behaviorally to prey odor (Piersanti et al., 2014). They have sensilla on the antennae that anatomically look like olfactory sensilla (Piersanti et al., 2010), and electrophysiological experiments with odorants confirm response of olfactory neurons (Rebora et al., 2012).

13.5.2 Contact Chemoreceptors–Gustatory Receptors

Gustatory receptors are taste receptors that respond to stimulus molecules in solution. Gustatory receptors function in numerous behaviors, including selection of food, avoidance of toxic and bad-tasting substances, courtship, mating, and oviposition. Olfactory receptors play a role in most of these behaviors as well. The ability to taste the food eaten is important in selection of what to eat and avoiding the feeding deterrents and toxic substances that many plants evolved to combat phytophagous animals. Chapman (2003) presented a general review of gustatory reception in insects, especially in relationship to feeding.

The availability of the complete genome of *Drosophila* has made it one of the most useful insects in the study of gustatory receptors. Montell (2009) presented a short review of gustatory receptors in *Drosophila*; the adult flies have 68 gustatory receptors spread over the proboscis, margins of the wings, legs, and ovipositor. The receptors are proteins with seven transmembrane domains (Slone et al., 2007; Zhang et al., 2011; Devambez et al., 2013). The number of genes encoding gustatory receptors varies widely across insects that have been studied: 68 in *Drosophila*, 13 in honeybees, 76 in some mosquitoes, and 65 in *Bombyx mori*, the silkworm moth (Sato et al., 2011). At present, the ligand activating most of the receptors has not been identified (Zhang et al., 2011). Most insects that have been studied have one or more receptors for sucrose; sucrose is almost always a feeding stimulant for insects (Slone et al., 2007; Sato et al., 2011). Receptors for trehalose (Kain et al., 2010), inositol, and some other sugars are known from a few insects. Some adult butterflies (e.g., Nymphalidae) have taste receptors on their walking legs and tarsi, enabling them to taste the surface of plants on which they rest or walk (Ômura et al., 2011). Briscoe et al. (2013) found a strong female bias in number and distribution of gustatory receptors in *Heliconius melpomene* and suggested that female oviposition and mating behavior is deriving the evolution of gustatory receptors in the *Heliconius* spp. of butterflies.

Phytophagous insects encounter many toxic substances in plants; many of these compounds taste bitter to humans, and possibly to insects (Weiss et al., 2011), and some of the compounds are very toxic. Two such feeding deterrent that have been investigated are caffeine and L-canavanine. Lee et al. (2009) found that two gustatory receptors were required for *Drosophila* to detect caffeine, and there is some evidence that two or more receptors may be required for detection of most substances, including sugars (Lee et al., 2009). L-Canavanine, present in some plants, is structurally similar to the amino acid, L-arginine, and if insects eat L-canavanine, it is incorporated into proteins that do not function properly; thus, it is toxic (Rosenthal, 2001). At least two receptor proteins are required for *Drosophila* to detect and avoid food containing L-canavanine (Lee et al., 2012; Devambez et al., 2013). Jorgensen et al. (2007) showed experimentally that adult *Heliothis virescens* could learn to associate the bitter compounds sinigrin and quinine (bitter tasting to humans and presumably to insects) in sugar solutions with a favored odor stimulus, resulting in a reduced proboscis extension reflex to the odor stimulus.

Polyphagous caterpillars that start feeding on one host plant where the mother laid eggs are reluctant, or may refuse, to feed upon a different, but equally suitable species of host plant. Apparently, their taste receptors have become conditioned, and/or mechanisms in the CNS have become imprinted upon the particular chemical taste of the first host plant. Abisgold and Simpson (1988) showed that the sensitivity of the taste receptors on the maxillary palps of *L. migratoria* exhibited reduced sensitivity (fewer spikes per second) after the locusts were fed a high-protein diet as compared with those fed a low-protein diet. The high-protein diet resulted in locusts spending longer quiescent periods after feeding and more often rejection of food.

Taste receptors on the tarsi and labellum of blowflies are housed in *S. trichodea*. There are no pore tubules in the cuticle covering the sensillum, the dendrites leading into the sensillum do not branch, and a ciliary region has not been observed. Usually, there is a single pore from about 0.25–0.5 nm in diameter near the tip of the hair; some have two pores. The labellar hairs of blowflies contain several neurons in each sensillum. The different neurons are sensitive to sugars, water, and anions (such as NaCl), and, in addition, one neuron usually acts as a mechanoreceptor. The unbranched terminals of the dendrites are bathed in a viscous fluid through which stimulating molecules must diffuse. It is convenient to study the response of the labellum sensilla by positioning a capillary electrode containing the stimulating substance (e.g., glucose) over the tip of the hair. This electrode also acts as the recording electrode, while the reference electrode is inserted into the body or head of the insect. Spikes generated in the axon portion of the receptor neurons near the base of the sensillum are conducted to the tip of the sensillum (by passive transmission) and can be recorded via the capillary electrode on an oscilloscope or other device. The number of spikes per sec is used as the indicator of receptor activity.

13.5.3 Specialists versus Generalists among Chemoreceptors

Among both taste receptors and olfactory receptors, there are some receptors specialized for detection of very specific chemical substances and others capable of responding to a wide variety of chemicals. These are commonly called "specialists" and "generalists," respectively. For example, each antenna of a male *Telea polyphemus* moth bears more than 60,000 sensilla containing about 150,000 sensory neurons. About 60%–70% of these neurons are specialized for detection of the female-produced sex pheromone, about 20% respond to other odors, and the remainder serve a variety of sensory functions. Even the specialists, however, are usually not absolutely specific. A few other chemicals in high concentration may also stimulate them. For example, specialists for 9-oxo-*trans*-2-decenoic acid (the sex pheromone) on the antenna of drone honeybees will also respond to caproic acid if it is presented at 10,000 times greater concentration than the pheromone (10^8 molecules/cc air for the pheromone and 10^{12} molecules/cc air for caproic acid).

Dethier (1971) discussed the coding and relaying of stimulus information into the insect brain in relation to specialist and generalist receptors. He described a strict specialist with its axon leading to the brain of an insect as an absolute labeled line. These, he noted, rarely if ever exist in the strictest sense. The information transmitted would be unambiguous but limited to identity of chemical and intensity of stimulus (concentration of chemical). Many absolutely labeled lines and, consequently, large sensory nerves (bundles of axons) and central ganglia would be necessary to accommodate a wide diversity of chemicals. Partially labeled lines, in which each receptor cell is capable of responding to several chemicals, would reduce the number of lines needed into the brain. Still greater capacity for information transmission is displayed by **across-fiber patterning** (Dethier, 1971), in which a receptor responds to several or many chemicals (stimuli) with differing response magnitudes. Thus, the brain could receive information about a specific chemical from a number of receptors and interpret the profile of responses received. The advantage of across-fiber patterning is that it allows only a few receptors to convey information about a large number of stimuli because each stimulus will result in a different profile of responses sent to the brain. Dethier (1971) considers this to be the way in which taste and olfactory stimuli are sensed and integrated in *M. sexta* caterpillars, which have only 48 taste receptors on the body and about 78 olfactory receptors. Even greater information coding can be achieved in across-fiber patterning by receptors that have different response latencies, rates of adaptation, after effects, and spontaneous activity (a stimulus may increase or inhibit spontaneous activity).

13.5.4 Stimulus–Receptor Excitation Coupling

How is the energy of a chemical stimulus transformed into the electrical energy of the neuron? Many theories (more than 30, according to Amoore et al., 1964) have been advanced. The **stereochemical theory** has gained the most support, and the isolation of pheromone-binding proteins and general odorant-binding proteins from insects and other organisms has contributed greatly to solidifying the stereochemical theory.

The stereochemical theory is most often associated with the work of Amoore (for review, see Amoore et al., 1964), although many others have contributed data to support the theory. Amoore's thesis is that the sense of smell (in humans) is based on the geometry of molecules, and he associated the seven primary odors (camphoraceous, musky, floral, pepperminty, ethereal, pungent, and putrid) with molecules having a particular shape. For example, molecules may be round, oblong, kite shaped, or have a positive charge (pungent) or a negative charge (putrid). The receptor site at the dendritic nerve endings should have a complementary shape or charge. Support for the stereochemical theory of odor perception has come from the study of some pheromones that exhibit chirality, and certain highly purified enantiomeric compounds, such as R-(–)-carvone and S-(+)-carvone. These are the organoleptic compounds in oil of spearmint and oil of caraway, respectively.

The two compounds have the same molecular formula and, therefore, are isomers of each other. Their mirror images are not superimposable upon each other, however, and they are called **enantiomers** of each other (see Chapter 19 for more details on enantiomers). The enantiomeric compounds R-(+)-limonene and S-(–)-limonene have the odors of orange and lemon, respectively, to humans; and S-(+)-amphetamine and R-(–)-amphetamine smell fecal and musty, respectively. Not all enantiomeric compounds, however, have distinctively different odors to humans.

The stereochemical theory proposes that a receptor site shaped for one of these molecules would not allow the opposite enantiomer to fit. If it smelled differently, it was because that enantiomer fits another receptor site. Further, support for the stereochemical theory comes from the actual isolation of some of the receptor molecules (proteins) at the receptor site in insect antennae (reviewed by Breer, 1997; Prestwich and Du, 1997).

13.6 REVIEW AND SELF-STUDY QUESTIONS

1. What is the physiological difference between a phasic and a tonic receptor response?
2. Describe the structure of a sensillum.
3. Is it always possible to determine the function of a sensory structure from its morphology? Why or why not?
4. If a seta from a receptor on the surface of the cuticle has no pores, how is it likely to function?
5. How does a chordotonal receptor differ from hair plate receptors?
6. What is a subgenual organ, where are they typically found, and what is their function?
7. What is the function of air sacs behind a tympanal organ?
8. Where on the body of insects have tympanal organs been found?
9. What is Johnston's organ, where is it found, and what is its structure?
10. Give some examples, and explain how IR receptors serve in the ecology and physiology of insects having IR receptors.
11. What are some of the functions that simple chordotonal organs serve?
12. Describe a hygroreceptor in insects. What is its function?
13. How do some beetles that lay their eggs in burned forest debris detect a forest fire?
14. Describe an IR receptor in insects.
15. What is the typical location of the cell body of an insect chemoreceptor?
16. What are the major differences in structure of olfactory and gustatory chemoreceptors?
17. Describe across-fiber patterning and receptors with multiple sensory functions, and explain how they are beneficial in the sensory system of an insect.
18. Explain the stereochemical theory of odor reception.

REFERENCES

Abisgold, J.D. and S.J. Simpson. 1988. The effect of dietary protein levels and haemolymph composition on the sensitivity of the maxillary palp chemoreceptors of locusts. *J. Exp. Biol.* 135: 215–229.

Altner, H. and R. Loftus. 1985. Ultrastructure and function of insect thermo- and hygroreceptors. *Annu. Rev. Entomol.* 30: 273–296.

Altner, H., R. Loftus, L. Schaller-Selzer, and H. Tichy. 1983. Modality-specificity in insect sensilla and multimodal input from body appendages. *Fortschr. Zool.* 28: 17–31.

Altner, H. and L. Prillinger. 1980. Ultrastructure of invertebrate chemo-, thermo- and hygroreceptors and its functional significance. *Int. Rev. Cytol.* 67: 69–139.

Amoore, J.E., J.W. Johnston, Jr., and M. Rubin. 1964. The stereochemical theory of odor. *Sci. Am.* 210: 42–49.

Apel, K.-H. 1989. Zur Verbreitung von *Melanophila acuminata* DEG. (Col., Buprestidae). *Entomol. Nach. Berlin* 33: 278–280.

Autrum, H. and W. Schneider. 1948. Vergleichende Untersuchungen über den Erschütterungssinn der Insekten. *Z. Vgl. Physiol.* 31: 77–88.

Bailey, W.J. 1991. *Acoustic Behaviour of Insects: An Evolutionary Perspective*. Chapman & Hall, London, U.K.

Barber, J.R., B.A. Chadwell, N. Garrett, B. Schmidt-French, and W.E. Conner. 2009. Naïve bats discriminate arctiid moth warning sounds but generalize their aposematic meaning. *J. Exp. Biol.* 212: 2141–2148.

Barber, J.R. and W.E. Conner. 2006. Tiger moth responses to a simulated bat attack: Timing and duty cycle. *J. Exp.Biol.* 209: 2637–2650.

Barber, J.R. and W.E. Conner. 2007. Acoustic mimicry in a predator–prey interaction. *Proc. Natl. Acad. Sci. USA* 104: 9331–9334.

Breer, H. 1997. Molecular mechanisms of pheromone reception in insect antennae, in R.T. Cardé and A.K. Minks (eds.), *Insect Pheromone Research: New Directions*. Chapman & Hall, New York, pp. 115–130.

Briscoe, A.D., A. Macias-Muñoz, K.M. Kozak, J.R. Walters, F. Yuan, G.A. Jamie, S.H. Martin et al. 2013. Female behavior drives expression and evolution of gustatory receptors in butterflies. *PLoS Genet.* 9(7): e1003620.

Brown, S.G., G.H. Boettner, and J.E. Yack. 2007. Clicking caterpillars: Acoustic aposematism in *Antheraea polyphemus* and other Bombycoidea. *J. Exp. Biol.* 210: 993–1005.

Burkhardt, D. 1960. Action potentials in the antennae of the blowfly (*Calliphora erythrocephala*) during mechanical stimulation. *J. Insect Physiol.* 4: 138–145.

Busnel, R.-G. 1963. *Acoustic Behavior of Animals*. Elsevier, Amsterdam, the Netherlands.

Chapman, R.F. 2003. Contact chemoreception in feeding by phytophagous insects. *Annu. Rev. Entomol.* 48: 455–484.

Conner, W.E. 1987. Ultrasound: Its role in the courtship of the arctiid moth, *Cycnia tenera*. *Esperientia* 43: 1029–1031.

Conner, W.E. 1999. 'Un chant d'appel amoureux': Acoustic communication in moths. *J. Exp. Biol.* 202: 1711–1723.

Conner, W.E. and A.J. Corcoran. 2012. Sound strategies: The 65-million-year-old battle between bats and insects. *Annu. Rev. Entomol.* 57: 21–39.

Corbière-Tichané, G. and R. Loftus. 1983. Antennal thermal receptors of the cave beetle *Speophyes lucidulus* Delar. *J. Comp. Physiol.* 153: 343–351.

Dethier, V.G. 1955. The physiology and histology of the contact chemoreceptors of the blowfly. *Q. Rev. Biol.* 30: 348–371.

Dethier, V.G. 1963. *The Physiology of Insect Senses*. Methuen, London, U.K.

Dethier, V.G. 1971. A surfeit of stimuli: A paucity of receptors. *Am. Sci.* 59: 706–715.

Dethier, V.G. and A. Gleperin. 1967. Hyperphagia in the blowfly. *J. Exp. Biol.* 47: 191–200.

Devambez, I., M.A. Agha, C. Mitri, J. Bockaert, M.-L. Parmentier, F. Marion-Poll, Y. Grau, and L. Soustelle. 2013. Gao is required for L-canavanine detection in Drosophila. *PLoS One* 8(5): e63484.

Evans, W.G. 1962. Notes on the biology and dispersal of *Melanophila* (Coleoptera: Buprestidae). *Pan-Pac. Entomol.* 38: 59–62.

Evans, W.G. 1966a. Perception of infrared radiation from forest fires by *Melanophila acuminata* De geer (Buprestidae, Coleoptera). *Ecology* 47: 1061–1065.

Evans, W.G. 1966b. Morphology of the infrared sense organ of *Melanophila acuminata* (Buprestidae, Coleoptera). *Ann. Entomol. Soc. Am.* 59: 873–877.

Ewing, A.W. 1989. *Arthropod Bioacoustics: Neurobiology and Behavior*. Comstock Publishing Association, Cornell University Press, Ithaca, NY.

Field, L.H. and T. Matheson. 1998. Chordotonal organs of insects. *Adv. Insect Physiol.* 27: 1–228.

Fonseca, P.J. and T. Correia. 2007. Effects of temperature on tuning of the auditory pathway in the cicada *Tettigetta josei* (Hemiptera, Tibicinidae). *J. Exp. Biol.* 210: 1834–1845.

Frazer, J.L. 1985. Nervous system: Sensory system, in M.S. Blum (ed.), *Fundamental of Insect Physiology*. John Wiley & Sons, New York, pp. 287–356.

French, A.S. 1988. Transduction mechanisms of mechanosensilla. *Annu. Rev. Entomol.* 33: 39–58.

Fullard, J.H., J.M. Radcliffe, and C.G. Christie. 2007. Acoustic feature recognition in the dogbane tiger moth, *Cycnia tenera*. *J. Exp. Biol.* 210: 2481–2488.

Fullard, J.H. and J.E. Yack. 1993. The evolutionary biology of insect hearing. *Trends Ecol. Evol.* 8: 248–252.

Galizia, C.G. and W. Rössler. 2010. Parallel olfactory systems in insects: Anatomy and Function. *Annu. Rev. Entomol.* 55: 399–420.

Gelperin, A. 1967. Stretch receptors in the foregut of a blowfly. *Science* 157: 208–210.

Griffin, D.R., F.A. Webster, and C.R. Michael. 1960. The echolocation of flying insects by bats. *Anim. Behav.* 8: 141–154.

Göpfert, M.C. and D. Robert. 2002. The mechanical basis of *Drosophila* audition. *J. Exp. Biol.* 205: 1119–1208.
Grosse-Wilde, E., L.S. Kuebler, S. Bucks, H. Vogel, D. Wicher, and B.S. Hansson. 2011. Antennal transcriptome of *Manduca sexta*. *Proc. Natl. Acad. Sci. USA* 108: 7449–7454.
Grosse-Wilde, E., R. Stieber, M. Forstner, J. Krieger, D. Wicher, and B.S. Hansson. 2010. Sex-specific odorant receptors of the tobacco hornworm *Manduca sexta*. *Front. Cell Neurosci.* 4: 1–7.
Hansson, B.S. and M.C. Stensmyr. 2011. Evolution of insect olfaction. *Neuron* 72: 698–711.
Hart, S. 1998. Beetle mania: An attraction to fire. *Bioscience* 48: 3–5.
Haskell, P.T. 1961. *Insect Sounds*. Quadrangle Books, Chicago, IL.
Hodgson, E.S. 1956. Electrophysiological studies of arthropod chemoreception. I. General properties of the labellar chemoreceptors of Diptera. *J. Cell. Comp. Physiol.* 48: 51–76.
Hodgson, E.S. 1958. Chemoreception in arthropods. *Annu. Rev. Entomol.* 3: 19–36.
Horridge, G.A. 1965. The Arthropoda: III Insecta, in T.H. Bullock and G.A. Horridge (eds.), *Structure and Function in the Nervous System of Invertebrates*. Freeman & Co., San Francisco, CA, pp. 1030–1055.
Howlett, N., K.L. Dauber, A. Shukla, B. Morton, J.I. Glendinning, E. Brent, C. Gleason et al. 2012. Identification of chemosensory receptor genes in *Manduca sexta* and knockdown by RNA interference. *BMC Genomics* 13: 211.
Hoy, R.R. 1989. Startle, categorical responses and attention in acoustic behavior of insects. *Annu. Rev. Neurosci.* 12: 355–375.
Hoy, R.R. and D. Robert. 1996. Tympanal hearing in insects. *Annu. Rev. Entomol.* 41: 433–450.
Jorgensen, K., M. Stranden, J.-C. Sandoz, R. Menzel, and H. Mustaparta. 2007. Effects of two bitter substances on olfactory conditioning in the moth *Heliothis virescens*. *J. Exp. Biol.* 210: 2563–2573.
Kain, P., F. Badsha, S.M. Hussain, A. Nair, G. Hasan, and V. Rodrigues. 2010. Mutants in phospholipid signaling attenuate the behavioral response of adult *Drosophila* to trehalose. *Chem. Senses* 35: 663–673.
Lapshin, D.N. and D.D. Vorontsov. 2013. Frequency tuning of individual auditory receptors in female mosquitoes (Diptera: Culicidae). *J. Insect Physiol.* 59: 828–839.
Lee, Y., M.J. Kang, J. Shim, C.U. Cheong, S.J. Moon, and C. Montell. 2012. Gustatory receptors required for avoiding the insecticide L-canavanine. *J. Neurosci.* 32: 1429–1435.
Lee, Y., S.J. Moon, and C. Montell. 2009. Multiple gustatory receptors required for the caffeine response in Drosophila. *Proc. Natl. Acad. Sci. USA* 106: 4495–4500.
Loftus, R. 1976. Temperature-dependent dry receptor on antenna of *Periplaneta*. Tonic response. *J. Comp. Physiol.* 111: 153–170.
Loftus, R. 1978. Peripheral thermal receptors, in M.A. Ali (ed.), *Sensory Ecology Review and Perspectives*. Plenum Press, New York, pp. 439–466.
Minet, J. and A. Surlykke. 2003. Auditory and sound producing organs, in N.P. Kristensen (ed.), *Handbook of Zoology*, vol. IV, Arthropoda: Insecta. Lepidoptera, Moths and Butterflies, vol. 2. W.G. deGruyter, Berlin, Germany, pp. 289–323.
Montell, C. 2009. A taste of the *Drosophila* gustatory receptors. *Curr. Opin. Neurobiol.* 19: 345–353.
Olberg, R.M. 2012. Visual control of pery-capture flight in dragonflies. *Curr. Opin. Neurobiol.* 22: 267–271.
Ômura, H., K. Honda, K. Asaoka, and T.A. Inoue. 2011. Divergent behavioral and electrophysiological taste responses in the mid-legs of adult butterflies, *Vanessa indica* and *Argyreus hyperbius*. *J. Insect Physiol.* 57: 118–126.
Piersanti, S., F. Frati, E. Conti, E. Gaino, M. Rebora, and G. Salerno. 2014. First evidence of the use of olfaction in Odonata behavior. *J. Insect Physiol.* 62: 26–31.
Piersanti, S., M. Rebora, and E. Gaino. 2010. A scanning electron microscope study of the antennal sensilla in adult zygoptera. *Odonatologica* 39: 235–241.
Prestwich, G.D. and G. Du. 1997. Pheromone-binding proteins, pheromone recognition, and signal transduction in moth olfaction, in R.T. Cardé and A.K. Minks (eds.), *Insect Pheromone Research: New Directions*. Chapman & Hall, New York, pp. 131–143.
Rebora, M., G. Salerno, S. Piersanti, A. Dell'Otto, and E. Gaino. 2012. Olfaction in dragonflies: An electrophysiological evidence. *J. Insect Physiol.* 58: 270–277.
Rees, C.J.C. 1968. The effect of aqueous solution of some 1:1 electrolytes on the electrical response of the type 1 ("salt") chemoreceptor cell in the labella of *Phormia*. *J. Insect Physiol.* 14: 1331–1364.
Roeder, K.D. 1962. The behavior of free flying moths in the presence of artificial ultrasonic pulses. *Anim. Behav.* 10: 300–304.
Roeder, K.D. 1964. Aspects of the noctuid tympanic nerve response having significance in the avoidance of bats. *J. Insect Physiol.* 10: 529–546.

Roeder, K.D. 1965. Moths and ultrasound. *Sci. Am.* 212: 94–102.
Roeder, K.D. 1966a. Acoustic sensitivity of the noctuid tympanic organ and its range for the cries of bats. *J. Insect Physiol.* 12: 843–859.
Roeder, K.D. 1966b. Auditory system of noctuid moths. *Science* 154(3756): 1515–1521.
Rosenthal, G.A. 2001. L-Canavanine: A higher plant insecticidal allelochemical. *Amino Acids* 21: 319–330.
Roth, L.M. 1948. A study of mosquito behavior. *Am. Midl. Natural.* 40: 265–352.
Sales, G. and D. Pye. 1974. *Ultrasonic Communication by Animals.* Chapman & Hall, London, U.K., pp. 71–97.
Sanderford, M.V. and W.E. Conner. 1990. Courtship sounds of the polka-dot wasp moth, *Syntomeida epilais.* *Naturwissenschaften* 77: 345–347.
Sanderford, M.V. and W.E. Conner. 1995. Acoustic courtship communication in *Syntomeida epilais* Wlk. (Lepidoptera: Arctiidae, Ctenuchinae). *J. Insect Behav.* 8(1): 19–31.
Sanderford, M.V., F. Coro, and W.E. Conner. 1998. Courtship behavior in *Empyreuma affinis* Roths. (Lepidoptera, Arctiidae, Ctenuchinae): Acoustic signals and tympanic response. *Naturwissenschaften* 85: 82–87.
Sato, K., M. Pellegrino, T. Nakagawa, T. Nakagawa, L.B. Vosshall, and K. Touhara. 2008. Insect olfactory receptors are heteromeric ligand-gated ion channels. *Nature* 452: 1002–1006.
Sato, K., K. Tanaka, and K. Touhara. 2011. Sugar-regulated cation channel formed by an insect gustatory receptor. *Proc. Natl. Acad. Sci. USA* 108: 11680–11685.
Schmitz, H., H. Bleckmann, and M. Mürtz. 1997. Infrared detection in a beetle. *Nature* 386: 773–774.
Schmitz, H., S. Trenner, M.H. Hofmann, and H. Bleckmann. 2000. The ability of *Rhodnius prolixus* (Hemiptera: Reduviidae) to approach a thermal source solely by its infrared radiation. *J. Insect Physiol.* 46: 745–751.
Slone, J., J. Daniels, and H. Amrein. 2007. Sugar receptors in *Drosophila.* *Curr. Biol.* 17: 1809–1816.
Spangler, H.G. 1988. Moth hearing, defense, and communication. *Annu. Rev. Entomol.* 33: 59–81.
Sueur, J., J.F.C. Windmill, and D. Robert. 2006. Tuning the drum: The mechanical basis for frequency discrimination in a Mediterranean cicada. *J. Exp. Biol.* 209: 4115–4128.
Vondran, T., K.-H. Apel, and H. Schmitz. 1995. The infrared receptor of *Melanophila acuminata* De Geer (Coleoptera: Buprestidae): Ultrastructural study of a unique insect thermoreceptor and its possible descent from a hair mechanoreceptor. *Tissue Cell* 27: 645–658.
Weiss, L.A., A. Dahanukar, J.Y. Kwon, D. Banerjee, and J.R. Carlson. 2011. The molecular and cellular basis of bitter taste in *Drosophila.* *Neuron* 69: 258–272.
Wicher, D., R. Schäfer, R. Bauernfeind, M.C. Stensmyr, R. Heller, S.H. Heinemann, and B.S. Hansson. 2008. *Drosophila* odorant receptors are both ligand-gated and cyclic–nucleotide–activated cation channels. *Nature* 452: 1007–1011.
Windmill, J.F.C., J.H. Fullard, and D. Robert. 2007. Mechanics of a 'simple' ear: Tympanal vibrations in noctuid moths. *J. Exp. Biol.* 210: 2637–2648.
Windmill, J.F.C., M.C. Göpfert, and D. Robert. 2005. Tympanal traveling waves in migratory locusts. *J. Exp. Biol.* 208: 157–168.
Windmill, J.F.C., J.C. Jackson, E.J. Tuck, and D. Robert. 2006. Keeping up with bats: Dynamic tuning in a moth. *Curr. Biol.* 16: 2418–2423.
Yack, J.E. and J.H. Fullard. 1993. What is an insect ear? *Ann. Entomol. Soc. Am.* 86: 677–682.
Yao, C.A. and J.R. Carlson. 2010. Role of G-proteins in odor-sensing and CO_2-sensing neurons in *Drosophila.* *J. Neurosci.* 30: 4562–4572.
Zacharuk, R.T. 1980. Ultrastructure and function of insect chemosensilla. *Annu. Rev. Entomol.* 25: 27–47.
Zhang, H.-J., A.R. Anderson, S.C. Trowell, A.-R. Luo, Z.-H. Xiang, and Q.-Y. Xia. 2011. Topological and functional characterization of an insect gustatory receptor. *PLoS One* 6(8): e24111.

CHAPTER 14

Vision

PREVIEW

Insects have several types of light receptors, including compound eyes, ocelli, stemmata, and simple dermal light receptors. Compound eyes form images, and while compound eyes of many insects are known to be sensitive to blue, green, and ultraviolet (UV) wavelengths, color vision has been demonstrated behaviorally in only a few insects. A rigorous test for color vision requires behavioral demonstration that an insect has discriminated between two colors (i.e., two wavelengths of light), and on this basis, color vision has been demonstrated in honeybees, some dipterans, and a few other insects. The visual process and visual cascade in insect compound eyes appears to be essentially the same as that in eyes of vertebrates. One exception is that rhodopsin does not split away from 11-*cis*-retinal in insect eyes after receiving a photon of light and becoming excited to the metarhodopsin state. By absorbing another photon of light, the metarhodopsin can be transformed back into rhodopsin, ready to repeat the visual process all over again. It has been demonstrated that a source of vitamin A or a carotenoid is needed by several species for visual acuity and normal structure of compound eyes. A number of different species detect and use plane-polarized light in behavioral orientation.

14.1 INTRODUCTION

The ultimate source of light and energy for life on earth is the sun, so it is not surprising that virtually all living organisms evolved some kind of response to light. Green plants, and a few types of bacteria, evolved mechanisms to capture light energy and use it to drive synthesis of organic molecules. Light causes phototropic movements of leaves and stems and timing of flowering in many plants, and wavelengths in the 600–700 nm range promote photosynthesis. Although animals did not evolve processes to convert light energy into synthesis of new chemical molecules in the way that plants did, they evolved physical structures and biochemical molecules that are sensitive to light. Light often influences sexual reproductive cycles, biological and seasonal rhythms, color changes in the skin, hormone secretion, and some chemical reactions in animals. Even the simplest plant and animal forms have pigments that enable them to respond to light. What is commonly called light is the visible part of the electromagnetic radiation from the sun that encompasses a wide spectrum from gamma rays and X-rays (<0.1 nm) to ultraviolet (UV), visible, infrared (IR), radio waves, and other longer wavelengths. Fortunately for living organisms, ozone in the stratosphere strongly absorbs short cosmic ray and UV wavelengths, which cause chemical changes in cells and deoxyribonucleic acid (DNA). The electromagnetic radiation that reaches the earth spans about 300–900 nm, with peak intensity at nearly 500 nm wavelength (Wolken, 1995).

Wavelengths around the peak of 500 nm can penetrate clear water to about 100 m, but other wavelengths are strongly absorbed by only a few meters of water. Little light penetrates deeper than 100 m. In spite of what might seem to be a vast covering of the earth by green plants that capture sunlight, relatively little of the solar radiation to the earth is captured, and most of it is reradiated out into space each night.

The light-sensitive receptors of insects, the **compound eyes**, **ocelli**, and **stemmata**, respond to light from about 350 nm (UV) into the red range at about 700 nm. All insects that have been studied have **UV receptors** in the eyes, but not all can detect the longer wavelengths of orange and red light. Some extraordinary parallels in the independent development of visual systems occurred during the evolution of animals, with image-forming eyes evolving in flatworms, annelid polychaetes, coelenterates, echinoderms, insects, arachnids, crustaceans, cephalopod mollusks, and vertebrates. Visual receptors, at least in terms of anatomy and structure, evolved independently in different groups of animals, possibly as many as seven times (Wolken, 1995). However, all animals use the same **chromophore, 11-*cis*-retinal** or a slightly modified molecule, and the **transmembrane protein opsin** in light reception. Opsin and the chromophore combine to form **rhodopsin**, the **visual pigment** in invertebrate and vertebrate photoreceptors. Species specificity is determined by the amino acid composition of opsin. Thus, even though the anatomical structure of eyes evolved independently, it appears that some form of rhodopsin was present in very ancient animal life and has been conserved through evolutionary time as has the basic structure of chlorophyll in plants. Compound eye structure evolved 500–600 million years ago in trilobites in the Cambrian period (Wolken, 1995). Such an eye may consist of a lens as simple as the cuticular covering over the light-sensitive receptor. Briscoe and Chittka (2001) speculate that UV, blue, and green photoreceptors probably existed in insects in the Devonian period about 300 million years ago.

Among invertebrates, crustaceans have the greatest diversity in eye structure, with structures possibly having evolved independently several times. Although some crustaceans have very simple eyes, many have compound eyes similar in structure and function to compound eyes in insects. Eye structure in the Mollusca ranges from simple eyecups of limpets to image-forming eyes with a lens in squid and octopus.

There are three types of visual receptors in insects: compound eyes, ocelli, and stemmata. Compound eyes are excellent motion detectors, responding to the movement of objects across many small facets, and in many insects, this may be their most important function. Ocelli are found on immatures and adults of some insects. The cuticular covering forms a single lens with photosensitive cells beneath it in the ocelli. Compound eyes, ocelli, and stemmata have the necessary structure of a lens to focus the light and photoreceptors for image formation, but the image in some cases (typically in ocelli) is not focused on the photoreceptor cells. It is not really possible to say what sort of image an insect sees because it really "sees" with the integrative centers in the brain. Does an insect see dozens or perhaps hundreds of small images at once with its compound eyes, or does the brain synthesize the incoming data into a single image? No one really knows. Ocelli probably function mainly in detecting the quality and intensity of light and its presence or absence. Stemmata, the visual receptors in larvae of many holometabolous insects, can focus an image on photoreceptors, but stemmata are very small, and it is likely that if an image is conveyed to the brain, it is probably fuzzy and poorly resolved. Immature grasshoppers, true bugs, cockroaches, and other hemimetabolous insects have compound eyes similar to adult compound eyes. Compound eyes, ocelli, and stemmata follow the same basic structural plan, with a lens to focus the light, light-sensitive cells, and axons from the photoreceptive cells projecting to the optic lobe of the brain. In all three types of eyes the photoreceptive cells have a rhabdomere region with many membrane layers where the visual pigment molecules are located. Color vision, form discrimination, and detection of plane-polarized light are well developed in some insects and play major roles in their behavior.

14.2 COMPOUND EYE STRUCTURE

The **compound eyes** are composed of multiple functional units called facets or **ommatidia** (*sing.*, **ommatidium**). Each ommatidium is composed of many cells and of functional parts that include the **dioptric structures**, the photosensitive cells containing the photosensitive pigments, and shielding cells that usually contain a variety of pigments as well. The small eyes of Thysanura contain only a few ommatidia (12 in *Lepisma* spp.), while the very large eyes of dragonflies contain as many as 10,000 ommatidia. Adult Collembola, Lepismatidae, Siphonaptera, and Strepsiptera do not have compound eyes, but instead have simple eyes similar to ocelli.

There is considerable diversity in the structural detail of ommatidia in different insect groups. Two major variations in structure are represented in Figure 14.1, the **photopic eye** of a dipteran, and in Figure 14.2, the **scotopic eye** of a moth. Goldsmith and Bernard (1974) recommend the terms "photopic" and "scotopic" as replacements for the older usage of **apposition** and **superposition**, respectively. Photopic eyes occur in diurnal insects, which are active during the day. The rhabdom extends from the cone to the basement membrane at the proximal limit of the eye.

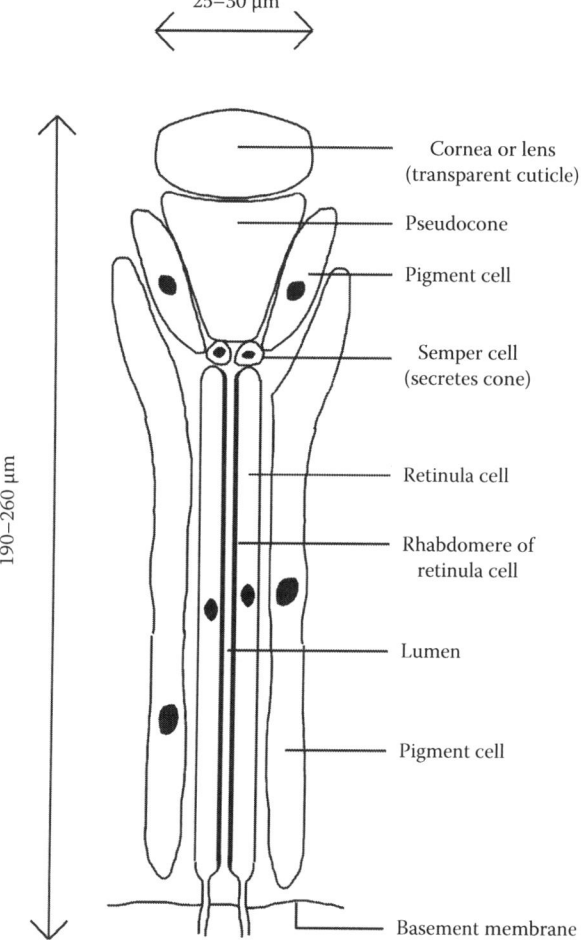

Figure 14.1 Diagrammatic representation of the structure of an ommatidium in the photopic compound eye of the tephritid *Anastrepha suspensa*. (Modified from Agee, H.R. et al., *Ann. Entomol. Soc. Am.*, 70, 359, 1977.)

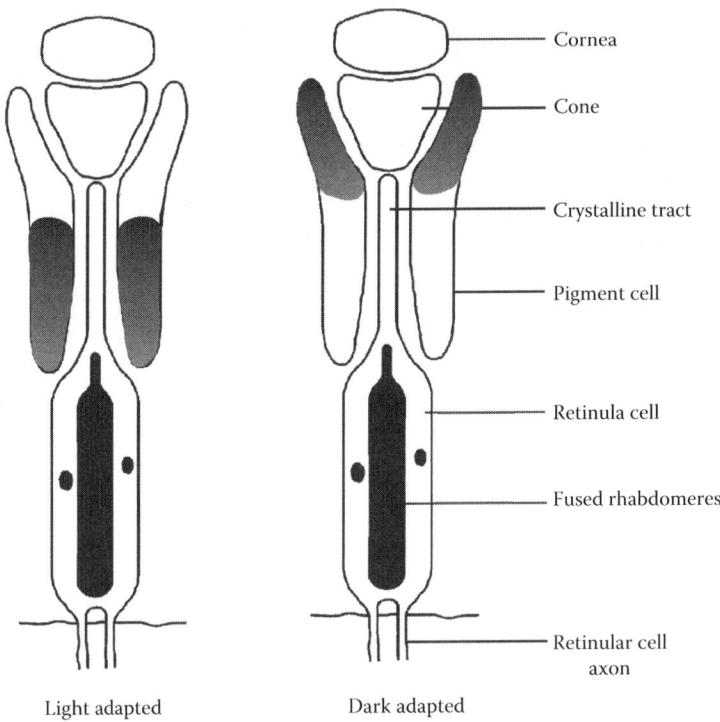

Figure 14.2 Diagrammatic illustration of the structure of an ommatidium in the scotopic eye of a nocturnal moth to show shielding pigment distribution in a light-adapted ommatidium and its distribution in a dark-adapted ommatidium. In the light-adapted eye, the dispersed pigment offers protection against light escaping from neighboring ommatidia, while in the dark-adapted state, migration of pigment to the periphery of the ommatidium allows the potential for light to enter from adjacent ommatidia. The latter condition may make for less sharp visual images, but probably allows better visual responses in dim light, which is likely an adaptation for night-flying moths.

The pigment in the shielding cells is uniformly dispersed, with little or no movement of it in the shielding cells. Thus, light can strike the photosensitive pigment in the rhabdom only by entering axially (i.e., directly from above after passing through the cornea and cone of the ommatidium). An image is focused on the **rhabdom** or **rhabdomeres** of retinula cells just below the level of the cone.

Scotopic eyes occur in nocturnal and crepuscular insects. The rhabdom is shorter than in photopic eyes and usually extends only about one-third of the distance from the basement membrane to the cone. The remaining distance to the cone may be filled with thin strands of the retinula cells forming the **crystalline tract** (Figure 14.2), or there may simply be a gap between the retinula cells and the cone. A major difference between photopic and scotopic eyes is that the pigment in shielding cells responds to the light intensity by migrating. At low light intensity, the pigments contract into the distal part of the cells (i.e., near the cone), allowing light from adjacent ommatidia to pass through and strike the rhabdom below. Pigment disperses throughout the shielding cells upon exposure to bright light or in response to higher environmental temperatures (Nordström and Warrant, 2000). This anatomical type of eye came to be called a superposition eye because it superimposed images (although likely not in perfect focus) from the visual field of several facets upon a single location. In light-adapted scotopic eyes, the dispersed shielding pigments tend to block light from adjacent facets (Figure 14.2), and this allows the scotopic eye to function more like a photopic eye, with an apposition image from the light coming into each ommatidium through its own cornea and cone. Migration of the pigment in bright light shields the rhabdomeres and may be a protective mechanism to prevent bright light from

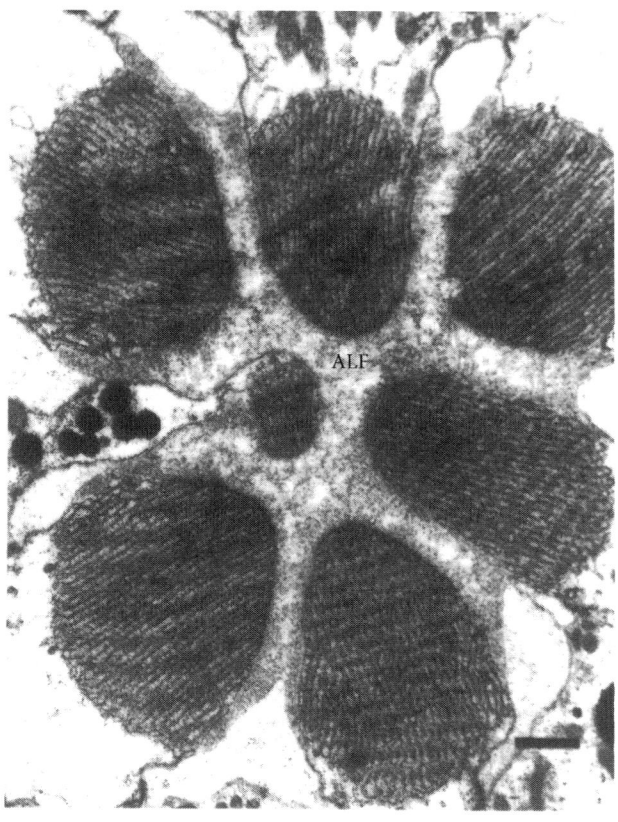

Figure 14.3 A transmission electron micrograph of the open rhabdom of a dipteran. (Photo courtesy of Dr. Clay Smith, University of Florida, Department of Ophthalmology, University of Florida Medical School, Gainesville, FL.)

bleaching the visual pigment. Experiments and observations suggest that scotopic eyes allow increased sensitivity in dim light with loss of some sharpness in image formation. Mantispids (Neuroptera, Mantispidae), however, have superposition compound eyes and are adept at capturing fast-moving insects with their raptorial forelegs. Their eyes have a frontal region of high acuity (Karl, 2013).

Another morphological variation in eye structure, although not correlated with photopic or scotopic eyes, is that rhabdomeres may be fused into a rhabdom in some insects (called a closed rhabdom) and remain unfused (open) in others (Figure 14.3).

14.3 DIOPTRIC STRUCTURES

Dioptric structures refract or bend light entering the eye and help focus an image. A variety of dioptric structures exist in different insect eyes, including cornea, cone, corneal nipples, crystalline tract, and layers of different density in one or more of these structures. All compound eyes and ocelli have a corneal covering that transmits and refracts the light passing through. The cornea is composed of transparent cuticle secreted by **corneagenous cells** (highly specialized and modified epidermal cells). Clearly, in order to allow effective vision, the cornea must transmit a large part of the light striking it, and about 90% of the light between 400 and 650 nm is transmitted by the cornea of *Manduca sexta*, while wavelengths shorter than 350 nm are strongly absorbed and almost no wavelengths shorter than 300 nm are transmitted.

The **cone** is another part of the dioptric apparatus in most insects. The light transmission characteristics of the crystalline cone in *M. sexta* are similar to the properties of the cornea. The cone also is variable in structure and is formed in several ways. In general, there are four cells, sometimes called Semper's cells, that form the cone. In insects without a cone (**acone eyes**), the cone cells themselves are transparent, but they are not greatly modified in shape nor do they secrete crystalline products as in insects with cones in the eye. Acone eyes are considered the most primitive type, and they occur in apterygote insects, but also in some Hemiptera, Coleoptera, and Diptera. **Eucone eyes** are common and occur in most orders of insects. In these, the cone cells contain a clear, hard intracellular secretion that fills most of the cell volume. The remaining small shell of living cytoplasm in the cone cells becomes pushed to the margin of the cells.

Pseudocone eyes are common in many Diptera, including the Cyclorrhapha (higher Diptera, such as the housefly and drosophilid and tephritid fruit flies), and in some Odonata. The pseudocone is a cavity below the cornea filled with a gelatinous or liquid secretion. The cytoplasm of the four cone cells is squeezed into a thin layer beneath the pseudocone.

Exocone eyes occur in some Coleoptera, most notably in fireflies and related beetles. Exocone eyes are not considered homologous to any of the preceding three types. In insects with exocone eyes, the cone is formed by an inward projection of the cornea and is cuticular in structure. The four Semper's cells form a short crystalline tract below the exocone, but they secrete no true crystalline core.

The compound eyes of some insects contain additional dioptric components much smaller in dimensions than the cone and cornea. These components include (1) a crystalline tract, (2) corneal nipples, (3) corneal layering, and (4) periodic layering of tracheolar structures, commonly referred to as a tapetum (Miller et al., 1968).

Crystalline tracts are found in some butterflies, moths, and fireflies. The diameter of a tract is an important parameter that determines whether the tract can act like a waveguide and transmit an image to the rhabdom below. The diameter varies from 2 µm in butterflies to 4 µm in *M. sexta* to 10 µm in some fireflies. At least in the moths and butterflies, the diameters measured are considered to be too small to allow an image to be transmitted (Miller et al., 1968). Other factors also are important to image transmission. The tracts of moths and butterflies can transmit light to the rhabdom by functioning as a light guide if the refractive index of the tract is greater than that of the surrounding medium so that the light is kept inside the tract and not diffused to the surrounding tissue. The presence of shielding cells is important in this respect. Finally, to be effective, the light will have to be brought to focus on the rhabdom. On the basis of both theoretical and experimental data, Miller et al. (1968) concluded that the tracts act as light guides in dark-adapted eyes of *M. sexta*, *Elpinor* spp. (a sphingid moth), *Cecropia* and *Polyphemus* (silk moths), and *Hobomok* spp. (a skipper).

Corneal nipples are present at the air interface of the eyes of many nocturnal Lepidoptera. The nipples probably function as antireflection devices, which in turn should increase the transmission of light into the eye, possibly by as much as 5%. The nipples also might benefit the insect by reducing the reflection of light to predators and parasites and by reducing or preventing internal reflection of light from the tapetum below the eye.

The **tapetum** is a thick mat of tracheoles at or near the base of the eye. The shiny surface of the air-filled tracheoles reflects light back into the upper parts of the eye and creates eye shine in moths and some other insects. The glow disappears in moths with light-adapted eyes because the shielding pigments move down in the pigment cells enough to absorb most of the light before it is reflected back. Very little detailed work has been done on the specific functions of the tapetum, but it may increase sensitivity of dark-adapted eyes and might allow increased sensitivity to contrast patterns (Miller et al., 1968). The reflection of light back into the distal parts of the eye are likely to create blurred images, but insects that have a large tapetum are those that are active at night when the light is dim anyway and not best suited for formation of sharp images.

14.4 CORNEAL LAYERING

The eyes of some dipterans, especially the Tabanidae (Figure 14.4), show patterns of bright and dark stripes, frequently in colors, in the cornea of the compound eyes. The patterns and colors are caused by the reflection of light from dense (refractive index 1.74) and thin layers (refractive index 1.40) in the corneal cuticle. Deerflies have as many as 20 layers in the cornea. Miller et al. (1968) suggested that the layers might serve as color contrast filters enabling fast-flying flies to more easily locate a host against the background.

14.5 RETINULA CELLS

The **photoreceptor cells** in all types of eyes of insects are called **retinula cells**, which are long slender cells extending over the greater part of the length of an ommatidium in photopic eyes, but are shorter in scotopic eyes. They are 60–100 μm long in *Drosophila melanogaster* (Wolken, 1957), about 100 μm long in *Periplaneta americana* (Wolken and Gupta, 1961), and from 150 to >200 μm long in *Anastrepha suspensa* (Agee et al., 1977).

Retinula cells are **primary receptors** (in contrast to secondary ones) because they send their axons directly into the optic lobe of the brain without synapsing. Eight is the most common number of retinula cells in each ommatidium, arranged in a circle in compound eyes. Ocelli and stemmata have smaller numbers of retinula cells. Some Lepidoptera have 10–11 retinula cells in each ommatidium of the compound eyes and honeybees have 9. Adult dragonflies have four retinula cells in the ommatidia of the dorsal part of the eye and six in the ventral part; dragonfly nymphs have eight retinula cells per ommatidium. In the most common situation, the eighth cell, or in honeybees, the ninth cell, is short and can be found only in the more proximal region of an ommatidium.

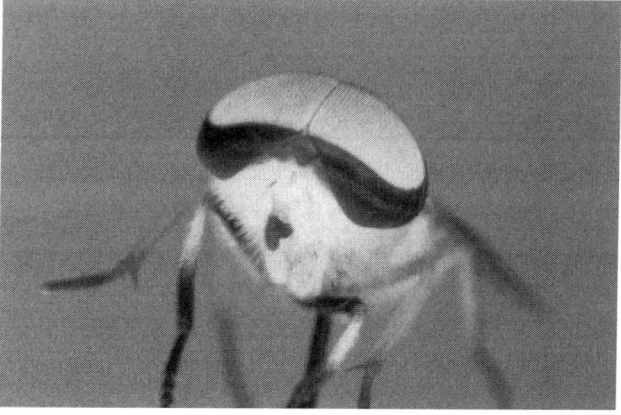

Figure 14.4 **(See color insert.)** A tabanid fly with a dark band across the lower part of the compound eyes. (Photo courtesy of Dr. Jerry Butler, University of Florida, Department of Entomology & Nematology, Gainesville, FL, Retired.)

14.6 RHABDOMERES

Each retinula cell contains a specialized region called the **rhabdomere** that is the site of the light-sensitive pigment. Perpendicular to the long axis of the retinula cell, the cell membrane is extended into thousands of **microvilli** or tubules, typically about 40–120 nm in diameter along all or part of the length of the cell, and these microvilli make up the rhabdomere. In *D. melanogaster*, there are about 60,000 microvilli per cell. Wolken (1968) estimated that there are about 80,000 in each rhabdomere of *P. americana*. The microvilli are the site of some 100 million molecules of rhodopsin per cell (Zuker, 1996). A rhabdomere extends over most or all of the length of a retinula cell and is usually oriented toward the circle formed by the group of retinula cells in an ommatidium. The diameter of the rhabdomere is about 1.2 µm in *D. melanogaster* (Wolken, 1957) and about 2 µm in *P. americana* (Wolken and Gupta, 1961). Rhabdomeres fuse at the center of the circle to form a closed or fused rhabdom in honeybees, Lepidoptera, and many other insects, but in some Diptera and Hemiptera, the rhabdomeres do not touch each other, leaving a central hollow space between them. In compound eyes, the microvilli typically are oriented perpendicular to the long axis of the retinula cell, but, in simpler eyes of some arthropods, the direction of microvilli is in line with the long axis of the retinula cell (Phillis and Cromroy, 1977).

14.7 ELECTRICAL ACTIVITY OF RETINULA CELLS

Resting potentials ranging from 25 to 70 mV have been recorded across the membranes of retinula cells. The inside of the cell is negative relative to the outside during resting conditions in the dark. In contrast to the polarizing effect of light on vertebrate eye receptors, light acts as a depolarizing stimulus for retinula cells in the compound eye. An electroretinogram (ERG) of electrical activity in response to stimuli can be recorded by placing one electrode on or into the eye and the reference electrode somewhere else in the head. An ERG is a summation of potentials from many retinula cells and possibly of electrical activity within the optic lobe. Agee (1977) described techniques and equipment for ERG measurements.

Illumination of the eye causes a slow or graded potential that increases with intensity of the stimulus. The slow potential is the receptor potential and it shows the typical characteristics of a graded potential. At higher intensities of illumination, there appears a transient "on" component that has a sharp threshold and a refractory period. Moreover, the "on" component can be abolished by high external K^+ ion concentration. Thus, the "on" component has the characteristics of a spike (action potential). Fuortes (1963) showed that both the transient response and the receptor potential (measured 900 ms after initiation of stimulus) are linear functions of the logarithm of light intensity. The transient "on" potential is a compound action potential of many individual axon responses. Increased illumination brings more and more axons into play and the "on" transient becomes larger.

14.8 NEURAL CONNECTIONS IN THE OPTIC LOBE

The **optic lobe** of the brain contains three large **neuropils** (Figure 14.5) in which synaptic connections occur. The most distal one, the **lamina ganglionaris**, is the first site where axons of the retinula cells synapse with **monopolar interneurons**. In dipterans, such as *Musca* and *Calliphora*, in which detailed studies have been made, groups of neurons collect into **neural cartridges** or **neuroommatidia** in the lamina ganglionaris (Figure 14.5). There are 3000 cartridges in each lobe corresponding to the 3000 ommatidia in the eye that send axons into the cartridges for synaptic connections in *Musca* and *Calliphora* (Braitenberg, 1972). The axons

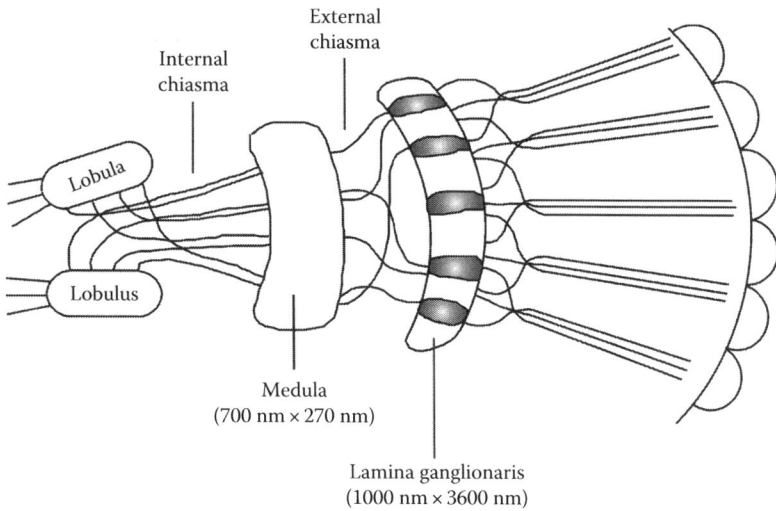

Figure 14.5 Basic structure of the optic lobe illustrating the four major optic neuropils and neural cartridges in the most distal neuropil, the lamina ganglionaris.

from the retinula cells decussate (cross) before entering the lamina so that axons from a single ommatidium do not synapse in the same neural cartridge and some retinula axons pass through the lamina without synapsing (Boschek, 1971).

Neural fibers leave the lamina, decussate, and enter the second optic ganglion—the **medulla**. The network of fibers between the lamina and medulla is called the external chiasma. Histological evidence indicates that within the medulla of *Musca* and *Calliphora*, there are regular arrays of repeating units receiving the input from the lamina (Braitenberg, 1972), but the anatomy of these units has not been studied in detail. Fibers leave the medulla, again decussate, and enter the **lobuli**—the third ganglionic mass of synapses. The lobuli consist of several, sometimes separated, masses of cell bodies and neuropil. In *Musca*, there are two masses known as the **lobula** and **lobulus**. Regular repeating units have not been found in the lobuli and little is known about synaptic connections or input/output interactions.

The large amount of crossing of fibers in the optic lobe of insects is not peculiar to insects, but it is a general feature of visual systems. Its function is not clearly understood, but it would appear to provide a great deal of backup security if small parts of either the external eye or brain suffered damage.

14.9 OCELLI

Ocelli have some of the same anatomical features as compound eyes, including a corneal lens, corneagenous cells that secrete the lens, retinula cells, and an ocellar nerve that leads into the protocerebrum (Goodman, 1970) (Figure 14.6). The retinula cells typically contain a rhabdomere region with microvilli. The corneal cuticle is probably transparent to various wavelengths of light, but few measurements have been made. Goldsmith and Ruck (1958) showed that the cornea from ocelli of the cockroach, *Blaberus craniifer*, was transparent to light from about 350–700 nm.

The **cornea** of an ocellus covers a large visual field and appears to form an image, but the image is focused beneath the layer of retinula cells and is not transmitted to the brain. The behavior of insects that have the compound eyes covered, but with functional ocelli, further indicates that ocelli do not convey an image to the brain. The retinula cells in ocelli are spontaneously active in the dark.

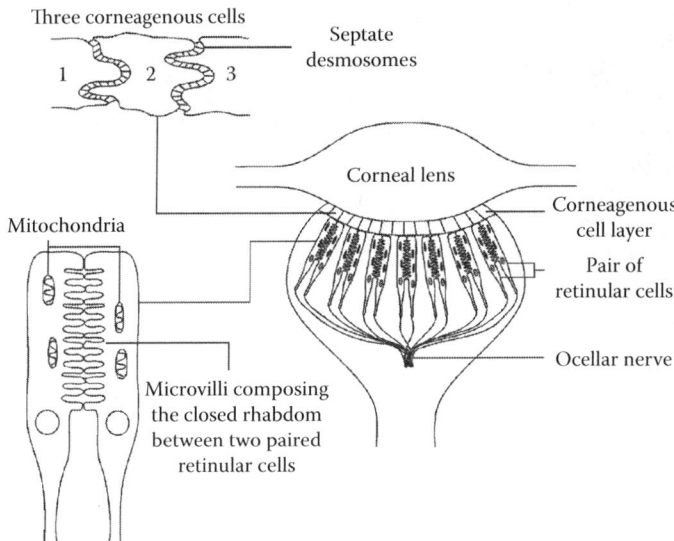

Figure 14.6 Diagram of the structure of an ocellus from a worker honeybee. (Modified from Toh, Y. and Kuwabara, M., *J. Morphol.*, 143, 285, 1974.)

Rather than depolarizing the ocellar retinula cells, light leads to a more stable or increased membrane potential and, hence, to a decrease or even cessation in spike activity sent into the brain. The ocelli can signal light on/off information and intensity of illumination and possibly in some insects may indicate the quality (wavelength) of light.

Ocelli in the cockroach, *P. americana*, contain a photosensitive pigment that has maximum sensitivity to light of 500 nm wavelength, while ocelli from honeybees contain receptors that show maximum sensitivity to light of 340 and 490 nm, probably indicative of two opsins (Goldsmith and Ruck, 1958). Rhabdomere structure of the ocellar retinula cells is similar to that in compound eyes. Honeybees have 3 ocelli and each contains about 800 retinula cells whose short axons converge on about eight second-order neurons (Toh and Kuwabara, 1974). Thus, the convergence ratio is about 100 retinula neurons to 1 second-order neuron. The second-order neurons from the three ocelli converge and form the ocellar nerve projecting to the posterior of the protocerebrum. The high convergence ratio is probably further evidence that an image is not transmitted.

14.10 LARVAL EYES: STEMMATA

Some larvae of holometabolous insects have **stemmata**, which vary in number and position on the head (Gilbert, 1994). Typically, larvae of Lepidoptera have six on each side of the head (Figure 14.7). Their general organization varies considerably among insect groups, but the most well-developed stemmata have an overlying transparent cuticle, a crystalline lens, and a few retinula cells with rhabdomere regions. Some stemmata lack a crystalline lens. In some larvae, the stemmata have two separate rhabdoms, a distal one nearest the overlying cornea and a proximal one below. Each of these rhabdoms is formed by extensions of microvilli from only a few cells, and, although the dioptric apparatus forms an image that falls on the rhabdomere surfaces, it seems likely that resolution of any image is poor.

Caterpillars frequently move the head from side to side, which may be a behavior that aids them in obtaining a wider field of view with small, multiple stemmata. Some sawfly larvae and tortricid caterpillars have been reported to sense plane-polarized light, but it is difficult to see how this could

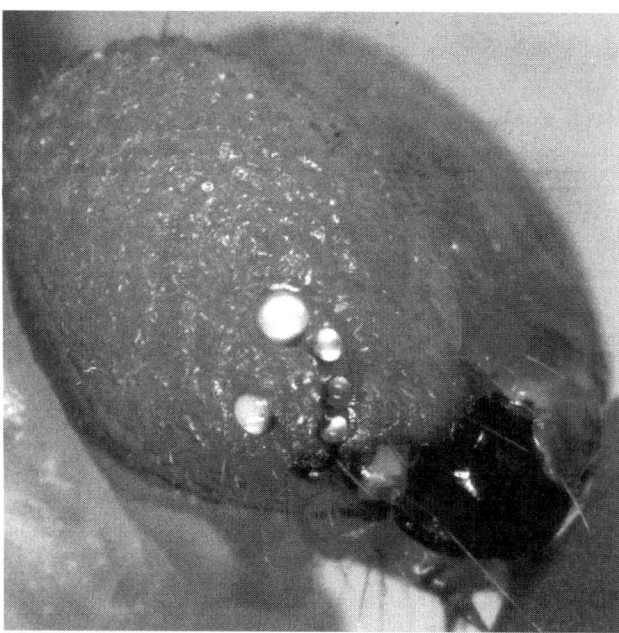

Figure 14.7 A photograph of the six stemmata (the large white circle and smaller ones nearby) on the head of the caterpillar of *Achlyodes mithridates*, a butterfly in the family Hesperiidae. (Photo courtesy of Dr. Andrei Sourakov, University of Florida and Florida Natural History Museum, Gainesville, FL.)

be adaptive to these larval insects. This, if true, needs to be verified and studied in more detail. Kral (2013) found that stemmata in mantispids (Neuroptera) may function in spatial vision because of their arrangement and structure.

Stemmata migrate inward along the developing optic nerve during metamorphosis of a larva into an adult and locate on the posterior surface of the adult optic lobe in a number of different insects (Gilbert, 1994). They retain their rhodopsins in this new location and may have new functional roles in adults (Gilbert, 1994). For example, the remnants of the larval stemmata in adult *Vanessa cardui*, the painted lady butterfly, are located on the ventral edge of the lamina ganglionaris and express messenger ribonucleic acids (mRNAs) for the UV-sensitive and green-sensitive rhodopsins found in both larval stemmata and adult compound eyes, but do not express a blue-sensitive RNA that is found in the compound eyes (Briscoe and White, 2005). The authors suggest that these adult stemmata might have some role in circadian rhythms, as has been shown in the eyelet photoreceptors of adult *Drosophila* (Regier et al., 2003). Eyelets are the adult form in *D. melanogaster* of larval Bolwig's organ, which is the homolog of stemmata in other insect larvae.

14.11 DERMAL LIGHT SENSE

There are small pockets of photosensitive cells in the cuticle of larvae of Diptera, which have no obvious eyes on the surface of the body. These cup-shaped pockets contain a few cells whose axons run into the short, stubby central nervous system (CNS). Tests of avoidance responses made with monochromatic light indicate that light of 540 nm is most effective in stimulating the photosensitive cells. The larvae try to avoid the light and attempt to crawl into the food or under other debris or cover. Many insect larvae can respond to light when the eyes are blackened or otherwise occluded, but very little information is available on the possible receptors involved.

14.12 CHEMISTRY OF INSECT VISION

The first event leading to a visual image is a **photochemical reaction**. Light quanta are absorbed by the pigment **rhodopsin** in rhabdomere microvilli. This leads to a chemical change in rhodopsin, which in turn leads to the opening of Na^+ and Ca^{2+} channels, depolarization, and electrical activity in the axons of the retinula cells. Thus, light energy is transduced into chemical energy by the pigment and, then, chemical energy is transduced into electrical energy by the neuron. Retinula cells are sensitive to one photon of light.

Rhodopsin is the name for the visual pigment in light-sensitive cells of all animals. It is composed of a protein, **opsin**, and a chromophore, **11-*cis*-retinal**, or a closely related molecule. In insect eyes, the chromophore (Figure 14.8) can be 11-*cis*-retinal or **3-hydroxy-11-*cis*-retinal** (Vogt, 1983; Vogt and Kirschfeld, 1984; Smith and Goldsmith, 1990). Rhodopsin is usually characterized on the basis of its maximum spectral sensitivity, for example, as a UV-sensitive rhodopsin, or blue sensitive, green sensitive, or red sensitive. Rhodopsin is a member of a family of **G-protein-coupled receptors** (Mizunami, 1994). **G-proteins** are intracellular proteins that are second messengers in a signaling process. The protein portion, opsin, is a transmembrane protein with seven α-helical domains (Figure 14.9). Opsin (and, of course, rhodopsin) is composed of varying numbers and sequences of amino acids, which give it species specificity and determine its spectral properties (Zhukovsky and Oprian, 1989). Retinal is held in a "pocket" of a portion of opsin that lies within the plasma membrane of the retinula cell. Retinal is held as a protonated Schiff base, having formed a covalent bond between the aldehyde group of retinal and the epsilon amino group of lysine residue 296 in transmembrane 7 of the protein (Stryer, 1975; Dratz and Hargrave, 1983; Zhukovsky et al., 1991). The proton of the Schiff base is counterbalanced by an induced negative charge on a glutamate residue in transmembrane 3 (Zhukovsky et al., 1991). Amino acid substitutions in opsin, especially in transmembrane 3 and transmembrane 6 domains, shift the distribution of electrons around the retinal chromophore and within the opsin portion of rhodopsin and alter the wavelength of maximum absorption of light. Briscoe and Bernard (2005) conclude that the difference between the slightly blue-shifted rhodopsin (R522, i.e., the λ_{max}) in the malachite butterfly, *Siproeta stelenes*, and R530 in the peacock butterfly, *Inachis io*, is due to the substitution of an amino acid at position S138A in the opsin of the peacock butterfly that cause a tuning shift toward the blue region.

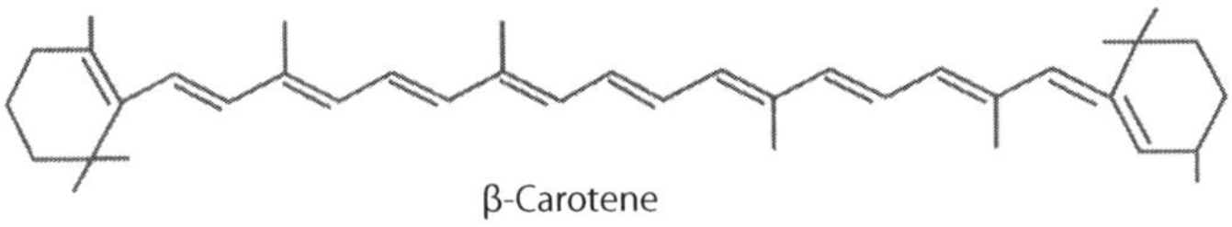

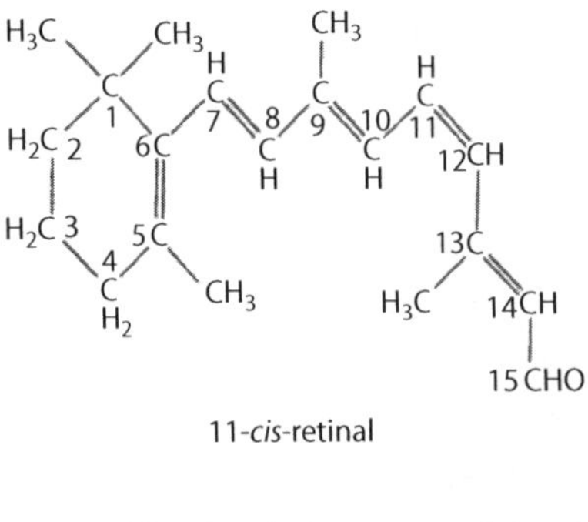

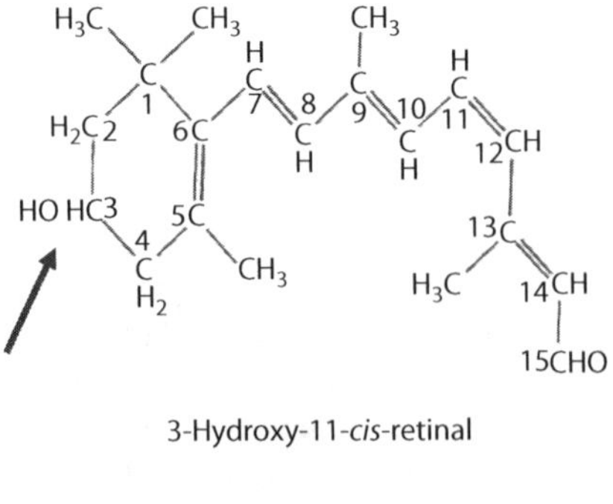

Figure 14.8 The chemical structure of the chromophores retinal and 3-hydroxy retinal found in insect eyes and β-carotene from which the compounds are derived.

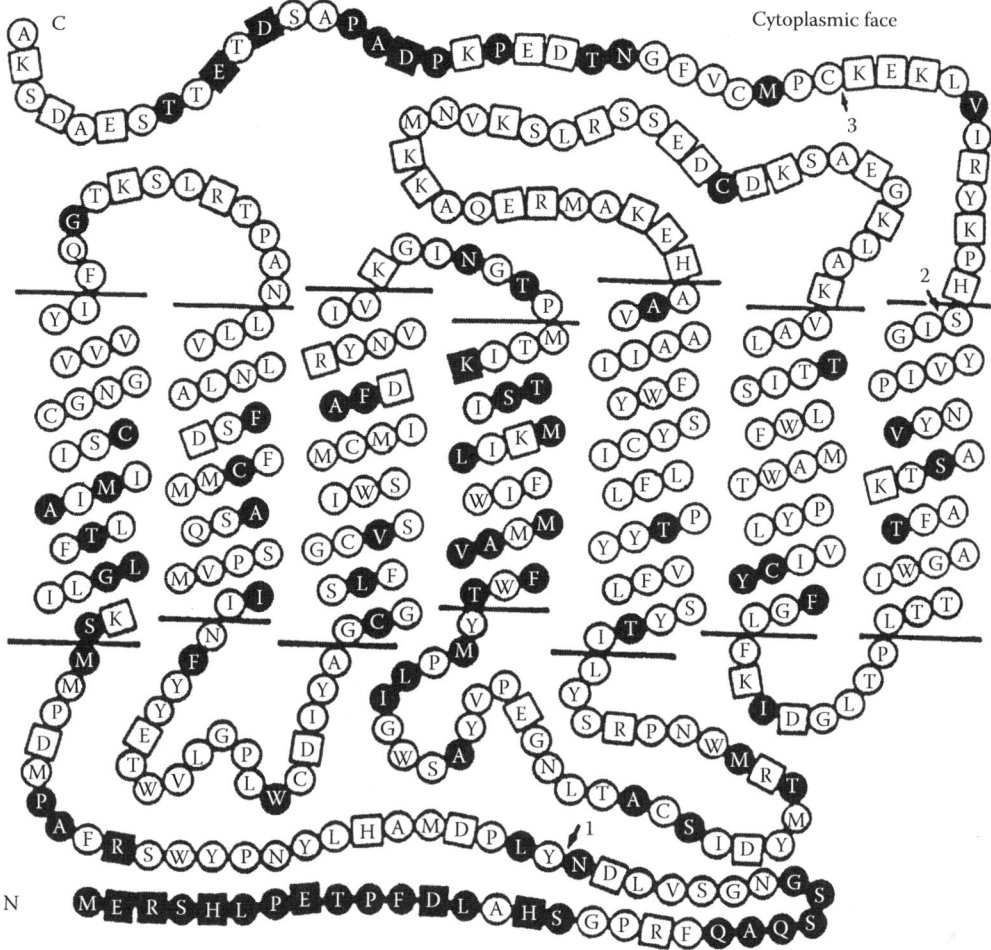

Figure 14.9 A diagram of the transmembrane nature of rhodopsin in the membranes of the rhabdomeres of retinular cells. (From Mizunami, M., *Adv. Insect Physiol.*, 25, 151, 1994. With permission.)

14.13 VISUAL CASCADE

When a photon of light is absorbed by rhodopsin, 11-*cis*-retinal is isomerized to 11-*trans*-retinal. This also initiates a conformational change in the protein portion of rhodopsin, and it becomes activated as **metarhodopsin**, a catalyst that activates an amplification cascade of reactions. A G-protein (also called **transducin**) binds to a loop of the metarhodopsin (Smith et al., 1991) at the inside surface of the plasma membrane (Figure 14.10). As is typical of G-proteins, transducin is composed of three subunits: Tα, Tβ, and Tγ. **Guanosine diphosphate (GDP)** is bound to the Tα subunit as a part of its structure. The G-protein–metarhodopsin complex is in an activated state enabling **guanosine triphosphate (GTP)** to replace GDP at the Tα subunit without an additional energy requirement. Transducin, now in a higher energy state, breaks away from metarhodopsin, and the α-subunit with bound GTP separates from the β–γ subunits. The Tα–GTP protein is the functional G-protein second messenger, and it activates phospholipase C, which cleaves **phosphatidylinositol bisphosphate (PIP₂)**, a component in the membranes of the microvilli, into **diacylglycerol** (DAG) and **inositol**

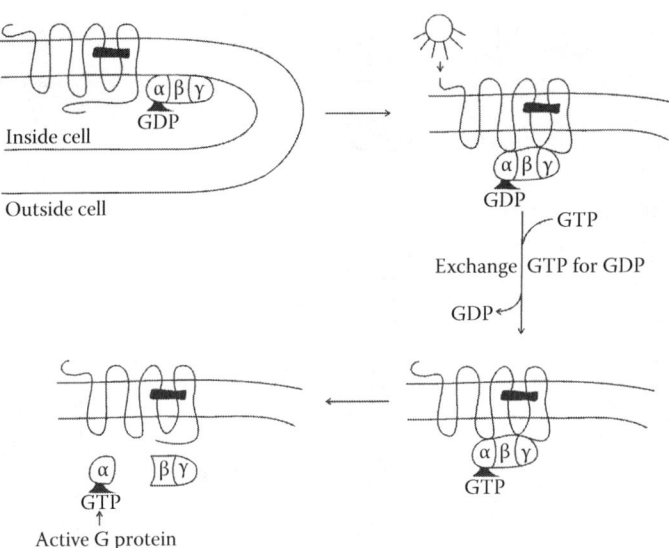

Figure 14.10 A diagram to illustrate the formation of active G-protein as the second messenger in the visual cascade when light activates rhodopsin in the rhabdomere membranes.

trisphosphate (IP_3). IP_3 promotes the release of calcium bound to smooth endoplasmic reticulum near the base of the microvilli in retinula cells (Spiegel et al., 1994). Free ionic calcium, Ca^{2+}, opens sodium channels that also admit more Ca^{2+} in the base of the microvilli. The inward movement of a few sodium ions generates a receptor (graded) potential (Hardie, 1991; Ranganathan et al., 1991; Hardie and Minke, 1994), and entry of more Ca^{2+} acts as an amplifying mechanism, resulting in still more sodium entry. If the receptor potential is strong enough, it gives rise to spikes in retinula axons. Liberated DAG activates protein kinase C, but its role in the visual process is not clear. It may be involved in the adaptation process in the eye. The β–γ subunits may also have some role in signal transduction, but that role, if any, also is not clear.

The visual signaling pathway is extremely fast and can be turned on and off many times per second. The cascade in *D. melanogaster* is the fastest G-protein cascade that has been measured thus far, and it takes only several tens of milliseconds to proceed from light activation of rhodopsin to the generation of a receptor potential (Zuker, 1996). The enzymatic nature of several steps of the visual cascade results in amplification of the signal by as much as 10^2–10^3-fold. Each metarhodopsin can catalyze GTP binding to as many as several hundred Tα subunits before it is deactivated (Yarfitz and Hurley, 1994), and similar amplification occurs in production of IP_3 and in release of Ca^{2+}. Thus, one photon light signal is turned into a barrage of neuronal impulses going into the lamina ganglionaris of the optic lobe.

14.14 REGULATION OF THE VISUAL CASCADE

How is the visual cascade turned off? Contrary to eye chemistry in vertebrates, one way is that insect metarhodopsin can absorb another photon of light and the metarhodopsin/11-*trans*-retinal complex is reconverted to rhodopsin with 11-*cis*-retinal, a process called **photoisomerization** (Smith and Goldsmith, 1991). Thus, rhodopsin and metarhodopsin are in a dynamic state and they cycle back and forth allowing the visual system to recover quickly for the next event. Metarhodopsin in vertebrate eyes loses its retinal chromophore on absorption of a photon of light, and a new rhodopsin molecule must be synthesized in the dark part of the eye. A second regulatory mechanism involves the binding of an inhibitory protein, called **arrestin**, to the metarhodopsin. The metarhodopsin–arrestin complex (MET–ARR) no longer binds G-protein, so the amplification cascade is prevented or stopped. Metarhodopsin is subsequently released from arrestin at the expense of adenosine triphosphate (ATP) input. There are also less well-known regulatory processes within the G-protein cascade and in activity of phospholipase C. Although constant resynthesis of 11-*cis*-retinal and rhodopsin after each visual stimulus is not necessary in insect compound eyes because of photoisomerization of metarhodopsin to rhodopsin, visual pigment molecules do wear out and new ones have to be synthesized by insects from time to time. Some insects shed the rhabdomere portions of the retinula cells and they have to synthesize new microvilli and rhodopsin. In order to synthesize new rhodopsin, 11-*trans*-retinol obtained from β-carotene must be enzymatically converted to 11-*trans*-retinal. A photoisomerase enzyme in eye tissues of honeybees (Smith and Goldsmith, 1990) is activated by absorbing a photon of light, and it catalyzes the isomerization of 11-*trans*-retinal to 11-*cis*-retinal, which is then combined with opsin.

14.15 COLOR VISION

Color vision is the ability to discriminate between two wavelengths of light, independent of intensity. Numerous reviews of color vision in insect, and particularly in butterflies, are available (Briscoe, 1998, 2000, 2001, 2002, 2008; Pichaud et al., 1999; Briscoe and Chittka, 2001; Briscoe et al., 2003; Bradley et al., 2009), and Briscoe and Chittka (2001) included phylogenies of insect groups with respect to color vision, data on ecological adaptation, and evolution of color vision, and they include a table of insects in which color vision has been documented. Nearly all species studied, with a few exceptions, have UV-, blue-, and green-sensitive photoreceptors, but red receptors (λ_{max} > 565 nm) have evolved only a few times, independently, in a few insects (Briscoe and Chittka, 2001; Frentiu and Briscoe, 2008).

Honeybees were the first insects in which perception of color was demonstrated with behavioral tests (see review by Srinivan, 2010). The German scientist and behaviorist Karl von Frisch (1964) trained honeybees to come to sugar water in a small dish placed on a sheet of blue paper lying on a table outdoors. After the bees had communicated the location to others in their hive (by the bee dance) and had recruited a regular stream of visitors to the dish, von Frisch replaced the blue paper with a clean one and an empty dish (the bees might have left an odor on the previous paper after alighting on it many times, and they might somehow smell the sugar solution or have added some olfactory cue to it by their feeding). He also made a checkerboard arrangement around the blue paper with gray papers of the same size as the blue one and graded in intensity from white to black. Each paper contained an empty dish. Von Frisch reasoned that colorblind bees would confuse the blue paper with one or more of the gray papers and probably would alight on the wrong dish or paper in their search for the sugar solution. They were not confused, however, but flew directly to the dish (now without sugar water) on the blue paper. He performed many variations of this experiment and found that the bees could be trained to come to sugar water on some other colored papers, but they could not distinguish red from black or dark-gray-colored papers. Subsequent work, including electrophysiological analysis of the spectral sensitivity of the honeybee compound eye, demonstrated that they have receptors with maximum sensitivity at 344 nm (UV), 436 nm (blue), and 544 nm (green). They do not have a red-sensitive receptor, which explains why von Frisch could not train them to discriminate red papers containing sugar water.

Red-sensitive photoreceptors have evolved sporadically only a few times in insects. There are documented cases in Odonata, Hymenoptera, Coleoptera, and Lepidoptera (Briscoe and Cittka, 2001). Red-sensitive photoreceptors have not been found in Blattaria, Orthoptera, Heteroptera, or Diptera (Briscoe, 2000). Red receptors are more common in Lepidoptera and have evolved at least four times, although some lepidopterans subsequently lost the red receptor (Swihart, 1967; Bernard,

1979; Briscoe and Chittka, 2001). The nymphalid butterfly, *Heliconius erato*, has rhodopsins with peak absorptions in the UV range, one in the blue-green range and one absorbing maximally at long wavelengths (long wavelength = the red region, $\lambda_{max} > 565$ nm). Contrary to what would be expected with only one long-wavelength receptor, these butterflies can discriminate colored light at 590, 620, and 640 nm (Zaccardi et al., 2006); thus, they can distinguish yellow orange from orange from orange red. The evidence is that different facets of *H. erato* contain the same rhodopsin, but filtering pigments near the rhabdom in different ommatidia probably tune the spectral sensitivity of the long-wavelength receptor, allowing discrimination at the different wavelengths. *Vanessa atalanta*, the Red Admiral butterfly, has the ability to distinguish red color from green and blue, but cannot distinguish the more subtle differences in the red part of the spectrum that *H. erato* can. The Japanese yellow swallowtail butterfly, *Papilio xuthus*, has photoreceptor cells in which rhodopsins absorb maximally in the UV at 360 nm, violet range at 400 nm (sensitivity to violet color may be because the receptor pigment is being tuned by filtering pigments; a rhodopsin with a spectral peak at 400 nm is not common in insects), blue range at 460 nm, green range at 520 nm, and red range at 600 nm. The butterflies were trained by Kinoshita et al. (1999) to feed on sugar water from dishes placed on colored discs of paper in the laboratory. The butterflies most easily learned to look for food on red and yellow colors, but training to other colors required more time, and they lost the ability to distinguish blue when the intensity of the color was reduced to 80% of the training intensity (intensity was reduced by placing neutral density filters over the color). Another swallowtail, *Papilio glaucus*, has rhodopsins with maximal absorption at approximately the same maxima as *P. xuthus*, except it does not have one absorbing in the violet range at 400 nm. The painted lady butterfly, *V. cardui*, has a UV-sensitive rhodopsin with peak absorption at 360 nm, a blue-sensitive one absorbing maximally at 470 nm, and a green-sensitive rhodopsin absorbing at 530 nm, but it does not have a red-sensitive rhodopsin (Briscoe and White, 2005).

Butterflies in the family Nymphalidae, genus *Heliconius*, are classic examples of evolution of mimicry, having undergone adaptive evolution in Mexico and Central and South America in which distantly related species have very similar color patterns on the wings (Hsu et al., 2001; Briscoe et al., 2010; Yuan et al., 2010; Bybee et al., 2012). Members of a species have to recognize each other, and evolution of genes in the eye appears to have tracked the evolution of wing colors. A gene duplication gave rise to the two UV-sensitive rhodopsins in *H. erato* at the same time that UV-reflecting yellow pigments based on 3-hydroxy-DL-kynurenine evolved in the wings of various species of *Heliconius* (Briscoe et al., 2010; Yuan et al., 2010; Bybee et al., 2012). These authors suggest that the second UV gene may enable the butterflies to discriminate among *Heliconius* species with the UV-reflecting and long-wavelength-reflecting yellow kynurenine derivative in the wings, while close relatives have yellow pigments that lack the same UV reflectance. *H. erato* and some other *Heliconius* species have four opsin-encoding genes, while most nymphalids have only three. The four genes in *H. erato* include *UVRh1* and *UVRh2* encoding two rhodopsins with λ_{max} in the UV range at 355 and 398, respectively, *BRh* encoding a rhodopsin with λ_{max} at 470 nm (blue), and *LWRh* encoding a rhodopsin with λ_{max} at 555 nm (long-wavelength range) (Yuan et al., 2010; Bybee et al., 2012). By contrast, most nymphalid butterflies have eyes that contain only one gene (*UVRh*) encoding a UV-sensitive rhodopsin with λ_{max} at 300–400 nm, a blue-sensitive rhodopsin with λ_{max} at 400–500 nm, and a long-wavelength-sensitive rhodopsin with λ_{max} at 500–600 nm. Another nymphalid butterfly, *Bicyclus anynana*, that occurs in Africa undergoes curious changes in wet vs. dry seasonal forms, driven by temperature, with concomitant plasticity in sexual dimorphism, phenotype (variation in wing color brightness and differences in which sex performs the most courting and choosing behavior), and opsin expression levels (Everett et al., 2012). Increasing temperature produced larger eyes in both sexes. But males always had larger eyes than females and ommatidia were larger in choosy dry season males. Dry season females, which were less choosy, had lower opsin transcript levels. Sison-Mangus et al. (2006) investigated the relationship between opsin evolution and wing color in *Lycaena rubidus*, a lycaenid butterfly. *L. rubidus* has four opsins with visual pigment peak absorbances of 360 nm (UV), 437 nm (blue), 500 nm (blue), and 568 nm (long wave [LW]). The 500 nm blue pigment is unusual in insects, and Sison-Mangus et al. (2006) traced its origin to a blue gene duplication event at the base of the polyommatine + thecline + lycaenine radiation; they further suggest that duplication of the blue opsin gene may have influenced the evolution of wing and body color of the blue lycaenid butterflies. Photographs of histological sections, a scanning electron microscope photograph, and distribution of opsin pigments in the eye can be viewed in the supplememtary material in the paper by Sison-Mangus et al. (2006) at http://jeb.biologists.org/cgi/content/full/209/16/3079/DCi (Sison-Mangus et al., 2006) (last viewed April 22, 2015).

Although relatively few moths have been investigated for color vision, two noctuid moths, *Spodoptera exempta* and *Mamestra brassicae*, have red-sensitive photoreceptors (Briscoe and Chittka, 2001). The tobacco hornworm moth, *M. sexta*, does not have a red-sensitive photoreceptor, but it has sensitivity to UV, blue, and green colors, with differential distribution of these receptors in three domains of the compound eyes, ventral, dorsal, and dorsal rim regions (White et al., 2003). Blue-sensitive photoreceptors predominate in the ventral part of the eye, and green-sensitive photoreceptors are about equally distributed in both ventral and dorsal regions. UV-sensitive photoreceptors are distributed in various parts of the eye, but predominate in the dorsal rim part of the eyes. The dorsal rim region of the compound eyes of many insects typically have predominately UV-sensitive photoreceptors, which seem to be involved in detection of plane-polarized light (see Section 14.18).

Both vertebrates and insects appear to have paralogously derived multiple red and green pigments (Briscoe, 2000); duplications of the green-sensitive opsin, and occasional substitution of amino acids at crucial sites in some of the duplicates, probably led to the evolution of red-sensitive rhodopsins in insects. It is known from genomic analyses and opsin sequencing data that diversity in the opsin gene family has occurred (Spaeth and Briscoe, 2005), although the insects analyzed often have only the typical sensitivity to UV, green, and blue wavelengths. The greatest diversity of opsin genes occurs in Lepidoptera (especially in the butterfly genus *Papilio*) and Diptera (*Anopheles gambiae*, the yellow fever mosquito) (Spaethe and Briscoe, 2004). *P. glaucus* has six opsins, three of which are long-wavelength opsins. *A. gambiae* has 12 opsin genes, the largest number known in insects so far, and 7 of them encode long-wavelength opsins (Hill et al., 2002). Chittka (1996) proposed that red pigments evolved from a very ancient class of green pigments, which are common in most insects, and there is bootstrap support for the idea that a *D. melanogaster* green-sensitive rhodopsin (Rh6) is ancestral to all the red-sensitive rhodopsins (Briscoe, 2000) and the red-sensitive rhodopsins in *P. glaucus* probably evolved from green ancestors in Lepidoptera. Frentiu et al. (2007) document that gene duplication has been a major mechanism for expanding the diversity of long-wavelength rhodopsins in butterflies, especially in the families Nymphalidae and Riodinidae. A blue gene duplication in the lycaenid butterfly *Polyommatus icarus* resulted in a green-sensitive rhodopsin (Sison-Mangus et al., 2008). Ancient gene duplications in the metazoan radiation gave rise to two genes (*cry1* and *cry2*) encoding cryptochromes, proteins that function in the circadian clock of metazoans (Yuan et al., 2007). CRY1 protein is photosensitive and present in *Drosophila*, while CRY2, present in nondrosophilid insects but not light sensitive, also is involved in the clock mechanism.

Little is known about the location of the visual pigments within the rhabdomeres, or whether more than one pigment might exist in the same ommatidium. Mote and Goldsmith (1970) believe, on the basis of intracellular recordings and dye-marked sites, that they recorded electrical activity from both a UV-sensitive retinula cell and a green-sensitive retinula cell within the same ommatidium of *P. americana*. The UV receptor was maximally sensitive to 365 nm, while the green receptor was maximally sensitive to 507 nm. Such experiments are technically difficult to do because of the small size of retinula cells and because the location of the recording electrode can be determined only by histological determination of dye location after the recording has been made and the experiment is over.

Menzel (1975) studied the spectral properties of eyes of *Formica polyctena*, the red wood ant in Europe, by taking advantage of the discovery that pigment migration in the retinula cells was light sensitive. In a fully dark-adapted eye (12 h darkness), the pigment was dispersed away from the rhabdom, while light adaptation (xenon light at 40,000 lux) caused the pigments to migrate toward the rhabdom (possibly adaptive as a shield against excessive light striking the rhabdom). Evaluations of pigment movements were made from a cross section through ommatidia viewed by electron microscopy. Each ommatidium in the central area of the eye contained two cells sensitive to UV (cells 1 and 5), and six cells sensitive to yellow light (cells 2, 3, 4, 6, 7, and 8). Although the UV-detecting cells were smaller, the sensitivity to UV was 20 times that to yellow light.

14.16 VISION IS IMPORTANT IN BEHAVIOR

Vision is clearly important in the ecology and behavior of many insects, enabling them to find mates, food resources, and oviposition sites. Olfaction may, of course, play a role in some or all of these actions. The tobacco hornworm, *M. sexta*, has UV, blue, and green receptors, but no red receptor. In flight tunnel tests, *M. sexta* moths responded best by flying upwind when a visual cue (a white paper flower) was presented with an olfactory stimulus (oil of bergamot, a known attractant for the moths). When the two stimuli were presented spatially or temporally separate from each other, the moths showed a response to each, with a stronger preference for the visual display, but less to either than to both presented together. Goyret et al. (2007) concluded from these experiments that the feeding behavior of *M. sexta* and possibly other nectar feeders is based on modality of stimulation as well as temporal and spatial perception of sensory stimulation. Omura and Honda (2005) showed with behavioral experiments that adult *Vanessa indica* butterflies depend mainly on the color of a flower and secondarily on the odor of the flower when foraging for nectar. Odor proved to be most important when the butterflies were presented with artificial flower models that were relatively unattractive to them, such as purple flower models. Monarch butterflies, *Danaus plexippus*, quickly learn to associate colors with nectar rewards and experimentally were demonstrated to discriminate color on the basis of wavelength rather than intensity, a criterion of true color vision, and they use lateral filtering pigments in the retinula cells to enhance long-wavelength colors (Blackiston et al., 2011). The authors suggest that color vision enhances the ability of monarchs to respond to nectar availability in milkweed plants of many different varieties, as well as other plants, as they migrate over long distances and changing food plants. The pine weevil, *Hylobius abietis*, used both visual and olfactory stimuli to locate pine seedlings for oviposition in experiments conducted by Björklund et al. (2005). Traps baited with a visual stimulus and spruce odor caught more beetles than traps baited with only one or the other stimulus. The effect of both stimuli presented together was additive in attraction, but when presented separately, the visual stimulus alone was as strong as the olfactory stimulus alone.

Arboreal ants, *Cephalotes atratus*, that fall or jump from tree branches orient their trajectories toward light-colored objects, which in their environment typically are tree trunks and lianas (Yanoviak and Dudley, 2006). The authors found that tree trunks had 2–10 times higher reflectance values than the surrounding vegetation in tropical forest environments.

14.17 NUTRITIONAL NEED FOR CAROTENOIDS IN INSECTS

All animals that have been critically studied, including some insects, require **β-carotene** or **vitamin A** or closely related **carotenoids** in the diet for normal vision. Carotenoids are readily synthesized by plants, but not by animals. There was a loss in visual sensitivity of 2.4–3.0 log units in electrophysiological tests of *Musca domestica* (Goldsmith et al., 1964) and *D. melanogaster*

(Zimmerman and Goldsmith, 1971) reared on β-carotene-free diets. Because the insects required so little carotene to satisfy their needs, the diet had to be very highly purified with respect to carotenoid pigments. In some cases, more than one generation had to be reared on the purified diet to exhaust the carotenoid pigments, which were passed through the eggs to the first or second generation. Histological changes in retinula cell structure and in underlying nervous tissue were demonstrated in *M. sexta* (Carlson et al., 1967) and in *Aedes aegypti* (Brammer and White, 1969) reared on carotenoid-free diets.

14.18 DETECTION OF PLANE-POLARIZED LIGHT

Many invertebrates are able to detect **plane-polarized** light in their surroundings. Light may be thought of in terms of particles or waves. In considering polarized light, it is best to mentally form a picture of light as a sine wave. Most of the light from the sun is not polarized, and the *e*-vector (vibration plane) is perpendicular to the direction of wave travel, with waves vibrating in every possible plane. A small percentage of the light becomes polarized by various molecules and small particles encountered in the atmosphere, and the *e*-vector of polarized waves vibrate in a specific plane. Light reflected from waxy and shiny surfaces, such as leaves, or other objects in the environment also has a polarized component. Both plane of vibration and degree of polarization from sunlight (as observed by an instrument or animal on the earth) vary with the position of the sun above the horizon (Figure 14.11) (Wehner, 1976). The direction of polarization is parallel to the horizon (only) along the path the sun takes toward the zenith and its path as it sinks in the west. At other positions above the horizon, the direction of polarization varies through all possible angles. The directions of polarization are opposite (e.g., +20° and −20°) at points separated by 180°. To further complicate the matter, the angle and degree of polarization varies at each elevation above the horizon. Clearly, using plane-polarized light as a compass is complicated.

Von Frisch demonstrated in the 1950s that honeybees use plane-polarized light for flight navigation. Houseflies, *Photuris pensylvanica* fireflies, Japanese beetles, and several species of ants orient

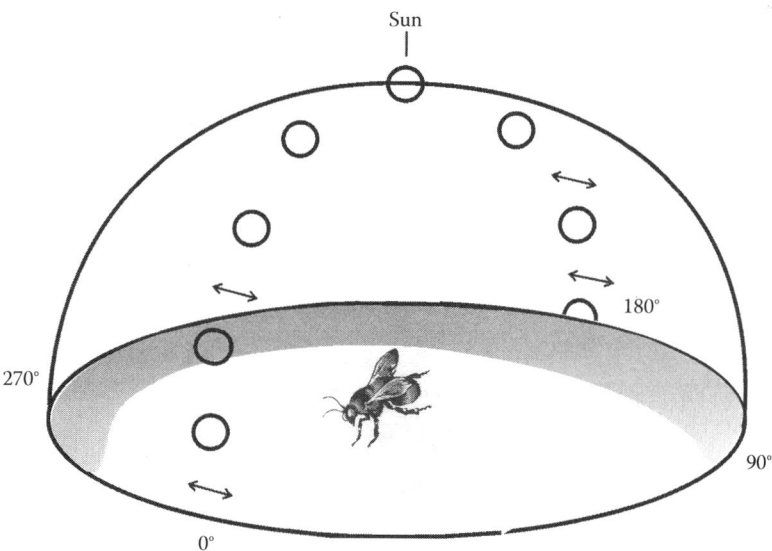

Figure 14.11 An illustration of the use of plane-polarized light by honeybees and other insects. (Modified from Wehner, R., *Sci. Am.*, 235, 106, 1976.)

to plane-polarized light under conditions that prevent them from using background reflections as orientation cues. Wehner (1976) concluded that an ant from the North African desert, *Cataglyphis bicolor*, utilized plane-polarized light to travel a direct path to its nest in the ground after having wandered, with many turns, up to 100 m from its nest in search of food. The cells sensitive to the plane of polarization are the UV-sensitive cells. Wehner determined this by holding a UV-absorbing shield over ants in the field, thereby causing them to wander aimlessly, unable to locate their underground nest. Menzel (1975) showed that *F. polyctena*, the red wood ant of northern Europe, uses polarized light as a compass.

Detection of the plane of polarized skylight and a time-compensated clock mechanism is clearly adaptive in the homing ability of bees and ants, but can plane-polarized light be used by insects that do not have homing behavior? Desert locusts, for example, do not show homing behavior, but they might use polarized light during migrations, and reflected polarized light from plant food sources may be processed by the polarization receptors. *Schistocerca gregaria* responds to polarized light, especially with two identified interneurons named TuTu1 and LoTu1 (Kinoshita et al., 2007) that respond to both polarized and ordinary light. Their response to plane-polarized light is based on blue-sensitive photoreceptors in the dorsal rim area of the compound eyes. LoTu1 neurons show approximately 2 log units greater sensitivity to polarized light than to nonpolarized light. The neurons from the dorsal rims of the compound eyes project to the brain along two different pathways (el Jundi and Homberg, 2010), one of which may have input to circadian time-compensated control of behavior.

The red swamp crayfish, not an insect, of course, is sensitive to plane-polarized light in behavioral tests. How the crayfish might benefit from detecting polarized light is not clear from experiments, but they may be able to detect transparent prey by the reflection of polarized light from their bodies and may detect predator fish by reflection of polarized light from their silvery scales. The level of polarization in the upper photic zone of water in which the crayfish forage for food is highest during crepuscular periods when the crayfish are actively feeding. Their compound eyes are similar to the compound eyes in insects, and the dorsal rim area of the eyes appears to be a site for detection of plane-polarized light (Tuthill and Johnsen, 2006). Monarch butterflies use a time-compensated sun compass during long migratory flights to Mexico, but behavioral experiments in a flight simulator that allowed the butterflies to take off in flight when exposed to a patch of naturally polarized light from the sky or to artificial polarizers or to the open sky did not indicate that the plane of polarization made any difference in their orientation (Stalleicken et al., 2005). When the dorsal rim area of the compound eyes was painted with black paint, they still used their time-compensated sun compass in orientation, but presumably could not detect the plane of polarization. The authors of these experiments concluded that the butterflies do not need polarized light cues to orient in their flight, but the ability to detect the *e*-vector of polarization might still be useful in some ecological way that these experiments did not probe.

Neotropical butterflies (family Nymphalidae) in Costa Rican tropical forests appear to use polarized light reflected from the shiny surfaces of leaves and the surface of insects to detect forage and oviposition sites and to identify conspecifics in the low light intensity of the forest foliage (Douglas et al., 2007). Polarized light may be useful in motion detection, particularly in dim light.

Mayflies (order Ephemeroptera) probably use reflected polarized light to identify water surfaces where they can lay their eggs. Unfortunately, they also can be fooled into laying them in the wrong place; mayflies in one site in Hungary (Kriska et al., 1998) were discovered laying masses of eggs on asphalt road surfaces near the stream from which they emerged. Measurements with instruments designed to measure polarized light indicated that the asphalt surface with the sun shining on it reflected plane-polarized light in much the same way that the sunlit water surface in a stream did. Thus, some mayflies were laying their eggs in an environment where they had no chance to hatch.

The dung beetles, *Scarabaeus zambesianus*, forage for fresh animal dung around sunset, a time when light intensity is low and the polarization pattern in the sky is the simplest of the day, with light of the entire sky polarized in one direction (Dacke et al., 2003). When a beetle locates fresh dung, it quickly makes a ball and rolls it away in a straight line, possibly an adaptive mechanism to avoid competition from other dung beetles and predators or parasites attracted to the fresh dung. Experiments with polarization filters that change the *e*-vector of polarization revealed that the beetles are sensitive to the *e*-vector and reorient the rolling direction in response to an experimental change in the *e*-vector. The dorsal rim area of the compound eyes has photoreceptor cells with large rhabdom surfaces, a lack of screening pigments in surrounding cells, and the microvilli in the rhabdoms oriented orthogonal to each other (perpendicular to each other), all features providing the best arrangement for detecting the contrast in *e*-vector of polarized light. The beetles cease foraging about 40–50 min after sunset when the degree of polarization at the zenith of the sky decreases from 45% to 5% within 15 min. The change in polarization, of course, might not be the sole factor involved in cessation of activity.

In the field cricket, *Gryllus campestris*, photoreceptor cells have orthogonally oriented microvilli in the dorsal rim area of the compound eyes with a blue-sensitive rhodopsin (λ_{max} about 440 nm). The cells show strong sensitivity to the *e*-vector of polarized light, and their input converges on polarization-sensitive neurons in the optic lobes of the brain. Input from nearly 200 ommatidia converge on the optic lobe neurons, which increases the signal-to-noise ratio and sensitivity to the *e*-vector (Labhart et al., 2001).

The precise way in which insects determine the plane of polarization and how some measure time lapse (necessary because the plane of polarization changes as the sun moves across the sky) is not known with certainty. Tentative explanations involve the twisting of retinula cells (Figure 14.12) and the orientation of rhodopsin molecules in the rhabdomere microvilli with respect to the plane of polarization (Menzel, 1975; Wehner et al., 1975; Wehner, 1976). In most of the insects that are known to detect plane-polarized light, the evidence suggests that UV-sensitive receptors in the dorsal rim of the compound eyes are the principal receptors of the *e*-vector of polarized light. Monarch butterflies use a time-compensated sun compass during migration from southern Canada and the northern United States to overwintering sites in central Mexico. Sauman et al. (2005) showed that the monarch UV-sensitive photoreceptors in the dorsal rim of the compound eyes were monochromatic and likely have input to the dorsal protocerebrum of the brain through cryptochrome-stained pathways as part of a circadian clock.

14.19 VISUAL ACUITY

Visual acuity is a measure of how well two close objects can be resolved. Put simply, higher visual acuity means better ability to see objects in the environment and to navigate, capture prey, and chase a potential mate in flight. Land (1997) has provided an excellent review of visual acuity and resolution in insect compound eyes with emphasis on the mathematics of eye structure, optics, and visual acuity.

Insect eyes, however, do not even come close to having the acuity and resolving power of human eyes. The principal factors that determine visual acuity of compound eyes include the angle between two adjacent ommatidia (Figure 14.13), the optical quality of the dioptric structures that focus the light, dimensions of the rhabdom, the light level, and speed of movement of an object across facets of the eye. The small size of facets of the compound eyes severely limits visual acuity, and larger facets increase visual acuity. The diameter of facets in compound eyes of many insects vary over different parts of the eyes. Smaller interommatidial angles allow greater distances at which objects, such as prey, predators, or host plants, can be resolved (Land, 1997). Dragonflies are among the insects that have the most acute vision, with an interommatidial angle as small as 0.24°. Most insects have considerably larger interommatidial angles of several degrees up to tens of degrees (Land, 1997). The very small nature of the lens in compound eyes severely limits resolving power

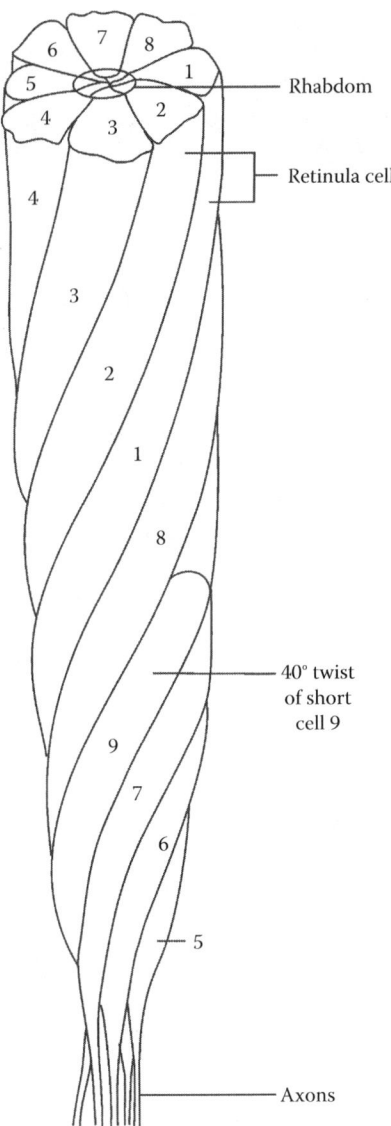

Figure 14.12 Ommatidial structure in the eye of the honeybee. About 5500 ommatidia occur in each compound eye. Eight of the retinular cells are elongated, and the ninth is short and confined to the base of the eye. The twisting of the cells may be involved in reception of polarized light; half of the ommatidia are twisted clockwise and half are twisted counterclockwise. The rhabdom is a closed or fused rhabdom in which the rhabdomeres from retinular cells touch each other. (Redrawn and modified from Wehner, R., *Sci. Am.*, 235, 106, 1976.)

because of diffraction. A human eye has much greater resolving power than a single facet of a compound eye because it is larger, has a larger opening to let light in, and has a single lens. Compound eyes are excellent motion detectors, but the fast movement of objects over the eyes causes any image to be blurred, just as movement of objects, or of the camera, causes blurring in photographs.

Some insects have variations that provide zones of greater acuity of vision (a **fovea**) in certain parts of the eye (Land, 1997; Kral, 2013). The fovea refers to the region in the human eye with the greatest density of cones (color and bright-light sensitive) where resolution is greatest when the eyes are focused directly on the object. An acute zone has evolved in the forward facing, and sometimes

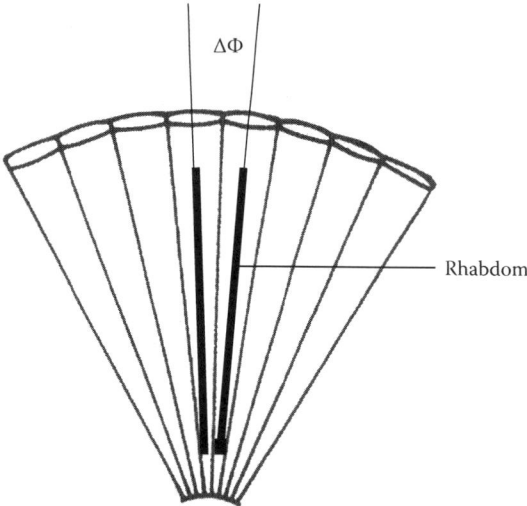

Figure 14.13 An illustration of the interommatidial angle; the smaller the angle, the better the insect can resolve objects at a distance.

upward looking, part of the compound eyes in some fast-flying insects, particularly those that capture prey in flight or chase flying potential mates (Figure 14.14). An insect in relatively straight line flight has a fairly stationary field of view straight in front, but highly blurred vision at the sides of the eyes as objects in the environment flash across the field of view. Bees, butterflies, and some acridid grasshoppers have an acute zone in the front of the compound eyes and better vertical acuity in a band around the equator of the eye. Male blowflies, *Calliphora erythrocephala*, drone honeybees, male hoverflies, some tabanid flies, and some other male insects that look for potential mates while flying have an acute zone that probably enables them to see the female better, particularly against the sky as a background. Both sexes of mantids, dragonflies, and robber flies have higher visual acuity near the forward part of the eye that likely enables them to see and capture prey more effectively.

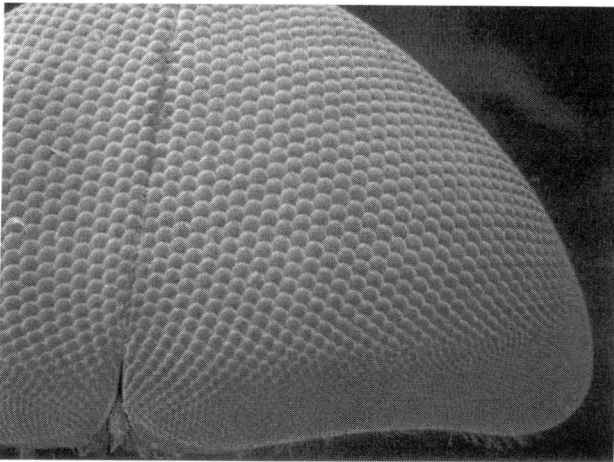

Figure 14.14 A scanning electron microscope picture of the eye of a horsefly showing larger facets at the front and upper part of the eye. (Photo courtesy of Dr. Jerry Butler, University of Florida, Department of Entomology & Nematology, Gainesville, FL, Retired.)

The fast-flying dragonfly, *Anax junius*, has 28,672 ommatidia per compound eye with the smallest known interommatidial angles, and they have an acute zone in the dorsal part of the eye with relatively large facets, as much as 62 μm across (Land, 1997). This gives the dragonfly the ability to catch mosquitoes and other small insects in flight. Another insect with good vision is the praying mantis, *Tenodera australasiae*. Facet diameters in the acute visual zone in the front of the eyes measure up to 50 μm across, and they have overlapping acute zones in the large binocular-looking eyes that enable them to determine distance of a prey object by binocular triangulation. They strike and capture the prey with the prothoracic pair of forelegs.

14.20 REVIEW AND SELF-STUDY QUESTIONS

1. What are the typical names for the three different types of light receptors (or eyes) of insects?
2. What is the name of the chromophore that is common to all animal eyes?
3. What is the difference between opsin and rhodopsin?
4. Where, in a compound eye, is the visual pigment located?
5. What are the major morphological and physiological differences between photopic and scotopic eyes?
6. What is the function of dioptric structures in insect eyes? What are some of the dioptric structures?
7. What kind of electrical activity has been recorded from insects' eyes?
8. In what major ways do ocelli differ from compound eyes?
9. Describe ways in which color vision has been shown to be important in insect behavior.
10. What are G-proteins?
11. Can insects see colors? If so, briefly describe some examples.
12. Describe the characteristics of plane-polarized light.
13. In general, what part of the compound eye in insects seems to be involved in detecting plane-polarized light?
14. Are the facets in the compound eyes of insects all of the same size? Explain answer.

REFERENCES

Agee, H.R. 1977. *Instrumentation and Techniques for Measuring the Quality of Insect Vision with the Electroretinogram*. U.S. Department of Agriculture, Washington, DC, ARS-S-162.

Agee, H.R., W.A. Phillips, and D.L. Chambers. 1977. The compound eye of the Caribbean fruit fly and the apple maggot fly. *Ann. Entomol. Soc. Am.* 70: 359–364.

Bernard, G.D. 1979. Red-absorbing visual pigment of butterflies. *Science* 203: 1125–1127. *BioEssays*. 30: 11–12.

Björklund, N., G. Nordlander, and H. Bylund. 2005. Olfactory and visual stimuli used in orientation to conifer seedlings by the pine weevil, *Hylobius abietis*. *Physiol. Entomol.* 30: 225–231.

Blackiston, D., A.D. Briscoe, and M.R. Weiss. 2011. Color vision and learning in the monarch butterfly, *Danaus plexippus* (Nymphalidae). *J. Exp. Biol.* 214: 509–520.

Boschek, C.B. 1971. On the fine structure of the peripheral retina and lamina ganglionaris of the fly *Musca domestica*. *Z. Zellforsch.* 118: 369–409.

Bradley, T.J., A.D. Briscoe, S.G. Brady, H.L. Contreras, B.N. Danforth, R. Dudley, D. Grimaldi et al. 2009. Episodes in insect evolution. *Integr. Comp. Biol.* 49: 590–606.

Braitenberg, V. 1972. I. Anatomy of the visual system. 1. Periodic structures and structural gradients in the visual ganglia of the fly, in R. Wehner (ed.), *Information Processing in the Visual Systems of Arthropods*. Springer-Verlag, New York, pp. 3–15.

Brammer, J.D. and R.H. White. 1969. Vitamin A deficiency: Effect on mosquito eye ultrastructure. *Science* 163: 821–823.

Briscoe, A.D. 1998. Molecular diversity of visual pigments in the butterfly *Papilio glaucus*. *Naturwissenschaften* 85: 33–35.

Briscoe, A.D. 2000. Six opsins from the butterfly *Papilio glaucus*: Molecular phylogenetic evidence for paralogous origins of red-sensitive visual pigments in insects. *J. Mol. Evol.* 51: 110–121.

Briscoe, A.D. 2001. Functional diversification of lepidopteran opsins following gene duplication. *Mol. Biol. Evol.* 18: 2270–2279.

Briscoe, A.D. 2002. Homology modeling suggests a functional role for parallel amino acid substitutions between bee and butterfly red- and green-sensitive opsins. *Mol. Biol. Evol.* 19: 983–986.

Briscoe, A.D. 2008. Reconstructing the ancestral butterfly eye: Focus on the opsins. *J. Exp. Biol.* 211: 1805–1813.

Briscoe, A.D. and G.D. Bernard. 2005. Eyeshine and spectral tuning of long wavelength-sensitive rhodopsins: No evidence for red-sensitive photoreceptors among five Nymphalini butterfly species. *J. Exp. Biol.* 208: 687–696.

Briscoe, A.D., G.D. Bernard, A.S. Szeto, L.M. Nagy, and R.H. White. 2003. Not all butterfly eyes are created equal: Rhodopsin absorption spectra, molecular identification, and localization of ultraviolet-,blue-, and green-sensitive rhodopsin-encoding mRNAs in the retina of *Vanessa cardui*. *J. Comp. Neurol.* 458: 334–349.

Briscoe, A.D., S.M. Bybee, G.D. Bernard, F. Yuan, M.P. Sison-Mangus, R.D. Reed, A.D. Warren, J. Llorente-Bousquets, and C.-C. Chiao. 2010. Positive selection of a duplicated UV-sensitive visual pigment coincides with wing pigment evolution in *Heliconius* butterflies. *Proc. Natl. Acad. Sci. USA* 107: 3628–3633.

Briscoe, A.D. and L. Chittka. 2001. The evolution of color vision in insects. *Annu. Rev. Entomol.* 46: 471–510.

Briscoe, A.D. and R.H. White. 2005. Adult stemmata of the butterfly *Vanessa cardui* express UV and green opsins mRNAs. *Cell Tissue Res.* 319: 175–179.

Bybee, S.M., F. Yuan, M.D. Ramstetter, J. Llorente-Bousquets, R.D. Reed, D. Orsorio, and A.D. Briscoe. 2012. UV photoreceptors and UV-yellow wing pigments in *Heliconius* butterflies allow a color signal to serve both mimicry and intraspecific communication. *Am. Nat.* 179: 38–51.

Carlson, S.D., H.R. Stevens III, J.S. Vandeberg, and W.E. Robbins. 1967. Vitamin A deficiency: Effect on retinal structure of the moth *Manduca sexta*. *Science* 158: 268–270.

Chittka, L. 1996. Does bee color vision predate the evolution of flower color? *Naturwissenschaften* 83: 136–138.

Dacke, M., P. Nordstrom, and C.H. Scholtz. 2003. Twilight orientation to polarized light in the crepuscular dung beetle *Scarabaeus zambesianus*. *J. Exp. Biol.* 206: 1535–1543.

Douglas, J.M., T.W. Cronin, T.-H. Chiou, and N.J. Dominy. 2007. Light habitats and the role of polarized iridescence in the sensory ecology of neotropical nymphalid butterflies (Lepidoptera: Nymphalidae). *J. Exp. Biol.* 210: 788–799.

Dratz, E.A. and P.A. Hargrave. 1983. The structure of rhodopsin and the rod outer segment disk membrane. *Trends Biochem. Sci.* 8: 128–131.

el Jundi, B. and U. Homberg. 2010. Evidence for the possible existence of a second polarization-vision pathway in the locust brain. *J. Insect Physiol.* 56: 971–979.

Everett, A., X. Tong, A.D. Briscoe, and A. Monteiro. 2012. Phenotypic plasticity in opsin expression in a butterfly compound eye complements sex role reversal. *BMC Evol. Biol.* 12: 232.

Frentiu, F.D., G.D. Bernard, M.P. Sison-Mangus, A. Van Zandt Brower, and A.D. Briscoe. 2007. Gene duplication is an evolutionary mechanism for expanding spectral diversity in the long-wavelength photopigments of butterflies. *Mol. Biol. Evol.* 24: 2016–2028.

Frentiu, F.D. and A.D. Briscoe. 2008. A butterfly eye's view of birds. *BioEssays* 30: 11–12.

Fuortes, M.G.F. 1963. Visual responses in the eye of the dragon fly. *Science* 142: 69–70.

Gilbert, C. 1994. Form and function of stemmata in larvae of holometabolous insects. *Annu. Rev. Entomol.* 39: 323–349.

Goldsmith, T.H., R.J. Barker, and C.F. Cohen. 1964. Sensitivity of visual receptors of carotenoid-depleted flies: A vitamin A deficiency in an invertebrate. *Science* 146: 65–67.

Goldsmith, T.H. and C.D. Bernard. 1974. The visual system of insects, in M. Rockstein (ed.), *The Physiology of Insecta*. Academic Press, New York, pp. 165–272.

Goldsmith, T.H. and P.R. Ruck. 1958. The spectral sensitivities of the dorsal ocelli of cockroaches and honey bees. *J. Gen. Physiol.* 41: 1171–1185.

Goodman, J.L. 1970. The structure and function of the insect dorsal ocellus. *Adv. Insect Physiol.* 7: 97–195.

Goyret, J., P.M. Markwell, and R.A. Raguso. 2007. The effect of decoupling olfactory and visual stimuli on the foraging behavior of *Manduca sexta*. *J. Exp. Biol.* 210: 1398–1405.

Hardie, R. 1991. Whole-cell recordings of the light induced current in dissociated *Drosophila* photoreceptors: Evidence for feedback by calcium permeating the light-sensitive channels. *Proc. Roy. Soc. Lond. B* 245: 203–210.

Hardie, R.C. and B. Minke. 1994. Spontaneous activation of light-sensitive channels in *Drosophila* photoreceptors. *J. Gen. Physiol.* 103: 389–407.

Hill, C.A., A.N. Fox, R.J. Pitts, L.B. Kent, P.L. Tan, M.A. Crystal, A. Cravchik, F.H. Collins, H.M. Robertson, and L.J. Zwiebel. 2002. G protein-coupled receptors in *Anopheles gambiae*. *Science* 298: 176–178.

Hsu, R., A.D. Briscoe, B.S.W. Chang, and N.E. Pierce. 2001. Molecular evolution of a long wavelength-sensitive opsin in mimetic *Heliconius* butterflies (Lepidoptera: Nymphalidae). *Biol. J. Linnean Soc.* 72: 435–449.

Kinoshita, M., K. Pfeiffer, and U. Homberg. 2007. Spectral properties of identified polarized-light sensitive interneurons in the brain of the desert locust *Schistocerca gregaria*. *J. Exp. Biol.* 210: 1350–1361.

Kinoshita, M., N. Shimada, and K. Arikawa. 1999. Colour vision of the foraging swallowtail butterfly *Papilio xuthus*. *J. Exp. Biol.* 202: 95–102.

Kral, K. 2013. Vision in the mantispid: A sit-and-wait and stalking predatory insect. *Physiol. Entomol.* 38: 1–12.

Kriska, G., G. Horvath, and S. Andrikovics. 1998. Why do mayflies lay their eggs *en masse* on dry asphalt roads? Water-imitating polarized light reflected from asphalt attracts Ephemeroptera. *J. Exp. Biol.* 201: 2273–2286.

Labhart, T., J. Petzold, and H. Helbling. 2001. Spatial integration in polarization-sensitive interneurons of crickets: A survey of evidence, mechanisms and benefits. *J. Exp. Biol.* 204: 2423–2430.

Land, M.F. 1997. Visual acuity in insects. *Annu. Rev. Entomol.* 42: 147–177.

Menzel, R. 1975. Polarization sensitivity in insect eyes with fused rhabdomes, in A.W. Snyder and R. Menzel (eds.), *Photoreceptor Optics*. Springer-Verlag, Berlin, Germany, pp. 372–387.

Miller, W.H., G.D. Bernard, and J.L. Allen. 1968. The optics of insect compound eyes. *Science* 162: 760–767.

Mizunami, M. 1994. The diversity of ocellar systems. *Adv. Insect Physiol.* 25: 151–265.

Mote, M.I. and T.H. Goldsmith. 1970. Compound eyes: Localization of two color receptors in the same ommatidium. *Science* 171: 1254–1255.

Nordström, P. and E.J. Warrant. 2000. Temperature-induced pupil movements in insect superposition eyes. *J. Exp. Biol.* 203: 685–692.

Omura, H. and K. Honda. 2005. Priority of color over scent during flower visitation by adult *Vanessa indica* butterflies. *Oecologia* 142: 588–596.

Phillis, W.A., III and H.L. Cromroy. 1977. The microanatomy of the eye of *Amblyomma americanum* (Acari: Ixodidae) and resultant implication of its structure. *J. Med. Entomol.* 13: 685–698.

Pichaud, F., A. Briscoe, and C. Desplan. 1999. Evolution of color vision. *Curr. Opin. Neurobiol.* 9: 622–627.

Ranganathan, R., G.L. Harris, C.F. Stevens, and C.S. Zuker. 1991. A *Drosophila* mutant defective in extracellular calcium-dependent photoreceptor deactivation and rapid desensitization. *Nature* (*London*) 354: 230–232.

Regier, D., R. Stanewsky, and C. Helfrich-Forster. 2003. Cryptochrome, compound eyes, Hofbauer-Buchner eyelets and ocelli play different roles in the entrainment and masking pathway of the locomotor activity rhythm in the fruit fly *Drosophila melanogaster*. *J. Biol. Rhythms* 18: 377–391.

Sauman, I., A.D. Bricoe, H. Zhu, D. Shi, O. Froy, J. Stalleicken, Q. Yuan, A. Casselman, and S.M. Reppert. 2005. Connecting the navigational clock to sun compass input in monarch butterfly brain. *Neuron* 46: 457–467.

Sison-Mangus, M.P., G.D. Bernard, J. Lampel, and A.D. Briscoe. 2006. Beauty in the eye of the beholder: The two blue opsins of lycaenid butterflies and the opsin gene-driven evolution of sexually dimorphic eyes. *J. Exp. Biol.* 209: 3079–3090.

Sison-Mangus, M.P., A.D. Briscoe, G. Zaccardi, H. Knüttel, and A. Kelber. 2008. The lycaenid butterfly *Polyommatus icarus* uses a duplicated blue opsin to see green. *J. Exp. Biol.* 211: 361–369.

Smith, D.P., R. Ranganathan, R.W. Hardy, J. Marx, T. Tsuchida, and C.S. Zuker. 1991. Photoreceptor deactivation and retinal degeneration mediated by a photoreceptor-specific protein kinase C. *Science* 254: 1478–1484.

Smith, W.C. and T.H. Goldsmith. 1990. Phyletic aspects of the distribution of 3-hydroxyretinal in the class Insecta. *J. Mol. Evol.* 30: 72–84.

Smith, W.C. and T.H. Goldsmith. 1991. The role of retinal photoisomerase in the visual cycle of the honey bee. *J. Gen. Physiol.* 97: 143–165.

Spaethe, J. and A.D. Briscoe. 2004. Early gene duplication and functional diversification of the opsin gene family in insects. *Mol. Biol. Evol.* 21: 1583–1594.

Spaethe, J. and A.D. Briscoe. 2005. Molecular characterization and expression of the UV opsin in bumblebees: Three ommatidial subtypes in the retina and a new photoreceptor organ in the lamina. *J. Exp. Biol.* 208: 2347–2361.
Spiegel, A.M., L. Teresa, Z. Jones, W.F. Simonds, and L.S. Weinstein. 1994. *G Proteins*. R.G. Landes Company, Austin, TX.
Srinivasan, M.V. 2010. Honey bees as a model for vision, perception, and cognition. *Annu. Rev. Entomol.* 55: 267–284.
Stalleicken, J., M. Mukhida, T. Labhart, R. Wehner, B. Frost, and H. Mouristen. 2005. Do monarch butterflies use polarized skylight for migratory orientation? *J. Exp. Biol.* 208: 2399–2408.
Stryer, L. 1975. *Biochemistry*. W.H. Freeman & Co., San Francisco, CA.
Swihart, S.L. 1967. Maturation of the visual mechanisms in the neotropical butterfly, *Heliconius sarae*. *J. Insect Physiol.* 13: 1679–1688.
Toh, Y. and M. Kuwabara. 1974. Fine structure of the dorsal ocellus of the worker honeybee. *J. Morphol.* 143: 285–306.
Tuthill, J.C. and S. Johnsen. 2006. Polarization sensitivity in the red swamp crayfish *Procambarus clarkii* enhances the detection of moving transparent objects. *J. Exp. Biol.* 209: 1612–1616.
Vogt, K. 1983. Is the fly visual pigment a rhodopsin? *Z. Naturforsch* 38C: 329–333.
Vogt, K. and K. Kirschfeld. 1984. Chemical identity of the chromophore of fly visual pigment. *Naturwissenschaften* 71: 211–213.
von Frisch, K. 1964. *Biology*. Harper & Row, New York.
Wehner, R. 1976. Polarized-light navigation by insects. *Sci. Am.* 235: 106–115.
Wehner, R., G.D. Bernard, and E. Geiger. 1975. Twisted and non-twisted rhabdomes and their significance for polarization detection in the bee. *J. Comp. Physiol.* 104: 225–245.
White, R.H., H. Xu, T.A. Münch, R.R. Bennett, and E.A. Grable. 2003. The retina of *Manduca sexta*: Rhodopsin expression, the mosaic of green-, blue- and UV-sensitive photoreceptors, and regional specialization. *J. Exp. Biol.* 206: 3337–3348.
Wolken, J.J. 1957. A comparative study of photoreceptors. *Trans. NY Acad. Sci.* 19: 315–327.
Wolken, J.J. 1968. The photoreceptors of arthropod eyes, in J.D. Carty and G.E. Newell (eds.), *Invertebrate Receptors*. Symposium Zoological Society of London No. 23, Academic Press, New York, pp. 113–133.
Wolken, J.J. 1995. *Light Detectors, Photoreceptors, and Imaging Systems in Nature*. Oxford University Press, New York.
Wolken, J.J. and P.D. Gupta. 1961. Photoreceptor structures of the retinal cells of the cockroach eye. IV. *Periplaneta americana* and *Blaberus giganteus*. *J. Biophys. Biochem. Cytol.* 9: 720–724.
Yanoviak, S.P. and R. Dudley. 2006. The role of visual cues in directed aerial descent of *Cephalotes atratus* workers (Hymenoptera: Formicidae). *J. Exp. Biol.* 209: 1777–1783.
Yarfitz, S. and J.B. Hurley. 1994. Minireview. Transduction mechanisms of vertebrate and invertebrate photoreceptors. *J. Biol. Chem.* 269: 14329–14332.
Yuan, F., A.D. Bernard, J. Le, and A.D. Briscoe. 2010. Contrasting modes of evolution of the visual pigments in *Heliconius* butterflies. *Mol. Biol. Evol.* 27: 2392–2405.
Yuan, Q., D. Metterville, A.D. Briscoe, and S.M. Reppert. 2007. Insect cryptochromes: Gene duplication and loss of define diverse ways to construct insect circadian clocks. *Mol. Biol. Evol.* 24: 948–955.
Zaccardi, G., A. Kelber, M.P. Sison-Mangus, and A.D. Briscoe. 2006. Color discrimination in the red range with only one long-wavelength sensitive opsin. *J. Exp. Biol.* 209: 1944–1955.
Zhukovsky, E.A. and D.D. Oprian. 1989. Effect of carboxylic acid side chains on the absorption maximum of visual pigments. *Science* 246: 928–930.
Zhukovsky, E.A., P.R. Robinson, and D.D. Oprian. 1991. Transducing activation by rhodopsin without a covalent bond to the 11-*cis*-retinal chromophore. *Science* 251: 558–560.
Zimmerman, W.F. and T.H. Goldsmith. 1971. Photosensitivity of the circadian rhythm and of visual receptors in carotenoid-depleted *Drosophila*. *Science* 171: 1167–1169.
Zuker, C.S. 1996. The biology of vision in *Drosophila*. *Proc. Natl. Acad. Sci. USA* 93: 571–576.

CHAPTER 15

Circulatory System

PREVIEW

The principal organ of whole-body circulation in insects is a tubular vessel lying just beneath the dorsal body wall that generally runs the length of the insect. The "blood," usually called hemolymph, typically enters the abdominal portion, the heart, through paired ostial openings, and is pumped anteriorly through the thoracic portion, the aorta. The aorta sends out branches in a few insects, but generally it is a simple tube with an open end in the head. It is not uncommon for the heartbeat to reverse in some insects, with the contraction wave beginning at the anterior and passing toward the posterior. The heartbeat is myogenic but modified by nervous input and by neurosecretions. Accessory pulsatile organs or hearts often occur at the base of appendages, such as antennae, legs, and wings, and promote circulation into the appendages. Hemolymph does not transport oxygen, but it is important in transport of nutrients and hormones and removal of waste products. It aids locomotion in caterpillars, serves as a hydrostatic force that aids in eversion of various organs and glandular tissues, and has several other important functions. Hemocytes, free cells in the hemolymph, show variability in structure and function in different insects. Coagulocytes initiate clotting of hemolymph at a wound or when hemolymph is withdrawn from the body. Hemolymph does not clot in some insects. Hemolymph typically contains relatively high concentrations of free amino acids, the disaccharide trehalose, and numerous other chemical substances that are often in transport from the site of synthesis to a site of utilization. The pH of the hemolymph usually is slightly acid but varies over a small range in different insects. Based on limited studies, it appears that insects regulate hemolymph pH by secretion of acid or base equivalents into the gut, excretion via the renal system and hindgut in some cases, and by increasing respiratory ventilation to control CO_2 content of the hemolymph.

15.1 INTRODUCTION: EMBRYONIC DEVELOPMENT OF THE CIRCULATORY SYSTEM AND HEMOCYTES

The heart is derived from mesodermal tissue in the developing embryo. The abdominal portion of the dorsal vessel is derived from cardioblasts, cells that fuse with each other, modify their shape, and form the dorsal tubular vessel. Eventually fusion with the cephalic portion of the vessel occurs. In some insects, the heart begins to beat only after dorsal closure. In other insects, the heartbeat may begin before the posterior part of the heart fuses with the anterior portion in the head. Embryonic hemocytes are present and circulate within the cavity of the embryonic heart as they travel anteriorly and are discharged from the cephalic open end in the developing embryo of the water strider, *Gerris paludum insularis* (Mori, 1996).

15.2 DORSAL VESSEL: HEART AND AORTA

The dorsal vessel (Figure 15.1) consists of two parts, the heart and the aorta. Sometimes, the entire vessel may be referred to in the literature as simply the heart. Although heart and aorta are terms borrowed from vertebrate anatomy, they have no real physiological meaning in describing the dorsal vessel of insects; the entire length of the dorsal vessel carries a wave of contraction. The **heart** is the abdominal portion of the vessel, but it may extend into the thorax in some insects. The major criteria for deciding when the heart ends and the aorta begins are the presence of alary muscles and incurrent ostia in the heart portion. The **aorta** does not have alary muscles and lacks incurrent ostia, but it may have excurrent ostia. Both alary muscles and incurrent ostia occur in the thoracic portion of the vessel in many Orthoptera and, consequently, this thoracic portion is still the heart and only a short aorta leads into the head.

Usually the entire dorsal vessel is a simple tube, but in Orthoptera and Dictyoptera, the heart has several pairs (four pairs in *Blaberus* spp.) of long diverticula (Figure 15.2) passing laterally to tergosternal muscles and fat body tissue. Although the posterior end of the heart is usually closed, it is open in immatures of crane flies (Tipulidae). In the Ephemeroptera (mayflies), the posterior end gives rise to three branches, each passing into one of the three caudal filaments. In some insects, a chambered effect is created by invaginations of ostia and the attachment of alary muscles, but division into chambers is not complete. The heart (Figure 15.3) contains both circular and longitudinal muscle fibers in most insects. The transverse tubules and longitudinal tubules of the sarcoplasmic reticulum are poorly developed in the dorsal vessel, and transverse (T) tubules may not penetrate at the Z bands as is typically the case in most muscles.

Although the abdominal heart lies just beneath the dorsal cuticle, the aorta in the thorax often meanders between the large masses of thoracic flight muscles. In Lepidoptera and Coleoptera, the dorsal vessel passes into the ventral region of the thorax but immediately rises toward the dorsal body wall (Figure 15.4). Near the dorsal wall, it makes a sharp turn and begins to descend in close contact with the ascending loop. At the apex of the dorsal loop, the aorta is joined with a mesothoracic pulsatile organ containing a pair of ostia. This accessory pump aids flow of hemolymph into the wings. Sections through the thorax in some insects show multiple images of the aorta because of the meandering path it takes.

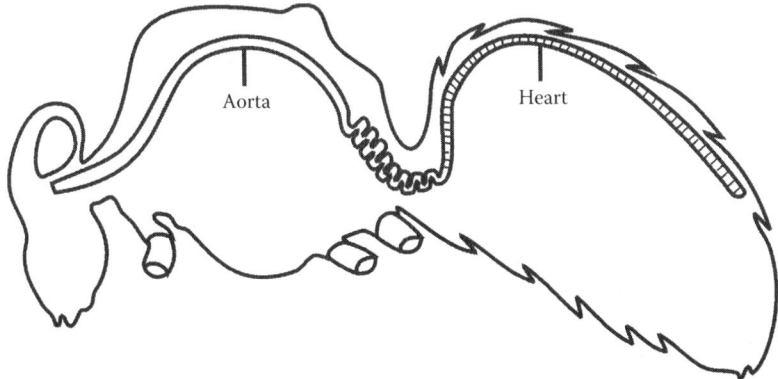

Figure 15.1 The dorsal vessel in a worker honeybee. The abdominal portion is called the heart and the thoracic portion is called the aorta. The distinction between heart and aorta is somewhat arbitrary, but the aorta has been defined by some authors as that part of the dorsal vessel that does not have incurrent ostia, although it may have excurrent ostia. By this definition, the heart sometimes extends into the thorax. (Drawing modified from Dadant & Sons (eds.), *The Hive and the Honey Bee*, Dadant & Sons, Hamilton, IL, 1975. Permission granted by acknowledging the source.)

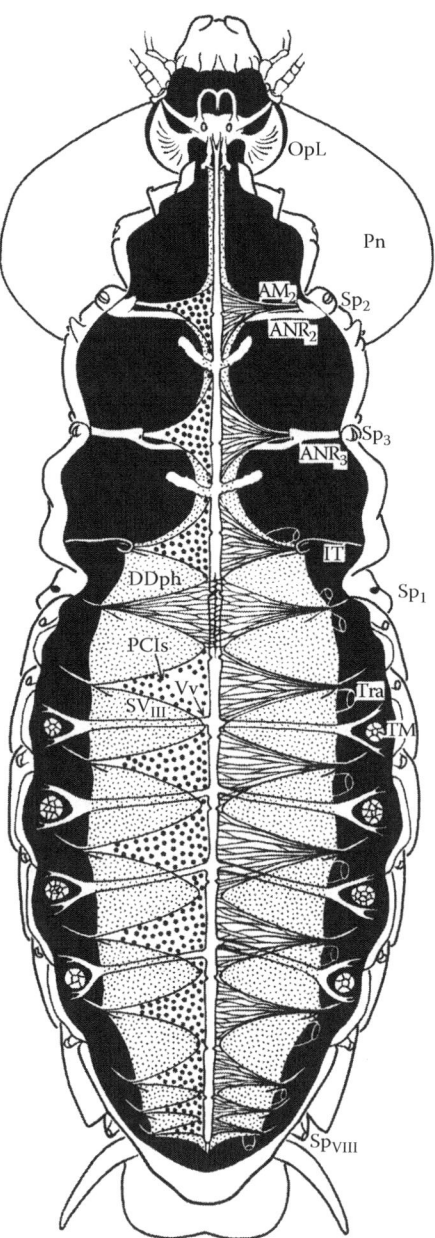

Figure 15.2 A diagram of the dorsal vessel, alary muscles, and branches of the dorsal vessel that pass out to the pleural region of the abdomen in a cockroach, *Blaberus* spp. (From Nutting, W.L., *J. Morphol.*, 89, 501, 1951. With permission.)

Hemolymph pressure is low and enters the dorsal vessel through incurrent openings called **ostia**, which may be guarded by valvular flaps in the abdominal portion of the vessel. Ostia usually exist in pairs. Incurrent ostia generally are confined to the abdomen, but sometimes there are incurrent ostia in the thoracic portion. Occasionally there are excurrent ostia in the abdomen and thorax, but most of the hemolymph exits from the open anterior end and flows posteriorly through the body cavity, called the hemocoel. There is no distinction of lymph and blood, as in vertebrates. The hemolymph is sometimes referred to as the extracellular fluid. The brain and head glands (corpora allata [CA],

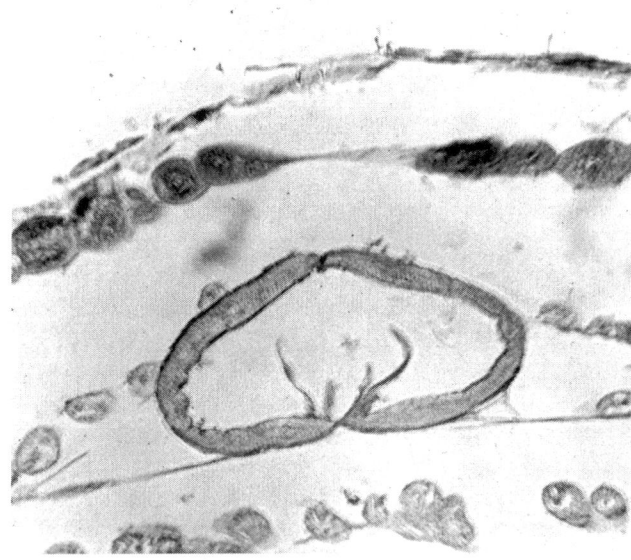

Figure 15.3 The illustration is a cross section of the heart (abdominal portion) in a honeybee. The dorsal diaphragm can be seen passing below the heart and extending laterally, and some pericardial cells are shown. (Photograph by the author.)

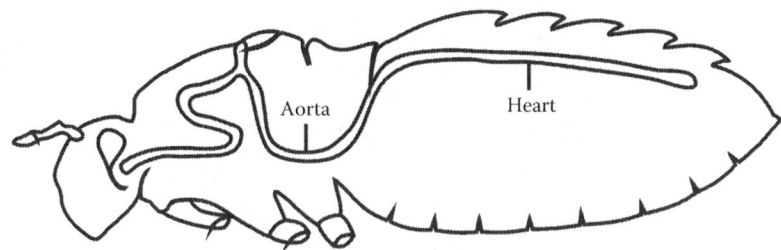

Figure 15.4 Loops of the dorsal aorta (the thoracic portion of the dorsal vessel) occur in the thorax of some insects; this drawing of the dorsal vessel in a sphingid moth shows an upward deflection of the dorsal vessel as it passes into the thorax and connects with an accessory pulsatile organ that helps pump hemolymph into the wings. The looping passage of the aorta among the large thoracic wing muscles helps transfer heat from the muscle to the hemolymph. The heat can be transferred to the abdomen as the hemolymph circulates from the anterior toward the abdomen. (Modified from Snodgrass, R.E., *Principles of Insect Morphology*, McGraw-Hill, New York, 1935; Heinrich, B., *The Thermal Warriors: Strategies of Insect Survival*, Harvard University Press, Cambridge, MA, 1996.)

corpora cardiaca [CC]) are bathed exceptionally well by the hemolymph. The continuous outflow at the head pushes hemolymph toward the posterior of the insect, where it flows around tissues and organs. It is aspirated into posterior incurrent ostia by relaxation (diastole) of the heart.

In most but not all insects, the abdomen is divided by the dorsal and the ventral diaphragm into three partitioned regions: the pericardial sinus, the perivisceral cavity, and the perineural sinus (Figure 15.5). The dorsal vessel lies on the upper surface of the dorsal diaphragm and is partially supported by it. The dorsal diaphragm consists of multiple layers of thin sheets of connective tissue, with small dorsal transverse muscle fibers, the alary muscles, enclosed within the sheets. The diaphragm is developed in grasshoppers as a nearly continuous sheet in the dorsal portion of the abdomen with hemolymph flow between the pericardial and perivisceral sinuses limited mostly

CIRCULATORY SYSTEM

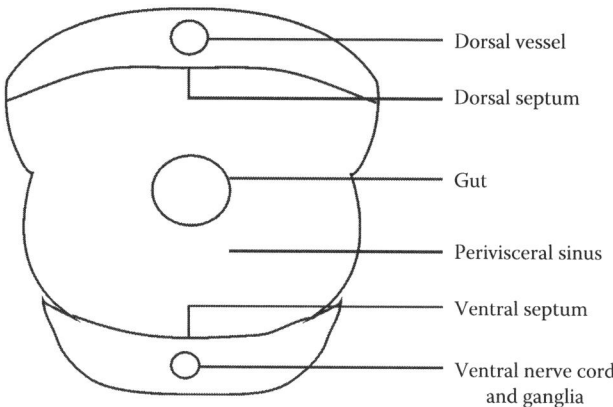

Figure 15.5 A diagram showing how the dorsal and ventral diaphragms partition the body into three regions: the dorsal pericardial sinus, the middle perivisceral sinus, and the ventral perineural sinus. The dorsal and ventral diaphragms are often fenestrated.

to the extreme posterior region of the abdomen where the diaphragm is not complete. In most other insects, the membrane is both fenestrated and incomplete laterally, so that hemolymph readily passes between the pericardial sinus and the perivisceral sinus. The dorsal diaphragm is present in most insects that have been examined, but in Hemiptera, it is greatly reduced and a series of muscles near the posterior end of the heart, along with tracheal connections, support the heart. The dorsal diaphragm usually does not extend into the thorax, although it extends into the thorax to a limited extent in some insects (e.g., some orthopterans). Loose clusters and strings of cells, called **pericardial cells**, often occur on the external surface of the heart, and they also are attached to the dorsal diaphragm at various places in the pericardial sinus. They phagocytize injected particles, such as India ink, some dyes, and other small particles, and are presumed to serve a protective phagocytic function.

The alimentary canal, reproductive organs, and some of the fat body lie in the central or perivisceral sinus below the dorsal diaphragm. There often is a ventral diaphragm separating the perineural sinus containing the ventral nerve cord and ganglia from organs in the perivisceral sinus, but the ventral diaphragm usually does not extend into the thorax. It is not even present in the abdomen of all insects. Generally, it is present in larvae and adults of Odonata, Orthoptera, Hymenoptera, Ephemeroptera, Lepidoptera, and Neuroptera. It is present only in adults of Mecoptera and lower Diptera, and not present in the higher Diptera (the Cyclorrhapha). The ventral diaphragm is fenestrated, allowing hemolymph to circulate.

Undulations of both the dorsal and ventral diaphragms due to intrinsic muscle fibers, alary muscle action, gut movements, ventilatory movements, and general body muscle action aid circulation of hemolymph and keep it mixed and moving within the body, especially in active insects. Studies on a wide variety of insects and developmental stages (Slama, 1999, and references therein) suggest that quiescent insects generate micropulses of pressure (coelopulses) in the circulatory system that aid circulation and breathing.

15.2.1 Alary Muscles

The **alary muscles** (see Figure 15.2), so named because of their general wing or delta shape in many insects, form part of the dorsal diaphragm. The muscles probably provide support for the heart. The muscle fibers typically fan out from a small point of origin on the lateral wall of the dorsum to a broad insertion on the heart in many insects, presenting the typical delta appearance.

In some insects (e.g., many grasshoppers), the origin and insertion are broad and the delta shape is not particularly evident. Some alary muscle fibers pass beneath the heart and extend from lateral side to side. In places, the fibers may also run parallel to the long axis of the heart for a short distance. The pairs of alary muscles tend to agree with the number of pairs of ostia. In addition to support, the alary muscle may assist in the expansion (diastole) of the heart after a contraction wave and, thus, aid in pulling hemolymph into the incurrent ostia. They are not necessary for diastole, however, as evidenced by severing them with little or no apparent effect on the heartbeat.

15.2.2 Ostia

Ostia are small, slit-like, paired openings in the dorsal vessel that allow hemolymph to enter or leave the vessel. Incurrent ostia allow hemolymph to enter during diastole and excurrent ones permit hemolymph to exit. Some Orthoptera have 12 pairs of incurrent ostia, 9 in the abdomen and 3 in the thorax, but most insects have fewer, with 2, 3, or 5 pairs of ostia being common. Ostia more commonly occur in the heart but may also occur in the aorta. Pairs of ostia are usually located laterally, with one on each side of the heart, but some are ventrally and dorsally located.

Ostial openings tend to occur at the base of shallow pockets or at deeper, funnel-shaped invaginations in the wall of the dorsal vessel, which often give the heart a chambered appearance. Incurrent and excurrent ostia may be difficult to distinguish (Jones, 1977). Excurrent ostia more often occur in the thoracic portion of the vessel but occur in the abdomen of some insects. Some ostia do not have developed valves that control hemolymph flow. Most incurrent ostia open with diastole, allowing hemolymph to be forced into the dorsal vessel by general body pressure and/or perhaps slight negative pressure inside the dorsal vessel. Some insects have valve flaps in incurrent ostia that open inward; hemolymph readily passes the valves into the lumen of the heart. During contraction, the valve flaps are forced together, preventing backflow, which may be their main function.

15.2.3 Heartbeat

The heartbeat is a wave of contractions (systole) generally originating at the posterior end of the heart and traveling anteriorly. The rate of contractions or beats is highly variable in different insects and varies with physiological conditions, temperature, species, stage of development, nervous activity, and neurosecretions. The rate may be as slow as 15 beats/min in the larva of *Lucanus cervus* (Coleoptera), to rates near or higher than 100 beats/min in several insects (see Beard, 1953; Jones, 1977, for large number of values). The beat is considered to be **myogenic**, originating in the muscle itself, although the heartbeat of many other invertebrates is neurogenic. The heart of the American cockroach, *Periplaneta americana*, has one of the most complex systems of innervation, but after careful removal of the lateral cardiac nerve cords, spontaneously active cardiac neurons, and lateral nerves, the heart still beats (Miller and Metcalf, 1968). When the heart is cut into numerous pieces after stripping all neurons away, each piece continues to express a beat, indicating that various parts of the dorsal vessel are capable of acting as a pacemaker. The pacemaker that usually dominates, however, is located at the posterior of the heart, and the contraction wave (systole) usually originates near the posterior of the heart and travels anteriorly. The contraction wave may move very slowly, only 1 or 2 mm/s, so that two or three contraction waves can be seen following each other. In other insects, the propagation rate may be so rapid that it appears as if the entire vessel is contracting simultaneously. In the mealworm, *Tenebrio molitor*, the conduction velocity for the contraction wave in the heart was measured at 14 mm/s, but only 1 mm/s in the aorta (Markou and Theophilidis, 2000).

One of the unusual and little understood features of insect hearts is that the beat can reverse and originate at or near the anterior and travel posteriorly. Marcello Malpighi (around 1669, quoted by Gerould, 1930) observed periodic reversals of the contraction wave, with systole beginning at the anterior of the heart and traveling posteriorly. Beat reversal has been observed in numerous insects in a number of orders (Gerould, 1933) and even prior to hatching (Davis, 1961). Contractions may begin occasionally at both ends and meet near the middle. Reversals are unpredictable, but usually the back-directed beats are slower, and duration of the reversal is usually brief. Certain experimental treatments, such as amputations of parts of the heart, lateral pressure on the dorsal vessel, and ligation near the middle of the heart may alter conduction rates and cause reversal of contraction waves in some insects. Smits et al. (2000) found that the contraction wave in larvae of the tobacco hornworm, *Manduca sexta*, consistently originates at the posterior end and passes anteriorly, with the dorsal vessel beating at about 34.8 beats/min. The heart rate is slower in pupae (21.5 beats/min) and irregular in rate, amplitude, and direction, with periods of cardiac arrest from a few seconds to as long as 20 min. Dorsal vessel contraction rates and direction of contraction wave are variable in adult moths, with fast-forward heart rates of 47.6 beats/min, slow-forward rates of 32.8 beats, and reversal of contraction wave (anterior origin) of 32.2 beats/min. Larval heart rate shows no increase in response to activity (induced by prodding), but adult heart rate rises from about 50 beats/min to as much as 223 beats/min in response to 1 min of prodding (Smits et al., 2000). Wasserthal (2012) described regular reversals of the heartbeat in the blowfly, *Calliphora vicina*, with a sequence of backward and forward pulses lasting about 34 s. The reverse pulse rate was higher than the forward rate.

15.2.4 Ionic Influences on Heartbeat

Ion influences on heartbeat are variable in different species and not well studied or understood. The resting potential of the heart in the American cockroach, *P. americana*, measures 40–70 mV, depending on technique, insect variability, and possibly other factors. Although potassium controls the resting potential of the American cockroach heart, the evidence suggests that potassium is not the controlling ion in other insects and, in some, more than one ion may be responsible for the resting potential. The ion or ions responsible for depolarization of the heart are not known for most insects. An action potential can be developed in isolated hearts of *P. americana* and the cecropia moth, *Hyalophora cecropia*, with sodium-free salines, but the identity of the ion(s) carrying the current is not known in these insects. Consistent with these observations, the heart of the American cockroach is insensitive to tetrodotoxin, a poison that completely blocks sodium channels in nerves, but does not stop the myogenic heartbeat. Calcium has been implicated in depolarization of the heart in some insects, but not others. During depolarization, overshoot potentials up to 20 mV have been recorded for some insects. Salines containing high concentrations of magnesium stop the heart of *P. americana*, a typical response of animal hearts to high magnesium ion concentrations. The heart of *H. cecropia*, however, cannot beat without magnesium in the saline, and this seems consistent with the fact that it is a lepidopteran, which typically contains high potassium and magnesium levels in its hemolymph as a phytophagous feeder.

15.2.5 Nerve Supply to the Heart

The heart receives innervation through lateral neurons from ventral ganglia in the abdomen (or from fused thoracic and abdominal ganglia in some advanced insects) and from a chain of cardiac neurons that lie alongside the heart in some insects, notably Odonata, Orthoptera, Dictyoptera, and Hemiptera. The American cockroach has spontaneously active cardiac ganglion (nerve) cells (Figure 15.6) that innervate the heart and maintain contact throughout the lateral cardiac chain running parallel with the dorsal vessel (Miller, 1968; Miller and Thomson, 1968). Spontaneously active neurosecretory cells also are located in the cardiac chain, especially near junctions with the lateral segmental nerves. The cardiac neurons in the cockroach are stretch-sensitive motor neurons.

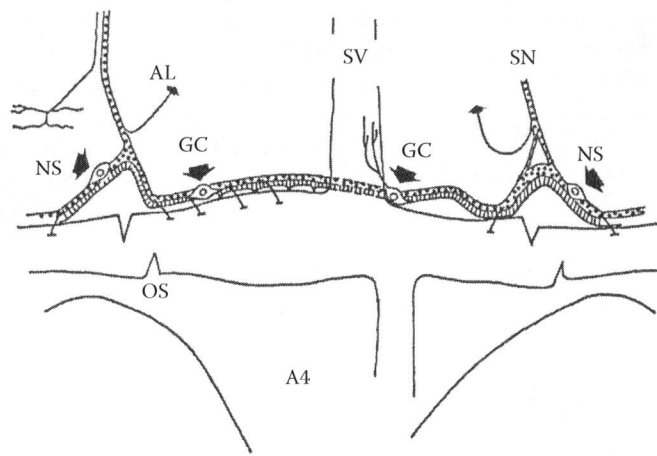

Figure 15.6 A diagram of one of the paired chains of cardiac neurons (consisting of cardiac motoneurons and neurosecretory neurons) that run along the heart in the abdomen of the cockroach *P. americana*. A4, fourth abdominal chamber; AL, alary muscles; GC, ganglion cell or motoneuron; NS, neurosecretory neuron; OS, incurrent ostial valve; SN, segmental nerves; SV, lateral segmental hemolymph vessel. (From Miller, T., *J. Insect Physiol.*, 14, 1265, 1968. With permission.)

They fire at increasing rates as the myocardium stretches in diastole and cease firing at each systolic contraction. They reinforce the simultaneous contraction of the dorsal vessel. Insects in the orders Diptera and Lepidoptera, and probably other orders, do not have cardiac neurons and a nerve chain paralleling the heart. The heart may still receive a nerve supply from the segmental ganglion.

15.2.6 Cardioactive Secretions

Various secretions and neurosecretory peptides act upon the heart to change the rate and amplitude of the heartbeat (Nässel, 1993). **Proctolin**, a neuropeptide produced in motoneurons, interneurons, and neurosecretory cells located at various places in different insects, typically in the brain, but sometimes in other ganglia, stimulates heart rate. Whether this is one of its functions in the normal course of insect life, however, is not clear. **Crustacean cardioactive peptide (CCAP)**, a cardioactive peptide composed of nine amino acids that was first isolated from a crab, has been isolated from *Locusta migratoria*, the migratory locust. It stimulates the heart, but it also is known to have physiological actions unrelated to the heart, so its role in heart action is unclear. One of several peptides isolated from the ventral ganglia of the tobacco hornworm, *M. sexta*, has the same amino acid sequence as CCAP but is usually designated as CAP_{2a} (Nässel, 1993). CCAP-immunoreactive neurons also have been detected in the brain of the mealworm, *T. molitor*. Another cardioactive peptide called **corazonin** has been isolated from the brain of the cockroach *P. americana*, and neurosecretory cells in the brain of a blowfly are positive to a corazonin antiserum. Numerous bioactive peptides have been isolated from insects (Nässel, 1993) and shown to have a variety of physiological effects, depending on the *in vitro* bioassays used, but their main or physiological role *in vivo* often remains unclear.

5-Hydroxytryptamine (serotonin) is a neurotransmitter in the nervous system, and it causes vigorous increase in heartbeat rate at very low concentrations in isolated or semi-isolated hearts of some insects, but not in others. Even in those in which it increases the heart rate, it seems to have little effect when injected into living insects, but this could be because it is quickly deactivated. Its status as a neurotransmitter at the cardiac neurons innervating the heart is at present uncertain. A number of other drugs and potential neurotransmitters, such as octopamine, dopamine, and tyramine, stimulate the semi-isolated hearts of some insects, but their role, if any, in heart function under normal physiological conditions is not clear.

15.3 ACCESSORY PULSATILE HEARTS

Accessory hearts have several shapes and variable morphology, but all are simple pulsatile, sac-like structures. They occur at a number of places in the body, but most commonly at the base of antennae and wings, and within the leg, usually near the femorotibial joint and in the dorsal region of the meso- and metathorax. They assist circulation of hemolymph into and through the appendages. Based on location, they are usually referred to as leg, antennal, or wing hearts. In most cases, they have no connections to the dorsal vessel. They merely aspirate hemolymph into a sinus cavity through ostial openings and pump it out. Less structured ones consist of little else than a pulsating muscle that aids movement of hemolymph. A mesothoracic accessory heart of *Bombyx mori* aids in pumping hemolymph into the wings. In *B. mori*, the metathoracic pulsatile heart is not directly connected to the aorta, although in some other insects it is connected to the aorta. The importance of wing expansion and wing circulation has been discussed previously.

Antennal hearts at the base of the antennae are common in many insects. Good circulation into the antennae is likely to be critical in supplying adequate nutrients to support the large number of sensory structures associated with the antennae. An accessory heart at the base of each antenna in the American cockroach, *P. americana*, is illustrated in Figure 15.7. A complex array of head accessory hearts occurs in *B. mori*, in which the aorta expands into a large sac on the anterior surface of the supraesophageal ganglion, the "brain." A short transverse tube arises from the sac and terminates in a pair of lateral ampullae, on each side of the head. Each functions as an accessory heart to pump hemolymph into the antenna via an antennal vessel running into the antenna, and to the optic lobe of the brain through a vessel that passes dorsally over the brain to the optic

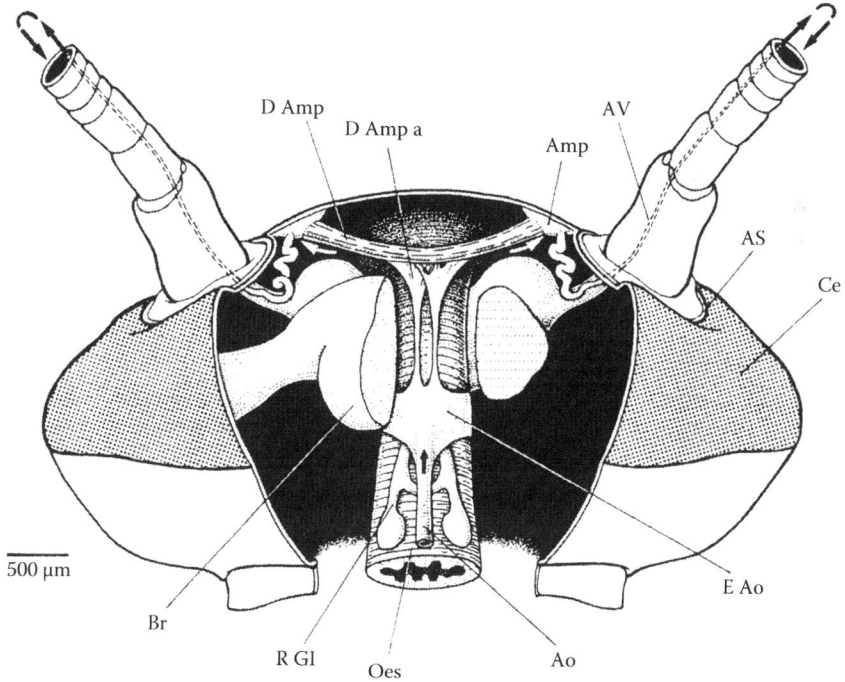

Figure 15.7 The accessory heart at the base of each antenna in *P. americana*, the American cockroach; Amp, ampulla of accessory heart; AV, Antennal vessel; Ao, aorta; AS, antennal sclerite; Br, brain; Ce, compound eye; D Amp, Amp a, dilator muscles of the ampulla; E Ao, enlarged anterior end of the aorta; Oes, esophagus; R Gl, corpora cardiaca and corpora allata, respectively. (From Pass, G., *J. Morphol.*, 185, 255, 1985. With permission.)

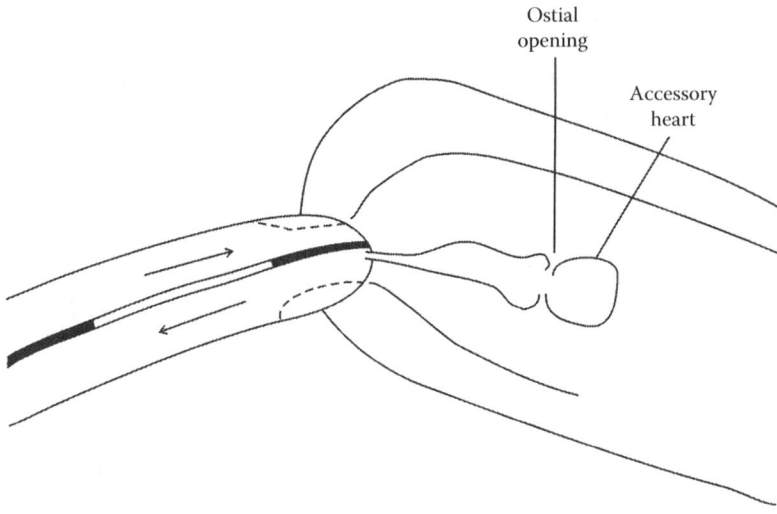

Figure 15.8 The accessory leg heart at the junction of the femur with the tibia in the beetle *Notonecta*. (Redrawn and modified from Weber, H., *Biologie der Hemipteren*, Springer, Berlin, Germany, 1930.)

lobe, and ends as an enlarged open sac. Antennal ampullae in some insects do not have direct connection with the aorta. Generally, antennal ampullae have simple attachments by tonofibrillae to the epidermis to keep them in place, and some are attached by tiny muscles to the pharynx. Some antennal ampullae have nervous connections, but detailed recordings of nervous control have not been conducted.

The structure of the antennal heart in certain earwigs (Dermaptera) may be indicative of early evolutionary history of accessory antennal hearts (Pass, 1988). The sac-like ampulla at the base of each antenna in earwigs is connected to an antennal blood vessel that runs to the apex of the antenna. The ampulla does not have a muscular wall, but it is compressed by a small independent muscle running across it like a belt. A valve-like structure near the origin of the antennal blood vessel prevents hemolymph from flowing back toward the ampulla. When the compression muscle relaxes, the natural elasticity of the ampulla assisted by the pull of elastic fibers attaching the ampulla to the wall of the head promotes diastole and filling with hemolymph via a ventral ostium.

Leg hearts (Figure 15.8) are common in some Odonata, Hemiptera, Homoptera, and higher Diptera. There may be a membranous septum between the ventral and dorsal regions of the tibia, and hemolymph is pumped into one channel and returns through the other, aided by muscle contractions in the leg that pump the septum.

15.4 HEMOCYTES

Hemocytes are blood cells. They change their appearance and shape (Figure 15.9) from time to time even in the same insect, and they can be distorted in shape by fixation, staining, and other procedures used in collecting and processing hemolymph. Procedures for examining and classifying hemocytes have not been standardized or agreed upon, and various classifications and morphological types have been published. Fixation, spreading, and drying of insect hemocytes on a glass slide, as is commonly done in vertebrate blood analysis prior to staining, tend to result in many bizarre and variable types due to distortion of cell shapes, probably by the drying process.

The electron microscope has been useful in hemocyte classification. The seven most common types of hemocytes found in insects are **prohemocytes**, **plasmatocytes**, **granulocytes**,

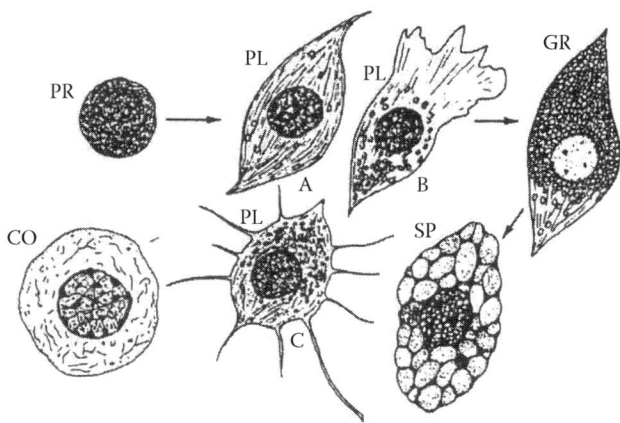

Figure 15.9 An illustration of the most common types of hemocytes from insect hemolymph. PR, prohemocyte; PL, plasmatocyte; GR, granulocyte; SP, spherulocyte; CO, coagulocyte (= hyaline hemocyte). Several different shapes of plasmatocytes are shown in A, B, and C. The arrows indicate transformations of cells that are believed to occur. (From Woodring, J.P., Circulatory systems, in: Blum, M.S. (ed.), *Fundamentals of Insect Physiology*, John Wiley & Sons, New York, pp. 5–57, 1985. With permission.)

spherulocytes, **adipohemocytes**, **oenocytoids**, and **coagulocytes** (Gupta, 1979a,b). Most insects that have been surveyed have prohemocytes, plasmatocytes, and granulocytes, but presence of the other types is variable. Only five types (prohemocytes, plasmatocytes, granulocytes, spherule cells or spherulocytes, and oenocytoids) were described from the pink bollworm, *Pectinophora gossypiella* (Lepidoptera: Gelechiidae) (Raina, 1976). Eight classes of hemocytes (prohemocytes, plasmatocytes, granular hemocytes, coagulocytes, crystal cells, spherule cells, oenocytoids, and thrombocytoids) were established by Lackie (1988) based on the classification scheme of Rowley and Ratcliffe (1981). Most insects will not have all eight of these classes, if indeed any have all eight. Application of monoclonal antibodies to the identification and classification of hemocytes has become a useful tool (Gillespie et al., 1997).

Prohemocytes are the smallest hemocytes and may be stem cells from which some other hemocytes may develop. Prohemocytes are known to divide and they may differentiate into plasmocytes, which, in turn, may give rise to granulocytes, and these may differentiate into spherulocytes. Although there is some evidence for this pattern of transformation, the evidence that they are the main source of hemocytes is not conclusive (Gupta, 1979a). Thus, prohemocytes may be one source of new hemocytes. The origin of other hemocytes is uncertain. Prohemocytes are typically round, 6–13 µm in diameter, with a relatively large nucleus (70%–80% of cell volume). The cells stain heavily with Wright's stain and the nucleus may not be easily discerned. They contain ribosomes and mitochondria but little endoplasmic reticulum and Golgi membranes. They are not mobile and do not participate in phagocytosis. The main function of hemocytes may be to divide and give rise to new hemocytes.

Plasmatocytes are small-to-large, polymorphic cells up to 40–50 µm in size, granular or agranular, and round-to-spindle shaped in wet suspensions, although they lose this shape when dried on a slide. They may be binucleate. "Young" plasmatocytes can be confused with prohemocytes (Gupta, 1979a). They contain lysosomal enzymes and are usually the most numerous of circulating cells (Lackie, 1988). They are phagocytic and participate in encapsulation, nodule formation, and wound healing (Ratcliffe and Götz, 1990).

Granulocytes are variable in size, spherical or oval, and up to 45 µm in size. The nucleus is usually small and the cytoplasm is granular. On the basis of histochemical tests, the granules are thought to be glycoproteins and mucopolysaccharides. Granulocytes may arise from plasmatocytes. The precise function of granulocytes is unproven, but some researchers have suggested that they serve storage and possibly secretory functions. They may be involved in cellular defensive functions in various insects and may be phagocytic in some insects, but in others, neither of these functions is established.

Spherulocytes are ovoid to round cells up to about 25 µm in length. They may contain few to many small spherical inclusions that stain for acid mucopolysaccharides (Ashhurst, 1982). Their function is unknown, but they may participate in phagocytosis.

Adipohemocytes may be small or large, spherical to oval, and contain lipid droplets. They might be plasmatocytes that are filled with lipids under certain physiological conditions (Gupta, 1979a).

Oenocytoids are variable in size, often large, may be binucleate, and lyse easily, but do not cause hemolymph coagulation when they lyse. They are nonphagocytic. Some evidence indicates that they contain prophenoloxidase, an inactive form of phenoloxidase (PO) (Lackie, 1988). Oenocytoids should not be confused with oenocytes, cells found among fat body cells and scattered among epidermal cells in many insects. Oenocytes are not blood cells.

Coagulocytes also have been called **hyaline hemocytes** (Grégoire, 1951) and cystocytes (Lackie, 1988). These cells rupture within seconds after injury or after taking a hemolymph sample from an insect and initiate the clotting process (Grégoire and Goffinet, 1979). The cells may contain granules. In phase contrast, they are nearly transparent, hence the name hyaline hemocytes. The hemolymph of some insects does not coagulate; for example, hemolymph of larval honeybees, *Apis mellifera*, does not coagulate.

The hemocytes in *Drosophila melanogaster* larvae do not correspond exactly to the hemocytes in Lepidoptera larvae. Crystal cells, plasmatocytes, and lamellocytes have been described from larval *D. melanogaster*. Ribeiro and Brehelin (2006) have attempted to harmonize the hemocytes of *Drosophila* with those in Lepidoptera, and their conclusions are summarized as follows. The hemocytes called plasmatocytes in *Drosophila* are not the equivalent of plasmatocytes in Lepidoptera, and Ribeiro and Brehelin propose calling them *Drosophila* plasmatocytes. They conclude that *Drosophila* lamellocytes show the most similarity to lepidopteran plasmatocytes, and they suggest keeping the name "lamellocytes." *Drosophila* does not have a true equivalent of lepidopteran granular hemocytes, but drosophila plasmatocytes (the new name) have more characteristics of lepidopteran granular hemocytes than they do to lepidopteran plasmatocytes. Cells called crystal cells in *Drosophila* have great similarity to lepidopteran oenocytoids, and Ribeiro and Brehelin suggest renaming them "oenocytoids." *Drosophila* oenocytoids (crystal cells) contain crystals of prophenoloxidase (Rizke and Rizke, 1980). *Drosophila* has no equivalent of lepidopteran spherule cells. Cells that look like prohemocytes are rare but present in *Drosophila* larvae (Lanot et al., 2001; Ribeiro and Brehelin, 2006). In summary, Ribeiro and Brehelin see lamellocytes, *Drosophila* plasmatocytes, oenocytoids, and rarely prohemocytes in *Drosophila* and plasmatocytes, granular hemocytes, oenocytoids, and Spherule cells in most Lepidoptera.

Good light microscope photos of prohemocytes, plasmatocytes, granulocytes, spherulocytes, and oenocytoids were presented by Arnold and Sohi (1974) from fresh hemolymph of the forest tent caterpillar, *Malacosoma disstria*. These authors found that only prohemocytes, plasmatocytes, and granulocytes could be maintained in cell cultures. Cultured prohemocytes and plasmatocytes divided by mitosis. Granulocytes divided, but the mechanism was not determined. The size of cells in culture was generally larger than those taken from fresh hemolymph. Hemocytes have been relatively easy to get into tissue or cell culture (Hink, 1976), and numerous cell lines have been developed.

15.4.1 Functions of Hemocytes

Hemocytes play major protective roles against invading microorganisms and the eggs or larvae of parasitoids by production of antibiotics, encapsulation, PO production and release, and phagocytosis. Recent reviews detail many aspects of the functions of hemocytes in immunity against invading microorganism (Tsakas and Marmaras, 2010; Siddiqui and Al-Khalifa, 2012). Er et al. (2011)

reviewed ways that parasitoids try to evade the hemocyte defenses of hosts. Hemocytes participate in wound healing by aggregating at the wound site, where some cells phagocytize cellular debris or foreign organisms, such as bacteria, and hemocytes activate PO, which is an antibacterial agent. Some authors have seen a parallel between the function of insect hemocytes in phagocytosis and mammalian phagocytic cells (Browne et al., 2013). Certain hemocytes, particularly coagulocytes and possibly granulocytes in some insects, participate in the coagulation of plasma to help plug a wound. Entrapment of hemocytes in the coagulum helps plug the wound. Some of the trapped cells may also play an active role in defense at the wound site.

Some hemocytes contain enzymes that aid in detoxification, including detoxification of some insecticides. Hemocytes participate in nodule formation and encapsulation of foreign objects (Gillespie et al., 1997). Invasions of bacteria may be attacked by nodule formation in which the bacteria are aggregated into nodules containing hemocytes and coagulum. Larger objects or invading parasites may be encapsulated. Plasmatocytes and granular hemocytes aggregate to form thin sheets of cells and plaster themselves around nodules, internal parasites, or other foreign objects. Sometimes encapsulated objects also become melanized by the action of the PO cascade of enzymatic action on tyrosine or other phenolic compounds. *Pseudoplusia includens*, a noctuid moth, attaches granular hemocytes directly to the target object, followed by multiple layers of plasmatocytes adhering to the inner layer of granular hemocytes, and finally a monolayer of granular hemocytes on the periphery of the capsule. The peripheral layer of granular hemocytes dies (apoptosis appears to be induced by substances released by the underlying plasmatocytes) leaving a basal lamina-like layer around the capsule that is important in successful capsule formation (Pech and Strand, 2000). Encapsulated objects often become attached to various tissues or organs in the body. The ability to encapsulate parasitoids seems to be dependent on an evolutionary race between the parasitoid and its host. A parasitoid that is not well adapted to a particular host is likely to have a high percentage of its eggs encapsulated, while a parasitoid well adapted to its host may avoid significant encapsulation. Similar to nodule formation, the encapsulated object is surrounded by many layers of plasmatocytes that form sheets of cells around the object. Often, but not always, the whole mass becomes melanized, a process that also may be toxic to the offending organism and assist in killing it.

Bacterial invasion of some insects results in synthesis and secretion into the hemolymph of combinations of several antibiotic peptides and proteins (see Chapter 16 for more details). These include lysozyme (14 kDa), the cecropins (4 kDa), the attacin/sarcotoxin II family of proteins (20–28 kDa), and the defensins (29–34 amino acids), a family of proline-rich antibacterial peptides (18–34 amino acids long) with a variety of names depending on the source (Gillespie et al., 1997). Fat body tissue is the usual site of synthesis of these proteins, but hemocytes and a variety of other tissues also contribute to the synthesis. Although combinations of the aforementioned are commonly secreted, not all are found in the same insect.

One final function of some hemocytes may be to form the basement membrane of some cells (Lackie, 1988), but this is a controversial issue that has not been conclusively resolved.

15.4.2 Hemocytopoietic Tissues and Origin of Hemocytes

During embryological development, hemocytes develop from mesodermal tissue. In larvae and adults of some insects, circulating hemocytes are known to divide and some may differentiate into other hemocytes. The rate of division of existing hemocytes in larvae of the wax moth, *Galleria mellonella*, seems to be sufficient to account for the numbers of circulating hemocytes (Jones and Liu, 1968). Some insects have **hemocytopoietic tissues** in which hemocytes are produced, for example, the cricket, *Gryllus bimaculatus* (Hoffmann, 1970; Hoffmann et al., 1979). Nutting (1951) described and illustrated structures in additional orthopteroid and related insects, but based upon dye injection experiments, he concluded that they were phagocytic tissues. Larvae of Lepidoptera have masses of loosely connected cells in a capsule located near the prothoracic spiracles that give rise to hemocytes. Hemocytes are released from the capsules through gaps in the covering surface. The organs disintegrate in pupae and release large numbers of hemocytes into the body. Small clusters of cells near the wing imaginal discs in the commercial silkworm, *B. mori*, and the tobacco hornworm, *M. sexta*, give rise to hemocytes during larval life (Nittono, 1964; Nardi et al., 2003; Wang et al., 2009; Nakahama et al., 2010). Hemocytes also arise from division of existing hemocytes in the circulating hemolymph in the silkworm (Tan et al., 2013). Similar structures without the capsule enclosure are reported to occur in some dipterous larvae, such as the housefly, *Musca domestica* (Arvy, 1954). The lymph gland in *D. melanogaster* gives rise to hemocytes (Jung et al., 2005; Crozier and Meister, 2007; Williams, 2007), with about 95% of the cells being plasmatocytes. A number of genes have been implicated in hemocyte proliferation in *Drosophila* (Minakhina and Steward, 2010; Gregorian et al., 2011; Kulkarni et al., 2011). Some Orthoptera (crickets) and some cockroaches (Dictyoptera) have complex hemocytopoietic organs composed of hollow sacs at the anterior end of the heart. Cells inside the sac are phagocytic, and some divide into stem cells that can further differentiate into several different types of hemocytes. Small groups of cells between pericardial cells located near the dorsal diaphragm are believed to produce hemocytes in *Locusta* (Orthoptera) and *Melolontha* (Coleoptera). Yamashita and Iwabuchi (2001) cultured prohemocytes from larvae of *B. mori* and found that more than 60% of the prohemocytes differentiated into plasmatocytes or granulocytes, and some granulocytes later differentiated into spherulocytes. Some prohemocytes divided into new prohemocytes. The authors of the study suggest that prohemocytes are the stem cells giving rise to plasmatocytes, granulocytes, and spherulocytes, but no oenocytoid cells were produced in culture, and they may come from a different stem cell line.

Probably insects are quite variable in how hemocytes are produced and whether they have hemocytopoietic tissues. In many insects there may be multiple origins of hemocytes (Arnold, 1974). Grigorian and Hartenstein (2013) present a review of hematopoiesis and tissues in which hemocytes are produced. In most insects, the source of new hemocytes is simply not known.

15.4.3 Number of Circulating Hemocytes

The number of hemocytes in the circulation of insects is quite variable from species to species and even within the same individual at different times depending on its physiological state. Sex, age, stage of development, and activity are known to influence the observed number of cells in some insects. Also, some hemocytes are sessile, attached to various tissue, at least for much of the time, but may be released into the circulation under certain circumstances (Figure 15.10). Measurements of hemocyte counts per microliter of hemolymph over time can be influenced by fluctuations in blood volume, which is not nearly so constant as in vertebrates.

Counting of hemocyte is usually done in a manner similar to the counting of red or white blood cells from a vertebrate with a standard hemocytometer-counting chamber. A measured volume of hemolymph is withdrawn and diluted to a known volume with a suitable diluent that does not lyse the cells, and the counting chamber is filled. One frequently used procedure in estimating cell counts is to heat kill the insect in water at 60°C for 2 min and then take the hemolymph sample; this is believed to put sessile cells into the circulation and fixes the hemocytes in their normal shape (Woodring, 1985). No really effective general anticoagulants for insect hemolymph are known, but rapid dilution with a saline solution tends to minimize coagulation, as does the heat treatment.

Hemocyte counts for a very large number of insects have been published (Jones, 1977). Different species contain widely differing numbers. For example, the American cockroach, *P. americana*, can have as many as 70,000–120,000 cells/µL hemolymph; the tobacco hornworm, *M. sexta*, has been reported to have 8,200 cells/µL; wax moth larvae, *G. mellonella*, have 35,000 cells/µL; and the blood-sucking bug, *Rhodnius prolixus*, can have from 300 to 5000 cells/µL hemolymph. The stage in the development of insects may have a dramatic effect on numbers of hemocytes in the hemolymph. Stoepler et al. (2013) found that the number of hemocytes increased at each developmental

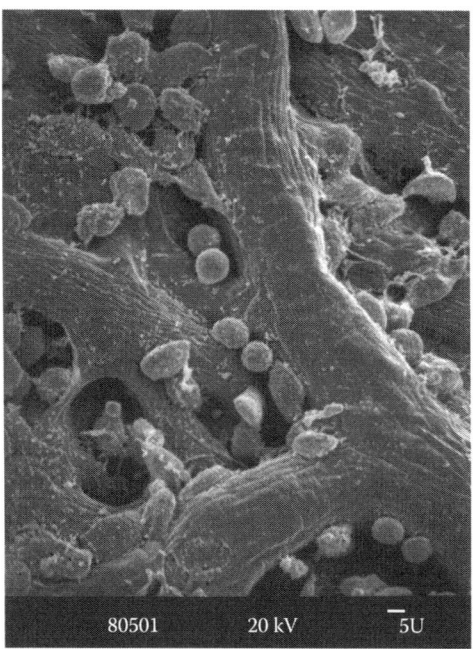

Figure 15.10 A scanning electron micrograph (SEM) of hemocytes that appear to be sessile and attached to the tissues. (SEM photograph by the author.)

stage when the slug caterpillar *Lithacodes fasciola* was challenged with injections of *Escherichia coli* bacteria. First and second instars of the Caribbean fruit fly, *Anastrepha suspensa*, and the housefly, *M. domestica*, have very few circulating hemocytes. Even large third instars have only a few thousand cells per microliter when they are only 1- and 2-day-old third instars, but as they approach larval maturity and prepare for pupariation, cells rapidly increase in circulation, culminating in up to 30,000 cells/µL of hemolymph (Figure 15.11).

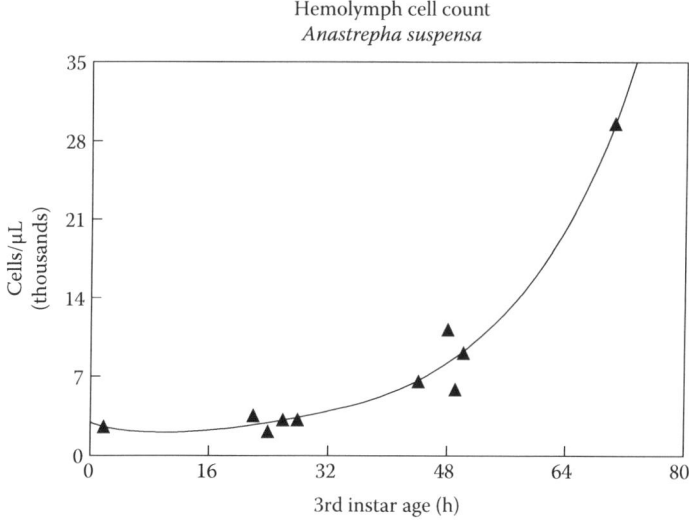

Figure 15.11 Age- and time-related increase in numbers of hemocytes in a tephritid fruit fly, *Anastrepha suspensa*. (Graph and data from the author.)

The absolute number and kind of hemocytes and, possibly, the temporal sequence of their appearance and increase in concentration may be important to the ability of an insect to defend itself from foreign invaders. For example, eggs of a braconid parasitoid, *Asobara tabida*, more often survive, hatch, and complete larvae development in *D. melanogaster*, than in a sister species, *D. simulans*, which usually encapsulates the parasitoid egg before it hatches. Early instars of *D. simulans* that successfully encapsulated the parasitoid eggs had several times more hemocytes than *D. melanogaster* larvae, which usually did not encapsulate the eggs. Individuals of *D. simulans* that successfully encapsulated eggs have greater numbers of hemocytes than those that did not encapsulate eggs (Eslin and Prevost, 1996). The authors of the study suggested that successful encapsulation and defense against parasitism may involve a physiological race between ability of the host to get hemocytes into circulation and ability of the parasitoid to locate and lay an egg in a very young host before many hemocytes are in circulation.

Some parasitoids may have evolved additional mechanisms to avoid encapsulation, such as mechanisms to destroy host hemocytes and/or inactivate other host defense mechanisms. For example, oviposition by the endoparasitoid, *Tranosema rostrale* (Hymenoptera: Ichneumonidae), into its host larvae of the spruce budworm, *Choristoneura fumiferana* (Lepidoptera: Tortricidae), results in up to 50% reduction in total hemocyte counts and some reduction of PO in the host after 3 days. Hemocyte reduction and reduced PO activity are caused by fluid from the calyx tissue of the ovaries that is injected at the same time as the egg. The mechanism(s) by which these actions are accomplished is not known (Doucet and Cusson, 1996). The ability to alter host total hemocytes and PO may be a mechanism that has evolved in successful parasitoids enabling parasitization of the host and avoidance of encapsulation or of otherwise being killed.

15.5 HEMOLYMPH

The circulating fluid in insects is called hemolymph. The hemolymph does not transport oxygen in insects (except as a small amount of oxygen dissolved in the aqueous nature of hemolymph). Hemolymph does transport a significant amount of carbon dioxide because it dissolves much more readily in an aqueous medium that oxygen does. The hemolymph is an important tissue in insects and serves numerous functions.

15.5.1 Functions of Hemolymph and Circulation

Circulation of hemolymph through the body of an insect serves a number of functions. The following are listed in no particular order:

1. Hemolymph is important as a sink for carbon dioxide (CO_2). CO_2 is soluble in hemolymph as the bicarbonate ion (HCO_3^-), and substantial amounts may be held in solution in the hemolymph of some insects. In diapausing *H. cecropia* pupae, for example, CO_2 in the gas phase builds up only slowly because most of the CO_2 goes into solution in the hemolymph. This allows the spiracles to remain closed for many minutes until CO_2 in the gas phase reaches a critical level and the spiracles open. Keeping the spiracles closed as much as possible prevents excessive water loss, a critical factor for the closed system pupa. Similar mechanisms have been demonstrated in other insects. Hemolymph does not have a role in transport of oxygen, and there is no pigment carrier for oxygen except in a very few species of insects. A limited amount of oxygen is present in solution in the hemolymph and this is probably used by cells, but it is a very small percentage of the oxygen that cells need. Oxygen is delivered by the tracheal system.
2. The circulatory system transports nutrients to cells and tissues. Hemolymph is a major storehouse of trehalose, the major insect sugar used for energy and, in particular, for flight energy. The main store of lipids utilized by moths in flight is the fat body, and rapid circulation is required to bring released lipids to the flight muscles located in the thorax.

3. The circulatory system delivers waste products, excess water absorbed from food, such as a blood meal or plant sap, and ingested allelochemicals or metabolites to excretory organs, or in some cases, to storage structures.
4. The circulatory system is a reservoir of fluid, nutrients, and enzymes. Some of the latter, such as lysozymes and POs, act as protective agents that can chemically modify potential toxicants, bacteria, parasitoid eggs and larvae, and other foreign invaders. Some large proteins, such as cecropins, with antibacterial and antifungal activity are induced and transported in the hemolymph after an insect is exposed to bacteria and fungi, or some of their chemical byproducts. Trehalose, some storage proteins, and amino acids are typical nutrients present in large quantity in hemolymph.
5. Hemolymph transports hormones from neurohemal organs to target tissues. Many neuropeptides, which have good water solubility, are transported in the hemolymph from their release points in the nervous system to target systems such as Malpighian tubules (diuretic hormone), prothoracic glands (PTTH), and pheromone glands (PBAN).
6. Hemolymph is a lubricant and hydraulic support that assists in maintaining body shape and movement, especially of soft-bodied forms such as caterpillars, and expansion of the wings in newly emerged adults. Pharate adults of dipteran Cyclorrhapha (houseflies, tephritid fruit flies, and related flies) utilize the hydraulic mechanism to force hemolymph into the ptilinum, a balloon-like structure on the front of the head, and, as it expands, it breaks open the old puparium. The new adult slowly wiggles out of the puparial case, again utilizing muscles, legs, and hydraulic action. When the fly is out, the ptilinum deflates, collapses inward, and its site is slowly sclerotized into a suture. Some insect glands are extruded by the hydraulic pressure of hemolymph. An example is the pair of osmeterial glands located behind the head on swallowtail butterfly caterpillars that are everted when the caterpillars are handled, probed, or attacked by predators. In late instars, an unpleasant odor (to humans) containing volatile derivatives of butyric acid and other compounds is emitted from the everted glands. The Caribbean fruit fly and some other tephritid fruit flies (Nation, 1989) contract muscles and force hemolymph toward the posterior of the body. The pressure balloons thin, lightly sclerotized lateral pouches at the sides of the abdomen and everts an anal pouch at the tip of the abdomen as part of their pheromone release behavior. Many other insects probably rely in part upon hydraulic pressure to expose pheromone glands and promote release of pheromone.
7. Pumping of hemolymph into the wings and, in some cases, secretion of plasticization factors are critical to proper expansion of the wings in newly emerged adults. Insects, such as moth and butterflies, typically rest quietly while the wings expand. The mesotergal and metatergal accessory hearts are important in directing hemolymph into the wings, but some insects have other accessory hearts within the wings. Even in mature, older adults of many insects, hemolymph flows through most of the veins of the wings, basically following a pattern of flowing into the wings at the anterior region of the wing and returning through the posterior wing veins. The flow may, however, reverse on occasion in some insects. Wings that are experimentally deprived of circulation become dry and brittle, and tracheae collapse and retract leaving gas bubbles behind in the wing veins. When parts of the wings are cut off, hemolymph does not ordinarily hemorrhage out of the wings, but alternative pathways of circulation through cross veins are established and some wing circulation continues. Three movies showing hemolymph flow and movement of sporozoites of the malaria parasite through the wing veins of mosquito adults can be viewed at http://jeb.biologists.org/cgi/content/full/208/16/3211/DC1 as supplementary material in the paper by Akaki and Dvorak (2005). (Last accessed April 22, 2015).
8. Hemolymph and the hemocytes provide protection from invading bacteria, eggs of parasitoids, and other foreign substances by biochemical "immune-type" reactions, phagocytosis and encapsulation by hemocytes, and wound healing by coagulation in some insects. Hemolymph does not coagulate in some insects.
9. The circulatory system is important in some insects as a means of heat transfer to prevent excessively high temperatures in the thorax from flight muscle activity or to hold heat in the thorax and allow thoracic temperature to warm above the ambient. Control of thoracic temperature is probably important to many insects in flight when the thoracic wing muscles generate much heat. Many insects do not initiate flight until the temperature in the thorax reaches some critical

temperature above the ambient. They warm the thorax by "wing whirring" before attempting to take flight. The tobacco hornworm moth, *M. sexta* (Lepidoptera: Sphingidae) (Heinrich, 1970, 1996), needs a thoracic temperature of about 38°C to begin flight. After 2 min of free flight at ambient temperatures ranging from 20°C to 30°C, its thoracic temperature is as high as 41°C–42°C, a temperature approaching the maximum the moth can tolerate in continuous flight. *M. sexta* moths will not fly continuously for 2 min at an ambient temperature of 35°C or higher, and those prodded into flight experience thoracic temperatures up to 43.3°C near the lethal point for many cells, especially neurons. The high temperatures in the thorax are the result of the intense muscular work by the flight muscles. In order to cool the thorax and prevent temperature from going too high, the moths circulate hemolymph from the hot thorax into the abdomen where little muscle activity is occurring during flight. Moreover, the abdomen is covered with a much thinner layer of scales (0.5 mm thick layer on abdomen as opposed to 2 mm thick on the thorax), and heat can escape by convection to the atmosphere more readily. When moths were prevented from circulating hemolymph into the abdomen by an experimentally placed ligature in the first segment of the abdomen, thoracic temperature in flying moths averaged about 23°C above ambient, and thoracic temperature increased directly with ambient temperature from 15°C to 23°C. When the heart was ligatured, the thorax soon overheated and moths ceased flying at temperatures above 23°C. The advantage of the physiological mechanisms that allow heat in the thorax either to accumulate or to be dissipated gives the moth the opportunity to fly at a wide range of ambient temperatures. More examples and details of the role of circulation in thermoregulation can be found in Heinrich (1996).

15.5.2 Hemolymph Volume

Hemolymph volume in insects is not constant. Dehydration, physiological stage of development, and other factors can cause large fluctuations in hemolymph volume. Volume may even fluctuate daily with food and water availability (Chapman, 1958). Several methods have been devised for measuring hemolymph volume. One of the simplest is blotting on paper as much fluid as can be removed from the insect and weighing the paper (Figure 15.12). Dye dilution methods following injection of a known volume of dye, and dilution of ^{14}C-labeled inulin, have been used (Levenbook, 1958, 1979; Wheeler, 1962, 1963; Wharton et al., 1965). Injection of any agent intended for use in dilution calculations may suffer from binding of the agent to tissues. Evans blue binds to the tissues in third instars of the blowfly, *C. vicina*, and calculation then overestimates hemolymph volume compared to the use of ^{14}C-labeled inulin (Levenbook, 1979). A micromethod has been devised for measurement of hemolymph volume in small insects (housefly) that have very little hemolymph in the body (Shatoury, 1966).

Fluctuation in hemolymph volume can distort measurements of the number of cells/µL and total concentration of any chemical component in the hemolymph. With the amaranth dye dilution method, Wheeler (1963) found that there is no change in absolute number of hemocytes/insect in *P. americana* just prior to a molt, but the number of cells/µL hemolymph increases due to a decrease in hemolymph volume. During ecdysis the total number of cells does not change, but number/µL decreases because of increase in hemolymph volume. About 24 h after ecdysis, there is a decrease in total number of cells.

15.5.3 Coagulation of Hemolymph

The hemolymph of many insects coagulates rapidly at the site of wounds or when withdrawn from the insect by capillary pipet, but in some insects, for example, in larvae of the honeybee, *A. mellifera*, the hemolymph does not coagulate. Hemolymph fails to coagulate in some Coleoptera, Hemiptera, some adult Lepidoptera, and many Diptera.

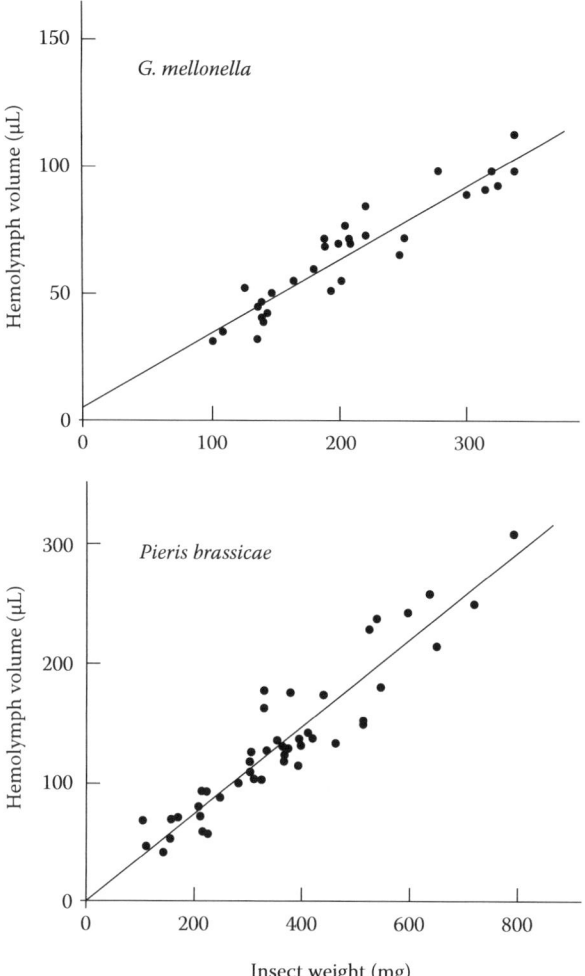

Figure 15.12 The graphs show regression lines of hemolymph volume on insect body weight for two lepidopterous caterpillars: wax moth larvae *Galleria mellonella*, and larvae of the cabbage butterfly *Pieris brassicae*. Each point represents the blood volume and body weight of one larva. (From Gagen, S.J. and Ratcliffe, N.A., *J. Invert. Pathol.*, 28, 17, 1976. With permission.)

Gregoire (1951) and Gregoire and Goffinet (1979) described coagulation from a very large number of insect species by observing with phase microscopy events occurring immediately after taking a hemolymph sample. Coagulation tends to be a continuous process initiated in all cases by rupture of a single type of hemocyte, the hyaline hemocytes, also called coagulocytes. Varying degrees of plasma clotting occur, from a general clotting to a limited reaction. A number of agents, including oxalate, citrate, magnesium sulfate, 2% methylene blue (a reducing agent), cocaine hydrochloride, sodium bisulfite, sodium thiosulfate, ethylenediaminetetraacetic acid (EDTA), and sodium hydrosulfite can prevent or reduce the clotting reaction in some insects. These reagents are not universally effective, however, and the mechanism by which they interfere with clotting has not been determined. Rapid dilution with a balanced saline solution is probably the easiest and most effective procedure to reduce or prevent clotting in most insects. Heparin, bee venom, and a number of other substances useful in vertebrate blood clotting usually are not effective in insects.

Three types of coagulation occur as follows:

1. Type I coagulation is initiated by immediate rupture of the hyaline hemocytes. Coagulation islands form around each ruptured hyaline hemocyte, and these gradually grow in area until many of them coalesce. This is the predominant type of coagulation in Orthoptera, Dermaptera, some Hemiptera, some Coleoptera, some Hymenoptera, some Homoptera, Neuroptera, Mecoptera, Trichoptera, and some Lepidoptera.
2. In Type II coagulation, there is an absence of coagulation islands and instead a pseudopodial meshwork develops from ruptured hyaline hemocytes. The meshwork gradually expands and traps other hemocytes within the net. This type of coagulation is found in some ground beetles (Carabidae), dragonfly (Odonata) nymphs, several Lepidoptera, and some Coleoptera (Scarabaeidae).
3. Type III coagulation is a combination of events taking place in Types I and II and is common in Homoptera, many Coleoptera, and Hymenoptera.

15.5.4 Hemolymph pH and Hemolymph Buffers

The great majority of insects have hemolymph that is slightly acid, but a few have hemolymph that is as alkaline as pH 7.5 or slightly greater. Hemolymph pH typically falls into a pH range of 6–7.5 (Buck, 1953). Measurement within the same species may vary by up to about 0.7 pH units. There is no strong correlation with sex, diet, stage of development, or taxonomic position. The greatest changes occur with metamorphosis; during the pupal stage, or sometimes just prior to pupation, there is a slight increase in acidity, but usually no more than about 0.3 pH unit.

Hemolymph pH stays reasonably stable during the life of an individual, which suggests that insects regulate hemolymph pH. Locusts and grasshoppers (and by extrapolation, possibly other insects) regulate hemolymph pH in at least two different ways: (1) by transfer of acid equivalents to the gut when acidosis is caused by a titratable acid (a natural situation might be accumulation of an organic acid from metabolism) and/or (2) by increased ventilation of the tracheal system when acidosis is caused by an increase of CO_2 in the hemolymph, as in vigorous muscular activity. Harrison et al. (1992) demonstrated that (fasting) desert locusts, *Schistocerca gregaria*, regulate extracellular pH after an experimental injection of HCl into the hemolymph. Recovery is mainly by transfer of about 75% of the acid equivalents from the hemolymph into the alimentary canal, with no evidence that the respiratory system aided the recovery (by eliminating CO_2). The experimentally injected locusts eventually eliminate ammonium urate, which may be a mechanism of compensation for extracellular acidosis (Harrison and Phillips, 1992).

A rise in body temperature results in a biphasic pattern of hemolymph pH regulation in two orthopterans, the two-striped grasshopper, *Melanoplus bivittatus*, and the locust, *Schistocerca nitens* (Harrison, 1988, 1989). Hemolymph pH remains constant at temperatures up to 25°C and then decreases about 0.017 pH units per degree Celsius above 25°C in both species. The two-striped grasshopper accumulates CO_2 in the hemolymph when forced to hop repeatedly over a 5-min period, and hemolymph pH decreases (Harrison et al., 1991). During repeated hopping and in recovery afterward, rates of gas exchange of O_2 and CO_2 increase, with a greater increase in rate of O_2 transfer than CO_2 during hopping, and a higher rate of CO_2 than O_2 exchange during recovery. The higher rate of O_2 exchange during hopping suggests that the jumping leg muscles work aerobically, like flight muscles, and that the oxygen functions as the ultimate acceptor for protons released during muscle metabolism. During a 2-min recovery period, the grasshoppers ventilate the tracheal system, rapidly flush excess CO_2 from the body, restore depleted O_2, and return hemolymph to normal resting values.

The western lubber grasshopper, *Taeniopoda eques*, shows a flexible ability to shift between excretion of excess acid and excess base equivalents depending on physiological condition in its homeostasis of hemolymph pH (Harrison and Kennedy, 1994). Such flexible ability seems adaptive in a highly polyphagous insect, such as *T. eques*, which may experience acidosis or alkalosis, depending on diet, allelochemicals in the food, type of proteins metabolized (e.g., proteins yielding basic, acid, or neutral amino acids), and other metabolic conditions (Harrison and Kennedy, 1994).

Buffering in the hemolymph of *S. gregaria* at 21°C, pH 7.31, when CO_2 is held constant is provided by (in milliequivalents per liter per pH unit) bicarbonate 20, protein 10, inorganic phosphate 1.6, organic phosphate 1.5, citrate 0.4, and the amino acid histidine 0.1 (Harrison et al., 1990). Thus, nearly 90% of the buffering capacity in hemolymph of these locusts is due to bicarbonate and proteins, with a small additional contribution by inorganic and organic phosphates. Amino acids, although in high concentrations in insect hemolymph, have dissociation constants well removed from typical hemolymph pH. The pK_1 is <3 for ionization of the carboxyl proton, and $pK_2 > 9$ for ionization of the amino group proton for nearly all the 20 amino acids that occur in insect hemolymph; only histidine has an ionizable proton from its ring structure with pK of 6.04 in the range of insect hemolymph.

15.5.5 Chemical Composition of Hemolymph

Hemolymph contains many dissolved inorganic and organic substances, colloidally suspended proteins, and lipoproteins. It is about 90% water and 10% solids. Probably most components contribute to the osmotic pressure and specific gravity of hemolymph. Specific gravity is usually slightly greater than 1. Hemolymph generally has an osmotic pressure, expressed in freezing point depression, of about 0.7 to slightly over 1°C. This is less than some Crustacea (*Cancer* [crab] and *Homarus* [lobster]) and slightly higher on average than that in humans. Osmotic pressure in a number of species, and in different developmental stages of the same species, is not correlated with age, sex, developmental stage, diet, or systematic position among orders (Buck, 1953).

In general, saline solutions that are equivalent to 0.9%–1.6% sodium chloride will encompass the range of osmotic pressures in insect hemolymph, but such a solution of NaCl will not be balanced ionically with respect to the composition of hemolymph. An isotonic saline developed for tissue perfusion of adult blowfly, *Phormia regina* (Diptera: Calliphoridae), contains 119 mM Na, 5.6 mM K, 2.4 mM Ca, 1 mM Mg, 97 mM glutamic acid, 44 mM glutamine, 97 mM proline, 48 mM alanine, and 26 mM glycine to give an osmotic pressure of 480 mOsM/kg. Some of the sodium is provided by NaOH and some by NaCl; the final pH is 7 (Chen and Friedman, 1975). A general purpose saline suitable for many insects can be made by dissolving 7.7 g NaCl, 0.36 g KCl, and 0.24 g $CaCl_2 \cdot 2H_2O$ in water to make a liter (Jones, 1977). If the saline is to be used for Lepidoptera and Coleoptera, the NaCl content should be reduced to 0.117 g and KCl increased to 7.46 g. Ruiz-Sanchez et al. (2007) give the composition of salines used successfully for *Drosophila* spp., *Acheta domesticus*, *R. prolixus*, *Aedes aegypti*, and *T. molitor*. A fixative and staining solution useful for staining and observing hemocytes from the American cockroach and other insects contains, in a final volume of 100 mL water, 0.5 g crystal violet, 1.09 g NaCl, 0.157 g KCl, 0.085 g $CaCl_2$, and 0.017 g $MgCl_2$, sufficient glacial acetic acid to make the pH 2.9 (Sarkaria et al., 1951). The acetic acid fixes the hemocytes in their natural shape and crystal violet lightly stains them for easier visualization. Numerous salines that have been recommended for use with Lepidoptera failed to adequately support normal heartbeat and neuromuscular transmission in several lepidopterans, but a new saline composed of 12–28 mM NaCl, 32–16 mM KCl ([Na^+] + [K^+] = 44 mM), 9 mM $CaCl_2$, 1.5 mM NaH_2PO_4, 1.5 M Na_2HPO_4, 18 mM $MgCl_2$, and 175 mM sucrose with pH 6.5 is satisfactory for *B. mori* and several other lepidopterans (Ai et al., 1995, and references therein).

15.5.5.1 Inorganic Ions

The ionic composition of hemolymph plasma is highly variable in different insects. Sodium, potassium, calcium, and magnesium are typical cations, and chloride, phosphate, amino acids, and sometimes bicarbonate are present as anions. Chloride and phosphate are the major anions in hemolymph of honeybee larvae, but these are not usually the major anions balancing the cations in hemolymph of most insects. Chloride accounts for about 7% of the anions in larvae of the horse bot *Gasterophilus intestinalis* (Diptera: Oestridae), 12% in larvae of the silk moth, *B. mori*, and up to 39% of anions in larvae of the southern armyworm, *Spodoptera eridania* (Lepidoptera: Noctuidae). Amino acids and organic acids account for a substantial part of the anions in some groups, particularly Neuroptera and Lepidoptera. Most insect have rather high levels of amino acids in the hemolymph, and some amino acids can contribute to the cation load.

Sodium and potassium concentrations and the Na/K ratio in hemolymph are variable. Plasma from Odonata, Diptera, carnivorous Coleoptera, Dictyoptera, and Orthoptera tends to have a relatively high sodium/potassium ratio (19.6 for *P. americana* cockroaches, 21.4 for *A. domesticus* crickets, 9.8 for *Schistocerca* locusts) and a calcium/magnesium ratio equal to about 1. Generally, this has been considered to approximate the early evolutionary condition in insects (Florkin and Jeuneaux, 1974).

Lepidoptera, phytophagous Coleoptera, Hemiptera, Homoptera, and Hymenoptera have a sodium/potassium ratio that is only a few multiples of 1, or even less than 1 (some as low as 0.3–0.1, or lower). Some researchers have speculated that evolution of a low Na/K ratio is related to evolution of phytophagous food habits, but the correlation is not strong and, if it did evolve this way, it is not strictly diet related in present-day insects. Carnivorous insects often maintain a Na/K ratio that is different from that of the phytophagous insects on which they feed. Phytophagous insects within some orders do not show Na/K ratios as low as those in Lepidoptera, but neither are they as high as in the Orthoptera or Dictyoptera. Force feeding of large doses of potassium chloride to *P. americana* depresses the Na/K ratio, but never to the level in Lepidoptera, indicating that some insects, at least, have the ability to regulate ion composition (Buck, 1953).

The very low Na/K hemolymph values in some insects would have consequences for nerve and muscle function were it not for barriers that protect cells. All cells (except hemocytes) are protected from direct contact with the hemolymph by a basement membrane on the hemolymph side, although few experimental data are available to assess how much it actually stops ion movements. The central nervous system and larger nerves especially are protected from direct hemolymph contact by the neurolemma (= perilemma) and perineurium, noncellular and cellular layers, respectively, that surround ganglia and large nerves. Individual neurons are protected by glial sheath cells.

Ion binding to macromolecules in hemolymph occurs in some insects (Weidler and Sieck, 1977). Macromolecules bind more than 20% of the sodium and magnesium, about 16% of the calcium, and about 10% of the chloride in whole hemolymph of the American cockroach, *P. americana*. Potassium is not bound. Similar binding data have been reported for other insects. Differential binding of ions could explain why hemolymph sampled from different body sites does not always have the same composition (Pichon, 1970).

15.5.5.2 Free Amino Acids

One of the interesting features of insect hemolymph is that it contains very large amounts of free amino acids, much more so than the body fluids of other animals. These amino acids contribute to the osmotic value of hemolymph and account for a substantial portion of the cations and anions of hemolymph. A good physiological explanation for such large quantities of amino acids in hemolymph is not available, but the hemolymph probably is a reservoir of amino acids for protein synthesis, much of which occurs in fat body cells.

15.5.5.3 Proteins

Hemolymph contains a wide variety of different proteins (reviewed by Kanost et al., 1990). Many protein bands can be detected by electrophoresis and these may change with physiological state, age, sex, and other factors. Some insects (larvae of some Diptera and Lepidoptera have been most intensively studied) synthesize large quantities of one or more storage proteins during late larval life. Calliphorin, named for its original source in *Calliphora erythrocephala* (Diptera: Calliphoridae), has a molecular weight of 540,000 and is composed of subunits of about 85,000. It accounts for up to 60% of the protein in mature blowfly larvae. It is synthesized by the fat body, released into the hemolymph, and finally reabsorbed back into the fat body and stored for use during pupation and adult development. Similar proteins have been described from other insects. Hemolymph proteins include many different enzymes, such as lysozyme, antibacterial and antifungal proteins, and transport proteins that carry hydrophobic substances, such as juvenile hormone, cholesterol, diglycerides, hydrocarbons, and other lipoidal substances through the aqueous medium of hemolymph.

One hemolymph enzyme, **PO**, plays an important role in sclerotization of the cuticle and in protecting the insect from foreign invaders (Ashida and Yamazaki, 1990; Nappi et al., 1991; Söderhäll and Aspán, 1993). It exists in the hemolymph and in hemocytes as a proenzyme that can be converted to active PO. PO converts a variety of phenols to quinones. The enzyme and some of its products are important in tanning the cuticle. PO activity often causes the hemolymph to melanize (darken) and eventually become dark brown or black at wound sites. When the hemolymph is withdrawn from many insects, it darkens. Active PO develops within minutes in tissue homogenates or drawn hemolymph in some insects, for example, in dipterous larvae. Other insects (lepidopterous caterpillars are an example) activate the prophenoloxidase more slowly over many minutes. PO may not be demonstrable in all stages of an insect; in a number of dipterans, it is only present in high titer in the last instar (Nation et al., 1995). Additional information on PO and other immune defense molecules is presented in Chapter 16.

15.5.5.4 Other Organic Constituents

Many organic compounds have been found in the hemolymph of insects. One of the main functions of circulating hemolymph is that of a major transport medium to move metabolic compounds synthesized in fat body cells or hormones synthesized in a variety of neurohemal organs to sites where they are needed. For example, triacylglycerols stored in fat body cells are released as diacylglycerols, and these are transported by lipoproteins (lipophorin) to thoracic flight muscles in Lepidoptera and some other insects.

Insect hemolymph is notable for the large concentration of trehalose in solution, from 0.5 g to as much as 2.5 g/100 mL in some cases (Woodring, 1985). Trehalose is the principal hemolymph sugar in most insects and is rapidly converted to glucose for immediate metabolism in glycolysis and the Krebs cycle. Ecdysteroids, juvenile hormone, PTTH, PBAN, AKH, HTH, diuretic hormone, and a host of other neuropeptides and hormones are transported from the site where they are released into the hemolymph to their targets. The titer of these in the hemolymph varies continuously with the physiological state of the insect.

Hemolymph is a major transporter of metabolic waste products for excretion or storage. Uric acid is transported by hemolymph from the main site of synthesis, the fat body, to the Malpighian tubules for excretion in many insects or for storage in various sites by some insects. Uric acid is poorly soluble in aqueous solutions, such as hemolymph, and its concentration in hemolymph is low (11.47 ± 0.99 mg/100 mL, mean $\pm$ SE, in last instars of the wax moth, *G. mellonella*) (Nation and Thomas, 1965). Hemolymph uric acid levels are variable in tobacco hornworm larvae, *M. sexta*, during development, and the concentration peaks near the middle of the last instar at slightly more than 33 mg/100 mL hemolymph (Buckner and Caldwell, 1980). Possibly, potassium and sodium urate account for much of the uric acid transported in hemolymph because the salts are much more soluble in an aqueous medium than the free acid. A possibility that has not been studied in detail is whether transport may be aided by binding of urates to proteins in the hemolymph.

15.6 RATE OF CIRCULATION

Not many measurements of the rate of circulation have been made. Dye injected into the posterior part of the abdomen of the cockroach, *P. Americana*, can be detected in the head in only 30 s but takes up to 8 min to reach the tarsus of the mesothoracic leg (Woodring, 1985). Similar results have been found with the large locusts. Some experiments with injected radioisotopes indicate that it takes 15–30 min for the isotope to be uniformly distributed in all parts of the body (Craig and Olson, 1951). Slow rates of complete mixing have consequences for determining blood volume by dye or isotope dilution, or for sampling any hemolymph component after some experimental procedure that is expected to influence concentration or distribution. For purposes of estimating blood volume, it has been recommended that at least 1 h be allowed for complete mixing of ^{14}C-inulin, and longer may be necessary for some insects (Woodring, 1985). Several successive samplings over time would be the best procedure to detect complete mixing. It is likely that in some insects the circulation rate is even slower than the aforementioned data indicate. On the other hand, based on the role of circulation in transporting nutrients as a source of flight energy, it is reasonable to assume that the rate of circulation is fairly rapid and efficient. Otherwise, flight could not continue for hours, as it does in some long-distance fliers (mostly the lipid burners). The action of the muscles themselves and the slight flexing action of the thorax during the wing cycle aid the circulation and move hemolymph more rapidly around the body than occurs in an insect at rest.

15.7 HEMOGLOBIN

Some species of chironomid larvae, *Chironomus tentans* and others (Diptera: Chironomidae); horse bot larvae, *G. intestinalis* (Diptera: Tendipendidae); and three bugs, *Buocnoa margaritacea*, *Anisops producta*, and *Macrocorixa geoffrey* (Hemiptera) have hemoglobin as a suspended colloid in the plasma of the hemolymph. The hemoglobins of the chironomids consists of as many as 12 monomers with molecular weights of about 15,900 each; each monomer may be coded by its own gene. In *C. tentans,* the hemoglobins account for up to 40% of the total proteins in hemolymph.

Insect hemoglobins have strong affinity for oxygen, and load to capacity at only a few mm Hg partial pressure of oxygen, consequently, insect hemoglobins remain fully saturated and do not release the oxygen for cell use unless the oxygen partial pressure is extremely low. At 7 mm Hg, *Chironomus* blood is still completely saturated with oxygen and half saturated at 4 mm Hg. Thus, how functional the hemoglobin is in supplying oxygen to tissues in the normal ecology of these insects is uncertain.

Whether the *Chironomus* hemoglobins exhibit a Bohr effect (unloading more readily in the presence of high tissue CO_2 concentration) is dependent upon pH (they do at pH 7.4–7.5), and analysis is complicated by the fact that more than one type of hemoglobin exists in the hemolymph. The large quantity in the hemolymph may have some effect as a pH buffer and help to provide a favorable pH for its unloading to tissues (Agosin, 1978).

Vertebrate hemoglobin exhibits a strong Bohr effect and, in the presence of high CO_2, such as in rapidly respiring tissues, it unloads oxygen more readily to the tissues, functionally a highly adaptive characteristic. If this effect does not occur in the *Chironomus* blood samples at physiological pH, oxygen tension must fall extremely low for the hemoglobin to be of much value to the insect. Nevertheless, experiments with *C. thummi* and *C. plumosus* suggest that the hemoglobin is beneficial under certain conditions in promoting quicker recovery from enforced anoxia and longer survival under anoxia and enables filter feeding behavior at oxygen partial pressures of 12–14 mm Hg. Species with hemoglobin are more active under conditions of experimentally falling partial pressure of oxygen than species of chironomids that do not have hemoglobin. Species with more hemoglobin in the hemolymph have a tendency to live in lakes with lower oxygen content than species that have little or no hemoglobin (Buck, 1953). Overall, chironomid larvae receive their oxygen supply most of the time by cutaneous respiration even if they do have hemoglobin, but it may improve their ability to survive short periods of very low oxygen tensions.

15.8 REVIEW AND SELF-STUDY QUESTIONS

1. Describe the location, structure, and function of the dorsal vessel in insects.
2. How is the circulatory system involved in heat transfer in *M. sexta* adult moths during flight?
3. Describe the flow of hemolymph into and out of the heart.
4. Is the heartbeat myogenic or neurogenic, and what does each mean?
5. What are accessory pulsatile organs and how are they involved in circulation?
6. Describe the major types of hemocytes.
7. What is a hemocytopoietic tissue? What is the location of such tissues?
8. List at least nine functions of hemolymph and circulation.
9. Does insect hemolymph coagulate? Explain briefly.
10. What are some of the buffering agents in insect hemolymph?
11. Give some of the general differences in the ion composition of insects.
12. How are cells in the insect body protected from ions and other chemicals in the open circulatory system?
13. What is the Bohr effect in hemoglobin? Is this important in insect hemoglobin?

REFERENCES

Agosin, M. 1978. Functional role of proteins, in M. Rockstein (ed.), *Biochemistry of Insects*. Academic Press, New York, pp. 93–203.

Ai, H., K. Kuwasawa, T. Yazawa, M. Kurokawa, M. Shimoda, and K. Kiguchi. 1995. A physiological saline for lepidopterous insects: Effects of ionic composition on heart beat and neuromuscular transmission. *J. Insect Physiol.* 41: 571–580.

Akaki, M. and J.A. Dvorak. 2005. A chemotactic response facilitates mosquito salivary gland infection by malaria sporozoites. *J. Exp. Biol.* 208: 3211–3218.

Arnold, J.W. 1974. The hemocytes of insects, in M. Rockstein (ed.), *Physiology of Insecta*, vol. 5. Academic Press, New York, pp. 201–254.

Arnold, J.W. and S.S. Sohi. 1974. Hemocytes of *Malacosoma disstria* Hübner (Lepidoptera: Lasiocampidae): Morphology of the cells in fresh blood and after cultivation in vitro. *Can. J. Zool.* 52: 481–485.

Arvy, L. 1954. Données sur la leucopoïèse chez *Musca domestica* L. *Proc. Roy. Ent. Soc., London A* 29: 39–41.

Ashhurst, D.E. 1982. Histochemical properties of the spherulocytes of *Galleria mellonella* L. (Lepidoptera: Pyralidae). *Int. J. Insect Morph. Embryol.* 11: 285–292.

Ashida, M. and H.I. Yamazaki. 1990. Biochemistry of the phenoloxidase system in insects: With special respect to its activation, in E. Ohnishi and H. Ishizaki (eds.), *Molting and Metamorphosis*. Japan Scientific Society Press, Tokyo, Japan; Springer-Verlag, Berlin, Germany, pp. 239–256.

Beard, R.L. 1953. Circulation, in K.D. Roeder (ed.), *Insect Physiology*. John Wiley & Sons, New York, pp. 232–272.

Browne, N., M. Heelan, and K. Kavanagh. 2013. An analysis of the structural and functional similarities of insect hemocytes and mammalian phagocytes. *Virulence* 4(7): 597–603.

Buck, J.B. 1953. Physical properties and chemical composition of insect blood, in K.D. Roeder (ed.), *Insect Physiology*. John Wiley & Sons, New York, pp. 147–190.

Buckner, J.S. and J.M. Caldwell. 1980. Uric acid levels during last larval instar of Manduca sexta, an abrupt transition from excretion to storage in fat body. *J. Insect Physiol.* 26: 27–32.

Chapman, R.F. 1958. A field study of the potassium concentration in the blood of the red locust, *Nomadacris septemfasciata* (Serv.), in relation to its activity. *Anim. Behav.* 6: 60–67.

Chen, A.C. and S. Friedman. 1975. An isotonic saline for the adult blowfly, *Phormia regina* and its application to perfusion experiments. *J. Insect Physiol.* 21: 529–536.

Craig, R. and N.A. Olson. 1951. Rate of circulation of the body fluid in adult Tenebrio molitor Linnaeus, *Anasa tristis* (de Geer), and *Murgantia histrionica* (Hahn). *Science* 113: 648–650.

Crozatier, M. and M. Meister. 2007. *Drosophila* haematopoiesis. *Cell. Microbiol.* 9(5): 1117–1126.

Dadant & Sons (eds.) 1975. *The Hive and the Honey Bee* (Revised edn.). Dadant & Sons, Publishers, Hamilton, IL, p. 740.

Davis, C.C. 1961. Periodic reversal of heart beat in the prolarva of a gyrinid. *J. Insect Physiol.* 7: 1–4.

Doucet, D. and M. Cusson. 1996. Role of calyx fluid in alterations of immunity in *Choristoneura fumiferana* larvae parasitized by *Tranosema rostrale*. *Comp. Biochem. Physiol.* 114A: 311–317.

Er, A., O. Sak, E. Ergin, F. Uckan, and D.B. Rivers. 2011. Venom-induced immunosuppression: An overview of hemocyte-mediated responses. *Psyche* 2011 (2011), Article ID 276376, 14 pages

Eslin, P. and G. Prevost. 1996. Variation in *Drosophila* concentration of haemocytes associated with different ability to encapsulate *Asobara tabida* larval parasitoid. *J. Insect Physiol.* 42: 549–555.

Florkin, M. and C. Jeuneaux. 1974. Hemolymph: Composition, in M. Rockstein (ed.), *The Physiology of Insecta*. Academic Press, New York, pp. 255–307.

Gagen, S.J. and N.A. Ratcliffe. 1976. Studies on the in vivo cellular reactions and fate of injected bacteria in *Galleria mellonella* and *Pieris brassicae* larvae. *J. Invert. Pathol.* 28: 17–24.

Gerould, J.H. 1930. History of the discovery of periodic reversal of heartbeat in insects. *Science* 71: 264–265.

Gerould, J.H. 1933. Orders with heartbeat reversal. *Biol. Bull., Woods Hole* 64: 424–431.

Gillespie, J.P., M.R. Kanost, and T. Trenczek. 1997. Biological mediators of insect immunity. *Annu. Rev. Entomol.* 42: 611–643.

Grégoire, C. and G. Goffinet. 1979. Controversies about the coagulocyte, in A.P. Gupta (ed.), *Insect Hemocytes*. Cambridge University Press, Cambridge, U.K., pp. 189–230.

Grégoire, C.H. 1951. Blood coagulation in arthropods. II. Phase contrast microscopic observations on hemolymph coagulation in sixty-one species of insects. *Blood* 6: 1173–1198.

Grigorian, M. and V. Hartenstein. 2013. Hematopoiesis and hematopoietic organs in arthropods. *Dev. Genes Evol.* 223: 103–115.

Grigorian, M., L. Mandal, M. Hakimi, I. Ortiz, and V. Hartenstein. 2011. The convergence of Notch and MAPK signaling specifies the blood progenitor fate in the *Drosophila* mesoderm. *Dev. Biol.* 353: 105–118.

Gupta, A.P. (ed.) 1979a. Arthropod hemocytes and phylogeny, in *Arthropod Phylogeny*. Van Nostrand Reinhold Co., New York, pp. 669–735.

Gupta, A.P. (ed.) 1979b. Hemocyte types: Their structures, synonymies, interrelationships and taxonomic significance, in *Insect Hemocytes*. Cambridge University Press, Cambridge, U.K., pp. 86–127.

Harrison, J.F. and M.J. Kennedy. 1994. In vivo studies of the acid-base physiology of grasshoppers: The effect of feeding state on acid-base and nitrogen excretion. *Physiol. Zool.* 67: 120–141.

Harrison, J.F. and J.E. Phillips. 1992. Recovery from acute haemolymph acidosis in unfed locusts. II. Role of ammonium and titratable acid excretion. *J. Exp. Biol.* 165: 97–110.

Harrison, J.F., J.E. Phillips, and T.T. Gleeson. 1991. Activity physiology of the two-striped grasshopper, *Melanoplus bivittatus*: Gas exchange, hemolymph acid-base status, lactate production, and the effect of temperature. *Physiol. Zool.* 64: 451–472.

Harrison, J.F., C.J.H. Wong, and J.E. Phillips. 1990. Haemolymph buffering in the locust *Schistocerca gregaria*. *J. Exp. Biol.* 154: 573–579.

Harrison, J.F., C.J.H. Wong, and J.E. Phillips. 1992. Recovery from acute hemolymph acidosis in unfed locusts. I. Acid transfer to the alimentary lumen is the dominant mechanism. *J. Exp. Biol.* 165: 85–96.

Harrison, J.M. 1988. Temperature effects on haemolymph acid-base status in vivo and in vitro in the two-striped grasshopper *Melanoplus bivittatus*. *J. Exp. Biol.* 140: 421–435.

Harrison, J.M. 1989. Temperature effects on intra-and extracellular acid-base status in the American locust, *Schistocerca nitens*. *J. Comp. Physiol. B* 158: 763–770.

Heinrich, B. 1970. Thoracic temperature stabilization by blood circulation in a free-flying moth. *Science* 168: 580–582.

Heinrich, B. 1996. *The Thermal Warriors: Strategies of Insect Survival*. Harvard University Press, Cambridge, MA.
Hink, W.F. 1976. A compilation of invertebrate cell lines and culture media, in K. Maramorosch (ed.), *Invertebrate Tissue Culture*. Academic Press, New York, pp. 319–369.
Hoffmann, J.A. 1970. Les organes hématopoïétiques de deux orthoptères: Locusta migratoria et *Gryllus bimaculatus*. *Z. Zellforsch.* 106: 451–472.
Hoffmann, J.A., D. Zachary, D. Hoffmann, and M. Brehélin. 1979. Postembryonic development and differentiation: Hemopoietic tissues and their functions in some insects, in A.P. Gupta (ed.), *Insect Hemocytes*. Cambridge University Press, Cambridge, U.K., pp. 29–82.
Jones, J.C. 1977. *The Circulatory System of Insects*. Charles C. Thomas, Springfield, IL.
Jones, J.C. and D.P. Liu. 1968. A quantitative study of mitotic divisions in haemocytes of *Galleria mellonella* larvae. *J. Insect Physiol.* 14: 1055–1061.
Jung, S.-H., C.J. Evans, C. Uemura, and U. Banerjee. 2005. The *Drosophila* lymph gland as a developmental model of hematopoiesis. *Development* 132: 2521–2533.
Kanost, M.R., J.K. Kawooya, J.H. Law, R.O. Ryan, M.C. Van Huesden, and R. Ziegler. 1990. Insect hemolymph proteins. *Adv. Insect Physiol.* 22: 299–396.
Kulkarni, V., R.J. Khadilkar, S.S. Magadi, and M.S. Inamdar. 2011. Asrij maintains the stem cell niche and controls differentiation during *Drosophila* lymph gland hematopoiesis. *PLoS One* 6 (11): e27667.
Lackie, A.M. 1988. Haemocyte behaviour. *Adv. Insect Physiol.* 21: 85–178.
Lanot, R., D. Zachary, F. Holder, and M. Meister. 2001. Postembryonic hematopoiesis in *Drosophila*. *Dev. Biol.* 230: 243–257.
Levenbook, L. 1958. Intracellular water of larval tissues of the southern armyworm as determined by the use of C14-inulin. *J. Cell. Comp. Physiol.* 52: 329–340.
Levenbook, L. 1979. Hemolymph volume during growth of *Calliphora vicina* larvae. *Ann. Entomol. Soc. Am.* 72: 454–455.
Markou, T. and G. Theophilidis. 2000. The pacemaker activity generating the intrinsic myogenic contraction of the dorsal vessel of *Tenebrio molitor* (Coleoptera). *J. Exp. Biol.* 203: 3471–3483.
Miller, T. 1968. Role of cardiac neurons in the cockroach heartbeat. *J. Insect Physiol.* 14: 1265–1275.
Miller, T. and R.L. Metcalf. 1968. Site of action of pharmacologically active compounds on the heart of *Periplaneta americana* L. *J. Insect Physiol.* 14: 383–394.
Miller, T. and W.W. Thomson. 1968. Ultrastructure of cockroach cardiac innervation. *J. Insect Physiol.* 14: 1099–1104.
Minakhina, S. and R. Steward. 2010. Hematopoietic stem cells in *Drosophila*. *Development* 137: 27–31.
Mori, H. 1996. Onset of embryonic heart movement in the water strider *Gerris paludum insularis* (Hemiptera: Gerridae). *Ann. Entomol. Soc. Am.* 89: 391–397.
Nakahara, Y., Y. Kanamori, M. Kiuchi, and M. Kamimura. 2010. Two hemocyte lineages exist in silkworm larval hematopoietic organ. *PLoS One* 5(7): e11816.
Nappi, A.J., Y. Carton, and F. Frey. 1991. Parasite induced enhancement of haemolymph tyrosinase in a selected immune reactive strain of *Drosophila melanogaster*. *Arch. Insect Biochem. Physiol.* 18: 159–168.
Nardi, J.B., B. Pilas, E. Ujhelyi, K. Garsha, and M.R. Kanost. 2003. Hematopoietic organs of *Manduca sexta* and hemocyte lineages. *Dev. Genes Evol.* 213: 477–491.
Nässel, D.R. 1993. Neuropeptides in the insect brain: A review. *Cell Tissue Res.* 273: 1–29.
Nation, J.L. 1989. The role of pheromones in the mating system of Anastrepha fruit flies, in A.S. Robinson and G. Hooper (eds.), *Fruit Flies, Their Biology, Natural Enemies and Control*, vol. 3A. Elsevier, Amsterdam, the Netherlands, pp. 189–205.
Nation, J.L., B. Smittle, and K. Milne. 1995. Radiation-induced changes in melanization and phenoloxidase in Caribbean fruit fly larvae (Diptera: Tephritidae) as the basis for a simple test of irradiation. *Ann. Entomol. Soc. Am.* 88: 201–205.
Nation, J.L. and K.K. Thomas. 1965. Quantitative studies on purine excretion in the greater wax moth, *Galleria mellonella*. *Ann. Entomol. Soc. Am.* 58: 883–885.
Nittono, Y. 1964. Formation of haemocytes near the imaginal wing disc in the silkworm, Bombyx mori L. *J. Sericultural Sci. Jap.* 33: 43–45.
Nutting, W.L. 1951. A comparative anatomical study of the heart and accessory structures of the orthopteroid insects. *J. Morphol.* 89: 501–597.
Pass, G. 1985. Gross and fine structure of the antennal circulatory organ in cockroaches (Blattodea, Insecta). *J. Morphol.* 185: 255–268.

Pass, G. 1988. Functional morphology and evolutionary aspects of unusual antennal circulatory organs in *Labidura riparia pallas* (Labiduridae), *Forficula auricularia* L. and *Chelidurella acanthopygia Géné* (Forficulidae) (Insecta: Dermaptera). *Int. J. Insect Morphol. Embryol.* 17: 103–112.

Pech, L.L. and M.R. Strand. 2000. Plasmatocytes from the moth *Pseudoplusia includens* induce apoptosis of granular cells. *J. Insect Physiol.* 46: 1565–1573.

Pichon, Y. 1970. Ionic content of haemolymph in the cockroach, *Periplaneta americana*: A critical analysis. *J. Exp. Biol.* 53: 195–209.

Raina, A.K. 1976. Ultrastructure of the larval hemocytes of the pink bollworm, *Pectinophora gossypiella* (Saunders) (Lepidoptera: Gelechiidae). *Int. J. Insect Morphol. Embryol.* 5: 187–195.

Ratcliffe, N.A. and P. Götz. 1990. Functional studies on insect haemocytes, including non-self recognition. *Res. Immunol.* 141: 919–922.

Ribeiro, C. and M. Brehelin. 2006. Insects hemocytes: What type of cell is that? *J. Insect Physiol.* 52: 417–429.

Rizki, R.M. and T.M. Rizke. 1980. Hemocyte responses to implanted tissues in *Drosophila melanogaster* larvae. *Roux's Arch. Dev. Biol.* 189: 207–213.

Rowley, A.F. and N.A. Ratcliffe. 1981. Insects, in N.A. Ratcliffe and A.F. Rowley (eds.), *Invertebrate Blood Cells*, vol. 2. Academic Press, New York, pp. 471–490.

Ruiz-Sanchez, E., M.C. Van Walderveen, A. Livingston, and M.J. O'Donnell. 2007. Transepithelial transport of salicylate by the Malpighian tubules of insects from different orders. *J. Insect Physiol.* 53: 1034–1045.

Sarkaria, D.S., S. Bettini, and R.L. Patton. 1951. A rapid staining method for clinical study of cockroach blood cells. *Can. Entomol.* 83: 329–332.

Shatoury, H.H. 1966. Micro-determination of insect blood volume. *Nature* 211: 317–318.

Siddiqui, M.I. and M.S. Al-Khalifa. 2012. Circulating haemocytes in insects: Phylogenic review of their types. *Pak. J. Zool.* 44: 1743–1750.

Slama, K. 1999. Active regulation of insect respiration. *Ann. Entomol. Soc. Am.* 92: 916–929.

Smits, A.W., W.W. Burggren, and D. Oliveras. 2000. Developmental changes in in vivo cardiac performance in the moth *Manduca sexta*. *J. Exp. Biol.* 203: 369–378.

Snodgrass, R.E. 1935. *Principles of Insect Morphology*. McGraw-Hill, New York.

Söderhäll, K. and A. Aspán. 1993. Prophenoloxidase activating system and its role in cellular communication, in J.P.N. Pathak (ed.), *Insect Immunity*. Oxford & IBH Publishing Co., New Delhi, India, pp. 113–129.

Stoepler, T.M., J.C. Castillo, J.T. Lill, and I. Eleftherianos. 2013. Hemocyte density increases with developmental stage in an immune-challenged forest caterpillar. *PLoS One* 8(8): e70978.

Tan, J., M. Xu, K. Zhang, X. Wang, S. Chen, T. Li, Z. Xiang, and H. Cui. 2013. Characterization of hemocytes proliferation in larval silkworm, *Bombyx mori*. *J. Insect Physiol.* 59: 595–603.

Tsakas, S. and V.J. Marmaras. 2010. Insect immunity and its signaling: An overview. *Inv. Surv. J.* 7: 228–238.

Wang, C.-L., Z.-X. Wang, M.M. Kariuki, Q.-Z. Ling, J. Kiguchi, and E.-J. Ling. 2009. Physiological functions of hemocytes newly emerged from the cultured hematopoietic organs in the silkworm, *Bombyx mori*. *Insect Sci.* 17: 7–20.

Wasserthal, L.T. 2012. Influence of periodic heartbeat reversal and abdominal movements on hemocoelic and tracheal pressure in resting blowflies *Calliphora vicina*. *J. Exp. Biol.* 215: 362–373.

Weber, H. 1930. *Biologie der Hemipteren*. J. Springer, Berlin, Germany.

Weidler, D.J. and G.C. Sieck. 1977. A study of ion binding in the hemolymph of *Periplaneta americana*. *Comp. Biochem. Physiol.* 56A: 11–14.

Wharton, R.A., M.L. Wharton, and J. Lola. 1965. Blood volume and water content of the male American cockroach *Periplaneta americana* L. Methods and influence of age and starvation. *J. Insect Physiol.* 11: 391–404.

Wheeler, R.E. 1962. Changes in hemolymph volume during the moulting cycle of *Periplaneta americana*. *Federation Proc.* 21: 123 (abstract).

Wheeler, R.E. 1963. Studies on the total haemocyte count and haemolymph volume in *Periplaneta americana* (L.) with special reference to the last molting cycle. *J. Insect Physiol.* 9: 223–235.

Williams, M.J. 2007. *Drosophila* hemopoiesis and cellular immunity. *J. Immunol.* 178: 4711–4716.

Woodring, J.P. 1985. Circulatory systems, in M.S. Blum (ed.), *Fundamentals of Insect Physiology*. John Wiley & Sons, New York, pp. 5–57.

Yamashita, M. and K. Iwabuchi. 2001. *Bombyx mori* prohemocyte division and differentiation in individual microcultures. *J. Insect Physiol.* 47: 325–331.

CHAPTER 16

Immunity

PREVIEW

The first line of defense in the resistance of insects to invasion of microorganisms is their external cuticular skeleton and cuticular lining of the fore- and hindgut, tracheal system, and parts of the reproductive system. When microorganisms get past the cuticular barrier, insects rapidly mount innate immune responses, including both cellular and humoral responses. Cellular responses include phagocytosis by hemocytes of small objects and encapsulation by layers of hemocytes of larger objects, such as parasitoid eggs or early instars of parasitoids. Virtually simultaneously with the cellular reactions, humoral responses begin with elaboration of pattern recognition proteins by epidermal cells, hemocytes, and fat body cells. The pattern recognition proteins serve several functions, but a major one is initiation of a cascade of metabolic reactions ending in the synthesis, mainly by fat body cells, of antibacterial and antifungal peptides that are released into the circulating hemolymph. The signaling pathways known as Toll, immune deficient (IMD), C-Jun N-terminal kinase (JNK), Janus kinase cascade (JAK), and signal transducers and activators of transcription (STAT) for eliciting synthesis of antibacterial and antifungal peptides are remarkably conserved, with variations, of course, from insects and other invertebrates to vertebrates, including humans. Additional components of responses include production of calcium-dependent (C-type) lectins that bind to particular carbohydrate sequences of invaders and may mark and clump them, control mechanisms involving serine proteinase inhibitors (serpins) to moderate cascading chemical reactions, and synthesis of phenoloxidase at wound sites and around encapsulated objects. Immune defense is not without the cost of energy and metabolic resources, and ecological tradeoffs with negative impact on fitness may occur, especially when insects encounter additional stresses, such as limited nutritional resources or adverse ecological conditions, when under bacterial or fungal attack. Parasitoids and parasites often elicit immune responses from their hosts, and hosts and parasites/parasitoids are in a continuing evolutionary race for survival. Evolutionarily, it may be more important for females under microbial attack to survive than males, and gender differences in immune responses and survival are known in some insects. Recent reviews of various aspects of the immune system in insects are available (Carton et al., 2008; Schmidt et al., 2010; Chambers and Schneider, 2012; Colinet et al., 2013; Merkling and Rij, 2013; Moreau, 2013).

16.1 INTRODUCTION

The first defense of insects against microbial organisms and fungi is the physical barrier presented by the tough sclerotized **cuticle** that covers the body, and thinner, more flexible cuticle that lines the tracheae, parts of the internal reproductive tract, foregut, and hindgut (see Chapters 2, 4, and 19 for details on cuticular surfaces of the body). Chemicals on the cuticular surface such as free fatty acids in some species (Gołębiowski et al., 2013) can be effective against some bacteria and fungi. When organisms succeed in getting past the cuticular barrier, insects rapidly mobilize **innate immune responses** to invading foreign organisms. Insects combat invading microbial organisms by several innate mechanisms including, (1) **phagocytosis** of small objects and **encapsulation** of larger objects with layers of hemocytes, (2) localized **coagulation of hemolymph** at wound sites, (3) **melanization reactions** at wound sites and usually at encapsulated objects, and (4) synthesis of **antimicrobial peptides**. Insects lack the complement system of acquired immunity with memory that occurs in vertebrates, although some experiments suggest that there is increased sensitivity and response to repeated challenges if the challenges are temporally close together (Schmid-Hempel, 2005).

Insect innate immune responses include both **cellular defenses** and **humoral defenses**. Cellular events are initiated by the cells that encounter the invading organisms, usually epithelial cells beneath the cuticle, hemocytes in the hemolymph, fat body cells, and epithelial cells lining the gut. These cells rapidly respond to the invasion by secreting **pattern recognition proteins** that have a variety of functions, including eliciting synthesis of antimicrobial peptides. Hemocyte proliferation occurs making increased numbers available to attack the invaders by phagocytosis, encapsulation, and nodule formation. Hemocytes release clotting agents in the hemolymph at wound sites, and **prophenoloxidase**, a zymogen circulating in the hemolymph, is activated to **phenoloxidase (PO)**, which promotes melanization of encapsulated objects and wound sites (Figure 16.1). Phenoloxidase action on phenolic compounds produces quinones and reactive oxygen and nitrogen compounds that are toxic to invading cells and to the host's own cells.

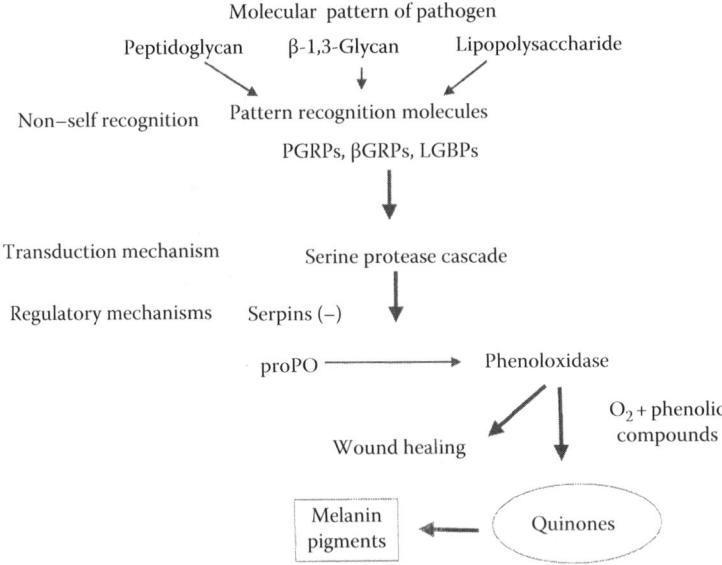

Figure 16.1 A conceptual scheme to illustrate non-self recognition by pattern recognition proteins and signal transduction through a serine protease cascade to activate prophenoloxidase to phenoloxidase and the production of melanin. PGRP, βGRP, and LGBPs are peptidoglycan recognition protein, betaglucan recognition protein, and liposaccharide recognition protein, respectively. Serpins are enzymatic proteins that have a negative effect by attacking the kinase cascade. proPO is prophenoloxidase.

Humoral responses usually are considered to be the elaboration of response agents that circulate in the hemolymph, which will include pattern recognition proteins. The pattern recognition proteins elaborated from epithelial cells, hemocytes, and fat body cells set in motion a cascade of enzymatic reactions in the cytoplasm that lead to the activation of nuclear genes that encode enzymes for the synthesis of antifungal and antibacterial peptides. Fat body cells are the principal sources of antibacterial and antifungal peptides, with some participation by hemocytes, and epithelial cells of the gut and epidermal cells below the cuticle. Additional humoral agents called serpins regulate cascades of reactions in the cytoplasm and help to localize the responses to site of fungal and bacterial invasion.

All of the cellular and humoral actions occur rapidly and nearly simultaneously in a challenged insect, although the humoral events initially lag slightly behind the cellular events (Schmid-Hempel and Ebert, 2003). In this chapter, these immune responses are discussed individually, but the reader should always keep in mind that the insect's response is a concert of cellular and humoral actions acting together.

16.2 PHYSICAL BARRIERS TO INVASION

The external **cuticle** of insects is a natural barrier to many microorganisms and fungi (Siva-Jothy et al., 2005), although some organisms have **chitinase** and protein-digesting enzymes that aid penetration through the cuticle. Probably the thickness of the cuticle is the main physical barrier, but in many insects there are numerous thin areas of cuticle, particularly at intersegmental boundaries between segments in the abdomen and in the lining of the tracheal system. Insect ingest many microorganisms with their food and, for some pathogens, the oral route is the main entry point into the insect. The foregut and hindgut have an attached cuticular lining on the lumen surface of the gut epithelial cells that protects the epithelial cells, but the midgut is more vulnerable. The midgut may have an unattached protective layer, the **peritrophic matrix** (Lehane, 1997) (see Chapter 2 for more details) composed of chitin and protein that protects the delicate brush border on midgut cells from harsh food particles as well as ingested microorganisms. Some insects, however, do not have a peritrophic matrix and yet their line has survived for hundreds of millions of years. The peritrophic matrix has holes or pores through which digestive enzymes secreted by the midgut cells pass into the food bolus and through which small digested molecules pass for absorption by midgut cells. Invading microorganism find these pores a potential site of entry to the midgut cells. In tsetse flies (*Glosina* spp., transmitters of the parasite for sleeping sickness to humans and nangana to animals), the pores are about 9 nm in diameter and probably are too small to allow invasion of the parasite through the pores. In some mosquitoes, the pores are up to 200 nm in size, and may allow the passage of some arboviruses, but not bacteria or other larger potential invaders. Some insects (notably some mosquitoes that are disease transmitters) secrete the peritrophic matrix only after food is ingested, and orally ingested microorganisms might contact the surface of midgut epithelial cells before the matrix is completely formed (Lehane, 2005). The mature ookinete stage of the malaria parasite produces the enzyme chitinase, which may aid in penetrating the midgut peritrophic matrix (Sieber et al., 1991; Shahabuddin, 1998). Penetration, however, may depend on how complete the secretion of the type I peritrophic matrix characteristic of the mosquito adults is and whether the blood meal of the mosquito female contains immature malarial gametocytes or mature ookinetes, which are the infective stage (Janse et al., 1985; Lehane, 2005). Billingsley and Rudin (1992) and Ponnudurrai et al. (1988) demonstrated that the type I peritrophic matrix of *Aedes aegypti* reduced the number of *Plasmodium gallinaceum* ookinetes that crossed the matrix when it was experimentally hardened by a chemical treatment that simulated the normal hardening of the matrix that occurs with time as the mosquito feeds on a host.

16.3 CELLULAR IMMUNE REACTIONS

Cellular reactions are initiated immediately upon the invasion of a foreign microorganism and involve direct attack of the foreign object by hemocytes in the circulating hemolymph (Tan et al., 2013). Figueiredo et al. (2006) found that **plasmatocytes** were the principal hemocytes involved in **phagocytosis** of yeast particles in *Rhodnius prolixus*, and also in the wax moth *Galleria mellonella* (Büyükgüzel et al., 2007) and plasmatocytes and granulocytes were the primary hemocytes involved in phagocytosis in a carabid ground beetle (Giglio and Giulianini, 2013). Browne et al. (2014) showed that a short-term boost to the immune system occurred in *Galleria* larvae when exposed to mild thermal or physical stress, with elevated expression of apolipophorin, arylophorin, and phenoloxidase, but the effect subsides within about 72 h (Browne et al., 2014). Wojda and Taszłow (2013) found increased transcription levels for the antimicrobial peptides cecropin, gallerimyucin, and galiomycin in the fat body of *Galleria* larvae exposed to a 30 min heat shock at 40°C.

The hemocyte load (i.e., having a large number of hemocytes) makes *Drosophila melanogaster* larvae more successful in recognizing and encapsulating the eggs of one of its parasitoid enemies, *Asobara tabida*. However, Gerritsma et al. (2013) found that variation in *D. melanogaster* populations collected from a wide-range of European sites in successfully encapsulating the eggs of *A. tabida* depended more upon the ratios among the different types of hemocytes than simple hemocyte load. *Drosophila* larvae have three types of hemocytes; plasmatocytes, lamellocytes, and crystal cells. Plasmatocytes phagocytize dead cells and microbial pathogens; lamellocytes encapsulate and sequester larger objects (such as the eggs of some parasitoids), and crystal cells produce phenoloxidase that is important in encapsulation of larger objects with melanization and wound healing. The proliferation of hemocytes in *Aedes aegypti* mosquitoes is stimulated by blood feeding and the insulin-like peptide 3, leading it to have more hemocytes to fight potential invading microbes (Castillo et al., 2011).

Melanization involves the action of **phenoloxidase** (PO) on phenolic compounds (tyrosine and dihydroxyphenylalanine, for example, Figures 16.1 and 16.2) to produce quinones that autopolymerize and produce **melanin** (Fujimoto et al., 1993; Chase et al., 2000). Activation of PO is one of the principal defenses of insects against bacterial and fungal invasion (Cerenius and Söderhäll, 2004; Piñera et al., 2013; Tokura et al., 2014). The thick layer of melanin and hemocytes formed around encapsulated organisms may help suffocate as well as be toxic to them. A **zymogen**, **prophenoloxidase**, is stored in cells, but mainly circulates in the open hemolymph system of insects. It is converted to active phenoloxidase by a series of serine proteases. Inhibitors (**serpins**, see Section 16.9) help regulate the active enzyme (see Figure 16.1) and serve to restrict its activity to the invasion site. Thus, excessive systemic damage is avoided by highly toxic and reactive compounds that result from PO action on a variety of substrate molecules (phenolic compounds). Additional limitation of toxic effects from phenolic and quinone compounds generated by the phenoloxidase reaction may be provided, at least in some insects, by amyloid fibers incorporated into the encapsulated object (Falabella et al., 2012) through action of an immune gene that encodes the protein of amyloid fibers (Di Lelio et al., 2014).

The *Drosophila* genome contains three genes coding proPO (Ross et al., 2003) and *Anopheles gambiae* has nine (Christophides et al., 2002). These different proPOs may have different functions (Cerenius and Söderhäll, 2004). Active PO is a heterodimer, with each subunit encoded by a different gene in *Bombyx mori* (Asano and Ashida, 2001a,b) and *Manduca sexta* (Jiang et al., 1997), but in *D. melanogaster* the active enzyme is a homodimer (Sezaki et al., 2001).

The serine protease cascade that converts proPO to active PO is activated by the binding of pattern recognition receptors to β-1,3-glycan in fungal invasion or to lipopolysaccharide or peptidoglycan in bacterial invasion (Figure 16.1), depending on whether the bacteria are Gram (+) or Gram (−). Simple damage to the tissues of an insect, such as cutting or puncturing the cuticle or other

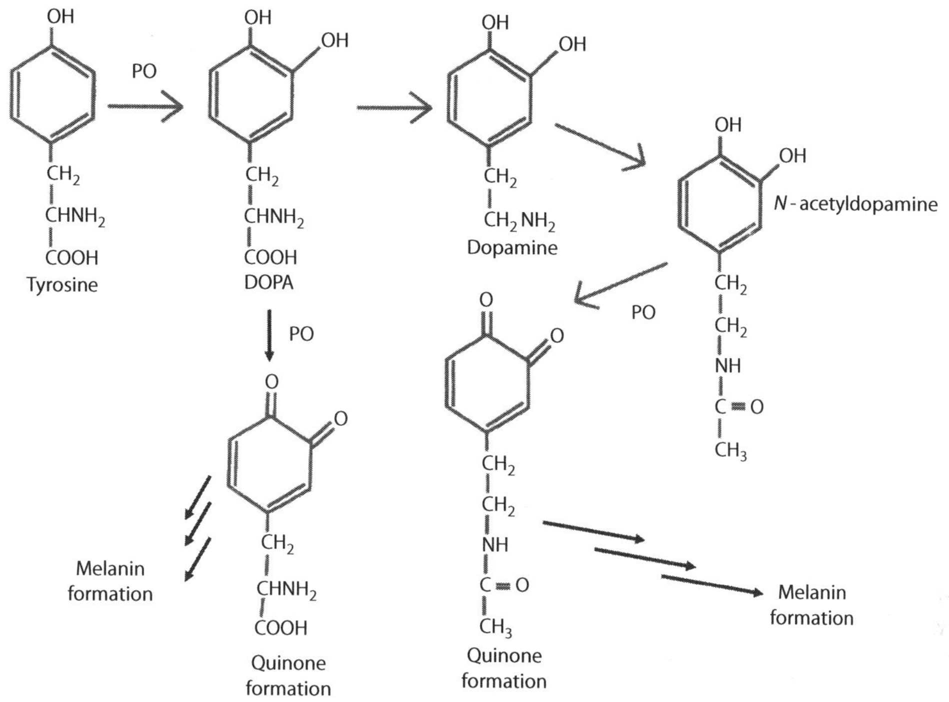

Figure 16.2 Some of the pathways that may be involved in melanin formation in insects during an immune response. Additional compounds with monohydroxyl or dihydroxyl groups are known to occur in insects and also may be involved. Key: PO, phenoloxidase; DOPA, dihydroxyphenylalanine.

tissue, also activates the proteinase cascade and serpins (De Gregorio et al., 2002; Ligoxygakis et al., 2002a,b). Thus, recognition of bacterial or fungal invasion elicits both positive and negative regulation of activation of active PO. Phenoloxidase also has a major function in tanning of the cuticle after a molt (see Chapter 4).

16.4 RECOGNITION OF NONSELF

Recognition of **nonself** by fat body cells, hemocytes, midgut epithelium, and cuticular epithelium is the first step in mounting a **humoral defense**. Insects may use the basement membrane that lies at the basal surface of all insect cells (except hemocytes) as an indicator of self against which they direct non-self reactivity (Siva-Jothy et al., 2005). When stimulated by invading microorganisms, epithelial cells in various parts of the body secrete **pattern recognition proteins** (**PRPs**, the designation used in this chapter, but also called **pattern recognition receptors** [**PRRs**] by some authors) (Werner et al., 2000). Some of the PRPs are released into the circulating hemolymph while others are attached to the surface of the cells that produce them. The PRPs recognize and bind to particular carbohydrate or carbohydrate-peptide linkages in the structure of invading fungi or bacteria by acting as receptors for characteristic bacterial or fungal wall components. The bacterial and fungal structures recognized by PRPs are called **pathogen-associated molecular patterns** or **PAMPs**. Identified PAMPs include **β-1,3-glucans** as a part of the fungal cell wall, and **lipopolysaccharides** (**LPS**) and **peptidoglycans** (**PGNs**) as part of the cell surface of bacteria. Binding of the PRPs to invading microbial cells marks them for destruction. Ferrandan et al. (2004) suggest that the open circulatory system of insects makes **PRRs** especially well suited to communicate rapidly the presence of nonself because the PPR-marked microorganisms are conveyed directly to hemocytes (which may encapsulate or phagocytize them) and to fat body cells (which are stimulated to synthesize antimicrobial peptides). Interestingly, some of the glucans may have glucanase activity, thus, participating in an enzymatic attack on a fungal pathogen. For example, a 40 kDa β1,3-glucan-binding protein isolated from the lepidopteran *Helicoverpa armigera* is secreted into the larval midgut when the larvae have fed upon a diet containing bacteria. The molecule has glucanase activity and accounts for most of the enzyme lysis of β1,3-glucans in the gut (Pauchet et al., 2009).

Drosophila has 13 genes coding for peptidoglycan recognition proteins (PGRPs; PGRP is a PRP, but some authors prefer the slightly more descriptive designation, PGRP). Some of the PGRPs are membrane-bound molecules while others circulate (Royet, 2004). In addition to PGRPs, *Drosophila* secretes Gram-negative binding proteins (GNBPs) that bind Gram (−) bacteria and β-glucan recognition proteins (βGRPs) that recognize the β-glucan structure of fungi (Ferrandan et al., 2004). Two characterized PGRPs are PGRP-LC and PGRP-LE, which specifically bind *meso*-diaminopimelic acid (DAP) in the peptidoglycan structure of Gram (−) bacteria. Another one designated as PGRP-SA (Wang et al., 2006) binds to the peptidoglycan structure in which lysine replaces DAP in the linking peptide, which is characteristic of Gram (+) bacteria. Two of the PGRPs are known to have immediate actions: PGRP-LC binding to Gram (+) bacteria initiates phagocytosis by hemocytes and PGRP-LE binding to Gram (−) bacteria results in phenoloxidase activation and melanin production.

Although having the complete genome of *D. melanogaster* stimulated and aided much of the early work on immunity, research was rapidly expanded to other insects. *Anopheles gambiae* mosquitoes are known to have seven PGRP genes. βGRP-1 and βGRP-2 have been isolated from hemolymph of larval *M. sexta* (Ma and Kanost, 2000), and βGRP-2 also occurs in the *Manduca* cuticle. Both of these pattern recognition proteins bind β-1,3-glucan, participate in initiating agglutination of Gram (−) and Gram (+) bacteria, and when bound to the PAMPs of the invading organism, they stimulate conversion of prophenoloxidase into the active phenoloxidase. Additional pattern recognition proteins have been identified from *B. mori*, *M. sexta*, and *Plodia interpunctella*. βGRP1 and βGRP2 from *M. sexta* and βGRP from *B. mori* are expressed in insect larvae before immune challenge, but are upregulated with challenge by bacteria and yeast (Ochiai and Ashida, 2000). The presence of a constitutively low level of some PRRs may serve to facilitate an immediate response to foreign invasion, and allow time for the induction of additional PRPs to assist in combating the infection (Fabrick et al., 2003). Fabrick et al. found that a βGRP isolated from *P. interpunctella* had broad binding capability and multiple recognition capability, binding to β-1,3-glucan, lipopolysaccharide, and lipoeichoic acid of fungi and bacteria, and it agglutinates yeast and bacteria.

Hemolin is also a carbohydrate-binding protein found in several lepidopterans (Rasmunson and Boman, 1979; Ladendorff and Kanost, 1990; Yu and Kanost, 1999; Lee et al., 2002; Kanost et al., 2004), but it has not been reported from *D. melanogaster* or *Anopheles gambiae*. Hemolin synthesis is induced when *M. sexta* is challenged by bacterial infection (Wang et al., 1995) and it binds to hemocytes (Ladendorff and Kanost, 1991; Zhao and Kanost, 1996). It may act as an opsonin that facilitates trapping of bacteria in aggregates of hemocytes and in nodule formation.

PRPs or PGRPs additionally have a critical role in initiating the cascade of reactions leading to the synthesis of antifungal and antibacterial peptides, as described in the next section.

16.5 SYNTHESIS OF ANTIFUNGAL AND ANTIBACTERIAL PEPTIDES

Antimicrobial peptides are vital components of the immune defense system of many organisms, including insects (Li et al., 2012). Three *defensin* genes were characterized from the braconid endoparasitoid *Cotesia vestalis* and designated as *CvDef1*, *CvDef2*, and *CvDef3* by Wang et al. (2013). Messenger RNA (mRNA) from Genes 1 and 3 was expressed in second and third instars, respectively, and gene 2 mRNA was present in all stages. Gene 1 was the most potent and wide ranging with antimicrobial action against both gram-positive and gram-negative bacteria. There are about 30 genes in the *Drosophila* genome that encode antimicrobial peptides (Royet, 2004), and antimicrobial peptides are known from other insects, including termites (Fefferman et al., 2007; Rosengaus et al., 2007), silkworm (Imamura et al., 2006), the mosquito *Anopheles gambiae* (Moita et al., 2006; Warr et al., 2006), the hemipteran *Triatoma* spp. (Araújo et al., 2006), the lepidopterans *Mamestra brassicae* (Lee et al., 2005) and *Galleria mellonella* (Cytryńska et al., 2006), locusts (Mullen and Goldsworthy, 2006), and other arthropods (Bulet et al., 2004; Zhou et al., 2006). Antimicrobial peptides are also important in the relationship between the insect host and its symbionts (Login et al., 2011; Login and Heddi, 2013; Vilcinskas, 2013).

PRPs described in the previous section are likely to be involved in different insects, but specific ones have been identified in only a few insects. Synthesis of the antimicrobial peptides involves activation of genes in the Toll, IMD, and JAK/STAT pathways in fat body cells. These pathways have been explored mostly in *Drosophila* (Agaisse and Perrimon, 2004) with the aid of specific mutants, but they probably operate in some form in all insects because the pathways are evolutionarily conserved and have been described from humans as well (Medzhitov et al., 1997; Rock et al., 1998; Pinheiro and Ellar, 2006).

Although high levels of antimicrobial peptides are synthesized in response to invasion of the insect body, constitutively low levels of some antimicrobial peptides (e.g., cecropin) have been reported (Junell et al., 2007).

16.6 TOLL PATHWAY FOR THE SYNTHESIS OF ANTIMICROBIAL PEPTIDES

Gram (+) bacteria and fungi each activate serine-based proteolytic cascades that converge on and activate the Toll pathway. Gram (+) bacteria induce fat body cells to synthesize the antimicrobial peptide drosomycin (possibly drosomysins) and fungi elicit the secretion of metchnikowin. Both antimicrobial compounds are synthesized after activation of the **Toll** pathway (Figures 16.3 and 16.4). The protein Toll and the Toll pathway are ancient parts of a conserved innate immune system that extends from lower invertebrates to humans, understandably with numerous variations and differences in detail along the evolutionary path. For example, the Toll protein is a transmembrane protein in fat body cells of insects, but in vertebrates it is a cytoplasmic component, not the transmembrane receptor, in the pathway.

Drosophila produces a number of PRPs, but only three peptidoglycan recognizing proteins, PGRP-SA, PGRP-LC, and PGRP-LE have been characterized functionally. PGRP-SA recognizes the peptidoglycan structure of Gram (+) bacteria and PGRP-LC and PGRP-LE recognize the peptidoglycan structure of Gram (−) bacteria. Gram (+) bacteria invasion of *Drosophila* results in the release of PGRP-SA and Osiris, another recognition molecule. The peptidoglycan recognition protein that binds an invading fungus in a *Drosophila* larva may be GNBP3.

Gram (+) bacteria and fungi activate different serine proteinase cascades in the hemolymph that result in the cleavage of Späetzle, an 82 kDa homodimer protein circulating in the hemolymph of *Drosophila* larvae (Ligoxygakis et al., 2002a). Most of the components of the cascades have not been identified, but one kinase is known to be Persephone that is elicited by fungi. Persephone, encoded by *Persephone*, cleaves Späetzle, the ligand that binds Toll and activates the Toll pathway. Späetzle is cleaved by an unknown kinase in response to Gram (+) invasion. *Drosophila* Toll is dimerized (probably as part of the cascade reactions) and binds two molecules of cleaved Späetzle (see Figure 16.3). Thus activated, Toll sets in motion an intracellular serine kinase cascade in the cytoplasm of fat body cells. Most of the components in this intracellular cascade have not been identified, but the proteins MyD88, Tube, and Pelle participate in *Drosophila* (Imler and Hoffmann, 2001; Steiner, 2004), and one of the kinases phosphorylates the Cactus domain of the Cactus-DIF (**dorsal immune factor**) complex (Ferrandan et al., 2004).

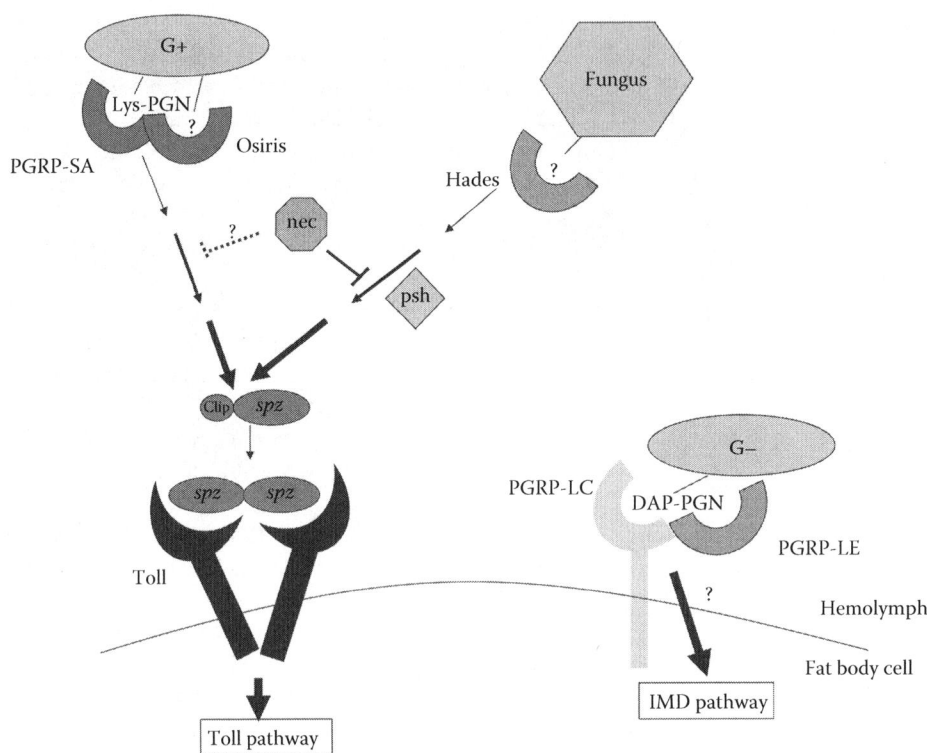

Figure 16.3 (See color insert.) The recognition of microbial infection in *Drosophila* leading to activation of the Toll and IMD pathways. Gram (+) bacteria are recognized by PGRP-SA, a peptidoglycan recognition protein that binds the lysine-type peptidoglycan of Gram (+) bacteria. Osiris is another peptidoglycan recognition protein that participates in the initial recognition reaction. A proteolytic kinase cascade is initiated that finally removes a clip domain from the dimer Späetzle, the ligand for Toll, a transmembrane protein in the fat body cell membrane. Fungal invasion also elicits recognition proteins, one of which may be Hades. A proteolytic cascade is activated, and one of the kinases, persephone, proteolytically removes the clip domain from Späetzle, which then binds Toll. Toll dimerizes, perhaps as a result of binding *spz*, and then binds two molecules of *spz*. Gram (−) bacteria, which have diaminopimelic acid (DAP)-type peptidoglycan in the outer coat, is recognized and bound by PGRP-LDC and possibly PGRP-LE, and the immune deficiency (IMD) pathway is activated. (From Leclerc, V. and Reichhart, J.-M., *Immunol. Rev.*, 198, 59, 2004. With permission.)

Cactus is an inhibitory protein that binds DIF and keeps DIF in the cytoplasm. After phosphorylation, Cactus releases DIF, which is translocated to the nucleus, and Cactus is degraded in the cytoplasm. **DIF** (also known as **Dorsal** in embryonic development) acts as a transcription factor in the nucleus for the gene *drosomysin*. *Drosomysin* expression results in the synthesis of drosomycin (Belvin and Anderson, 1996; Imler and Hoffmann, 2001; Ferrandan et al., 2004; Royet, 2004). Drosomysin is secreted into the circulating hemolymph where it attacks the invading microorganisms. The Toll path in *Drosophila* is an active area of investigation and has several intracellular branches as shown in Figure 16.4.

Identified genes that function in the Toll pathway include *Späetzle*, *Toll*, *Tube*, *Pelle*, and *Cactus* operating in the antifungal response of *Drosophila*. The genes *Späetzle*, *Toll*, and *Cactus* are expressed in the adult fat body and are upregulated when *Drosophila* is immune challenged. Mutants in which these genes fail to function are very susceptible to fungal infection (Lemaitre et al., 1996). Although 9 Toll genes are present in the *Drosophila* genome (Tauszig et al., 2000) and 10 in the genome of *A. gambiae* (Christophides et al., 2002), functions for most of the Toll receptors

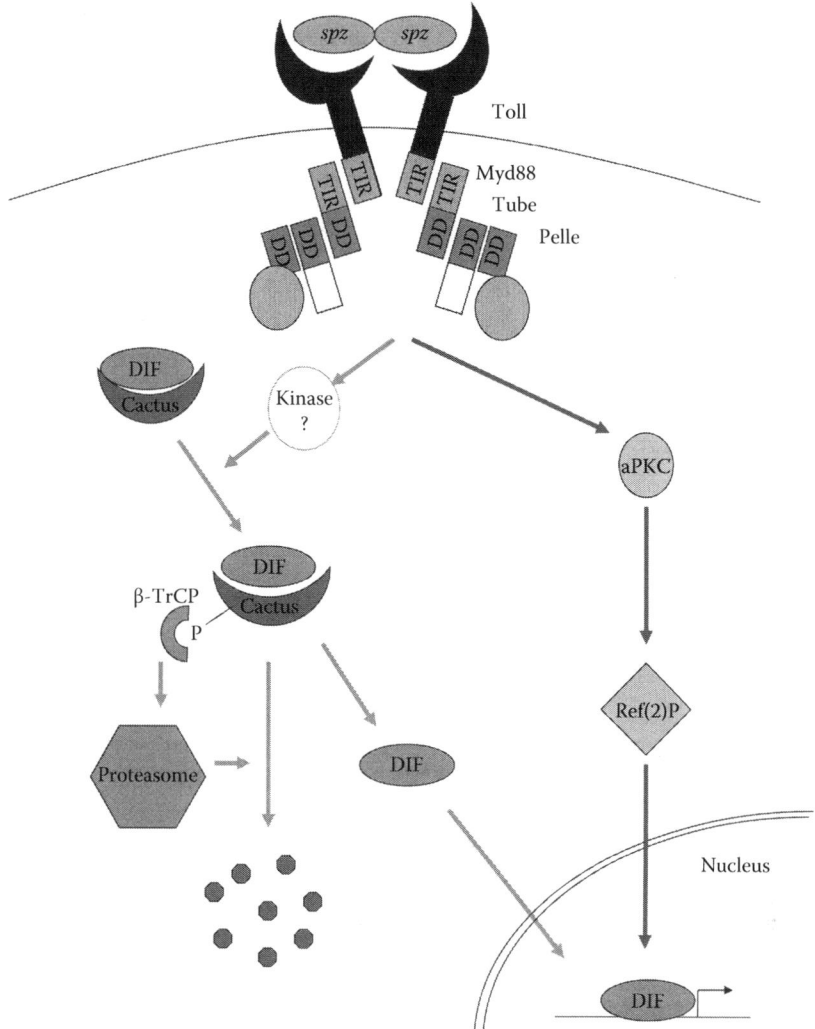

Figure 16.4 (See color insert.) An illustration of selected details of the intracellular Toll pathway that culminates in the synthesis of antimicrobial peptide drosomysin. Each of the Toll receptors recruits several additional proteins including MYD88, Tube, and Pelle, and an intracellular proteolytic kinase cascade is activated. At least one of the kinases in the cascade hydrolyzes the Cactus domain from the Cactus-DIF complex in the cell cytoplasm. Cactus in then further degraded by enzymes, and DIF is translocated to the nucleus where it acts as a transcription factor for *drosomyosin*. Expression of *drosomyosin* results in synthesis of the antimicrobial peptide Drosomyosin and its secretion into the circulating hemolymph where it attacks the invading microoganisms. (From Leclerc, V. and Reichhart, J.-M., *Immunol. Rev.*, 198, 59, 2004. With permission.)

are not known. The *Drosophila* genome encodes five homologs of Späetzle, but functions for those other than the Toll ligand is uncertain; they might possibly serve as ligands for other Toll receptors (Imler and Hoffmann, 2001; Parker et al., 2001) and/or they have developmental functions because they are expressed in the embryo (Gerttula et al., 1988) and during metamorphosis in appropriately timed sequences (Eldon et al., 1994; Kambris et al., 2002). Additional details on Toll and the Toll pathway can be found in reviews by Imler and Hoffmann (2001), Hoffmann (2003), Ferrandan et al. (2004), Leclerc and Reichhart (2004), Christophides et al. (2004), Steiner (2004), Bangham et al. (2006), and Pinheiro and Ellar (2006).

16.7 IMD PATHWAY FOR THE SYNTHESIS OF ANTIMICROBIAL PEPTIDES

Gram (−) bacteria and some Gram (+) bacilli induce the synthesis of the antibacterial peptides, diptericin, drosocin, cecropins, and attacins. These peptides are synthesized by fat body cells after genes in the nucleus have been turned on by signaling through the **IMD** pathway (Figure 16.5). IMD stands for immune deficiency, so named because earlier discovery (Lemaitre et al., 1995) of a mutation in the gene *imd* reduced the resistance of *Drosophila* to Gram (−) bacterial infection. Gram (−) bacteria likely induce pattern recognition receptors, but

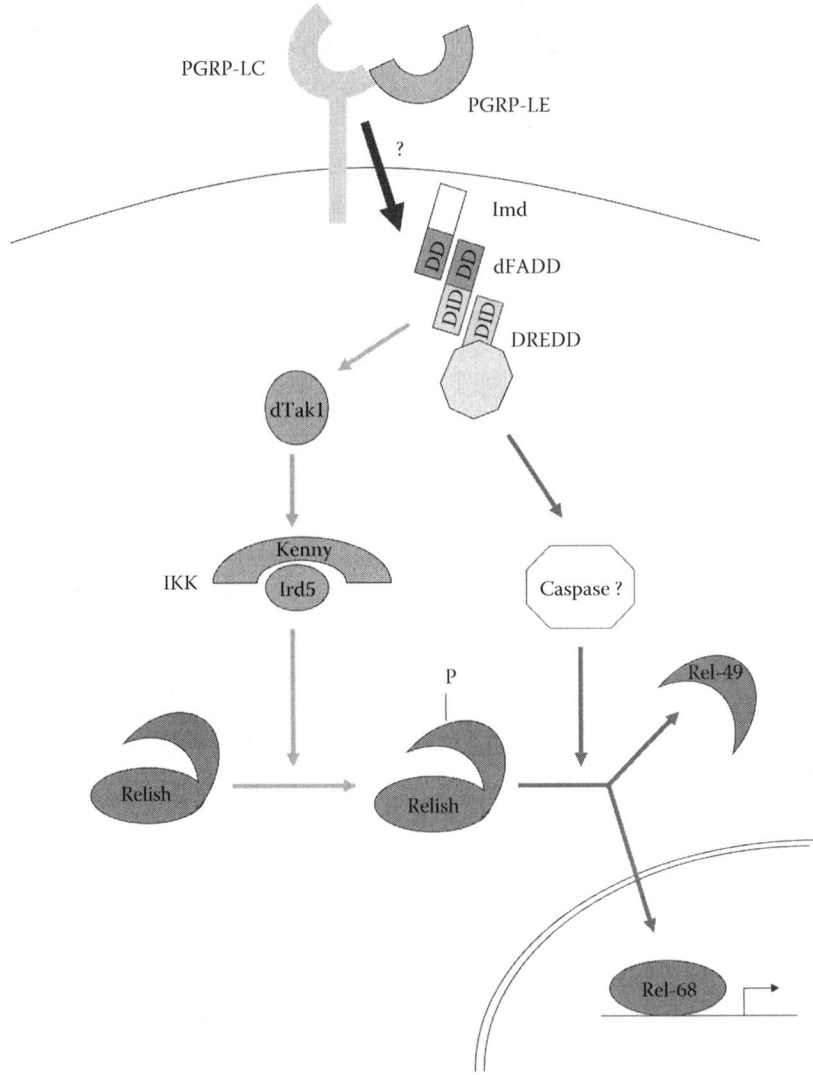

Figure 16.5 (See color insert.) The IMD pathway results in the synthesis of an antimicrobial peptide, diptericin, effective against Gram (−) bacteria. Several factors are recruited when the IMD receptor is activated and two intracellular proteolytic cascades are initiated. One cascade results in phosphorylation of Relish. The second cascade cleaves Relish into an inhibitory molecule, Rel-49, which remains in the cytoplasm, while Rel-68 is translocated to the nucleus where it serves as a transcription factor for *diptericin*. The antimicrobial peptide diptericin is then synthesized and secreted into the hemolymph. (From Leclerc, V. and Reichhart, J.-M., *Immunol. Rev.*, 198, 59, 2004. With permission.)

they have not been identified. A serine proteinase cascade activates the transmembrane IMD receptor. PGRP-LE seems to be involved near or at the fat body cell membrane, possibly acting in conjunction with the IMD receptor. PGRP-LC and PGRP-LE recognize peptidoglycan (PGN) with diaminopimelic acid (DAP) in the short cross-linking peptide bridges in the peptidoglycan molecule. Binding of DAP–PGN activates the IMD pathway in the cell cytoplasm. The intracellular portion of the IMD pathway involves an additional cascade of intracellular proteolytic kinase reactions that lead to the movement of **Relish** (Rel) into the nucleus. Rel is a transcription factor, and *Diptericin* is transcribed and translated in the cytoplasm with the synthesis of **diptericin** and related antibacterial peptides. These peptides are exported into the circulating hemolymph to attack the invading bacteria. The IMD pathway controls the transcription of several hundred genes in *Drosophila*, but only a few are known to function in immune defense.

The IMD pathway has a branch in the cytoplasm downstream of the IMD membrane receptor named the **JNK** path (Delaney et al., 2006). JNK is a c-Jun N-terminal kinase cascade that leads to the activation of cytoskeletal genes important in tissue repair. Thus, the upstream activation of IMD receptor may allow linking of antimicrobial defense with tissue repair due to the invasion or damage from invasion. Additional details about the IMD pathway can be found in reviews by Hoffmann (2003) and Ferrandan et al. (2004).

16.8 C-TYPE LECTINS

C-type lectins are calcium-dependent proteins that bind to particular carbohydrate sequences and, thus, function as PRPs (Kanost et al., 2004). Four C-type lectins found in the hemolymph of *M. sexta* and named immunolectins (IMLs) bind to bacterial LPS, and IML-1 and IML-2 stimulate activation of proPO in plasma (Yu et al., 1999, 2002; Yu and Kanost, 2000). IML-2 appears to have activity against Gram (−) bacteria (Yu and Kanost, 2003). IMLs may help in localizing PO response to the surface of invading bacteria (Yu et al., 2003) and, thus, avoid widespread action of phenoloxidase that might be detrimental to the tissues of the insect.

16.9 SERPINS

Serpin is a coined name that stands for **(ser)**ine **(p)**roteinase **(in)**hibitor. Serpins act like a brake to regulate the cascade of reactions (see Figure 16.1) occurring in the immune response and in numerous other cascade reactions involving serine proteinases (Pelte et al., 2006). Serpins are proteins about 45 kDa in size forming a family of serine proteinase inhibitors that function in both invertebrates and mammals (Kanost, 1999). **Necrotic** (Nec) is a serpin that regulates Toll activation by inhibiting a serine proteinase involved in the cleavage of Späetzle. They perform their inhibition of serine proteinases by trapping and distorting the proteinase enzyme in a covalently linked serpin–proteinase complex, which is then targeted for destruction.

Manduca sexta synthesizes a mixture of 12 serpins, differing mainly in the carboxyl-terminal reactive site loop that is critical to the mechanism of inhibition of serine proteinases. Serpin-1 is not upregulated in response to challenge and ecdysteroid negatively regulates serpin-1 in fat body cells (Kanost et al., 1995). Serpin-2 from *M. sexta* is upregulated by bacterial challenge (Gan et al., 2001), but its specific target proteinase is unknown. Serpins 3–5 are present in fat body cells and are upregulated by challenge. Kanost et al. (2004) suggest that serpin-3 might help prevent excessive melanization in the hemocoel of insects by localizing proPO activation to the site of infection (i.e., serpin-3 might inhibit the prophenoloxidase-activating protein, a serine proteinase that is upregulated in an infection). The sequenced genomes of the mosquitoes *A. gambiae*, *Aedes aegypti*, and *Culex quinquefasciatus* contain 18, 23, and 31 serpin genes, respectively (Gulley et al., 2013). The serpins can cause lysis of the malarial parasite and play a major role in the activation of phenoloxidase, and possibly function in the activation of the Toll pathway in the mosquitoes.

16.10 ECOLOGY, BEHAVIOR, AND IMMUNITY

The environment, and especially the social environment of colony insects, the internal symbionts environment, nutrition, and reproductive status all have important consequences and potential tradeoffs for insect immunity (Figure 16.6). Schmid-Hempel (2005), Ponton et al. (2011, 2013), Chambers and Schneider (2012), and Bauerfeind and Fischer (2014) reviewed interrelationships between nutrition, ecology, and immunity in insects. The interaction between pathology and microbial numbers, and the influence of environmental parameters such as diet, temperature, humidity, crowding, and particularly intraspecies contacts in social insects are important factors in whether insects become infected with pathogens and whether they survive. The primary and secondary symbionts that many insects harbor often provide some benefits against infection with viruses and other pathogenic microbes (Oliver et al., 2014). The burying beetle *Nicrophorus vespilloides*, is a member of a group of insects with behavioral mechanisms (Cotter and Kilmer, 2010; Cotter et al., 2013; Urbański et al., 2014) that require balancing individual vs. social immunity, and differences in its immune response varies with developmental stage, parental care, and with a rapidly changing food supply. A benefit of the social environment in a colony of bumblebees is that the gut microbiota of bumblebees protects them from *Crithidia bombi*, a virulent parasite (Koch and Schmid-Hempel, 2011). Bauerfeind and Fischer (2014) investigated the effects of climate change upon the immune responses of the butterfly, *Pieris napi*, and found that butterflies reared at 25°C (as contrasted to those reared at 17°C) had reduced number of hemocytes and reduced phenoloxidase activity; the authors suggested that higher temperatures do not always increase the competency of the immune system.

Evolutionists are interested in how defenses are selected and how they adapt insects for success in their ecological setting. Synthesis of the new compounds involved in an immune response is likely to be costly and may divert energy away from other activities, with possible

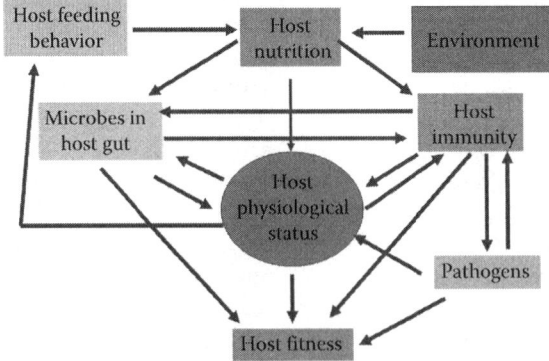

Figure 16.6 (See color insert.) A scheme to show some of the factors affecting the host physiological status, host immunity, and host fitness. The interactions are multiple and complex. Environmental influences may act directly upon most of the other components illustrated in this diagram, but to avoid excessive lines, it is shown interacting through host nutrition, which in turn has an influence on all the other components. (Based on Ponton, F. et al., *PloS Pathog.*, 7(12), e1002223, 2011; Ponton, F. et al., *J. Insect Physiol.*, 59, 130, 2013; Chambers, M.C. and Schneider, D.S., *Curr. Opin. Immunol.*, 198, 116, 2012; Bauerfeind, S.S. and Fischer, K., *J. Insect Physiol.*, 64, 14, 2014.)

tradeoffs operating during the evolution of defensive mechanisms that negatively impact fitness of insects. Benefits of defense are generally obvious, such as survival of an attack, continued growth, development, and success in reproduction. The costs of defense are less clear-cut, but may involve reduced function in some other biological process, especially when nutritional resources are limited. Nutritional resources may be limited for many insects at certain times, and especially when an insect feeds on one or only a few food sources. For example, *Rhodnius prolixus*, a blood feeder, takes only one blood meal in each instar as an immature, and adult females need blood to produce eggs. Feder et al. (1997) found that *Rhodnius* was less able to resist challenge with bacteria if fed only plasma instead of whole blood, which suggested that the plasma was not nutritionally as good as whole blood. On plasma, the bugs had reduced production of antimicrobial peptides, reduced nodule formation, and synthesized less lysozyme than those getting whole blood. Addition of α-ecdysone aided these plasma-fed bugs during infection, and although the mechanism by which α-ecdysone might work is not clear, it might signal ecdysteroid-related control of some defense mechanisms. The quality of the protein fed to *D. melanogaster* in the larval stage increased the expression of genes encoding antimicrobial peptides (Fellous and Lazzaro, 2010). Alaux et al. (2010) found that the quality of proteins (from mixed-floral pollens, as opposed to pollens from a single flora source) led to increased immunocompetency in honeybees, *Apis mellifera*. The mixed-pollens-fed bees had higher levels of glucose oxidase activity, higher hemocyte numbers, and higher levels of phenoloxidase. Mounting an immune response is reduced in the mealworm, *Tenebrio molitor*, if it is deprived of food for short periods of time (Siva-Jothy and Thompson, 2002). Welburn et al. (1989) found that lectins in the midgut of various species of *Glossina* (tsetse flies that feed on vertebrate blood) were less able to kill trypanosomes when the insects were fed with modified blood. Vogelweith et al. (2013) found that larger larvae of the vine moth, *Eupoecilia ambiguella*, were more able to resist a bacterial challenge than smaller larvae, and they suggest that often overlooked size variation in individuals may be a factor in variable immunocompetency among individuals, populations, and species.

16.11 COST OF DEFENSE

Schmid-Hempel (2003) suggests that evolution of an effective immune defense can result in some other fitness-related parameter becoming reduced because of "pleiotropic effects or genetic covariance" or, conversely, that strong selective pressure for a non-immune-related fitness trait may reduce preparedness in immune defense. Yan et al. (1997) demonstrated that mosquito host lines that resisted infection with the malarial agent had decreased fecundity. *D. melanogaster* selected for parasitoid resistance were less competitive in feeding, especially under crowded conditions and scarcity of food (Kraaijeveld and Godfray, 1997). Intense sexual selection in dung flies resulted in reduced immune response in the flies, indicating a microevolutionary tradeoff between the reproduction fitness vs. immune response (Hosken, 2003). Hoang (2001) showed that *D. melanogaster* was less able to resist desiccation and starvation when having to mount an immune response, and Fellowes et al. (1999) found reduced fitness measured as reduced fecundity in *D. melanogaster* that had been challenged with, and successfully defended, attacks from the braconid parasitoid, *Asobara tabida*, and some other insects have shown similar reduced immune response following mating and oviposition (Siva-Jothy et al., 1998). The costs of short-term food stress in the larval stage was demonstrated as a carryover effect after metamorphosis in adults of the damselfly *Lestes viridis*, which had reduced levels of phenoloxidase and hemocyte number (Block and Stoks, 2008), and in another damselfly, *Coenagrion puella* (Campero et al., 2008). Both groups of authors attributed the reduced immune function to the allocation of limited nutrients to growth of adult size and vigor, factors influencing reproductive success.

An age-dependent tradeoff in cotton boll moth males between immunity and allocation of body resources to sperm production was documented by McNamara et al. (2013).

16.12 COEVOLUTIONARY RACE BETWEEN PARASITOID ESCAPE MECHANISMS AND HOST DEFENSE MECHANISMS

Many, perhaps most, insects have **parasitoids** that attack them by laying their eggs in or on them, and the hatching larvae feed upon host hemolymph and tissues. If the eggs are laid inside the host, then the host may be able to encapsulate the egg or larva upon hatching. Parasitoids, in their own evolutionary race to survive, evolved ways to suppress host immunity. Usually this has occurred in specific host–parasitoid evolutionary relationships, and the parasitoid may not be able to suppress the defenses of all potential hosts. Some parasitoids that are infected with polydnaviruses (PDVs) rely upon these viral symbionts to escape the host immune response. The ichneumonid parastoid *Hyposoter didymator* is often protected by its PDA symbiont, but Doremujs et al. (2013) found that proteins on the egg chorion were sufficient to allow them to evade encapsulation by hemocytes of *Spodoptera frugiperda*, the fall armyworm, and the PDA was not required. However, the parasitoid larvae did require injection of the virus in the calyx fluid in order to escape encapsulation. Gomes et al. (2003) showed that the mobilization of prophenoloxidase in *R. prolixus* was suppressed by oral infection with *Trypanosoma rangeli*. Humoral immune defenses of wax moth larvae, *Galleria mellonella*, a convenient host for rearing large populations of nematodes for biological control purposes, were suppressed by the nematode *Steinernema feltiae* (Brivio et al., 2002). Rather than suppressing the host immune system, injection of the nematode *Heterorhabditis bacteriophora* into wild-type adult *D. melanogaster* stimulates the upregulation of genes in the Toll, Imd, and JAK/STAT in the host, as well as other gene pathways (Castillo et al., 2013). *Pimpla hypochondriaca* (a hymenopteran parasitoid) injects a toxin with its egg that suppresses the immune function of hemocytes of larval *Lacanobia oleracea*, the tomato moth (Richards and Parkinson, 2000). Edwards et al. (2006) found that the ectoparasitic wasp, *Eulophus pennicornis*, uses instar-specific endocrine disruption strategies to suppress the development of *L. oleracea*. Growth and development of *H. virescens* are inhibited by members of the CsIV cys-motif gene family from polydnavirus (Fath-Goodin et al., 2006). Uka et al. (2006) describe an interesting case of one parasitoid adversely affecting the development of a polyembryonic parasitoid in the host.

The ectoparasitic mite, *Varroa destructor*, parasitizes honeybees, *A. mellifera*, and can eventually kill a colony of bees if not controlled. Yang and Cox-Foster (2005) have presented evidence that the mites somehow suppress the immune system of honeybees and allow greater expression of the picorna-like deformed wing virus (DWV). They suggest that the demise of honeybee colonies that has recently hit the apicultural industry may be caused by the Varroa mite's suppression of immunity when exposed to microbial attack. Some of the currently used acaracides against *Varroa* mites raised the expression of antimicrobial peptides in the bees; for example, flumethrin increased the expression of hymenoptaecin and abaecin, thus, possibly aiding in the control of *Varroa* (Garrido et al., 2013). The entomopathogen *Bacillus sphaericus* is an effective control agent for *Culex* species of mosquitoes that are vectors of serious disease organisms to humans, but some mosquitoes have become resistant to the bacterium. Guo et al. (2013) showed that resistance in the mosquitoes is associated with the loss of a 60-kDa toxin receptor in the microvilli of the larval midgut. The lack of the receptor, which has α-glucosidase activity, apparently prevents binding of the toxin and/or release of toxic subunits by enzymatic action (Guo et al., 2013). In the silkworm, *Bombyx mori*, evolution leading to the expansion of multiple copies of glucose-methanol-choline (GMC) oxidoreductases genes is associated with increased immunity (Sun et al., 2012).

16.13 AUTOIMMUNE CONSEQUENCES OF SOME DEFENSE REACTIONS

Some of the immune reactions have potential for autotoxicity to the insect itself, a subject area recently reviewed by Zuk and Stoehr (2002), Schmid-Hempel (2003), and Rolff and Siva-Jothy (2003). Some of the reactive oxygen and nitrogen radicals generated by phenoloxidase activity may damage the host's own tissues is some cases. Nevertheless, melanin, a product of the action of phenoloxidase, may be beneficial as a scavenger of reactive free radicals (Schmidt-Hempel, 2005). Green et al. (2003) found mutants of *D. melanogaster* that experienced necrotic tissues after microbial challenge, apparently because the mutants were not able to mobilize serpin inhibitors of the prophenoloxidase activation cascade. Thus, it appeared that phenoloxidase activity was not localized and excessive tissue damage occurred. The potential for autoimmune reactions is another potential cost of activating the immune system. Rahman et al. (2006) present evidence that lipopharin particles may assemble into cage-like coagulation products that protect tissues and cells from some pathogens and PO products.

16.14 GENDER DIFFERENCES IN IMMUNE RESPONSES

Male insects typically invest less in the next generation and their role is usually simply related to mating, although some males offer a nuptial gift to the female and some provide nutrients in the seminal fluid. Many male insects have less immune capacity than females (Schmid-Hempel, 2005, and references therein). Meylaers et al. (2007) found that male adults of the wax moth, *Galleria mellonella,* were smaller than females when the larvae were immune challenged. There were also stage-specific effects, with pupae having the most effective immune response, the larvae next, and adults with the lowest immune response in the wax moth. Kageyama et al. (2007) found that infection of *D. melanogaster* with spiroplasms had an adverse effect on the longevity of male flies, and could kill them early or late, depending on maternal host age when infected. Male *D. melanogaster* that were allowed to repeatedly mate were found to have reduced immune capability (McKean and Nunney, 2001), but Ryder and Siva-Jothy (2000) and Siva-Jothy (2000) found positive correlation between sexually selected traits and immune function. Vigorous courtship behavior may be indicative of general vigor and the fact that the individual has escaped from parasites, predators, and pathogens (Contreras-Garduno et al., 2007, and references therein). Territorial males of the American ruby spot damselfly, *Hetaerina americana,* have larger red spot areas on the wings, greater phenoloxidase activity, more hydrolytic enzyme activity, and higher survival after challenge with *Serratia marcescens* than males with smaller spots (Contreras-Garduno et al., 2007).

16.15 CONCLUSIONS

Insect immunity, and indeed invertebrate immunity (Nappi and Ottaviani, 2000; Loker et al., 2004), is an active area of investigation. The availability of the genome of many insects and other genome projects underway will facilitate the identification of the immune genes and identification of the details of the pathways for antibacterial and antifungal peptides (Evans et al. 2006). The recognition that the TOLL, IMD, and JAK/STAT pathways are present in both insects and vertebrates, including humans, provides added incentives to insect investigations, in which insect immune defenses can be a model for study of vertebrate immune reactions. Furthermore, because of the tragedy of so many deaths from malaria, understanding the immune system of *A. gambiae* and how the malaria organism avoids the mosquito's system is of great importance.

16.16 REVIEW AND SELF-STUDY QUESTIONS

1. What are the differences between innate immunity and the complement system of acquired immunity?
2. What are the five major systems of innate immunity in insects?
3. What is phenoloxidase? What are some of its important functions?
4. What are pattern recognition proteins and what is/are their function(s)?
5. What are PAMPs?
6. What cells are involved in the synthesis of antifungal and antibacterial peptides?
7. How is the Toll pathway different in insects and in vertebrates?
8. What is the IMD pathway?
9. What is/are the roles of DIF or Dorsal in immunity and in embryonic development?
10. What are serpins? What words have been used to coin the word "serpin?"
11. What are C-type lectins?
12. In what ways may an insect be ecologically and physiologically challenged by mounting an immune defense to an invading organism?
13. Is there any possibility of an insect suffering an autoimmune reaction?
14. Are there gender differences in the immune response of insects? From an evolutionary point of view, how are gender differences explained?

REFERENCES

Agaisse, H. and N. Perrimon. 2004. The roles of JAK/STAT signaling in *Drosophila* immune responses. *Immunol. Rev.* 198: 72–82.

Alaux, C., F. Ducloz, D. Crauser, and Y. Le Conte. 2010. Diet effects on honeybee immunocompetence. *Biol. Lett.* 6: 562–565.

Araújo, C.A.C., P.J. Waniek, P. Sock, C. Mayer, A.M. Jansen, and G.A. Schaub. 2006. Sequence characterization and expression patterns of defensin and lysozyme encoding genes from the gut of the reduviid bug *Triatoma brasiliensis*. *Insect Biochem. Mol. Biol.* 36: 547–560.

Asano, T. and M. Ahida. 2001a. Cuticular pro-phenoloxidase of the silkworm, *Bombyx mori*. *J. Biol. Chem.* 276: 11100–11112.

Asano, T. and M. Ahida. 2001b. Transepithelially transported pro-phenoloxidase in the cuticle of the silk worm, *Bombyx mori*. *J. Biol. Chem.* 276: 11113–11125.

Bangham, J., F. Jiggins, and B. Lemaitre. 2006. Insect immunity: The post-genomic era. *Immunity* 25: 1–5.

Bauerfeind, S.S. and K. Fischer. 2014. Integrating temperature and nutrition—Environmental impacts on an insect immune system. *J. Insect Physiol.* 64: 14–20.

Belvin, M.P. and K.V. Anderson. 1996. A conserved signaling pathway: The *Drosophila* toll-dorsal pathway. *Annu. Rev. Cell Dev. Biol.* 12: 393–416.

Billingsley, P.F. and W. Rudin. 1992. The role of the mosquito peritrophic membrane in bloodmeal digestion and infectivity of *Plasmodium* species. *J. Parasit.* 78: 430–440.

Brivio, M., M. Pagani, and S. Restelli. 2002. Immune suppression of *Galleria mellonella* (Insecta, Lepidoptera) humoral defenses induced by *Steinernema feltiae* (Nematoda, Rhabditida). Involvement of the parasite cuticle. *Exp. Parasitol.* 101: 149–156.

Browne, N., C. Surlis, and K. Kavanagh. 2014. Thermal and physical stresses induce a short-term immune priming effect in *Galleria mellonella*. *J. Insect Physiol.* 63: 21–26.

Bulet, P., R. Stöcklin, and L. Menin. 2004. Anti-microbial peptides: From invertebrates to vertebrates. *Immunol. Rev.* 198: 169–184.

Büyükgüzel, E., H. Tunaz, D. Stanley, and K. Büyükgüzel. 2007. Eicosanoids mediate *Galleria mellonella* cellular immune response to viral infection. *J. Insect Physiol.* 53: 99–105.

Campero, M., M. De Block, F. Ollevier, and R. Stoks. 2008. Correcting the short-term effect of food deprivation in a damselfly: Mechanisms and costs. *J. Anim. Ecol.* 77: 66–73.

Carton, Y., M. Poirié, and A.J. Nappi. 2008. Insect immune resistance to parasitoids. *Insect Sci.* 15: 67–87.

Castillo, J., M.R. Brown, and M.R. Strand. 2011. Blood feeding and insulin-like peptide 3 stimulate proliferation of hemocytes in the mosquito *Aedes aegypti*. *PLoS Pathog.* 7(10): e1002274.

Castillo, J.C., U. Shokal, and I. Eleftherianos. 2013. Immune gene transcription in *Drosophila* adult flies infected by entomopathogenic nematode and their mutualistic bacteria. *J. Insect Physiol.* 59: 179–185.

Cerenius, L. and K. Söderhäll. 2004. The prophenoloxidase-activating system in invertebrates. *Immunol. Rev.* 198: 116–126.

Chambers, M.C. and D.S. Schneider. 2012. Pioneering immunology: Insect style. *Curr. Opin. Immunol.* 24: 10–14.

Chase, M., K. Raina, J. Bruno, and M. Sugumaran. 2000. Purification, characterization and molecular cloning of prophenoloxidase from *Sarcophaga bullata*. *Insect Biochem. Mol. Biol.* 30: 953–967.

Christophides, G.K., D. Viachou, and F.C. Kafatos. 2004. Comparative and functional genomics of the innate immune system in the malaria vector *Anopheles gambiae*. *Immunol. Rev.* 198: 127–148.

Christophides, G.K., E. Zdobnov, C. Barillas-Mury, E. Birney, S. Blanton, C. Blass, P.T. Brey et al. 2002. Immunity-related genes and gene families in *Anopheles gambiae*. *Science* 298: 159–165.

Colinet, D., H. Mathé-Hubert, R. Allemand, and J.-L. Gatti. 2013. Variability of venom components in immune suppressive parasitoid wasps: From a phylogenetic to a population approach. *J. Insect Physiol.* 59: 205–212.

Contreras-Garduño, J., H. Lanz-Mendoza, and A. Córdoba-Aguilar. 2007. The expression of a sexually selected trait correlate with different immune defense components and survival in males of the American ruby spot. *J. Insect Physiol.* 53: 612–621.

Cotter, S.C. and R.M. Kilmer. 2010. Sexual division of antibacterial defense in breeding burying beetles, *Nicrophorus vespilloides*. *J. Anim. Ecol.* 79: 35–43.

Cotter, S.C., J.E. Littlefair, P.J. Grantham, and R.M. Kilmer. 2013. A direct physiological trade-off between personal and social immunity. *J. Anim. Ecol.* 82: 846–853.

Cytryńska, M., A. Zdybicka-Barabas, and T. Jakubowicz. 2006. Studies on the role of protein kinase A in humoral immune response of *Galleria mellonella* larvae. *J. Insect Physiol.* 52: 744–753.

De Block, M. and R. Stoks. 2008. Short-term larval food stress and associated compensatory growth reduce adult immune function in a damselfly. *Ecol. Entomol.* 33: 796–801.

De Gregorio, E., S.-J. Han, W.-J. Lee, M.-J. Baek, T. Osaki, S.-I. Kawabata, B.-L. Lee, S. Iwanaga, B. Lemaitre, and P.T. Brey. 2002. An immune-responsive serpin regulates the melanization cascade in *Drosophila*. *Dev. Cell* 3: 581–592.

Delaney, J.R., S. Stöven, H. Uvell, K.V. Anderson, Y. Engström, and M. Mlodzik. 2006. Cooperative control of *Drosophila* immune responses by the JNK and NF-kappa B signaling pathways. *EMBO J.* 12: 3068–3077.

Di Lelio, I., P. Varricchio, G. Di Prisco, A. Marinelli, V. Lasco, A. Caccia, M. Casartelli, B. Giordana, R. Rao, S. Gigliotti, and F. Pennacchio. 2014. Functional aanalysis of an immune gene of *Spodoptera littoralis* by RNAi. *J. Insect Physiol.* 64: 90–97.

Dorémus, T., V. Jouan, S. Urbach, F. Cousserans, P. Wincker, M. Ravallec, E. Wajnberg, and A.-N. Volkoff. 2013. *Hyposoter didymator* uses a combination of passive and active strategies to escape from the *Spodoptera frugiperda* cellular immune response. *J. Insect Physiol.* 59: 500–508.

Edwards, J.P., H.A. Bell, N. Audsley, G.C. Marris, A. Kirkbride-Smith, G. Bryning, C. Frisco, and M. Cusson. 2006. The ectoparasitic wasp *Eulophus pennicornis* (Hymenoptera: Eulophidae) uses instar-specific endocrine disruption strategies to suppress the development of its host *Lacanobia oleracea* (Lepidoptera: Noctuidae). *J. Insect Physiol.* 52: 1153–1162.

Eldon, E., S. Kooyer, D. D'Evelyn, M. Duman, P. Lawinger, and J. Botas. 1994. The *Drosophila* 18 wheeler gene is required for morphogenesis and has striking similarities to Toll. *Development* 120: 885–899.

Evans, J.D., K. Aronstein, Y.P. Chen, C. Hetru, J.-L. Imler, H. Jiang, M. Kanost, G.J. Thompson, Z. Zou, and D. Hultmark. 2006. Immune pathways and defense mechanisms in honey bees *Apis mellifera*. *Insect Mol. Biol.* 15: 645–656.

Fabrick, J.A., J.E. Baker, and M.R. Kanost. 2003. cDNA cloning, purification, properties, and function of a β-1,3-glucan recognition protein from a pyralid moth, *Plodia interpunctella*. *Insect Biochem. Mol. Biol.* 33: 579–594.

Falabella, P., L. Riviello, M. Pascale, L. Di Lelio, G. Tettamanti, A. Grimaldi, C. Iannone et al. 2012. Functional amyloids in insect immune response. *Insect Biochem. Mol. Biol.* 42: 203–211.

Fath-Goodin, A., T.A. Gill, S.B. Martin, and B.A. Webb. 2006. Effect of *Campoletis sonorensis* ichnovirus *cys*-motif proteins on *Heliothis virescens* larval development. *J. Insect Physiol.* 52: 576–585.

Feder, D., C.B. Mello, E.S. Garcia, and P. Azambuja. 1997. Immune response in *Rhodnius prolixus*: Influence of nutrition and ecdysone. *J. Insect Physiol.* 43: 513–519.

Fefferman, N.H., J.F.A. Traniello, R.B. Rosengaus, and D.V. Calleri II. 2007. Disease prevention and resistance in social insects: Modeling the survival consequences of immunity, hygienic behavior, and colony organization. *Behav. Ecol. Sociobiol.* 61: 565–577.

Fellous, S. and B.P. Lazzaro. 2010. Larval food quality affects adult (but not larval) immune gene expression independent of effect on general condition. *Mol. Ecol.* 19: 1462–1468.

Fellowes, M.D.E., A.R. Kraaijeveld, and H.C.J. Godfray. 1999. The relative fitness of *Drosophila melanogaster* (Diptera, Drosophilidae) that have successfully defended themselves against the parasitoid *Asobara tabida* (Hymenoptera: Braconidae). *J. Evol. Biol.* 12: 123–128.

Ferrandon, D., J.-L. Imler, and J.A. Hoffman. 2004. Sensing infection in *Drosophila*: Toll and beyond. *Sem. Immunol.* 16: 45–53.

Figueiredo, M.B., D.P. Castro, N.F.S. Nogueira, E.S. Garcia, and P. Azambuja. 2006. Cellular immune response in *Rhodnius prolixus*: Role of ecdysone in hemocyte phagocytosis. *J. Insect Physiol.* 52: 711–716.

Fujimoto, K., K. Masuda, N. Asada, and E. Ohnishi. 1993. Purification and characterization of prophenoloxidase from pupae of *Drosophila melanogaster*. *J. Biochem.* 113: 285–291.

Gan, H., Y. Wang, H. Jiang, K. Mita, and M.R. Kanost. 2001. A bacteria-induced, intracellular serpin in granular hemocytes of *Manduca sexta*. *Insect Biochem. Mol. Biol.* 31: 887–898.

Garrido, P.M., K. Antúnez, M. Martín, M.P. Porrini, P. Zunino, and M.J. Eguaras. 2013. Immune-related gene expression in nurse honeybees (*Apis mellifera*) exposed to synthetic acaricides. *J. Insect Physiol.* 59: 113–119.

Gerritsma, S., A. de Haan, L. van de Zande, and B. Wertheim. 2013. Natural variation in differentiated hemocytes is related to parasitoid resistance in *Drosophila melanogaster*. *J. Insect Physiol.* 59: 148–158.

Gerttula, S., Y.S. Jin, and K.V. Anderson. 1988. Zygotic expression and activity of the *Drosophila* Toll gene, a gene required maternally for embryonic dorsal-ventral pattern formation. *Genetics* 119: 123–133.

Giglio, A. and P.G. Giulianini. 2013. Phenoloxidase activity among developmental stages and pupal cell types of the ground beetle *Carabus* (*Chaetocarabus*) *lefebvrei* (Coleoptera, Carabidae). *J. Insect Physiol.* 59: 466–474.

Gołębiowski, M., M. Cerkowniak, M.I. Boguś, E. Włóka, M. Dawgul, W. Kamysz, and P. Stepnowski. 2013. Free fatty acids in the cuticular and internal lipids of *Calliphora vomitoria* and their antimicrobial activity. *J. Insect Physiol.* 59: 416–429.

Gomes, S.A.O., D. Feder, E.S. Garcia, and P. Azambuja. 2003. Suppression of the prophenoloxidase system in *Rhodnius prolixus* orally infected with *Trypanosoma rangeli*. *J. Insect Physiol.* 49: 829–837.

Green, C., G. Brown, T.R. Dafforn, J.-M. Reichart, T. Morley, D. Lomas, and D. Grubb. 2003. *Drosophila necrotic* mutations mirror disease-associated variants of human serpins. *Development* 130: 1473–1478.

Guo, Q.-Y., Q.-X. Cai, J.-P. Yan, X.-M. Hu, D.-S. Zheng, and A.-H. Yuan. 2013. Single nucleotide deletion of *cqm1* gene results in the development of resistance to *Bacillus sphaericus* in *Culex quinquefasciatus*. *J. Insect Physiol.* 59: 967–973.

Gulley, M.M., X. Zhang, and K. Michel. 2013. The roles of serpins in mosquito immunology and physiology. *J. Insect Physiol.* 59: 138–147.

Hoang, A. 2001. Immune response to parasitism reduces resistance of *Drosophila melangaster* to desiccation and starvation. *Evolution* 55: 2353–2358.

Hoffmann, J.A. 2003. The immune response of *Drosophila*. *Nature* 426: 33–38.

Hosken, D. 2003. Sex and death: Microevolutionary trade-offs between reproductive and immune investment in dung flies. *Curr. Biol.* 11: R379–R380.

Imamura, M., Y. Nakahara, T. Kanda, T. Tamura, and K. Taniai. 2006. A transgenic silkworm expressing the immune-inducible cecropin B-GFP reporter gene. *Insect Biochem. Mol. Biol.* 36: 429–434.

Imler, J-L. and J.A. Hoffmann. 2001. Toll receptors in innate immunity. *Trends Cell Biol.* 11: 304–311.

Janse, C.J., R.J. Rouwenhorst, P.F.J. van der Klooster, H.J. van der Kay, and J.P. Overdulve. 1985. Development of *Plasmodium berghei* ookinetes in the midgut of *Anopheles* atroparvus mosquitoes and in vitro. *J. Parasitol.* 91: 219–225.

Jiang, H., Y. Wang, C. Ma, and M.R. Kanost. 1997. Subunit composition of pro-phenoloxidase from *Manduca sexta*: Molecular cloning of subunit proPO-P1. *Insect Biochem. Mol. Biol.* 27: 835–850.

Junell, A., H. Uvell, L. Pick, and Y. EngstrÖm. 2007. Isolation of regulators of *Drosophila* immune defense genes by a double screen in yeast. *Insect Biochem. Mol. Biol.* 37: 202–212.

Kageyama, D., H. Anbutsu, M. Shimada, and T. Fukatsu. 2007. Spiroplasms infection causes either early or late male killing in *Drosophila*, depending on maternal host age. *Naturwissenschaften* 94: 333–337.

Kambris, Z., J.A. Hoffmann, J.-L. Imler, and M. Capovilla. 2002. Tissue and stage-specific expression of the Tolls in *Drosophila* embryos. *Gene Expr. Patterns* 2: 311–317.

Kanost, M.R. 1999. Serine proteinase inhibitors in arthropod immunity. *Dev. Comp. Immunol.* 23: 291–301.

Kanost, M.R., J. Jiang, and Z.-Q. Yu. 2004. Innate immune responses of a lepidopteran insect, *Manduca sexta*. *Immunol. Rev.* 198: 97–105.

Kanost, M.R., S.V. Prasad, Y. Huang, and E. Willott. 1995. Regulation of serpin gene-1 in *Manduca sexta*. *Insect Biochem. Mol. Biol.* 25: 285–291.

Koch, H. and P. Schmid-Hempel. 2011. Socially transmitted gut microbiota protect bumble bees against an intestinal parasite. *Proc. Natl. Acad. Sci. USA* 108: 19288–19292.

Kraaijeveld, A.R. and H.J.C. Godfray. 1997. Trade-off between parasitoid resistance and larval competitive ability in *Drosophila melanogaster*. *Nature* 389: 278–280.

Ladendorff, N.E. and M.R. Kanost. 1990. Isolation and characterization of bacteria-induced protein P4 from hemolymph of *Manduca sexta*. *Arch. Insect Biochem. Physiol.* 15: 33–41.

Ladendorff, N.E. and M.R. Kanost 1991. Bacteria-induced protein P4 (hemolin) from *Manduca sexta*: A member of the immunoglobulin superfamily which can inhibit hemocyte aggregation. *Arch. Insect Biochem. Physiol.* 18: 285–300.

Leclerc, V. and J.-M. Reichhart. 2004. The immune response of *Drosophila melanogaster*. *Immunol. Rev.* 198: 59–71.

Lee, K.Y., F.M. Horodyski, A.P. Valaitis, and D.L. Denlinger. 2002. Molecular characterization of the insect immune protein hemolin and its high induction during embryonic diapause in the gypsy moth, *Lymantria dispar*. *Insect Biochem. Mol. Biol.* 32: 1457–1467.

Lee, M., C.S. Yoon, J. Yi, J.R. Cho, and H.S. Kim. 2005. Cellular immune responses and FAD-glucose dehydrogenase activity of *Mamestra brassicae* (Lepidoptera: Noctuidae) challenged with three species of entomopathogenic fungi. *Physiol. Entomol.* 30: 287–292.

Lehane, M.J. 1997. Peritrophic matrix structure and function. *Annu. Rev. Entomol.* 42: 525–550.

Lehane, M.J. 2005. *The Biology of Blood-Sucking in Insects*, 2nd edn. Cambridge University Press, Cambridge, U.K.

Lemaitre, B., E. Kromer-Metzger, L. Michaut, E. Nicolas, M. Meister, P. Georgel, J.-M. Reichhart, and J.A. Hoffmannn. 1995. A novel mutation, *immune deficiency*, defines two distinct control pathways in the *Drosophila* host defense. *Proc. Natl. Acad. Sci. USA* 92: 9465–9469.

Lemaitre, B., E. Nicolas, L. Michaut, J.-M. Reichhart, and J.A. Hoffmann. 1996. The dorsoventral regulatory gene cassette *spätzle/Toll/cactus* controls the potent antifungal response in *Drosophila* adults. *Cell* 86: 973–983.

Li, Y., Q. Xiang, Q. Zhang, Y. Huang, and Z. Su. 2012. Overview on the recent study of antimicrobial peptides: Origins, functions, relative mechanisms and application. *Peptides* 37: 207–215.

Ligoxygakis, P., N. Pelte, J.A. Hoffmann, and J.-M. Reichhart. 2002a. Activation of *Drosophila* Toll during fungal infection by a blood serine protease. *Science* 297: 114–116.

Ligoxygakis, P., N. Pelte, C. Ji, V. Leclerc, B. Duvic, M. Belvin, H. Jiang, J.A. Hoffmann, and J.-M. Reichhart. 2002b. A serpin mutant links Toll activation to melanization in the host defense of *Drosophila*. *EMBO J.* 21: 6330–6337.

Login, F.H., S. Balmand, A. Vallier, C. Vincent-Monégat, A. Vigneron, M. Weiss-Gayet, D. Rochat, and A. Heddi. 2011. Antimicrobial peptides keep insect endosymbionts under control. *Science* 334: 362–365.

Login, F.H. and A. Heddi. 2013. Insect immune system maintains long-term resident bacteria through a local response. *J. Insect Physiol.* 59: 232–239.

Loker, E.S., C.M. Adema, S.-M. Zhang, and T.B. Kepler. 2004. Invertebrate immune systems—Not homogenous, not simple, not well understood. *Immunol. Rev.* 198: 10–24.

Ma, C. and M.R. Kanost. 2000. A β-1,3 glycan-recognition protein from an insect, *Manduca sexta*, agglutinates microorganisms and activates the phenoloxidase cascade. *J. Biol. Chem.* 275: 7505–7514.

McKean, K.A. and L. Nunney. 2001. Increased sexual activity reduces male immune function in *Drosophila melanogaster*. *Proc. Natl. Acad. Sci. USA* 98: 7904–7909.

McNamara, K.B., E. van Lieshout, T.M. Jones, and L.W. Simmons. 2013. Age-dependent trade-offs between immunity and male, but no female, reproduction. *J. Anim. Ecol.* 82: 235–244.

Medzhitov, R., P. Preston-Hurlburt, and C.A. Janeway, Jr. 1997. A human homologue of the *Drosophila* Toll protein signals activation of adaptive immunity. *Nature* 388: 394–397.

Merkling, S.K. and R.P. van Rij. 2013. Beyond RNAi: Antiviral defense strategies in *Drosophila* and mosquito. *J. Insect Physiol.* 59: 159–170.

Meylaers, K., D. Freitak, and L. Schoofs. 2007. Immunocompetence of *Galleria mellonella*: Sex- and stage-specific differences and the physiological cost of mounting an immune response during metamorphosis. *J. Insect Physiol.* 53: 146–156.

Moita, L.F., G. Vriend, V. Mahairaki, C. Louis, and F.C. Kafatos. 2006. Integrins of *Anopheles gambiae* and a putative role of a new β integrin, BINT2, in phagocytosis of *E. coli*. *Insect Biochem. Mol. Biol.* 36: 282–290.

Moreau, S.J.M. 2013. "It stings a bit but it cleans well": Venoms of Hymenoptera and their antimicrobial potential. *J. Insect Physiol.* 59: 186–204.

Mullen, L.M. and G.J. Goldsworthy. 2006. Immune responses of locusts to challenge with the pathogenic fungus *Metarhizium* or high doses of laminarin. *J. Insect Physiol.* 52: 389–398.

Nappi, A.J. and E. Ottaviani. 2000. Cytotoxicity and cytotoxic molecules in invertebrates. *BioEssays* 22: 469–480.

Ochiai, M. and M. Ashida. 2000. A pattern recognition protein for β-1,3 glycan. *J. Biol. Chem.* 275: 4995–5002.

Oliver, K.M., A.H. Smith, and J.A. Russell. 2014. Defensive symbiosis in the real world—Advancing ecological studies of heritable, protective bacteria in aphids and beyond. *Funct. Ecol.* 28: 341–355.

Parker, J.S., K. Mizuguchi, and N.J. Gay. 2001. A family of proteins related to Spaetzle, the toll receptor ligand, is encoded in the *Drosophila* genome. *Proteins* 45: 71–80.

Pauchet, Y., D. Freitak, H.M. Heidel-Fischer, D.G. Heckel, and H. Vogel. 2009. Immunity or digestion: Glucanase activity in a glucan-binding protein family from Lepidoptera. *J. Biol. Chem.* 284: 2214–2224.

Pelte, N., A.S. Robertson, Z. Zou, D. Belorgey, T.R. Dafforn, H. Jiang, D. Lomas, J.-M. Reichart, and D. Grubb. 2006. Immune challenge induces N-terminal cleavage of the *Drosophila* serpin Necrotic. *Insect Biochem. Mol. Biol.* 36: 37–46.

Piñera, A.V., H.M. Charles, T.A. Dinh, and K.A. Killian. 2013. Maturation of the immune system of the male house cricket, *Acheta domesticus*. *J. Insect Physiol.* 59: 752–760.

Pinheiro, V.B. and D.J. Ellar. 2006. How to kill a mocking bug? *Cell. Microbiol.* 8: 545–557.

Ponnudurai, T., P.F. Billingsley, and W. Rudin. 1988. Differential infectivity of *Plasmodium* for mosquitoes. *Parasitol. Today* 4: 319–321.

Ponton, F., K. Wilson, S.C. Cotter, D. Raubenheimer, and S.J. Simpson. 2011. Nutritional immunology: A multi-dimensional approach. *PloS Pathog.* 7(12): e1002223.

Ponton, F., K. Wilson, A.J. Holmes, S.C. Cotter, D. Raubenheimer, and S.J. Simpson. 2013. Integrating nutrition and immunology: A new frontier. *J. Insect Physiol.* 59: 130–137.

Rahman, M.M., G. Ma, H.L.S. Roberts, and O. Schmidt. 2006. Cell-free immune reactions in insects. *J. Insect Physiol.* 52: 754–762.

Rasmunson, T. and H.G. Boman. 1979. Insect immunity. V. Purification and some properties of immune protein P4 and hemolymph of *Hyalophora cecropia* pupae. *Insect Biochem.* 9: 259–264.

Richards, E.H. and N.M. Parkinson. 2000. Venom from the endoparasitoid wasp *Pimpla hypochondriaca* adversely affects the morphology, viability, and immune functions of hemocytes from larvae of the tomato moth, *Lacanobia oleracea*. *J. Invertb. Pathol.* 76: 33–42.

Rock, F.L., G. Hardiman, J.C. Timans, R.A. Kastelein, and J.F. Bazan. 1998. A family of human receptors structurally related to *Drosophila* Toll. *Proc. Natl. Acad. Sci. USA* 95: 588–593.

Rolff, J. and M.T. Siva-Jothy. 2003. Invertebrate ecological immunity. *Science* 301: 472–475.

Rosengaus, R.B., T. Cornelisse, K. Guschanski, and J.F.A. Traniello. 2007. Inducible immune proteins in the dampwood termite *Zootermopsis angusticollis*. *Naturwissenschaften* 94: 25–33.

Ross, J., H. Jang, M.R. Kanost, and Y. Wang. 2003. Serine proteases and their homologs in the *Drosophila melanogaster* genome: An initial analysis of sequence conservation and phylogenetic relationships. *Gene* 304: 117–131.

Royet, J. 2004. *Drosophila melanogaster* innate immunity: An emerging role for peptidoglycan recognition proteins in bacteria detection. *Cell. Mol. Life Sci.* 61: 537–546.

Ryder, J. and M.T. Siva-Jothy. 2000. Male calling song provides a reliable signal of immune function in a cricket. *Proc. Roy. Soc. London Sci. Ser. B* 263: 1171–1175.

Schmid-Hempel, P. 2003. Variation in immune defense as a question of evolutionary ecology. *Proc. Roy. Soc. London Sci. Ser. B* 270: 357–366.

Schmid-Hempel, P. 2005. Evolutionary ecology of insect innate defenses. *Annu. Rev. Entomol.* 50: 529–551.

Schmid-Hempel, P. and D. Ebert. 2003. On the evolutionary ecology of specific immune defense. *Trends Ecol. Evol.* 18: 27–32.

Schmidt, O., K. Sönderhäll, U. Theopold, and I. Faye. 2010. Role of adhesion in arthropod immune recognition. *Annu. Rev. Entomol.* 55: 485–504.

Sezaki, H., N. Kawamoto, and N. Asada. 2001. Effect of ionic concentration on the higher-order structure of prophenol oxidase in *Drosophila melanogaster*. *Biochem. Genet.* 39: 83–92.

Shahabuddin, M. 1998. *Plasmodium* ookinete development in the mosquito midgut: A case of reciprocal manipulation. *J. Parasitol.* 116: S83–S93.

Sieber, K.-P., M. Huber, D. Kaslow, S.M. Banks, M. Torii, M. Aikawa, and L.H. Miller. 1991. The peritrophic membrane as a barrier: Its penetration by *Plasmodium gallinaceum* and the effect of a monoclonal antibody to ookinetes. *Exp. Parasitol.* 72: 145–156.

Siva-Jothy, M.T. 2000. A mechanistic link between parasite resistance and expression of sexually selected trait in a damselfly. *Proc. Roy. Soc. London Sci. Ser. B* 267: 2523–2527.

Siva-Jothy, M.T., Y. Moret, and J. Rolff. 2005. Insect immunity: An evolutionary ecology perspective. *Adv. Insect Physiol.* 32: 1–48.

Siva-Jothy, M.T. and J.J.W. Thompson 2002. Short-term nutrient deprivation affects immune function. *Physiol. Entomol.* 27: 206–212.

Siva-Jothy, M.T., Y. Tsubaki, and R.E. Hooper. 1998. Decreased immune response as a proximate cost of copulation and oviposition in a damselfly. *Physiol. Entomol.* 23: 274–277.

Steiner, H. 2004. Peptidoglycan recognition proteins: On and off switches for innate immunity. *Immunol. Rev.* 198: 8–96.

Sun, W., Y.-H. Shen, W.-J. Yang, Y.-F. Cao, Z.-H. Xiang, and Z. Zhang. 2012. Expansion of the silkworm GMC oxidoreductase genes is associated with immunity. *Insect Biochem. Mol. Biol.* 42: 935–945.

Tan, J., M. Xu, K. Zhang, X. Wang, S. Chen, T. Li, Z. Xiang, and H. Cui. 2013. Characterization of hemocytes proliferation in larval silkworm *Bombyx mori*. *J. Insect Physiol.* 59: 595–603.

Tauszig, S., E. Jouanguy, J.A. Hoffmann, and J.-L. Imler. 2000. Toll-related receptors and the control of antimicrobial peptide expression in *Drosophila*. *Proc. Natl. Acad. Sci. USA* 97: 10520–10525.

Tokura, A., G.S. Fu, M. Sakamoto, H. Endo, S. Tanaka, S. Kikuta, H. Tabunoki, and R. Sato. 2014. Factors functioning in nodule melanization of insects and their mechanisms of accumulation in nodules. *J. Insect Physiol.* 60: 40–49.

Uka, D., T. Hiraoka, and K. Iwabuchi. 2006. Physiological suppression of the larval parasitoid *Glyptapanteles pallipes* by the polyembyonic parasitoid *Copidosoma floridanum*. *J. Insect Physiol.* 52: 1137–1142.

Urbański, A., E. Czarniewska, E. Baraniak, and G. Rosiński. 2014. Developmental changes in cellular and humoral responses of the burying beetle *Nicrophorus vespilloides* (Coleoptera, Silphidae). *J. Insect Physiol.* 60: 98–103.

Vilcinskas, A. 2013. Evolutionary plasticity of insect immunity. *J. Insect Physiol.* 59: 123–129.

Vogelweith, F., D. Thiery, Y. Moret, and J. Moreau. 2013. Immunocompetence increases with larval body size in a phytophagous moth. *Physiol. Entomol.* 38: 219–225.

Wang, L., A.N.R. Weber, M.L. Atilano, S.R. Filipe, N.J. Gay, and P. Ligoxygakis. 2006. Sensing of gram-positive bacteria in *Drosophila*: GNBP1 is needed to process and present peptidolgycan to PGRP-SA. *EMBO J.* 25: 5005–5014.

Wang, Y., E. Willott, and M.R. Kanost. 1995. Organization and expression of the hemolin gene, a member of the immunoglobulin superfamily in an insect, *Manduca sexta*. *Insect Mol. Biol.* 4: 113–123.

Wang, Z.-Z., M. Shi, W. Zhao, Q.-L. Bian, C.-Y. Ye, and X.-X Chen. 2013. Identification and characterization of *defensin* genes from the endoparasitoid wasp *Cotesia vestalis* (Hymenoptera: Braconidae). *J. Insect Physiol.* 59: 1095–1103.

Warr, E., L. Lambrechts, J.C. Koella, C. Burgouin, and G. Dimopoulos. 2006. *Anopheles gambiae* immune responses to Sepahadex beads: Involvement of anti-*Plasmodium* factors in regulating melanization. *Insect Biochem. Mol. Biol.* 36: 769–778.

Welburn, S.C., I. Maudlin, and D.S. Ellis. 1989. Rate of trypanosome killing by lectins in midgets of different species and strains of *Glossina*. *Med. Vet. Entomol.* 3: 77–82.

Werner, T., G. Liu, D. Kang, S. Ekengren, H. Steiner, and D. Hultmark. 2000. A family of peptidoglycan recognition proteins in the fruit fly *Drosophila melanogaster*. *Proc. Natl. Acad. Sci. USA* 97: 13772–13777.

Wojda, I. and P. Taszłow. 2013. Heat shock affects host-pathogen interaction in *Galleria mellonella* infected with *Bacillus thuringiensis*. *J. Insect Physiol.* 59: 894–905.

Yan, G., D.W. Severson, and B.M. Christensen. 1997. Costs and benefits of mosquito refractoriness to malaria parasites: Implications for genetic variability of mosquitoes and genetic control of malaria. *Evolution* 51: 441–450.

Yang, X. and D.L. Cox-Foster. 2005. Impact of an extoparasite on the immunity and pathology of an invertebrate: Evidence for host immunosuppression and viral amplification. *Proc. Natl. Acad. Sci. USA* 102: 7470–7475.

Yu, Q.-Y. and M.R. Kanost 1999. Developmental expression of *Manduca sexta* hemolin. *Arch. Insect Biochem. Physiol.* 42: 198–212.

Yu, Q.-Y. and M.R. Kanost. 2003. *Manduca sexta* lipopolysaccharide-specific immulectin-2 protects larvae from bacterial infection. *Dev. Comp. Immunol.* 27: 189–196.

Yu, X.-Q., H. Gan, and M.R. Kanost. 1999. Immulectin, an inducible C-type lectin from an insect, *Manduca sexta*, stimulates activation of plasma prophenol oxidase. *Insect Biochem. Mol. Biol.* 29: 585–597.

Yu, X.-Q., H. Jiang, Y. Wang, and M.R. Kanost. 2003. Nonproteolytic serine proteinase homologs are involved in prophenoloxidase activation in the tobacco hornworm *Manduca sexta*. *Insect Biochem. Mol. Biol.* 33: 197–208.

Yu, X.-Q. and M.R. Kanost. 2000. Immulectin-2, a lipopolysaccharide-specific lectin from an insect, *Manduca sexta*, is induced in response to Gram-negative bacteria. *J. Biol. Chem.* 275: 37373–37381.

Yu, X.-Q., Y. Zhu, C. Ma, J.A. Fabrick, and M.R. Kanost. 2002. Pattern recognition proteins in *Manduca sexta* plasma. *Insect Biochem. Mol. Biol.* 32: 1287–1293.

Zhao, L. and M.R. Kanost. 1996. In search of a function for hemolin, a hemolymph protein from the immunoglobulin superfamily. *J. Insect Physiol.* 42: 73–79.

Zhou, J., M. Ueda, R. Umemiya, B. Battsetseg, D. Boldbaatar, X. Xuan, and K. Fujisaki. 2006. A secreted cystatin from the tick *Haemaphysalis longicornis* and its distinct expression patterns in relation to innate immunity. *Insect Biochem. Mol. Biol.* 36: 527–535.

Zuk, M. and A.M. Stoehr. 2002. Immune defense and host life history. *Am. Nat.* 160: S9–S22.

CHAPTER 17

Respiration

PREVIEW

Respiration, used in the sense that it means breathing and gas exchange, is a function of the tracheal system, a tubular network that originates at spiracular openings on the body surface and radiates to all parts of the insect body. Spiracular valves at the body surface often can be closed to reduce water loss from the system. Large longitudinal and transverse tracheae may be up to 0.2 mm in diameter, and branches from these become smaller in diameter as they penetrate between cells and tissues and finally terminate as tracheoles. Tracheoles are blind tubules less than 1 μm in diameter. Often, especially in active tissues that demand rapid gas exchange, the tracheoles push against and indent cell membranes, like pushing a finger into a soft balloon, until they terminate within a few micrometers of mitochondria. The tracheal system is very efficient for insects and delivers oxygen to flight muscles in sufficient amounts so that they conduct aerobic metabolism even during flight. In some very small insects, simple diffusion of gases through the system of tracheal tubules may suffice, but most insects actively ventilate the system by muscular pumping motions. Many insects demonstrate discontinuous ventilation in which the spiracles are tightly closed or flutter almost imperceptibly for varying periods interspersed with bursts of open spiracles. The cuticular lining of tracheae is shed at each molt, but the lining of tracheoles is not molted. Aquatic insects have a tracheal system that is essentially the same in structure as that of terrestrial insects. Many adaptations to an aquatic life occur, however, in the tracheal system, including compressible gas bubbles (gas gills), incompressible gas films (plastrons), and thin cuticular flaps (gills) that take oxygen from water by cutaneous diffusion. Insect eggs often have plastrons as a part of the eggshell that aid gas exchange for the developing embryo. In addition to gas exchange, the tracheal system has been adapted in some insects to serve nonrespiratory functions, such as an attachment site for endocrine cells, participation in sound reception and sound production as a source of hissing sounds, and delivery of a distasteful froth. In all insects, the vast network of tracheae and tracheoles tie cells and tissues together.

17.1 INTRODUCTION

Insects breathe by delivering air through a network of small tubes to within a few micrometers of mitochondria in cells. These tubes, the **tracheae**, arise at openings on the sides of the body, the **spiracles**. In most insects, there are interconnecting longitudinal and transverse tracheal trunks and, sometimes, large air sacs that tie together the entire system (Figure 17.1). Larger tracheal tubes send branches that become smaller in diameter as they ramify to all tissues, with the smallest diameter **tracheoles** (tubes less than 1 µm diameter) (Figure 17.2) touching most cells and even indenting some cells. The airflow may be tidal or directed, depending on the insect and its physiological state. The interconnections make directed airflow possible, in which air enters through one or more anterior spiracles, gets pumped through the body by muscular ventilatory movements, and is expelled through one or more posterior spiracles. Such a directed flow is more efficient than tidal inflow and outflow from the same spiracles because the system is constantly flushed and incoming air is not mixed with used air. In some very simple tracheal systems, the tracheae arising from each spiracle are independent of other tracheae and spiracles and only tidal flow is possible.

Whitten (1972) has provided a review of comparative anatomy of the tracheal system among insect groups, and various aspects of the physiology of gas exchange and breathing have been reviewed by Miller (1966a), Slama (1994), Hadley (1994), Snyder et al. (1995), and Lighton (1996). Dyby (1998) developed a novel method for infiltrating the tracheal system of newly hatched insects for effective visualization. Recent use of synchrotron X-ray techniques reveals details of the internal anatomy of the tracheal system without invasive techniques or harm to the insects (Socha et al., 2007; Socha and De Carlo, 2008; Westneat et al., 2008; Greenlee et al., 2009; Socha et al., 2010; Waters et al., 2013).

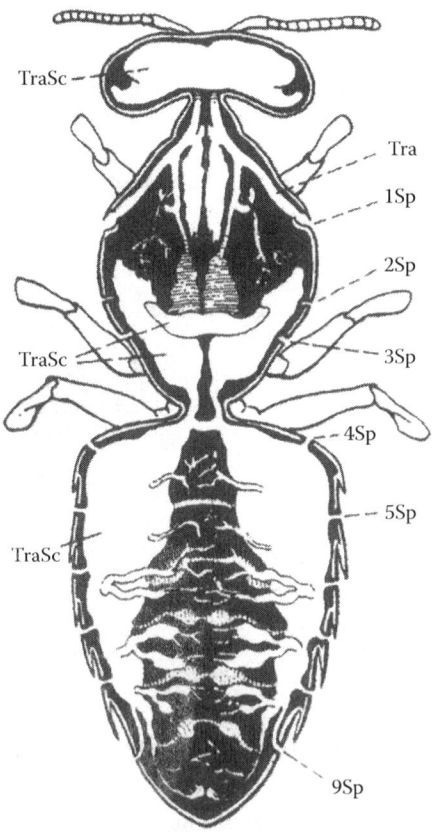

Figure 17.1 Large longitudinal tracheal sacs in the body of a honeybee. Sp, spiracle openings in the pleural region of the body; Tra, trachea; TraSc, tracheal sac. (From Dadant & Sons (eds.), *The Hive and the Honey Bee*, Dadant & Sons Publishers, Hamilton, IL, 1975. Permission granted with acknowledgment of source.)

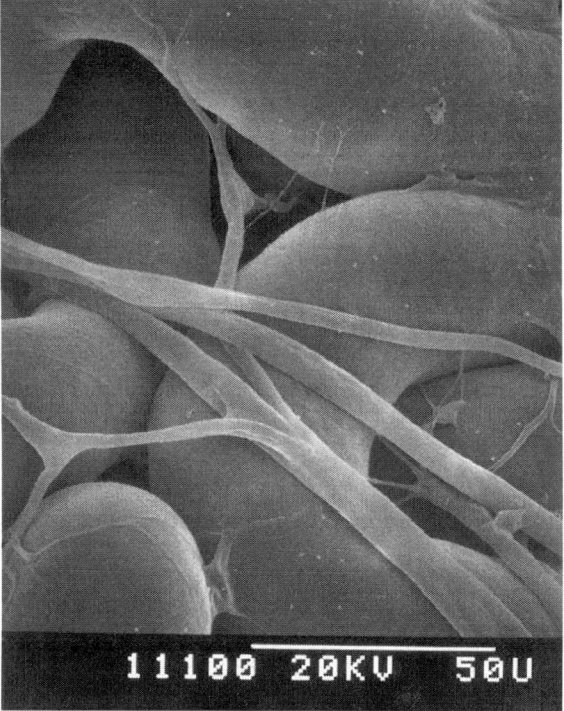

Figure 17.2 Scanning Electron Microscope (SEM) view of a tracheal tube branching into many fine tracheoles that disappear beneath the surface of cells in the salivary gland of a male Caribbean fruit fly *Anastrepha suspensa*. (scanning electron microscope (SEM) photograph by the author.)

17.2 STRUCTURE OF THE TRACHEAL SYSTEM

17.2.1 Tracheae and Tracheole Structure

The tracheal system develops from embryonic ectodermal tissue. Tracheae and tracheoles have an epicuticular lining, composed primarily of a cuticulin layer that is continuous with the external cuticle. Larger tracheal trunks have an endocuticle layer that gives more strength to the tubular structure. A hydrophobic substance is secreted on the lumen surface of tracheae that helps prevent water from entering the tracheae and reduces evaporative water loss from the humid, extensive internal tracheal surfaces.

The major distinction between tracheae and tracheoles is one of size. Tubes down to about 1 μm in size are called **tracheae** (*sing.*, **trachea**), while those smaller than 1 μm are **tracheoles**. Tracheoles typically have fluid in their terminal endings, and the fluid level seems to be related to metabolic demand for oxygen delivery to cells near the endings. Wigglesworth (1935) suggested that the change in fluid level and, thus, the open air path when the fluid recedes or is absorbed into the tissues is probably caused by changes in the osmotic pressure between tracheolar fluid and the intracellular medium, and Förster and Woods (2012) suggest that active ion transport is most likely involved and that the major ions are K^+, H^+, and bicarbonate. The lower limit for effective tracheolar diameter is limited by the mean free path of diffusing oxygen molecules, which is about 0.072 µm at 300 K (27°C) and 1 atm, according to Pickard (1974). The smallest tracheoles that have been observed are slightly smaller than 0.2 µm or are 2–3 times the limiting diameter. Tracheoles in the lantern of some fireflies are very specialized in structure, with stiff, reinforcing material in the tracheole to help it resist folding or collapse under what appear to be conditions in which there are rapid and strong osmotic changes across the tracheolar cell membrane when nerve impulses signal the change in permeability (Ghiradella, 1978). Some aspects of the mechanical strength of the tracheal tubes in the American cockroach, *Periplaneta americana*, have been measured (Webster et al., 2011).

Oxygen delivery to photocytes, the cells in fireflies that produce light, is the basis for the ability of fireflies to use light as a channel of communication. Timmins et al. (2001) demonstrated that normobaric hyperoxia (i.e., change in partial pressure of oxygen from the normal 21 to 101 kPa) resulted in constant glowing rather than the normal intermittent flashing typical of fireflies. The finding supports the theory that gating of oxygen access to the photocytes is the basis for firefly flashing, and the authors further suggested that the control or gating of oxygen likely relies on modulating the level of fluid in tracheoles supplying the photocytes.

Thickened, tight spirals of the cuticular intima, the **taenidia**, strengthen tracheae and tracheoles, provide elasticity, and help the tubes resist compression and collapse (Figures 17.3 and 17.4). Even tracheoles have taenidial reinforcements, contrary to older reports published prior to availability of the electron microscope. In larger tracheae, the taenidia are up to 450 nm in width and

Figure 17.3 SEM cutaway view of the inside of a large trachea of a mole cricket *Scapteriscus borellii*, showing the origin of two smaller tracheae and taenidial windings. (SEM photograph by the author.)

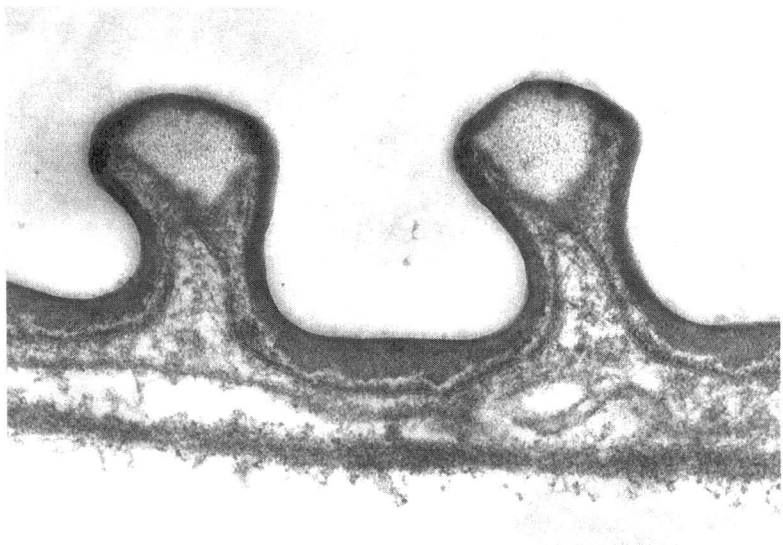

Figure 17.4 A lateral view of taenidia in a trachea from a termite *Cubitermes fungifaber*. (From Bordereau, C., *Int. J. Insect Morphol. Embryol.*, 4, 431, 1975. With permission. Photomicrograph Courtesy of Prof. Dr. C. Bordereau, retired, in France.)

are spaced about 300 nm apart, but, in tracheoles, the taenidia are smaller (50–80 nm in width) and are spaced farther apart than their width. Within the taenidial folds, there is a component, probably similar to procuticle, that adds strength to the taenidia (Bordereau, 1975). The cuticular intima is as thick as 200 nm at the taenidial spirals and as thin as 10–40 nm between spirals. The micelles of the cuticulin layer are oriented so that their long axis is parallel to the long axis of a trachea or tracheole in intertaenidial areas but perpendicular to the long axis (i.e., circular in orientation) within the taenidial thickenings. The orientation of the cuticulin micelles lends strength to the tubes.

When tracheae first form, they have smooth walls, but the taenidia soon appear. Locke (1958a) proposed that taenidial formation is the result of expansion and buckling of the tracheal wall. With a mathematical model, he accounted for the frequency of taenidia, tube-wall thickness, and orientation of the cuticulin micelles within the taenidia and intertaenidial regions. The buckling hypothesis may explain taenidial formation, but to date, experimental proof for it is lacking, and the buckling theory is not universally accepted.

17.2.2 Spiracle Structure and Function

The openings of the tracheal system at the body surface are called **spiracles**. Spiracles usually occur on the pleural surfaces of the body, typically one on each side of each segment, but numerous variations have evolved. Most terrestrial insects can close the spiracles as an adaptation for water conservation. The simplest closing mechanism consists of folds of the integument, which can be pulled together over the spiracular opening by a closer muscle. Many insects have valves or flaps of cuticle (Figures 17.5 and 17.6) that rotate and close over the spiracle. Sometimes an opener muscle is present, but in most cases opening occurs because of elasticity of the cuticle when the closer muscle relaxes. The spiracle on the prothorax of Orthoptera has both closer and opener muscles, and although the spiracle opens partly by natural elasticity after relaxation of the closer muscle, the opener muscle can cause the spiracle to open more widely for increased ventilation (Miller, 1960a).

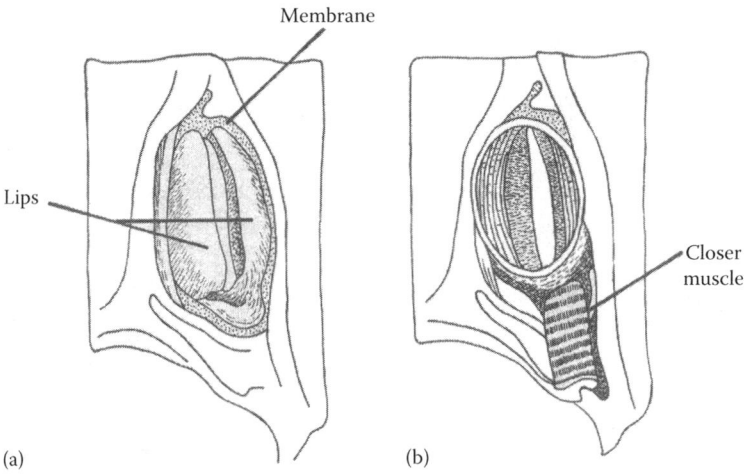

Figure 17.5 The second thoracic spiracle of a grasshopper illustrating (a) outer and (b) inner views of spiracular valves. (Modified from Snodgrass, R.E., *Principles of Insect Morphology*, McGraw-Hill, New York, 1935.)

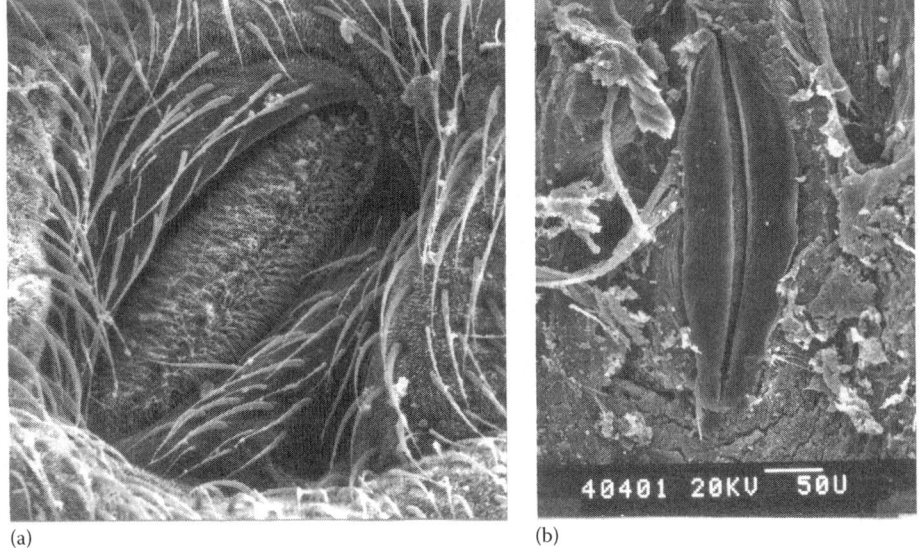

Figure 17.6 SEMs of the (a) external and (b) internal views of the valves that close the spiracular opening of a mole cricket *Scapteriscus vicinus*. The many fine setae on the valve and around the spiracle surface may trap a film of air and act as a plastron or compressible gill when the cricket gets flooded in its underground tunnel. (SEM photographs by the author.)

The openings may be simple unguarded pores, but frequently there are additional structural details associated with the openings. Typically, a ring of sclerotized cuticle, the peritreme, around the spiracular opening provides reinforcement. In many insects there is a slightly enlarged chamber or atrium just inside the spiracular opening from which tracheae branch in a variety of directions. The atrium may contain various structural adaptations to filter the air, such as dust-catching setae or hairs in the atrium space (Figure 17.7). Many terrestrial Diptera, Coleoptera, Lepidoptera, and some aquatic insects have thin perforated partitions, called sieve plates, in the

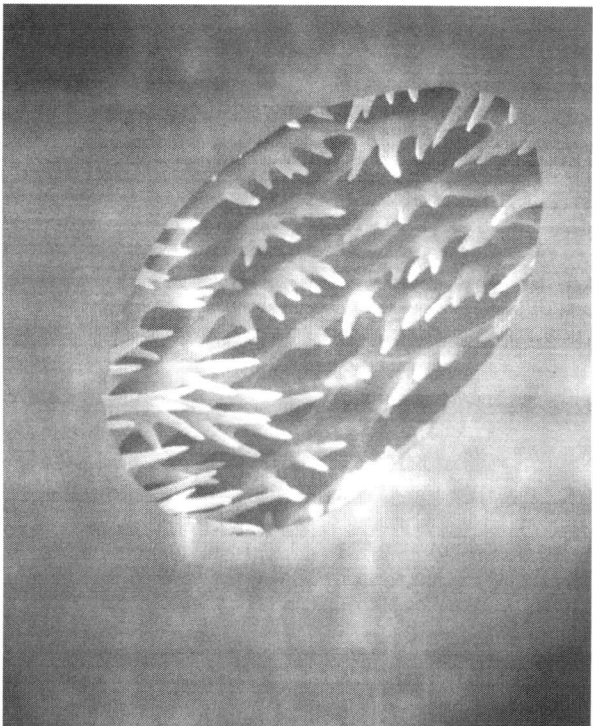

Figure 17.7 Setae inside the external opening of a spiracle of the black turpentine beetle *Dendroctonus terebrans* that probably act as a dust and particle filter to protect the tracheal system. (SEM photograph by the author.)

atrium that act as a filter to keep dust particles out of the tracheae. In aquatic insects, it aids in excluding water. A "felt chamber," a dense mat of very fine hairs and setae, occurs just inside the spiracle of some dipterous insects.

The muscles associated with the spiracles are innervated by a branch of the median nerve from the ganglion in the same segment or from the ganglion in the anterior segment (Case, 1957). Repetitive action potentials from the median nerve cause contracture of the closer muscle and close the spiracle. The closer muscle of *Hyalophora cecropia* has a myogenic rhythm leading to slow (graded) pacemaker potentials that give rise to spike discharges (Van Der Kloot, 1963). Experimental increments of hyperpolarization of the muscle membrane potential slows the rate by which the pacemaker potentials depolarize the membrane, and strong hyperpolarization stops the ability of the graded potentials to exceed the threshold for action potentials. Prolonged exposure of the closer muscle to CO_2 causes sufficient hyperpolarization that the pacemaker potentials are ineffective and the muscle relaxes and the spiracle opens. Anesthetization of insects with CO_2 is a common practice in laboratory experiments, and in most insects this treatment causes the spiracles to open and results in significant loss of water if the insects are kept anesthetized very long and/or not given access to water after recovery.

A bilateral pair of spiracles occurred on each thoracic segment, and one pair on each abdominal segment in the evolutionary primitive arrangement. Apparently, no openings evolved on the head. Some existing Diplura have 12 spiracles, but most insects have fewer (Edwards, 1953). Some insects have only one or two functional spiracles, and a few, particularly some aquatic ones, have no functional spiracles at all, and tracheae do not open to the outside. Gas exchange between tracheae and the environment in this latter type of system occurs by diffusion through a thin cuticle.

17.2.3 Tracheal Epithelium

The tracheal system is not an acellular system of tubes and tubules; every part of the system contains living cells. **Tracheal epithelial cells**, derived like cuticular epidermal cells from embryonic ectoderm, surround tracheal tubes and tracheoles throughout all parts of the system. The epithelium cells are flattened, like pieces of a ribbon that might be cut and pasted around a tube. These epithelial cells secrete the cuticulin lining and the hydrophobic compounds at the lumen surface of the tubules. In larger tracheae, the cells secrete a new cuticle lining on their apical side (i.e., toward the old cuticle) prior to each molt and the old lining of tracheae is shed at each molt.

17.2.4 Development of New Tracheoles

The formation of new tracheae and tracheoles varies with respiratory demand of tissues (Locke, 2001). Active larvae of the tenebrionid mealworm, *Tenebrio molitor*, show a large increase in number and diameter of tracheae during exposure to 10% oxygen atmosphere, and occluding the spiracles induces growth of new tracheae and tracheoles, particularly on muscles and gut that have a high demand for gas exchange (Locke, 1958b). The process by which new tracheoles form is primarily based upon the work of Margaret Keister (1948), who patiently observed events occurring in the cells of *Sciara coprophila* Lintner, a mycetophilid fly whose larvae are so transparent that the internal organs can be observed with transmitted light. The initial event in development of a new tracheole begins when a tracheal epithelial cell begins to grow out, usually in a triangular shape, toward a group of cells that (apparently) need better gas exchange (Figure 17.8). It may be that metabolites from tissue cells induce the tracheal epithelial cell to grow a process out toward

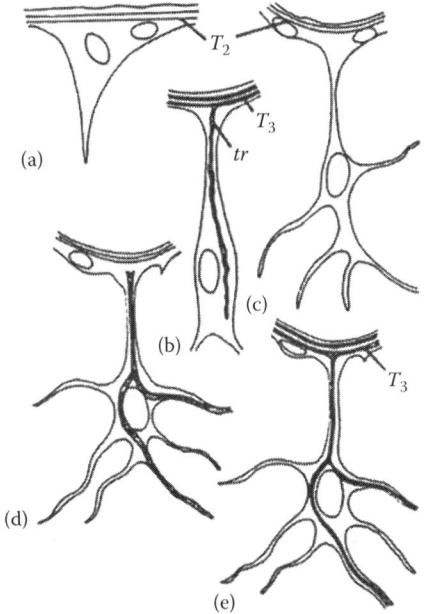

Figure 17.8 Stages in the growth of new tracheoles in *S. coprophila*, a fungus fly (Diptera). (a) Enlargement of a tracheal epithelium cell to become a tracheoblast and growth of a pseudopod toward the area needing more oxygen. (b) Beginning of a tracheole within the elongated epithelium cell. (c) A tracheoblast that has developed several cytoplasmic filaments before formation of a tracheole has begun. (d) Tracheole that has formed in a tracheoblast with multiple cytoplasmic extensions. (e) The tracheole will only receive an open contact with the preexisting tracheal trunk at the next molt. (From Keister, M., *J. Morphol.*, 83, 373, 1948. With permission.)

the tissue. A tracheal epithelial cell that initiates such growth has been called, by various authors, a tracheoblast, stellate cell, transition cell, and tracheal end cell. Such a cell may be at the end of an existing small trachea, but not necessarily so. As it grows, the tracheoblast tends to become spindle shaped, and a single unbranched tracheole may form within it at this stage. More often, however, the cell develops multiple finger-like projections (hence the name stellate cell) before a tracheole forms, and then tracheole branches may form in the several fingers. A fluid-filled, linear channel begins to appear in the cytoplasm of the tracheoblast and gradually becomes longer as the tracheole grows. The tracheole does not grow from the larger trachea, but grows toward it, the tracheoblast never completely having detached from its home base on the surface of a larger trachea. The new tracheole only becomes air filled and functional (in the *Sciara* model), however, at the next molt.

Wigglesworth (1954) observed essentially the same processes occurring in the cuticular epithelium of the hemipteran, *Rhodnius prolixus*. Columns of epithelial cells grew out from an existing tracheal tube, and new tracheae and tracheoles developed within the outgrowing cells. New tracheae and tracheoles rapidly grew toward regions made artificially deficient in gas exchange by cutting existing small tracheae to a section of cuticular epithelium and into transplanted organs. Wigglesworth (1959, 1981) observed that tracheoles were pulled by cytoplasmic strands radiating from the cells needing oxygen (Figure 17.9). In some cases, the tips of tracheoles migrated as much as a millimeter in distance.

Tracheoles also move in cultured *Galleria mellonella* wing discs and in organ transplants in Diptera. Tracheole migration is ecdysone dependent in *Galleria* and may be related to the presence of large numbers of microtubules in cultured wing discs. Microtubule formation is disrupted and tracheole migration is prevented by 10^{-8} M vinblastine or 10^{-5} M colchicine in the culture medium (Oberlander, 1980).

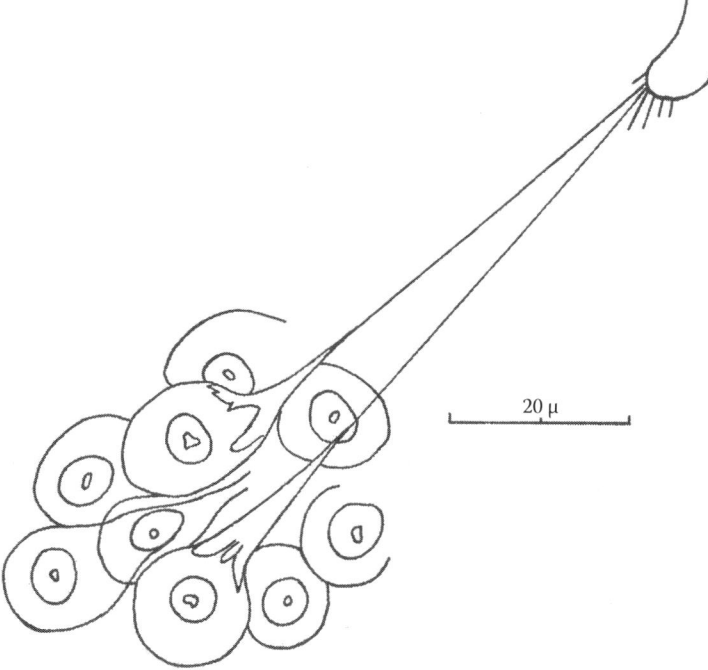

Figure 17.9 Contractile strands from epidermal cells pulling a tracheole into an oxygen-deficient region in the epidermis of *R. prolixus*. (Redrawn and modified from Wigglesworth, V.B., *J. Exp. Biol.*, 36, 632, 1959.)

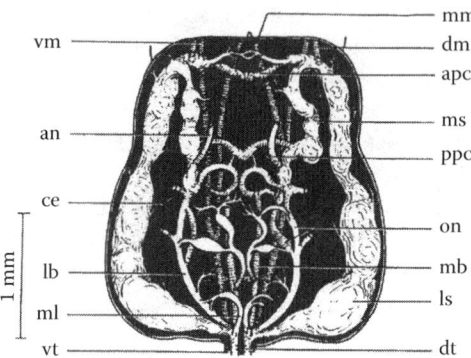

Figure 17.10 Tracheal supply and large air sacs in the head of ant *Camponotus pennsylvanicus*: an, branch to antenna; apc, anterior commissure ventral to pharynx; ce, branches to compound eye; dm, dorsal trachea to mandible; dt, dorsal trachea from thorax; lb, trachea lateral to brain; ls, large lateral air sac; mb, trachea medial to brain; ml, loop around muscle; mm, trachea to mouth parts; ms, small median air sac; on, trachea encircling optic nerve; ppc, posterior commissure ventral to pharynx; vm, ventral trachea to mandible; vt, ventral trachea from thorax. (From Keister, M., *Ann. Entomol. Soc. Am.*, 56, 336, 1963. With permission.)

17.2.5 Air Sacs

Dilations of both primary and secondary tracheae occur in many insects and are called **air sacs** (Figures 17.1 and 17.10). They are variable in size but frequently are large in flying insects, such as honeybees, cicadas, many adult Diptera, and some scarab and buprestid beetles. The intima of air sacs may contain typical taenidia, but these may be reduced or irregular. With *Drosophila* as a model, Weis-Fogh (1964a) showed that the air sacs provide a large surface in contact with flight muscles for exchange of gases. The rhythmical squeezing action of working flight muscles pumps the air sacs like a bellows and increases the flow of air through the system. Air sacs sometimes collapse as growing tissues fill the body space, and by collapsing they make room for the new tissue or organ, with little change in the general body shape. Air sacs serve a hydrostatic function in some aquatic insects and allow more freedom in vibration of the tympanic membrane in some sound-producing insects. They may increase hemolymph concentration of solutes without necessarily increasing total solute and reduce hemolymph volume by restricting space that the volume of circulating hemolymph must serve.

17.2.6 Molting of Tracheae

Prior to a molt, tracheal epithelial cells secrete a new cuticulin layer on their apical surface (the surface toward the lumen of the trachea). Often a fluid layer then separates the two cuticulin layers, the old and the new. The ecdysis of the external cuticle pulls the old, air-filled tubes out, leaving behind the newly formed, fluid-filled ones. The new tracheal tube fills with air rapidly, and although details are scarce, it seems likely that the fluid in the new system is actively reabsorbed into the surrounding tissue, thus making way for the air. Wigglesworth (1959) found that tracheoles, which are never shed in *Rhodnius*, become cemented to the new tracheae at each molt by a ring of adhesive material.

17.3 TRACHEAL SUPPLY TO TISSUES AND ORGANS

The heart and dorsal musculature usually are aerated by tracheal branches in each segment from the dorsal longitudinal trunks. The visceral and internal reproductive organs receive tracheal branches from the lateral trunks. The ventral nerve cord and ventral musculature are supplied by branches from

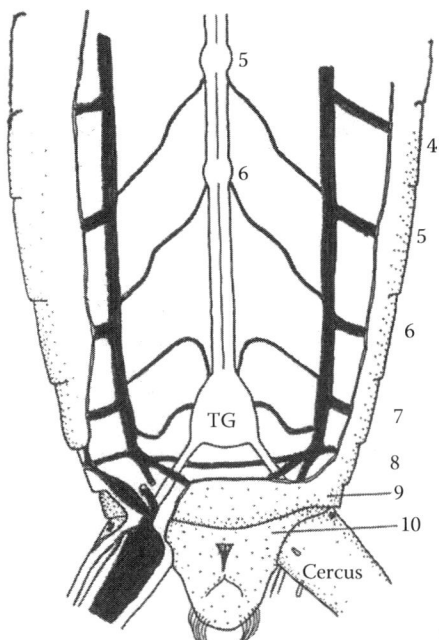

Figure 17.11 Tracheae supplying abdominal ganglia in the abdomen of the house cricket *Acheta domesticus*. (From Longley, A. and Edwards, J.S., *J. Morphol.*, 159, 233, 1979. With permission.)

the ventral trunks or by branches off ventral transverse connectives (Figure 17.11). The legs and wings are supplied by tracheae from the thoracic spiracles, while the head and associated structures are generally supplied with branches from the first spiracle and the dorsal longitudinal trunk.

The only cells or tissues not having direct connections with tracheoles are the hemocytes, the cells that circulate in the hemolymph. Locke (1998) has shown that anoxia causes structural changes in hemocytes and causes sessile hemocytes to be released from the surface of various tissues where they may be attached. Under the anoxic conditions, hemocytes accumulate on very-thin-walled tufts of tracheae near the last pair of abdominal spiracles and in the **tokus** compartment at the tip of the abdomen (Figure 17.12). Thus, the highly branched tracheal system of the last segment and the tokus appear to serve the function of a lung for aeration of hemocytes.

17.3.1 Adaptations of Tracheae to Supply Flight Muscles

Working muscles, especially flight muscles, have a very high demand for oxygen and there is an extensive tracheal supply to flight muscles enabling them to avoid an oxygen debt even with prolonged flight. For example, an adult blowfly, *Lucilia sericata*, requires about 33–50 µL O_2/min/g body weight at rest, but within seconds of taking flight, it increases its oxygen consumption to as much as 1625 µL/min/g body weight. Most of this oxygen is used by flight muscles. This rate of oxygen use is from 30 to 50 times the maximum rate of O_2 used by active vertebrate leg or heart muscle per unit volume. Flight muscles typically have an extensive tracheal supply.

Primary tracheae that originate at a thoracic spiracle may (1) pass through the core of a muscle, giving rise to numerous branches (centroradial system), or (2) run along its surface with branches penetrating deep into the interior of the muscle (lateroradial system), or (3) the tracheae may expand into air sacs that lie on top of the muscle (laterolinear system) (Weis-Fogh, 1964a). The latter arrangement is common in small insects. Some insects, such as the locust, *Schistocerca gregaria*, have each of the three systems in some wing muscles. Secondary tracheae branch from the primary

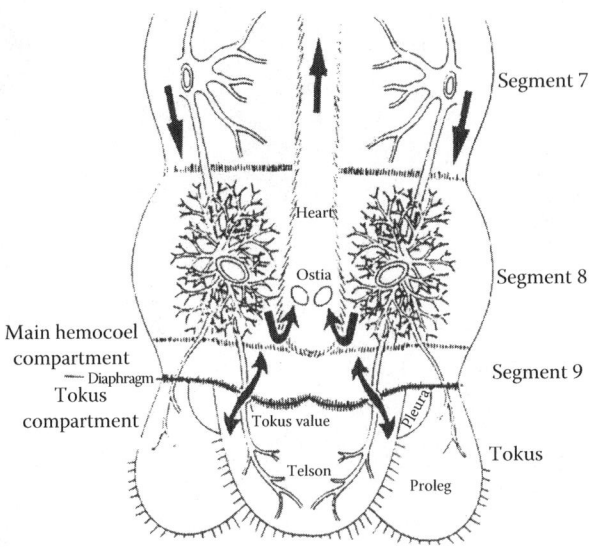

Figure 17.12 The tracheal cluster at the eighth abdominal spiracle in lepidopterous larvae and tracheal extensions into the tokus at the tip of the abdomen, sites where hemocytes accumulate and become aerated. (From Locke, M., *J. Insect Physiol.*, 44, 1, 1998. With permission.)

tracheae and radiate into the spaces between muscle fibers, and tertiary tracheae branch from the secondaries. In the tergosternal muscle (a flight muscle) of *Aeshna* spp. dragonflies, tertiary tracheae branch from the secondary tracheae at about 20 μm intervals, and the tertiary tracheae eventually narrow and each branches into 20–30 tracheoles. In metabolically active tissues and, especially in flight muscles, tracheoblasts push against the sarcolemma of a muscle fiber and indent the fiber like a finger pushed into a balloon to become **intracellular** tracheoles. Despite the term "intracellular," the tracheoles are not inside the cells they penetrate because they still maintain their own cellular epithelium, and the host-cell membrane, though indented, remains intact. Functionally, intracellular tracheoles bring gas exchange capability within a few tenths of a micrometer of mitochondria.

The tracheal system is very efficient, delivering O_2 through an air path branching into smaller and smaller tracheoles that often end within a few micrometers of mitochondria. The tissues of insects typically maintain constant rates of O_2 consumption when atmospheric levels of O_2 are as low as a few kPa (1 Torr = 1 mmHg = 133.322 Pa; 7.5006 mmHg = 1 kPa). This resting safety margin may have evolved in response to several selective pressures, including (1) a relatively low resting metabolic rate that can undergo up to 50-fold increase during intense activity, such as flight; (2) muscular pumping mechanisms that ventilate the tracheal system when demand increases; and (3) regulation of tracheal conductance (quantity of gas transferred divided by the partial pressure gradient) (Greenlee and Harrison, 1998). Although the efficiency of the tracheal system enables working flight muscle to perform aerobically at normal atmospheric O_2 levels, the safety margin is reduced so that hypoxic gas mixtures of 5–10 kPa O_2 decrease flight metabolic rate in the dragonfly *Erythemis* (*Mesothemis*) *simplicicollis* (Harrison and Lighton, 1998).

17.4 VENTILATION AND DIFFUSION OF GASES WITHIN THE SYSTEM

Movement of O_2 and CO_2 through tracheal tubes is promoted by active ventilation in most insects. Simple diffusion may suffice in very small insects, but in all insects, body movements and muscle and gut movements serve to pump the tracheal system and move gases within it and aid gas exchange within the system.

17.4.1 Simple Diffusion Is Usually Not Adequate

The idea that simple diffusion may suffice has a long history. In the early part of the twentieth century, August Krogh concluded that the process of diffusion alone was sufficient to supply oxygen demands of small insects, but some of his assumptions were questioned, and Weis-Fogh reexamined the problem. In brief, Weis-Fogh (1964b) confirmed Krogh's earlier conclusions that diffusion is, indeed, adequate in very small insects, but not in larger ones. **Even in small insects, there probably is pumping ventilation by muscle action and body movements**. Weis-Fogh (1964b) characterized the area in a microscopic cross section of tissue that is occupied by tracheal tubes as the "hole fraction" value, and he found values ranging from 10^{-1} to 10^{-2} for the secondary tracheal supply to wing muscles and 10^{-2} to 10^{-3} for the tertiary tracheae and tracheoles to wing muscles. Mathematical expressions (19 different ones, in all) were derived to describe diffusion in different tracheal branching systems. The simplest formula for calculating the pressure necessary to cause oxygen to diffuse from the spiracles to the site of use applies to those systems in which the cross-sectional area of all tracheae is constant from the last point of branching as in *Cossus* (Lepidoptera) and *Rhodnius* (Hemiptera) systems and can be calculated as

$$\Delta p = \frac{mL^2}{2aP}$$

where
 Δp is the pressure necessary to cause oxygen to diffuse over the distance L (cm) to satisfy the metabolic rate
 m, in mL O_2/g tissue/min that is used
 P is a permeability constant for O_2 diffusion in an air path (about 12 mL/min/cm^2/atm/cm)
 The hole fraction is represented by a

Different patterns of branching called for more complex treatments, for example, in systems such as those in *Schistocerca* and *Aeshna* in which tracheal cross-sectional area decreases with branching.

Weis-Fogh concluded that if an insect uses about 25% of the oxygen entering the spiracles, there would be enough pressure drop from spiracle to tracheole in small insects to allow oxygen to diffuse at a rate that could supply a few mL O_2/g tissue/h. He concluded, however, that diffusion would not be sufficient to supply larger amounts of oxygen or to supply it over distances involved in larger insects and that active ventilation of the tracheal system was necessary. The fact that as much as 10% of the muscle tissue consists of tracheae and that they are ventilated facilitates aerobic metabolism.

17.4.2 Active Ventilation of Tracheae

Insects as large as, or larger than, *Drosophila melanogaster* actively ventilate the tracheal system by muscle and body movements (Weis-Fogh, 1967; Lewis et al., 1973; Sláma, 1988, 1994, 1999; Lighton and Wehner, 1993; Lighton, 1996). Abdominal pumping may occur by dorsoventral compression of the abdomen or by longitudinal telescoping of abdominal segments, and pumping of large tracheae and air sacs is created by action of the large flight musculature in the thorax when an insect is flying. Moreover, simply the change in the shape of the thorax during flight acts as a pumping mechanism. Finally, some insects seem to get ventilation from suction-ventilation pressure differences in the large tracheae in the thorax with differential opening and closing of thoracic spiracles. Komai (1998) found that a relatively large hawk moth, *Agrius convolvuli*, could meet resting oxygen demand by diffusion, but flight muscle pumping was needed during flight to supply enough oxygen. A bumblebee, *Bombus hypocrite hypocrita*, used abdominal pumping even at rest to kept the $_pO_2$ high at 8.5–9.2 kPa. Prior to taking flight, abdominal pumping elevated $_pO_2$, but it fell during flight to a mean level of 6.36 kPa (Komai, 2001). Abdominal pumping is common in large insects at rest and continues during flight. In some fast-flying insects, such as the wasp, *Vespa crabro*, and possibly in other hymenopterans, abdominal pumping alone is sufficient to supply the wing muscles during flight. A somewhat unusual mechanism of rhythmic extension and retraction of the proboscis during flight provides ventilatory flow of air in flying *D. melanogaster* with little or no abdominal pumping detected (Lehmann and Heymann, 2005). Westneat et al. (2003) used X-ray imaging with photography to clearly demonstrate the alternate compression and relaxation of large tracheae by flight muscles in flying insects. When cold, *Manduca sexta* adults shiver the flight muscles during warm up for flight causing a tidal in-and-out flow in thoracic spiracles, but once in flight, there is unidirectional flow, with air entering at the mesothoracic spiracles, and positive pressure in the mesoscutellar air sacs forcing air toward the posterior. Used air and carbon dioxide are released from the posterior thoracic spiracles. The downstroke of the wings in flight slightly increases the volume of the thorax and increases the volume of the thoracic air sacs, creating a slight negative pressure that sucks air in at the mesothoracic spiracles. The openings to the posterior thoracic spiracles are compressed by the downward movement of the wing hinge, but during the upstroke, the thoracic air sacs are compressed and the posterior spiracles are open so that expiration occurs (Wasserthal, 2001). If this is difficult to visualize, a review of the wing and thoracic movements in Chapter 11 may be helpful.

Sláma (1988, 1994, 1999, 2000) recorded periodically repeated **extracardiac miniature pulsations** in hemocoelic pressure mediated from a center in the mesothoracic ganglion (Figures 17.13 and 17.14). Such pulsations, called **coelopulses** by Sláma, appear to be common in many, and perhaps all, insects, including larvae, pupae, and adults. Coelopulses enable insects to breathe through selected spiracles and promote a unidirectional ventilatory stream of air through the body. In pupae of *T. molitor*, the extracardiac coelopulses produce 30–90 µm movements of the

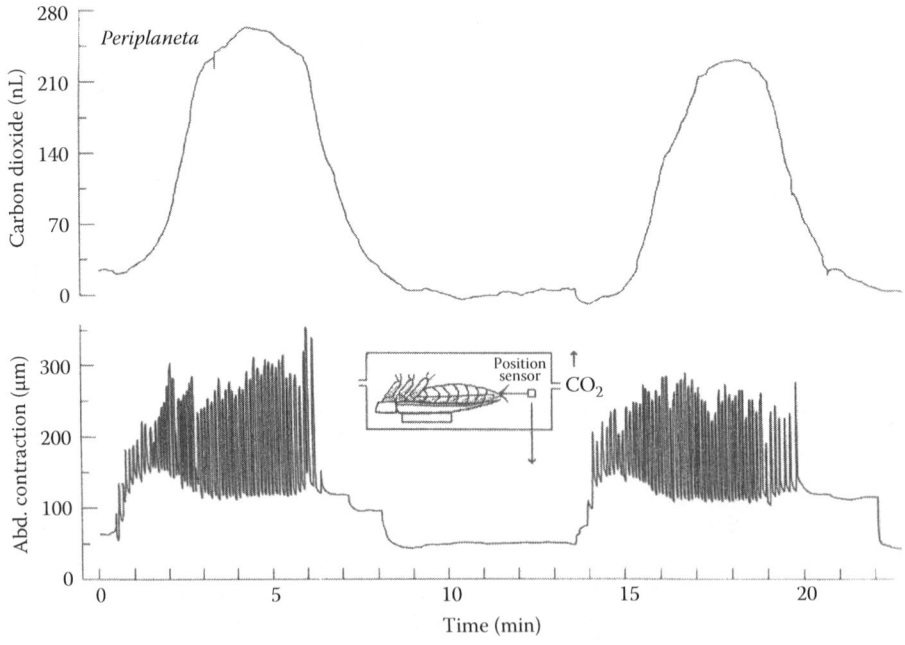

Figure 17.13 A recording of the CO_2 release (top) and extracardiac hemocoelic pulsations (coelopulses) of the abdomen (bottom) in an immobilized American cockroach *P. americana* at 25°C. (From Slama, K., *Ann. Entomol. Soc. Am.*, 92, 916, 1999. With permission.)

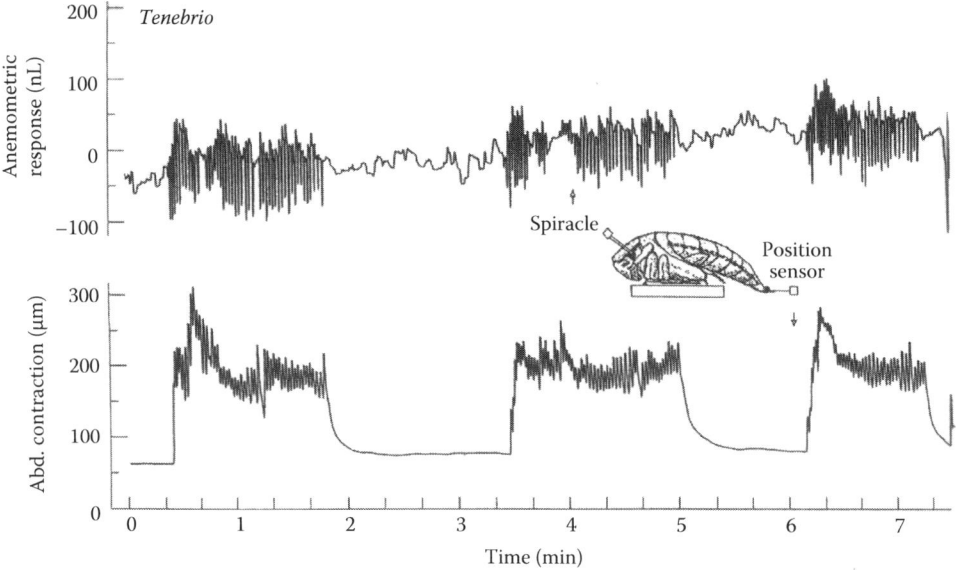

Figure 17.14 A recording of the outbursts of tracheal gas (top) synchronized with coelopulses in hemocoelic pressure (bottom, monitored as longitudinal contractions of abdominal segments) of a pupa of the mealworm *T. molitor* at 23°C. (From Slama, K., *Ann. Entomol. Soc. Am.*, 92, 916, 1999. With permission.)

abdomen, preventing hemolymph stagnation, promoting flow of hemolymph around organs, and probably pumping small tracheoles that aid air movements (Sláma, 2000).

In dragonflies and locusts, thoracic muscular pumping is important for ventilation of wing muscles and experimentally can be demonstrated to be sufficient in the absence of abdominal mechanisms. Although abdominal pumping in the locust is vigorous, abdominal action is not necessary to support flight, as demonstrated by selective injury to the nervous system that blocks or greatly reduces abdominal pumping without altering tethered flight. Thoracic pumping alone can adequately ventilate the muscles and body as a result of thoracic movements and wing action during flight.

Thoracic pumping tends to create a tidal flow that moves air in and out of the same spiracles, while abdominal pumping promotes a directed flow in through thoracic spiracles and out through abdominal spiracles. Abdominal pumping also can create a tidal flow in some insects. At rest, abdominal pumping by the locust, *S. gregaria*, can move 40 L of air/kg body weight/h in through the thoracic spiracles and out through the abdominal ones. As expected the volume of air ventilated during flight goes up (180 L/kg body weight/h for the first 5 min of flight, falling to 150 L/kg/h). As much as 16%–17% of the airflow goes to the head, ventral thorax, and ventral abdomen where it mainly serves nervous system demand, and about 39% goes to the pterothorax to support flight muscle demand. The remaining 44% goes to other systems in the body (Miller, 1960a,b).

Each abdominal ganglion of *Schistocerca* is capable of initiating ventilatory movements for its segment, but the rhythm that synchronizes the overall movements originates in the metathoracic ganglion (Bustami and Hustert, 2000). The isolated metathoracic ganglion shows efferent discharges consistent with continuous ventilation during activity of the locusts and discontinuous ventilation during quiescent periods. The rhythm can be altered by CO_2 concentration and to a lesser extent by hypoxia. Local perfusion of the head or thoracic ganglia with gas mixtures indicates that each region can alter the rhythm of ventilation. A nonflying locust is capable of additional ventilatory mechanisms involving longitudinal telescoping of abdomen, protraction and retraction of head ("neck ventilation"), and protraction and retraction of prothorax ("prothoracic ventilation") (Miller, 1960b). These mechanisms do not come into play under normal resting conditions, but do operate for short intervals following periods of great activity, such as after flight. Groenewald et al. (2012) found that *S. gregaria* locusts pump the tracheal system by dorsoventral contractions and longitudinal telescoping actions during discontinuous gas exchange and suggest that such pumping actions may serve several functions including mixing of gases in the closed tracheal system and possibly promoting water reabsorption from the tips of tracheoles.

Dragonflies (*Aeshna* spp.) and wasps (*V. crabro*) vigorously ventilate the tracheal system during flight. In sustained horizontal flight, dragonflies and wasps use about 20 L O_2/kg body weight/h (Weis-Fogh, 1967). Dragonflies supply the wing muscles entirely from thoracic pumping. Wasps, however, have a very hard, rigid pterothorax that allows little thoracic pumping, but abdominal ventilatory movements increase in amplitude and frequency (to 180/min) during flight.

The way (or ways) in which insects regulate ventilation is not well understood, and diverse mechanisms may be involved depending upon the activity state of the insect, as is the case in grasshoppers (reviewed by Harrison, 1997). *Romalea guttata* and *Schistocerca americana*, two large grasshoppers, regulate O_2 level (at about 18 kPa) and CO_2 level (2 kPa) in the large, longitudinal tracheal trunks by adjustments in ventilation rate. Experimentally elevating tracheal $_pO_2$ above normal or decreasing tracheal $_pCO_2$ below normal values decreases ventilatory rate (Gulinson and Harrison, 1996). The ventilation rate is not altered by experimentally changing hemolymph pH, but elevating hemolymph HCO_3^- increases ventilation. In contrast to the influence of tracheal gas tension on resting ventilation, postexercise increase in ventilatory rate in *S. americana* and *Melanoplus differentialis* (a grasshopper) is not due to the normal rise in internal $_pCO_2$ that accompanies intense muscle activity (Krolikowski and Harrison, 1996). The authors suggest that the reason is because CO_2 receptors are located centrally in thoracic and head ganglia and, thus, they are too far from the site of CO_2 production in working muscles for rapid response to CO_2 levels. Krolikowski and Harrison (1996) suggest humoral mechanisms, nerve activation, neuromodulators, and ionic or metabolic products from working muscle as possible mechanisms that could be involved in controlling activity-related increase in ventilation rate.

Does the tracheal system set an upper limit on the size of insects? Some fossil insects were much larger than present-day insects, but they probably lived in an environment richer in oxygen, perhaps as much as 35% O_2 compared to about 20% today. Greenlee et al. (2007) tested the safety margin for gas exchange in a series of grasshoppers ranging in size from 0.6 g to more than 8 g by exposing them to a series of decreasing oxygen partial pressures ranging from normal oxygen atmospheric partial pressure down to 0 kPa. In this experiment, CO_2 emission rate scaled with body mass to the power 0.92 ± 0.07. During hypoxia, ventilatory activity and tidal volume increased, with ventilatory frequency approximately doubling. In the size range of grasshoppers studied, the authors did not find any effect of body size on safety margin for gas exchange, and they concluded that larger insects compensate for potential diffusion limitation at the tracheole endings by matching tracheal conductance to tissue requirements. In earlier experiments, Rascón and Harrison (2005) and Harrison et al. (2006) found that during intense activity, such as flight or terrestrial running or jumping, there may be decreasing safety margin in a reduced oxygen atmosphere, but under normoxic conditions, the tracheal system is capable of providing for adequate gas exchange in the tissues. The safety margin actually increased with development from instar to instar in *S. americana*, the American locust, due to increased abdominal ventilatory pumping (Greenlee and Harrison, 2004a; Kirkton et al., 2005; Harrison et al., 2013), although tracheal conductance decreased 20%–33% near the time of each molt to the next instar, probably because of compression of the tracheal system by increased tissue growth (Greenlee and Harrison, 2004b; Harrison et al., 2005), and similar effects occurred as a caterpillar of *M. sexta* approached a molt (Greenlee and Harrison, 2005). This may be a common occurrence in very active insects as they approach a molt because the tracheal system cannot be enlarged to keep up with tissue growth until the molt occurs. Some reduction occurs in capability to deliver oxygen to tissues, and especially the jumping leg muscles of *S. americana* during intermolt periods, as the tissues grow but the tracheal system must await molting to enlarge (Kirkton et al., 2012). The safety margin for oxygen delivery (Greenlee et al., 2007) does not change independent of body size across a number of locust species, and major tracheae in the thorax of *M. sexta* increased isometrically with body mass as the caterpillar molted to larger instars, but tracheae in the head did not scale with mass, which suggests a larger safety factor for gas exchange in the head (Greenlee et al., 2013). Callier and Nijhout (2011) and Kirkton et al. (2012) suggest that reduced oxygen delivery capacity may be an important trigger for secretion of the hormonal cascade for molting as an insect gets larger and its tracheal system becomes less efficient. In conclusion, a tracheal system certainly could not support the gas exchange of a mammal or other large animal, but even the largest insects living today seem to have an ample safety margin in the ability of the tracheal system to provide adequate gas exchange in the tissues, although some reduced capacity may occur during intermolt periods.

17.4.3 Diffusion from Tracheoles to Mitochondria

The evidence suggests that there is little or no difference in the permeability of larger tracheae and tracheoles to O_2, but only tracheoles usually will be close enough to mitochondria for O_2 to diffuse across the aqueous path to the mitochondria in the quantities needed. In its final path from tracheole to mitochondria, O_2 must move by diffusion, crossing tracheole wall, cell membranes, cell cytoplasm, and mitochondrial membranes. This is the slowest part of the pathway for the final delivery of oxygen for metabolism. The diffusion of O_2 in a water-filled path is about one million times less rapid than its movement in an air path, thus the closer the tracheole can get to the site of mitochondria, the higher the rate of O_2 consumption that can be supported. Tracheoles need to be within about 10 µm of a mitochondrion in order to deliver sufficient O_2 to support active metabolism. The evolution of intracellular tracheoles seems to be an adaption to support a high rate of metabolism in large, active cells. Many insect cells are typically 30–60 µm diameter, or even larger in the case of the most active fibrillar muscle fibers, but tracheoles often indent these muscle fibers and may even touch and indent mitochondria (Afzelius and Gonnert, 1972).

17.5 DISCONTINUOUS GAS EXCHANGE

Some insects are able to keep the spiracles tightly closed and/or "apparently closed" for a high percentage of the time. Gas exchange occurs in three periods named the **open, flutter**, and **closed** periods (often designated as O, F, and C, respectively) because of the action of spiracles over the duration of a cycle. This functional pattern, variously known as **discontinuous release of CO_2, passive suction ventilation, discontinuous ventilation cycle**, and as the **discontinuous gas exchange cycle (DGC)** (Lighton and Garrigan, 1995), has been known to occur in some insects for more than half a century (reviewed by Slàma, 1988, 1994; Lighton, 1996). The earliest studies were in diapausing pupae of Lepidoptera, and Schneiderman and Williams (1955) cannulated the spiracles of large cecropia pupae and made the first analyses of intratracheal air. Initially, DGC was thought to be limited to quiescent insects in a depressed state of metabolism, but DGC patterns have been observed in a number of different insects in various states of activity, including ants, the cockroach (*P. americana*), a number of adult tenebrionid beetles, the locust *S. gregaria*, the lubber grasshopper (*R. guttata*), and adults of additional species (Punt, 1950; Lighton, 1988, 1990, 1991, 1994, 1996; Lighton and Wehner, 1993; Lighton et al., 1993b; Hadley, 1994; Slàma, 1999, 2000; Chown and Holter, 2000; Duncan and Newton, 2000; Vogt and Appel, 2000). DGC behavior also occurs in some arthropods other than insects (Lighton et al., 1993a; Lighton and Duncan, 1995).

During DGC, accumulated CO_2 is discharged periodically in episodic bursts during brief intervals when spiracles are open (Figure 17.15). After a burst, the spiracles are closed for some period of time that varies from species to species. During the closed interval, oxygen is used by tissues and intratracheal oxygen tension falls. Most insects that exhibit DGC allow the spiracular valve to flutter with an amplitude often imperceptible to the unaided eye during a portion of a cycle. The fluttering (F) phase allows small amounts of O_2 to be sucked into the tracheal system by the slight negative pressure arising from O_2 consumption by the tissues. The F phase is usually considered to involve convective transfer of O_2. This has given rise to the name "passive suction ventilation" that sometimes is applied to the process. The slight internal vacuum retards the outward loss of water vapor and CO_2 during the fluttering phase, and the low influx of O_2 lengthens the time to full opening or

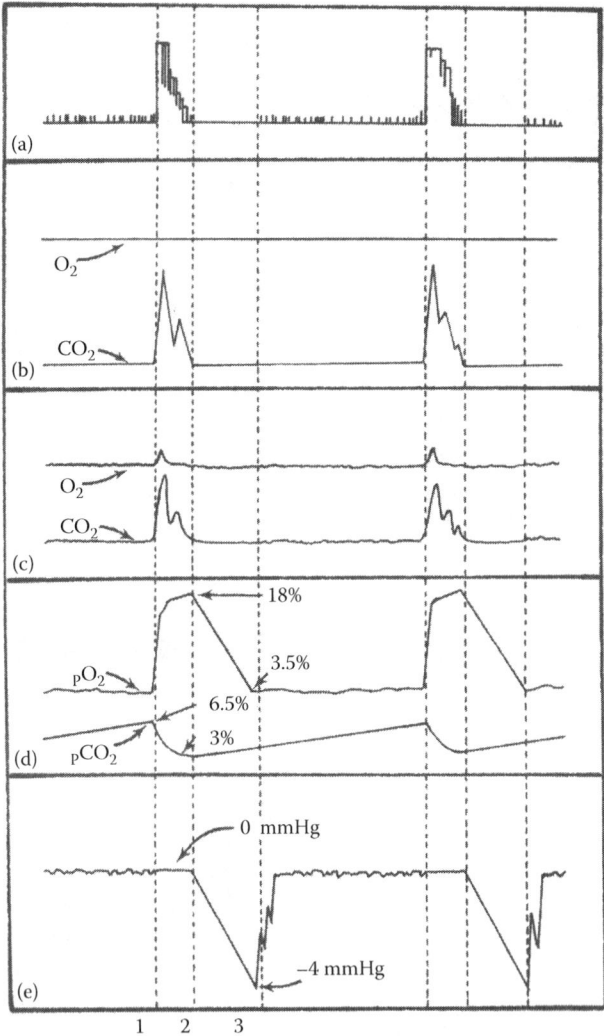

Figure 17.15 Correlation of events occurring during the respiratory cycle in a cecropia pupa. A cycle is divided into major phases in which the spiracle is (1) fluttering, (2) open, or (3) tightly closed. (a) Spiracular opening (rise above the baseline) and closing (return to baseline). (b) Gas exchange measured by manometric methods. (c) Gas exchange measured by diaferometric method. (d) Tracheal gas composition. (e) Intratracheal pressure. (From Levy, R.I. and Schneiderman, H., *J. Insect Physiol.*, 12, 465, 1966b. With permission.)

"burst" of spiracles. CO_2 is produced by tissues even when the spiracles are closed, but the high solubility of CO_2 in aqueous solutions enables insects to accumulate bicarbonate ion, HCO_3^-, in the hemolymph according to the following reactions:

$$CO_2 + H_2O \rightarrow H_2CO_3 \rightarrow H^+ + HCO_3^-$$

Buffering capacity of hemolymph aids in solubilizing CO_2 as bicarbonate ions (HCO_3^-), and this keeps gaseous CO_2 from building up rapidly in the tracheal system. At some point, probably different for different insects, the relationship between gaseous O_2 and CO_2 and bicarbonate in solution reaches an equilibrium at which tracheal tension of CO_2 and O_2, and/or pH change in the hemolymph, triggers spiracle opening and release of CO_2 from the hemolymph as a gas. Slàma (1999) identified specially modified tracheal sacs near abdominal spiracles two to five in pupae of *G. mellonella* that he calls carboniferous tracheae that selectively extract dissolved CO_2 from the hemolymph and release it as respiratory gas. Low oxygen tension (Burkett and Schneiderman, 1967, 1974; Lighton, 1996) acts centrally on the ganglion controlling a spiracle to trigger fluttering of spiracular valves, while CO_2 acts directly on the closer muscle of *H. cecropia*. Slàma (1994) presented evidence that the neural center controlling DGC cycles in the lacewing *Chrysoperla carnea* (Neuroptera) is located in thoracic ganglia. During quiescent periods at 15°C, the western lubber grasshopper, *Taeniopoda eques*, can go up to 40 min between bursts, during which partial pressure of CO_2 builds to 2.26 kPa with little acidification of the hemolymph (Harrison et al., 1995). These data suggest that when CO_2 tension in tracheae reaches a threshold between 2 and 2.9 kPa, opening of the spiracle is triggered.

Diapausing *H. cecropia* pupae can tolerate very low oxygen tensions in the tracheae, as low as 5% O_2, before fluttering is triggered. Continued fluttering of the spiracular valve keeps O_2 in the intratracheal air of a cecropia pupa at about 3.5% until the next burst (see Figure 17.15). This low level of O_2 is sufficient to support the slow metabolic processes of the diapausing pupa, but it would not likely support an active insect. Intervals between bursts of CO_2, or spiracle opening, increase as the diapausing pupa sinks deeper into the diapause state and requires less O_2 (Schneiderman and Williams, 1955; Levy and Schneiderman, 1966a,b). Discontinuous respiration continues with longer and longer intervals between bursts down to −5°C (Burkett and Schneiderman, 1974), but the tracheal valves freeze closed at lower temperatures, and any additional gas exchange has to occur through the cuticle. Thus, discontinuous respiration is highly functional and adaptive for cecropia pupae, which spend the winter buried beneath litter and soil.

It has often been assumed that the functional benefit to an insect of discontinuous ventilation is conservation of water, and some evidence supports this theory (Williams et al., 2010). In diapausing pupae that must pass a long winter under the soil, leaf litter, or other pupation site, water conservation seems quite necessary and a reasonable driving mechanism for the evolution of discontinuous ventilation because the pupa is a closed (no food or water intake) system. Most adult insects that exhibit discontinuous ventilation do so only intermittently, and the rest of the time they ventilate the system continuously. Surprisingly, some insects do not discontinuously ventilate under conditions that might be expected (based on assumptions) to promote the behavior. For example, the lubber grasshopper, *R. guttata*, which discontinuously ventilates at times, tends to ventilate continuously when dehydrated, a physiological condition in which it presumably has a great need to conserve water (Hadley and Quinlan, 1993; Quinlan and Hadley, 1993). No good explanation has been provided for such behavior. The ant, *Camponotus vicinus*, which exhibits discontinuous gas exchange, actually loses more water than CO_2 during the period when the spiracles are open (Lighton and Garrigan, 1995). The harvester ant, *Pogonomyrmex rugosus*, and similar desert ants have a relatively high percentage (up to 13%) of body water loss through the tracheal system (Lighton et al., 1993b), even though they exhibit discontinuous ventilation at times. Whether the somewhat higher rate of water loss from the tracheal system in these ants is a significant stress for them may depend on how much of the time the ants exhibit discontinuous gas exchange, as well as how much access they have to food with high water content and exposure to environmental extremes. Although it seems somewhat intuitive that water loss might be high from the tracheal system, actual measurements in some insects indicate that about 90% of total body water loss occurs through the cuticle, with only 2%–5% typically lost from the tracheal system (Hadley, 1994). Cyclic release of CO_2 occurs in drywood termites, *Incisitermes minor* (Hagen); Formosan subterranean termites, *Coptotermes formosanus* Shiraki; and Eastern subterranean termites, *Reticulitermes flavipes* (Kollar), but the termites do not exhibit a strict DGC pattern, and water loss through the respiratory system is less than 10% of the total daily water loss (Shelton and Appel, 2000, 2001). Populations of the grasshopper, *Melanoplus sanguinipes*, collected in California at several geographic and altitudinal sites show behavior leading to discontinuous gas exchange, but DGC decreases at elevated temperatures when it might be expected to be more pronounced if respiratory water conservation were important to the grasshoppers. The quantity and melting point of cuticular lipids, however, show strong correlation with lower rates of water loss (Rourke, 2000).

Gibbs and Johnson (2004) devised a mathematical and graphic method to separate cuticular water loss from tracheal system water loss. *Karoophasma biedouwensis*, the heelwalker and a member of the recently discovered wingless family of (and presumably ancient) insects, shows cyclic gas exchange with lack of a flutter period (Chown et al., 2006). About 70% of its water loss came from cuticular surfaces and about 29% from the respiratory system. Thus, as intuitive as the water loss hypothesis for evolution of respiratory patterns may seem, the evidence indicates that most insects lose water from the cuticle far in excess of what they lose from the tracheal system.

An alternative to the water conservation theory was put forth by Lighton and Berrigan (1995) and amplified by Lighton (1998), who suggested that DGC may occur primarily in insects that experience **hypoxic** (low O_2) and **hypercapnic** (high CO_2) conditions (such as ants living underground) and that it is not necessarily essential to reducing water loss through the tracheal system. Chappell and Rogowitz (2000) also concluded that the hypoxic and hypercapnic environment in which eucalyptus-boring cerambycid beetles live may have had more influence on evolution of discontinuous ventilation than water conservation. Nespolo et al. (2007) support the hypoxic/hypercapnic theory with data from the Chilean red cricket, which lives in burrows in the ground. Marais et al. (2005) reviewed the data on 118 species from 8 orders in which gas exchange patterns have been studied, and they concluded that the data show no pattern of association with subterranean or nonsubterranean lifestyle nor with winged or wingless species, and discontinuous and cyclic gas exchange patterns have evolved in species that live in both mesic and xeric habitats. Chown and Holter (2000) propose still another explanation for DGC, suggesting that the periodic nature of DGC may be the result of two feedback systems (i.e., sensors detecting low O_2 and those detecting high CO_2) interacting during times of minimal demand, and cite Kauffman (1993), who has described varying effects of interacting feedback systems, ranging from a single steady state to cyclic behavior. Hetz and Bradley (2005) suggest that discontinuous respiratory cycles aid insects in avoiding oxygen toxicity. Terblanche et al. (2008) cannulated the spiracles of *Samia cynthia* and concluded that DGE is maintained and regulated in response to various environmental situations, including preventing oxidative damage and water loss, but Matthews et al. (2012) found no evidence for DGE as a defense against high oxygen levels in *Locusta migratoria*.

In a review of discontinuous gas exchange, Chown (2011) concluded that the evolutionary driving mechanism for DGE was energy saving in metabolic rate when the brain transferred control of the spiracles to segmental ganglia in the thorax and abdomen, leading to discontinuous respiration based upon peripheral CO_2 sensing and central O_2 sensing. Thus, DGE might serve a variety of ecological functions in different insects and in different ecological environments, with reduced energy demand.

In summary, a definitive evolutionary mechanism for evolution of discontinuous gas exchange cycling may still be debatable, but the mechanism is fairly widespread in many insects, including larvae, pupae, and large and small, winged and wingless adults. Water conservation, ecological niche occupied by insects, interactions of sensory mechanisms, and potential oxygen toxicity and energy savings may be factors.

17.6 WATER BALANCE DURING FLIGHT

A flying insect can lose large amounts of water due to evaporation from the tracheal surfaces as large volumes of air are ventilated through the system. Physiological and behavioral adaptations, particularly in long-distance flyers, such as those that migrate, help prevent desiccation. One adaptation is the use of fat as a flight fuel. Lepidoptera, Orthoptera, and some other groups mobilize triacylglycerols from the fat body, transport the molecules to flight muscles, and oxidize them for energy during flight. In *S. gregaria* about 7 g fat/kg body weight/h is burned during flight, resulting in the production of 8.1 g H_2O/kg body weight/h (Weis-Fogh, 1967). This gain of metabolic water helps offset the loss from evaporation. Water loss is related also to relative humidity and temperature of the air. At moderate to high relative humidity and at temperatures between 25°C and 30°C, a flying *S. gregaria* locust can stay in water balance during sustained flight, but at very low relative humidity, water will be lost and this is intensified at higher temperatures. Lehmann et al. (2000) suggest that small dipterans, such as the smaller species of *Drosophila*, face high risk of desiccation during flight. They determined that during hovering flight, the average water loss in four species of *Drosophila* is 67.3 ± 36.9 µL/g/h. If the flies metabolize glycogen during flight, they can produce 0.56 mg H_2O/mg glycogen metabolized (Schmidt-Nielsen, 1997). Based on the amount of CO_2 produced during hovering flight (32.4 ± 5.1 mL/g/h) and a factor of 1.19 mg glycogen/mL CO_2 produced, the authors estimate that on average, the four species produce 21.6 ± 3.4 µL metabolic water/g body mass/h, or about 41.7% of the total water loss during hovering flight.

Behavioral mechanisms, such as flight at night when temperature is usually lower and relative humidity usually higher and flight at higher altitudes where temperatures are lower and prevailing winds can blow insects along, can aid insects in reducing water loss. Some insects may make other behavioral adjustments, including being quiescent, seeking shade, and refusing to take flight. For example, tobacco hornworm adult moths will not fly at air temperatures of about 43°C, and this protects them from overheating and dehydration (Heinrich, 1970, 1996), either of which would be especially detrimental to the nervous system and probably other organ systems.

17.7 GAS EXCHANGE IN AQUATIC INSECTS

The tracheal system of most aquatic insects is structurally the same as that of terrestrial insects, that is, open spiracles and an extensive network of tracheae and tracheoles. These aquatic insects breathe air by frequently coming to the surface. Water is prevented from entering the system by the hydrophobic surface of the tracheae and, in some cases, by closed spiracles. Many aquatic larvae have a metapneustic system, with only one functional pair of spiracles near the tip of the abdomen (Keilin, 1944). When the insect comes to the surface, it does so with the posterior end upper most and only the tip of the abdomen bearing the spiracles is held above the surface. Some have a siphon at the tip of the abdomen that is pushed above the surface of the water for gas exchange. During submergence, the spiracles are kept closed. Probably a small vacuum is created within the tracheal system due to O_2 use while submerged, as in discontinuous release of CO_2, and this aids the intake of oxygen when the larva or pupa comes to the surface. Larvae of *Glossina* spp. (tsetse flies, Diptera: Glossinidae) develop inside the uterus of the mother, an environment in which they get their oxygen through a pair of posterior spiracles after the air enters the vulva of the mother (Zdárek et al., 1996).

Some aquatic insects living under water must be able to break the surface film of water to get air from the surface. Typically they have structural adaptations of the body (**hydrofuges**) that are difficult to wet and these facilitate surfacing and hanging at the surface. The hydrofuge repels water, is not easily wetted, and tends to buoy the insect at the surface of the water. Tufts of long hairs surround the siphon in mosquito larvae, and when the mosquito larva comes to the surface, the ring of hydrofuge hairs spread out in a circle on top of the water, keeping the tip of the siphon just above the water level. When the larva submerges, the hydrofuge hairs tend to collapse inward on each other and form a small dome over the tip of the siphon. Oily secretions from cuticular glands coat the hydrofuge regions or hairs and aid in maintaining their hydrofuge nature. Larvae of *D. melanogaster* often live in a wet environment, and three unicellular glands associated with each of the two posterior spiracles (Jarial and Engstrom, 1995) appear to secrete oily substances on to the cuticle around the spiracles.

17.7.1 Compressible Gas Gills

A large number of aquatic insects submerge with a bubble or film of air enclosing one or more spiracles. These gas bubbles or films of air have been called gas gills, and they may be compressible and require replenishing periodically, or they may be incompressible and enable the insect to stay submerged indefinitely. **Compressible gas gills**, a bubble or film of air somewhere on the body, are widespread among aquatic members of the Coleoptera (Dryopoidae and Hydrophilidae), Hemiptera (the genera *Gerris* and *Velia*), and Lepidoptera (some arctiids and pyralids). Compressible gills slowly collapse as the oxygen is used, although additional oxygen is usually extracted from the water before the bubble must be renewed at the surface (Tsubaki et al., 2006). Air stores in compressible gas gills may be carried in a fine network of hydrofuge hairs. Sometimes there is a fine, dense set of hairs nearest the body surface from which the gas volume is very slowly used and a set of longer, larger hairs over these, from which the gas is rapidly used. Alternatively, a gas gill may be carried beneath the elytra as in dytiscid diving beetles or as a gas bubble on the posterior part of the abdomen. In addition to serving a respiratory function, gas gills also provide buoyancy and a hydrostatic function. When the insect comes to the surface to renew the air store, the buoyancy of the remaining bubble allows the part of the body carrying the bubble to come to the top first. Thus, the air in the bubble is restored with a minimum of exposure of the insect at the surface. Immediately after an insect has surfaced, any air bubble on its body should contain approximately 21% O_2 and 79% N_2, in equilibrium with the atmosphere. As the insect submerges, O_2 will be used for metabolic processes and O_2 tension in the air bubble will fall, while the partial pressure of the nitrogen (pN_2) will rise. In well-aerated water, the gas composition is approximately 33% O_2, 64% N_2, and 3% CO_2. Since oxygen is ordinarily in higher concentration in the water than in the air bubble, O_2 from the water will diffuse into the gas bubble and N_2 in the bubble will begin to diffuse out into the water. Equilibrium pressure in the bubble cannot be attained because the insect is continually using O_2, but this dynamic exchange allows the insect to gain up to 13 times the quantity of O_2 originally carried within the bubble. Eventually, the insect must surface and renew the bubble. The N_2 in the bubble plays a crucial role because its relatively low solubility in water causes it to move out of the bubble more slowly than O_2 enters the bubble. The N_2 remaining in the bubble prevents it from collapsing, giving oxygen from the water a chance to diffuse in. Beetles allowed to fill the bubble at the surface with pure O_2 cannot stay submerged as long as normal because the bubble shrinks rapidly as oxygen is used by the insect and there is little or no nitrogen to keep the bubble expanded. The time that an insect can stay submerged with a compressible gas gill depends upon the size of the air bubble, activity of the insect, and O_2 tension in the water. *Dytiscus* diving beetles were able to stay submerged from 3 to 5 min with a bubble that initially contained 19.5% O_2, but then dropped to 2%, requiring them to surface for replenishment (Wigglesworth, 1972).

17.7.2 Incompressible Gas Gills: A Plastron

Incompressible gills do not collapse and oxygen can continue to be extracted indefinitely from the water into the gill (if the water is well aerated), allowing the insect to live underwater. Incompressible gas gills also are called **plastrons**. A plastron consists of any extensive physical meshwork, either of fine hairs or setae, or a meshwork of small pores and channels in the cuticular surface of some insects and eggs that can hold a volume of air and can present a large water–air interface. When the meshwork is extensive enough, a constant film of air can be held that takes oxygen from the surrounding water, provided that the water is well aerated. Plastrons are common adaptations of insects living in aquatic arrangements and can take many physical forms. A plastron contains a film of air so tightly guarded by a dense network of nonwettable hairs or meshwork of pores that even though the gas equilibrium may change due to O_2 use, water cannot invade the air space. The minimum amount of water–air interface in a meshwork necessary to enable it to function as a plastron has not been determined, but Hinton (1964) suggested that a water–air interface to weight ratio of 15,000 μm^2/mg weight was sufficient to qualify as a plastron. He based this conclusion upon the water–air interface/mg ratio found in the pupa of the fly, *Eutanyderus wilsoni*. This fly has the poorest ratio known for an insect obviously adapted for living in water. Most insects with a plastron have a water–air interface of from 10^5 to 10^6 μm^2/mg weight. The thickness of the hair pile will obviously help determine the efficiency of the plastron. *Aphelocheirus aestivalis* (Hemiptera), for example, has from 2×10^8 to 2.5×10^8 hairs/cm^2 forming the plastron. Insects that have 10^6 to 10^8 hairs/cm^2 generally have a very efficient plastron and usually can stay submerged for months. Some Coleoptera and Lepidoptera also have very efficient plastrons and often have eight or nine spiracles that open into the plastron air space. Agents that lower the surface tension of the water (e.g., soap and alcohols) will cause wetting of the plastron and failure to retain the air space. High pressure, if applied over long enough periods of time, will cause wetting of the plastron, but these high pressures are not likely to occur in the natural habitat. A plastron can work in reverse and extract O_2 from the insect tissue and pass it into water that is very low in O_2 content, as might occur in cases of severe pollution. Insects utilizing a plastron usually live in well-aerated water of streams, lake edges, and intertidal zones.

17.7.3 Use of Aquatic Plants as Air Source

Some insects with a hydrofuge are able to capture and utilize the gas bubbles released by aquatic plants, either by inserting a part of the body bearing a spiracle into the plant tissue (Figures 17.16 and 17.17) or by biting into the air spaces of the plant. Some species of Diptera, Coleoptera,

Figure 17.16 Mosquito larva, *Mansonia* spp., with its postabdominal spiracles inserted in a branch of water lettuce, *Pistia stratiotes*. (Photograph by Tom Loyless, Courtesy of D.O. Deonier, deceased.)

478   INSECT PHYSIOLOGY AND BIOCHEMISTRY

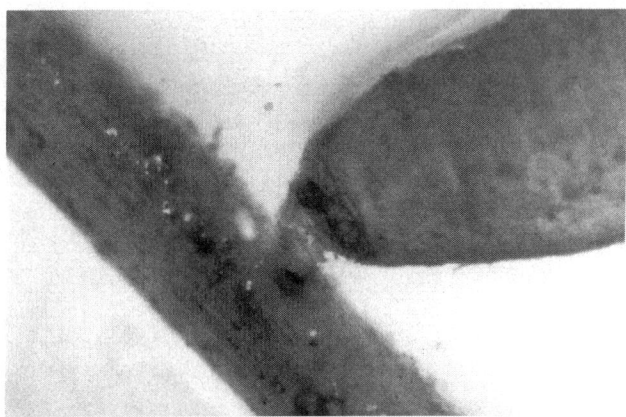

Figure 17.17 A puparium of *Notiphila carinata* (Diptera: Ephydridae) attached to the root of a water weed. The larva and pupa live underwater and obtain their air from the host plant. The larva inserts a pointed, root-piercing spiracle into the plant just before it pupariates, and the pupa obtains its air supply from the plant. (Photograph Courtesy of D.O. Deonier, deceased.)

and Lepidoptera independently evolved modifications for piercing aquatic plants for air. Dipterous larvae in the family Ephydridae mine leaves of aquatic or semiaquatic plants (Deonier, 1993). One species, *Hydrellia pakistanae* Deonier, the Indian hydrilla leaf miner, is a small fly that lays its eggs on leaves of the aquatic weed hydrilla (Buckingham, 1994). The larvae mine the leaves, which usually are underwater much or all of the time, and the larvae get their air either from the plant or by cutaneous diffusion through the cuticle. Before pupariation, a larva leaves the leaf and anchors itself to the stem of the weed by inserting its paired anal spines into the stem, and the pupa gets its air from the plant through the terminal spiracles. Pupae do not survive if mechanically removed from the plant (Cuda, 2001, personal communication). Remarkably, small hymenopteran wasps (*Trichopria columbiana*) hunt and lay their eggs on the pupa under the water. The small wasps enter the water and crawl downward along a hydrilla stem searching for pupae, and upon finding one it inserts an egg through the cuticle of the pupa. The larva of the wasp develops in the pupa and usually has its posterior end oriented with the posterior of the pupa that is inserted into the plant, suggesting that the wasp larva also gets its oxygen from the plant. The adult wasp can stay submerged for several hours, and perhaps much longer in its hunt for a pupa (Deonier, 1971), but its mechanism for obtaining air while submerged in search of a host pupa is not known. It may be able to trap a film of air on the body when it enters the water. Another dipteran, *Notiphila riparia*, similarly pierces submerged plants as a larva and a pupa in order to obtain an air supply.

Certain coleopterans (the Donaciinae) live in the mud around the roots of aquatic plants, and larvae penetrate the plant roots with a pointed, posterior siphon. The thick mud gives them a resistant surface to push against in inserting the siphon. A spiracle at the end of the siphon allows gas entry. Larvae also bite into the root before pupation and construct a pupal case over the lesion. Air probably continues to be released from the lesion into the pupal case to support the pupa. Some lepidopterous larvae in the genus *Hyrocampa* also insert a respiratory siphon into the air spaces of aquatic plants, while some other lepidopterans bite into the plant.

17.7.4 Cutaneous Respiration: Closed Tracheal System in Some Aquatic Insects

A closed tracheal system without functional spiracles is present in some aquatic insects. The lack of functional spiracles eliminates any chance that water will enter the system, but oxygen also must diffuse through the cuticle, a breathing mechanism called **cutaneous respiration**.

Internally, these insects still have an extensive tracheal system like terrestrial insects. Often in larger, more active insects, there are tracheal gills that greatly increase the cuticular surface for gas exchange with the water. Relatively large larvae of Trichoptera, Plecoptera, Odonata, and some Lepidoptera utilize cutaneous respiration that is facilitated by extensive elaborations of thin hair-like or flap-like cuticular extensions from the body surface called **tracheal gills**.

Cuticular extensions from the body surface called tracheal gills (Figure 17.18) occur in several orders of insects and are highly variable in structure and in location on the body. The cuticle of insects with tracheal gills is very thin, and usually large numbers of small tracheae and tracheoles lie just beneath the cuticle. Larvae of Trichoptera, Odonata, and Coleoptera display a highly structured arrangement of tracheoles that are uniform in size and spacing in the gills as an adaptation to utilize optimal functional efficiency with a minimum of tracheoles (Wichard and Komnick, 1974). The tracheoles run parallel to the gill surface and are just beneath the thin cuticle. These characteristics appear to be highly adaptive for trapping O_2 that diffuses through the cuticle. Three caudal gills are characteristic of larvae of the Zygoptera, while tufts of thin gill filaments are located on the head, thorax, abdomen, and coxae of some Plecoptera. In Plecoptera, the gills have a tube-like shape with the center cavity filled with hemolymph that is continuous with the hemocoel of the body (Wichard and Komnick, 1974). The cuticle is thin (0.2–1.2 µm thick). The tracheoles have a diameter less than 1 µm, and they indent the epidermal cells so that they lie immediately beneath the body surface (Figure 17.19). The tracheoles are not uniform in size, but vary in diameter from 0.2 to 1.0 µm, and spacing and distribution are less uniform than in Trichoptera. Filamentous gills occur on the abdomen of some Trichoptera, Diptera, and Lepidoptera and on the thorax and abdomen of

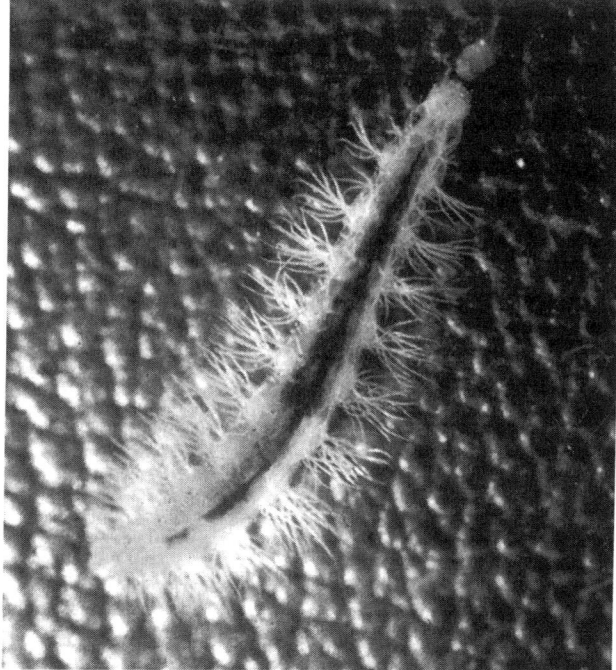

Figure 17.18 Filamentous tracheal gills on the larva of *P. seminealis* (Pyralidae: Lepidoptera) that lives its larval life underwater and obtains air by cutaneous respiration primarily through the tracheal gills (Photograph Courtesy of Dale Habeck. Professor, Entomology and Nematology, University of Florida [deceased].)

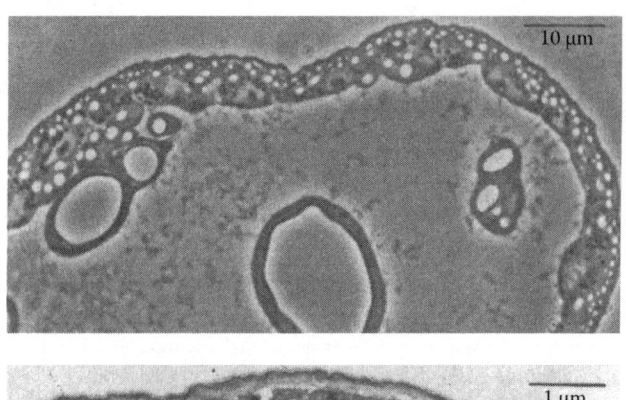
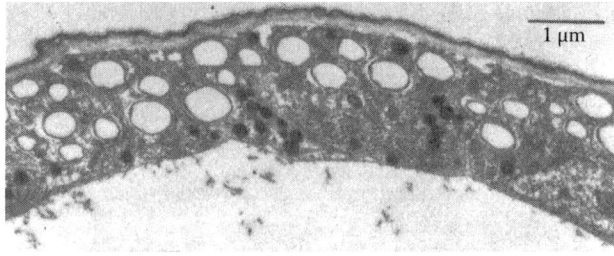

Figure 17.19 Top: Cross section through a gill filament of a *Perla* species (Perlidae: Plecoptera) showing many small tracheoles and a few larger tracheae just beneath the thin cuticle of the gill filament. Bottom: Higher magnification to show more clearly the tracheoles (tubules less than 1 μm in diameter) in a cross section of a gill filament. (From Wichard, W. and Komnick, H., *J. Insect Physiol.*, 20, 2397, 1974. With permission.)

a few Coleoptera. Tracheal gills generally are larval structures, but some Trichoptera pupae have them, and they persist as atrophied, probably nonfunctional, structures in some adults of Trichoptera and Plecoptera.

Pupae of the simuliids (dipterans) develop plastrons on the gill surface at the last larval molt. The plastron fills with air shortly before the larval–pupal ecdysis and, after ecdysis, the plastron expands into its functional appearance before the cuticle hardens. The entire structure of the gill, except a small area at the base, bears a plastron in gills of *Simulium ornatum* (Miller, 1966b).

After O_2 diffuses across the thin cuticle of the typical gill surface and enters the tracheae, it probably is distributed to different body tissues much as in terrestrial insects by diffusion and by body and muscular movements. Some researchers have raised questions about the respiratory function of tracheal gills since some insects from which the gills were removed did not show much, if any, change in behavior or in O_2 consumption. Probably in these insects, there is a high level of general cutaneous respiration that is assisted by the gills.

Movement of water over the gill and body surface of aquatic insects is important in maintaining a fresh supply of oxygenated water in contact with the body, and most use undulations of the body and/or movements of the gills themselves to create ventilatory currents of water. Larvae of some dragonflies (Anisoptera) draw water into the rectum by elastic expansion of the body as dorsoventral compressor muscles relax. Typically there are six main gill folds as extensions of the cuticular intima in the anterior part of the rectum that extract O_2 from the water. The water is pumped out by dorsoventral compression of the abdomen. The rate of ventilation varies with several factors including the O_2 content of the water. About 85% of the water in the rectum is renewed during each pumping cycle, and 25–50 cycles/min have been recorded. Larvae will also come to the surface and ventilate the rectum with air when oxygen content of the water is very low.

17.8 RESPIRATION IN ENDOPARASITIC INSECTS

Many hymenopterans and dipterans are parasitic on other insects and have been little studied with respect to respiration, perhaps for the obvious reasons that they are usually small and hidden in the body of the host. Cutaneous respiration is probably very important. The first instar often has a fluid-filled tracheal system, necessitating cutaneous respiration. Although air displaces the fluid at the molt into the second instar, the mechanism for clearing the fluid from the system is not known; possibly, the fluid is reabsorbed into the tissues of the larva. The spiracles become functional only just before the larva is ready to leave the host to pupate. Many chalcid wasps and tachinid flies that hatch from eggs laid on the surface of the host insect eat their way into the host. They have a metapneustic system and orient the posterior pair of spiracles at the body surface of the host so that they breathe air directly. Chalcid wasps remain in contact with the hollow pedicel of the egg from which they hatched at the host's integumental surface. Some tachinids stimulate the host's integument to become invaginated into a sheath around the parasite, leaving it with an air opening at the host's surface. The larvae of botflies, *Hypoderma* spp., migrate to the skin of the vertebrate host where it bores a tiny opening to the surface through which gas exchange occurs. In earlier instars that are migrating, respiration is presumably cutaneous.

17.9 RESPIRATORY PIGMENTS

Only a few insects have **respiratory pigments** in the hemolymph or cells. *Chironomus* spp. larvae have a small **hemoglobin** molecule composed of two chains with a molecular weight of 31,400. The hemoglobin is present in the hemolymph, but it is not in the hemocytes. It has a high affinity for oxygen and is 50% saturated at a $_pO_2$ of 0.6 mmHg at 17°C. Its loading curve is not shifted by CO_2 tension and temperature as in the case of vertebrate hemoglobin. The implication of such strong affinity of the hemoglobin for O_2 is that it will unload O_2 to the tissues only at a site of extremely low oxygen deficit. Moreover, the expected high CO_2 at such a site will not promote unloading as in the case of vertebrate hemoglobin. Its principal function may be to aid recovery from anaerobic conditions and to provide limited O_2 to some critical tissues, such as the nervous system.

Hemoglobin occurs within certain cells of *Gasterophilus* spp. (horse bots) and in larvae and adults of some beetles (Notonectidae, *Anisops*, and *Buenoa* spp.) (Mill, 1974). In contrast to the situation in *Chironomus* larvae, the hemoglobin of the beetle *Anisops pellucens* is only 50% saturation at a $_pO_2$ of 28 mmHg at 24°C, giving it much more functional potential. The beetle may get up to 75% of the O_2 consumed during a normal dive from its hemoglobin (Miller, 1966b). Although CO_2 tension causes little tendency to unload, increased temperature does shift the curve to the right and increases unloading of O_2 in actively working tissues such as muscles.

17.10 RESPIRATION IN EGGS AND DEVELOPING EMBRYOS

The embryo developing inside the egg must obtain sufficient oxygen for development. Contradictory to an intuitive approach, the majority of aquatic and semiaquatic insects lay eggs with no special respiratory structures incorporated into the shell, while eggs of a majority of terrestrial insects contain special structures for respiration, including an extensive, inner chorionic meshwork that can function as a plastron when the egg is submerged in well-aerated water (Hinton, 1969). Some eggs have special respiratory structures, but even in these, the eggs do not have a large enough water–air interface per unit weight to enable effective function of the air space as a plastron for very long. Plastrons on eggs laid in a terrestrial environment may help prevent an O_2 deficiency if the eggs are subjected to only short periods of wetting from dew, rain, or temporary flooding. A 4-mm diameter raindrop falling on an egg can exert up to about 30 cmHg pressure, but for only a fraction of a second, so plastrons of eggs exposed directly to rainfall usually do not become wet.

Gas exchange in eggs with no special respiratory structure occurs by simple diffusion through interstices in the eggshell (Hinton, 1969). Developing eggs of *M. sexta* experienced O_2 limitation even at normoxic conditions (normoxia equals 21 kPa at sea level), and metabolic rates of eggs were depressed at higher temperatures under hyperoxic conditions (Woods and Hill, 2004). These authors speculate that the need to conserve water in the egg resulted in a trade-off with oxygen diffusion through the rather impermeable eggshell. They also suggest that the higher O_2 level (as much as 35%) in the air during the late Carboniferous may have been a factor in larger eggs and larger insects during the time and that subsequent evolution led to smaller eggs and smaller insects.

The eggs of many terrestrial insects frequently are laid in wet environments, including decaying organic matter, animal manure, fruits, leaves, and stems of plants. To be effective, the plastron must resist the action of naturally occurring surfactants that often are present in animal dung and decaying flesh. Eggs laid in sites containing natural surfactants resist wetting better than those eggs lain in sites where surface-active agents are less likely to be encountered. Clearly this greater resistance to wetting in the presence of surface-active agents is an adaptation to the egg environment. The eggshell, the **chorion**, may contain many small, twisting tubules called **aeropyles** that connect the inner part of the eggshell with the outside air. The aeropyles present little open surface to the air interface (e.g., 536 μm^2 in eggs of the hemipteran *R. prolixus*), and water loss from the egg is not increased greatly.

Some eggs have the plastron elevated on a **respiratory horn** (Hinton, 1961) that may be up to 10–13 mm long, with the plastron covering most of the surface of the horn. When the egg is submerged completely in water, the length of the longer horns make the gradient for gas exchange favorable only in the proximal part of the horn. It may be that long horns are useful as a conduit to atmospheric air when the egg is not too deep in water.

17.11 NONRESPIRATORY FUNCTIONS OF THE TRACHEAL SYSTEM

Tracheae serve the functions of connective tissue, typically tying cells and tissues together. Organs, such as Malpighian tubules, frequently are tied to each other and to other structures, such as the gut, by tracheae. Tracheae are important as a structural base for at least two important endocrine tissues: (1) the diffuse cluster of cells making up the prothoracic glands that are attached to the prothoracic tracheae near the prothoracic spiracle in larvae of Lepidoptera and (2) the epitracheal glands attached to the ventral surface of the major ventrolateral tracheal trunk connection to each spiracle in lepidopterous larvae (Žitňan et al., 1996).

Air sacs that back sound-producing organs in insects, particularly in cicadas (Homoptera) that produce very loud sounds, are important to loudness or sound production and its modulation (Claridge, 1985). A large air sac backs up to the tymbal located on each side of the first abdominal segment of male cicadas. The size of the air sac varies with different species, and its size and tuning are partly responsible for the species-specific quality of the sound produced by male cicadas. In many cicada species, the air sacs are broadly tuned to resonate over a range of frequencies, spanning the natural vibration frequency of the tymbal (Pringle, 1954). Some cicada species are able to damp the air sac resonance and produce complex pulses of sound. Most cicada calls generally have a characteristic sound frequency varying from 4 to 7 kHz (Pringle, 1954), but an Australian species, *Cystosoma saundersii*, is able to call and resonate at about 1 kHz because of a very large air sac in this species. The interconnected prothoracic tracheae extending across the prothorax and into the prothoracic legs of crickets act as a resonator of sounds. They aid the cricket

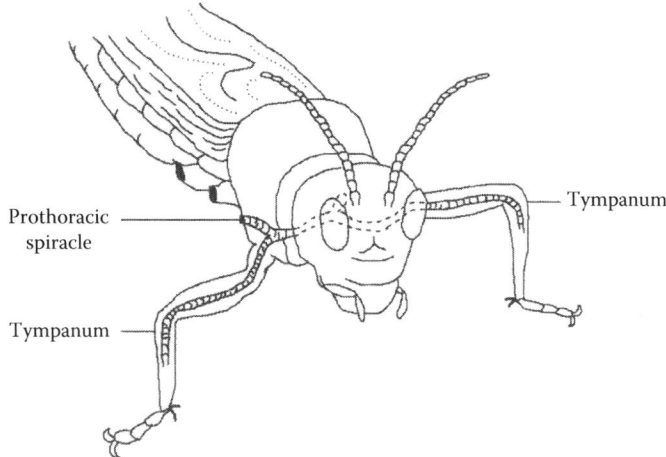

Figure 17.20 Drawing of the interconnected prothoracic leg tracheae forming a pathway between the two tympanic organs located just below the tibia of the prothoracic legs of the cricket *Teleogryllus commodus*. The large tracheal pathway acts as a sounding board to increase sensitivity and allows increased directional sensitivity to sound waves impinging upon the two tympanal membranes. (Drawing modified from Hill, K.G. and Boyan, G.S., *Nature* (*Lond.*), 262, 390, 1976.)

in discriminating direction of sounds reaching the tympanal membrane located just below the joint of the femur with the tibia (i.e., the knee joint) on each prothoracic leg (Figure 17.20). Hissing cockroaches expel air forcefully from certain spiracles and produce a hissing sound when they are disturbed. The sound is apparently a warning signal intended to scare away a potential predator or parasite. Lubber grasshoppers, *R. guttata*, release a yellow-colored foam containing quinones from spiracle 4 on the side of the attack when an ant attacks or when stimulated by probing in the laboratory (Roth and Stay, 1958).

17.12 REVIEW AND SELF-STUDY QUESTIONS

1. What are the differences between tracheae and tracheoles?
2. What functions do air sacs play in the body of some insects?
3. Describe the structure of taenidia in tracheae.
4. Briefly describe how a new tracheole is formed in *S. coprophila*, the model for new tracheal development.
5. In what ways are tracheae adapted morphologically for supplying large amounts of air to thoracic muscles involved in flight?
6. Describe the tokus and its function in some insects.
7. Describe the active ventilation of tracheae. When might simple diffusion of gases be sufficient for insects?
8. What morphological adaptations are there for gas exchange within the body tissues?
9. What are coelopulses and what is their function?
10. What is discontinuous gas exchange, and how might it benefit an insect?
11. Describe gas exchange in a typical aquatic insect?
12. What is an incompressible gas gill? What is its typical structure?
13. Describe the way in which some insects use aquatic plants as a source of oxygen.
14. Describe how filamentous tracheal gills enable the larva of *Parapoynx seminealis*, a pyralid moth larva, to live its entire larval life under water? How does it breathe?
15. What is an aeropyle?
16. What are some nonrespiratory functions of the tracheal system?

REFERENCES

Afzelius, B.A. and N. Gonnert. 1972. Intramitochondrial tracheoles in flight muscle from the hornet *Vespa crabro*. *J. Submicrosc. OSC Cytol.* 4: 16.

Bordereau, C. 1975. Croissance des Trachées au cours de L'evolution de la physogastrie chez la reine des termites superieurs (Isoptera: Termitidae). *Int. J. Insect Morphol. Embryol.* 4: 431–465.

Buckingham, G.R. 1994. Biological control of aquatic weeds, in D. Rosen, F.D. Bennett, and J.L. Capinera (eds.), *Pest Management in the Subtropics*, vol. 1. Intercept Limited, Andover, U.K., pp. 413–480.

Burkett, B.N. and H.A. Schneiderman. 1967. Control of spiracles in silk moths by oxygen and carbon dioxide. *Science* 156: 1604–1606.

Burkett, B.N. and H.A. Schneiderman. 1974. Discontinuous respiration in insects at low temperatures: Intratracheal changes and spiracular valve behavior. *Biol. Bull.* 147: 294–310.

Bustami, H.P. and R. Hustert. 2000. Typical ventilatory pattern of the intact locust is produced by the isolated CNS. *J. Insect Physiol.* 46: 1285–1293.

Callier, V. and H.F. Nijhout. 2011. Control of body size by oxygen supply reveals size-dependent and size-independent mechanisms of molting and metamorphosis. *Proc. Natl. Acad. Sci. USA* 108: 14664–14669.

Case, J.F. 1957. The median nerves and cockroach spiracular function. *J. Insect Physiol.* 1: 85–94.

Chappell, M.A. and G.L. Rogowitz. 2000. Mass, temperature and metabolic effects on discontinuous gas exchange cycles in eucalyptus-boring beetles (Coleoptera: Cerambycidae). *J. Exp. Biol.* 203: 3809–3820.

Chown, S.L. 2011. Discontinuous gas exchange: New perspectives on evolutionary origins and ecological implications. *Func. Ecol.* 25(6): 1163–1168. doi:10.1111/j.1365-2435.2011.01879x.

Chown, S.L. and P. Holter. 2000. Discontinuous gas exchange cycles in *Aphodius fossor* (Scarabaeidae): A test of hypotheses concerning origins and mechanisms. *J. Exp. Biol.* 203: 397–403.

Claridge, M.F. 1985. Acoustic signals in the Homoptera: Behavior, taxonomy, and evolution. *Annu. Rev. Entomol.* 30: 297–317.

Cuda, J. Personal communication, October 2001. Professor of Entomology and Nematology, University of Florida, Gainesville, FL.

Dadant & Sons (eds.). 1975. *The Hive and the Honey Bee*, revised ed. Dadant & Sons, Publishers, Hamilton, IL.

Deonier, D.L. 1971. A systematic and ecological study of Nearctic *Hydrellia* (Diptera: Ephydridae). *Smithson. Contrib. Zool.* 68: 11–47.

Deonier, D.L. 1993. A critical taxonomic analysis of the *Hydrellia pakistanae* species group (Diptera: Ephydridae). *Insecta Mundi* 7: 133–158.

Duncan, F.D. and R.D. Newton. 2000. The use of the anaesthetic, enflurane, for determination of metabolic rates and respiratory parameters in insects, using the ant, *Camponotus maculatus* (Fabricius) as the model. *J. Insect Physiol.* 46: 1529–1532.

Dyby, S.D. 1998. Method for visualizing the tracheal system of newly hatched insects. *Ann. Entomol. Soc. Am.* 91: 350–352.

Edwards, G.A. 1953. Respiratory systems, in K.D. Roeder (ed.), *Insect Physiology*. John Wiley & Sons, New York, pp. 55–95.

Förster, T.D. and H.A. Woods. 2012. Mechanisms of tracheal filling in insects. *Biol. Rev.* 88(1): 1–14. doi:10.1111/j.1469-185X.2012.00233.x.

Ghiradella, H. 1978. Reinforced tracheoles in three firefly lanterns: Further reflections on specialized tracheoles. *J. Morphol.* 157: 281–300.

Gibbs, A.G. and R.A. Johnson. 2004. The role of discontinuous gas exchange in insects: The chthonic hypothesis does not hold water. *J. Exp. Biol.* 207: 3477–3482.

Greenlee, K.J. and J.F. Harrison. 1998. Acid-base and respiratory responses hypoxia in the grasshopper *Schistocerca americana*. *J. Exp. Biol.* 201: 2843–2855.

Greenlee, K.J. and J.F. Harrison. 2004a. Development of respiratory function in the American locust *Schistocerca americana*. I. Across-instar effects. *J. Exp. Biol.* 207: 497–508.

Greenlee, K.J. and J.F. Harrison. 2004b. Development of respiratory function in the American locust *Schistocerca americana*. II. Within-instar effects. *J. Exp. Biol.* 207: 509–517.

Greenlee, K.J. and J.F. Harrison. 2005. Respiratory changes throughout ontogeny in the tobacco hornworm caterpillar, *Manduca sexta*. *J. Exp. Biol.* 208: 1385–1392.

Greenlee, K.J., J.R. Henry, S.D. Kirkton, M.K. Westneat, K. Fezzaa, W.-K. Lee, and J.F. Harrison. 2009. Synchrotron imaging of the grasshopper tracheal system: Morphological and physiological components of tracheal hypermetry. *Am. J. Physiol. Regul. Integr. Comp. Physiol.* 297: R1343–R1350.

Greenlee, K.J., C. Nebeker, and J.F. Harrison. 2007. Body size-independent safety margins for gas exchange across grasshopper species. *J. Exp. Biol.* 210: 1288–1296.

Greenlee, K.J., J.J. Socha, H.B. Eubanks, P. Pedersen, W.-K. Lee, and S.D. Kirkton. 2013. Hypoxia-induced compression in the tracheal system of the tobacco hornworm caterpillar, *Manduca sexta*. *J. Exp. Biol.* 216: 2293–2301.

Groenewald, B., S.K. Hetz, S.L. Chown, and J.S. Terblanche. 2012. Respiratory dynamics of discontinuous gas exchange in the tracheal system of the desert locust, *Schistocerca gregaria*. *J. Exp. Biol.* 215: 2301–2307.

Gulinson, S.L. and J.F. Harrison. 1996. Control of resting ventilation rate in grasshoppers. *J. Exp. Biol.* 199: 379–389.

Hadley, N.F. 1994. Ventilatory patterns and respiratory transpirations in adult terrestrial insects. *Physiol. Zool.* 67: 175–184.

Hadley, N.F. and M.C. Quinlan. 1993. Discontinuous CO_2 release in the eastern lubber grasshopper *Romalea guttata* and its effect on respiratory transpiration. *J. Exp. Biol.* 177: 169–180.

Harrison, J., M.R. Frazier, J.R. Henry, A. Kaiser, C.J. Klok, and B. Rascón. 2006. Responses of terrestrial insects of hypoxia or hyperoxia. *Respir. Physiol. Neurobiol.* 154: 4–17.

Harrison, J.F. 1997. Ventilatory mechanism and control in grasshoppers. *Am. Zool.* 37: 73–81.

Harrison, J.F., N.F. Hadley, and M.C. Quinlan. 1995. Acid-base status and spiracular control during discontinuous ventilation in grasshoppers. *J. Exp. Biol.* 198: 1755–1763.

Harrison, J.F., J.J. Lafreniere, and K.J. Greenlee. 2005. Ontogeny of tracheal dimensions and gas exchange capacities in the grasshopper, *Schistocerca americana*. *Comp. Biochem. Physiol. Part A* 141: 372–380.

Harrison, J.F. and J.R.B. Lighton. 1998. Oxygen-sensitive flight metabolism in the dragonfly *Erythemis simplicicollis*. *J. Exp. Biol.* 201: 1739–1744.

Harrison, J.F., J.S. Waters, A.J. Cease, J.M. VandenBrooks, V. Callier, C.J. Klok, K. Schaffer, and J.J. Socha. 2013. How Locusts Breathe. *Physiol.* 28: 18–27. doi:10.1152/physiol.00043.2012.

Heinrich, B. 1970. Thoracic temperature stabilization by blood circulation in a free-flying moth. *Science* 168: 580–582.

Heinrich, B. 1996. *The Thermal Warriors: Strategies of Insect Survival*. Harvard University Press, Cambridge, MA, 221pp.

Hetz, S. and T.J. Bradley. 2005. Insect breathe discontinuously to avoid oxygen toxicity. *Nature* 433: 516–519.

Hill, K.G. and G.S. Boyan. 1976. Directional hearing in crickets. *Nature (London)* 262: 390–391.

Hinton, H.E. 1961. The structure and function of the eggshell in the Nepidae (Hemiptera). *J. Insect Physiol.* 7: 224–257.

Hinton, H.E. 1964. The respiratory efficiency of the spiracular gill of *Simulium*. *J. Insect Physiol.* 10: 73–80.

Hinton, H.E. 1969. Respiratory systems of insect egg shells. *Annu. Rev. Entomol.* 14: 343–368.

Jarial, M.S. and L. Engstrom. 1995. Fine structure of the spiracular glands in larval *Drosophila melanogaster* (Meig.) (Diptera: Drosophilidae). *Int. J. Insect Morphol. Embryol.* 24: 1–12.

Kauffman, S.A. 1993. *The Origins of Order, Self Organization and Selection in Evolution*. Oxford University Press, Oxford, U.K.

Keilin, D. 1944. Respiratory system and respiratory adaptations in larvae and pupae of Diptera. *Parasitology* 36: 1–66.

Keister, M. 1948. The morphogenesis of the tracheal system of *Sciara*. *J. Morphol.* 83: 373–423.

Keister, M. 1963. The anatomy of the tracheal system of *Camponotus pennsylvanicus* (Hymenoptera: Formicidae). *Ann. Entomol. Soc. Am.* 56: 336–340.

Kirkton, S.D., L.E. Hennessey, B. Duffy, M.M. Bennett, W.-K. Lee, and K.J. Greenlee. 2012. Intermolt development reduces oxygen delivery capacity and jumping performance in the American locust (*Schistocerca americana*). *J. Comp. Physiol. B* 182: 217–230.

Kirkton, S.D., J. Niska, and J.F. Harrison. 2005. Ontogenetic effects on aerobic and anaerobic metabolism during jumping in the American locust, *Schistocerca americana*. *J. Exp. Biol.* 208: 3003–3012.

Komai, Y. 1998. Augmented respiration in a flying insect. *J. Exp. Biol.* 201: 2359–2366.

Komai, Y. 2001. Direct measurement of oxygen partial pressure in a flying bumblebee. *J. Exp. Biol.* 204: 2999–3007.

Krolikowsi, K. and J.F. Harrison. 1996. Haemolymph acid-base status, tracheal gas levels and the control of post-exercise ventilation rate in grasshoppers. *J. Exp. Biol.* 199: 391–399.

Lehmann, F.-O., M.H. Dickinson, and J. Staunton. 2000. The scaling of carbon dioxide release and respiratory water loss in flying fruit flies (*Drosophila* spp.). *J. Exp. Biol.* 203: 1613–1624.

Lehmann, F.-O. and N. Heymann. 2005. Unconventional mechanisms control cyclic respiratory gas release in flying *Drosophila*. *J. Exp. Biol.* 208: 3645–3654.

Levy, R.I. and H. Schneiderman. 1966a. Discontinuous respiration in insects. II. The direct measurement and significance of changes in tracheal gas composition during the respiratory cycle of silkworm pupae. *J. Insect Physiol.* 12: 83–104.

Levy, R.I., and H. Schneiderman. 1966b. Discontinuous respiration in insects. IV. Changes in intratracheal pressure during the respiratory cycle of silkworm pupae. *J. Insect Physiol.* 12: 465–492.

Lewis, G.W., P.L. Miller, and P.S. Mills. 1973. Neuromuscular mechanisms of abdominal pumping in the locust. *J. Exp. Biol.* 59: 149–168.

Lighton, J.R.B. 1988. Simultaneous measurement of oxygen uptake and carbon dioxide emission during discontinuous ventilation in the tok-tok beetle, *Psammodes striatus*. *J. Insect Physiol.* 34: 361–367.

Lighton, J.R.B. 1990. Slow discontinuous ventilation in the Namib dune-sea ant, *Camponotus detritus* (Hymenoptera, Formicidae). *J. Exp. Biol.* 151: 71–82.

Lighton, J.R.B. 1991. Ventilation in Namib Desert tenebrionid beetles: Mass scaling, and evidence of a novel quantitized flutter phase. *J. Exp. Biol.* 159: 249–268.

Lighton, J.R.B. 1994. Discontinuous ventilation in terrestrial insects. *Physiol. Zool.* 67: 142–162.

Lighton, J.R.B. 1996. Discontinuous gas exchange in insects. *Annu. Rev. Entomol.* 41: 309–324.

Lighton, J.R.B. 1998. Notes from underground: Towards ultimate hypotheses of cyclic, discontinuous gas-exchange in tracheate arthropods. *Am. Zool.* 38: 483–491.

Lighton, J.R.B. and D. Berrigan. 1995. Questioning paradigms: Caste-specific ventilation in harvester ants, *Messor pergandei* and *M. julianus* (Hymenoptera: Formicidae). *J. Exp. Biol.* 198: 521–530.

Lighton, J.R.B. and F.D. Duncan. 1995. Standard and exercise metabolism and the dynamics of gas exchange in the giant red velvet mite, *Dinothrombium magnificum*. *J. Insect Physiol.* 41: 877–884.

Lighton, J.R.B., L. Fielden, and Y. Rechav. 1993a. Characterization of discontinuous ventilation in a non-insect, the tick *Amblyomma marmoreum* (Acari: Ixodidae). *J. Exp. Biol.* 180: 229–245.

Lighton, J.R.B. and D. Garrigan. 1995. Ant breathing: Testing regulation and mechanism hypotheses with hypoxia. *J. Exp. Biol.* 198: 1613–1620.

Lighton, J.R.B., D. Garrigan, F.D. Duncan, and R.A. Johnson. 1993b. Respiratory water loss during discontinuous ventilation in female alates of the harvester ant, *Pogonomyrmex rugosus*. *J. Exp. Biol.* 179: 233–244.

Lighton, J.R.B. and R. Wehner. 1993. Ventilation and respiratory metabolism in the thermophilic desert ant, *Cataglyphis bicolor* (Hymenoptera: Formicidae). *J. Comp. Physiol.* 163: 12–17.

Locke, M. 1958a. The formation of tracheae and tracheoles in *Rhodnius prolixus*. *Q. J. Microscop. Sci.* 99: 29–46.

Locke, M. 1958b. The coordination of growth in the tracheal system of insects. *Q. J. Microscop. Sci.* 99: 373–391.

Locke, M. 1998. Caterpillars have evolved lungs for hemocyte gas exchange. *J. Insect Physiol.* 44: 1–20.

Locke, M. 2001. The Wigglesworth lecture: Insects for studying fundamental problems in biology. *J. Insect Physiol.* 47: 495–507.

Longley, A. and J.S. Edwards. 1979. Tracheation of abdominal ganglia and cerci in the house cricket *Acheta domesticus* (Orthoptera: Gryllidae). *J. Morphol.* 159: 233–244.

Marais, E., C.J. Klok, J.S. Terblanche, and S.L. Chown. 2005. Insect gas exchange patterns: A phylogenetic perspective. *J. Exp. Biol.* 208: 4495–4507.

Matthews, P.G.D., E.P. Snelling, R.S. Seymour, and C.R. White. 2012. A test of the oxidative damage hypothesis for discontinuous gas exchange in the locust *Schistocerca migratoria*. *Biol. Lett.* 8: 682–684.

Mill, P.J. 1974. Respiration: Aquatic insects, in M. Rockstein (ed.), *The Physiology of Insecta*, vol. VI, 2nd ed. Academic Press, New York, pp. 403–467.

Miller, P.L. 1960a. Respiration in the desert locust. I. The control of ventilation. *J. Exp. Biol.* 37: 224–236.

Miller, P.L. 1960b. Respiration in the desert locust. III. Ventilation and the spiracles during flight. *J. Exp. Biol.* 37: 264–278.

Miller, P.L. 1966a. The regulation of breathing in insects. *Adv. Insect Physiol.* 3: 279–354.

Miller, P.L. 1966b. The function of haemoglobin in relation to the maintenance of neutral buoyancy in *Anisops pellucens* (Notonectidae, Hemiptera). *J. Exp. Biol.* 44: 529–543.

Nespolo, R.F., P. Artacho, and L.E. Castaneda. 2007. Cyclic gas-exchange in the Chilean red cricket: Inter-individual variation and thermal dependence. *J. Exp. Biol.* 210: 668–675.

Oberlander, H. 1980. Morphogenesis in tissue culture: Control by ecdysteroids, in M. Locke and D.S. Smith (eds.), *Insect Biology in the Future.* Academic Press, New York, pp. 423–438.

Pickard, W.F. 1974. Transition regime diffusion and the structure of the insect tracheolar system. *J. Insect Physiol.* 20: 947–956.

Pringle, J.W.S. 1954. A physiological analysis of cicada song. *J. Exp. Biol.* 31: 525–556.

Punt, A. 1950. The respiration in insects. *Physiol. Comp.* 2: 59–74.

Quinlan, M.C. and N.F. Hadley. 1993. Gas exchange, ventilatory patterns, and water loss in two lubber grasshoppers: Quantifying cuticular and respiratory transpiration. *Physiol. Zool.* 66: 628–642.

Rascón, B. and J.F. Harrison. 2005. Oxygen effects on metabolic rate and flight behavior in the American locust *Schistocerca americana. J. Insect Physiol.* 51: 1193–1199.

Roth, L.M. and B. Stay. 1958. The occurrence of para-quinones in some arthropods, with emphasis on the quinone-secreting tracheal glands of *Diploptera punctata* (Blattaria). *J. Insect Physiol.* 1: 305–318.

Rourke, B.C. 2000. Geographic and altitudinal variation in water balance and metabolic rate in a California grasshopper, *Melanoplus sanguinipes. J. Exp. Biol.* 203: 2699–2712.

Schmidt-Nielsen, K. 1997. *Animal Physiology.* Cambridge University Press, Cambridge, MA.

Schneiderman, H.A. and C.M. Williams. 1955. An experimental analysis of the discontinuous respiration of the Cecropia silkworm. *Biol. Bull.* 109: 123–143.

Shelton, T.G. and A.G. Appel. 2000. Cyclic CO_2 release and water loss in the western drywood termite (Isoptera: Kalotermitidae). *Ann. Entomol. Soc. Am.* 93: 1300–1307.

Shelton, T.G. and A.G. Appel. 2001. Carbon dioxide release in *Coptotermes formosanus* Shiraki and *Reticulitermes flavipes* (Kollar): Effects of caste, mass, and movement. *J. Insect Physiol.* 47: 213–224.

Sláma, K. 1988. A new look at insect respiration. *Biol. Bull. Woods Hole* 175: 289–300.

Sláma, K. 1994. Regulation of respiratory acidemia by the autonomic nervous system (coelopulse) in insects and ticks. *Physiol. Zool.* 67: 163–174.

Sláma, K. 1999. Active regulation of insect respiration. *Ann. Entomol. Soc. Am.* 92: 916–929.

Sláma, K. 2000. Extracardiac vs. cardiac haemocoelic pulsations in pupae of the mealworm (*Tenebrio molitor* L.). *J. Insect Physiol.* 46: 977–992.

Snodgrass, R.E. 1935. *Principles of Insect Morphology.* McGraw-Hill, New York.

Snyder, G.K., B. Sheafor, D. Scholnick, and C. Farrelly. 1995. Gas exchange in the insect tracheal system. *J. Theor. Biol.* 172: 199–207.

Socha, J.J. and F. De Carlo. 2008. Use of synchrotron tomography to image naturalistic anatomy in insects. *Proc. SPIE* 7078: 70780A. doi:10.1117/12.795210.

Socha, J.J., M.W. Eestneat, J.F. Harrison, J.S. Waters, and W.-K. Lee. 2007. Real-time phase-contrast x-ray imaging: A new technique for the study of animal form and function. *BMC Biol.* 5: 6. doi:10.1186/1741-7007-5-6.

Socha, J.J., T.D. Förster, and K.J.Greenlee. 2010. Issues of convection in insect respiration: Insights from synchrotron Z-ray imaging and beyond. *Respir. Physiol. Neurobiol.* 173(Supplement): S65–S73.

Terblanche, J.S., E. Marais, S.K. Hetz, and S.L. Chown. 2008. Control of discontinuous gas exchange in *Samia cynthia*: Effects of atmospheric oxygen, carbon dioxide and moisture. *J. Exp. Biol.* 211: 3272–3280.

Timmins, G.S., F.J. Robb, C.M. Wilmot, S.K. Jackson, and H.M.Swartz. 2001. Firefly flashing is controlled by gating oxygen to light-emitting cells. *J. Exp. Biol.* 204: 2795–2801.

Tsubaki, Y., C. Kato, and S. Shintani. 2006. On the respiratory mechanism during underwater oviposition in a damselfly, *Calopteryx cornelia* Selys. *J. Insect Physiol.* 52: 499–505.

Van der Kloot, W.G. 1963. The electrophysiology and the nervous control of the spiracular muscle of pupae of the giant silkmoths. *Comp. Biochem. Physiol.* 9: 317–333.

Vogt, J.T. and A.G. Appel. 2000. Discontinuous gas exchange in the fire ant, *Solenopsis invicta* Buren: Caste differences and temperature effects. *J. Insect Physiol.* 46: 403–416.

Wasserthal, L.T. 2001. Flight-motor-driven respiratory air flow in the hawkmoth *Manduca sexta. J. Exp. Biol.* 204: 2209–2220.

Waters, J.S., W.-K. Lee, M.W. Westneat, and J.J. Socha. 2013. Dynamics of tracheal compression in the horned passalus beetle. *Am. J. Physiol. Regul. Integr. Comp. Physiol.* 304: R621–R627.

Webster, M.R., R. De Vita, J.N. Twigg, and J.J. Socha. 2011. Mechanical properties of tracheal tubes in the American cockroach (*Periplaneta americana*). *Smart. Mater. Struct.* 20: doi:10.1088/0964-1726/20/9/094017.

Weis-Fogh, T. 1964a. Functional design of the tracheal system of flying insects as compared with the avian lung. *J. Exp. Biol.* 41: 207–226.

Weis-Fogh, T. 1964b. Diffusion in insect wing muscle, the most active tissue known. *J. Exp. Biol.* 41: 229–256.

Weis-Fogh, T. 1967. Respiration and tracheal ventilation in locusts and other flying insects. *J. Exp. Biol.* 47: 561–587.

Westneat, M.W., O. Betz, R.W. Blob, K. Fezzaa, W.J. Cooper, and W. Lee. 2003. Tracheal respiration in insects visualized with synchrotron x-ray imaging. *Science* 299: 558–560.

Westneat, M.W., J.J. Socha, and W.-K. Lee. 2008. Advances in biological structure, function, and physiology using synchrotron x-ray imaging. *Annu. Rev. Physiol.* 70: 119–142.

Whitten, J.M. 1972. Comparative anatomy of the tracheal system. *Annu. Rev. Entomol.* 17: 373–402.

Wichard, W. and H. Komnick. 1974. Structure and function of the respiratory epithelium in the tracheal gills of stonefly larvae. *J. Insect Physiol.* 20: 2397–2406.

Wigglesworth, V.B. 1935. Regulation of respiration in the flea, *Xenopsylla cheopsis*, Roths (Pulicidae). *Proc. R. Soc. Lond. B* 118: 397–419.

Wigglesworth, V.B. 1954. Growth and regeneration in the tracheal system of an insect, *Rhodnius prolixus* (Hemiptera). *Q. J. Microscop. Sci.* 95: 125–137.

Wigglesworth, V.B. 1959. The role of the epidermal cells in the "migration" of tracheoles in *Rhodnius prolixus* (Hemiptera). *J. Exp. Biol.* 36: 632–640.

Wigglesworth, V.B. 1972. *The Principles of Insect Physiology*, 7th ed. Chapman & Hall, London, U.K.

Wigglesworth, V.B. 1981. The natural history of insect tracheoles. *Physiol. Entomol.* 6: 121–128.

Williams, C.M., S.L. Pelini, J.J. Hellmann, and B.J. Sinclair. 2010. Intra-individual variation allows an explicit test of the hygric hypothesis for discontinuous gas exchange in insects. *Biol. Lett.* 6: 274–277.

Woods, H.A. and R.L. Hill. 2004. Temperature-dependent oxygen limitation in insect eggs. *J. Exp. Biol.* 207: 2267–2276.

Zdárek, J., F. Weyda, M.M.B. Chimtawi, and D.L. Denlinger. 1996. Functional morphology and anatomy of the polypneustic lobes of the last larval instar of tsetse flies, *Glossina* spp. (Diptera: Glossinidae). *Int. J. Insect Morphol. Embryol.* 25: 235–248.

Žitňan, D., T.G. Kingan, J.L. Hermesman, and M.E. Adams. 1996. Identification of ecdysis-triggering hormone from an epitracheal endocrine system. *Science* 271: 88–91.

CHAPTER 18

Excretion

PREVIEW

The excretory system consists of two organ systems working together: the Malpighian tubules and the hindgut. The Malpighian tubules typically arise at the junction of the mid- and hindgut. Protons secreted into the Malpighian tubule lumen by a membrane proton pump provide the driving force for urine formation. Potassium ions (K^+) enter Malpighian tubule cells from the hemolymph side through potassium channels, and then they are secreted into the tubule lumen by a membrane antiporter mechanism that exchanges hydrogen ions (H^+) for K^+. Fluid from the hemolymph follows the osmotic gradient created by K^+ movement across tubule cells and carries dissolved substances into the tubule lumen. The tubules transfer the accumulated urine to the hindgut, where selective reabsorption by the ileum and rectum retains necessary substances within the body, while allowing waste products and excesses of useful substances to be voided with the fecal wastes. Specializations in hindgut epithelial cells facilitate reabsorption processes. The excretory system plays a major role in homeostasis of hemolymph, cells, and tissues by helping control levels of electrolytes, water, acid–base equivalents, and nitrogen metabolites. Homeostasis is challenged by food habits, habitat, and metabolic state of the insect. Some excretory products may be stored in the body or cuticle where they may offer protection from predation and parasitism. Insects have a high surface-to-volume ratio and challenge their excretory system to conserve water rather than excrete it. Some insects take large liquid meals, such as vertebrate blood, plant phloem sap, or xylem sap, each food consisting of more water than needed, so they rapidly excrete the water and concentrate the nutrients. A few insects (mainly Lepidoptera and Coleoptera) have a cryptonephridial system of Malpighian tubules in which the distal ends of the tubules are held closely to the surface of the rectum in many loops and folds. The loops are more extensive in insects that live in very dry environments, apparently aiding in water reabsorption.

18.1 INTRODUCTION

Excretion can be defined broadly as any process that eliminates the interaction of harmful substances with cells and tissues. Even useful substances, such as glucose, amino acids, and certain ions, can be harmful if present in excess amounts. Nitrogenous metabolites, ions, water, and ingested chemicals are substances that an insect may need to excrete. Allelochemicals from eaten plant tissues may be excreted from the body or stored in some inert location in the body. For example, *Manduca sexta*, the tobacco hornworm, efficiently excretes nicotine ingested with its host food, the leaves of tobacco plants (Baldwin, 1991), while *Danaus plexippus*, the monarch butterfly, stores cardenolides from milkweed in the cuticle of the larva and adult, where they act as chemical protection from predators (Brower, 1969).

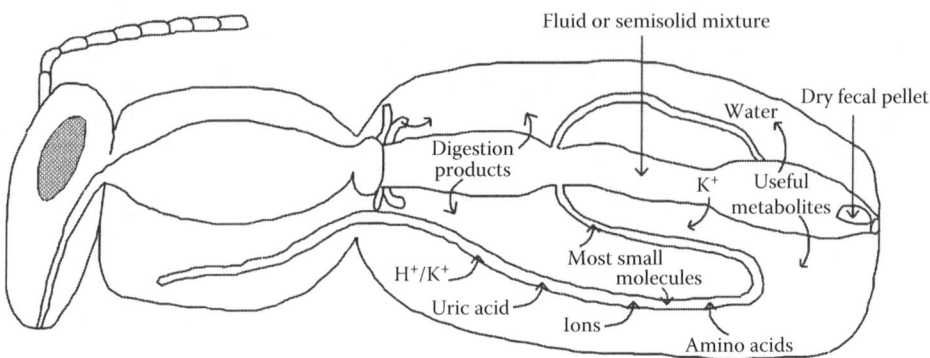

Figure 18.1 A generalized scheme of excretion showing the collection of fluid in the Malpighian tubules (promoted by active secretion of protons followed by an antiporter exchange of K^+ for H^+ in the tubule lumen), and extensive reabsorption of water, K^+, and useful substances from the hindgut, primarily the rectum.

Maddrell (1971) defined **storage excretion** to include materials with future potential use such as glycogen as a storage form of glucose or amino acids stored as proteins. He viewed **deposit excretion** as waste material of no further use that needed to be removed from harmful interaction with the tissues. The distinction between stored and deposit excretion, however, is very subtle and not agreed on by everyone.

In insects, both the **Malpighian tubules** and the **hindgut** function together as excretory organs. The Malpighian tubules collect a filtrate from the hemolymph and pass this primary urine to the hindgut. Additional components are secreted into the excreta by the hindgut, and some substances are reabsorbed into the hemolymph (Figure 18.1). The term **excreta** describes the material actually eliminated from the anus by insects because it is a mixture of undigested materials passing through the gut, substances acted upon and possibly modified by bacterial action in the gut, and urinary materials from the Malpighian tubules. Insect excretion has been reviewed frequently and extensively. The biology of the Malpighian tubules has been reviewed frequently (Maddrell, 1977, 1980; Phillips, 1981; Bradley, 1985; Spring, 1990; Nicolson, 1993; Pannabecker, 1995; O'Donnell, 2008; Beyenbach et al., 2010). Selective reabsorption in the hindgut has been reviewed by Phillips et al. (1986). Bursell (1967) and Cochran (1975, 1985a,b) summarized the overall function of the excretory system in insects.

18.2 MALPIGHIAN TUBULES

Malpighian tubules, the first of the two systems involved in excretion, are long, tubular structures, usually arising at the junction of the mid- and hindgut and terminating blindly in the hemocoel. Some variations in the gross morphology of Malpighian tubule systems are shown in Figure 18.2. The tubules vary in number from 2 to more than 100 in various insect species. Collembola, Aphididae, and some Thysanura lack Malpighian tubules altogether, and other cells and glands take over the functions of excretion. In some members of the Lepidoptera and Coleoptera, the distal ends of the tubules are embedded in the wall of the rectum (see later section). This arrangement, called **cryptosolenic** or **cryptonephridial** tubules, appears to be a modification that aids water conservation. Tracheal connections to Malpighian tubules are numerous and are indicative of a high metabolic demand for oxygen. A small spiral muscle (Figure 18.3) frequently runs along the surface of a tubule, promoting coiling movements that assist proximal flow of fluid and increase hemolymph in contact with the tubule. Several structural types of tubules may occur in the same insect (Figure 18.4).

EXCRETION

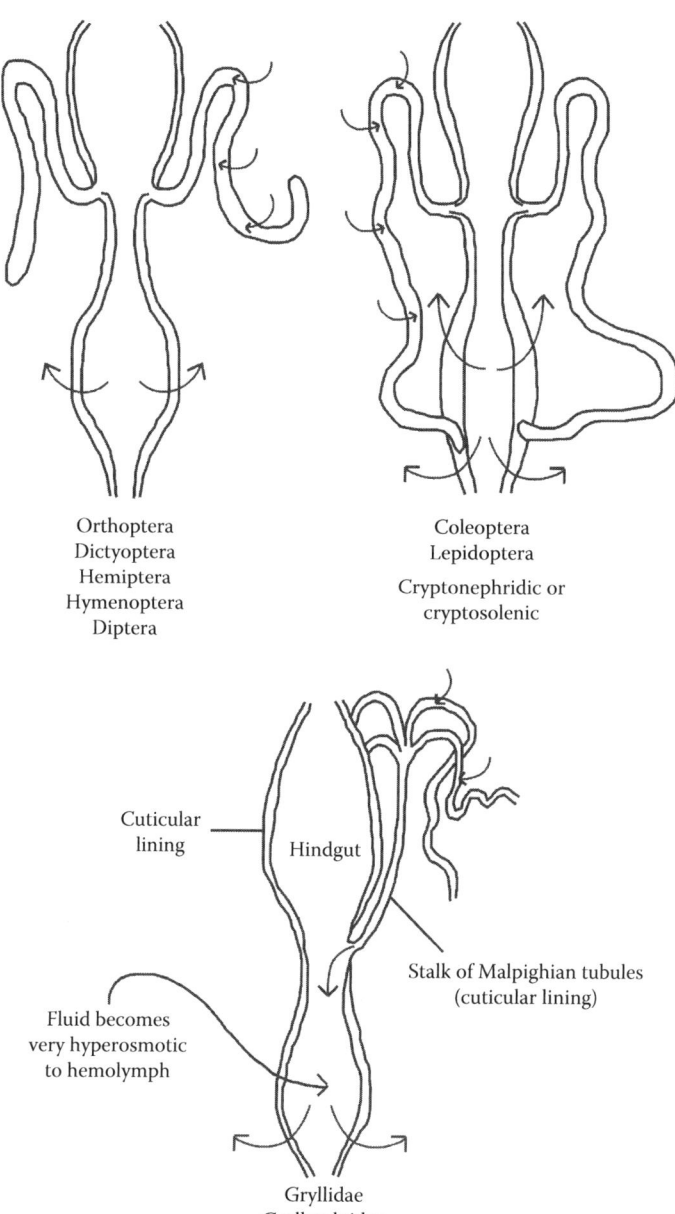

Figure 18.2 Variations in the Malpighian tubule systems of insects. The majority of insects have Malpighian tubules that originate at the junction of the mid- and hindgut and terminate as blind tubules in various regions of the hemocoel. Cryptosolenic tubules in which the distal ends of the tubules lie on top of the terminal part of the hindgut occur in most Lepidoptera and Coleoptera. The Malpighian tubules arise from a short, cuticle-lined stalk in gryllid crickets and mole crickets.

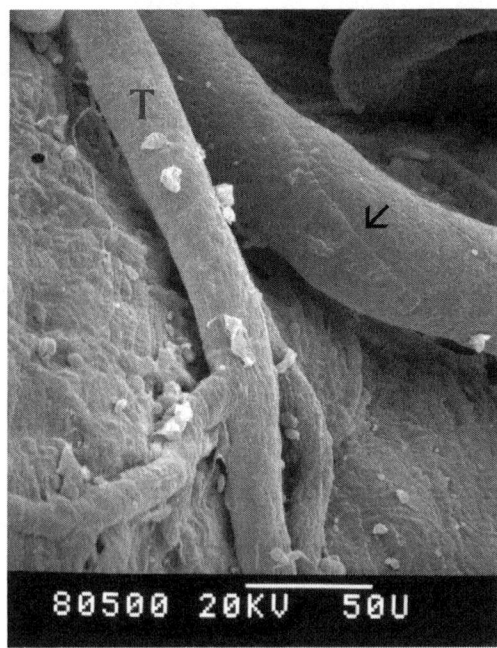

Figure 18.3 A scanning electron microscope (SEM) photo of the small muscle (arrow) that often spirals along the length of a Malpighian tubule of some insects, in this case, the cricket *Gryllus assimilis*. Contraction of the muscle throws the tubules into tight coils in some insects. The muscle probably serves to keep the tubule moving through the hemolymph. T, a tracheal tube supplying air. The bar is 50 μm. (SEM photo by the author.)

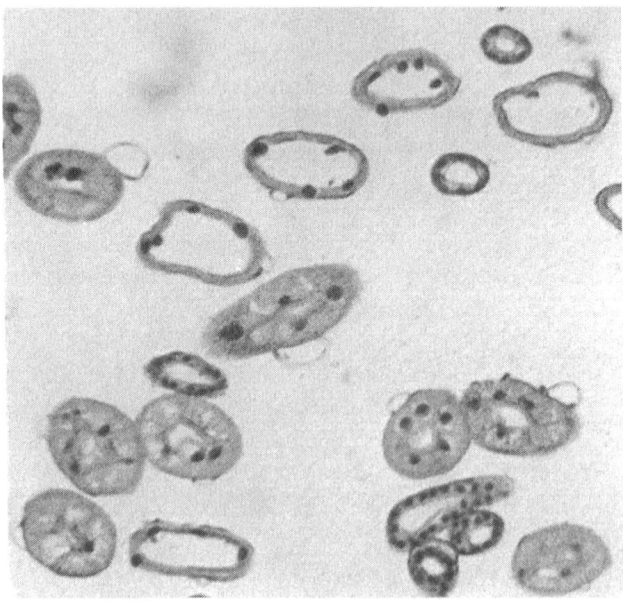

Figure 18.4 A photograph from a histological preparation showing the three types of Malpighian tubules in a mole cricket *Scapteriscus vicinus*. (Photo by the author.)

Malpighian tubules are not only important in excretion. They have many functions in insects (Dow, 2009), including detoxification (Leader and O'Donnell, 2005; O'Donnell, 2009), metabolic functions, and a role in immunity (Davies et al., 2012), and now that the genomes of *Drosophila melanogaster*, *Anopheles gambiae*, and *Apis mellifera* are known, they may serve for organotypic study of human genes (reviewed and discussed by Dow and Davies, 2006).

18.3 MALPIGHIAN TUBULE CELLS

A single layer of epithelial cells (usually comprising two to five cells) surrounds the lumen of a tubule. Several different cell types have been identified, but their specific functions have not been elucidated in many cases. Some, but not all, tubule cells have a **brush border** of **microvilli** on the apical surface, and these have been called **type 1** or **principal tubule cells** (Figure 18.5). They are very important in ion and water regulation. Smaller, **type II stellate cells** are usually present and they play a role in chloride and water flux across the Malpighian tubule, as well (Dow, 2012). Multipotent stem cells that can differentiate into several types of cells in the Malpighian tubules were recently identified in the tubules of *Drosophila* (Singh and Hou, 2009). Cells in the distal half of the tubules in *Rhodnius prolixus* have a brush border on the apical surface of the cells and are involved in formation of the primary urine. Cells in the proximal half of the tubules have a relatively smooth apical surface over which reabsorption occurs, probably by energy-requiring mechanisms (Wigglesworth, 1931).

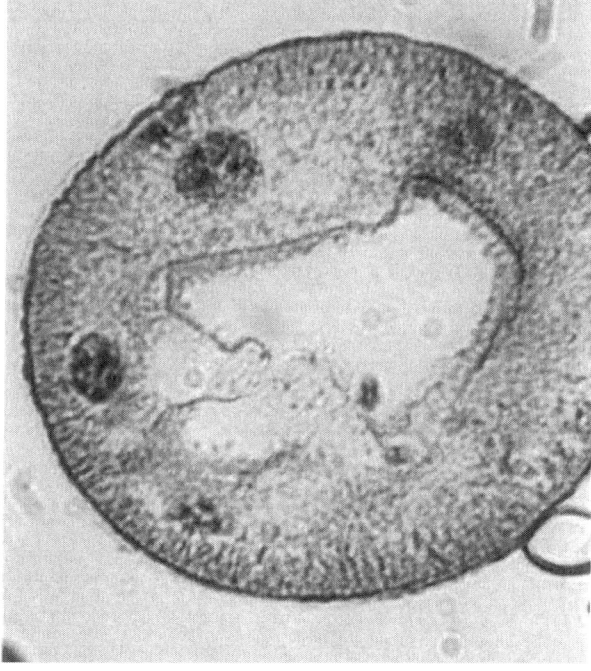

Figure 18.5 A cross section through the primary type of Malpighian tubule in *S. vicinus* showing microvilli on the lumen surface of the cell. (Photo by the author.)

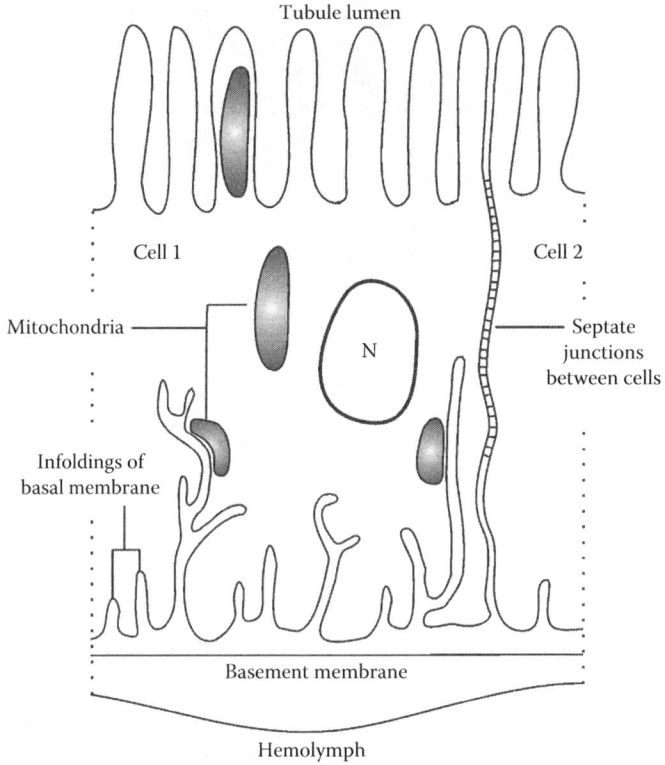

Figure 18.6 The drawing illustrates the general structure of a Malpighian tubule cell from the proximal tubule segment of the last instar of *D. melanogaster* showing extensive basal infoldings, a relatively short path across the narrow cell, and long microvilli on the apical (luminal) surface of the cells. The microvilli often contain large mitochondria, as shown in this illustration.

Tubule cells are thin, sheetlike cells that wrap around the tubule lumen. Water and hemolymph components entering at the basal side of the cells have only a short distance to traverse to reach the apical surface where they may be secreted, or diffuse, into the tubule lumen. Malpighian tubule cells are characterized by extensive infolding of the basal membrane (the membrane on the hemolymph side), creating many twisting channels that reach 5–10 μm or more into the cell (O'Donnell et al., 1985) (Figure 18.6). Potassium ions in the hemolymph enter Malpighian tubule cells through potassium ion gates in these infolded channels in the basal membrane (Nicolson and Isaacson, 1990), and water and dissolved substances in the aqueous medium of the hemolymph follow the osmotic gradient (Figure 18.7). Extensive membrane surfaces presented by both the basal and apical faces of Malpighian tubule cells and large mitochondria are indicative of specialization for active transport as well as passive diffusion.

18.4 FORMATION OF PRIMARY URINE IN MALPIGHIAN TUBULES

The primary urine formed in the lumen of the Malpighian tubules is a **filtrate of the hemolymph** (Ramsay, 1953, 1955a,b, 1956, 1958), and it contains most of the small ions and molecules (sugars, amino acids, ions, as well as other components) that occur in the hemolymph. The urine/hemolymph concentration ratio for many of the filtered substances approaches unity, indicating passive movement across the tubule cell membranes, but some components are actively secreted and their urine/hemolymph ratio is always greater than 1.

EXCRETION

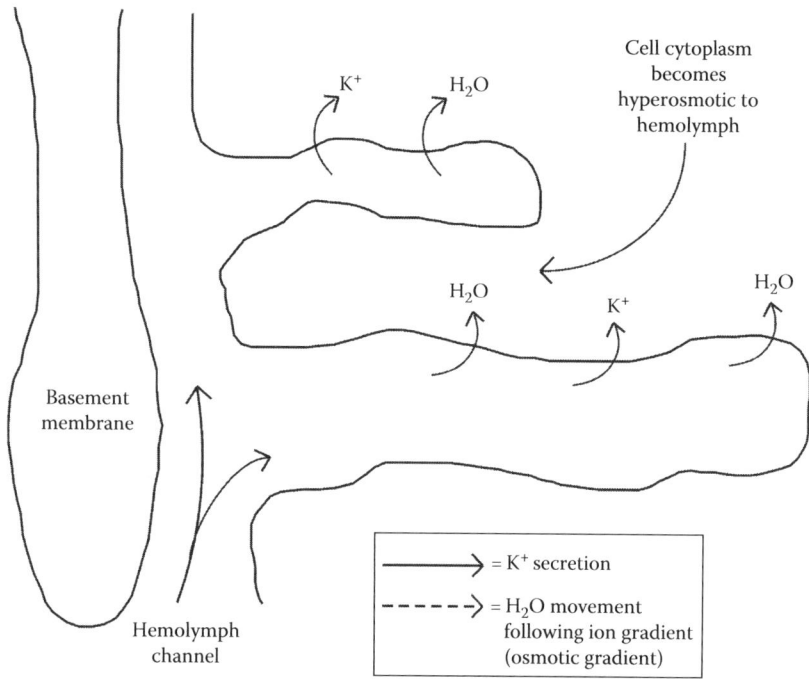

Figure 18.7 A diagram to illustrate the influx of K+ through potassium ion channels in the basal infoldings of Malpighian tubule cells and the passive movement of water and dissolved solutes following the microosmotic gradients setup by K+ movement.

18.5 PROTON PUMP AS A DRIVING MECHANISM FOR URINE FORMATION AND HOMEOSTASIS

Urine formation in Malpighian tubules relies upon a **proton pump**. Proton pumps are ubiquitous in all eukaryotic cells (Huss and Wieczorek, 2009). Such pumps are particularly important not only in the Malpighian tubules but also in the midgut, salivary glands (Baumann and Walz, 2012), and rectum. The pump is located in the apical membrane (the side facing the lumen) of Malpighian tubule cells and actively secretes protons, H+, into the tubule lumen against an electrochemical gradient (Wieczorek et al., 2000; Hopkin et al., 2001; Weng et al., 2003; Muench et al., 2009; Harvey and Xiang, 2012; Tiburcy et al., 2013). The pump consists of a V_1 complex of proteins in the cytoplasm of principal cells (the primary cells in Malpighian tubules) but also in stellate cells (Dow, 2012) of the Malpighian tubules. An ion channel is formed by the transmembrane V_o complex embedded in the lipid bilayer of the apical membrane. The pump causes the tubule lumen to become positive (as much as +30 mV or more in some insects) to the hemolymph and creates highly variable gradients in pH across the apical membrane of principal cells. The proton gradient provides the energy for an **antiporter mechanism** that exchanges K+ for H+ across the apical membrane (Forgac, 1989; Maddrell and O'Donnell, 1992; Weltens et al., 1992; Zhang et al., 1993; Wieczorek et al., 2000; Xiang et al., 2012). The net result is that K+ is secreted into the tubule lumen and concentrated against an electrochemical gradient. Three genes encoding the inward rectifier channels in *Drosophila*, *ir*, *irk2*, and *irk3*, have been identified in principal cells (Evans et al., 2005). In some insects that take blood meals rich in Na+ (e.g., mosquito adults, *R. prolixus*, and tsetse fly adults), Na+ is actively transported by the pump mechanism (Gámez et al., 2012). The pump is probably regulated by dissociation/reassociation of its components (Kane and Parra, 2000; Baumann and Walz, 2012) and likely is under genetic regulation at the transcriptional and posttranscriptional levels (Wieczorek et al., 1999).

Cations, such as K^+ and/or Na^+ and probably anions, must enter the Malpighian tubule cells (at the basolateral surface) from the hemolymph in order for secretion into the lumen at the apical face of cells to continue. Entry of ions at the basolateral membrane surface, however, is not well understood. Secretion of cations (H^+, Na^+, and K^+) across the apical membrane appears to be electrically coupled with Cl^- transport (Beyenbach, 1995; Dijkstra et al., 1995) in the basolateral membrane of tubule cells, providing for balance and steady-state conditions between entry of cations from the hemolymph across the basolateral membrane and secretion across the apical membrane (Beyenbach et al., 2000b). Piermarni et al. (2013) describe three inward-rectifying K^+ channels (Kir) that are expressed in the Malpighian tubules of *Aedes aegypti* and Raphemot et al. (2013) provide some experimental evidence that inhibition of these Kir channels might be a viable way to cause renal failure and control of *A. aegypti* and possibly other mosquitoes that are serious disease vectors.

Chloride ions are transported by a paracellular pathway (i.e., between adjacent cells) and transcellular in some insects such as *A. aegypti* (Beyenbach et al., 2000a,b, 2009; Beyenbach and Piermarini, 2011; Beyenbach, 2012). A diuretic hormone, aedeskinin-III, increases the paracellular conductance of Cl^- in *A. aegypti* (Miyauchi et al., 2013). The rapid increase in Cl^- movement also increases the transepithelial secretion of Na^+ and K^+ as well as water (Beyenbach, 2012). In tubules of *D. melanogaster*, chloride ions move by a transcellular pathway (O'Donnell et al., 1998). There is evidence that the route for Cl^- transport may vary within the same organism in response to variable physiological conditions (Dijkstra et al., 1995) and/or hormonal stimulation (Yu and Beyenbach, 2001).

Based on measurement of transepithelial potentials across Malpighian tubules cells and lumen, Ianowski and O'Donnell (2001) suggest a stoichiometry of $Na^+:K^+:2Cl^-$ cotransport across the basolateral membrane of tubule cells in *R. prolixus*.

Although the secretion of K^+ to the tubule lumen has been known for half a century (Ramsay, 1953, 1955a,b), a cellular/molecular explanation was not known until discovery of the proton pump. The formation of urine volume is highly dependent upon K^+ concentration in the bathing hemolymph or saline, but fluid formation stops even when K^+ concentration in the bathing saline is high if the H^+ pump is inhibited (Bertram et al., 1991; Weltens et al., 1992). Molecules, such as sugars, amino acids, and some allelochemicals in the hemolymph, follow the osmotic gradient created by the transport of K^+ and other ions. A large number of ion and solute (protein-based) carriers have been identified in Malpighian tubule cells and in a number of other cell types (e.g., alimentary cells, fat body) (Carpenter et al., 2012; Hirata et al., 2012; Linser et al., 2012). In addition to secreting K^+, Malpighian tubule cells in some insects also secrete Na^+, some other ions, and organic molecules. One organic molecule, the amino acid proline, is actively secreted into the urine by tubules of the desert locust, *Schistocerca gregaria*, and, after passage into the hindgut, it is used as an energy source for adenosine triphosphate (ATP) production to fuel ion pumps in the hindgut epithelium (Phillips et al., 1994).

The process driven by the proton pump has been called a **standing gradient process** (Berridge and Oschman, 1969). Although it probably accounts for most of the urine formed, there may be additional processes by which substances enter the tubule lumen. Wessing and Eichelberg (1975) suggested that there might be a number of mechanisms operating in various insects to account for some components in the urine, and they presented electron micrograph evidence, which they interpreted as indicative of more than one process operating in tubule cells of *D. melanogaster* (Wessing and Eichelberg, 1978). Additional processes may include transport of substances enclosed in vesicles (Riegel, 1966; Linton and O'Donnell, 2000), free movement of substances through the cell cytoplasm, and passage of substances between adjacent cells by movement through the intercellular space. None of these seems likely to be mutually exclusive of the others, and several mechanisms might operate in the same insect.

The rates of urine formation and ion secretion are controlled by diuretic peptide hormones and certain nonpeptide compounds such as 5-hydroxytryptamine (5-HT or serotonin) (reviewed by O'Donnell and Spring, 2000). When maximally stimulated, the Malpighian tubules can secrete a volume of fluid equal to their cell volume every 10 s, a record rate of secretion in biology (Maddrell, 1991; Evans et al., 2005). The peptide compounds fall into two major classes, those similar to vertebrate corticotropin-releasing factor, called CFC-related peptides, and smaller kinins. The CFC-type peptides range in size from 30 to 46 amino acids, while the kinins are smaller and comprise 6 to 15 amino acids. Although both types stimulate urine formation, they act through different mechanisms. CFC peptides (and 5-HT) stimulate adenylate cyclase and raise the level of cyclic adenosine monophosphate (cAMP), while the kinins that have been studied to this point activate the Ca^{2+}-signaling pathway. Malpighian tubule secretion rate typically is controlled by the interaction of several of these compounds. Diuretic factors may inhibit synergism (i.e., greater activity in combination than the additive effects of each alone) or alternatively, effects of multiple compounds may only be additive, but cation and anion pathways are controlled separately by different second messengers. In one case, an inhibitory interaction is known. In the blood-feeding hemipteran *R. prolixus*, two factors, 5-HT and diuretic hormone (DH), act synergistically to stimulate urine formation in the tubules, but a third peptide, cardioacceleratory peptide 2b (CAP_{2b}), acts as an antidiuretic hormone when part of the mixture. Its effects are mediated through stimulation of cyclic guanosine monophosphate (cGMP) that then inhibits the action of 5-HT. Synergistic action of these hormones probably benefits an insect by reducing quantities of hormones released and ensuring that all tubules respond rapidly. In insects that take a large fluid meal, such as mosquitoes, some hemipterans, and phloem and xylem feeders, synergistic action of hormones may compensate for hemolymph dilution as water from the large meal is absorbed into the hemolymph.

The primary urine formed by the Malpighian tubules is **isosmotic** or sometimes slightly **hyposmotic** to the hemolymph. Malpighian tubules are not capable of producing primary urine that is appreciably **hyperosmotic** to the hemolymph. The proximal tubules may modify urine by reabsorption of some substances (e.g., in *R. prolixus*), but many insects transfer the tubule fluid to the hindgut with few or no changes in its chemical composition or volume. The hindgut then proceeds to concentrate waste products by reabsorbing water and useful substances.

The use of isolated tubules (Figures 18.8 and 18.9), a technique originally devised by Ramsay (1954), continues to be an important research technique for elucidating physiology of

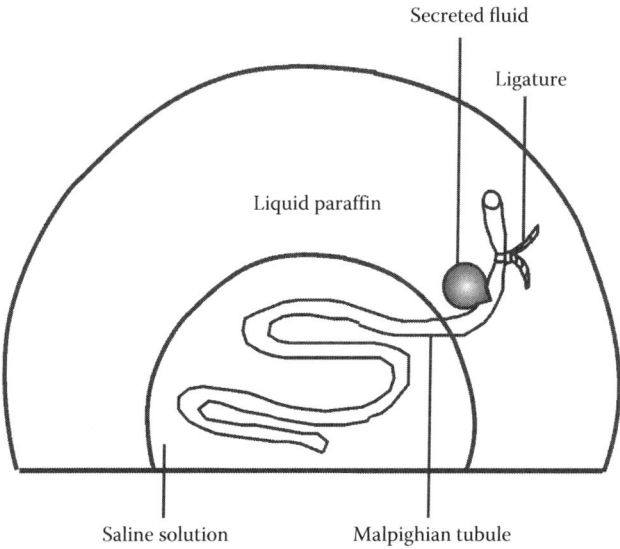

Figure 18.8 An illustration of the isolated Malpighian tubule technique for studying urine formation in response to hormones, ions, inhibitors, or other agents dissolved in the bathing saline. (Drawing modified from Ramsay, J.A., *J. Exp. Biol.*, 31, 104, 1954.)

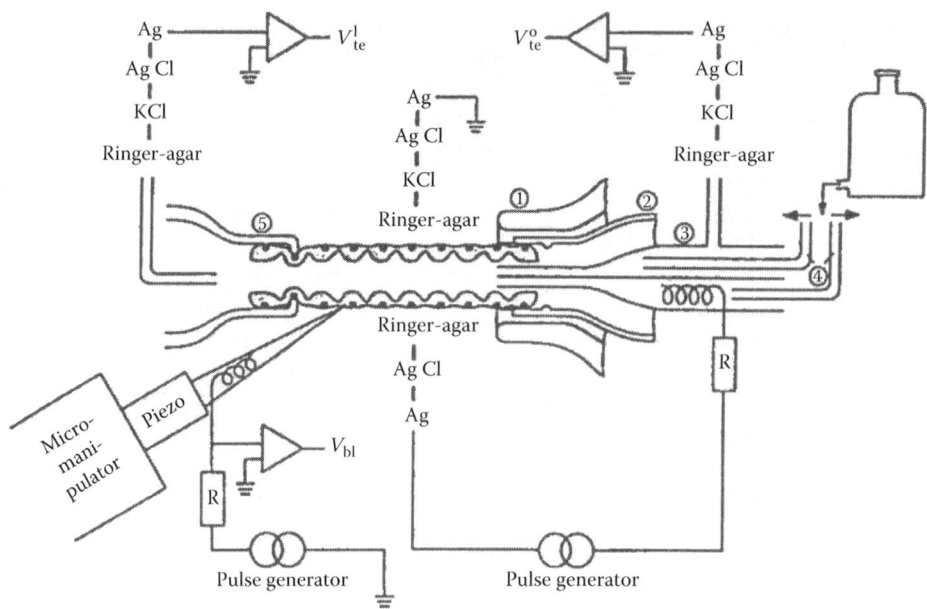

Figure 18.9 Arrangement for experimental perfusion of an isolated Malpighian tubule. (From Leyssens, A. et al., *J. Insect Physiol.*, 38, 431, 1992. With permission.)

the tubules (Nicolson and Hanrahan, 1986; Isaacson et al., 1989; Hegarty et al., 1991; Leyssens et al., 1992, 1993). Isolated tubules secrete a droplet of urine that can be measured volumetrically by assuming the droplet has the dimensions of a sphere. Typical secretion rates measured in nanoliters per minute (nL/min) over a period of several hours are shown in Figure 18.10. With current ultrasensitive techniques, sufficient fluid can be recovered for microchemical analyses (Beyenbach, 1995, and references therein).

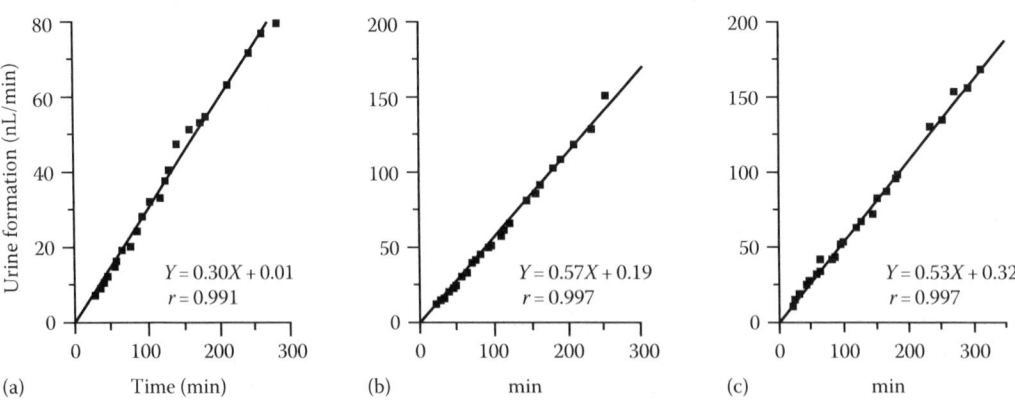

Figure 18.10 The cumulative formation of primary urine by an isolated Malpighian tubule. (a–c) show results from three separate tubules. (From Hegarty, J.L. et al., *Am. J. Physiol.*, 261, C521, 1991. With permission.)

18.6 SELECTIVE REABSORPTION IN THE HINDGUT

18.6.1 Anatomical Specialization of Hindgut Epithelial Cells

The hindgut is the second system that completes the excretion process by selectively reabsorbing some substances into the hemolymph, leaving others in the lumen, and actively secreting some substances into the hindgut lumen. The rectal cuticular lining has greater permeability than the cuticular lining on foregut cells, and the epithelial cells of the hindgut are specialized for both active secretion and active reabsorption. Phillips and Dockrill (1968), who removed and tested the permeability of the cuticular lining from the hindgut of *S. gregaria*, found that molecules with a molecular weight (MW) of 300–500 Da crossed the membrane slowly and molecules having a radius larger than about 0.5–0.6 nm penetrated very slowly or not at all. Glucose (0.42 nm radius, MW 180 Da) penetrated readily, while trehalose (0.52 nm radius, MW 342 Da) penetrated much more slowly. Although this should not be taken as the norm for all insects, it is probably similar in other insects.

In the rectum, small groups of cells are variously called the **rectal cells**, **rectal pad cells**, or **rectal papillae cells** in different insects. These groups of cells have special modifications for reabsorption. In Diptera, four to six fingerlike papillae (Gupta and Berridge, 1966; Hopkins, 1967) are attached to the wall of the rectum and project into the rectal lumen (Figure 18.11). The chitinous lining on the luminal surface of the papillae is continuous with the lining on the inner wall of the rectum. The cells of a rectal papilla are large, usually cuboidal cells that surround a central channel in the papilla that opens into the hemolymph space through a valve (Figure 18.12). Fluid that crosses the rectal papillae cells and enters the central channel is returned to the hemolymph. A small tracheal trunk and a nerve pass into the central cavity and the tracheal trunk branches into many finer tracheae and tracheoles, suggesting a high demand for oxygen for performance of

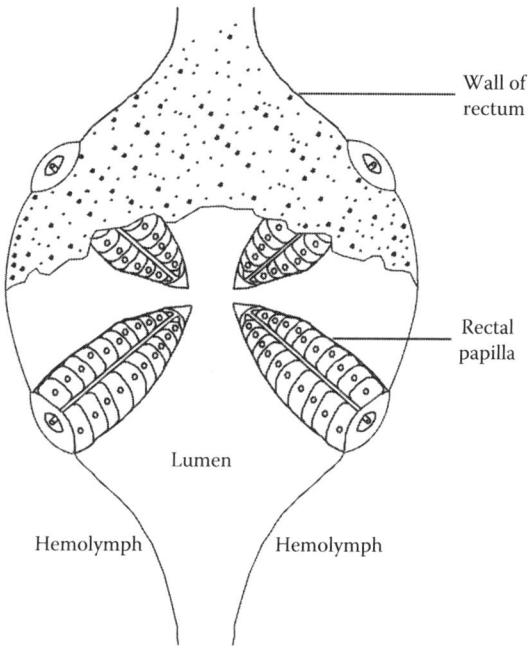

Figure 18.11 A diagram of the rectal papillae in the rectum of adult dipterans and adult siphonapterans (fleas).

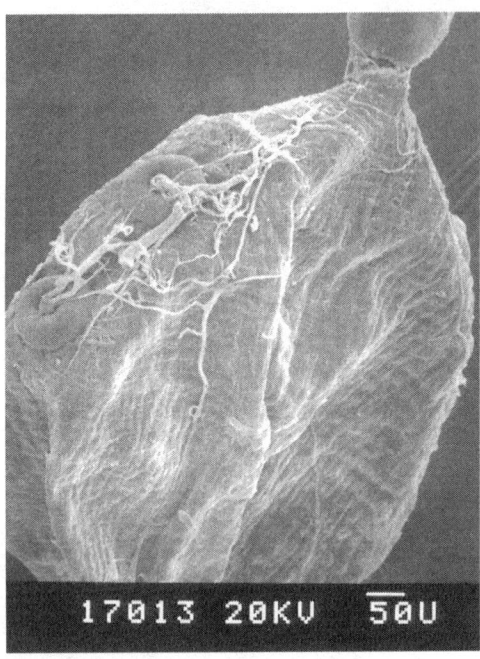

Figure 18.12 An SEM of the hemolymph side of the rectum of the tephritid fruit fly *Anastrepha suspensa* showing the exit to the hemolymph of two of the four rectal papillae. Note the extensive tracheal network and the large trachea entering each papilla indicating a rich oxygen supply. (SEM photo by the author.)

metabolic and secretory work. In rectal papillae cells of the mosquito, *A. aegypti*, the lateral cell membranes are elongated into extensive inward-directed folds of membranes lying very close to each other and projecting nearly to the basal and apical surfaces of the cells (Hopkins, 1967). These elaborate membrane folds create many membrane-bounded channels and spaces within cells in papillae. Cell nuclei are large and prominent. The apical cell membrane (facing the lumen of the rectum) of rectal papillae cells also is greatly infolded, and large mitochondria are usually associated closely with the intercellular channels created by the extensive membrane folding.

Rectal pad cells are common in many insects, and typically are enlarged, columnar to cuboidal cells arranged in six clusters separated by smaller, squamous cells between the "pads" of absorptive cells. The rectal pad cells in the cockroach, *Periplaneta americana* (Phillips, 1981), have highly folded cell membranes at the surface of the rectal lumen that present 10–20 times the surface area of smooth membranes. Mitochondria are located near and within the apical folds and often occur in compact stacks in conjunction with the highly infolded lateral membranes. The extensive infoldings of membranes create many intracellular channels that collect fluid (with dissolved solutes) from the rectum and direct it toward channels between adjacent cells (intercellular channels). The intercellular channels lead to the basal membrane of pad cells where water and useful solutes reenter the hemolymph. The membranes of the intercellular channels are straight and smooth; thus, they present relatively little surface for back diffusion into the rectal pad cells. Using a micropuncture technique to withdraw minute amounts of fluid from various regions of the rectal pad cells of the American cockroach, *P. americana*, Wall and Oschman (1970) found that fluid from the basal sub-epithelial sinus is hyposmotic to rectal lumen fluid. This was interpreted to mean that its ion load had been reduced by the reabsorption of K^+ into the cells for recycling.

18.6.2 Secretion and Reabsorption in the Ileum

The ileum is the most anterior part of the hindgut, occurring just posterior to the origin of the Malpighian tubules in most insects. The most detailed studies of ileal function have been conducted in the desert locust, *S. gregaria*. In the locust, the **ileum** is a major site for isosmotic fluid reabsorption, for active Na^+ and Cl^- reabsorption, and for active secretion of proline as an energy source to support metabolic processes (Phillips et al., 1988, 1994; Audsley et al., 1992a,b). The driving mechanism for ion and water reabsorption in the ileum is an electrogenic Cl^- pump (Phillips et al., 1986, 1988). A neuropeptide, the **ion transport peptide (ITP)**, isolated from the fused corpus cardiacum of *S. gregaria* stimulates Na^+, Cl^-, and water reabsorption and promotes passive reabsorption of K^+ by electrical coupling (Audsley et al., 1992a,b; Phillips et al., 1994; Harrison, 1995; Meredith et al., 1996). ITP needs a second messenger, which is probably cAMP based on observations that exogenously applied cAMP stimulates ion and fluid reabsorption. Some uncertainties still exist in the way that ITP acts upon the ileum. Although ITP inhibits H^+ secretion (i.e., inhibits formation of NH_4^+) in the ileum, cAMP stimulates NH_4^+ formation (Audsley et al., 1992b). The gene encoding ITP has been investigated by Meredith et al. (1996), who prepared a cDNA encoding a peptide with 130 residues that may be a propeptide of ITP, but the mechanism by which the active ITP is derived from this prohormone was not elucidated. The net result from movement of the excretory contents through the ileum of the desert locust is that the volume of fluid is reduced as Na^+, Cl^-, K^+, and fluid are reabsorbed (the ions by active mechanisms and fluid following the osmotic gradients).

The ileum plays a major role in **acid–base balance** (see Section 18.7.3) by secretion of H^+ into the lumen, formation of NH_4^+, and reabsorption of HCO_3^-. Metabolism of reabsorbed nonessential amino acids (alanine, asparagine, glutamine, serine, and proline) present in the urine releases energy for ATP synthesis, an essential power source needed to drive the active reabsorption of Na^+ and Cl^-. About 80% of the ammonia produced in the epithelial cells from metabolism of these amino acids is transported (mechanism not clarified) into the lumen (Phillips et al., 1994) where it is excreted as NH_4^+.

18.6.3 Reabsorption in the Rectum

The **rectum** is the final and major site for reabsorption of ions, water, and nutrients, and it is capable of reabsorbing fluid against strong osmotic gradients, ultimately producing in many insects a very concentrated, **hyperosmotic** excreta. The driving mechanism for cation and water reabsorption, as in the ileum, is an electrogenic Cl^- pump under the influence of a neuropeptide hormone, **chloride transport stimulating hormone**, from the corpora cardiaca (CC). It acts on the rectal epithelium to promote active Cl^- absorption (Phillips, 1964; Phillips et al., 1986, 1994; Harrison, 1995). The pump provides the energy for K^+ reabsorption. The exact route followed by K^+ as it crosses the rectal epithelial cells to reenter the hemolymph or Malpighian tubules varies in different insects depending on the anatomy of the rectum and the cells involved in the process. Water from the rectal lumen and dissolved solutes follows the osmotic gradient created by ion absorption. The result is that water is reabsorbed into the epithelial cells against an increasingly strong concentration gradient in the rectal lumen. The excreta in the rectum become very pasty or even dry in many insects as water is reabsorbed. The rectal epithelial cells actively reabsorb amino acids from the lumen and metabolize them (primarily proline) to produce ATP needed to energize the pump. Proline is metabolized within mitochondria by the proline dehydrogenase pathway (Chamberlin and Phillips, 1982, 1983). Thus, active ion secretion and electrogenic pumps play a major role in the formation of primary urine in the Malpighian tubules and then in reclaiming water from the hindgut. Several different hormones are responsible for regulating the different functions.

18.7 ROLE OF THE EXCRETORY SYSTEM IN MAINTAINING HOMEOSTASIS

Maintenance of constancy of the internal environment of cells, tissues, and organisms is the process of **homeostasis**. The excretory system plays a major role in eliminating metabolic wastes and toxins acquired with or from the food, typically by increasing the basal rate of fluid secretion, and transport of the toxin, into the lumen of Malpighian tubules (Rheault et al., 2006; Ruiz-Sanchez and O'Donnell, 2006, 2007a,b; Ruiz-Sanchez et al., 2007; Maddrell, 2009). Dynamic changes in salt, water, acid–base, and nitrogen amounts occur from time to time in all organisms as a result of food ingested, environmental conditions, and metabolism. Regulatory mechanisms that respond rapidly to these changes are necessary to preserve the integrity of cells and tissues. For example, herbivores ingest relatively large amounts of potassium and little sodium with their plant-based diet, while blood feeders, such as some heteropterans and mosquitoes, ingest relatively large amounts of sodium (mostly as sodium chloride) and little potassium with their food. Ingestion of plant phloem or xylem sap results in an excessive intake of water, and usually more sugar and some amino acids than needed. Nitrogen metabolites from proteins, amino acids, and purines must be disposed of by all cells. Blood-feeding insects usually take especially large meals when the opportunity presents itself and ionic homeostasis must be rapidly rectified (reviewed by Coast, 2009).

The female yellow-fever mosquito *A. aegypti* can ingest a blood meal equal to as much as twice its body weight, resulting in a serious impediment to flight. Hence, it begins urine excretion even while still feeding (Beyenbach, 2003), rapidly getting rid of unneeded NaCl and water while retaining the red blood cells rich in proteins and potassium in the alimentary canal for later digestion. Several peptide hormones are secreted by the nervous system into the hemolymph to promote the rapid excretory process (Jagge and Pietrantonio, 2008; Beyenback et al., 2010; Miyauchi et al., 2013).

Rhodnius prolixus is another insect that has been a major research animal for nearly 100 years; it was a favorite of V.B. Wigglesworth and subsequently of many of his students and colleagues. *Rhodnius* takes only one blood meal in each instar and spends the remainder of the instar digesting it and growing toward the next molt. It may ingest up to 10 times its unfed body weight (Paluzzi et al., 2013), ballooning the abdomen and presenting a major challenge to homeostasis. As in mosquitoes, hormones are secreted and excretion of excess water and NaCl begins almost immediately. Stretch receptors in the expanded abdomen respond by triggering secretion of serotonin (5-hydroxytryptamine, 5HT) and a corticotropin-releasing factor-related peptide named RhoprCRF, or designated as RhoprDH in some publications) (Te Brugge et al., 2011; Paluzzi et al., 2012, 2013). Excretion in *Rhodnius* presents a beautiful model of the intertwining of physiological systems; even during diuresis to remove excess water and ion, especially Na^+, antidiuretic peptides begin to be secreted from the central nervous system (CNS) (Orchard and Paluzzi, 2009; Paluzzi and Orchard, 2010; Paluzzi et al., 2010, 2012), to slow down diuresis and eventually stop it. RhoprCAPA-α2 inhibits fluid secretion stimulated by 5-HT and inhibits absorption of fluids from the midgut into the hemolymph (Paluzzi and Orchard, 2009; Ianowski et al., 2010; Paluzzi et al., 2012). Two genes have been identified that encode RhoprCAPA-α2 (Paluzzi and Orchard, 2010).

18.7.1 Electrolyte Homeostasis

Electrolyte homeostasis and the necessity of ion and solute transport across cell membranes have been reviewed (Beyenbach, 1995; Coast, 2009; Dircksen, 2009), with special emphasis on *Drosophila* and blood-feeding insects such as the mosquito *A. aegypti* and the heteropteran *R. prolixus*. Davies and Terhzaz (2009) reviewed calcium signaling as a common second messenger in movement of ions and solutes in and out of cells.

Larval and adult *A. aegypti* live in different habitats, have different food habits, and control Malpighian tubule function by different hormones. Adult female mosquitoes need a blood meal in order to mature each batch of eggs, but with the blood comes a large salt (NaCl) load that must be excreted. Sodium excretion is an active process and occurs in Malpighian tubule cells of the adult mosquito in response to stimulation from the **mosquito natriuretic peptide** (**MNP**) released from the CC (Wheelock et al., 1988; Beyenbach, 1995, 2003). A proton pump coupled with a H^+–Na^+ exchange mechanism secretes sodium into the tubule lumen. The pump appears to work as previously described for Malpighian tubules in general, except that Na^+, rather than K^+, is the principal ion exchanged for protons pumped into the tubule lumen.

Prior to a blood meal, urine forms slowly in isolated tubules from *A. aegypti* at about 0.4 nL/min, and the measured **transcellular resistance** (**Rc**) across the tubule cells is high, keeping cation and water movement low. Feeding on a blood meal stimulates the release of MNP, and cAMP is produced as a second messenger at the inner surface of the basolateral membrane of tubule cells. cAMP acts selectively to open Na^+ channels in the basolateral membrane. As Na^+ enters the tubule cell from the hemolymph, the Rc falls to about 40% of its prefeeding value (Wheelock et al., 1988). Movement of water into tubule cells follows the osmotic gradient. Urine flow rates as high as 2.8 nL/min in hormone stimulated tubules are promoted by the apical membrane H^+-pump coupled with H^+–Na^+ exchange. The ion flux generated by MNP and cAMP is specifically an increase in secretion of sodium. Potassium movement is not influenced. The voltage in the lumen of the Malpighian tubules increases from about +52 mV in unfed mosquitoes to about +70 mV in fed mosquitoes (lumen positive to hemolymph in both cases).

The large chloride load from the blood meal also must be excreted, and chloride (Cl^-) moves from hemolymph to tubule lumen in a passive transport pathway between the cells (called the **paracellular pathway**) (Pennabecker et al., 1993). The permeability of the paracellular pathway is increased by leucokinin VIII, a neuropeptide (Wang et al., 1996), although it is not known whether this or a similar hormone is secreted by the mosquito.

Larval *A. aegypti* live in freshwater and, in response to an increase in salinity, they secrete **5-hydroxytryptamine** into the hemolymph, leading to an increase in cAMP formation in the Malpighian tubules (Clark and Bradley, 1993). 5-Hydroxytryptamine and cAMP stimulate fluid and ion (Na^+ and K^+) secretion rates in isolated larval tubules, but the urine is not concentrated with respect to the ions (Clark and Bradley, 1996). The blood-feeding hemipteran *R. prolixus* also secretes Na^+ (and K^+) into the lumen of its Malpighian tubules. Hematophagous behavior may have driven the evolution of Na^+ secretion by Malpighian tubule cells, thus enabling blood feeders to regulate ion homeostasis after a large, salty meal.

Beyenbach (1995) reviews three potential physiological processes through which *A. aegypti* may regulate (control) rates of ion and fluid excretion. These processes are (1) the proton pump that supplies energy for Na^+ and K^+ secretion to the tubule lumen, (2) the resistance Rc across the tubule cells that control ion channels in the basolateral membrane, and (3) the resistance of the passive transport pathway for chloride movement. Regulation of the proton pump has not been demonstrated in mosquito Malpighian tubules (Beyenbach, 1995). However, Rc and basolateral ion channels are regulated by the natriuretic peptide and cAMP in mosquitoes. The passive transport pathway between tubule cells may be a function of extracellular secretion of a leucokinin-type peptide in adult mosquitoes.

18.7.2 Water Homeostasis

Water homeostasis is very important to insects because they have a high surface-to-volume ratio and their food often has variable water content. Water homeostasis is particularly important to honeybees, both individually and in the social environment of the hive where there is high water turnover and water evaporation from the stored nectar in the hive (Nicolson, 2009). Although many physiological systems and behaviors are related to water homeostasis, ridding the body of excess water is primarily a function of the excretory system. Water excretion and retention are regulated by hormones. **Diuretic hormones** promote fluid formation and rapid excretion by the Malpighian tubules, while the currently known **antidiuretic hormones** act upon the hindgut (with one exception) and promote water reabsorption. The exception is an antidiuretic hormone demonstrated from water-deprived, dehydrated house crickets, *Acheta domesticus*, that inhibits fluid formation by the Malpighian tubules without action upon the hindgut (Spring et al., 1988).

18.7.2.1 Diuretic Hormones

Currently, more than 20 insect diuretic hormones are known. All are neuropeptides. They control the formation of fluid by Malpighian tubules and fluid transport across the midgut and hindgut in insects in the orders Orthoptera, Lepidoptera, Diptera, Dictyoptera, and Coleoptera (Wheeler and Coast, 1990; Coast et al., 2011; Vanderveken and O'Donnell, 2014). Serotonin (5-hydroxytryptamine), which is not a neuropeptide, also stimulates urine formation (Barrett and Orchard, 1990; Maddrell et al., 1991). Diuresis reduces the flight load in those insects that take large blood meals at one time or ingest plant sap in large quantities, clearly an adaptive physiological mechanism because the excess water in not needed and critical nutrients in the liquid food need to be concentrated and conserved. Diuresis in beetles that live in very dry environments is not so intuitively obvious, but may have important functions as explained later.

Rhodnius, the first insect studied in detail with respect to a neuropeptide diuretic hormone (Maddrell, 1963, 1964a,b, 1966), takes one large blood meal in each instar and within hours rapidly excretes a large volume of urine. This action leaves concentrated proteins from the blood meal in the midgut and rids the body of excess water and Na^+. Within minutes after feeding starts, the large volume of blood ingested greatly distends the abdomen and activates stretch receptors located in the tergosternal muscles near the lateral edge of abdominal segments two to seven. Rapid circulation of hemolymph is induced, caused at least in part by vigorous peristaltic movements of the alimentary canal. Diuresis starts within 3 min after feeding, initiated by secretion of 5-HT and a neuropeptide diuretic hormone synthesized in large neurosecretory cells (NSCs) in the fused mesothoracic ganglion (Figure 18.13). The neuropeptide is released from a series of enlarged axonal endings of abdominal nerves originating from the mesothoracic ganglion (Maddrell, 1966). The hormone is transported to the Malpighian tubules, the target tissue, by rapid circulation of hemolymph. By the time diuresis is over, *Rhodnius* has lost 40% of its freshly fed weight. The osmotic concentration of its hemolymph first falls due to the dilution effect of absorbing so much water from the midgut, but after diuresis, the osmotic concentration is about the same as before feeding. Malpighian tubules and possibly other tissues rapidly destroy the diuretic hormone, and sustained release of the hormone is necessary to maintain the excretion of a large volume of fluid. *Triatoma infestans*, the vector of the Chagas disease organism, also takes large blood meals and then needs to get rid of large volumes of water. The bugs produce an allatotropin-like peptide that has a diuretic effect and aids the excretion of excess water (Santini and Ronderos, 2007).

Diuresis is also under similar hormonal control in the cotton stainer, *Dysdercus fasciatus* (Hemiptera), which takes plant sap. Medial NSCs of the brain are the principal source of the hormone in *D. fasciatus*, but some activity is present also in CC and in the mesothoracic ganglion. Extracts from median NSC accelerated urine flow rate from the normal value of $3.1 \text{ mm}^3 \times 10^{-3}$ to $9.87 \text{ mm}^3 \times 10^{-3}$ per minute (Berridge, 1966).

A diuretic neuropeptide (**MAS-DH**) with 41 amino acid residues has been isolated from *M. sexta* (Kataoka et al., 1989). It has some sequence similarity to **corticotropin-releasing factor** (**CRF**) and **urotensin I**, two vertebrate neuropeptides with hormonal activity, and to a toxin, **sauvagine**, from the skin of a South American tree frog. A receptor for MAS-DH has been characterized from the Malpighian tubules of fifth instars of tobacco hornworms (Reagan et al., 1993). MAS-DH binds to the receptor rapidly and in a reversible manner (Lehmberg et al., 1991). A diuretic neuropeptide isolated

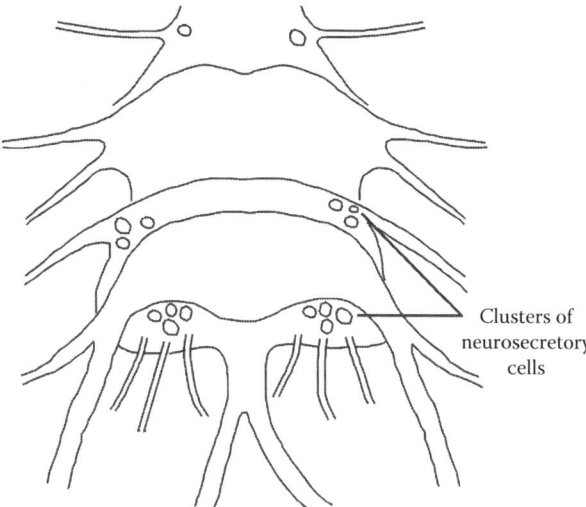

Figure 18.13 The drawing illustrates the large thoracic ganglion containing neurosecretory cells that are the synthesis site of the diuretic hormone in *R. prolixus*. The neuropeptide is released from swollen axons arising from the posterior of the ganglion. The complex ganglion contains fused abdominal ganglia. (Drawing modified from Maddrell, S.H.P., *Exp. Biol.*, 40, 247, 1963.)

from *Locusta migratoria* (Mordue and Morgan, 1985; Proux et al., 1987; Lehmberg et al., 1991, 1993) consists of two antiparallel 9-amino acid peptides joined by two disulfide bonds (Figure 18.14).

Malpighian tubules from the cricket *A. domesticus* are stimulated to secrete fluid by tissue extracts (Spring and Hazelton, 1987; Coast, 1988) of CC, corpora allata (CA) (Coast, 1989), and some other parts of the CNS (Coast and Wheeler, 1990). Tubules in this cricket have three physiologically and morphologically different segments (Kim and Spring, 1992), and different extracts may act on different segments by different mechanisms.

The beetle, *Tenebrio molitor*, lives in dry grain and grain products and needs no water other than that already present in the food and that derived from metabolism. Surprisingly, it produces a diuretic hormone (Nicolson, 1991). Why would an insect living in a water-impoverished environment evolve a diuretic hormone? Nicolson (1991) suggests that its function is to act as a **clearance hormone** to flush the Malpighian tubules and hindgut. The tubule fluid may be passed into the hindgut to help move the very dry food residue through the gut and may help promote a countercurrent flow in the midgut. Fluid reaching the rectal region is nearly all reabsorbed by the cryptonephridic tubules and returned to the hemolymph.

Another tenebrionid beetle, *Onymacris plana*, living in the desert also has a diuretic hormone, and it, too, may have a flushing function. Nicolson and Hanrahan (1986) found that an isolated Malpighian tubule from *O. plana* typically produces about 3 nL/min/tubule, but with stimulation by a diuretic hormone from the CC, the tubule forms 40–60 nL, or sometimes up to 100 nL/min/tubule. They speculated that a flushing function might be beneficial to the insects by aiding in removal of plant allelochemicals eaten with the food. The rectum reabsorbs the water and conserves it for reuse.

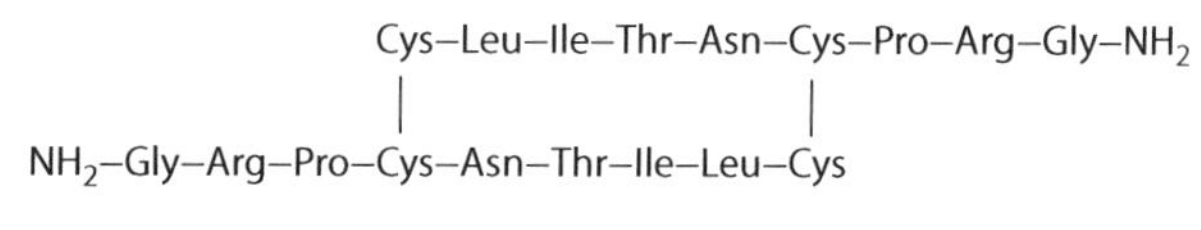

$$\begin{array}{c} \text{Cys–Leu–Ile–Thr–Asn–Cys–Pro–Arg–Gly–NH}_2 \\ |\qquad\qquad\qquad\qquad\qquad| \\ \text{NH}_2\text{–Gly–Arg–Pro–Cys–Asn–Thr–Ile–Leu–Cys} \end{array}$$

Figure 18.14 The structure of the diuretic neuropeptide from *L. migratoria*, consisting of two antiparallel 9-amino acid peptides joined by two disulfide bonds.

D. melanogaster regulates fluid secretion with at least four types of diuretic peptide hormones, including DH_{44} and DH_{31}, kinin neuropeptides, and capa neuropeptides that act on different cell types in Malpighian tubules (Pollock et al., 2004; Johnson et al., 2005, and references therein). The capa peptide family and their amino acid sequences include CAPA1 and CAPA 2 from *Drosophila* (GANMGLYAFPRVamide and ASGLVAFPRVamide, respectively); *Ang*CAPA-QGL and *Ang*CAPA-GPT from *A. gambiae* (QGLVPFPRVamide and GPTVGLFAFPRVamide, respectively); and CAP2b from *M. sexta* (PyroELYAFPRVamide) (Pollock et al., 2004, and references therein). DH_{44} and DH_{31} increase secretion rates by elevating levels of cAMP in principal cells of the tubules (Johnson et al., 2005). The leucokinin neuropeptide regulates Cl⁻ transport in stellate cells by increasing intracellular calcium, while capa peptides raise fluid transport by upregulating the signaling messengers calcium, nitric oxide, and cGMP in principal cells of Malpighian tubules. The mechanism of action of these neuropeptides is not the same in all of relatively few insects that have been studied to date. Although the capa neuropeptides occur in several orders of insects and nitric oxide synthase, a prerequisite for capa peptide function, is known to occur in Malpighian tubules of some dipterans, lepidopterans, and orthopterans, the physiological role of the peptides is not uniform. Capa neuropeptides from dipterans and from *M. sexta* do not stimulate fluid secretion nor activate nitric oxide/cGMP signaling in *S. gregaria* or *L. migratoria* (Pollock et al., 2004). In the mosquito, *A. aegypti*, leucokinin stimulates Malpighian tubule secretion rates, not across stellate cells but by a paracellular pathway through septate junctions between principal cells (Yu and Beyenbach, 2004).

18.7.2.2 Antidiuretic Hormones

Generally, insects that feed on dry solid food probably need to conserve water rather than excrete it; thus, evolution of an **antidiuretic hormone** might be expected. Mills (1967) and Goldbard et al. (1970) found evidence for both an antidiuretic and a diuretic hormone in *P. americana*. The antidiuretic hormone promotes water conservation and ordinarily is the predominate hormone. A diuretic hormone was demonstrated in *P. americana* only by depriving male cockroaches of water for 3 days, after which the thirsty insects drank enough water at one time to stimulate the release of the diuretic hormone, thus temporarily increasing water excretion. Ligation experiments indicated the diuretic hormone was released from the posterior part of the abdomen, and extracts of various tissues indicated that the terminal abdominal ganglion was the source. Fluid reabsorption across the cryptonephritic complex of larval *M. sexta* has been demonstrated with an antidiuretic factor extracted from the brain/CC/CA complex of the caterpillars by Liao et al. (2000), who suggest that the factor involves cAMP increase and activation of a Cl⁻ pump in the cryptonephritic system.

18.7.3 Acid–Base Homeostasis

The excretory system is important in maintaining **acid–base balance** of body fluids and tissues (Harrison, 2001). Acidosis or alkalosis may be experienced by an insect depending on various foods, the presence of certain types of chemical compounds in plants eaten, type of proteins metabolized (whether proteins yield a high proportion of acidic, basic, or neutral amino acids), and metabolic conditions such as exercise (e.g., flight) that produce acids in the tissues (Harrison and Kennedy, 1994). The western lubber grasshopper, *Taeniopoda eques*, exhibits flexibility in shifting between excretions of excess acid or excess base equivalents (Harrison and Kennedy, 1994) depending on need. This flexibility and regulatory ability in a highly polyphagous insect seems adaptive, and many other insects may show similar ability (Harrison and Kennedy, 1994).

Acid–base regulation has been most thoroughly studied in the desert locust, *S. gregaria* (see Harrison and Kennedy, 1994; Phillips et al., 1994, and references therein). Locusts experimentally injected with HCl excrete most of the acid equivalents by secretion of protons by the hindgut epithelium. The Malpighian tubules participate only marginally. Secretion of H⁺ and formation of ammonium ions (NH_4^+) in the ileum is a principal mechanism for excreting excess acid equivalents (Harrison, 1994; Harrison and Kennedy, 1994; Phillips et al., 1994). The ileum is a major site of **ammoniagenesis**, the formation of ammonia from precursors, in locusts in which hindgut cells specifically metabolize amino acids and glucose for energy (Peach and Phillips, 1991). The excess nitrogen from amino groups is incorporated into formation of ammonia. These metabolites are present in the lumen of the ileum, having come with urine formed in the Malpighian tubules. Ammonia and urate are about equal in concentration in the fluid within the Malpighian tubules, but because of ileal and rectal secretion of ammonia/ammonium ions, about half of the total nitrogen excreted by desert locusts is ammonia nitrogen (Harrison and Phillips, 1992). Some of the ammonia reacts with uric acid in the hindgut to form ammonium urate and spares sodium and potassium, which also react with uric acid to form sodium and potassium urate. Thus, excretion of total ammonia nitrogen serves several functions in locusts (Harrison and Phillips, 1992; Phillips et al., 1994; Harrison, 1995), including the following:

1. Ammonium urate allows the insects to conserve Na⁺, an ion that is not high in the food of locusts.
2. Conversion of ammonia (NH_3) to ammonium (NH_4^+) in the ileal cells is equivalent to removal of protons (H⁺), and excretion of ammonia is more than sufficient to explain the recovery of hemolymph pH after a load of HCl is injected into the hemocoel.
3. Excretion of ammonia by locusts conserves water (because of precipitation of not very soluble ammonium urate salt).
4. Increasing nitrogen excretion by 25% more than excretion of only sodium or potassium urate.

18.7.4 Nitrogen Homeostasis

Harrison (1995) has addressed the fact that nitrogen, although known to be a growth-limiting nutrient for some insects, is nevertheless excreted in several forms by insects. Excess nitrogen that must be excreted may come from ingestion of proteins leading to an imbalance of amino acids; they save essential amino acids and may metabolize the nonessential ones as energy sources. Insects also ingest nitrogen in the nucleic acids of their food. Excess protein nitrogen is excreted as **uric acid** (a purine) (Figure 18.15), as purines related to uric acid, as **ammonia** or **ammonium salts**, and in several other (usually minor) forms. Nitrogen from nucleic acids also is excreted as uric acid (or as related purine metabolites or metabolites of uric acid). Typically, only trace or small amounts of

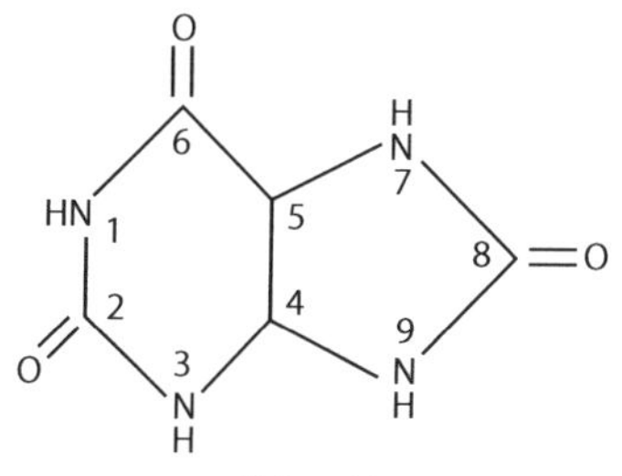

Uric acid

Figure 18.15 The structure of uric acid, the principal nitrogenous excretory product for most insects. The source of nitrogen atoms has not been determined in insects, but they almost certainly come from the amino nitrogen of amino acids that are metabolized. Carbon at position 2 is derived from the carbon of formate, carbon 4 from the carboxyl carbon of glycine, carbon 5 from the alpha carbon of glycine, carbon 6 from carbon dioxide derived from the carboxyl carbon of glycine, and carbon 8 from formate. (Drawing modified from Barrett, F.M. and Friend, W.G., J. Insect Physiol., 16, 121, 1970.)

urea are excreted. The complete sequence of enzymes in the ureotelic pathway for synthesis of urea has not been found in insects, but arginase, a primary enzyme in the pathway, is active in fat body of the abdomen and thorax throughout the life cycle of *A. aegypti* mosquitoes. Although uric acid is the primary excretory product, small amounts of urea are excreted (von Dungern and Briegel, 2001). Bursell (1967) and Cochran (1975) provided thorough reviews of early literature on nitrogen excretory products in insects.

18.7.4.1 Ammonia Excretion

Ammonia is a product of protein and amino acid metabolism. Free ammonia cannot be stored in tissues or cells because it is a very strong base influencing pH and, in its free form, it is very toxic to all cells. It must be rapidly excreted or transformed into a less toxic compound. If water is readily available for dilution, ammonia can be excreted as the free base or as an ammonium salt. Animals that excrete ammonia as their primary nitrogenous waste product are described as **ammonotelic**. Weihrauch et al. (2012) reviewed ammonia production and excretion in insects.

Ammonia is a major excretory product for larval stages of some Diptera that live in very wet environments. Larvae of *Calliphora erythrocephala*, the common blowfly; *Wohlfahrtia vigil*, a sarcophagid fly (Brown, 1936); *Phormia regina*, a blowfly; and *Lucilia cuprina*, the sheep ked (Hitchcock and Haub, 1941) excrete ammonia into wet surroundings that dilute it to nontoxic levels. *Lucilia sericata*, another blowfly, excretes up to 15-fold more ammonia than uric acid (Brown, 1938). Although uric acid is synthesized, most of it is stored in the tissues (storage excretion). Allantoin, a breakdown product of uric acid, is also excreted by larvae of *L. sericata*. In these dipterans, ammonia excretion ceases at pupariation, and the adults excrete uric acid and, in a few cases, some allantoin.

Staddon (1955, 1959) found that most of the nitrogen excreted by the aquatic larva of the neuropteran, *Sialis lutaria*, and the odonate, *Aeshna cyanea*, is ammonia. Although some aquatic insects excrete substantial amounts of ammonia, many synthesize and excrete uric acid (or a further metabolic derivative of uric acid).

Some terrestrial insects excrete the majority of their excretory nitrogen as ammonia or ammonium salts. The American cockroach, *P. americana*, excretes ammonia as a major excretory product (Mullins and Cochran, 1972), but the precise mechanism(s) of its excretion has not been elucidated. Ammonia and ammonium nitrogen can account for 10%–46% of the total nitrogen excreted by the desert locust, *S. gregaria*, according to Harrison (1995), who cautions that previous studies on the distribution of nitrogen in excreta of terrestrial insects may have missed more labile ammonia and ammonium nitrogen by the methods and techniques employed. Ammonia nitrogen is rapidly lost from fecal pellets, especially if they are dried prior to analysis. In some cases, ammonium urate may be lost because it is poorly soluble unless excretory material is extracted with a large volume of aqueous solvent (Harrison, 1995). Fecal pellets should be collected within minutes after excreted, deposited in acid solution, and kept frozen until analysis (Harrison, 1995).

Most animals, including insects, synthesize ammonia into less toxic compounds such as urea (mammals) or uric acid (birds, reptiles, insects, and Dalmatian dogs). Enzymes involved in amino acid metabolism and ammonia production include amino transaminases (or transferases), glutamic acid and alanine dehydrogenases, L- and D-amino acid oxidases, adenosine deaminase, and monoamine oxidase. All of these enzymes have been detected in a number of insects (Cochran, 1975).

The **amino transaminases** are widely distributed in insect tissues and enable the amino group of one amino acid to be transferred to a ketoacid, thereby forming a new amino acid. Transaminase reactions provide a way to interconvert nonessential amino acids and to make amino acids available to enzymes that deaminate with release of ammonia such as glutamic dehydrogenase and alanine dehydrogenase. Ammonia formed in these reactions is rapidly excreted or converted into less toxic compounds. The α-ketoglutaric acid and pyruvate formed can be metabolized through the Krebs cycle as an energy source (Cochran, 1975).

Ammonia production also may come from turnover and replacement of an insect's own nucleic acids as well as from the metabolism of nucleic acids ingested with food. The enzymes adenosine deaminase, guanine deaminase, and adenine deaminase, all of which give rise to ammonia from metabolism of nucleic acids or their derived products, have been reported from various insects.

18.7.4.2 Uric Acid Synthesis and Excretion

Uric acid (Figure 18.15) is synthesized in insects from protein nitrogen as well as from nucleic acid nitrogen. The major portion is synthesized from protein nitrogen simply because insects ingest relatively much more protein nitrogen (or amino acid nitrogen) in their diet than nucleic acid nitrogen (Cochran, 1975). In birds (and presumably in insects) synthesis of 1 mol of uric acid from NH_3 requires the expenditure of 8 mols of ATP (Cochran, 1975). The 8 ATP/uric acid formed ignore other costs that may be incurred such as transport across cell membranes and maintenance of enzymatic machinery. Thus, excretion of uric acid as the main product of protein metabolism is energetically costly. Its advantages are that it rids the body of four nitrogen atoms, which make up 33.3% of the MW of uric acid (MW of uric acid = 168.11 Da, 4 N = 56), and it is very insoluble in water. It often reaches concentrations that promote precipitation from solution in the Malpighian tubules and hindgut and, as a precipitate, it does not contribute to osmotic values across cells that line the tubules or hindgut.

The fat body is the primary site for uric acid synthesis. Barrett and Friend (1970) found that glycine contributes a carboxyl carbon to position 4 and an α-carbon to position 5 during synthesis of uric acid in *R. prolixus*. Formate contributed carbons to positions 2 and 8. The labeled carboxyl carbon of glycine was converted rapidly in *Rhodnius* to $^{14}CO_2$, and much of the $^{14}CO_2$ formed gave rise to labeled carbon 6 in uric acid. The α-carbon of glycine also can contribute to the synthesis of formate, so that ultimately some or even much of carbon-8 of uric acid also might come from glycine. Although the origin of the nitrogen atoms in uric acid from *Rhodnius* was not determined, it seems reasonably certain that the nitrogen atoms are derived from NH_3 resulting from metabolism of proteins, amino acids, and nucleic acids (Barrett and Friend, 1970).

The final steps in the synthesis of uric acid involve the conversion of hypoxanthine and xanthine to uric acid (Figure 18.16). The enzyme that catalyzes the two-step conversion is **xanthine dehydrogenase** (Irzykiewicz, 1955). Its counterpart in vertebrates (birds and reptiles) is xanthine oxidase, a true oxidase since molecular oxygen can accept the protons removed from hypoxanthine and xanthine. The insect enzyme does not work without NAD^+ or FAD^+ (or a synthetic acceptor such as methylene blue) as the proton acceptor, so it is a true dehydrogenase. Transfer of electrons from reduced cofactors through the electron transport system could yield six (NADH) or four ($FADH_2$) ATP/uric acid molecules formed. Evolution of xanthine dehydrogenase in insects, instead of xanthine oxidase, may have occurred as a potential way to recover some of the costs of urate synthesis (Cochran, 1975).

Uric acid and related uricotelic compounds are admirably adaptive excretory products for animals that live with water stress and that need to conserve water. These compounds have limited solubility, and when they crystallize from solution, they reduce the osmotic work required to reabsorb water from the rectum. Uric acid is the least soluble of the compounds (6 mg/100 mL of water) followed by allantoin (60 mg/100 mL), hypoxanthine (70 mg/100 mL), and xanthine (260 mg/100 mL) (Bursell, 1967). Uric acid may crystallize as free uric acid and/or as a salt. Data presented by Harrison (1995) show that ammonium urate is much less soluble than either sodium or potassium urate, and it has the advantage that the NH_4^+ rids the body of additional nitrogen and acid equivalents.

Storage excretion of uric acid, or deposition in various parts of the body, is common in cockroaches and some other insects. Crystals of uric acid may occur in the hemolymph, and in other tissues, especially in fat body. Cochran (1973) found high levels of crystalline uric acid in the fat bodies of 14 species of cockroaches. Male cockroaches deposit uric acid in accessory glands associated

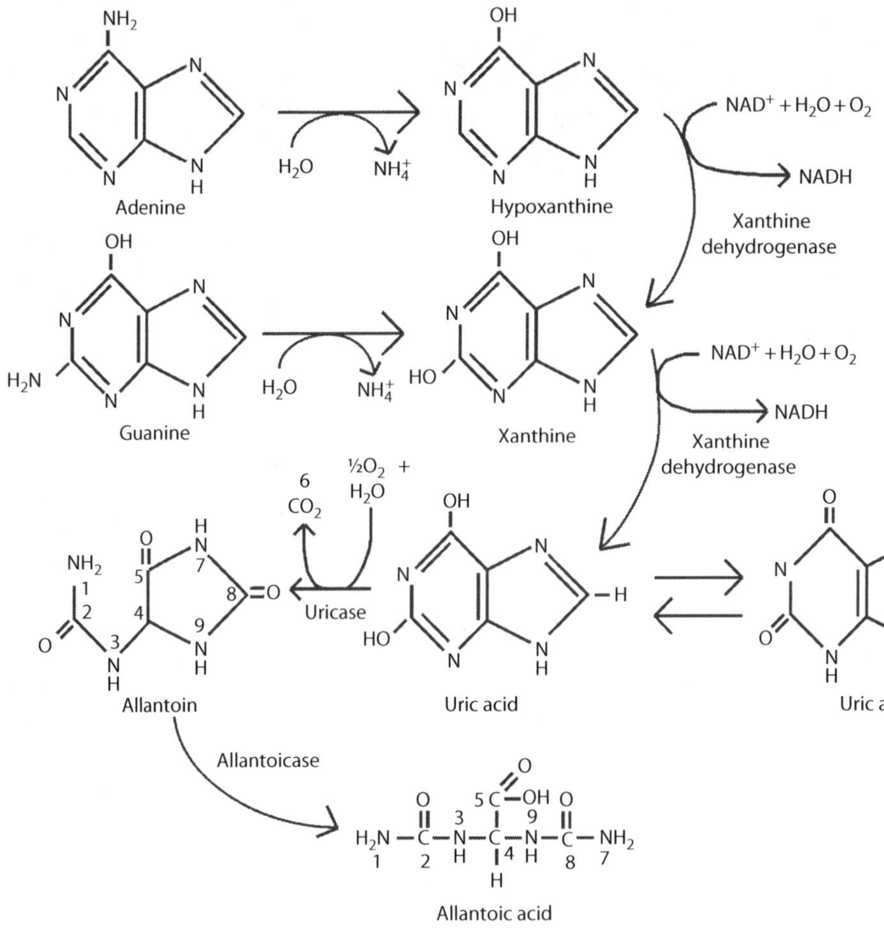

Figure 18.16 The final stages in the synthesis of uric acid and various metabolic breakdown products of uric acid that some insects excrete.

with their reproductive tract and deposit urates from the glands on the outside of the sperm packet, or spermatophore, that they produce and insert into the female at mating. Mullins and Keil (1980) found that labeled uric acid of male *Blattella germanica* cockroaches could be recovered in female German cockroaches and in their oothecae after mating. They suggested that the urates represented a nitrogen source for the female and a **paternal investment** by the male in her progeny.

Paternal investment behavior is also shown by the tropical cockroach, *Xestoblatta hamata*, in which females feed upon urates deposited in the genital chamber by males during mating. Females transfer the urates to their terminal oocytes and the ootheca (Schal and Bell, 1982). Under both field and laboratory conditions, male *X. hamata* choose high-protein foods, which are known to result in greater production of urates in cockroaches (Haydak, 1953), and they feed opportunistically upon substances containing uric acid such as bird droppings.

The Malpighian tubules of the American cockroach, *P. americana*, do not clear uric acid from the hemolymph, and the cockroaches do not excrete uric acid with the fecal wastes (except in small amounts under crowded conditions, which may be because they have cannibalized another cockroach and ingested uric acid that simply passes through the gut) (Mullins and Cochran, 1972). Most of the nitrogen excreted by American cockroaches is ammonia nitrogen (Mullins and Cochran, 1972).

Razet (1966) found that many insects excrete small to large percentages of their excretory nitrogen as **allantoin**, a breakdown product of uric acid catalyzed by the enzyme **uricase**. Occurrence of uricase is widespread in insects and in their tissues, but no particular advantage is known for the conversion of uric acid to allantoin nor for it as an excretory product. A few insects excrete some **allantoic acid**, an oxidation product of allantoin, but the enzyme **allantoicase** is not widespread in insects and, if present at all, allantoic acid is a very minor excretory product.

Animals that excrete most or all of their excretory nitrogen as uric acid are described as **uricotelic**. Occasionally, it has been questioned whether insects really fit the definition of uricotelism (discussed in Cochran, 1975) because some insects do not excrete most of their nitrogen as uric acid. Bursell (1970) proposed extending the definition of uricotelism to include excretion of allantoin and allantoic acid, since both are derived from further metabolism of uric acid. Cochran (1975) concurred with the broader definition and extended it further to include excretion of the uric acid precursors hypoxanthine, xanthine, and guanine excreted by some insects (Morita, 1958; Mitchell et al., 1959; Nation and Patton, 1961; Nation, 1963; Mitlin and Vickers, 1964; Nation and Thomas, 1965). Thus, Bursell (1970) and Cochran (1975) concluded that, as a group, insects should still be considered uricotelic in excretion even though some excrete a variety of nitrogenous compounds and a few excrete relatively little or no uric acid.

18.8 CRYPTONEPHRIDIAL SYSTEMS

Many families of Coleoptera, Lepidoptera, and some sawfly larvae (Hymenoptera) have an arrangement of Malpighian tubules in which the distal ends of the tubules are enveloped within a membrane and held close to the surface of the rectum (Figure 18.17). This arrangement is known as a **cryptosolenic** or **cryptonephridial** tubule system. It appears to be an arrangement that enables very efficient conservation of water. Insects living in the driest habitats and eating very dry food have the most extensive development and network of cryptonephridial tubules.

Cryptonephridial tubules do not penetrate the lumen of the rectum, but lie on the outer surface of the rectum, encased within a **perinephric chamber** bounded by the **perinephric membrane** (Figure 18.18). The perinephric membrane is composed of thin, elongated cells that seal the tubules from the hemocoel and hemolymph at the initial point of contact with the gut (Saini, 1964). The tubules do not terminate immediately after contacting the rectum, but typically are thrown into many loops and convolutions, with segments running radially around the rectum as well as looping anteriorly and posteriorly along the length of the rectum (Saini, 1964). The perinephric membrane follows the various convolutions and turns, always enclosing the tubules like a blanket. Several layers of tubules may lie on the rectum in those insects that live in the driest environments (Saini, 1964). A small **perirectal space** occurs between the epithelial cell layer of the rectum and the innermost layer of tubules within the perinephric chamber.

The food of insects, particularly the food of Coleoptera, is variable in water content. Some Coleoptera are phytophagous, eating food with relatively high water content, while others feed upon

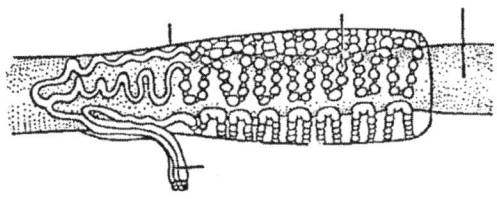

Figure 18.17 The cryptosolenic or cryptonephridial tubule system characteristic of most Lepidoptera and Coleoptera. (From Grimstone, A.V. et al., *Phil. Trans. R. Soc. Ser. B*, 253, 343, 1968. With permission.)

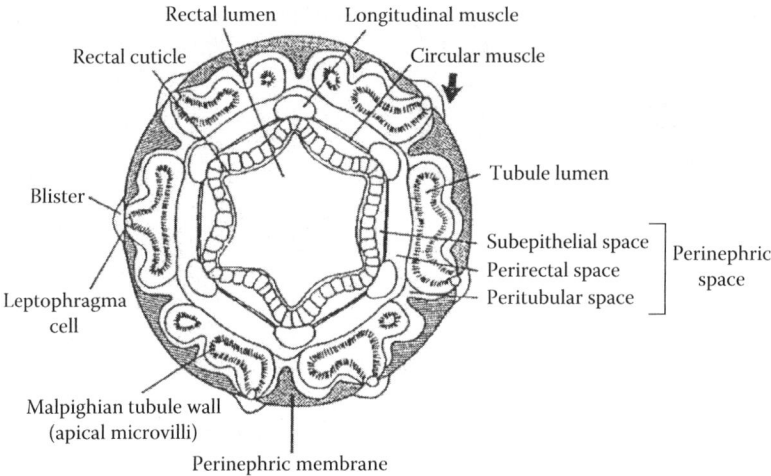

Figure 18.18 Cross-sectional view of the rectum with cryptonephridial tubules from the yellow mealworm, *T. molitor*. (From Grimstone, A.V. et al., *Phil. Trans. R. Soc. Ser. B*, 253, 343, 1968. With permission.)

dry stored grain products or similar dry food materials. In phytophagous Coleoptera, the posterior rectal region containing the fecal pellet just prior to its being expelled is smaller and has fewer convolutions of the tubules than in Coleoptera that feed upon dry food (Saini, 1964). Coleoptera living in very dry environments, such as *T. molitor* and other grain-infesting beetles and weevils, have a large cryptonephridial system enabling them to extract more water from the fecal pellet.

At frequent points in most of the Coleoptera, a cryptonephridial tubule in the outermost layer makes contact with the outer perinephric membrane through a single, highly modified cell of the tubule wall called the **leptophragma cell**. These points of contact when a tubule is separated from the hemolymph only by the thin cell membrane of the leptophragma cell and the very thin (at this particular point) perinephric membrane are called **leptophragmata**. Only two families of beetles, Ptinidae and Anobiidae, do not have leptophragmata, but none of the Lepidoptera studied by Saini (1964) have them.

In *T. molitor* and some other beetles, the thin perinephric membrane is expanded into a ***boursouflure***, a French word meaning a blister (Figure 18.19) above each leptophragma cell. The exact function of the *boursouflure* is uncertain. Ramsay (1964) and Grimstone et al. (1968) proposed that it may be a site of active secretion of ions from the hemolymph into the perinephric tubules, thus creating high osmotic values in the cryptonephridic tubules to aid passive movement of water down the osmotic gradient from rectal lumen to the tubule lumen. There is some evidence (Maddrell, 1971) that a high-molecular-weight compound (probably proteinaceous) is secreted into the perirectal space of *T. molitor* and that it first absorbs water from the rectum, with the water passed on to the tubules.

Cryptonephridial tubules of Lepidoptera have neither leptophragmata nor leptophragma cells. Thus, regardless of the function of leptophragma cells in Coleoptera, they are not essential to the system in Lepidoptera. As in the Coleoptera, the extent of layering of cryptonephridial loops on the rectum is correlated with the habitat and food eaten by larvae (Saini, 1964). In larvae feeding upon green plant matter, the innermost layer of tubules extends only about one-third the length of the anterior portion of the rectum, and there is only a single layer of tubules not packed very close together in the posterior half of the anterior rectum. In those larvae living in very dry conditions and eating dry food (e.g., Galleriidae, Phycitidae, and Tineidae), the convoluted tubules are closely packed together in both the inner and outer layers and extend the entire length of the anterior rectum. There is no cryptonephridial system of tubules in the aquatic Lepidoptera, *Parapoynx* (*Nymphula*)

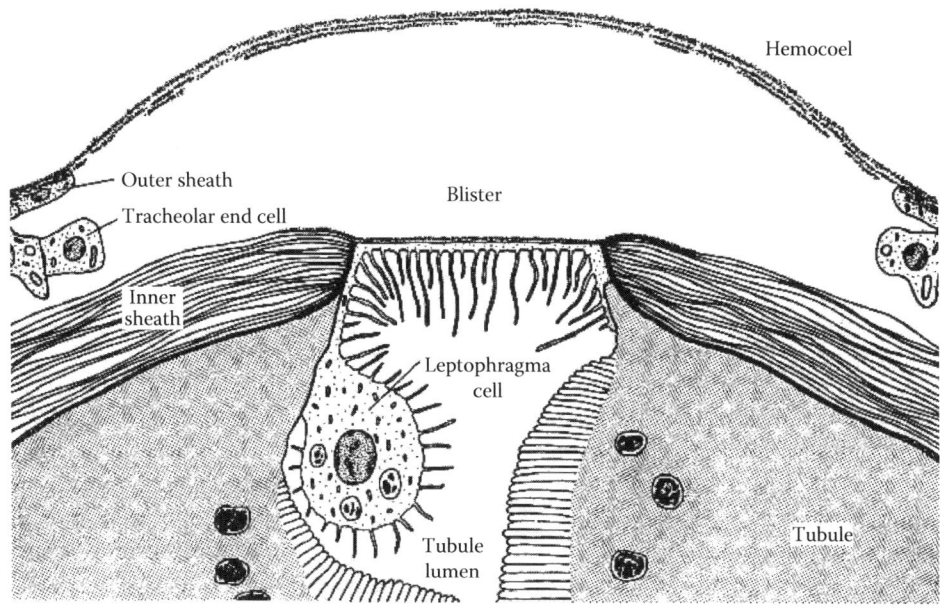

Figure 18.19 Diagram of the *boursouflure* and underlying leptophragma cell and cryptonephridial tubule cells in a larva of *T. molitor*. (From Grimstone, A.V. et al., *Phil. Trans. R. Soc. Ser. B*, 253, 343, 1968. With permission.)

stratiotata and *Cataclysta lemnata* (Pyraustidae). In those insects, the distal ends of the Malpighian tubules merely lie on the rectum in association with some fat body tissue and tracheae. Although the cryptonephric complex has been well studied anatomically, physiological studies are sparse. In a recent study, Azuma et al. (2012) describe two water-specific **aquaporins** (transmembrane protein guarded channels) in the cryptonephritic system of *Bombyx mori*. One of the aquaporins is expressed at the apical plasma membrane of the colon and rectal epithelial cells and the other in the basal plasma membrane of those same cells. It appears that these two aquaporins channel water reabsorption in the silkworm. Aquaporins, although only recently discovered in biological systems, appear to be universal in biological tissues (reviewed by Spring et al., 2009; Drake et al., 2010).

18.9 SELF-STUDY AND REVIEW QUESTIONS

1. In what ways do the Malpighian tubules and hindgut function together in excretion?
2. What are cryptonephritic tubules in insects? Which groups of insects have them?
3. What is the molecular driving mechanism in the formation of Malpighian tubule fluid?
4. Where, other than in the Malpighian tubules, is a proton pump an important functional mechanism in insects?
5. What are some functions of cations and anions in the excretory process?
6. Typically, what is the tonicity of the fluid formed in Malpighian tubules?
7. What are rectal papillae and what is/are their function(s)? What groups of insects have them?
8. What type of water reabsorptive cells do cockroaches have in the hindgut?
9. What is the meaning of the word "homeostasis?"
10. What are diuretic and antidiuretic hormones in insects? Explain a function for each type.
11. How is the diuretic process an important physiological adaptation for *R. prolixus*?
12. How might a diuretic hormone be functionally important in an insect like *T. molitor* that lives in a very dry environment?

13. Explain or show by a formula how the excretion of ammonium ions is a mechanism for acid control in an insect.
14. What are some of the chemical forms in which insects excrete excess nitrogen?
15. How is uric acid excretion especially adaptive for many insects?
16. Explain the structure and function of a cyptonephridic excretory system in insects.
17. What are aquaporins and how do they function in insects?

REFERENCES

Audsley, N., C. McIntosh, and J.E. Phillips. 1992a. Isolation of a neuropeptide from locust corpus cardiacum which influences ileal transport. *J. Exp. Biol.* 173: 261–274.

Audsley, N., C. McIntosh, and J.E. Phillips. 1992b. Actions of ion-transport peptide from locust corpus cardiacum on several hindgut transport processes. *J. Exp. Biol.* 173: 275–288.

Azuma, M., T. Nagae, M. Maruyama, N. Kataoka, and S. Miyake. 2012. Two water-specific aquaporins at the apical and basal plasma membranes of insect epithelia: Molecular basis for water recycling through the cryptonephric rectal complex of lepidopteran larvae. *J. Insect Physiol.* 58: 523–533.

Baldwin, I.T. 1991. Damage-induced alkaloids in wild tobacco, in D.W. Tallamy and M.J. Raupp (eds.), *Phytochemical Induction by Herbivores*. Wiley Interscience, New York, pp. 47–69.

Barrett, F.M. and W.G. Friend. 1970. Uric acid synthesis in *Rhodnius prolixus*. *J. Insect Physiol.* 16: 121–129.

Barrett, M. and I. Orchard. 1990. Serotonin-induced elevation of cyclic AMP levels in the epidermis of the blood-sucking bug *Rhodnius prolixus*. *J. Insect Physiol.* 36: 625–634.

Baumann, O. and B. Walz. 2012. The blowfly salivary gland—A model system for analyzing the regulation of plasma membrane V-ATPase. *J. Insect Physiol.* 58: 450–458.

Berridge, M.J. 1966. The physiology of excretion in the cotton stainer, *Dysdercus fasciatus* Signoret. IV. Hormonal control of excretion. *J. Exp. Biol.* 44: 553–566.

Berridge, M.J. and J.L. Oschman. 1969. A structural basis for fluid secretion by Malpighian tubules. *Tissue Cell* 1: 247–272.

Bertram, G., L. Schleithoff, P. Zimmermann, and A. Wessing. 1991. Bafilomycin A_1 is a potent inhibitor of urine formation by Malpighian tubules of *Drosophila hydei*: Is a vacuolar-type ATPase involved in ion and fluid secretion? *J. Insect Physiol.* 37: 201–209.

Beyenbach, K.W. 1995. Mechanism and regulation of electrolyte transport in Malpighian tubules. *J. Insect Physiol.* 41: 197–207.

Beyenbach, K.W. 2003. Transport mechanisms of diuresis in Malpighian tubules of insects. *J. Exp. Biol.* 206: 3845–3856.

Beyenbach, K.W. 2012. A dynamic paracellular pathway serves diuresis in mosquito Malpighian (renal) tubules. *Ann. N. Y. Acad. Sci.* 1258: 166–176.

Beyerbach, K.W., D.J. Aneshansley, T.L. Pannabecker, R. Masia, D. Gray, and M.-J. Yu. 2000a. Oscillations of voltage and resistance in Malpighian tubules of *Aedes aegypti*. *J. Insect Physiol.* 46: 321–333.

Beyenbach, K.W., S. Baumgart, K. Lau, P.M. Piermarini, and S. Zhang. 2009. Signaling to the apical membrane and to the paracellular pathway: Changes in the cytosolic proteome of *Aedes* Malpighian tubules. *J. Exp. Biol.* 212: 329–340.

Beyenbach, K.W., T.L. Pannabecker, and W. Nagel. 2000b. Central role of the apical membrane H^+-ATPase in electrogenesis and epithelial transport in Malpighian tubules. *J. Exp. Biol.* 203: 1459–1468.

Beyenbach, K.W. and P.M. Piermarini. 2011. Transcellular and paracellular pathways of transepithelial fluid secretion in Malpighian (renal) tubules of the yellow fever mosquito *Aedes aegypti*. *Acta Physiol.* 202: 387–407.

Beyenback, K.W., H. Skaer, and J.A.T. Dow. 2010. The developmental, molecular, and transport biology of Malpighian tubules. *Annu. Rev. Entomol.* 55: 351–374.

Bradley, T.J. 1985. The excretory system: Structure and physiology, in G.A. Kerkut and L.I. Gilbert (eds.), *Comparative Insect Physiology, Biochemistry, and Pharmacology*, vol. 4. Pergamon Press, New York, pp. 421–465.

Brower, L.P. 1969. Ecological chemistry. *Sci. Am.* 220: 22–29.

Brown, A.W.A. 1936. The excretion of ammonia and uric acid during the larval life of certain muscoid flies. *J. Exp. Biol.* 13: 131–139.

Brown, A.W.A. 1938. The nitrogen metabolism of an insect (*Lucilia sericata* Meig.) I. Uric acid, allantoin, and uricase. *Biochem. J.* 32: 895–902.
Bursell, E. 1967. The excretion of nitrogen in insects. *Adv. Insect Physiol.* 4: 33–67.
Bursell, E. 1970. *An Introduction to Insect Physiology*. Academic Press, London, U.K.
Carpenter, V.K., L.L. Drake, S.E. Aguirre, D.P. Price, S.D. Rodriguez, and I.A. Hansen. 2012. SLC7 amino acid transporters of the yellow fever mosquito *Aedes aegypti* and their role in fat body TOR signaling and reproduction. *J. Insect Physiol.* 58: 513–522.
Chamberlin, M.E. and J.E. Philips. 1982. Regulation of hemolymph amino acid levels and active secretion of proline by Malpighian tubules of locusts. *Can. J. Zool.* 60: 2745–2752.
Chamberlin, M.E. and J.E. Philips. 1983. Oxidative metabolism in the locust rectum. *J. Comp. Physiol. B* 151: 191–198.
Clark, T.M. and T.J. Bradley. 1993. Short term changes in hemolymph properties of larval *Aedes aegypti* in response to physiological challenges affect Malpighian tubule secretion rates in vitro. *Am. Zool.* 33: 43A.
Clark, T.M. and T.J. Bradley. 1996. Stimulation of Malpighian tubules from larval *Aedes aegypti* by secretagogues. *J. Insect Physiol.* 42: 593–602.
Coast, G.M. 1988. Fluid secretion by single isolated Malpighian tubules of the house cricket, *Acheta domesticus*, and their response to diuretic hormone. *Physiol. Entomol.* 13: 381–391.
Coast, G.M. 1989. Stimulation of fluid secretion by single isolated Malpighian tubules of the house cricket, *Acheta domesticus*. *Physiol. Entomol.* 14: 21–30.
Coast, G.M. 2009. Neuroendocrine control of ionic homeostasis in blood-sucking insects. *J. Exp. Biol.* 212: 378–386.
Coast, G.M., R.J. Nachman, and J. Lopez. 2011. The control of Malpighian tubule secretion in a predacious hemipteran insect, the spined soldier bug *Podisus maculiventris* (Heteroptera, Pentatomidae). *Peptides* 32: 493–499.
Coast, G.M. and C.H. Wheeler. 1990. The distribution and relative potency of diuretic peptides in the house cricket, *Acheta domesticus*. *Physiol. Entomol.* 15: 13–21.
Cochran, D.G. 1973. Comparative analysis of excreta from twenty cockroach species. *Comp. Biochem. Physiol.* 46A: 409–419.
Cochran, D.G. 1975. Excretion in insects, in D.J. Candy and B.A. Kilby (eds.), *Insect Biochemistry and Function*. Chapman & Hall, London, U.K., pp. 177–281.
Cochran, D.G. 1985a. Nitrogen excretion in cockroaches. *Annu. Rev. Entomol.* 30: 29–49.
Cochran, D.G. 1985b. Nitrogen excretion, in G.A. Kerket and L.I. Gilbert (eds.), *Comprehensive Insect Physiology, Biochemistry and Pharmacology*, vol 4. Pergamon Press, New York, pp. 467–506.
Davies, S.A. and S. Terhzaz. 2009. Organellar calcium signalling mechanisms in *Drosophila* epithelial function. *J. Exp. Biol.* 212: 387–400.
Davies, S.-A., C. Overend, S. Sebastian, M. Cundall, P. Cabrero, J.A.T. Dow, and S. Terhzaz. 2012. Immune and stress response "cross-talk" in the *Drosophila* Malpighian tubule. *J. Insect Physiol.* 58: 488–497.
Dijkstra, S., A. Leyssens, E. van Kerkhove, W. Zeiske, and P. Steels. 1995. A cellular pathway for Cl⁻ during fluid secretion in ant Malpighian tubules: Evidence from ion-sensitive microelectrode studies? *J. Insect Physiol.* 41: 695–703.
Dircksen, H. 2009. Insect ion transport peptides are derived from alternatively spliced genes and differentially expressed in the central and peripheral nervous system *J. Exp. Biol.* 212: 401–412.
Dow, J.A.T. 2009. Insights into the Malpighian tubule from functional genomics. *J. Exp. Biol.* 212: 435–445.
Dow, J.A.T. 2012. The versatile stellate cell—More than just a space filler. *J. Insect Physiol.* 58: 467–472.
Dow, J.A.T. and S.A. Davies. 2006. The Malpighian tubule: Rapid insights from post-genomic biology. *J. Insect Physiol.* 52: 365–378.
Drake, L.L., D.Y. Boudko, O. Marinotti, V.K. Carpenter, A.L. Dawe, and I.A. Hansen. 2010. The aquaporin gene family of the yellow fever mosquito, *Aedes aegypti*. *PLoS One* 5, e15578.
Evans, J.M., A.K. Allan, S.A. Davies, and J.A.T. Dow. 2005. Sulphonylurea sensitivity and enriched expression implicate inward rectifier K⁺ channels in *Drosophila melanogaster* renal function. *J. Exp. Biol.* 208: 3771–3783.
Forgac, M. 1989. Structure and function of vacuolar class of ATP-driven proton pumps. *Physiol. Rev.* 69: 765–796.
Gámez, A.D., A.M. Gutiérrez, R. García, and G. Whittembury. 2012. Recent experiments towards a model for fluid secretion in *Rhodnius* upper Malpighian tubules (UMT). *J. Insect Physiol.* 58: 543–550.

Goldbard, G.A., J.R. Sauer, and R.R. Mills. 1970. Hormonal control of excretion in the American cockroach. II. Preliminary purification of a diuretic and an antidiuretic hormone. *Comp. Gen. Pharmacol.* 1: 82–86.

Grimstone, A.V., A.M. Mullinger, and J.A. Ramsay. 1968. Further studies on the rectal complex of the mealworm, *Tenebrio molitor* L. (Coleoptera, Tenebrionidae). *Phil. Trans. R. Soc. Ser. B* 253: 343–382.

Gupta, B.J. and M.J. Berridge. 1966. Fine structural organization of the rectum in the blowfly, *Calliphora erythrocephala* (Meig.) with special reference to connective tissue, tracheae and neurosecretory innervation in the rectal papillae. *J. Morph.* 120: 23–82.

Harrison, J.F. 1994. Respiratory and ionic aspects of acid-base regulation in insects: An introduction. *Physiol. Zool.* 67: 1–6.

Harrison, J.F. 1995. Nitrogen metabolism and excretion in locusts, in P.J. Walsh and R. Wright (eds.), *Nitrogen Metabolism and Excretion.* CRC Press, Boca Raton, FL, pp. 119–131.

Harrison, J.F. 2001. Insect acid-base physiology. *Annu. Rev. Entomol.* 46: 221–250.

Harrison, J.F. and M.J. Kennedy. 1994. In vivo studies of the acid-base physiology of grasshoppers: The effect of feeding state on acid-base and nitrogen excretion. *Physiol. Zool.* 67: 120–141.

Harrison, J.F. and J.E. Phillips. 1992. Recovery from acute haemolymph acidosis in unfed locusts II. Role of ammonium and titratable acid excretion. *J. Exp. Biol.* 165: 97–110.

Harvey, W.R. and M.A. Xiang. 2012. K^+ pump: From caterpillar midgut to human cochlea. *J. Insect Physiol.* 58: 590–598.

Haydak, M.H. 1953. Influence of the protein level of the diet on the longevity of cockroaches. *Ann. Entomol. Soc. Am.* 46: 547–560.

Hegarty, J.L., B. Zhang, T.L. Pannabecker, D.H. Petzel, M.D. Baustian, and K.W. Beyenbach. 1991. Dibutyryl cAMP activates bumetanide-sensitive electrolyte transport in Malpighian tubules. *Am. J. Physiol.* 261: C521–C529.

Hirata, T., A. Czapar, L. Brin, A. Haritonova, D.P. Bondeson, P. Linser, P. Cabrero, J. Thompson, J.A.T. Dow, and M.F. Romero. 2012. Ion and solute transport by Prestin in *Drosophila* and *Anopheles*. *J. Insect Physiol.* 58: 563–569.

Hitchcock, F.A. and J.G. Haub. 1941. The interconversion of foodstuffs in the blowfly (*Phormia regina*) during metamorphosis. I. Respiratory metabolism and nitrogen excretion. *Ann. Entomol. Soc. Am.* 34: 17–25.

Hopkin, R., J.H. Anstee, and K. Bowler. 2001. An investigation into the effects of inhibitors of fluid production by *Locusta* Malpighian tubule Type I cells on their secretion and elemental composition. *J. Insect Physiol.* 47: 359–367.

Hopkins, C.R. 1967. The fine-structural changes observed in the rectal papillae of the mosquito *Aedes aegypti*, L. and their relation to the epithelial transport of water and inorganic ions. *J. R. Micros. Soc.* 86: 235–252.

Huss, M. and H. Wieczorek. 2009. Inhibitors of V-ATPases: Old and new players. *J. Exp. Biol.* 212: 341–346.

Ianowski, J.P. and M.J. O'Donnell. 2001. Transepithelial potential in Malpighian tubules of *Rhodnius prolixus*: Lumen-negative voltages and the triphasic response to serotonin. *J. Insect Physiol.* 47: 411–421.

Ianowski, J.P., J.-P. Paluzzi, V.A. Te Brugge, and I. Orchard. 2010. The antidiuretic neurohormone RhoprCAPA-2 downregulates fluid transport across the anterior midgut in the blood-feeding insect *Rhodnius prolixus*. *American J. Physiol.-Regul. Integr. Comp. Physiol.* 298: R548–R557.

Irzykiewicz, H. 1955. Xanthine oxidase of the clothes moth, *Tineola bisselliella*, and some other insects. *Aust. J. Biol. Sci.* 8: 369–377.

Isaacson, L.C., S.W. Nicolson, and D.W. Fisher. 1989. Electrophysiological and cable parameters of perfused beetle Malpighian tubules. *Am. J. Physiol.* 257: R1190–R1198.

Jagge, C.L. and P.V. Pietrantonio. 2008. Diuretic hormone 44 receptor in Malpighian tubules of the mosquito *Aedes aegypti*: Evidence for transcriptional regulation paralleling urination. *Insect Mol. Biol.* 17: 413–426.

Johnson, E.C., O.T. Shafer, J.S. Trigg, J. Park, D.A. Schooley, J.A. Dow, and P.H. Taghert. 2005. A novel diuretic hormone receptor in *Drosophila*: Evidence for conservation of CGRP signaling. *J. Exp. Biol.* 208: 1239–1246.

Kane, P.M. and K.J. Parra. 2000. Assembly and regulation of the yeast vacuolar H^+-ATPase. *J. Exp. Biol.* 203: 81–87.

Kataoka, H., R.G. Troetschler, J.P. Li, S.J. Kramer, R.L. Carney, and D.A. Schooley. 1989. Isolation and identification of a diuretic hormone from the tobacco hornworm, *Manduca sexta*. *Proc. Natl. Acad. Sci. USA* 86: 2976–2980.

Kim, I.S. and J.H. Spring. 1992. Excretion in the house cricket: Relative contribution of distal and mid-tubule to diuresis. *J. Insect Physiol.* 38: 373–381.

Leader, J.P. and M.J. O'Donnell. 2005. Transepithelial transport of fluorescent p-glycoprotein and MRP2 substrates by insect Malpighian tubules: Confocal microscopic analysis of secreted fluid droplets. *J. Exp. Biol.* 208: 4363–4376.

Lehmberg, E., R.B. Ota, K. Furuya, D.S. King, S.W. Applebaum, H.-J. Ferenz, and D.A. Schooley. 1991. Identification of a diuretic hormone of *Locusta migratoria*. *Biochem. Biophy. Res. Commun.* 179: 1036–1041.

Lehmberg, E., D.A. Schooley, H.-J. Ferenz, and S.W. Applebaum. 1993. Characteristics of *Locusta migratoria* diuretic hormone. *Arch. Insect Biochem. Physiol.* 22: 133–140.

Leyssens, A., P. Steels, E. Lohrmann, R. Weltens, and E. Van Kerkhove. 1992. Intrinsic regulation of K^+ transport in Malpighian tubules (*Formica*): Electrophysiological Evidence. *J. Insect Physiol.* 38: 431–446.

Leyssens, A., S.-L. Zhang, E. Van Kerkhove, and P. Steels. 1993. Both dinitrophenol and Ba^{2+} reduce KCl and fluid secretion in Malpighian tubules of *Formica polyctena*: The role of the apical H^+ and K^+ concentration gradient. *J. Insect Physiol.* 39: 1061–1073.

Liao, A., N. Audsley, and D.A. Schooley. 2000. Antidiuretic effects of a factor in brain/corpora cardiaca/corpora allata extract on fluid reabsorption across the cryptonephric complex of *Manduca sexta*. *J. Exp. Biol.* 203: 605–615.

Linser, P.J., M.N. Oviedo, T. Hirata, T.J. Seron, K.E. Smith, P.M. Piermarini, and M.F. Romero. 2012. Slc4-like anion transporter of the larval mosquito alimentary canal. *J. Insect Physiol.* 58: 551–562.

Linton, S.M. and M.J. O'Donnell. 2000. Novel aspects of the transport of organic anions by the Malpighian tubules of *Drosophila melanogaster*. *J. Exp. Biol.* 203: 3575–3584.

Maddrell, S. 2009. Insect homeostasis: Past and future. *J. Exp. Biol.* 212: 446–451.

Maddrell, S.H.P. 1963. Excretion in the blood-sucking bug, *Rhodnius prolixus* Stål. I. The control of diuresis. *J. Exp. Biol.* 40: 247–256.

Maddrell, S.H.P. 1964a. Excretion in the blood-sucking bug, *Rhodnius prolixus* Stål. II. The normal course of diuresis and the effect of temperature. *J. Exp. Biol.* 41: 163–176.

Maddrell, S.H.P. 1964b. Excretion in the blood-sucking bug, *Rhodnius prolixus* Stål. III. The control of the release of the diuretic hormone. *J. Exp. Biol.* 41: 459–472.

Maddrell, S.H.P. 1966. The site of release of the diuretic hormone in *Rhodnius*—A new neurohaemal system in insects. *J. Exp. Biol.* 45: 499–508.

Maddrell, S.H.P. 1971. The mechanisms of insect excretory systems. *Adv. Insect Physiol.* 8: 199–331.

Maddrell, S.H.P. 1977. Insect Malpighian tubules, in B.L. Gupta, R.B. Moreton, J.L. Oschman, and B.J. Wall (eds.), *Transport of Ions and Water in Animals*. Academic Press, London, U.K., pp. 541–569.

Maddrell, S.H.P. 1980. Characteristics of epithelial transport in insect Malpighian tubules, in F. Bonner and A. Kleinzeller (eds.), *Current Topics in Membranes and Transport*, vol. 14. Academic Press, New York, pp. 427–463.

Maddrell, S.H.P. 1991. The fastest fluid-secreting cell known: The upper Malpighian tubule cell of *Rhodnius*. *BioEssays* 13: 357–362.

Maddrell, S.H.P., W.S. Herman, R.L. Mooney, and J.A. Overton. 1991. 5-Hydroxytryptamine: A second diuretic hormone in *Rhodnius prolixus*. *J. Exp. Biol.* 156: 557–566.

Maddrell, S.H.P. and M.J. O'Donnell. 1992. Insect Malpighian tubules: V-ATPase action in ion and fluid transport. *J. Exp. Biol.* 172: 417–430.

Meredith, J., M. Ring, A. Macins, J. Marschall, N.N. Cheng, D. Theilmann, H.W. Brock, and J.E. Phillips. 1996. Locust ion transport peptide (ITP): Primary structure cDNA and expression in a baculovirus system. *J. Exp. Biol.* 199: 1053–1061.

Mills, R.R. 1967. Hormonal control of excretion in the American cockroach. I. Release of a diuretic hormone from the terminal abdominal ganglion. *J. Exp. Biol.* 46: 35–41.

Mitchell, H.K., E. Glassman, and E. Hadorn. 1959. Hypoxanthine in *rosy2* and *maroon-like* mutants of *Drosophila melanogaster*. *Science* 129: 268–269.

Mitlin, N. and D.H. Vickers. 1964. Guanine in the excreta of the boll weevil. *Nature* 203: 1403–1404.

Miyauchi, J.T., P.M. Piermarini, J.D. Yang, D.M. Gilligan, and K.W. Beyenbach. 2013. Roles of PKC and phospho-adducin in transepithelial fluid secretion by Malpighian tubules of the yellow fever mosquito. *Tissues Barriers* 11: 1.e23120.

Mordue, W. and P.J. Morgan. 1985. Chemistry of peptide hormones, in G.A. Kerkut and L.I. Gilbert (eds.), *Comprehensive Insect Physiology, Biochemistry and Pharmacology*, vol. 7. Pergamon Press, New York, pp. 153–183.

Morita, T. 1958. Purine catabolism in *Drosophila melanogaster*. *Science* 128: 1135.

Muench, S.P., M. Huss, C.F. Song, C. Phillips, H. Wieczorek, J. Trinick, and M.S. Harrison. 2009. Cryo-electron microscopy of the vacuolar ATPase motor reveals its mechanical and regulatory complexity. *J. Mol. Biol.* 386: 989–999.

Mullins, D.E. and D.G. Cochran. 1972. Nitrogen excretion in cockroaches: Uric acid is not a major product. *Science* 177: 699–701.

Mullins, D.E. and C.B. Keil. 1980. Paternal investment of urates in cockroaches. *Nature* 283: 567–569.

Nation, J.L. 1963. Identification of xanthine in excreta of the greater wax moth, *Galleria mellonella* (L). *J. Insect Physiol.* 9: 195–200.

Nation, J.L. and R.L. Patton. 1961. A study of nitrogen excretion in insects. *J. Insect Physiol.* 6: 299–308.

Nation, J.L. and K.K. Thomas. 1965. Quantitative studies on purine excretion in the greater wax moth, *Galleria mellonella*. *Ann. Entomol. Soc. Am.* 58: 983–885.

Nicolson, S.W. 1991. Diuresis or clearance: Is there a physiological role for the "diuretic hormone" of the desert beetle *Onymacris*? *J. Insect Physiol.* 37: 447–452.

Nicolson, S.W. 1993. The ionic basis of fluid secretion in insect Malpighian tubules: Advances in the last ten years. *J. Insect Physiol.* 39: 451–458.

Nicolson, S.W. 2009. Water homeostasis in bees, with the emphasis on sociality. *J. Exp. Biol.* 212: 429–434.

Nicolson, S.W. and S.A. Hanrahan. 1986. Diuresis in a desert beetle? Hormonal control of the Malpighian tubules of *Onymacris plana* (Coleoptera: Tenebrionidae). *J. Comp. Physiol. B* 156: 407–413.

Nicolson, S.W. and L. Isaacson. 1990. Patch clamp of the basal membrane of beetle Malpighian tubules: Direct demonstration of potassium channels. *J. Insect Physiol.* 36: 877–884.

O'Donnell, M. 2008. Insect excretory mechanisms, in S.J. Simpson (ed.), *Advances in Insect Physiology*, vol. 35. Elsevier, Amsterdam, the Netherlands, pp. 1–105.

O'Donnell, M.J. 2009. Too much of a good thing: How insects cope with excess ions or toxins in the diet. *J. Exp. Biol.* 212: 363–372.

O'Donnell, M.J., S.H.P. Maddrell, H.B. Skaer, and J.B. Harrison. 1985. Elaborations of the basal surface of the cells of the Malpighian tubules of an insect. *Tissue Cell* 17: 865–881.

O'Donnell, M.J., M.R. Rheault, S.A. Davies, P. Rosay, B.J. Harvey, S.H.P. Maddrell, K. Kaiser, and J.A.T. Dow. 1998. Hormonally controlled chloride movement across *Drosophila* tubules is via ion channels in stellate cells. *Am. J. Physiol.* 43: R1039–R1049.

O'Donnell, M.J. and J.H. Spring. 2000. Modes of control of insect Malpighian tubules: Synergism, antagonism, cooperation and autonomous regulation. *J. Insect Physiol.* 46: 107–117.

Orchard, I. and J.-P. Paluzzi. 2009. Diuretic and antidiuretic hormones in the blood gorging bug *Rhodnius prolixus*. *Ann. N. Y. Acad. Sci.* 1163: 501–503.

Paluzzi, J.P. and I. Orchard. 2010. A second gene encodes the anti-diuretic hormone in the insect, *Rhodnius prolixus*. *Mol. Cell. Endocrinol.* 317: 53–63.

Paluzzi, J.-P., Y. Park, R.J. Nachman, and I. Orchard. 2010. Isolation, expression analysis and functional characterization of the first antidiuretic hormone receptor in insects. *Proc. Natl. Acad. Sci. USA* 107: 10290–10295.

Paluzzi, J.-P., C. Yeung, and M.J. O'Donnell. 2013. Investigations of the signaling cascade involved in diuretic hormone stimulation of Malpighian tubule fluid secretion in *Rhodnius prolixus*. *J. Insect Physiol.* 59: 1179–1185.

Paluzzi, J.-P.V., W. Naikkhwah, and M.J. O'Donnell. 2012. Natriuresis and diuretic synergism in *R. prolixus* upper Malpighian tubules is inhibited by the anti-diuretic hormone, RhoprCAPA-α2. *J. Insect Physiol.* 58: 534–542.

Pannabecker, T. 1995. Physiology of the Malpighian tubule. *Annu. Rev. Entomol.* 40: 493–510.

Pannabecker, T.L., T.K. Hayes, and K.W. Beyenbach. 1993. Regulation of epithelial shunt conductance by the peptide leucokinin. *J. Membr. Biol.* 132: 63–76.

Peach, J.L. and J.E. Phillips. 1991. Metabolic support of chloride-dependent short-circuit current across the locust (*Schistocerca gregaria*) ileum. *J. Insect Physiol.* 37: 255–260.

Phillips, J.E. 1964. Rectal absorption in the desert locust, *Schistocerca gregaria* Forskal II. The nature of the excretory process. *J. Exp. Biol.* 41: 68–80.

Phillips, J.E. 1981. Comparative physiology of insect renal function. *Am. J. Physiol.* 241: R241–R257.
Phillips, J.E., N. Audsley, R. Lechleitner, B. Thomson, J. Meredith, and M. Chamberlin. 1988. Some major transport mechanisms of insect absorptive epithelia. *Comp. Biochem. Physiol.* 90A: 643–650.
Phillips, J.E. and A.A. Dockrill. 1968. Molecular sieving of hydrophilic molecules by the rectal intima of the desert locust (*Schistocerca gregaria*). *J. Exp. Biol.* 48: 521–532.
Phillips, J.E., J. Hanrahan, M. Chamberlin, and B. Thomson. 1986. Mechanisms and control of reabsorption in insect hindgut. *Adv. Insect Physiol.* 19: 329–422.
Phillips, J.E., R.E. Thomson, N. Audsley, J.L. Peach, and A.P. Stagg. 1994. Mechanisms of acid-base transport and control in locust excretory system. *Physiol. Zool.* 67: 95–119.
Piermarini, P.M., M.F. Rouhier, M. Schepel, C. Kosse, and K.W. Beyenbach. 2013. Cloning and functional characterization of inward-rectifying potassium (Kir) channels from Malpighian tubules of the mosquito *Aedes Aegypti*. *Insect Biochem. Mol. Biol.* 43: 75–90.
Pollock, V.P., J. McGettigan, P. Cabrero, I.M. Maudlin, J.A.T. Dow, and S.A. Davies. 2004. Conservation of capa peptide-induced nitric oxide signaling in Diptera. *J. Exp. Biol.* 207: 4135–4145.
Proux, J.P., C.A. Miller, J.P. Li, R.L. Carney, A. Girardie, M. Delaage, and D.A. Schooley. 1987. Identification of an arginine vasopressin-like diuretic hormone from *Locusta migratoria*. *Biochem. Biophys. Res. Comm.* 149: 180–186.
Ramsay, J.A. 1953. Active transport of potassium by the Malpighian tubules of insects. *J. Exp. Biol.* 30: 358–369.
Ramsay, J.A. 1954. Active transport of water by the Malpighian tubules of the stick insect, *Dixippus morosus* (Orthoptera, Phasmidae). *J. Exp. Biol.* 31: 104–113.
Ramsay, J.A. 1955a. The excretory system of the stick insect *Dixippus morosus* (Orthoptera, Phasmidae). *J. Exp. Biol.* 32: 183–199.
Ramsay, J.A. 1955b. The excretion of sodium, potassium and water by the Malpighian tubules of the stick insect, *Dixippus morosus* (Orthoptera, Phasmidae). *J. Exp. Biol.* 32: 200–216.
Ramsay, J.A. 1956. Excretion by the Malpighian tubules of the stick insect, *Dixippus morosus* (Orthoptera, Phasmidae): Calcium, magnesium, chloride, phosphate and hydrogen ions. *J. Exp. Biol.* 33: 697–709.
Ramsay, J.A. 1958. Excretion by the Malpighian tubules of the stick insect, *Dixippus morosus* (Orthoptera, Phasmidae): Amino acids, sugars and urea. *J. Exp. Biol.* 35: 871–891.
Ramsay, J.A. 1964. The rectal complex of the mealworm *Tenebrio molitor* L. (Coleoptera, Tenebrionidae). *Phil Trans. R. Soc. Ser. B* 248: 279–314.
Raphemot, R., M.F. Rouhier, C.R. Hopkins, R.D. Gogliotti, K.M. Lovell, R.M. Hine, D. Ghosalkar et al. 2013. Eliciting renal failure in mosquitoes with a small-molecule inhibitor of inward-rectifying potassium channels. *PLoS One* 8(5): e64905.
Razet, P. 1966. Les Éléments terminaux du catabolisme azoté chez les insectes. *Annee Biol.* 5: 43–73.
Reagan, J.D., J.P. Li, R.L. Carney, and S.J. Kramer. 1993. Characterization of a diuretic hormone receptor from the tobacco hornworm, *Manduca sexta*. *Arch. Insect Biochem. Physiol.* 23: 135–145.
Rheault, M., J.S. Plaumann, and M.J. O'Donnell. 2006. Tetraethylammonium and nicotine transport by the Malpighian tubules of insects. *J. Insect Physiol.* 52: 487–498.
Riegel, J.A. 1966. Micropuncture studies of formed body secretion by the excretory organs of the crayfish, frog, and stick insect. *J. Exp. Biol.* 44: 379–385.
Ruiz-Sanchez, E. and M.J. O'Donnell. 2006. Characterization of salicylate uptake across the basolateral membrane of the Malpighian tubules of *Drosophila melanogaster*. *J. Insect Physiol.* 52: 920–928.
Ruiz-Sanchez, E. and M.J. O'Donnell. 2007a. Characterization of transepithelial transport of salicylate by the Malpighian tubules of *D. melanogaster* and the effects of changes in fluid secretion rate. *Physiol. Entomol.* 32: 157–166.
Ruiz-Sanchez, E. and M.J. O'Donnell. 2007b. Effects of chronic exposure to dietary salicylate on elimination and renal excretion of salicylate by *Drosophila melanogaster* larvae. *J. Exp. Biol.* 210: 2464–2471.
Ruiz-Sanchez, E., M.C. van Walderveen, A. Livingston, and M.J. O'Donnell. 2007. Transepithelial transport of salicylate by the Malpighian tubules of insects from different orders. *J. Insect Physiol.* 53: 1034–1045.
Saini, R.S. 1964. Histology and physiology of the cryptonephridial system of insects. *Trans. R. Entomol. Soc. Lond.* 116: 347–392.
Santini, M.S. and J.R. Rondereos. 2007. Allatotropin-like peptide released by Malpighian tubules induces hindgut activity associated with diuresis in the Chagas disease vector *Triatoma infestans* (Klug). *J. Exp. Biol.* 210: 1986–1991.

Schal, C. and W.J. Bell. 1982. Ecological correlates of paternal investment of urates in a tropical cockroach. *Science* 218: 170–172.
Singh, S.R. and S.X. Hou. 2009. Multipotent stem cells in the Malpighian tubules of adult *Drosophila melanogaster*. *J. Exp. Biol.* 212: 413–423.
Spring, J.H. 1990. Endocrine regulation of diuresis in insects. *J. Insect Physiol.* 36: 13–22.
Spring, J.H. and S.R. Hazelton. 1987. Excretion in the house cricket (*Acheta domesticus*): Stimulation of diuresis by tissue homogenates. *J. Exp. Biol.* 129: 63–81.
Spring, J.H., A.M. Morgan, and S.R. Hazelton. 1988. A novel target for antidiuretic hormone in insects. *Science* 241: 1096–1098.
Spring, J.H., S.R. Robichaux, and J.A. Hamlin. 2009. The role of aquaporins in excretion in insects. *J. Exp. Biol.* 212: 358–362.
Staddon, B.W. 1955. The excretion and storage of ammonia by the aquatic larva of *Sialis lutaria* (Neuroptera). *J. Exp. Biol.* 32: 84–94.
Staddon, B.W. 1959. Nitrogen excretion in nymphs of *Aeshna cyanea* (Müll.) (Odonata, Anisoptera). *J. Exp. Biol.* 36: 566–574.
Te Brugge, V., J.-P. Paluzzi, D.A. Schooley, and I. Orchard. 2011. Identification of the elusive peptidergic diuretic hormone in the blood-feeding bug *Rhodnius prolixus*. A CRF-related peptide. *J. Exp. Biol.* 214: 371–381.
Tiburcy, F., K.W. Beyenbach, and H. Wieczorek. 2013. Protein kinase A-dependent and–independent activation of the V-ATPase in Malpighian tubules of *Aedes aegypti*. *J. Exp. Biol.* 216: 881–891.
Vanderveken, M. and M.J. O'Donnell. 2014. Effects of diuretic hormone 31, drosokinin, and allatostatin A on transepithelial K$^+$ transport and contraction frequency in the midgut and hindgut of larval *Drosophila melanogaster*. *Arch. Insect Biochem. Physiol.* 85: 76–93.
von Dungern, P. and H. Briegel. 2001. Protein catabolism in mosquitoes: Ureotely and uricotely in larval and imaginal *Aedes aegypti*. *J. Insect Physiol.* 47: 131–141.
Wall, B.J. and J.L. Ochsman. 1970. Water and solute uptake by rectal pads of *Periplaneta americana*. *Am. J. Physiol.* 218: 1208–1215.
Wang, S., A.B. Rubenfeld, T.K. Hayes, and K.W. Beyenbach. 1996. Leucokinin increases paracellular permeability in insect Malpighian tubules. *J. Exp. Biol.* 199: 2537–2542.
Weihrauch, D., A. Donini, and M.J. O'Donnell. 2012. Ammonia transport by terrestrial and aquatic insects. *J. Insect Physiol.* 58: 473–487.
Weltens, R., A. Leyssens, S.L. Zhang, E. Lohrmann, P. Steels, and E. Van Kerkhove. 1992. Unmasking of the apical electrogenic H pump in isolated Malpighian tubules (*Formica polyctena*) by use of barium. *Cell. Physiol. Biochem.* 2: 101–116.
Weng, X.-H., M. Huss, H. Wieczorek, and K.W. Beyenbach. 2003. The V-type H$^+$-ATPase in Malpighian tubules of *Aedes aegypti*: Localization and activity. *J. Exp. Biol.* 206: 221–2219.
Wessing, A. and D. Eichelberg. 1975. Ultrastructural aspects of transport and accumulation of substances in the Malpighian tubules. *Fortsch. Zool.* 23: 148–172.
Wessing, A. and D. Eichelberg. 1978. Malpighian tubules, rectal papillae and excretion, in M. Ashburner and T.R.F. Wright (eds.), *The Genetics and Biology of Drosophila*, vol. 2c. Academic Press, London, U.K., pp. 1–42.
Wheeler, C.H. and G.M. Coast. 1990. Assay and characterization of diuretic factors in insects. *J. Insect Physiol.* 36: 23–34.
Wheelock, G.D., D.H. Petzel, J.D. Gillet, K.W. Beyenbach, and H.H. Hagedorn. 1988. Evidence for hormonal control of diuresis after a blood meal in the mosquito *Aedes aegypti*. *Arch. Insect Biochem. Physiol.* 7: 75–89.
Wieczorek, H., G. Grüber, W.R. Harvey, M. Huss, and H. Merzendorfer. 1999. The plasma membrane H$^+$ V-ATPase from tobacco hornworm midgut. *J. Bioenerg. Biomembr.* 31: 67–74.
Wieczorek, H., G. Grüber, W.R. Harvey, M. Huss, H. Merzendorfer, and W. Zeiske. 2000. Structure and regulation of insect plasma membrane H$^+$ V-ATPase. *J. Exp. Biol.* 203: 127–135.
Wigglesworth, V.B. 1931. The physiology of excretion in a blood-sucking insect, *Rhodnius prolixus* (Hemiptera, Reduviidae). III. The mechanism of uric acid excretion. *J. Exp. Biol.* 8: 443–451.
Xiang, M.A., P.J. Linser, D.A. Price, and W.R. Harvey. 2012. Localization of two NA$^+$ or K$^+$-H$^+$ antiporters, AgNHA1 and AgNHA2, in *Anopheles gambiae* larval Malpighian tubules and the functional expression of AgNHA2 in yeast. *J. Insect Physiol.* 58: 570–579.

Yu, M.-J. and K.W. Beyenbach. 2001. Leucokinin and the modulation of the shunt pathway in Malpighian tubules. *J. Insect Physiol.* 47: 263–276.

Yu, M.-J. and K.W. Beyenbach. 2004. Effects of leucokinin-VIII on *Aedes* Malpighian tubule segments lacking stellate cells. *J. Exp. Biol.* 207: 519–526.

Zhang, S.-L., A. Leyssens, E. Van Kerkhove, R. Weltens, W. Van Driessche, and P. Steels. 1993. Electrophysiological evidence for the presence of an apical H-ATPase in Malpighian tubules of *Formica polyctena*: Intracellular and luminal pH measurements. *Pflug. Archs.* 426: 443–451.

CHAPTER **19**

Semiochemicals

PREVIEW

Semiochemicals are chemicals produced and released for communication functions. The communication may be intraspecific, and the semiochemicals are then called pheromones. Semiochemicals also function interspecifically, between species, and these sometimes are called allelochemicals. Pheromones often are characterized by the type of behavior elicited in the receiving organism, such as sex attraction, alarm, trail following, and numerous other categories. Interspecific semiochemical classification is usually based on the nature of who benefits from the chemical message. Interspecific semiochemicals include allomones benefiting the sender, kairomones benefiting the receiver, and synomones benefiting both sender and receiver. The same chemical substance may serve more than one function, such as a sex pheromone that attracts a potential mate and predator or parasitoid, and thus functions as a pheromone and a kairomone. Social insects have evolved a further elaboration of semiochemical function in which one chemical may elicit several different behaviors within the colony depending on social context and/or when the chemical is released. For example, queen substance in honeybees controls colony unity and behavior, suppresses ovary development in genetically female worker bees, and serves in the proper context as a sex pheromone to attract male bees to mate with the queen. This multiple functionality is called pheromone parsimony. Several thousand semiochemicals have been identified, and they comprise a large variety of chemical structures, functional groups, and molecular variations including geometric and positional isomers and chirality. Research continues to focus on chemical identification of new semiochemicals and on understanding receptor physiology and the nervous processes involved in responding to a semiochemical. Sex pheromones are usually blends of several chemicals. Sometimes the opposite sex will respond to the major chemical, or to a partial blend, but in some species, only the full blend in the correct proportions will attract a potential mate. Knowledge of what components and blend proportions will attract becomes particularly important in attempting to formulate pheromones for practical application in traps for population monitoring. Information processing of odor plumes is an active area of research. Odor plumes tend to be discontinuous pulses of chemical in the air, and their discontinuities appear to be important to the detection process by the receiving insect. Research shows that the male tobacco hornworm moth detects (with antennal receptors) pheromone quality, quantity, and pulse rate of the pheromone plume several times per second. Sex pheromones often function as species-isolating mechanisms, and populational and geographic differences suggest evolution in progress in pheromone blends based upon what currently is believed to be a single species in several well-studied cases. Pheromone production in insects is regulated by hormones, with the pheromone biosynthesis activating neuropeptide (PBAN) as one important hormone. Juvenile hormone (JH) may regulate pheromone production in some insects. Much of the research on pheromones has been driven by the desire to use pheromones in the control of insect populations. The most effective uses currently are as monitoring tools for population appearance or increase and in mating disruption techniques.

19.1 INTRODUCTION

Communication through exchange of chemical signals is undoubtedly the oldest language known. It probably evolved with life itself, first as a means of intracellular communication, later as a means of intercellular exchange of signals, and finally as a vehicle for communication between organisms. **Semiochemicals** are signaling chemicals produced by one organism and received by another organism as a message. Semiochemicals elicit changes in the behavior or physiology of the receiving organism. Many examples of chemical communication can be found in living organisms ranging from the simplest unicellular plants, microorganisms, and animals. Recent studies have shown how important chemical signals from microorganisms (Davis et al., 2013; Six, 2013, Raffa, 2014) are to insects, and there is a long history of the importance of semiochemicals in interactions between plants and insects (Bell and Cardé, 1984; Cardé and Bell, 1995; Nishida et al., 1997; Musser et al., 2012; Gu and Yang, 2013; Milet-Pinheiro et al., 2013).

The secretion of hormones and "second messengers" are examples of internal secretions that influence internal physiology, biochemistry, and behavior of the producing organism, but these are not called semiochemicals.

19.2 CLASSES OF SEMIOCHEMICALS

Semiochemicals have been defined both as **intraspecific** agents influencing the physiology or behavior of members of the same species as that of the producer (**pheromones**) and as **interspecific** agents influencing a different species than that of the producer (**allomones and kairomones**) (Regnier, 1971). Recently, some workers have suggested use of the term "**infochemical**" as a subcategory of semiochemicals, with redefinition of the types of infochemicals based upon a context-specific rather than a chemical-specific basis (Dicke and Sabelis, 1988; Vet and Dicke, 1992). The term "pheromone" was coined by Karlson and Luscher (1959) from two Greek words, *pherein*, meaning to carry, and *horman*, meaning to excite. In the early 1960s when only a few pheromones had been identified, it was thought that a pheromone would be a single specific chemical. The sensitivity and sophistication of the work at that time generally resulted in identification of the major component in what later turned out to be blends in all the insects from which the first few pheromones were identified. Nearly all sex pheromones, and many other types of pheromones, are blends of components, and the total blend generally is considered to be the pheromone. The blend or ratio of components allows more information to be encoded and allows greater discrimination among closely related species, which often use the same, or some of the same, components.

The roles of semiochemicals in the lives of living organisms are often unexpected, surprising, and manipulative, as when a pheromone modulates preflight warm-up behavior in a male moth (Crespo et al., 2012), possibly enabling it to get into flight and beat competitors to a pheromone-releasing female, or a predator of aphids modulates the production of the aphid alarm pheromone (Joachim et al., 2013). A male that has attracted a *Nasonia vitripennis* female with his male-produced pheromone can terminate her response to other male pheromone by feeding her an oral secretion during courtship (Ruther and Hammerl, 2014). The male-produced sex pheromone hydroxydanaidal, derived from pyrrolizidine alkaloids by male *Utetheisa ornatrix* moths (family Arctiidae), and plant-derived alkaloids are used by the female moth to evaluate the fitness of male moths, resulting in a model for sexual selection (Eisner and Meinwald, 1995).

Pheromones have been identified from several thousand species of insects in many families from most orders (Roelofs, 1995; El-Sayed, 2014). Lists of Lepidoptera from which compounds have been identified as pheromonal components are available electronically El-Sayed, 2014, last accessed April 22, 2015. A pheromone that has an effect upon the physiology or biochemistry of an animal, such as suppression of ovary development in honeybees or stimulation of maturation in the desert locust, is called a **primer** pheromone because it "primes the pump" by requiring a finite period of time for its action to be effective. It causes relatively slow changes in the physiology, and usually behavior, of the animal. In contrast, a sex pheromone that attracts a potential mate is called a **releaser** pheromone because the pheromone almost instantly releases some behavior such as upwind searching behavior or attempted mating.

Allomones, **kairomones**, and **synomones** influence behavior or physiology of members of a different species than that of the producer. They are interspecific semiochemicals and sometimes they also are called allelochemicals because they act between species. Whittaker and Feeny (1971) used the term "allelochemical" to describe "chemicals significant to organisms of a species different from their source, for reasons other than food…"

Defensive secretions are **allomones** and they benefit the producer. An example of an allomone is the defensive spray of a bombardier beetle directed against attacking ants. A recent review (Berenbaum, 1995) includes a discussion of the theory of defensive secretions in both plants and animals. On the other hand, a **kairomone** often works to the detriment of the producer, and to the benefit of the receiver. An example is the odor of a prey, whether rabbit or lepidopterous caterpillar, that leads a parasite or predator to seek and find it. The host odors that attract a phytophagous insect to its food plant are often called kairomones. **Synomones** are mutually beneficial to producer and receiver and operate in cases of mutualism or commensalism. Semiochemicals are used by insects in a variety of ways. Some well-known cases have been described in which a chemical can play more than one of the roles noted earlier. The sex pheromone of *Dendroctonus frontalis*, a bark beetle, not only attracts a potential mate, but may attract a clerid predator, *Thanasimus dubius*. Thus, it functions as a pheromone to the beetles and as a kairomone to the predator. Many parasitoids and predators evidently are in an evolutionary race with their hosts to home in on semiochemical components, while the hosts tend to modify the chemicals or blend they release (reviewed by Vet and Dicke, 1992).

19.3 IMPORTANCE OF THE OLFACTORY SENSE IN INSECTS

Examples of nearly incredible sensitivity to pheromonal chemicals occur in some insects, and especially among representatives of moths (Lepidoptera). Most moths mate in low light intensity during some part of the evening or night hours. Although the eyes of moths are adapted for low light intensity, they probably rely less upon vision than on olfaction, and certainly, olfaction is important for long-distance orientation. Perhaps because of feeding and mating activity in dim light, their ability to detect chemicals in the air has evolved to a very high degree of sensitivity.

The antennae of many male Lepidoptera are plumose, with thousands of small hairs containing pheromone-sensitive sensory neurons (Figure 19.1). Although female moths usually produce a sex pheromone that attracts males, many male butterflies (Andersson et al., 2007) and some male moths release pheromone components from a variety of glandular structures, but often from hair pencils (Nishida et al., 1996; Honda et al., 2006) that can be extruded from pouches at the tip of the abdomen. Figures 19.2 and 19.3 show, respectively, hair pencils of the male moth, *Heliothis virescens*, and those of the male danaine butterfly, *Idea leuconoe*.

The male commercial silk moth, *Bombyx mori*, is a good example of an insect with very high sensitivity to its sex pheromone and was the first insect from which a sex pheromone component was chemically identified (Butenandt et al., 1959). The male has about 17,000 olfactory receptors on its antennae and about 50% of these are tuned to detect the sex pheromone (Schneider, 1974). The advantage of having a very large number of receptors tuned to the pheromone is that sensitivity to low concentrations in the air is greatly increased. These male pheromone receptors fire spontaneously, with each antenna sending about 1600 impulses/s into the brain even when not stimulated by pheromone. According to established information theory in biological systems, in order for the pheromone stimulus to be perceived in the brain, the firing rate must increase by about three times the square root of the noise (noise in this case is the spontaneous rate). Thus, three times the square root of the noise is equal to about 120 firings/s, and an antenna must increase its firing rate to about 1720 impulses/s. Electrophysiological recordings from the antenna indicate that when a *single* pheromone molecule strikes a receptor, there is a receptor response. When a total of about 200 molecules of the sex pheromone simultaneously strike the antenna of the male moth (i.e., 200 receptor neurons are activated), the male moth responds by searching upwind for the source of the chemical. The male of the commercial variety of silkworm, however, has lost the ability to fly because it is too heavy. When it detects the sex pheromone of the female, motor output to the wing muscles causes the wings to start vibrating, but it merely walks upwind in a zigzag path toward the source of the pheromone.

Figure 19.1 The plumose antenna of a gypsy male moth. There are thousands of pheromone receptors on the tiny hairs of the antennae. (Photo courtesy of USDA, Washington, DC.)

Figure 19.2 Hair pencils everted from the tip of the abdomen by a male *H. virescens* in response to a pheromone source. The hair pencils release male pheromone components that act upon the female. (Photo courtesy of Dr. Peter Teal, USDA, Gainesville, FL.)

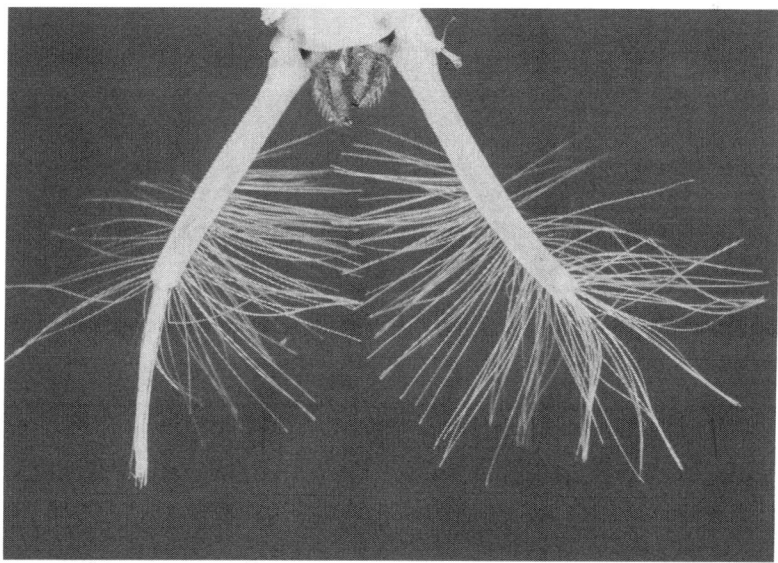

Figure 19.3 **(See color insert.)** Hair pencils of a male giant danaine butterfly, *Idea leuconoe*. Pheromonal components are released from the hair pencils that influence female response. (A previously unpublished photograph, courtesy of Dr. Ritsuo Nishida, Kyoto, Japan.)

19.4 ACTIVE SPACE CONCEPT

The **active space** is the physical space in which the concentration of a pheromone is high enough to cause a behavioral effect in the receiving individual, and it has been mathematically modeled (Bossert and Wilson, 1963; Wilson and Bossert, 1963). The active space is influenced by the sensitivity of the receiver, the quantity of chemical produced and released by the sender per unit time, the volatility of the chemical(s) involved, and environmental factors such as wind velocity and temperature. The female silkworm moth contains only about 0.01 μg pheromone in her body, but if the total were released instantly and uniformly distributed through air moving at about 1.6 km/h, the active space for the male response system theoretically would extend for 4560 m (2.8 miles) downwind in a swath 215 m wide and 108 m high. Any male within that active space, or wandering into its periphery, should be excited to fly upwind. If uniformly distributed, the female can release enough chemical in that space to potentially attract about 1 billion moths.

The possibility of a male detecting a single female up to 4.5 km distant stretches the imagination. Almost certainly such uniform distribution of pheromone never exists in nature because of changes in wind current and direction and the presence of buildings, trees, and other objects between the female and the male that disrupt the airflow and create turbulence. In addition, plants and other objects adsorb pheromone, reducing the amount in the air. Nevertheless, the success of small insects in locating each other under semidark conditions, as most moths do, is impressive.

19.5 PHEROMONES CLASSIFIED ACCORDING TO BEHAVIOR ELICITED

Pheromones are often described or classified by the behavior elicited from the receiver (Shorey, 1973). There are sex pheromones, aggregation pheromones, alarm pheromones, egg-laying pheromones, brood-tending pheromones, recruitment pheromones, trail-following pheromones, and territory-marking pheromones, to name a few. Either sex, or both, may produce one or several pheromones.

Trail-following pheromones have been widely studied. Ants, for example, have evolved highly sensitive mechanisms for trail following (Wilson, 1962; Wilson and Bossert, 1963; Janssen et al., 1997; Kern et al., 1997). They live underground where it is always dark, and in many species, their eyes are small and their visual sense is not acute. However, their olfactory sense is very keen, and they use the sense of smell to follow a chemical trail. Each ant reinforces the trail as it follows it. The trail chemicals come from a gland, Dufour's gland, near the tip of the abdomen. If the trail gets old—no ants pass over it and the trail is not reinforced with new chemical deposition—then the trail vanishes. The trail is easy to demonstrate by allowing a column of ants to establish a trail over a glass slide or strip of paper placed on the ground. When the trail is well established, the slide or paper can be turned 90°, whereupon the trail is completely disrupted. Ants arriving at the point where the trail is interrupted wander around searching for the trail. If the trail is reestablished after a few minutes by turning the slide again to its original position, the ants pick up the trail and continue. This simple experiment has been the basis for some of the behavioral tests to isolate and chemically identify the trail pheromone components. Other insects that use trail pheromones include termites (Matsumura et al., 1968; Grace et al., 1988; Reinhard and Kaib, 1995), some caterpillars (Capinera, 1980; Fitzgerald and Costa, 1986; Fitzgerald, 1993; Fitzgerald and Underwood, 1998), and bumblebees (Svensson and Bergström, 1977; Bergman and Bergström, 1997).

19.6 PHEROMONE PARSIMONY

Pheromone parsimony refers to the fact that the same pheromonal compound, sometimes synergized by additional compounds, can serve multiple functions depending on ecological and behavioral contexts (Blum, 1996). The phenomenon is prevalent in social insects (Hölldobler et al., 2013; Rottler et al., 2013; Ayasse and Jarau, 2014), but examples have been found in nonsocial insects such as scarab beetles (Vuts et al., 2014) and cerambycid beetles (Hanks and Millar, 2013; Mitchell et al., 2013). The alarm pheromones of social insects often also serve, in the proper context, as defensive allomones, attractants, trail pheromones, antimicrobial agents, and as releasers of several additional behavioral actions. The pheromones produced and released by the queen of a colony of social insects frequently have several functions depending on the social context. For example, the principal queen substance from the queen honeybee, *Apis mellifera*, is (*E*)-9-oxo-2-decenoic acid (9-ODA). It releases behavior in worker bees that makes them form a cluster around the queen (called a queen retinue) and lick 9-ODA and other pheromonal components from her body. The workers subsequently spread the components throughout the colony by communal feeding. Distributed in this way, 9-ODA acts in conjunction with several other queen-produced components to suppress ovary development of workers (a primer function) and inhibits construction of queen cells in which new queens might be reared. 9-ODA also releases mating behavior in male bees (drones), but only when the drones are flying and when the queen substance is released from several meters up in the air, the normal site for mating by honeybees (Gary, 1962). Drones will attempt to mate with a variety of objects, including inanimate queen mimics and dead queens, if 9-ODA is released from them and if they are displayed in the air, for example, on a flagpole. In the colony, drones exhibit no mating behavior.

19.7 CHEMICAL CHARACTERISTICS OF SEMIOCHEMICALS

To have sufficient volatility, airborne pheromones have to be relatively small molecules. In general, the molecular weight must not be much over 200, to get the volatility needed. Not all pheromones are airborne; some insect pheromones are contact pheromones and some crustaceans have waterborne pheromones. No particular chemical structure is used exclusively by insects as a pheromone, and low-molecular-weight acids, esters, alcohols, aldehydes, ketones, epoxides, lactones, hydrocarbons, terpenes, and sesquiterpenes are common components in pheromones. A few large molecules serve as pheromones, but they are predominately contact pheromones. For example, (*Z*)-9-tricosene (a hydrocarbon composed of 23 carbons) is the sex pheromone of the housefly, *Musca domestica*. It has very low volatility. Male houseflies are stimulated to make contact with any small dark object, such as a knot in a black shoelace, which was used as one of the bioassay devices for the pheromone. They detect the pheromone, if it is present, after landing on the object and then attempt to mate with it. Even larger cuticular hydrocarbons on the surface of female tsetse flies serve as contact pheromones. In general, a male tsetse fly cannot distinguish a female from a male until it contacts the body surface of the fly. The chemoreceptors may be located on the tarsi of the male.

Pheromones often serve as species-isolating mechanisms, and the receptors on the antennae of the opposite sex must be tuned to the pheromone of its species. There are several ways that species specificity is coded in pheromones. One of the most common adaptations is the use of a blend of two or more chemical components in the pheromone as well as variations in different blend ratios when the same components are incorporated into the pheromone. Many of the sex pheromone components that have been identified from moths are acetates and alcohols with a backbone of 14–16 carbons with one or more double bonds in the molecule. The position of the double bond offers many variations on each backbone. Still further specificity can be encoded in (E) or (Z) configuration of the groups or atoms on the carbons at the double bond. Finally, **chirality** in molecules provides another way that insects can specify their pheromone signal (Mori, 1984; Silverstein, 1988). The word "**chiral**" is derived from the Greek word for hand. Chiral compounds have a 3D shape analogous to human hands. Although the right hand is a mirror image of the left hand, the two hands are not perfectly superimposable upon each other with respect to matching shape when both palms are down or both up. All objects, including chemical molecules, have a mirror image, but like human hands, the mirror image of a chiral compound cannot be superimposed upon its structure. The mirror image of ethanol, for example, can be superimposed on its structure, but 2-butanol can exist in two 3D shapes that cannot be superimposed upon each other (Figure 19.4). Nonsuperimposability of mirror images gives rise to enantiomerism and distinct optical activity for the **enantiomers**. One feature that makes for chirality is when a carbon atom is attached to four different groups; such an arrangement produces a stereo center. A molecule with one stereo center is a chiral molecule. Many of the most common biochemical compounds in living organisms are chiral such as amino acids and glucose. Most enzymes that utilize amino acids to synthesize proteins accept only L-amino acids, and D-glucose is the form synthesized into glycogen, trehalose, and chitin.

A molecule may have more than one stereo center, but this gets more complicated, and such a molecule may or may not be chiral. It is not uncommon, however, for pheromone molecules to have more than one stereo center. Other atoms in addition to carbon may also be a source of chirality in a molecule, a situation not often relevant to pheromones. The designation of the two enantiomers of a chiral compound as R or S is based upon the Cahn–Ingold–Prelog system (simple sugars and amino acids have been excepted from the system because of the long-term usage of D- and L-designations). Chemists assign a priority value to chemical groups, such as –OH, –CH$_3$, and –CH$_2$CH$_3$, based on certain priority rules (Table 19.1) (Cahn, 1964a,b).

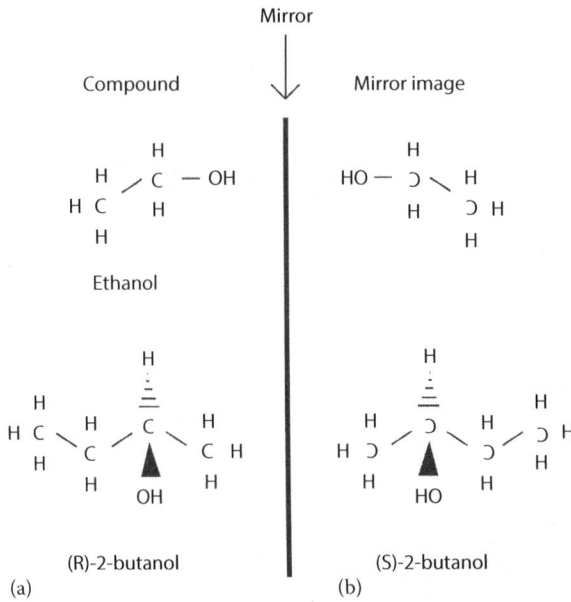

Figure 19.4 The concept of chirality. (a) The mirror images of ethanol can be superimposed upon each other because ethanol does not have a stereo center. (b) The mirror images of 2-butanol are not superimposable, and the carbon bearing the –OH group is a stereo center. On the left is the R enantiomer, and on the right is the S enantiomer.

Table 19.1 Priority in the Cahn–Ingold–Prelog System for Determining the Absolute Configuration of a Molecule Bearing One of the More Common Groups Found in Insect Pheromones

Atom or Group	Priority
Hydrogen	1
Methyl	2
Ethyl	3
n-Propyl	4
n-Butyl	5
n-Pentyl	6
n-Hexyl	7
Iso-pentyl	8
Iso-butyl	9
Allyl	10
Iso-propyl	14
Vinyl	15
Cyclohexyl	17
Acetyl	36
Carboxy	38
Amino	43
Hydroxy	57
Fluoro	68
Chloro	74

Sources: Data from Cahn, R.S., *J. Chem. Educ.*, 41, 116, 1964a; Cahn, R.S., *J. Chem. Educ.*, 41, 508, 1964b.

The molecule is viewed by looking at it so that the group of lowest priority is behind the molecule or farthest from the eye (Figure 19.5). Then the decreasing priority order of the remaining groups is noted and, if the order decreases clockwise around the chiral carbon, the designation is *R*, from the Latin word *rectus* for right. If the order decreases counterclockwise, the designation is *S*, from Latin for *sinister* meaning left.

Enantiomers are optically active; each rotates the plane of polarized light, but in different directions, either clockwise, designated dextrorotatory (+), or counterclockwise, designated levorotatory (−). The direction of rotation of plane-polarized light is completely independent of the *R* or *S* designation. Thus, a particular molecule might be *R*-(+)-, or it might be *R*-(−)-. Whatever it turns out to be, its enantiomer will have the opposite designation and rotation of plane-polarized light.

For chirality in a pheromone to provide specificity, a receptor has to recognize the difference in the two enantiomers. The opposite enantiomer of a chiral pheromone component may have no effect at the sensory neuron, may stimulate, or may inhibit the response to the natural enantiomer. A good example of enantiomeric differentiation at the receptor site occurs in the Japanese beetle, *Popillia japonica* Newman (family Scarabaeidae). Tumlinson et al. (1977) identified (*R,Z*)-5-(1-decenyl) dihydro-2(3H)-furanone as the pheromone produced by female beetles and showed inhibition by the *S,Z* enantiomer. Mirror images of the enantiomers are shown in Figure 19.6. A **racemic mixture** (i.e., equal mixture of the *R* and *S* configurations) of the synthetic (*Z*) isomer is completely inactive in field tests, and as little as 1% of the unnatural synthetic (*S,Z*) enantiomer mixed with the natural (*R,Z*) enantiomer substantially reduces male response. When a trap contains 5% of the (*S,Z*) enantiomer, capture of beetles is reduced to the level of an empty trap. One can only guess at what happens at the receptor level in this case because the receptor has not been isolated. One possibility is that the (*S,Z*) enantiomer binds rapidly and perhaps preferentially to the

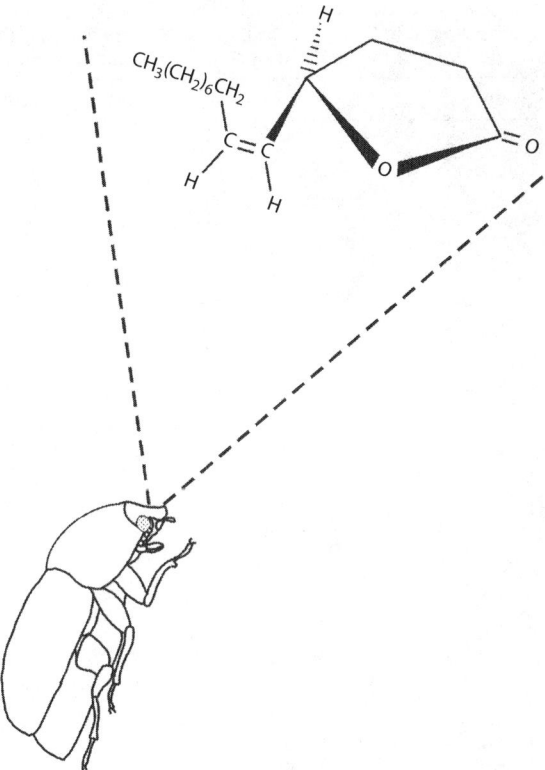

Figure 19.5 An illustration to show how one determines whether a molecule has the R or S configuration. In the drawing, a male Japanese beetle *P. japonica* views the chiral pheromone produced by a female as a chemist would view it, with the group of lowest priority behind the molecule. The natural enantiomer of the Japanese beetle pheromone has the R configuration because the priority of the groups on the chiral carbon decreases in priority in a clockwise direction.

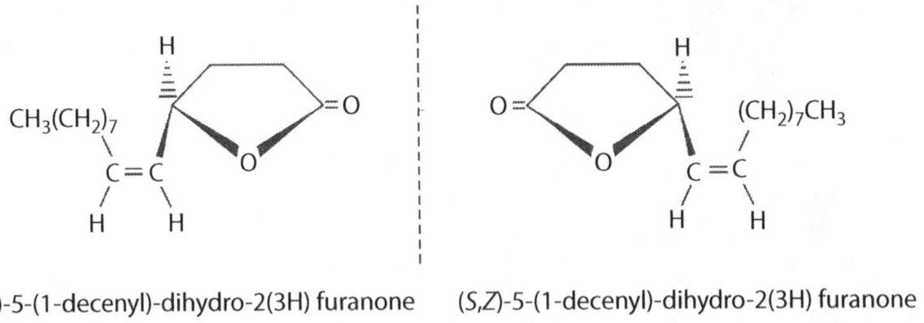

(R,Z)-5-(1-decenyl)-dihydro-2(3H) furanone   (S,Z)-5-(1-decenyl)-dihydro-2(3H) furanone

Figure 19.6 The R and S enantiomers of the Japanese beetle pheromone depicted as mirror images of each other. The natural enantiomer that the males respond to is (R,Z)-(1-decenyl)-dihydro-2(3H)-furanone.

receptor and cannot be removed. Thus, it might block binding of the natural enantiomer. However, other explanations are possible, including physiological actions within the brain of the insect.

Grain beetles in the family Cucujidae illustrate the probable utilization of pheromone chirality in evolution and isolation of species (Table 19.2) (Oehlschlager et al., 1987). The pheromonal compounds produced by males during feeding act as aggregation pheromones; they attract both sexes. Starved males are not good producers. Possibly one evolutionary factor

Table 19.2 Similarities and Differences in the Pheromone Blend Components and Enantiomeric Composition of Blend Components in Closely Related Grain Beetles in the Family Cucujidae

Species	Pheromone
Cryptolestes ferrugineus	(S,Z)-3-dodecen-11-olide
Oryzaephilus mercator	(R,Z)-3-dodecen-11-olide
	(R,Z,Z)-3,6-dodecadien-11-olide
O. surinamensis	(R,Z,Z)-3,6-dodecadien-11-olide
	(R,Z,Z)-5,8-tetradecadien-13-olide
	(Z,Z)-3,6-dodecadienolide
C. turcicus	(R,Z,Z)-5,8-tetradecadien-13-olide, 85%
	(S,Z,Z)-5,8-tetradecadien-13-olide, 15%
	(R,Z)-5-tetradecen-13-olide, 33%
	(S,Z)-5-tetradecen-13-olide, 67%
C. pusillus	(S,Z)-5-tetradecen-13-olide

Source: Data from Oehlschlager, A.C. et al., *J. Chem. Ecol.*, 13: 1543, 1987.

operating in the selection of these pheromones is that they may indicate a food source, and both sexes could benefit by responding to the pheromone. Mixed sexes of each species are attracted best to blends of compounds produced by feeding males of their own species.

19.8 INSECT RECEPTORS AND THE DETECTION PROCESS

Pheromones are generally detected by olfactory receptors located mostly on the antennae. Many male Lepidoptera have pheromone receptors specialized for reception of the sex pheromone. These receptors are relatively insensitive to other chemicals. A sex pheromone receptor on the antenna typically consists of one or two nerve cells housed within a seta or fine "hair" on the antenna. The whole structure is called the sensillum (Figure 19.7). There often are many thousands of such sensilla on each antenna of a male moth. Each seta has microscopic pores along its length through which the airborne molecules enter the sensillum and make contact with the sensory neuron.

Kaissling (1987) postulated that the following six steps occur as a part of the process of semiochemical detection.

1. **Adsorption** of an odor molecule by sensory hairs (setae) on the antennae.
2. **Penetration** of the molecule through pores in the setal wall.
3. **Receptor binding** of the molecule and transport to the sensory nerve endings. It is believed that molecules adsorb to the cuticular surface, and then move into a pore, although an occasional molecule may hit a pore directly. Processes that may promote movement of adsorbed molecules into a pore are unknown.
4. **Membrane alteration**, probably opening of sodium channels.
5. **Receptor potential generation**, a graded potential, followed by spikes in the axon hillock region.
6. **Inactivation** of the odor molecule and removal. Inactivation clears the receptor so that it can respond again.

19.8.1 Pheromone-Binding Proteins

After a molecule enters a pore, it must cross the **sensillum liquor** (lymph, extracellular fluid) in order to reach the dendritic nerve endings. Most pheromonal molecules are lipid soluble, and the sensillum liquor is an aqueous medium that does not readily dissolve lipids. Thus, the molecule

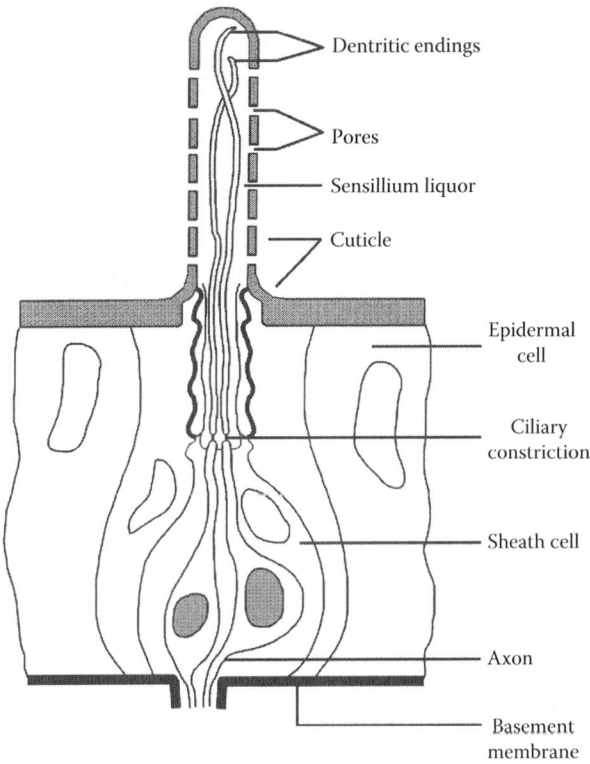

Figure 19.7 Basic structure of an olfactory receptor with pores and dendritic nerve endings in the cuticular hair.

usually must bind with a specific binding protein, which then transports it across or through the sensillum liquor. Pheromones bind to **pheromone-binding proteins** (**PBPs**) (Vogt and Riddiford, 1981; Klein, 1987; Vogt, 1987), a subset of a larger group of **odorant-binding proteins** (**OBPs**) or **general OBPs** (**GOBPs**) known from both vertebrates and invertebrates. A number of PBPs and GOBPs have been isolated and sequenced (Prestwich and Du, 1997).

GOBPs of Lepidoptera bind a variety of odorant molecules associated with food, habitat, and oviposition substrates (Breer et al., 1990a; Vogt et al., 1991a,b). GOBPs may occur in other insects, perhaps in males in some cases as well as in females, but currently little research has been done in groups other than Lepidoptera.

PBPs are synthesized in male Lepidoptera just prior to adult emergence and are localized in the extracellular fluid (the sensillum liquor) of pheromone responsive sensilla on the male antennae (Vogt et al., 1989). Auxiliary cells associated with the receptor cell appear to be responsible for synthesis (Steinbrecht et al., 1992). By binding the (usually) hydrophobic pheromone molecule, the PBP aids in transport of the pheromone through the aqueous sensillum lymph so that it makes contact with specific receptor proteins in the dendritic nerve endings (Vogt and Riddiford, 1981; Prestwich, 1993b; Breer, 1997; Prestwich and Du, 1997). One current hypothesis is that the PBP may attach directly to the dendritic membrane where subsequent activation of a G-protein-coupled cascade of events results in a receptor potential in the dendrite (Prestwich and Du, 1997). Specific binding proteins may serve as filters that protect the dendritic endings from many other chemical molecules of the air that also enter sensilla pores (Prestwich and Du, 1997). Some evidence suggests that PBPs also may be involved in destruction of the pheromone after signal transduction (Vogt et al., 1985; Prestwich, 1993a).

Prestwich and Du (1997) determined the active site for pheromone binding in a binding protein isolated from *Antheraea polyphemus* by using photoaffinity labeling. The major component of the *A. polyphemus* sex pheromone, $6(E),11(Z)$-hexadecadienyl acetate, has only one binding site, but a second pheromonal component, $4(E),9(Z)$-tetradecadienyl acetate, can bind in two slightly different ways (Du et al., 1994).

19.8.2 Signal Transduction and Receptor Response

At the dendritic ending, the pheromone probably combines with a receptor protein in the dendritic membrane, although a specific protein receptor has not been identified in an insect as yet. G-proteins, cyclic adenosine monophosphate (cAMP), and inositol trisphosphate (IP_3) are involved in the odor transduction process in several vertebrates (Buck and Axel, 1991; Ngai et al., 1993; Ressler et al., 1993), and some or all of these may mediate and amplify the pheromone signal at the dendritic ending in insects. The concentration of cAMP, however, is low in antennae, and its concentration is not stimulated by pheromone (Breer et al., 1990a; Ziegelberger et al., 1990), so it seems an unlikely participant. High doses of pheromone elevate cyclic guanosine monophosphate (cGMP) in the antennae, but only slowly and after a long delay (Boekhoff et al., 1993), so it also seems unlikely to mediate the fast responses needed. Ishida et al. (2012) found an aquaporin expressed in the olfactory neurons of the blowfly, *Phormia regina*. They speculated that the aquaporin functions to maintain homeostasis in the sensillum liquor.

Some evidence suggests that IP_3 may be a participant in pheromone response. Phospholipase C, the enzyme that hydrolyses phosphatidylinositol bisphosphate (PIP_2) to IP_3 and diacylglycerol (DAG), has high activity in the antenna (Boekhoff et al., 1990a; Breer et al., 1990a). Moreover, IP_3 shows a rapid phasic increase followed by a tonic decline in male antennae after sex pheromone application (Boekhoff et al., 1990b, 1993). Kinetic measurements following pheromone application indicate that IP_3 reaches a stimulus-dependent maximum in about 50 ms and declines to the basal level within a few hundred milliseconds (Breer et al., 1990b). Such a time course is consistent with the observed ability of some insects to resolve several pheromone pulses/s (Marion-Poll and Tobin, 1992). The slow increase and sustained elevation of cGMP and experiments indicating that elevated cGMP modify the response to pheromone may mean that it is involved in the adaptation of pheromone receptors exposed to high and sustained pheromone levels (Breer, 1997).

Electrophysiological recordings from single cells and from the whole antenna have been used to study the response to pheromone components and in bioassay of potential pheromone components during pheromone identification. The technically more easily accomplished recordings from the antenna have been used more frequently. The procedure requires the mounting of an antenna between two electrodes. The antenna can be left attached to the head of the insect, but frequently, it is severed from the head and mounted. The response is called the electroantennogram (EAG), and it is a summed potential from many receptors responding simultaneously or in rapid sequence (Schneider, 1957; Boeckh et al., 1965). The technique has been effectively used in pheromone identification (Roelofs, 1984) and in combination with gas chromatographic mass spectrometer (GC-MS) technique by splitting the column effluent, sending part to a flame ionization detector (FID) or ion trap and part to the antenna (Arn et al., 1975; Cossé et al., 1995). In such a situation, the antenna is called an electroantennogram detector (EAD) (Figures 19.8 and 19.9). Portable EAG devices have been designed and built to measure pheromone concentrations in the field (Baker and Haynes, 1989; Sauer et al., 1992; Karg and Sauer, 1995; Rumbo et al., 1995; Leal et al., 1997).

Single-cell recordings usually have been made extracellularly by carefully placing a glass capillary electrode filled with saline over a single hair on the antenna. After a relatively long latency (15 ms or longer), slow graded potentials with superimposed spikes can be obtained when pheromone is pulsed over the antenna (Kaissling, 1986). Single sensillum recordings also have been adapted for use as a GC-MS detector (Wadhams, 1982).

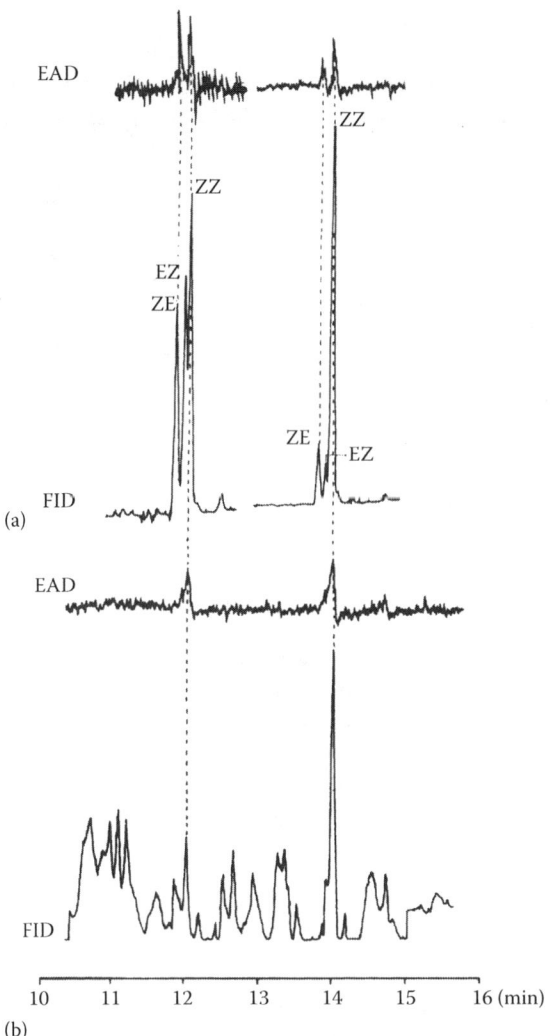

Figure 19.8 Use of the EAG response as an EAD in conjunction with an FID in a GC to recognize behaviorally active peaks in a gas chromatogram. (a) The EAD responses were made by the antenna of a male *Idea aversata* geometrid moth to the effluent from a GC. The GC FID response is shown to a mixture of Z7,E9-, E7,Z9-, and Z7,Z9-dodecadienyl acetate and a mixture of Z9,E11-, E9,Z11-, and Z9,Z11-tetradecadienyl acetate. (b) Response of the male antenna to female gland extract is shown as the EAD and the FID response to the gland extract. (From Zhu, J. et al., *J. Chem. Ecol.*, 22, 1505, 1996. With permission.)

Axons from sex pheromone receptor neurons located peripherally project into the antennal lobe (AL) of the deutocerebrum as labeled lines (pheromone-specific neurons), while general food odors, host odors, and environmental odors are transmitted by across-fiber patterning (Masson and Mustaparta, 1990). Dethier (1972) described the concept of labeled lines and across-fiber patterning lines from studies with taste receptors. It currently appears that an odor is defined in the deutocerebrum by an across-glomeruli pattern based on input that the glomeruli receive from peripheral receptors (Todd and Baker, 1997).

Pheromone receptors on the antennae of some insects (and perhaps on most insects) are spontaneously active and display a constant low level of firing. Contact with a pheromone

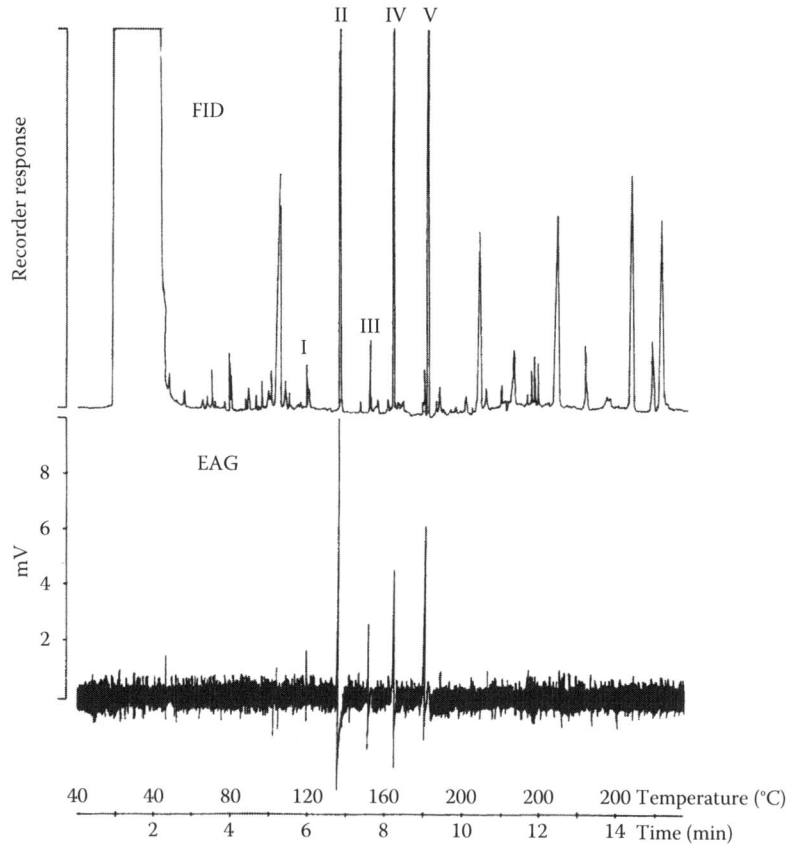

Figure 19.9 Application of the EAD and FID to study of female Mediterranean fruit fly responses to male-released volatiles. The lower trace shows EAG (or EAD) responses by female antennal receptors to some of the volatiles released by "calling" Mediterranean fruit fly males. The upper trace shows FID responses to male-released volatiles injected into a GC. Female antennal receptors respond strongly to only some of the male-released volatiles, and the EAG-active ones are labeled I, II, III, IV, and V. (From Cossé, A.A. et al., *J. Chem. Ecol.*, 21, 1823, 1995. With permission.)

component to which a receptor is sensitive usually results in an increase in firing. In some cases, a receptor is differentially sensitive to a particular component of a pheromone blend, and it typically responds to low concentrations of that component by a large increase in rate of firing. Exposure of a specialist receptor to large amounts of other blend components also may cause an increase in firing to nonspecific components, but usually at a much lower rate, that is, it is less sensitive to the other components (Almaas and Mustaparta, 1990; Berg et al., 1995; Berg and Mustaparta, 1995).

Most insects with pheromone receptors appear to have more than one type of receptor, each being sensitive to one or more of the blend components of the pheromone. It may be that when a responding insect flies upwind in a plume of pheromone composed of several components, each of its several receptor types responds to one component while ignoring other components in the blend. Thus, the nerve activity going into the deutocerebrum and other parts of the brain is via labeled lines, and the brain presumably has to integrate the input as the relative activities of different receptor neurons (Mustaparta, 1997). Few neurophysiological details are available to substantiate or refute this concept.

19.8.3 Pheromone Inactivation and Clearing of the Receptor

When pheromone molecules have made contact with the dendritic endings, it is important to destroy or inactivate them in order to clear the receptor active site and allow the receptor to be sensitive to incoming pheromone. Although only a few studies are available, the evidence indicates that enzymes attack the pheromone and destroy it. An esterase in *A. polyphemus* and an aldehyde oxidase from *Manduca sexta* have been identified (Vogt and Riddiford, 1981; Vogt et al., 1985; Klein, 1987; Rybczynski et al., 1989). In less than 0.5 s, the esterase in *M. sexta* antennae can destroy a million pheromone molecules, and this seems consistent with the rapid changes in upwind or casting behavior that males make in a changing pheromone plume (Vogt, 1987). The pheromone-destroying enzymes also may aid an insect by destroying small amounts of pheromone slowly leaking into the pores after adsorption on the antennal surface (Breer, 1997). Such a slow, persistent pheromone leak might create a high background noise in the system.

19.8.4 Do Insects Smell the Blend or Just the Major Components?

Pheromone specialists have been divided over the issue of whether the responding insects smell and respond to only one or perhaps two major components in the pheromone blend, or whether the entire blend is necessary for response. Actually, the behavior of male moths in this respect is quite variable depending on the species. It appears that different species utilize different strategies to locate a mate. In some the major component may be sufficient for the complete behavioral response and mating, while in others a more complex or complete blend is necessary. Males of the red-banded leafroller, *Argyrotaenia velutinana*, and Oriental fruit moth, *Grapholita molesta*, are typical of insects that need the blend. Very few males of the red-banded leafroller moth fly upwind toward a pheromone source containing only the major pheromonal component. Significantly more males of the Oriental fruit moth give behavioral displays (wing fanning) up to 60 m away from a pheromone source when the source contains the correct proportions of a three-component pheromone blend as opposed to when the source contains only the major component (Linn et al., 1987). Alternatively, a high percentage of the males of a number of other insects (cabbage looper, *T. ni*; cotton bollworms, *Helicoverpa armigera*, *H. zea*, and *Heliothis virescens*) take flight, fly the typical zigzag pattern in a pheromone plume, and may contact the source and attempt mating when exposed to only the major component of the blend (Kehat and Dunkelblum, 1990; Vickers et al., 1991; Mayer and McLaughlin, 1992).

A model (reviewed by Christensen, 1997) based on *M. sexta* diversity in neuronal input and output in the macroglomerular complex (MGC) provides insight into possible neuronal explanations of the blend versus major component response. Males usually do not fly upwind when exposed to only one component; they need the correct blend. Nevertheless, in *M. sexta*, some olfactory receptors on the antennae of males respond best when exposed to the major component of the blend (component A), and others respond best to a second component (B). Similar responses are obtained from some receptors on the antennae of male *H. zea*. The input neurons synapse in different glomeruli in *M. sexta* and in the same glomeruli in *H. zea*. The outputs from the glomeruli go to higher brain centers through local and projection neurons through at least four pathways (A output only, B output only, A or B output, and blend output). Thus, higher centers in the protocerebrum receive a variety of inputs, depending on exposure of antennal receptors. The majority of the output interneurons in *M. sexta* respond best to a blend. The majority of the output neurons in *H. zea* respond strongly to the major component (A, in the model), but some respond strongly to either the A component or to the blend. *H. zea* males also respond behaviorally to the major component and to the blend. Detailed study of many more species is needed.

19.9 INFORMATION CODING AND PROCESSING

19.9.1 Structure of Odor Plumes

A female moth typically releases pheromone in pulses and a filamentous plume snakes out from the female (Figures 19.10 and 19.11). There are frequent changes in pheromone concentration within the plume (Murlis and Jones, 1981). Pulsed pheromone released at a rate of 3 pulses/s is more effective in causing male orientation and upwind flight than continuous release (Kaissling, 1986). A male insect responding to the female-produced pheromone must detect, process, initiate flight commands, and clear the sensory receptors rapidly in order to respond effectively to the rapid changes that occur in a pheromone plume. The typical response of a flying male insect to a pheromone plume is **optomotor anemotaxis**, or upwind flight (Kennedy, 1940; Kennedy and Marsh, 1974; Marsh et al., 1978; Vickers, 2000; Cardé and Willis, 2008; Girling et al., 2013). Once in flight, its own movement through the air prevents an insect from determining wind direction

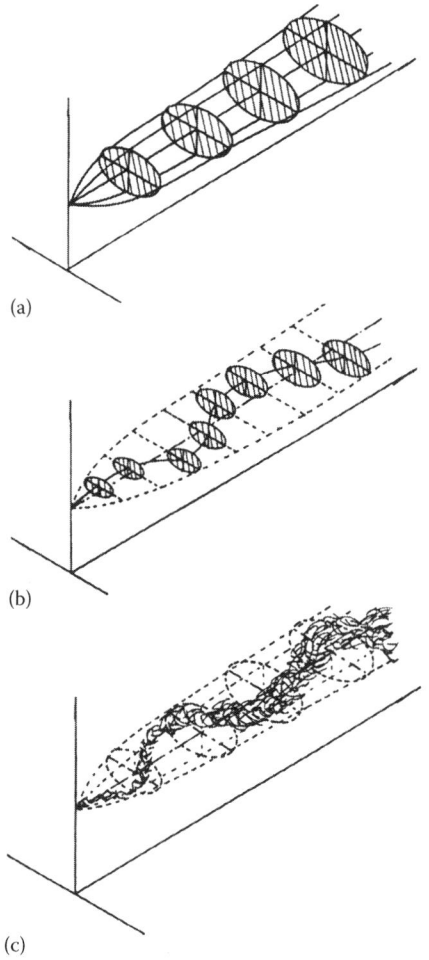

Figure 19.10 A schematic diagram to illustrate the structure of a pheromone plume in the air: (a) a time-averaged approach, (b) a more realistic meandering plume, and (c) the discontinuous, filamentous structure of a real plume. (From Murlis, J. et al., *Annu. Rev. Entomol.*, 37, 505, 1992. With permission.)

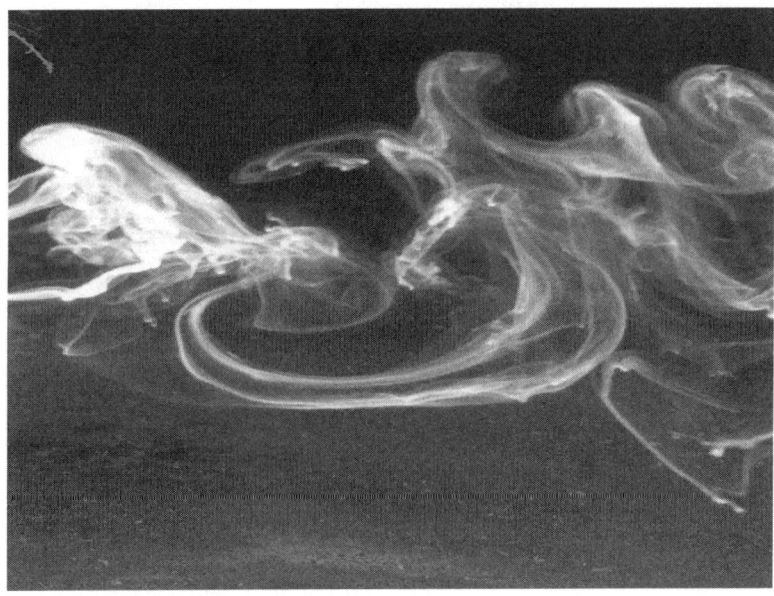

Figure 19.11 (See color insert.) A smoke plume released into the air in an orchard in California. The eddies of smoke show the likely eddies in a volatile chemical, such as a pheromone, or another type of semiochemical, released into the air. (Photo courtesy of Dr. Ring Cardé; From *J. Chem. Ecol.* Cover Photo, Vol. 39(9), 2013. With permission.)

except by its visual displacement over the ground. Flight directly into the wind direction causes the image received by the eyes to move in line with the body axis, while a crosswind causes the insect to experience lateral drift. Usually a male flies in a zigzag pattern upwind in response to the sex pheromone. Its behavior changes from upwind flight to casting from side to side in as little as 0.5 s when it loses the plume (Kennedy et al., 1980), indicating that its sensory and nervous system can detect and respond rapidly.

Male silkworm moths, *B. mori*, are too heavy to fly, and they walk upwind in a zigzag pattern while vibrating the wings at 40–50 Hz (Kanzaki and Shibuya, 1986; Kanzaki, 1997). Loudon and Koehl (2000) determined that the wings are flapped through a stroke angle of 90°–110° at about 40 Hz, directing an unsteady flow of air (average speed of 0.3–0.4 m/s) toward the antennae. Airflow over the antennae is about 15 times faster than that produced by walking and it is 560 times faster through the spaces between the sensory hairs. They fail to move upwind in a steady plume of pheromone and only respond when the stimulus is pulsed. Many species of flying moths exhibit similar behavior and resort to casting from side to side without forward movement or come to rest on some convenient substrate in a steady plume of pheromone (Kramer, 1997). Experiments have demonstrated that discontinuous pulsing of the pheromone in the plume is much more important to upwind flight than the concentration of pheromone in the plume (Kramer, 1986).

Attempts to explain the mechanism underlying zigzag and casting behavior have led to many experiments and numerous arguments. One idea has been that the male moth initiates flight upwind, but cannot steer a perfect course, so sooner or later it comes to the edge of the active space. It then makes a turn back into the active space. Experiments show, however, that males of some species make turns in clean air with no pheromone and within a (presumably) uniform cloud of pheromone (Kennedy, 1983). Such observations led to the idea that perhaps zigzag flight is not an essential component of optomotor anemotaxis, and an internal turn generator that operates independently of the odor plume has been proposed (Wright, 1958; Kennedy and Marsh, 1974; Baker et al., 1984).

A second explanation is the **flight imprecision model** (Mafra-Neto and Cardé, 1995) based on experimental data from gypsy moths and a computer simulation program (Preiss and Kramer, 1986a,b). Data from these experiments and flight simulations lend support to the idea that the turns are course corrections caused by the inability of the moth to head straight upwind. Presumably it would fly directly upwind if it could, and when its movement relative to the ground (substrate) indicates lateral drift, it corrects course.

A third possibility is that zigzag flight is caused by blend quality that does not perfectly mimic the natural blend released by a calling female (Witzgall and Arn, 1990, 1991; Witzgall, 1997). For example, several different investigators have observed that males of *Lobesia botrana*, *G. molesta*, and *Eupoecilia ambiguella* are able to fly nearly straight upwind without the zigzag movement in response to a calling female. They fly the zigzag pattern, however, when exposed to synthetic pheromone, which is presumed, at best, to be slightly different from the blend released by the female.

Regardless of the arguments about mechanisms operating during upwind orientation, the microstructure of the pheromone plume controls the optomotor anemotaxis response. Pheromone does not disperse in the natural environment of insects as a uniform cloud; it travels as a plume of discrete filaments or eddies ranging in size from less than a centimeter to many meters (Murlis et al., 1992; Murlis 1997; Girling et al., 2013) interspersed with pheromone-free air gaps. The gaps become larger and the pheromone filaments smaller at greater distances from the source. Thus, the signal gets broken into discrete stimuli lasting many milliseconds and reoccurring several times per second as a moth flies upwind. The peak pulses of pheromone are greater than the time-averaged mean concentration by up to a factor of 10, and the differential becomes greater with increasing distance from the point source (Murlis, 1997). Studies with *Cadra cautella*, the almond moth, show that males surge upwind with each turbulent pulse of pheromone they encounter. Five pulses/s of pheromone released into the air result in rapid, straight-line, or nearly straight flight with few or shallow zigzags, but less than 1 pulse/s causes slow movement upwind and wide zigzag excursions (Mafra-Neto and Cardé, 1994, 1995). The conclusion (although not universally accepted as the only explanation of zigzag flight) is that zigzags result from low pheromone filament encounter rate combined with a (presumed) counterturning program. Zigzag flight is not a necessity of optomotor anemotaxis and straight upwind flight can occur when there is a high rate of pheromone filament encounters (Cardé and Mafra-Neto, 1997).

When a moth emerges into clean air devoid of pheromone or host-odor plume, it begins casting (zigzagging) from side to side, with little or no forward movement. Under natural conditions, males may lose the plume due to sudden wind shifts or by flight out of a plume because the upwind direction is not aligned with the long axis of the plume. The adaptive value of casting is that it may maximize encounters with the plume after it is lost (David et al., 1983). Those who believe that zigzag flight is due to an internal turn generator and widely spaced pheromone filaments believe that casting simply may be a manifestation of zigzag flight without any significant encounters with a pulse of pheromone. Hence, an internal generator in the central nervous system (CNS) repeatedly generates turns (Kuenen and Cardé, 1994). Casting or coming to rest on the substrate occurs in very high concentrations of pheromone as well, presumably because the antennal receptors are not receiving pulsed stimuli.

19.9.2 Pheromone Signal Processing

Signal processing has been studied intensively in only a few insects. Males of *M. sexta* have about 10^5 sensilla with single walls and pores that house about 3×10^5 receptor neurons. Each sensillum typically contains two sensory neurons, one that is sensitive to (*E,Z*)-10,12-hexadecadienal (*EZ*-10,12–16:AL) and the other sensitive to (*E,E,Z*)-10,12,14-hexadecatrienal (*E,E,Z*-10,12,14–16:AL), two main components of the eight C16 aldehydes in the *M. sexta* female-produced pheromone (Tumlinson et al., 1989). All eight components are important and give the best results in the field (Tumlinson et al., 1994), but detailed neurophysiological data are available only for the two main components. The antennal receptor neurons respond to pheromone aldehydes by opening Na^+, K^+, and Ca^{2+} channels, which are not ligand (pheromone) gated, but mobilized through the second messenger system of G-proteins (Stengl et al., 1992).

Axons from olfactory receptors on the antennae pass through the antennal nerve and enter the large AL of the deutocerebrum. In moths, the antennal nerve breaks into two branches as it enters the AL (Hansson, 1997). One branch carries axons from nonpheromone olfactory receptors to **glomeruli** in a mechano- and taste-sensitive region of the AL, while the second branch carries axons from pheromone receptors to glomeruli in the **Macroglomerular Complex (MGC)**. In the MGC, incoming axons synapse with AL interneurons that interconnect parts of the AL (local interneurons) or pass to other parts of the brain including the protocerebrum (projection interneurons). Similarly organized olfactory glomeruli occur in a wide variety of organisms in which olfaction is very important, including vertebrates and other (noninsect) invertebrates (Hildebrand, 1995, 1996; Christensen, 1997).

In all Lepidoptera that have been investigated, the MGC is a sex-specific region found only in the AL of males. In male *M. sexta* (Camazine and Hildebrand, 1979; Rospars and Hildebrand, 1992; Christensen et al., 1993), the large globular glomerulus near the point at which the antennal nerve enters the AL is called the **cumulus** (because of its resemblance to the cumulus cloud shape), and the ring or donut-shaped area beneath it is called the **toroid** (Hansson, 1997). Similar glomeruli varying in shape and size are known also in a number of other male moths, including *B. mori*, *A. polyphemus*, *Agrotis segetum*, *T. ni*, *Spodoptera littoralis*, *H. zea*, and *H. virescens*. All have one large glomerulus at the entrance of the antennal nerve into the AL, and several smaller satellite glomeruli beneath the large one.

Females in some other insect groups that receive information about male-produced sex pheromones may have a similar structure in the deutocerebrum (Anton and Hansson, 1994), but studies are limited. Glial cells invest the glomeruli and provide protection. In addition to glomeruli, the AL contains lateral and medial groups of cell bodies of the interneuron associated with the glomeruli. The cell bodies of the primary olfactory neurons are located peripherally in the antennae, as is characteristic of sensory neurons in insects.

M. sexta males can detect the pheromone quality, quantity of pheromone, and the frequency of pulses in pheromone plumes. Some pheromone receptor neurons in the male act as pheromone generalists that respond to either of the two aldehyde components or to the total pheromone blend, while other pheromone specialists discriminate between the two aldehydes and respond differently to them (Christensen and Hildebrand, 1990). Thus, the response either to aldehyde or to a blend of both is sent to the MGC as information about the blend, that is, its quality.

A subset of pheromone-specialist neurons provides further discrimination by responding in opposite ways to the two aldehydes. For example, some of the neurons are stimulated by (*E,Z*)-10,12-hexadecadienal, resulting in excitation, but exposure to (*E,E,Z*)-10,12,14-hexadecatrienal inhibits these same neurons. The opposite scenario can also occur, that is, (*E,E,Z*)-10,12,14-hexadecatrienal may stimulate some neurons while (*E,Z*)-10,12-hexadecadienal inhibits them. A blend of aldehyde components results in a unique mixture of inhibitory and excitatory responses depending on the mixing of these two input channels. These pheromone specialists respond and recover rapidly enough to detect the natural intermittent pheromone release by the female at frequencies of about 10 pheromone plumes or pulsed releases/s.

In some moth species (*A. segetum*, *S. littoralis*, and *M. sexta*) receptor neurons responding to different pheromone components synapse in different glomeruli in the MGC. For example, receptor neurons of male *M. sexta* that respond when one of the sex pheromone components [(*E,Z*)-10,12-hexadecadienal] is blown onto the antenna have terminal arborizations in the toroid, while receptors responding to (*E,E,Z*)-10,12,14-hexadecatrienal, a second component, terminate in the cumulus

(Hansson et al., 1991). In some other species (*H. virescens* and *A. polyphemus*), receptor neurons tuned to different pheromone components synapse in the same MGC glomeruli, with some neurons also possibly projecting to a second glomerulus (Hansson, 1997).

JH has been shown to play a role in nervous system regulation of certain behaviors and nervous system structure in a few insects (Gadenne and Anton, 2000, and references therein). In allatectomized mature male *Agrotis ipsilon*, the proportion of low threshold AL interneurons sensitive to the female sex pheromone is lower than in intact males. Injection of JH restores (in allatectomized males) or induces (in intact males) a larger proportion of low threshold interneurons and increases the specificity of *A. ipsilon* males for its own female-produced blend compared to the very similar blend from a closely related species (Gadenne and Anton, 2000).

A number of examples are known of male moths that have receptor neurons on the antennae sensitive to one or more components that inhibit pheromone response. These inhibitory components may serve as isolating mechanisms for closely related species that share blend components. In all cases known, the receptor neurons that detect inhibitory compounds synapse in a glomerulus that does not receive input from pheromone components (Hansson, 1997).

In the AL, labeled lines and across-fiber patterning occur together because most receptor neurons make synaptic contacts with many different local interneurons (Christensen et al., 1993). Male and female insects have dimorphic numbers of receptors sensitive to sex pheromone on the antennae. Moths and cockroaches in particular have large numbers of receptors on the male antennae that respond mainly to the sex pheromone produced by females. Sex-specific pheromone receptors do not occur (or do not occur in large numbers) on the antennae of female moths and cockroaches.

19.10 HORMONAL CONTROL OF PHEROMONE SYNTHESIS AND RELEASE

Pheromone production in some insects, and perhaps in most if not all Lepidoptera, is under hormonal control (see Holman et al., 1990; Cardé and Minks, 1997, for reviews). Early evidence for possible hormonal control of pheromone production and/or release came from studies on the cockroaches, *Byrsotria fumigata* and *Pycnoscelus surinamensis* (Barth, 1964, 1965). Allatectomized cockroaches do not produce sex pheromone, leading to the hypothesis that probably JH from the corpora allata is a pheromonotropic hormone in cockroaches.

A polypeptide hormone that controls the synthesis of the sex pheromone in moths has been named **PBAN** (Raina and Klun, 1984) and was isolated first from the subesophageal ganglion (SEG) of the moth *H.* (formerly *Heliothis*) *zea* and determined to consist of 33 amino acids (MW = 3900) (Raina et al., 1989). PBAN is now known from several different sources, and that from *H. zea* is currently called **Hez-PBAN**; PBAN from *B. mori* is called **Bom-PBAN-I** (Raina and Gäde, 1988; Kitamura et al., 1989); and a second PBAN isolated from *B. mori* is called **Bom-PBAN-II** (Kitamura et al., 1990). PBAN isolated from the gypsy moth, *Lymantria dispar*, is called **Lyd-PBAN** (Masler et al., 1994). The PBANs belong to a class of peptides called **pyrokinins** and those isolated so far have 33–34 amino acid residues. All PBANs share, with other pyrokinins, a common C-terminal sequence of five amino acids Phe-X-Pro-Arg-Leu-NH2 (X can be Gly, Ser. Thr, or Val) that is required for biological activity. Substitution in the X position is extremely critical to pheromonotropic activity; putting glycine in the X position causes loss of activity, whereas a molecule containing threonine in position X is active (Abernathy et al., 1995).

Pyrokinin-type peptides, including PBAN, may have several different functions in insects (Choi et al., 2001, 2009, 2011, 2012, Wang et al., 2013). **Myotropic** activity of a pyrokinin-type peptide has been recorded in an *in vitro* cockroach hindgut assay (Nachman et al., 1986) and another (Bom-DH) functions as an **egg diapause hormone** in *B. mori* (Imai et al., 1991). Some PBAN peptides have myotropic activity *in vitro* in the cockroach hindgut assay (Nachman and

Holman, 1991). Some degree of cross-reactivity in the different bioassays is common and is probably due to the characteristic C-terminal peptide sequence. Some of the cockroach and locust pyrokinins also stimulate sex pheromone synthesis in *B. mori* females (Fónagy et al., 1992; Kuniyoshi et al., 1992).

Teal et al. (1996) suggested that there may be a number of neuropeptides involved with pheromone production and that some of the neuropeptides may have other physiological functions as well. They cite as evidence the fact that genes encoding for PBAN in *H. zea* and in *B. mori* encode for propheromones. In *B. mori* (Kawano et al., 1992), the propheromones give rise to Bom-PBAN, Bom-DH, a peptide with homology to Pss-Pt (a pheromonotropic peptide from the army worm, *Pseudaletia separata*) and other peptides.

19.10.1 Mode of Action of PBAN

How the PBAN signal is transposed into a cellular signal for pheromone synthesis in the pheromone gland cells is an area of active investigation. A PBAN receptor (PBANR) has been demonstrated as a G-protein-coupled membrane protein with 7-transmembrane domains in the cell membrane of pheromone gland cells (Choi et al., 2003; Hull et al., 2004). Binding of PBAN by its receptor results in the second messenger cAMP and an influx of calcium into the pheromone biosynthetic cells (Rafaeli and Jurenka, 2003). Hull et al. (2005) found that the C-terminus of the *Bombyx* PBANR is different from that of the *H. zea* PBANR by an additional 67 amino acids and contains specific amino acid residues that are essential for internalization of the *Bombyx* receptor. Although cAMP is involved as a second messenger in some insects, there is no evidence for involvement of cAMP in pheromone gland cells of *B. mori* (Hull et al., 2007; Matsumoto et al., 2007).

In the red-banded leafroller, *A. velutinana*, PBAN regulates pheromone biosynthesis by increasing the supply of octadecanoyl and hexadecanoyl fatty acids needed for pheromone biosynthesis, although the exact mechanism by which it does this is not clear (Tang et al., 1989). PBAN regulates the $\Delta 11$-desaturase (Roelofs and Jurenka, 1997), an enzyme widely distributed in Lepidoptera. The enzyme introduces a double bond into the pheromone precursor in *Mamestra brassicae* (Bestmann et al., 1989) and *Chrysodeixis chalcites* (Alstein et al., 1989). In two lepidopterans, *B. mori* and *S. littoralis*, PBAN influences the reduction of the fatty acid to the alcohol precursor of the pheromone (Martinez et al., 1990; Arima et el., 1991). Fang et al. (1992) were unable to determine the step or steps influenced by PBAN in *M. sexta* females, but they did determine that injection of PBAN during the photophase stimulated pheromone production and that putative fatty acid precursors for potential conversion to the aldehyde pheromone were present, but not changed by PBAN injection.

PBAN may control pheromone biosynthesis through regulation of fatty acid biosynthesis in *H. zea*, or it may control a step prior to fatty acid synthesis (Jurenka et al., 1991). The sex pheromone of *H. zea* is secreted from glandular cells in the intersegmental membrane located between the eighth and ninth abdominal segments (Jefferson et al., 1969). The major component (about 92%) of the pheromone is (Z)-11-hexadecenal (Klun et al., 1979) derived from fatty acid metabolism. The females normally synthesize and release the pheromone during the scotophase period, and only then can it be detected in the glandular tissue; little or no pheromone occurs in the gland during the photophase (Raina et al., 1986; Teal and Tumlinson, 1989). The hormone is detectable in the hemolymph only during the time when the pheromone is being produced (Raina and Klun, 1984). Injection of exogenous PBAN into a female moth can cause pheromone synthesis independent of the photoperiod. As little as 0.06 pmol of Hez-PBAN injected into a female moth stimulated pheromone biosynthesis, and 2 pmol stimulated maximum pheromone synthesis; higher doses produced no additional increase in synthesis.

PBAN is produced in the **Suboesophagal Ganglion (SEG)** of *H. zea* and *B. mori*. In studies with *H. zea*, Raina et al. (1989) found that the PBAN is released from the SEG and transported by the hemolymph to the pheromone gland, its target. Teal et al. (1989) have evidence that PBAN may be transported through the ventral nerve cord to the **terminal abdominal ganglion (TAG)**, where it causes the release of some second messenger (unidentified as yet) that acts on the pheromone gland cells. They propose that the target for PBAN is the TAG and that nerves from the TAG must be intact to the pheromone gland in order for extracts of brain SEG, applied to the TAG, to ultimately elicit pheromone synthesis. Ma and Roelofs (1995) showed in the female European corn borer, *Ostrinia nubilalis*, that PBAN was synthesized in three sets of neurosecretory cells (NSC) in the SEG and released from the corpora cardiaca (CC). Even though PBAN immunoreactivity was present throughout the ventral nerve cord, complete removal of the nerve cord did not alter female response to exogenous PBAN. In the gypsy moth, *L. dispar*, PBAN-immunoreactive material can be detected in the SEG and cells in each segmental ganglion (Golubeva et al., 1997), and transection of the ventral nerve cord disrupts pheromone production in females. It thus appears that origin, transport, and release mechanism of PBAN vary with species. A similar hormone, a sort of generic PBAN, probably functions in most, if not all, Lepidoptera.

19.11 BIOSYNTHESIS OF PHEROMONES

Most of the insects that have been studied synthesize their pheromonal components from small metabolic pool precursor molecules. Some insects modify precursors obtained in the food. For example, scolytid bark beetles use one or more of the terpenes in the host tree, sometimes aided by their complement of symbionts, as a pheromone precursor (Six, 2013; Raffa, 2014) (Figure 19.12), or some pheromonal components may be synthesized *de novo* from metabolic precursors (Song et al., 2014). Males of *Bactrocera dorsalis* fruit flies are strongly attracted to certain Hawaiian flowers and feed upon phenylpropanoid compounds on the petals (Figure 19.13a and b) and use the compounds to make a pheromone that attracts female flies (Nishida et al., 1997). Many moths use monounsaturated alcohols, acetates, or aldehydes with 12–16 carbons as pheromonal components. The starting material for synthesis of these pheromonal components is a saturated fatty acid synthesized from the acetate pool. Radiolabeled tracer studies show that common fatty acid starting materials are stearic (C18:COOH), palmitic (C16:COOH), myristic (C14:COOH), and lauric (C12:COOH) acids.

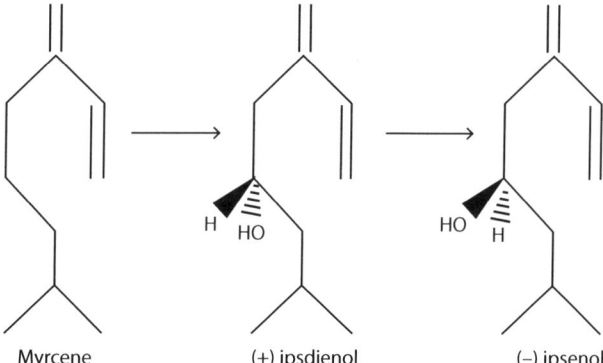

Figure 19.12 Some bark beetles use a naturally occurring compound in their host tree phloem resin as a precursor to synthesize pheromonal components. For example, *Ips confusus* uses naturally occurring myrcene to synthesize ipsdienol and ipsenol.

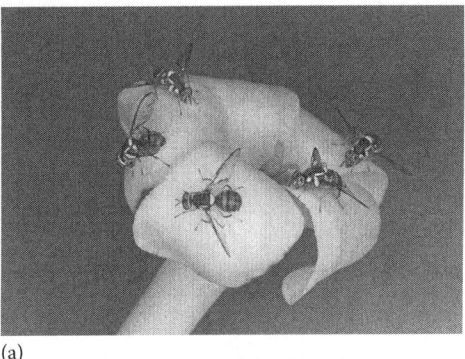

(a) (b)

Figure 19.13 (See color insert.) (a) Males of the Oriental fruit fly, *B. dorsalis*, collect naturally occurring phenylpropanoid compounds from the petals of the Hawaiian lei flower, *Fagraea berteriana*, and use the compounds to make a male-produced pheromone, *trans*-coniferyl alcohol, that attracts female flies. (From Nishida, H. et al., *J. Chem. Ecol.*, 22, 949, 1997. With permission. Color photograph courtesy of Dr. Ritsuo Nishida.) (b) *B. dorsalis* on a flower of the Hawaiian rainbow shower tree, a *Cassia* sp. that releases methyl eugenol, a pheromone precursor for the Oriental fruit fly males. (Photograph courtesy of Ethel Villalobos and Todd Shelly, USDA, Waimanalo, HI.)

Depending on the number of carbons in the pheromone component, one or more of these fatty acids is chain shortened in the pheromone gland by β-oxidation. A Δ11-desaturase enzyme (Figure 19.14) introduces a double bond, and *E* and *Z* isomers can be produced, and a different desaturase enzyme was discovered recently (Hagström et al., 2014), and others may yet be found. The moths reduce monounsaturated fatty acid intermediates to alcohols, acetates, or aldehydes to produce their pheromonal components (Roelofs and Bjostad, 1984; Morse and Meighen, 1986; Teal and Tumlinson, 1986; Bjostad et al., 1987; Roelofs and Wolf, 1988).

Some moths use diunsaturated pheromonal components, and two double bonds can be introduced by the Δ11-desaturase acting before and after chain shortening. Deuterium-labeled palmitic and myristic acids were used to demonstrate production of (*E*)-11-14 acetate and (*E,E*)-9,11 acetate, major components of the pheromone of *Epiphyas postvittana* (Bellas et al., 1983).

$$
\begin{array}{ccccccc}
18C & \xrightarrow{-2C} & 16C & \xrightarrow{-2C} & 14C & \xrightarrow{-2C} & 12C \\
\downarrow & & \downarrow & & \downarrow & & \downarrow \\
\Delta 11\text{-}18C & \xrightarrow{-2C} & \Delta 11\text{-}16C & \xrightarrow{-2C} & \Delta 11\text{-}14C & \xrightarrow{-2C} & \Delta 11\text{-}12C \\
\downarrow -2C & & \downarrow -2C & & \downarrow -2C & & \\
\Delta 9\text{-}16C & \xrightarrow{-2C} & \Delta 9\text{-}14C & \xrightarrow{-2C} & \Delta 9\text{-}12C & & \\
\downarrow -2C & & \downarrow -2C & & & & \\
\Delta 7\text{-}14C & \xrightarrow{-2C} & \Delta 7\text{-}12C & & & & \\
\downarrow -2C & & & & & & \\
\Delta 5\text{-}12C & & & & & &
\end{array}
$$

Figure 19.14 Biosynthetic pathways for *de novo* synthesis of pheromone components by female moths from precursors in the general metabolic pool. (Modified from Roelofs, W.L. and Wolf, W.A., *J. Chem. Ecol.*, 14, 2019, 1988.)

In addition to the Δ11-desaturase, a Δ10-desaturase has been identified in the New Zealand leafroller, *Planotortrix excessana* (Foster and Roelofs, 1988), and a Δ9-desaturase in the brown-headed leafroller, *Ctenopseutis obliquana*. Roelofs and Wolf (1988) speculate that in the early evolutionary stages of these Tortricidae, the β-oxidation step was used to shorten oleic and/or palmitoleic acids from which shorter chain pheromone components were synthesized. Later, evolution of Δ10- and Δ11-desaturases made it possible for the moths to biosynthesize a wide range of unique pheromonal components and may have paved the way for evolution of many different species-specific blends.

19.12 GEOGRAPHICAL AND POPULATION DIFFERENCES AND EVOLUTION OF PHEROMONE BLENDS

It is clear that closely related species in several groups of insects have evolved closely related pheromone blends, but how the evolution took place has not been elucidated. Leal (1999) found close similarity of pheromone enantiomers for two scarab beetles, and related chemical structures have been demonstrated in a group of scarab beetles (Leal, 1997). Just as there is variation in virtually all biological attributes of organisms, it seems reasonable to assume that there may be some biological variability in pheromone blends and in response patterns of the opposite sex to blend differences. Blend producers and blend responders that are too far off the mark probably seldom mate, but the genes for variability may be kept in the population, or lie dormant, for a long time. If a pheromone blend changes or shows variability in a population, then one expects variability in the responders. If populations are separated and random or "saltatorial" shifts (Roelofs et al., 2002) in pheromone blends occur, it could eventually lead to new species.

One species displaying blend differences among several populations is *Ips pini*, the pine engraver. This insect is a major tree killer in the Great Lakes area of the United States, and in California and Idaho. Males locate a new host tree, bore into the phloem layer, feed, and release as the principal component of its pheromone the compound 2-methyl-6-methylene-2,7-octadien-4-ol, also known as ipsdienol. California and Idaho populations are attracted to *R*-(−)-ipsdienol, and attraction is reduced if *S*-(+)-ipsdienol is present. On the other hand, New York populations are most attracted to a 50:50 blend of *R*-(−)-ipsdienol and *S*-(+)-ipsdienol. Beetles from Wisconsin respond preferentially to a 75:25 mixture of the *S*-(+):*R*(−). The beetles also show regional differences in response to a minor pheromonal component, lanierone (Miller et al., 1997). What could be promoting such populational changes?

One possible answer is predation pressure. Two major predators of the pine engraver are the adult beetles, *T. dubius* (family Cleridae) and *Cylistix cylindrica* (family Histeridae). These beetles use the pine engraver pheromone as a kairomone, leading them to their prey, which they aggressively attack and eat. Both predators show strong preference for attraction to a mixture of 25% *S*-(+):75% *R*-(−)-ipsdienol. Populations may be evolving aggregation pheromone blends that are less effective in attracting their predators.

For individuals to successfully mate, there must be concomitant evolution in pheromone blend with evolution in responder receptors. The turnip moth, *Agrotis segetum*, seems to provide a good example of coevolution between chemical pheromone and receptor. There are three populations of the moth, one in France, one in Sweden, and one in the general area of Armenia/Bulgaria. The pheromone is multicomponent, but female moths in the French population produce a large amount of (Z)-5-decenyl acetate, and males have many receptors on the antennae that respond electrophysiologically to this component. Moths in the Swedish population produce less (Z)-5-decenyl acetate, and males have a smaller population of receptors on their antennae that respond to the compound. Finally, female moths in the Armenian/Bulgarian population produce very little (Z)-5-decenyl acetate, and males have very few receptors for it. Clearly, evolution of changes in female production of

(Z)-5-decenyl acetate has been correlated with changes in receptor specificity in males. Mechanisms driving the changes have not been elucidated (Hansson et al., 1990).

In New York state, there are three races of the European corn borer, *Ostrinia nubilalis*: a bivoltine Z race, a univoltine Z race, and a bivoltine E race. Males in the Z races respond maximally to a blend of (Z)- and (S)-11-tetradecenyl acetate in a ratio of 97:3 Z/E, while males in the E race respond to a blend of 1:99 Z/E. A close relative *O. furnacalis*, the Asian corn borer, produces a pheromone blend of 2:1 (Z)- and (E)-12-tetradecenyl acetates. Roelofs et al. (2002) showed that both species are capable of chain-shortening fatty acids and both have (and probably have had for millions of years) the genes capable of expressing the Δ11-desaturase and the Δ14-desaturase enzyme needed to put the double bond into the pheromone molecule. The gene for the Δ11-desaturase is not known to be expressed in the Asian corn borer, and Roelofs et al. (2002) speculate that the Δ14-desaturase has not been expressed in the European corn borer until recently. Phelan (1992, 1997) argued for an asymmetric tracking model that would predict selection of males for variation in response specificity, which may be applicable if mutations arise in a pheromone synthetic pathway (Haynes, 1997) or if a pheromone blend undergoes "saltatory" shifts in composition (Roelofs et al., 2002). In a follow-up study, Linn et al. (2003) found support for the asymmetric tracking model in the response of 1%–5% of European corn borer males able to respond to blends produced by the Asian corn borer (and also to their own European corn borer blends).

19.13 PRACTICAL APPLICATIONS OF PHEROMONES

Much of the stimulus for identification of pheromones has come from the expectation that pheromones would have practical application in population management of insects. The most widespread and successful use of pheromones has been in **monitoring** insect emergence (in the spring or summer) and population buildup. Deployed in traps, pheromones can indicate the presence of pest insects, timing of emergence, flights, and movement into a crop. When large numbers are caught, the decision may be made to apply control measures such as a pesticide.

Direct control with pheromones also is possible in some cases, including (1) **mass trapping**, (2) **lure and kill**, and (3) **mating disruption**. Mass trapping and luring large aggregations to trap trees where they can be killed by conventional insecticides has been tried with limited success in population control of bark beetles (Borden, 1997).

Disruption of mating has been one of the successful applications of pheromone to direct insect control of a number of lepidopteran species, although it, like other control procedures, sometimes fails. Current successes and some reasons for failure in Lepidoptera have been reviewed by Cardé and Minks (1995), Arn and Louis (1997), Sanders (1997), Staten et al. (1997), and Suckling and Karg (1997). Mating disruption currently is more costly than conventional insecticide treatments if environmental issues are not considered. Disruption of mating in tortricid moths in Switzerland and Germany costs two to four times as much as conventional insecticides. Reducing the cost of pheromone, more effective delivery systems (Figure 19.15) that release pheromone in controlled, pulsed amounts and only when the target insect is flying (Shorey et al., 1996; Shorey and Gerber, 1996; Baker et al., 1997), and application of the least amount of pheromone that will work are ways to reduce costs (Arn and Louis, 1997). Disruption of bark beetle aggregations with inhibitors of the aggregation response has enjoyed some success and promises to have a future (Borden, 1997).

Successful mating under field conditions involves several behavioral actions, such as long-range orientation (upwind flight in response to pheromone) followed by close-range courtship (possibly also involving pheromone, vision, and other sensory modes such as mechanoreception). Disruption tactics may target any or combinations of these behaviors. Upwind flight in theory and practice has been more amenable to disruption (Sanders, 1997) than close-range behaviors.

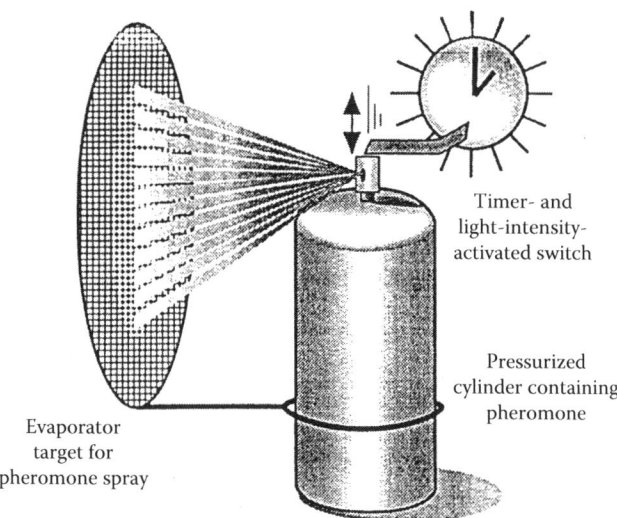

Figure 19.15 One design for a puffer device that disperses pheromone from a pressurized canister at timed intervals. (From Shorey, H.H. et al., *Environ. Entomol.*, 25, 446, 1996. With permission.)

19.13.1 Mechanisms Operating in Mating Disruption

How mating disruption occurs is poorly understood and may occur by one or more of the following mechanisms:

1. Sensory fatigue, which can be divided into the component parts of adaptation of receptors at the periphery of the insect (the antenna) and/or habituation in the CNS
2. Competition between natural and synthetic sources (false trail following)
3. Camouflage of the natural pheromone trail
4. Use of blend imbalance and antagonists that stop the response to pheromone

19.13.1.1 Sensory Fatigue

Males of many moth species show failure to successfully find and mate with a female in wind tunnel tests when exposed to high concentrations of pheromone. Males also preconditioned by periodic exposure to pheromone exhibit reduced response and/or fail to find and attempt mating with a female when exposed to pheromone in subsequent wind tunnel tests. Preexposure to pulses of pheromone are more effective than exposure to a constant concentration of pheromone in creating subsequent failure or low response (Kuenen and Baker, 1981).

19.13.1.2 False Trail Following

Pheromone that is widely dispersed in the environment may cause male moths to follow the synthetic pheromone plume as opposed to the plume from a female. In order for this mechanism to work, the synthetic pheromone sources must be at least as attractive as the natural female, and there must be enough artificial sources to make it improbable that a male will find a female by chance. Presumably, males will spend virtually all their energy and time in following false trails. To make the artificial pheromone sources as attractive as the natural pheromone usually requires that the complete pheromone blend and rate of release by the female must be known and used in the synthetic pheromone sources (Minks and Cardé, 1988). In theory and practice, the method works best at low population density when the artificial sources greatly outnumber the feral females (Webb et al., 1990; Howell et al., 1992). For disruption to work against *L. botrana* in grape vineyards, population density should be no more than four pairs per 10 m^2 (Feldhege et al., 1995).

19.13.1.3 Camouflage of Natural Pheromone Plume

Camouflage of natural pheromone trails is similar to the previous method, but it is predicated on the assumption that a male moth cannot detect the true pheromone trail if it is surrounded by pheromone or in a pheromone fog (Sanders, 1997). In such a situation, the male would either remain at rest, as in some wind tunnel tests, or spend all its time casting back and forth in search of discontinuous pheromone filaments, which, if located, signal it to fly upwind. It is probably not possible in a field situation to have a uniform fog of pheromone because of wind eddies and air disturbances due to vegetation breaking up the movement of pheromone. Male moths may successfully find significant numbers of feral females.

19.13.1.4 Pheromone Antagonists and Imbalanced Blends

In a few cases, the use of pheromone antagonists and incomplete blends show promise in field tests. For example, mating disruption of some lepidopterans has been achieved with incomplete blends. Disruption of mating in the navel orange worm was based on the presence of inhibitors of the natural pheromone (Curtis et al., 1987), and the pea moth, *Cydia nigricana*, is inhibited by a pheromone blend containing attraction inhibitors (Bengtsson et al., 1994). The female tortricid moth, *E. ambiguella*, produces (Z)-9-dodecenyl acetate, and males are inhibited if a synthetic blend contains more than about 0.1% of the *E* isomer. A technical grade of Z9-12:AC that has been used successfully in mating disruption contains a small percentage of the *E* isomer (Arn and Louis, 1997). Disruption of mating in *E. ambiguella* is the most important method of control of this moth in grape vineyards in Switzerland and parts of Germany (Arn and Louis, 1997). Aggregations of some species of bark beetles can be reduced by applications of several available inhibitors of the aggregation response, and these inhibitors may have good potential for protection of valuable specimens and ornamental trees (Borden, 1997).

Any of the methods in which pheromones are used to control a population have the potential to select for survival and reproduction of those (possibly few) individuals that respond to variable blend ratios, different release rates, or that somehow compensate for possible inhibitors.

19.14 REVIEW AND SELF-STUDY QUESTIONS

1. What are allelochemicals? Name some types of allelochemicals.
2. How are pheromones defined?
3. What is the difference between a releaser pheromone and a primer one?
4. Define kairomone. Define allomone. Give an example from insects of each.
5. The olfactory sense in most Lepidoptera seems to be extraordinarily sensitive. How can you explain the evolution of such a sensitive system?
6. What is the active space concept?
7. Name five different types of pheromones based upon the behavior influenced.
8. What is pheromone parsimony? Give an example.
9. What are pheromone enantiomers?
10. How are R and S enantiomers described or characterized?
11. What is a racemic mixture of a pheromone?
12. Would it be less expensive to put a racemic mixture into a trap or a pure enantiomer in most cases? Why?
13. What is the general structure of a pheromone (or semiochemical) receptor?

14. What is an EAG? What information can it give to an experimenter?
15. If you wanted to identify a sex pheromone from an insect that had not been studied before, would you begin by making an extract of the two sexes and take it to a chemist for him or her to identify the chemicals in the extract? Why or why not?
16. What is the difference between an EAG and an EAD?
17. What is an optomotor anemotaxis response in insects?
18. What does the pheromone plume of an insect look like?
19. What is PBAN?
20. How can/are pheromones used in a practical way to manage insect populations?

REFERENCES

Abernathy, R.L., R.J. Nachman, P.E.A. Teal, O. Yamashita, and J.H. Tumlinson. 1995. Pheromonotropic activity of naturally occurring pyrokinin insect neuropeptides (FXPRLamine) in *Helicoverpa zea*. *Peptides* 16: 215–219.

Almaas, T.J. and H. Mustaparta. 1990. Pheromone reception in tobacco budworm moth, *Heliothis virescens*. *J. Chem. Ecol.* 16: 1331–1347.

Altstein, M., M. Harel, and E. Dunkelblum. 1989. Effect of a neuroendocrine factor on sex pheromone biosynthesis in the tomato looper, *Chrysodeixis chalcites* (Lepidoptera: Noctuidae). *Insect Biochem.* 19: 645–649.

Andersson, J., A.-K. Borg-Karlson, N. Vongvanich, and C. Wiklund. 2007. Male sex pheromone release and female mate choice in a butterfly. *J. Exp. Biol.* 210: 964–970.

Anton, S. and B.S. Hansson. 1994. Central processing of sex pheromone, host odour, and oviposition deterrent information by interneurons in the antennal lobe of the female *Spodoptera littoralis* (Lepidoptera: Noctuidae). *J. Comp. Neurol.* 350: 199–214.

Arima, R., K. Takahara, T. Kadoshima, F. Numazaki, T. Ando, M. Uchiyama, H. Nagasawa, A. Kitamura, and A. Suzuki. 1991. Hormonal regulation of pheromone biosynthesis in the silkworm moth, *Bombyx mori* (Lepidoptera: Bombycidae). *Appl. Entomol. Zool.* 26: 137–147.

Arn, H. and F. Louis. 1997. Mating disruption in European vineyards, in R.T. Cardé and A.K. Minks (eds.), *Insect Pheromone Research. New Directions*. Chapman & Hall, New York, pp. 377–382.

Arn, H., E. Stadler, and S. Rauscher. 1975. The electroantennographic detector—A selective and sensitive tool in the gas chromatographic analysis of insect pheromones. *Z. Naturforsch.* 30c: 722–725.

Ayasse, M. and S. Jarau. 2014. Chemical ecology of bumble bees. *Annu. Rev. Entomol.* 59: 299–319.

Baker, T.C. and K.F. Haynes. 1989. Field and laboratory electroantennographic measurements of pheromone plume structure correlated with oriental fruit moth behavior. *Physiol. Entomol.* 14: 1–12.

Baker, T.C., A. Mafra-Neto, T. Dittl, and M.E. Rice. 1997. A novel controlled-release device for disrupting sex pheromone communication in moths, in P. Witzgall and H. Arn (eds.), *Technology Transfer in Mating Disruption*, IOBC wprs Bulletin, vol. 20(1), Avignon, France.

Baker, T.C., M.A. Willis, and P.L. Phelan. 1984. Optometer anemotaxis polarizes self-steered zigzagging in flying moths. *Physiol. Entomol.* 9: 365–376.

Barth, R.H., Jr. 1964. The mating behavior of *Byrsotria fumigata*. *Behavior* 23: 1–30.

Barth, R.H., Jr. 1965. Endocrine control of a chemical communication system. *Science* 149: 882–883.

Bell, W.J. and R.T. Cardé, 1984. *Chemical Ecology of Insects*. Chapman & Hall, New York.

Bellas, T.E., R.J. Bartell, and A. Hill. 1983. Identification of two components of the sex pheromone of the moth, *Epiphyas postvittana* (Lepidoptera, Tortricidae). *J. Chem. Ecol.* 9: 503–512.

Bengtsson, M., G. Karg, P.A. Kirsch, J. Löfqvist, A. Sauer, and P. Witzgall. 1994. Mating disruption of pea moth *Cydia nigricana* F. (Lepidoptera: Tortricidae) by a repellent blend of sex pheromone and attraction inhibitors. *J. Chem. Ecol.* 20: 871–887.

Berenbaum, M.R. 1995. The chemistry of defense: Theory and practice. *Proc. Natl. Acad. Sci. USA* 92: 2–8.

Berg, B.G. and H. Mustaparta. 1995. The significance of major pheromone components and interspecific signals as expressed by receptor neurons in the oriental tobacco budworm moth, *Helicoverpa assulta*. *J. Comp. Physiol. A* 177: 683–694.

Berg, B.G., J.H. Tumlinson, and H. Mustaparta. 1995. Chemical communication in heliothine moths. IV. Receptor neuron responses to pheromone compounds and formate analogues in the tobacco budworm moth *Heliothis virescens*. *J. Comp. Physiol. A* 177: 527–534.

Bergman, P. and G. Bergström. 1997. Scent marking, scent origin, and species specificity in male premating behavior of two Scandinavian bumblebees. *J. Chem. Ecol.* 23: 1235–1251.

Bestmann, H.J., M. Herrig, A.B. Attygalle, and M. Hupe. 1989. Regulatory steps in sex pheromone biosynthesis in *Mamestra brassicae* L. (Lepidoptera: Noctuidae). *Experientia* 45: 778–781.

Bjostad, L.B., W.A. Wolf, and W.L. Roelofs. 1987. Pheromone biosynthesis in lepidopterans: Desaturation and chain shortening, in G.D. Prestwich and G.J. Blomquist (eds.), *Pheromone Biochemistry*. Academic Press, New York, pp. 77–120.

Blum, M.S. 1996. Semiochemical parsimony in the Arthropoda. *Annu. Rev. Entomol.* 41: 353–374.

Boekhoff, I., K. Raming, and H. Breer. 1990a. Pheromone-induced stimulation of inositol-trisphosphate formation in insect antennae is mediated by G-proteins. *J. Comp. Physiol.* 160: 99–103.

Boekhoff, I., E. Seifert, S. Goggerle, M. Lindemann, B.W. Kruger, and H. Breer. 1993. Pheromone-induced second-messenger signaling in insect antennae. *Insect Biochem. Mol. Biol.* 23: 757–762.

Boekhoff, I., J. Strotmann, K. Raming, E. Tareilus, and H. Breer. 1990b. Odorant-sensitive phospholipase C in insect antennae. *Cell. Signal.* 2: 49–56.

Borden, J.H. 1997. Disruption of semiochemical-mediated aggregation in bark beetles, in R.T. Cardé and A.K. Minks (eds.), *Insect Pheromone Research: New Directions*. Chapman & Hall, New York, pp. 421–438.

Bossert, W.H. and E.O. Wilson. 1963. The analysis of olfactory communication among animals. *J. Theoret. Biol.* 5: 443–469.

Breer, H. 1997. Molecular mechanisms of pheromone reception in insect antennae, in R.T. Cardé and A.K. Minks (eds.), *Insect Pheromone Research: New Directions*. Chapman & Hall, New York, pp. 115–130.

Breer, H., I. Boekhoff, J. Strotmann, K. Raming, and E. Tareilus. 1990a. Molecular elements of olfactory signal transduction in insect antennae, in D. Schild (ed.), *Information Processing of Chemical Sensory Stimuli in Physiological and Artificial Systems*. Springer-Verlag, Berlin, Germany, pp. 77–86.

Breer, H., I. Boekhoff, and E. Tarelius. 1990b. Rapid kinetics of second messenger formation in olfactory transduction. *Nature* 345: 65–68.

Buck, L. and R. Axel. 1991. A novel multigene family may encode odorant receptors: A molecular basis for odor recognition. *Cell* 65: 175–187.

Butenandt, A., R. Beckmann, D. Stamm, and E. Hecker. 1959. Über den Sexual-Lockstoff des Seidenspinners *Bombyx mori*. Reindarstellung und Konstitutionsermittlung. *Z. für Naturforschung* 14b: 283–284.

Cahn, R.S. 1964a. An introduction to the sequence rule. *J. Chem. Educ.* 41: 116–125.

Cahn, R.S. 1964b. Errata. *J. Chem. Educ.* 41: 508.

Camazine, S.M. and J.G. Hildebrand. 1979. Central projections of antennal sensory neurons in mature and developing *Manduca sexta*. *Soc. Neurosci. Abstr.* 5: 155.

Capinera, J.L. 1980. A trail pheromone from the silk produced by larvae of the range caterpillar *Hemileuca oliviae* (Lepidoptera: Saturniidae) and observations on aggregation behavior. *J. Chem. Ecol.* 3: 655–644.

Cardé, R. and A. Mafra-Neto. 1997. Mechanisms of flight of male moths to pheromone, in R.T. Cardé and A.K. Minks (eds.), *Insect Pheromone Research: New Directions*. Chapman & Hall, New York, pp. 275–290.

Cardé, R.T. and W.J. Bell. 1995. *Chemical Ecology of Insects 2*. Chapman & Hall, New York, 433pp.

Cardé, R.T. and A.K. Minks. 1995. Control of moth pests by mating disruption: Successes and constraints. *Annu. Rev. Entomol.* 40: 559–585.

Cardé, R.T. and A.K. Minks (eds.). 1997. *Insect Pheromone Research: New Directions*. Chapman & Hall, New York.

Cardé, R.T. and M.A. Willis. 2008. Navigational strategies used by insect to find distant, wind-borne sources of odor. *J. Chem. Ecol.* 34: 854–866.

Choi, M.Y., E.J. Fuerst, A. Rafaeli, and R. Jurenka. 2003. Identification of a G protein-coupled receptor for pheromone biosynthesis activating neuropeptide from pheromone glands of the moth *Helicoverpa zea*. *Proc. Natl. Acad. Sci. USA* 100: 9721–9726.

Choi, M.-Y., A. Rafaeli, and R.A. Jurenka. 2001. Pyrokinin/PBAN-like peptides in the central nervous system of *Drosophila melanogaster*. *Cell Tissue Res.* 306: 459–465.

Choi, M.-Y., A. Raina, and R.K. Vander Meer. 2009. PBAN/pyrokinin peptides in the central nervous system of the fire ant, *Solenopsis invicta*. *Cell Tissue Res.* 335: 431–439.

Choi, M.-Y. and R.K. Vander Meer. 2012. Molecular structure and diversity of PBAN/pyrokinin family peptides in ants. *Front. Endocrinol.* 3: 1–8. doi:10.3389/fendo.2012.00032.

Choi, M.-Y., R.K. Vander Meer, D. Shoemaker, and S.M. Valles. PBAN gene architecture and expression in the fire ant, *Solenopsis invicta. J. Insect Physiol.* 57: 161–165.

Christensen, T.A. 1997. Anatomical and physiological diversity in the central processing of sex-pheromone information in different moth species, in R.T. Cardé and A.K. Minks (eds.), *Insect Pheromone Research: New Directions.* Chapman & Hall, New York, pp. 184–193.

Christensen, T.A. and J.G. Hildebrand. 1990. Representation of sex-pheromonal information in the insect brain, in K.B. Døving (ed.), *Proceedings of X International Symposium on Olfaction and Taste.* Graphic Communication Systems, Oslo, Norway, pp. 142–150.

Christensen, T.A., B.R. Waldrop, I.D. Harrow, and J.G. Hildebrand. 1993. Local interneurons and information processing in the olfactory glomeruli of the moth *Manduca sexta. J. Comp. Physiol. A* 173: 385–399.

Cossé, A.A., J.L. Todd, J.G. Millar, L.A. Martínez, and T.C. Baker. 1995. Electroantennographic and coupled gas chromatographic-electroantennographic responses of the Mediterranean fruit fly, *Ceratitis capitata*, to male-produced volatiles and mango odor. *J. Chem. Ecol.* 21: 1823–1836.

Crespo, J.G., F. Goller, and N.J. Vickers. 2012. Pheromone mediated modulation of pre-flight warm-up behavior in male moths. *J. Exp. Biol.* 215: 2203–2209.

Curtis, C.F., J.D. Clark, D.A. Carlson, and J.A. Coffelt. 1987. A pheromone mimic: Disruption of mating communication in the navel orangeworm, *Amyelois transitella*, with Z,Z-1,12,14-heptadecatriene. *Ent. Exp. Appl.* 44: 249–255.

David, C.T., J.S. Kennedy, and A.R. Ludlow. 1983. Finding a sex pheromone source by gypsy moths, *Lymantria dispar*, released in the field. *Nature* 303: 804–806.

Davis, T.S., T.L. Crippen, R.W. Hofstetter, and J.K. Tomberlin. 2013. Microbial volatile emissions as insect semiochemicals. *J. Chem. Ecol.* 39: 840–859.

Dethier, V.G. 1972. A surfeit of stimuli: A paucity of receptors. *Am. Sci.* 59: 706–715.

Dicke, M. and M.W. Sabelis. 1988. Infochemical terminology: Based on cost-benefit analysis rather than on origin of compounds? *Funct. Ecol.* 2: 131–139.

Du, G., C.-S. Ng, and G.D. Prestwich. 1994. Odorant binding by a pheromone binding protein: Active site mapping by photoaffinity labeling. *Biochemistry* 33: 4812–4819.

Eisner, T. and J. Meinwald. 1995. The chemistry of sexual selection. *Proc. Natl. Acad. Sci. USA* 92: 50–55.

El-Sayed, A.M. 2014. The Pherobase: Database of pheromone and semiochemicals. http://www.pherobase.com, (last accessed April 22, 2015).

Fang, N., J.H. Tumlinson, P.E.A. Teal, and H. Oberlander. 1992. Fatty acyl pheromone precursors in the sex pheromone gland of female hornworm moths, *Manduca sexta* (L.). *Insect Biochem. Mol. Biol.* 22: 621–631.

Feldhege, M., F. Louis, and H. Schmutterer. 1995. Untersuchungen uber Falterabundanzen des Bekreuzten Traubenwicklers *Lobesia botrana* Sciff. im Weinbau. *Anz. Schaldlingsk., Pflanzenschutz, Umweltschutz* 68: 85–91.

Fitzgerald, T.D. 1993. Trail and arena marking by caterpillars of *Archips cerasivoranus* (Lepidoptera: Tortricidae). *J. Chem. Ecol.* 19: 1479–1489.

Fitzgerald, T.D. and J.T. Costa. 1986. Trail-based communication and foraging behavior of young colonies of the forest tent caterpillar *Malacosoma disstria* Hubn. (Lepidoptera: Lasiocampidae). *Ann. Entomol. Soc. Am.* 79: 999–1007.

Fitzgerald, T.D. and D.L.A. Underwood. 1998. Communal foraging behavior and recruitment communication in *Gloveria* sp. *J. Chem. Ecol.* 24: 1381–1396.

Fónagy, A., L. Schoofs, S. Matsumoto, A. De Loof, and T. Mitsui. 1992. Functional cross-reactivities of some locustamyotropins and *Bombyx* pheromone biosynthesis activating neuropeptide. *J. Insect Physiol.* 38: 651–657.

Foster, S.P. and W.L. Roelofs. 1988. Sex pheromone biosynthesis in the leafroller moth *Planotortrix excessana* by $\Delta 10$ desaturation. *Arch. Insect Biochem. Physiol.* 8: 1–9.

Gadenne, C. and S. Anton. 2000. Central processing of sex pheromone stimuli is differentially regulated by juvenile hormone in a male moth. *J. Insect Physiol.* 46: 1195–1206.

Gary, N. 1962. Chemical mating attractants in the queen honey bee. *Science* 136: 773–774.

Girling, R.D., B.S. Higbee, and R.T. Cardé. 2013. The plume also rises: Trajectories of pheromone plumes issuing from point sources in an orchard canopy at night. *J. Chem. Ecol.* 39: 1150–1160.

Golubeva, E., T.G. Kingan, M.B. Blackburn, E.P. Masler, and A.K. Raina. 1997. The distribution of PBAN (pheromone biosynthesis activating neuropeptide)-like immunoreactivity in the nervous system of the gypsy moth, *Lymantria dispar*. *Arch. Insect Bochem. Physiol.* 34: 391–408.

Grace, J.K., D.L. Wood, and G.W. Frankie. 1988. Trail-following behavior of *Reticulitermes hesperus* banks (Isoptera: Rhinotermitidae). *J. Chem. Ecol.* 14: 653–667.

Gu, D. and D.-R. Yang. 2013. Utilisation of chemical signals by inquiline wasps in entering their host figs. *J. Insect Phyisol.* 59: 1065–1068.

Hagström, Å.K., J. Albre, L.K. Tooman, A.H. Thirmawithana, J. Corcoran, C. Löfstedt, and R.D. Newcomb. 2014. A novel fatty acyl desaturase from the pheromone glands of *Ctenopseustis obliquana* and *C. herana* with specific Z-5-desaturase activity on myristic acid. *J. Chem. Ecol.* 40: 63–70.

Hanks, L.M. and J.G. Millar. 2013. Field bioassays of cerambycid pheromones reveal widespread parsimony of pheromone structures, enhancement by host plant volatiles, and antagonism by components from heterospecifics. *Chemoecology* 23: 21–44.

Hansson, B.S. 1997. Antennal lobe projection patterns of pheromone-specific olfactory receptor neurons in moths, in R.T. Cardé and A.K. Minks (eds.), *Insect Pheromone Research: New Directions*. Chapman & Hall, New York, pp. 164–183.

Hansson, B.S., T.A. Christensen, and J.G. Hildebrand. 1991. Functionally distinct subdivisions of the macroglomerular complex in the antennal lobe of the male sphinx moth *Manduca sexta*. *J. Comp. Neurol.* 312: 264–278.

Hansson, B.S., M. Toth, C. Löfstedt, G. Szos, M. Subchev, and J. Lofqvist. 1990. Pheromone variation among eastern European and a western population of the turnip moth *Agrotis segetum*. *J. Chem. Ecol.* 16: 1611–1622.

Haynes, K. 1997. Genetics of pheromone communication in the cabbage looper moth, *Trichoplusia ni*, in R.T. Cardé and A.K. Minks (eds.), *Insect Pheromone Research: New Directions*. Chapman & Hall, New York, pp. 525–534.

Hildebrand, J.G. 1995. Analysis of chemical signals by nervous systems. *Proc. Natl. Acad. Sci. USA* 92: 67–74.

Hildebrand, J.G. 1996. Olfactory control of behavior in moths: Central processing of odor information and the functional significance of olfactory glomeruli. *J. Comp. Physiol A* 178: 5–19.

Hölldobler, B., N.J.R. Plowes, R.A. Johnson, U. Nishshanka, C. Liu, and A.B. Attygalle. 2013. Pygidial gland chemistry and potential alarm-recruitment function in column foraging, but not solitary, Nearctic *Messor* harvesting ants (Hymenoptera: Formicidae: Myrmicinae). *J. Insect Physiol.* 59: 863–869.

Holman, G.M., R.J. Nachman, and M.S. Wright. 1990. Insect neuropeptides. *Annu. Rev. Entomol.* 35: 201–217.

Honda, Y., K. Honda, and H. Ômura. 2006. Major components in the hairpencil secretion of a butterfly, *Euploea mulciber* (Lepidoptera, Danaidae): Their origins and male behavioral responses to pyrrolizidine alkaloids. *J. Insect Physiol.* 52: 1043–1053.

Howell, J.F., A.L. Knight, T.R. Unruh, D.F. Brown, J.L. Krysan, C.R. Sell, and P.A Kirsch. 1992. Control of codling moth in apple and pear with sex pheromone-mediated mating disruption. *J. Econ. Entomol.* 85: 918–925.

Hull, J.J., R. Kajigaya, K. Imai, and S. Matsumoto. 2007. The *Bombyx mori* sex pheromone biosynthetic pathway is not mediated by cAMP. *J. Insect Physiol.* 53: 782–793.

Hull, J.J., A. Ohnishi, and S. Matsumoto. 2005. Regulatory mechanisms underlying pheromone biosynthesis activating neuropeptide (PBAN)-induced internalization of the *Bombyx mori* PBAN receptor. *Biochem. Biophys. Res. Com.* 334: 69–78.

Hull, J.J., A. Ohnishi, K. Moto, Y. Kawasaki, R. Kurata, M.G. Suzuki, and S. Matsumoto. 2004. Cloning and characterization of the pheromone biosynthesis activating neuropeptide receptor from the silkmoth, *Bombyx mori*: Significance of the carboxyl terminus in receptor internalization. *J. Biol. Chem.* 279: 51500–51507.

Imai, K., T. Konna, Y. Nakazawa, T. Komiya, M. Isobe, K. Koga, T. Goto et al. 1991. Isolation and structure of diapause hormone of the silkworm moth, *Bombyx mori*. *Proc. Jpn. Acad.* 67: 98–101.

Ishida, Y., T. Nagae, and M. Azuma. 2012. A water-specific aquaporin is expressed in the olfactory organs of the blowfly, *Phormia regina*. *J. Chem. Ecol.* 38: 1057–1061.

Janssen, E., B. Hölldobler, F. Kern, H.-J. Bestmann, and K. Tsuji. 1997. Trail pheromone of myrmicine ant *Pristomyrmex pungens*. *J. Chem. Ecol.* 23: 1025–1034.

Jefferson, R.N., H.H. Shorey, and R.E. Rubin. 1969. Sex pheromones of noctuid moths. XVI. The morphology of the female sex pheromone glands of eight species. *Ann. Entomol. Soc. Am.* 61: 861–865.

Joachim, C., E. Hatano, A. David, M. Kunert, C. Linse, and W.W. Weisser. 2013. Modulation of aphid alarm pheromone emission of pea aphid prey by predators. *J. Chem. Ecol.* 39: 773–782.

Jurenka, R.A., E. Jacquin, and W.L. Roelofs. 1991. Control of the pheromone biosynthetic pathway in *Helicoverpa zea* by the pheromone biosynthesis activating neuropeptide. *Arch. Insect Biochem. Physiol.* 17: 81–91.

Kaissling, K.-E. 1986. Chemo-electrical transduction in insect olfactory receptors. *Annu. Rev. Neurosci.* 9: 121–145.

Kaissling, K.-E. 1987. Transduction processes in olfactory receptors of moths, in J.H. Law (ed.), *Molecular Entomology*. Alan R. Liss, New York, pp. 33–43.

Kanzaki, R. 1997. Pheromone processing in the lateral accessory lobes of the moth brain: Flip-flopping signals related to zigzagging upwind walking, in R.T. Cardé and A.K. Minks (eds.), *Insect Pheromone Research: New Directions*. Chapman & Hall, New York, pp. 291–303.

Kanzaki, R. and T. Shibuya. 1986. Descending protocerebral neurons related to the mating dance of the male silkworm moth. *Brain Res.* 377: 378–382.

Karg, G. and A.E. Sauer. 1995. Spatial distribution of pheromone in vineyards treated for mating disruption of the grape vine moth *Lobesia botrana* measured with electroantennograms. *J. Chem. Ecol.* 21: 1299–1314.

Karlson, P. and M. Luscher. 1959. Pheromones: A new term for a class of biologically active substances. *Nature* 183: 55.

Kawano, T., H. Kataoka, H. Nagasawa, A. Isogai, and A. Suzuki. 1992. cDNA cloning and sequence determination of the pheromone biosynthesis activating neuropeptide of the silkworm, *Bombyx mori*. *Biochem. Biophys. Res. Comm.* 189: 221–226.

Kehat, M. and E. Dunkelblum. 1990. Behavioral responses of male *Heliothis armigera* (Lepidoptera: Noctuidae) moths in a flight tunnel to combinations of components identified from female sex pheromone glands. *J. Insect Behav.* 3: 75–83.

Kennedy, J.S. 1940. The visual responses of flying mosquitoes. *Proc. Zool. Soc. Lond.* A109: 221–242.

Kennedy, J.S. 1983. Zigzagging and casting as a programmed response to wind-borne odour: A review. *Physiol. Entomol.* 8: 109–120.

Kennedy, J.S., A.R. Ludlow, and C.J. Sanders. 1980. Guidance system used in moth sex attraction. *Nature* 288: 475–477.

Kennedy, J.S. and D. Marsh. 1974. Pheromone-regulated anemotaxis in flying moths. *Science* 184: 999–1001.

Kern, F., R.W. Klein, E. Janssen, H.-J. Bestmann, A.B. Attygalle, D. Schäfer, and U. Maschwitz. 1997. Mullein, a trail pheromone component of the ant *Lasius fuliginosus*. *J. Chem. Ecol.* 23: 779–792.

Kitamura, A., H. Nagasawa, H. Kataoka, T. Ando, and A. Suzuki. 1990. Amino acid sequence of pheromone-biosynthesis-activating neuropeptide-II (PBAN-II) of the silkworm, *Bombyx mori*. *Agric. Biol. Chem.* 54: 2495–2497.

Kitamura, A., H. Nagasawa, H. Kataoka, T. Inoue, S. Matsumoto, T. Ando, and A. Suzuki. 1989. Amino acid sequence of pheromone-biosynthesis-activating neuropeptide (PBAN) of the silkworm, *Bombyx mori*. *Biochem. Biophys. Res. Comm.* 163: 520–526.

Klein, U. 1987. Sensillum-lymph proteins from antennal olfactory hairs of the moth *Antheraea polyphemus* (Saturniidae) *Insect Biochem.* 17: 1193–1204.

Klun, J.A., J.R. Plimmer, B.A. Bierl-Leonhardt, A.N. Sparks, and O.L. Chapman. 1979. Trace chemicals: The essence of sexual communication systems in *Heliothis* species. *Science* 204: 1328–1330.

Kramer, E. 1986. Turbulent diffusion and pheromone-triggered anemotaxis, in T.L. Payne, M.C. Birch, and C.E.J. Kennedy (eds.), *Mechanisms in Insect Olfaction*. Clarendon Press, Oxford, U.K., pp. 59–67.

Kramer, E. 1997. A tentative intercausal nexus and its computer model on insect orientation in windborne pheromone plumes, in R.T. Cardé and A.K. Minks (eds.), *Insect Pheromone Research: New Directions*. International Thompson Publishing, New York.

Kuenen, L.P.S. and T.C. Baker. 1981. Habituation versus sensory adaptation as the cause of reduced attraction following pulsed and constant sex pheromone preexposure by *Trichoplusia ni*. *J. Insect Physiol.* 27: 721–726.

Kuenen, L.P.S. and R.T. Cardé. 1994. Strategies for recontacting a lost pheromone plume: Casting and upwind flight in the male gypsy moth. *Physiol. Entomol.* 19: 15–29.

Kuniyoshi, H., R.A. Nagasawa, A. Suzuki, R.J. Nachman, and G.M. Holman. 1992. Cross-reactivity between pheromone biosynthesis activating neuropeptide (PBAN) and myotropic pyrokinin insect peptides. *Biosci. Biotech. Biochem.* 56: 167–168.

Leal, W.S. 1997. Evolution of sex pheromone communication in plant-feeding scarab beetles, in R.T. Cardé and A.K. Minks (eds.), *Insect Pheromone Research: New Directions*. International Thompson Publishing, New York, pp. 505–513.

Leal, W.S. 1999. Enantiomeric anosmia in scarab beetles. *J. Chem. Ecol.* 25: 1055–1066.

Leal, W.S., J.I.L. Moura, J.M.S. Bento, E.F. Vilela, and P.B. Pereira. 1997. Electrophysiological and behavioral evidence for a sex pheromone in the wasp *Bephratelloides pomorum* congeneric to a parthenogenetic species. *J. Chem. Ecol.* 23: 1281–1289.

Linn, C.E., Jr., M. O'Connor, and W. Roelofs. 2003. Silent genes and rare males: A fresh look at pheromone response specificity in the European corn borer moth, *Ostrinia nubilalis*. *J. Insect Sci.* 3: 15.

Linn, C.E., Jr., M.G. Campbell, and W.L. Roelofs. 1987. Pheromone components and active spaces: What do moths smell and when do they smell it? *Science* 237: 650–652.

Loudon, C. and M.A.R. Koehl. 2000. Sniffing by a silkworm moth: Wing fanning enhances air penetration through the pheromone interception by antennae. *J. Exp. Biol.* 203: 2977–2990.

Ma, P.W.K. and W.L. Roelofs. 1995. Sites of synthesis and release of PBAN-like factor in the female European corn borer, *Ostrinia nubilalis*. *J. Insect Physiol.* 41: 339–350.

Mafra-Neto, A. and R.T. Cardé. 1994. Fine-scale structure of pheromone plumes modulates upwind orientation of flying moths. *Nature* 369: 142.

Mafra-Neto, A. and R.T. Cardé. 1995. Effect of the fine-scale structure of pheromone plumes: Pulse frequency modulates activation and upwind flight of almond moth males. *Physiol. Entomol.* 20: 229–242.

Marion-Poll, F. and T.R. Tobin. 1992. Temporal coding of pheromone pulses and trains in *Manduca sexta*. *J. Comp. Physiol.* 171: 505–512.

Marsh, D., J.S. Kennedy, and A.R. Ludlow. 1978. An analysis of anemotactic zigzagging flight in male moths stimulated by pheromone. *Physiol. Entomol.* 3: 221–240.

Martinez, T., G. Fabrias, and F. Camps. 1990. Sex pheromone biosynthetic pathway in *Spodoptera littoralis* and its activation by a neurohormone. *J. Biol. Chem.* 265: 1381–1387.

Masler, E.P., A.K. Raina, R.M. Wagner, and J.P. Kochansky. 1994. Isolation and identification of a pheromonotropic neuropeptide from the brain-subesophageal ganglion complex of *Lymantria dispar*: A new member of the PBAN family. *Insect Biochem. Mol. Biol.* 24: 829–836.

Masson, C. and H. Mustaparta. 1990. Chemical information processing in the olfactory system of insects. *Physiol. Rev.* 70: 199–245.

Matsumoto, S., J.J. Hull, A. Ohnishi, K. Moto, and A. Fónagy. 2007. Molecular mechanisms underlying sex pheromone production in the silkmoth, *Bombyx mori*: Characterization of the molecular components involved in bombykol biosynthesis. *J. Insect Physiol.* 53: 752–759.

Matsumura, F., H.C. Coppel, and A. Tai. 1968. Isolation and identification of termite trail-following pheromone. *Nature* 219: 963–964.

Mayer, M.S. and J.R. McLaughlin. 1992. Discrimination of female sex pheromone by male *Trichoplusia ni* (Hubner). *Chem. Senses* 16: 699–710.

Milet-Pinheiro, P., M. Ayasse, H.E.M. Dobson, C. Schlindwein, W. Francke, and S. Dötterl. 2013. The chemical basis of host-plant recognition in a specialized bee pollinator. *J. Chem. Ecol.* 39: 1347–1360.

Miller, D.R., K.E. Gibson, K.F. Raffa, S.J. Seybold, S.A. Teale, and D.L. Wood. 1997. Geographic variation in response of pine engraver, *Ips pini*, and associated species to pheromone, Lanierone. *J. Chem. Ecol.* 23: 2013–2031.

Minks, A.K. and R.T. Cardé. 1988. Disruption of pheromone communication in moths: Is the natural blend really more efficacious? *Ent. Exp. Appl.* 49: 25–36.

Mitchell, R.F., J.G. Millar, and L.M. Hanks. 2013. Blends of *R*-3-hydroxyhexan-2-one and alkan-2-ones identified as potential pheromones produced by three species of cerambycid beetles. *Chemoecology* 23: 121–127.

Mori, K. 1984. The significance of chirality: Methods for determining absolute configuration and optical purity of pheromones and related compounds, in H.E. Hummel and T.A. Miller (eds.), *Techniques in Pheromone Research*. Springer-Verlag, New York, pp. 323–370.

Morse, D. and E.A. Meighen. 1986. Pheromone biosynthesis and role of functional groups in pheromone specificity. *J. Chem. Ecol.* 12: 335–351.

Murlis, J. 1997. Odor plumes and the signal they provide, in R.T. Cardé and A.K. Minks (eds.), *Insect Pheromone Research: New Directions*. International Thompson Publishing, New York, pp. 221–231.

Murlis, J., J.S. Elkinton, and R.T. Cardé. 1992. Odor plumes and how insects use them. *Annu. Rev. Entomol.* 37: 505–532.

Murlis, J. and C.D. Jones. 1981. Fine scale structure of odour plumes in relation to insect orientation to distant pheromone and other attractant sources. *Physiol. Entomol.* 6: 71–86.

Musser, R.O., S.M. Hum-Musser, H.K. Lee, B.L. DesRochers, S.A. Williams, and H. Vogel. 2012. Caterpillar labial saliva alters tomato plant gene expression. *J. Chem. Ecol.* 38: 1387–1401.

Mustaparta, H. 1997. Olfactory coding mechanisms for pheromone and interspecific signal information in related moth species, in R.T. Cardé and A.K. Minks (eds.), *Insect Pheromone Research: New Directions*. International Thompson Publishing, New York, pp. 144–163.

Nachman, R.J. and G.M. Holman. 1991. Myotropic insect neuropeptide families from the cockroach *Leucophaea maderae*: Structure-activity relationships, in J.J. Menn, T.J. Kelly, and E.P. Masler (eds.), *Insect Neuropeptides: Chemistry, Biology and Action*. American Chemical Society, Washington, DC, pp. 194–214.

Nachman, R.J., G.M. Holman, and B.J. Cook. 1986. Active fragments and analogs of the insect neuropeptide leucopyrokinin: Structure-activity studies. *Biochem. Biophys. Res. Comm.* 137: 936–942.

Ngai, J., A. Chess, M.M. Dowling, N. Necles, E.R. Macgno, and R. Axel. 1993. Coding of olfactory information: Topography of odorant expression in the catfish olfactory epithelium. *Cell* 72: 667–680.

Nishida, R., S. Schultz, C.S. Kim, H. Fukami, Y. Kuwahara, K. Honda, and N. Hayashi. 1996. Male sex pheromone of a giant danaine butterfly, *Idea leuconoe*. *J. Chem. Ecol.* 22: 949–972.

Nishida, R., T.E. Shelly, and K.Y. Kaneshiro. 1997. Acquisition of female-attracting fragrance by males of Oriental fruit fly from a Hawaiian lei flower, *Fagraea berteriana*. *J. Chem. Ecol.* 23: 2275–2285.

Oehlschlager, A.C., G.G.S. King, H.D. Pierce, Jr., A.M. Pierce, K.N. Slessor, J.G. Millar, and J.H. Borden. 1987. Chirality of macrolide pheromones of grain beetles in the genera *Oryzaephilus* and *Cryptolestes* and its implications for species specificity. *J. Chem. Ecol.* 13: 1543–1554.

Phelan, P.L. 1992. Evolution of sex pheromones and the role of asymmetric tracking, in B.D. Roitberg and M.B. Isman (eds.), *Insect Chemical Ecology: An Evolutionary Approach*. Chapman & Hall, New York, pp. 265–314.

Phelan, P.L. 1997. Evolution of mate-signaling in moths: Phylogenetic considerations and predictions from the asymmetric tracking hypothesis, in J.C. Choe and B.J. Crespi (eds.), *The Evolution of Mating Systems in Insects and Arachnids*. Cambridge University Press, Cambridge, U.K., pp. 240–256.

Preiss, R. and E. Kramer. 1986a. Pheromone-induced anemotaxis in simulated free flight, in T.L. Payne, M.C. Birch, and C.E.J. Kennedy (eds.), *Mechanisms in Insect Olfaction*. Clarendon Press, Oxford, U.K., pp. 69–79.

Preiss, R. and E. Kramer. 1986b. Mechanism of pheromone orientation in flying moths. *Naturwissenschaften* 73: 555–557.

Prestwich, G.D. 1993a. Bacterial expression and photoaffinity labeling of a pheromone binding protein. *Protein Sci.* 2: 420–428.

Prestwich, G.D. 1993b. Chemical studies of pheromone receptors in insects. *Arch. Insect Biochem. Physiol.* 22: 75–83.

Prestwich, G.D. and G. Du. 1997. Pheromone-binding proteins, pheromone recognition, and signal transduction in moth olfaction, in R.T. Cardé and A.K. Minks (eds.), *Insect Pheromone Research: New Directions*. Chapman & Hall, New York, pp. 131–143.

Rafaeli, A. and R.A. Jurenka. 2003. PBAN regulation of pheromone biosynthesis in female moths, in G. Blomquist and R. Vogt (eds.), *Insect Pheromone Biochemistry and Molecular Biology*. Elsevier, Amsterdam, the Netherlands, pp. 107–136.

Raffa, K.F. 2014. Terpenes tell different tales at different scales: Glimpses into the chemical ecology of conifer—Bark beetle—Microbial interactions. *J. Chem. Ecol.* 40: 1–20.

Raina, A.K. and G. Gäde. 1988. Insect peptide nomenclature. *Insect Biochem.* 18: 785–787.

Raina, A.K., H. Jaffe, T.G. Kempe, P. Keim, R.W. Blacher, H.M. Fales, C.T. Riley, J.A. Klun, R.L. Ridgway, and D.K. Hayes. 1989. Identification of a neuropeptide hormone that regulates sex pheromone production in female moths. *Science* 244: 796–798.

Raina, A.K. and J.A. Klun. 1984. Brain factor control of sex pheromone production in the female corn earworm moth. *Science* 225: 531–533.

Raina, A.K., J.A. Klun, and E.A. Stadelbacker. 1986. Diel periodicity and effect of age and mating on female sex pheromone titer in *Heliothis zea* (Lepidoptera: Noctuidae). *Ann. Entomol. Soc. Am.* 79: 128–131.

Regnier, F.E. 1971. Semiochemicals—Structure and function. *Biol. Reprod.* 4(3): 309–326.

Reinhard, J. and M. Kaib. 1995. Interaction of pheromones during food exploitation by the termite *Schedorhinotermes lamanianus*. *Physiol. Entomol.* 20: 266–272.

Ressler, K.J., S.L. Sullivan, and L.B. Buck. 1993. A zonal organization of odorant receptor gene expression in the olfactory epithelium. *Cell* 73: 597–609.

Roelofs, W.L. 1984. Electroantennogram assays: Rapid and convenient screening procedures for pheromones, in H.E. Hummel and T.A. Miller (eds.), *Techniques in Pheromone Research*. Springer, New York, pp. 131–159.

Roelofs, W.L. 1995. Chemistry of sex attraction. *Proc. Natl. Acad. Sci. USA* 92: 44–49.

Roelofs, W.L. and L.B. Bjostad. 1984. Biosynthesis of lepidopteran pheromones. *Bioorg. Chem.* 12: 279–298.

Roelofs, W.L. and R.A. Jurenka. 1997. Interaction of PBAN with biosynthetic enzymes, in R.T. Cardé and A.K. Minks (eds.), *Insect Pheromone Research: New Directions*. Chapman & Hall, New York, pp. 42–45.

Roelofs, W.L., W. Liu, G. Hao, H. Jiao, A.P. Rooney, and C.E. Linn, Jr. 2002. Evolution of moth sex pheromones via ancestral genes. *Proc. Natl. Acad. Sci. USA* 99: 13621–13626.

Roelofs, W.L. and W.A. Wolf. 1988. Pheromone biosynthesis in Lepidoptera. *J. Chem. Ecol.* 14: 2019–2031.

Rospars, J.P. and J.G. Hildebrand. 1992. Anatomical identification of glomeruli in the antennal lobes of the male sphinx moth *Manduca sexta*. *Cell Tissue Res.* 270: 205–227.

Rottler, A.-M., S. Schulz, and M. Ayasse. 2013. Wax lipids signal nest identity in bumblebee colonies. *J. Chem. Ecol.* 39: 67–75.

Rumbo, E.R., D.M. Suckling, and G. Karg. 1995. Measurement of airborne pheromone concentrations using electroantennograms: Interactions between environmental volatiles and pheromone. *J. Insect Physiol.* 41: 465–471.

Ruther, J. and T. Hammerl. 2014. An oral male courtship pheromone terminates the response of *Nasonia vitripennis* females to the male-produced sex attractant. *J. Chem. Ecol.* 40: 56–62.

Rybczynski, R., J. Reagan, and M. Lerner. 1989. A pheromone-degrading aldehyde oxidase in the antennae of the moth *Manduca sexta*. *J. Neurosci.* 9: 1341–1353.

Sanders, C.J. 1997. Mechanisms of mating disruption in moths, in R.T. Cardé and A.K. Minks (eds.), *Insect Pheromone Research: New Directions*. Chapman & Hall, New York, pp. 333–346.

Sauer, A.E., G. Karg, U.T. Koch, J.J. De Kramer, and R. Milli. 1992. A portable EAG system for the measurement of pheromone concentrations in the field. *Chem. Senses* 17: 543–553.

Schneider, D. 1957. Electrophysiologische Untersuchungten von Chemo- und Mechanorezeptroren der Antenne des Seidenspinsers *Bombyx mori* L. *Z. Vgl. Physiol.* 40: 8–41.

Schneider, D. 1974. The sex-attractant receptor of moths. *Sci. Am.* July: 28–35.

Shorey, H.H. 1973. Behavioral responses to insect pheromones. *Annu. Rev. Entomol.* 18: 349–380.

Shorey, H.H. and R.G. Gerber. 1996. Use of puffers for disruption of pheromone communication of codling moths (Lepidoptera: Tortricidae) in walnut orchards. *Environ. Entomol.* 25: 1401–1405.

Shorey, H.H., C.B. Sisk, and R.G. Gerber. 1996. Widely separated release sites for disruption of sex pheromone communication in two species of Lepidoptera. *Environ. Entomol.* 25: 446–451.

Silverstein, R.M. 1988. Chirality in insect communication. *J. Chem. Ecol.* 14: 1981–2004.

Six, D.L. 2013. The bark beetle holobiont: Why microbes matter. *J. Chem. Ecol.* 39: 989–1002.

Song, M., A. Gorzalski, T.T. Nguyen, X. Liu, X. Jeffrey, G.J. Blomquist, and C. Tittiger. 2014. *exo*-Brevicomin biosynthesis in the fat body of the mountain pine beetle, *Dendroctonus ponderosae*. *J. Chem. Ecol.* 40: 181–189.

Spencer, K.C. 1988. *Chemical Mediation of Coevolution*. Academic Press, Inc., San Diego, CA, 609pp.

Staten, R.R., O. El-Lissy, and L. Antilla. 1997. Successful area-wide program to control pink bollworm by mating disruption, in R.T. Cardé and A.K. Minks (eds.), *Insect Pheromone Research: New Directions*. Chapman & Hall, New York, pp. 383–396.

Steinbrecht, R.A., M. Ozaki, and G. Ziegelberger. 1992. Immunocytochemical localization of pheromone-binding protein in moth antennae. *Cell Tissue Res.* 270: 287–302.

Stengl, M., H. Hatt, and H. Breer. 1992. Peripheral processes in insect olfaction. *Annu. Rev. Physiol.* 54: 665–681.

Suckling, D.M. and G. Karg. 1997. Mating disruption of the lightbrown apple moth: Portable electroantennogram equipment and other aspects, in R.T. Cardé and A.K. Minks (eds.), *Insect Pheromone Research: New Directions*. Chapman & Hall, New York, pp. 411–420.

Svensson, B.G. and G. Bergström. 1977. Volatile marking secretions from the labial gland of north European *Pyrobombus* D.T. males (Hymenoptera, Apidae). *Insectes Soc.* 24: 213–224.

Tang, J.D., R.E. Carlton, R.A. Jurenka, W.A. Wolf, P.L. Phelan, L. Sreng, and W.L. Roelofs. 1989. Regulation of pheromone biosynthesis by a brain hormone in two moth species. *Proc. Nat. Acad. Sci. USA* 86: 1806–1810.

Teal, P.E.A., R.L. Abernathy, R.J. Nachman, N. Fang, J.A. Meredith, and J.H. Tumlinson. 1996. Pheromone biosynthesis activating neuropeptides: Functions and chemistry. *Peptides* 17: 337–344.

Teal, P.E.A. and J.H. Tumlinson. 1986. Terminal steps in pheromone biosynthesis by *Heliothis virescens* and *H. zea. J. Chem. Ecol.* 12: 353–366.

Teal, P.E.A. and J.H. Tumlinson. 1989. Neuronal induction of pheromone biosynthesis by *Heliothis zea* (Broddie) during the photophase. *Can. Entomol.* 121: 43–46.

Teal, P.E.A., J.H. Tumlinson, and H. Oberlander. 1989. Neural regulation of sex pheromone biosynthesis in *Heliothis* moths. *Proc. Natl. Acad. Sci. USA* 86: 2488–2492.

Todd, J.L. and T.C. Baker. 1997. The cutting edge of insect olfaction *Am. Entomol.* Fall: 174–182.

Tumlinson, J.H., M.M. Brennan, R.E. Doolittle, E.R. Mitchell, A. Brabham, B.E. Mazomenos, A.H. Baumhover, and D.M. Jackson. 1989. Identification of a pheromone blend attractive to *Manduca sexta* (L.) males in a wind tunnel. *Arch. Insect Biochem. Physiol.* 10: 255–271.

Tumlinson, J.H., M.G. Glein, R.E. Doolittle, T.L. Ladd, and A.T. Proveaux. 1977. Identification of the female Japanese beetle sex pheromone: Inhibition of male response by an enantiomer. *Science* 197: 789–792.

Tumlinson, J.H., E.R. Mitchell, R.E. Doolittle, and D.M. Jackson. 1994. Field tests of synthetic *Manduca sexta* sex pheromone. *J. Chem. Ecol.* 20: 579–591.

Vet, L.E.M. and M. Dicke. 1992. Ecology of infochemical use by natural enemies in a tritrophic context. *Annu. Rev. Entomol.* 37: 141–172.

Vickers, N.J. 2000. Mechanisms of animal navigation in odor plumes. *Biol. Bull.* 198: 203–212.

Vickers, N.J., T.A. Christensen, H. Mustaparta, and T.C. Baker. 1991. Chemical communication in heliothine moths III. Flight behavior of male *Helicoverpa zea* and *Heliothis virescens* in response to varying ratios of intra- and interspecific sex pheromone components. *J. Comp. Physiol. A* 169: 275–280.

Vogt, R.G. 1987. The molecular basis of pheromone reception: Its influence on behavior, in G.D. Prestwich and G.L. Blomquist (eds.), *Pheromone Biochemistry*. Academic Press, Orlando, FL, pp. 385–431.

Vogt, R.G., A.C. Kohne, J.T. Dubnau, and G.D. Prestwich. 1989. Expression of pheromone binding proteins during antennal development in the gypsy moth *Lymantria dispar. J. Neurosci.* 9: 3332–3346.

Vogt, R.G., J.D. Prestwich, and M.R. Lerner. 1991a. Odorant-binding-protein subfamilies associate with distinct classes of olfactory receptor neurons in insects. *J. Neurobiol.* 22: 74–84.

Vogt, R.G. and L.M. Riddiford. 1981. Pheromone binding and activation by moth antennae. *Nature* 293: 161–163.

Vogt, R.G., L.M. Riddiford, and G.D. Prestwich. 1985. Kinetic properties of a pheromone degrading enzyme: The sensillar esterase of *Antheraea polyphemus. Proc. Natl. Acad. Sci. USA* 82: 8827–8831.

Vogt, R.G., R. Rybcynski, and M.R. Lerner. 1991b. Molecular cloning and sequencing of general odorant-binding proteins GOBP1 and GOPB2 from the tobacco hawk moth *Manduca sexta*: Comparison with other insect OBPs and their signal peptides. *J. Neurosci.* 11: 2972–2984.

Vuts, J., Z. Imrei, M.A. Birkett, J.A. Pickett, C.H. Woodcock, and M. Tóth. 2014. Semiochemistry of the Scarabaeoidea. *J. Chem. Ecol.* 40: 190–210.

Wadhams, L.J. 1982. Coupled gas chromatography-single cell recording: A new technique for use in the analysis of insect pheromone. *Z. Naturforschung* 37c: 947–952.

Wang, H.L., C.H. Zhao, G. Szöcs, S.P. Chinta, S. Schulz, and C. Löfstedt. 2013. Biosynthesis and PBAN-regulated transport of pheromone polyenes in the winter moth, *Operophtera brumata. J. Chem. Ecol.* 39: 790–796.

Webb, R.E., B.A. Leonhardt, J.R. Plimmer, K.M. Tatman, V.K. Boyd, D.L. Cohen, C.P. Schwalbe, and L.W. Douglass. 1990. Effect of racemic disparlure released from grids of plastic ropes on mating success of gypsy moth (Lepidoptera: Lymantriidae) as influenced by dose and population density. *J. Econ. Entomol.* 83: 910–916.

Whittaker, R.H. and P.P. Feeny. 1971. Allelochemics: Chemical interactions between species. *Science* 171: 757–770.

Wilson, E.O. 1962. Chemical communication among workers of the fire ant *Solenopsis saevissima* (Fr. Smith). 1. The organization of mass-foraging. *Anim. Behav.* 10: 134–147.

Wilson, E.O. and W.H. Bossert. 1963. Chemical communication among animals. *Recent Prog. Horm. Res.* 19: 673–716.

Witzgall, P. 1997. Modulation of pheromone-mediated flight in male moths, in R.T. Cardé and A.K. Minks (eds.), *Insect Pheromone Research: New Directions*. International Thompson Publishing, New York, pp. 265–290.

Witzgall, P. and H. Arn. 1990. Direct measurement of the flight behavior of male moths to calling females and synthetic sex pheromones. *Z. Naturforsch.* 45c: 1067–1069.

Witzgall, P. and H. Arn. 1991. Recording flight tracks of *Lobesia botrana* in the wind tunnel, in I. Hrdy (ed.), *Insect Chemical Ecology*. SPB Academic Publishers, The Hague, the Netherlands, pp. 187–193.

Wright, R.H. 1958. The olfactory guidance of flying insects. *Can. Entomol.* 90: 81–89.

Zhu, J., N. Ryrholm, H. Ljungberg, B.S. Hansson, D. Hall, D. Reed, and C. Lofstedt. 1996. Olefinic acetates, Δ-9,11–14:OAc and Δ-7,9–12:OAc used as sex pheromone components in three geometrid moths, *Idaea aversata*, *I. straminata*, and *I. biselata* (Geometridae, Lepidoptera). *J. Chem. Ecol.* 22: 1505–1526.

Ziegelberger, G., M.J. Van den Berg, K.-E. Kaissling, S. Klump, and J.E. Schultz. 1990. Cyclic GMP levels and guanylate cyclase activity in pheromone-sensitive antennae of the silkmoths *Antheraea polyphemus* and *Bombyx mori*. *J. Neurosci.* 10: 1217–1225.

CHAPTER 20

Reproduction

PREVIEW

Insects display great diversity in modes of reproduction. Most insects reproduce in the adult stage by laying eggs, but a few produce gametes and reproduce during an immature stage, a process known as pedogenesis. Some insects (e.g., aphids, some flies) give birth to live young. Although sexual reproduction by union of male and female gametes is typical, certain insects reproduce some or all of the time by laying unfertilized eggs (parthenogenesis). Oocytes accumulate yolk and cytoplasm and develop in egg chambers or follicles in the ovarioles within the ovary. Ovarioles may contain nurse cells (meroistic ovarioles) in several different configurations relative to the developing oocyte, or there may be no special nurse cells (panoistic ovarioles). Nurse cells provide nutrients and gene products for the developing oocyte. Presumably, similar components are provided by other cells in panoistic ovarioles. In most females, developing oocytes incorporate into the yolk large glycolipoproteins called **vitellogenins (Vgs)** that are synthesized in fat body cells and transported to the ovaries by hemolymph. Higher Diptera evolved small protein–lipid complexes (called yolk proteins) for incorporation in the yolk. Yolk proteins are synthesized in both fat body and follicular epithelial cells under the influence of a different set of genes than the Vgs of other insects. In most insects, several hormones regulate oogenesis, synthesis of yolk proteins, and additional aspects of reproduction such as pheromone production, mating, and oviposition. The hormones controlling these processes are not the same in all groups of insects. When maturation of the oocyte is nearly complete, a vitelline membrane is secreted, followed by secretion of the eggshell, or chorion. Eggs are fertilized after the chorion is put on as the egg passes down the median oviduct and past the opening to the spermatheca where sperm are stored. Sperm enter through the micropyle, a twisting channel through the chorion. Male insects produce sperm in the testes. Males typically transfer sperm to the female tract by insertion of the aedeagus into the reproductive tract of the female, or by incorporating the sperm into a spermatophore, a protein sac, that may be inserted into the opening of the female's reproductive tract or formed there as mating occurs. A spermatophore is usually viewed as an investment of protein nutrition by the male to the next generation. There are at least three chromosomal systems for gender determination in insects; in some insects the male is heterogametic, while in other species the female is heterogametic. Hymenoptera and some coccids (Homoptera) have haploid males and diploid females. The ratio of sex chromosome to autosomes and/or the presence of sex-determining genes are two mechanisms known to determine gender in most insects.

20.1 INTRODUCTION

With the diversity in insect life histories and ecology, it should not be surprising to find great diversity in the details of mating behavior, endocrine regulation of egg development, and pheromone production and in the physical structures associated with reproduction. This chapter deals primarily with internal reproductive structures and the physiology of gamete production and yolk deposition in the developing eggs.

20.2 FEMALE REPRODUCTIVE SYSTEM

20.2.1 Structure of Ovaries

An excellent review of the functional and comparative anatomy of the ovarian system has been presented by Bonhag (1958), the principal source for the information presented here and for details on model types of ovarioles discussed below. The general anatomy of the female internal organs consisting of the **paired ovaries, lateral oviducts, common oviduct, spermatheca,** and **accessory glands** is illustrated in Figure 20.1. The internal reproductive structures of both sexes are located dorsally to the alimentary tract. Each ovary consists of one to many **ovarioles**, with each ovariole containing a string of "egg-shaped" egg chambers called **follicles**. A follicle is separated from the preceding one by a constriction and a bit of interfollicular tissue. Each ovariole is typically enclosed in an epithelial sheath of variable structure in different insects. Striated muscle fibers are often associated with this outer epithelial sheath.

The number of ovarioles per ovary varies in different species, and even to some extent within a species. Viviparous Diptera have one (tsetse fly *Glossina* spp.) or two (*Melophagus* spp. and *Hippobosca* spp.); the American cockroach, *Periplaneta americana,* has eight; *Drosophila*

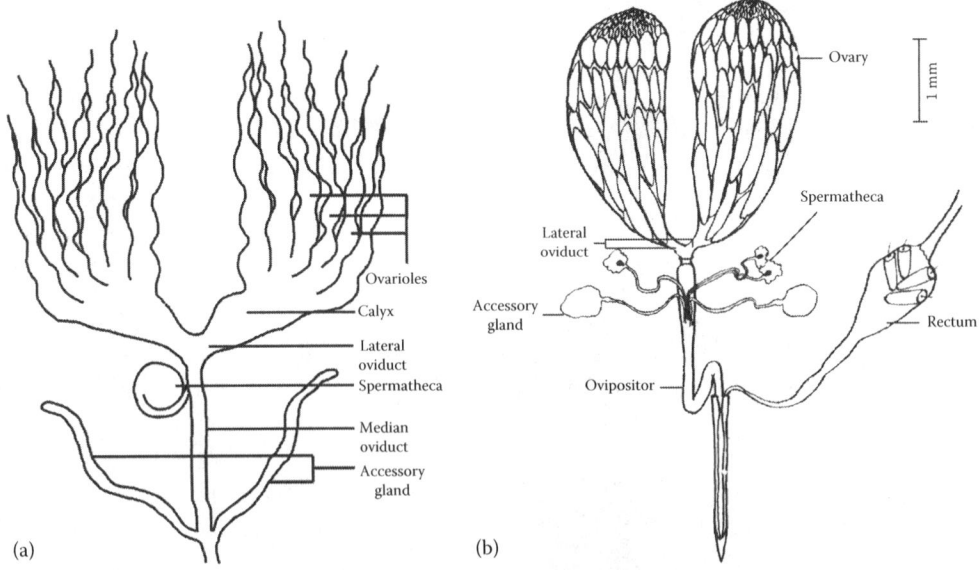

Figure 20.1 (a) The internal reproductive structures of a female milkweed bug, *Oncopeltus fasciatus*. (b) Internal structures of a female Caribbean fruit fly, *Anastrepha suspensa*. There are three spermathecae in most of the tephritid fruit flies. (a: Drawing by the author; b: Drawing modified from Dodson, G., *Fla. Entomol.*, 61, 231, 1978.)

melanogaster has from 10 to 30; the blowfly, *Calliphora erythrocephala*, has about 100; and termite queens (Isoptera) may have up to 2000. Just one ovary containing one ovariole occurs in some aphids, and Collembola have sac-like ovaries that do not contain ovarioles.

Insect ovaries and the ovarioles are classified into two main types depending on whether there are nurse cells associated with the developing egg (**meroistic ovaries**) or no nurse cells (**panoistic ovaries**) (Figure 20.2). Panoistic ovaries are considered to be the earliest type to evolve and occur in present-day Thysanura, Odonata, Plecoptera, Dictyoptera, and Isoptera. They evolved secondarily in Ephemeroptera, Orthoptera, and Siphonaptera (Bonhag, 1958). Meroistic ovaries occur in most of the Holometabola (except Siphonaptera, noted above), and in Hemiptera, Dermaptera, Psocoptera, Anoplura, and Mallophaga among the Hemimetabola. Meroistic ovaries can also be divided into two types, **polytrophic** and **telotrophic**, depending upon whether nurse cells are located in one of two ways, as follows:

1. In polytrophic ovarioles, each oocyte is closely associated with nurse cells in its follicle or an adjacent follicle. Most of the Holometabola, some Coleoptera (Adephaga), Dermaptera, Psocoptera, Anoplura, and Mallophaga have polytrophic ovarioles (Figure 20.2). The nurse cells may be present in the follicle containing the developing oocyte as in higher Diptera or they may occupy a separate follicle adjacent to that of the developing oocyte, as in the honeybee, *Apis mellifera*, and other Hymenoptera.
2. In acrotrophic or telotrophic ovarioles, the nurse cells are located at the distal apex of the ovariole, in the germarial region, and long, connecting nutritive chords (Figure 20.2) extend from the nurse cells to each developing oocyte. Hemiptera and some Coleoptera (polyphaga) have the telotrophic arrangement (Bonhag, 1955, 1958).

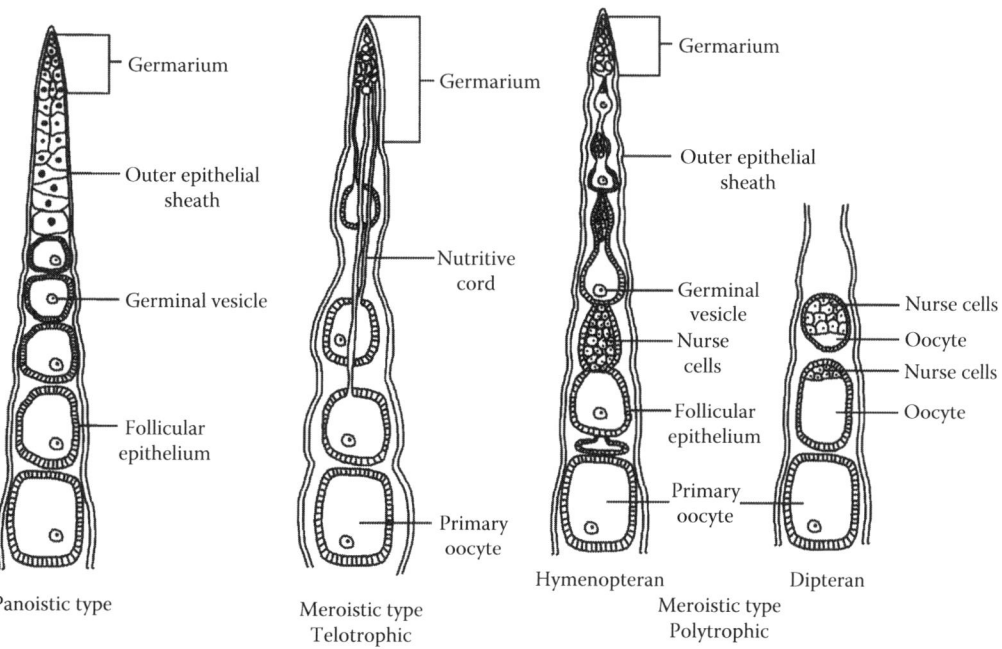

Figure 20.2 Major types of ovary structure in insects. The panoistic ovary is typical of Orthoptera and Dictyoptera with no nurse cells. Meroistic telotrophic ovaries have nurse cells in the germarial region and cytoplasmic strands extend to the developing oocytes. Coleoptera and Hemiptera have telotrophic ovaries. Meroistic ovaries may be polytrophic, as for example, in Hymenoptera and higher Diptera. In polytrophic ovaries the nurse cells occur in an adjacent follicle (Hymenoptera) or in the follicle with the developing oocyte (higher Diptera). In all cases, the nurse cells pass nutrients and gene products (mRNAs) to the developing oocyte. (Drawing by the author.)

Each ovariole terminates distally in a thin, slightly elastic filament that attaches to the dorsal diaphragm or dorsal cuticle. Frequently, terminal filaments from all ovarioles fuse into a suspensory ligament that is likewise attached. Proximally, each ovariole connects with the lateral oviduct, and the two lateral oviducts join the common or medial oviduct as a passage for the eggs to the outside.

20.2.1.1 Panoistic Ovarioles

Panoistic ovarioles do not have nurse cells. The panoistic ovary in *Thermobia domestica* (Packard), the firebrat (Thysanura), is a well-studied example of a panoistic ovary. Each ovary is composed of only five ovarioles, with each ovariole enclosed in a thin, largely membranous, epithelial sheath containing occasional nuclei. The distal germarial region contains oogonia that can undergo mitosis to produce additional oogonia. Proximal to the germarium, and in the area nearest the germarium, are young oocytes not yet arranged in single file, although more proximally they become arranged into a single string of developing oocytes. Many small prefollicular nuclei are present in a common cytoplasm, but later these acquire cell boundaries and arrange themselves around each developing oocyte as a follicular epithelium. Interfollicular tissue separates each follicle from the next above it. The follicular epithelial cells secrete a noncellular, thick membranous tunica between themselves and the outside epithelial sheath. The terminal oocyte, the most proximal one, sequesters yolk and increases in size to become a mature oocyte. The follicular epithelial cells secrete the chorion, the eggshell. Fertilization occurs after the eggshell is in place as the egg passes down the median oviduct and past spermatheca where sperm are stored. The final maturation divisions of the egg nucleus do not take place in most eggs until they are ovulated.

20.2.1.2 Telotrophic Ovarioles

There are nurse cells located in the germarial region of telotrophic ovarioles, and long cytoplasmic cords or strands extend from the germarium to the developing oocytes. Nutrients and maternal gene products pass down the cytoplasmic cords. The milkweed bug *Oncopeltus fasciatus* has been intensively studied as a model of an insect with a telotrophic ovary (Bonhag, 1955, 1958). The germarial region at the distal end of each of the seven ovarioles in each ovary is primarily occupied by the apical trophic tissue. These cells send long cytoplasmic strands to each developing oocyte through which nutrients and genetic messages pass to the oocytes. Proximal to the apical trophic tissue are the young oocytes, all of which are produced during the nymphal stage. No additional oocytes are produced in the milkweed bug during the adult stage (Wick and Bonhag, 1955).

Each ovariole is covered by an inner envelope composed of a single layer of cells and an outer thin syncytial epithelial sheath. As the terminal oocyte matures, these two outer layers become stretched very thin and generally are evident only in the region around the interfollicular tissue dividing each follicle from the preceding one. A single layer of follicular epithelium is arranged around each oocyte and, as the oocyte enlarges and grows toward maturity, the follicular epithelial cells first become large, binucleate rounded cells, and finally become squamous. Prior to the egg moving down the oviduct, the **chorion** is secreted by the follicular epithelium.

Detailed studies of the telotrophic ovary in the yellow mealworm, *Tenebrio molitor*, have been made as a model for the polyphaga group of Coleoptera. Many of the basic features are the same as in the milkweed bug, but there are some differences. The publication by Schlottman and Bonhag (1956) may be consulted for more details.

20.2.1.3 Polytrophic Ovarioles

In polytrophic ovarioles, the nurse cells are located in the egg chambers, either in the same egg chamber with the oocyte or in an adjacent chamber. The earwig, *Anisolabis maritima,* has five ovarioles in each ovary, with each ovariole enclosed in a syncytial outer epithelial sheath. Each ovariole consists of a terminal filament, a germarium, and a string of egg chambers. Each follicle contains one developing oocyte and one **trophocyte** or nurse cell. Initially, the trophocyte is the larger of the two cells in a follicle, but this changes as the oocyte matures. The cells of the follicular epithelium divide by mitosis to accommodate the need for more cells to surround the growing oocyte during the previtellogenesis growth, but later during vitellogenesis the follicular epithelial cells mainly grow and change shape, becoming more thin and squamous as they stretch to cover the enlarging oocyte.

The number of nurse cells or trophocytes varies in different insects. Earwigs are somewhat unusual in having just one trophocyte for each oocyte. Many insects have multiple trophocytes per oocyte: Lepidoptera typically have five nurse cells per oocyte; the diving beetle, *Dytiscus marginalis,* has 15 trophocytes per oocyte; the gyrinid beetle, *Dinuetes nigrior,* has seven trophocytes per oocyte; *Drosophila* and higher Diptera have 15 nurse cells per oocyte; and the honeybee, *Apis mellifera,* has 48 nurse cells occupying the follicle preceding the oocyte. In advanced polytrophic ovarioles in which the trophocytes occupy a separate follicle adjacent to the oocyte, an oocyte process typically extends into the nutritive follicle through which nutrients and maternal gene products are passed to the oocyte.

Multiple trophocytes and a cell destined to become the oocyte are produced by mitotic division of an oogonium to produce two cells, with successive mitotic division of these to produce four, and so on to give a group of sister cells, all diploid in chromosome number. For example, in *Drosophila,* eight divisions produce 16 cells, and 15 become trophocytes and one becomes the oocyte. In *Drosophila* (and probably other insects), interconnecting cytoplasmic strands, often called ring canals, allow nutrients and gene products to pass from nurse cells to the developing oocyte (Chapter 1, Figure 1.8).

The nurse cells of some insects amplify rRNA genes so that they produce a large complement of ribosomes for the egg (Kafatos et al., 1985), but only a few copies are put into the oocytes of other insects (Schäfer and Kunz, 1987). In the polytrophic ovary of dipterans, including *D. melanogaster,* rRNA is not amplified, but ribosomes are supplied by the highly polyploid nurse cells. The oocyte also usually receives some nutrients from the layer of follicle cells surrounding it.

Typically, several oocytes in various stages of development occur in each ovariole. For example, in the housefly oogonial division begins during early pupation and the first egg chamber is formed before emergence of the adult. At the beginning of oviposition, several days after emergence, there is one mature egg in each ovariole, and secondary oocytes already in various stages of development.

20.2.1.4 Oviposition

In some orders of insects there are no special external structures for oviposition. An egg ready to be laid passes through the lateral oviducts and into the common oviduct, and is deposited on some substrate. In others, the last segments of the abdomen are elongated into an ovipositor. The egg then passes out from the ovipositor and is deposited on the surface of a leaf or other substrate. In some insects the ovipositor is hardened and pointed and can be inserted inside another insect, such as in a parasitoid that inserts its egg into a host insect, or in the case of tephritid fruit flies into a fruit where

Figure 20.3 (See color insert.) A female olive fruit fly probing a green olive fruit, perhaps about to lay an egg beneath the skin of the olive fruit. (Photo courtesy of Dr. Hanife Genc, Çanakkale Onsekiz Mart Üniversitesi, Çanakkale, Turkey.)

the egg is deposited (Figure 20.3). The ovipositor of many insects have been shown to have mechanoreceptors and chemoreceptors that provide sensory information about the substrate they probe with the ovipositor. Tephritid fruit flies often probe a fruit without, or before, eventually laying an egg, apparently receiving information about the external surface and internal milieu.

20.2.2 Nutrients for Oogenesis

The availability of nutrients during oogenesis is a major limiting factor in the ability of an insect to successfully reproduce (Wheeler, 1996). In addition, mating (Gillott and Friedel, 1977) and physical activity, such as flight, influence the physiological availability of nutrients in many insects. Mating is a stimulus that induces mobilization of reserves in some females. The male may make nutrient contributions to the female during courtship and mating (Boggs, 1990) by offering nuptial gifts of food, and nutrients may be obtained from seminal fluid and a spermatophore.

Oogenesis, the formation of eggs, requires incorporation of relatively large amounts of protein and lipids and, thus, is an energy-intensive activity in most insects. Insects that live only short lives as adults typically accumulate nutrients for oogenesis during their larval stage. Adults with longer lives often have a period of preoviposition development of the ovaries and usually require nitrogenous foods for maximum growth of ovaries and egg production. Among Diptera, the terms **autogenous**, the ability to develop a first set of eggs without an exogenous nitrogen source, and **anautogenous**, the need for a protein or nitrogen source as an adult to develop eggs, are used to describe the requirements for a blood meal to provide nutrients for egg development. Spielman (1971) suggests that some individuals in all populations are likely to show autogeny. Autogenous individuals have been found in populations of two anautogenous higher dipterans (*Sarcophaga bullata* and *Musca domestica*) (Robbins and Shortino, 1962; Baxter et al., 1973). Some species of mosquitoes are autogenous while others are anautogenous. The autogenous species can mature one set of eggs without a nitrogen source as an adult, but subsequent egg development depends on taking a blood meal. Anautogenous species of mosquitoes need a blood meal in order to develop the first and each subsequent set of eggs. Some parasitic insects produce small (20–200 µm), almost yolk-less eggs that are deposited in a host (another insect) where the developing embryo absorbs nutrients from the host through a thin chorion shell (Flanders, 1942; Fisher, 1971; Wheeler, 1996).

Physical activity, and especially flight, which demands so much energy, can compete with the ovary for nutrients (Lorenz, 2007). Hawkmoths (Lepidoptera: Sphingidae) are very active fliers, hovering in flight as well as flying backwards, with very high oxygen demands in flight (up to 148 times resting level, Bartholomew and Casey, 1978). Research by von Arx et al. (2013) showed that the female hawkmoth *Hyles lineata* lived twice as long and laid more fertile eggs when allowed access to sucrose after mating compared with controls that were allowed only water. Thus, they suggested that evolution should strongly select behavior for continuing foraging for nectar during and between oviposition events. The energy demands during long migratory flights generally results in inhibition of oogenesis or when lengthy flights must be taken to locate a new host or suitable oviposition site (Wheeler, 1996). A new brood of scolytid bark beetles (e.g., the genera *Ips* and *Dendroctonus*) emerging from the log or tree in which they have developed often must fly some distance in seeking a new suitable tree to colonize. When a suitable one has been selected, the wing muscles of some females degenerate, making nutrients available quickly for oogenesis. After the mating flight, queen ants break off their wings (Fletcher and Blum, 1981), and the wing muscles degenerate to make nutrients available for the first oogenesis cycle. Lorenz (2007) found a similar oogenesis flight syndrome in female crickets, *Gryllus bimaculatus*, in that muscle mass increased at days 2 and 3 in the new adult, a time also coinciding with maximum tendency to flight. Between days 2 and 3 and up to day 10, the ovary developed with vitellogenesis and the flight muscles progressively underwent histolysis.

20.2.3 Hormonal Regulation of Ovary Development and Synthesis of Egg Proteins

Hormones control ovary growth, synthesis of **Vg** by fat body cells, and uptake of Vg by the developing oocytes (Hagedorn, 1985; Adams and Filipi, 1988). In some groups, only **juvenile hormone (JH)** produced by the corpora allata appears to be involved in regulating reproductive biology, while in others both **JH** and **ecdysone**, produced by the ovaries (Hagedorn et al., 1975), are important, and, in still other groups, **JH**, **ecdysteroids**, and additional hormones are known. Briefly, hormonal regulation in different groups is as follows:

- *Dictyoptera* (*Cockroaches*): **JH** has numerous pleiotropic actions on the development and reproduction of cockroaches, including maturation of gonads, production of attractant and courtship pheromones, "calling" behavior and pheromone release, and sexual receptivity (Schal et al., 1997). It appears to be the only hormone that is involved in controlling fat body synthesis of Vgs ovary growth, and uptake of the Vgs (Figure 20.4) in the German cockroach, *Blatella germanica* and *Leucophaea maderae*. Mating, high-quality nutrition, social interactions, and the presence of vitellogenic ovaries influence JH synthesis by the corpora allata (Schal et al., 1997). One or more additional factors may be involved in the decline of Vg synthesis in *B. germanica* late in the gonatropic cycle because JH levels remain high while Vg production is declining (Martín et al., 1995).
- *Orthoptera*: **JH** is the principal hormone that stimulates fat body cells and the ovary, and the **adipokinetic hormone** (from the corpora cardiaca) inhibits Vg mRNA translation at the end of an egg production cycle in *Locusta migratoria* (Bownes, 1986). The evidence suggest that JH controls the Vg gene because JH analogs promote transcription of a gene coding for Vg in the fat body (Glinka and Wyatt, 1996). JH also regulates uptake of the Vgs by the developing oocytes, and the mature ovary appears to have feedback to the brain and/or corpora, which regulates down the production of JH until the primary set of eggs is laid. JH III is the major JH in several gryllid crickets (*Acheta domesticus*, *Teleogryllus commodus*, and *Gryllus bimaculatus*) (Strambi et al., 1997). The ovaries of *A. domesticus* synthesize ecdysone and conjugate most of it with fatty acids to form ecdysone 22-fatty acyl esters (Whiting et al., 1997). The precise role of the ecdysone or fatty acyl esters is not known. Migratory locusts, *Locusta migratoria*, and many other grasshoppers dig an oviposition

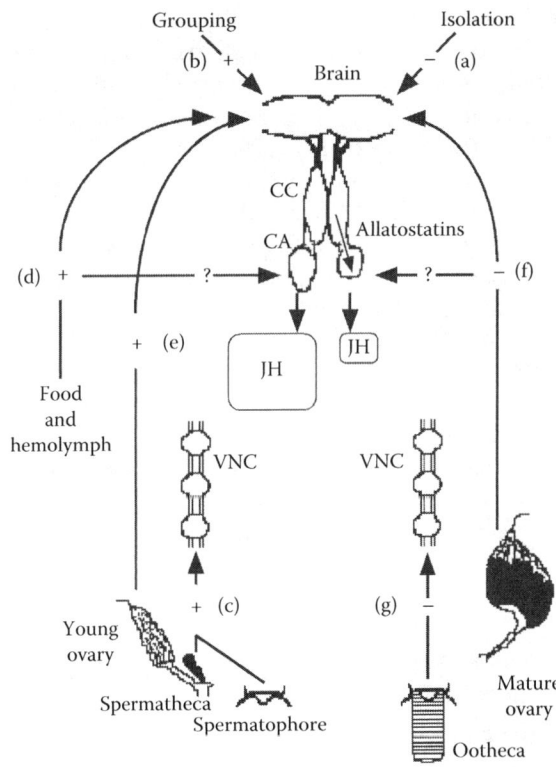

Figure 20.4 A model to suggest external and internal regulation of juvenile hormone (JH) synthesis in female German cockroaches, *Blattella germanica*. Inhibitory regulators are indicated by a minus sign and positive regulators by a plus sign. JH synthesis is inhibited by social isolation, a mature ovary, or an ootheca (an egg case carried for some time by a female and finally deposited before eggs hatch). Social interaction, mating, food availability, and a young ovary stimulate JH synthesis. (From Schal, C. et al., *Arch. Insect Biochem. Physiol.*, 35, 405, 1997. With permission.)

hole in the soil with the abdomen and lay eggs in the hole. During digging, eggs ready to be laid are held in the lateral oviducts by rhythmic contractions of lower part of the lateral oviducts, while at the same time contractions of the upper part of the lateral oviducts tends to push eggs toward the common oviduct for deposition. The abdomen is extended into the hole as digging continues (Ayali and Lange, 2010); the sclerotized appendages at the tip of the abdomen (called valves) open and close as they work their way into the soil and pull abdominal segments apart (Rose et al., 2000). In order for egg laying to begin, octopamine relaxes the lateral and common oviducts to enable passage of eggs when the hole is dug (Wong and Lange, 2014).

- *Diptera*: Diptera exhibit complex control of reproduction involving multiple hormones. Hormonal control of reproduction in mosquitoes has been reviewed by Klowden (1997) and in the higher Diptera (Cyclorrhapa = Muscamorpha) by Yin and Stoffolano (1997). Female mosquitoes, and some other dipterans, are hematophagous, that is, foragers on blood. Some mosquitoes, such as *Aedes aegypti* and *Culex pipiens pipiens*, are anautogenous and must have a blood meal in order to mature the first set of eggs. Other mosquitoes, including *A. taeniorhynchus*, *A. atropaplus*, and *Culex pipiens molestus*, are autogenous. They can mature one set of eggs by using stored reserves from their larval life, but require a blood meal for subsequent egg development. Hormonal controls appear to be similar in both kinds of mosquitoes (Klowden, 1997). Autogeny and anautogeny occurs also in the cyclorrhaphous Diptera.

In anautogenous mosquitoes, hormonal control of ovary and egg development is conveniently divided into **previtellogenic**, **vitellogenic**, and **postvitellogenic** stages (Figure 20.5). Soon after an

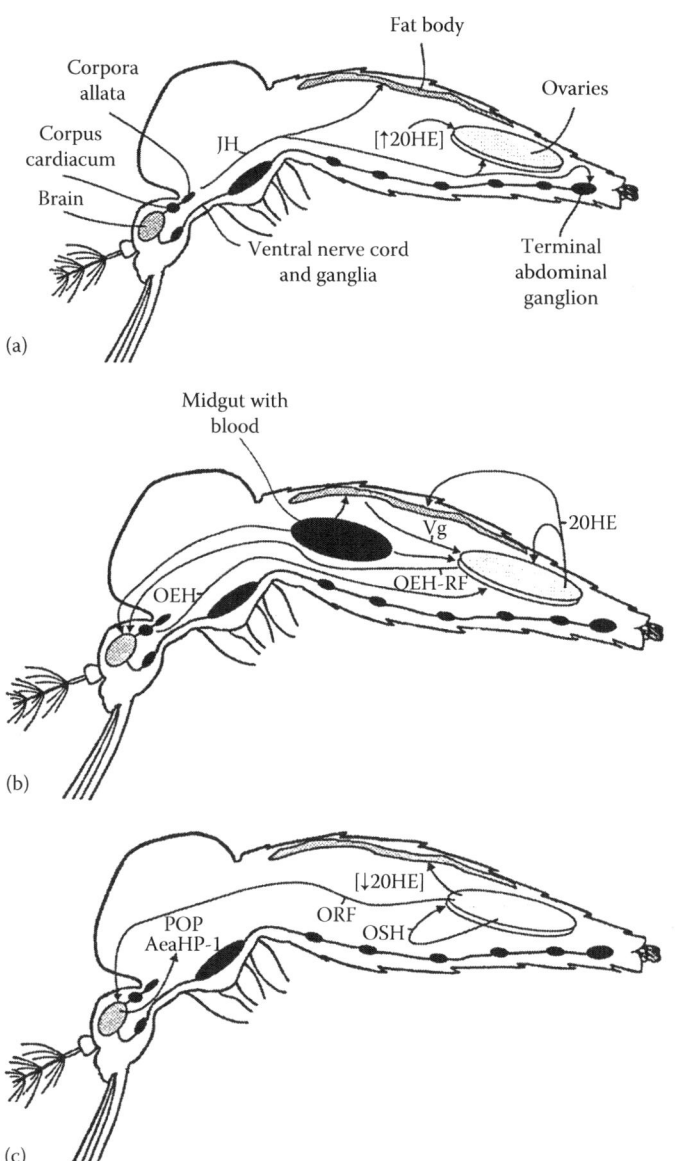

Figure 20.5 (a) Hormonal control of previtellogenesis in mosquitoes. Paired corpora allata (CA) produce juvenile hormones (JHs) prior to a blood meal, and JH acts on fat body and ovaries to make them competent to respond to later hormones and secrete vitellogenin. JH also acts on the terminal abdominal ganglion and mediates mating acceptance. Circulating 20-hydroxyecdysone left from the pupal–adult molt initiates follicle formation in the ovary. (b) Hormonal action during vitellogenesis. The blood meal is digested and nutrients are incorporated into the fat body. Ovarian ecdysteroidogenic hormone-releasing factor (OEH-RF) is released from the ovaries. OEH-RF causes the corpora cardiaca (CC) to release OEH, which stimulates competent ovaries to synthesize ecdysone. Ecdysone is converted to 20-hydroxyecdysone (20-HE) by the fat body and other tissues, and it stimulates the fat body to synthesize and release vitellogenin (Vg). 20-HE also stimulates the separation of additional follicles from the germarium. Vg is taken up by oocytes. (c) Postvitellogenic hormonal controls. The fat body ceases to produce Vg in response to falling titers of 20-HE. An oostatic hormone (OSH) is produced by growing primary follicles and inhibits development of secondary follicles. Hormones that inhibit host-seeking behavior and promote preoviposition behavior are released by the ovarian releasing factor. (From Klowden, M.J., *Arch. Insect Biochem. Physiol.*, 35, 491, 1997. With permission.)

adult female mosquito emerges, the **previtellogenic stage** begins with JH secretion from the corpora allata (CA). In *A. aegypti*, JH III is the only JH known (Baker et al., 1983). The stimulus for JH secretion in not known, but it is not the taking of the blood meal itself because secretion occurs prior to the meal. JH has at least three actions:

1. It makes females receptive to mating.
2. It stimulates previtellogenic growth of ovaries.
3. It prepares the fat body so that it is competent for responding to later hormones and secretion of Vgs.

Follicles begin to separate from the growing previtellogenic ovary, possibly under the influence of residual 20-hydroxyecdysone remaining from the pupal to adult transformation (Whisenton et al., 1989). Soon after emerging, females seek a blood meal, which provides proteins, other nutrients, and initiates the **vitellogenic stage**. With availability of nutrients, the ovary, stimulated by JH, releases **corpora cardiaca stimulating factor** (**CCSF**), a neuropeptide probably produced somewhere in the nervous system (before the blood meal) and stored in the young ovary. The target cells for CCSF are in the corpora cardiaca (CC), which release **egg development neurohormone** (**EDNH**) (Hagedorn et al., 1979), recently given the more descriptive name of **ovarian ecdysteroidogenic hormone I** (**OEH**). OEH (MW 8803) is a polypeptide comprising 86 amino acids that is synthesized in brain medial neurosecretory cells and stored in the CC. There is evidence that both humoral and nervous stimuli are important in causing the release of OEH from the CC (Klowden, 1987). Follicular epithelium cells in the ovary respond to OEH by producing and releasing **ecdysone** into the circulating hemolymph. Ecdysone is converted rapidly to **20-hydroxyecdysone** by many types of cells throughout the body, including the target fat body cells. Fat body cells respond to 20-hydroxyecdysone by synthesizing Vgs (but only if the fat body has previously been exposed to JH; see Step 3 for JH stated earlier). A receptor for 20-hydroxyecdysone is expressed in fat body and ovary (Cho et al., 1995). **JH** also stimulates Vg synthesis (Wu et al., 1987; Wyatt et al., 1987; Bradfield et al., 1989; Hagedorn, 1989).

The postvitellogenic stage terminates Vg production in the fat body when the primary oocytes complement is approaching or has reached maturity (one mature egg per ovariole). The ovaries (the exact site is not established) release an **oostatic hormone (OSH)**, called the **trypsin modulating oostatic factor** (**TMOF**) in *A. aegypti* by Borovsky (1982, 1988) and simply **OSH** by Klowden (1997). The oostatic hormone stops the uptake of yolk by secondary oocytes, thus stopping their growth until the primary set of eggs is laid. Borovsky (1988, 2003) and Borovsky et al. (1990, 1994, 2006) present evidence that the function of TMOF is to stop the synthesis of late trypsin enzyme (see Chapter 2 for role of early and late trypsin in digestion) in midgut cells, interrupting blood meal digestion and denying secondary oocytes nutrients. The overall function of the oostatic hormone is to keep the ovary and abdomen from becoming over-distended by too many eggs growing to maturity at the same time. Females now seek an appropriate site for oviposition. Hormones inhibiting host-seeking behavior and stimulating preoviposition-searching behavior are activated by an ovarian releasing factor (see Figure 20.5c). Exactly how the inhibition of trypsin synthesis is released when eggs are laid is not clear.

Alternative mechanisms for the action of oostatic hormones may exist. Oostatic hormone activity has been demonstrated in *A. atropalpus*, an autogenous mosquito that does not need to feed on blood for the first set of eggs nor seems likely to depend on trypsin enzyme activity in the midgut to mature the first set of eggs (Kelly et al., 1984, 1986). Both an oostatic hormone (Adams et al., 1968) and an ecdysteroidogenin hormone (Adams et al., 1997) have been demonstrated in the housefly, *Musca domestica*. In response to purified extracts of the ecdyteroidogenin, housefly ovaries *in vitro* produced approximately 500 pg each of ecdysone, 20-hydroxyecdysone, and makisterone A.

JH plays a major role in the reproduction of **cyclorrhaphous Diptera**, but Yin and Stoffolano (1997) suggest that dipterans are too diverse a group (with diversity even within the two major divisions, the Cyclorrhapha and the Nematocera) to expect JH (and by extrapolation other hormones) to play a common role. Adult life history and nutrition determine in large part the way in which JH and the neuroendocrine system exert their control over reproduction (Yin and Stoffolano, 1997). Juvenile hormone III bisepoxide (**JHB$_3$**) as well as **JH III** and **methyl farnesoate** have been identified in a number of cyclorrhaphous larval and adult dipterans. In the blowfly, *Phormia regina* (Meigen), JH influences mating behavior in both sexes, fat body development, ovary growth and development, and pinocytotic uptake of Vg by oocytes (Yin et al., 1989), but JH is synthesized by the CA only at low levels until a protein meal is taken by the flies (Yin et al., 1995). Biosynthesis of Vg is primarily controlled by **ecdysteroids** released from the ovaries (Yin et al., 1990). In contrast to mosquitoes, there presently is no published evidence for an OEH to stimulate ecdysteroid synthesis or for the necessity of the ovaries to be exposed first to JH to make them competent for further function (Yin and Stoffolano, 1997). Liver-fed flies produce approximately normal levels of Vg but oocytes do not sequester it when JH production is suppressed by precocene II treatment (Yin et al., 1989), but precocene-treated flies sequester Vg and mature eggs if they are rescued by treating them with methoprene, a JH mimic (Stoffolano et al., 1992). Examples of specific hormonal roles in different groups of insects are as follows:

- **Apterygota**: *Thermobia domestica*, the firebrat, represents the present-day success of a very early evolutionary group of insects. These insects continue to secrete **ecdysteroids** and **molt as adults**. JH is implicated in control of ovary development and oogenesis as indicated by allatectomy or treatment with precocenes, either of which prevents egg development. Both procedures also prevent the secretion of JH, and it is likely that JH is the main hormone controlling vitellogenesis. Fat body and ovaries are involved in producing the two large Vg molecules that go into the eggs (Rousset and Bitsch, 1993), but details relating to precise endocrine controls on synthesis are not available. JH III is the major JH of *Euborellia annulipes* (Lucas) (Dermaptera), another group that evolved early with strong sister group relationships with Dictyoptera, Isoptera, and Mantoidea, but methyl farnesoate is also present in the CA and the medium in which glands are incubated (Rankin et al., 1997).
- **Coleoptera**: Engelmann (1983) has shown that **JH** is responsible for Vg synthesis in the Colorado potato beetle *Leptinotarsa decemlineata*, the yellow mealworm *Tenebrio molitor*, and in some other beetles. The peptide Neb-colloostatin is important in physiology of the ovary in *T. molitor* (Czarniewska et al., 2014).
- **Hemiptera**: JH promotes Vg synthesis and uptake by the oocytes of *Rhodnius prolixus* (Davey, 1993, 1997), and the milkweed bug *Oncopeltus fasciatus*, *Pyrrhocoris apterus*, and *Triatoma protracta* (Engelmann, 1983).
- **Lepidoptera**: Ramaswamy et al. (1997) and Bellés (1998) suggest evolution of flexibility in the hormonal control of vitellogenesis in Lepidoptera. **Ecdysteroids** control vitellogenesis in those species that start vitellogenesis in the larval or early pupal stages, with progression to a combination of ecdysteroids and **JH** in those that start vitellogenesis prior to emergence in the pharate adult stage, and finally only **JH** controls egg protein synthesis in those species that begin vitellogenesis after adult emergence.

The cecropia moth, *Hyalophora cecropia*, initiates Vg synthesis early in the prepupal stage, and the silkmoth, *Bombyx mori*, begins Vg synthesis in the early pupal stage. In these two moths JH does not seem to have a role, and 20-hydroxyecdysone stimulates Vg synthesis (Tsuchida et al., 1987). 20-Hydroxyecdysone stimulates Vg synthesis in the gypsy moth, in which Vg synthesis begins late in the last instar, and experimental treatment with JH inhibits Vg synthesis (Fescemyer et al., 1992). Some moths, including some pyralid moths, use falling ecdysteroid concentrations to induce Vg synthesis, and eggs mature before eclosion of adults. In the fall armyworm, *Spodoptera frugiperda*, both JH and ecdysteroids promote Vg synthesis, but sequestering of Vg by the oocytes is under JH control (Sorge et al., 2000). Only JH induces Vgs synthesis in the monarch butterfly, *Danaus plexippus* (Pan and Wyatt, 1971); the moth, *Heliothis virescens* (Zeng et al., 1997); and in a number of other lepidopterans that begin vitellogenesis after emergence of the adult (Cusson et al., 1994). Males transfer JH to female *H. virescens* during mating (Park et al., 1998; Ramaswamy et al., 2000), and mating itself stimulates the CA in females to synthesize JH II (and small amounts of JH I and III) and causes inhibition of JH esterase that could potentially destroy JH transferred or synthesized.

20.3 VITELLOGENINS AND YOLK PROTEINS

20.3.1 Biochemical Characteristics of Vitellogenins and Yolk Proteins

The egg yolk is rich in proteins and lipids. Sex-limited proteins present in the hemolymph, which are incorporated into developing eggs, were discovered initially in *Hyalophora cecropia*, the cecropia silkmoth (Telfer, 1954), and since have been found in many different groups of insects. Some early work suggested the proteins were sex-specific and found only in females, but later research has shown varying, but small, amounts of the same proteins in males of some species, including some lepidopterans, hemipterans, orthopterans, and honeybees. In some males, egg proteins can be induced with hormone treatments (Shirk et al., 1983; Wyatt, 1991).

Although synthesis of the egg proteins occurs in both fat body (Keely, 1985) and the follicular cells of the ovary in some insects, the major source of proteins is the fat body in most insects. The proteins are called **Vg** while they are being produced by the fat body and during transport to the ovaries by the hemolymph (Pan et al., 1969), but after incorporation into developing eggs, the proteins are known as **vitellins (Vns)**.

Insect Vgs typically are large **glycolipoproteins**, from 400 to 600 kDa, that are composed of small (40–60 kDa) and large (120–200 kDa) subunits (Kunkel and Nordin, 1985; Borovsky and Whitney, 1987; Shirk, 1987; Yano et al., 1994a,b; Tufail and Takeda, 2008). Vgs contain 7%–15% lipids, consisting primarily of phospholipids and diacylglycerol (Raikhel and Dhadialla, 1992) and apoproteins (Zeng et al., 1997). Vgs usually exist as **dimers**, but **monomers** are known from the cockroach, *Nauphoeta cinereae* (Imboden et al., 1987). Genes controlling synthesis of Vgs have been identified and cloned from a number of insects, including *Locusta migratoria*, *Aedes aegypti*, *Anopheles gambiae*, and *Drosophila melanogaster* (Bownes, 1986; Wyatt, 1991, and references therein). The apoproteins of insect and vertebrate Vgs diverged from a superfamily of proteins controlled by genes with an ancient heritage (Speith et al., 1985; Borovsky and Whitney, 1987; Blumenthal and Zucker-Aprison, 1997).

Egg proteins of Diptera fall into two classes that split along the lines of the two suborders of Diptera. Lower Diptera in the suborder Nematocera, including mosquitoes and some other dipterans, have Vgs similar in structure to those of other insects, that is, large glycolipoproteins composed of small and large subunits. A major evolutionary shift in gene control of yolk proteins occurred in the higher Diptera (suborder Cyclorrhapha), and the proteins that go into the yolk are not homologous with the Vgs of other insects (Romans et al., 1995). Consequently, these proteins are not called Vgs, and instead are called **yolk proteins (YPs)**. As noted in Section 20.2.3, a cascade of hormones is often involved in controlling the synthesis of Vgs and YPs, with JH and ecdysteroids playing important roles. In nonfeeding moths, synthesis of Vgs and their uptake by oocytes appear to be controlled by ecdysteroids during prepupal, pupal, or pharate adult development (depending on the species), and experimental addition of exogenous JH inhibits Vg production (Satyanarayana et al., 1994, and references therein).

20.3.2 Yolk Proteins of Higher Diptera

Known **YPs** in higher Diptera are **small polypeptides** (Figure 20.6) composed of small subunits. For example, in *D. melanogaster* and some other higher dipterans, the yolk protein is composed of three subunits called YP1, YP2, and YP3 (46, 45, and 44 kDa, respectively) (Figure 20.7), and each is coded by single-copy genes on the X chromosome (Bownes et al., 1993). The numbers of YPs differ in several different *Drosophila* spp., but in all investigated, the YPs are small polypeptides. Small YPs have been found in a number of other higher Diptera, including blowflies, flesh flies, houseflies, and several tephritid fruit flies (Huybrechts and DeLoof, 1982; DeBianchi et al., 1985; Handler and Shirk, 1988; Rina and Savakis, 1991; Martínez and Bownes, 1994).

The genetic controls of YPs and Vgs are different. The YPs of *D. melanogaster* are under control of a family of genes different from that controlling the more widespread Vgs (Bownes et al., 1993). The *Drosophila* genes have greater sequence similarity to genes controlling mammalian triacylglycerol lipase than to Vg genes of other insects (Baker, 1988; Bownes et al., 1988; Terpstra and Geert, 1988), and the YP genes and vertebrate lipase genes may have evolved from ancestral progenitors (Kirchgessner et al., 1989). **YPs** may **sequester ecdysteroids** and make them available to the developing embryo as the proteins are digested during embryogenesis. Bownes et al. (1988) found that degradation of the YPs from *Drosophila* releases ecdysteroid in proportion to the degree of enzyme attack by protease and esterase. Ecdysteroids may have multiple functions in the embryo, but one function that seems likely is to promote the secretion of a cuticle. Some embryos secrete and molt more than one cuticle during embryogenesis (Bownes et al., 1988).

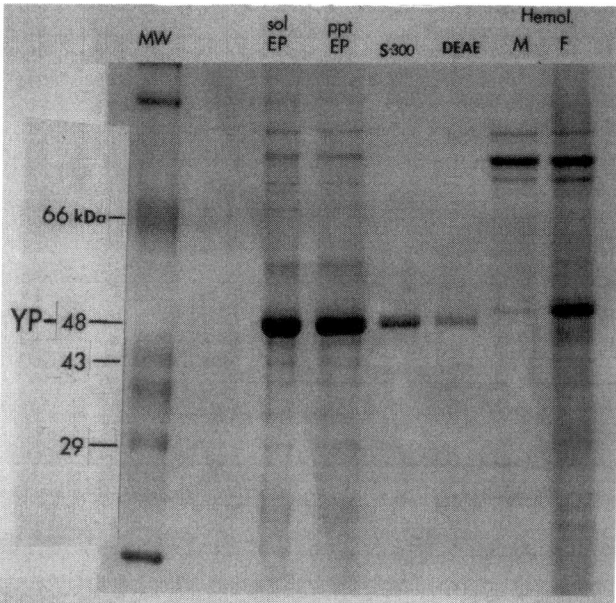

Figure 20.6 Resolution and identification of yolk protein (YP) from eggs of the Caribbean fruit fly, *Anastrepha suspensa*, on 10% SDS-PAGE stained for protein with Coomassie Blue. Key: Lane a, molecular mass standards; lane b, soluble proteins from oviposited eggs; lane c, ammonium sulfate precipitated egg proteins; lane d, combined YP fractions from S-300 separation; lane e, combined YP fractions from DEAE separation; lane f, 5 μL hemolymph from 3- to 4-day-old males; lane g, 5 μL hemolymph from 3- to 4-day-old females. (Photo courtesy of Al Handler and Paul Shirk, USDA, Gainesville, FL.)

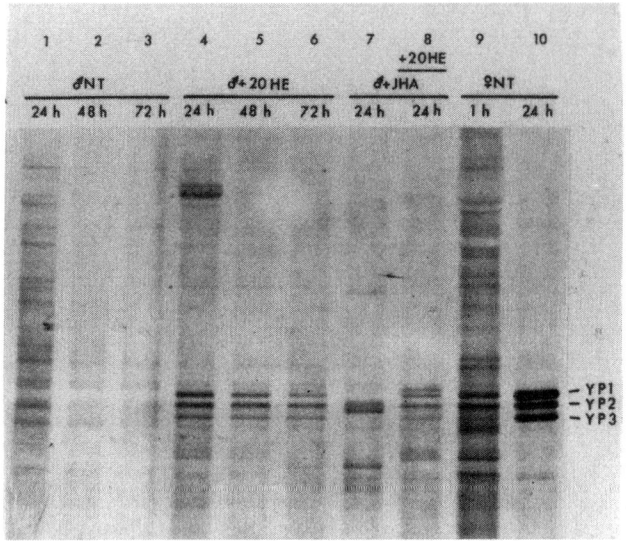

Figure 20.7 Yolk proteins (YP) 1, 2, and 3 in normal adult males and females and hormonally stimulated males of *Drosophila melanogaster*. Hormonally stimulated flies were injected with 0.3 μL of 0.1 mM 20-hydroxyecdysone (20-HE) or were topically treated with 0.16 μg ZR-515 juvenile hormone mimic in acetone 8 h prior to the collection of hemolymph. Hemolymph was collected at times indicated in the lanes mentioned and subjected to gel electrophoresis on 0.1% sodium dodecyl sulfate (SDS), 9%–12% polyacrylamide slab gels. Lanes 1, 2, and 3 (from left) represent males with no treatment; lanes 4, 5, and 6 represent males treated with 20-HE; lane 7 shows hemolymph from ZR-515-treated male; lane 8 represents ZR-515 + 20-HE-treated males; and lanes 9 and 10 represent normal untreated females. Bands in lane 7 that nearly match YPs also appeared in untreated 24-h-old males (lane 1). These polypeptides do not migrate exactly with the YPs and are not immunoprecipitable and probably are not YPs. They may be synthesized by remaining larval fat body cells because they are not present in 2- and 3-day-old males (lanes 2 and 3) when larval fat body has disappeared. (Photo courtesy of Paul Shirk, USDA, Gainesville, FL.)

20.4 SEQUESTERING OF VITELLOGENINS AND YOLK PROTEINS BY OOCYTES

20.4.1 Patency of Follicular Cells

Oocytes take up proteins through channels between the follicular cells (Figure 20.8) (Telfer, 1961, 1965; Davey, 1981; Raikhel and Dhadialla, 1992). **JH** acts with a membrane receptor to promote Vg uptake by promoting widening of the intercellualr spaces in the follicular epithelium. A Na/K-ATPase is activated and the cells shrink. **Phosphatidylinositol** and **protein kinase C** are involved (Ilenchuk and Davey, 1987; Davey, 1993). The opening of spaces between follicular cells is called **patency**. Egg proteins and experimentally added dyes readily pass through the spaces between follicle cells when patency has occurred. Patency is inhibited by ouabain and metabolic poisons that stop or reduce Na^+/K^+-ATPase activity, and by colchicine and cytochalasin B that inhibit cytoskeletal elements such as microtubules (Davey, 1981). At the end of vitellogenesis, new junctions between follicular cells seal the interfollicle cell channels and protein uptake rapidly falls (Rubenstein, 1979; Koller et al., 1989).

The Vgs and YPs bind to specific receptors at the surface of the oocyte plasma membrane between and at the base of microvilli. The receptor–protein complex tends to sink inward at the oocyte surface, forming a pit, with a **clathrin protein coat** on the cytoplasmic side (Raikhel, 1984,

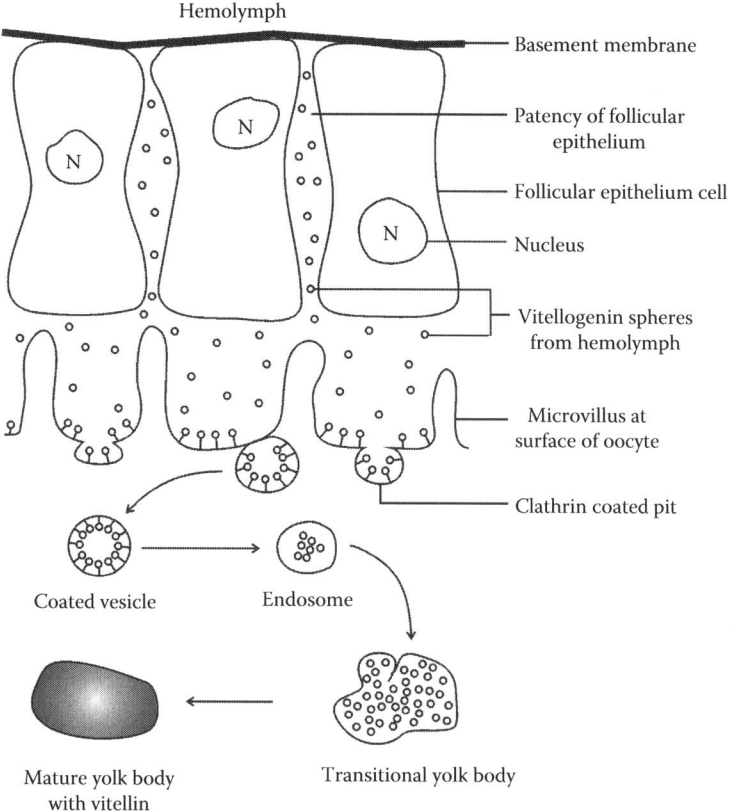

Figure 20.8 A schematic diagram to illustrate the uptake of vitellogenin by a developing oocyte. The proteinaceous vitellogenin (Vg) is mainly synthesized in fat body cells. It is transported by the hemolymph and passes between spaces in follicular epithelial cells that have shrunk (shrinking of the cells is hormonally induced and is called patency). Vg is bound to receptors on the surface of the developing oocyte. The membrane with bound vitellogenin invaginates and pinches off as small vesicles. Within the vesicles, the vitellogenin is released from the receptors and the receptor molecules probably return to the membrane where they bind more vitellogenin. Vesicles filled with vitellogenin are called endosomes and they coalesce as transitional yolk bodies that finally become mature yolk bodies containing vitellin, as the protein is called after it is stored in the yolk bodies. (Drawing modified from Raikhel, A.S. and Dhadialla, T.S., *Annu. Rev. Entomol.*, 37, 217, 1992.)

1987; Raikhel and Dhadialla, 1992). The pits continue to invaginate, close up, and become pinched off as small, coated vesicles (150–190 nm diameter) inside the oocyte. Coated vesicles have been isolated from the ovaries of *Locusta migratoria* locusts (Röhrkasten and Ferenz, 1987) and the clathrin heavy chains have a molecular weight of 180,000.

The clathrin coat soon dissociates from the vesicles containing the receptor–protein molecules, and the vesicles are then called **endosomes**. Bound egg proteins dissociate from the receptor within the endosome, probably as a result of ATP-dependent acidification within the endosome (Stynen et al., 1988). The clathrin molecules and receptors probably recycle to the oocyte surface for reuse.

Endosomes coalesce into a larger transitional **yolk body**, and the egg proteins, now called Vns, begin to crystallize. Additional Vns are added to the transitional yolk body until it becomes a mature yolk body (see Figure 20.8). Generally, Vgs and Vns have the same immunological properties, chemical composition, and physical properties. Some Vns derived from YPs are sulfated, but it is not known whether this is a general property of Vns (Baeuerle and Huttner, 1985; Dhadialla and Raikhel, 1990).

20.4.2 Egg Proteins Produced by Follicular Cells

Proteins produced in the ovary, usually by follicular cells lining the follicle, do not enter the hemolymph, but are passed directly into the developing oocytes. In several *Drosophila* species that have been studied, the proportion of the several YPs synthesized in the fat body and in the follicular cells varies with species, but follicular cells are a major source of some YPs. **Paravitellogenin** is a 70 kDa protein produced in the follicular cells of *H. crecropia* and taken into the oocyte (Bast and Telfer, 1976; Telfer et al., 1981). A similar, perhaps homologous, protein making up to 25% of the egg proteins is produced in the follicular cells of *B. mori* (Sato et al., 1990). Proteins produced in the follicular cells of Indian meal moths, *Plodia interpunctella*, and in several mosquitoes are incorporated into eggs (Borovsky and Van Handel, 1980; Bean et al., 1988).

20.4.3 Proteins in Addition to Vitellogenin and Yolk Proteins in the Egg

Developing eggs often contain varying amounts of proteins that are sequestered from the hemolymph in addition to Vgs and YPs. They are not usually considered to be Vgs because they are present in small amounts and/or do not have the general structure of Vgs. **Lipophorin**, a general lipid transport protein of insects, and **microvitellin**, a small female-specific protein, are present in eggs of *H. cecropia* (Telfer and Pan, 1988) and *Manduca sexta* (Law, 1989), respectively. In addition, *M. sexta* eggs contain **insecticyanin**, a blue biliprotein, giving the eggs a pale blue color. A group of storage proteins of about 30 kDa are synthesized during larval stages of both sexes of the commercial silk moth, *B. mori*, and become the dominant hemolymph proteins during larval life and persist into the pupal stage (Izumi et al., 1981). Substantial quantities are taken into the developing eggs, but they usually are not considered to be Vgs. One of these is a glycoprotein with a molecular weight of 55,000. It contains 2% carbohydrate (mannose and traces of amino sugars) and 4% lipid. It is the second major protein in the yolk and is used early in embryogenesis, although some of the original Vns are still present at hatching (Irie and Yamashita, 1983). The hemolymph of the female migratory locust, *L. migratoria*, has a low concentration of a 21-kDa monomer whose synthesis is stimulated in the fat body by JH treatment. The protein is female specific and taken into developing oocytes (Zhang and Wyatt, 1990). In most cases, little is known about the fate of these proteins once they enter the oocyte.

20.5 FORMATION OF THE VITELLINE MEMBRANE

Near the termination of vitellogenesis, a thin protein sheet, the **vitelline membrane**, is secreted at the inner surface of the follicle cells (Raikhel and Dhadialla, 1992). The vitelline membrane in *D. melanogaster* is composed of numerous proteins ranging from 14 to 130 kDa (Fargnoli and Waring, 1982) that are encoded by a family of genes (Wyatt, 1991).

20.6 CHORION

The eggshell, the **chorion**, is composed of a number of sclerotized proteins. It contains no chitin. With very few exceptions, it is not mineralized like the eggshell of birds. The chorion is placed on the egg while it is still in the ovary and before fertilization. Sperm, which are released from the spermatheca as the egg passes down the common oviduct, have to enter the egg through a small, usually twisted channel, the **micropyle**, which passes through the various layers of the chorion. More than one micropyle channel is not uncommon; although most Diptera have only one, *Locusta* has 35–43 openings. Follicular epithelial cells secrete chorionic proteins on the outer surface of the vitelline membrane, thus enclosing it on the inside of the chorion. The follicular epithelial cells lay down proteins sequentially, indicative of the sequential expression of a superfamily of genes, to produce a laminar structure. In wild silk moths, *Antheraea polyphemus*, chorion formation requires about 2 days and more than 100 low molecular weight proteins are secreted (Lecanidou et al., 1986). A large gene family controls the secretion of chorion proteins in *A. polyphemus* without gene amplification (Kafatos et al., 1987). By contrast, in *D. melanogaster* only about 20 proteins are secreted under control of a small family of genes (Waring and Mahowald, 1979). The single-copy genes are amplified 20- to 80-fold in the follicle cells about 15 h prior to transcription (Spradling and Mahowald, 1980). These multiple gene copies (after transcription) allow the follicle cells to secrete a large amount of chorionic proteins in a short time, and the chorion is completed in about 5 h (Hammond and Laird, 1985). The proteins become sclerotized to produce a tough, water-impermeable covering for the egg and developing embryo. When secretion of the chorion is complete, the old follicular epithelial cells are sloughed off as the egg passes into the median oviduct. The chorion does not contain chitin and, except in a few dipterans, no significant quantity of minerals. Intricate surface sculpturing is characteristic of many insect eggs. Although hormonal control of choriogenesis in *D. melanogaster* has not been demonstrated, recent work has shown that the DNA site to which a chorion gene transcription factor, CF1, binds has part of the sequence of the ecdysone response element. This suggests the possibility of hormonal control (Shea et al., 1990; Wyatt, 1991).

20.7 GAS EXCHANGE IN EGGS

Many eggs have a porous gas-filled meshwork near the inner (yolk) side. In some cases, this is a **plastron**, a surface that is not easy to wet. Several channels called aeropyles connect the meshwork or plastron to the external surface of the egg. The function of such hard-to-wet structures is to supply oxygen to the developing embryo if the egg becomes submerged under water for some time, or when the natural site for laying the eggs is in wet decaying organic matter, animal manure, fruits, or similar plant tissues. When eggs have a plastron surface, tests have shown that the plastron surface resists the wetting action of raindrops, which can exert up to about 30 cm Hg pressure for about a millisecond. When there is an egg plastron, it is usually a part of the chorion itself, but some eggs have the plastron surface on **respiratory horns** or filaments protruding from the egg. These might give a submerged egg the opportunity to have the plastron surface above the fluid medium if it were not deep.

20.8 MALE REPRODUCTIVE SYSTEM

In many species, males make elaborate courtship displays directed at females, often in conjunction with production of sounds or pheromone, or offering a nuptial gift of food. Although it has been shown that females often are more choosy than males, and males make much less investment in reproduction than females, recent studies suggest that male investment is not trivial in some cases (Papadopoulos et al., 2010; Edward and Chapman, 2011, 2013; Nandy et al., 2012; Wegener et al., 2013). Now Khan and Prasad (2013) show that when male *Drosophila melanogaster* are presented with females infected with pathogenic bacteria, they preferentially mated with sham-infected females, providing more evidence for male choice. Male *Drosophila* use auditory and olfactory senses (Bretman et al., 2011) and vision to detect the presence of male rivals during courtship of females and then increase seminal fluid proteins (Fedorka et al., 2011) and the number of sperm in the ejaculate (Garbaczewska et al., 2013), possibly mechanisms to try to keep the female from mating again soon, and to ensure that their sperm will fertilize eggs even if the female mates again before all eggs are laid (Parker and Pizzari, 2010).

Courtship in *Drosophila* males is an elaborate ritual involving numerous sensory stimuli and motor actions. The gene *fruitless* (*fru*) is necessary for courtship display in male *Drosophila*, and the gene is spliced differently in males and females (Demir and Dickson, 2005; supplemental data and several movies of courtship are on the Internet at http://www.cell.com/cgi/content/full/121/5/785/DC1). Demir and Dickson (2005) constructed alleles of *fru* with male or female splicing, and show that male splicing of *fru* is essential and sufficient for male courtship behavior. Moreover, male splicing of *fru* generates typically male behavior in females that are otherwise normal females. These females courted females and also males that were genetically engineered to produce female pheromones. The authors demonstrated that *fruitless* is a switch gene, a gene that is necessary and sufficient for a complex behavior, such as male courtship. Tobback et al. (2012) showed that there is involvement of clock genes (*period* and *timeless*) in the reproduction of male *Schistocerca gregaria*; use of RNAi to silence those genes interfered with successful reproduction.

The internal organs of the male reproductive system are the **paired testes**, the **vas deferens**, the **accessory glands**, and the **ejaculatory duct** (Figure 20.9). All parts of the system may produce secretions that aid the transfer of sperm to the female (Happ, 1992). Each testis generally consists of a number of tubes or **follicles** in which **spermatozoa** are matured (Figure 20.10). Follicles may vary from 1 to greater than 100 follicles, and may be incompletely separated from each other, such as lepidopteran testis, or the testes may consist of several lobes, each with several follicles. In Diptera, the testes consist of a simple, elongated and undivided sac (French and Hoopingarner, 1965). **Zones of maturation** stages of sperm exist along the length of a typical follicle. The distal part of a testicular tubule is a **germarium** in which repeated mitotic divisions give rise to undifferentiated, diploid, **spermatogonia**. In a **growth zone** (Zone I), the spermatogonia divide by mitosis into many diploid **spermatocytes** enclosed within a cyst or capsule of somatic cells. All spermatocytes within a sac or cyst generally arise from the same spermatogonial cell and their development is synchronized. The spermatocytes may undergo more mitotic divisions; there are five to eight divisions in Acrididae and seven in *Melanoplus*, but eventually in the "zone of maturation" (Zone II) **meiosis** and haploid **spermatids** are produced. A spermatid has completed its meiotic divisions, but is an immature sperm. According to Jones (1978), meiosis in *Schistocerca gregaria* males depends on the presence of ecdysteroids. Normally four sperm are produced from each spermatocyte. In Zone III, **the region of transformation**, the mature sperm develop.

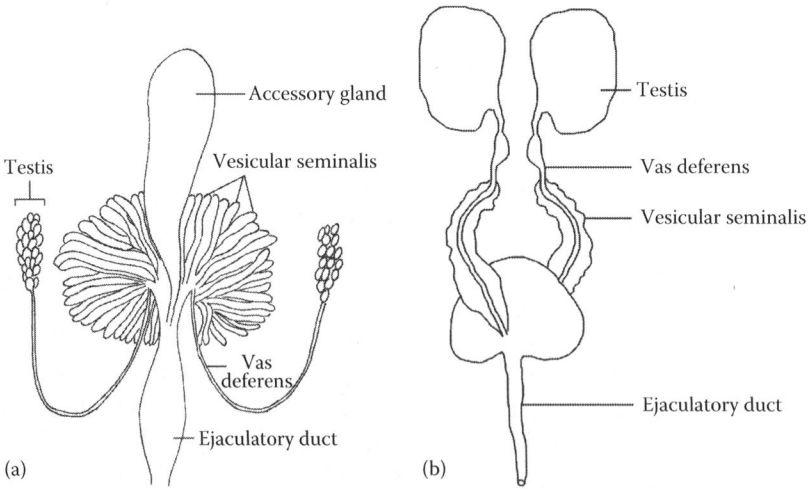

Figure 20.9 An illustration of the male reproductive system from (a) the American cockroach, *Periplaneta americana* and (b) the milkweed bug, *Oncopeltus fasciatus*. (Drawing by the author.)

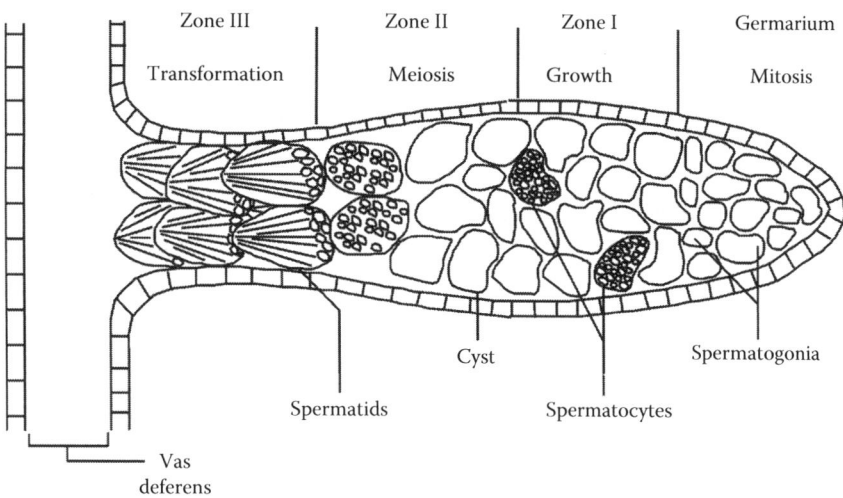

Figure 20.10 An illustration of zones of maturation of spermatozoa that can be observed in the testes of some insects. (Drawing by the author.)

Insect spermatozoa tend to be very long (300 μm in *Rhodnius prolixus*) and have a slender head region, probably as an evolutionary adaptation to the necessity to navigate the micropyle. Usually the mature sperm remain bundled together in Zone III.

Many insects contain mature sperm in the late pupal stage, while others may require several days as an adult to mature sperm. In *R. prolixus* and grasshoppers, the accessory glands in males are influenced in their development by secretions from the corpora allata. In contrast to the situation in most vertebrates, sperm survival within the genital tract of female insects may be prolonged for weeks, months, or even years. Honeybee queens have been known to lay fertilized eggs after several years (8–9 years in one reported case), and queen ants were reported to contain viable sperm after 15 years. Spermatozoa survive in female *R. prolixus* for about a month, and for about 10 weeks in *Schistocerca gregaria*.

For a very long period of time, it has been thought that spermatogenesis in insects is not under hormonal control, but since the 1970s and later, it is now clear that ecdysteroids are synthesized in the testis of flies, crickets, mosquitoes, and various lepidopterans, and that ecdysteroids stimulate spermatogenesis in several insects (Wagner et al., 1997, and numerous references therein). These observations bring spermatogenesis in insects more in line with known hormonal controls of spermatogenesis in other animals. Wagner et al. (1997) recently described an ecdysiotropic peptide from the brain of gypsy moths that stimulates synthesis of ecdysteroids (at reasonable hormonal levels of about 10^{-13} to 10^{-15} M) in testes of larval and pupal gypsy moth males.

20.8.1 Apyrene and Eupyrene Sperm of Lepidoptera

Most Lepidoptera produce two types of sperm, **apyrene** sperm without a nucleus and nucleated **eupyrene** sperm. Only the latter type can fertilize an egg. The two types of sperm are produced by major differences in the meiotic process (Garvey et al., 2000). Sperm dichotomy appeared early in the evolution of Lepidoptera, but apyrene sperm are not present in the Micropterigidae, one of the most primitive families of Lepidoptera (Sonnenschein and Hauser, 1990; Hamon and Chauvin, 1992). Eupyrene sperm are usually packaged into bundles, while apyrene sperm are dissociated as single, but immobile cells. Both types of sperm are incorporated into the **spermatophore** that is formed in the female bursa copulatrix at the time of mating by secretions from the male.

Maturation division (to produce the haploid number of chromosomes) of the eupyrene sperm and development of motility in both types of sperm occur in the spermatophore of some insects (Osanai et al., 1987, 1989). He et al. (1995) demonstrated a correlation between decrease of apyrene sperm in the spermatheca of an army worm, *Pseudaletia separata*, and remating patterns, and suggested that the number of apyrene sperm in the spermatheca may influence female remating.

Male *B. mori* have an **endopeptidase**, called **initiatorin**, in secretions of the posterior segment of the ejaculatory duct (Osanai et al., 1989) that is important in the activation of both apyrene and eupyrene sperm and in the maturation of the eupyrene sperm. Initiatorin is a **serine endoprotease** that is active at pH 9.2. It digests the surface coat of apyrene sperm most easily, and these sperm become motile before the euprene sperm are completely freed from their bundles. Their vigorous movements serve to stir the viscous contents of the spermatophore, aiding the liberation of the eupyrene sperm and facilitating metabolic reactions that promote eupyrene sperm maturation. In addition, initiatorin converts an inactive **procarboxypeptidase** secreted by the ampulla, the region of the ejaculatory duct where the vasa deferentia and ducts from the accessory glands converge and empty, into an active **carboxypeptidase**. The carboxypeptidase digests proteins and liberates arginine and other amino acids (Kasuga et al., 1987). Arginine is subsequently converted to glutamate, which is metabolized by the sperm to support motility (Aigaki et al., 1987). By virtue of its production in the terminal portion of the ejaculatory duct, initiatorin is kept away from the sperm and the procarboxypeptidase until ejaculation at mating. Similar processes likely occur in other Lepidoptera.

20.8.2 Male Accessory Glands

Many males have **accessory glands** associated with the reproductive tract. The accessory glands, which have varied morphology in different insects (Chen, 1984; Davey, 1985), empty into the ejaculatory duct. Their secretions are used to form the **spermatophore** in some insects, or if no spermatophore is formed, the secretions are added to the sperm prior to transfer to the female. Some of the secretory products may stimulate contractions in the reproductive tract of females, thus aiding movement of sperm into the spermathecae of the female (Avila et al., 2011).

Accessory glands are present in *Drosophila* species and are called **paragonial glands**. The glands synthesize more than 85 proteins (Stumm-Zollinger and Chen, 1985; Coulthart and Singh, 1988). One of the proteins, sometimes called a "**sex peptide**," is passed at mating to the female, which is then much less receptive to remating for 6–9 days (Chen et al., 1988). The peptide also stimulates oviposition (Chen, 1984) and increased food intake, produces more concentrated excreta (Apger-McGlaughon and Wolfner, 2013), and slows down intestinal transit, which may increase nutrient absorption (Cognigni et al., 2011). In some insects at least (e.g., *Lygus herperus* bugs) several male-derived compounds inserted with the spermatophore induce both short- and longer-term refractoriness in the female after mating. In *L. hesperus*, the lateral and medial accessory glands produce products that inhibit remating in females as do some of the proteins incorporated into the spermatophore (Brent and Hull, 2014). Genes controlling the synthesis of a number of the accessory gland proteins have been identified (reviewed by Happ, 1992). Accessory glands are lacking in some insects; for example, they are not present in some dipterans, including the housefly (*Musca domestica*) and stable fly (*Stomoxys calcitrans*).

20.8.3 Transfer of Sperm

Some insects transfer packets or bundles of sperm to the female reproductive tract by insertion of the **aedeagus** into the reproductive tract of the female. Many insects produce a **spermatophore** that contains the sperm and is transferred to the female. Accessory glands secrete **spermatophorins**, proteins that form the spermatophore. Mealworm adults (*Tenebrio molitor*) contain up to eight cell types in the wall of the accessory glands that secrete different proteins in a sequential manner so that specific layers of the spermatophore are formed (Grimnes et al., 1986; Happ, 1987; Shinbo et al., 1987). Spermatophorin production is stimulated in the mealworm by **20-hydroxyecdysone** (Yaginuma et al., 1988), but **JH** stimulates production in the hemipteran, *R. prolixus* (Gold and Davey, 1989). In *R. prolixus*, the spermatophore consists of a pear-shaped mass of transparent mucoprotein in a sol or gel state depending on the pH. The semen is contained in a slit inside the jelly mass. The protein jelly is secreted in the accessory glands and is first fluid at the pH (about 7) in the glands. As the fluid moves down the reproductive system, the pH decreases to about 5.5 in the bulbous ejaculatorius and intromittent organ. This is evidently at or near the isoelectric point of the secretion and it gels. Spermatozoa are released from the spermatophore after it is deposited in the bursa copulatrix of the female *Rhodnius*. Mechanical abrasion of the spermatophore or the action of proteolytic enzymes, or both, may play a role in releasing sperm. Although it appears that sperm are moved to the spermathecae to be stored without active participation from the female, contractions in the oviducts of the female induced by secretions from the male probably help to force sperm toward the spermathecae. Formation of the spermatophore results in loss of protein from the male, but the effects of this or consequences for the male usually have not been evaluated. The protein content of the spermatophore has often been viewed as male investment in the next generation. In at least one cricket, the loss of protein during spermatophore formation and transfer amounted to 40% of the body weight.

20.9 GENDER DETERMINATION

Insects have at least three chromosomal systems for gender determination, with variations existing within the types (reviewed by Lauge, 1985; Wyatt, 1991). In **type 1**, probably the most primitive mechanism, the male is heterogametic, or **maleXY** and **femaleXX**. Type 1 occurs in many different groups, including *D. melanogaster* and other Diptera. A variation within this type is the loss of the Y chromosome, so that the male is **XO**, as found in Odonata, Orthoptera, and among some groups in some orders. The female is heterogametic (**femaleZW**, **maleZZ**) in **type 2**, found in Lepidoptera and the closely related Trichoptera. In the heterogametic sex (females in Lepidoptera), a poorly understood process called **chromosome dosage compensation** regulates or equalizes expression of the homogametic sex chromosome and autosomal genes in the heterogametic sex. Compensation that is not complete (Mank, 2013) can lead to sex-biased gene expression, and current research suggests that dosage compensation patterns are variable across species and sex-determination systems (Mank et al., 2011), including some Lepidoptera. Smith et al. (2014) present evidence for complete dosage compensation in two lepidopterans, *Bombyx mori* and *Manduca sexta*, and they suggest that more lepidopterans should be examined. **Type 3**, in which **females are diploid** while **males are haploid**, is present in Hymenoptera, and in some coccids in the order Homoptera.

Variations in the types include loss or suppression of chromosomes during the early cleavage stages in embryogenesis resulting in only the germ cells retaining a complete set of chromosomes; subsequently, the sex of individuals is controlled by differences in the incomplete chromosome sets retained by somatic cells. In a few insects, gender can be determined by prevailing temperature, as in subarctic mosquitoes, and, in some gall midges, by available nutrition (Nöthiger and Steinmann-Zwicky, 1985, 1987).

The genetic mechanisms by which the several chromosomal patterns lead to gender determination are variable and poorly known for all except a few insects. Two broad mechanisms are known: the **ratio of sex chromosomes to autosomes** and the presence of **sex-determining genes**. Gender in *D. melanogaster* is determined by the ratio of sets of sex chromosomes to autosomes, or X:A (A = autosome set). Although the Y chromosome carries genes for factors necessary for the production of motile sperm, it does not carry gender-determining genes. The ratio, 2X:2A = 1, as in normal females (or the ratio of 3X:2A in aneuploids), results in female phenotype. Conversely, a ratio of 0.5, as in normal males (1X:2A), or a smaller ratio (1X:3A), produces male phenotypes. Intermediate ratios are known that result in mosaic intersexes in which the individual contains both male and female cells. Although vast genetic information in *D. melanogaster* has resulted in better understanding of gender determination and early development, especially as controlled by genes, these fruit flies cannot be considered representative of insects in general. Chromosomal ratios are ultimately expressed in specific gene actions, and mechanisms are not well understood in insects. Specific **gender determining genes** control the phenotype of some insects (Baker and Sakai, 1976).

In *Drosophila*, the expression of the chromosome ratio is related to the expression of a set of genes involved in somatic gender determination. The X:A ratio is fixed irreversibly at the blastoderm stage in the embryo (Sanchez and Nöthiger, 1983) by genes (*sisterless-a* and *sisterless-b*, and possibly others) coding for certain proteins that probably act as transcription factors (Torres and Sanchez, 1989). The genes and/or protein products are involved in determining the X:A ratio. The proteins are crucial for neural tube formation, and may be expressed prior to the establishment of dosage compensation, so that they exert a greater effect in a female (2X) than in a male (1X), at a time when gender determination is being established (Torres and Sanchez, 1989; Hodgkin, 1990). Another gene, *daughterless* (*da*), known to be required in females (but not males), acts synergistically in unknown ways with the *sisterless-a* and -*b* genes in determining femaleness, possibly in allowing a female-specific expression of a main regulatory on/off gender determination gene, *Sex-lethal* (*Sxl*). Once expression of *Sex-lethal* begins, its activity is maintained by positive feedback from its own gene products in females, and it sets in motion a cascade of gene actions (Baker and Belote, 1983, and briefly reviewed in Wyatt, 1991) that leads to differentiation into a female. In males, *Sex-lethal* does not seem to be involved in causing maleness, but it does regulate a set of genes controlling dosage compensation of the X chromosome, so that the male, with only one X, realizes twofold expression of the X-linked genes (Gergen, 1987; Hazelrigg, 1987). Maintenance of female sexual behavior patterns requires sustained expression of at least one of the cascade genes, *transformer* (*tra+*) in the adult female. When genetic females with certain *tra* genotypes are reared at high temperature (29°C) or when adult females are transferred from 16°C to 29°C, they display male courtship behavior (Belote and Baker, 1987; Wyatt, 1991). Presumably at 29°C, the *tra* gene cannot be expressed properly and female behavior is abnormal.

Nöthiger and Steinmann-Zwicky (1985) have proposed a unifying model for gender determination that depends on a primary signal that is monitored by a key gene whose activity state, off or on, controls gender differentiation genes. When the key gene is "on," a female or male can be determined depending on a second "male/female" switch mechanism. When the key gene is "off," a female cannot be produced and only male development is possible.

20.10 REVIEW AND SELF-STUDY QUESTIONS

1. Briefly indicate the major differences in the ovary structure of insects.
2. What is the function of the spermatheca? Can an insect have more than one spermatheca? Give an example.
3. What are the major physiological differences between autogenous and anautogenous mosquitoes?
4. What are the differences between vitellogenin and vitellin?
5. Describe some of the variations in hormonal regulation of ovary development and formation of eggs in insects.
6. JH has several functions in the reproductive process in mosquitoes. What are these functions?
7. What is TMOF? What function does it serve?
8. What is/are the differences between vitellogenins and yolk proteins?
9. What is meant by the patency of follicular cells?
10. What are endosomes?

11. What is a yolk body?
12. Are all of the proteins in insect eggs called vitellin? Explain variations.
13. Describe the chemical structure of the chorion.
14. What is the micropyle? What is an aeropyle?
15. What is a plastron and how does it function in insects?
16. Are spermatids haploid or diploid in a cockroach?
17. What are spermatocytes and are they haploid or diploid?
18. What are the differences between apyrene and eupyrene sperm in Lepidoptera?
19. What are the variations in gender determination and the genetic structure of gender in various groups of insects?

REFERENCES

Adams, T.S. and P.A. Filipi. 1988. Vitellin and vitellogenin concentrations during oogenesis in the first gonotrophic cycle of the house fly, *Musca domestica. J. Insect Physiol.* 29: 723–733.

Adams, T.S., J.W. Gerst, and E.P. Masler. 1997. Regulation of ovarian ecdysteroid production in the housefly, *Musca domestica. Arch. Insect Biochem. Physiol.* 35: 135–148.

Adams, T.S., A.M. Hintz, and J.G. Pomonis. 1968. Oostatic hormone production in houseflies, *Musca domestica*, with developing ovaries. *J. Insect Physiol.* 14: 983–993.

Aigaki, T., H. Kasuga, and M. Osanai. 1987. A specific endopeptidase, BAEE esterase, in the glandula prostatica of the male reproductive system of the silkworm, *Bombyx mori. Insect Biochem.* 17: 323–328.

Apger-McGlaughon, J. and M.F. Wolfner. 2013. Post-mating change in excretion by mated *Drosophila melanogaster* females is a long-term response that depends upon sex peptide and sperm. *J. Insect Physiol.* 59: 1024–1030.

Avila, F.W., L.K. Sirot, B.A. Laflamme, C.D. Rubinstein, and M.F. Wolfner. 2011. Insect seminal fluid proteins: Identification and function. *Annu. Rev. Entomol.* 56: 21–40.

Ayali, A. and A.B. Lange. 2010. Rhythmic behaviour and pattern-generating circuits in the locust: Key concepts and recent updates. *J. Insect Physiol.* 56: 834–843.

Baeuerle, P.A. and W.B. Huttner. 1985. Tyrosine sulfation of yolk proteins 1, 2, and 3 in *Drosophila melanogaster. J. Biol. Chem.* 260: 6434–6439.

Baker, B.S. and J.M. Belote. 1983. Sex determination and dosage compensation in *Drosophila melanogaster. Annu. Rev. Genet.* 17: 345–393.

Baker, F.C., H.H. Hagedorn, D.A. Schooley, and G. Wheelock. 1983. Mosquito juvenile hormone: Identification and bioassay activity. *J. Insect Physiol.* 29: 465–470.

Baker, M.E. 1988. Is vitellogenin an ancestor of apolipoprotein B-100 of human LDL and human lipoprotein lipase? *Biochem. J.* 255: 1057–1060.

Baker, R.H. and R.K. Sakai. 1976. Male determining factor on chromosome 3 in the mosquito, *Culex tritaeniorhynchus. J. Hered.* 67: 289–294.

Bartholomew, G.A. and T.M. Casey. 1978. Oxygen consumption of moths during rest, pre-flight warm-up, and flight in relation to body size and wing morphology. *J. Exp. Biol.* 76: 11–25.

Bast, R.E. and W.H. Telfer. 1976. Follicle cell protein synthesis and its contribution to the yolk of the cecropia moth oocyte. *Dev. Biol.* 52: 83–97.

Baxter, J.A., A.M. Mjeni, and P.E. Morrison. 1973. Expression of autogeny in relation to larval population density of *Sarcophaga bullata* Parker (Diptera: Sarcophagidae). *Can. J. Zool.* 51: 1189–1193.

Bean, D.W., P.D. Shirk, and V.J. Brookes. 1988. Characterization of yolk proteins from the eggs of the Indian meal moth, *Plodia interpunctella. Insect Biochem.* 18: 199–210.

Bellés, X. 1998. Endocrine effectors in insect vitellogenesis, in G.M. Coast and S.G. Webster (eds.), *Recent Advances in Arthropod Endocrinology*. Cambridge University Press, Cambridge, U.K., pp. 71–90.

Belote, J.M. and B.S. Baker. 1987. Sexual behavior: Its genetic control during development and adulthood in *Drosophila melanogaster. Proc. Natl. Acad. Sci. USA* 84: 8026–8030.

Blumenthal, T. and E. Zucker-Aprison. 1997. Evolution and regulation of vitellogenin genes, in J.D. O'Connor (ed.), *Molecular Biology of Invertebrate Development*. Alan R. Liss, New York, pp. 3–19.

Boggs, C.L. 1990. A general model of the role of male-donated nutrients in female insects' reproduction. *Am. Nat.* 136: 598–617.

Bonhag, P.F. 1955. Histochemical studies of the ovarian nurse tissues and oocytes of the milkweed bug, *Oncopeltus fasciatus* (Dallas). *J. Morphol.* 96: 381–439.

Bonhag, P.F. 1958. Ovarian structure and vitellogenesis in insects. *Annu. Rev. Entomol.* 3: 137–160.

Borovsky, D. 1982. Release of egg development neurosecretory hormone in *Aedes aegypti* and *Aedes taeniarhynchus* induced by an ovarian factor. *J. Insect Physiol.* 28: 311–316.

Borovsky, D. 1988. Oostatic hormone inhibits biosynthesis of midgut proteolytic enzymes and egg development in mosquitoes. *Arch. Insect Biochem. Physiol.* 7: 187–210.

Borovsky, D. 2003. Trypsin-modulating oostatic factor: A potential new larvicide for mosquito control. *J. Exp. Biol.* 206: 3869–3875.

Borovsky, D., D.A. Carlson, P.R. Griffin, J. Shabanowitz, and D.F. Hunt. 1990. Mosquito oostatic factor: A novel decapeptide modulating trypsin-like enzyme biosynthesis in the midgut. *Fed. Am. Soc. Exp. Bio. (FASEB) J.* 4: 3015–3020.

Borovsky, D., C.A. Powell, J.K. Nayar, J.E. Blalock, and T.K Hayes. 1994. Characterization and localization of mosquito-gut receptors for trypsin modulating oostatic factor using a complementary peptide and immunocytochemistry. *Fed. Am. Soc. Exp. Bio. (FASEB) J.* 8: 350–355.

Borovsky, D., S. Rabindran, W.O. Dawson, C.A. Powell, D.A. Iannotti, T.J. Morris, J. Shabanowitz, D.F. Hunt, H.L. DeBondt, and A. DeLoof. 2006. Expression of *Aedes* trypsin-modulating oostatic factor on the virion of TMV: A potential larvicide. *Proc. Natl. Acad. Sci. USA* 103: 18963–18968.

Borovsky, D. and E. van Handel. 1980. Synthesis of ovary-specific proteins in mosquitoes. *Intl. J. Invertebr. Reprod.* 2: 153–163.

Borovsky, D. and P.L. Whitney. 1987. Biosynthesis, purification, and characterization of *Aedes aegypti* vitellin and vitellogenin. *Arch. Insect Biochem. Physiol.* 4: 81–99.

Bownes, M. 1986. Expression of the genes coding for vitellogenin (yolk protein). *Annu. Rev. Entomol.* 32: 507–531.

Bownes, M., E. Ronaldson, D. Mauchline, and A. Martínez. 1993. Regulation of vitellogenesis in *Drosophila*. *Int. J. Insect Morphol. Embryol.* 22: 349–367.

Bownes, M., A. Shirras, M. Blair, J. Collins, and A. Coulson. 1988. Evidence that insect embryogenesis is regulated by ecdysteroids released from yolk proteins. *Proc. Natl. Acad. Sci. USA* 84: 1554–1557.

Bradfield, J.Y., R.L. Berlin, and L.L. Keeley. 1989. Constrasting modulations of gene expression by a juvenile hormone analog. *Insect Biochem.* 20: 105–111.

Brent, C.S. and J.J. Hull. 2014. Characterization of male-derived factors inhibiting female sexual receptivity in *Lygus hesperus*. *J. Insect Physiol.* 60: 104–110.

Bretman, A., J.D. Westmancoat, M.J.G. Gage, and T. Chapman. 2011. Males use multiple, redundant cues to detect mating rivals. *Current Biol.* 21: 617–622.

Chen, P.S. 1984. The functional morphology and biochemistry of insect male accessory glands and their secretions. *Annu. Rev. Entomol.* 29: 233–255.

Chen, P.S., E. Stumm-Zollinger, T. Aigaki, J. Balmer, M. Bienz, and P. Böhlen. 1988. A male accessory gland peptide that regulates reproductive behavior of female *D. melanogaster*. *Cell* 54: 291–298.

Cho, W.-L., M.Z. Kapitskaya, and A.S. Raikhel. 1995. Mosquito ecdysteroid receptor: Analysis of the cDNA and expression during vitellogenesis. *Insect Biochem. Mol. Biol.* 25: 19–27.

Cognigni, P., A.P. Bailey, and I. Miguel-Aliaga. 2011. Enteric neurons and systemic signals couple nutritional and reproductive status with intestinal homeostasis. *Cell Metab.* 13: 92–104.

Coulthart, M.B. and R.S. Singh. 1988. Differing amounts of genetic polymorphism in testes and male accessory glands of *Drosophila melanogaster* and *Drosophila simulans*. *Biochem. Genet.* 26: 153–164.

Cusson, M., C.G. Yu, K. Carruthers, G.R. Wyatt, S.S. Tobe, and J.N. McNeil. 1994. Regulation of vitellogenin production in armyworm moths, *Pseudaletia unipuncta*. *J. Insect Physiol.* 40: 129–136.

Czarniewska, E., G. Rosiński, E. Gabała, and M. Kuczer. 2014. The natural insect peptide Neb-colloostatin induces ovarian atresia and apoptosis in the mealworm *Tenebrio molitor*. *BMC Dev. Biol.* 14: 4.

Davey, K.G. 1981. Hormonal control of vitellogenin uptake in *Rhodnius prolixus* Stål. *Am. Zool.* 21: 701–705.

Davey, K.G. 1985. The male reproductive tract, in G.A. Kerkut and L.I. Gilbert (eds.), *Comprehensive Insect Physiology, Biochemistry and Pharmacology*, vol. 1. Pergamon Press, Oxford, U.K., pp. 1–14.

Davey, K.G. 1993. Hormonal control of egg production in *Rhodnius prolixus*. *Am. Zool.* 33: 397–402.

Davey, K.G. 1997. Hormonal controls on reproduction in female Heteroptera. *Arch. Biochem. Physiol.* 35: 443–453.

DeBianchi, A.G., M. Coutinho, S.D. Pereira, O. Marinotti, and H.J. Targa. 1985. Vitellogenin and vitellin of *Musca domestica*. Quantification and synthesis by fat bodies and ovaries. *Insect Biochem.* 15: 77–84.

Demir, E. and B.J. Dickson. 2005. *fruitless* splicing specifies male courtship behavior in *Drosophila*. *Cell* 121: 785–794.
Dhadialla, T.S. and A.S. Raikhel. 1990. Biosynthesis of mosquito vitellogenin. *J. Biol. Chem.* 265: 9924–9933.
Dodson, G. 1978. Morphology of the reproductive system in *Anastrepha suspensa* (Loew) and notes on related species. *Fla. Entomol.* 61: 231–239.
Edward, D.A. and T. Chapman. 2011. The evolution and significance of male mate choice. *TREE* 26: 647–654.
Edward, D.A. and T. Chapman. 2013. Variation in male mate choice in *Drosophila melanogaster*. *PLoS One* 8(2): e56299.
Engelmann, F. 1983. Vitellogenesis controlled by juvenile hormone, in R.G.H. Downer and H. Laufer (eds.), *Endocrinology of Insects*. Alan R. Liss, New York, pp. 259–270.
Fargnoli, J. and G.L. Waring. 1982. Identification of vitelline membrane proteins in *Drosophila melanogaster*. *Dev. Biol.* 92: 306–314.
Fedorka, K.M., W.E. Winterhalter, and B. Ware. 2011. Perceived sperm competition intensity influences seminal fluid protein production prior to courtship and mating. *Evolution* 65: 584–590.
Fescemyer, H.W., E.P. Masler, R.E. Davis, and T.J. Kelly. 1992. Vitellogenin synthesis in female larvae of the gypsy moth, *Lymantria dispar* (L.): Suppression by juvenile hormone. *Comp. Biochem. Physiol.* 103B: 533–542.
Fisher, R.C. 1971. Aspects of the physiology of endoparasitic Hymenoptera. *Biol. Rev.* 46: 243–278.
Flanders, S.E. 1942. Oosorption and ovulation in relation to oviposition in the parasitic Hymenoptera. *Ann. Entomol. Soc. Am.* 35: 251–266.
Fletcher, D.J.C. and M.S. Blum. 1981. Pheromonal control of dealation and oogenesis in virgin queen fire ants. *Science* 212: 73–75.
French, A. and R. Hoopingarner. 1965. Gametogenesis in the housefly, *Musca domestica*. *Ann. Entomol. Soc. Am.* 58: 650–657.
Garbaczewska, M., J.-C. Billeter, and J.D. Levine. 2013. *Drosophila melanogaster* males increase the number of sperm in their ejaculate when perceiving rival males. *J. Insect Physiol.* 59: 306–310.
Garvey, L.K., G.M. Gutierrez, and H.M. Krider. 2000. Ultrastructure and morphogenesis of apyrene and eupyrene spermatozoa in the gypsy moth (Lepidoptera: Lymantriidae). *Ann. Entomol. Soc. Am.* 93: 1147–1155.
Gergen, J.P. 1987. Dosage compensation in *Drosophila*: Evidence that *daughterless* and *Sex-lethal* control X chromosome activity at the blastoderm stage of embryogenesis. *Genetics* 117: 177–185.
Gillott, C. and T. Friedel. 1977. Fecundity-enhancing and receptivity-inhibiting substances produced by male insects: A review. *Adv. Invertebr. Reprod.* 1: 199–218.
Glinka, A.V. and G.R. Wyatt. 1996. Juvenile hormone activation of gene transcription in locust fat body. *Insect Biochem. Mol. Biol.* 26: 13–18.
Gold, S.M.W. and K.G. Davey. 1989. The effect of juvenile hormone on protein synthesis in the transparent accessory gland of male *Rhodnius prolixus*. *Insect Biochem.* 19: 139–143.
Grimnes, K.A., C.S. Bricker, and G.M Happ. 1986. Ordered flow of secretion from accessory glands to specific layers of a spermatophore of mealworm beetles: Demonstration with a monoclonal antibody. *J. Exp. Zool.* 240: 275–286.
Hagedorn, H. 1985. The role of ecdysteroids in reproduction, in G.A. Kerkut and L.I. Gilbert (eds.), *Comprehensive Insect Physiology, Biochemistry and Pharmacology*, vol. 8. Pergamon Press, Oxford, U.K., pp. 205–261.
Hagedorn, H. 1989. Physiological roles of hemolymph ecdysteroids in the adult insect, in J. Koolman (ed.), *Ecdysone, from Chemistry to Mode of Action*. Thieme, Stuttgart, Germany, pp. 279–289.
Hagedorn, H.H., J.D. O'Connor, M.S. Fuchs, B. Sage, D.A. Schlaeger, and M.K. Bohm. 1975. The ovary as a source of an ecdysone in an adult mosquito. *Proc. Natl. Acad. Sci. USA* 72: 3255–3259.
Hagedorn, H.H., J.P. Shapiro, and K. Hanaoka. 1979. Ovarian ecdysone secretion is controlled by a brain hormone in an adult mosquito. *Nature* 282: 92–94.
Hammond, M.P. and C.D. Laird. 1985. Chromosome structure and DNA replication in nurse and follicle cells of *Drosophila melanogaster*. *Chromosoma* 91: 267–278.
Hamon, C. and G. Chauvin. 1992. Ultrastructural analysis of spermatozoa of *Korscheltellus lupulinus* L. (Lepidoptera: Hepialidae) and *Micropterix calthella* L. (Lepidoptera: Micropterigidae). *Int. J. Insect Morphol. Embryol.* 21: 149–160.
Handler, A.M. and P. Shirk. 1988. Identification and analysis of the major yolk polypeptides from the Caribbean fruit fly, *Anastrepha suspensa* (Loew). *Arch. Insect Biochem. Physiol.* 9: 91–106.

Happ, G.M. 1987. Accessory gland development in mealworm beetles, in J.L. Law (ed.), *Molecular Entomology*, vol. 49. Alan R. Liss, New York, pp. 433–442.

Happ, G.M. 1992. Maturation of the male reproductive system and its endocrine regulation. *Annu. Rev. Entomol.* 37: 303–320.

Hazelrigg, T. 1987. The *Drosophila white* gene: A molecular update. *Trends Genet.* 3: 43–47.

He, Y., T. Tanaka, and T. Miyata. 1995. Eupyrene and apyrene sprerm and their numerical fluctuations inside the female reproductive tract of the army worm *Pseudaletia separata*. *J. Insect Physiol.* 41: 689–694.

Hodgkin, J. 1990. Sex determination compared in *Drosophila* and *Caenorhabditis*. *Nature (Lond.)* 244: 721–728.

Huybrechts, R. and A. DeLoof. 1982. Similarities in vitellogenin and control of vitellogenin synthesis within the genera *Sarcophaga*, *Calliphora*, *Phormia* and *Lucilia* (Diptera). *Comp. Biochem. Physiol.* 72B: 339–344.

Ilenchuk, T.T. and K.G. Davey. 1987. Effects of various compounds on Na/K-ATPase activity, JH I binding capacity and patency response in follicles of *Rhodnius prolixus*. *Insect Biochem.* 17: 1085–1088.

Imboden, H., R. König, P. Ott, A. Lustig, U. Kämpfer, and B. Lanzrein. 1987. Characterization of the native vitellogenin and vitellin of the cockroach, *Nauphoeta cineraea*, and comparison with other species. *Insect Biochem.* 17: 353–365.

Irie, K. and O. Yamashita. 1983. Egg-specific protein in the silkworm, *Bombyx mori*: Purification, properties, localization and titre changes during oogenesis and embryogenesis. *Insect Biochem.* 13: 71–80.

Izumi, S., J. Fujie, S. Yamada, and S. Tomino. 1981. Molecular properties and biosynthesis of major plasma proteins in *Bombyx mori*. *Biochim. Biophys. Acta* 670: 222–229.

Jones, T. 1978. The blood/germ cell barrier in male *Schistocerca gregaria*: The time of its establishment and factors affecting its formation. *J. Cell. Sci.* 31: 145–163.

Kafatos, F.C., W. Orr, and C. Delidakis. 1985. Developmentally regulated gene amplification. *Trends Genet.* 1: 301–306.

Kafatos, F.C., N. Spoerel, S.A. Mitsialis, H.T. Nguyen, C. Romano, J.R. Lingappa, B.D. Mariani, G.C. Rodakis, R. Lecanidou, and S.G. Tsitilou. 1987. Developmental control and evolution in the chorion gene families of insects. *Adv. Genet.* 24: 223–242.

Kasuga, H., R. Aigaki, and M. Osanai. 1987. System for supply of free arginine in the spermatophore of *Bombyx mori*. Arginine-liberating activities of contents of male reproductive glands. *Insect Biochem.* 17: 317–322.

Keely, L.L. 1985. Physiology and biochemistry of the fat body, in G.A. Kerkut and L.I. Gilbert (eds.), *Comprehensive Insect Physiology, Biochemistry and Pharmacology*, vol. 3. Pergamon Press, Oxford, U.K., pp. 211–248.

Kelly, T.J., M.J. Birnbaum, C.W. Woods, and A.B. Borkovec. 1984. Effects of house fly oostatic hormone on egg development neurosecretory hormone action in *Aedes atropalpus*. *J. Exp. Zool.* 229: 491–496.

Kelly, T.J., E.P. Masler, M.B. Schwartz, and S.B. Haught. 1986. Inhibitory effects of oostatic hormone on ovarian maturation and ecdysteroid production in Diptera. *Insect Biochem.* 16: 273–279.

Khan, I. and N.G. Prasad. 2013. Male *Drosophila melanogaster* show adaptive mating bias in response to female infection status. *J. Insect Physiol.* 59: 1017–1023.

Kirchgessner, T.G., J.-C. Chuat, C. Heinzmann, J. Etienne, S. Guilhot, K. Svenson, D. Ameis et al. 1989. Organization of the human lipoprotein lipase gene and evolution of the lipase gene family. *Proc. Natl. Acad. Sci. USA* 86: 9647–9651.

Klowden, M.J. 1987. Distention-mediated egg maturation in the mosquito, *Aedes aegypti*. *J. Insect Physiol.* 33: 83–87.

Klowden, M.J. 1997. Endocrine aspects of mosquito reproduction. *Arch. Insect Biochem. Physiol.* 35: 491–512.

Koller, C.N., T.S. Dhadialla, and A.S. Raikhel. 1989. A study of receptor-mediated endocytosis of vitellogenin in mosquito oocytes. *Insect Biochem.* 19: 693–702.

Kunkel, J.G. and J.H. Nordin. 1985. Yolk proteins, in G.A. Kerkut and L.I. Gilbert (eds.), *Comprehensive Insect Physiology, Biochemistry and Pharmacology*, vol. 1. Pergamon Press, Oxford, U.K., pp. 84–111.

Lauge, G. 1985. Sex determination: Genetic and epigenetic factors, in G.A. Kerkut and L.I. Gilbert (eds.), *Comprehensive Insect Physiology, Biochemistry and Pharmacology*, vol. 1. Pergamon Press, Oxford, U.K., pp. 295–318.

Law, J.H. 1989. Egg proteins other than vitellin, in *Fifth International Congress of Invertebrate Reproduction*, Nagoya, Japan. (Abstract, p. 45.)

Lecanidou, R., G.C. Rodakis, T.H. Eickbush, and F.C. Kafatos. 1986. Evolution of the silk moth chorion gene superfamily: Gene families CA and CB. *Proc. Natl. Acad. Sci. USA* 83: 6514–6518.

Lorenz, M.W. 2007. Oogenesis-flight syndrome in crickets: Age-dependent egg production, flight performance, and biochemical composition of the flight muscles in adult female *Gryllus bimaculatus*. *J. Insect Physiol.* 53: 819–832.

Mank, J.E. 2013. Sex chromosome dosage compensation: Definitely not for everyone. *Trends Genet.* 29: 677–683.

Mank, J.E., D.J. Hosken, and N. Wendell. 2011. Some inconvenient truths about sex chromosome dosage compensation and the potential role of sexual conflict. *Evolution* 65: 2133–2144.

Martín, D., M.-D. Piulachs, and X. Bellés. 1995. Patterns of haemolymph vitellogenin and ovarian vitellin in the German cockroach, and the role of juvenile hormone. *Physiol. Entomol.* 20: 59–65.

Martínez, A. and M. Bownes. 1994. The sequence and expression pattern of the *Calliphora erythrocephala* yolk protein A and B genes. *J. Mol. Evol.* 38: 336–351.

Nandy, B. and N.G. Prasad. 2011. Reproductive behavior and fitness components in male *Drosophila melanogaster* are nonlinearly affected by the number of male co-inhabitants early in adult life. *J. Insect Sci.* 11: 67.

Nöthiger, R. and M. Steinmann-Zwicky. 1985. A single principle for sex determination in insects. *Cold Spring Harb. Symp. Quant. Biol.* 50: 615–621.

Nöthiger, R. and M. Steinmann-Zwicky. 1987. Genetics of sex determination in eukaryotes, in W. Hennig (ed.), *Structure and Function of Eukaryotic Chromosomes, Results and Problems in Cell Differentiation*, vol. 14. Springer-Verlag, Berlin, Germany, pp. 271–300.

Osanai, M., T. Aigaki, and H. Kasuga. 1987. Arginine degradation cascade as an energy-yielding system for sperm maturation in the spermatophore of the silkworm, *Bombyx mori*, in H. Mohri (ed.), *New Horizons in Sperm Research*. Japanese Scientific Society Press, Tokyo, Japan, pp. 185–195.

Osanai, M., H. Kasuga, and T. Aigaki. 1989. Induction of motility of apyrene spermatozoa and dissociation of eupyrene sperm bundles of the silkworm, *Bombyx mori* by initiatorin and trypsin. *Invert. Reprod. Dev.* 15: 97–103.

Pan, M.-L., W.J. Bell, and W.H. Telfer. 1969. Vitellogenic blood protein synthesis by insect fat body. *Science* 165: 393–394.

Pan, M.-L. and G.R. Wyatt. 1971. Juvenile hormone induces vitellogenin synthesis in the monarch butterfly. *Science* 174: 503–505.

Papadopoulos, N.T., P. Liedo, H. Müller, J. Wang, and J.R. Carey. 2010. Cost of reproduction in male medflies: The primacy of sexual courting in extreme longevity reduction. *J. Insect Physiol.* 56: 283–287.

Park, Y.I., S. Shu, S.B. Ramaswamy, and A. Srinivasan. 1998. Mating in *Heliothis virescens*: Transfer of juvenile hormone during copulation by male to female and stimulation of biosynthesis of endogenous juvenile hormone. *Arch. Insect Biochem. Physiol.* 38: 100–107.

Parker, G. and T. Pizzari. 2010. Sperm competition and ejaculate economics. *Biol. Rev.* 85: 897–934.

Raikhel, A.S. 1984. The accumulative pathway of vitellogenin in the mosquito oocyte; a high resolution immuno- and cytochemical study. *J. Ultrastruct. Res.* 87: 285–302.

Raikhel, A.S. 1987. Monoclonal antibodies as probes for processing of yolk protein in the mosquito; a high-resolution immunolocalization of secretory and accumulative pathways. *Tissue Cell* 19: 515–529.

Raikhel, A.S. and T.S. Dhadialla. 1992. Accumulation of yolk proteins in insect oocytes. *Annu. Rev. Entomol.* 37: 217–251.

Ramaswamy, S.B., S. Shu, G.N. Mbata, A. Rachinsky, Y.I. Park, L. Crigler, S. Donald, and A. Srinivasan. 2000. Role of juvenile hormone-esterase in mating-stimulated egg development in the moth *Heliothis virescens*. *Insect Biochem. Mol. Biol.* 30: 785–791.

Ramaswamy, S.B., S. Shu, Y.I. Park, and F. Zeng. 1997. Dynamics of juvenile hormone-mediated gonadotropism in the Lepidoptera. *Arch. Insect Biochem. Physiol.* 35: 539–558.

Rankin, S.M., J. Chambers, and J.P. Edwards. 1997. Juvenile hormone in earwigs: Roles in oogenesis, mating, and maternal behaviors. *Arch. Insect Biochem. Physiol.* 35: 427–442.

Rina, M. and C. Savakis. 1991. A cluster of vitellogenin genes in the Mediterranean fruit fly, *Ceratitis capitata*: Sequence and structural conservation in dipteran yolk proteins and their genes. *Genetics* 127: 769–780.

Robbins, W.E. and T.J. Shortino. 1962. Effect of cholesterol in the larval diet on ovarian development in the adult housefly. *Nature* 194: 502–503.

Röhrkasten, A. and H.-J. Ferenz. 1987. Coated vesicles from locust oocytes: Isolation and characterization. *Int. J. Invertebr. Reprod. Dev.* 12: 341–346.

Romans, P., Z. Tu, Z. Ke, and H. Hagedorn. 1995. Analysis of a vitellogenin gene of the mosquito, *Aedes aegypti*, and comparisons to vitellogenins from other organisms. *Insect Biochem. Mol. Biol.* 25: 939–958.

Rose, U., G. Seebohm, and R. Hustert. 2000. The role of internal pressure and muscle activation during locust oviposition. *J. Insect Physiol.* 46: 69–80.

Rousset, A. and C. Bitsch. 1993. Comparison between endogenous and exogenous yolk proteins along an ovarian cycle in the firebrat *Thermobia domestica* (Insects, Thysanura). *Comp. Biochem. Physiol.* 104B: 33–44.

Rubenstein, E.C. 1979. The role of an epithelial occlusion zone in the termination of vitellogenesis in *Hyalophora cecropia* ovarian follicles. *Dev. Biol.* 71: 115–127.

Sanchez, L. and R. Nöthiger. 1983. Sex determination and dosage compensation in *Drosophila melanogaster*: Production of male clones in XX females. *EMBO J.* 2: 485–491.

Sato, Y., S. Inagaki, and O. Yamashita. 1990. Egg-specific protein in the silkworm, *Bombyx mori*: Gene structure, expression and post-translational modification, in M. Hoslin and O. Yamashita (eds.), *Advances in Invertebrate Reproduction*. Elsevier, Amsterdam, the Netherlands, pp. 91–95.

Satyanarayana, K., J.Y. Bradfield, G. Bhaskaran, and K.H. Dahm. 1994. Stimulation of vitellogenin production by methoprene in prepupae and pupae of *Manduca sexta*. *Arch. Insect Biochem. Physiol.* 25: 21–37.

Schäfer, M. and W. Kunz. 1987. Ribosomal gene amplification does not occur in the oocytes of *Locusta migratoria*. *Dev. Biol.* 120: 43–52.

Schal, C., G.L. Holbrook, J.A.S. Bachmann, and V.L. Sevala. 1997. Reproductive biology of the German cockroach, *Blattella germanica*: Juvenile hormone as a pleiotropic master regulator. *Arch. Insect Biochem. Physiol.* 35: 405–426.

Schlottman, L. and P. Bonhag. 1956. Histology of the ovary of the adult mealworm *Tenebrio molitor* L. (Coleoptera, Tenebrionidae). *Univ. Calif. Publ. Entomol. Berkeley* 11: 351–394.

Shea, M.J., D.L. King, M.J. Conboy, B.D. Mariani, and F.C. Kafatos. 1990. Proteins that bind to *Drosophila* chorion *cis*-regulatory elements: A new C_2H_2 zinc finger protein and a C_2C_2 steroid receptor-like component. *Genes Dev.* 4: 1128–1140.

Shinbo, H., T. Yaginuma, and G.M. Happ. 1987. Purification and characterization of a proline-rich secretory protein on an insect spermatophore. *J. Biol. Chem.* 262: 4794–4799.

Shirk, P.D. 1987. Comparison of yolk production in seven pyralid moth species. *Int. J. Invertebr. Reprod. Dev.* 11: 173–188.

Shirk, P.D., P. Minoo, and J.H. Postlethwait. 1983. 20-Hydroxyecdysone stimulates the accumulation of translatable yolk polypeptide gene transcript in adult male *Drosophila melanogaster*. *Proc. Natl. Acad. Sci. USA* 80: 186–190.

Smith, G., Y.-R. Chen, G.W. Blissard, and A.D. Briscoe. 2014. Complete dosage compensation and sex-biased gene expression in the moth *Manduca sexta*. *Genome Biol. Evol.* 6: 536–537.

Sonnenschein, M. and Ch.L. Hauser. 1990. Presence of only eupyrene spermatozoa in adult males of the genus *Microtpterix* Hubner and its phylogenetic significance (Lepidoptera: Zeugloptera, Micropterigidae). *Int. J. Insect Morphol. Embryol.* 19: 269–276.

Sorge, D., R. Nauen, S. Range, and K.H. Hoffmann. 2000. Regulation of vitellogenesis in the fall armyworm, *Spodoptera frugiperda* (Lepidoptera: Noctuidae). *J. Insect Physiol.* 46: 969–976.

Speith, J., K. Denison, S. Kirtland, J. Cand, and T. Blumenthal. 1985. The *C. elegans* vitellogenin genes: Short sequence repeats in the promoter regions and homology to the vertebrate genes. *Nucleic Acids Res.* 13: 5283–5295.

Spielman, A. 1971. Bionomics of autogenous mosquitoes. *Annu. Rev. Entomol.* 16: 231–248.

Spradling, A.C. and A.P. Mahowald. 1980. Amplification of genes for chorion proteins during oogenesis in *Drosophila melanogaster*. *Proc. Natl. Acad. Sci. USA* 77: 1096–1100.

Stoffolano, J.G., Jr., M.-F. Li, B.-X. Zou, and C.-M. Yin. 1992. Vitellogenin uptake, not synthesis, is dependent on juvenile hormone in adults of *Phormia regina* (Meigen). *J. Insect Physiol.* 11: 839–845.

Strambi, A., C. Strambi, and M. Cayre. 1997. Hormonal control of reproduction and reproductive behavior in crickets. *Arch. Insect Biochem. Physiol.* 35: 393–404.

Stumm-Zollinger, E. and P.S. Chen. 1985. Protein metabolism of *Drosophila melanogaster* male accessory glands. 1. Characterization of secretory proteins. *Insect Biochem.* 15: 375–383.

Stynen, D., R.I. Woodruff, and W.H. Telfer. 1988. Effect of ionophores on vitellogenin uptake by *Hyalophora* oocytes. *Arch. Insect Biochem. Physiol.* 8: 261–276.

Telfer, W.H. 1954. Immunological studies of insect metamorphosis. II. The role of a sex-limited blood protein in egg formation by the cecropia silkworm. *J. Gen. Physiol.* 37: 539–558.
Telfer, W.H. 1961. The route of entry and localization of blood proteins in the oocytes of saturniid moths. *J. Biophys. Biochem. Cytol.* 9: 747–759.
Telfer, W.H. 1965. The mechanism and control of yolk formation. *Annu. Rev. Entomol.* 10: 161–184.
Telfer, W.H. and M.-L. Pan. 1988. Adsorptive endocytosis of vitellogenin, lipophorin, and microvitellogenin during yolk formation in *Hyalophora*. *Arch. Insect Biochem. Physiol.* 9: 339–355.
Telfer, W.H., E. Rubinstein, and M.-L. Pan. 1981. How the ovary makes yolk in *Hyalophora*, in F. Sehnal, A. Zabza, J. Menn, and B. Cymborowski (eds.), *Regulation of Insect Development and Behavior*. Wroclaw Technical University Press, Wroclaw, Poland, pp. 637–654.
Terpstra, P. and A.B. Geert. 1988. Homology of *Drosophila* yolk proteins and the triacylglycerol lipase family. *J. Mol. Biol.* 202: 663–665.
Tobback, J., B. Boerjan, H.P. Vandersmissen, and R. Huybrechts. 2012. Male reproduction is affected by RNA interference of *period* and *timeless* in the desert locust *Schistocerca gregaria*. *Insect Biochem. Mol Biol.* 42: 109–115.
Torres, M. and Sanchez, L. 1989. The *scute* (T4) gene acts as a numerator element of the X: A signal that determines the state of activity of *Sex-lethal* in *Drosophila*. *EMBO J.* 8: 3079–3086.
Tsuchida, K., M. Nagata, and A. Suzuki. 1987. Hormonal control of ovarian development in the silkworm, *Bombyx mori*. *Arch. Insect Biochem. Physiol.* 5: 167–177.
Tufail, M. and M. Takeda. 2008. Molecular characteristics of insect vitellogenins. *J. Insect Physiol.* 54: 1447–1458.
von Arx, M., K.S. Sullivan, and R.A. Raguso. 2013. Dual fitness benefits of post-mating sugar meals for female hawkmoths (*Hyles lineata*). *J. Insect Physiol.* 59: 458–465.
Wagner, R.M., M.J. Loeb, J.P. Kochansky, D.B. Gelman, W.R. Lusby, and R.A. Bell. 1997. Identification and characterization of an ecdysiotropic peptide from brain extracts of the gypsy moth, *Lymantria dispar*. *Arch. Insect Biochem. Physiol.* 34: 175–189.
Wahli, W. 1988. Evolution and expression of vitellogenin genes. *Trends Genet.* 4: 227–232.
Waring, G.L. and A.P. Mahowald. 1979. Identification and time of synthesis of chorion proteins in *Drosophila melanogaster*. *Cell* 16: 599–607.
Wegener, B.J., D.M. Stuart-Fox, M.D. Norman, and B.B.M. Wong. 2013. Strategic male mate choice minimizes ejaculate consumption. *Behav. Ecol.* 24: 668–671.
Wheeler, D. 1996. The role of nourishment in oogenesis. *Annu. Rev. Entomol.* 41: 407–431.
Whisenton, L.R., J.T. Warren, M.K. Manning, and W.E. Bollenbacher. 1989. Ecdysteroid titers during pupal-adult development of *Aedes aegypti*: Basis for a sexual dimorphism in the rate of development. *J. Insect Physiol.* 35: 67–73.
Whiting, P., S. Sparks, and L. Dinan. 1997. Endogenous ecdysteroid levels and rates of ecdysone acylation by intact ovaries in vitro in relation to ovarian development in adult female house crickets, *Acheta domesticus*. *Arch. Insect Biochem. Physiol.* 35: 279–299.
Wick, J.R. and P.F. Bonhag. 1955. Postembryonic development of the ovaries of *Oncopeltus fasciatus* (Dallas). *J. Morphol.* 96: 31–60.
Wong, R. and A.B. Lange. 2014. Octopomine modulates a central pattern generator associated with egg-laying in the locust, *Locusta migratoria*. *J. Insect Physiol.* 63: 1–8.
Wu, S.-J., J.-Z. Zhang, and M. Ma. 1987. Monitoring the effects of juvenile hormones and 20-hydroxyecdysone on yolk polypeptide production of *Drosophila melanogaster* with enzyme immunoassay. *Physiol. Entomol.* 12: 355–361.
Wyatt, G.R. 1991. Gene regulation in insect reproduction. *Invertebr. Reprod. Dev.* 20: 1–35.
Wyatt, G.R., K.E. Cook, H. Firko, and T.S. Dhadialla. 1987. Juvenile hormone action on locust fat body. *Insect Biochem.* 17: 1071–1074.
Yaginuma, T., H. Kai, and G.M. Happ. 1988. 20-Hydroxyecdysone accelerates the flow of cells into the G_1 phase and the S phase in a male accessory gland of a mealworm pupa. *Dev. Biol.* 126: 173–181.
Yano, K., M.T. Sakurai, S. Izumi, and S. Tomino. 1994a. Vitellogenin gene of the silkworm, *Bombyx mori*: Structure and sex-dependent expression. *FEBS Lett.* 356: 207–211.
Yano, K., M.T. Sakurai, S. Watabe, S. Izumi, and S. Tomino. 1994b. Structure and expression of mRNA for vitellogenin in *Bombyx mori*. *Biochimica Biophysica Acta* 1218: 1–10.

Yin, C-M. and J.G. Stoffolano, Jr. 1997. Juvenile hormone regulation of reproduction in the cyclorrhaphous Diptera with emphasis on oogenesis. *Arch. Insect Biochem. Physiol.* 35: 513–537.

Yin, C-M., B.X. Zou, M.G. Jiang, M.F. Li, W.H. Qin, T.L. Potter, and J.G. Stoffolano. 1995. Identification of juvenile hormone III bisepoxide (JHB_3), juvenile hormone III and methyl farnesoate secreted by the corpus allatum of *Phormia regina* (Meigen), in vitro and function of JHB_3 either applied alone or as a part of a juvenoid blend. *J. Insect Physiol.* 41: 473–479.

Yin, C.-M., B.-X. Zou, and J.G. Stoffolano, Jr. 1989. Precocene II treatment inhibits terminal oöcyte development but not vitellogenin synthesis and release in the black blowfly, *Phormia regina* (Meigen). *J. Insect Physiol.* 36: 375–382.

Yin, C.-M., B.-X. Zou, S.-X. Yi, and J.G. Stoffolano, Jr. 1990. Ecdysteroid activity during oögenesis in the black blowfly, *Phormia regina* (Meigen*). J. Insect Physiol.* 36: 375–382.

Zeng, E., S. Shu, and S.B. Ramaswamy. 1997. Vitellogenin and egg production in the moth, *Heliothis virescens*. *Arch. Biochem. Physiol.* 34: 287–300.

Zhang, J.-Z. and G.A. Wyatt. 1990. A new member of a low molecular weight hemolymph protein family from *Locusta migratoria*(abstract), in H.H. Hagedorn, J.C. Hildebrand, M.G. Kidwell, and J.H. Law (eds.), *Molecular Insect Science*. Plenum Press, New York, pp. 385.

CHAPTER 21

Insect Symbioses

PREVIEW

Symbiosis is the living together of any two different species. All insects, indeed all organisms, have microorganisms within and on their surface, and even a few microorganisms have their own associated microorganisms. An animal (or plant) and all the microbes associated with it comprise its holobiont. Thus, a study of symbiosis could include any organism, any insect, or any human. In this chapter, the emphasis is on symbiotic relationships with selected groups of insects, namely, leaf-cutter ants, termites, bark beetles, aphids, and tsetse flies, and the symbiotic relationship of one of the most ubiquitous microorganisms common in many insects, *Wolbachia*. Several hundred species of ants culture fungi growing in their nests, and some species of leaf-cutter ants have been fungus farming for millions of years. Moreover, the fungus-culturing ants may be associated with beneficial and harmful bacteria and sometimes nematodes. Some termites and over 3000 species of bark and ambrosia beetles also culture fungi growing in their nests and galleries. The fungal–insect associations are often mutualistic; the insects derive nutrients from the fungi, and the fungi have a protected place to grow and often get distributed to new locations with the movement of their insect hosts. Termites that do not depend directly upon fungi for nutrients have vast numbers of bacteria or flagellated protists or both in their body and/or alimentary canal that digest the wood that is the primary food of many termites. Glucose and related compounds from microorganismal action on the wood particles is then available to the termites. Aphids may have a variety of facultative symbionts, but all except two families of aphids have an obligate symbiont, *Buchnera*, again a mutualistic association. *Buchnera* synthesizes some amino acids and perhaps some vitamins, nutrients that tend to be in low supply in the plant phloem sap that aphids consume, and the aphids pass *Buchnera* on to their offspring. The symbiotic relationship is ancient. Tsetse flies can transmit the trypanosome that causes sleeping sickness in humans and Nagana in cattle in large areas in Africa. The flies have lived with an obligate symbiont, *Wigglesworthia glossinidia*, for more than 50 million years, and may contain several other facultative symbionts in addition to the disease-causing trypanosomes. *Wolbachia* are generally facultative intracellular bacteria that have been estimated to infect upward of 40% of all insects. *Wolbachia* often manipulate the behavior, physiology, and reproduction of their host. The relatively recent outbreak of bedbug problems has led to the discovery that *Wolbachia* may be in a mutualistic, obligate relationship with bedbugs. In face of increasing concern about emerging pathogens in agriculture and human health, the delineation of symbionic relationships in all organism is receiving attention, and understanding insect symbiotic relationships is important to conserving beneficial insects and potentially in controlling economically important pest such as mosquitoes and other agricultural pests.

21.1 INTRODUCTION

Symbiosis, a term invented by Albert Frank in 1877 to apply to nonparasitic interactions involving microorganisms, was broadened later to include any two different species living together (Six, 2012). Probably, every multicellular creature has microorganisms in it, on it, and around it. Gilbert et al. (2012) noted (paraphrased here) that plants and animals have never been just individuals, but all evolved with a host of associated microorganisms. An animal (or plant) and all the microbes associated with it comprise its **holobiont** (Margulis and Fester, 1991; Six, 2013). Symbiotic relationships span the range of mutualism, commensalism, antagonism, or amensalism (amensalism is advantageous to one organism but the other is unaffected). Microorganisms often play important, and unsuspected, roles in insect–insect and insect–plant interactions (Clark et al., 2010; Douglas, 2013). For example, *Drosophila melanogaster* raised in a microbe-free environment were slower in developing, suffered metabolic abnormalities, and had reduced immunity reactions (Ryu et al., 2008; Storelli et al., 2011; Ridley et al., 2012). Moreover, the insulin-like peptide genes that regulate hemolymph glucose levels were unable to control glucose, allowing it to rise to abnormal levels; the situation could be corrected by adding acetic acid to the diet, suggesting that acetic acid from bacterial metabolism influences insulin-like peptide action (Shin et al., 2011). Research is clarifying that microorganisms contribute to health and ecology of most multicellular organisms (Duron and Hurst, 2013; McFall-Ngai et al., 2013). Some of the microorganisms are harmless, some helpful, some toxic, and even deadly to the host and/or to other organisms to which the microbes are sometimes transferred (e.g., the malaria parasite transferred to humans by the bite of a mosquito).

Symbionts may be external symbionts or internal as **endosymbionts**. Symbionts may be **obligate (primary) symbionts** in which the host and symbiont are necessary partners, and the symbiont typically is transmitted vertically from the mother to the offspring. Contrastingly, symbionts may be **facultative (secondary)**, present in only a portion of the host population, and often living in the gut. They may be transmitted horizontally from individual to individual, or vertically from mother to offspring (Oliver et al., 2010). Some primary symbionts have become entirely dependent upon the host because they have undergone great reduction in their genome and it cannot be independent (McCutcheon and Moran, 2012), and some insects survive only poorly, or not at all, without their primary symbiont(s).

Symbionts may interact with each other in relationship to the host. For example, the secondary symbiont *Rickettsia insecticola* in pea aphids offers protection to the aphid against the entomopathogenic fungus *Pandora neoaphidis* (Farrai et al., 2012). Aphids infected with the secondary symbiont *Serratia symbiotica* keep more bacteriocytes and primary symbiont cells when exposed to high temperatures than when *S. symbiotica* is not present (Wernegreen, 2012).

Provision of nutrients is a benefit that some hosts receive from their symbionts. For example, *Buchnera* provides aphids with certain essential amino acids, and *Wigglesworthia* makes the vitamin B1, thiamine, available to tsetse flies (Snyder et al., 2010). Ants in the subfamily Myrmicinae, tribe Attini, many bark beetles, and some termites practice "agriculture" (Mueller et al., 2005; Schultz and Braddy, 2008) because they have evolved an obligate symbiosis with fungi that they culture in their underground nests or, in the case of bark beetles, in their tunnels in the bark or wood of trees. These three groups of insects transfer their symbionts to new colonies and nests; thus, the symbionts gain by being propagated (most of them cannot live outside their host), and the host insect gains food and vital nutrients such as essential amino acids, vitamins, and sterols (the latter of which insects cannot synthesize). Microorganisms provide essential amino acids and better balance of amino acids to periodical cicadas feeding on plant phloem sap (Christensen and Fogel, 2011), provide sterols (which insects cannot synthesize) (Thompson et al., 2013), and may provide enzymes important in host metabolism (De Fine Licht Biedermann, 2012). Still another host benefit is aid in detoxification of plant toxins and insecticides (Kikuchi et al., 2012; Koch et al., 2012; Priya et al., 2012).

Intracellular symbionts are often housed in special host cells called **mycetocytes** (if the symbionts are fungi) or **bacteriocytes** (if they are bacteria) (Buchner, 1965; Ishikawa, 1989); however, "mycetocyte" is often used regardless of the nature of the symbiont (Ishikawa, 2003).

Mycetocytes are often aggregated into **mycetomes** and bacteriocytes into **bacteriosomes** (Buchner, 1965; Douglas, 1989). The mycetomes may be located intracellularly within fat body cells, and both intracellularly and extracellular in other locations in some insects (e.g., in the midgut of stinkbugs [Dong et al., 2011] or yeasts in the fat body of some planthoppers [Nikoh et al., 2011]). Intracellular symbionts in mycetomes are typically enveloped in two membranes of their own and then surrounded by a membrane of the host cell. Many intracellular bacteria have become dependent upon their insect host because they have lost many of the genes of free-living bacteria; their genome is often small (<1 Mb) (Shigenobu et al., 2000; McCutcheon et al., 2009; McCutcheon and Moran, 2010; Nikoh et al., 2011; Sloan and Moran, 2012).

Numerous books on symbioses in insects are available (Bucher, 1965; Krishna and Weesner, 1970; Anderson et al., 1984; Smith and Douglas, 1987; O'Neill et al., 1997; Bourtzis and Miller, 2003, 2006, 2009; Douglas, 2010). My intention in this chapter is to describe selected insect symbioses. Even in the limited cases I have chosen, the literature is voluminous. I apologize to those authors who have special interests in a symbiotic relationship, which is not included here. This chapter will include symbiosis in leaf-cutting ants, termites, bark beetles, aphids, tsetse flies, and the widespread symbiont *Wolbachia*.

21.2 SYMBIOSES AMONG LEAF-CUTTING ANTS, FUNGI, AND BACTERIA

Ants are one of the most abundant animal groups, making up to 80% of the arthropods in tropical rain forests (Hulldobler and Wilson, 1990; Kautz et al., 2013). There are more than 230 species of **fungus-culturing ants** in 12 genera, mostly in tropical rain forests. **Leaf-cutting ants** in the genera *Atta* and *Acromyrmex*, a monophyletic clade of neotropical ants, exhibit the largest and most complex fungus-farming colonies (Scott et al., 2010; Mueller, 2012; Kautz et al., 2013). The ants bring fresh leaves or parts of leaves into their subterranean nests and add them to **fungal gardens** that subsequently serve as their main food source (Figure 21.1). The youngest newly emerged workers tend the garden (Poulson et al., 2002). The fungi are propagated when the ants establish a new nest (Schultz and Brady, 2008). A queen on a mating flight carries a pellet of the fungal garden in her buccal cavity and subsequently establishes a new nest and garden (Weber, 1982; Fernández-Marin et al., 2004; Poulson and Boomsma, 2005). The fungi benefit by being protected within the ants' nest, by having a substrate on which to grow, and by constant infusion of new material to the cultures. The relationships are ancient, probably evolving about 50 MYA, with evolution to bringing fresh leaves into the nest probably 40 MYA (Schulz and Brady, 2008; Mikheyev et al., 2010; Cafaro et al., 2011).

Mature colonies of *Atta* ants may contain millions of workers in hundreds of interconnecting tunnels with **fungal chambers** where their food is grown (Hölldobler and Wilson, 1990). The ants are major foragers in their tropical environment; it has been estimated that colonies of *Atta colombica* in Panama harvested about 250 kg (dry weight basis) of leaves, fruit, nuts, and flowers per colony on an annual basis (Herz et al., 2007). The ants chew the fresh material brought in, mix it with ant feces, and inoculate it with fungal mycelia (Scott et al., 2010). Much of the cellulose in the fungal garden is digested by the fungi, and the smaller compounds such as starch and pectin that result are eaten by the ants; lignin is not appreciably digested (Scott et al., 2010; De Fine Licht, 2013).

Leaf-cutting ants in the genera *Atta* and *Acromyrmex* use fresh leaf fragments to add to their fungal gardens in their in-ground nests. The fungus that grows is the main food of the ant larvae; adult workers use the sugary sap from the leaves they cut. The fungus will die at high temperatures (>about 33°C) and at humidity below about 60% (Quinlan and Cherrett, 1978; Powell and Stradling, 1986), so the ants position the nests deep in the ground, have multiple gardens in their numerous

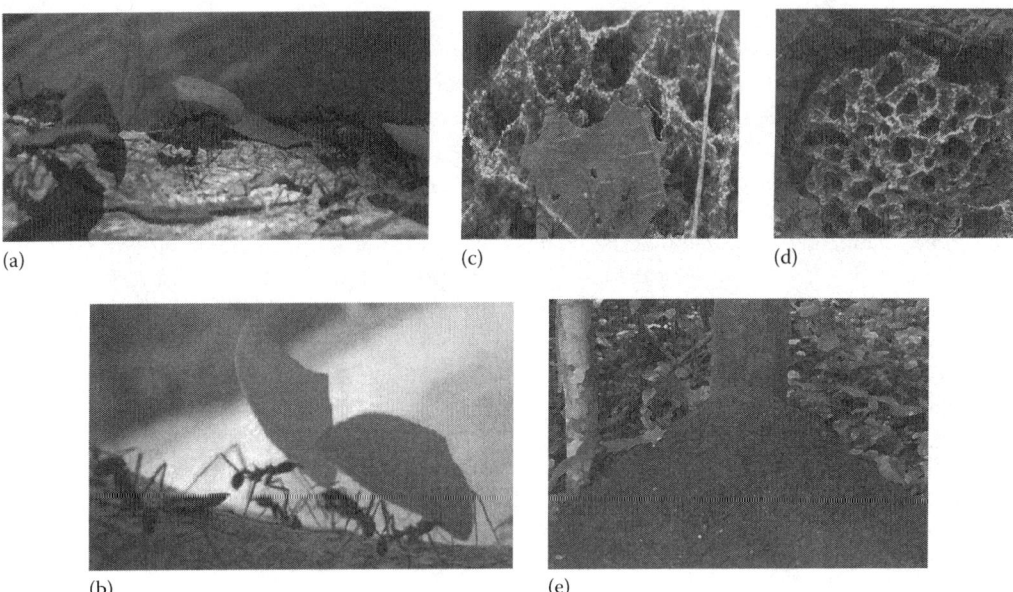

Figure 21.1 **(See color insert.)** Leaf-cutter ants, nest, and dump. (a and b) Fresh plant material being harvested by foraging workers returning to the nest. (c) An active nest comb with fungal tissue and leaf shown. (d) Older nest comb. (e) Older substrate and spent fungal materials are removed from the bottom of the garden and transported to a refuse dump. Older workers manipulate material on the dump presumably to facilitate degradation of the material. (a, d, e: From Scott, J.J. et al., *PLoS ONE* 5, e9922, 2010. With permission; b, c: From Aylward, F.O. et al., *ISME J.*, 6, 1688, 2012. With permission; all photos courtesy of J. Scott.)

tunnels, and relocate fungal gardens when microclimate conditions dictate (Bollazzi and Roces, 2002, 2010a,b; Bollazzi et al., 2008). The fungal gardens are subject to attack by various parasites, including nematodes and a parasitic fungus, genus ***Escovopsis***. The ants defend and protect their fungal cultures by fungal grooming, weeding (removal of invasive fungi), adding secretions from their metapleural glands, and keeping the nest area clean (Currie and Stuart, 2001; Poulsen and Currie, 2006; Yek and Mueller, 2011). Most ants have only a small range of fungal cultivars. The fungal systems grown by specific groups of ants have been described in anthropomorphic terms as various forms of agriculture (Schultz and Brady, 2008) as follows: (1) "lower agriculture" in which Palaeoattini and Neoattini ants (considered the most basal of the attine ants) grow fungi in the family Lepiotaceae, (2) "coral fungus agriculture" developed by some Palaeoattini ants by cultivating fungi from the family Pterulaceae, (3) "yeast culture" by the *Cyphomyrmex* genus of ants (Piattini), (4) "higher agriculture," and (5) "leaf-cutter agriculture" in which *Atta* and *Acromyrmex* ants only use fresh plant material to cultivate their fungi.

The fungal gardens of leaf-cutting ants are threatened by various foreign fungi and organisms. One of the best studied ones is the parasitic fungus *Escovopsis* sp., which, if it gets establish in the fungal garden, can destroy the garden and possibly cause death of the ants. Ants protect their fungal gardens by removing contaminating fungi and nematodes from the gardens and by cleaning the fresh leaf material brought into the nest before it is added to the fungal garden. They manipulate the leaf surface in their mouth and remove contaminating materials to their infrabuccal pocket (Little et al., 2006) (sometimes, the ants are said to "lick" the leaf surface and fungal garden). Worker ants routinely groom themselves and are groomed by other colony members (allogrooming) as another protective mechanism guarding their fungal gardens (Currie and Stuart, 2001; Morelos-Juárez et al., 2010), and the metapleural glands secrete 3-hydroxydecanoic acid as an antifungal compound (Bot et al., 2002; Fernández-Marin et al., 2006; Schoenian et al., 2011).

The fungal gardens, and the ants themselves, host a great abundance of bacteria. The active gardens as well as the voluminous refuse contain a diversity of biomass-degrading bacterial colonies, metabolic profiles of which correspond to the state of degradation of the plant matter (Scott et al., 2010). The exoskeleton of ants in particular is often covered by a **bacterial biofilm** dominated by **actinomycetes**, such as *Streptomyces* and *Pseudonocardia* (Currie et al., 2003). Some of the bacterial associates on the ant exoskeleton produce antifungal compounds (Poulsen and Currie, 2006), and some are also vertically transmitted to new nests, which results in biogeographical concordance between the ant population structure and specific bacterial strains (Caldera and Currie, 2012). The exact role of bacteria in the functioning of this symbiotic system is under intensive research. It remains to be clarified whether the presence of bacteria is an adaptive feature of the ants (Currie et al., 2003; Barke et al., 2010) or whether the diverse communities are highly specific commensals (Sen et al., 2009). The bacteria appear to be benefited by being carried on specialized morphological structures on the cuticle of most of the attine ants, and they may receive nourishment from specialized glandular structures that appear to secrete substances that support the growth of the bacteria (Poulsen and Currie, 2006).

A variety of antibiotics are produced by various microorganisms associated with leaf-cutter ants (Barke et al., 2010; Seipke et al., 2011). A biofilm of bacteria on the cuticle of worker ants, including *Pseudonocardia* sp. and *Streptomyces*, produces antibiotics that are active against the parasitic fungus *Escovopsis* (Cafaro et al., 2011; Schoenian et al., 2011). *Streptomyces* produced antimycin A1–A4, valinomycins, and actinomycins. Valinomycin was found on the cuticle of *Acromyrmex echinatior* workers, and along with actinomycins, in the waste deposits from *A. echinatior* and *A. niger* leaf-cutting ants. Although the suppression of *Escovopsis* and other invading fungi in bioassays suggests a protective function in the fungus garden, the actinomycins also tended to inhibit the growth of *Leucoagaricus gongylophorus*, the principal food fungus. Both *Escovopsis* and the extracuticular bacteria appear to have been present in leaf-cutting ant gardens for millions of years (Currie et al., 2003). Not all ant nests have the same strain of *Pseudonocardia* bacteria; groups of ants have become associated with different free-living strains of *Pseudonocardia* during evolution (Cafaro et al., 2011). The production of antifungal compounds may be a general feature of the actinomycetes on the ant body. For example, bacteria on the bodies of immature worker ants in *Acromyrmex subterraneus subterraneus* display activity against the entomopathogenic fungus *Metarhizium anisopliae* (Mattoso et al., 2012).

The ants remove waste from their fungal gardens and their living area by transporting such wastes to specific, multiple dump areas (Hart and Ratnieks, 2001; Haeder et al., 2009; Lacerda et al., 2010; Scott et al., 2010; Farji-Brener and Tadey, 2012) well isolated from their fungal gardens. Waste or dump areas (Figure 21.1) are established within the subterranean nests of some leaf-cutting ants, but well separated from the fungal gardens, while other species of ants establish aboveground dump areas (Moser, 1963; Hart and Ratnieks, 2002; Scott et al., 2010). Because the dumps are highly contaminated with organisms that could kill the fungal gardens and perhaps even the ants themselves, they employ a division of labor that minimizes contact with fungal gardens and dump areas by the same individuals (Bot et al., 2001; Hart and Ratnieks, 2001, 2002; Hart et al., 2002a,b; Fernández-Marin et al., 2003; Ballari et al., 2007).

21.3 BIOLOGY OF TERMITES

Termites have a highly developed social behavior, typically living in large colonies. Until recently, termites were treated as their own order Isoptera, but phylogenetic systematics places them unambiguously as a highly derived family within cockroaches (Blattodea: Termitidae). This classification helps to better understand the origin of termites and their evolutionary origin: their closest relatives are **social cockroaches** *Cryptocercus* that also live inside wood and digest cellulose (Inward et al., 2007; Brune, 2011). The family Termitidae comprises at least 2761 described species (Eggleton, 2011; Konig et al., 2013) in several hundred genera, but many species are yet to be described and named. The largest populations are in subtropical and tropical areas.

The whole diversity of termites is often informally divided into two groups: the lower and **higher termites**, based on evolution, phylogeny, gut symbionts, food habits, and other aspects of their biology and ecology. **Lower termites** (about 29% of species) have flagellated protists and bacteria in their gut, while the higher termites lack protists and have only bacteria in the gut (with a few exceptions, Rohrmann and Rossman, 1980). Lower termites primarily feed upon the lignocellulose in wood; some higher termites also feed upon wood, but many have diverse feeding habits including groups that feed upon fungus gardens, grass, animal manure, soil, and humus (Konig et al., 2013).

The capacity to digest **lignocellulose** represents a major ecological advantage. Ljungdahl and Eriksson (1985) estimated that plants synthesize 136×10^{15} g of dry material each year, with cellulose and lignin comprising the major components of the plant synthetic activity, typically designated as lignocellulose. Carbon is cycled through the environment by synthetic activities of plants and animals and subsequent degradation of biological material; a significant part of organic matter is also mineralized as deposits that then enter the much longer geological carbon cycle. Bacteria, fungi, and various invertebrates are major players in the breakdown and recycling of carbon from plant and animal synthetic activity.

21.3.1 Symbionts in Termites

The symbionts in both lower and higher termites are located within the alimentary canal (Figure 21.2), with the exception of *Mastotermes darwiniensis*, a lower termite with intracellular bacteria housed in mycetocytes within the fat body, as well as in the alimentary canal (Jucci, 1952). This species lives in northern Australia and is an important economic pest (also considered to be the most primitive of living termites). Although some bacteria live in the fore- and midgut, most of the bacteria and protozoa are located in the hindgut, mostly in enlarged portions called the paunch.

Symbionts in the lower termites include eubacteria, archaea, and eukaryotic flagellate protozoans and yeast (Konig et al., 2013). The most abundant bacterial symbionts of termites are classified as spirochetes. These spiral or undulate and mobile prokaryotes make up one of the largest groups of bacteria. Although spirochetes are found in many other types of invertebrates, the specific ones in termites are not common in other organisms. The role of spirochetes in the gut of termites is not well known, and whether they are beneficial to the termites or not, they at least do not seem to be harmful to the termites.

The flagellated protists (Figure 21.3) in the hindgut of lower termites have been most often studied, but few of them can be cultured outside the termite gut. They are of interest because they are uncommon in other organisms and they clearly participate in the digestion of lignocellulose, although it is now well established that termites themselves produce cellulases independent of their flagellated protists. Many of the protists in the lower termites incorporate bacteria into their own cytoplasm, and other bacteria are attached to the protist's outside cell surface. In both lower and higher termites, cellulose digestion occurs in the enlarged hindgut, and glucose released from cellulose digestion is converted by symbionts to **short-chain fatty acids**, a process called **acetogenesis** (particularly, acetate, propionate, and butyrate). These acetates are the buffered forms of the free acids (acetic, propionic, and butyric acids). The fatty acids (the acetates) are the major source of carbohydrate nutrients for the termites and are metabolized for energy (Odelson and Breznak, 1983) and carbon source for synthesis of new body tissues (Blomquist et al., 1979; Prestwich et al., 1981; Mauldin, 1982; Guo et al., 1991).

Figure 21.2 Lower and higher termites and a comparison of their gut structure. (a) *Coptotermes formosanus*, lower termites. (b) *Termes comis*, wood- and humus-feeding higher termites. (c) The alimentary canal of *C. formosanus*. (d) The alimentary canal of *T. comis*. Labels: F, foregut; M, midgut; P1, first paunch; P3, third paunch; P4, fourth paunch; and P5, fifth paunch. Bars = 1 mm. (From Hongoh, Y., *Cell. Mol. Life Sci.*, 68, 1311, 2011. With permission.)

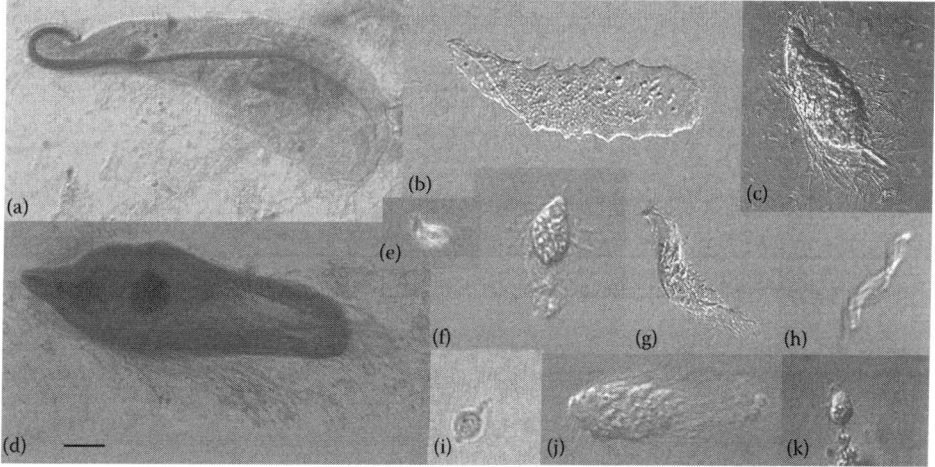

Figure 21.3 Protist species from the gut of the termite *Reticulitermes flavipes*. (a) *Pyrsonympha vertens*, (b) *P. major*, (c) *Spirotrichonympha flagellata*, (d) *Trichonympha agilis*, (e) *Trichomitus trypanoides*, (f) *Spironympha kofoidi*, (g) *Dinenympha fimbriata*, (h) *D. gracilis*, (i) *Monocercomonas* sp., (j) *Holomastigotes elongatum*, and (k) *Microjoenia fallax*. Scale bar = 10 μm (photographs taken with a Leica microscope at 400× magnification and images acquired with an AxioCam digital camera). (From Lewis, J.L. and Forschler, B.T., *Ann. Entomol. Soc. Am.*, 97, 1242, 2004. With permission.)

Carbon dioxide, hydrogen, and methane are end products of symbiont action on cellulose. Under anaerobic conditions in some parts of the termite hindgut, some hydrogen and carbon dioxide are fermented to methane (CH_4). Although only the protozoan *Trichomitopsis termopsidis* has been extensively studied in cultures (Odelson and Breznak, 1985a,b), it is likely that most of the flagellated protists participate in similar digestion (Brune and Stingl, 2005; Ohkuma, 2008). Reduction of CO_2 and H_2 to acetate was first described by Breznak and Switzer (1986) and has been confirmed by other researchers (Pester and Brune, 2007, and references therein). Some workers have suggested that the spirochete bacteria in the termite gut may have a role in acetogenesis (Pester and Brune, 2007).

The very small chewing mandibles of termites require them to take small bites of wood, and the action of the mandibles and the proventriculus at the junction of the fore- and midgut grind wood particles into extremely small fragments (20–100 μm diameter) (Brune and Ohkuma, 2011). At least some of the lower termites produce endoglucanases in their salivary gland and midgut that digest amorphous cellulose (Wheeler et al., 2007, 2010; Zhou et al., 2007, 2010; Tartar et al., 2009; Scharf et al., 2010; Watanabe and Tokuda, 2010).

21.3.2 Lignocellulose Structure

Plant cell walls mainly contain the polysaccharides cellulose and hemicellulose embedded in a pectin matrix (Palin and Geitmann, 2012; Pettolino et al., 2012). Secondary cell walls have polysaccharides covalently linked with ester and ether linkages with a lignin polymer to form lignocellulose, composed of 28%–50% cellulose, 20%–30% hemicelluloses, and 18%–30% lignin (Thompson, 1983; Breznak and Brune, 1994). Lignin is composed of phenylpropanoid subunits linked by ether or carbon-to-carbon bonds and subsequently linked to cellulose, much of which is highly crystalline (structured, as opposed to amorphous), making lignocellulose very difficult to digest. Several cellulase enzymes acting in a concert of hydrolytic actions are required to digest it. Lignin cannot be released from lignocellulose without some degradation of the lignin structure.

At least three types of **cellulolytic enzymes** are involved in lignocellulose digestion: **endoglucanases** (endo-1,4-β-glucanase, EC 3.2.1.4) hydrolyze 1,4-β bonds in cellulose chains; **cellobiohydrolases** (exo-1,4-β-glucanase, EC 3.2.1.91) break cellobiosyl units at the nonreducing end of cellulose chains; and **β-glucosidases** (EC 3.2.1.21) break off glucosyl units from the nonreducing ends of cello-oligosaccharides (Watanabe and Tokuda, 2010).

Most termites that have been treated to remove their symbionts usually cannot live more than a few weeks or months (a few have survived for a year or more), so it seems clear that exocellobiohydrolases and endo-β-1,4-glucanases produced by their symbionts are necessary for efficient and complete digestion of more crystallized lignocellulose. The flagellated protozoans in the hindgut phagocytize small wood particles into digestive vacuoles within themselves, where more depolymerization of the crystalline cellulose occurs. The engulfed wood particles clearly retard the passage of food through the gut, allowing more time for digestion. Although some of the protozoans are common among lower termites, each termite species seems to have some flagellate protozoans that are specific to the particular termite species, which is probably the result of their long coevolution (Brune and Ohkuma, 2011).

Contrary to earlier ideas, the gut of termites is not totally anaerobic. The hindgut is actually a system of different microbial populations in a gradient of oxygen and hydrogen (Brune and Friedrich, 2000), with populations of aerobic and anaerobic bacteria in the small volume of the hindgut. Oxygen deficiency is typically measured as a low (negative) **redox potential**, and values as low as −50 to −270 mV have been measured in different species of termites (Bignell and Anderson, 1980; Veivers, 1980, 1982; Bignell, 1884).

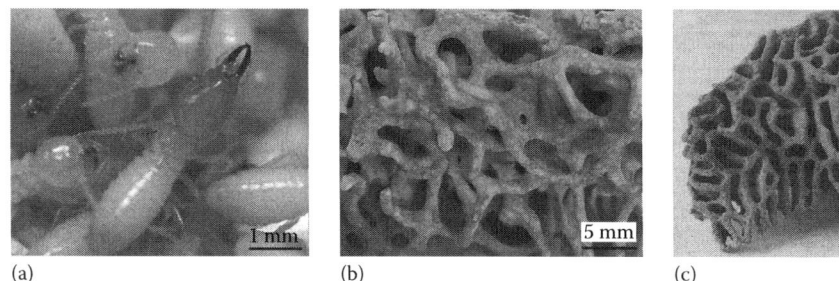

Figure 21.4 (See color insert.) (a) The subterranean termite *Coptotermes formosanus*, (b) nest material, and (c) *Termitomyces* fungal comb from the termite *Macrotermes bellicosus*. (From Chouvenc, T. et al., *Proc. R. Soc. Lond. B*, 280, 20131885, 2013. With permission, Photos courtesy of Thomas Chouvenc, University of Florida, Ft. Lauderdale, FL.)

21.3.3 Nitrogen Metabolism

Wood is very poor in available nitrogen, and this deficiency is overcome in several ways by the symbionts in termites. Uric acid, the end product of nitrogen metabolism in termites, may be stored in the termite fat body in **urocytes**, special cells that accumulate uric acid. While in the urocytes, the uric acid cannot be recycled by the symbionts, but termites are cannibalistic and eat the bodies of dead nest mates, releasing uric acid into the alimentary canal, where it is recycled by gut symbionts to acetate, CO_2, and ammonia. **Nitrogen fixation** from the air also is widespread in termites (Breznak et al., 1973, Breznak, 2000), but is less important in fungus cultivating and soil feeding higher termites because in fungus feeders, nitrogen is made available by the fungi to the termites, and soil feeders usually get adequate nitrogen from the soil and humus in the diet (Eggleton and Tayasu, 2001). Tayasu (1998) using isotope analyses showed that 30%–60% of the nitrogen in the bodies of *Neotermes koshunensis* workers came from nitrogen fixation. Proctodeal trophallaxis (ingestion of feces from one individual by another) is also important in replenishing the symbionts in the gut of newly molted individuals (which lose most or all of their symbionts because the lining of the hindgut is shed at each molt) and in nitrogen recycling through feeding upon the cells of dead symbionts (Brune and Ohkuma, 2011).

21.3.4 Fungal Culture

Higher termites (subfamily Macrotermitinae) have an intimate symbiotic relationship with **basidiomycete fungi** in the genus *Termitomyces* (Chouvenc et al., 2013). Termites cultivate the fungus within the termite nest on plant material that is undergoing partial digestion by mycelia of the fungi (Figure 21.4). The termites make constant additions to the fungal mass by deposition of termite feces containing partially digested plant substance. Termites feed upon older parts of the fungal mass (reviewed by Wood and Thomas, 1989; Aanen and Boomsma, 2006).

21.4 BARK AND AMBROSIA BEETLES AND THEIR SYMBIONTS

Bark beetles (about 3700 species) and **ambrosia beetles** (about 3500 species) are a group of highly modified weevils in two subfamilies, Scolytinae and Platypodinae (Wood, 1982; Bright, 1993; Marvaldi et al., 2002; Kasson et al., 2013) in the beetle family Curculionidae. The common names and the scientific classification are not exactly overlapping: the term "bark beetle" refers to the subfamily Scolytinae, but it is also an ecological term for beetles colonizing tree phloem. The term "ambrosia beetle" is also primarily ecological, not taxonomic, and includes all of 11 or more groups within Scolytinae and all Platypodinae that evolved fungus farming.

The group includes some of the most important forest insects. According to Raffa (2014), "Bark beetles that attack living conifers likely exert stronger ecological impacts ..." in forest ecosystems than any other insect group by influencing nutrient recycling, forest structure, fire, biodiversity, and other ecological events. Fortunately, only a few of the bark beetle species attack and kill living trees. The vast majority of both bark and ambrosia beetles colonize dead or dying trees, many others are seed feeders, and several unusual species are even able to colonize herbs or occupy live tree tissues as parasites. All ambrosia beetles and some bark beetles depend upon fungal symbionts for part of their nutrition. In addition to the fungi cultivated by the beetles themselves, there are other microorganisms associated with bark beetles, all forming the **bark beetle holobiont** (Six, 2013). Neither the beetles nor other living organisms are ever simple "individuals" as Gilbert et al. (2012) stressed. The interactions of the beetles, fungi, bacteria, and host plant chemicals are complex (Adams et al., 2009, 2011; Raffa, 2014).

21.4.1 Ambrosia Beetles

Ambrosia beetles (Figure 21.5) tunnel into the woody part of trees, the sapwood, which is nutritionally poor, and they depend upon their associated fungi for various nutrients. Nearly all ambrosia beetles have a **mycangium**, a special cuticular structure in which fungal spores are stored and transported as the beetles move to colonize a new tree. A mycangium (Figure 21.6) is an invagination of the integument that is lined with secretory glands or secretory cells and is specialized for receipt and transport of fungi (Batra, 1963; Levieux et al., 1991). **Mycangia** have

Figure 21.5 **(See color insert.)** (a) The ambrosia beetle *Xylosandrus crassiusculus* in its gallery with fungal garden. (b) *X. crassiusculus* gallery with brood in pupal stage. (c) Characteristic frass appearance of *X. crassiusculus* attack on a tree. All photos courtesy of Jiri Hulcr, University of Florida. (a and c: Unpublished photos; b: From Six, D.L., *Insects*, 3, 339, 2012. With permission.)

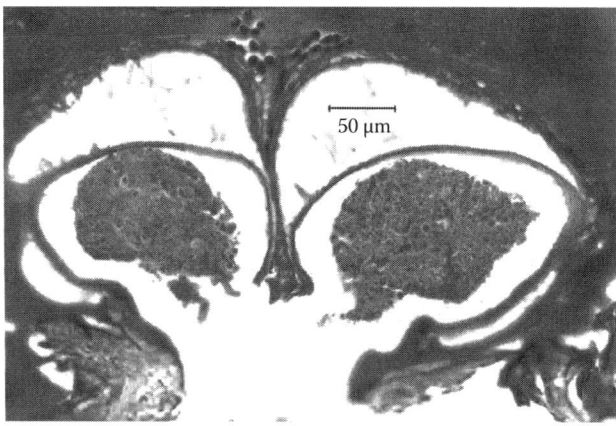

Figure 21.6 (See color insert.) Transverse section through the paired mandibular mycangia of a female ambrosia *Euwallacea validus* beetle showing novel *Fusarium* and *Raffaelea* nutritional symbionts. The preparation was paraffin embedded, sectioned, and stained with Harris hematoxylin and eosin–phloxine. (From Kasson, M.T. et al., *Fungal Genet. Biol.*, 56, 147, 2013. With permission, photo courtesy of Mike Kasson.)

evolved independently many times in Scolytinae. Mycangia are present in nearly all ambrosia beetles and in many bark beetles. Mycangia occur in various places on the external cuticle of beetles and sometimes differ between the sexes (Six and Paine, 1999; Farrell et al., 2001). Sometimes, any cuticular structure that consistently serves to transport fungi is called a mycangium even if secretory gland or cells are not present. Thus, pits, setal brushes, and saclike structures are sometimes called mycangia. When there are secretory glands or cells, the secretion may provide support for the fungi or spores and may protect from antagonistic fungi (Six, 2003).

The beetles are dependent upon the fungi as food (Six, 2012) obtaining vitamins, amino acids, and sterols (Norris et al., 1969; Kok et al., 1970; Kok, 1979; Nasir and Noda, 2003) from the fungi, and fungi secrete enzymes that aid in the digestion of the garden materials (De Fine Licht and Biedermann, 2012). Females lay their eggs in the galleries constructed by adult beetles and, at least in some ambrosia beetle, are the primary caretakers for the fungal gardens. If a female in the gallery dies or is removed, the fungal gardens tend to be overgrown by contaminating fungi and bacteria, and death of the larvae often ensues (Norris, 1979; Biedermann and Taborsky, 2011; Six, 2012). Larvae pupate within the galleries, and when new adult beetles emerge, they carry fungi with them to new trees. After a suitable new host tree has been located, adults construct galleries and begin a new generation with new fungal gardens on the walls of the gallery (Six, 2003).

In some ambrosia beetles, both sexes disperse and carry fungi to new host trees (Kirkendall, 1983), while in others, only the female disperses to a new host tree. The galleries after construction are defended by the beetles, thus reducing resource competition. The association between host beetles and their symbionts is mutualistic—beetles and symbionts benefit. The beetles supplement the nutrient-poor sapwood of trees by feeding upon the fungi, while the life cycle of the fungus continues by being transported from tree to tree by the beetles (Mueller et al., 2005). Recently, Hulcr and Cognato (2010) discovered a novel activity of some ambrosia beetles in which there has been repeated evolution of stealing of one beetle's fungal crop by another beetle that tunnels into the brood galleries where much of the fungal growth occurs. They called this activity **mycocleptism** and reported that it has evolved independently in several beetle clades (Figure 21.7). In addition, the red bay ambrosia beetle, *Xyleborus glabratus*, is attracted to volatiles from its (tree killing) fungus *Raffaelea lauricola* (Hulcr et al., 2011), suggesting an area of chemical ecology for greater exploration (Biedermann and Kaltenpoth, 2014).

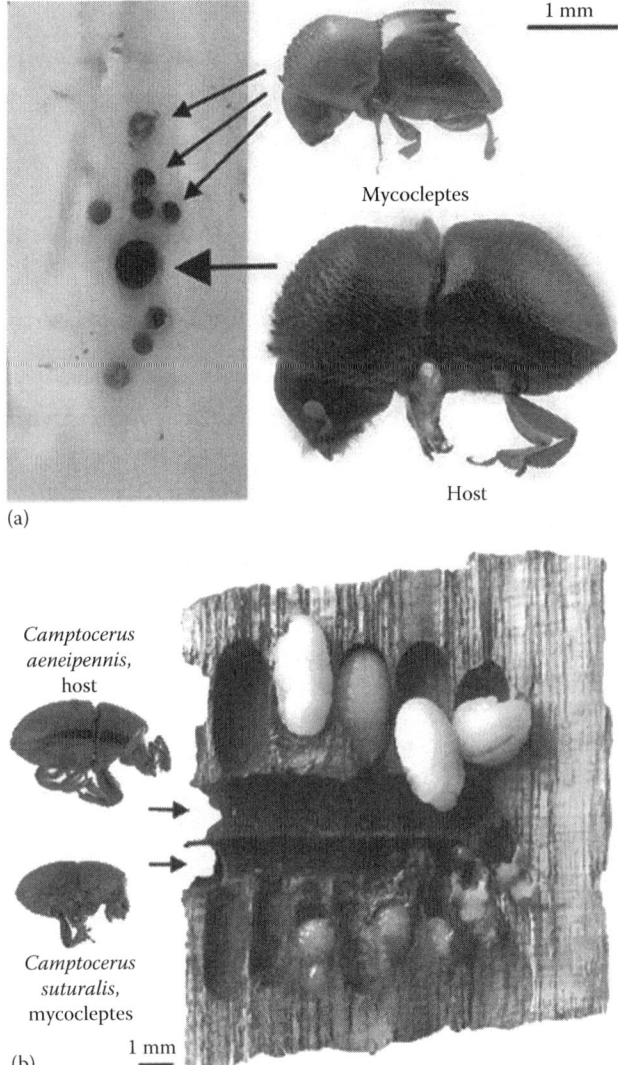

Figure 21.7 **(See color insert.)** (a) Multiple small entrances by the mycocleptic beetle *Diuncus duodecimspinatus* around the larger gallery entrance of its host or provider species, *Hadrodemius globus*, in a branch of a *Ficus* sp. in Papua New Guinea. (b) Detrimental effect of mycocleptic beetle *Camptocerus suturalis* on the gallery of its host, *Camptocerus aeneipennis* in Guyana. The bottom half of the host gallery has been destroyed and replaced by a gallery and brood of the mycocleptic *C. suturalis*. Typically, in healthy galleries of *C. aeneipennis*, the larval chambers line both sides of the maternal gallery. Photos in a from Jiri Hulcr, University of Florida, and in b from Sarah M. Smith, Michigan State University. (From Hulcr, J. and Cognato, A.I., *Evolution*, 64, 3205, 2010. With permission.)

21.4.2 Bark Beetles

Bark beetles construct their galleries in the phloem layer of trees, which is still low in nutrients, but better than the sapwood. They chew some of the phloem as they excavate their galleries, probably deriving some soluble nutrients and resulting in production of substantial amounts of **frass** that is pushed to the end of their galleries. Some, but not all, bark beetles feed upon fungi growing in their galleries. For ambrosia beetles and bark beetles (and many other types of insects), it is likely that a soluble form of nitrogen and a suitable sterol are the most limiting factors in growth (Hodges et al., 1968; Ayres et al., 2000). Nitrogen, particularly in the form of essential amino acids, is low in sapwood and phloem, and fungal feeding is beneficial (Cook et al., 2010). Insect cannot condense the carbon skeleton of squalene into the sterol rings, so a dietary sterol is required by all insects. Many phytophagous insects are able to modify various plant sterols into cholesterol for incorporation into cell membrane structure and for synthesizing the molting hormone, ecdysone (see Chapter 5, **Hormones and Development**, for more details). **Ergosterol** is commonly produced by many fungi and is a sterol that insects readily convert into cholesterol (Norris et al., 1969; Morales-Ramos et al., 2000; Bentz and Six, 2006).

21.4.3 Fungal Role in Supplementing Limited Nutrients in Wood and Phloem

Fungus feeding has enable ambrosia and bark beetles to live upon nutrient-poor tree resources. The nitrogen content of loblolly pine trees, a host to several bark beetles, is about 0.38%, requiring the beetles to consume a large amount of food or supplement their diet with richer sources of nitrogen to support the typical insect body content of 6%–10% nitrogen (Hodges et al., 1968; Fajer, 1989; Slansky and Feeny, 1997; Ayres et al., 2000).

A study with the bark beetles *Ips grandicollis* and *Dendroctonus frontalis* (Ayres et al., 2000) related tunnel length with fungal feeding. *I. grandicollis* beetles do not have a mycangium for transporting fungi, although some fungi may get transported on their body surface as they move to a new tree (Rumbold, 1931; Stone and Simpson, 1990); the beetles construct long galleries in the phloem, chewing the phloem as they lengthen their galleries and pushing the frass to the end of the gallery. They do not construct feeding chambers and seem not to require fungi to supplement their diet (Six and Wingfield, 2011; Giordano et al., 2013). The large amount of phloem chewed as they make the long galleries seems to serve their nutritional needs. *D. frontalis* beetles have a mycangium for transporting fungi; in a new tree, they construct only short galleries that have terminal feeding chambers in which the fungi grow. The beetles spend much of their time in the feeding chamber feeding upon the fungus. The larvae of *I. grandicollis* chewed their way through 79% more phloem than did *D. frontalis* larvae (Ayres et al., 2000), suggesting that without fungal culture, more phloem has to be chewed and longer tunnels constructed to get needed nutrients. *D. frontalis* beetles that develop with fungi are larger than those beetles that develop without mycangial fungi, and larval survival is higher and galleries are shorter when mutualistic fungi are present in the galleries than when fungi are absent (Barras, 1970; Six and Paine, 1998; Ayres et al., 2000). Newly enclosed beetles typically feed upon fungal spores that grow upon the walls of the pupal chambers before they actually emerge from the galleries and fly to new host trees (Whitney, 1971; Six and Paine, 1998), possibly gaining important nutrition for reproduction.

21.4.4 Evolution of Fungal Feeding in Bark Beetles

The evolution of fungus feeding has two possible explanations. Mueller et al. (2001, 2005) suggested that bark beetles became associated with vectoring fungi and later began to obtain nutrition from the fungi, leading to requirement for the fungi as a food source. Six (2012) suggests that beetles began to incorporate fungi into their mixed diet and later became specialized for feeding almost exclusively on the fungi. It is possible that both types of evolution occurred in the bark beetle–fungus associations (Six, 2012). Once fungal feeding behavior evolved, no reversals have been known to occur in ambrosia beetles (Farrell et al., 2001), and although some beetles have lost the mycangial mechanism for transporting fungi, some (by mycocleptism) continue to feed on fungi, albeit not their own (Huler and Cognato, 2010).

In addition to providing food, some of the microorganisms associated with bark beetles may play additional roles in the life of beetles. Some yeasts associated with *Dendroctonus ponderosae* are able to convert **trans-verbenol**, an aggregating pheromone, to **verbenone**, an **antiaggregating pheromone** (Hunt and Borden, 1990). The aggregating pheromone is produced by the beetles and aids in recruiting additional beetles to attack trees where beetles already have galleries (*D. ponderosae* is one of the few bark beetles that can kill trees when it makes mass attacks), and verbenone, also produced by the beetles as well as by the yeasts, acts as an antiaggregating pheromone to limit the attack, presumably when resource competition becomes a pressure.

Some fungi and other organisms found in beetle galleries appear to be detrimental to bark beetles. The fungus *Ophiostoma minus*, for example, may be detrimental to development of *D. frontalis*; larvae developing in galleries in which this fungus was prevalent often failed to survive (Hemingway et al., 1977; Lombardero et al., 2003; Hofstetter et al., 2006). Some bacteria (Nair et al., 2002; Adams et al., 2008) may inhibit the growth of beneficial fungi that grow in the beetle's tunnels. Volatiles from the yeast, *Ogataea pini* (Davis and Hofstetter, 2011; Davis et al., 2011), stimulated growth of beneficial fungi of *Dendroctonus brevicomis* and helped control the growth of the pathogen *Beauveria bassiana*. *Dendroctonus rufipennis* produces an oral secretion that inhibits some of the fungi in its galleries. Some gallery fungi may inhibit harmful ones (Scott et al., 2008). Even the volatile compounds released by fungi may impact bark beetles by influencing the host plant, beetle predators, and other microbial symbionts associated with the beetles (Davis et al., 2013; Groenhagen et al., 2013). Bacteria and fungi associated with bark beetles can reduce the concentration of some of the plant-defensive compounds induced after bark beetles attack a tree (Boone et al., 2013; Hammerbacher et al., 2013).

Bark beetles probably evolved about 100 MYA (Cognato and Grimaldi, 2009; Kirejtshuk et al., 2009; Jordahl et al., 2011) with their ancestral hosts, the conifers. The principal fungi associated with bark beetles probably evolved earlier than the beetles, perhaps as early as 170 MYA. Various members of the Scolytinae transferred to angiosperms several times and, after a time, returned to conifers. Shifts to angiosperms were periods of increased species diversity, while reversals to conifers resulted in low diversity (Farrell et al., 2001). Fungal feeding behavior seems to have evolved several different times, at least eight times according to some authorities (Farrell et al., 2001; Huler, et al., 2007; Six, 2012), and fungal feeding by ambrosia beetles evolving with their shifts to angiosperms (Cognato and Grimaldi, 2009).

21.4.5 Bacteria as Part of the Bark Beetle Holobiont

Bacteria occur in and on bark beetles, and in their galleries, but it is not clear from the limited research available what roles, if any, they play in most cases. The mycangia of ambrosia beetles are a rich source of bacteria (Hulcr et al., 2012). Some bacteria may provide nutritional benefits to beetles. For example, *Rahnella aquatilis*, a nitrogen fixer found in all stages of *Dendroctonus rhizophagus* and in other bark beetles, may provide nitrogen to beetles (Morales-Jimenez et al., 2012). Detoxification of the phenolics and terpenoids found in pine tree resin may aid *D. ponderosae* whose gut is rich in several genera of bacteria with genes for terpene degradation (Deguistine et al., 2011, Clark et al., 2012; Adams et al., 2013; Boone et al., 2013; Wang et al., 2013).

21.4.6 Anthropogenic Effects upon Bark Beetles and Their Symbionts

Climate change will affect all living organisms and ecosystems. Bark beetles, their symbionts, and their forest ecosystems have already experienced, and will continue to experience, new challenges from climate change (Six, 2009; Six et al., 2011b; Addison et al., 2013; Six, 2013). Diamond et al. (2012) present data to support the idea that organisms living at lower latitudes are likely to be more adversely affected by climate warming than species living at higher latitudes. The rationale behind this hypothesis is that organisms living at lower latitudes are likely to have lower tolerance for warming because they may already be living at near the upper temperature limit for life processes. Their data from the study of ants in tropical environments support the hypothesis: they are more at risk than ants at higher latitudes. Natural enemies that help control damaging populations of bark beetles may be impacted by climate change (Adams and Six, 2008; Boone et al., 2008). According to Six (2013), some of the larger, more severe outbreaks of bark beetles, including range and host expansion (Bentz et al., 2010; Logan et al., 2010; Safranyik et al., 2010) are likely the effects of climate change. Some fungal symbionts that grow best at cooler temperatures may be impacted by substantially warmer climate in some regions (Hofstetter et al., 2006a; Addison et al., 2013). Movement of bark beetles into new habitats in new places in the world may present new problems for preservation of forest ecosystems. Introduction of exotics has long been a worldwide problem, and some bark beetles have moved into regions where they never existed before. For example, *Dendroctonus valens*, the red turpentine beetle, does not kill trees and is not a major pest in its native habitat in the United States, but its introduction into China has caused high mortality in pine trees (Sun et al., 2013) where it is associated with a more virulent tree-killing fungus (Lu et al., 2010). Reduced forest ecosystems, introduction of exotic forest insects and diseases, and climate change are likely to have major impacts on bark beetles and their holobiont (Six, 2009; Addison et al., 2013; Six et al., 2013).

21.5 *BUCHNERA* IN APHIDS

Buchner (1966) showed that all aphids except Phylloxeridae and Adelgidae contained **coccoid bacteria** in mycetocytes within the aphid cells. Only in the 1990s were these primary bacteria in aphids identified as γ-**proteobacteria** related to *Escherichia coli* (Munson et al., 1991), and the genus given the name ***Buchnera***. *Buchnera* are ordinarily located in mycetocytes (also called bacteriocytes) in the body cavity (hemocoel) of the aphid (Munson et al., 1991; Moran and Buchmann, 1994). Mycetocytes increase in size as the bacteria multiply, creating very large cells. A pea aphid of 4 mg might contain 100 mycetocytes (mean 85 µm diameter) with each housing as many as 23,500 *Buchnera* cells (Wilkinson and Douglas, 1998; Douglas, 2003). Mittler (1971) demonstrated the importance of the bacteria to aphids by showing that aphids treated with the antibiotic tetracycline lost their symbionts and grew very poorly and could not produce offspring. The fossil record of aphids in amber indicates an ancient association of *Buchnera* and aphids, and sequences of 16S rDNA of several species of aphids indicate that the association may go back as far as 200–250 MYA (Moran and Buchmann, 1994).

The phloem and xylem sap that aphids feed upon is low in nitrogen and especially low in the essential amino acids that insects need, and *Buchnera* provides the essential amino acids needed for growth of the aphids (Douglas, 1993, 1998; Sasaki and Ishikawa, 1995). In addition, *Buchnera* probably provides its aphid host with the vitamin B2, riboflavin (Nakabachi and Ishikawa, 1999), possibly some growth factors (Fares et al., 2004), and shares several metabolic pathways with aphids in a nutritionally mutualistic relationship (Russell et al., 2013). Aphids that are starved cannibalize their bacteriocytes, but when well fed again, the bacteriocytes recover their numbers. Aphids and

Buchnera both benefit from this mutualistic association; aphids receive nutrients and *Buchnera* is propagated into the next generation of aphids.

Buchnera is transmitted to offspring **transovarially** through the eggs of the mother. The mycetocytes associate with the germarial region in the ovaries, and *Buchnera* are shed from the mycetocytes as the embryos in viviparous aphids develop to the blastoderm stage or as eggs begin vitellogenesis in oviparous aphids. *Buchnera* travel between follicle cells surrounding the developing embryos or eggs and finally are incorporated into the mycetocytes as the embryos or eggs develop (Douglas, 2003).

An adult pea aphid could have as many as 100 mycetocytes, with *Buchnera* cells averaging 2.5 μm and occupying 60% of the volume of the mycetocyte (Whitehead and Douglas, 1993). Such density of bacteria could make 2.2×10^6 cells available to transmit, but typically only 50–500 cells are transmitted. It is not known why so few are transmitted, but the small number transmitted theoretically could result in genomic deterioration of *Buchnera*, and Rispe and Moran (2000) demonstrated that genomic decay has occurred in *Buchnera*. The genome of *Buchnera* from pea aphids *Acyrthosiphon pisum* (Shigenobu et al., 2000) is small, 0.45–0.64 Mb (Wernegreen et al., 2000; Lawrence et al., 2001; Gil et al., 2002). Independent survival of *Buchnera* outside aphids is not possible because genes for phospholipid synthesis required for cell structure and genes controlling the Tricarboxylic Acid (TCA) (metabolic) cycle are not present. Nevertheless, the long evolutionary development between aphids and *Buchnera* is attested by the fact that the symbionts still have the genes for synthesis of essential amino acids that are so important to the aphids (Shigenobu et al., 2000). Only genes most essential to *Buchnera* in its present lifestyle are retained. Aphids deprived of bacteria have low body protein content (20% depressed), although lipid levels are normal, and some amino acids are as much as 40% elevated (Prosser and Douglas, 1991). However, phenylalanine and methionine, two essential amino acids for insects, are particularly low in the symbiont-free aphids (Prosser and Douglas, 1991; Douglas, 1996). Plant phloem sap contains 10%–25% of the essential amino acids (Douglas, 1993), but animal proteins contain about 50% essential amino acids. Growth and reproduction of aphids with *Buchnera* on synthetic diets mimicking phloem sap with low or no essential amino acids is normal, but *Buchnera*-free aphids grow poorly and do not reproduce (Mittler, 1971; Douglas et al., 2001) on such diets.

21.6 TSETSE FLY SYMBIONTS

Tsetse flies for ages have plagued parts of Africa as agents of human disease (**sleeping sickness**) and a related cattle disease (**Nagana**), reviewed by Weiss et al. (2012). Tsetse flies (Diptera: Glossinidae) harbor the protozoan ***Trypanosoma brucei***, which is the disease agent that can be transferred to humans and animals. The flies also have the primary bacterial symbiont **Wigglesworthia glossinidia** and the secondary bacterial symbiont **Sodalis glossinidius** and sometimes a third bacterium **Wolbachia pipientis**. *Sodalis* and *Wigglesworthia* belong to the gammaproteobacteria; *Wolbachia* belongs to the alphaproteobacteria. Tsetse flies are blood feeders (both sexes) and can transmit *Trypanosoma brucei* to humans and animals while feeding. Harboring the trypanosomes is not without cost to the flies; female reproductive fitness is impacted by increasing the larval incubation period (Hu et al., 2008). The generic name "*Wigglesworthia*" was established in honor of Sir V. B. Wigglesworth, who worked with vectors of tropical diseases and is often considered the father of insect physiology as a scientific discipline.

Wigglesworthia has been found in every fly examined, and its relationship with flies goes back an estimated 50–80 MYA. The genome of *Wigglesworthia* is small (about 700 kb), and it is no longer able to survive outside the tsetse fly (Akman et al., 2002). Bacteriocytes containing *Wigglesworthia* form a bacteriome organ in the anterior part of the midgut. Extracellular *Wigglesworthia* also are in the milk gland lumen where they probably get transmitted to future fly embryos. The exact function of *Wigglesworthia* to the flies is not fully known, but *Wigglesworthia* does have genes that could enable it to produce several critical vitamins, including riboflavin, dihydrofolate, thiamine, and genes for lipoic acid, intermediates in biotin synthesis, and pyridoxine (vitamin B6) (Akman et al., 2002). Thus, it is likely that *Wigglesworthia* aids flies by producing some vitamins and possibly other components that the blood meal of the flies does not provide an adequate quantity.

Sodalis is found in laboratory colonies of tsetse flies and in some feral populations as a secondary facultative symbiont (Rio et al., 2003). It seems likely to be a recent associate with the flies and may be transmitted horizontally from individual to individual. Some insects besides tsetse flies have infections with related species of *Sodalis*. In contrast to *Wigglesworthia*, *Sodalis* has a large 4.5-Mb genome (Toh et al., 2006). The role of *Sodalis* to the flies is not clear. In spite of its large genome, it has lost the ability to synthesize the vitamin thiamin, even though the vitamin is required for its survival. It may get this vitamin from the flies and/or from *Wigglesworthia* synthesis (Snyder et al., 2010). Data on whether *Sodalis* influences the life span of tsetse flies are conflicting; Toh et al. (2006) claimed that it does, while Weiss et al. (2006) did not find such an effect. Also in conflict are ideas that it might have an influence on susceptibility of the flies to trypanosome infection. However, in experiments in which flies were treated with ampicillin, the load of *Sodalis* in the tsetse flies was still high, while *Wigglesworthia* appeared to be gone, and these tsetse flies were highly susceptible to infection with trypanosomes (Pais et al., 2008). Some populations of *Glossina palpalis gambiensis* tsetse flies have genetically different populations of *Sodalis*, and these flies exhibited a tendency to show a relationship between challenge with trypanosomes and certain populations of *Sodalis*, so *Sodalis* may be important in the infection of tsetse with trypanosomes in some situation or populations.

Female tsetse flies larviposit a mature larva that is ready to pupariate, and it pupariates within about 30 min after deposition by burrowing into the detritus or soil where it is deposited. Only a single egg matures in the ovary of the mother. The embryo and the larva are nourished within the body of the female with secretion from a modified accessory gland (called the milk gland) that is connected to the uterus. The gland has tubules that extend into the abdominal cavity. A female is capable of giving birth to a total of 8–10 progeny over her lifetime. *Wigglesworthia* and *Sodalis* are transmitted to the new progeny through the secretions of the milk gland. Pais et al. (2008) found that treating females with ampicillin removed *Wigglesworthia*, but not *Sodalis*, and the females could still reproduce. There appeared to be no effect on longevity of the treated females, but they were less able to digest the blood meal they fed upon, and they were highly susceptible to infection with trypanosomes, suggesting that their immune system may have been affected. The IMD pathway (refer to the Chapter 16, **Immunity**, for details) is involved in immunity of tsetse flies, and it plays a major role in infection of tsetse with trypanosomes (Nayduch and Aksoy, 2007; Wang et al., 2009). Hu and Aksoy (2006) showed that immunocompromised flies had a higher midgut parasite level than normal flies. Diptericin and attacin (bacteriocides produced by insects) had trypanocidal activity in lab and host tsetse midgut tissue (Hao et al., 2001; Hu and Aksoy, 2005).

In spite of the similarity of the phylogenetic analysis of ribosomal proteins in *Sodalis* showing similarity to pathogenic bacteria, *Sodalis* appears to be nonpathogenic to tsetse flies. Weiss et al. (2008) showed that *Sodalis* has several modifications that probably account for its lack of pathogenicity, and Toh et al. (2006) found modifications in the outer component of the cell wall of *Sodalis* that are different from its close homologue *E. coli*.

21.7 WOLBACHIA

Wolbachia are intracellular bacteria in the α-proteobacteria group. *Wolbachia* infects various invertebrates including mites, crustaceans, nematodes, and insects and is especially common in insects. Zug and Hammerstein (2012) estimated that as many as 40% may be infected. *Wolbachia* typically is a facultative symbiont, but recently, Brownlie et al. (2009) found evidence for a mutualistic relationship between *Wolbachia* and *D. melanogaster* by influencing iron metabolism in flies on iron-restricted and iron-overloaded diets in the laboratory. Similar effects were found in field-collected flies. Hosokawa et al. (2010) and Nikoh et al. (2014) studied a *Wolbachia* strain localized in bacteriosomes in the bedbug *Cimex lectularius* that appears to be an obligate nutritional mutualist. Elimination of *Wolbachia* from the bedbugs resulted in retarded growth and sterility of the insects. In these bedbugs, *Wolbachia* was transmitted to offspring vertically through somatic stem cells to the oocytes in the ovary of the mother. Nikoh et al. (2014) showed that the *Wolbachia* strain in bedbugs synthesized the vitamin B7, biotin.

Wolbachia manipulates its host's behavior and biology, causing cytoplasmic incompatibility (CI) in many species, thelytokous parthenogenesis in haplodiploid species (Southamer et al., 1990), male death in some insects (Hurst et al., 1999), and feminization of genetic males in some insects and other arthropods. There are a number of recent reviews of *Wolbachia* and host biology (Werren, 1997; Stouthamer et al., 1999; Bourtzis et al., 2003; Dedeine et al., 2003; Dobson, 2003; Huigens and Stouthamer, 2003; Bouchon et al., 2009; Narita and Kageyama, 2009).

Wolbachia was first described as a rickettsia-like bacterial infection in the house mosquito *Culex pipiens* (Hertig and Wolbach, 1924). The bacteria were irregular in shape, with both rod and coccoid forms. The bacteria were (and are) enclosed within three membranes, an outer layer from the host cell, inside that the cell wall of the bacteria, and inside that the plasma membrane of the bacteria. Hertig in 1936 named the bacteria **Wolbachia pipientis** in honor of his colleague, Wolbach, and *pipientis* because the bacteria were first discovered in *C. pipiens* (Hertig, 1936). The bacteria are often simply referred to as **Wolbachia** because while there are clearly different strains, which some have attempted to name as different *Wolbachia* species, the only name officially recognized is *W. pipientis*. *Wolbachia* are not rickettsia bacteria, but those types of bacteria are its closest relatives. *Wolbachia* is an ancient symbiont in insects, with genetic lineage analyses suggesting a common ancestor for *Wolbachia* some 25–100 MYA (O'Neill et al., 1992; Rousset et al., 1992; Werren et al., 1995).

Initially, it was thought that the bacteria in mosquitoes were more or less harmless to the mosquitoes, but later work revealed incompatibility effects in as many as 17 *Culex* crossing types, leading Laven (1967) to coin the term **cytoplasmic incompatibility (CI)**. Males infected with *Wolbachia* could not produce progeny with uninfected females. The World Health Organization (WHO) became interested and promoted field trials releasing cytoplasmically incompatible males of *Culex quinquefasciatus*, the primary vector of human filariasis in Southeast Asia. Males that were incompatibility with females (it was not known at this time exactly why they were incompatible) were released into a small isolated population of the mosquitoes in Burma, and egg rafts were periodically collected and monitored for hatching, which soon dropped from 85.7% to 0% (Levan, 1967). However, due to various political problems, WHO did not continue the work nor extend it to other species of mosquitoes at the time. Yen and Barr (1971, 1973), after studying the original work by Hertig and Wolbach (1924), proposed that infection with *W. pipientis* was the causative agent of CI. CI has been demonstrated in a number of insect orders including Coleoptera, Hymenoptera, Diptera, Heteroptera, Orthoptera, and Lepidoptera. Currently, *Wolbachia* is being investigated as a possible control for the mosquito *Aedes aegypti*, the principal vector of dengue fever (Bull and Turelli, 2013).

21.7.1 Cytoplasmic Incompatibility Inducing Effect of *Wolbachia*

Matings between *Wolbachia*-infected males and uninfected females fail to produce progeny due to CI, but mating of infected females and uninfected males or matings when both sexes in the same population are infected result in normal progeny (Yen and Barr, 1971). CI may be unidirectional or bidirectional. Bidirectional CI typically happens when both sexes contain different strains of *Wolbachia*, and then both sexes are incompatible with each other. *Wolbachia* is thought to interfere in ways as yet not understood with the physiology of male sperm and/or male chromosomes and that an infected oocyte (from an infected female) can rescue sperm from an infected male (Dobson, 2003). In some cases, it has been shown that sperm from *Wolbachia*-infected males will not condense in the cytoplasm of an uninfected egg, but will condense in the cytoplasm of an infected egg (Breeuwer and Werren, 1990; O'Neill and Karr, 1990). In *Culex* species of mosquitoes, all individuals in feral populations are infected, but only matings of mosquitoes from the same population result in viable offspring. Probably, different populations of these mosquitoes are infected with different strains of *Wolbachia*, and incompatibility occurs between different strains. It has been suggested that such incompatibility between different populations could, over long periods of time, lead to speciation of sympatric populations.

Nearly always, *Wolbachia* is inherited only through the female passing the bacteria on to her eggs in the ovary, and the bacteria are quickly incorporated into pole cells shortly after fertilization in *Drosophila* (Hadfield and Axton, 1999). *Wolbachia* multiply in the nurse cells and move into oocytes by the cytoplasmic bridges from nurse cells to oocytes (Zchori-Fein et al., 1998). Vertical transmission is expected to select for reduced adverse effect to the host because the parasite is dependent upon the host for transmission and its future. Although males may be infected with *Wolbachia*, it is a dead end for *Wolbachia* because males cannot pass on the infection. In some situations, there is clear evidence of lateral transmission from individual to individual, but the best examples of lateral transmission are in isopod crustaceans. Nevertheless, when Huigens et al. (2000) allowed *Trichogramma kaykai* wasps to feed on butterfly eggs that were infected with *Wolbachia*, 40% of the wasps produced female offspring with infection, suggesting horizontal transfer.

In a population that includes infection and uninfected sexes with *Wolbachia*, infected females have a reproductive advantage in that they can mate successfully with infected or uninfected males, but uninfected females can successfully mate only with uninfected males. *Wolbachia* acts in a number of ways to benefit itself and its female hosts, including decreasing male fitness by CI, killing of males (Hurst et al., 1999), inducing **parthenogenesis** (Stouthamer et al., 1993), and **feminization** of males (Rousset et al., 1992). All of these effects on males increase the population success of females at the expense of males and, at the same time, increase the chances for *Wolbachia* to be passed on to the next generation of hosts by **vertical transmission** to the host's eggs. Parthenogenesis in a bisexual population is expected to increase the spread of *Wolbachia* if parthenogenetic females produce more daughters, which, of course, will also be infected with *Wolbachia* (Stouthamer et al., 2001). Such a process can lead to complete parthenogenesis in a population, and it cannot be reversed by removal of the *Wolbachia* infection (Zchori-Fein et al., 1992). Infections that kill males can become selected in a population in which there are strong antagonistic sibling interactions and in cases where there is strong within species competition for resources. *Wolbachia*-infected females may get additional nutritional benefits from cannibalizing their unhatched male siblings (Majerus et al., 2000; von der Schulenburg et al., 2000, 2001; Zakharov et al., 2000) and have less competition for food and other resource if males are not part of the population.

Wolbachia infection potentially can lead to speciation in several ways, one of which is the possibility of genetic drift in the infected females, leading to populations that are genetically incompatible due to deleterious gene combinations. Another possible mechanism may be formation of postzygotic reproductive barriers created by bidirectional CI (Bordenstein et al., 2001; Wade, 2001). Finally, some have speculated that unidirectional CI might lead to reproductive isolation of two taxa (Shoemaker et al., 1999).

21.7.2 Parthenogenesis Inducing Effect of *Wolbachia*

Haplodiploidy, a natural mixture of parthenogenetic and sexual reproduction, is common in many Hymenoptera. When there are haploid males and diploid females in the population, the situation is called arrhenotoky; when all eggs, fertilized or not, develop into females, it is called **thelytoky**. *Wolbachia*-induced parthenogenesis has been mainly found in Hymenoptera; for example, Grenier et al. (1998) showed that parthenogenetic reproduction in *Trichogramma* wasps was caused by *Wolbachia*. In some populations of insects that have only females, males can be induced by antibiotic treatment (Stouthamer et al., 1990). Dedeine et al. (2001) showed, however, that a female braconid wasp cured of *Wolbachia* could not reproduce, and egg production in *Asobara tabida* completely depends upon infection with *Wolbachia*.

21.7.3 Feminizing Strains of *Wolbachia*

Strains of *Wolbachia* have been discovered in several insects, most notably, in certain Lepidoptera that cause **feminization of genetic males**. In the Asian corn borer *Ostrinia furnacalis*, treatment with antibiotics cures the feminization, and the evidence is that *Wolbachia* is the causative agent (Kageyama et al., 1998). In the butterfly *Eurema hecabe* (Figure 21.8) infected with two strains of *Wolbachia*, genetic males are changed into functional females, and all female broods are produced. Addition of an antibiotic to the diet during various stages of larval life did not completely reverse the feminization process, but did result in some individuals with intersex phenotypes (Narita et al., 2007).

Figure 21.8 (See color insert.) Feminization of the butterfly *Eurema hecabe* infected with two different strains of *Wolbachia* (wHecCl2 and wHecFem2). Butterfly caterpillars were infected with the two strains of *Wolbachia*, and then as different instars fed a diet containing an antibiotic to suppress, but not eliminate, the feminization effect of the male genotype. (a and b) Adult *E. hecabe* emerged with deformed wings from infection with the double strain of *Wolbachia*, when fed the antibiotic diet from third to fourth instar. (c) Failure of an adult butterfly infected with the double strain of *Wolbachia* to escape from the pupal case, when fed the antibiotic diet from first to fourth instar. (d and e) Normal adult *E. hecabe* females pale in ground color and without sex brands, from a nontreated line of insects singly infected with wHecCl2; (f and g) normal adult male *E. hecabe* bright in color with sex brands (arrow) from a nontreated insect line singly infected with wHecCl2. Bars = 10 mm. (From Narita, S. et al., *Appl. Environ. Microbiol.*, 73, 4332, 2007. With permission.)

21.8 REVIEW AND SELF-STUDY QUESTIONS

1. Define the current usage of the term "symbiosis."
2. What does it mean to say that some insects practice "agriculture"? Elaborate several examples.
3. What are mycetocytes, bacteriosomes, and mycetomes?
4. Do some symbionts live freely in the gut or among other tissues of insects? Explain with some examples.
5. Is the development of the ovaries and eggs under the same hormonal control in all insects? Explain answer.
6. Describe an ant fungal garden by explaining what it comprises, how it is constructed, and how ants benefit from a garden.
7. Briefly describe an *Atta* sp. ant colony and its feeding behavior.
8. Elaborate upon the role of fungi in producing antibiotics in a leaf-cutter ant society.
9. Describe the role of a refuse "dump" in a leaf-cutter colony. How is it maintained so that it does not contaminate the main fungal gardens of the colony?
10. What is current evidence for the relationship between termites and cockroaches?
11. What are some of the principal differences in biology of the so-called "lower" and "higher" termites?
12. Where are the symbionts located in termites?
13. How is the glucose that is released from symbiont digestion of cellulose made nutritionally available to termites?
14. Describe the biology of the hindgut of termites.
15. What, if any, are the differences between the terms "bark beetles" and "ambrosia beetles," and what are the major differences in their galleries and attack on trees?
16. What is a bark beetle mycangium?
17. Describe the relationship between *Buchnera* and aphids.
18. What human and animal pathogen is vectored by tsetse flies?
19. Is the symbiont *W. glossinidia* a secondary or primary symbiont in tsetse flies? Where is this symbiont located in tsetse flies?
20. Briefly describe the reproductive biology of tsetse fly females.
21. What is the scientific classification of *Wolbachia*?
22. Explain the origin of the name "*Wolbachia*."
23. How can various strains of *Wolbachia* affect insect biology?
24. Discuss the various aspects of CI induced by *Wolbachia*.
25. What is the evidence that *Wolbachia* can be both a facultative or obligatory symbiont?

REFERENCES

Aanen, D.K. and J.J. Boomsma. 2006. The evolutionary origin and maintenance of the mutualistic symbiosis between termites and fungi, in K. Bourtzis and T.A. Miller (eds.), *Insect Symbiosis*, vol. 2. CRC Press, Boca Raton, FL, pp. 79–95.

Adams, A.S., F.O. Aylward, S.M. Adams, N. Erbilgen, B.H. Aukema, C.R. Currie, G. Suen, and K.F. Raffa. 2013. Mountain pine beetles colonizing historical and naïve host trees are associated with a bacterial community highly enriched in genes contributing to terpene metabolism. *Appl. Environ. Microbiol.* 79: 3468–3475.

Adams, A.S., C.K. Boone, J. Buhlmann, and K.F. Raffa. 2011. Response of bark beetle associated-bacteria to host monoterpenes and their relationship to insect life histories. *Environ. Entomol.* 37: 808–817.

Adams, A.S., C.R. Currie, Y. Cardoza, K.D. Klepzig, and K.F. Raffa. 2009. Effects of symbiotic bacteria and tree chemistry on the growth and reproduction of bark beetle fungal symbionts. *Can. J. Forest Res.* 39: 1133–1147.

Adams, A.S. and D.L. Six. 2008. Detection of host habitat by parasitoids using cues associated with mycangial fungi of the mountain pine beetle, *Dendroctonus ponderosae*. *Can. Entomol.* 140: 124–127.

Adams, A.S., D.L. Six, S. Adams, and W. Holben. 2008. In vitro interactions among yeasts, bacteria and the fungal symbionts of the mountain pine beetle, *Dendroctonus ponderosae*. *Microb. Ecol.* 56: 460–466.

Addison, A., J.A. Powell, D.L. Six, M. Moore, and B.J. Bentz. 2013. The role of temperature variability in stabilizing the mountain pine beetle-fungus mutualism. *J. Theor. Biol.* 335: 40–50. doi:10.1016/j.jtbi.2013.06.012.

Akman, L., A. Yamashita, H. Watanabe, K. Oshima, T. Shiba, M. Hattori, and S. Aksoy. 2002. Genome sequence of the endocellular obligate symbiont of tsetse flies, *Wigglesworthia glossinidia*. *Nat. Genet.* 32: 402–407.

Anderson, J.M., A.D.M. Rayer, and D.W.H. Walton. 1984. *Invertebrate-Microbial Interactions*. Cambridge University Press, Cambridge, U.K.

Aylward, F.O., K.E. Burnum, J.J. Scott, G. Suen, S.G. Tringe, S.M. Adams, K.W. Barry et al. 2012. Metagenomic and metaproteomic insights into bacterial communities in leaf-cutter ant fungus gardens. *ISME J.* 6: 1688–1701.

Ayres, M.P., R.T. Wilkens, J.J. Ruel, MJ. Lombardero, and E. Vallery. 2000. Nitrogen budgets of phloem-feeding bark beetles with and without symbiotic fungi (Coleoptera: Scolytidae). *Ecology* 81: 2198–2210.

Ballari, S., A.G. Farji-Brener, and M. Tadey. 2007. Waste management in the leaf-cutting ant *Acromyrmex lobicornis*: Division of labor, aggressive behavior, and location of external refuse dumps. *J. Insect Behav.* 20: 87–98.

Barke, J., R.F. Seipke, S. Grueschow, D. Heavens, N. Drou, M.J. Bibb, R.J.M. Goss, D.W. Yu, and M.I. Hutchings. 2010. A mixed community of Actinomycetes produce multiple antibiotics for the fungus farming ant *Acromyrmex octospinosus*. *BMC Biol.* 8: 109

Barras, S.J. 1970. Antagonism between *Dendroctonus frontalis* and the fungus *Ceratocystis minor*. *Ann. Entomol. Soc. Am.* 63: 1187–1190.

Batra, L.R. 1963. Ecology of ambrosia fungi and their dissemination by beetles. *Trans. Kans. Acad. Sci.* 66: 213–236.

Bentz, B.J., J. Regniere, C.J. Fettig, E.M. Hansen, J.L. Hayes, J.A. Hick, R.G. Kelsey, J.F. Negron, and S.J. Seybold. 2010. Climate change and bark beetles in western United States and Canada; direct and indirect effects. *Bioscience* 60: 602–613.

Bentz, B.J. and D.L. Six. 2006. Ergosterol content of four fungal symbionts associated with *Dendroctonus ponderosae* and *D. rufipennis* (Coleoptera: Curculionidae, Scolytinae). *Ann. Entomol. Soc. Am.* 99: 189–194.

Biedermann, P.H.W. and M. Kaltenpoth. 2014. New synthesis: The chemistry of partner-choice in insect-microbe mutualisms. *J. Chem. Ecol.* 40: 99.

Biedermann, P.H.W. and M. Taborsky. 2011. Larval helpers and age polyethism in ambrosia beetles. *Proc. Natl. Acad. Sci. USA* 108: 17064–17069.

Bignell, D.E. 1984. Direct potentiometric determination of redox potentials of the gut contents in the termites *Zootermopsis nevadensis* and *Cubitermes severus* and in three other arthropods. *J. Insect Physiol.* 30: 169–174.

Bignell, D.E. and J.M. Anderson. 1980. Determination of pH and oxygen status in the guts of lower and higher termites. *J. Insect Physiol.* 26: 183–188.

Blomquist, C.L., R.W. Howard, and C.A. McDaniel. 1979. Biosynthesis of cuticular hydrocarbons of the termite *Zootermopsis angusticollis* (Hagen). Incorporation of propionate into dimethylalkanes. *Insect Biochem.* 9: 371–374.

Bollazzi, M.J., J. Kronenbitter, and F. Roces. 2008. Soil temperature, digging behaviour, and the adaptive value of nest depth in South American species of *Acromyrmex* leaf-cutting ants. *Oecologia* 158: 165–175.

Bollazzi, M.J. and F. Roces. 2002. Thermal preference for fungus culturing and brood location by workers of the thatching grass-cutting ant *Acromyrmex heyeri*. *Insect. Soc.* 49: 153–157.

Bollazzi, M.J. and F. Roces. 2010a. Control of nest water losses through building behavior in leaf-cutting ants. *Insect. Soc.* 57: 267–273.

Bollazzi, M.J. and F. Roces. 2010b. The thermoregulatory function of thatched nests in the South American grass-cutting ant *Acromyrmex heyeri*. *J. Insect Sci.* 10: 154.

Boone, C.K., K. Keefover-Ring, A.C. Mapes, A.S. Adams, J. Bohlmann, and K.F. Raffa. 2013. Bacteria associated with a tree-killing insect reduce concentrations of plant defense compounds. *J. Chem. Ecol.* 39: 1003–1006.

Boone, C.K., D.L. Six, Y. Zheng, and K.F. Raffa. 2008. Parasitoids and dipteran predators exploit volatiles from microbial symbionts to locate bark beetles. *Environ. Entomol.* 37: 150–161.

Bordenstein, S.R., F.P. O'Hara, and J.H. Warren. 2001. *Wolbachia*-induced incompatibility precedes other hybrid incompatibilities in *Nasonia*. *Nature* 409: 707–710.

Bot, A., C.R. Currie, A.G. Hart, and J.J. Boomsma. 2001. Waste management in leaf-cutting ants. *Ethol. Ecol. Evol.* 13: 225–237.

Bot, A.N.M., D. Ortius-Lechner, K. Finster, R. Maile, and J.J. Boomsma. 2002. Variable sensitivity of fungi and bacteria to compounds produced by the metapleural glands of leaf-cutting ants. *Insect. Soc.* 49: 363–370.

Bouchon, D., R. Cordaux, and P. Greve. 2009. Feminizing *Wolbachia* and the evolution of sex determination in isopods, in K. Bourtzis and T.A. Miller (eds.), *Insect Symbiosis*, vol. 3. CRC Press, Boca Raton, FL, pp. 273–294.

Bourtzis, K., H.R. Braig, and T.L. Karr. 2003. Cytoplasmic incompatibility, in K. Bourtzis and T.A. Miller (eds.), *Insect Symbiosis*, vol. 1. CRC Press, Boca Raton, FL, 347pp.

Bourtzis, K. and T.A. Miller. 2003. *Insect Symbiosis*, vol. 1. CRC Press, Boca Raton, FL, 347pp.

Bourtzis, K. and T.A. Miller. 2006. *Insect Symbiosis*, vol. 2. CRC Press, Boca Raton, FL, 276pp.

Bourtzis, K. and T.A. Miller. 2009. *Insect Symbiosis*, vol. 3. CRC Press, Boca Raton, FL, 408pp.

Breeuwer, J.A. and J.H. Werren. 1990. Microorganism Associated with chromosome destruction and reproductive isolation between two insect species. *Nature* 346: 558–560.

Breznak, J.A. 2000. Ecology of prokaryotic microbes in the guts of wood- and litter-feeding termites, in T. Abe, D.E. Bignell, and M. Hagashi (eds.), *Termites: Evolution, Sociality, Symbiosis, Ecology*. Kluwer Academic Publishers, Dordrecht, the Netherlands, pp. 209–231.

Breznak, J.A., W.J. Brill, J.W. Mertins, and H.C. Coppel. 1973. Nitrogen fixation in termites. *Nature* 244: 577–580.

Breznak, J.A. and A. Brune. 1994. Role of microorganisms in the digestion of lignocellulose by termites. *Annu. Rev. Entomol.* 39: 453–487.

Breznak, J.A. and J.M. Switzer. 1986. Acetate synthesis from H_2 plus CO_2 by termite gut microbes. *Appl. Environ. Microbiol.* 52: 623–630.

Bright, D.E. 1993. Systematics of bark beetles, in T.D. Schowalter and G.M. Phillips (eds.), *Beetle-Pathogen Interactions in Conifer Forests*, Academic Press, New York, pp. 23–33.

Brownlie, J.C., B.N. Cass, M. Riegler, J.J. Witsenburg, I. Iturbe-Ormaetxe, E.A. McGraw, and S.L. O'Neill. 2009. Evidence for metabolic provisioning by a common invertebrate endosymbiont, *Wolbachia pipientis*, during periods of nutritional stress. *PLoS Pathog.* 5(4), e1000368.

Brune, A. 2011. Microbial symbioses in the digestive tract of lower termites, in E. Rosenberg and U. Gophna (eds.), *Beneficial Microorganisms in Multicellular Life Forms*. Springer-Verlag, Berlin, Germany, pp. 3–25.

Brune, A. and M. Friedrich. 2000. Microecology of the termite gut: Structure and function on a microscale. *Curr. Opin. Microbiol.* 3: 263–269.

Brune, A. and M. Ohkuma. 2011. Role of the termite gut microbiota in symbiotic digestion, in D.E. Bignell, Y. Roisin, and N. Lo (eds.), *Biology of Termites: A Modern Synthesis*. Springer Science+Business Media B.V, Dordrecht, the Netherlands, pp. 439–475.

Brune, A. and U. Stingl. 2005. Prokaryotic symbionts of termite gut flagellates: Phylogenetic and metabolic implications of a tripartite symbiosis, in J. Overmann (ed.), *Molecular Basis of Symbiosis*. Springer-Verlag, Berlin, Germany, pp. 39–60.

Buchner, P. 1965. *Endosymbiosis of Animals with Plant Microorganisms*. Interscience Publishers, New York.

Buchner, P. 1966. *Endosymbiosis of Animals with Plant Micro-Organisms*. John Wiley & Sons, London, U.K.

Bull, J.J. and M. Turelli. 2013. *Wolbachia* versus dengue. *Evol. Med. Public Health*, pp. 197–207. doi: 10.1093/emph/eot018.

Cafaro, M.J., M. Poulsen, A.E.F. Little, S.L. Price, N.M. Gerardo, D. Wong, A.E. Stuart, B. Larget, P. Abbot, and C.R. Currie. 2011. Specificity in the symbiotic association between fungus-growing ants and protective *Pseudonocardia* bacteria. *Proc. R. Soc. Lond. B*: 278: 1814–1822.

Caldera, E.J. and C.R. Currie. 2012. The population structure of antibiotic-producing bacterial symbionts of *Apterostigma dentigerum* ants: Impacts of coevolution and multipartite symbiosis. *Am. Nat.* 180(5): 604–617.

Chouvenc, T., C.A. Efstathion, M.L. Elliott, and N.-Y. Su. 2013. Extended disease resistance emerging from the faecal nest of a subterranean termite. *Proc. R. Soc. Lond. B* 280: 20131885.

Christensen, H. and M.L. Fogel. 2011. Feeding ecology and evidence for amino acid synthesis in the periodical cicada (Magicicada). *J. Insect Physiol.* 57: 211–219.

Clark, E.L., D.P.W. Huber, and A.L. Carroll. 2012. The legacy of attack: Implications of high phloem resin monoterpene levels in lodgepole pines following mass attack by mountain pine beetle, *Dendroctonus ponderosae* Hopkins. *Environ. Entomol.* 41: 392–398.

Clark, E.L., A.J. Karley, and S.F. Hubbard. 2010. Insect endosymbionts: Manipulators of insect herbivore trophic interactions. *Protoplasma* 244: 25–51.

Cognato, A.I. and D. Grimaldi. 2009. 100 million years of morphological conservation in bark beetles (Coleoptera: Curculionidae: Scolytinae). *Syst. Entomol.* 34: 93–100.

Cook, S.S., B.M. Shirley, and P. Zambino. 2010. Nitrogen concentration in mountain pine beetle larvae reflects nitrogen status of tree host and two fungal associates. *Environ. Entomol.* 39: 821–826.

Currie, C.R. and A.E. Stuart. 2001. Weeding and grooming of pathogens in agriculture by ants. *Proc. R. Soc. Lond. B* 268: 1033–1039.

Currie, C.R., B. Wong, A.E. Stuart, T.R. Schultz, S.A. Rehner, U.G. Muller, G.-H. Sung, J.W. Spatafora, and N.A. Straus. 2003. Ancient tripartite coevolution in the attine ant-microbe symbiosis. *Science* 299: 386–388.

Davis, S.T. and R.W. Hofstetter. 2011. Interactions between the yeast *Orgataea pini* and filamentous fungi associated with the western pine beetle. *Microb. Ecol.* 61: 626–634.

Davis, T.S., T.L. Crippen, R.W. Hofstetter, and J.K. Tomberlin. 2013. Microbial volatile emissions as insect semiochemicals. *J. Chem. Ecol.* 39: 840–859.

Davis, T.S., R.W. Hofstetter, J.T. Foster, N.E. Foote, and P. Keim. 2011. Interactions between the yeast *Orgataea pini* and filamentous fungi associated with the western pine beetle. *Microb. Ecol.* 61: 626–634.

De Fine Licht, H.H. and P.H.W. Biedermann. 2012. Patterns of functional enzyme activity in fungus framing ambrosia beetles. *Front. Zool.* 9: 13.

De Fine Licht, H.H., M. Schiøtt, A. Rogowska-Wrzesinska, S. Nygaard, P. Roepstorff, and J.J. Boomsma. 2013. Laccase detoxification mediates the nutritional alliance between leaf-cutting ants and fungus-garden symbionts. *Proc. Natl. Acad. Sci. USA* 110(2): 583–587.

Dedeine, F., C. Bandi, M. Boulétreau, and L.H. Kramer. 2003. Insights into *Wolbachia* obligatory symbiosis, in K. Bourtzis and T.A. Miller (eds.), *Insect Symbiosis*, vol. 1. CRC Press, Boca Raton, FL, pp. 267–282.

Dedeine, F., F. Vavre, F. Fleury, B. Loppin, E. Hochberg, and M. Boulétreau. 2001. Removing symbiotic *Wolbachia* bacteria specifically inhibits oogenesis in a parasitic wasp. *Proc. Natl. Acad. Sci. USA* 98: 6247–6252.

Diamond, S.E., D.M. Sorger, J. Hulcr, S.L. Pelini, I. Del Toro, C. Hirsch, E. Oberg, and R.R. Dunn. 2012. Who like it hot? A global analysis of the climatic, ecological, and evolutionary determinants of warming tolerance in ants. *Glob. Chang. Biol.* 18: 448–456.

Diguistini, S., Y. Wang, N. Liao, G. Taylor, P. Tanguay, N. Feau, B. Henrissat et al. 2011. Genome and transcriptome analyses of the mountain pine beetle-fungal symbiont *Grosmannia clavigera*, a lodgepole pine pathogen. *Proc. Natl. Acad. Sci. USA* 108: 2504–2509.

Dobson, S.L. 2003. *Wolbachia pipientis*: Impotent by association, in K. Bourtzis and T.A. Miller (eds.), *Insect Symbiosis*, vol. 1. CRC Press, Boca Raton, FL, pp. 199–215.

Dong, S., K. Pang, X. Bai, X. Yu, and P. Hao. 2011. Identification of two species of yeast-like symbiotes in the brown planthopper, *Nilaparvata lugens. Curr. Microbiol.* 63: 1133–1138.

Douglas, A.E. 1989. Mycetocyte symbiosis in insects. *Biol. Rev.* 64: 409–434.

Douglas, A.E. 1993. The nutritional quality of phloem sap utilized by natural aphid populations. *Ecol. Entomol.* 18: 31–38.

Douglas, A.E. 1996. Reproductive failure and the amino acid pools in pea aphids (*Acyrthosiphon pisum*) lacking symbiotic bacteria. *J. Insect Physiol.* 42: 247–255.

Douglas, A.E. 1998. Nutritional interactions in insect-microbial symbioses. *Annu. Rev. Entomol.* 43: 17–37.

Douglas, A.E. 2003. *Buchnera* bacteria and other symbionts of aphids, in K. Bourtis and T.A. Miller (eds.), *Insect Symbiosis*, vol. 1. CRC Press, Boca Raton, FL, pp. 23–38.

Douglas, A.E. 2010. *The Symbiotic Habit*. Princeton University Press, Princeton, NJ.

Douglas, A.E. 2013. Microbial brokers of insect-plant interactions revisited. *J. Chem. Ecol.* 39: 952–961.

Douglas, A.E., L.B. Minto, and T.L. Wilkinson. 2001. Quantifying nutrient production by the microbial symbiosis in an aphid. *J. Exp. Biol.* 204: 349–358.

Duron, O. and G.D.D. Hurst. 2013. Arthropods and inherited bacteria: From counting the symbionts to understanding how symbionts count. *BMC Biol.* 11: 45.

Eggleton, P. 2011. An Introduction to termites: Biology, taxonomy and functional morphology, in D.E. Bignell, Y. Roisin, and N. Lo (eds.), *Biology of Termites: A Modern Synthesis*, 2nd ed. Springer, New York, pp. 1–26.

Eggleton, P. and I. Tayasu. 2001. Feeding groups, lifetypes and the global ecology of termites. *Ecol. Res.* 16: 941–960.

Fajer, E.D. 1989. The effects of enriched CO_2 atmospheres on plant insect herbivore interactions: Growth responses of larvae of the specialist butterfly, *Junonia coenia* (Lepidoptera: Nymphalidae). *Oecologia* 81: 514–520.

Fares, M.A., M.X. Ruiz-González, A. Moya, S.F. Elena, and E. Barrio. 2004. GroEL buffers against deleterious mutations. *Nature* 417: 398.

Farji-Brener, A.G. and M. Tadey. 2012. Trash to treasure: Leaf-cutting ants repair nest-mound damage by recycling refuse dump materials. *Behav. Ecol.* 23: 1195–1202.

Farrell, B.D., A.S. Sequiera, B.C. O'Meara, B.B. Normark, J.H. Chung, and B.H. Jordahl. 2001. The evolution of agriculture in beetles (Curculionidae: Scolytidae and Platypodidae). *Evolution* 55: 2011–2017.

Fernández-Marín, H., J.K. Zimmerman, S.A. Rehner, and W.T. Wcislo. 2006. Active use of the metapleural glands by ants in controlling fungal infection. *Proc. R. Soc. Lond. B* 273: 1689–1695.

Fernández-Marín, H., J.K. Zimmerman, and W.T. Wcislo. 2003. Nest-founding in *Acromyrmex octospinosus* (Hymenoptera, Formicidae, Attini): Demography and putative prophylactic behaviors. *Insect. Soc.* 50: 304–308.

Fernández-Marín, H., J.K. Zimmerman, and W.T. Wcislo. 2004. Ecological traits and evolutionary sequence of nest establishment in fungus-growing ants (Hymenoptera, Formicidae, Attini). *Biol. J. Linn. Soc.* 81: 39–48.

Ferrari, J., J.A. West, S. Via, and H.C.J. Godfray. 2012. Population genetic structure and secondary symbionts in host-associated populations of the pea aphid complex. *Evolution* 66: 375–390.

Gil, R., B. Sabater-Munoz, A. Latorre, F.J. Silva, and A. Moya. 2002. Extreme genome reduction in *Buchnera* spp. toward the minimal genome needed for symbiotic life. *Proc. Natl. Acad. Sci. USA* 99: 4454–4458.

Gilbert, S.F., J. Sapp, and A.I. Tauber. 2012. A symbiotic view of life: We have never been individuals. *Q. Rev. Biol.* 87: 325–341.

Giordano, L., M. Garbelotto, G. Nicolotti, and P. Gonthier. 2013. Characterization of fungal communities associated with the bark beetle *Ips typographus* varies depending on detection method, location, and beetle population levels. *Mycol. Prog.* 12: 127–140.

Grenier, S., B. Pintureau, A. Heddi, F. Lassablière, C. Jager, C. Louis, and C. Khatchadourian. 1998. Successful horizontal transfer of *Wolbachia* symbionts between *Trichogramma* wasps. *Proc. R. Soc. Lond. B* 265: 1441–1445.

Groenhagen, U., R. Baumgartner, A. Bailly, M.A. Gardiner, L. Eberl, S. Shulz, and L. Weisskopf. 2013. Production of bioactive volatiles by different *Burkholderia ambifaria* strains. *J. Chem. Ecol.* 39: 892–906.

Guo, L., D.R. Quilici, I. Chase, and G.L. Blomquist. 1991. Gut tract microorganisms supply the precursors for methyl-branched hydrocarbon biosynthesis in the termite. *Zootermopsis nevadensis. Insect Biochem.* 21: 327–333.

Hadfield, S.J. and J.M. Axton. 1999. Germ cells colonized by endosymbiotic bacteria. *Nature* 402: 482.

Haeder, S., R. Wirth, H. Herz, and D. Spiteller. 2009. Candicidin-producing *Streptomyces* support leaf-cutting ants to protect their fungus garden against the pathogenic fungus *Escovopsis. Proc. Natl. Acad. Sci. USA* 106: 4742–4746.

Hammerbacher, A., A. Schmidt, N. Wadke, L.P. Wright, B. Schenider, J. Bohlmann, W.A. Brand, T.M. Fenning, J. Gershenzon, and C. Paetz. 2013. A common fungal associate of the spruce bark beetle metabolizes the stilbene defenses of Norway spruce. *Plant Physiol.* 162: 1324–1336.

Hao, Z.R., I. Kasumba, M.J. Lehane, W.C. Gibson, J. Kwon, and S. Aksoy. 2001. Tsetse immune responses and trypanosome transmission: Implications for the development of tsetse-based strategies to reduce trypanosomiasis. *Proc. Natl. Acad. Sci. USA* 98: 12648–12653.

Hart, A., C. Anderson, and F. Ratnieks. 2002a. Task partitioning in leaf-cutting ants. *Acta Ethologica* 5: 1–11.

Hart, A., A. Bot, and M. Brown. 2002b. A colony-level response to disease control in a leaf-cutting ant. *Naturwissenschaften* 89: 275–277.

Hart, A. and F. Ratnieks. 2001. Task partitioning, division of labor and nest compartmentalization collectively isolate hazardous waste in the leaf-cutting ant *Atta cephalotes. Behav. Ecol. Sociobiol.* 49: 367–392.

Hart, A. and F. Ratnieks. 2002. Waste management in the leaf-cutting ant *Atta cephalotes. Behav. Ecol.* 13: 224–231.

Hemingway, R.W., G.C. McGraw, and S.J. Barras. 1977. Polyphenols in *Ceratocystis minor*-infected *Pinus taeda*: Fungal metabolites, phloem and zylem phenols. *Agric. Food. Chem.* 25: 717–722.

Hertig, M. 1936. The rickettsia, *Wolbachia pipientis* (gen. et sp. n.) and associated inclusions of the mosquito, *Culex pipiens. Parasitology* 28: 453–486.

Hertig, M. and S.B. Walbach. 1924. Studies on rickettsia-like microorganisms in insects. *J. Med. Res.* 44: 329–374.

Herz, H., W. Beyschlag, and B. Hölldobler. 2007. Herbivory rate of leaf-cutting ants in a tropical moist forest in Panama at the population and ecosystem scales. *Biotropica* 39: 482–488.

Hodges, J.E., S.J. Barras, and J. Maudlin. 1968. Amino acids in the inner bark of loblolly pine as affected by the southern pine beetle and associated organisms. *Can. J. Bot.* 46: 1467–1472.

Hofstetter, R.W., J. Cronin, D.D. Klepzig, J.C. Moser, and M.P. Ayres. 2006. Antagonisms, mutualisms and commensalisms affect outbreak dynamics of the southern pine beetle. *Oecologia* 147: 679–691.

Hölldobler, B. and E.O. Wilson. 1990. *The Ants.* Springer, Berlin, Germany.

Hongoh, Y. 2011. Toward the functional analysis of uncultivable, symbiotic microorganisms in the termite gut. *Cell. Mol. Life Sci.* 68: 1311–1325.

Hosokawa, T., R. Koga, Y. Kikuchi, X.-Y. Meng, and T. Fukatsu. 2010. *Wolbachia* as a bacteriocyte-associated nutritional mutualist. *Proc. Natl. Acad. Sci. USA* 107(2): 769–774.

Hu, C. and S. Aksoy. 2006. Innate immune responses regulate trypanosome parasite infection of the tsetse fly *Glossina morsitans morsitans*. *Mol. Microbiol.* 60: 1194–1204.

Hu, C., R.V. Rio, J. Medlock, L.R. Haines, D. Nayduch, A.F. Savage, N. Guz et al. 2008. Infections with immunogenic trypanosomes reduce tsetse reproductive fitness: Potential impact of different parasite strains on vector population structure. *PLoS Negl. Trop. Dis.* 2: e192.

Hu, Y. and S. Aksoy. 2005. An antimicrobial peptide with trypanocidal activity characterized from *Glossina morsitans morsitans*. *Insect Biochem. Mol. Biol.* 35: 105–115

Huigens, M.E., R.F. Luck, R.G.H. Klaassen, M.F.P.M. Maas, M.J.T.N. Timmermans, and R. Stouthamer. 2000. Infectious parthenogenesis. *Nature* 405: 178–179.

Huigens, M.E. and R. Stouthamer. 2003. Parthenogenesis associated with *Wolbachia*, in K. Bourtzis and T.A. Miller (eds.), *Insect Symbiosis*, vol. 1. CRC Press, Boca Raton, FL, pp. 247–266.

Hulcr, J. and A.I. Cognato. 2010. Repeated evolution of crop theft in fungus-farming ambrosia beetles. *Evolution* 64: 3205–3212.

Hulcr, J., M. Kolarick, and L. Kirkendall. 2007. A new record of a fungus-beetle symbiosis in *Scolytodes* bark beetles (Scolytine, Curculionidae, Coleoptera). *Symbiosis* 43: 151–159.

Hulcr, J., R. Mann, and L.L. Stelinski. 2011. The scent of a partner: Ambrosia beetles are attracted to volatiles from their fungal symbionts. *J. Chem. Ecol.* 37: 1374–1377.

Hulcr, J., N.R. Roundtree, S.E. Diamond, L.L. Stelinski, N. Fierer, and R.R. Dunn. 2012. Mycangia of ambrosia beetles host communities of bacteria. *Invertebr. Microbiol.* 64: 784–793.

Hunt, D.W.A. and J.H. Borden. 1990. Conversion of verbenols to verbenone by yeasts isolated from *Dendroctonus ponderosae* (Coleoptera: Scolytidae). *J. Chem. Ecol.* 16: 1385–1397.

Hurst, G.D.D., F.M. Jiggins, H.H.G. von der Schulenburg, D. Bertrand, S.A. West, I.I. Goriacheva, I.A. Zakharov, J.H.Werren, R. Stouthamer, and M.E.N. Majerus. 1999. Male-killing *Wolbachia* in two species of insect. *Proc. R. Soc. Lond. B* 26: 735–740.

Inward, D., G. Beccaloni, and P. Eggleton. 2007. Death of an order: A comprehensive molecular phylogenetic study confirms that termites are eusocial cockroaches. *Biol. Lett.* 3: 331–335.

Ishikawa, H. 1989. Biochemical and molecular aspects of endosymbiosis in insects. *Int. Rev. Cytol.* 116: 1–45.

Ishikawa, H. 2003. Insect symbiosis: An introduction, in K. Bourtzis and T.A. Miller (eds.), *Insect Symbiosis*. CRC Press, Boca Raton, FL, pp. 1–21.

Jordahl, B.H., A. Sequeira, and A. Cognato. 2011. The age and phylogeny of wood boring weevils and the origin of subsociality. *Mol. Phylogenet. Evol.* 59: 708–724.

Jucci, C. 1952. Symbiosis and phylogenesis in the Isoptera. *Nature* 169: 837.

Kageyama, D., A. Hoshizaki, and Y. Ishikawa. 1998. Female biased sex ratio in Asian corn borer, *Ostrinia furnacalis*: Evidence for the occurrence of feminizing bacteria in an insect. *Heredity* 81: 311–316.

Kasson, M.T., K. O'Donnell, A.P. Rooney, S. Sink, R.C. Ploetz, J.N. Ploetz, J.L. Konkol et al. 2013. An inordinate fondness for *Fusarium*: Phylogenetic diversity of fusaria cultivated by ambrosia beetles in the genus *Euwallacea* on avocado and other plant hosts. *Fungal Genet. Biol.* 56: 147–157.

Kautz, S., B.E.R. Rubin, J.A. Russel, and C.S. Moreau. 2013. Surveying the microbiome of ants: Comparing 454 pyrosequencing with traditional methods to uncover bacterial diversity. *Appl. Environ. Microbiol.* 79(2): 525–534.

Kikuchi, Y., M. Hayatsu, T. Hosokawa, A. Nagayama, K. Tago, and T. Fakatsu. 2012. Symbiont-mediated insecticide resistance. *Proc. Natl. Acad. Sci. USA* 109(22): 8618–8622.

Kirejtshuk, A.G., D. Azar, R.A. Beaver, M.Y. Mandelshtam, and A. Nel. 2009. The most ancient bark beetle known; A new tribe, genus and species from Lebanese amber (Coleoptera, Curculionidae, Scolytinae). *Syst. Entomol.* 34: 101–112.

Kirkendall, L.R. 1983. The evolution of mating systems in bark and ambrosia beetles (Coleoptera: Scolytidae and Platypodidae). *Zool. J. Linnaean Soc.* 77: 293–352.

Koch, H., G. Cisarovsky, and P. Schmid-Hempel. 2012. Ecological effects on gut bacterial communities in wild bumblebee colonies. *J. Anim. Ecol.* 81: 1202–1210.

Kok, L.T. 1979. Lipids of ambrosia fungi and the life of mutualistic beetles, in L. Batra (ed.), *Insect-Fungus Symbiosis: Nutrition, Mutualism, and Commensalism*. John Wiley & Sons, Hoboken, NJ, pp. 33–52.

Kok, L.T., D.M. Norris, and H.M. Chu. 1970. Sterol metabolism as a basis for a mutualistic symbiosis. *Nature* 225: 661–662.

König, H., L. Li, and J. Fröhlich. 2013. The cellulolytic system of the termite gut. *Appl. Microbiol. Biotechnol.* 97: 7943–7962.

Krishna, K. and R.M. Weesner. 1970. *Biology of Termites II*. Academic Press, London, U.K.

Lacerda, F.G., T.M.C. Della Lucia, O.L. Periera, L. Pternelli, and M.R. Tótola. 2010. Mortality of *Atta sexdens* in contact with colony waste from different plant sources. *Bull. Entomol. Res.* 100: 99–103.

Laven, H. 1967. Eradication of *Culex pipiens fatigans* through cytoplasmic incompatibility. *Nature* 216: 383–384.

Lawrence, J.G., R.W. Hendrix, and S. Casjens. 2001. Where are the pseudogenes in bacterial genomes? *Trends Microbiol.* 9: 535–540.

Levieux, J., P. Cassier, L. Guillaumin, and A. Roques. 1991. Structures implicated in the transportation of pathogenic fungi by the European bark beetle, *Ips sexdentatus* Boerner: Ultrastructure of a mycangium. *Can. Entomol.* 123: 245–254.

Lewis, J.L. and B.T. Forschler. 2004. Protist communities from four castes and three species of *Reticulitermes* (Isoptera: Rhinotermitidae). *Ann. Entomol. Soc. Am.* 97: 1242–1251.

Little, A.E.F., T. Murakami, U.G. Mueller, and C.R. Currie. 2006. Defending against parasites: Fungus-growing ants combine specialized behaviours and microbial symbionts to protect their fungus gardens. *Biol Lett.* 2: 12–16.

Ljungdahl, L.G. and K.-E. Eriksson. 1985. Ecology of microbial cellulose degradation. *Adv. Microbiol. Ecol.* 8: 237–299.

Logan, J.A., W.W. Macfarlane, and L. Wilcox. 2010. White bark pine vulnerability to climate-driven mountain pine beetle disturbance in the Greater Yellowstone Ecosystem. *Ecol. Appl.* 20: 895–902.

Lombardero, M.J., M.P. Ayres, R.W. Hofstetter, J.C. Moser, and K.D. Klepzig. 2003. Strong indirect interactions of *Tarsonemus* mites (Acarina: Tarsonemidae) and *Dendroctonus frontalis* (Coleoptera: Scolytidae). *Oikos* 102: 243–252.

Lu, M., M.J. Wingfield, N.E. Gillete, S.R. Mori, and J. Sun. 2010. Complex interactions among pine hosts and fungi vectored by an invasive bark beetle. *PLoS One* 2: e1302.

Majerus, M.E.N., J. Hinrich, G.V.D. Schulenburg, and I.A. Zakharov. 2000. Multiple causes of male-killing in a single sample of the two-spot ladybird, *Adalia bipunctata* (Coleoptera: Coccinellidae) from Moscow. *Heredity* 84: 605–609.

Margulis, L. and R. Fester. 1991. *Symbiosis as a Source of Evolutionary Innovation: Speciation and Morphogenesis*. MIT Press, Boston, MA.

Marvaldi, A.E., A.S. Sequiera, C.W. O'Brien, and B.W. Farell. 2002. Molecular and morphological phylogenetics of weevils (Coleoptera, Curculionoidea): Do niche shifts accompany diversification? *Syst. Biol.* 51: 761–785.

Mattoso, T.C., D.D.O. Moreira, and R.I. Samuels. 2012. Symbiotic bacteria on the cuticle of the leaf-cutting ant *Acromyrmex subterraneus subterraneus* protect workers from attack by entomopathogenic fungi. *Biol. Lett.* 8: 461–464.

Mauldin, J.K. 1982. Lipid synthesis from [^{14}C]-acetate by two subterranean termites, *Reticulitermes flavipes* and *Coptotermes formosanus*. *Insect Biochem.* 12: 193–199.

McCutcheon, J.P., B.R. McDonald, and N.A. Moran. 2009. Convergent evolution of metabolic roles in bacterial co-symbionts of insects. *Proc. Natl. Acad. Sci. USA* 106: 15394–15399.

McCutcheon, J.P. and N.A. Moran. 2010. Functional convergence in reduced genomes of bacterial symbionts spanning 200 My of evolution. *Genome Biol. Evol.* 2: 708–718.

McCutcheon, J.P. and N.A. Moran. 2012. Extreme genome reduction in symbiotic bacteria. *Nat. Rev. Microbiol.* 10: 13–26.

McFall-Ngai, M., M.G. Hadfield, T.C. Bosch, H.V. Carey, T. Domazet-Lošo, A.E. Douglas, N. Dubilier et al. 2013. Animals in a bacterial world, a new imperative for the life sciences. *Proc. Natl. Acad. Sci. USA* 110(9): 3229–3236.

Mikheyev, A.S., U.G. Mueller, and P. Abbot. 2010. Comparative dating of attine ant and lepiotaceous cultivar phylogenies reveals coevolutionary synchrony and discord. *Am. Nat.* 175: 126–133.

Mittler, T.E. 1971. Dietary amino acid requirement of the aphid *Myzus persicae* affected by antibiotic uptake. *J. Nutr.* 101: 1023–1028.

Morales-Jimenez, J., G. Zuniga, H.C. Ramirez-Saad, and C. Hernandez-Rodriguez. 2012. Gut-associated bacteria throughout the life cycle of the bark beetle *Dendroctonus rhizophagus* Thomas and Bright (Curculionidae: Scolytinae) and their cellulolytic activities. *Microb. Ecol.* 64: 268–278.

Morales-Ramos, J.A., M.G. Rojas, H. Sittertz-Bhatkar, and G. Saldana. 2000. Symbiotic relationship between *Hypothenemus hampei* (Coleoptera: Scolytidae) and *Fusarium solani* (Moniliales: Tuberculariaceae. *Ann. Entomol. Soc. Am.* 93: 541–547.

Moran, N.A. and P. Baumann. 1994. Phylogenetics of cytoplasmically inherited microorganisms of arthropods. *Trends Ecol. Evol.* 9: 15–20.

Morelos-Juárez, C., T.N. Walker, J.F.S. Lopes, and W.O.H. Hughes. 2010. Ant farmers practice proactive personal hygiene to protect their fungal crop. *Curr. Biol.* 20: R553–R554.

Moser, J.C. 1963. Contents and structure of *Atta texana* nest in summer. *Ann. Entomol. Soc. Am.* 56: 286–291.

Mueller, U.G. 2012. Symbiont recruitment versus ant-symbiont coevolution in the attine ant-microbe symbiosis. *Curr. Opin. Microbiol.* 15: 269–277.

Mueller, U.G., N. Gerardo, D. Aanen, D.L. Six, and T. Schultz. 2005. The evolution of agriculture in insects. *Annu. Rev. Ecol. Evol. Syst.* 36: 563–595.

Mueller, U.G., T.R. Schultz, C.R. Currie, R.M.M. Adams, and D. Mallock. 2001. The origin of attine ant-fungus mutualism. *Q. Rev. Biol.* 76: 169–197.

Munson, M.A., P. Baumann, M.A. Clark, L. Baumann, N.A. Moran, D.J. Voegtlin, and B.C. Campbell. 1991. Evidence for the establishment of aphid-eubacterium endosymbiosis in an ancestor of four aphid families. *J. Bacteriol.* 173: 6321–6324.

Nair, J.R., G. Singh, and V. Sekar. 2002. Isolation and characterization of a novel bacillus strain from coffee phyllosphere showing anti-fungal activity. *J. Appl. Microbiol.* 93: 772–780.

Nakabachi, A. and H. Ishikawa. 1999. Provision of riboflavin to the host aphid, *Acyrthosiphon pisum*, by endosymbiotic bacteria, *Buchnera*. *J. Insect Physiol.* 45: 1–6.

Narita, S. and D. Kageyama. 2009. *Wolbachia*-induced sex reversal in Lepidoptera, in K. Bourtzis and T.A. Miller (eds.), *Insect Symbiosis*, vol. 3. CRC Press, Boca Raton, FL, pp. 295–319.

Narita, S., D. Kageyama, M. Nomura, and T. Fukatsu. 2007. Unexpected mechanism of symbiont-induced reversal of insect sex: Feminizing *Wolbachia* continuously acts on the butterfly *Eurema hecabe* during larval development. *Appl. Environ. Microbiol.* 73: 4332–4341.

Nasir, H. and H. Noda. 2003. Yeast-like symbiotes as a sterol source in anobiid beetles (Coleoptera: Anobiidae): Possible metabolic pathways from fungal sterols of 7-dehydrocholesterol. *Arch. Insect Biochem. Physiol.* 52: 175–182.

Nayduch, D. and S. Aksoy. 2007. Refractoriness in tsetse flies (Diptera: Glossinidae) may be a matter of timing. *J. Med. Entomol.* 44: 660–665.

Nikoh, N., T. Hosokawa, M. Moriyama, K. Oshima, M. Hattori, and T. Fukatsu. 2014. Evolutionary origin of insect–*Wolbachia* nutritional mutualism. *Proc. Natl. Acad. Sci. USA* 111(28): 10257–10262.

Nikoh, N., T. Hosokawa, K. Oshima, M. Hattori, and T. Fukatsu. 2011. Reductive evolution of bacterial genome in insect gut environment. *Genome Biol. Evol.* 3: 702–714.

Norris, D.M. 1979. The mutualistic fungi of the Xyleborini beetles, in L. Batra (ed.), *Insect-Fungus Symbiosis: Nutrition, Mutualism, and Commensalism*. John Wiley & Sons, Hoboken, NJ, pp. 53–64.

Norris, D.M., J.K. Baker, and H.M. Chu. 1969. Symbiotic interrelationships between microbes and ambrosia beetles III. Ergosterol as the source of sterol to the insect. *Ann. Entomol. Soc. Am.* 62: 413–414.

O'Neill, S.L., R. Giordano, A.M. Colbert, T.L. Karr, and H.M. Robinson. 1992. 16S rRNA phylogenetic analysis of the bacterial endosymbionts associated with cytoplasmic incompatibility in insects. *Proc. Natl. Acad. Sci. USA* 89: 2699–2702.

O'Neill, S.L. and Karr, T.L. 1990. Bidirectional cytoplasmic incompatibility between conspecific populations of *Drosophila simulans*. *Nature* 348: 178–180.

O'Neill, S.L, J.H. Werren, and A.A. Hoffmann. 1997. *Influential Passengers: Inherited Microorganisms and Arthropod Reproduction*. Oxford University Press, Oxford, U.K.

Odelson, D.A. and J.A. Breznak. 1983. Volatile fatty acid production by the hind-gut microbiota of xylophagous termites. *Appl. Environ. Microbiol.* 45: 1602–1613.

Odelson, D.A. and J.A. Breznak. 1985a. Nutrition and growth characteristics of *Trichomitopsis termopsidis*, a cellulolytic protozoan from termites. *Appl. Environ. Microbiol.* 49: 614–621.

Odelson, D.A. and J.A. Breznak. 1985b. Cellulase and other polymer-hydrolyzing activities of *Trichomitopsis termopsidis*, a symbiotic protozoan from termites. *Appl. Environ. Microbiol.* 49: 622–626.

Ohkuma, M. 2008. Symbiosis of flagellates and prokaryote in the gut of lower termites. *Trends Microbiol.* 16: 345–352.

Oliver, K.M., P.H. Degnan, G.R. Burke, and N.A. Moran. 2010. Facultative symbionts in aphids and the horizontal transfer of ecologically important traits. *Annu. Rev. Entomol.* 55: 247–266.

Pais, R., C. Lohs, Y. Wu, J. Wang, and S. Aksoy. 2008. The obligate mutualist *Wigglesworthia glossinidia* influences reproduction, digestion. and immunity processes of its host, the tsetse fly. *Appl. Environ. Microbiol. Lett.* 74: 5965–5974.

Palin, R. and A. Geitmann. 2012. The role of pectin in plant morphogenesis. *Biosystems* 109: 397–402.

Pester, M. and A. Brune. 2007. Hydrogen is the central free intermediate during lignocellulose degradation by termite gut symbionts. *ISME J.* 1: 551–565.

Pettolino, F.A., C. Walsh, G.B. Fincher, and A. Bacic. 2012. Determining the polysaccharide composition of plant cell walls. *Nat. Protoc.* 7: 1590–1607.

Poulsen, M. and J.J. Boomsma. 2005. Mutualistic fungi control crop diversity in fungus-growing ants. *Science* 307: 741–744.

Poulsen, M., A.N.M. Bot, M.G. Nielsen, and J.J. Boomsma. 2002. Mutualistic bacteria and a possible trade-off between alternative defense mechanisms in *Acromyrmex* leaf-cutting ants. *Insect. Soc.* 49: 15–19.

Poulsen, M. and C.R. Currie. 2006. Complexity of insect-fungal associations: Exploring the influence of microorganisms on the attine ant-fungus symbiosis, in K. Bourtzis and T.A. Miller (eds.), *Insect Symbiosis*, vol. 2. CRC Press, Taylor & Francis Group, Boca Raton, FL, pp. 57–77.

Powell, R. and D. Stradling. 1986. Factors influencing the growth of *Attamyces bromatificus*, a symbiont of attine ants. *Trans. Br. Mycol. Soc.* 87: 205–213.

Prestwich, G.D., R.W. Jones, and M.S. Collins.1981. Terpene biosynthesis by nasute termite soldiers (Isoptera: Nasutitermitinae). *Insect Biochem.* 11: 331–336.

Priya, N.G., A. Ojha, M.K. Kajla, A. Raj, and R. Rajagopal. 2012. Host plant induced variation in gut bacteria of *Helicoverpa armigera*. *PLoS One* 7: e30768.

Prosser, W.A. and A.E. Douglas. 1991. The aposymbiotic aphid: An analysis of chlortetracycline-treated pea aphid. *Acyrthosiphon pisum*. *J. Insect Physiol.* 37: 713–719.

Quinlan, R.J. and J.M. Cherrett. 1978. Aspects of the symbiosis of the leaf-cutting ant *Acromyrmex octospinosus* and its food fungus. *Ecol. Entomol.* 3: 221–230.

Raffa, K.F. 2014. Terpenes tell different tales at different scales: Glimpses into chemical ecology of conifer-bark beetle-microbial interactions. *J. Chem. Ecol.* 40: 1–20.

Ridley, E.V., A.C. Wong, S. Westmiller, and A.E. Douglas. 2012. Impact of the resident microbiota on the nutritional phenotype of *Drosophila melanogaster. PLoS One* 7: e36765.

Rio, R.V, C. Lefèvre, A. Heddi, and S. Aksoy. 2003. Comparative genomics of insect-symbiotic bacteria: Influence of host environment on microbial genome composition. *Appl. Environ. Microbiol.* 69: 6825–6832.

Rispe, C. and N.A. Moran. 2000. Accumulation of deleterious mutations in endosymbionts: Muller's ratchet with two levels of selection. *Am. Nat.* 156: 425–441.

Rohrmann, G.F. and A.Y. Rossman. 1980. Nutrient strategies of *Macrotermes ukuzii* (Isoptera: Termitidae). *Pedobiologia* 20: 61–73.

Rousset, F., D. Bouchon, B. Pintureau, P. Juchault, and M. Solignac. 1992. *Wolbachia* endosymbionts responsible for various alterations of sexuality in arthropods. *Proc. R. Soc. Lond. B* 250: 91–98.

Rumbold, C.T. 1931. Two blue-staining fungi associated with bark beetle infestation of pines. *J. Agric. Res.* 43: 847–873.

Russell, C.W., S. Bouvaine, P.D. Newell, and A.E. Douglas. 2013. Shared metabolic pathways in a coevolved insect-bacterial symbiosis. *App. Environ. Microbiol.* 79(19): 6117–6123.

Ryu, J.H., S.H. Kim, H.Y. Lee, J.Y. Bai, Y.D. Nam, J.W. Bac, D.G. Lee, S.C. Shin, E.M. Ha, and W.J. Lee. 2008. Innate immune homeostasis by the homeobox gene caudal and commensal-gut mutualism in *Drosophila*. *Science* 319: 777–782.

Safranyik, L., A.L. Carroll, J. Regniere, D.W. Langor, W.G. Riel, T.L. Shore, B. Peter, B.J. Cooke, V.G. Nealis, and S.W. Taylor. 2010. Potential for range expansion of mountain pine beetle into the northern boreal forest of North America. *Can. Entomol.* 142: 415–442.

Sasaki, T. and H. Ishikawa. 1995. Production of essential amino acids from glutamate by mycetocyte symbionts of the pea aphid *Acyrthosiphon pisum*. *J. Insect Physiol.* 41: 81–86.

Scharf, M.E., E.S. Kovaleva, S. Jadhao, J.H. Campbell, G.W. Buchman, and D.G. Boucias. 2010. Functional and translational analyses of a beta-glucosidase gene (glycosyl hydrolase family 1) isolated from the gut of the lower termite *Reticulitermes flavipes*. *Insect Biochem. Mol. Biol.* 40: 611–620.

Schoenian, I., M. Spiteller, M. Ghaste, R. Wirth, H. Herz, and D. Spiteller. 2011. Chemical basis of the synergism and antagonism in microbial communities in the nests of leaf-cutting ants. *Proc. Natl. Acad. Sci. USA* 108: 1955–1960.

Schultz, T.R. and S.G. Brady. 2008. Major evolutionary transitions in ant agriculture. *Proc. Natl. Acad. Sci. USA* 105: 5435–5440.

Scott, J.J., K.J. Budsberg, G. Suen, D.L. Wixon, T.C. Balser, and R.C. Currie. 2010. Microbial community structure of leaf-cutter ant fungus gardens and refuse dumps. *PLoS One* 5: e9922.

Seipke, R.F., J. Barke, C. Brearley, L. Hill, D.W. Yu, R.J. Goss, and M.I. Hutchings. 2011. A single *Streptomyces* symbiont makes multiple antifungals to support the fungus farming ant *Acromyrmex octospinosus*. *PLoS One* 6: e22028

Sen, R., H.D. Ishak, D. Estrada, S.E. Dowd, E. Hong, and U.G. Mueller. 2009. Generalized antifungal activity and 454-screening of *Pseudonocardia* and *Amycolatopsis* bacteria in nests of fungus-growing ants. *Proc. Natl. Acad. Sci. USA* 106: 17805–17810

Shigenobu, S., H. Watanabe, M. Hattori, Y. Sakaki, and H. Ishikawa. 2000. Genome sequence of the endocellular bacterial symbiont of aphids *Buchnera* sp. APS. *Nature* 407: 81–86.

Shin, S.C., S.-H. Kim, H. You, B. Kim, A.C. Kim, K.-A. Lee, J.-H. Yoon, J.-H. Ryu, and W.-J. Lee. 2011. *Drosophila* microbiome modulates host developmental and metabolic homeostasis via insulin signaling. *Science* 334: 670–674.

Shoemaker, D.D., V. Katju, and J. Jaenike. 1999. *Wolbachia* and the evolution of reproductive isolation between *Drosophila recens* and *Drosophila subquinaria*. *Evolution* 53: 11576–1164.

Six, D.L. 2003. Bark beetle-fungus symbioses, in K. Bourtzis and T.A. Miller (eds.), *Insect Symbiosis*. CRC Press, Boca Raton, FL, pp. 97–114.

Six, D.L. 2009. Climate change and mutualism. *Nat. Microbiol. Rev.* 7: 686.

Six, D.L. 2012. Ecological and evolutionary determinants of bark beetle-fungus symbioses. *Insects* 3: 339–366.

Six, D.L. 2013. Bark beetle holobiont: Why microbes matter. *J. Chem. Ecol.* 39: 989–1002.

Six, D.L. and T.D. Paine. 1998. Effects of mycangial fungi and host tree species on progeny survival and emergence of *Dendroctonus ponderosae* (Coleoptera: Scolytidae). *Environ. Entomol.* 27: 1393–1401.

Six, D.L. and T.D. Paine. 1999. A phylogenetic comparison of ascomycete mycangial fungi and *Dendroctonus* bark beetles (Coleoptera: Scolytidae). *Ann. Entomol. Soc. Am.* 92: 159–166.

Six, D.L, M. Poulsen, A.K. Hansen, M.J. Wingfield, J. Roux, P. Eggleton, B. Slippers, and T.D. Paine. 2011. Anthropogenic effects on insect-microbial symbioses in forest and savanna ecosystems. *Symbiosis* 53: 101–121.

Six, D.L. and M.J. Wingfield. 2011. The role of phytopathogenicity in bark beetle-fungus symbioses: A challenge to the classic paradigm. *Annu. Rev. Entomol.* 56: 255–272.

Slansky, F. and P. Feeny. 1997. Stabilization of the rate of nitrogen accumulation by larvae of the cabbage butterfly on wild and cultivated plants. *Ecol. Monogr.* 47: 209–228.

Sloan, D.B. and N.A. Moran. 2012. Endosymbiotic bacteria as a source of carotenoids in whiteflies. *Biol. Lett.* 8: 986–989.

Snyder, A.K., J.W. Deberry, L. Runyen-Janecky, and R.V. Rio. 2010. Nutrient provisioning facilitates homeostasis between tsetse fly (Diptera: Glossinidae) symbionts. *Proc. R. Soc. Lond. B* 277: 2389–2397.

Stone, C. and J.A. Simpson. 1990. Species associations in *Ips grandicollis* galleries in *Pinus taeda*. *N. Z. J. For. Sci.* 20: 75–96.

Storelli, G., A. Defaye, B. Erkosar, P. Hols, J. Royet, and F. Leulier. 2011. *Lactobacillus plantarum* promotes *Drosophila* systemic growth by modulating hormonal signals through TOR-dependent nutrient sensing. *Cell Metab.* 14: 403–414.

Stouthamer, R., J.A.J. Breeuwer, and G.D.D. Hurst. 1999. *Wolbachia pipientis*: Microbial manipulator of arthropod reproduction. *Annu. Rev. Microbiol.* 53: 71–102.

Stouthamer, R., J.A.J. Breeuwer, R.F. Luck, and J.H. Werren. 1993. Molecular identification of microorganisms associated with parthenogenesis. *Nature* 361: 66–68.

Stouthamer, R., R.F. Luck, and W.D. Hamilton. 1990. Antibiotics cause parthenogenetic *Trichogramma* to revert to sex. *Proc. Natl. Acad. Sci. USA* 87: 2424–2427.

Stouthamer, R., M. van Tilborg, J.H. de Jong, L. Nunney, and R.F. Luck. 2001. Selfish element maintains sex in natural populations of a parasitoid wasp. *Proc. R. Soc. Lond. B* 268: 617–622.

Sun, J., L. Min, N.E. Gillette, and M.J. Wingfield. 2013. Red turpentine beetle: Innocuous native becomes tree killer in China. *Annu. Rev. Entomol.* 58: 293–311.

Tartar, A., M.M. Wheeler, X. Zhou, M.R. Coy, D.G. Boucias, and M.E. Scharf. 2009. Parallel metatranscriptome analyses of host and symbiont gene expression in the gut of the termite *Reticulitermes flavipes*. *Biotechnol. Biofuels*, 19 pp. http://www.biotechnologyforbiofuels.com/2/1/25 (last accessed April 22, 2015).

Taysu, I. 1998. The use of carbon and nitrogen isotope ratios in termite research. *Ecol. Res.* 13: 377–387.

Thompson, B.M., R.J. Grebenok, S.T. Behmer, and D.S. Gruner. 2013. Microbial symbionts shape the sterol profile of the xylem-feeding woodwasp, *Sirex noctilio*. *J. Chem. Ecol.* 39: 129–139.

Thompson, N.S. 1983. Hemicellulose as a biomass resource, in E.J. Soltes (ed.), *Wood and Agricultural Residues. Research on Use for Feed, Fuels, and Chemicals*. Academic Press, New York, pp. 101–119.

Toh, H., B.L. Weiss, S.A. Perkin, A.Yamashita, K. Oshima, M. Hattori, and S. Aksoy. 2006. Massive genome erosion and functional adaptations provide insights into the symbiotic lifestyle of *Sodalis glossinidius* in the tsetse host. *Genome Res.* 16: 149–156.

Veivers, P.C., R.W. O'Brien, and M. Slaytor. 1980. The redox state of the gut of termites. *J. Insect Physiol.* 26: 75–77.

Veivers, P.C, R.W. O'Brien, and M. Slaytor. 1982. Role of bacteria in maintaining the redox potential in the hindgut of termites and preventing entry of foreign bacteria. *J. Insect Physiol.* 28: 947–951.

von der Schulenburg, J.H., G.D.D. Hurst, M.E. Huigens, M.M.M. van Meer, F.M. Jiggins, and M.E.N. Majerus. 2000. Molecular evolution and phylogenetic utility of *Wolbachia* ftsZ and wsp gene sequences with special reference to the origin of male-killing. *Mol. Biol. Evol.* 17: 584–600.

von der Schulenburg, J.H.G., M. Habig, J.J.Sloggett, K.M. Wsbberley, D. Bertrand, G.D.D. Hurst, and M.E.N. Majerus. 2001. Incidence of male-killing *Rickettsia* spp. (a-proteobacteria) in the ten-spot ladybird beetle *Adalia decempunctata* L. (Coleoptera: Coccinellidae). *Appl. Environ. Microbiol.* 67: 270–277.

Wade, M.J. 2001. Infectious speciation. *Nature* 409: 675–677.

Wang, J., Y. Wu, G.Yang, and S. Aksoy. 2009. Interactions between mutualist *Wigglesworthia* and tsetse peptidoglycan recognition protein (PGRP-LB) influence trypanosome transmission. *Proc. Natl. Acad. Sci. USA* 106: 12133–12138.

Wang, Y., L. Lim, S. Diguistini, G. Robertson, J. Bohlmann, and C. Breuil. 2013. A specialized ABC efflux transporter GcABC-G1 confers monoterpene resistance to *Grosmannia clavigera*, a bark beetle-associated fungal pathogen of pine. *New Phytol.* 197: 886–898.

Watanabe, H. and G. Tokuda. 2010. Cellulolytic systems in insects. *Annu. Rev. Entomol.* 55: 609–632.

Weber, N.A. 1982. Fungus ants, in H.R. Hermann (ed.), *Social Insects*, Vol. IV. Academic Press, London, U.K., pp. 255–363.

Weiss, B.L., R. Mouchotte, R.V. Rio, Y.N. Wu, Z. Wu, A. Heddi, and S. Aksoy. 2006. Inter-specific transfer of bacterial endosymbionts between tsetse species: Infection establishment and effect on host fitness. *Appl. Environ. Microbiol.* 72: 7013–7021.

Weiss, B.L., J. Wang, G.M. Attardo, and S. Aksoy. 2012. Host and symbiont adaptations provide tolerance to beneficial microbes *Sodalis* and *Wigglesworthia* symbioses in tsetse flies, in E. Zchori-Fein and K. Bourtzis (eds.), *Manipulative Tenants, Bacteria Associated with Arthropods*. CRC Press, Taylor & Francis Group, Boca Raton, FL, pp. 175–189.

Weiss, B.L., Y. Wu, J.J. Schwank, N.S. Tolwinski, and S. Aksoy. 2008. An insect symbiosis is influenced by bacterium-specific polymorphisms in outer-membrane protein A. *Proc.Natl.Acad. Sci. USA* 105: 15088–15093.

Wernegreen, J.J. 2012. Mutualism meltdown in insects: Bacteria constrain thermal adaptation. *Curr. Opin. Microb.* 15: 255–262.

Wernegreen, J.J., H. Ochman, I.B. Jones, and N.A. Moran. 2000. Decoupling of genome size and sequence divergence in a symbiotic bacterium. *J. Bacteriol.* 182: 3867–3869.

Werren, J.H. 1997. Biology of *Wolbachia. Annu. Rev. Entomol.* 42: 587–609.

Werren, J.H., W. Zhang, and L.R. Guo. 1995. Evolution and phylogeny of *Wolbachia*: Reproductive parasites of arthropods. *Proc. R. Soc. Lond. B* 261: 55–63.

Wheeler, M.M., M.R. Tarver, M.R. Coy, and M.E. Scharf. 2010. Characterization of four esterase genes and esterase activity from the gut of the termite *Reticulitermes flavipes. Arch. Insect Biochem. Physiol.* 73: 30–48.

Wheeler, M.M., X. Zhou, M.E. Scharf, and F.M. Oi. 2007. Molecular and biochemical markers for monitoring dynamic shifts of cellulolytic protozoa in *Reticulitermes flavipes. Insect Biochem. Mol. Biol.* 37: 1366–1374.

Whitehead, L.F. and A.E. Douglas. 1993. Populations of symbiotic bacteria in the parthenogenetic pea aphid (*Acyrthosiphon pisum*) symbiosis. *Proc. R. Soc. Lond. B* 254: 29–32.

Whitney, H.S. 1971. Association of *Dendroctonus ponderosae* (Coleoptera: Scolytidae) with blue-stain fungi and yeasts during brood development in lodgepole pine. *Can. Entomol.* 103: 1495–1503.

Wilkinson, T.L. and A.E. Douglas. 1998. Host cell allometry and regulation of the symbiosis between pea aphids, *Acyrthosiphon pisum,* and bacteria, *Buchnera. J. insect Physiol.* 44: 629–635.

Wood, S.L. 1982. The bark and ambrosia beetles of North and Central America (Coleoptera: Scolytidae), a taxonomic monograph, *Great Basin Natural Memoir 6.* 1359pp.

Wood, T.G. and R.J. Thomas. 1989. The mutualistic association between Macrotermitinae and Termitomyces, in N. Wilding, N.M. Collins, P.M. Hammond, and J.F. Weber (eds.), *Insect-Fungus Interactions.* Academic Press, New York, pp. 69–92.

Yek, S.-H. and U.G. Mueller. 2011. The metapleural gland of ants. *Biol. Rev.* 91: 201–224

Yen, J.H. and A.R. Barr. 1971. New hypothesis of the cause of cytoplasmic incompatibility in *Culex pipiens* L. *Nature* 232: 657–658.

Yen, J.H. and A.R. Barr. 1973. The etiological agent of cytoplasmic incompatibility in *Culex pipiens. J. Invertebr. Pathol.* 22: 242–250.

Zakharov, I.A., I.I. Goryacheva, E.V. Shaikevich, J.H. Schulenburg, and M.E.N. Majerus. 2000. *Wolbachia*, a new bacterial agent causing sex-ratio bias in the two-spot ladybird *Adalia bipunctata* L. *Russ. J. Genet.* 36: 385–388.

Zchori-Fein, E., R.T. Roush, and M.S. Hunter. 1992. Male production by antibiotic treatment in *Encarsia formosa* (Hymenoptera:Aphelinidae), an asexual species. *Experentia* 48: 102–105.

Zchori-Fein, E., R.T. Roush, and D. Rosen. 1998. Distribution of Parthenogenesis-inducing symbionts in ovaries and eggs of *Aphytis* (Hymenoptera: Aphelinidae). *Curr. Microbiol.* 36: 1–8.

Zhou, X., E.S. Kovaleva, D. Wu-Scharf, J.H. Campbell, G.W. Buchman, D.G. Boucias, and M.E. Scharf. 2010. Production and characterization of a recombinant beta-1,4-endoglucanase (Glycohydrolase family 9) from the termite *Reticulitermes flavipes. Arch. Insect Biochem. Physiol.* 74: 147–162.

Zhou, X., J.A. Smith, F.M. Oi, P.G. Koehler, G.W. Bennett, and M.E. Scharf. 2007. Correlation of cellulose gene expression and cellulolytic activity throughout the gut of the termite *Reticulitermes flavipes. Gene* 395: 29–39.

Zug, R. and P. Hammerstein. 2012. Still a host of hosts for *Wolbachia*: Analysis of recent data suggests that 40% of terrestrial arthropod species are infected. *PLoS One* 7(6): e38544.

Index

A

Abdominal ganglia, 255, 257–258, 265–266, 336
Absolute refractory period, 295
Accessory glands, 562, 578, 580
Acetogenesis, 596
Acetylcholine (ACh), 286, 300–303
Acheta domesticus, 9, 272, 323, 465, 504, 567
Achetakinins, 279
Acid–base homeostasis, 506–507
Acone eyes, 382
Acron, 16, 18
Across-fiber patterning, 353, 371, 536
Actin
    active state, 322
    alary muscles, 330
    F actin, 320
    G actin, 320
    heart muscles, 329
    myofibrils, 310
    myosin head, release of, 322–323
    sliding filament theory, 313
    visceral muscles, 329
Actin binding (TnI), 320
Actinomycetes, 595
Action potential
    chloride ions, 292
    cockroach giant axon, 287–288
    local circuit theory, 298–299
    Na^+ channels, 285
    Nernst equation, 294
    sodium activation, 294–296
    synchronous bursts, 112
Active space, 528
Adipohemocytes, 415–416
Adipokinetic hormone (AKH), 230, 245, 247, 277, 567
Adsorption, semiochemical detection, 533
Adult diapause, 214
Aedeagus, 580
Aedes aegypti, 14, 44, 50, 436
Aeropyles, 482, 577
Agrotis segetum, 547
Air pressure receptors, 363
AKH, *see* Adipokinetic hormone (AKH)
Alary muscles, 330 406–407, 409–410
Allatostatins (ASTs), 59, 159, 188, 280
Allatotropins (ATs), 159, 280
Allelochemicals, 525
Allomones, 524–525
α-D-glucopyranosyl-α-D-glucopyranoside, 226
Amacrine neurons, 260
Ambrosia beetles, 599–602, 604
Aminopeptidases, 50
AMMC, *see* Antennal mechanosensory and motor center (AMMC)
Ammonotelic, 508–510
Amnion, 9–11
Amnionic cavity, 10, 13
Amnioserosa, 10–11
α-Amylase, 48, 51
Anatrepsis, 11–13
Anautogenous, 566
Animal origin, 34–36
ANT-C, *see* Antennapedia complex (ANT-C)
Antennal lobe (AL)
    in brain, 260–262
    glomeruli, 262–263
Antennal mechanosensory and motor center (AMMC), 260–261
Antennapedia (Antp), 20
Antennapedia complex (ANT-C), 20
Antiaggregating pheromone, 604
Antidiuretic hormones, 504, 506
Antimicrobial peptides, 434, 438–439
Antiporter mechanism, 42, 495
Aphids
    Buchnera in, 605–606
    green vetch, 196
    light-sensitive periods, physiology, 196
    minerals, 86–87
    symbiotic relationships, 591–593
Apocrine secretion, 47
Apolysial space, 108
Apolysis, 107–108, 314
Apoptosis, 24
Approximate digestibility (AD), 88–90
Apterygota, 2, 5–7, 255, 265, 571
Apyrene sperm, 579–580
Aquaglyceroporins, 58
Aquaporins, 58, 513
Arginine, 79–80, 580
Arginine phosphate, 226
Argyrotaenia velutinana, 89
Arrestin, 391
Arresting factor, 218
Arylphorins, 217
Ascorbic acid, 84
Aseptic conditions, 77
L-aspartate, 316
Aspartic acid, 49
L-aspartic acid, 296, 300, 304
Association neurons, 271
Asynchronous muscles
    contraction frequency, 317
    control muscles, 317
    direct and indirect muscles, 319
    dorsal longitudinal and ventral, 317
    myofibrils, 319–321
    rhythmical input, 319
    tymbal muscles, 319
    wingbeats, 319
Auditory receptors, 363
Autogenous, 566, 568, 570

Axenic conditions, 77
Axenic culture, 87
Axonal process, 276–277, 286
Axon, sensory and motor neuron, 269

B

Bacterial biofilm, 595
Bacteriocytes, 593, 605–606
Bacteriosomes, 593, 608
Bactrocera dorsalis, 545–546
Ball and chain model, 291
Bark beetles
    anthropogenic effects, 605
    fungal feeding, evolution of, 603–604
    holobiont, 600, 604
    pheromones, 545
    symbionts, 599, 603
Basalar muscles, 316–317, 337, 340, 346
Basidiomycete fungi, 599
BBB, *see* Blood–brain barrier (BBB)
Bicoid protein, 16–18
Bicyclus anynana, 76, 392
Bioassay
    ecdysteroids, *Calliphora* assay, 151
    for PTTH activity, 141–143
Biological rhythms
    circadian clock
        molecular basis for, 183–186
        peripheral organs and tissues, 188–189
        in reproduction, 190–193
        in social behavior of honeybees, 189–190
    circadian rhythms, 182
        characteristics of, 182–183
        hormone secretion, circadian regulation of, 186–188
    clock genes, 186
    clock models
        external coincidence model, 197
        hourglass model, 196
        internal coincidence model, 197
        resonance model, 197–198
    hotoperiodic response, 193–195
    photoperiodic rhythms, characteristics of, 182–183
Bioluminescence, 189
Bithorax complex (BX-C), 20
Blastoderm
    Apterygota, 5–7
    blastokinesis and extraembryonic membranes, 11
    cytoplasmic strands, 5
    development, stage in, 6
    gastrulation, 9–10
    germ band
        elongation, 10–11
        formation, 9
    Hemimetabola, 7–8
    Holometabola, 8–9

Blastokinesis
    anatrepsis and katatrepsis, 13–14
    Oncopeltus fasciatus development, 11–12
    serosa and/or amnion, 14–15
Blastomeres, 5–6, 19
Blood–brain barrier (BBB), 268
Bombyx mori
    accessory heart, 413
    amino acids, 54
    diapause, 212
    embryogenesis, 2
    metamorphosis, 137
    stage-specific proteins, 126
    20-hydroxyecdysone, 124
    Vg synthesis, 571
Bouligand helicoids, 120
Brain
    CNS, 256–259
    deutocerebrum
        AL, 261–263
        AMMC, 260–261
        chemosensory neurons, 260–261
        mechanosensory neurons, 260
        tritocerebrum, 263
    hormone, 137
    NSC and PTTH
        bioassay for, 141–143
        mode of action, 145–146
        secretion, stimuli for, 143
        source and chemistry, 141
        stretch receptors, brain activation, 143–144
        tissue regulation and hemolymph levels, 144
        tobacco hornworm, 144
    protocerebrum, 259–260
    supraesophageal ganglion, 258
Brain hormone, *see* Molting, hormone
Brain neurosecretory cells
    neuropeptide secretion, 138
    and PTTH
        bioassay for, 141–143
        mode of action, 145–146
        secretion, stimuli for, 143
        source and chemistry, 141
        stretch receptors, brain activation, 143–144
        tissue regulation and hemolymph levels, 144
        tobacco hornworm, 144
Brush border, 44
Buccal cavity, 36
Buchnera, aphids, 605–606
Bursicon, 110, 113, 116, 138
BX-C, *see Bithorax complex (BX-C)*

C

Cahn–Ingold–Prelog system, 530–531
Calliphora assay, 151
cAMP, *see* Cyclic adenosine monophosphate (cAMP)
Campesterol, 148–150
CAMs, *see* Cell adhesion molecules (CAMs)
Cap cell, 358–359

Carbohydrate-digesting enzymes, 48–49
Carbohydrates
    digesting enzymes, 48–49
    hormones controlling, 230
    nutrition, requirements, 81
    resources
        glycogen, 229–230
        trehalose, 226–229
Carboxypeptidase, 51, 580
Cardiac sphincter, 37
Cardioactive peptide, 280
Carnitine, 84, 246–248
Carnivorous feeding, 34
Carotene, 84
Carotenoids, 394–395
Cascade of gene action, 16, 19
Cathepsins, 49–50
CCAP, see Crustacean cardioactive peptide (CCAP)
CCSF, see Corpora cardiaca stimulating factor (CCSF)
Cell adhesion molecules (CAMs), 22
Cellobiohydrolases, 598
Cell-to-cell interaction, 15
Cellular defenses, 434
Cellulolytic enzymes, 598
Cement, 101–103
Central body complex, 260
Central command (COM) neurons, 274
Central nervous system (CNS)
    AKH, 278
    ATs and ASTs, 280
    brain
        deutocerebrum, 260–263
        oxygen and glucose supply, 266–267
        protocerebrum, 259–260
        structure of, 256–257
        supraesophageal ganglion, 258
    cardioactive peptide, 280
    FMRFamides, 278–279
    giant axons, 273
    nerve cord, 256–258
    NSC, 276–278
    PBAN, 280
    pigment-dispersing factor, 279
    proctolin, 278
    tachykinins, 279
    vasopressin, 279–280
    ventral ganglia
        abdominal, 265–266
        lateral nerves, 266
        mesothoracic ganglia, 264
        metathoracic ganglia, 264
        oxygen and glucose supply, 266–267
        prothoracic ganglia, 264
        structure, 256–258
Centrifugal neurons (CNs), 262
Centrolecithal, 3
Chemiosmotic hypothesis, 238, 240
Chemoreceptors
    gustatory receptors, 369–370
    olfactory sensilla, 368–369

Phormia regina, labellum of, 354–355
specialists *vs.* generalists, 371
stimulus–receptor excitation coupling, 371–372
Chiral, 530–531
Chirality, 530–531
Chitin
    biosynthesis of, 121–123
    bouligand helicoids, 120
    α-chitin, 119
    β-chitin, 120
    γ-chitin, 120
    ^{13}C-NMR, determination, 118
    cuticle, chemical composition, 116
    enzymatic hydrolysis, 116
    inertness and insolubility of, 119
    thoracic cuticle, freeze-fractured break, 121
α-Chitin, 118–119
β-Chitin, 120
Chitinase, 48, 108–109, 435
Cholesterol, 82–83, 146–149
Chordotonal sensillum, 358–359
Chorion, 3, 482, 561, 564, 576–577
Chromophore, 378, 388
Chromosomal puffs, 164–166
Chromosome dosage compensation, 581
Chymotrypsin, 49–50
Chymotrypsin-like endoproteinases, 50
Ciliary constriction, 367–368
Circadian clocks
    diapause, 215
    external coincidence model, 197
    hourglass model, 196
    internal coincidence model, 197
    molecular basis for, 183–186
    peripheral organs and tissues, 188–189
    in reproduction, 190–193
    resonance model, 197–198
    in social behavior of honeybees, 189–190
Circadian rhythms
    characteristics of, 182–183
    clocks (*see* Circadian clocks)
    diapause, 215
    hormone secretion, circadian regulation of, 186–188
Circulatory system
    accessory pulsatile hearts, 413–414
    circulation rate, 428
    dorsal vessel, heart and aorta
        abdominal portion, 406
        alary muscles, 407, 409–410
        in *Blaberus* spp., 406–407
        cardioactive secretions, 412
        circular and longitudinal muscle
            fibers, 406, 408
        heartbeat, 410–411
        ionic influences, 411
        loops, 406, 408
        nerve supply, 411–412
        ostia, 407, 410
        pericardial sinus, 408–409
        perineural sinus, 408–409

perivisceral cavity, 408–409
thoracic portion, 406
embryonic development, 405
hemocytes
adipohemocytes, 416
appearance and shape changes, 414–415
coagulocytes, 416
functions, 416–417
granulocytes, 415–416
hemocytopoietic tissues, 417–418
number, 418–420
oenocytoids, 416
origin of, 418
plasmatocytes, 415
prohemocytes, 414
spherulocytes, 416
hemoglobin, 428–429
hemolymph (see Hemolymph)
Clap and fling wing motion, 342
Clathrin protein coat, 574
Climate change, 605
Clock genes, 183, 186, 192–193, 215
Close junctions, 104–105
Close-packed fibers, 319
CNS, see Central nervous system (CNS)
Coagulation of hemolymph, 422–424, 434
Coagulocytes, 415–416
Coccoid bacteria, 605
Coleoptera, 61, 478, 491, 511–512, 571
Columnar cells, 40–41
Columnar neurons, 260
Common oviduct, 562
Compound eyes
Cambrian period, 378
dorsal rim area of, 396–397
Drosophila, 215
facets of, 397
lens, 378
ommatidia (see Ommatidia)
tabanid fly, 383
Corazonin, 110, 113, 280, 412
Cornea
corneal layering, 383
dioptric structures, 381–382
ocellus, 385
Corneagenous cells, 381
Corneal layering, 383
Corpora allata (CA)
immunostaining, 188
and JH
assays for, 157
downstream transcription factors, 163–164
Galleria wax test, 158
glandular source and chemistry, 155–157
growth regulators and compounds, 161
mode of action and receptors, 162–163
molting and metamorphosis, 137
previtellogenesis, hormonal control of, 569
Corpora cardiaca stimulating factor (CCSF), 570
Corpora pedunculata, 259

Corticotropin-releasing factor (CRF), 504
Countercurrent circulation, 53–54
Crop, 36–38, 55, 58–59
Crustacean cardioactive peptide (CCAP), 110, 113, 280, 412
Crustecdysone, 148
Cryptoglossa verrucosa, 127
Cryptonephridial tubules, 490, 511–512
Cryptosolenic tubules, 490–491
Crystalline tract, 380, 382
Crystallite, 119–120
Cumulus, 542
Cuticle
integument
chitin (see Chitin)
cuticular proteins, 122–125
formation, 108–109
lipids, 126–129
mineralization of, 129
protective functions, 126
resilin, 125
stage-specific differences, 126
secretion, embryo, 23
serosal, formation, 14
Cuticular epidermis, 189
Cuticular epithelium, 103, 463
Cuticular proteins
chemical composition, 122–125
protective functions of, 126
stage-specific differences in, 126
Cuticulin envelope, 101
Cyclic adenosine monophosphate (cAMP), 145–146, 229, 501, 503, 544
Cyclic GMP (cGMP), 112
Cyclorrhaphous Diptera, 571
Cydia nigricana, 550
Cysteine, 49, 167
Cytoplasmic glycerol-3-phosphate dehydrogenase, 234
Cytoplasmic incompatibility (CI), 608–609

D

Dance communication, 189
Dauer, 141, 143
Dendrites, 269–270, 370
Dendritic processes, 286
de novo synthesis, 546–547
Depolarization, 289, 291, 294, 298, 411
Deposit excretion, 490
Dermal light sense, 387
Desmosomes, 9
Deterrents, feeding, 92
Deutocerebrum
across-glomeruli pattern, 536
AL, 261–263, 542
AMMC, 260–261
brain, consists of, 258–259
chemosensory neurons, 260–261
mechanosensory neurons, 260
tritocerebrum, 263

Devonian period, 378
DGC, see Discontinuous gas exchange (DGC)
Diacylglycerol (DAG), 54, 224, 389
Diapause
    behavior and physiology, 193
    daily and seasonal biological clocks, 215
    dormancy and, 207
    in embryonic stage, 208
    environmental conditions, 208
    and gene expression, 215–216
    hormonal control
        adult diapause, 214
        embryonic diapause, 212–213
        larval diapause, 213
        pupal diapause, 213–214
    molecular studies of, 217–218
    nutrient accumulation for, 216–217
    nutrients storage and conservation, 216–217
    phases of
        induction and preparation, 210–211
        initiation and maintenance, 211–212
        termination, 212
    Pieris napi, 192
    quiescence and, 207–208
    seasonal morph, 195
    states of, 207–208
    survival strategy, 208–210
Dictyoptera, 59, 418, 563, 567
DIF, see Dorsal immune factor (DIF)
Digestion
    absorption of, 53–54
    digestive enzymes
        carbohydrate, 48–49
        control mechanisms, 47
        holocrine secretion, 47
        lipid, 49
        merocrine secretion, 47
        microapocrine secretion, 47–48
        protein, 49–50
        proteinase inhibitors, 51
    digestive system morphology and physiology
        Coleoptera, 61
        Dictyoptera, 59
        Diptera, 62–63
        Hemiptera, 60–61
        Homoptera, 61
        Hymenoptera, 62
        Isoptera, 59–60
        Lepidoptera, 63–64
        Orthoptera, 58–59
    foregut, 36–38
    glycocalyx, 45
    gut
        pH, 55–57
        TEP and redox potential, 55
    hematophagy, 57–58
    hindgut, 36, 39–40
    insects, 33–34
    midgut, 36, 39
        columnar cells, 40–41
        countercurrent circulation, 53–54
        goblet cells, 41–44
        hormonal influence, 51–53
        microvilli/brush border of, 44
        regenerative cells, 40–41
    peritrophic matrix
        components, 46
        functions of, 46–47
        type I, 45
        type II, 45
    plant and animal disease organisms, spread control, 64–65
    population management, potential target, 64–65
    solid *vs.* liquid diet, 34–36
Digestive enzymes
    carbohydrate, 48–49
    control mechanisms, 47
    lipid, 49
    protein, 49–50
    proteinase inhibitors, 51
    secretion of, 47
    ways of, secretion, 47–48
3,4-Dihydroxy benzoic acid, 116
Dimerization, 169
Dimers, 163, 572
Dioptric structures, 379, 381–383
Diphenoloxidase, 116
Diploid, 581
Diploptera punctata, 158, 280
Diptera
    digestive system, 62–63
    gut pH, 57
    imaginal discs, 24
    ovary development, hormonal regulation of, 567
    tympanal organs, 361
    YPs of, 573–574
Direct wing muscles, 317, 337
Discontinuous gas exchange (DGC)
    closed period, 471–472
    definitive evolutionary mechanism, 474–475
    DGC cycle, 471
    diapausing pupae, 473
    discontinuous release of CO_2, 471
    discontinuous ventilation cycle, 471
    evolutionary driving mechanism, 474
    feedback systems, 474
    fluttering phase, 472
    flutter period, 471–472
    gas exchange measurement, 472
    hypoxic and hypercapnic theory, 474
    intratracheal pressure, 472
    low oxygen tension, 473
    open period, 471–472
    passive suction ventilation, 471–472
    spiracular opening and closing, 472
    tightly/apparently closed, 471
    tracheal gas composition, 472
Disulfide bonds, 141, 505
Dityrosine, 125

Diuretic hormones, 504–506
DLM, see Dorsal longitudinal muscles (DLM)
DNA-binding domain, 166
Dormancy, 207
Dorsal immune factor (DIF), 439–440
Dorsal longitudinal muscles (DLM), 316–317, 336–339
Dorsal motor neuropil, 266
Dorsal organ, 6–7
Dorsal protein, 19
Dorsal unpaired median (DUM)
    motoneurons, 270–271
Dorsal–ventral, 19
Dorsal vessel
  heart and aorta
    abdominal portion, 406
    alary muscles, 407, 409–410
    in *Blaberus* spp., 406–407
    cardioactive secretions, 412
    circular and longitudinal muscle fibers, 406, 408
    heartbeat, 410–411
    ionic influences, 411
    loops, 406, 408
    nerve supply, 411–412
    ostia, 407, 410
    pericardial sinus, 408–409
    perineural sinus, 408–409
    perivisceral cavity, 408–409
    thoracic portion, 406
  honeybee, body outline, 34
Dorsoventral muscles (DVMs), 317, 336–339
Downstroke
  dragonflies and damselflies, 340–341
  lift forces, 342–344
  pitch control and wings twisting, 346
  wings, 338
Drosophila
  accessory glands, 580
  antimicrobial peptides, 439–441
  bicoid gene and anterior determination in, 16–18
  CCAP, 113
  cell death, 24
  circadian oscillations, 188
  clock model, 182–184
  cry-d gene, 186, 189
  developmental stages of, 2–3
  DLM and DVM, 337–338
  dorsal longitudinal flight muscle, 310
  dumpy gene, 314
  ecdysone receptor, 166
  ecdysteroid secretion, 140
  flight mechanics, 336
  gender determination, 582
  genetic control, development, 15
  gustatory receptors, 370
  hemocytes, 416
  imaginal discs (*see* Imaginal discs)
  insulin signaling pathway, 159
  JH, 162–163
  male reproductive system, 577–578
  Malpighian tubules, 189
  neuronal network, 185
  neuropil, 267–268
  neurotransmitters, 260
  nutrient imbalance, 76
  PGRP, 438
  polytrophic ovarioles, 565
  ponasterone, 167
  potassium ion channels, 291–292
  PRPs, 439
  puff patterns, 164–165
  reproduction, circadian clock, 190–191
  retinula cells, 383
  segmentation genes, 19
  simple chordotonal organs, 364
  stage-specific proteins, 126
  taste receptors, 188
  water balance, 475
  YPs, 573–574
DUM motoneurons, see Dorsal unpaired median (DUM) motoneurons
Dumpy protein, 314
DVMs, see Dorsoventral muscles (DVMs)
Dyad/triad junctions, 310
Dysdercus fasciatus, 35, 149, 152, 504

E

Early trypsin, 51–52
Ecdysial membrane, 108
Ecdysiostatin, 52
Ecdysis
  Hyalophora cecropia, VNC, 111–112
  larval monarch, 111
  motor program controlling, 274
  new cuticle, waterproofing, 113–114
  phases, 110
  post-ecdysis wing expansion, 113–114
  ptilinum, 109–110
  PTTH, 110
  sclerotization, 114–116
  shedding, old cuticle, 112–113
Ecdysone
  biosynthesis of, 146–148
  conversion of, 148
  cuticle, 110
  degradation of, 153–154
  EREs, 167
  ovary development, hormonal regulation of, 567
  receptor, 166
  RIA, 151–152
β-Ecdysone, 148
Ecdysone response elements (EREs), 167
Ecdysteroid-binding domain, 166
Ecdysteroids
  gene level, mode of action of
    chromosomal puffs, 164–166
    differential tissue and cell response, 168–170
    ecdysteroid receptor isolation, 166–168
  hormone control development, 138–140

ovary development, hormonal regulation of, 567
prothoracic glands and
    assays, tissues and cell cultures, 153
    Calliphora assay for, 151
    ecdysone (*see* Ecdysone)
    molting hormone, molecular diversity in, 148–151
    parasitoids dependence, 155
    physicochemical techniques, 152–153
    secreation, 146
    virus degradation of, 155
Ecdysterone, 148, 164
Eclosion hormone (EH), 110, 112–113
Eclosion rhythm, 191, 197
EcR, 166–169, 217
Ectoderm, 10
Ectodermal layer, 10
Edysis-triggering hormone (ETH), 110
Efficiency of conversion (ECI), 88
Egg(s)
    cortex, 8
    diapause hormone, 543
    gas exchange in, 577
    proteins, synthesis of
        Apterygota, 571
        Coleoptera, 571
        Dictyoptera, 567
        Diptera, 568
        Hemiptera, 571
        JH, 567–568, 570–571
        Lepidoptera, 571
        Orthoptera, 567–568
        stages, 568–570
    structure, 3
Egg development neurohormone (EDNH), 570
Ejaculatory duct, 578, 580
Electroantennogram (EAG), 188, 192, 535
Electrolyte homeostasis, 502–503
Electron transport system, 237–240
Electrotonus, 286–287
11-*cis*-retinal, 378–389
Embryo
    anterior and posterior, 17–18
    cuticle secretion in, 23
    development, 3, 5, 481–482
    formation of, 2
    metamerization, 11
Embryogenesis
    Apterygota, 5–7
    blastoderm, 5–6
    blastokinesis and extraembryonic membranes
        Aedes aegypti, 14
        anatrepsis, 11
        germ band stage embryo model, 11, 13
        katatrepsis, 11–13
        Oncopeltus fasciatus development, 11–13
        serosa/amnion, 14
        Tc-zen1 and *Tc-zen2*, 14–15
    divisions, 2
    Drosophila, developmental stages of, 2–3

egg, fertilization, and zygote formation, 3–4
energid, 4–5
gastrulation, 9–10
genetic control
    acron and telson, 18
    cell-to-cell interaction, 15
    classification, 15
    Drosophila, Bicoid gene and anterior determination in, 16–18
    in *Drosophila melanogaster*, 15
    posterior group genes and posterior pattern formation, 18
    regional localization, 15
germ band
    elongation, 10–11
    formation, 9
hatching, 24
Hemimetabola, 7–8
Holometabola, 8–9
homeotic genes
    Antp, 20
    gene complexes, 20
    homeobox, 21
imaginal discs
    definition, 1
    of *Drosophila melanogaster*, 25–27
    IDGF, 26–27
    JH levels, 25
    small clumps of, 24
    in tephritid fruit fly, 24
organogenesis
    apoptosis, 24
    cell movements, 23–24
    embryo, cuticle secretion in, 23
    gut development, 22–23
    malpighian tubules, 23
    neurogenesis, 21–22
    oenocytes, 23
    tracheal system, 23
    segmentation genes, 19–20
    zygotic nucleus divisions, 4
Embryonic diapause, 212–213
Enantiomers, 372, 530–531
Endocuticle, 103
Endoderm, 10
Endoectoperitrophic countercurrent flow, 53, 58–59, 61–62
Endogenous daily rhythm, 193
Endoglucanases, 49, 598
Endopeptidase, 580
Endoproteases, 49
Endosomes, 575
Endosymbionts, 592
Energid, 5
Entrained circadian rhythm, 182, 184
Environmental stimuli, 143
Epidermis, 24, 103
EPSPs, *see* Excitatory Post Synaptic Potentials (EPSPs)
Ergosterol, 82, 603
Escovopsis, 594–595
Esophageal valves, 37–38

Essential amino acids
    classic deletion method, 80
    insects, number of, 79
    isotope labeling technique, 81
Eucone eyes, 382
Eupoecilia ambiguella, 445, 541
Eupyrene sperm, 579–580
Excitatory Post Synaptic Potentials (EPSPs), 286, 300–301, 304
Excreta, 490
Excretion
    cryptonephridial systems, 513–515
    definition, 489
    hindgut function, 490
    homeostasis (*see* Homeostasis)
    malpighian tubules
        cells, 493–495
        cryptosolenic/cryptonephridial tubules, 490
        detoxification and immunity, 493
        function, 490
        proton pump (*see* Urine formation)
        small spiral muscle, 490, 492
        storage and deposit, 490
        structural types of, 490, 492
        variations, 490–491
    selective reabsorption
        anatomical specialization, 499–500
        in ileum, 501
        in rectum, 501
Exocone eyes, 382
Exocytosis, 47–48, 61, 106
Exopeptidases, 49–50
Exoskeleton muscles, 313–314
External coincidence model, 197
Extracellular signal-regulated kinases (ERKs), 217–218
Extraembryonic membranes
    Aedes aegypti, 14
    anatrepsis, 11
    germ band stage embryo model, 11, 13
    katatrepsis, 11–13
    Oncopeltus fasciatus development, 11–13
    serosa/amnion, 14
    Tc-zen1 and *Tc-zen2*, 14–15
Extraoral digestion, 37

F

Facilitated diffusion, 54, 227
Facultative symbionts, 591, 607
F1 ATP synthase complex, 240
Fatty acids
    activation of, 247–248
    lipases hydrolyze, 217
    metabolism, 245–246
    polyunsaturated, 83–84
Female reproductive system
    oogenesis, nutrients for, 566–567
    ovaries (*see* Ovaries)
Feminization, 608–610
Fertilization, 3, 564, 576, 609

Fibrillar muscles, *see* Asynchronous muscles; Indirect flight muscles
5-hydroxytryptamine (5-HT), 412, 503
Flavin adenine dinucleotide (FAD), 233
Flexor burst generator (FBG) neurons, 274
Flight imprecision model, 541
FMRFamides, 278–279
Follicles, 562, 570, 578
Follicular cells
    patency of, 574–575
    YPs, 576
Fovea, 398
β-Fructofuranosidase, 48
Fungal gardens, 593–595, 601
Fungus-culturing ants, 593

G

GABA, *see* γ-aminobutyric acid (GABA)
β-Galactosidase, 48
Galleria wax test, 157–158
γ-aminobutyric acid (GABA), 260, 271, 286, 300–301, 316–317
Ganglion mother cells (GMCs), 21
Gap genes, 19–20
Gap junctions, 104–105
Gas
    chromatography, 152–153
    exchange
        aquatic plants, 477–478
        compressible gas gills, 476
        cutaneous respiration, 478
        in eggs, 577
        gill folds, 480
        hydrofuges, 476
        incompressible gas gills, 477
        Perla species, 479–480
        tracheal gills, 479
    ventilation and diffusion
        abdominal pumping, 469
        coelopulses, 468–469
        "hole fraction" value, 467
        humoral mechanisms, 470
        ionic/metabolic products, 470
        neck ventilation, 469–470
        nerve activation, 470
        neuromodulators, 470
        prothoracic ventilation, 470
        pumping mechanism, 467
        safety margin, 470–471
        thoracic muscular pumping, 469
        tracheoles, 471
Gastric caeca, 39, 46, 53, 59
Gastrula, 9–10
Gastrulation, 9–10
Gender determination, 581–582
Gender determining genes, 582
Gene expression
    behavior/clock, 189
    diapause and, 215–216

General odorant-binding proteins (GOBPs), 534
Gene transcription, 52, 140, 154, 166, 577
Germarium, 578
Germ band
　elongation, 10–11
　formation of, 9
Germ cell expressed (*gce*) gene, 162–163
Giant axons, 271, 273
Glial cells, 272–273, 301
Glomeruli, 262–263, 542
Glucose
　anaerobic fermentation, 57
　biosynthetic pathway, 228
β-Glucosidases, 48, 116, 598
Gluphisia septentrionis, 85–86
Glutamate receptors (GluR), 317
Glycerol-3-phosphate shuttle, 231–235
Glycocalyx, 45
Glycogen
　biosynthetic pathway, 228
　carbohydrate resources, 226
　fat body, 224
　phosphorylase, 229
　storage and synthesis, 229–230
　synthetase, 230
Glycolipoproteins, 572
Glycolysis
　glucose, metabolism of, 230–232
　glycerol-3-phosphate shuttle and NAD^+ regeneration, 232–235
　significance and control of, 235
α-Glycosidases, 48
β-Glycosidases, 48
β-(1-4)-glycosidic linkages, 116
L-glutamic acid, 286, 300–301, 316–317
GMCs, *see* Ganglion mother cells (GMCs)
Gnotobiotic culture, 87
Goblet cells, 41–44
Goldman constant field equation, 294
Golgi complex, 40, 48, 106
G-protein-coupled receptors (G-proteins), 388, 390
Graded membrane response, 286–288
Granulocytes, 414–418
Green vetch aphids, 196
Growth cone, 22
Growth zone, 578
Guanosine diphosphate (GDP), 389
Guanosine triphosphate (GTP), 389
Gustatory function, 356–357
Gut
　development of, 22–23
　fates, glucose absorbtion, 227
　foregut, 36–38
　hindgut, 36, 39–40
　midgut
　　columnar cells, 40–41
　　countercurrent circulation, 53–54
　　gastric caeca, 39
　　goblet cells, 41–44
　　hormonal influence, 51–53
　　microvilli/brush border of, 44
　　regenerative cells, 40–41
　pH, 55–57
　plant and animal disease organisms, spread control, 64–65
　population management, potential target, 64–65
　structure and function, 34–36
　TEP and redox potential, 55

H

Hair plates, 358
Haplodiploidy, 609
Haploid
　males, 581
　spermatids, 578
Hatching, 24
Heartbeat, 410–411
Heart muscles, 329–330
Helicoverpa zea, 51, 77, 81, 280
Hematophagy, 57–58
Hemidesmosomes, 104
Hemimetabola, 2, 7–8, 11, 13, 24
Hemiptera, 60–61, 571
Hemocytes
　adipohemocytes, 416
　appearance and shape changes, 414–415
　coagulocytes, 416
　functions, 416–417
　granulocytes, 415–416
　hemocytopoietic tissues, 417–418
　number, 418–420
　oenocytoids, 416
　origin of, 418
　plasmatocytes, 415
　prohemocytes, 414
　spherulocytes, 416
Hemocytopoietic tissues, 417–418
Hemolymph
　buffer, 425
　chemical composition
　　free amino acids, 426
　　inorganic ions, 426
　　organic constituents, 427–428
　　proteins, 427
　coagulation of, 422–424
　definition, 420
　functions, 420–422
　pH, 424–425
　volume, 422–423
Heterodimers, 169–170
Hexamerins, 217
High-performance liquid chromatography (HPLC), 152–153
Hindgut, 36, 39–40
Holidic diets, 87
Holobiont, 592, 605
Holocrine secretion, 40, 47
Holometabola, 2, 8–9
HOM-C, *see* Homeotic complex (HOM-C)

Homeobox, 11, 21, 28
Homeodomain, 17
Homeostasis
    acid–base balance, 506–507
    electrolyte, 502–503
    eliminating metabolic wastes and toxins, 502
    female yellow-fever mosquito *A. aegypti*, 502
    nitrogen
        allantoin, 511
        ammonia excretion, 508
        paternal investment behavior, 510
        uric acid, 507, 509–510
        uricotelic, 511
    regulatory mechanisms, 502
    Rhodnius prolixus, 502
    and urine formation (*see* Urine formation)
    water
        antidiuretic hormones, 506
        diuretic hormones, 504–506
Homeotic complex (HOM-C), 20
Homeotic genes
    Antp, 20
    gene complexes, 20
    homeobox, 21
    identity and sequence determination, 15
Homodimers, 169
Homoptera, 61
Honeybee
    circadian clock, 189–190
    color perception, 391
    energy demands, 225
    structure, 34
Honey stopper, 37–38
Hormonal regulation, 213, 567
Hormonal stimulation, 47, 168
Hormones
    brain, 141
    carbohydrate metabolism, controlling, 230
    corpora allata and JH
        assays for, 157
        downstream transcription factors, 163–164
        Galleria wax test, 158
        glandular source and chemistry, 155–157
        growth regulators and compounds, 161
        hormone development, 137–140
        mode of action and receptors, 162–163
        tissue regulation and hemolymph
            levels, 157–161
    ecdysone–gene interaction, 170–171
    ecdysteroids
        assays, tissues and cell cultures, 153
        Calliphora assay for, 151
        chromosomal puffs, 164–166
        description, 146
        differential tissue and cell response, 168–170
        ecdysone (*see* Ecdysone)
        ecdysteroid receptor isolation, 166–168
        hormone development, 137–140
        molting hormone, molecular diversity in, 148–151
        parasitoids dependence, 155
        physicochemical techniques, 152–153
        virus degradation of, 155
    molting and metamorphosis, control of, 136–137
    molting process, possible timer gene in, 170
    ovaries, regulation of
        Apterygota, 571
        Coleoptera, 571
        Dictyoptera, 567
        Diptera, 568
        Hemiptera, 571
        JH, 567–568, 570–571
        Lepidoptera, 571
        Orthoptera, 567–568
        stages, 568–570
    pheromones, control of
        egg diapause hormone, 543
        Lepidoptera, 543
        mode of action, 544–545
        myotropic activity, 543
        neuropeptides, 544
        PBAN, 543
    prothoracic glands
        assays, tissues and cell cultures, 153
        Calliphora assay for, 151
        description, 146
        ecdysone (*see* Ecdysone)
        molting hormone, molecular diversity in, 148–151
        parasitoids dependence, 155
        physicochemical techniques, 152–153
        virus degradation of, 155
    PTTH
        bioassay for, 141–143
        hormone development, 137–140
        mode of action, 145–146
        secretion, stimuli for, 143
        source and chemistry, 141
        stretch receptors, brain activation, 143–144
        tissue regulation and hemolymph levels, 144
        tobacco hornworm, 144
Hormone secretion, circadian regulation of, 186–188
Hourglass model, 196
Humoral defenses, 434–435
Hunchback, 17–19
Hyaline hemocytes, *see* Coagulocytes
Hydrocarbons, 128–129, 247
Hydrogen bonding, 119
Hygroreceptors, 364–365
Hymenoptera, 62
Hyperglycemic hormone (HGH), 229
Hyperpolarization, 296, 298, 300, 316, 354, 461
Hypertrehalosemic hormone (HTH), 229–230
Hypocerebral ganglion, 263
Hypodermis, 103

I

Idea leuconoe, 526–527
IDGF, *see* Imaginal disc growth factor (IDGF)

Imaginal disc growth factor (IDGF)
  Diaprepes abbreviatus, 26–27
  phylogenetic tree, 27
Imaginal discs
  definition, 1
  of *Drosophila melanogaster*, 25–27
  IDGF, 26–27
  JH levels, 25
  small clumps of, 24
  in tephritid fruit fly, 24
IMD pathway, 442–443, 607
Immunity
  antifungal and antibacterial peptides, 438–439
  antimicrobial peptides
    IMD pathway, 442–443
    Toll pathway, 439–441
  autoimmune reactions, 447
  cellular immune reactions, 436–437
  cost of defense, 445–446
  C-type lectins, 443
  gender differences, 447
  host
    defense mechanisms, 446
    fitness, 444–445
    nutrition, 444
    physiological status, 444
  innate mechanisms, 434
  nonself recognition, 437–438
  parasitoid escape mechanisms, 446
  physical barriers, 435
  serpins, 443–444
  tough sclerotized cuticle, 433
Incompressible gas gills, 477
Indirect flight muscles, 317–318, 337
Infrared (IR) receptors, 366–368
Ingluvial ganglia, 263
Initiatorin, 580
Inka cells, 110, 112, 114
Inositol trisphosphate (IP_3), 389–390
Insect(s)
  carotenoids, 394–395
  damselfly flight, 340–341
  dragonfly flight, 340–341
  flight behavior, 348
  hovering flight, 344–346
  light-sensitive receptors
    compound eyes, 378
    ocelli, 378
    stemmata, 378
  muscles
    asynchronous muscles, 339–340
    direct flight, 337–338
    indirect flight, 337
    metabolic activity, wing muscles, 347
    pleuroalar muscle, 337–338
    power output, flight muscles, 346–347
    synchronous, 339
  nicotinic and muscarinic cholinergic receptors, 302–303
  symbiosis (*see* Symbiosis)
  thoracic structure and wing hinges, 336–337
  wingbeat frequency, 339–340
  wings (*see* Wings)
Insecticyanin, 576
Insulin/insulin-like growth factor, 144
Integument
  composition, 100
  cuticle, chemical composition
    chitin (*see* Chitin)
    cuticular proteins, 122–125
    lipids, 126–129
    protective functions, cuticle proteins, 126
    resilin, 125
    stage-specific differences, cuticle proteins, 126
  cuticular surfaces, atmospheric water on, 129
  ecdysis, 109
    Hyalophora cecropia, VNC, 111–112
    larval monarch, 111
    new cuticle, waterproofing, 113–114
    phases, 110
    post-ecdysis wing expansion, 113–114
    ptilinum, 109–110
    PTTH, 110
    sclerotization, 114–116
    shedding, old cuticle, 112–113
  mineralization, cuticles, 129
  molting
    apolysial space, 108
    apolysis, 106–107
    cuticle ecdysis, 107
    cuticle secretion, 107
    fluid secretion, 108
    molting fluid, reabsorption of, 109
    new cuticle formation, 108–109
    preparation for, 107
  structure of
    basement membrane, 103
    close junctions, 104
    cuticulin envelope, 101
    DPY module, 106
    dumpy (*dp*) gene, 106
    epicuticle, 101–102
    gap junctions, 104
    golgi complex, 106
    hemidesmosomes, 104
    junctional contacts, 104–105
    layers, 100–101
    oenocytes, 106
    pore canals and wax channels, 103
    procuticle, 102–103
    RER, 106
    septate junctions, 104–105
    SER, 106
    tight junctions, 104–105
Intermediary metabolism
  carbohydrate metabolism, hormones controlling, 230
  fat body, 224–225
  insect flight, energy demands for, 225–226

intense muscular activity
    electron transport system, 237–240
    glycolysis (see Glycolysis)
    krebs cycle, 235–237
    lipids (see Mobilization)
    proline, 240–244
    storage (see Carbohydrates, resources)
Intermediate germ band, 9
Internal coincidence model, 196–197
Internuncials, 271
Intersegmental interneurons, 271
Intracellular tracheoles, 312, 466, 471
Ionic influences, 411
Isoptera, 59–60

J

Japyx solifugus (Diplura), 6
JH-binding proteins (JHBPs), 160–161, 187
JH bisepoxide (JHB3), 155–157
JH epoxide hydrolases (JHEHs), 160
Johnston's organ, 261, 355, 358–359, 363–364
Junctional contacts, 104, 272, 314–315
Juvenile hormone (JH)
    ATs, 280
    corpora allata and
        assays for, 157
        downstream transcription factors, 163–164
        Galleria wax test, 158
        glandular source and chemistry, 155–157
        growth regulators and compounds, 161
        mode of action and receptors, 162–163
        tissue regulation and hemolymph levels, 157–161
    diapause, 211
    disc formation, 25
    hormonal regulation of ovary development, 567–568, 570
    hormone control development, 137–140
    hormone secretion, circadian regulation of, 187
    level of, 25
    ovary growth, 567
    protein deprivation, 78
    PTTH interplay, ecdysteroids, and, 137–140
    sperm transfer, 580
    trypsin, 110

K

Kairomones, 524–525, 547
Katatrepsis, 11–14
Kinases, 145–146, 439–441
Krebs cycle
    carbohydrate metabolism regulation, 237
    control of, 237
    in insect mitochondria, 235–237
Krüppel homolog 1 (*Kr-h1*), 163

L

Lamina ganglionaris, 271, 384–385, 387, 390
Larval diapause, 213

Lateral nerves, 112, 263, 266, 270, 276
Lateral oviducts, 562, 564–565, 568
Late trypsin, 51–52, 570
Leading-edge vortex (LEV), 341
Leaf-cutting ants, 149, 151, 593–595
Lepidoptera, 63–64
    apyrene and eupyrene sperm of, 579–580
    GOBPs, 534
    mating disruption, 548
    MGC, 542
    ovary development, hormonal regulation of, 571
    PBAN, 544
    PBPs, 535
    pheromones
        hormonal control, 543
        olfactory receptors, 534
        olfactory sense, 526
Leptophragma cell, 512–513
Leucine, 52, 54, 79–81, 91, 114
Leucokinins, 277, 279, 503, 506
Ligand gated, 290–291, 369
Lignocellulose, 596, 598
Lipases, 49, 55, 59, 245
Lipid-digesting enzymes, 49
Lipids
    cuticle, 126–129
    mobilization and
        AKH, 244–245
        fat body, 243–244
        fatty acid activation, 247–248
        lipophorin, lipid transport, 247
        mitochondria entry, 247–248
        β-oxidation, 246–248
    nutrition, requirement, 82
Lipophorin, 54, 128, 160, 247, 576
Liquid food, 34–36
Local circuit theory, 298–299
Local Interneurons (LNs), 262, 271, 542–543
Local potential, 286–287
Locustatachykinins, 277, 279
Long germ band, 9
Lure and kill, 548
Lysine, 50, 54, 79–81, 91, 114, 388, 438

M

Macroglomerular complex (MGC), 262, 538, 542
Makisterone A, 148, 150, 153
Makisterone C, 149, 153
Male reproductive system
    accessory glands, 580
    apyrene and eupyrene sperm, Lepidoptera, 579–580
    in *Drosophila*, 577
    internal organs, 578
    sperm transfer, 580–581
    zones of maturation, 578–579
Malpighian tubules, 23, 189
    cells
        extensive basal infoldings, 494–495
        type I/principal tubule cells, 493

INDEX

type II stellate cells, 493
water and hemolymph components, 494
cryptosolenic/cryptonephridial tubules, 490
detoxification and immunity, 493
function, 490
small spiral muscle, 490, 492
structural types of, 490, 492
variations, 490–491
Manduca sexta, 154
AKH, 230
antennal transcriptome, 369
capa neuropeptides, 506
cardioacceleratory peptides, 277
CCAP, 412
cone, 382
C-type lectins, 443
ecdysis, 110, 146
esterase, 538
glomeruli, 262
hemocytes, 418
hemolymph uric acid, 427
insecticyanin, 576
JH, 25, 157
JHBP, 160
males, 542
Mas-AST, 280
MAS-DH, 504
molting fluid, 109
nicotine excretion, 489
NSC, 278
olfactory receptors, 538
PBAN, 544
PTTH secretion, 138–141, 144
self-select nutritional components, 77
serpin, 443
taste and olfactory stimuli, 371
vision, 394
vitamins, 84
Mass trapping, 548
Master clock, 184–185
Maternal genes, 15–16, 18–20, 565
Mating disruption
direct insect control, 548
operating mechanisms
blend imbalance and antagonists, 550
camouflage of natural pheromone trails, 550
false trail following, 549–550
sensory fatigue, 549
pheromone, 548–549
population appearance, 524
Mechanoreceptors
AMMC, 260–261
chordotonal organs, 364
chordotonal sensillum, 358–359
hair plates, 324, 358
hermoreceptors and hygroreceptors, 365–366
IR receptors, 366–368
Johnston's organ, 364
ovipositor, 566
proprioceptors, 355

sensillum, 357
sensory neurons, 270, 273
subgenual organ, 359–360
tactile hair, 357–358
tympanal organs (*see* Tympanal organs)
Meconium, 109
Medulla, 385
Meiosis, 578
Melanization reaction, 434, 436
Membrane alteration, semiochemical detection, 533
Membrane ion channels, 240
ball and chain model, 291
Cl^- concentration, 289
ion homeostasis, 289
large negatively charged organic ions, 289
ligand gated, 290–291
mesaxon, 289–290
pores, 289
potassium channels, 289, 291–292
sodium channels, 289, 291
transmembrane proteins, 289–290
voltage gated, 290
Meridic diets, 87
Merocrine secretion, 47
Meroistic ovaries, 9, 563
Mesaxon, 289–290, 292, 294–296
Mesocuticle, 102–103
Mesothoracic ganglia, 264, 468, 504
Metabolic water, 238, 248, 475
Metalloproteinases, 49
Metamerization, 11, 19
Metamorphosis, 136, 138, 424
Hemimetabola, 2, 13
Holometabola, 2
hormonal control of, 136–137
JH, 162–164
nutrient-dependent hormonal factor, 25
prothoracic glands, 146
sensory neurons, 270
stemmata migration, 387
Metathoracic ganglia, 264–265, 275–276, 336
Methionine, 79, 81, 157
Methoprene-tolerant gene (*Met*), 162–163
Methyl farnesoate, 161, 571
MGC, *see* Macroglomerular complex (MGC)
Microapocrine secretion, 47
Micropyle, 314, 576, 579
Microvilli, 384, 397
columnar cells, 40, 48, 60
epidermal cells, 109, 122
follicular cells, 574
goblet cells, 41
malpighian tubule cells, 493
of midgut cells, 44
peritrophic matrix, 45
posterior midgut cells, 62
Microvitellin, 576
Middle integrative neuropil, 266

Midgut, 36–38
  anterior and posterior, 23
  aquaporins, 58
  coleoptera, 61
  columnar cells, 40–41
  countercurrent circulation, 53–54
  dictyoptera, 59
  digestive enzymes, 47–50
  diptera, 62–63
  diuretic hormones, 504
  endoderm, 10, 39
  gastric caeca, 39
  glucose absorption, 224, 227
  goblet cells, 41–44
  hemiptera, 60
  hormonal influence, 51–53
  huckebein, 18
  hymenoptera, 62
  Leishmania, transmission of, 64
  Lepidoptera, 63
  mesendoderm, 10
  microvilli or brush border of, 44
  orthoptera, 58–59
  peritrophic matrix, 45–46, 435
  pH, 55–57
  proton pump, 495
  regenerative cells, 40–41
  toxin, 64
  visceral muscles, 329
  yolk cells, 7
Midline precursor cells (MPCs), 21
Minerals, of cuticles, 85–87, 129
Minimum cost of transport (MCOT), 323
Mitochondrial glycerol-3-phosphate dehydrogenase, 234, 238
Mitogen-activated protein kinases (MAP kinases), 218
Mobilization, 237
  and lipids
    AKH, 244–245, 278
    fat body, 243–244
    fatty acid activation, 247–248
    lipophorin, lipid transport, 247
    mitochondria entry, 247–248
    β-oxidation, 246–248
  PDP1, 225
  prophenoloxidase, 446
Molting, 136
  hormone
    control of, 136–138
    ecdysone, 603
    inactivation, 154
    molecular diversity, 148–151
    N-acetyldopamine, 114
    PTTH, 277
    β-sitosterol, 152
    sterol, 82
  integument (*see* Integument)
  JH, 136, 162
  possible timer gene in, 170

Molybdenum (Mo), 85
Monomers, 428, 572
Monopolar interneurons, 384
Morphogen, 15–16, 19
Mosquito natriuretic peptide (MNP), 503
Motoneurons, 22, 276, 336
  controls walking, 274
  giant axons, 273
  input and output synapses, 269
  lateral nerves, 266
  proctolin, 412
  somata, 270–271
  tritocerebrum, 263
  ventilatory movements, 276
Motor programs
  neuroanatomy
    controls walking, 274
    ecdysis, 110, 274
    motor pattern, rhythmic breathing, 274–276
  VNC, 111
Multiple ecdysone receptors, 168
Multiterminal innervation, 314–316
Muscles
  asynchronous muscles
    contraction frequency, 317
    control muscles, 317
    direct and indirect muscles, 319
    dorsal longitudinal and ventral, 317
    myofibrils, 319–321
    rhythmical input, 319
    tymbal muscles, 319
    wingbeats, 319
  ATP, 310
  exoskeleton, 313–314
  insect, 337–338
    asynchronous muscles, 339–340
    direct flight, 337–338
    indirect flight, 337
    metabolic activity, wing muscles, 347
    pleuroalar muscle, 337–338
    power output, flight muscles, 346–347
    synchronous, 339
  jumping, adaptation, 325–326
  locomotion, 323
  macro- and micro-structure
    intracellular tracheoles, 312
    nuclei, 310
    sliding filament theory, 313
    SR, 310–311
  nerve–muscle junctions, 316–317
  nonskeletal muscle
    alary, 330
    heart, 329–330
    visceral, 329
  polyneuronal and multiterminal innervation, 314–316
  proteins and contraction physiology, 320–323
  running, walking, and survival, 323–325
  stridulatory, 328–329

synchronous muscles
   close packed, 319
   contraction frequency, 317
   control muscles, 317
   function of, 318
   neuronal repolarization, 318
   skeletal, 314, 318
   SR and T system, 311, 318
   tubular, 318
tymbal, 327–328
visceral, 310
Mycangia, 600–601, 604
Mycangium, 600, 603
Mycetocytes, 593, 605–606
Mycetomes, 78, 87, 593
Mycocleptism, 601, 604
Myofibrils
  active state, 323
  fibrillar muscles, 319–320
  sarcomeres
    actin and myosin, 310
    A band, 312
    H and I zone, 312–313
    Z line, 312
  slab-like, 318
  sliding filament theory, 313
  thin filaments, 313
Myosin, 310
  actin, 322–323
  alary muscles, 330
  crossbridges, 320
  growth cone lamellae, 22
  heart muscles, 329
  M line, 312
  shape of, 320–321
  sliding filament theory, 313
  thick filaments, 312
  visceral muscles, 329

N

N-acetyl-β-D-glucosamine, 116
N-acetyl-β-D-glucosaminidase, 108
N-acetyldopamine, 114–115
N-acetylglucosamine, 108
Nagana, 606
Nauphoeta cinerea, 138–139, 572
N-β-alanyldopamine, 114
Neb-TMOF, 52
Negative feedback mechanism, 229
Nernst equation, 198, 293–295
Nerve–muscle junctions, 316–317
Nerve supply, 316, 329, 411–412
Neural cartridges, 271, 384–385
Neurites, 269–271
Neuroanatomy
  building blocks of
    axon, 269
    glial cells, 272–273
    interneurons, 271
    motoneurons, 270–271
    sensory neurons, 270
    soma, 269
  CNS (*see* Central nervous system (CNS))
  hemolymph–CNS barrier, 268–269
  motor programs
    controlling walking, 274
    ecdysis, 274
    motor pattern, rhythmic breathing, 274–276
  neuropil, 258, 267–268
Neuroblasts (NBs), 21
Neurogenesis, 21–22
Neurolemma, 268–269
Neuromeres, 265–266
Neuroommatidia, 384
Neurophysiology
  axonal, 286
  dendritic processes, 286
  soma, 286
  stimulus response
    action potential, 294–296
    electrotonus, 286–387
    graded membrane response, 286–288
    local potential, 286
    membrane ion channels (*see* Membrane ion channels)
    resting potential, 292–294
    sodium inactivation and repolarization, 296–297
    spike potential, 288
    voltage clamp technique, 297–298
  synapse
    acetylcholine-mediated, 301–303
    electric transmission, 304
    EPSP, 298–301
    neuromuscular junctions, 304
    nicotinic and muscarinic cholinergic receptors, 302–303
Neuropil, 255, 267–268, 384
Neurosecretory cells (NSC), 116, 276–278
  brain, PTTH
    bioassay for, 141–143
    mode of action, 145–146
    secretion, stimuli for, 143
    source and chemistry, 141
    stretch receptors, brain activation, 143–144
    tissue regulation and hemolymph levels, 144
    tobacco hornworm, 144
  neuropeptides, 256
  SEG, 545
Nicotinamide adenine dinucleotide (NAD^+), 232–235
Nitrogen
  homeostasis, 507–508
  metabolism, 599
  source, 78–79
  termites, 599
Nonneuronal cells (NNCs), 21
NSC, *see* Neurosecretory cells (NSC)

Nutrients
    for oogenesis, 566–567
    requirements, 77–78
    storage, 224
Nutrition
    balance of, 76–77
    evaluating nutritional quality, 87–88
    feeding deterrents, 92
    food intake and utilization, 88–90
    growth and reproduction, 75–76
    phagostimulants, 90–91
    requirements
        essential amino acids, 79–81
        lipids, 82
        minerals, 85–87
        polyunsaturated fatty acids, 83–84
        proteins and amino acids, 78
        sterols, 82–83
        vitamins, 84–85
    self-select dietary components, 77
    stems, 76
    techniques, 87

O

Obligate (primary) symbionts, 592
Oblique dorsal muscles, 337
OBPs, see Odorant-binding proteins (OBPs)
Ocelli, 378, 385–386
(Z,Z)-9,12 Octadecadienoic acid, 83
(Z,Z,Z)-6,9,12-Octadecatrienoic acid, 83
Octopamine, 159, 260, 412
Odorant-binding proteins (OBPs), 369, 534
Odor plumes, 523, 539–541
Oenocytes, 23, 106, 128, 243
Oenocytoids, 416
Olfactory receptors, 356, 369
Olfactory sense, 525–527
Oligidic diets, 87
Ommatidia
    dioptric structures, 379
    photopic eye, 379–380
    scotopic eye, 379–381
Oncopeltus fasciatus, 11–12
    blastokinesis, 11–12
    female reproductive structures, 562
    insect growth regulators and compounds, 161
    male reproductive system, 578
    molting hormone, 148
    thoracic ganglion, 265
Oocytes
    maturation divisions of, 4
    vitellogenins sequestering and YPs
        follicular cells, patency of, 574–575
        paravitellogenin, 576
        proteins, 576
Oogenesis, 566–567
Oostatic hormone (OSH), 570
Opsin, 378, 388, 393

Optomotor anemotaxis, 539–541
Organogenesis
    apoptosis, 24
    cell movements, 23–24
    embryo, cuticle secretion in, 23
    gut development, 22–23
    malpighian tubules, 23
    neurogenesis, 21–22
    oenocytes, 23
    tracheal system, 23
Orthoptera, 58–59, 567–568
Ostia, 407, 410
Ostrinia nubilalis, 548
Ovarian ecdysteroidogenic hormone I (OEH), 570
Ovaries
    hormonal regulation of
        Apterygota, 571
        Coleoptera, 571
        Dictyoptera, 567
        Diptera, 568
        Hemiptera, 571
        JH, 567–568, 570–571
        Lepidoptera, 571
        Orthoptera, 567–568
        stages, 568–570
    structure of
        internal, 562
        oviposition, 565–566
        panoistic ovarioles, 564
        telotrophic ovarioles, 564
Ovarioles, see Ovaries
Overshoot potential, 294–296
Oviparae, 196
Oviposition, 565–566

P

Paired ovaries, 562
Paired testes, 578
Pair-rule genes, 19–20
Panoistic ovaries, 9, 563–564
Panoistic ovarioles, 564
Paracellular pathway, 496, 503
Paracrine control, 47
Paragonial glands, 580
Paranotal lobe theory, 335
Parasegments, 11, 19–21
Parasitoids, 155, 210, 416–417, 420, 446
Paravitellogenin, 576
Pars intercerebralis, 116, 159, 260, 277
Parthenogenesis, 609–610
Patency, follicular cells, 574–575
Paternal investment behavior, 510
Pathogen-associated molecular patterns (PAMPs), 437–438
Pattern recognition proteins (PRPs), 434–435, 437–439
Pattern recognition receptors (PRRs), 434, 436–437
PBAN, see Pheromone biosynthesis-activating neuropeptide (PBAN)

PBPs, *see* Pheromone-binding proteins (PBPs)
Penetration, semiochemical detection, 533
Pericardial cells, 409
Pericardial sinus, 408–409
Perineural sinus, 408–409
Perineurium, 268
Peripheral clocks, 184
Periplaneta americana
    accessory heart, 413
    acetylcholine, 800
    action potential, 294–295
    bursicon, 116
    corazonin, 412
    extracardiac hemocoelic pulsations, 468
    feeding deterrents, 92
    giant axons, 273
    gut
        pH, 55
        TEP, 55
    heartbeat, 410
    hemocyte counts, 418
    male reproductive system, 578
    motor programs, 274–275
    Na/K ratio, 426
    nicotinic ACh receptors, 302
    ocelli, 386
    regenerative cells, 40
    resting potential, heart, 411
    retinula cells, 383
    rhabdomere, 384
    sclerotization, 116–117
    subgenual organ, 360
    synapses, 300
    TAG, 265
    thermoreceptors and hygroreceptors, 365
    vitamins, 84
Periplasm, 3, 8
Peritrophic matrix (PM)
    components, 46
    countercurrent flow, 53–54
    functions of, 46–47
    glycocalyx material, 45
    Hemiptera, 60–61
    invasion, physical barriers, 435
    type I, 45, 58, 63
    type II, 45, 62
Peritrophic membrane, *see* Peritrophic matrix (PM)
Perivisceral cavity, 408–409
Phagocytosis, 434, 436
Pharate, 107, 109, 114, 221
Phasic receptors, 355
Phenoloxidase (PO), 434, 436–437, 447
Phenylalanine, 79–81, 606
Pheromone-binding proteins (PBPs), 533–535
Pheromone biosynthesis-activating neuropeptide (PBAN), 188, 280
    Bom-PBAN-I, 543
    Bom-PBAN-II, 543
    Hez-PBAN, 543
    Lyd-PBAN, 543
    mode of action, 544–545
Pheromones
    active space, 528
    biosynthesis, 545–547
    chemical characteristics
        Cahn–Ingold–Prelog system, 530–531
        chirality, 530
        enantiomers, 530–531
        Grain beetles, 532–533
        R and S configurations, 531–532
    classification, 528
    destroy/inactivate, 538
    geographical differences, 547–548
    hormonal control
        egg diapause hormone, 543
        Lepidoptera, 543
        mode of action, 544–545
        myotropic activity, 543
        neuropeptides, 544
        PBAN, 543
    in male tobacco hornworm, 523
    in odor plumes, 523
    olfactory receptors, 533
    olfactory sense, 526–527
    parsimony, 523, 529
    population differences, 547–548
    practical application
        luring large aggregations, 548
        mass trapping, 548
        mating (*see* Mating disruption)
        successful mating, 548
    reception, 192
    signal processing, 541–543
    signal transduction and receptor response, 535–537
    smell and responds, 538
Phosphagen, 226
Phosphatidylinositol bisphosphate (PIP_2), 389
Phosphoglucose isomerase (PGI), 323–324
Phospholipase C (PLC), 145, 349, 535
Photochemical reaction, 388
Photoinducible response, 197
Photoisomerization, 390–391
Photoperiodic clock, 183, 193–195
Photoperiodic responses, 182
Photophase, 144, 146, 182
Photopic eye, 379–380
Photoreceptor cells, 383, 397
Phytophagous, 34, 46, 48, 148, 370
Pigment-dispersing factors, 279
Plane-polarized light, 393, 395–397, 531
Planotortrix excessana, 547
Plasma membrane plaques, 108–109
Plasmatocytes, 415–418, 436
Plastrons, *see* Incompressible gas gills
Pleural wing process, 336–338, 340
Pleuroalar muscle, 337–338
Polar body, 3–4

Polyhydric alcohols, 211
Polyneuronal innervation, 314–316
Polytene chromosomes, 164–165
Polytrophic ovarioles, 563, 565
Polyunsaturated fatty acids, nutrition requirement, 83–84
Pontine neurons, 260
Pore canals, 103
Post–ecdysis, 110
Postvitellogenic stages, 570
Prandial control, 47
Prediapause, 210–211
Pre-ecdysis I, 110–111
Pre-ecdysis II, 110, 112
Pre-ecdysis-triggering hormone (PETH), 110
Previtellogenic stage, 568, 570
Procarboxypeptidase, 580
Proctolin, 278, 329, 412
Procuticle, 99, 101–103, 109
Programmed cell death, 24
Prohemocytes, 414–416, 418
Proline dehydrogenase, 240
Prophenoloxidase, 434, 436, 446
Proprioceptors, 355
Proteinase inhibitors, 51
Proteinases, 49–51, 108
Protein-digesting enzymes, 49–50
Protein kinase C, 574
γ-Proteobacteria, 605–606
Prothoracic glands (PGL)
   in American cockroach nymph, 142
   and ecdysteroids
      assays, tissues and cell cultures, 153
      Calliphora assay for, 151
      ecdysone, biosynthesis of, 146–148
      ecdysone, conversion of, 148
      ecdysone, degradation of, 153–154
      ecdysone, RIA, 151–152
      molting hormone, molecular diversity in, 148–151
      parasitoids dependence, 155
      physicochemical techniques, 152–153
      virus degradation of, 155
Prothoracicotropic hormone (PTTH)
   brain neurosecretory cells and
      bioassay for, 141–143
      mode of action, 145–146
      secretion, stimuli for, 143
      source and chemistry, 141
      stretch receptors, brain activation, 143–144
      tissue regulation and hemolymph levels, 144
      Tobacco Hornworm, 144
   hormone control development, 137–140
   hormone secretion, circadian regulation of, 186–187
   pupal diapause, 213–214
Protocephalon, 10–11
Protocerebrum, 141–142, 258–260
Proton ATPase pump, 41–43
Proventriculus, 34–35, 37–38
Pseudocone eyes, 382
Ptilinum, 109–110
Pupal diapause, 213–214

Q

Quiescence, 99, 207–208, 216
Quinone sclerotization, 114, 116
Quinone tanning, 114–116, 120

R

Radioimmunoassay (RIA), 144, 151–152, 157, 245
Reactive oxygen species (ROS), 145, 193
Redox potential, 55, 59, 239, 598
Regenerative cells, 40–41, 48, 60, 62
Regional localization, 15
Relative growth rate (R.G.R.), 88
Reproduction
   chorion, 576–577
   circadian clock, 190–193
   female reproductive system
      oogenesis, nutrients for, 566–567
      ovaries (*see* Ovaries)
   gas exchange in eggs, 577
   gender determination, 581–582
   male reproductive system
      accessory glands, 580
      apyrene and eupyrene sperm, Lepidoptera, 579–580
      in *Drosophila*, 577
      internal organs, 578
      sperm transfer, 580–581
      zones of maturation, 578–579
   oocytes (*see* Oocytes)
   vitelline membrane formation, 576
   vitellogenins and YPs
      biochemical characteristics of, 572
      higher Diptera, 573–574
Resilin, 125, 326, 347
Resonance model, 196–198
Respiration
   DGC
      closed period, 471–472
      cycle, 471
      definitive evolutionary mechanism, 474–475
      diapausing pupae, 473
      discontinuous release of CO_2, 471
      discontinuous ventilation cycle, 471
      evolutionary driving mechanism, 474
      feedback systems, 474
      fluttering phase, 472
      flutter period, 471–472
      gas exchange measurement, 472
      hypoxic and hypercapnic theory, 474
      intratracheal pressure, 472
      low oxygen tension, 473
      open period, 471–472
      passive suction ventilation, 471–472
      spiracular opening and closing, 472
      tightly/apparently closed, 471
      tracheal gas composition, 472

in eggs and embryo development, 481–482
in endoparasitic insects, 481
gas exchange
  aquatic plants, 477–478
  compressible gas gills, 476
  cutaneous respiration, 478
  gill folds, 480
  hydrofuges, 476
  incompressible gas gills, 477
  Perla species, 479–480
  tracheal gills, 479
gas ventilation and diffusion
  abdominal pumping, 469
  coelopulses, 468–469
  "hole fraction" value, 467
  humoral mechanisms, 470
  ionic or metabolic products, 470
  neck ventilation, 469–470
  nerve activation, 470
  neuromodulators, 470
  prothoracic ventilation, 470
  pumping mechanism, 467
  safety margin, 470–471
  thoracic muscular pumping, 469
  tracheoles, 471
large longitudinal tracheal sacs, 455–456
respiratory pigments, 481
synchrotron X-ray techniques, 456
tracheal supply
  abdominal ganglia, 464–465
  abdominal spiracle, 465–466
  flight muscles, 465–466
tracheal system
  air sacs, 456, 464
  buckling hypothesis, 459
  mole cricket *Scapteriscus borellii*, 458
  molting, 464
  nonrespiratory functions, 482–483
  spiracles, 459–461
  termite *Cubitermes fungifaber*, 458–459
  tracheae, 457
  tracheal epithelial cells, 462
  tracheole, 457–458
  tracheoles development, 462–463
tracheal tube branching, 456–457
water balance, 475
Resting potential, 292–295, 411
Retinula cells, 380, 383–386, 388
Rhabdom/rhabdomeres, 379–382, 384, 386, 394, 397–398
Rhodnius prolixus, 50, 463, 497, 502, 579, 581
  accessory glands, 579
  CA, 137
  CCAP, 113
  gut structure and food habits, 35
  Hemiptera, 60
  IR, 367
  JH, 571, 581
  proteolytic enzyme, 50
  PTTH, 143
  spermatozoa, 579

Rhodopsin, 378, 387–393
Ring gland, 151, 153, 156, 166, 218
Rough endoplasmic reticulum (RER), 40, 106

S

Sarcomeres
  alary muscles, 330
  heart muscles, 329
  length, 312, 319
  Z bands, 312
Sarcophaga bullata, 128, 214, 566
Sarcoplasmic reticulum (SR), 310–311
Schistocerca americana, 470–471
Sclerotization, 99–103, 113–117, 124, 427
β-Sclerotization, 115
Scolopale cell, 358–359, 361
Scolopidium, 358–359
Scotophase, 144, 146, 182, 184, 186–187, 196–197, 544
Scotopic eye, 379–381, 383
Segmentation genes, 15–17, 19–20
Segment polarity genes, 19–20, 27
Self-select dietary components, 77
Semiochemicals, 524
  active space, 528
  allomones, 523, 525
  chemical characteristics, 529–533
  chemical signals, exchange, 524
  information coding and processing
    odor plumes, 539–541
    pheromone signal processing, 541–543
  insect receptors and detection process
    components, 538
    pheromone-binding proteins, 533–535
    pheromone inactivation and receptor clearing, 538
    signal transduction and receptor response, 535–537
    steps, 533
  insects, olfactory sense in, 525–527
  interspecific classification, 523
  kairomones, 523, 525
  pheromones (*see* Pheromones)
  synomones, 523, 525
Sensillum, 355–359, 365–366, 369–370, 535, 541
Sensillum liquor, 369, 533–535
Sensory neurons, 270
  axons, 354
  characteristic, 354
Sensory receptors
  adaptation, 354–355
  chemoreceptors, 354–355
    gustatory receptors, 369–370
    olfactory sensilla, 368–369
    specialists *vs.* generalists, 371
    stimulus–receptor excitation coupling, 371–372
  external or internal environment, 355
  functional classification, 355–356
    lack of pores, 357
    multiple pores, 356–357
    single pore, 357

mechanoreceptors
  chordotonal organs, 364
  chordotonal sensillum, 358–359
  hair plates, 358
  hermoreceptors and hygroreceptors, 365–366
  IR receptors, 366–368
  Johnston's organ, 364
  subgenual organ, 359–360
  tactile hair, 357–358
  tympanal organs (*see* Tympanal organs)
 phasic receptors, 355
 repetitive discharge, 354
 sensillum, 355
Septate junctions, 104–105, 506
Sequester ecdysteroids, 573
(ser)ine (p)roteinase (in)hibitor (serpin), 443–444
Serosa, 7, 9–15, 23
Sesquiterpenoid, 155
Sex-determining genes, 581
Sex peptide, 580
Sex pheromone, 188, 192, 362, 371, 525–526, 529, 533, 535, 542–544
Short-chain fatty acids, 40, 57, 59, 596
Short germ band, 9, 11, 13
Simple chordotonal organs, 364
Skeletal muscles, 231, 234, 302, 310, 313–319, 329–330, 347
Skipped bisepoxide, 157
Sleeping sickness, 535, 606
Slow potentials, 287, 292, 296, 384
Small polypeptides, 116, 141, 573
Smooth endoplasmic reticulum (SER), 23, 106, 390
Social organization, 189–190
Sodalis glossinidius, 606
Sodium activation, 294–295
Sodium inactivation, 296
Solid food, 47, 54, 506
Soma, 185, 262, 269–273, 276, 286, 291, 357
Somata, 260, 262, 267, 270–273, 276
Sorbitol, 211, 218
Sparing sterol, 82
Spermatheca, 562, 564, 576, 580–581
Spermatids, 578
Spermatocytes, 4, 578
Spermatogonia, 578
Spermatophore, 510, 566, 579–581
Spermatozoa, 4, 578–579, 581
Sperm, maturation divisions of, 4
Spherulocytes, 415–416, 418
Spike potential, 288, 294
Spodoptera littoralis, 154, 190–192, 542, 544
Stemmata, 378, 383, 386–387
Stereochemical theory, 371–372
Sterols, 82–83, 147–149, 603
Stomatogastric nervous system, 263, 329
Stretch receptors, 107, 143–144, 355, 502, 504
Striated border, 44
Stridulatory muscle, 318, 327–329
Subalar muscles, 317, 337, 340, 346
Subesophageal ganglion (SEG), 258–263, 273, 278–279

Subgenual organ, 359–360, 363, 367
Sun-compass orientation, 188–189, 396
Supernumerary larvae, 137
Supraesophageal ganglion, 258, 413
Symbionts, 592
  ambrosia beetles, 599–602
  anthropogenic effects, 605
  bark beetles, 599, 603
  evolution, fungal feeding, 603–604
  nutrients, fungal role in, 603
  in termites, 596–598
  tsetse flies, 606–607
Symbiosis, 592–593
  antibiotics, 595
  bacterial biofilm, 595
  Buchnera in aphids, 605–606
  fungal gardens, 593–595
  leaf-cutting ants, 593–594
  symbionts
    ambrosia beetles, 599–602
    anthropogenic effects, 605
    bark beetles, 599, 603
    evolution, fungal feeding, 603–604
    external/internal symbionts, 592
    nutrients, fungal role in, 603
    tsetse flies, 606–607
  termites
    fungal culture, 599
    higher termites, 596
    lignocellulose structure, 598
    lower termites, 596
    nitrogen metabolism, 599
    social cockroaches *Cryptocercus*, 595–596
    symbionts in, 596–598
  Wolbachia
    cytoplasmic incompatibility, 608–609
    feminizing strains of, 610
    host's behavior and biology, 608
    infection, 607
    parthenogenesis, 609–610
    strain localization, 608
Synchronous muscles
  close packed (microfibrillar and mosaic fibers), 319
  contraction frequency, 317
  control muscles, 317
  function of, 318
  neuronal repolarization, 318
  skeletal, 314, 318
  SR and T system, 311, 318
  tubular, 318
Synomones, 525

T

Tabanidae, 383
Tachykinins, 260, 279
Tactile hair, 355–359
TAG, *see* Terminal abdominal ganglion (TAG)
Tanning, 113–116, 120, 427, 437
Tapetum, 382–383

Task-related plasticity, 190
Taste receptors, 77, 92, 188, 357, 369–371, 536
Telotrophic ovarioles, 564
Telson, 11, 16, 18
Terminal abdominal ganglion (TAG), 265, 545, 569
Termites
    fungal culture, 599
    higher termites, 596
    lignocellulose structure, 598
    lower termites, 596
    nitrogen metabolism, 599
    social cockroaches *Cryptocercus*, 595–596
    symbionts in, 596–598
Thecogen, 357, 366
Thelytoky, 610
Thermoreceptors, 364–365
Thin-layer chromatography (TLC), 152–153
Third axillary muscles, 317, 337
3-hydroxy-11-*cis*-retinal, 388
Threonine, 52, 79–81, 543
Tight junctions, 104–105, 272
TIMELESS (TIM) proteins, 184
Time-linked memory, 189
Timer gene, 170, 215
Tissue-specific factors, 170
Toll pathway, 439–441, 444
Tonic receptors, 143, 355
Tonofibrillae, 313–314, 414
Tormogen, 357, 366
Toroid, 542
Tracheal system
    air sacs, 464
    cutaneous respiration, 478–480
    epithelium, 462
    flight muscles, 465–466
    fluid-filled, 481
    gas exchange, 470–471
    molting, 464
    new tracheoles development, 462–464
    nonrespiratory functions, 482–483
    spiracle structure and function, 459–461
    tracheae and tracheole structure, 457–459
    water loss, 473–474
Transcellular resistance (Rc), 503
Transcription factor, 11, 21, 163–164
Transcripts (mRNA), 16, 19
Transducin, 389
Transepithelial potential (TEP), 44, 54–55, 496
trans-verbenol, 604
α-Trehalase, 48
Trehalose resources, 226–229
Triacylglycerols (TAGs), 211, 217, 265, 545
Tribolium castaneum, 13–14, 84, 113, 162
Trichogen, 357, 366
Triosephosphate isomerase, 232
Tritocerebrum, 258–259, 263
Trityrosine, 125
Trophocyte, 225, 565
Troponin, 320–322
Trypanosoma brucei, 606

Trypsin, 48–52, 55, 61–63, 108
Trypsin modulating oostatic factor (TMOF), 51, 570
Tryptophan, 50, 52, 79, 81, 114, 125
Tsetse flies, 240, 243–244, 529, 606–607
Tubular muscle, 318–319
24-epi-Makisterone A, 149, 151
20-hydroxy-24-α-methyl ecdysone, 148, 150
20-hydroxy-24-β-methyl ecdysone, 149, 151
20-hydroxyecdysone, 148, 570, 580
Tymbal muscles, 314, 316, 318–319, 327–328
Tympanal organs
    acoustical sense, 362
    components, 360–361
    echolocating insectivorous bats, 361–362
    locations, 360
    polka-dot moth, 362–363
    scolopidia, 361
    supplementary material, 363
    vibrations, 361
    warning signals, 362–363
Type I peritrophic matrix, 45, 58–59, 63, 435
Type I/principal tubule cells, 493
Type II peritrophic matrix, 45
Type II stellate cells, 493

U

UDP-glucose-glycogen transglycosylase, 230
Ultraspiracle (USP), 163, 169, 216–217
Univoltine life cycle, 208
Unsteady forces, 342
Upstroke, wings, 338, 344, 346
Uric acid, 89, 427, 507–511, 599
Uridine 5'-diphosphate (UDP)-glucosyl transferase, 155
Urine formation
    antiporter mechanism, 495
    cations, 496
    chloride ions, 496
    cumulative formation, 498
    diuretic factor, 497
    hemolymph filtrate, 494
    hyperosmotic, 497
    hyposmotic, 497
    isolated tubule technique, 497–498
    isosmotic, 497
    standing gradient process, 496
    transmembrane V_o complex embedded, 495
Urocytes, 599
UV receptors, 378, 393

V

Valine, 52, 79–81
Vas deferens, 578
Vasopressin, 279–280
Ventral sensory neuropil, 266
Verbenone, 604
Vertical transmission, 609
Virginoparae, 196
Visceral muscles, 310, 329

Vision
- blue photoreceptors, 378, 391, 393
- carotenoids, 394–395
- chromophore, 378
- compound eyes
  - Cambrian period, 378
  - lens, 378
  - ommatidia (*see* Ommatidia)
- corneal layering, 383
- dark-adapted eye, 394
- dioptric structures, 381–383
- electromagnetic radiation, 377–378
- 11-*cis*-retinal, 378
- green photoreceptors, 378, 393
- insects behavior, 394
- light, 377–378
- ocelli, 378, 385–386
- optic lobe, neural connections in, 384–385
- photochemical reaction, 388
- plane-polarized light, 395–397
- red receptors, 391–393
- retinula cells, 383–384
- rhabdomere, 378, 384
- rhodopsin, 388–389
- stemmata, 378, 386–387
- transmembrane protein opsin, 378
- UV receptors, 378, 392–393
- visual acuity
  - acute zones, 400
  - definition, 397
  - fovea, 398–399
  - horsefly eye, 399
  - interommatidial angles, 400
  - ommatidia, 397–398
  - praying mantis, 400
- visual cascade
  - GDP, 389
  - G-protein cascade, 390
  - GTP, 389–390
  - metarhodopsin, 389–390
  - receptor potential, 390
  - regulation of, 390–391
  - Tα-GTP protein, 389–390
- wing colors, butterflies, 392–393

Vitamins, 84–85
Vitelline membrane, 3, 576–577
Vitellins (Vns), 572, 575
Vitellogenic stage, 570
Vitellogenins (Vgs), 561, 567, 572
- sequestering and YPs
  - follicular cells, patency of, 574–575
  - paravitellogenin, 576
  - proteins, 576
- and YPs
  - biochemical characteristics of, 572
  - higher Diptera, 573–574

Vitellophages, 5, 7–8
Voltage clamp technique, 297–298
Voltage gated, 290–292, 298–299

W

Water homeostasis, 503–504
Wax blooms, 102, 127–128
Wax channels, 103
Wigglesworthia glossinidia, 606
Winged morphs, 193, 195–196
Wings
- downstroke, 338
- large interneurons, 336
- lift forces
  - clap and fling wing motion, 342
  - drag and delayed stall, 343–345
  - dynamically scaled models, 342
  - steady-state aerodynamic calculations, 341–342
- paranotal lobe theory, 335
- pitch and twisting, 346
- upstroke, 338

Wolbachia, 607–608
- cytoplasmic incompatibility, 608–609
- feminizing strains of, 610
- parthenogenesis, 609–610
- *W. pipientis*, 606, 608

Y

Yolk body, 575
Yolk proteins (YPs)
- biochemical characteristics of, 572
- follicular cells, patency of, 574–575
- higher Diptera, 573–574
- paravitellogenin, 576
- proteins, 576

Z

Zinc fingers, 167–168
Zones of maturation, 578–579
Zygote, 3, 6, 8, 15
Zygotic genes, 15–19
Zymogen, 47, 108, 434